14th EDITION

MATHEMATICAL IDEAS

Charles D. Miller

Vern E. Heeren
American River College

John Hornsby
University of New Orleans

Christopher Heeren
American River College

AND

Margaret L. Morrow
Pittsburgh State University of New York
for the chapter on Graph Theory

Jill Van Newenhizen
Lake Forest College
for the chapter on Voting and Apportionment

Director, Portfolio Management: Anne Kelly
Courseware Portfolio Specialist: Marnie Greenhut
Courseware Portfolio Specialist Assistant: Richard Feathers
Content Producer: Patty Bergin
Managing Producer: Karen Wernholm
Media Producer: Nicholas Sweeny
TestGen Associate Content Producer: Rajinder Singh
Product Manager, Content Development: Bob Carroll
Marketing Manager: Erin Kelly
Marketing Assistant: Brooke Imbornone
Field Marketing Manager: Andrew Noble
Field Marketing Assistant: Rosemary Morton
Manager, Senior Author Support/Technology Specialist: Joe Vetere
Rights and Permissions: Gina Cheselka
Manufacturing Buyer: Carol Melville
Text Design, Production Coordination, Composition, and Illustrations:
  Pearson CSC
Cover Design: Jerilyn Bockorick/Pearson CSC

Library of Congress Cataloging-in-Publication Data
Names: Miller, Charles D. (Charles David), 1942-1986, author. | Heeren, Vern
  E., author. | Hornsby, John, 1949- author. | Heeren, Christopher, author.
Title: Mathematical ideas.
Description: 14th edition / Charles D. Miller, Vern E. Heeren (American River
  College), John Hornsby (University of New Orleans), Christopher Heeren
  (American River College). | Boston : Pearson, [2020] | Includes indexes.
Identifiers: LCCN 2018041190| ISBN 9780134995588 (student edition) | ISBN
  0134995589 (student edition)
Subjects: LCSH: Mathematics--Textbooks.
Classification: LCC QA39.3 .M55 2020 | DDC 510--dc23
LC record available at https://lccn.loc.gov/2018041190

ISBN 13: 978-0-13-499558-8
ISBN 10: 0-13-499558-9

3 2020

*To Sweet Baby Jess—we love you—*PPJ

*To my beloved wife, Carole, and to our three sons, their wives, and our ten grandchildren, each one a continuing inspiration in my life—*VERN

*To my children—you continually inspire me with your courage and creativity—*CHRIS

CONTENTS

The Real Numbers and Their Representations 227

The Basic Concepts of Algebra 297

*Online section can be found in the MyLab Math course or at www.pearsonhighered.com/mathstatsresources

8   Graphs, Functions, and Systems of Equations and Inequalities   375

Geometry 461

Counting Methods 549

Personal Financial Management 731

Graph Theory 801

Voting and Apportionment 865

NOTE: Trigonometry module and Metrics module available in MyLab Math or online at www.pearsonhighered.com/mathstatsresources.

PREFACE

After thirteen editions and over five decades, *Mathematical Ideas* continues to be one of the most popular texts in liberal arts mathematics education. We are proud to present the fourteenth edition of a text that offers non-STEM students a practical coverage that connects mathematics to the world around them. It is a flexible text that has evolved alongside changing trends but remains steadfast to its original objectives.

Mathematical Ideas is written with a variety of students in mind. It is well suited for several courses, including those geared toward the aforementioned liberal arts audience and survey courses in mathematics or finite mathematics. Students taking these courses will pursue careers in nursing and healthcare, the construction trades, communications, hospitality, information technology, criminal justice, retail management and sales, computer programming, political science, school administration, and myriad other fields. Accordingly, we have chosen to continue showcasing how the math in this course will be relevant in this wide array of career options. Chapter openers continue to focus on particular careers commonly anticipated by students.

- Chapter openers address how the chapter topics can be applied within the context of work and future careers.

- We made sure to retain the hundreds of examples and exercises from the previous edition that pertain to these interests.

- Every chapter also contains from one to three of the popular ***When Will I Ever Use This?*** features that help students connect mathematics to the workplace.

Interesting and mathematically pertinent movie and television applications and references are still interspersed throughout the chapters.

Ample topics are included for a two-term course, yet the variety of topics and flexibility of sequence make the text suitable for shorter courses as well. Our main objectives continue to be comprehensive coverage, appropriate organization, clear exposition, an abundance of examples, and well-planned exercise sets with numerous applications.

New and Expanded in This Edition

- Career applications have taken on greater prominence, especially through the inclusion of additional *When Will I Ever Use This?* feature boxes.

- Many of the learning objectives that begin each section have been reworded for clarity and specificity.

- All exercise sets now begin with **Concept Check** questions to confirm that students truly understand concepts before diving right into homework. Exercise sets have once again been updated, and there are over 1000 new or modified exercises, many with an emphasis on career applications.

- Most **chapter summaries** have been reorganized to improve readability. They still include the following components:

  ○ A list of **Key Terms** for each section of the chapter

  ○ **New Symbols** with definitions, to clarify newly introduced symbols

  ○ **Test Your Word Power** questions that enable students to test their knowledge of new vocabulary

  ○ A **Quick Review** that gives a brief summary of concepts covered in the chapter, along with examples illustrating those concepts

- Actual screen captures from the **TI-84 Plus C** model calculator are now included.

- The presentation has been made more uniform whenever clarity for the reader could be served.

- The general style has been freshened with more pedagogical use of color, new photos and art, and opening of the exposition.

- A new **Workbook with Investigations and Integrated Review Worksheets** accompanies the text. The workbook contains worksheets for every objective in the text, Investigation activities students can do in or out of class on their own or in a group, and worksheets to complement the Integrated Review content in the MyLab Math course.

- **NEW online resources**

  ○ **StatCrunch** applets and activities are available as indicated by the StatCrunch icon in the margin. Data sets will also be available for ease of use in the StatCrunch program.

  ○ **Group review modules in Learning Catalytics Modules** have been developed to complement each chapter's review test for use as classroom review. Using this student engagement, assessment, and classroom intelligence system gives instructors real-time feedback on student learning before the test. Instructions on how to use the Learning Catalytics Chapter Test modules are provided in the MyLab Math course.

  ○ The **Corequisite MyLab Math Support Course** takes the traditional integrated review content a step further to provide one complete solution for instructors considering a corequiste model. By mapping relevant algebra objectives from the *Mathematical Ideas* text to particular chapters and sections, students have access to the foundational support they need to successfully complete the course. The Corequisite Support Course includes over 30 review topics and enables students to access material for both their credit and support course with a single access code.

  ○ **Section lecture videos** will cover the section objectives with a modern, clear, and engaging approach. The videos will incorporate the use of animations, applets, and StatCrunch to demonstrate concepts, making use of resources available to students in the MyLab Math course.

  ○ The **Animations Library** will help students explore concepts more deeply, encouraging them to visualize and interact with concepts such as Venn diagrams, probability, logic, debt, graph theory, geometry, statistics, and more. An icon in the text indicates where the animations can aid in understanding.

  ○ **Mindset** materials are included in the **Video & Resource Library: Skills for Success** section of the MyLab Math course, along with Math-Reading Connections, College Success Modules, and Professionalism Tools.

Overview of Chapters

- **Chapter 1 (The Art of Problem Solving)** introduces the student to inductive reasoning, pattern recognition, and problem-solving techniques. We continue to provide exercises based on the monthly Calendar from *Mathematics Teacher* and have added new ones throughout this edition. The chapter opener recounts the solving of the Rubik's cube by a college professor. The *When Will I Ever Use This?* feature shows how estimation techniques may be used by a group home employee charged with holiday grocery shopping.

- **Chapter 2 (The Basic Concepts of Set Theory)** includes updated examples and exercises on surveys. The chapter opener suggests ways in which set theory can apply to exploring career opportunities in nursing and other health-related careers. One *When Will I Ever Use This?* feature applies Venn diagrams to a genetic brain disease study. Another applies survey techniques to the allocation of work crews in the building trade.

- **Chapter 3 (Introduction to Logic)** introduces the fundamental concepts of inductive and deductive logic. The chapter opener connects logic with fantasy literature, and new exercises further illustrate this relationship. *For Further Thought* addresses common fallacies in everyday life. One *When Will I Ever Use This?* feature connects circuit logic to the design and installation of home monitoring systems. Another shows a pediatric nurse applying a logical flowchart and truth tables to a child's vaccination protocol.

- **Chapter 4 (Numeration Systems)** covers historical numeration systems, including Egyptian, Roman, Chinese, Babylonian, Mayan, Greek, and Hindu-Arabic systems. A connection between base conversions in positional numeration systems and Web design is suggested in the new chapter opener and illustrated in the *When Will I Ever Use This?* feature and a new example.

- **Chapter 5 (Number Theory)** presents an introduction to the prime and composite numbers, the Fibonacci sequence, and a cross section of related historical developments. The largest currently known prime numbers of various categories are identified, and recent progress on Goldbach's conjecture and the twin prime conjecture are noted. The chapter opener and one *When Will I Ever Use This?* feature apply cryptography and modular arithmetic to criminal justice in the context of cyber security. Another such feature shows how a health practicioner might use the concept of least common denominator in determining proper drug dosage.

- **Chapter 6 (The Real Numbers and Their Representations)** introduces some of the basic concepts of real numbers, their various forms of representation, and operations of arithmetic with them. The chapter opener and *When Will I Ever Use This?* feature discuss applying percents to tipping and retail pricing markup.

- **Chapter 7 (The Basic Concepts of Algebra)** can be used to present the basics of algebra (linear and quadratic equations, applications, exponents, polynomials, and factoring) to students for the first time, or as a review of previous courses. One *When Will I Ever Use This?* feature applies proportions to an automobile owner's determination of fuel mileage, and another relates inequalities to computation of the score needed to maintain a certain grade point average.

- **Chapter 8 (Graphs, Functions, and Systems of Equations and Inequalities)** is the second of our two algebra chapters. It continues with graphs, equations, and applications of linear, quadratic, exponential, and logarithmic functions and models, along with systems of equations. The chapter opener shows how an automobile owner can use a linear graph to relate price per gallon, amount purchased, and total cost. One *When Will I Ever Use This?* feature discusses a "Eureka moment" pertaining to pumping gasoline, and another connects logarithms with the interpretation of earthquake reporting in the news.

- **Chapter 9 (Geometry)** covers elementary plane geometry, transformational geometry, basic geometric constructions, non-Euclidean geometry (including projective geometry), and chaos and fractals. The chapter opener and one *When Will I Ever Use This?* feature connect geometric volume formulas to a video game developer's job of designing the visual field of a game screen. Another *When Will I Ever Use This?* feature relates right triangle geometry to a forester's determination of safe tree-felling parameters, and a third connects perimeter and circumference formulas to a pest control specialist's task of determining the quantity of pesticide needed to prevent termite infestation.

- **Chapter 10 (Counting Methods)** focuses on elementary counting techniques, in preparation for the probability chapter. The chapter opener relates how a restaurateur used counting methods to help design the sales counter signage in a new restaurant. One *When Will I Ever Use This?* feature describes an entrepreneur's use of probability and sports statistics in designing a game and in building a successful company based on it. Another explains the categories of hands in 5-card poker. The binomial theorem is now discussed in more detail.

- **Chapter 11 (Probability)** covers the basics of probability, odds, and expected value. The chapter opener relates to the professions of weather forecaster, actuary, baseball manager, and corporate manager, applying probability, statistics, and expected value to interpreting forecasts, determining insurance rates, selecting optimum strategies, and making business decisions. One *When Will I Ever Use This?* feature shows how a tree diagram helps a decision maker provide equal chances of winning to three players in a game of chance. A second such feature shows how knowledge of probability can help some TV game show contestants to choose a strategy that will boost their odds of winning! Many additional exercises have been added to Section 11.3.

- **Chapter 12 (Statistics)** is an introduction to statistics that focuses on the measures of central tendency, dispersion, and position and discusses the normal distribution and its applications. The chapter opener and two *When Will I Ever Use This?* features connect probability and graph construction and interpretation to how a psychological therapist may motivate and carry out treatment for alcohol and tobacco addiction. A third such feature applies normal curves to analyze worker skills in an electronics assembly plant.

- **Chapter 13 (Personal Financial Management)** provides the student with the basics of the mathematics of finance as applied to inflation, consumer debt, and home mortgages. We also include a section on investing, with emphasis on stocks, bonds, and mutual funds. Tables, examples, and exercises have been updated to reflect current interest rates and investment returns. Margin notes feature smart apps for financial calculations. In response to reviewer requests, added emphasis has been placed on the use of financial calculators in examples and exercises. This chapter includes two *When Will I Ever Use This?* features. The first explores the cost-effectiveness of solar energy, applying chapter topics essential for a sales representative in the solar energy industry. The second shows how a financial planner might apply several topics addressed in the chapter to compare the cost of renting a house to the cost of buying one. The chapter opener connects the time value of money to how a financial planner can help clients make wise financial decisions.

- **Chapter 14 (Graph Theory)** covers the basic concepts of graph theory and its applications. The chapter opener shows how a writer can apply graph theory to the analysis of poetic rhyme. One *When Will I Ever Use This?* feature connects graph theory to how a postal or delivery service manager could determine the most efficient delivery routes. Another tells of a unique use by an entrepreneur who developed a business based on finding time-efficient ways to navigate theme parks.

- **Chapter 15 (Voting and Apportionment)** deals with issues in voting methods and apportionment of representation, topics that have become increasingly popular in liberal arts mathematics courses. To illustrate the important work of a political consultant, the chapter opener connects different methods of analyzing votes. One *When Will I Ever Use This?* feature relates voting methods to the functioning of governing boards. Another gives an example of how understanding apportionment methods can help in the work of a school administrator. A new margin note and a new *For Further Thought* feature address the increasingly contentious issues related to the electoral college and gerrymandering, respectively.

Course Outline Considerations

Chapters in the text are, in most cases, independent and may be covered in the order chosen by the instructor. The few exceptions are as follows:

- Chapter 6 contains some material dependent on the ideas found in Chapter 5.

- Chapter 6 should be covered before Chapter 7 if student background so dictates.

- Chapters 7 and 8 form an algebraic "package" and should be covered in sequential order.

- A thorough coverage of Chapter 11 depends on knowledge of Chapter 10 material, although probability can be covered without teaching extensive counting methods by avoiding the more difficult exercises.

Features of the Fourteenth Edition

Chapter Openers   In keeping with the career theme, chapter openers address a situation related to a particular career. Some are new to this edition, and some include a problem that the reader is asked to solve. We hope that you find these chapter openers useful and practical.

ENHANCED! Varied Exercise Sets   We continue to present a variety of exercises that integrate drill, conceptual, and applied problems, and there are over 1000 new or modified exercises in this edition. The text contains a wealth of exercises to provide students with opportunities to practice, apply, connect, and extend the mathematical skills they are learning. We have updated the exercises that focus on real-life data and have retained their titles for easy identification. Several chapters are enriched with new applications, particularly Chapters 6, 7, 8, 11, 12, and 13. We continue to use graphs, tables, and charts where appropriate. Many of the graphs are presented in a style similar to that seen by students in today's print and electronic media.

UPDATED! Emphasis on Real Data in the Form of Graphs, Charts, and Tables   We continue to use up-to-date information from magazines, newspapers, and the Internet to create real applications that are relevant and meaningful.

Problem-Solving Strategies   Special paragraphs labeled "Problem-Solving Strategy" relate the discussion of problem-solving strategies to techniques that have been presented earlier.

For Further Thought   These entries, following the exercise sets of many sections, encourage students to share their reasoning processes among themselves to gain a deeper understanding of key mathematical concepts.

ENHANCED! When Will I Ever Use This?   These features in each chapter connect chapter topics to career or workplace situations (and answer that age-old question)!

ENHANCED! Margin Notes   This popular feature is a hallmark of this text and has been retained and updated where appropriate. These notes are interspersed throughout the text and are drawn from various sources, such as the lives of mathematicians, historical vignettes, anecdotes on mathematics textbooks of the past, newspaper and magazine articles, and current research in mathematics.

Optional Graphing Technology   We continue to provide graphing calculator screens to show how technology can be used to support results found analytically. It is not essential, however, that a student have a graphing calculator to study from this text. The technology component is optional.

ENHANCED! Chapter Summaries   Extensive summaries at the end of each chapter include Key Terms, New Symbols with definitions, Test Your Word Power vocabulary checks, and a Quick Review that provides a brief summary of concepts (with examples) covered in the chapter.

Chapter Tests   Each chapter concludes with a chapter test so that students can check their mastery of the material.

Online Resources

- **NEW! StatCrunch** applets and activities are available as indicated by the StatCrunch icon in the margin. Data sets will also be available for ease of use in the Stat-Crunch program.

- ***Trigonometry, Metrics, and Magic Squares** content that was previously in the text is now found in the MyLab Math course, including the assignable MyLab Math questions.

- ***Extensions** cover topics such as Infinite Sets and their Cardinalities, Logic Problems, Sudoku, Z, and Modern Cryptography. Extensions include instruction, exercises, and homework problems in the MyLab Math course.

- **NEW! Group chapter review modules are available in Learning Catalytics** The team-based assessment format facilitates both individual review outside of class and group interaction during class. As students enter responses using smartphones or other connected devices, instructors receive immediate feedback, allowing them to focus on concepts that need attention.

- **ENHANCED! Integrated Review** content in the MyLab Math course uses newly developed worksheets including key terms, summaries, notes, guided problems, and practice problems. This MyLab™ includes a full suite of supporting Integrated Review resources for the Mathematical Ideas course, including pre-made, assignable (and editable) quizzes to assess the prerequisite skills needed for each chapter, and personalized remediation for any gaps in skills that are identified. Each student, therefore, receives just the help that he or she needs—no more, no less.

- **NEW! Corequisite MyLab Math** takes the traditional integrated review content a step further to provide one complete solution for instructors considering a corequiste model. By mapping relevant algebra objectives from the *Mathematical Ideas* text to particular chapters and sections, students have access to the foundational support they need to successfully complete the course. The Corequisite Support Course includes over 30 review topics and enables students to access material for both their credit and support course with a single access code.

- **NEW!** The **Animations Library** will help students explore concepts more deeply, encouraging them to visualize and interact with concepts such as Venn diagrams, probability, logic, debt, graph theory, geometry, statistics, and more. An Animation icon in the text indicates where the animations can aid in understanding.

---

*Online material can be found in the MyLabMath course or at www.pearsonhighered.com/mathstatsresources

- Culinary math review, basic math in baking, methods to calculate drug doses, pharmacology mathematics review, carpentry, and nursing modules are all available in the MyLab Math course to further support the connection to careers and the workplace.

- **NEW! Mindset** materials are included in the **Video & Resource Library: Skills for Success** section of the MyLab Math course, along with Math-Reading Connections, College Success Modules, and Professionalism Tools.

- **Video Program**

  - **NEW! Section lecture videos** will cover the section objectives with a modern, clear, and engaging approach. The videos will incorporate the use of animations, applets, and StatCrunch to demonstrate concepts, making use of resources available to students in the MyLab Math course.

  - **Interactive Concept Check videos with assignable MyLab Math questions** These videos walk students through a concept and then ask them to answer a question within the video. If students answer correctly, the concept is summarized. If a student selects one of the two incorrect answers, the video continues focusing on why students probably selected that answer and works to correct that line of thinking and explain the concept. Then students get another chance to answer a question to prove mastery.

  - **When Will I Ever Use This? videos** bring the ideas in the *When Will I Ever Use This?* feature to life in a fun, memorable way.

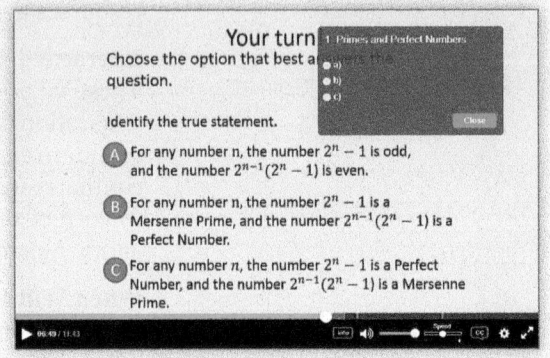
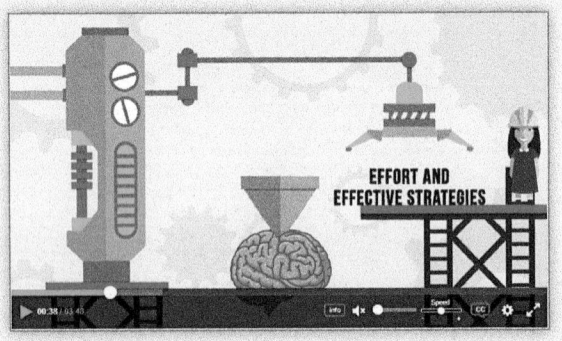

Resources for Success

Instructor Resources

Annotated Instructor's Edition

When possible, answers are on the page with the exercises. Longer answers are in the back of the book.

Online Supplements

The following instructor material is available for download from Pearson's Instructor Resource Center (www.pearsonhighered.com/irc) or within the text's MyLab Math course.

Instructor's Solutions Manual

This Manual includes fully worked solutions to all text exercises.

PowerPoint® Lecture Slides

These fully editable lecture slides present key concepts and definitions from the text. Fully accessible PowerPoint lecture slides are also available. The **Image Resource Library**, found in the Instructor Resources section of the MyLab Math course, includes art from the text that instructors can use when editing the existing lecture slides or for developing their own slides or worksheets.

Instructor's Testing Manual

This manual includes tests with answer keys for each chapter of the text.

TestGen®

Enables instructors to build, edit, print, and administer tests using a computerized bank of questions developed to cover all the objectives of the text.

Student Resources

New! Workbook with Investigations and Integrated Review Worksheets

ISBN: 0-13-499726-3 / 978-0-13-499726-1

Worksheets for every objective in the text include key terms, summaries, a notes section, guided problems and practice problems. Also included are Investigations students can do in or out of class, on their own or in a group. Worksheets to support the integrated review content in the MyLab Math course are also included.

Student Solutions Manual

ISBN: 0-13-499736-0 / 978-0-13-499736-0

This manual provides detailed worked-out solutions to odd-numbered exercises.

ACKNOWLEDGMENTS

We wish to thank the following reviewers for their helpful comments and suggestions for this and previous editions of the text. (Reviewers of the fourteenth edition are noted with an asterisk.)

H. Achepohl, *College of DuPage*

Shahrokh Ahmadi, *Northern Virginia Community College*

*Mark Alexander, *Kapiolani Community College*

Richard Andrews, *Florida A&M University*

Cindy Anfinson, *Palomar College*

Erika Asano, *University of South Florida, St. Petersburg*

Elaine Barber, *Germanna Community College*

Anna Baumgartner, *Carthage College*

James E. Beamer, *Northeastern State University*

Brad Beauchamp, *Vernon College*

Elliot Benjamin, *Unity College*

Jaime Bestard, *Barry University*

Joyce Blair, *Belmont University*

*Andrea Blanchard, *Waldorf University*

Gus Brar, *Delaware County Community College*

Roger L. Brown, *Davenport College*

Douglas Burke, *Malcolm X College*

John Busovicki, *Indiana University of Pennsylvania*

Ann Cascarelle, *St. Petersburg Junior College*

Kenneth Chapman, *St. Petersburg Junior College*

Gordon M. Clarke, *University of the Incarnate Word*

M. Marsha Cupitt, *Durham Technical Community College*

James Curry, *American River College*

Rosemary Danaher, *Sacred Heart University*

*Dr. Erin Davis, *University of Evansville*

Ken Davis, *Mesa State College*

Nancy Davis, *Brunswick Community College*

George DeRise, *Thomas Nelson Community College*

Catherine Dermott, *Hudson Valley Community College*

*Robert Diaz, *Fullerton College*

Greg Dietrich, *Florida Community College at Jacksonville*

Vincent Dimiceli, *Oral Roberts University*

Qiang Dotzel, *University of Missouri, St. Louis*

Diana C. Dwan, *Yavapai College*

Laura Dyer, *Belleville Area College*

Jan Eardley, *Barat College*

Joe Eitel, *Folsom College*

Azin Enshai, *American River College*

Gayle Farmer, *Northeastern State University*

Michael Farndale, *Waldorf College*

Gordon Feathers, *Passaic County Community College*

*Angela Fishman, *Penn State University*

Thomas Flohr, *New River Community College*

*Janet Frewing, *Norco College*

Bill Fulton, *Black Hawk College—East*

Anne Gardner, *Wenatchee Valley College*

Justin M. Gash, *Franklin College*

*Kim Ghiselin, *State College of Florida, Manatee-Sarasota*

Donald Goral, *Northern Virginia Community College*

Glen Granzow, *Idaho State University*

Larry Green, *Lake Tahoe Community College*

Arthur D. Grissinger, *Lock Haven University*

Don Hancock, *Pepperdine University*

Denis Hanson, *University of Regina*

Marilyn Hasty, *Southern Illinois University*

Shelby L. Hawthorne, *Thomas Nelson Community College*

Jeff Heiking, *St. Petersburg Junior College*

Laura Hillerbrand, *Broward Community College*

*Jennifer Huri, *Spokane Falls Community College*

Corinne Irwin, *University of Texas at Austin*

Neha Jain, *Northern Virginia Community College*

Jacqueline Jensen, *Sam Houston State University*

Emanuel Jinich, *Endicott College*

Frank Juric, *Brevard Community College–Palm Bay*

Karla Karstens, *University of Vermont*

Najam Khaja, *Centennial College*

Hilary Kight, *Wesleyan College*

*Mary Margaret Kittle, *Coastal Carolina University*

Barbara J. Kniepkamp, *Southern Illinois University at Edwardsville*

*James Kornmeyer, *Seminole State College*

Suda Kunyosying, *Shepherd College*

Yu-Ju Kuo, *Indiana University of Pennsylvania*

Stephane Lafortune, *College of Charleston*

Pam Lamb, *J. Sargeant Reynolds Community College*

John Lattanzio, *Indiana University of Pennsylvania*

John W. Legge, *Pikeville College*

Dawn Locklear, *Crown College*

Bin Lu, *California State University, Sacramento*

Leo Lusk, *Gulf Coast Community College*

Sherrie Lutsch, *Northwest Indian College*

Rhonda Macleod, *Florida State University*

*James Magee, *Diablo Valley College*

Andrew Markoe, *Rider University*

Darlene Marnich, *Point Park College*

Victoria Martinez, *Okaloosa Walton Community College*

Chris Mason, *Community College of Vermont*

Mark Maxwell, *Maryville University*

Carol McCarron, *Harrisburg Area Community College*

Delois McCormick, *Germanna Community College*

Daisy McCoy, *Lyndon State College*

Cynthia McGinnis, *Okaloosa Walton Community College*

Vena McGrath, *Davenport College*

Robert Moyer, *Fort Valley State University*

Shai Neumann, *Brevard Community College*

Barbara Nienstedt, *Gloucester County College*

Chaitanya Nigam, *Gateway Community-Technical College*

Vladimir Nikiforov, *University of Memphis*

Vicky Ohlson, *Trenholm State Technical College*

Jean Okumura, *Windward Community College*

Stan Perrine, *Charleston Southern University*

Mary Anne Petruska, *Pensacola State College*

Bob Phillips, *Mesabi Range Community College*

Kathy Pinchback, *University of Memphis*

Fatima Prioleau, *Borough of Manhattan Community College*

Priscilla Putman, *New Jersey City University*

Scott C. Radtke, *Davenport College*

Doraiswamy Ramachandran, *California State University, Sacramento*

John Reily, *Montclair State University*

Beth Reynolds, *Mater Dei College*

Shirley I. Robertson, *High Point University*

Andrew M. Rockett, *CW Post Campus of Long Island University*

Kathleen Rodak, *St. Mary's College of Ave Maria University*

Cynthia Roemer, *Union County College*

Lisa Rombes, *Washtenaw Community College*

Abby Roscum, *Marshalltown Community College*

Catherine A. Sausville, *George Mason University*

D. Schraeder, *McLennan Community College*

Wilfred Schulte, *Cosumnes River College*

Melinda Schulteis, *Concordia University*

Gary D. Shaffer, *Allegany College of Maryland*

Doug Shaw, *University of North Iowa*

*Kate Sims-Drew, *Prairie State College*

Jane Sinibaldi, *York College of Pennsylvania*

Nancy Skocik, *California University of Pennsylvania*

Larry Smith, *Peninsula College*

Marguerite Smith, *Merced College*

Charlene D. Snow, *Lower Columbia College*

H. Jeannette Stephens, *Whatcom Community College*

Suzanne J. Stock, *Oakton Community College*

Dawn M. Strickland, *Winthrop University*

Dian Thom, *McKendree College*

Claude C. Thompson, *Hollins University*

Mark Tom, *College of the Sequoias*

Ida Umphers, *University of Arkansas at Little Rock*

Karen Villarreal, *University of New Orleans*

Dr. Karen Walters, *Northern Virginia Community College*

Wayne Wanamaker, *Central Florida Community College*

David Wasilewski, *Luzerne County Community College*

William Watkins, *California State University, Northridge*

*Jonathan Weisbrod, *Rowan College at Burlington County*

Alice Williamson, *Sussex County Community College*

Susan Williford, *Columbia State Community College*

Tom Witten, *Southwest Virginia Community College*

Fred Worth, *Henderson State University*

Rob Wylie, *Carl Albert State College*

Henry Wyzinski, *Indiana University Northwest*

A project of this magnitude cannot be accomplished without the help of many other dedicated individuals. Marnie Greenhut served as portfolio manager for this edition. Carol Merrigan provided excellent production supervision. Anne Kelly and Patty Bergin of Pearson gave us their unwavering support.

Terry McGinnis gave her usual excellent behind-the-scenes guidance. Thanks go to Dr. Margaret L. Morrow of Plattsburgh State University and Dr. Jill Van Newenhizen of Lake Forest College, who wrote the material on graph theory and voting/apportionment, respectively. Paul Lorczak, Hal Whipple, Beverly Fusfield, and Jack Hornsby did an outstanding job of accuracy- and answer-checking. And finally, we thank our loyal users over these many editions for making this book one of the most successful in its market.

Vern E. Heeren
John Hornsby
Christopher Heeren

Vern Heeren grew up in the Sacramento Valley of California. After earning a Bachelor of Arts degree in mathematics, with a minor in physics, at Occidental College, and completing his Master of Arts degree in mathematics at the University of California, Davis, he began a 38-year teaching career at American River College, teaching math and a little physics. He coauthored *Mathematical Ideas* in 1968 with office mate Charles Miller, and he has enjoyed researching and revising it over the years. It has been a joy for him to complete the fourteenth edition, along with long-time coauthor John Hornsby, and with son Christopher.

These days, besides pursuing his mathematical interests, Vern enjoys spending time with his wife Carole and their family, exploring the wonders of nature at and near their home in central Oregon.

John Hornsby joined the author team of Margaret Lial, Charles Miller, and Vern Heeren in 1988. In 1990, the sixth edition of *Mathematical Ideas* became the first of nearly 150 titles he has coauthored for Scott Foresman, HarperCollins, Addison-Wesley, and Pearson in the years that have followed. His books cover the areas of developmental and college algebra, precalculus, trigonometry, and mathematics for the liberal arts. He is a native and resident of New Roads, Louisiana.

Christopher Heeren is a native of Sacramento, California. While studying engineering in college, he had an opportunity to teach a math class at a local high school, and this sparked both a passion for teaching and a change of major. He received a Bachelor of Arts degree and a Master of Arts degree, both in mathematics, from California State University, Sacramento. Chris has taught mathematics at the middle school, high school, and college levels, and he currently teaches at American River College in Sacramento. He has a continuing interest in using technology to bring mathematics to life. When not writing, teaching, or preparing to teach, Chris enjoys spending time with his lovely wife Heather and their three children.

1 The Art of Problem Solving

Professor Terry Krieger, of Rochester (Minnesota) Community College, shares his thoughts about why he decided to become a mathematics teacher. He is an expert at the Rubik's Cube. Here, he explains how he mastered this classic puzzle.

From a very young age I always enjoyed solving problems, especially problems involving numbers and patterns. There is something inherently beautiful in the process of discovering mathematical truth. Mathematics may be the only discipline in which different people, using wildly varied but logically sound methods, will arrive at the same correct result—not just once, but every time! It is this aspect of mathematics that led me to my career as an educator. As a mathematics instructor, I get to be part of, and sometimes guide, the discovery process.

I received a Rubik's Cube as a gift my junior year of high school. I was fascinated by it. I devoted the better part of three months to solving it for the first time, sometimes working 3 or 4 hours per day on it.

There was a lot of trial and error involved. I devised a process that allowed me to move only a small number of pieces at a time while keeping other pieces in their places. Most of my moves affect only three or four of the 26 unique pieces of the puzzle. What sets my solution apart from those found in many books is that I hold the cube in a consistent position and work from the top to the bottom. Most book solutions work upward from the bottom.

I worked on the solution so much that I started seeing cube moves in my sleep. In fact, I figured out the moves for one of my most frustrating sticking points while sleeping. I just woke up knowing how to do it.

The eight corners of the cube represented a particularly difficult challenge for me. Finding a consistent method for placing the corners appropriately took many, many hours. To this day, the amount of time that it takes for me to solve a scrambled cube depends largely on the amount of time that it takes for me to place the corners.

When I first honed my technique, I was able to consistently solve the cube in 2 to 3 minutes. My average time is now about 65 seconds. My fastest time is 42 seconds.

1.1 SOLVING PROBLEMS BY INDUCTIVE REASONING

OBJECTIVES

1 Distinguish between inductive and deductive reasoning.

2 Recognize that inductive reasoning may not always lead to valid conclusions.

Characteristics of Inductive and Deductive Reasoning

The development of mathematics can be traced to the Egyptian and Babylonian cultures (3000 B.C.–A.D. 260) as a necessity for counting and problem solving. To solve a problem, a cookbook-like recipe was given, and it was followed repeatedly to solve similar problems. By observing that a specific method worked for a certain type of problem, the Babylonians and the Egyptians concluded that the same method would work for any similar type of problem. Such a conclusion is called a *conjecture*. A **conjecture** is an educated guess based on repeated observations of a particular process or pattern.

The method of reasoning just described is called *inductive reasoning*.

INDUCTIVE REASONING

Inductive reasoning is characterized by drawing a general conclusion (making a conjecture) from repeated observations of specific examples. The conjecture may or may not be true.

In testing a conjecture obtained by inductive reasoning, it takes only one example that does not work to prove the conjecture false. Such an example is called a **counterexample.**

Inductive reasoning provides a powerful method of drawing conclusions, but there is no assurance that the observed conjecture will always be true. For this reason, mathematicians do not accept a conjecture as an absolute truth until it is formally proved using methods of *deductive reasoning*. Deductive reasoning characterized the development and approach of Greek mathematics, as seen in the works of Euclid, Pythagoras, Archimedes, and others. During the classical Greek period (600 B.C.–A.D. 450), general concepts were applied to specific problems, resulting in a structured, logical development of mathematics.

Explaining Number Tricks Using
Deductive Reasoning

DEDUCTIVE REASONING

Deductive reasoning is characterized by applying general principles to specific examples.

We now look at examples of these two types of reasoning. In this chapter, we often refer to the **natural, or counting, numbers.**

$$1, 2, 3, \ldots \quad \text{Natural (counting) numbers}$$

Ellipsis points

The three dots (*ellipsis points*) indicate that the numbers continue indefinitely in the pattern that has been established. The most probable rule for continuing this pattern is "Add 1 to the previous number," and this is indeed the rule that we follow.

Now consider the following list of natural numbers:

$$2, 9, 16, 23, 30.$$

What is the next number of this list? What is the pattern? After studying the numbers, we might see that

$$2 + 7 = 9, \quad \text{and} \quad 9 + 7 = 16.$$

Do we add 16 and 7 to get 23? Do we add 23 and 7 to get 30? Yes. It seems that any number in the given list can be found by adding 7 to the preceding number, so the next number in the list would be

$$30 + 7 = \mathbf{37}.$$

We set out to find the "next number" by reasoning from observation of the numbers in the list. We may have jumped from these observations to the general statement that any number in the list is 7 more than the preceding number. This is an example of inductive reasoning.

By using inductive reasoning, we concluded that 37 was the next number. Suppose the person making up the list has another answer in mind. The list of numbers

$$2, 9, 16, 23, 30$$

actually gives the dates of Mondays in June if June 1 falls on a Sunday. The next Monday after June 30 is July 7. With this pattern, the list continues as

$$2, 9, 16, 23, 30, 7, 14, 21, 28, \ldots.$$

See the calendar in **Figure 1.** The correct answer would then be 7. The process used to obtain the rule "add 7" in the preceding list reveals a main flaw of inductive reasoning.

We can never be sure that what is true in a specific case will be true in general. Inductive reasoning does not guarantee a true result, but it does provide a means of making a conjecture.

We now review some basic notation. Throughout this book, we use *exponents* to represent repeated multiplication.

$$\text{Base} \rightarrow \mathbf{4}^3 = 4 \cdot 4 \cdot 4 = 64 \quad \text{4 is used as a factor 3 times.}$$

Exponent

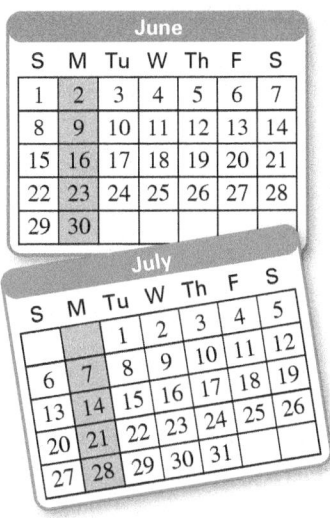

Figure 1

EXPONENTIAL EXPRESSION

If a is a number and n is a counting number $(1, 2, 3, \dots)$, then the exponential expression a^n is defined as follows.

$$a^n = \underbrace{a \cdot a \cdot a \cdot \ldots \cdot a}_{n \text{ factors of } a}$$

The number a is the **base** and n is the **exponent.**

With deductive reasoning, we use general statements and apply them to specific situations. For example, a basic rule for converting feet to inches is to multiply the number of feet by 12 in order to obtain the equivalent number of inches. This can be expressed as a formula.

$$\text{Number of inches} = 12 \times \text{number of feet}$$

This general rule can be applied to any specific case. For example, the number of inches in 3 feet is $12 \times 3 = 36$ inches.

Reasoning through a problem usually requires certain *premises*. A **premise** can be an assumption, law, rule, widely held idea, or observation. Then reason inductively or deductively from the premises to obtain a **conclusion.** The premises and conclusion make up a **logical argument.**

EXAMPLE 1 **Identifying Premises and Conclusions**

Identify each premise and the conclusion in each of the following arguments. Then tell whether each argument is an example of inductive or deductive reasoning.

(a) Our house is made of brick. Both of my next-door neighbors have brick houses. Therefore, all houses in our neighborhood are made of brick.

(b) All keyboards have the symbol @. I have a keyboard. My keyboard has the symbol @.

(c) Today is Tuesday. Tomorrow will be Wednesday.

Solution

(a) The premises are "Our house is made of brick," and "Both of my next-door neighbors have brick houses." The conclusion is "Therefore, all houses in our neighborhood are made of brick." Because the reasoning goes from specific examples to a general statement, the argument is an example of inductive reasoning (although it may very well give a faulty conclusion).

(b) Here the premises are "All keyboards have the symbol @" and "I have a keyboard." The conclusion is "My keyboard has the symbol @." This reasoning goes from general to specific, so deductive reasoning was used.

(c) There is only one premise here, "Today is Tuesday." The conclusion is "Tomorrow will be Wednesday." The fact that Wednesday immediately follows Tuesday is being used, even though this fact is not explicitly stated. Because the conclusion comes from general facts that apply to this special case, deductive reasoning was used.

While inductive reasoning may, at times, lead to false conclusions, in many cases it does provide correct results if we look for the most *probable* answer.

Fibonacci (1170–1250) discovered the sequence named after him in solving a problem on rabbits. Fibonacci ("son of Bonaccio") is one of several names for **Leonardo of Pisa**. His father managed a warehouse in present-day Bougie (or Bejaia), in Algeria. Thus it was that Leonardo Pisano studied with a Moorish teacher and learned the "Indian" numbers that the Moors and other followers of Mohammed brought with them in their westward drive.

Fibonacci wrote books on algebra, geometry, and trigonometry.

EXAMPLE 2   Predicting the Next Number in a List

Use inductive reasoning to determine the *probable* next number in each list below.

(a) 5, 9, 13, 17, 21, 25, 29   **(b)** 1, 1, 2, 3, 5, 8, 13, 21   **(c)** 2, 4, 8, 16, 32

Solution

(a) Each number in the list is obtained by adding 4 to the previous number. The probable next number is $29 + 4 = 33$. (This is an example of an *arithmetic sequence*.)

(b) Beginning with the third number in the list, 2, each number is obtained by adding the two previous numbers in the list. That is,

$$1 + 1 = 2, \qquad 1 + 2 = 3, \qquad 2 + 3 = 5,$$

and so on. The probable next number is $13 + 21 = 34$. (These are the first few terms of the *Fibonacci sequence*.)

(c) It appears here that to obtain each number after the first, we must double the previous number. Therefore, the probable next number is $32 \times 2 = 64$. (This is an example of a *geometric sequence*.)

EXAMPLE 3   Predicting the Product of Two Numbers

Consider the list of equations. Predict the next multiplication fact in the list.

$$37 \times 3 = 111$$
$$37 \times 6 = 222$$
$$37 \times 9 = 333$$
$$37 \times 12 = 444$$

Solution

The left side of each equation has two factors, the first 37 and the second a multiple of 3, beginning with 3. Each product (answer) consists of three digits, all the same, beginning with 111 for 37×3. Thus, the next multiplication fact would be

$$37 \times 15 = 555, \quad \text{which is indeed true.}$$

Pitfalls of Inductive Reasoning

There are pitfalls associated with inductive reasoning. A classic example involves the maximum number of regions formed when chords are constructed in a circle. When two points on a circle are joined with a line segment, a *chord* is formed.

Locate a single point on a circle. Because no chords are formed, a single interior region is formed. See **Figure 2(a)** on the next page. Locate two points and draw a chord. Two interior regions are formed, as shown in **Figure 2(b)**. Continue this pattern. Locate three points, and draw all possible chords. Four interior regions are formed, as shown in **Figure 2(c)**. Four points yield 8 regions and five points yield 16 regions. See **Figures 2(d) and 2(e)**.

The results of the preceding observations are summarized in **Table 1**. The pattern formed in the column headed "Number of Regions" is the same one we saw in **Example 2(c)**, where we predicted that the next number would be 64. It seems here that for each additional point on the circle, the number of regions doubles.

Table 1

Number of Points	Number of Regions
1	1
2	2
3	4
4	8
5	16

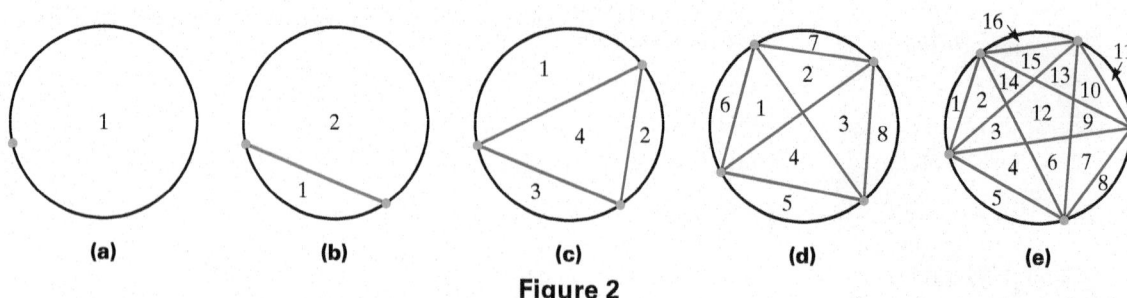

Figure 2

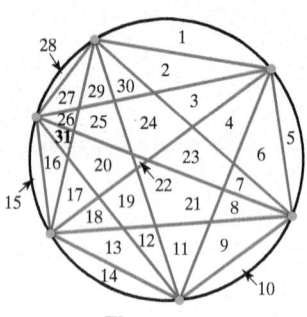

Figure 3

A reasonable inductive conjecture would be that for six points, 32 regions would be formed. But as **Figure 3** indicates, there are *only 31 regions*. The pattern of doubling ends when the sixth point is considered. Adding a seventh point would yield 57 regions. The numbers obtained here are

$$1, 2, 4, 8, 16, 31, 57.$$

For *n* points on the circle, the number of regions is given by the expression

$$\frac{n^4 - 6n^3 + 23n^2 - 18n + 24}{24}.$$

1.1 EXERCISES

CONCEPT CHECK *Fill in the blank with the correct response.*

1. The first five natural numbers are _____.

2. "If today is Friday, then tomorrow is _____," is an example of valid deductive reasoning.

3. The most *probable* next number in the list 1, 4, 7, 10, 13 is ____.

4. "Donna's first three children were named Anna, Bernard, and Cary. Therefore, her next child's name will start with the letter ____," is an example of inductive reasoning.

5. If the natural numbers between 1 and 8 inclusive are multiplied by 9, the first digit of the product increases by 1 each time, and the second digit decreases by 1 each time, as shown in the following list. Inductive reasoning suggests that the product 9×9 is equal to ____.

$$1 \times 9 = 09$$
$$2 \times 9 = 18$$
$$3 \times 9 = 27$$
$$4 \times 9 = 36$$
$$5 \times 9 = 45$$
$$6 \times 9 = 54$$
$$7 \times 9 = 63$$
$$8 \times 9 = 72$$

6. See **Example 2(b).** The number following 34 in the Fibonacci sequence is ____.

Determine whether the reasoning is an example of deductive or inductive reasoning.

7. The next number in the pattern 2, 4, 6, 8, 10 is 12.

8. My dog barked and woke me up at 1:02 a.m., 2:03 a.m., and 3:04 a.m. So he will bark again and wake me up at 4:05 a.m.

9. To find the perimeter *P* of a square with side of length *s*, I can use the formula $P = 4s$. So the perimeter of a square with side of length 7 inches is $4 \times 7 = 28$ inches.

10. A company charges a 10% re-stocking fee for returning an item. So when I return a radio that cost $150, I will get only $135 back.

11. If a mechanic says that it will take seven days to repair your SUV, then it will actually take ten days. The mechanic says, "I figure it'll take exactly one week to fix it, ma'am." Then you can expect it to be ready ten days from now.

12. If you take your medicine, you'll feel a lot better. You take your medicine. Therefore, you'll feel a lot better.

13. It has rained every day for the past seven days, and it is raining today as well. So it will also rain tomorrow.

14. Rhonda's first five children were boys. If she has another baby, it will be a boy.

15. The 2000 movie *Cast Away* stars Tom Hanks as the only human survivor of a plane crash, stranded on a tropical island. He estimates that his distance from where the plane lost radio contact is about 400 miles (a radius), and then he uses the formula for the area of a circle,

$$\text{Area} = \pi \, (\text{radius})^2,$$

to determine that a search party would have to cover an area of over 500,000 square miles to look for him and his "pal" Wilson.

16. If the same number is subtracted from both sides of a true equation, the new equation is also true. I know that $11 + 18 = 29$. Therefore, $(11 + 18) - 13 = 29 - 13$.

17. If you build it, they will come. You build it. Therefore, they will come.

18. All men are mortal. Socrates is a man. Therefore, Socrates is mortal.

19. It is a fact that every student who ever attended Southern University was accepted into graduate school. Because I am attending Southern, I can expect to be accepted into graduate school, too.

20. For the past 57 years, a rare plant has bloomed in Colombia each summer, alternating between yellow and green flowers. Last summer, it bloomed with green flowers, so this summer it will bloom with yellow flowers.

21. In the sequence 5, 10, 15, 20, 25, . . . , the most probable next number is 30.

22. (This anecdote is adapted from a story by Howard Eves in *In Mathematical Circles*.) A scientist had a group of 100 fleas, and one by one he would tell each flea "Jump," and the flea would jump. Then with the same fleas, he yanked off their hind legs and repeated "Jump," but the fleas would not jump. He concluded that when a flea has its hind legs yanked off, it cannot hear.

23. Discuss the differences between inductive and deductive reasoning. Give an example of each.

24. Give an example of inductive reasoning with a faulty conclusion.

Determine the most probable next term in each of the following lists of numbers.

25. 6, 9, 12, 15, 18

26. 13, 18, 23, 28, 33

27. 3, 12, 48, 192, 768

28. 32, 16, 8, 4, 2

29. 3, 6, 9, 15, 24, 39

30. $\dfrac{1}{3}, \dfrac{3}{5}, \dfrac{5}{7}, \dfrac{7}{9}, \dfrac{9}{11}$

31. $\dfrac{1}{2}, \dfrac{3}{4}, \dfrac{5}{6}, \dfrac{7}{8}, \dfrac{9}{10}$

32. 1, 4, 9, 16, 25

33. 1, 8, 27, 64, 125

34. 2, 6, 12, 20, 30, 42

35. 4, 7, 12, 19, 28, 39

36. 27, 21, 16, 12, 9

37. Construct a list of numbers similar to that in **Exercise 25** such that the most probable next number in the list is 60.

38. Construct a list of numbers similar to that in **Exercise 36** such that the most probable next number in the list is 8.

Use the list of equations and inductive reasoning to predict the next equation, and then verify your conjecture.

39.
$(9 \times 9) + 7 = 88$
$(98 \times 9) + 6 = 888$
$(987 \times 9) + 5 = 8888$
$(9876 \times 9) + 4 = 88,888$

40.
$(1 \times 9) + 2 = 11$
$(12 \times 9) + 3 = 111$
$(123 \times 9) + 4 = 1111$
$(1234 \times 9) + 5 = 11,111$

41.
$3367 \times 3 = 10,101$
$3367 \times 6 = 20,202$
$3367 \times 9 = 30,303$
$3367 \times 12 = 40,404$

42.
$15,873 \times 7 = 111,111$
$15,873 \times 14 = 222,222$
$15,873 \times 21 = 333,333$
$15,873 \times 28 = 444,444$

43.
$11 \times 11 = 121$
$111 \times 111 = 12,321$
$1111 \times 1111 = 1,234,321$

44.
$34 \times 34 = 1156$
$334 \times 334 = 111,556$
$3334 \times 3334 = 11,115,556$

45.
$$3 = \frac{3(2)}{2}$$
$$3 + 6 = \frac{6(3)}{2}$$
$$3 + 6 + 9 = \frac{9(4)}{2}$$
$$3 + 6 + 9 + 12 = \frac{12(5)}{2}$$

46.
$$2 = 4 - 2$$
$$2 + 4 = 8 - 2$$
$$2 + 4 + 8 = 16 - 2$$
$$2 + 4 + 8 + 16 = 32 - 2$$

47.
$$5(6) = 6(6 - 1)$$
$$5(6) + 5(36) = 6(36 - 1)$$
$$5(6) + 5(36) + 5(216) = 6(216 - 1)$$
$$5(6) + 5(36) + 5(216) + 5(1296) = 6(1296 - 1)$$

48.
$$3 = \frac{3(3 - 1)}{2}$$
$$3 + 9 = \frac{3(9 - 1)}{2}$$
$$3 + 9 + 27 = \frac{3(27 - 1)}{2}$$
$$3 + 9 + 27 + 81 = \frac{3(81 - 1)}{2}$$

49.
$$\frac{1}{2} = 1 - \frac{1}{2}$$
$$\frac{1}{2} + \frac{1}{4} = 1 - \frac{1}{4}$$
$$\frac{1}{2} + \frac{1}{4} + \frac{1}{8} = 1 - \frac{1}{8}$$
$$\frac{1}{2} + \frac{1}{4} + \frac{1}{8} + \frac{1}{16} = 1 - \frac{1}{16}$$

50.
$$\frac{1}{1 \cdot 2} = \frac{1}{2}$$
$$\frac{1}{1 \cdot 2} + \frac{1}{2 \cdot 3} = \frac{2}{3}$$
$$\frac{1}{1 \cdot 2} + \frac{1}{2 \cdot 3} + \frac{1}{3 \cdot 4} = \frac{3}{4}$$
$$\frac{1}{1 \cdot 2} + \frac{1}{2 \cdot 3} + \frac{1}{3 \cdot 4} + \frac{1}{4 \cdot 5} = \frac{4}{5}$$

Legend has it that the great mathematician Carl Friedrich Gauss (1777–1855) at a very young age was told by his teacher to find the sum of the first 100 counting numbers.

While his classmates toiled at the problem, Carl simply wrote down a single number and handed the correct answer in to his teacher. The young Carl explained that he observed that there were 50 pairs of numbers that each added up to 101. (See below.) So the sum of all the numbers must be $50 \times 101 = 5050$.

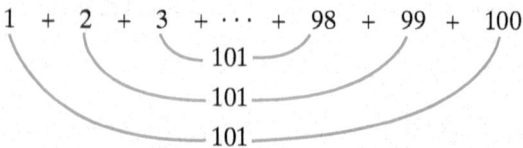

50 sums of 101 = 50 × 101 = 5050

Use the method of Gauss to find each sum.

51. $1 + 2 + 3 + \cdots + 200$ **52.** $1 + 2 + 3 + \cdots + 400$

53. $1 + 2 + 3 + \cdots + 800$ **54.** $1 + 2 + 3 + \cdots + 2000$

55. Modify the procedure of Gauss to find the sum $1 + 2 + 3 + \cdots + 175$.

56. Explain in your own words how the procedure of Gauss can be modified to find the sum $1 + 2 + 3 + \cdots + n$, where n is an odd natural number. (When an odd natural number is divided by 2, it leaves a remainder of 1.)

57. Modify the procedure of Gauss to find the sum $2 + 4 + 6 + \cdots + 100$.

58. Use the result of **Exercise 57** to find the sum $4 + 8 + 12 + \cdots + 200$.

Solve each problem.

59. What is the most probable next number in this list?

$$12, 1, 1, 1, 2, 1, 3$$

(*Hint:* Think about a clock with chimes.)

60. What is the next term in this list?

$$O, T, T, F, F, S, S, E, N, T$$

(*Hint:* Think about words and their relationship to numbers.)

61. Choose any three-digit number with all different digits, and follow these steps.

(a) Reverse the digits, and subtract the smaller from the larger. Record your result.

(b) Choose another three-digit number and repeat this process. Do this as many times as it takes for you to see a pattern in the different results you obtain. (*Hint:* What is the middle digit? What is the sum of the first and third digits?)

(c) Write an explanation of this pattern.

62. Choose any number, and follow these steps.

    (a) Multiply by 2.   **(b)** Add 6.

    (c) Divide by 2.   **(d)** Subtract the number you started with.

    (e) Record your result.

    Repeat the process, except in Step (b), add 8. Record your final result. Repeat the process once more, except in Step (b), add 10. Record your final result.

    (f) Observe what you have done. Then use inductive reasoning to explain how to predict the final result.

63. Complete the following.

$142{,}857 \times 1 = \underline{\hspace{1cm}}$ $142{,}857 \times 2 = \underline{\hspace{1cm}}$

$142{,}857 \times 3 = \underline{\hspace{1cm}}$ $142{,}857 \times 4 = \underline{\hspace{1cm}}$

$142{,}857 \times 5 = \underline{\hspace{1cm}}$ $142{,}857 \times 6 = \underline{\hspace{1cm}}$

What pattern exists in the successive answers? Now multiply 142,857 by 7 to obtain an interesting result.

64. Refer to **Figures 2(b)–(e)** and **Figure 3.** Instead of counting interior regions of the circle, count the chords formed. Use inductive reasoning to predict the number of chords that would be formed if seven points were used.

1.2   AN APPLICATION OF INDUCTIVE REASONING: NUMBER PATTERNS

OBJECTIVES

1 Recognize arithmetic and geometric sequences.

2 Apply the method of successive differences to predict the next term in a sequence.

3 Recognize number patterns.

4 Use sum formulas.

5 Recognize triangular, square, and pentagonal numbers.

Number Sequences

An ordered list of numbers such as

$$3, 9, 15, 21, 27, \ldots$$

is called a *sequence*. A **number sequence** is a list of numbers having a first number, a second number, a third number, and so on. These are the **terms of the sequence.**

    The sequence that begins

$$5, 9, 13, 17, 21, \ldots$$

is an *arithmetic sequence,* or *arithmetic progression.* In an **arithmetic sequence,** each term after the first is obtained by adding the same number, the **common difference,** to the preceding term. To find the common difference, choose any term after the first and subtract from it the preceding term. If we choose

$$9 - 5 \quad \text{(the second term minus the first term),}$$

for example, we see that the common difference is 4. To find the term following 21, we add 4 to get

$$21 + 4 = 25.$$

    The sequence that begins

$$2, 4, 8, 16, 32, \ldots$$

is a *geometric sequence,* or *geometric progression.* In a **geometric sequence,** each term after the first is obtained by multiplying the preceding term by the same number, the **common ratio.** To find the common ratio, choose any term after the first and divide it by the preceding term. If we choose

$$\frac{4}{2} \quad \text{(the second term divided by the first term),}$$

for example, we see that the common ratio is 2. To find the term following 32, we multiply by 2 to get

$$32 \cdot 2 = 64.$$

StatCrunch Identifying Arithmetic and Geometric Sequences

| **EXAMPLE 1** | **Identifying Arithmetic and Geometric Sequences** |

For each sequence, determine whether it is an *arithmetic sequence,* a *geometric sequence,* or *neither.* If it is either arithmetic or geometric, give the next term.

(a) $5, 10, 15, 20, 25, \ldots$ **(b)** $3, 12, 48, 192, 768, \ldots$ **(c)** $1, 4, 9, 16, 25, \ldots$

Solution

(a) If we choose *any* term after the first term, and subtract the preceding term, we find that the common difference is 5.

$$10 - 5 = 5 \qquad 15 - 10 = 5 \qquad 20 - 15 = 5 \qquad 25 - 20 = 5$$

Therefore, this is an arithmetic sequence. The next term is $25 + 5 = 30$.

(b) If any term after the first is divided by the previous term, we find the common ratio 4.

$$\frac{12}{3} = 4 \qquad \frac{48}{12} = 4 \qquad \frac{192}{48} = 4 \qquad \frac{768}{192} = 4$$

This is a geometric sequence. The next term is $768 \cdot 4 = 3072$.

(c) Although there is a pattern here (the terms are the squares of the first five counting numbers), there is neither a common difference nor a common ratio. This is neither an arithmetic nor a geometric sequence.

Successive Differences

Some sequences present more difficulty than our earlier examples when we are making a conjecture about the next term. Often the **method of successive differences** may be applied in such cases. Consider the sequence

$$2, 6, 22, 56, 114, \ldots.$$

Because the next term is not obvious, subtract the first term from the second term, the second from the third, the third from the fourth, and so on.

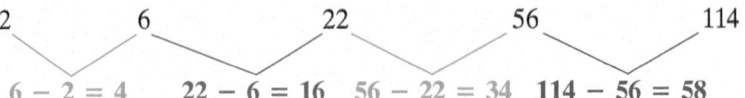

Now repeat the process with the sequence 4, 16, 34, 58, and continue repeating until the difference is a constant value, as shown in line (4).

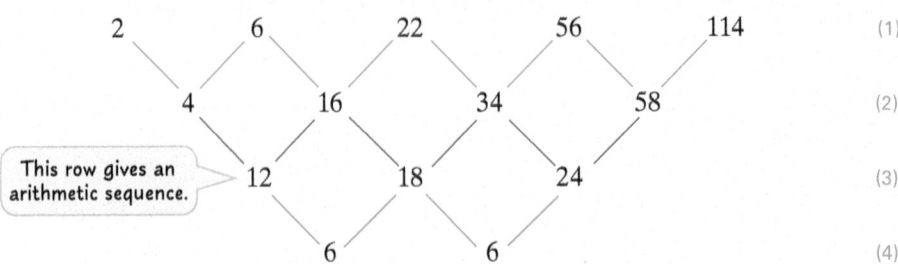

Once a line of constant values is obtained, simply work "backward" by adding until the desired term of the given sequence is obtained. Thus, for this pattern to continue, another 6 should appear in line (4), meaning that the next term in line (3) would have to be $24 + 6 = 30$. The next term in line (2) would be $58 + 30 = 88$. Finally, the next term in the given sequence would be $114 + 88 = 202$.

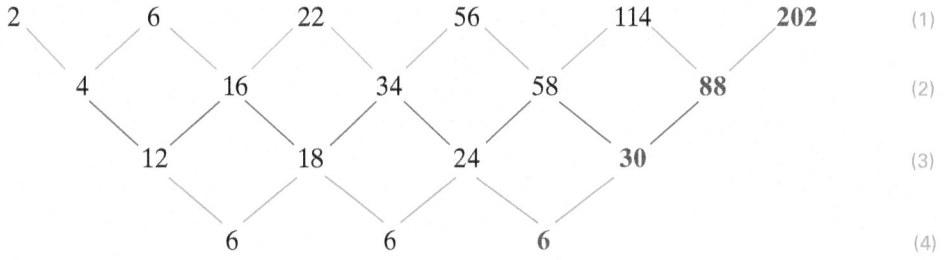

StatCrunch Successive Differences

EXAMPLE 2   **Using Successive Differences**

Use successive differences to determine the next number in each sequence.

(a) 14, 22, 32, 44, . . .   **(b)** 5, 15, 37, 77, 141, . . .

Solution

(a) Subtract a term from the one that follows it, and continue until a pattern is observed.

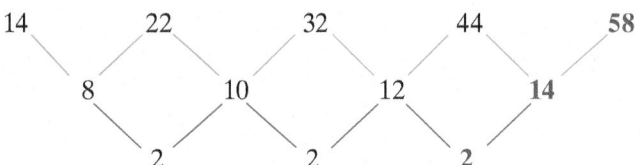

Once the row of 2s was obtained and extended, we were able to obtain

$$12 + 2 = 14, \quad \text{and} \quad 44 + 14 = 58 \quad \text{as shown above.}$$

The next number in the sequence is **58**.

(b) Proceed as before to obtain the following diagram.

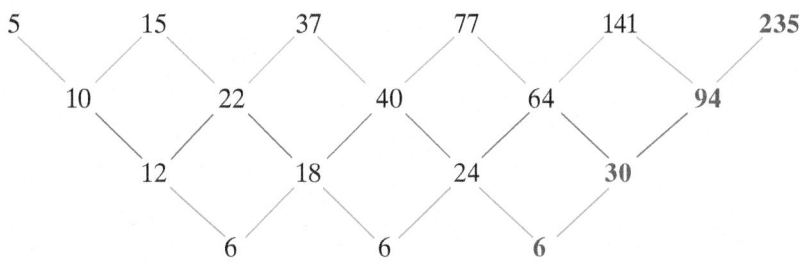

The numbers in the "diagonal" at the far right were obtained by adding:

$$24 + 6 = 30,$$

$$64 + 30 = 94,$$

and $$141 + 94 = \textbf{235}.$$

The next number in the sequence is **235**.

The method of successive differences does not always work. For example, try it on the Fibonacci sequence in **Example 2(b)** of **Section 1.1** and notice what happens. The resulting sequence is also a Fibonacci-type sequence, beginning with 0, 1.

$$0, 1, 1, 2, 3, 5, 8, \ldots$$

Continuing this way never does lead to a sequence of constant values.

Number Patterns and Sum Formulas

Observe the number pattern at the right. In each case, the left side of the equation is the indicated sum of consecutive odd counting numbers beginning with 1, and the right side is the square of the number of terms on the left side. Inductive reasoning would suggest that the next line in this pattern is as follows.

$$1 = 1^2$$
$$1 + 3 = 2^2$$
$$1 + 3 + 5 = 3^2$$
$$1 + 3 + 5 + 7 = 4^2$$
$$1 + 3 + 5 + 7 + 9 = 5^2$$

$$1 + 3 + 5 + 7 + 9 + 11 = 6^2 \quad \text{Each side simplifies to 36.}$$

We cannot conclude that this pattern will continue indefinitely, because observation of a finite number of examples does *not* guarantee that the pattern will continue. However, mathematicians have proved that this pattern does indeed continue indefinitely, using a method of proof called **mathematical induction.** (See any comprehensive college algebra text.)

Any even counting number may be written in the form $2k$, where k is a counting number. It follows that the kth odd counting number is written $2k - 1$. For example, the third odd counting number, 5, can be written $2(3) - 1$.

SUM OF THE FIRST n ODD COUNTING NUMBERS

If n is any counting number, then the following is true.
$$1 + 3 + 5 + \cdots + (2n - 1) = n^2$$

EXAMPLE 3 Predicting the Next Equation in a List

Several equations are given illustrating a suspected number pattern. Determine what the next equation would be, and verify that it is indeed a true statement.

(a)
$$1^2 = 1^3$$
$$(1 + 2)^2 = 1^3 + 2^3$$
$$(1 + 2 + 3)^2 = 1^3 + 2^3 + 3^3$$
$$(1 + 2 + 3 + 4)^2 = 1^3 + 2^3 + 3^3 + 4^3$$

(b)
$$1 = 1^3$$
$$3 + 5 = 2^3$$
$$7 + 9 + 11 = 3^3$$
$$13 + 15 + 17 + 19 = 4^3$$

(c)
$$1 = \frac{1 \cdot 2}{2}$$
$$1 + 2 = \frac{2 \cdot 3}{2}$$
$$1 + 2 + 3 = \frac{3 \cdot 4}{2}$$
$$1 + 2 + 3 + 4 = \frac{4 \cdot 5}{2}$$

(d)
$$12,345,679 \times 9 = 111,111,111$$
$$12,345,679 \times 18 = 222,222,222$$
$$12,345,679 \times 27 = 333,333,333$$
$$12,345,679 \times 36 = 444,444,444$$
$$12,345,679 \times 45 = 555,555,555$$

Notice that there is no 8 here.

Solution

(a) The left side of each equation is the square of the sum of the first n counting numbers. The right side is the sum of their cubes. The next equation would be
$$(1 + 2 + 3 + 4 + 5)^2 = 1^3 + 2^3 + 3^3 + 4^3 + 5^3.$$

Each side simplifies to 225, so the pattern is true for this equation.

(b) The left sides of the equations contain the sum of odd counting numbers, starting with the first (1) in the first equation, the second and third (3 and 5) in the second equation, the fourth, fifth, and sixth (7, 9, and 11) in the third equation, and so on. Each right side contains the cube (third power) of the number of terms on the left side. Following this pattern, the next equation would be

$$21 + 23 + 25 + 27 + 29 = 5^3,$$

which can be verified by computation.

(c) The left side of each equation gives the indicated sum of the first n counting numbers, and the right side is always of the form

$$\frac{n(n+1)}{2}.$$

For the pattern to continue, the next equation would be

$$1 + 2 + 3 + 4 + 5 = \frac{5 \cdot 6}{2}.$$

Because each side simplifies to 15, the pattern is true for this equation.

(d) In each case, the first factor on the left is 12,345,679 and the second factor is a multiple of 9 (that is, 9, 18, 27, 36, 45). The right side consists of a nine-digit number, all digits of which are the same (that is, 1, 2, 3, 4, 5). For the pattern to continue, the next equation would be as follows.

$$12{,}345{,}679 \times 54 = 666{,}666{,}666$$

Verify that this is a true statement.

The patterns established in **Examples 3(a) and 3(c)** can be written as follows.

SPECIAL SUM FORMULAS

For any counting number n, the following are true.

$$(1 + 2 + 3 + \cdots + n)^2 = 1^3 + 2^3 + 3^3 + \cdots + n^3$$

$$1 + 2 + 3 + \cdots + n = \frac{n(n+1)}{2}$$

We can provide a general deductive argument showing how the second equation is obtained. Let S represent the sum $1 + 2 + 3 + \cdots + n$. This sum can also be written as

$$S = n + (n-1) + (n-2) + \cdots + 1.$$

Write these two equations as follows.

$$S = 1 \qquad + 2 \qquad + 3 \qquad + \cdots + n$$
$$\underline{S = n \qquad + (n-1) + (n-2) + \cdots + 1}$$
$$2S = (n+1) + (n+1) + (n+1) + \cdots + (n+1) \qquad \text{Add the corresponding sides.}$$

$$2S = n(n+1) \qquad \text{There are } n \text{ terms of } n+1.$$

$$S = \frac{n(n+1)}{2} \qquad \text{Divide both sides by 2.}$$

Figurate Numbers

Pythagoras and his Pythagorean brotherhood studied numbers of geometric arrangements of points, such as **triangular numbers, square numbers,** and **pentagonal numbers. Figure 4** illustrates the first few of each of these types of numbers.

The **figurate numbers** possess numerous interesting patterns. For example, every square number greater than 1 is the sum of two consecutive triangular numbers. ($9 = 3 + 6$, $25 = 10 + 15$, and so on.)

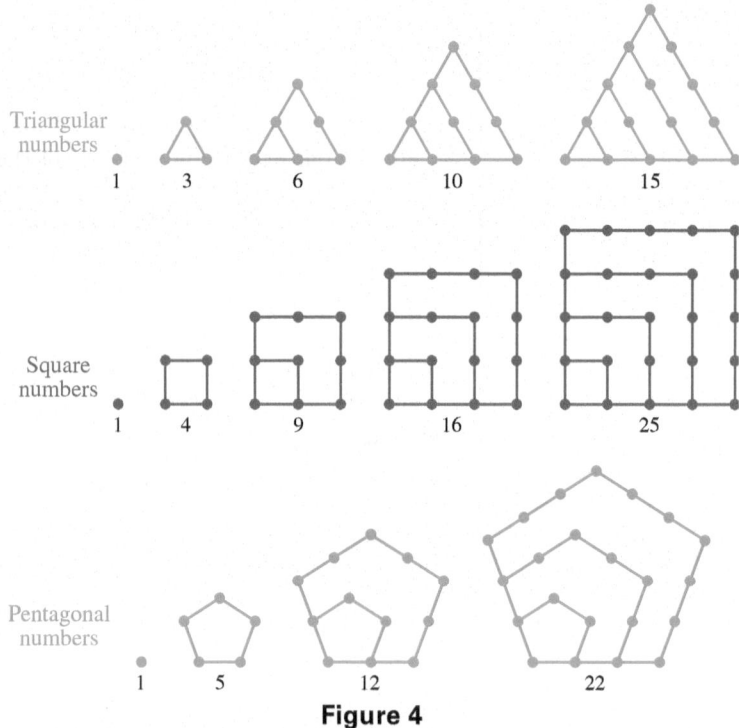

Figure 4

Every pentagonal number can be represented as the sum of a square number and a triangular number. (For example, $5 = 4 + 1$ and $12 = 9 + 3$.) Many other such relationships exist.

In the expression T_n, n is called a **subscript.** T_n is read **"T sub *n*,"** and it represents the triangular number in the *n*th position in the sequence. For example,

$$T_1 = 1, \quad T_2 = 3, \quad T_3 = 6, \quad \text{and} \quad T_4 = 10.$$

S_n and P_n represent the *n*th square and pentagonal numbers, respectively.

FORMULAS FOR TRIANGULAR, SQUARE, AND PENTAGONAL NUMBERS

For any natural number *n*, the following are true.

The *n*th triangular number is given by $T_n = \dfrac{n(n + 1)}{2}$.

The *n*th square number is given by $S_n = n^2$.

The *n*th pentagonal number is given by $P_n = \dfrac{n(3n - 1)}{2}$.

EXAMPLE 4 **Using the Formulas for Figurate Numbers**

Use the formulas to find each of the following.

(a) the seventh triangular number   **(b)** the twelfth square number

(c) the sixth pentagonal number

Solution

(a) $T_7 = \dfrac{n(n+1)}{2} = \dfrac{7(7+1)}{2} = \dfrac{7(8)}{2} = \dfrac{56}{2} = 28$ Formula for a triangular number, with $n = 7$

(b) $S_{12} = n^2 = 12^2 = 144$ Formula for a square number, with $n = 12$

$12^2 = 12 \cdot 12$

Inside the brackets, multiply first and then subtract.

(c) $P_6 = \dfrac{n(3n-1)}{2} = \dfrac{6[3(6)-1]}{2} = \dfrac{6(17)}{2} = 51$ Formula for a pentagonal number, with $n = 6$

EXAMPLE 5 **Illustrating a Figurate Number Relationship**

Show that the sixth pentagonal number is equal to the sum of 6 and 3 times the fifth triangular number.

Solution

From **Example 4(c),** $P_6 = 51$. The fifth triangular number is 15. Thus,

$$51 = 6 + 3(15) = 6 + 45 = 51.$$

The general relationship examined in **Example 5** can be written as follows.

$$P_n = n + 3 \cdot T_{n-1} \quad (n \geq 2)$$

(The equation above is read "P sub-n is equal to n plus 3 times T sub-n minus 1, n is greater than or equal to 2.")

EXAMPLE 6 **Predicting the Value of a Pentagonal Number**

The first five pentagonal numbers are

$$1, 5, 12, 22, 35.$$

Use the method of successive differences to predict the sixth pentagonal number.

Solution

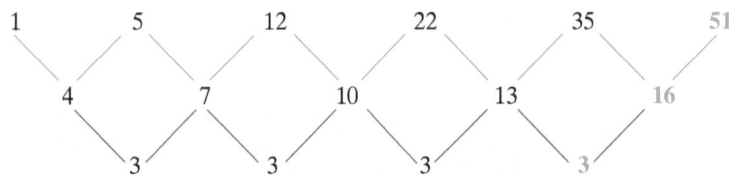

After the second line of successive differences, we work backward to find that the sixth pentagonal number is 51, which was also found in **Example 4(c).**

1.2   EXERCISES

CONCEPT CHECK *Fill in the blank with the correct response.*

1. If 5, 8 are two adjacent terms in an arithmetic sequence (in that order), then the next term is ____ .

2. If 5, 25 are two adjacent terms in a geometric sequence (in that order), then the next term is ____ .

3. The expression for finding the sum $1 + 2 + 3 + 4 + 5 + 6$ without actually performing the addition is

$$\frac{6(\underline{})}{2}.$$

4. The expression for finding the sum $1 + 3 + 5 + 7 + 9 + 11$ without actually performing the addition is

_____ .

(*Hint:* There are six terms to be added here.)

For each sequence, determine whether it is an arithmetic *sequence, a* geometric *sequence, or* neither. *If it is either arithmetic or geometric, give the next term in the sequence.*

5. 6, 16, 26, 36, 46, . . .

6. 8, 16, 24, 32, 40, . . .

7. 5, 15, 45, 135, 405, . . .

8. 2, 12, 72, 432, 2592, . . .

9. 1, 8, 27, 81, 243, . . .

10. 2, 8, 18, 32, 50, . . .

11. 256, 128, 64, 32, 16, . . .

12. 4096, 1024, 256, 64, 16, . . .

13. 1, 3, 4, 7, 11, . . .

14. 0, 1, 1, 2, 3, . . .

15. 12, 14, 16, 18, 20, . . .

16. 10, 50, 90, 130, 170, . . .

Use the method of successive differences to determine the next number in each sequence.

17. 1, 4, 11, 22, 37, 56, . . .

18. 3, 14, 31, 54, 83, 118, . . .

19. 6, 20, 50, 102, 182, 296, . . .

20. 1, 11, 35, 79, 149, 251, . . .

21. 0, 12, 72, 240, 600, 1260, 2352, . . .

22. 2, 57, 220, 575, 1230, 2317, . . .

23. 5, 34, 243, 1022, 3121, 7770, 16,799, . . .

24. 3, 19, 165, 771, 2503, 6483, 14,409, . . .

Solve each problem.

25. Refer to **Figures 2 and 3** in **Section 1.1.** The method of successive differences can be applied to the sequence of interior regions,

$$1, 2, 4, 8, 16, 31,$$

to find the number of regions determined by seven points on the circle. What is the next term in this sequence? How many regions would be determined by eight points? Verify this using the formula given at the end of that section.

26. The 1952 film *Hans Christian Andersen* stars Danny Kaye as the Danish writer of fairy tales. In a scene outside a schoolhouse window, Kaye sings a song to an inchworm. "Inchworm" was written for the film by the composer Frank Loesser and has been recorded by many artists, including Paul McCartney and Kenny Loggins. It was once featured on an episode of *The Muppets* and sung by Charles Aznavour.

As Kaye sings the song, the children in the school room are heard chanting addition facts:

$$2 + 2 = 4, \quad 4 + 4 = 8, \quad 8 + 8 = 16, \quad \text{and so on.}$$

(a) Use patterns to state the next addition fact (as heard in the movie).

(b) If the children were to extend their facts to the next four in the pattern, what would those facts be?

Several equations are given illustrating a suspected number pattern. Determine what the next equation would be, and verify that it is indeed a true statement.

27. $(1 \times 9) - 1 = 8$
$(21 \times 9) - 1 = 188$
$(321 \times 9) - 1 = 2888$

28. $(1 \times 8) + 1 = 9$
$(12 \times 8) + 2 = 98$
$(123 \times 8) + 3 = 987$

29. $999,999 \times 2 = 1,999,998$
$999,999 \times 3 = 2,999,997$

30. $101 \times 101 = 10,201$
$10,101 \times 10,101 = 102,030,201$

31. $3^2 - 1^2 = 2^3$
$6^2 - 3^2 = 3^3$
$10^2 - 6^2 = 4^3$
$15^2 - 10^2 = 5^3$

32. $1 = 1^2$
$1 + 2 + 1 = 2^2$
$1 + 2 + 3 + 2 + 1 = 3^2$
$1 + 2 + 3 + 4 + 3 + 2 + 1 = 4^2$

33. $2^2 - 1^2 = 2 + 1$
$3^2 - 2^2 = 3 + 2$
$4^2 - 3^2 = 4 + 3$

34. $1^2 + 1 = 2^2 - 2$
$2^2 + 2 = 3^2 - 3$
$3^2 + 3 = 4^2 - 4$

35. $1 = 1 \times 1$
$1 + 5 = 2 \times 3$
$1 + 5 + 9 = 3 \times 5$

36. $1 + 2 = 3$
$4 + 5 + 6 = 7 + 8$
$9 + 10 + 11 + 12 = 13 + 14 + 15$

Use the formula $S = \frac{n(n+1)}{2}$ to find each sum.

37. $1 + 2 + 3 + \cdots + 300$

38. $1 + 2 + 3 + \cdots + 500$

39. $1 + 2 + 3 + \cdots + 675$

40. $1 + 2 + 3 + \cdots + 825$

Use the formula $S = n^2$ to find each sum. (Hint: To find n, add 1 to the last term and divide by 2.)

41. $1 + 3 + 5 + \cdots + 101$

42. $1 + 3 + 5 + \cdots + 49$

43. $1 + 3 + 5 + \cdots + 999$

44. $1 + 3 + 5 + \cdots + 301$

Solve each problem.

45. Use the formula for finding the sum
$$1 + 2 + 3 + \cdots + n$$
to discover a formula for finding the sum
$$2 + 4 + 6 + \cdots + 2n.$$

46. State in your own words the following formula discussed in this section.
$$(1 + 2 + 3 + \cdots + n)^2 = 1^3 + 2^3 + 3^3 + \cdots + n^3$$

47. Explain how the following diagram geometrically illustrates the formula $1 + 3 + 5 + 7 + 9 = 5^2$.

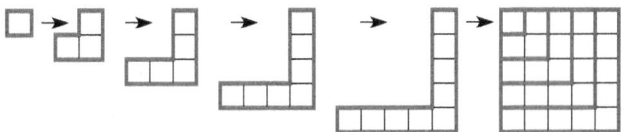

48. Explain how the following diagram geometrically illustrates the formula $1 + 2 + 3 + 4 = \frac{4 \times 5}{2}$.

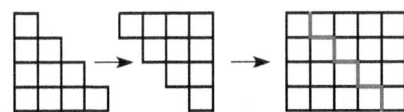

49. Use patterns to complete the table below.

Figurate Number	1st	2nd	3rd	4th	5th	6th	7th	8th
Triangular	1	3	6	10	15	21		
Square	1	4	9	16	25			
Pentagonal	1	5	12	22				
Hexagonal	1	6	15					
Heptagonal	1	7						
Octagonal	1							

50. The first five triangular, square, and pentagonal numbers can be obtained using sums of terms of sequences as shown below.

Triangular	Square	Pentagonal
$1 = 1$	$1 = 1$	$1 = 1$
$3 = 1 + 2$	$4 = 1 + 3$	$5 = 1 + 4$
$6 = 1 + 2 + 3$	$9 = 1 + 3 + 5$	$12 = 1 + 4 + 7$
$10 = 1 + 2 + 3 + 4$	$16 = 1 + 3 + 5 + 7$	$22 = 1 + 4 + 7 + 10$
$15 = 1 + 2 + 3 + 4 + 5$	$25 = 1 + 3 + 5 + 7 + 9$	$35 = 1 + 4 + 7 + 10 + 13$

Notice the successive differences of the added terms on the right sides of the equations. The next type of figurate number is the **hexagonal** number. (A hexagon has six sides.) Use the patterns above to predict the first five hexagonal numbers.

51. Eight times any triangular number, plus 1, is a square number. Show that this is true for the first four triangular numbers.

52. Divide the first triangular number by 3 and record the remainder. Divide the second triangular number by 3 and record the remainder. Repeat this procedure several more times. Do you notice a pattern?

53. Repeat **Exercise 52,** but instead use square numbers and divide by 4. What pattern is determined?

54. Exercises 52 and 53 are specific cases of the following: When the numbers in the sequence of n-agonal numbers are divided by n, the sequence of remainders obtained is a repeating sequence. Verify this for $n = 5$ and $n = 6$.

55. Every square number can be written as the sum of two triangular numbers. For example, $16 = 6 + 10$. This can be represented geometrically by dividing a square array of dots with a line as shown.

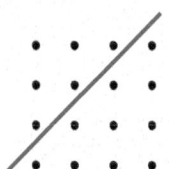

The triangular arrangement above the line represents 6, the one below the line represents 10, and the whole arrangement represents 16. Show how the square numbers 25 and 36 may likewise be geometrically represented as the sum of two triangular numbers.

56. A fraction is in *lowest terms* if the greatest common factor of its numerator and its denominator is 1. For example, $\frac{3}{8}$ is in lowest terms, but $\frac{4}{12}$ is not.

(a) For $n = 2$ to $n = 8$, form the fractions

$$\frac{n\text{th square number}}{(n + 1)\text{st square number}}.$$

(b) Repeat part (a) with triangular numbers.

(c) Use inductive reasoning to make a conjecture based on your results from parts (a) and (b), observing whether the fractions are in lowest terms.

*In addition to the formulas for T_n, S_n, and P_n, the following formulas are true for **hexagonal** numbers (H), **heptagonal** numbers (Hp), and **octagonal** numbers (O).*

$$H_n = \frac{n(4n - 2)}{2}, \quad Hp_n = \frac{n(5n - 3)}{2}, \quad O_n = \frac{n(6n - 4)}{2}$$

Use these formulas to find each of the following.

57. the sixteenth square number

58. the eleventh triangular number

59. the ninth pentagonal number

60. the seventh hexagonal number

61. the tenth heptagonal number

62. the twelfth octagonal number

63. Observe the formulas given for H_n, Hp_n, and O_n, and use patterns and inductive reasoning to predict the formula for N_n, the nth **nonagonal** number. (A nonagon has nine sides.) Then use the fact that the sixth nonagonal number is 111 to further confirm your conjecture.

64. Use the result of **Exercise 63** to find the tenth nonagonal number.

Use inductive reasoning to answer each question.

65. If you add two consecutive triangular numbers, what kind of figurate number do you get?

66. If you add the squares of two consecutive triangular numbers, what kind of figurate number do you get?

67. Square a triangular number. Square the next triangular number. Subtract the smaller result from the larger. What kind of number do you get?

68. Choose a value of n greater than or equal to 2. Find T_{n-1}, multiply it by 3, and add n. What kind of figurate number do you get?

In an arithmetic sequence, the nth term a_n is given by the formula

$$a_n = a_1 + (n - 1)d,$$

where a_1 is the first term and d is the common difference. Similarly, in a geometric sequence, the nth term is given by

$$a_n = a_1 \cdot r^{n-1}.$$

Here r is the common ratio. Use these formulas to determine the indicated term in the given sequence.

69. the eleventh term of $2, 6, 10, 14, \ldots$

70. the sixteenth term of $5, 15, 25, 35, \ldots$

71. the 21st term of $19, 39, 59, 79, \ldots$

72. the 36th term of $8, 38, 68, 98, \ldots$

73. the 101st term of $\frac{1}{2}, 1, \frac{3}{2}, 2, \ldots$

74. the 151st term of $0.75, 1.50, 2.25, 3.00, \ldots$

75. the eleventh term of $2, 4, 8, 16, \ldots$

76. the ninth term of $1, 4, 16, 64, \ldots$

77. the 12th term of $1, \frac{1}{2}, \frac{1}{4}, \frac{1}{8}, \ldots$

78. the 10th term of $1, \frac{1}{3}, \frac{1}{9}, \frac{1}{27}, \ldots$

79. the 8th term of $40, 10, \frac{5}{2}, \frac{5}{8}, \ldots$

80. the 9th term of $10, 2, \frac{2}{5}, \frac{2}{25}, \ldots$

*The mathematical array of numbers known as **Pascal's triangle** consists of rows of numbers, each of which contains one more entry than the previous row. The first six rows are shown here.*

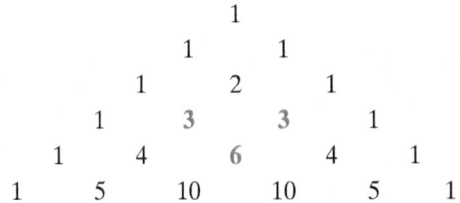

Refer to this array to answer the following.

81. Each row begins and ends with a 1. Discover a method whereby the other entries in a row can be determined from the entries in the row immediately above it. (*Hint:* See the entries in color above.) Find the next three rows of the triangle, and prepare a copy of the first nine rows for later reference.

82. Find the sum of the entries in each of the first eight rows. What is the pattern that emerges? Predict the sum of the entries in the ninth row, and confirm your prediction.

83. The first six rows of the triangle are arranged "flush left" here.

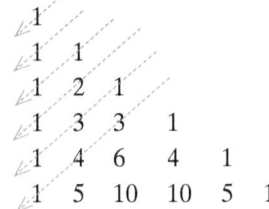

Add along the blue diagonal lines. Write these sums in order from left to right. What sequence is this?

84. Find the values of the first four powers of the number 11, starting with 11^0, which by definition is equal to 1. Predict what the next power of 11 will equal by observing the rows of Pascal's triangle. Confirm your prediction by actual computation.

FOR FURTHER THOUGHT

**Dattathreya Kaprekar
(1905–1986)**

Kaprekar Constants

Take any four-digit number whose digits are all different. Arrange the digits in decreasing order, and then arrange them in increasing order. Now subtract. Repeat the process, called the **Kaprekar routine,** until the same result appears.

For example, suppose that we choose a number whose digits are

1, 5, 7, and 9, such as 1579.

```
  9751     8721     7443
 -1579    -1278    -3447
 ─────    ─────    ─────
  8172     7443     3996

  9963     6642     7641
 -3699    -2466    -1467
 ─────    ─────    ─────
  6264     4176     6174
```

Note that we have obtained the number 6174, and the process will lead to 6174 again. The number 6174 is called a **Kaprekar constant.** This number 6174 will always be generated eventually if this process is applied to such a four-digit number.

For Group or Individual Investigation

1. Apply the Kaprekar routine to a four-digit number of your choice, in which the digits are all different. How many steps did it take for you to arrive at 6174?

2. Apply the Kaprekar routine, starting with a *three*-digit number of your choice whose digits are all different. You should arrive at a particular three-digit number that has the same property described for 6174. What is this three-digit number?

3. Applying the Kaprekar routine to a five-digit number does not reach a single repeating result. Instead, it reaches one of the following ten numbers and then cycles repeatedly through a subset of these ten numbers.

53,955	59,994
61,974	62,964
63,954	71,973
74,943	75,933
82,962	83,952

(a) Start the routine with the five-digit number 45,986, and determine which one of the ten numbers above is reached first.

(b) Start with a five-digit number of your own, and determine which one of the ten numbers is eventually reached first.

1.3   STRATEGIES FOR PROBLEM SOLVING

OBJECTIVES

1 Use George Polya's four-step method of problem solving.

2 Apply various strategies for solving problems.

A General Problem-Solving Method

In the first two sections of this chapter we stressed the importance of pattern recognition and the use of inductive reasoning in solving problems. Probably the most famous study of problem-solving techniques was developed by George Polya (1888–1985), among whose many publications was the modern classic *How to Solve It.* In this book, Polya proposed a four-step method for problem solving.

George Polya, author of the classic *How to Solve It*, died at the age of 97 on September 7, 1985. A native of Budapest, Hungary, he was once asked why there were so many good mathematicians to come out of Hungary at the turn of the century. He theorized that it was because mathematics is the cheapest science. It does not require any expensive equipment, only pencil and paper.

POLYA'S FOUR-STEP METHOD FOR PROBLEM SOLVING

Step 1 **Understand the problem.** You cannot solve a problem if you do not understand what you are asked to find. The problem must be read and analyzed carefully. You may need to read it several times. After you have done so, ask yourself, "*What must I find?*"

Step 2 **Devise a plan.** There are many ways to attack a problem. Decide what plan is appropriate for the particular problem you are solving.

Step 3 **Carry out the plan.** Once you know how to approach the problem, carry out your plan. You may run into "dead ends" and unforeseen roadblocks, but be persistent.

Step 4 **Look back and check.** Check your answer to see that it is reasonable. *Does it satisfy the conditions of the problem? Have you answered all the questions the problem asks? Can you solve the problem a different way and come up with the same answer?*

In Step 2 of Polya's problem-solving method, we are told to devise a plan. Here are some strategies that may prove useful.

Problem-Solving Strategies

- Make a table or a chart.
- Look for a pattern.
- Solve a similar, simpler problem.
- Draw a sketch.
- Use inductive reasoning.
- Write an equation and solve it.
- If a formula applies, use it.

- Work backward.
- Guess and check.
- Use trial and error.
- Use common sense.
- Look for a "catch" if an answer seems too obvious or impossible.

Using a Table or Chart

EXAMPLE 1 Solving Fibonacci's Rabbit Problem

A man put a pair of rabbits in a cage. During the first month the rabbits produced no offspring but each month thereafter produced one new pair of rabbits. If each new pair thus produced *reproduces* in the same manner, how many pairs of rabbits will there be at the end of 1 year? (This problem is a famous one in the history of mathematics and first appeared in *Liber Abaci*, a book written by the Italian mathematician Leonardo Pisano (also known as Fibonacci) in the year 1202.)

Solution

Step 1 **Understand the problem.** We can reword the problem as follows:

How many pairs of rabbits will the man have at the end of one year if he starts with one pair, and they reproduce in the following way? During the first month of life, each pair produces no new rabbits, but each month thereafter each pair produces one new pair.

Step 2 **Devise a plan.** Because there is a definite pattern to how the rabbits will reproduce, we can construct **Table 2** on the next page.

1, 1, 2, 3, 5, 8, 13, 21, ···

In the 2003 movie *A Wrinkle in Time*, young Charles Wallace, played by David Dorfman, is challenged to identify a particular sequence of numbers. He correctly identifies it as the **Fibonacci sequence.**

StatCrunch   Problem Solving—
Using a Table

Animation
→

Scheduling Classes Using
an Organized List

Table 2

Month	Number of Pairs at Start	Number of New Pairs Produced	Number of Pairs at End of Month
1ˢᵗ			
2ⁿᵈ			
3ʳᵈ			
4ᵗʰ			
5ᵗʰ			
6ᵗʰ			
7ᵗʰ			
8ᵗʰ			
9ᵗʰ			
10ᵗʰ			
11ᵗʰ			
12ᵗʰ			

The answer will go here.

Step 3 **Carry out the plan.**   At the start of the first month, there is only one pair of rabbits. No new pairs are produced during the first month, so there is $1 + 0 = 1$ pair present at the end of the first month. This pattern continues. In **Table 3,** we add the number in the first column of numbers to the number in the second column to get the number in the third.

Table 3

Month	Number of Pairs at Start	+	Number of New Pairs Produced	=	Number of Pairs at End of Month	
1ˢᵗ	1		0		1	$1 + 0 = 1$
2ⁿᵈ	1		1		2	$1 + 1 = 2$
3ʳᵈ	2		1		3	$2 + 1 = 3$
4ᵗʰ	3		2		5	.
5ᵗʰ	5		3		8	.
6ᵗʰ	8		5		13	.
7ᵗʰ	13		8		21	
8ᵗʰ	21		13		34	
9ᵗʰ	34		21		55	.
10ᵗʰ	55		34		89	.
11ᵗʰ	89		55		144	.
12ᵗʰ	144		89		**233**	$144 + 89 = 233$

The answer is the final entry.

NUMB3RS   On January 23, 2005, the CBS television network presented the first episode of *NUMB3RS*, a show focusing on how mathematics is used in solving crimes. David Krumholtz plays Charlie Eppes, a brilliant mathematician who assists his FBI agent brother (Rob Morrow).

In the first-season episode "Sabotage" (2/25/2005), one of the agents admits that she was not a good math student, and Charlie uses the **Fibonacci sequence** and its relationship to nature to enlighten her.

The sequence shown in color in **Table 3** is the Fibonacci sequence, mentioned in **Example 2(b)** of **Section 1.1.**

There will be **233** pairs of rabbits at the end of one year.

Step 4 **Look back and check.**   Go back and make sure that we have interpreted the problem correctly. Double-check the arithmetic. We have answered the question posed by the problem, so the problem is solved.

Working Backward

EXAMPLE 2 Determining Money for a Craft Fair

For three consecutive days, Ronnie set up a booth at a crafts fair. On the first day, he tripled the amount of money he brought with him, but spent $12 at another booth. The second day he brought the money with him, doubled it, and spent $40 at another booth. On the third day, he again brought the money with him, quadrupled it, and spent nothing. He ended up with $224. How much did he bring with him the first day?

Solution

This problem asks us to find Ronnie's starting amount. Since we know his final amount, the method of working backward can be applied.

Because his final amount was $224 and this represents four times the amount he started with on the third day, we *divide* $224 by 4 to find that he started the third day with $56. Before he spent $40 the second day, he had this $56 plus the $40 he spent, giving him $96.

The $96 represented double what he started with, so he started with $96 *divided by* 2, or $48, the second day. Repeating this process once more for the first day, before his $12 expenditure he had $48 + $12 = $60, which represents triple what he started with. Now divide by 3 to find that he started with

$$\$60 \div 3 = \$20. \quad \text{Answer}$$

CHECK Observe the following equations.

> ***First week:*** $(3 \times \$20) - \$12 = \$60 - \$12 = \$48$
> ***Second week:*** $(2 \times \$48) - \$40 = \$96 - \$40 = \$56$
> ***Third week:*** $(4 \times \$56) = \$224$ His final amount

Using Trial and Error

Recall that $5^2 = 5 \cdot 5 = 25$. That is, 5 squared is 25. Thus, 25 is called a **perfect square.** Other perfect squares are 1, 4, 9, 16, 25, 36, and so on.

EXAMPLE 3 Finding Augustus De Morgan's Birth Year

The mathematician Augustus De Morgan lived in the nineteenth century. He made the following statement: *"I was x years old in the year x^2."* In what year was he born?

Solution

We must find the year of De Morgan's birth. The problem tells us that he lived in the nineteenth century, which is another way of saying that he lived during the 1800s. One year of his life was a *perfect square,* so we must find a number between 1800 and 1900 that is a perfect square. Use trial and error.

$$42^2 = 42 \cdot 42 = 1764$$
$$43^2 = 43 \cdot 43 = 1849 \quad \text{1849 is between 1800 and 1900.}$$
$$44^2 = 44 \cdot 44 = 1936$$

The only natural number whose square is between 1800 and 1900 is 43, because $43^2 = 1849$. Therefore, De Morgan was 43 years old in 1849. The final step in solving the problem is to subtract 43 from 1849 to find the year of his birth.

$$1849 - 43 = 1806 \quad \text{He was born in 1806.}$$

CHECK Look up De Morgan's birth date in a mathematics history book, such as *An Introduction to the History of Mathematics,* Sixth Edition, by Howard W. Eves.

Augustus De Morgan was an English mathematician and philosopher who served as professor at the University of London. He wrote numerous books, one of which was *A Budget of Paradoxes.* His work in **set theory** and **logic** led to laws that bear his name and are covered in other chapters.

Guessing and Checking

As mentioned above, $5^2 = 25$. The inverse procedure for squaring a number is called taking the **square root.** We indicate the positive square root using a **radical symbol** $\sqrt{\ }$. Thus, $\sqrt{25} = 5$. Also,

$$\sqrt{4} = 2, \quad \sqrt{9} = 3, \quad \sqrt{16} = 4, \quad \text{and so on.} \quad \text{Square roots}$$

The next problem deals with a square root and dates back to Hindu mathematics, circa 850.

EXAMPLE 4 Finding the Number of Camels

One-fourth of a herd of camels was seen in the forest. Twice the square root of that herd had gone to the mountain slopes, and 3 times 5 camels remained on the riverbank. What is the numerical measure of that herd of camels?

Solution

The numerical measure of a herd of camels must be a counting number. Because the problem mentions

"one-fourth of a herd" and "the square root of that herd,"

the number of camels must be both a multiple of 4 and a perfect square, so only whole numbers are used. The least counting number that satisfies both conditions is 4.

We write an equation where x represents the numerical measure of the herd, and then substitute 4 for x to see whether it is a solution.

One-fourth of the herd	$+$	Twice the square root of that herd	$+$	3 times 5 camels	$=$	The numerical measure of the herd.

$$\frac{1}{4}x \quad + \quad 2\sqrt{x} \quad + \quad 3\cdot5 \quad = \quad x$$

$$\frac{1}{4}(4) + 2\sqrt{4} + 3\cdot5 \overset{?}{=} 4 \quad \text{Let } x = 4.$$

$$1 + 4 + 15 \overset{?}{=} 4 \quad \sqrt{4} = 2$$

$$20 \neq 4$$

Because 4 is not the solution, try 16, the next perfect square that is a multiple of 4.

$$\frac{1}{4}(16) + 2\sqrt{16} + 3\cdot5 \overset{?}{=} 16 \quad \text{Let } x = 16.$$

$$4 + 8 + 15 \overset{?}{=} 16 \quad \sqrt{16} = 4$$

$$27 \neq 16$$

Because 16 is not a solution, try 36.

$$\frac{1}{4}(36) + 2\sqrt{36} + 3\cdot5 \overset{?}{=} 36 \quad \text{Let } x = 36.$$

$$9 + 12 + 15 \overset{?}{=} 36 \quad \sqrt{36} = 6$$

$$36 = 36$$

Thus, 36 is the numerical measure of the herd.

CHECK "One-fourth of 36, plus twice the square root of 36, plus 3 times 5" gives

$$9 + 12 + 15, \quad \text{which equals} \quad 36.$$

Considering a Similar, Simpler Problem

EXAMPLE 5 Finding the Units Digit of a Power

The digit farthest to the right in a counting number is called the *ones* or *units* digit, because it tells how many ones are contained in the number when grouping by tens is considered. What is the ones (or units) digit in 2^{4000}?

Solution

Recall that 2^{4000} means that 2 is used as a factor 4000 times.

$$2^{4000} = \underbrace{2 \times 2 \times 2 \times \cdots \times 2}_{4000 \text{ factors}}$$

To answer the question, we examine some smaller powers of 2 and then look for a pattern. We start with the exponent 1 and look at the first twelve powers of 2.

$$2^1 = 2 \qquad 2^5 = 32 \qquad 2^9 = 512$$
$$2^2 = 4 \qquad 2^6 = 64 \qquad 2^{10} = 1024$$
$$2^3 = 8 \qquad 2^7 = 128 \qquad 2^{11} = 2048$$
$$2^4 = 16 \qquad 2^8 = 256 \qquad 2^{12} = 4096$$

Notice that in any one of the four rows above, the ones digit is the same all the way across the row. The final row, which contains the exponents 4, 8, and 12, has the ones digit 6. Each of these exponents is divisible by 4, and because 4000 is divisible by 4, we can use inductive reasoning to predict that the ones digit in 2^{4000} is **6**.

(*Note:* The ones digit for any other power can be found if we divide the exponent by 4 and consider the remainder. Then compare the result to the list of powers above. For example, to find the ones digit of 2^{543}, divide 543 by 4 to get a quotient of 135 and a remainder of 3. The ones digit is the same as that of 2^3, which is 8.)

Drawing a Sketch

EXAMPLE 6 Connecting the Dots

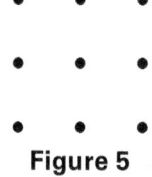

Figure 5

An array of nine dots is arranged in a 3 × 3 square, as shown in **Figure 5**. Is it possible to join the dots with exactly four straight line segments if you are not allowed to pick up your pencil from the paper and may not trace over a segment that has already been drawn? If so, show how.

Solution

Figure 6 shows three attempts. In each case, something is wrong. In the first sketch, one dot is not joined. In the second, the figure cannot be drawn without picking up your pencil from the paper or tracing over a line that has already been drawn. In the third figure, all dots have been joined, but you have used five line segments as well as retraced over the figure.

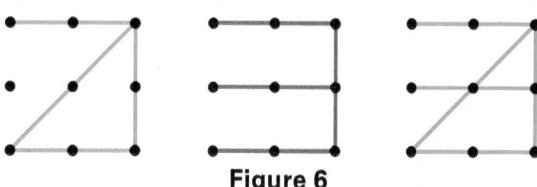

Figure 6

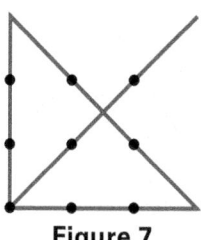

Figure 7

The conditions of the problem can be satisfied, as shown in **Figure 7**. We "went outside of the box," which was not prohibited by the conditions of the problem. This is an example of creative thinking—we used a strategy that often is not considered at first.

Using Common Sense

Problem-Solving Strategies

Some problems involve a "catch." They seem too easy or perhaps impossible at first because we tend to overlook an obvious situation. Look carefully at the use of language in such problems. And, of course, never forget to use common sense.

EXAMPLE 7 Determining Coin Denominations

Two currently minted United States coins together have a total value of $1.05. One is not a dollar. What are the two coins?

Solution

Our initial reaction might be, "The only way to have two such coins with a total of $1.05 is to have a nickel and a dollar, but the problem says that one of them is not a dollar." This statement is indeed true. The one that is not a dollar is the nickel, and the *other* coin is a dollar! So the two coins are a dollar and a nickel.

1.3 EXERCISES

One of the most popular features of the journal Mathematics Teacher, *published by the National Council of Teachers of Mathematics, is the monthly calendar. It provides an interesting, unusual, or challenging problem for each day of the month. Some of these exercises, and others to follow in this text, are chosen from these calendars and designated "MT calendar problem." The authors thank the many contributors for permission to use these problems.*

Use the various problem-solving strategies to solve each problem. In many cases there is more than one possible approach, so be creative.

1. **Digits of a Year** The year 2013 contains four digits whose values are consecutive integers (0, 1, 2, 3). How many years after 2013 will this event occur next? (*MT* calendar problem)

2. **Final Digits of a Power of 5** What are the last two digits of 5^{2007}? (*MT* calendar problem)

3. **Making Change** Given nine nickels and five pennies, how many different sums of money less than 50 cents can be formed using one or more coins? (*MT* calendar problem)

4. **Taking Steps** In how many ways can you walk up a stairway that has 7 steps if you take only 1 or 2 steps at a time? (*MT* calendar problem)

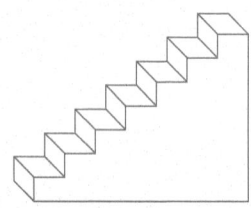

5. **Number Trick** Take a two-digit number, represented by ab, such that neither a nor b is zero and a is not equal to b. Reverse the digits and add the result to the original number. Divide this sum by $a + b$. What is the quotient? (*MT* calendar problem)

6. **Favorite Foods** Violet, Kathy, Lucy, Charles, and John are at a barbecue with their favorite foods: hot dogs, hamburgers, meatless pasta salad, veggie burgers, and ribs. Charles and Kathy are carnivores, but neither likes hamburgers. Violet is a vegetarian. The women dislike ribs. Lucy loves pasta salad. If no two people have the same favorite food, match each of the people to her or his food. (*MT* calendar problem)

7. **Broken Elevator** A man enters a building on the first floor and runs up to the third floor in 20 seconds. At this rate, how many seconds would it take for the man to run from the first floor up to the sixth floor? (*MT* calendar problem)

8. **Saving Her Dollars** Every day Sally saved a penny, a dime, and a quarter. What is the least number of days required for her to save an amount equal to an integral (counting) number of dollars? (*MT* calendar problem)

9. *Do You Have a Match?* Move 4 of the matches in the figure to create exactly 3 equilateral triangles. (An *equilateral triangle* has all three sides the same length.) (*MT* calendar problem)

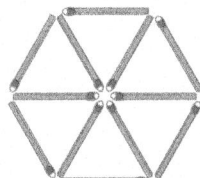

10. *Sudoku* Sudoku is an $n \times n$ puzzle that requires the solver to fill in all the squares using the integers 1 through n. Each row, column, and subrectangle contains exactly one of each number. Complete the $n \times n$ puzzle. (*MT* calendar problem)

			4
2			
	1		
		1	

11. *Break This Code* Each letter of the alphabet is assigned an integer, starting with A = 0, B = 1, and so on. The numbers repeat after every seven letters, so that G = 6, H = 0, and I = 1, continuing on to Z. What two-letter word is represented by the digits 16? (*MT* calendar problem)

12. *A Real Problem* We are given the following sequence:

PROBLEMSOLVINGPROBLEMSOLVINGPROB...

If the pattern continues, what letter will be in the 2012th position? (*MT* calendar problem)

13. *How Old Is Mommy?* A mother has two children whose ages differ by 5 years. The sum of the squares of their ages is 97. The square of the mother's age can be found by writing the squares of the children's ages one after the other as a four-digit number. How old is the mother? (*MT* calendar problem)

14. *An Alarming Situation* You have three alarms in your room. Your cell phone alarm is set to ring every 30 minutes, your computer alarm is set to ring every 20 minutes, and your clock alarm is set to ring every 45 minutes. If all three alarms go off simultaneously at 12:34 p.m., when is the next time that they will go off simultaneously? (*MT* calendar problem)

15. *Laundry Day* Every Monday evening, a mathematics teacher stops by the dry cleaners, drops off the shirts that he wore for the week, and picks up his previous week's load. If he wears a clean shirt every day, including Saturday and Sunday, what is the minimum number of shirts that he needs to own? (*MT* calendar problem)

16. *Pick an Envelope* Three envelopes contain a total of six bills. One envelope contains two $10 bills, one contains two $20 bills, and the third contains one $10 and one $20 bill. A label on each envelope indicates the sum of money in one of the other envelopes. It is possible to select one envelope, see one bill in that envelope, and then state the contents of all of the envelopes. Which envelope should you choose? (*MT* calendar problem)

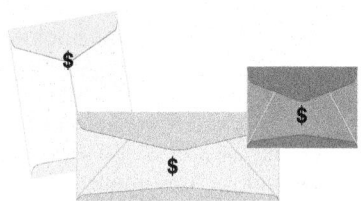

17. *Class Members* A classroom contains an equal number of boys and girls. If 8 girls leave, twice as many boys as girls remain. What was the original number of students present? (*MT* calendar problem)

18. *Give Me a Digit* Given a two-digit number, make a three-digit number by putting a 6 as the rightmost digit. Then add 6 to the resulting three-digit number, and remove the rightmost digit to obtain another two-digit number. If the result is 76, what is the original two-digit number? (*MT* calendar problem)

19. *Missing Digit* Look for a pattern and find the missing digit x.

3	2	4	8
7	2	1	3
8	4	x	5
4	3	6	9

(*MT* calendar problem)

20. *Abundancy* An integer $n > 1$ is **abundant** if the sum of its proper divisors (positive integer divisors smaller than n) is greater than n. Find the smallest abundant integer. (*MT* calendar problem)

21. *Cross-Country Competition* The schools in an athletic conference compete in a cross-country meet to which each school sends three participants. Erin, Katelyn, and Iliana are the three representatives from one school. Erin finished the race in the middle position; Katelyn finished after Erin, in the 19th position; and Iliana finished 28th. How many schools took part in the race? (*MT* calendar problem)

22. *Gone Fishing* Four friends go fishing one day and bring home a total of 11 fish. If each person caught at least 1 fish, then which of the following *must* be true?

    A. One person caught exactly 2 fish.

    B. One person caught exactly 3 fish.

    C. One person caught fewer than 3 fish.

    D. One person caught more than 3 fish.

    E. Two people each caught more than 1 fish.

    (*MT* calendar problem)

23. *Bookworm Snack* A 26-volume encyclopedia (one for each letter) is placed on a bookshelf in alphabetical order from left to right. Each volume is 2 inches thick, including the front and back covers. Each cover is $\frac{1}{4}$ inch thick. A bookworm eats straight through the encyclopedia, beginning inside the front cover of volume A and ending after eating through the back cover of volume Z. How many inches of book did the bookworm eat? (*MT* calendar problem)

24. *You Lie!* Max, Sam, and Brett were playing basketball. One of them broke a window, and the other two saw him break it. Max said, "I am innocent." Sam said, "Max and I are both innocent." Brett said, "Max and Sam are both innocent." If only one of them is telling the truth, who broke the window? (*MT* calendar problem)

25. *Catwoman's Cats* If you ask Batman's nemesis, Catwoman, how many cats she has, she answers with a riddle: "Five-sixths of my cats plus seven." How many cats does Catwoman have? (*MT* calendar problem)

26. *Pencil Collection* Bob gave four-fifths of his pencils to Barbara; then he gave two-thirds of the remaining pencils to Bonnie. If he ended up with ten pencils for himself, with how many did he start? (*MT* calendar problem)

27. *Adding Gasoline* The gasoline gauge on a van initially read $\frac{1}{8}$ full. When 15 gallons were added to the tank, the gauge read $\frac{3}{4}$ full. How many more gallons are needed to fill the tank? (*MT* calendar problem)

28. *Gasoline Tank Capacity* When 6 gallons of gasoline are put into a car's tank, the indicator goes from $\frac{1}{4}$ of a tank to $\frac{5}{8}$. What is the total capacity of the gasoline tank? (*MT* calendar problem)

29. *Number Pattern* What is the relationship between the rows of numbers?

    18, 38, 24, 46, 42
    8, 24, 8, 24, 8

    (*MT* calendar problem)

30. *Locking Boxes* You and I each have one lock and a corresponding key. I want to mail you a box with a ring in it, but any box that is not locked will be emptied before it reaches its recipient. How can I safely send you the ring? (Note that you and I each have keys to our own lock but not to the other lock.) (*MT* calendar problem)

31. *Unfolding and Folding a Box* An unfolded box is shown below.

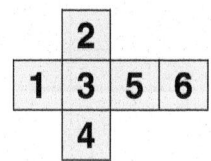

    Which figure shows the box folded up? (*MT* calendar problem)

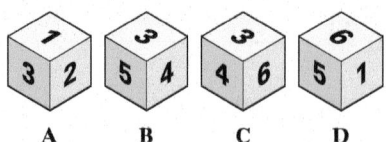

    A   B   C   D

32. *Unknown Number* Cindy was asked by her teacher to subtract 3 from a certain number and then divide the result by 9. Instead, she subtracted 9 and then divided the result by 3, giving an answer of 43. What would her answer have been if she had worked the problem correctly? (*MT* calendar problem)

33. *Labeling Boxes* You are working in a store that has been very careless with the stock. Three boxes of socks are each incorrectly labeled. The labels say *red socks*, *green socks*, and *red and green socks*. How can you relabel the boxes correctly by taking only one sock out of one box, without looking inside the boxes? (*MT* calendar problem)

34. *Vertical Symmetry in States' Names* (If a vertical line is drawn through the center of a figure, and the left and right sides are reflections of each other across this line, the figure is said to have vertical symmetry.) When HAWAII is spelled with all capital letters, each letter has vertical symmetry. Find the name of a state whose capital letters all have both vertical and horizontal symmetry. (*MT* calendar problem)

35. *Sum of Hidden Dots on Dice* Three dice with faces numbered 1 through 6 are stacked as shown. Seven of the eighteen faces are visible, leaving eleven faces hidden on the back, on the bottom, and between dice. The total number of dots not visible in this view is ____.

A. 21

B. 22

C. 31

D. 41

E. 53

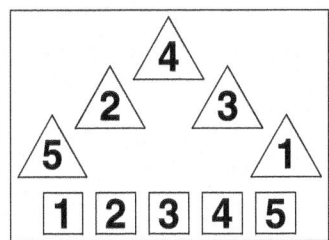

(*MT* calendar problem)

36. *Mr. Green's Age* At his birthday party, Mr. Green would not directly tell how old he was. He said,

> "If you add the year of my birth to this year, subtract the year of my tenth birthday and the year of my fiftieth birthday, and then add my present age, the result is eighty."

How old was Mr. Green? (*MT* calendar problem)

37. *Matching Triangles and Squares* How can you connect each square with the triangle that has the same number? Lines cannot cross, enter a square or triangle, or go outside the diagram. (*MT* calendar problem)

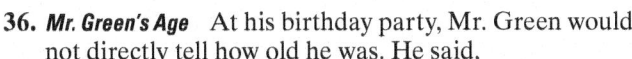

38. *Age of the Bus Driver* Today is your first day driving a city bus. When you leave downtown, you have twenty-three passengers. At the first stop, three people exit and five people get on the bus. At the second stop, eleven people exit and eight people get on the bus. At the third stop, five people exit and ten people get on. How old is the bus driver? (*MT* calendar problem)

39. *Difference Triangle* Balls numbered 1 through 6 are arranged in a **difference triangle.** Note that in any row, the difference between the larger and the smaller of two successive balls is the number of the ball that appears below them. Arrange balls numbered 1 through 10 in a difference triangle. (*MT* calendar problem)

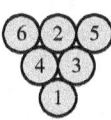

40. *Clock Face* By drawing two straight lines, divide the face of a clock into three regions such that the numbers in the regions have the same total. (*MT* calendar problem)

41. *Alphametric* If *a*, *b*, and *c* are digits for which

$$
\begin{array}{r}
7\ a\ 2 \\
-4\ 8\ b \\
\hline
c\ 7\ 3,
\end{array}
$$

then $a + b + c =$ ____.

A. 14   **B.** 15   **C.** 16   **D.** 17   **E.** 18

(*MT* calendar problem)

42. *Perfect Square* Only one of these numbers is a perfect square. Which one is it? (*MT* calendar problem)

329,476   389,372   964,328

326,047   724,203

43. *Sleeping on the Way to Grandma's House* While traveling to his grandmother's for Christmas, George fell asleep halfway through the journey. When he awoke, he still had to travel half the distance that he had traveled while sleeping. For what part of the entire journey had he been asleep? (*MT* calendar problem)

44. *Buckets of Water* You have brought two unmarked buckets to a stream. The buckets hold 7 gallons and 3 gallons of water, respectively. How can you obtain exactly 5 gallons of water to take home? (*MT* calendar problem)

45. *Counting Puzzle (Rectangles)* How many rectangles are in the figure? (*MT* calendar problem)

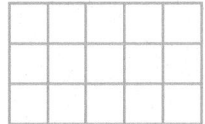

46. *Digit Puzzle* Place each of the digits

1, 2, 3, 4, 5, 6, 7,   and   8

in separate boxes so that boxes that share common corners do not contain successive digits. (*MT* calendar problem)

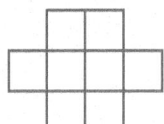

47. Palindromic Number (*Note:* A **palindromic number** is a number whose digits read the same left to right as right to left. For example, 383, 12321, and 9876789 are palindromic.) The odometer of the family car read 15951 when the driver noticed that the number was palindromic. "Curious," said the driver to herself. "It will be a long time before that happens again." But 2 hours later, the odometer showed a new palindromic number. (*Author's note:* Assume it was the next possible one.) How fast was the car driving in those 2 hours? (*MT* calendar problem)

48. How Much Is That Doggie in the Window? A man wishes to sell a puppy for $11. A customer who wants to buy it has only foreign currency. The exchange rate for the foreign currency is as follows: 11 round coins = $15, 11 square coins = $16, 11 triangular coins = $17. How many of each coin should the customer pay? (*MT* calendar problem)

49. Final Digits of a Power of 7 What are the final two digits of 7^{1997}? (*MT* calendar problem)

50. Units Digit of a Power of 3 If you raise 3 to the 324th power, what is the units (ones) digit of the result?

51. Summing the Digits When $10^{50} - 50$ is expressed as a single whole number, what is the sum of its digits? (*MT* calendar problem)

52. Frog Climbing up a Well A frog is at the bottom of a 20-foot well. Each day it crawls up 4 feet, but each night it slips back 3 feet. After how many days will the frog reach the top of the well?

53. Units Digit of a Power of 7 What is the units digit in 7^{491}?

54. Money Spent at a Bazaar Christine bought a book for $10 and then spent half her remaining money on a train ticket. She then bought lunch for $4 and spent half her remaining money at a bazaar. She left the bazaar with $8. How much money did she start with?

55. Matching Socks A drawer contains 20 black socks and 20 white socks. If the light is off and you reach into the drawer to get your socks, what is the minimum number of socks you must pull out in order to be sure that you have a matching pair?

56. Counting Puzzle (Squares) How many squares are in the figure?

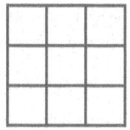

57. Perfect Number A **perfect number** is a counting number that is equal to the sum of all its counting number divisors except itself. For example, 28 is a perfect number because its divisors other than itself are 1, 2, 4, 7, and 14, and $1 + 2 + 4 + 7 + 14 = 28$. What is the least perfect number?

58. Counting Puzzle (Triangles) How many triangles are in the figure?

59. Growth of a Lily Pad A lily pad grows so that each day it doubles its size. On the twentieth day of its life, it completely covers a pond. On what day was the pond half covered?

60. Naming Children Becky's mother has three daughters. She named her first daughter Penny and her second daughter Nichole. What did she name her third daughter?

61. *High School Graduation Year of Author* One of the authors of this book graduated from high school in the year that satisfies these conditions: (1) The sum of the digits is 23; (2) The hundreds digit is 3 more than the tens digit; (3) No digit is an 8. In what year did he graduate?

62. *Interesting Property of a Sentence* Comment on an interesting property of this sentence: "A man, a plan, a canal, Panama." (*Hint:* See **Exercise 47.**)

63. *Adam and Eve's Assets* Eve said to Adam, "If you give me one dollar, then we will have the same amount of money." Adam then replied, "Eve, if you give me one dollar, I will have double the amount of money you are left with." How much does each have?

64. *Missing Digits Puzzle* In the addition problem below, some digits are missing, as indicated by the blanks. If the problem is done correctly, what is the sum of the missing digits?

$$
\begin{array}{r}
\_\ 3\ 5 \\
8\ \_\ 6 \\
+\ 1\ 4\ \_ \\
\hline
\_\ 4\ 0\ 8
\end{array}
$$

65. *Missing Digits Puzzle* Fill in the blanks so that the multiplication problem below uses all digits 0, 1, 2, 3, ..., 9 exactly once and is worked correctly.

$$
\begin{array}{r}
\_\ 0\ 2 \\
\times\quad\ \_\ 3\ \_ \\
\hline
\_\quad 5,\ \_\ \_\ \_
\end{array}
$$

66. *Magic Square* A **magic square** is a square array of numbers that has the property that the sum of the numbers in any row, column, or diagonal is the same. Fill in the square below so that it becomes a magic square, and all digits 1, 2, 3, ..., 9 are used exactly once.

6		8
	5	
		4

67. *Magic Square* Refer to **Exercise 66.** Complete the magic square below so that all counting numbers 1, 2, 3, ..., 16 are used exactly once, and the sum in each row, column, or diagonal is 34.

6			9
	15		14
11		10	
16		13	

68. *Decimal Digit* What is the 100th digit in the decimal representation for $\frac{1}{7}$?

69. *Pitches in a Baseball Game* What is the minimum number of pitches that a baseball player who pitches a complete game can make in a regulation 9-inning baseball game?

70. *Weighing Coins* You have eight coins. Seven are genuine and one is a fake, which weighs a little less than the other seven. You have a balance scale, which you may use only three times. Tell how to locate the bad coin in three weighings. (Then show how to detect the bad coin in only *two* weighings.)

71. *Geometry Puzzle* When the diagram shown is folded to form a cube, what letter is opposite the face marked Z?

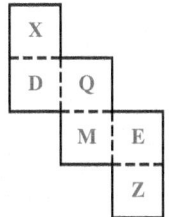

72. *Geometry Puzzle* Draw the following figure without picking up your pencil from the paper and without tracing over a line you have already drawn.

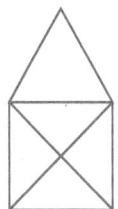

73. *Geometry Puzzle* Repeat **Exercise 72** for this figure.

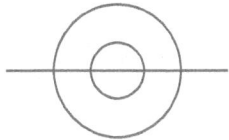

74. *Books on a Shelf* Volumes 1 and 2 of *The Complete Works of Wally Smart* are standing in numerical order from left to right on your bookshelf. Volume 1 has 450 pages and Volume 2 has 475 pages. Excluding the covers, how many pages are between page 1 of Volume 1 and page 475 of Volume 2?

75. *Paying for a Mint* Brian has an unlimited number of cents (pennies), nickels, and dimes. In how many different ways can he pay 15¢ for a chocolate mint? (For example, one way is 1 dime and 5 pennies.)

76. *Teenager's Age* A teenager's age increased by 2 gives a perfect square. Her age decreased by 10 gives the square root of that perfect square. She is 5 years older than her brother. How old is her brother?

77. *Area and Perimeter* Triangle *ABC* has sides 10, 24, and 26 cm long. A rectangle that has an area equal to that of the triangle is 3 cm wide. Find the perimeter of the rectangle. (*MT* calendar problem)

78. *Making Change* In how many different ways can you make change for a half dollar using currently minted U.S. coins, if cents (pennies) are not allowed?

79. *Ages* James, Dan, Jessica, and Cathy form a pair of married heterosexual couples. Their ages are 36, 31, 30, and 29. Jessica is married to the oldest person in the group. James is older than Jessica but younger than Cathy. Who is married to whom, and what are their ages?

80. *Final Digit* What is the last digit of $49{,}327^{1783}$? (*MT* calendar problem)

81. *Geometry Puzzle* What is the maximum number of small squares in which we may place crosses ($\times$) and not have any row, column, or diagonal completely filled with crosses? Illustrate your answer.

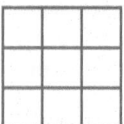

82. *Making Change* Webster has some pennies, dimes, and quarters in his pocket. When Josefa asks him for change for a dollar, Webster discovers that he cannot make the change exactly. What is the largest possible total value of the coins in his pocket? (*MT* calendar problem)

1.4 NUMERACY IN TODAY'S WORLD

OBJECTIVES

1 Use a calculator for routine mathematical operations.

2 Use estimation techniques.

3 Interpret information by reading circle, bar, and line graphs.

4 Use writing skills to convey information about mathematics.

The familiar term *literacy* applies to language in the same way that the term **numeracy** applies to mathematics. It is virtually impossible to function in the world today without understanding fundamental number concepts. The basic ideas of calculating, estimating, interpreting data from graphs, and conveying mathematics via language and writing are among the skills required to be "numerate."

Calculation

The search for easier ways to calculate and compute has culminated in the development of hand-held calculators and computers. For the general population, a calculator that performs the operations of arithmetic and a few other functions is sufficient. These are known as **four-function calculators.** Students who take higher mathematics courses (engineers, for example) usually need the added power of **scientific calculators. Graphing calculators,** which actually plot graphs on small screens, are also available. *Always refer to your owner's manual if you need assistance in performing an operation with your calculator. If you need further help, ask your instructor or another student who is using the same model.*

Today's smartphones routinely include a calculator application (app). For example, Apple's iPhone has an app that serves as a four-function calculator when the phone is held vertically, but it becomes a scientific calculator when held horizontally. Furthermore, graphing calculator apps are available at little or no cost. Although it is not necessary to have a graphing calculator to study the material presented in this text, we occasionally include graphing calculator screens to support results obtained or to provide supplemental information.*

---

*Because it is one of the most popular graphing calculators, we use screens from the TI-84 Plus C from Texas Instruments.

The screens that follow illustrate some common entries and operations.

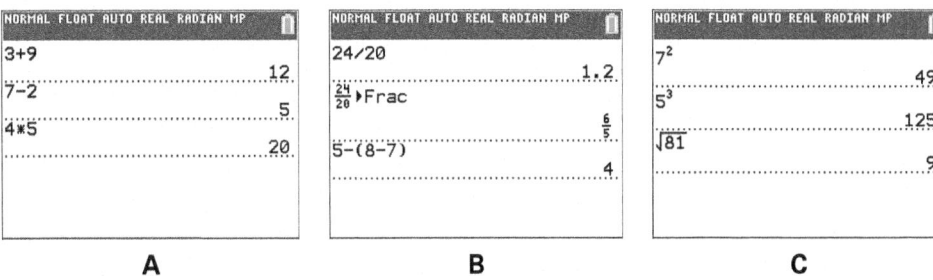

A   B   C

Screen A illustrates how two numbers can be added, subtracted, or multiplied. Screen B shows how two numbers can be divided, how a quotient can be converted into a fraction in lowest terms, and how parentheses can be used in a computation. Screen C shows how a number can be squared, how it can be cubed, and how its square root can be taken.

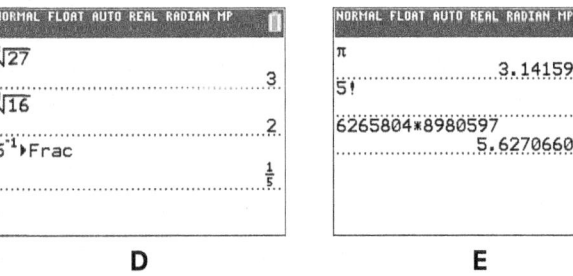

D   E

Shown here is an example of a **calculator app** from a smartphone. Since the introduction of hand-held calculators in the early 1970s, the methods of everyday arithmetic have been drastically altered. One of the first consumer models available was the Texas Instruments SR-10, which sold for nearly $150 in 1973. It could perform the four operations of arithmetic and take square roots, but it could do very little more.

Screen D shows how other roots (cube root and fourth root) can be found, and how the reciprocal of a number can be found using −1 as an exponent. Screen E shows how π can be accessed with its own special key, how a *factorial* (as indicated by !) can be found, and how a result might be displayed in *scientific notation*. (The "E13" following 5.62706606 means that this number is multiplied by 10^{13}. This answer is still only an approximation, because the product $6{,}265{,}804 \times 8{,}980{,}597$ contains more digits than the calculator can display.)

We repeat the list of problem-solving strategies from **Section 1.3.**

Problem-Solving Strategies

- Make a table or a chart.
- Look for a pattern.
- Solve a similar, simpler problem.
- Draw a sketch.
- Use inductive reasoning.
- Write an equation and solve it.
- If a formula applies, use it.
- Work backward.
- Guess and check.
- Use trial and error.
- Use common sense.
- Look for a "catch" if an answer seems too obvious or impossible.

As Terry Krieger points out in the chapter opener, "mathematics may be the only discipline in which different people, using wildly varied but logically sound methods, will arrive at the same correct result," with respect to solving problems. Sometimes more than one strategy can be used in a particular situation.

In **Example 1,** we present a type of problem that has been around for thousands of years in various forms and is solved by observing a pattern within a table. One ancient form of this problem deals with doubling a kernel of corn for each square on a checkerboard. It illustrates an example of exponential growth, covered in more detail in a later chapter.

The popular **TI-84 Plus C** graphing calculator is shown here.

EXAMPLE 1 Calculating a Sum That Involves a Pattern

Lucille Ball starred in *The Lucy Show,* which aired on CBS in the 1960s. She worked for Mr. Mooney (Gale Gordon), who was very careful with his money.

In the September 26, 1966, show "*Lucy, the Bean Queen,*" Mr. Mooney refused to lend her $1500 to buy furniture because he claimed she did not know the value of money. He explained that if she saved so that she would have one penny on Day 1, two pennies on Day 2, four pennies on Day 3, and so on, she would have more than enough money to buy her furniture after only nineteen days. Use a calculator to verify this fact.

Solution

A calculator will help in constructing **Table 4.**

Table 4

Day Number	Accumulated Savings on That Day	Day Number	Accumulated Savings on That Day
1	$0.01	11	$10.24
2	0.02	12	20.48
3	0.04	13	40.96
4	0.08	14	81.92
5	0.16	15	163.84
6	0.32	16	327.68
7	0.64	17	655.36
8	1.28	18	1310.72
9	2.56	19	2621.44
10	5.12		

Lucy will have accumulated $2621.44 on Day 19, enough to buy the furniture.

Estimation

Although calculators can make life easier when it comes to computations, many times we need only estimate an answer to a problem, and in these cases, using a calculator may not be necessary or appropriate.

EXAMPLE 2 Estimating the Number of Birdhouses

A birdhouse for swallows can accommodate up to 8 nests. How many birdhouses would be necessary to accommodate 58 nests?

Solution

If we divide 58 by 8 either by hand or with a calculator, we get 7.25. Can this possibly be the desired number? Of course not, because we cannot consider fractions of birdhouses. Do we need 7 or 8 birdhouses?

To provide nesting space for the nests left over after the 7 birdhouses (as indicated by the decimal fraction), we should plan to use 8 birdhouses. In this problem, we must round our answer *up* to the next counting number.

In the introduction to his book *Innumeracy: Mathematical Illiteracy and Its Consequences*, Temple University professor **John Allen Paulos** writes,

Innumeracy, an inability to deal comfortably with the fundamental notions of number and chance, plagues far too many otherwise knowledgeable citizens. . . .

[W]e were watching the news and the TV weathercaster announced that there was a 50 percent chance of rain for Saturday and a 50 percent chance for Sunday, and concluded that there was therefore a 100 percent chance of rain that weekend. . . .

[U]nlike other failings which are hidden, mathematical illiteracy is often flaunted. "I can't even balance my checkbook." "I'm a people person, not a numbers person." Or "I always hated math."

EXAMPLE 3   Approximating Average Number of Yards per Carry

In 2017, Marshawn Lynch of the Oakland Raiders carried the football a total of 207 times for 891 yards (Data from www.nfl.com). Approximate his average number of yards per carry that year.

Solution

Because we are told only to find Lynch's approximate average, we can say that he carried about 200 times for about 900 yards, and his average was therefore about $\frac{900}{200} = 4.5$ yards per carry. (A calculator shows that his average to the nearest tenth was 4.3 yards per carry. Verify this.)

Interpretation of Graphs

In a **circle graph,** or **pie chart,** a circle is used to indicate the total of all the data categories represented. The circle is divided into sectors, or wedges (like pieces of a pie), whose sizes show the relative magnitudes of the categories. The sum of all the fractional parts must be 1 (for one whole circle).

EXAMPLE 4   Interpreting Information in a Circle Graph

In late 2015, there were about 3300 million (3.3 billion) Internet users worldwide. The circle graph in **Figure 8** shows the approximate shares of these users living in various regions of the world.

(a) Which region had the largest share of Internet users? What was that share?

(b) Estimate the number of Internet users in Asia.

(c) How many actual Internet users were there in Asia?

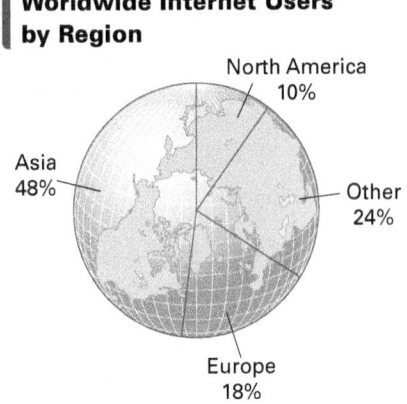

Worldwide Internet Users by Region

North America 10%
Asia 48%
Other 24%
Europe 18%

Data from www.internetworldstats.com

Figure 8

Solution

(a) In the circle graph, the sector for Asia is the largest, so Asia had the largest share of Internet users, 48%.

(b) A share of 48% can be rounded up to 50%. Then find 50% of 3300 million by finding $\frac{1}{2}$ of 3300, or 1650. There were *about* 1650 million users in Asia. (This estimate is a bit high, because we rounded up.)

(c) To find the actual number of users, find 48% of 3300 million. We do this by multiplying 0.48×3300 million.

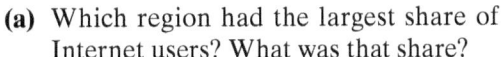

$$0.48 \times 3300 \text{ million} = 1584 \text{ million}$$

48%   of   total   Actual number of users in Asia

A **bar graph** is used to show comparisons. It consists of a series of bars (or simulations of bars) arranged either vertically or horizontally. In a bar graph, values from two categories are paired with each other (for example, years with dollar amounts).

EXAMPLE 5   Interpreting Information in a Bar Graph

The bar graph in **Figure 9** shows annual per-capita spending on health care in the United States for the years 2009 through 2014.

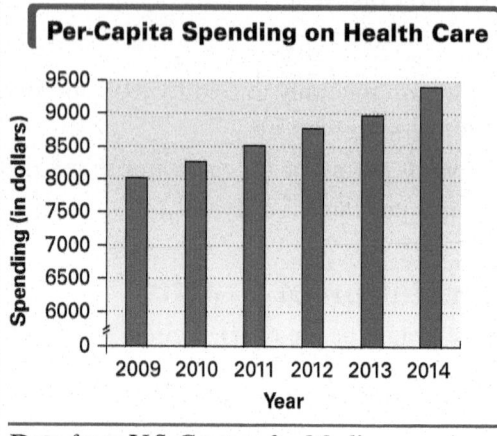
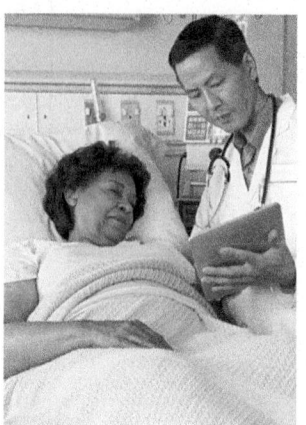

Data from U.S. Centers for Medicare and Medicaid Services.

Figure 9

(a) In what years was per-capita health care spending greater than $8500?

(b) Estimate per-capita health care spending in 2011 and 2014.

(c) Describe the change in per-capita spending as the years progressed.

Solution

(a) Locate 8500 on the vertical axis and follow the line across to the right. Four years—2011, 2012, 2013, and 2014—have bars that extend above the line for 8500, so per-capita health care spending was greater than $8500 in those years.

(b) Locate the top of the bar for 2011 and move horizontally across to the vertical scale to see that it is just above 8500. Per-capita health care spending for 2011 was a bit over $8500.

Similarly, follow the top of the bar for 2014 across to the vertical scale to see that it lies a little above 9400, so per-capita health care spending in 2014 was about $9400.

(c) As the years progressed, per-capita spending on health care increased steadily, from about $8000 in 2009 to about $9400 in 2014.

A **line graph** is used to show changes or trends in data over time. To form a line graph, we connect a series of points representing data with line segments.

EXAMPLE 6   Interpreting Information in a Line Graph

The line graph in **Figure 10** on the next page shows average prices of a gallon of regular unleaded gasoline in the United States for the years 2008 through 2016.

(a) Between which years did the average price of a gallon of gasoline increase?

(b) What was the general trend in the average price of a gallon of gasoline from 2012 through 2016?

(c) Estimate the average price of a gallon of gasoline in 2011 and in 2016. About how much did the price decrease between 2011 and 2016?

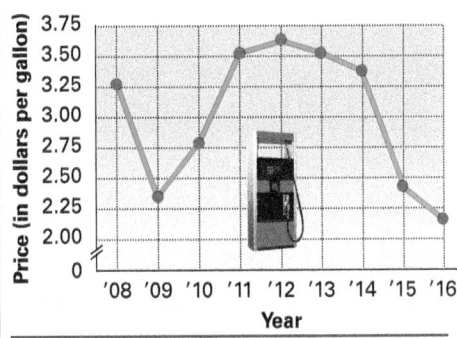

Average U.S. Gasoline Prices

Data from *Energy Prices and Taxes*,
International Energy Agency.
Figure 10

Solution

(a) The line segments between 2009 and 2010, 2010 and 2011, and 2011 and 2012 rise from left to right, so the average price of a gallon of gasoline increased from 2009 to 2012.

(b) The line segments fall from 2012 to 2016, so the average price of a gallon of gasoline decreased over those years.

(c) Move up from 2011 on the horizontal scale to the point plotted for 2011. This point is just above the line on the vertical scale for $3.50, which is an estimate for the price in 2011.

Similarly, locate the point plotted for 2016. Moving across to the vertical scale, the graph indicates that the price for a gallon of gasoline in 2016 was about $2.15. Between 2011 and 2016, the average price of a gallon of gasoline decreased by about

$$\$3.50 - \$2.15 = \$1.35.$$

Communicating Mathematics through Language Skills

Research has indicated that the ability to express mathematical observations in writing can serve as a positive force in one's continued development as a mathematics student. The implementation of writing in the mathematics class can use several approaches.

One way of using writing in mathematics is to keep a **journal** in which you spend a few minutes explaining what happened in class that day. The journal entries may be general or specific, depending on the topic covered, the degree to which you understand the topic, your interest level at the time, and so on. Journal entries are usually written in informal language.

Although journal entries are for the most part informal writings in which the student's thoughts are allowed to roam freely, entries in **learning logs** are typically more formal. An instructor may pose a specific question for a student to answer in a learning log. In this text, we include in each exercise set some exercises that require written answers that are appropriate for answering in a learning log.

EXAMPLE 7   **Writing an Answer to a Conceptual Exercise**

(See Section 1.1, Exercise 23.)
Write a short paragraph to answer this exercise.

Discuss the differences between inductive and deductive reasoning. Give an example of each.

Mathematical writing takes many forms. One of the most famous author/mathematicians was **Charles Dodgson** (1832–1898), who used the pen name **Lewis Carroll.**

Dodgson was a mathematics lecturer at Oxford University in England. Queen Victoria told Dodgson how much she enjoyed *Alice's Adventures in Wonderland* and how much she wanted to read his next book; he is said to have sent her *Symbolic Logic,* his most famous mathematical work.

The *Alice* books made Carroll famous. Late in life, however, Dodgson shunned attention and denied that he and Carroll were the same person, even though he gave away hundreds of signed copies to children and children's hospitals.

Solution

Here is one possible response.

> *Deductive reasoning occurs when you go from general ideas to specific ones. For example, I know that I can multiply both sides of $\frac{1}{2}x = 6$ by 2 to get $x = 12$, because I can multiply both sides of any equation by whatever I want (except 0). Inductive reasoning goes the other way. If I have a general conclusion from specific observations, that's inductive reasoning. Example – in the numbers 4, 8, 12, 16, and so on, I can conclude that the next number is 20, since I always add 4 to get the next number.*

Many professors in mathematics survey courses require students to write short term papers. A list of important mathematicians, philosophers, and scientists follows.

Abel, N.	Copernicus, N.	Kepler, J.	Pascal, B.
Al-Khowârizmi	De Morgan, A.	Kronecker, L.	Plato
Archimedes	Descartes, R.	Leibniz, G.	Pythagoras
Aristotle	Euler, L.	L'Hôspital, G.	Ramanujan, S.
Babbage, C.	Fermat, P.	Lobachevsky, N.	Riemann, G.
Bernoulli, Jakob	Fibonacci	Mandelbrot, B.	Russell, B.
Bernoulli, Johann	Galileo	Napier, J.	Somerville, M.
Cantor, G.	Galois, E.	Nash, J.	Tartaglia, N.
Cardano, G.	Gauss, C.	Newton, I.	Whitehead, A.
	Hilbert, D.	Noether, E.	Wiles, A.

The following topics in the history and development of mathematics can also be used for term papers.

Babylonian and Egyptian mathematics	Historical methods of computation
The origin of zero	Pascal's triangle
Plimpton 322	Women in mathematics
The Rhind papyrus	Unsolved problems in mathematics
Origins of the Pythagorean theorem	The four-color theorem
The regular (Platonic) solids	The proof of Fermat's Last Theorem
The Golden Ratio (Golden Section)	The search for large primes
The three famous construction problems of the Greeks	The co-inventors of calculus
The history of π	The role of the computer in the study of mathematics
Euclid and his *Elements*	Mathematics and music
Early Chinese and Hindu mathematics	Police mathematics
Magic squares	Goldbach's conjecture
Figurate numbers	The development of graphing calculators
The Fibonacci sequence	Multicultural mathematics
The Cardano/Tartaglia controversy	

The motto **"Publish or perish"** implies that a scholar in pursuit of an academic position must publish in a journal in his or her field. There are numerous such journals in mathematics research and/or education. The National Council of Teachers of Mathematics publishes *Teaching Children Mathematics, Mathematics Teaching in the Middle School, Mathematics Teacher, Journal for Research in Mathematics Education,* and *Mathematics Teacher Educator.* Refer to the Web site https://www.nctm.org to access these journals, or consult print copies in your local library.

WHEN WILL I **EVER** USE THIS **?**

Suppose that you are an employee of a group home, and you must purchase the necessary items for a traditional Thanksgiving meal for the residents. Your budget is $80. At the supermarket, you do not want to approach the checkout counter and not have enough for the purchase.

As you shop, you can get a fairly accurate idea of what the total will be by mentally rounding each item (up or down) and keeping a running total.

Item	Actual Cost	Estimate
18-lb turkey	$26.82	$30
12-pack of dinner rolls	2.29	2
15-oz container of margarine	2.79	3
40-oz can of yams	3.34	3
28-oz can of cranberry sauce	3.97	4
26-oz can of cream of mushroom soup	2.99	3
14-oz bag of herb stuffing	3.59	4
1-lb bag of pecans	8.79	9
50-oz can of green beans	2.59	3
28-oz can of onion flakes	2.19	2
22-bag pack of tea bags	4.29	4

Before reaching the checkout counter, look at the items in the basket (move from the top to the bottom in the third column) and add, rounding to simplify the computation. Here is one of many ways this can be done.

"30 plus 2 gives 32; the next three items total 10, so 32 plus 10 equals 42, which I will round down to 40; plus 3 gives 43; plus 4 gives 47, which I will round up to 50; I will round 9 to 10 and add 50 to 10 to give 60; the final three items total 9, which I will round down to 8 (because I rounded up just before) and add to 60 to get 68; 68 is about 70."

Before tax is added, the items total about $70. The sales tax is 8.75%, so round this to 10%. To take 10% of a number, just move the decimal point one place to the left. In this case, the tax will be about $7, so add 7 to 70 to get an approximate total of $77. It looks like $80 will cover the total cost.

Now, as the cashier rings up the purchase, the screen shows that the actual total cost is $63.65, the sales tax is $5.57, and the grand total is $69.22. The estimate is a bit high in this example (because most of the roundoffs were "upward"), but it is "in the ballpark"—that is, the estimate is close enough for our purpose. **Happy Thanksgiving.**

1.4 EXERCISES

CONCEPT CHECK *Fill in the blank with the correct answer. Use a calculator in each case.*

1. Add any two positive numbers. The result is a _____ number.
(negative/positive)

2. Divide any number by $\frac{1}{2}$. The result is the same as multiplying the number by ____.

3. Divide 0 by any number other than 0. The numerical result is ____ .

4. Divide any number by 0. The result is _____ .

Perform the indicated operations, and give as many digits in your answer as shown on your calculator display. (The number of displayed digits may vary depending on the model used.)

5. $39.7 + (8.2 - 4.1)$

6. $2.8 \times (3.2 - 1.1)$

7. $\sqrt{5.56440921}$

8. $\sqrt{37.38711025}$

9. $\sqrt[3]{418.508992}$

10. $\sqrt[3]{700.227072}$

11. 2.67^2

12. 3.49^3

13. 5.76^5

14. 1.48^6

15. $\dfrac{14.32 - 8.1}{2 \times 3.11}$

16. $\dfrac{12.3 + 18.276}{3 \times 1.04}$

17. $\sqrt[5]{1.35}$

18. $\sqrt[6]{3.21}$

19. $\dfrac{\pi}{\sqrt{2}}$

20. $\dfrac{2\pi}{\sqrt{3}}$

21. $\sqrt[4]{\dfrac{2143}{22}}$

22. $\dfrac{12,345,679 \times 72}{\sqrt[3]{27}}$

23. $\dfrac{\sqrt{2}}{\sqrt[3]{6}}$

24. $\dfrac{\sqrt[3]{12}}{\sqrt{3}}$

Perform each calculation and observe the answers. Then fill in the blank with the appropriate response.

25. $\boxed{5.6^0}$; $\boxed{\pi^0}$; $\boxed{2^0}$; $\boxed{120^0}$

Raising a nonzero number to the power 0 gives a result of _____.

26. $\boxed{1^2}$; $\boxed{1^3}$; $\boxed{1^{-3}}$; $\boxed{1^0}$

Raising 1 to any power gives a result of _____.

27. $\boxed{\frac{1}{7}}$; $\boxed{\frac{1}{-9}}$; $\boxed{\frac{1}{3}}$; $\boxed{\frac{1}{-8}}$

The sign of the reciprocal of a number is _____ the sign of the number.
(the same as/different from)

28. $\boxed{\sqrt{-3}}$; $\boxed{\sqrt{-4}}$; $\boxed{\sqrt{-10}}$

Taking the square root of a negative number gives _____ on a basic calculator.

29. $\boxed{-3 * -4 * -5}$; $\boxed{-3 * -4 * -5 * -6 * -7}$;

$\boxed{-3 * -4 * -5 * -6 * -7 * -8 * -9}$

Multiplying an *odd* number of negative numbers gives a _____ product.
(positive/negative)

30. $\boxed{-3 * -4}$; $\boxed{-3 * -4 * -5 * -6}$;

$\boxed{-3 * -4 * -5 * -6 * -7 * -8}$

Multiplying an *even* number of negative numbers gives a _____ product.
(positive/negative)

Solve each problem.

31. Choose any number consisting of five digits. Multiply it by 9 on your calculator. Now add the digits in the answer. If the sum is more than 9, add the digits of this sum, and repeat until the sum is less than 10. Your answer will always be 9. Repeat the exercise with a number consisting of six digits. Does the same result hold?

32. Use your calculator to *square* the following two-digit numbers ending in 5: 15, 25, 35, 45, 55, 65, 75, 85. Write down your results, and examine the pattern that develops. Then use inductive reasoning to predict the value of 95^2. Write an explanation of how you can mentally square a two-digit number ending in 5.

33. Find the decimal representation of $\frac{1}{6}$ on your calculator. Following the decimal point will be a 1 and a string of 6s. The final digit will be a 7 if your calculator *rounds off* or a 6 if it *truncates*. Which kind of calculator do you have?

34. Choose any three-digit number and enter the digits into a calculator. Then enter them again to get a six-digit number. Divide this six-digit number by 7. Divide the result by 13. Divide the result by 11. What is interesting about your answer? Explain why this happens.

35. Choose any digit except 0. Multiply it by 429. Now multiply the result by 259. What is interesting about your answer? Explain why this happens.

36. Refer to **Example 1.** If Lucy continues to double her savings amount each day, on what day will she become a millionaire?

Give an appropriate counting number answer to each question. (Find the least counting number that will work.)

37. Pages to Store Trading Cards A plastic page designed to hold trading cards will hold up to 9 cards. How many pages will be needed to store 563 cards?

38. Drawers for DVDs A sliding drawer designed to hold DVD cases has 20 compartments. If Chris wants to house his collection of 408 Disney DVDs, how many such drawers will he need?

39. Containers for African Violets A gardener wants to fertilize 800 African violets. Each container of fertilizer will supply up to 60 plants. How many containers will she need to do the job?

40. Fifth-Grade Teachers Needed False River Academy has 155 fifth-grade students. The principal has decided that each fifth-grade teacher should have a maximum of 24 students. How many fifth-grade teachers does the principal need?

Use estimation to determine the choice closest to the correct answer.

41. Price per Acre of Land To build a "millennium clock" on Mount Washington in Nevada that would tick once each year, chime once each century, and last at least 10,000 years, the nonprofit Long Now Foundation purchased 80 acres of land for $140,000. Which one of the following is the closest estimate to the price per acre?

A. $1000 **B.** $2000 **C.** $4000 **D.** $11,200

42. Time of a Round Trip The distance from Seattle, Washington, to Springfield, Missouri, is 2009 miles. About how many hours would a round trip from Seattle to Springfield and back take a bus that averages 50 miles per hour for the entire trip?

A. 60 **B.** 70 **C.** 80 **D.** 90

43. People per Square Mile Baton Rouge, Louisiana, has a population of 230,058 and covers 76.9 square miles. About how many people per square mile live in Baton Rouge?

A. 3000 **B.** 300 **C.** 30 **D.** 30,000

44. Revolutions of Mercury The planet Mercury takes 88.0 Earth days to revolve around the sun once. Pluto takes 90,824.2 days to do the same. When Pluto has revolved around the sun once, about how many times will Mercury have revolved around the sun?

A. 100,000 **B.** 10,000 **C.** 1000 **D.** 100

45. Reception Average In 2017, Keenan Allen of the Los Angeles Chargers caught 102 passes for 1393 yards. His approximate number of yards gained per catch was ____.

A. $\dfrac{1}{14}$ **B.** 0.07 **C.** 142,086 **D.** 14

46. Area of the Sistine Chapel The Sistine Chapel in Vatican City measures 40.5 meters by 13.5 meters.

Which is the closest approximation to its area?

A. 110 meters **B.** 55 meters

C. 110 square meters **D.** 600 square meters

Visitors to the U.S. In a recent year, approximately 60 million people from other countries visited the United States. The circle graph shows the distribution of these international visitors by country or region. Use the graph to answer Exercises 47–50 on the following page.

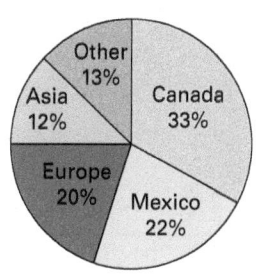

International Travelers to the United States

Other 13%
Asia 12%
Canada 33%
Europe 20%
Mexico 22%

Data from U.S. Dept. of Commerce.

47. What share was from Other regions?

48. What share was from Europe or Canada?

49. How many people (in millions) were from Europe?

50. How many more people (in millions) were from Mexico than from Asia?

Milk Production *The bar graph shows total U.S. milk production (in billions of pounds) for the years 2010 through 2016. Use the bar graph to answer the following.*

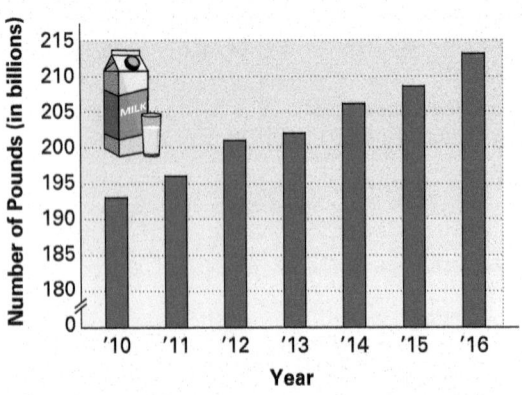

U.S. Milk Production

Data from U.S. Dept. of Agriculture.

51. In what years was U.S. milk production greater than 200 billion pounds?

52. In what years was U.S. milk production about the same?

53. Estimate U.S. milk production in 2010 and 2016.

54. Describe the change from 2010 to 2016.

Unemployment Rate *The line graph below shows the unemployment rate in the U.S. civilian labor force for the years 2009 through 2015. Use the graph to answer the following.*

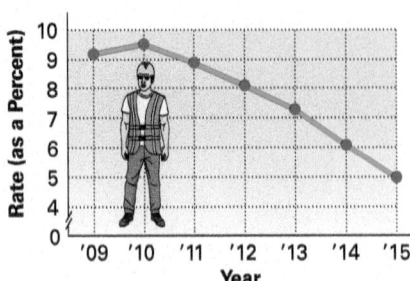

Unemployment Rate

Data from Bureau of Labor Statistics.

55. Between which years did the unemployment rate increase?

56. What was the general trend in the unemployment rate between 2010 and 2015?

57. Estimate the unemployment rate in 2010 and that in 2015. About how much did the rate decrease between 2010 and 2015?

58. Which year was the first shown to have an unemployment rate less than 8%?

Use effective writing skills to address the following.

59. ***Mathematics Web sites*** The Web sites below provide a list of mathematics-related topics. Investigate, choose a topic that interests you, and report on it according to the guidelines provided by your instructor.

http://mathworld.wolfram.com

world.std.com/~reinhold/mathmovies.html

http://www.surrey.ac.uk/department-mathematics/

http://www.cut-the-knot.org

http://www.ics.uci.edu/~eppstein/recmath.html

http://www.nctm.org/mathforum

http://rosettacode.org

http://plus.maths.org

http://mathlair.allfunandgames.ca/recreational.php

http://www.khanacademy.org/math/math-for-fun-and-glory

60. ***The Simpsons*** The longest-running animated television series is *The Simpsons*, which began in 1989. The Web site www.cs.appstate.edu/~sjg/simpsonsmath/ explores the occurrence of mathematics in the episodes on a season-by-season basis. Watch several episodes and elaborate on the mathematics found in them.

61. ***Donald in Mathmagic Land*** One of the most popular mathematical films of all time is *Donald in Mathmagic Land,* a 1959 Disney short. Spend an entertaining half-hour watching this film, and write a report on it according to the guidelines provided by your instructor.

62. ***Mathematics in Hollywood*** A theme of mathematics-related scenes in movies and television is found throughout this text. Prepare a report on one or more such scenes, and determine whether the mathematics involved is correct or incorrect. If it is correct, show why. If it is incorrect, find the correct answer.

Chapter 1   SUMMARY

KEY TERMS

1.1

conjecture
inductive reasoning
counterexample
deductive reasoning
natural (counting) numbers
base
exponent
premise
conclusion
logical argument

1.2

number sequence
terms of a sequence
arithmetic sequence
common difference
geometric sequence
common ratio
method of successive
    differences
mathematical
    induction

triangular, square, and
    pentagonal numbers
figurate number
subscript
Kaprekar constant

1.3

perfect square
square root
radical symbol

1.4

numeracy
four-function calculator
scientific calculator
graphing calculator
circle graph (pie chart)
bar graph
line graph
journal
learning log

TEST YOUR WORD POWER

See how well you have learned the vocabulary in this chapter.

1. A **conjecture** is
   A. a statement that has been proved to be true.
   B. an educated guess based on repeated observations.
   C. an example that shows that a general statement is false.
   D. an example of deductive reasoning.

2. An example of a **natural number** is
   A. 0.   B. $\frac{1}{2}$.   C. −1.   D. 1.

3. An **arithmetic sequence** is
   A. a sequence that has a common difference between any two successive terms.
   B. a sequence that has a common sum of any two successive terms.
   C. a sequence that has a common ratio between any two successive terms.
   D. a sequence that can begin 1, 1, 2, 3, 5. . . .

4. A **geometric sequence** is
   A. a sequence that has a common difference between any two successive terms.
   B. a sequence that has a common sum of any two successive terms.
   C. a sequence that has a common ratio between any two successive terms.
   D. A sequence that can begin 1, 1, 2, 3, 5, ...

5. The symbol T_n, which uses the **subscript** n, is read
   A. "T to the nth power."   B. "T times n."
   C. "T of n."   D. "T sub n."

ANSWERS
1. B   2. D   3. A   4. C   5. D

QUICK REVIEW

Concepts   *Examples*

1.1 SOLVING PROBLEMS BY INDUCTIVE REASONING

Inductive Reasoning
Inductive reasoning is characterized by drawing a general conclusion (making a conjecture) from repeated observations of specific examples. The conjecture may or may not be true.

Consider the following:
When I square the first twenty numbers ending in 5, the result always ends in 25. Therefore, I make the conjecture that this happens in the twenty-first case.

A general conclusion from inductive reasoning can be shown to be false by providing a single counterexample.

This is an example of inductive reasoning because a general conclusion follows from repeated observations.

Concepts	*Examples*

Deductive Reasoning
Deductive reasoning is characterized by applying general principles to specific examples.

Consider the following:

The formula for finding the perimeter P of a rectangle with length L and width W is $P = 2L + 2W$. Therefore, the perimeter P is

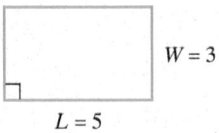

$$2(5) + 2(3) = 16.$$

This is an example of deductive reasoning because a specific conclusion follows from a mathematical formula that is true in general.

1.2 AN APPLICATION OF INDUCTIVE REASONING: NUMBER PATTERNS

Sequences
A number sequence is a list of numbers having a first number, a second number, a third number, and so on, which are called the terms of the sequence.

The Fibonacci sequence

$$1, 1, 2, 3, 5, 8, \ldots$$

has fourth term 3.

Arithmetic Sequence
In an arithmetic sequence, each term after the first is obtained by adding the same number, called the common difference.

The arithmetic sequence that begins

$$2, 4, 6, 8$$

has common difference $4 - 2 = 2$, and the next term in the sequence is $8 + 2 = 10$.

Geometric Sequence
In a geometric sequence, each term after the first is obtained by multiplying by the same number, called the common ratio.

The geometric sequence that begins

$$4, 20, 100, 500$$

has common ratio $\frac{20}{4} = 5$, and the next term in the sequence is $500 \times 5 = 2500$.

Method of Successive Differences
The next term in a sequence can sometimes be found by computing successive differences between terms until a pattern can be established.

The sequence that begins

$$7, 15, 25, 37$$

has the following successive differences.

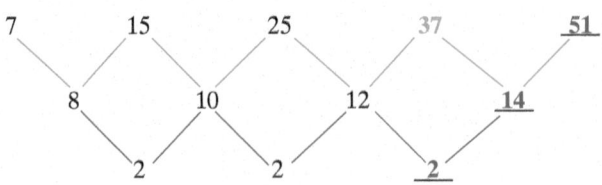

The next term in the sequence is $37 + 14 = 51$.

Figurate Numbers
Figurate numbers, such as triangular and square numbers, can be represented by geometric arrangements of points.

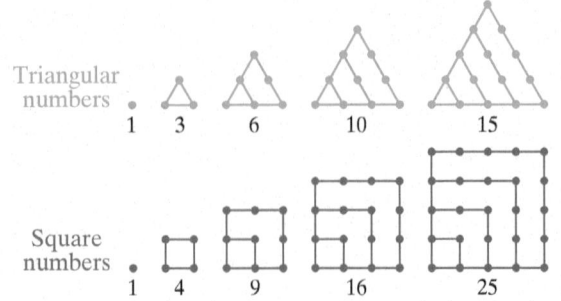

Concepts   *Examples*

1.3 STRATEGIES FOR PROBLEM SOLVING

Polya's Four-Step Method for Problem Solving

Step 1   Understand the problem.

Step 2   Devise a plan.

Step 3   Carry out the plan.

Step 4   Look back and check.

Problem-Solving Strategies

- Make a table or a chart.
- Look for a pattern.
- Solve a similar, simpler problem.
- Draw a sketch.
- Use inductive reasoning.
- Write an equation and solve it.
- If a formula applies, use it.
- Work backward.
- Guess and check.
- Use trial and error.
- Use common sense.
- Look for a "catch" if an answer seems too obvious or impossible.

Determine the ones, or units, digit in 7^{350}.

We can observe a pattern in the table of simpler powers of 7. (Use a calculator.)

$7^1 = 7$	$7^5 = 16,807$	$7^9 = 40,353,607$
$7^2 = 49$	$7^6 = 117,649$	$\ldots$
$7^3 = 343$	$7^7 = 823,543$	$\ldots$
$7^4 = 2401$	$7^8 = 5,764,801$	$\ldots$

The ones digit appears in a pattern of four digits over and over: 7, 9, 3, 1, 7, 9, 3, 1, ... If the exponent is divided by 4, the remainder helps predict the ones digit. If we divide the exponent 350 by 4, the quotient is 87 and the remainder is 2, just as it is in the second row above for 7^2 and 7^6, where the units digit is 9. So the units digit in 7^{350} is 9.

How many ways are there to make change equivalent to one dollar using only nickels, dimes, and quarters? You do not need at least one coin of each denomination. (*MT* calendar problem)

If we start with 4 quarters, there is 1 way to make change for a dollar.

If we start with 3 quarters, there are 3 ways.

If we start with 2 quarters, there are 6 ways.   Thus, there are $1 + 3 + 6 + 8 + 11 = 29$ ways in all.

If we start with 1 quarter, there are 8 ways.

If we start with 0 quarters, there are 11 ways.

1.4 NUMERACY IN TODAY'S WORLD

In practical applications, it is often convenient to simply approximate to get an idea of an answer to a problem.

Circle graphs (pie charts), bar graphs, and line graphs are used in today's media to illustrate data in a compact way.

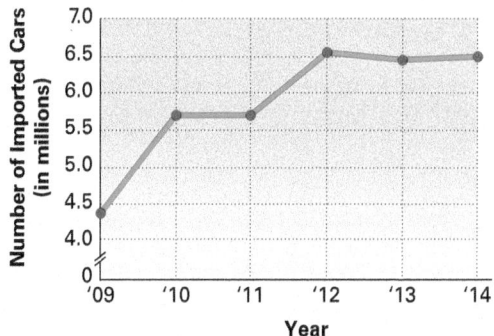

Passenger Cars Imported into the United States

Data from U.S. Census Bureau.

Over which two consecutive years did the number of passenger cars imported decrease?

The graph falls from left to right between 2012 and 2013. In all other cases, the graph either stays the same or rises between consecutive years.

Chapter 1 TEST

Decide whether the reasoning involved is an example of inductive *or* deductive *reasoning.*

1. Michelle is a sales representative for a publishing company. For the past 16 years, she has exceeded her annual sales goal, primarily by selling mathematics texts. Therefore, she will also exceed her annual sales goal this year.

2. For all natural numbers n, n^2 is also a natural number. For example, 176 is a natural number. Therefore, 176^2 is a natural number.

Solve each problem.

3. Use the list of equations and inductive reasoning to predict the next equation, and then verify your conjecture.

   $65,359,477,124,183 \times 17 = 1,111,111,111,111,111$
   $65,359,477,124,183 \times 34 = 2,222,222,222,222,222$
   $65,359,477,124,183 \times 51 = 3,333,333,333,333,333$

4. Use the method of successive differences to find the next term in the sequence

   $$3, 11, 31, 69, 131, 223, \ldots.$$

5. Find the sum $1 + 2 + 3 + \cdots + 250$.

6. Consider the following equations, where the left side of each is an octagonal number.

   $$1 = 1$$
   $$8 = 1 + 7$$
   $$21 = 1 + 7 + 13$$
   $$40 = 1 + 7 + 13 + 19$$

   Use the pattern established on the right sides to predict the next octagonal number. What is the next equation in the list?

7. Use the result of **Exercise 6** and the method of successive differences to find the first eight octagonal numbers. Then divide each by 4 and record the remainder. What is the pattern obtained?

8. Describe the pattern used to obtain the terms of the Fibonacci sequence below. What is the next term?

   $$1, 1, 2, 3, 5, 8, 13, 21, \ldots.$$

Use problem-solving strategies to solve each problem.

9. *Building a Fraction* Each of the four digits 2, 4, 6, and 9 is placed in one of the boxes to form a fraction. The numerator and the denominator are both two-digit whole numbers. What is the smallest value of all the common fractions that can be formed? Express your answer as a common fraction. (*MT* calendar problem)

   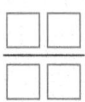

10. *Counting Puzzle (Rectangles)* How many rectangles of any size are in the figure shown? (*MT* calendar problem)

11. *Units Digit of a Power of 9* What is the units digit (ones digit) in the decimal representation of 9^{1997}? (*MT* calendar problem)

12. *Counting Puzzle (Triangles)* How many triangles are in this figure? (*MT* calendar problem)

    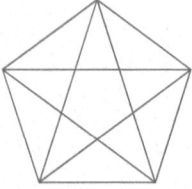

13. *Missing Digit* A digit is placed in each empty square so that each row and each column contain the digits 1, 2, 3, 4, 5. What digit is placed in the square at the bottom right corner? (*MT* calendar problem)

	5	4		
1	3			
		5	3	
2		3	1	

14. *Shrinkage* Dr. Small is 36 inches tall, and Ms. Tall is 96 inches tall. If Dr. Small shrinks 2 inches per year and Ms. Tall grows $\frac{2}{3}$ of an inch per year, how tall will Ms. Tall be when Dr. Small disappears altogether? (*MT* calendar problem)

15. *Units Digit of a Sum* Find the units digit (ones digit) of the decimal numeral representing the number

$$11^{11} + 14^{14} + 16^{16}.$$

(*MT* calendar problem)

16. Based on your knowledge of elementary arithmetic, describe the pattern that can be observed when the following operations are performed.

$$9 \times 1, \quad 9 \times 2, \quad 9 \times 3, \ldots, \quad 9 \times 9$$

(*Hint:* Add the digits in the answers. What do you notice?)

Use your calculator to evaluate each of the following. Give as many decimal places as the calculator displays.

17. $\sqrt{98.16}$ **18.** 3.25^3

19. *Basketball Scoring Results* The Division I record for NCAA Women's Basketball free throws in a career is held by Kelsey Plum of Washington. In her career (2014–2017), she made 912 free throws in a total of 1036 attempts. This means that for every 100 attempts, she made *approximately* _____ of them.

A. 1   **B.** 9   **C.** 99   **D.** 90

20. *Degrees Earned* The line graph shows the percent of 4-year college students who earned a degree within 5 years. Use the graph to answer the following.

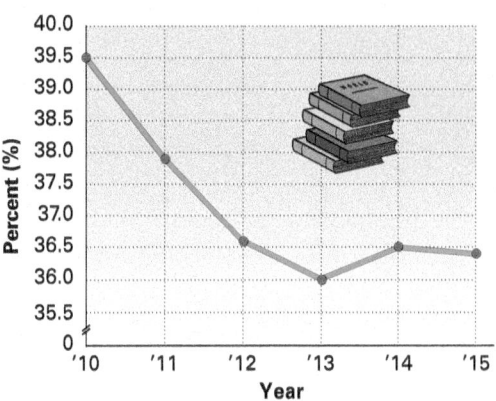

Percents of Students Earning a Degree within 5 Years

Data from ACT.

(a) Between the years 2010 and 2013, what was the general trend in the percent of students who earned a degree within 5 years?

(b) Estimate the percent of students earning a degree within 5 years in 2010 and 2013.

(c) About how much did the percent decrease between 2010 and 2013?

2 The Basic Concepts of Set Theory

The job outlook in the United States is good—and improving—for all types of nursing careers, as well as other health-related categories. Students may be considering groups of different training programs; lists of certificate or degree objectives; an array of employment opportunities; the pros and cons of different opportunities relative to upward mobility, stress level, and flexibility; and many others. The properties of these arrays, groups, and collections can be better understood by treating them all as *sets* (the mathematical term) and applying the methods presented in this chapter. (See, for example, the opening discussion of the fourth section of this chapter.)

Refer to the table on the next page. To three significant figures, how many jobs are predicted for 2026 for RNs? For LPNs? (Turn 3 pages ahead for the answers.)

	Typical Entry-Level Education	2016 Median Annual Pay	Number of Jobs in 2016	Job Outlook, 2016–2026
Registered Nurses, RNs	Bachelor's Degree	$68,450	2,955,200	+14.8%
Licensed Practical (or Vocational) Nurses, LPNs (or LVNs)	Postsecondary Nondegree Award	$44,000	724,500	+12.2%

Data from www.bls.gov

2.1   SYMBOLS AND TERMINOLOGY

OBJECTIVES

1   Use three methods to designate sets.

2   Recognize important categories of numbers, and determine cardinal numbers of sets.

3   Distinguish between finite and infinite sets.

4   Determine whether two sets are equal.

The basic ideas of set theory were developed by the German mathematician **Georg Cantor** (1845–1918) in about 1875. Cantor created a new field of theory and at the same time continued the long debate over infinity that began in ancient times. He developed counting by one-to-one correspondence to determine how many objects are contained in a set. Infinite sets differ from finite sets by not obeying the familiar law that the whole is greater than any of its parts.

Designating Sets

A **set** is a collection of objects. The objects belonging to the set are the **elements,** or **members,** of the set. Sets are designated using the following three methods:

(1) *word description*,   (2) the *listing method*,   and   (3) *set-builder notation*.

The set of even counting numbers less than 10	Word description
$\{2, 4, 6, 8\}$	Listing method
$\{x \mid x \text{ is an even counting number less than } 10\}$	Set-builder notation

The set-builder notation above is read

"the set of all x such that x is an even counting number less than 10."

Set-builder notation uses the algebraic idea of a *variable*. (Any symbol would do, but just as in other algebraic applications, the letter x is a common choice.)

Variable representing an
element in general
↓
$$\{x \mid x \text{ is an even counting number less than } 10\}$$
↑
Criteria by which an element
qualifies for membership in the set

Sets are commonly given names (usually capital letters), such as E for the set of all letters of the English alphabet.

$$E = \{a, b, c, d, e, f, g, h, i, j, k, l, m, n, o, p, q, r, s, t, u, v, w, x, y, z\}$$

The listing notation can often be shortened by establishing the pattern of elements included and using ellipsis points to indicate a continuation of the pattern.

$$E = \{a, b, c, d, \ldots, x, y, z\}, \quad \text{or} \quad E = \{a, b, c, d, e, \ldots, z\}$$

The set containing no elements is called the **empty set,** or **null set.** The symbol $\varnothing$ is used to denote the empty set, so $\varnothing$ and $\{ \ \}$ have the same meaning. We do *not* denote the empty set with the symbol $\{\varnothing\}$ because this notation represents a set with one element (that element being the empty set).

StatCrunch Listing Elements of Sets

EXAMPLE 1 **Listing Elements of Sets**

Give a complete listing of all the elements of each set.

(a) the set of counting numbers between eight and thirteen

(b) $\{5, 6, 7, \ldots, 13\}$

(c) $\{x \mid x$ is a counting number between 4 and 5$\}$

Solution

(a) This set can be denoted $\{9, 10, 11, 12\}$. (Note that the word "*between*" excludes the endpoint values.)

(b) This set contains the element 5, then 6, then 7, and so on, with each element obtained by adding 1 to the previous element in the list. This pattern stops at 13, so a complete listing is

$$\{5, 6, 7, 8, 9, 10, 11, 12, 13\}.$$

(c) There are no counting numbers between 4 and 5, so this is the empty set:

$$\{ \ \}, \quad \text{or} \quad \varnothing.$$

For a set to be useful, it must be *well defined*. The set E of the letters of the English alphabet is well defined. For example, it is clear that the letter h is an element of E, whereas the Greek letter θ (theta) is not an element of E.

However, given the set G of all good singers, and a particular singer, Adilah, it may not be possible to say whether

Adilah is an element of G or   Adilah is *not* an element of G.

The problem is the word "good." How good is good? Because we cannot necessarily decide whether a given singer belongs to set G, set G is not well defined.

The fact that the letter h is an element of set E is denoted by using the symbol $\in$.

$$h \in E \quad \text{This is read "h is an element of set } E."$$

The letter θ is not an element of E. To show this, $\in$ with a slash mark is used.

$$\theta \notin E \quad \text{This is read "}\theta \text{ is not an element of set } E."$$

Many other mathematical symbols also have their meanings negated by use of a **slash mark.** The most common example is

$$\neq, \quad \text{which means} \quad \text{"does not equal"} \quad \text{or} \quad \text{"is not equal to".}$$

StatCrunch Identiying Elements of Sets

EXAMPLE 2 **Applying the Symbol $\in$**

Decide whether each statement is *true* or *false*.

(a) $4 \in \{1, 2, 5, 8, 13\}$ **(b)** $0 \in \{0, 1, 2, 3\}$ **(c)** $\frac{1}{6} \notin \left\{\frac{1}{3}, \frac{1}{4}, \frac{1}{5}\right\}$

Solution

(a) Because 4 is *not* an element of the set $\{1, 2, 5, 8, 13\}$, the statement is *false*.

(b) Because 0 is indeed an element of the set $\{0, 1, 2, 3\}$, the statement is *true*.

(c) This statement says that $\frac{1}{6}$ is not an element of the set $\left\{\frac{1}{3}, \frac{1}{4}, \frac{1}{5}\right\}$, which is *true*.

Sets of Numbers and Cardinality

Important categories of numbers are summarized below.

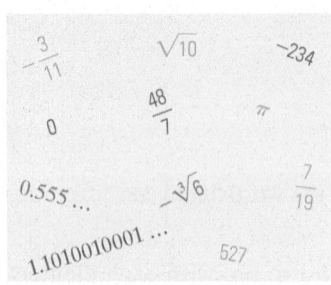

Most concepts in this chapter will be illustrated using the **sets of numbers** shown at the right, not only to solidify understanding of these sets of numbers but also because all these sets are precisely, or "well," defined and therefore provide clear illustrations.

SETS OF NUMBERS

Natural numbers (or counting numbers) $\{1, 2, 3, 4, \ldots\}$

Whole numbers $\{0, 1, 2, 3, 4, \ldots\}$

Integers $\{\ldots, -3, -2, -1, 0, 1, 2, 3, \ldots\}$

Rational numbers $\left\{\frac{p}{q} \mid p \text{ and } q \text{ are integers, and } q \neq 0\right\}$

(*Examples:* $\frac{3}{5}, -\frac{7}{9}, 5, 0$. Any rational number may be written as a terminating decimal number, such as 0.25, or as a repeating decimal number, such as 0.666 . . . .)

Real numbers $\{x \mid x \text{ is a number that can be expressed as a decimal}\}$

Irrational numbers $\{x \mid x \text{ is a real number and } x \text{ cannot be expressed as a quotient of integers}\}$

(*Examples:* $\sqrt{2}, \sqrt[3]{4}, \pi$. Decimal representations of irrational numbers are neither terminating nor repeating.)

The number of elements in a set is called the **cardinal number,** or **cardinality,** of the set. The symbol

$$n(A), \quad \text{which is read "} n \text{ of } A,\text{"}$$

represents the cardinal number of set A. If elements are repeated in a set listing, they should not be counted more than once when determining the cardinal number of the set.

StatCrunch Finding Cardinal Numbers

EXAMPLE 3   Finding Cardinal Numbers

Find the cardinal number of each set.

(a) $K = \{3, 9, 27, 81\}$ **(b)** $M = \{0\}$ **(c)** $B = \{1, 1, 2, 3, 2\}$

(d) $R = \{7, 8, \ldots, 15, 16\}$ **(e)** $\varnothing$

Solution

(a) Set K contains four elements, so the cardinal number of set K is 4, and $n(K) = 4$.

(b) Set M contains only one element, 0, so $n(M) = 1$.

(c) Do not count repeated elements more than once. Set B has only three *distinct* elements, so $n(B) = 3$.

(d) Although only four elements are specifically listed, the ellipsis points indicate that there are additional elements in the set. Counting them all, we find that there are ten elements, so $n(R) = 10$.

(e) The empty set, $\varnothing$, contains no elements, so $n(\varnothing) = 0$.

Finite and Infinite Sets

If the cardinal number of a set is a particular whole number (0 or a counting number), as in all parts of **Example 3,** we call that set a **finite set.** Given enough time, we could finish counting all the elements of any finite set and arrive at its cardinal number.

Some sets, however, are so large that we could never finish the counting process. The counting numbers themselves form such a set. Whenever a set is so large that its cardinal number is not found among the whole numbers, we call that set an **infinite set.**

A close-up of a camera lens shows the **infinity symbol**, ∞, defined in this case as any distance greater than 1000 times the focal length of a lens.

The sign was invented by the mathematician John Wallis in 1655. Wallis used **1**/∞ to represent an infinitely small quantity.

EXAMPLE 4 **Designating an Infinite Set**

Designate all odd counting numbers by the three common methods of set notation.

Solution

The set of all odd counting numbers   Word description

$\{1, 3, 5, 7, 9, \dots\}$ Listing method

$\{x \mid x \text{ is an odd counting number}\}$ Set-builder notation

Equality of Sets

SET EQUALITY

Set A is **equal** to set B provided the following two conditions are met:

1. Every element of A is an element of B, and
2. Every element of B is an element of A.

Two sets are equal if they contain exactly the same elements, regardless of order.

$$\{a, b, c, d\} = \{a, c, d, b\} \quad \text{Both sets contain exactly the same elements.}$$

Repetition of elements in a set listing does not add new elements.

$$\{1, 0, 1, 5, 3, 3\} = \{0, 1, 3, 5\} \quad \text{Both sets contain exactly the same elements.}$$

EXAMPLE 5 **Determining Whether Two Sets Are Equal**

Are $\{-4, 3, 2, 5\}$ and $\{-4, 0, 3, 2, 5\}$ equal sets?

Solution

Every element of the first set is an element of the second. However, 0 is an element of the second and not of the first. The sets do not contain exactly the same elements.

$$\{-4, 3, 2, 5\} \neq \{-4, 0, 3, 2, 5\} \quad \text{The sets are not equal.}$$

Cardinalities of Infinite Sets Two sets are **equivalent** if they have the *same number* of elements. (See **Exercises 85–88**.) Georg Cantor extended the idea of equivalence to infinite sets, used one-to-one correspondence to establish equivalence, and showed that, surprisingly, the natural numbers, the whole numbers, the integers, and the rational numbers are all equivalent. The elements of any one of these sets will match up, one-to-one, with those of any other, with no elements left over in either set. All these sets have cardinal number $\aleph_0$ (which is read *aleph null*).

However, the irrational numbers and the real numbers, though equivalent to one another, are of a higher infinite order than the sets mentioned above. Their cardinal number is denoted **c** (representing the *continuum* of points on a line). In other words, there are different sizes of infinities!

EXAMPLE 6 **Determining Whether Two Sets Are Equal**

Decide whether each statement is *true* or *false*.

(a) $\{3\} = \{x \mid x \text{ is a counting number between 1 and 5}\}$

(b) $\{x \mid x \text{ is a negative whole number}\} = \{y \mid y \text{ is a number that is both rational and irrational}\}$

(c) $\{(0, 0), (1, 1), (2, 4)\} = \{(x, y) \mid x \text{ is a natural number less than 3, and } y = x^2\}$

Solution

(a) The set on the right contains *all* counting numbers between 1 and 5, namely 2, 3, and 4, while the set on the left contains *only* the number 3. Because the sets do not contain exactly the same elements, they are not equal. The statement is *false*.

(b) No whole numbers are negative, so the set on the left is ∅. By definition, if a number is rational, it cannot be irrational, so the set on the right is also ∅. Because each set is the empty set, the sets are equal. The statement is *true*.

(c) The first listed ordered pair in the set on the left has x-value 0, which is not a natural number. Therefore, the ordered pair $(0, 0)$ is not an element of the set on the right, even though the relationship $y = x^2$ is true for $(0, 0)$. Thus the sets are not equal. The statement is *false*.

2.1 EXERCISES

CONCEPT CHECK *Write* true *or* false *for each statement.*

1. $n(\varnothing) = 0$

2. Every whole number is a natural number.

3. The set of all even integers less than 6 is the same as $\{0, 2, 4\}$.

4. $\{x \mid x \text{ is an odd counting number}\} = \{1, 3, 5, 7, 9, \ldots\}$

CONCEPT CHECK *Match each set in Group I with the appropriate description in Group II.*

I

5. $\{1, 3, 5, 7, 9\}$

6. $\{x \mid x \text{ is an even integer greater than 4 and less than 6}\}$

7. $\{\ldots, -4, -3, -2, -1\}$

8. $\{\ldots, -5, -3, -1, 1, 3, 5, \ldots\}$

9. $\{2, 4, 8, 16, 32\}$

10. $\{\ldots, -4, -2, 0, 2, 4, \ldots\}$

11. $\{2, 4, 6, 8, 10\}$

12. $\{2, 4, 6, 8\}$

II

A. the set of all even integers

B. the set of the five least positive integer powers of 2

C. the set of even positive integers less than 10

D. the set of all odd integers

E. the set of all negative integers

F. the set of odd positive integers less than 10

G. $\varnothing$

H. the set of the five least positive integer multiples of 2

List all the elements of each set. Use set notation and the listing method to describe the set.

13. the set of all counting numbers less than or equal to 6

14. the set of all whole numbers between 8 and 18

15. the set of all whole numbers not greater than 4

16. the set of all natural numbers between 4 and 14

17. $\{6, 7, 8, \ldots, 14\}$

18. $\{3, 6, 9, 12, \ldots, 30\}$

19. $\{2, 4, 8, \ldots, 256\}$

20. $\{90, 87, 84, \ldots, 69\}$

21. $\{x \mid x \text{ is an even whole number less than 11}\}$

22. $\{x \mid x \text{ is an odd integer between } -8 \text{ and 7}\}$

Denote each set by the listing method. There may be more than one correct answer.

23. the set of U.S. Great Lakes

24. $\{x \mid x \text{ is a negative multiple of 6}\}$

25. $\{x \mid x \text{ is the reciprocal of a natural number}\}$

26. $\{x \mid x \text{ is a positive integer power of 4}\}$

Denote each set by set-builder notation, using x as the variable. There may be more than one correct answer.

27. the set of all rational numbers

28. the set of all even natural numbers

29. $\{1, 3, 5, \ldots, 75\}$

30. $\{35, 40, 45, \ldots, 95\}$

Give a word description for each set. There may be more than one correct answer.

31. $\{-9, -8, -7, \ldots, 7, 8, 9\}$

32. $\left\{\dfrac{1}{2}, \dfrac{2}{3}, \dfrac{3}{4}, \ldots\right\}$

33. $\{\text{Alabama, Alaska, Arizona}, \ldots, \text{Wisconsin, Wyoming}\}$

34. $\{\text{Alaska, California, Hawaii, Oregon, Washington}\}$

Identify each set as finite *or* infinite.

35. $\{2, 4, 6, \ldots, 932\}$

36. $\{6, 12, 18\}$

37. $\{x \mid x \text{ is a natural number greater than 50}\}$

38. $\{3, 6, 9, \ldots\}$

39. $\{x \mid x \text{ is a rational number}\}$

40. $\{x \mid x \text{ is a rational number between 0 and 1}\}$

Find n(A) for each set.

41. $A = \{0, 1, 2, 3, 4, 5, 6, 7\}$

42. $A = \{-3, -1, 1, 3, 5, 7, 9\}$

43. $A = \{2, 4, 6, \ldots, 1000\}$

44. $A = \{0, 1, 2, 3, \ldots, 2000\}$

45. $A = \{a, b, c, \ldots, z\}$

46. $A = \{x \mid x \text{ is a vowel in the English alphabet}\}$

47. $A =$ the set of integers between -20 and 20

48. $A =$ the set of sanctioned U.S. senate seats

49. $A = \left\{\dfrac{1}{3}, \dfrac{2}{4}, \dfrac{3}{5}, \dfrac{4}{6}, \ldots, \dfrac{27}{29}, \dfrac{28}{30}\right\}$

50. $A = \left\{\dfrac{1}{2}, -\dfrac{1}{2}, \dfrac{1}{3}, -\dfrac{1}{3}, \ldots, \dfrac{1}{10}, -\dfrac{1}{10}\right\}$

Write a short answer for each problem.

51. Although x is a consonant, why can we write "x is a vowel in the English alphabet" in **Exercise 46?**

52. Explain how **Exercise 49** can be answered without actually listing and then counting all the elements.

Identify each set as well defined *or* not well defined.

53. $\{x \mid x \text{ is a real number}\}$

54. $\{x \mid x \text{ is a good worker}\}$

55. $\{x \mid x \text{ is a difficult person}\}$

56. $\{x \mid x \text{ is a counting number less than 2}\}$

Fill each blank with either $\in$ or $\notin$ to make each statement true.

57. $3 \_\_ \{2, 4, 5, 7\}$ **58.** $-4 \_\_ \{4, 7, 8, 12\}$

59. $8 \_\_ \{3, 8, 12, 18\}$ **60.** $0 \_\_ \{-2, 0, 5, 9\}$

61. $8 \_\_ \{10 - 2, 10\}$ **62.** $\{6\} \_\_ \{5 + 1, 6 + 1\}$

Write a short answer for each problem.

63. Is the statement $\{0\} = \varnothing$ *true*, or is it *false*? Explain.

64. The statement

$$3 \in \{9 - 6, 8 - 6, 7 - 6\}$$

is true even though the *symbol* 3 does not appear in the set. Explain.

Write true *or* false *for each statement.*

65. $3 \in \{2, 5, 6, 8\}$ **66.** $i \in \{i, h, o, p\}$

67. $c \in \{c, d, a, b\}$ **68.** $2 \in \{-2, 5, 8, 9\}$

69. $\{k, c, r, a\} = \{k, c, a, r\}$ **70.** $\{e, h, a, n\} = \{a, h, e, n\}$

71. $\{5, 8, 9\} = \{5, 8, 9, 0\}$ **72.** $\{3, 7\} = \{3, 7, \varnothing\}$

73. $\{4\} \in \{\{3\}, \{4\}, \{5\}\}$ **74.** $4 \in \{\{3\}, \{4\}, \{5\}\}$

75. $\{x \mid x \text{ is a natural number less than 3}\} = \{1, 2\}$

76. $\{x \mid x \text{ is a natural number greater than 10}\} = \{11, 12, 13, \ldots\}$

77. $\{x \mid x \text{ is a positive whole number}\} = \{0, 1, 2, 3, \ldots\}$

78. $\{x \mid x \text{ is an integer greater than 2}\} = \{\ldots, -5, -4, -3, 3, 4, 5, \ldots\}$

79. $\{x \mid x \text{ is a positive integer}\} = \{0, 1, 2, 3, \ldots\}$

80. $\{x \mid x \text{ is a nonpositive whole number}\} = \{0\}$

Write true *or* false *for each statement.*

Let $A = \{2, 4, 6, 8, 10, 12\}$, $B = \{2, 4, 8, 10\}$, and $C = \{4, 10, 12\}$.

81. $4 \in A$ **82.** $10 \in B$

83. $4 \notin C$ **84.** $10 \notin A$

85. Every element of C is also an element of A.

86. Every element of C is also an element of B.

87. Every element of A is also an element of B.

88. There are exactly two elements of A that are not elements of B.

Write a short answer for each problem.

89. The human mind likes to create collections. Why do you suppose this is so? In your explanation, use one or more particular "collections," mathematical or otherwise.

90. Explain the difference between a well-defined set and a set that is not well defined. Give examples, and use terms introduced in this section.

*Two sets are **equal** if they contain identical elements. Two sets are **equivalent** if they contain the same number of elements (but not necessarily the same elements). For each condition, give an example or explain why it is impossible.*

91. two sets that are neither equal nor equivalent

92. two sets that are equal but not equivalent

93. two sets that are equivalent but not equal

94. two sets that are both equal and equivalent

Solve each problem.

95. ***Hiring Nurses*** A medical organization plans to hire three nurses from the pool of applicants shown.

Name	Certification
Brock	RN
Heather	RN
Boris	LVN
Natalie	LVN
Taylor	RN

Show all possible sets of hires that would include

(a) two RNs and one LVN.

(b) one RN and two LVNs.

(c) no LVNs.

96. ***Burning Calories*** Candice likes cotton candy, each serving of which contains 220 calories. To burn off unwanted calories, Candice participates in her favorite activities, shown in the next column, in increments of 1 hour and never repeats a given activity on a given day.

Activity	Symbol	Calories Burned per Hour
Volleyball	v	160
Golf	g	260
Canoeing	c	340
Swimming	s	410
Running	r	680

(a) On Monday, Candice has time for no more than two hours of activities. List all possible sets of activities that would burn off at least the number of calories obtained from three cotton candies.

(b) Assume that Candice can afford up to three hours of time for activities on Wednesday. List all sets of activities that would burn off at least the number of calories in five cotton candies.

(c) Candice can spend up to four hours in activities on Saturday. List all sets of activities that would burn off at least the number of calories in seven cotton candies.

2.2   VENN DIAGRAMS AND SUBSETS

OBJECTIVES

1 Use Venn diagrams to depict set relationships.

2 Determine the complement of a set within a universal set.

3 Determine whether one set is a subset of another.

4 Distinguish between a subset and a proper subset.

5 Determine the number of subsets of a given set.

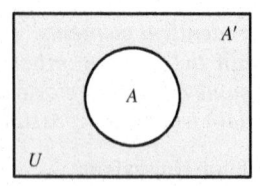

The entire region bounded by the rectangle represents the universal set *U*, and the portion bounded by the circle represents set *A*.

Figure 1

Venn Diagrams

In most discussions, there is either a stated or an implied **universe of discourse.** The universe of discourse includes all things under discussion at a given time. For example, if the topic of interest is what courses to offer at a vocational school, the universe of discourse might be all students at the school, or the board of trustees of the school, or the members of a local overseer board, or the members of a state regulatory agency, or perhaps all these groups of people.

In set theory, the universe of discourse is called the **universal set,** typically designated by the letter **U**. The universal set might change from one discussion to another.

Venn diagrams show one or more (overlapping) sets within the universal set. See **Figure 1** and **Figures 10, 11,** and **12** in the next section. (Euler diagrams show specific relationships between sets, as in **Figure 2** on the next page. They are used to analyze logical arguments in a later chapter.)

Complement of a Set

In **Figure 1** the colored region inside *U* and outside the circle is labeled *A'* (read "*A* prime"). This set, called the *complement* of *A*, contains all elements of *U* that are not contained in *A*.

THE COMPLEMENT OF A SET

For any set *A* within a universal set *U*, the **complement** of *A*, written *A'*, is the set of elements of *U* that are not elements of *A*. That is,

$$A' = \{x \mid x \in U \text{ and } x \notin A\}.$$

StatCrunch Finding Complements

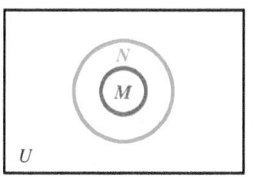

John Venn (1834–1923) improved upon the pictorial representation of sets, which appeared in earlier work by Gottfried Leibniz and Leonhard Euler. Venn diagrams are used in set theory and probability and Venn made use of them in developing the symbolic logic of George Boole. Leibniz, Euler, and Boole are profiled in a later chapter.

Figure 2

EXAMPLE 1 **Finding Complements**

Find each of the following sets.

Let $U = \{a, b, c, d, e, f, g, h\}$, $M = \{a, b, e, f\}$, and $N = \{b, d, e, g, h\}$.

(a) M' **(b)** N'

Solution

(a) Set M' contains all the elements of set U that are *not* in set M. Because set M contains a, b, e, and f, these elements will be disqualified from belonging to set M'.

$$M' = \{c, d, g, h\}$$

(b) Set N' contains all the elements of U that are not in set N, so $N' = \{a, c, f\}$.

Consider the complement of the universal set, denoted U'. The set U' is found by selecting all the elements of U that do *not* belong to U. There are no such elements, so there can be no elements in set U'. This means that for any universal set U,

$$U' = \varnothing.$$

Now consider the complement of the empty set, denoted $\varnothing'$. The set $\varnothing'$ includes all elements of U that do *not* belong to $\varnothing$. All elements of U qualify, because none of them belongs to $\varnothing$. Therefore, for any universal set U,

$$\varnothing' = U.$$

Subsets of a Set

Suppose that we are given the universal set $U = \{1, 2, 3, 4, 5\}$, while $A = \{1, 2, 3\}$. Every element of set A is also an element of set U. Because of this, set A is called a *subset* of set U, written

$$A \subseteq U.$$

("A is not a subset of set U" would be written $A \nsubseteq U$.)

An Euler diagram showing that set M is a subset of set N is shown in **Figure 2**.

SUBSET OF A SET

Set A is a **subset** of set B if every element of A is also an element of B. This is written $A \subseteq B$.

EXAMPLE 2 **Determining If One Set Is a Subset of Another**

Write $\subseteq$ or $\nsubseteq$ in each blank to make a true statement.

(a) $\{3, 4, 5, 6\}$ _____ $\{3, 4, 5, 6, 8\}$ **(b)** $\{1, 2, 6\}$ _____ $\{2, 4, 6, 8\}$

(c) $\{5, 6, 7, 8\}$ _____ $\{6, 5, 8, 7\}$

Solution

(a) Because every element of $\{3, 4, 5, 6\}$ is also an element of $\{3, 4, 5, 6, 8\}$, the first set is a subset of the second, so $\subseteq$ goes in the blank.

$$\{3, 4, 5, 6\} \subseteq \{3, 4, 5, 6, 8\}$$

(b) $\{1, 2, 6\} \nsubseteq \{2, 4, 6, 8\}$ 1 does not belong to $\{2, 4, 6, 8\}$.

(c) $\{5, 6, 7, 8\} \subseteq \{6, 5, 8, 7\}$

As **Example 2(c)** suggests, every set is a subset of itself.

$$B \subseteq B, \quad \text{for any set } B.$$

> This is a restatement, in terms of subsets, of the definition given earlier.

SET EQUALITY (ALTERNATIVE DEFINITION)

Suppose A and B are sets. Then $A = B$ if $A \subseteq B$ and $B \subseteq A$ are both true.

Proper Subsets

Suppose that we are given the following sets.

$$B = \{5, 6, 7, 8\} \quad \text{and} \quad A = \{6, 7\}$$

A is a subset of B, but A is not all of B. There is at least one element in B that is not in A. (Actually, in this case there are two such elements, 5 and 8.) In this situation, A is called a *proper subset* of B, written $A \subset B$.

> Notice the similarity of the subset symbols, $\subset$ and $\subseteq$, to the inequality symbols from algebra, $<$ and $\leq$.

PROPER SUBSET OF A SET

Set A is a **proper subset** of set B if $A \subseteq B$ and $A \neq B$. This is written $\mathbf{A \subset B}$.

> **One-to-one correspondence** was employed by Georg Cantor to establish many controversial facts about infinite sets. For example, the correspondence
>
> $$\{1, 2, 3, 4, \ldots, n \ldots\}$$
> $$\updownarrow \updownarrow \updownarrow \updownarrow \quad \updownarrow$$
> $$\{2, 4, 6, 8, \ldots, 2n, \ldots\},$$
>
> which can be continued indefinitely without leaving any elements in either set unpaired, shows that the counting numbers and the even counting numbers are equivalent (have the same number of elements).
>
> To many in Cantor's day (and perhaps today), such results seemed logically contradictory. The first set may seem to have twice as many elements as the second. The symbol below, two arrows pointed at each other, is commonly used in mathematics to indicate "Contradiction."
>
> $$\Longrightarrow\!\!\Longleftarrow$$

EXAMPLE 3 Determining Subsets and Proper Subsets

Decide whether $\subset$, $\subseteq$, or both could be placed in each blank to make a true statement.

(a) $\{5, 6, 7\}$ _____ $\{5, 6, 7, 8\}$ **(b)** $\{a, b, c\}$ _____ $\{a, b, c\}$

Solution

(a) Every element of $\{5, 6, 7\}$ is contained in $\{5, 6, 7, 8\}$, so $\subseteq$ could be placed in the blank. Also, the element 8 belongs to $\{5, 6, 7, 8\}$ but not to $\{5, 6, 7\}$, making $\{5, 6, 7\}$ a proper subset of $\{5, 6, 7, 8\}$. Thus $\subset$ could also be placed in the blank.

(b) The set $\{a, b, c\}$ is a subset of $\{a, b, c\}$. Because the two sets are equal, $\{a, b, c\}$ is not a proper subset of $\{a, b, c\}$. Only $\subseteq$ may be placed in the blank.

Set A is a subset of set B if every element of set A is also an element of set B. Alternatively, we say that set A is a subset of set B if there are no elements of A that are not also elements of B. Thus, the empty set is a subset of any set.

$$\varnothing \subseteq B, \quad \text{for any set } B.$$

This is true because it is not possible to find any element of $\varnothing$ that is not also in B. (There are no elements in $\varnothing$.) The empty set $\varnothing$ is a proper subset of every set except itself.

$$\varnothing \subset B \quad \text{if } B \text{ is any set other than } \varnothing.$$

Every set (except $\varnothing$) has at least two subsets, $\varnothing$ and the set itself.

EXAMPLE 4 Listing All Subsets of a Set

StatCrunch Listing All Subsets of a Set

Find all possible subsets of each set.

(a) $\{7, 8\}$ **(b)** $\{a, b, c\}$

Solution

(a) By trial and error, the set $\{7, 8\}$ has four subsets: $\varnothing, \{7\}, \{8\}, \{7, 8\}$.

(b) Trial and error leads to eight subsets for $\{a, b, c\}$:

$$\varnothing, \{a\}, \{b\}, \{c\}, \{a, b\}, \{a, c\}, \{b, c\}, \{a, b, c\}.$$

Counting Subsets

In **Example 4,** the subsets of $\{7, 8\}$ and the subsets of $\{a, b, c\}$ were found by trial and error. An alternative method involves drawing a **tree diagram,** a systematic way of listing all the subsets of a given set. See **Figure 3**.

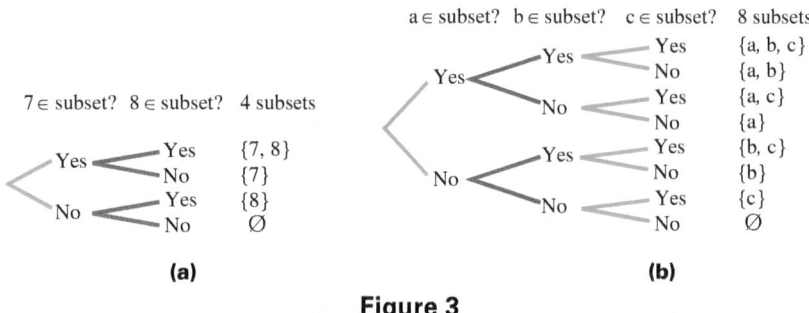

(a)   **(b)**

Figure 3

In **Example 4,** we determined the number of subsets of a given set by making a list of all such subsets and then counting them. The tree diagram method also produced a list of all possible subsets.

Problem-Solving Strategies

In order to extend the observations from above to a general formula for the number of subsets of a set, we will now apply three strategies.

- Make a table or chart.
- Look for a pattern.
- Use inductive reasoning.

Begin with the set containing the least number of elements possible—the empty set. This set, $\varnothing$, has only one subset, $\varnothing$ itself. Next, a set with one element has only two subsets, itself and $\varnothing$. These facts, together with those obtained in **Example 4** for sets with two and three elements, are summarized here.

Number of elements	0	1	2	3
Number of subsets	1	2	4	8

This chart suggests that as the number of elements of the set increases by one, the number of subsets doubles. If so, then the number of subsets in each case might be a power of 2. Since every number in the second row of the chart is indeed a power of 2, add this information to the chart.

Number of elements	0	1	2	3
Number of subsets	$1 = 2^0$	$2 = 2^1$	$4 = 2^2$	$8 = 2^3$

This chart shows that the number of elements in each case is the same as the exponent on the base 2. Inductive reasoning gives the following generalization.

NUMBER OF SUBSETS

The number of subsets of a set with n elements is 2^n.

Powers of 2

$2^0 = 1$

$2^1 = 2$

$2^2 = 2 \cdot 2 = 4$

$2^3 = 2 \cdot 2 \cdot 2 = 8$

$2^4 = 2 \cdot 2 \cdot 2 \cdot 2 = 16$

$2^5 = 32$

$2^6 = 64$

$2^7 = 128$

$2^8 = 256$

$2^9 = 512$

$2^{10} = 1024$

$2^{11} = 2048$

$2^{12} = 4096$

$2^{15} = 32,768$

$2^{20} = 1,048,576$

$2^{25} = 33,554,432$

$2^{30} = 1,073,741,824$

Because the set itself is one of the 2^n subsets, we must subtract 1 from this value to obtain the number of proper subsets of a set containing n elements.

NUMBER OF PROPER SUBSETS

The number of proper subsets of a set with n elements is $2^n - 1$.

As shown earlier, although inductive reasoning is a good way of *discovering* principles or arriving at a *conjecture*, it does not provide a proof that the conjecture is true in general. The two formulas above are true, by observation, for $n = 0, 1, 2$, and 3. (For a general proof, see **Exercise 67.**)

EXAMPLE 5 Finding Numbers of Subsets and Proper Subsets

Find the number of subsets and the number of proper subsets of each set.

(a) $\{3, 4, 5, 6, 7\}$ **(b)** $\{1, 2, 3, 4, 5, 9, 12, 14\}$

Solution

(a) This set has 5 elements and $2^5 = 2 \cdot 2 \cdot 2 \cdot 2 \cdot 2 = 32$ subsets. Of these,

$$2^5 - 1 = 32 - 1 = 31 \text{ are proper subsets.}$$

(b) This set has 8 elements. There are

$$2^8 = 256 \text{ subsets and } 255 \text{ proper subsets.}$$

2.2 EXERCISES

CONCEPT CHECK *Write* true *or* false *for each statement.*

1. A Venn diagram usually involves a rectangle and one or more circles (or ovals).

2. If U is the universal set, then $U' = \varnothing$.

3. The set $\varnothing$ has the same number of subsets as it has proper subsets.

4. If A and B are sets, then $A \subseteq B$ and $A \subset B$ cannot both be true.

CONCEPT CHECK *Match each set or sets in Column I with the appropriate description in Column II.*

I	II
5. $\{p\}, \{q\}, \{p,q\}, \varnothing$	**A.** the proper subsets of $\{p,q\}$
6. $\{p\}, \{q\}, \varnothing$	**B.** the complement of $\{c,d\}$, if $U = \{a, b, c, d\}$
7. $\{a, b\}$	**C.** the complement of U
8. $\varnothing$	**D.** the subsets of $\{p,q\}$

Insert $\subseteq$ *or* $\nsubseteq$ *in each blank to obtain a true statement.*

9. $\{-2, 0, 2\}$ ____ $\{-2, -1, 1, 2\}$

10. $\{M, W, F\}$ ____ $\{S, M, T, W, Th\}$

11. $\{2, 5\}$ ____ $\{0, 1, 5, 3, 7, 2\}$

12. $\{a, n, d\}$ ____ $\{r, a, n, d, y\}$

13. $\varnothing$ ____ $\{a, b, c, d, e\}$

14. $\varnothing$ ____ $\varnothing$

15. $\{-5, 2, 9\}$ ____ $\{x \mid x \text{ is an odd integer}\}$

16. $\left\{1, 2, \dfrac{9}{3}\right\}$ ____ the set of rational numbers

Decide whether $\subset$, $\subseteq$, *both, or neither can be placed in each blank to make the statement true.*

17. $\{P, Q, R\}$ _____ $\{P, Q, R, S\}$

18. $\{red, blue, yellow\}$ _____ $\{yellow, blue, red\}$

19. $\{9, 1, 7, 3, 5\}$ _____ $\{1, 3, 5, 7, 9\}$

20. $\{S, M, T, W, Th\}$ _____ $\{W, E, E, K\}$

21. $\varnothing$ _____ $\{0\}$

22. $\varnothing$ _____ $\varnothing$

23. $\{0, 1, 2, 3\}$ _____ $\{1, 2, 3, 4\}$

24. $\left\{\dfrac{5}{6}, \dfrac{9}{8}\right\}$ _____ $\left\{\dfrac{6}{5}, \dfrac{8}{9}\right\}$

Write true *or* false *for each statement. U is the universal set.*

Let $U = \{a, b, c, d, e, f, g\}$, $A = \{a, e\}$,
$B = \{a, b, e, f, g\}$, $C = \{b, f, g\}$, $D = \{d, e\}$.

25. $A \subset U$

26. $C \not\subset U$

27. $D \subseteq B$

28. $D \not\subseteq A$

29. $A \subset B$

30. $B \subseteq C$

31. $\varnothing \not\subset A$

32. $\varnothing \subseteq D$

33. $D \not\subseteq B$

34. $A \not\subseteq B$

35. There are exactly 6 subsets of C.

36. There are exactly 31 subsets of B.

37. There are exactly 3 proper subsets of A.

38. There are exactly 4 subsets of D.

39. The diagram below correctly represents the relationship among sets A, D, and U.

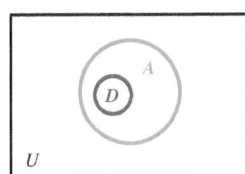

40. The diagram below correctly represents the relationship among sets B, C, and U.

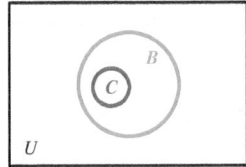

Find **(a)** *the number of subsets and* **(b)** *the number of proper subsets of each set.*

41. $\{a, b, c, d, e, f\}$

42. the set of days of the week

43. $\{x \mid x$ is an odd integer between -4 and $6\}$

44. $\{x \mid x$ is an even whole number less than $4\}$

Let $U = \{1, 2, 3, 4, 5, 6, 7, 8, 9, 10\}$ *and find the complement of each set.*

45. U

46. $\varnothing$

47. $\{1, 2, 3, 4, 6, 8\}$

48. $\{2, 5, 9, 10\}$

Vacationing in Florida *Terry is planning a trip with her two sons to Florida. In weighing her options concerning whether to fly or drive from their home in Iowa, she has listed the following considerations.*

Fly to Florida	Drive to Florida
Higher cost	Lower cost
Educational	Educational
More time to see the sights in Florida	Less time to see the sights in Florida
Cannot visit friends along the way	Can visit friends along the way

Refer to the table for the following exercises.

49. Find the smallest universal set U that contains all listed considerations of both options.

Let F represent the set of considerations of the flying option and let D represent the set of considerations of the driving option. Use the universal set from ***Exercise 49.***

50. Give the set F'.

51. Give the set D'.

Find the set of considerations common to both sets.

52. F and D

53. F' and D'

54. F and D'

Meeting in the Conference Room *Anh, Bruce, Corey, Dino, and Eladia, members of an architectural firm, plan to meet in the company conference room to discuss the project coordinator's plans for their next project. Denoting these five people by* A, B, C, D, *and* E, *list all the possible sets of this group in which the given number of them can gather.*

55. five people

56. four people

57. three people

58. two people

59. one person

60. no people

61. Find the total number of ways that members of this group can gather. (*Hint:* Find the total number of sets in your answers to **Exercises 55–60.**)

62. How does your answer to **Exercise 61** compare with the number of subsets of a set of five elements? Interpret the answer to **Exercise 61** in terms of subsets.

Solve each problem.

63. **Selecting a Club Delegation** The twenty-five members of the mathematics club must send a delegation to a meeting for student groups at their school. The delegation can include as many members of the club as desired, but at least one member must attend. How many different delegations are possible? (*MT* calendar problem)

64. **Selecting a Club Delegation** In **Exercise 63,** suppose ten of the club members say they do not want to be part of the delegation. Now how many delegations are possible?

65. **Selecting Bills** Suppose you have the bills shown here.

(a) How many sums of money can you make using nonempty subsets of these bills?

(b) Repeat part (a) without the condition "nonempty."

66. **Selecting Coins** The photo shows a group of obsolete U.S. coins, consisting of one each of the half dollar, quarter, dime, nickel and penny. Repeat **Exercise 65,** replacing "bill(s)" with "coin(s)."

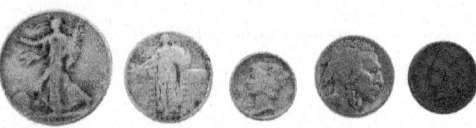

67. **Number of Subsets** In discovering the expression (2^n) for the number of subsets of a set with n elements, we observed that for the first few values of n, increasing the number of elements by one doubles the number of subsets. Here, we can prove the formula in general by showing that the same is true for any value of n. Assume set A has n elements and s subsets. Now add one additional element, say e, to the set A. (We now have a new set, say B, with $n + 1$ elements.) Divide the subsets of B into those that do not contain e and those that do.

(a) How many subsets of B do not contain e? (*Hint:* Each of these is a subset of the original set A.)

(b) How many subsets of B do contain e? (*Hint:* Each of these would be a subset of the original set A, with the additional element e included.)

(c) What is the total number of subsets of B?

(d) What do you conclude?

68. **Subsets and Elements** Explain why $\varnothing$ is both a subset and an element of $\{\varnothing\}$.

2.3 SET OPERATIONS

OBJECTIVES

1 Determine intersections of sets.

2 Determine unions of sets.

3 Determine the difference of two sets.

4 List and make use of ordered pairs.

5 Determine Cartesian products of sets.

6 Analyze sets and set operations with diagrams.

7 Apply De Morgan's laws for sets.

Intersection of Sets

Two candidates, Aiko and Dorian, are running for a seat on the city council. A voter deciding for whom she should vote recorded the campaign promises, each given a code letter, made by the candidates.

Honest Aiko	Determined Dorian
Spend less money, *m*	Spend less money, *m*
Emphasize traffic law enforcement, *t*	Crack down on crooked politicians, *p*
Increase service to suburban areas, *s*	Increase service to the city, *c*

The only promise common to both candidates is promise m, to spend less money. Suppose we take each candidate's promises to be a set. Aiko's promises give the set $\{m, t, s\}$, while Dorian's promises give $\{m, p, c\}$. The common element m belongs to the *intersection* of the two sets, as shown in color in the Venn diagram in **Figure 4** on the next page.

$$\{m, t, s\} \cap \{m, p, c\} = \{m\}$$ ∩ represents set intersection.

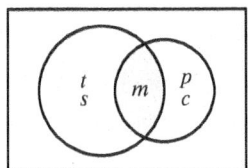

Figure 4

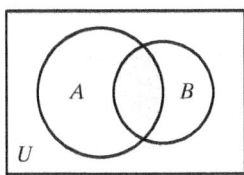

$A \cap B$

Figure 5

StatCrunch Finding Intersections

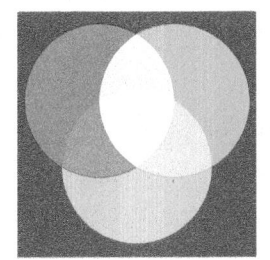

White light can be viewed as the intersection of the three primary colors.

The intersection of two sets is itself a set.

INTERSECTION OF SETS

The **intersection** of sets A and B, written $A \cap B$, is the set of elements common to both A and B.

$$A \cap B = \{x \mid x \in A \text{ and } x \in B\}$$

Form the intersection of sets A and B by taking only the elements included in both sets, as shown in color in **Figure 5**.

EXAMPLE 1   **Finding Intersections**

Find each intersection.

(a) $\{3, 4, 5, 6, 7\} \cap \{4, 6, 8, 10\}$ (b) $\{9, 14, 25, 30\} \cap \{10, 17, 19, 38, 52\}$

(c) $\{5, 9, 11\} \cap \varnothing$

Solution

(a) The elements common to both sets are 4 and 6.

$$\{3, 4, 5, 6, 7\} \cap \{4, 6, 8, 10\} = \{4, 6\}$$

(b) These two sets have no elements in common.

$$\{9, 14, 25, 30\} \cap \{10, 17, 19, 38, 52\} = \varnothing$$

(c) There are no elements in $\varnothing$, so there can be no elements belonging to both $\{5, 9, 11\}$ and $\varnothing$.

$$\{5, 9, 11\} \cap \varnothing = \varnothing$$

Examples 1(b) and 1(c) show two sets that have no elements in common. Sets with no elements in common are called **disjoint sets.** (See **Figure 6**.) A set of dogs and a set of cats would be disjoint sets.

Sets A and B are disjoint if $A \cap B = \varnothing.$

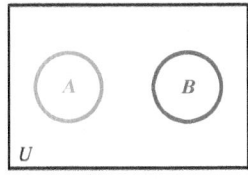

Disjoint sets

Figure 6

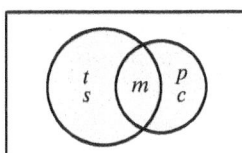

Figure 7

Union of Sets

Referring again to the lists of campaign promises, suppose a pollster wants to summarize the types of promises made by the candidates. The pollster would need to study *all* the promises made by *either* candidate, or the set

$$\{m, t, s, p, c\}.$$

This set is the *union* of the sets of promises, as shown in color in the Venn diagram in **Figure 7**.

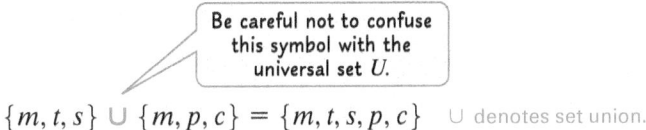

Be careful not to confuse this symbol with the universal set U.

$\{m, t, s\} \cup \{m, p, c\} = \{m, t, s, p, c\}$ ∪ denotes set union.

Again, the union of two sets is a set.

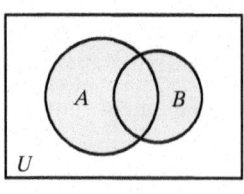

$A \cup B$

Figure 8

UNION OF SETS

The **union** of sets A and B, written $A \cup B$, is the set of all elements belonging to either A or B.

$$A \cup B = \{x \mid x \in A \text{ or } x \in B\}$$

Form the union of sets A and B by first taking every element of set A and then also including every element of set B that is not already listed. See **Figure 8.**

StatCrunch Finding Unions

EXAMPLE 2 Finding Unions

Find each union.

(a) $\{2, 4, 6\} \cup \{4, 6, 8, 10, 12\}$ **(b)** $\{a, b, d, f, g, h\} \cup \{c, f, g, h, k\}$

(c) $\{3, 4, 5\} \cup \varnothing$

Solution

(a) Start by listing all the elements from the first set, 2, 4, and 6. Then list all the elements from the second set that are not in the first set, 8, 10, and 12. The union is made up of *all* these elements.

$$\{2, 4, 6\} \cup \{4, 6, 8, 10, 12\} = \{2, 4, 6, 8, 10, 12\}$$

(b) $\{a, b, d, f, g, h\} \cup \{c, f, g, h, k\} = \{a, b, c, d, f, g, h, k\}$

(c) Because there are no elements in $\varnothing$, the union of $\{3, 4, 5\}$ and $\varnothing$ contains only the elements 3, 4, and 5.

$$\{3, 4, 5\} \cup \varnothing = \{3, 4, 5\}$$

Recall from the previous section that A' represents the *complement* of set A. *Set A' is formed by taking every element of the universal set U that is not in set A.*

StatCrunch Finding Intersections and Unions of Complements

EXAMPLE 3 Finding Intersections and Unions of Complements

Find each set. Let

$$U = \{1, 2, 3, 4, 5, 6, 9\}, \quad A = \{1, 2, 3, 4\}, \quad B = \{2, 4, 6\}, \quad \text{and} \quad C = \{1, 3, 6, 9\}.$$

(a) $A' \cap B$ **(b)** $B' \cup C'$ **(c)** $A \cap (B \cup C')$ **(d)** $(B \cup C)'$

Solution

(a) First identify the elements of set A', the elements of U that are not in set A.

$$A' = \{5, 6, 9\}$$

Now, find $A' \cap B$, the set of elements belonging both to A' and to B.

$$A' \cap B = \{5, 6, 9\} \cap \{2, 4, 6\} = \{6\}$$

(b) $B' \cup C' = \{1, 3, 5, 9\} \cup \{2, 4, 5\} = \{1, 2, 3, 4, 5, 9\}$

(c) First find the set inside the parentheses.

$$B \cup C' = \{2, 4, 6\} \cup \{2, 4, 5\} = \{2, 4, 5, 6\}$$

Now, find the intersection of this set with A.

$$A \cap (B \cup C') = A \cap \{2, 4, 5, 6\}$$
$$= \{1, 2, 3, 4\} \cap \{2, 4, 5, 6\}$$
$$= \{2, 4\}$$

(d) $B \cup C = \{2, 4, 6\} \cup \{1, 3, 6, 9\} = \{1, 2, 3, 4, 6, 9\}$, so

$$(B \cup C)' = \{5\}.$$

Comparing **Examples 3(b) and 3(d),** we see that, interestingly, $(B \cup C)'$ is not the same as $B' \cup C'$. This fact will be investigated further later in this section.

EXAMPLE 4   **Describing Sets in Words**

Describe each set in words.

(a) $A \cap (B \cup C')$ **(b)** $(A' \cup C') \cap B'$

Solution

(a) This set might be described as "the set of all elements that are in A, and also are in B or not in C."

(b) One possibility is "the set of all elements that are not in A or not in C, and also are not in B."

Difference of Sets

Suppose that $A = \{1, 2, 3, \ldots, 10\}$ and $B = \{2, 4, 6, 8, 10\}$. If the elements of B are excluded (or taken away) from A, the set $C = \{1, 3, 5, 7, 9\}$ is obtained. C is called the *difference* of sets A and B.

DIFFERENCE OF SETS

The **difference** of sets A and B, written $A - B$, is the set of all elements belonging to set A and not to set B.

$$A - B = \{x \,|\, x \in A \text{ and } x \notin B\}$$

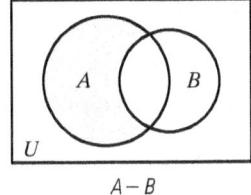

$A - B$

Figure 9

Assume a universal set U containing both A and B. Then, because $x \notin B$ has the same meaning as $x \in B'$, the set difference $A - B$ can also be described as

$$\{x \,|\, x \in A \text{ and } x \in B'\}, \quad \text{or} \quad A \cap B'.$$

Figure 9 illustrates the idea of set difference. The region in color represents $A - B$.

StatCrunch Finding Set Differences

EXAMPLE 5   **Finding Set Differences**

Find each set.

$$\text{Let} \quad U = \{1, 2, 3, 4, 5, 6, 7\}, \quad A = \{1, 2, 3, 4, 5, 6\},$$
$$B = \{2, 3, 6\}, \quad \text{and} \quad C = \{3, 5, 7\}.$$

(a) $A - B$ **(b)** $B - A$ **(c)** $(A - B) \cup C'$

$U = \{1, 2, 3, 4, 5, 6, 7\}$

$A = \{1, 2, 3, 4, 5, 6\}$

$B = \{2, 3, 6\}$

$C = \{3, 5, 7\}$

Solution

(a) To find $A - B$, begin with set A and exclude any elements found also in set B.

$$A - B = \{1, 2, 3, 4, 5, 6\} - \{2, 3, 6\} = \{1, 4, 5\}$$

(b) To be in $B - A$, an element must be in set B and not in set A. But all elements of B are also in A. Thus,

$$B - A = \varnothing.$$

(c) From part (a), $A - B = \{1, 4, 5\}$. Also, $C' = \{1, 2, 4, 6\}$.

$$(A - B) \cup C' = \{1, 2, 4, 5, 6\}$$

The results in **Examples 5(a) and 5(b)** illustrate that, in general,

$$A - B \neq B - A.$$

Ordered Pairs

In any set that contains several elements, the order in which the elements appear is not relevant. For example,

$$\{n, o\} = \{o, n\}.$$

However, there are many instances in mathematics where, when two objects are paired, the order in which the objects are written is important. This leads to the idea of *ordered pair*. When writing ordered pairs, use parentheses rather than braces, which are reserved for writing sets.

ORDERED PAIRS

In the **ordered pair** (a, b), a is called the **first component** and b is called the **second component**. In general, $(a, b) \neq (b, a)$.

Two ordered pairs (a, b) and (c, d) are **equal** provided that their first components are equal and their second components are equal.

$$(a, b) = (c, d) \quad \text{provided that} \quad a = c \quad \text{and} \quad b = d.$$

EXAMPLE 6   Determining Equality of Sets and of Ordered Pairs

Decide whether each statement is *true* or *false*.

(a) $(3, 4) = (5 - 2, 1 + 3)$ **(b)** $\{3, 4\} \neq \{4, 3\}$ **(c)** $(7, 4) = (4, 7)$

Solution

(a) Because $3 = 5 - 2$ and $4 = 1 + 3$, the first components are equal and the second components are equal. The statement is *true*.

(b) Because these are sets and not ordered pairs, the order in which the elements are listed is not important. Because these sets are equal, the statement is *false*.

(c) The ordered pairs $(7, 4)$ and $(4, 7)$ are not equal because their corresponding components are not equal. The statement is *false*.

Cartesian Product of Sets

A set may contain ordered pairs as elements. If A and B are sets, then each element of A can be paired with each element of B, and the results can be written as ordered pairs. The set of all such ordered pairs is called the *Cartesian product* of A and B, which is written $A \times B$ and read **"A cross B."** The name comes from that of the French mathematician René Descartes (1596–1650).

CARTESIAN PRODUCT OF SETS

The **Cartesian product** of sets A and B is defined as follows.

$$A \times B = \{(a, b) \mid a \in A \text{ and } b \in B\}$$

EXAMPLE 7   **Finding Cartesian Products**

Let $A = \{1, 5, 9\}$ and $B = \{6, 7\}$. Find each set.

(a) $A \times B$ **(b)** $B \times A$

Solution

(a) Pair each element of A with each element of B. Write the results as ordered pairs, with the element of A written first and the element of B written second. Write as a set.

$$A \times B = \{(1, 6), (1, 7), (5, 6), (5, 7), (9, 6), (9, 7)\}$$

(b) Because B appears first in the product, this set will consist of ordered pairs that have their components interchanged when compared to those in part (a).

$$B \times A = \{(6, 1), (7, 1), (6, 5), (7, 5), (6, 9), (7, 9)\}$$

The order in which the ordered pairs themselves are listed is not important. For example, another way to write $B \times A$ in **Example 7(b)** would be

$$\{(6, 1), (6, 5), (6, 9), (7, 1), (7, 5), (7, 9)\}.$$

From **Example 7** it can be seen that, in general,

$$A \times B \neq B \times A,$$

because they do not contain exactly the same ordered pairs. However, each set contains the same *number* of elements, six. Furthermore,

$$n(A) = 3, \quad n(B) = 2, \quad \text{and} \quad n(A \times B) = n(B \times A) = 6.$$

Because $3 \cdot 2 = 6$, one might conclude that the cardinal number of the Cartesian product of two sets is equal to the product of the cardinal numbers of the sets. In general, this conclusion is correct.

CARDINAL NUMBER OF A CARTESIAN PRODUCT

If $n(A) = a$ and $n(B) = b$, then the following is true.

$$n(A \times B) = n(B \times A) = n(A) \cdot n(B) = n(B) \cdot n(A) = ab = ba$$

EXAMPLE 8 Finding Cardinal Numbers of Cartesian Products

Find $n(A \times B)$ and $n(B \times A)$ from the given information.

(a) $A = \{a, b, c, d, e, f, g\}$ and $B = \{2, 4, 6\}$ **(b)** $n(A) = 24$ and $n(B) = 5$

Solution

(a) Because $n(A) = 7$ and $n(B) = 3$, $n(A \times B)$ and $n(B \times A)$ both equal $7 \cdot 3$, or 21.

(b) $n(A \times B) = n(B \times A) = 24 \cdot 5 = 5 \cdot 24 = 120$

An **operation** is a rule or procedure by which one or more objects are used to obtain another object. The most common operations on sets are summarized in the following box.

SET OPERATIONS

Let A and B be any sets within a universal set U.

The **complement** of A, written A', is

$$A' = \{x \mid x \in U \text{ and } x \notin A\}.$$

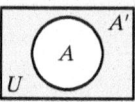

The **intersection** of A and B is

$$A \cap B = \{x \mid x \in A \text{ and } x \in B\}.$$

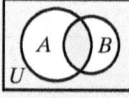

The **union** of A and B is

$$A \cup B = \{x \mid x \in A \text{ or } x \in B\}.$$

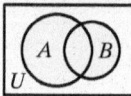

The **difference** of A and B is

$$A - B = \{x \mid x \in A \text{ and } x \notin B\}.$$

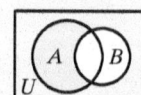

The **Cartesian product** of A and B is

$$A \times B = \{(x, y) \mid x \in A \text{ and } y \in B\}.$$

Venn Diagrams

More on Venn Diagrams

It is often helpful to use numbers in Venn diagrams, as in **Figures 10, 11, and 12,** depending on whether the discussion involves one, two, or three (distinct) sets, respectively. In each case, the numbers are neither elements nor cardinal numbers, but simply arbitrary labels for the various regions within the diagram.

In **Figure 11,** region 3 includes the elements belonging to both A and B, while region 4 includes those elements (if any) belonging to B but not to A. How would you describe region 7 in **Figure 12?**

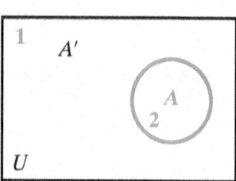

Figure 10

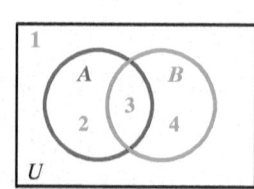

Figure 11

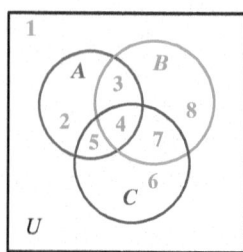

Figure 12

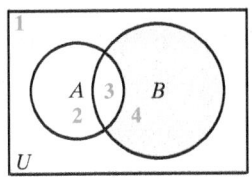

$A' \cap B$, or $B - A$

Figure 13

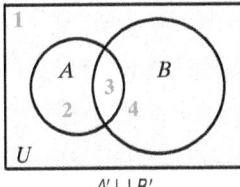

$A' \cup B'$

Figure 14

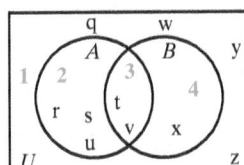

Figure 15

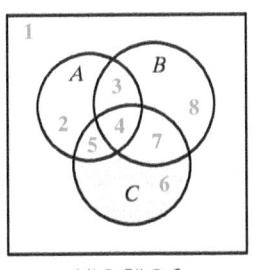

$(A' \cap B') \cap C$

Figure 16

EXAMPLE 9   Shading Venn Diagrams to Represent Sets

Draw a Venn diagram similar to **Figure 11,** and shade the region or regions representing each set.

(a) $A' \cap B$ **(b)** $A' \cup B'$

Solution

(a) See **Figure 11.** Set A' contains all the elements outside of set A—in other words, the elements in regions 1 and 4. Set B contains the elements in regions 3 and 4. The intersection of sets A' and B is made up of the elements in the region common to (1 and 4) and (3 and 4), which is region 4. Thus, $A' \cap B$ is represented by region 4, shown in color in **Figure 13.** This region can also be described as $B - A$.

(b) Again, set A' is represented by regions 1 and 4, and B' is made up of regions 1 and 2. The union of A' and B', the set $A' \cup B'$, is made up of the elements belonging to the union of (1 and 4) with (1 and 2)—that is, regions 1, 2, and 4, shown in color in **Figure 14.**

EXAMPLE 10   Locating Elements in a Venn Diagram

Place the elements of the sets in their proper locations in a Venn diagram.

Let $U = \{q, r, s, t, u, v, w, x, y, z\}$, $A = \{r, s, t, u, v\}$,   and $B = \{t, v, x\}$.

Solution

Because $A \cap B = \{t, v\}$, elements t and v are placed in region 3 in **Figure 15.** The remaining elements of A—that is, r, s, and u—go in region 2. The figure shows the proper placement of all other elements.

EXAMPLE 11   Shading a Set in a Venn Diagram

Shade the set $(A' \cap B') \cap C$ in a Venn diagram similar to the one in **Figure 12.**

Solution

Work first inside the parentheses. Set A' is made up of the regions outside set A, or regions 1, 6, 7, and 8. Set B' is made up of regions 1, 2, 5, and 6. The intersection of these sets is given by the overlap of regions 1, 6, 7, 8 and 1, 2, 5, 6, or regions 1 and 6.

For the final Venn diagram, find the intersection of regions 1 and 6 with set C. Set C is made up of regions 4, 5, 6, and 7. The overlap of regions 1, 6 and 4, 5, 6, 7 is region 6, the region shown in color in **Figure 16.**

EXAMPLE 12   Verifying a Statement Using a Venn Diagram

Suppose $A, B \subseteq U$. Is the statement $(A \cap B)' = A' \cup B'$ true for every choice of sets A and B?

Solution

Use the regions labeled in **Figure 11.** Set $A \cap B$ is made up of region 3, so $(A \cap B)'$ is made up of regions 1, 2, and 4. These regions are shown in color in **Figure 17(a)** on the next page.

To identify set $A' \cup B'$, proceed as in **Example 9(b).** The result, shown in **Figure 14,** is repeated in **Figure 17(b)** on the next page.

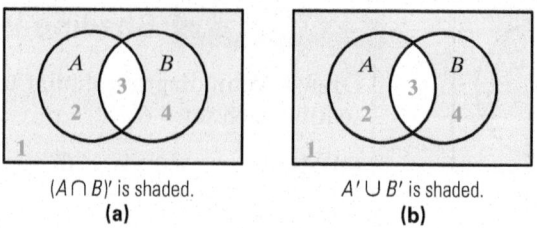

(A ∩ B)′ is shaded.   A′ ∪ B′ is shaded.
        (a)   (b)

Figure 17

The fact that the same regions are in color in both Venn diagrams suggests that

$$(A \cap B)' = A' \cup B'.$$

De Morgan's Laws

Venn Diagrams

The result of **Example 12** can be stated in words.

> *The complement of the intersection of two sets is equal to the union of the complements of the two sets.*

Interchanging the words "intersection" and "union" produces another true statement.

> *The complement of the union of two sets is equal to the intersection of the complements of the two sets.*

Both of these "laws" were established by the British logician Augustus De Morgan (1806–1871). They are stated in set symbols as follows.

DE MORGAN'S LAWS FOR SETS

For any sets A and B, where $A, B \subseteq U$,

$$(A \cap B)' = A' \cup B' \quad \text{and} \quad (A \cup B)' = A' \cap B'.$$

The Venn diagrams in **Figure 17** strongly suggest the truth of the first of De Morgan's laws. They provide a *conjecture*. Actual proofs of De Morgan's laws would require methods used in more advanced courses in set theory.

EXAMPLE 13 Describing Venn Diagram Regions Using Symbols

For the Venn diagrams, write several symbolic descriptions of the region in color, using $A, B, C, \cap, \cup, -$, and $'$ as necessary.

(a) 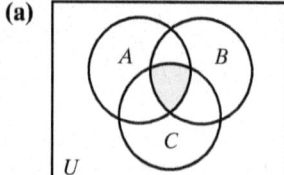   (b)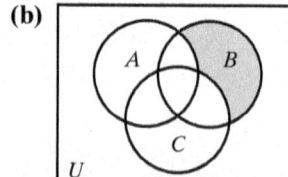

Solution

(a) The region in color can be described as belonging to all three sets, A and B and C. Therefore, the region corresponds to

$$(A \cap B) \cap C, \quad \text{or} \quad A \cap (B \cap C), \quad \text{or} \quad A \cap B \cap C.$$

(b) The region in color is in set B and is not in A and is not in C. Because it is not in A, it is in A', and similarly it is in C'. The region can be described as

$$B \cap A' \cap C', \quad \text{or} \quad B - (A \cup C), \quad \text{or} \quad B \cap (A \cup C)'.$$

WHEN WILL I **EVER** USE THIS?

The Venn diagram below reflects relationships among 5000 specific human genes. The information helps guide research into three brain-related disorders:

Alzheimer's disease, multiple sclerosis (MS), and stroke.

For example, we see that

36 genes are implicated in Alzheimer's and MS but not in stroke.

1. Make similar statements for each of the following numbers in the diagram.

(a) 96 (b) 14 (c) 708 (d) 784

2. State in words what is represented by each sum.

(a) 36 + 14 (b) 318 + 96 + 2701

(c) 36 + 96 + 343 (d) 36 + 96 + 343 + 14

(e) 318 + 708 + 2701

(f) 318 + 708 + 2701 + 784

(g) 318 + 708 + 2701 + 36 + 96 + 343 + 14

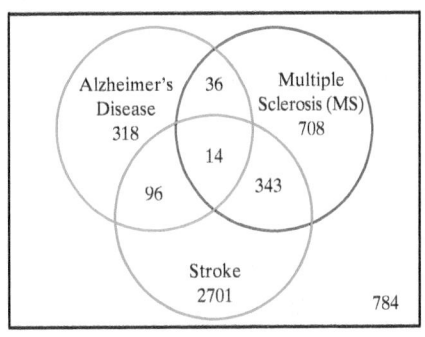

Data from http://www.pasteur.gr

Answers: **1.** (a) 96 genes are implicated in Alzheimer's and stroke but not in MS. (b) 14 genes are implicated in all three of these disorders. (c) 708 genes are implicated in MS but in neither of the other two. (d) 784 genes are not implicated in any of these three disorders. **2.** (a) 36 + 14 is the number of genes implicated in both Alzheimer's and MS. (b) 318 + 96 + 2701 is the number of genes implicated in Alzheimer's or stroke but not in MS. (c) 36 + 96 + 343 is the number of genes implicated in exactly two of these disorders. (d) 36 + 96 + 343 + 14 is the number of genes implicated in at least two of these disorders. (e) 318 + 708 + 2701 is the number of genes implicated in exactly one of these disorders. (f) 318 + 708 + 2701 + 784 is the number of genes implicated in no more than one of these disorders. (g) 318 + 708 + 2701 + 36 + 96 + 343 + 14 is the number of genes implicated in at least one of these disorders.

2.3 EXERCISES

CONCEPT CHECK *Write* true *or* false *for each statement.*

1. $(3, 2) = (5 - 2, 1 + 1)$

2. $(2, 13) = (13, 2)$

3. $\{6, 3\} = \{3, 6\}$

4. $\{(5, 9), (4, 8), (4, 2)\} = \{(4, 8), (5, 9), (2, 4)\}$

CONCEPT CHECK *Match each term in Group I with the appropriate description from A–F in Group II. Assume that A and B are sets.*

I

5. the intersection of A and B

6. the union of A and B

7. the difference of A and B

8. the complement of A

9. the Cartesian product of A and B

10. the difference of B and A

II

A. the set of elements in A that are not in B

B. the set of elements common to both A and B

C. the set of elements in the universal set that are not in A

D. the set of elements in B that are not in A

E. the set of ordered pairs such that each first component is from A and each second component is from B, with every element of A paired with every element of B

F. the set of elements that are in A or in B or in both A and B

Perform the indicated operations, and designate each answer using the listing method.

> Let $U = \{a, b, c, d, e, f, g\}$, $X = \{a, c, e, g\}$,
> $Y = \{a, b, c\}$, and $Z = \{b, c, d, e, f\}$.

11. $X \cap Y$ 12. $X \cup Y$

13. $Y \cup Z$ 14. $Y \cap Z$

15. X' 16. Y'

17. $X' \cap Y'$ 18. $X' \cap Z$

19. $X \cup (Y \cap Z)$ 20. $X \cap (Y \cup Z)$

21. $X - Y$ 22. $Y - X$

23. $(Z \cup X')' \cap Y$ 24. $(Y \cap X') \cup Z'$

25. $X \cap (X - Y)$ 26. $Y \cup (Y - X)$

27. $X' - Y$ 28. $Y' - (X \cap Z)$

Describe each set in words.

29. $A \cup (B' \cap C')$ 30. $(A \cap B') \cup (B \cap A')$

31. $(C - B) \cup A$ 32. $(A' \cap B') \cup C'$

Adverse Effects of Tobacco and Alcohol   *The table lists some common adverse effects of prolonged tobacco and alcohol use.*

Tobacco	Alcohol
Emphysema, e	Liver damage, l
Heart damage, h	Brain damage, b
Cancer, c	Heart damage, h

Let T be the set of listed effects of tobacco and A be the set of listed effects of alcohol. Find each set.

33. the smallest possible universal set U that includes all the effects listed

34. $T - A$ 35. $T \cup A$ 36. $T \cap A'$

An accountant is sorting tax returns in her files that require attention in the next week.

Let U = the set of all tax returns in the file,

     A = the set of all tax returns with itemized deductions,

     B = the set of all tax returns showing business income,

     C = the set of all tax returns filed in 2018,

     D = the set of all tax returns selected for audit.

Give a word description of each of the following sets.

37. $C - A$ 38. $D \cup A'$

39. $(A \cup B) - D$ 40. $(C \cap A) \cap B'$

Assume that A and B represent any two sets. Identify each statement as either always true *or* not always true.

41. $(A \cap B) \subseteq A$ 42. $A \subseteq (A \cap B)$

43. $n(A \cup B) = n(A) + n(B)$

44. $n(A \cup B) = n(A) + n(B) - n(A \cap B)$

In each problem, use your results in parts (a) and (b) to answer part (c).

> Let $U = \{1, 2, 3, 4, 5\}$, $X = \{1, 3, 5\}$, $Y = \{1, 2, 3\}$,
> and $Z = \{3, 4, 5\}$.

45. **(a)** Find $X \cup Y$.   **(b)** Find $Y \cup X$.

     (c) State a conjecture.

46. (a) Find $X \cap Y$.   **(b)** Find $Y \cap X$.

(c) State a conjecture.

47. (a) Find $X \cup (Y \cup Z)$.   **(b)** Find $(X \cup Y) \cup Z$.

(c) State a conjecture.

48. (a) Find $X \cap (Y \cap Z)$.   **(b)** Find $(X \cap Y) \cap Z$.

(c) State a conjecture.

Find $A \times B$ and $B \times A$, for A and B defined as follows.

49. $A = \{d, o, g\}, \quad B = \{p, i, g\}$

50. $A = \{3, 6, 9\}, \quad B = \{6, 8\}$

Use the given information to find $n(A \times B)$ and $n(B \times A)$.

51. $n(A) = 35$ and $n(B) = 6$

52. $n(A) = 8$ and $n(B) = 5$

Find the cardinal number specified.

53. If $n(A \times B) = 72$ and $n(A) = 12$, find $n(B)$.

54. If $n(A \times B) = 300$ and $n(B) = 30$, find $n(A)$.

Place the elements of these sets in the proper locations in the given Venn diagram.

55. Let $U = \{a, b, c, d, e, f, g\}$,
   $A = \{b, d, f, g\}$,
   $B = \{a, b, d, e, g\}$.

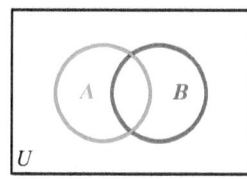

56. Let $U = \{5, 6, 7, 8, 9, 10, 11, 12, 13\}$,
   $M = \{5, 8, 10, 11\}$,
   $N = \{5, 6, 7, 9, 10\}$.

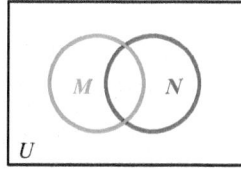

Use a Venn diagram similar to the one shown below to shade each set in Exercises 57–64.

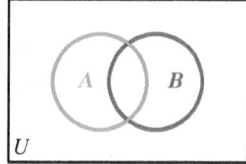

57. $A' \cup B$ **58.** $A' \cap B'$

59. $B \cap A'$ **60.** $A \cup B$

61. $B' \cap B$ **62.** $A' \cup A$

63. $B' \cup (A' \cap B')$ **64.** $(A - B) \cup (B - A)$

Place the elements of the sets in the proper locations in a Venn diagram similar to the one shown below.

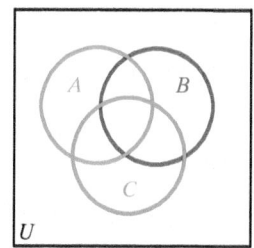

65. Let $U = \{m, n, o, p, q, r, s, t, u, v, w\}$,
   $A = \{m, n, p, q, r, t\}$,
   $B = \{m, o, p, q, s, u\}$,
   $C = \{m, o, p, r, s, t, u, v\}$.

66. Let $U = \{1, 2, 3, 4, 5, 6, 7, 8\}$,
   $A = \{1, 3, 5, 7\}$,
   $B = \{1, 3, 4, 6\}$,
   $C = \{1, 4, 5, 6, 7, 8\}$.

Use a Venn diagram similar to the one above to shade each set.

67. $(A \cap B) \cap C$ **68.** $(A' \cap B) \cap C$

69. $(A' \cap B') \cap C$ **70.** $(A \cap C') \cap B$

71. $(A \cap B') \cap C'$ **72.** $(A \cap B)' \cup C$

Write a description of each shaded area. Use the symbols

$$A, \; B, \; C, \; \cap, \; \cup, \; -, \; and \; '.$$

Different answers are possible.

73.
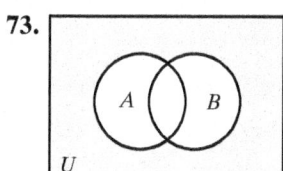

74.
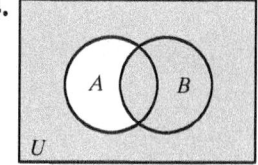

75.
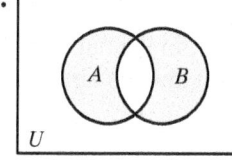

76.
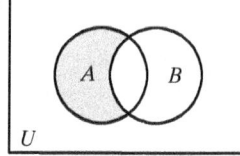

77.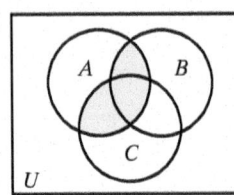

78.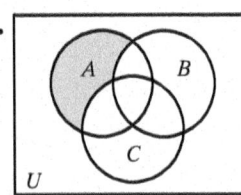

Suppose A and B are sets. Describe the conditions under which each statement would be true.

79. $A = A - B$

80. $A = B - A$

81. $A = A - \varnothing$

82. $A \cap \varnothing = \varnothing$

83. $A \cup B = A$

84. $A \cap B = B$

Draw two Venn diagrams to decide whether the statement is always true or not always true.

85. $(A \cap B) \subseteq A$

86. $(A \cup B) \subseteq A$

87. If $A \subseteq B$, then $A \cup B = A$.

88. If $A \subseteq B$, then $A \cap B = A$.

Write a short answer for each problem.

89. Explain why, if A and B are sets, it is not necessarily true that $n(A - B) = n(A) - n(B)$. Give a counterexample.

90. The five set operations listed in this section are applied to subsets of U (that is, to A and/or B). Is the result always a subset of U also? Explain why or why not.

FOR FURTHER THOUGHT

Comparing Properties

The arithmetic operations of addition and multiplication, when applied to numbers, have some familiar properties discussed in detail later in this text. If a, b, and c are *real numbers*, then the **commutative property of addition** says that the order of the numbers being added makes no difference:

$$a + b = b + a.$$

(Is there a **commutative property of multiplication?**) The **associative property of addition** says that when three numbers are added, the grouping used makes no difference:

$$(a + b) + c = a + (b + c).$$

(Is there an **associative property of multiplication?**) The number 0 is called the **identity element for addition** since adding it to any number does not change that number:

$$a + 0 = a.$$

(What is the **identity element for multiplication?**) Finally, the **distributive property of multiplication over addition** says that

$$a(b + c) = ab + ac.$$

(Is there a distributive property of addition over multiplication?)

For Group or Individual Investigation

Now consider the operations of union and intersection, applied to sets. By recalling definitions, trying examples, or using Venn diagrams, answer the following questions.

1. Is set union commutative? Set intersection?

2. Is set union associative? Set intersection?

3. Is there an identity element for set union? If so, what is it? How about set intersection?

4. Is set intersection distributive over set union? Is set union distributive over set intersection?

2.4 SURVEYS AND CARDINAL NUMBERS

OBJECTIVES

1. Analyze survey results.
2. Apply the cardinal number formula.
3. Interpret information from tables.

Surveys

As suggested in the chapter opener, the techniques of set theory can be applied to many different groups of people (or objects). The problems that are addressed sometimes require analyzing known information about certain subsets to obtain cardinal numbers of other subsets. This "known information" is quite often (though not always) obtained by conducting a survey.

Surveying Media Users

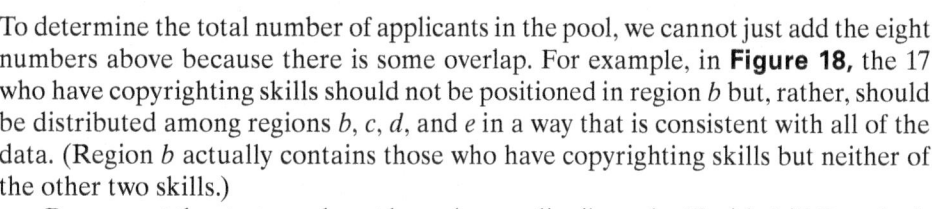

Problem-Solving Strategies

In this section, we apply three problem-solving strategies.
- Make a table or chart—to organize known data.
- Draw a sketch—in this case, a Venn diagram.
- Apply a formula—the cardinal number formula.

Suppose a Human Resources (HR) director for a marketing firm is evaluating a pool of job applicants, with general emphasis on digital and social networking skills, and specifically in copyrighting, graphic design, and link building. The applicants' resumes show the following information.

17 have copyrighting skills (C).

18 have graphic design skills (G).

15 have link-building skills (L).

9 are skilled in both C and G.

5 are skilled in both C and L.

6 are skilled in both L and G.

2 have all three skills.

4 applicants have none of these three skills.

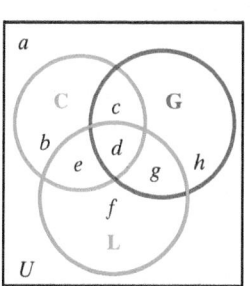

Figure 18

To determine the total number of applicants in the pool, we cannot just add the eight numbers above because there is some overlap. For example, in **Figure 18,** the 17 who have copyrighting skills should not be positioned in region b but, rather, should be distributed among regions b, c, d, and e in a way that is consistent with all of the data. (Region b actually contains those who have copyrighting skills but neither of the other two skills.)

Because, at the start, we do not know how to distribute the 17 with skill C, we look first for some more manageable data. The smallest total listed, the 2 who have all three skills, can be placed in region d (the intersection of the three sets), as shown in **Figure 19.** The 4 who have none of the three skills must go into region a. Then, the 9 who have skills C and G must go into regions c and d. Because region d already contains 2 applicants, we must place

$$9 - 2 = 7 \quad \text{in region } c.$$

Because 5 have skills C and L (regions d and e), we place

$$5 - 2 = 3 \quad \text{in region } e.$$

Now that regions c, d, and e contain 7, 2, and 3, respectively, we must place

$$17 - 7 - 2 - 3 = 5 \quad \text{in region } b.$$

By similar reasoning, all regions are assigned their correct numbers in **Figure 19.**

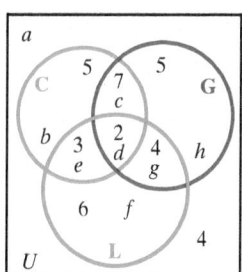

Figure 19

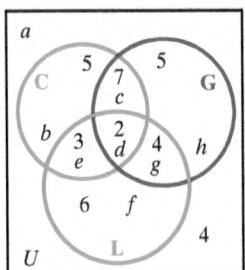

Figure 19 (repeated)

EXAMPLE 1 **Analyzing a Survey**

Using the data on applicants' job skills, as summarized in **Figure 19,** answer each question.

(a) How many applicants are skilled in copyrighting only?

(b) How many applicants have exactly two of these three skills?

(c) How many applicants are in the pool?

Solution

(a) An applicant with skill C only does not have skill G and does not have skill L. These applicants are inside the regions for C and outside the regions for G and L. Region b is the appropriate region in **Figure 19,** and we see that five applicants have skill C only.

(b) The applicants in regions c, e, and g have exactly two of the three skills. The total number of such applicants is

$$7 + 3 + 4 = 14.$$

(c) Each applicant has been placed in exactly one region of **Figure 19,** so the total number in the pool is the sum of the numbers in all eight regions:

$$4 + 5 + 7 + 5 + 3 + 2 + 4 + 6 = 36.$$

Cardinal Number Formula

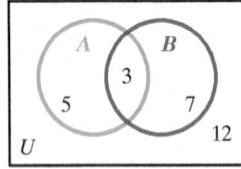

Figure 20

If the numbers shown in **Figure 20** are the cardinal numbers of the individual regions, then

$$n(A) = 5 + 3 = 8, \quad n(B) = 3 + 7 = 10, \quad n(A \cap B) = 3,$$

and

$$n(A \cup B) = 5 + 3 + 7 = 15.$$

Notice that $n(A \cup B) = n(A) + n(B) - n(A \cap B)$ because $15 = 8 + 10 - 3$. This relationship is true for any two sets A and B.

CARDINAL NUMBER FORMULA

For any two sets A and B, the following is true.

$$n(A \cup B) = n(A) + n(B) - n(A \cap B)$$

The cardinal number formula can be rearranged to find any one of its four terms when the others are known.

EXAMPLE 2 **Applying the Cardinal Number Formula**

Find $n(A)$ if $n(A \cup B) = 22, n(A \cap B) = 8,$ and $n(B) = 12.$

Solution

We solve the cardinal number formula for $n(A)$.

Subtract $n(B)$ and add $n(A \cap B)$ to each side.

$$n(A) = n(A \cup B) - n(B) + n(A \cap B)$$

$$= 22 - 12 + 8$$

$$= 18$$

EXAMPLE 3   Analyzing Data in a Report

Scott, who leads a group of technical agents investigating illegal activities on social networking sites, reported the following information.

T = the set of agents following patterns on Twitter

F = the set of agents following patterns on Facebook

L = the set of agents following patterns on LinkedIn

$$n(T) = 13 \quad n(T \cap F) = 9 \quad\quad n(T \cap F \cap L) = 5$$
$$n(F) = 16 \quad n(F \cap L) = 10 \quad n(T' \cap F' \cap L') = 3$$
$$n(L) = 13 \quad n(T \cap L) = 6$$

How many agents are in Scott's group?

Solution

The data supplied by Scott are reflected in **Figure 21.** The sum of the numbers in the diagram gives the total number of agents in the group.

$$3 + 3 + 1 + 2 + 5 + 5 + 4 + 2 = 25$$

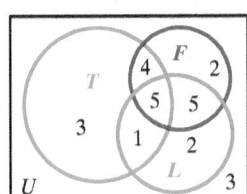

Figure 21

Tables

Sometimes information appears in a table rather than a Venn diagram, but the basic ideas of union and intersection still apply.

EXAMPLE 4   Analyzing Data in a Table

Melanie, the officer in charge of the cafeteria on a military base, wanted to know whether the beverage that enlisted men and women preferred with lunch depended on their age. On a given day, Melanie categorized her lunch patrons according to age and preferred beverage, recording the results in a table.

		Beverage			
		Cola (*C*)	Iced Tea (*I*)	Sweet Tea (*S*)	Totals
	18–25 (*Y*)	45	10	35	90
Age	26–33 (*M*)	20	25	30	75
	Over 33 (*O*)	5	30	20	55
	Totals	70	65	85	220

Using the letters in the table, find the number of people in each set.

(a) $Y \cap C$ **(b)** $O' \cup I$

Solution

(a) The set Y includes all personnel represented across the top row of the table (90 in all), while C includes the 70 down the left column. The intersection of these two sets is just the upper left entry, 45 people.

(b) The set O' excludes the bottom row, so it includes the first and second rows. The set I includes the middle column only. The union of the two sets represents

$$45 + 10 + 35 + 20 + 25 + 30 + 30 = 195 \text{ people.}$$

WHEN WILL I **EVER** USE THIS?

Suppose you run a small construction company, building a few "spec" homes at a time. This week's work will require the following jobs:

Hanging drywall (D),

Installing roofing (R),

Doing electrical work (E).

You will assign 9 workers, with job skills as described here.

6 of the 9 can do D

4 can do R

7 can do E

5 can do both D and E

4 can do both R and E

4 can do all three

1. Construct a Venn diagram to decide how many of your workers have

   (a) exactly two of the three skills

   (b) none of the three skills

   (c) no more than one of the three skills

   (d) at least two of the three skills.

2. Which skill is common to the greatest number of workers, and how many possess that skill?

Answers: **1. (a)** 1 **(b)** 1 **(c)** 4 **(d)** 5 **2.** Electrical work; 7

2.4 EXERCISES

CONCEPT CHECK *Let A and B be any two sets. Write* true *or* false *for each statement.*

1. $n(A \cup B) = n(A) + n(B)$

2. $n(A \cap B) = n(A) \times n(B)$

3. $n(A \cup B) + n(A \cap B) = n(A) + n(B)$

4. $n(A \cup B)$ cannot be less than $n(A \cap B)$

CONCEPT CHECK *Use the numerals representing cardinalities in the Venn diagrams to give the cardinality of each set specified.*

5.
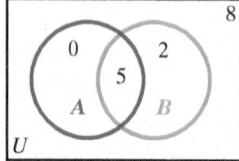

(a) $A \cap B$

(b) $A \cup B$

(c) $A \cap B'$

(d) $A' \cap B$

(e) $A' \cap B'$

6.
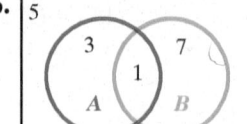

(a) $A \cap B$

(b) $A \cup B$

(c) $A \cap B'$

(d) $A' \cap B$

(e) $A' \cap B'$

7.
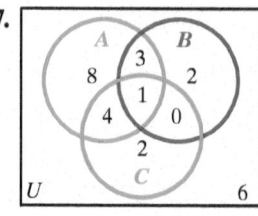

(a) $A \cap B \cap C$

(b) $A \cap B \cap C'$

(c) $A \cap B' \cap C$

(d) $A' \cap B \cap C$

(e) $A' \cap B' \cap C$

(f) $A \cap B' \cap C'$

(g) $A' \cap B \cap C'$

(h) $A' \cap B' \cap C'$

8.

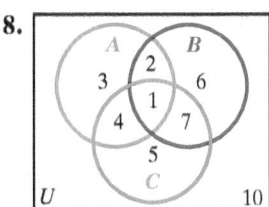

(a) $A \cap B \cap C$

(b) $A \cap B \cap C'$

(c) $A \cap B' \cap C$

(d) $A' \cap B \cap C$

(e) $A' \cap B' \cap C$

(f) $A \cap B' \cap C'$

(g) $A' \cap B \cap C'$

(h) $A' \cap B' \cap C'$

In each case, make use of an appropriate formula.

9. Find the value of $n(A \cup B)$ if $n(A) = 12, n(B) = 14$, and $n(A \cap B) = 5$.

10. Find the value of $n(A \cup B)$ if $n(A) = 16, n(B) = 28$, and $n(A \cap B) = 5$.

11. Find the value of $n(A \cap B)$ if $n(A) = 20, n(B) = 12$, and $n(A \cup B) = 25$.

12. Find the value of $n(A \cap B)$ if $n(A) = 20, n(B) = 24$, and $n(A \cup B) = 30$.

13. Find the value of $n(A)$ if $n(B) = 35, n(A \cap B) = 15$, and $n(A \cup B) = 55$.

14. Find the value of $n(B)$ if $n(A) = 20, n(A \cap B) = 6$, and $n(A \cup B) = 30$.

Draw a Venn diagram and use the given information to fill in the number of elements in each region.

15. $n(A) = 19, \ n(B) = 13, \ n(A \cup B) = 25,$
$n(A') = 11$

16. $n(U) = 43, \ n(A) = 25, \ n(A \cap B) = 5,$
$n(B') = 30$

17. $n(A') = 25, \ n(B) = 28, \ n(A' \cup B') = 40,$
$n(A \cap B) = 10$

18. $n(A \cup B) = 15, \ n(A \cap B) = 8, \ n(A) = 13,$
$n(A' \cup B') = 11$

19. $n(A) = 57, \ n(A \cap B) = 35, \ n(A \cup B) = 81,$
$n(A \cap B \cap C) = 15, \ n(A \cap C) = 21, \ n(B \cap C) = 25,$
$n(C) = 49, \ n(B') = 52$

20. $n(A) = 24, \ n(B) = 24, \ n(C) = 26, \ n(A \cap B) = 10,$
$n(B \cap C) = 8, \ n(A \cap C) = 15, \ n(A \cap B \cap C) = 6,$
$n(U) = 50$

21. $n(A) = 15, \ n(A \cap B \cap C) = 5, \ n(A \cap C) = 13,$
$n(A \cap B') = 9, \ n(B \cap C) = 8, \ n(A' \cap B' \cap C') = 21,$
$n(B \cap C') = 3, \ n(B \cup C) = 32$

22. $n(A \cap B) = 21, \ n(A \cap B \cap C) = 6, \ n(A \cap C) = 26,$
$n(B \cap C) = 7, \ n(A \cap C') = 20, \ n(B \cap C') = 25,$
$n(C) = 40, \ n(A' \cap B' \cap C') = 2$

Use a Venn diagram to solve each problem.

23. ***Winter Break Shopping*** A total of 120 high school students were surveyed about their winter break shopping habits. The results revealed that 65% shop for new clothes, 50% shop for school supplies, and 40% shop for clothes and school supplies. How many students shop for clothes or school supplies over the break? (*MT* calendar problem)

24. ***Internet Browsers*** Among 240 Microsoft Windows users, 65% use the browser Google Chrome, 35% use Internet Explorer, and 95% use at least one of the two. How many Windows users use both Chrome and Internet Explorer?

25. ***Writing and Producing Music*** Joe Long worked on 9 music projects last year.

Joe Long, Bob Gaudio, Tommy DeVito, and Frankie Valli
The Four Seasons

He wrote and produced 3 projects.

He wrote a total of 5 projects.

He produced a total of 7 projects.

(a) How many projects did he write but not produce?

(b) How many projects did he produce but not write?

26. ***Compact Disc Collection*** Gitti is a fan of the music of Paul Simon and Art Garfunkel. In her collection of 25 compact discs, she has the following.

5 on which both Simon and Garfunkel sing

7 on which Simon sings

8 on which Garfunkel sings

15 on which neither Simon nor Garfunkel sings

(a) How many of her compact discs feature only Paul Simon?

(b) How many of her compact discs feature only Art Garfunkel?

(c) How many feature at least one of these two artists?

(d) How many feature at most one of these two artists?

27. *Student Response to Classical Composers*   The 65 students in a classical music lecture class were polled, with the following results.

> 37 like Wolfgang Amadeus Mozart.
>
> 36 like Ludwig van Beethoven.
>
> 31 like Franz Joseph Haydn.
>
> 14 like Mozart and Beethoven.
>
> 21 like Mozart and Haydn.
>
> 14 like Beethoven and Haydn.
>
> 8 like all three composers.

How many of these students like:

(a) exactly two of these composers?

(b) exactly one of these composers?

(c) none of these composers?

(d) Mozart, but neither Beethoven nor Haydn?

(e) Haydn and exactly one of the other two?

(f) no more than two of these composers?

28. *Financial Aid for Students*   At a western college, half of the 48 mathematics majors were receiving federal financial aid.

> 5 had Pell Grants.
>
> 14 participated in the College Work Study Program.
>
> 4 had TOPS scholarships.
>
> 2 had TOPS scholarships and participated in Work Study.
>
> Those with Pell Grants had no other federal aid.

How many of the 48 math majors had:

(a) no federal aid?

(b) more than one of these three forms of aid?

(c) federal aid other than these three forms?

(d) a TOPS scholarship or Work Study?

(e) exactly one of these three forms of aid?

(f) no more than one of these three forms of aid?

29. *Animated Movies*   A middle school counselor, attempting to correlate school performance with leisure interests, found the following information for a group of students.

> 34 had seen *Despicable Me 3*.
>
> 29 had seen *Coco*.
>
> 26 had seen *The Boss Baby*.
>
> 16 had seen *Despicable Me 3* and *Coco*.
>
> 12 had seen *Despicable Me 3* and *The Boss Baby*.
>
> 10 had seen *Coco* and *The Boss Baby*.
>
> 4 had seen all three of these films.
>
> 5 had seen none of the three films.

(a) How many students had seen *The Boss Baby* only?

(b) How many had seen exactly two of the films?

(c) How many students were surveyed?

30. *Nursing Student Education Preferences*   A group of attendees at an educational seminar, all of whom desire to become registered nurses (RNs), are asked their preferences among the three traditional ways to become an RN, and the following information is produced.

> 23 would consider pursuing the Bachelor of Science in Nursing (BSN).
>
> 16 would consider pursuing the Associate Degree in Nursing (ADN).
>
> 7 would consider a Diploma program (Diploma).
>
> 10 would consider both the BSN and the ADN.
>
> 5 would consider both the BSN and the Diploma.
>
> 3 would consider both the ADN and the Diploma.
>
> 2 would consider all three options.
>
> 4 are looking for an option other than these three.

How many attendees:

(a) like the BSN option only?

(b) like exactly two of these three options?

(c) were surveyed altogether?

31. *Orchestral Skills*   In an orchestra, 23 people can play stringed instruments, 18 can play brass, and 12 can play percussion. Further, 10 of the performers can play both strings and brass, whereas 6 can play both strings and percussion. If no one can play all three types of instruments, what are the maximum and minimum numbers of people in the orchestra? (*MT* calendar problem) (*Hint:* Consider expressing the numbers in some of the regions of your diagram in terms of a single unknown, say x.)

32. *Student Goals*   Sofia, an educational researcher, interviewed graduating seniors on a college campus to find out the main goals of today's students.

> Let W = the set of those who want to be wealthy,
>
> F = the set of those who want to raise a family,
>
> E = the set of those who want to become experts in their fields.

Sofia's findings are summarized here.

$$n(W) = 160 \qquad n(E \cap F) = 90$$
$$n(F) = 140 \qquad n(W \cap F \cap E) = 80$$
$$n(E) = 130 \qquad n(E') = 95$$
$$n(W \cap F) = 95 \qquad n[(W \cup F \cup E)'] = 10$$

Find the total number of students interviewed.

33. *Hospital Patient Symptoms* A survey was conducted among 75 patients admitted to a hospital cardiac unit during a two-week period.

Let $\;B =$ the set of patients with high blood pressure,

$\quad\;\; C =$ the set of patients with high cholesterol levels,

$\quad\;\; S =$ the set of patients who smoke cigarettes.

The survey produced the following data.

$$n(B) = 47 \qquad\quad n(B \cap S) = 33$$
$$n(C) = 46 \qquad\quad n(B \cap C) = 31$$
$$n(S) = 52 \qquad\quad n(B \cap C \cap S) = 21$$
$$n[(B \cap C) \cup (B \cap S) \cup (C \cap S)] = 51$$

Find the number of these patients who:

(a) had either high blood pressure or high cholesterol levels, but not both.

(b) had fewer than two of the indications listed.

(c) were smokers but had neither high blood pressure nor high cholesterol levels.

(d) did not have exactly two of the indications listed.

(e) had at least one but not all three of the indications listed.

(f) had none of the indications listed.

34. *Song Themes* It was once said that country-western songs emphasize three basic themes: love, prison, and trucks. A survey of the local country-western radio station produced the following data.

    12 songs about a truck driver who is in love while in prison

    13 about a prisoner in love

    28 about a person in love

    18 about a truck driver in love

    3 about a truck driver in prison who is not in love

    2 about people in prison who are not in love and do not drive trucks

    8 about people who are out of prison, are not in love, and do not drive trucks

    16 about truck drivers who are not in prison

(a) How many songs were surveyed?

Find the number of songs about:

(b) truck drivers.

(c) prisoners.

(d) truck drivers in prison.

(e) people not in prison.

(f) people not in love.

(g) a prisoner in love.

(h) exactly two of the basic themes listed.

Solve each problem.

35. *Army Housing* A study of U.S. Army housing trends categorized personnel as commissioned officers (C), warrant officers (W), or enlisted (E), and categorized their living facilities as on-base (B), rented off-base (R), or owned off-base (O). One survey yielded the following data.

		Facilities			
		B	**R**	**O**	**Totals**
	C	12	29	54	95
Personnel	**W**	4	5	6	15
	E	374	71	285	730
Totals		390	105	345	840

Find the number of personnel in each of the following sets.

(a) $W \cap O$

(b) $C \cup B$

(c) $R' \cup W'$

(d) $(C \cup W) \cap (B \cup R)$

(e) $(C \cap B) \cup (E \cap O)$

(f) $B \cap (W \cup R)'$

36. Basketball Positions Sakda runs a youth basketball program. On the first day of the season, 60 young women showed up and were categorized by age level and by preferred basketball position, as shown in the following table.

	Position Guard (G)	Position Forward (F)	Position Center (N)	Totals
Junior High (J)	9	6	4	19
Age Senior High (S)	12	5	9	26
College (C)	5	8	2	15
Totals	26	19	15	60

Using the set labels (letters) in the table, find the number of players in each of the following sets.

(a) $J \cap G$

(b) $S \cap N$

(c) $N \cup (S \cap F)$

(d) $S' \cap (G \cup N)$

(e) $(S \cap N') \cup (C \cap G')$

(f) $N' \cap (S' \cap C')$

Write a short answer for each problem.

37. Could the information of **Example 4** have been presented in a Venn diagram similar to those in **Examples 1 and 3?** If so, construct such a diagram. Otherwise, explain the essential difference of **Example 4.**

38. Explain how a cardinal number formula can be derived for the case where *three* sets occur. Specifically, give a formula relating $n(A \cup B \cup C)$ to

$$n(A), \ n(B), \ n(C), \ n(A \cap B), \ n(A \cap C),$$

$$n(B \cap C), \ \text{and} \ n(A \cap B \cap C).$$

Illustrate with a Venn diagram.

Chapter 2 SUMMARY

KEY TERMS

2.1

set
elements
members
empty (null) set
natural (counting)
  numbers
whole numbers

integers
rational numbers
real numbers
irrational
  numbers
cardinal number
  (cardinality)
finite set
infinite set

2.2

universal set
Venn diagram
complement
  (of a set)
subset (of a set)
proper subset (of a set)
tree diagram

2.3

intersection (of sets)
disjoint sets
union (of sets)
difference (of sets)
ordered pairs
Cartesian product (of sets)
operation (on sets)

NEW SYMBOLS

$\emptyset$	empty set (or null set)		$\not\subseteq$	is not a subset of
$\in$	is an element of		$\subset$	is a proper subset of
$\notin$	is not an element of		$\not\subset$	is not a proper subset of
$n(A)$	cardinal number of set A		$\cap$	intersection (of sets)
$\aleph_0$	aleph null		$\cup$	union (of sets)
U	universal set		$-$	difference (of sets)
A'	complement of set A		(a, b)	ordered pair
$\subseteq$	is a subset of		$\times$	Cartesian product

TEST YOUR WORD POWER

See how well you have learned the vocabulary in this chapter.

1. In an **ordered pair,**
   A. the components are always numbers of the same type.
   B. the first component must be less than the second.
   C. the components can be any kinds of objects.
   D. the first component is a subset of the second component.

2. The **complement** of a set
   A. contains only some, but not all, of that set's elements.
   B. contains the same number of elements as the given set.
   C. always contains fewer elements than the given set.
   D. cannot be determined until a universal set is given.

3. Any **subset** of set A
  A. must have fewer elements than A.
  B. has fewer elements than A only if it is a proper subset.
  C. is an element of A.
  D. contains all the elements of A, plus at least one additional element.

4. Examples of **operations on sets** are
  A. union, intersection, and subset.
  B. Cartesian product, difference, and proper subset.
  C. complement, supplement, and intersection.
  D. complement, union, and Cartesian product.

5. The **set difference $A - B$** must have cardinality
  A. $n(A) - n(B)$.
  B. less than or equal to $n(A)$.
  C. greater than or equal to $n(B)$.
  D. less than $n(A)$.

6. The **cardinal number formula** says that
  A. $n(A \cup B) = n(A) + n(B) - n(A \cap B)$.
  B. $n(A \cup B) = n(A) + n(B) + n(A \cap B)$.
  C. $n(A \cup B) = n(A) + n(B)$.
  D. $n(A \cup B) = n(A) \cdot n(B)$.

ANSWERS
1. C **2.** D **3.** B **4.** D **5.** B **6.** A

QUICK REVIEW

Concepts | *Example*

2.1 SYMBOLS AND TERMINOLOGY

Designating Sets
Sets are designated using the following methods:

(1) word descriptions,

(2) the listing method,

(3) set-builder notation.

Designate the set containing 1, 3, and 5 in three ways.

The set of odd counting numbers less than 7

$\{1, 3, 5\}$

$\{x \mid x$ is an odd counting number and $x < 7\}$

All are equal.

The **cardinal number** of a set is the number of elements it contains.

Let $A = \{10, 20, 30, \ldots, 80\}$.
$$n(A) = 8$$

Equal sets have exactly the same elements.

$\{a, e, i, o, u\} = \{i, o, u, a, e\}, \quad \{q, r, s, t\} \neq \{q, p, s, t\}$

2.2 VENN DIAGRAMS AND SUBSETS

Sets are normally discussed within the context of a designated **universal set, U.**

The **complement** of a set A contains all elements in U that are not in A.

Set B is a **subset** of set A if every element of B is also an element of A.

Set B is a **proper subset** of A if $B \subseteq A$ and $B \neq A$.

If a set has cardinal number n, then it has
$$2^n \text{ subsets} \quad \text{and} \quad 2^n - 1 \text{ proper subsets.}$$

Let $U = \{x \mid x$ is a whole number$\}$
and $A = \{x \mid x$ is an even whole number$\}$.

$$A' = \{x \mid x \text{ is an odd whole number}\}$$

Let $U = \{2, 3, 5, 7\}$, $A = \{3, 5, 7\}$, and $B = \{5\}$.
$$A' = \{2\} \quad \text{and} \quad B \subseteq A$$

For the sets A and B given above,
$$B \subset A.$$

If $D = \{x, y, z\}$, then $n(D) = 3$.

D has $2^3 = 8$ subsets and
$$2^3 - 1 = 8 - 1 = 7 \text{ proper subsets.}$$

Concepts	Example

(2.3) SET OPERATIONS

Intersection of Sets

$$A \cap B = \{x \mid x \in A \text{ and } x \in B\}$$

Union of Sets

$$A \cup B = \{x \mid x \in A \text{ or } x \in B\}$$

Difference of Sets

$$A - B = \{x \mid x \in A \text{ and } x \notin B\}$$

Cartesian Product of Sets

$$A \times B = \{(a, b) \mid a \in A \text{ and } b \in B\}$$

Cardinal Number of a Cartesian Product

If $n(A) = a$ and $n(B) = b$, then

$$n(A \times B) = n(A) \cdot n(B) = ab.$$

Perform the indicated operations.

$$\{1, 2, 7\} \cap \{5, 7, 9, 11\} = \{7\}$$

$$\{20, 40, 60\} \cup \{40, 60, 80\} = \{20, 40, 60, 80\}$$

$$\{5, 6, 7, 8\} - \{1, 3, 5, 7\} = \{6, 8\}$$

$$\{1, 2\} \times \{30, 40, 50\}$$
$$= \{(1, 30), (1, 40), (1, 50), (2, 30), (2, 40), (2, 50)\}$$

Let $A = \{2, 4, 6, 8\}$ and $B = \{3, 7\}$.

$$n(A \times B) = n(A) \cdot n(B) = 4 \cdot 2 = 8$$

Verify this by listing the elements of $A \times B$.

$$\{(2, 3), (2, 7), (4, 3), (4, 7), (6, 3), (6, 7), (8, 3), (8, 7)\}$$

There are indeed 8 elements.

Numbering the regions in a Venn diagram facilitates identification of various sets and relationships among them.

Refer to the Venn diagram.

A consists of regions 2, 3, 4, and 5.

$A \cap B$ consists of regions 3 and 4.

$B \cup C$ consists of regions 3, 4, 5, 6, 7, and 8.

$(A \cup B) \cap C'$ consists of regions 2, 3, and 8.

$A \cap B \cap C$ consists of region 4.

$(A \cup B \cup C)'$ consists of region 1.

$A - C$ consists of regions 2 and 3.

De Morgan's Laws

For any sets A and B, where $A \subseteq U$ and $B \subseteq U$, the following are true.

$$(A \cap B)' = A' \cup B'$$

$$(A \cup B)' = A' \cap B'$$

Let $U = \{1, 2, 3, 4, \ldots, 9\}$, $A = \{2, 4, 6, 8\}$, and $B = \{4, 5, 6, 7\}$.

Then $A \cap B = \{4, 6\}$,

so $(A \cap B)' = \{1, 2, 3, 5, 7, 8, 9\}$.

Also, $A' = \{1, 3, 5, 7, 9\}$ and $B' = \{1, 2, 3, 8, 9\}$, — Same

so $A' \cup B' = \{1, 2, 3, 5, 7, 8, 9\}$.

Using sets A and B given above,

$$A \cup B = \{2, 4, 5, 6, 7, 8\},$$

so $(A \cup B)' = \{1, 3, 9\}$.

Also, $A' = \{1, 3, 5, 7, 9\}$ and $B' = \{1, 2, 3, 8, 9\}$, — Same

so $A' \cap B' = \{1, 3, 9\}$.

2.4 SURVEYS AND CARDINAL NUMBERS

Cardinal Number Formula

For any two sets A and B, the following is true.

$$n(A \cup B) = n(A) + n(B) - n(A \cap B)$$

Enter known facts in a Venn diagram to find desired facts.

Given $n(A) = 15, n(B) = 12$, and $n(A \cup B) = 25$.

To find $n(A \cap B)$, first solve the formula for that term.

$$n(A \cap B) = n(A) + n(B) - n(A \cup B)$$
$$= 15 + 12 - 25$$
$$= 2$$

Given $n(A) = 12, n(B) = 27, n(A \cup B) = 32$, and $n(U) = 50$, find $n(A - B)$ and $n[(A \cap B)']$.

$$n(A \cap B) = n(A) + n(B) - n(A \cup B)$$
$$= 12 + 27 - 32$$
$$= 7$$

The cardinalities of all regions can now be entered in a Venn diagram, starting with 7 in the center.

Observe that $n(A - B) = 5$ and

$$n[(A \cap B)'] = 18 + 5 + 20 = 43.$$

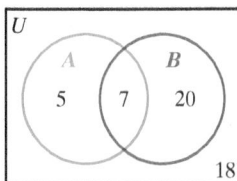

Chapter 2 TEST

In Exercises 1–14, let

$$U = \{a, b, c, d, e, f, g, h\}, \quad A = \{a, b, c, d\},$$
$$B = \{b, e, a, d\}, \quad and \quad C = \{e, a\}.$$

Find each set.

1. $A \cup C$ **2.** $B \cap A$

3. B' **4.** $A - (B \cap C')$

Write true *or* false *for each statement.*

5. $e \in A$ **6.** $C \subseteq B$

7. $B \subset (A \cup C)$ **8.** $c \notin C$

9. $n[(A \cup B) - C] = 4$ **10.** $\varnothing \not\subset C$

11. $A \cap B'$ is equivalent to $B \cap A'$

12. $(A \cup B)' = A' \cap B'$

Evaluate each of the following.

13. $n(B \times A)$

14. the number of proper subsets of B

Give a word description for each set.

15. $\{-3, -1, 1, 3, 5, 7, 9\}$

16. {Sun., Mon., Tue., ..., Sat.}

Express each set in set-builder notation.

17. $\{-1, -2, -3, -4, \ldots\}$

18. $\{24, 32, 40, 48, \ldots, 88\}$

Place $\subset$, $\subseteq$, *both, or neither in each blank to make a true statement.*

19. $\varnothing$ _____ $\{x \mid x$ is a counting number between 20 and 21$\}$

20. $\{3, 5, 7\}$ _____ $\{4, 5, 6, 7, 8, 9, 10\}$

Shade each set in an appropriate Venn diagram.

21. $X \cup Y'$ **22.** $X' \cap Y'$

23. $(X \cup Y) - Z$

24. $[(X \cap Y) \cup (X \cap Z)] - (Y \cap Z)$

25. State De Morgan's laws for sets in words rather than symbols.

Facts about Inventions   *The table lists ten inventions, together with other pertinent data.*

Invention	Date	Inventor	Nation
Adding machine	1642	Pascal	France
Baking powder	1843	Bird	England
Electric razor	1917	Schick	U.S.
Fiber optics	1955	Kapany	England
Geiger counter	1913	Geiger	Germany
Pendulum clock	1657	Huygens	Holland
Radar	1940	Watson-Watt	Scotland
Telegraph	1837	Morse	U.S.
Thermometer	1593	Galileo	Italy
Zipper	1891	Judson	U.S.

Let U = the set of all ten inventions,

   A = the set of items invented in the United States,

and T = the set of items invented in the twentieth
        century.

List the elements of each set.

26. $A \cap T$ **27.** $(A \cup T)'$ **28.** $T' - A'$

Solve each problem.

29. The numerals in the Venn diagram indicate the number of elements in each particular region. Determine the number of elements in each set.

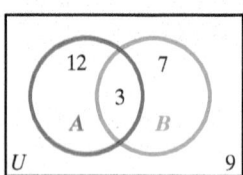

(a) $A \cup B$ **(b)** $A \cap B'$ **(c)** $(A \cap B)'$

30. *Financial Aid to College Students*   Three sources of financial aid are government grants, private scholarships, and the colleges themselves. Susan, a financial aid director of a private college, surveyed the records of 100 sophomores and found that:

   49 receive government grants.

   55 receive private scholarships.

   43 receive aid from the college.

   23 receive government grants and private scholarships.

   18 receive government grants and aid from the college.

   28 receive private scholarships and aid from the college.

   8 receive help from all three sources.

How many of the students in the survey:

(a) have government grants only?

(b) have scholarships but not government grants?

(c) receive financial aid from only one of these sources?

(d) receive aid from exactly two of these sources?

(e) receive no financial aid from any of these sources?

(f) receive no aid from the college or the government?

To capture the imagination of their readers, authors of fantasy literature (such as J. R. R. Tolkien, J. K. Rowling, and Lewis Carroll) have long used magical imagery, including riddles. Such a riddle confronts a hero named Humphrey, who finds himself trapped in a magical maze with only two exits.

One door to freedom opens wide,
The other naught but dungeon hides.
Twin watchers, one by truth is bound,
The other never truth will sound.

The watcher here shall speak when he
Spies a man who would be free.
Name him Lying Troll or Truthful,
Rightly named he shall be useful.

If he moves, then be not late,
Choose a door and seal thy fate.

As Humphrey approaches the troll, it speaks:

"If truly Truthful Troll I be,
then go thou east, and be thou free."

Logic, the topic of this chapter, is useful to authors in the creation of such puzzles—and also to readers in solving them.

3.1 STATEMENTS AND QUANTIFIERS

OBJECTIVES

1 Distinguish between statements and non-statements.

2 Compose negations of statements.

3 Translate between words and symbols.

4 Interpret statements with quantifiers and form their negations.

5 Find truth values of statements involving quantifiers and number sets.

Statements

This section introduces the study of **symbolic logic,** which uses letters to represent statements, and symbols for words such as *and, or, not.* Logic is used to determine the **truth value** (that is, the truth or falsity) of statements with multiple parts.

Many kinds of sentences occur in ordinary language, including factual statements, opinions, commands, and questions. Symbolic logic discusses only statements involving facts. A **statement** is a declarative sentence that is either true or false, but not both simultaneously.

Text messaging provides a means of communication.
$12 + 6 = 13$

> Statements
> Each is either true or false.

Access the file.
Did the Astros win the World Series?
Kyrie Irving is a better basketball player than James Harden.
This sentence is false.

> Not statements
> Each cannot be identified as being either true or false.

Of the sentences that are not statements, the first is a command, and the second is a question. The third is an opinion. "This sentence is false" is a paradox: If we assume it is true, then it is false, and if we assume it is false, then it is true.

A **compound statement** may be formed by combining two or more statements. The statements comprising a compound statement are called **component statements.** Various **logical connectives,** or simply **connectives,** such as *and, or, not,* and *if . . . then,* can be used in forming compound statements. (Although a statement such as "Today is not Tuesday" does not consist of two component statements, for convenience it is considered compound, because its truth value is determined by noting the truth value of a different statement, "Today is Tuesday.")

EXAMPLE 1 Deciding Whether a Statement Is Compound

Decide whether each statement is compound. If so, identify the connective.

(a) Lord Byron wrote sonnets, and the poem exhibits iambic pentameter.

(b) You can pay me now, or you can pay me later.

(c) If it's on the Internet, then it must be true.

(d) My accountant works for Ernst and Young.

Solution

(a) This statement is compound, with component statements "Lord Byron wrote sonnets" and "the poem exhibits iambic pentameter." The connective is *and.*

(b) The connective here is *or.* The statement is compound.

(c) The connective here is *if…then,* discussed in more detail later in the chapter. The statement is compound.

(d) Although the word "and" is used in this statement, it is not used as a *logical* connective. It is part of the name of the company. The statement is not compound.

Negations

The sentence "Anthony Mansella has a red truck" is a statement. The **negation** of this statement is "Anthony Mansella does not have a red truck." ***The negation of a true statement is false, and the negation of a false statement is true.***

Gottfried Leibniz (1646–1716) was a wide-ranging philosopher and a universalist who tried to patch up Catholic–Protestant conflicts. He promoted cultural exchange between Europe and the East. Chinese ideograms led him to search for a universal symbolism. He was an early inventor of **symbolic logic.**

EXAMPLE 2 **Forming Negations**

Form the negation of each statement.

(a) That city has a mayor.   **(b)** The moon is not a planet.

Solution

(a) To negate this statement, we introduce *not* into the sentence: "That city does not have a mayor."

(b) The negation is "The moon is a planet."

One way to detect incorrect negations is to check truth values. ***A negation must have the opposite truth value from the original statement.***

The next example uses some of the inequality symbols in **Table 1.** In the case of an inequality involving a variable, the negation must have the opposite truth value for *any* replacement of the variable.

```
NORMAL FLOAT AUTO REAL RADIAN MP
TEST LOGIC
1:=
2:≠
3:>
4:≥
5:<
6:≤
```

The TEST menu of the TI-84 Plus C calculator allows the user to test the truth value of statements involving =, ≠, >, ≥, <, and ≤. If a statement is true, it returns a 1. If a statement is false, it returns a 0.

Table 1 Inequality Symbols

Symbolism	Meaning	Examples	
$a < b$	a is less than b	$4 < 9$	$\frac{1}{2} < \frac{3}{4}$
$a > b$	a is greater than b	$6 > 2$	$-5 > -11$
$a \leq b$	a is less than or equal to b	$8 \leq 10$	$3 \leq 3$
$a \geq b$	a is greater than or equal to b	$-2 \geq -3$	$-5 \geq -5$

```
NORMAL FLOAT AUTO REAL RADIAN MP
4<9
                              1.
4>9
                              0.
```

$4 < 9$ is true, as indicated by the 1.
$4 > 9$ is false, as indicated by the 0.

EXAMPLE 3 **Negating Inequalities**

Give a negation of each inequality. Do *not* use a slash symbol.

(a) $x < 9$ **(b)** $7x + 11y \geq 77$

Solution

(a) The negation of "x is less than 9" is "x is *not* less than 9." Because we cannot use "not," which would require writing $x \nless 9$, phrase the negation as "x is greater than or equal to 9," or

$$x \geq 9.$$

(b) The negation, with no slash, is

$$7x + 11y < 77.$$

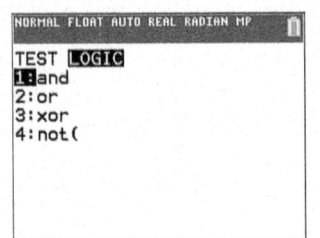

NORMAL FLOAT AUTO REAL RADIAN MP

TEST **LOGIC**
1: and
2: or
3: xor
4: not(

The LOGIC menu of the TI-84 Plus C calculator allows the user to test the truth value of statements involving *and*, *or*, *exclusive or*, and *not*.

Symbols

The study of logic uses symbols. Statements are represented with letters, such as *p*, *q*, or *r*. Several symbols for connectives are shown in **Table 2**.

Table 2 Symbols for Connectives

Connective	Symbol	Type of Statement
and	$\wedge$	Conjunction
or	$\vee$	Disjunction
not	$\sim$	Negation

The symbol ~ represents the connective *not*. If *p* represents the statement "Barack Obama was president in 2014," then ~*p* represents "Barack Obama was *not* president in 2014."

EXAMPLE 4 **Translating from Symbols to Words**

Let *p* represent the statement "Informatics is a growing field," and let *q* represent "Critical care will always be in demand." Translate each statement from symbols to words.

(a) $p \vee q$ **(b)** $\sim p \wedge q$ **(c)** $\sim (p \vee q)$ **(d)** $\sim (p \wedge q)$

Solution

(a) From the table, $\vee$ symbolizes *or*. Thus, $p \vee q$ represents

Informatics is a growing field, or critical care will always be in demand.

(b) Informatics is not a growing field and critical care will always be in demand.

(c) It is not the case that informatics is a growing field or critical care will always be in demand. (This is usually translated as **"Neither *p* nor *q*."**)

(d) It is not the case that informatics is a growing field and critical care will always be in demand.

Quantifiers

Quantifiers are used to indicate *how many* cases of a particular situation exist. The words *all, each, every,* and *no(ne)* are **universal quantifiers,** while words and phrases such as *some, there exists,* and *(for) at least one* are **existential quantifiers.**

Be careful when forming the negation of a statement involving quantifiers. A statement and its negation must have opposite truth values in all possible cases. Consider this statement.

Every month in the list has 31 days.

Many people would write the negation of this statement as "No months in the list have 31 days" or "Every month in the list does not have 31 days." But neither of these is correct. To see why, look at the three lists below.

List 1:   January, March, May

List 2:   August, September, October

List 3:   February, April, June, November

These lists contain all possibilities that need to be considered. In List 1, *every* month has 31 days. In List 2, *some* months have 31 days (and some do not). In List 3, *no* months have 31 days.

Aristotle, the first to systematize the logic we use in everyday life, appears above in a detail from the painting *The School of Athens,* by Raphael. He is shown debating a point with his teacher **Plato.**

Consider **Table 3.** Keep in mind that "some" means "at least one (and possibly all)."

Table 3 Truth Value as Applied to:

	List 1	List 2	List 3
(1) Every month in the list has 31 days. (Given)	T	F	F
(2) No month in the list has 31 days. (Possible negation)	F	F	T
(3) Every month in the list does not have 31 days. (Possible negation)	F	F	T
(4) Some months in the list do not have 31 days. (Possible negation)	F	T	T

Negation

The negation of the given statement (1) must have opposite truth values in *all* cases. It can be seen that statements (2) and (3) do not satisfy this condition (for List 2), but statement (4) does. It may be concluded that the correct negation for "Every month in the list has 31 days" is "Some months in the list do not have 31 days." Other ways of stating the negation include the following.

Not every month in the list has 31 days.

It is not the case that every month in the list has 31 days.

At least one month in the list does not have 31 days.

Table 4 shows how to find the negation of a statement involving quantifiers.

Table 4 Negations of Quantified Statements

Statement	Negation
All do.	Some do not. (Equivalently: Not all do.)
Some do.	None do. (Equivalently: All do not.)

The negation of the negation of a statement is simply the statement itself. For instance, the negations of the statements in the Negation column of **Table 4** are simply the corresponding original statements in the Statement column.

EXAMPLE 5 **Forming Negations of Quantified Statements**

Form the negation of each statement.

(a) Some cats have fleas.   **(b)** Some cats do not have fleas.

(c) No cats have fleas.

Solution

(a) Because *some* means "at least one," the statement "Some cats have fleas" is really the same as "At least one cat has fleas." The negation of this is

"No cats have fleas."

(b) The statement "Some cats do not have fleas" claims that at least one cat, somewhere, does not have fleas. The negation of this is

"All cats have fleas."

(c) The negation is "Some cats have fleas." —— Avoid the incorrect answer "All cats have fleas."

Quantifiers and Number Sets

Earlier we introduced sets of numbers.

SETS OF NUMBERS

Natural numbers (or counting numbers) $\{1, 2, 3, 4, \ldots\}$

Whole numbers $\{0, 1, 2, 3, 4, \ldots\}$

Integers $\{\ldots, -3, -2, -1, 0, 1, 2, 3, \ldots\}$

Rational numbers $\left\{\frac{p}{q} \middle| p \text{ and } q \text{ are integers, and } q \neq 0\right\}$

Real numbers $\{x \mid x \text{ is a number that can be written as a decimal}\}$

Irrational numbers $\{x \mid x \text{ is a real number and } x \text{ cannot be written as a quotient of integers}\}$

EXAMPLE 6 Deciding Whether Quantified Statements Are True or False

Decide whether each statement involving a quantifier is *true* or *false*.

(a) There exists a whole number that is not a natural number.

(b) Every integer is a natural number.

(c) Every natural number is a rational number.

(d) There exists an irrational number that is not real.

Solution

(a) Because there is such a whole number (it is 0), this statement is true.

(b) This statement is false, because we can find at least one integer that is not a natural number. For example, −1 is an integer but is not a natural number.

(c) Because every natural number can be written as a fraction with denominator 1, this statement is true.

(d) In order to be an irrational number, a number must first be real. Because we cannot give an irrational number that is not real, this statement is false. (Had we been able to find at least one, the statement would have been true.)

3.1 EXERCISES

CONCEPT CHECK *Fill in the blank with the correct response.*

1. If the blanks are filled in, then this sentence will be an example of a(n) _____ _____.

2. The statements $x < 5$ and $x \geq 5$ are _____ of each other.

3. The statement "Some batters strike out" contains a(n) _____ quantifier.

4. Not every whole number is a(n) _____ number.

Decide whether each is a statement or is not a statement.

5. February 2, 2009, was a Monday.

6. The ZIP code for Oscar, Louisiana, is 70762.

7. Listen, my children, and you shall hear of the midnight ride of Paul Revere.

8. Did you silence your cell phone?

9. $5 + 9 \neq 14$ and $4 - 1 = 12$

10. $5 + 9 \neq 12$ or $4 - 2 = 5$

11. Some numbers are positive.

12. Grover Cleveland was president of the United States in 1885 and 1897.

13. Accidents are the main cause of deaths of children under the age of 7.

14. It is projected that in the United States between 2016 and 2026, there will be over 100,000 job openings per year for carpenters, with median annual salaries of about $44,000.

15. Where are you going tomorrow?

16. Behave yourself and sit down.

Decide whether each statement is compound.

17. Tomorrow is Saturday.

18. Jing is younger than 18 years of age, and so is her friend Shu-fen.

19. I read the *Detroit Free Press,* and I follow the Detroit Pistons.

20. My brother got married in Copenhagen.

21. If Lorri sells her quota, then Michelle will be happy.

22. If Bobby is a politician, then Mitch is a crook.

23. Jay's wife loves Ben and Jerry's ice cream.

24. The sign on the back of the car read "Canada or bust!"

Write a negation for each statement.

25. Her aunt's name is Hermione.

26. The trash needs to be collected.

27. Some books are longer than this book.

28. No rain fell in southern California today.

29. No computer repairman can play blackjack.

30. All students present will get another chance.

31. Everybody loves somebody sometime.

32. Some people have all the luck.

Give a negation for each inequality. Do not use a slash symbol.

33. $x > 10$

34. $x < -6$

35. $x \geq 2$

36. $x \leq 19$

Write a short answer for each problem.

37. Try to negate the sentence "The exact number of words in this sentence is ten" and see what happens. Explain the problem that arises.

38. Explain why the negation of "$x \geq 5$" is not "$x \leq 5$."

Let p represent the statement "She has green eyes" *and let q represent the statement* "He is 60 years old." *Translate each symbolic compound statement into words.*

39. $\sim p$

40. $\sim q$

41. $p \wedge q$

42. $p \vee q$

43. $\sim p \vee q$

44. $p \wedge \sim q$

45. $\sim p \vee \sim q$

46. $\sim p \wedge \sim q$

47. $\sim (\sim p \wedge q)$

48. $\sim (p \vee \sim q)$

Let p represent the statement "Tyler collects DVDs" *and let q represent the statement* "Josh is an art major." *Convert each compound statement into symbols.*

49. Tyler collects DVDs and Josh is not an art major.

50. Tyler does not collect DVDs or Josh is not an art major.

51. Tyler does not collect DVDs or Josh is an art major.

52. Josh is an art major and Tyler does not collect DVDs.

53. Neither Tyler collects DVDs nor Josh is an art major.

54. Either Josh is an art major or Tyler collects DVDs, and it is not the case that both Josh is an art major and Tyler collects DVDs.

Solve each problem.

55. Quantifiers are often misused in everyday language. Suppose you hear that a local electronics chain is having a 40% off sale, and the radio advertisement states "All items are not available in all stores." Do you think that, literally translated, the ad really means what it says? What do you think is really meant? Explain your answer.

56. Repeat **Exercise 55** for the following: "All people don't have the time to devote to maintaining their vehicles properly."

Refer to the groups of images of mathematicians labeled A, B, and C, and identify by letter the group or groups that satisfy the given statements involving quantifiers.

A

B

C

57. All pictures have frames.

58. No picture has a frame.

59. At least one picture does not have a frame.

60. Not every picture has a frame.

61. At least one picture has a frame.

62. No picture does not have a frame.

63. All pictures do not have frames.

64. Not every picture does not have a frame.

Decide whether each statement involving a quantifier is true *or* false.

65. Every whole number is an integer.

66. Every integer is a whole number.

67. There exists a natural number that is not an integer.

68. There exists an integer that is not a natural number.

69. All rational numbers are real numbers.

70. All irrational numbers are real numbers.

71. Some rational numbers are not integers.

72. Some whole numbers are not rational numbers.

73. Each whole number is a positive number.

74. Each rational number is a positive number.

Solve each problem.

75. Explain the difference between the statements "All students did not pass the test" and "Not all students passed the test."

76. The statement "For some real number x, $x^2 \geq 0$" is true. However, your friend does not understand why, because he claims that $x^2 \geq 0$ is true for *all* real numbers x (and not *some*). How would you explain his misconception to him?

77. Write the following statement using "every": There is no one here who has not made mistakes before.

78. Only one of these statements is true. Which one is it?

   A. For some real number x, $x \not< 0$.

   B. For all real numbers x, $x^3 > 0$.

   C. For all real numbers x less than 0, x^2 is also less than 0.

   D. For some real number x, $x^2 < 0$.

Symbolic logic also uses symbols for quantifiers. The symbol for the existential quantifier is

$$\exists \quad (a\ rotated\ \mathrm{E}),$$

and the symbol for the universal quantifier is

$$\forall \quad (an\ inverted\ \mathrm{A}).$$

The statement "For some x, p is true" can be symbolized

$$(\exists x)(p).$$

The statement "For all x, p is true" can be symbolized

$$(\forall x)(p).$$

The negation of $(\exists x)(p)$ *is*

$$(\forall x)(\sim p),$$

and the negation of $(\forall x)(p)$ *is*

$$(\exists x)(\sim p).$$

79. Refer to **Example 5.** If we let c represent "cat" and f represent "The cat has fleas," then the statement "Some cats have fleas" can be represented by $(\exists c)(f)$. Use symbols to express the negation of this statement.

80. Use symbols to express the statements for parts (b) and (c) of **Example 5** and their negations. Verify that the symbolic expressions translate to the negations found in the text.

3.2   TRUTH TABLES AND EQUIVALENT STATEMENTS

OBJECTIVES

1 Find the truth value of a conjunction.

2 Find the truth value of a disjunction.

3 Find the truth values for compound mathematical statements.

4 Construct truth tables for compound statements.

5 Determine equivalence of statements.

6 Use De Morgan's laws to find negations of compound statements.

Conjunctions

Truth values of compound statements are found by using truth values of component statements. To begin, we must decide on truth values of the **conjunction** *p and q,* symbolized *p* ∧ *q.* Here, the connective *and* implies the idea of "both." The following statement is true, because each component statement is true.

> Monday immediately follows Sunday, and March immediately follows February.
>
> True

On the other hand, the following statement is false, even though part of the statement (Monday immediately follows Sunday) is true.

> Monday immediately follows Sunday, and March immediately follows January.
>
> False

For the conjunction p ∧ q to be true, both p and q must be true. This result is summarized by a table, called a **truth table,** which shows truth values of *p* ∧ *q* for all four possible combinations of truth values for the component statements *p* and *q*.

TRUTH TABLE FOR THE CONJUNCTION   *p* and *q*

p and q		
p	*q*	*p* ∧ *q*
T	T	T
T	F	F
F	T	F
F	F	F

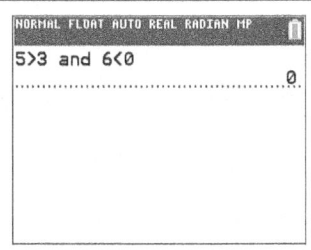

NORMAL FLOAT AUTO REAL RADIAN MP

5>3 and 6<0
                                                    0.

The calculator returns a 0 for

5 > 3 *and* 6 < 0,

indicating that the statement is false.

EXAMPLE 1   Finding the Truth Value of a Conjunction

Let *p* represent "5 > 3" and let *q* represent "6 < 0." Find the truth value of *p* ∧ *q*.

Solution

Here *p* is true and *q* is false. The second row of the conjunction truth table shows that *p* ∧ *q* is false in this case.

In some cases, the logical connective *but* is used in compound statements.

> He wants to go to the mountains but she wants to go to the beach.

Here, *but* is used in place of *and* to emphasize the contrast between the component statements. We consider this statement as we would consider the conjunction using the word *and*. The truth table for the conjunction, given above, would apply.

Disjunctions

In ordinary language, the word *or* can be ambiguous. The expression "this or that" can mean either "this or that or both," or "this or that but not both." For example, consider the following statement.

> I will paint the wall or I will paint the ceiling.

This statement probably means: "I will paint the wall or I will paint the ceiling or I will paint both."

On the other hand, consider the following statement.

> I will wear my glasses or my contact lenses.

It probably means "I will wear my glasses, or I will wear my contacts, but I will not wear both."

The symbol ∨ represents the first *or* described. That is,

$$p \lor q \text{ means "} p \text{ or } q \text{ or both."} \quad \text{Disjunction}$$

With this meaning of *or*, $p \lor q$ is called the **inclusive disjunction,** or just the **disjunction,** of p and q. In everyday language, the disjunction implies the idea of "either." For example, consider the following disjunction.

> I have a quarter or I have a dime.

It is true whenever I have either a quarter, a dime, or both. This disjunction could be false only if I had neither coin. *The disjunction $p \lor q$ is false only if both component statements are false.*

TRUTH TABLE FOR THE DISJUNCTION p or q

p or q

p	q	$p \lor q$
T	T	T
T	F	T
F	T	T
F	F	F

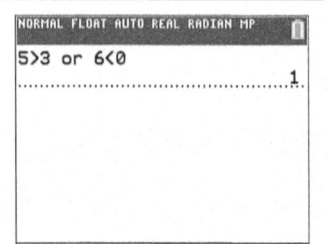

The calculator returns a 1 for

$$5 > 3 \text{ or } 6 < 0,$$

indicating that the statement is true.

EXAMPLE 2 Finding the Truth Value of a Disjunction

Let p represent "$5 > 3$" and let q represent "$6 < 0$." Find the truth value of $p \lor q$.

Solution

Here, as in **Example 1,** p is true and q is false. The second row of the disjunction truth table shows that $p \lor q$ is true.

The symbol $\geq$ is read **"is greater than or equal to,"** while $\leq$ is read **"is less than or equal to."** If a and b are real numbers, then $a \leq b$ is true if $a < b$ or $a = b$. See **Table 5**.

Negations

The **negation** of a statement p, symbolized **$\sim p$,** must have the opposite truth value from the statement p itself. This leads to the truth table for the negation.

Table 5 True Inequalities

Statement	Reason It Is True
$8 \geq 8$	$8 = 8$
$3 \geq 1$	$3 > 1$
$-5 \leq -3$	$-5 < -3$
$-4 \leq -4$	$-4 = -4$

TRUTH TABLE FOR THE NEGATION not p

not p

p	$\sim p$
T	F
F	T

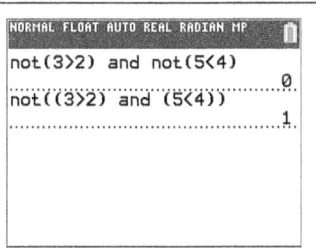

Order of operations in logic is similar to that of mathematics.

Logic: ( ), ∧, ∨

Math: ( ), ×, +

Example 3 shows the use of parentheses to prioritize steps in determining the truth value of the expression $\sim p \wedge (q \vee \sim r)$.

The calculator screen above shows that, in both mathematics and logic, parentheses matter!

EXAMPLE 3   Finding the Truth Value of a Compound Statement

Suppose p is true, q is false, and r is false. What is the truth value of the compound statement $\sim p \wedge (q \vee \sim r)$?

Solution

Here parentheses are used to group q and $\sim r$ together. Work first inside the parentheses. Because r is false, $\sim r$ will be true. Because $\sim r$ is true and q is false, the truth value of $q \vee \sim r$ is T, as shown in the third row of the *or* truth table.

Because p is true, $\sim p$ is false, and the truth value of $\sim p \wedge (q \vee \sim r)$ is found in the third row of the *and* truth table. The statement

$$\sim p \wedge (q \vee \sim r) \quad \text{is false.}$$

We can use a shortcut symbolic method that involves replacing the statements with their truth values, letting T represent a true statement and F represent a false statement.

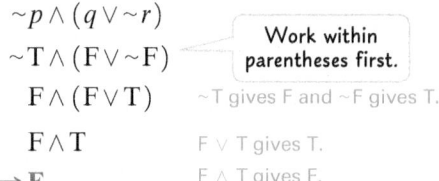

$$\sim p \wedge (q \vee \sim r)$$
$$\sim T \wedge (F \vee \sim F) \quad \text{Work within parentheses first.}$$
$$F \wedge (F \vee T) \quad \text{~T gives F and ~F gives T.}$$
$$F \wedge T \quad \text{F ∨ T gives T.}$$

The compound statement is false. → **F**   F ∧ T gives F.

Mathematical Statements

We can use truth tables to determine the truth values of compound mathematical statements.

EXAMPLE 4   Deciding Whether Compound Mathematical Statements Are True or False

Let p represent the statement $3 > 2$, q represent $5 < 4$, and r represent $3 < 8$. Decide whether each statement is *true* or *false*.

(a) $\sim p \wedge \sim q$ **(b)** $\sim (p \wedge q)$ **(c)** $(\sim p \wedge r) \vee (\sim q \wedge \sim p)$

Solution

(a) Because p is true, $\sim p$ is false. By the *and* truth table, if one part of an "and" statement is false, the entire statement is false.

$$\sim p \wedge \sim q \quad \text{is false.}$$

(b) For $\sim (p \wedge q)$, first work within the parentheses. Because p is true and q is false, $p \wedge q$ is false by the *and* truth table. Next, apply the negation. The negation of a false statement is true.

$$\sim (p \wedge q) \quad \text{is true.}$$

(c) Here p is true, q is false, and r is true. This makes $\sim p$ false and $\sim q$ true. By the *and* truth table, $\sim p \wedge r$ is false, and $\sim q \wedge \sim p$ is also false. By the *or* truth table,

$$(\sim p \wedge r) \vee (\sim q \wedge \sim p) \quad \text{is false.}$$

(Alternatively, see **Example 8(b)**.)

$$\begin{array}{ccc} \downarrow & & \downarrow \\ F & \vee & F \end{array}$$

Example 4(a) explains why

$\sim (3 > 2) \wedge \sim (5 < 4)$

is false. The calculator returns a 0. For a true statement such as

$\sim [(3 > 2) \wedge (5 < 4)]$

from **Example 4(b)**, it returns a 1.

When a quantifier is used with a conjunction or a disjunction, we must be careful in determining the truth value, as shown in the following example.

George Boole (1815–1864) grew up in poverty. His father, a London tradesman, gave him his first mathematics lessons and taught him to make optical instruments. Boole was largely self-educated. At 16 he worked in an elementary school and by age 20 had opened his own school. He studied mathematics in his spare time. He died of lung disease at age 49.

    Boole's ideas have been used in the design of computers, telephone systems, and search engines.

p	q	Compound Statement
T	T	
T	F	
F	T	
F	F	

Exploring Truth Values for Compound Statements

EXAMPLE 5   **Deciding Whether Quantified Mathematical Statements Are True or False**

Decide whether each statement is *true* or *false*.

(a) For some real number x, $x < 5$ and $x > 2$.

(b) For every real number x, $x > 0$ or $x < 1$.

(c) For all real numbers x, $x^2 > 0$.

Solution

(a) Because "some" is an *existential* quantifier, we need only find one real number x that makes both component statements true, and $x = 3$ is such a number. The statement is true by the *and* truth table.

(b) No matter which real number might be tried as a replacement for x, at least one of the two statements

$$x > 0, \quad x < 1$$

will be true. Because an "or" statement is true if one or both component statements are true, the entire statement as given is true.

(c) Because "for all" is a *universal* quantifier, we need only find one case in which the inequality is false to make the entire statement false. Can we find a real number whose square is not positive (that is, not greater than 0)? Yes, we can—0 is such a number. In fact 0 is the *only* real number whose square is not positive. This statement is false.

Truth Tables

In the preceding examples, the truth value for a given statement was found by going back to the basic truth tables. It is generally easier to first create a complete truth table for the given statement itself. Then final truth values can be read directly from this table.

    In this text we use the standard format shown in the margin for listing the possible truth values in compound statements involving two component statements.

EXAMPLE 6   **Constructing a Truth Table**

Consider the statement $(\sim p \wedge q) \vee \sim q$.

(a) Construct a truth table.

(b) Suppose both p and q are true. Find the truth value of the compound statement.

Solution

(a) As shown below, begin by listing all possible combinations of truth values for p and q. Then list the truth values of $\sim p$, which are the opposite of those of p. Use the "$\sim p$" and "q" columns, along with the *and* truth table, to find the truth values of $\sim p \wedge q$. List them in a separate column.

Next include a column for $\sim q$.

p	q	$\sim p$	$\sim p \wedge q$	$\sim q$
T	T	F	F	F
T	F	F	F	T
F	T	T	T	F
F	F	T	F	T

Finally, make a column for the entire compound statement. To find the truth values, use *or* to combine $\sim p \wedge q$ with $\sim q$ and refer to the *or* truth table.

p	q	$\sim p$	$\sim p \wedge q$	$\sim q$	$(\sim p \wedge q) \vee \sim q$
T	T	F	F	F	F
T	F	F	F	T	T
F	T	T	T	F	T
F	F	T	F	T	T

(b) Look in the first row of the final truth table above, where both p and q have truth value T. Read across the row to find that the compound statement is false.

EXAMPLE 7 **Constructing a Truth Table**

Construct the truth table for $p \wedge (\sim p \vee \sim q)$.

Solution

p	q	$\sim p$	$\sim q$	$\sim p \vee \sim q$	$p \wedge (\sim p \vee \sim q)$
T	T	F	F	F	F
T	F	F	T	T	T
F	T	T	F	T	F
F	F	T	T	T	F

If a compound statement involves three component statements p, q, and r, we will use the following standard format in setting up the truth table.

p	q	r	**Compound Statement**
T	T	T	
T	T	F	
T	F	T	
T	F	F	
F	T	T	
F	T	F	
F	F	T	
F	F	F	

EXAMPLE 8 **Constructing a Truth Table**

Consider the statement $(\sim p \wedge r) \vee (\sim q \wedge \sim p)$.

(a) Construct a truth table.

(b) Suppose p is true, q is false, and r is true. Find the truth value of this statement.

Solution

(a) There are three component statements: p, q, and r. The truth table thus requires eight rows to list all possible combinations of truth values of p, q, and r. The final truth table can be found in much the same way as the ones earlier.

p	q	r	$\sim p$	$\sim p \wedge r$	$\sim q$	$\sim q \wedge \sim p$	$(\sim p \wedge r) \vee (\sim q \wedge \sim p)$
T	T	T	F	F	F	F	F
T	T	F	F	F	F	F	F
T	F	T	F	F	T	F	F
T	F	F	F	F	T	F	F
F	T	T	T	T	F	F	T
F	T	F	T	F	F	F	F
F	F	T	T	T	T	T	T
F	F	F	T	F	T	T	T

(b) By the third row of the truth table in part (a), the compound statement is false.

Emilie, Marquise du Châtelet
(1706–1749) participated in the scientific activity of the generation after Newton and Leibniz. Educated in science, music, and literature, she was studying mathematics at the time (1733) she began a long intellectual relationship with the philosopher **François Voltaire** (1694–1778). She and Voltaire competed independently in 1738 for a prize offered by the French Academy on the subject of fire. Although du Châtelet did not win, her dissertation was published by the academy in 1744.

EXAMPLE 9   **Using Inductive Reasoning**

Suppose that n is a counting number, and a logical statement is composed of n component statements. How many rows will appear in the truth table for the compound statement?

Solution

We examine some of the earlier truth tables in this section. The truth table for the negation has one statement and two rows. The truth tables for the conjunction and the disjunction have two component statements, and each has four rows. The truth table in **Example 8(a)** has three component statements and eight rows.

Summarizing these in **Table 6** (seen in the margin) reveals a pattern encountered earlier. Inductive reasoning leads us to the conjecture that if a logical statement is composed of n component statements, it will have 2^n rows. This can be proved using more advanced concepts.

Table 6

Number of Statements	Number of Rows
1	$2 = 2^1$
2	$4 = 2^2$
3	$8 = 2^3$

The result of **Example 9** is reminiscent of the formula for the number of subsets of a set having n elements.

NUMBER OF ROWS IN A TRUTH TABLE

A logical statement having n component statements will have 2^n rows in its truth table.

Alternative Method for Constructing Truth Tables

After making a reasonable number of truth tables, some people prefer the shortcut method shown in **Example 10,** which repeats **Examples 6 and 8.**

EXAMPLE 10   **Constructing Truth Tables**

Construct the truth table for each compound statement.

(a) $(\sim p \wedge q) \vee \sim q$ **(b)** $(\sim p \wedge r) \vee (\sim q \wedge \sim p)$

Solution

(a) Start by inserting truth values for $\sim p$ and for q. Then use the *and* truth table to obtain the truth values for $\sim p \wedge q$.

p	q	$(\sim p$	$\wedge$	$q)$	$\vee$	$\sim q$
T	T	F				T
T	F	F				F
F	T	T				T
F	F	T				F

p	q	$(\sim p$	$\wedge$	$q)$	$\vee$	$\sim q$
T	T	F	F		T	
T	F	F	F		F	
F	T	T	T		T	
F	F	T	F		F	

Ada Lovelace (1815–1852) was born Augusta Ada Byron. Her talents as a mathematician and logician led to her work with **Charles Babbage** (1791–1871) on his Analytical Engine, the first programmable mechanical computer. Lovelace's notes on this machine include what is regarded by many as the first computer program and reveal her visionary belief that computing machines would have applications beyond numerical calculations.

Now disregard the two preliminary columns of truth values for $\sim p$ and for q, and insert truth values for $\sim q$. Finally, use the *or* truth table.

p	q	($\sim p \wedge q$)	$\vee$	$\sim q$
T	T	F		F
T	F	F		T
F	T	T		F
F	F	F		T

p	q	($\sim p \wedge q$)	$\vee$	$\sim q$
T	T	F	F	F
T	F	F	T	T
F	T	T	T	F
F	F	F	T	T

These steps can be summarized as follows.

p	q	($\sim p$	$\wedge$	q)	$\vee$	$\sim q$
T	T	F	F	T	F	F
T	F	F	F	F	T	T
F	T	T	T	T	T	F
F	F	T	F	F	T	T
		①	②	①	④	③

The circled numbers indicate the order in which the various columns of the truth table were found.

(b) Work as follows.

p	q	r	($\sim p$	$\wedge$	r)	$\vee$	($\sim q$	$\wedge$	$\sim p$)
T	T	T	F	F	T	F	F	F	F
T	T	F	F	F	F	F	F	F	F
T	F	T	F	F	T	F	T	F	F
T	F	F	F	F	F	F	T	F	F
F	T	T	T	T	T	T	F	F	T
F	T	F	T	F	F	F	F	F	T
F	F	T	T	T	T	T	T	T	T
F	F	F	T	F	F	T	T	T	T
			①	②	①	⑤	③	④	③

The circled numbers indicate the order.

Equivalent Statements and De Morgan's Laws

Two statements are **equivalent** if they have the same truth value in *every* possible situation. The columns of the two truth tables that were the last to be completed will be the same for equivalent statements.

EXAMPLE 11 Deciding Whether Two Statements Are Equivalent

Are the following two statements equivalent?

$$\sim p \vee \sim q \quad \text{and} \quad \sim (p \wedge q)$$

Solution

Construct a truth table for each statement.

p	q	$\sim p \vee \sim q$
T	T	F
T	F	T
F	T	T
F	F	T

p	q	$\sim (p \wedge q)$
T	T	F
T	F	T
F	T	T
F	F	T

Because the truth values are the same in all cases, as shown in the columns in color, the statements $\sim p \vee \sim q$ and $\sim (p \wedge q)$ are equivalent.

$$\sim p \vee \sim q \equiv \sim (p \wedge q) \quad \text{The symbol } \equiv \text{ denotes equivalence.}$$

In the same way, the statements $\sim p \wedge \sim q$ and $\sim (p \vee q)$ are equivalent. We call these equivalences *De Morgan's laws*.

DE MORGAN'S LAWS FOR LOGICAL STATEMENTS

For any statements p and q, the following equivalences are valid.

$$\sim(p \vee q) \equiv \sim p \wedge \sim q \qquad \sim(p \wedge q) \equiv \sim p \vee \sim q$$

Compare **De Morgan's Laws** for logical statements with the set theoretical version.

EXAMPLE 12 Applying De Morgan's Laws

Find a negation of each statement by applying De Morgan's laws.

(a) I laughed or I cried.   **(b)** She won't try and he will succeed.

(c) $\sim p \vee (q \wedge \sim p)$

Solution

(a) If p represents "I laughed" and q represents "I cried," then the compound statement is symbolized $p \vee q$. The negation of $p \vee q$ is $\sim(p \vee q)$. By one of De Morgan's laws, this is equivalent to $\sim p \wedge \sim q$, or, in words,

I didn't laugh and I didn't cry.

This negation is reasonable—the original statement says that I laughed or cried. The negation says that I did neither.

(b) From De Morgan's laws, $\sim(p \wedge q) \equiv \sim p \vee \sim q$, so the negation becomes

She will try or he won't succeed.

(c) Negate both component statements and change $\vee$ to $\wedge$.

$$\sim[\sim p \vee (q \wedge \sim p)] \equiv p \wedge \sim(q \wedge \sim p) \qquad \text{Apply De Morgan's law.}$$

$$\equiv p \wedge (\sim q \vee \sim(\sim p)) \qquad \text{Apply De Morgan's law again.}$$

$$\equiv p \wedge (\sim q \vee p)$$

A truth table will show that the statements

$$\sim p \vee (q \wedge \sim p) \quad \text{and} \quad p \wedge (\sim q \vee p) \quad \text{are negations of each other.}$$

3.2   EXERCISES

CONCEPT CHECK *Fill in the blank with the correct response.*

1. The _____ $p \vee q$ is false only if p is
(conjunction/disjunction)
_____ and q is _____.

2. The _____ $p \wedge q$ is true only if p is
(conjunction/disjunction)
_____ and q is _____.

3. The statement $6 \geq 2$ is true because _____.

4. The compound statement $8 \geq 8$ is a _____ and is true because _____.

5. If a compound statement has two component statements, its truth table will have _____ rows.

6. The negation of a disjunction may be expressed as a conjunction of negations using _____ _____.

Answer each question.

7. If q is false, what must be the truth value of the statement $(p \wedge \sim q) \wedge q$?

8. If q is true, what must be the truth value of the statement $q \vee (q \wedge \sim p)$?

9. If $p \vee (q \wedge \sim q)$ is true, what must be the truth value of p?

10. If $p \wedge \sim(q \vee r)$ is true, what must be the truth value of the component statements?

11. If $\sim(p \vee q)$ is true, what must be the truth values of the component statements?

12. If $\sim(p \wedge q)$ is false, what must be the truth values of the component statements?

Let p represent a true statement and let q represent a false statement. Find the truth value of each compound statement.

13. $\sim p$

14. $\sim q$

15. $p \vee q$

16. $p \wedge q$

17. $p \vee \sim q$

18. $\sim p \wedge q$

19. $\sim p \vee \sim q$

20. $p \wedge \sim q$

21. $\sim(p \wedge \sim q)$

22. $\sim(\sim p \vee \sim q)$

23. $\sim[\sim p \wedge (\sim q \vee p)]$

24. $\sim[(\sim p \wedge \sim q) \vee \sim q]$

Let p represent a true statement, and let q and r represent false statements. Find the truth value of each compound statement.

25. $(p \wedge r) \vee \sim q$

26. $(q \vee \sim r) \wedge p$

27. $p \wedge (q \vee r)$

28. $(\sim p \wedge q) \vee \sim r$

29. $\sim(p \wedge q) \wedge (r \vee \sim q)$

30. $(\sim r \wedge \sim q) \vee (\sim r \wedge q)$

31. $\sim[(\sim p \wedge q) \vee r]$

32. $\sim[r \vee (\sim q \wedge \sim p)]$

33. $\sim[\sim q \vee (r \wedge \sim p)]$

34. $\sim(p \vee q) \wedge \sim(p \wedge q)$

Let p represent the statement $16 < 8$, let q represent the statement $5 \not> 4$, and let r represent the statement $17 \leq 17$. Find the truth value of each compound statement.

35. $p \wedge r$

36. $p \vee \sim q$

37. $\sim q \vee \sim r$

38. $\sim p \wedge \sim r$

39. $(p \wedge q) \vee r$

40. $\sim p \vee (\sim r \vee \sim q)$

41. $(\sim r \wedge q) \vee \sim p$

42. $\sim(p \vee \sim q) \vee \sim r$

Give the number of rows in the truth table for each compound statement.

43. $p \vee \sim r$

44. $p \wedge (r \wedge \sim s)$

45. $(\sim p \wedge q) \vee (\sim r \vee \sim s) \wedge r$

46. $[(p \vee q) \wedge (r \wedge s)] \wedge (t \vee \sim p)$

47. $[(\sim p \wedge \sim q) \wedge (\sim r \wedge s \wedge \sim t)] \wedge (\sim u \vee \sim v)$

48. $[(\sim p \wedge \sim q) \vee (\sim r \vee \sim s)]$
  $\quad \vee [(\sim m \wedge \sim n) \wedge (u \wedge \sim v)]$

CONCEPT CHECK *Answer each of the following.*

49. If the truth table for a certain compound statement has 64 rows, how many distinct component statements does the statement have?

50. Is it possible for the truth table of a compound statement to have exactly 54 rows? Why or why not?

Construct a truth table for each compound statement.

51. $\sim p \wedge q$

52. $\sim p \vee \sim q$

53. $(q \vee \sim p) \vee \sim q$

54. $(p \wedge \sim q) \wedge p$

55. $(p \vee \sim q) \wedge (p \wedge q)$

56. $(\sim p \wedge \sim q) \vee (\sim p \vee q)$

57. $(\sim p \wedge q) \wedge r$

58. $(\sim r \vee \sim p) \wedge (\sim p \vee \sim q)$

59. $(\sim p \wedge \sim q) \vee (\sim r \vee \sim p)$

60. $r \vee (p \wedge \sim q)$

61. $\sim(\sim p \wedge \sim q) \vee (\sim r \vee \sim s)$

62. $(\sim r \vee s) \wedge (\sim p \wedge q)$

Use one of De Morgan's laws to write the negation of each statement.

63. You can pay me now or you can pay me later.

64. I am not going or she is going.

65. It is summer and there is no snow.

66. $\frac{1}{2}$ is a positive number and -9 is less than zero.

67. I said yes but she said no.

68. Dan tried to sell the software, but he was unable to do so.

69. $6 - 1 = 5$ and $9 + 13 \neq 7$

70. $8 < 10$ or $5 \neq 2$

Identify each statement as true *or* false.

71. For every real number x, $x < 14$ or $x > 6$.

72. For every real number x, $x > 9$ or $x < 9$.

73. There exists an integer n such that $n > 0$ and $n < 0$.

74. For some integer n, $n \geq 3$ and $n \leq 3$.

Solve each problem.

75. Solve the truth table for *exclusive disjunction*. The symbol $\underline{\vee}$ represents "one or the other is true, but not both."

p	q	$p \underline{\vee} q$
T	T	
T	F	
F	T	
F	F	

Exclusive disjunction

76. Attorneys sometimes use the phrase "and/or." This phrase corresponds to which usage of the word *or:* inclusive or exclusive disjunction?

Decide whether each compound statement is true *or* false. ($\underline{\vee}$ *is the exclusive disjunction of* **Exercise 75.**)

77. $3 + 1 = 4 \underline{\vee} 2 + 5 = 7$

78. $3 + 1 = 4 \underline{\vee} 2 + 5 = 10$

79. $3 + 1 = 6 \underline{\vee} 2 + 5 = 7$

80. $3 + 1 = 12 \underline{\vee} 2 + 5 = 10$

Solve each problem.

81. *Whose Picture Am I Looking At?* Raymond Smullyan (1919–2017) authored several books on recreational logic. *What Is the Name of This Book?* includes the following puzzle, which has been around for many years.

The puzzle tells us that a man was looking at a portrait. Someone asked him, "Whose picture are you looking at?" He replied,

> "Brothers and sisters, I have none, but this man's father is my father's son."

("This man's father" means, of course, the father of the man in the picture.) Whose picture was the man looking at?

82. *Choose a Door* In his book *The Lady or the Tiger and Other Logic Puzzles*, Raymond Smullyan proposed the following problem. It is taken from the classic Frank Stockton short story, in which a prisoner must make a choice between two doors: behind one is a beautiful lady, and behind the other is a hungry tiger.

What if each door has a sign, and the man knows that only one sign is true?

The sign on Door 1 reads:

IN THIS ROOM THERE IS A LADY AND IN THE OTHER ROOM THERE IS A TIGER.

The sign on Door 2 reads:

IN ONE OF THESE ROOMS THERE IS A LADY AND IN ONE OF THESE ROOMS THERE IS A TIGER.

With this information, the man is able to choose the correct door. Can you?

83. De Morgan's law

$$\sim(p \vee q) \equiv \sim p \wedge \sim q$$

can be stated verbally, "The negation of a disjunction is equivalent to the conjunction of the negations." Give a similar verbal statement of

$$\sim(p \wedge q) \equiv \sim p \vee \sim q.$$

84. (a) Build truth tables for

$$p \vee (q \wedge r) \quad \text{and} \quad (p \vee q) \wedge (p \vee r).$$

Decide whether it can be said that "OR distributes over AND." Explain.

(b) Build truth tables for

$$p \wedge (q \vee r) \quad \text{and} \quad (p \wedge q) \vee (p \wedge r).$$

Decide whether it can be said that "AND distributes over OR." Explain.

(c) Describe how the logical equivalences developed in parts (a) and (b) are related to the set-theoretical equations

$$X \cup (Y \cap Z) = (X \cup Y) \cap (X \cup Z)$$

and $X \cap (Y \cup Z) = (X \cap Y) \cup (X \cap Z).$

3.3 THE CONDITIONAL AND CIRCUITS

In his April 21, 1989, five-star review of **Field of Dreams,** the *Chicago Sun-Times* movie critic Roger Ebert gave an explanation of why the movie has become an American classic.

There is a speech in this movie about baseball that is so simple and true that it is heartbreaking. And the whole attitude toward the players reflects that attitude. Why do they come back from the great beyond and play in this cornfield? Not to make any kind of vast, earthshattering statement, but simply to hit a few and field a few, and remind us of a good and innocent time.

The photo above was taken in 2007 in Dyersville, Iowa, at the actual scene of the filming. The carving "Ray Loves Annie" in the bleacher seats can be seen in a quick shot during the movie. It has weathered over time.

What famous **conditional statement** inspired Ray to build a baseball field in his cornfield?

Conditionals

"If truly Truthful Troll I be,
then go thou east, and be thou free."

This of course is the statement uttered by the troll on the second page of this chapter. A more modern paraphrase would be, "*If* I am the troll who always tells the truth, *then* the door to the east is the one that leads to freedom."

The troll's utterance is an example of a conditional statement. A **conditional** statement is a compound statement that uses the connective *if . . . then.*

If I read for too long, *then* I get tired.

If looks could kill, *then* I would be dead.   Conditional statements

If he doesn't get back soon, *then* you should go look for him.

In each of these conditional statements, the component coming after the word *if* gives a condition (but not necessarily the only condition) under which the statement coming after *then* will be true. For example, "If it is over 90°, then I'll go to the mountains" tells one possible condition under which I will go to the mountains—if the temperature is over 90°.

The conditional is written with an arrow and symbolized as follows.

$$p \rightarrow q \qquad \text{If } p, \text{ then } q.$$

We read $p \rightarrow q$ as "**p implies q**" or "**If p, then q.**" In the conditional $p \rightarrow q$, the statement p is the **antecedent,** while q is the **consequent.**

The conditional connective may not always be explicitly stated. That is, it may be "hidden" in an everyday expression. For example, consider the following statement.

Quitters never win.

It can be written in *if . . . then* form as

If you're a quitter, *then* you will never win.

As another example, consider this statement.

It is difficult to study when you are distracted.

It can be written

If you are distracted, *then* it is difficult to study.

In the quotation "If you aim at nothing, you will hit it every time," the word "then," though not stated explicitly, is understood from the context of the statement. "You aim at nothing" is the antecedent, and "you will hit it every time" is the consequent.

The conditional truth table is more difficult to define than the tables in the previous section. To see how to define the conditional truth table, imagine you have bought a used car (with financing from the car dealer), and the used-car salesperson says,

If you fail to make your payment on time, *then* your car will be taken.

Let p represent "You fail to make your payment on time," and let q represent "Your car will be taken." There are four combinations of truth values for the two component statements.

You Lie! (or Do You?) Refer to the four possibilities in the used car scenario. Granted, the T for Case 3 is less obvious than the F for Case 2. However, the laws of symbolic logic permit only one of two truth values. Because no lie can be established in Case 3, we give the salesperson the benefit of the doubt. Likewise, *any* conditional statement is declared to be true whenever its antecedent is false.

As we consider these four possibilities, it is helpful to ask,

"Did the salesperson lie?"

If so, then the conditional statement is considered false. Otherwise, the conditional statement is considered true.

Possibility	Failed to Pay on Time?	Car Taken?	
1	Yes	Yes	p is T, q is T.
2	Yes	No	p is T, q is F.
3	No	Yes	p is F, q is T.
4	No	No	p is F, q is F.

The four possibilities are as follows.

1. In the first case, assume you failed to make your payment on time, and your car *was* taken (p is T, q is T). The salesperson told the truth, so place T in the first row of the truth table. (We do not claim that your car was taken *because* you failed to pay on time. It may be gone for a completely different reason.)

2. In the second case, assume that you failed to make your payment on time, and your car was *not* taken (p is T, q is F). The salesperson lied (gave a false statement), so place an F in the second row of the truth table.

3. In the third case, assume that you paid on time, but your car was taken anyway (p is F, q is T). The salesperson did *not* lie. She only said what would happen if you were late on payments, not what would happen if you paid on time. (The crime here is stealing, not lying.) Since we cannot say that the salesperson lied, place a T in the third row of the truth table.

4. Finally, assume that you made timely payment, and your car was not taken (p is F, q is F). This certainly does not contradict the salesperson's statement, so place a T in the last row of the truth table.

The completed truth table for the conditional is defined as follows.

TRUTH TABLE FOR THE CONDITIONAL   **If p, then q**

If p, then q

p	q	$p \rightarrow q$
T	T	T
T	F	F
F	T	T
F	F	T

The use of the conditional connective in no way implies a cause-and-effect relationship. Any two statements may have an arrow placed between them to create a compound statement. Consider this example.

If I pass mathematics, then the sun will rise the next day.

It is true, because the consequent is true. (See the special characteristics following **Example 1** on the next page.) There is, however, no cause-and-effect connection between my passing mathematics and the rising of the sun. The sun will rise no matter what grade I get.

```
NORMAL FLOAT AUTO REAL RADIAN MP

PROGRAM:PARITY
:Prompt N
:If N/2=int(N/2)
:Then
:Disp "EVEN"
:Else
:Disp "ODD"
:End
:
```

```
NORMAL FLOAT AUTO REAL RADIAN MP

prgmPARITY
N=?7
ODD
                        Done
```

```
NORMAL FLOAT AUTO REAL RADIAN MP

prgmPARITY
N=?6
EVEN
                        Done
```

Conditional statements are useful in writing programs. The short program in the first screen determines whether an integer is even or odd. Notice the lines that begin with *If* and *Then*.

EXAMPLE 1 Finding the Truth Value of a Conditional

Given that p, q, and r are all false, find the truth value of the following statement.

$$(p \rightarrow \sim q) \rightarrow (\sim r \rightarrow q)$$

Solution

Using the shortcut method explained in **Example 3** of the previous section, we can replace p, q, and r with F (since each is false) and proceed as before, using the negation and conditional truth tables as necessary.

$$
\begin{array}{cccl}
(p \rightarrow \sim q) & \rightarrow & (\sim r \rightarrow q) & \\
(F \rightarrow \sim F) & \rightarrow & (\sim F \rightarrow F) & \\
(F \rightarrow T) & \rightarrow & (T \rightarrow F) & \text{Use the negation truth table.} \\
T & \rightarrow & F & \text{Use the conditional truth table.} \\
& \mathbf{F} & &
\end{array}
$$

The statement $(p \rightarrow \sim q) \rightarrow (\sim r \rightarrow q)$ is false when p, q, and r are all false.

SPECIAL CHARACTERISTICS OF CONDITIONAL STATEMENTS

1. $p \rightarrow q$ is false only when the antecedent is *true* and the consequent is *false*.

2. If the antecedent is *false*, then $p \rightarrow q$ is automatically *true*.

3. If the consequent is *true*, then $p \rightarrow q$ is automatically *true*.

EXAMPLE 2 Determining Whether Conditionals Are True or False

Write *true* or *false* for each statement. Here T represents a true statement, and F represents a false statement.

(a) $T \rightarrow (7 = 3)$ **(b)** $(8 < 2) \rightarrow F$ **(c)** $(4 \neq 3 + 1) \rightarrow T$

Solution

(a) Because the antecedent is true, while the consequent, $7 = 3$, is false, the given statement is false by the first point mentioned above.

(b) The antecedent is false, so the given statement is true by the second observation.

(c) The consequent is true, making the statement true by the third characteristic of conditional statements.

EXAMPLE 3 Constructing Truth Tables

Construct a truth table for each statement.

(a) $(\sim p \rightarrow \sim q) \rightarrow (\sim p \wedge q)$ **(b)** $(p \rightarrow q) \rightarrow (\sim p \vee q)$

Solution

(a) Insert the truth values of $\sim p$ and $\sim q$. Find the truth values of $\sim p \rightarrow \sim q$.

p	q	$\sim p$	$\sim q$	$\sim p \rightarrow \sim q$
T	T	F	F	T
T	F	F	T	T
F	T	T	F	F
F	F	T	T	T

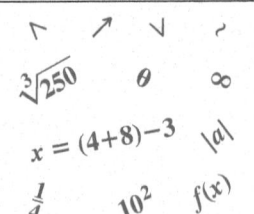

The importance of **symbols** was emphasized by the American philosopher-logician **Charles Sanders Peirce** (1839–1914), who asserted the nature of humans as symbol-using or sign-using organisms. Symbolic notation is half of mathematics, Bertrand Russell once said.

Next use $\sim p$ and q to find the truth values of $\sim p \wedge q$.

p	q	$\sim p$	$\sim q$	$\sim p \rightarrow \sim q$	$\sim p \wedge q$
T	T	F	F	T	F
T	F	F	T	T	F
F	T	T	F	F	T
F	F	T	T	T	F

Finally, find the truth values of $(\sim p \rightarrow \sim q) \rightarrow (\sim p \wedge q)$.

p	q	$\sim p$	$\sim q$	$\sim p \rightarrow \sim q$	$\sim p \wedge q$	$(\sim p \rightarrow \sim q) \rightarrow (\sim p \wedge q)$
T	T	F	F	T	F	F
T	F	F	T	T	F	F
F	T	T	F	F	T	T
F	F	T	T	T	F	F

(b) For $(p \rightarrow q) \rightarrow (\sim p \vee q)$, follow steps similar to the ones above.

p	q	$p \rightarrow q$	$\sim p$	$\sim p \vee q$	$(p \rightarrow q) \rightarrow (\sim p \vee q)$
T	T	T	F	T	T
T	F	F	F	F	T
F	T	T	T	T	T
F	F	T	T	T	T

As the truth table in **Example 3(b)** shows, the statement

$$(p \rightarrow q) \rightarrow (\sim p \vee q)$$

is always true, no matter what the truth values of the components. Such a statement is called a **tautology.** Several other examples of tautologies are

$$p \vee \sim p, \quad p \rightarrow p, \quad \text{and} \quad (\sim p \vee \sim q) \rightarrow \sim (p \wedge q). \quad \text{Tautologies}$$

The truth tables in **Example 3** also could have been found by the alternative method shown in the previous section.

Writing a Conditional as a Disjunction

p	q	$p \rightarrow q$
T	T	T
T	F	F
F	T	T
F	F	T

Recall that the truth table for the conditional (repeated in the margin) shows that $p \rightarrow q$ is false only when p is true and q is false. But we also know that the disjunction is false for only one combination of component truth values, and it is easy to see that $\sim p \vee q$ will be false only when p is true and q is false, as the following truth table indicates.

p	q	$\sim p$	$\vee$	q
T	T	F	T	T
T	F	F	F	F
F	T	T	T	T
F	F	T	T	F

Thus we see that the disjunction $\sim p \vee q$ is equivalent to the conditional $p \rightarrow q$.

WRITING A CONDITIONAL AS A DISJUNCTION

$p \rightarrow q$ is equivalent to $\sim p \vee q.$

Negation Temptation It is tempting to think that the negation of a conditional is another conditional. To avoid this misconception, remember that a conditional can be stated as a disjunction, having only one "false" in its truth table's final column. Its negation, then, has only one "true" value, so it may be expressed as a conjunction, but not as a disjunction (and thus not as a conditional).

p	q	$p \to q$	$\sim(p \to q)$
T	T	T	F
T	F	F	T
F	T	T	F
F	F	T	F

p	q	$\sim p \lor q$	$p \land \sim q$
T	T	T	F
T	F	F	T
F	T	T	F
F	F	T	F

We now know that

$$p \to q \equiv \sim p \lor q,$$

so the negation of the conditional is

$$\sim(p \to q) \equiv \sim(\sim p \lor q).$$

Applying De Morgan's law to the right side of the above equivalence gives the negation as a conjunction. The truth table is shown in the margin note.

NEGATION OF $p \to q$

The negation of $p \to q$ is $p \land \sim q$.

EXAMPLE 4 Determining Negations

Determine the negation of each statement.

(a) If I'm hungry, I will eat.   **(b)** All ponies love apples.

Solution

(a) If p represents "I'm hungry" and q represents "I will eat," then the given statement can be symbolized by $p \to q$. The negation of $p \to q$, as shown earlier, is $p \land \sim q$, so the negation of the statement is

I'm hungry and I will not eat.

(b) First, we must restate the given statement in *if . . . then* form.

If it is a pony, then it loves apples.

Based on our earlier discussion, the negation is

It is a pony and it does not love apples.

As seen in **Example 4,** the negation of a conditional statement is written as a conjunction.

EXAMPLE 5 Determining Statements Equivalent to Conditionals

Write each conditional as an equivalent statement without using *if ... then*.

(a) If the Indians win the pennant, then Johnny will go to the World Series.

(b) If it's Borden's, it's got to be good.

Solution

(a) Because the conditional $p \to q$ is equivalent to $\sim p \lor q$, let p represent "The Indians win the pennant" and q represent "Johnny will go to the World Series." Restate the conditional as

The Indians do not win the pennant or Johnny will go to the World Series.

(b) If p represents "it's Borden's" and if q represents "it's got to be good," the conditional may be restated as

It's not Borden's or it's got to be good.

Figure 1

Series circuit

Figure 2

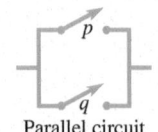

Parallel circuit

Figure 3

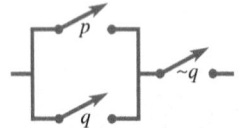

Figure 4

Switch Postion	Current	Truth Value
Closed	flows	T
Open	stops	F

Circuits

One of the first nonmathematical applications of symbolic logic was seen in the master's thesis of Claude Shannon in 1937. Shannon showed how logic could be used to design electrical circuits. His work was soon applied by computer designers to make simplifications at the development stage, significantly reducing assembly costs.

To see how Shannon's ideas work, consider the electrical switch shown in **Figure 1.** We assume that current will flow through this switch when it is closed but not when it is open.

Figure 2 shows two switches connected in *series.* In such a circuit, current will flow only when both switches are closed. Note how closely a series circuit corresponds to the conjunction $p \wedge q$. We know that $p \wedge q$ is true only when both p and q are true.

A circuit corresponding to the disjunction $p \vee q$ can be found by drawing a *parallel* circuit, as in **Figure 3.** Here, current flows if either p or q is closed or if both p and q are closed.

The circuit in **Figure 4** corresponds to the statement $(p \vee q) \wedge {\sim}q$, which is a compound statement involving both a conjunction and a disjunction.

Simplifying an electrical circuit depends on the idea of equivalent statements. Recall that two statements are equivalent if they have the same truth table final column. The symbol $\equiv$ is used to indicate that the two statements are equivalent. Some equivalent statements are shown in the following box.

EQUIVALENT STATEMENTS USED TO SIMPLIFY CIRCUITS

$$p \vee (q \wedge r) \equiv (p \vee q) \wedge (p \vee r) \qquad p \vee p \equiv p$$
$$p \wedge (q \vee r) \equiv (p \wedge q) \vee (p \wedge r) \qquad p \wedge p \equiv p$$
$$p \rightarrow q \equiv {\sim}q \rightarrow {\sim}p \qquad {\sim}(p \wedge q) \equiv {\sim}p \vee {\sim}q$$
$$p \rightarrow q \equiv {\sim}p \vee q \qquad {\sim}(p \vee q) \equiv {\sim}p \wedge {\sim}q$$

If T represents any true statement and F represents any false statement, then

$$p \vee \text{T} \equiv \text{T} \qquad p \vee {\sim}p \equiv \text{T}$$
$$p \wedge \text{F} \equiv \text{F} \qquad p \wedge {\sim}p \equiv \text{F}.$$

Circuits can be used as models of compound statements, with a closed switch corresponding to T (current flowing) and an open switch corresponding to F (current not flowing).

EXAMPLE 6 Simplifying a Circuit

Simplify the circuit of **Figure 5.**

Solution

At the top of **Figure 5,** p and q are connected in series, and at the bottom, p and r are connected in series. These are interpreted as the compound statements $p \wedge q$ and $p \wedge r$, respectively. These two conjunctions are connected in parallel, as indicated by the figure treated as a whole.

Write the disjunction of the two conjunctions.

$$(p \wedge q) \vee (p \wedge r)$$

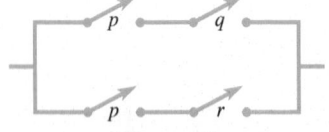

Figure 5

Figure 6

(Think of the two switches labeled "p" as being controlled by the same lever.) By one of the pairs of equivalent statements in the preceding box,

$$(p \wedge q) \vee (p \wedge r) \equiv p \wedge (q \vee r),$$

which has the circuit of **Figure 6**. This circuit is logically equivalent to the one in **Figure 5,** and yet it contains only three switches instead of four—which might well lead to a large savings in manufacturing costs.

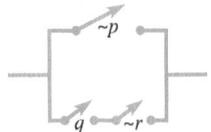

Figure 7

EXAMPLE 7   Drawing a Circuit for a Conditional Statement

Draw a circuit for $p \rightarrow (q \wedge \sim r)$.

Solution

From the list of equivalent statements in the box, $p \rightarrow q$ is equivalent to $\sim p \vee q$. This equivalence gives $p \rightarrow (q \wedge \sim r) \equiv \sim p \vee (q \wedge \sim r)$, which has the circuit diagram in **Figure 7**.

WHEN WILL I **EVER** USE THIS?

Suppose you are a home-monitoring and control system designer. A home-owner wants the capability of controlling his home air-conditioning system remotely. You will need to install a control module for the AC unit that receives signals from your customer's smartphone.

1. Draw a circuit that will allow the AC unit to turn on when both a master switch and the control module are activated, and when either (or both) of two thermostats is (or are) triggered.

2. Write a logical statement for the circuit.

   Since the master switch (m) and the control module (c) both need to be activated, they must be connected in series, followed by two thermostats T_1 and T_2 in parallel (because only one needs to be triggered). The circuit is shown at right, and the corresponding logical statement is

   $$m \wedge c \wedge (T_1 \vee T_2).$$

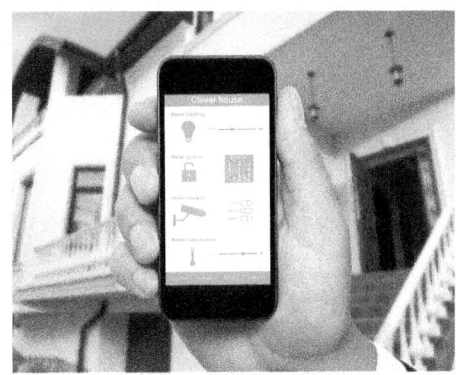

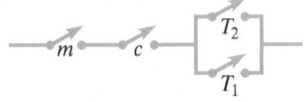

3.3 EXERCISES

CONCEPT CHECK *Write* true *or* false *for each statement.*

1. There are multiple ways in which a conditional may be false.

2. A conditional may be expressed as a disjunction.

3. Conditional statements *must* be expressed using the *if ... then* connective.

4. "If p then q" implies that the occurrence of p causes q to be true.

CONCEPT CHECK *Fill in the blank with the correct response.*

5. The negation of a conditional _____ a conditional.
                                    (is / is not)

6. The negation of a conditional is a(n) _____.
                                    (conjunction / disjunction)

7. A(n) _____ circuit corresponds to a conjunction.

8. A(n) _____ circuit corresponds to a disjunction.

Rewrite each statement using the if . . . then *connective. Rearrange the wording or add words as necessary.*

9. You can do it if you just believe.

10. It must be bad for you if it's sweet.

11. Every even integer divisible by 5 is divisible by 10.

12. No perfect square integers have units digit 2, 3, 7, or 8.

13. No grizzly bears live in California.

14. No guinea pigs get lonely.

15. Surfers can't stay away from the beach.

16. Running Bear loves Little White Dove.

Decide whether each statement is true *or* false.

17. If the antecedent of a conditional statement is false, the conditional statement is true.

18. If the consequent of a conditional statement is true, the conditional statement is true.

19. If p is false, then $(p \land (q \to r)) \to q$ is true.

20. If q is true, then $\sim p \to (q \lor r)$ is true.

21. The negation of "If pigs fly, I'll believe it" is "If pigs don't fly, I won't believe it."

22. The statements "If it flies, then it's a bird" and "It does not fly or it's a bird" are logically equivalent.

Tell whether each conditional is true (T) *or* false (F).

23. $T \to (7 < 3)$ **24.** $F \to (4 \neq 8)$

25. $F \to (5 \neq 5)$ **26.** $(8 \geq 8) \to F$

27. $(5^2 \neq 25) \to (8 - 8 = 16)$

28. $(5 = 12 - 7) \to (9 = 0)$

Write a short answer for each problem.

29. Explain why the statement "If $3 > 5$, then $4 < 6$" is true.

30. Explain how to determine the truth value of a conditional statement.

Let s represent "she sings for a living," let p represent "he fixes cars," and let m represent "they collect classics." Express each compound statement in words.

31. $\sim m \to p$ **32.** $p \to \sim m$

33. $s \to (m \land p)$ **34.** $(s \land p) \to m$

35. $\sim p \to (\sim m \lor s)$ **36.** $(\sim s \lor \sim m) \to \sim p$

Let c represent "I take my car," let s represent "it is sunny," and let g represent "the gym is open." Write each compound statement in symbols.

37. If I take my car, then the gym is open.

38. If I do not take my car, then it is not sunny.

39. The gym is open, and if it is sunny then I do not take my car.

40. I take my car, or if the gym is open then it is sunny.

41. It is sunny if the gym is open.

42. I'll take my car if it is not sunny.

Find the truth value of each statement. Assume that p and r are false, and q is true.

43. $\sim r \to q$ **44.** $q \to p$

45. $\sim p \to (q \land r)$ **46.** $(\sim r \lor p) \to p$

47. $\sim q \to (p \land r)$ **48.** $(\sim p \land \sim q) \to (p \land \sim r)$

49. $(p \rightarrow \sim q) \rightarrow (\sim p \wedge \sim r)$

50. $[(p \rightarrow \sim q) \wedge (p \rightarrow r)] \rightarrow r$

Solve each problem.

51. Explain why we know that

$$[r \vee (p \vee s)] \rightarrow [(p \vee q) \vee \sim p]$$

is true, even if we are not given the truth values of p, q, r, and s.

52. Construct a true statement involving a conditional, a conjunction, a disjunction, and a negation (not necessarily in that order) that consists of component statements p, q, and r, with all of these component statements false.

Construct a truth table for each statement. Identify any tautologies.

53. $\sim q \rightarrow p$

54. $(\sim q \rightarrow \sim p) \rightarrow \sim q$

55. $(\sim p \rightarrow q) \rightarrow p$

56. $(p \wedge q) \rightarrow (p \vee q)$

57. $(p \vee q) \rightarrow (q \vee p)$

58. $(\sim p \rightarrow \sim q) \rightarrow (p \wedge q)$

59. $[(r \vee p) \wedge \sim q] \rightarrow p$

60. $[(r \wedge p) \wedge (p \wedge q)] \rightarrow p$

61. $(\sim r \rightarrow s) \vee (p \rightarrow \sim q)$

62. $(\sim p \wedge \sim q) \rightarrow (s \rightarrow r)$

CONCEPT CHECK *Answer each of the following.*

63. What is the minimum number of Fs that must appear in the final column of a truth table for us to be assured that the statement is not a tautology?

64. If all truth values in the final column of a truth table are F, how can we easily change the statement so that it becomes a tautology?

Write the negation of each statement. Remember that the negation of $p \rightarrow q$ is $p \wedge \sim q$.

65. If that is an authentic diamond, I'll be surprised.

66. If Muley Jones hits that note, he will shatter glass.

67. "If you talk in your sleep don't mention my name." *Elvis Presley*

68. "If I die young, bury me in satin." *The Band Perry*

69. If the bullfighter doesn't get going, he's going to get gored.

70. If you don't say "I do," then you'll regret it for the rest of your life.

Write each statement as an equivalent statement that does not use the if . . . then *connective. Remember that*

$$p \rightarrow q \quad \text{is equivalent to} \quad \sim p \vee q.$$

71. If you call, I will answer.

72. If you need a friend, I will be there.

73. If they turn the ball over one more time, they'll lose.

74. If you can just hang on, you'll make it.

75. All champions have had their challenges.

76. Every dog has its day.

Use truth tables to decide which of the pairs of statements are equivalent.

77. $p \rightarrow q$; $\sim p \vee q$ **78.** $\sim(p \rightarrow q)$; $p \wedge \sim q$

79. $p \rightarrow q$; $\sim q \rightarrow \sim p$ **80.** $p \rightarrow q$; $q \rightarrow p$

81. $q \rightarrow \sim p$; $p \rightarrow \sim q$

82. $\sim(p \vee q) \rightarrow r$; $(p \vee q) \vee r$

Write a logical statement representing each of the following circuits. Simplify each circuit when possible.

83.

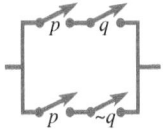

84.

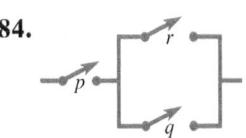

85.

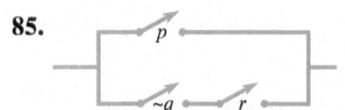

86.

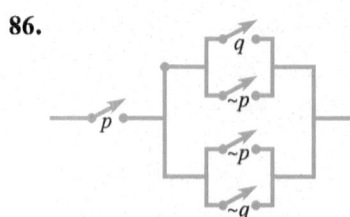

87.

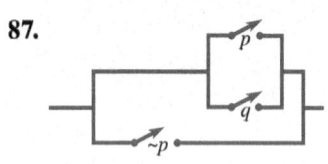

88.

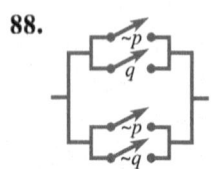

Draw circuits representing the following statements as they are given. Simplify if possible.

89. $p \wedge (q \vee {\sim}p)$ **90.** $({\sim}p \wedge {\sim}q) \wedge {\sim}r$

91. $(p \vee q) \wedge ({\sim}p \wedge {\sim}q)$

92. $({\sim}q \wedge {\sim}p) \vee ({\sim}p \vee q)$

93. $[(p \vee q) \wedge r] \wedge {\sim}p$

94. $[({\sim}p \wedge {\sim}r) \vee {\sim}q] \wedge ({\sim}p \wedge r)$

95. ${\sim}q \rightarrow ({\sim}p \rightarrow q)$ **96.** ${\sim}p \rightarrow ({\sim}p \vee {\sim}q)$

Write a short answer for each problem.

97. Refer to **Figures 5 and 6** in **Example 6**. Suppose the cost of the use of one switch for an hour is $0.06. When one uses the circuit in **Figure 6** rather than the circuit in **Figure 5,** what is the savings for a year of 365 days, assuming that the circuit is in continuous use?

98. Explain why the circuit shown will always have exactly one open switch. What does this circuit simplify to?

3.4   THE CONDITIONAL AND RELATED STATEMENTS

Converse, Inverse, and Contrapositive

Many mathematical properties and theorems are stated in *if . . . then* form. Any conditional statement $p \rightarrow q$ is made up of an antecedent p and a consequent q. If they are interchanged, negated, or both, a new conditional statement is formed. Suppose that we begin with a conditional statement.

If you stay, then I go.   Conditional statement

By interchanging the antecedent ("you stay") and the consequent ("I go"), we obtain a new conditional statement.

If I go, then you stay.   Converse

This new conditional is called the **converse** of the given conditional statement.

By negating both the antecedent and the consequent, we obtain the **inverse** of the given conditional statement.

If you do not stay, then I do not go.   Inverse

If the antecedent and the consequent are both interchanged *and* negated, the **contrapositive** of the given conditional statement is formed.

If I do not go, then you do not stay.   Contrapositive

Alfred North Whitehead (1861–1947) and Bertrand Russell worked together on *Principia Mathematica*. During that time, Whitehead was teaching mathematics at Cambridge University and had written *Universal Algebra*. In 1910 he went to the University of London, exploring not only the philosophical basis of science but also the "aims of education" (as he called one of his books). It was as a philosopher that he was invited to Harvard University in 1924. Whitehead died at the age of 86 in Cambridge, Massachusetts.

These three related statements for the conditional $p \rightarrow q$ are summarized below.

RELATED CONDITIONAL STATEMENTS

Conditional Statement	$p \rightarrow q$	(If p, then q.)
Converse	$q \rightarrow p$	(If q, then p.)
Inverse	$\sim p \rightarrow \sim q$	(If not p, then not q.)
Contrapositive	$\sim q \rightarrow \sim p$	(If not q, then not p.)

Notice that the inverse is the contrapositive of the converse.

EXAMPLE 1 **Determining Related Conditional Statements**

Determine each of the following, given the conditional statement

If I am running, then I am moving.

(a) the converse   **(b)** the inverse   **(c)** the contrapositive

Solution

(a) Let p represent "I am running" and q represent "I am moving." Then the given statement may be written $p \rightarrow q$. The converse, $q \rightarrow p$, is

If I am moving, then I am running.

The converse is not necessarily true, even though the given statement is true.

(b) The inverse of $p \rightarrow q$ is $\sim p \rightarrow \sim q$. Thus the inverse is

If I am not running, then I am not moving.

Again, this is not necessarily true.

(c) The contrapositive, $\sim q \rightarrow \sim p$, is

If I am not moving, then I am not running.

The contrapositive, like the given conditional statement, is true.

Contrapositive Thinking   Some students find it a challenge to see that the conditional

$$p \rightarrow q$$

is equivalent to the statement

q is necessary for p.

It is often helpful to consider the contrapositive form

$$\sim q \rightarrow \sim p,$$

which says, "If q is not true, then p is not true." This statement makes clear that q is necessary for p.

Example 1 shows that the converse and inverse of a true statement need not be true. They *can* be true, but they need not be. The relationships between the related conditionals are shown in the truth table that follows.

Equivalent

Equivalent

		Conditional	**Converse**	**Inverse**	Contrapositive
p	q	$p \rightarrow q$	$q \rightarrow p$	$\sim p \rightarrow \sim q$	$\sim q \rightarrow \sim p$
T	T	T	T	T	T
T	F	F	T	T	F
F	T	T	F	F	T
F	F	T	T	T	T

Bertrand Russell (1872–1970) was a student of Whitehead's before they wrote the *Principia*. Like his teacher, Russell turned toward philosophy. His works include a critique of Leibniz, analyses of mind and of matter, and a history of Western thought.

Russell became a public figure because of his involvement in social issues. Deeply aware of human loneliness, he was "passionately desirous of finding ways of diminishing this tragic isolation." During World War I he was an antiwar crusader, and he was imprisoned briefly. Again in the 1960s he championed peace. He wrote many books on social issues, winning the Nobel Prize for Literature in 1950.

As this truth table shows,

1. *A conditional statement and its contrapositive always have the same truth value,* making it possible to replace any statement with its contrapositive without affecting the logical meaning.

2. *The converse and inverse always have the same truth value.*

> ### EQUIVALENCES
>
> A conditional statement and its contrapositive are equivalent. Also, the converse and the inverse are equivalent.

EXAMPLE 2   Determining Related Conditional Statements

For the conditional statement $\sim p \to q$, write each of the following.

(a) the converse   **(b)** the inverse   **(c)** the contrapositive

Solution

(a) The converse of $\sim p \to q$ is $q \to \sim p$.

(b) The inverse is $\sim(\sim p) \to \sim q$, which simplifies to $p \to \sim q$.

(c) The contrapositive is $\sim q \to \sim(\sim p)$, which simplifies to $\sim q \to p$.

Alternative Forms of "If p, then q"

The conditional statement "If p, then q" can be stated in several other ways in English. Consider this statement.

> If you take Tylenol, then you will find relief from your symptoms.

It can also be written as follows.

> Taking Tylenol is sufficient for relieving your symptoms.

According to this statement, taking Tylenol is enough to relieve your symptoms. Taking other medications or using other treatment techniques *might* also result in symptom relief, but at least we *know* that taking Tylenol will. Thus $p \to q$ can be written "p is sufficient for q." Knowing that p has occurred is sufficient to guarantee that q will also occur.

On the other hand, consider this statement, which has a different structure.

> Fresh ingredients are necessary for making a good pizza.   (*)

This statement claims that fresh ingredients are one condition for making a good pizza. But there may be other conditions (such as a working oven, for example). The statement labeled (*) could be written as

> If you want good pizza, then you need fresh ingredients.

As this example suggests, $p \to q$ is the same as "q is necessary for p." In other words, if q doesn't happen, then neither will p. Notice how this idea is closely related to the idea of equivalence between a conditional statement and its contrapositive.

Kurt Gödel (1906–1978) is widely regarded as the most influential mathematical logician of the twentieth century. He proved by his Incompleteness Theorem that the search for a set of axioms from which all mathematical truths could be proved was futile. In particular, "the vast structure of the *Principia Mathematica* of Whitehead and Russell was inadequate for deciding all mathematical questions."

After the death of his friend **Albert Einstein** (1879–1955), Gödel developed paranoia, and his life ended tragically when, convinced he was being poisoned, he refused to eat, essentially starving himself to death.

COMMON TRANSLATIONS OF $p \rightarrow q$

The conditional $p \rightarrow q$ can be translated in any of the following ways, none of which depends on the truth or falsity of $p \rightarrow q$.

If p, then q.	p is sufficient for q.
If p, q.	q is necessary for p.
p implies q.	All p are q.
p only if q.	q if p.

Example: If you live in Alamogordo, then you live in New Mexico.   Statement

You live in New Mexico if you live in Alamogordo.
You live in Alamogordo only if you live in New Mexico.
Living in New Mexico is necessary for living in Alamogordo.
Living in Alamogordo is sufficient for living in New Mexico.
All residents of Alamogordo are residents of New Mexico.
Being a resident of Alamogordo implies residency in New Mexico.

Common translations

EXAMPLE 3   Rewording Conditional Statements

Rewrite each statement in the form "If p, then q."

(a) You'll starve if you don't eat.

(b) Go to the doctor only if your temperature exceeds 101°F.

(c) Everyone at the game had a great time.

Solution

(a) If you don't eat, then you'll starve.

(b) If you go to the doctor, then your temperature exceeds 101°F.

(c) If you were at the game, then you had a great time.

EXAMPLE 4   Translating from Words to Symbols

Let p represent "A triangle is equilateral," and let q represent "A triangle has three sides of equal length." Write each of the following in symbols.

(a) A triangle is equilateral if it has three sides of equal length.

(b) A triangle is equilateral only if it has three sides of equal length.

Solution

(a) $q \rightarrow p$ **(b)** $p \rightarrow q$

Animation
→
Exploring Truth Values for Compound Statements

Biconditionals

The compound statement *p if and only if q* (often abbreviated *p iff q*) is called a **biconditional.** It is symbolized $p \leftrightarrow q$ and is interpreted as the conjunction of the two conditionals $p \rightarrow q$ and $q \rightarrow p$.

Principia Mathematica, the title chosen by Whitehead and Russell, was a deliberate reference to *Philosophiae naturalis principia mathematica,* or "mathematical principles of the philosophy of nature," Isaac Newton's epochal work of 1687. Newton's *Principia* pictured a kind of "clockwork universe" that ran via his Law of Gravitation. Newton independently invented the calculus, unaware that Leibniz had published his own formulation of it earlier.

Using symbols, the conjunction of the conditionals $p \rightarrow q$ and $q \rightarrow p$ is written $(q \rightarrow p) \wedge (p \rightarrow q)$ so that, by definition,

$$p \leftrightarrow q \equiv (q \rightarrow p) \wedge (p \rightarrow q). \quad \text{Biconditional}$$

The truth table for the biconditional $p \leftrightarrow q$ can be determined using this definition.

TRUTH TABLE FOR THE BICONDITIONAL p **if and only if** q

p **if and only if** *q*

p	q	$p \leftrightarrow q$
T	T	T
T	F	F
F	T	F
F	F	T

A biconditional is true when both component statements have the same truth value. It is false when they have different truth values.

EXAMPLE 5   **Determining Whether Biconditionals Are True or False**

Determine whether each biconditional statement is *true* or *false.*

(a) $6 + 8 = 14$ if and only if $11 + 5 = 16$

(b) $3 = 7$ if and only if $8 < 5$

(c) Mars is a moon if and only if Jupiter is a planet.

Solution

(a) Both $6 + 8 = 14$ and $11 + 5 = 16$ are true. By the truth table for the biconditional, this biconditional is true.

(b) Both component statements are false, so by the last line of the truth table for the biconditional, this biconditional statement is true.

(c) Because the first component is false, and the second is true, this biconditional statement is false.

Summary of Truth Tables

Truth tables have been derived for several important types of compound statements.

SUMMARY OF BASIC TRUTH TABLES

1. $\sim p,$ the **negation** of p, has truth value opposite that of p.

2. $p \wedge q,$ the **conjunction,** is true only when both p and q are true.

3. $p \vee q,$ the **disjunction,** is false only when both p and q are false.

4. $p \rightarrow q,$ the **conditional,** is false only when p is true and q is false.

5. $p \leftrightarrow q,$ the **biconditional,** is true only when both p and q have the same truth value.

3.4 EXERCISES

CONCEPT CHECK *Fill in the blank with the correct response.*

1. Given the conditional $p \rightarrow q$,

  (a) the statement $\sim q \rightarrow \sim p$ is the _____.

  (b) the statement $q \rightarrow p$ is the _____.

  (c) the statement $\sim p \rightarrow \sim q$ is the _____.

2. Provide the missing column titles in the truth table.

p	q	___	___	___	___
T	T	T	T	T	T
T	F	F	T	F	F
F	T	F	T	T	F
F	F	T	F	T	F

3. A conditional and its _____ always have the same truth values.

4. The inverse and the _____ of a conditional are logically equivalent.

For each given conditional statement (or statement that can be written as a conditional), write **(a)** *the converse,* **(b)** *the inverse, and* **(c)** *the contrapositive in* if . . . then *form. In some of the exercises, it may be helpful to first restate the given statement in* if . . . then *form.*

5. If beauty were a minute, then you would be an hour.

6. If you lead, then I will follow.

7. If it ain't broke, don't fix it.

8. If I had a nickel for each time that happened, I would be rich.

9. If you build it, he will come.

10. Where there's smoke, there's fire.

11. Walking in front of a moving car is dangerous to your health.

12. Vegetables contain micronutrients.

13. Birds of a feather flock together.

14. A rolling stone gathers no moss.

Use symbols to write the **(a)** *converse,* **(b)** *inverse, and* **(c)** *contrapositive for each conditional statement.*

15. $p \rightarrow \sim q$ **16.** $\sim p \rightarrow q$

17. $\sim p \rightarrow \sim q$ **18.** $\sim q \rightarrow \sim p$

19. $p \rightarrow (q \vee r)$ (*Hint:* Use one of De Morgan's laws as necessary.)

20. $(r \wedge \sim q) \rightarrow p$ (*Hint:* Use one of De Morgan's laws as necessary.)

CONCEPT CHECK *Answer each of the following.*

21. Discuss the equivalences that exist among a conditional statement, its converse, its inverse, and its contrapositive.

22. State the contrapositive of "If the square of a natural number is odd, then the natural number is odd." The two statements must have the same truth value. Use several examples and inductive reasoning to decide whether both are true or both are false.

Write each statement in if . . . then *form.*

23. Legs of 3 and 4 imply a hypotenuse of 5.

24. "This is a leap year" implies that next year is not.

25. All whole numbers are rational numbers.

26. No irrational numbers are rational.

27. Two coats of paint are necessary to cover the graffiti.

28. Surgery is necessary to correct the overbite.

29. Doing logic puzzles is sufficient for driving me crazy.

30. Being in Kalispell is sufficient for being in Montana.

31. No whole numbers are not integers.

32. No integers are irrational numbers.

33. Employment will improve only if the economy recovers.

34. The economy will recover only if employment improves.

35. The Cubs will win the pennant when their pitching improves.

36. The grass will be greener when we're on the other side.

37. A rectangle is a parallelogram with perpendicular adjacent sides.

38. A square is a rectangle with two adjacent sides equal.

39. The square of a three-digit number whose units digit is 5 will end in 25.

40. An integer whose units digit is 0 or 5 is divisible by 5.

41. A triangle with two perpendicular sides is a right triangle.

42. A parallelogram is a four-sided figure with opposite sides parallel.

CONCEPT CHECK *Answer each the following.*

43. One of the following statements is not equivalent to all the others. Which one is it?

   A. *r* only if *s*.   **B.** *r* implies *s*.

   C. If *r*, then *s*.   **D.** *r* is necessary for *s*.

44. Many students have difficulty interpreting *necessary* and *sufficient*. Use the statement "Being in Vancouver is sufficient for being in North America" to explain why "*p* is sufficient for *q*" translates as "if *p*, then *q*."

45. Use the statement "To be an integer, it is necessary that a number be rational" to explain why "*p* is necessary for *q*" translates as "if *q*, then *p*."

46. Explain why the statement "A week has eight days if and only if October has forty days" is true.

Identify each statement as true *or* false.

47. $6 = 9 - 3$ if and only if $8 + 2 = 10$.

48. $3 + 1 \neq 7$ if and only if $8 \neq 8$.

49. $8 + 7 \neq 15$ if and only if $3 \times 5 \neq 8$.

50. $6 \times 2 = 18$ if and only if $9 + 7 \neq 16$.

51. Theodore Roosevelt was president if and only if Franklin D. Roosevelt was not president.

52. McDonald's sells Whoppers if and only if Apple manufactures iPhones.

Two statements that can both be true about the same object are **consistent.** *For example, "It is green" and "It weighs 60 pounds" are consistent statements. Statements that cannot both be true about the same object are called* **contrary.** *"It is a Nissan" and "It is a Mazda" are contrary. Label each pair of statements as either* contrary *or* consistent.

53. Michael Jackson is alive. Michael Jackson is dead.

54. That car is used. That same car costs more than $10,000.

55. This number is a whole number. This same number is irrational.

56. This number is negative. This same number is a natural number.

57. This number is an integer. This same number is a rational number.

58. This number is positive. This same number is irrational.

Trolling for Freedom *Refer to the Chapter Opener. Humphrey the hero deduced from the riddle that there are twin trolls who take shifts guarding the magical doors. One of them tells only truths, the other only lies. The troll standing guard says,*

> *"If truly Truthful Troll I be, then go thou east and be thou free."*

Humphrey needs to decide which troll he is addressing, call him by name, and tell him which door he would like opened.

   There are two things Humphrey must get right: the name of the troll and the proper door. If he misidentifies the troll, no door will be opened. If he correctly names the troll and picks the wrong door, he will be confined to the dungeon behind it.

59. Because the troll either always lies or always tells the truth, Humphrey knows that

   (1) if the troll is Truthful Troll, then the conditional statement he uttered is true, and

   (2) if the conditional statement he uttered is true, then he is Truthful Troll.

   Let *p* represent "the troll is Truthful Troll" and let *q* represent "the door to the east leads to freedom." Express the statements in (1) and (2) in symbolic form.

60. The conjunction of the answers from **Exercise 59** is a biconditional that must be true.

   (a) Build a truth table for this biconditional.

   (b) Use the fact that it *must* be true to solve Humphrey's riddle.

3.5 ANALYZING ARGUMENTS WITH EULER DIAGRAMS

OBJECTIVES

1 Define logical arguments.

2 Use Euler diagrams to analyze arguments with universal quantifiers.

3 Use Euler diagrams to analyze arguments with existential quantifiers.

Logical Arguments

With inductive reasoning we observe patterns to solve problems. Now we study how deductive reasoning may be used to determine whether logical arguments are valid or invalid.

   A logical argument is made up of **premises** (assumptions, laws, rules, widely held ideas, or observations) and a **conclusion.** Recall that *deductive* reasoning involves drawing specific conclusions from given general premises. When reasoning from the premises of an argument to obtain a conclusion, we want the argument to be *valid*.

Leonhard Euler (1707–1783) won the Academy prize and edged out du Châtelet and Voltaire. That was a minor achievement, as was the invention of "Euler circles" (which antedated Venn diagrams). Euler was the most prolific mathematician of his generation despite eventual blindness that forced him to dictate from memory.

VALID AND INVALID ARGUMENTS

An argument is **valid** if the fact that all the premises are true forces the conclusion to be true. An argument that is not valid is **invalid.** It is called a **fallacy.**

"Valid" and "true" do not have the same meaning—an argument can be valid even though the conclusion is false (see Example 4), or invalid even though the conclusion is true (see Example 6).

Arguments with Universal Quantifiers

Several techniques can be used to check whether an argument is valid. One such technique is based on **Euler diagrams.**

Leonhard Euler (pronounced "Oiler") was one of the greatest mathematicians who ever lived. He is immortalized in mathematics history with the important irrational number e, named in his honor. This number appears throughout mathematics.

EXAMPLE 1   **Using an Euler Diagram to Determine Validity**

Is the following argument valid?

> No accidents happen on purpose.
> Spilling the beans was an accident.
> The beans were not spilled on purpose.

Solution

To begin, draw regions to represent the first premise. Because no accidents happen on purpose, the region for "accidents" goes outside the region for "things that happen on purpose," as shown in **Figure 8.**

The second premise, "Spilling the beans was an accident," suggests that "spilling the beans" belongs in the region representing "accidents." Let x represent "spilling the beans." **Figure 9** shows that "spilling the beans" is not in the region for "things that happen on purpose." If both premises are true, the conclusion that the beans were not spilled on purpose is also true. The argument is valid.

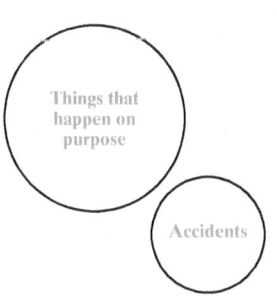

Figure 8

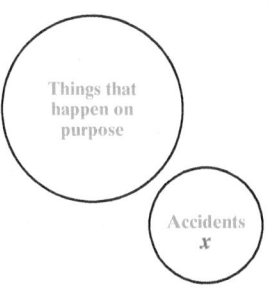

Figure 9

EXAMPLE 2   **Using an Euler Diagram to Determine Validity**

Is the following argument valid?

> All rainy days are cloudy.
> Today is not cloudy.
> Today is not rainy.

Solution

The first premise is reflected in **Figure 10.** Because "Today is *not* cloudy," place an x for "today" *outside* the region for "cloudy days." See **Figure 11.** Placing the x outside the region for "cloudy days" forces it also to be located outside the region for "rainy days." Thus, if the two premises are true, then it is also true that today is not rainy. The argument is valid.

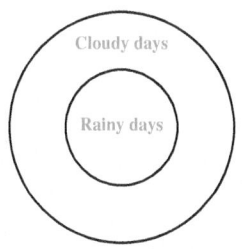

Figure 10

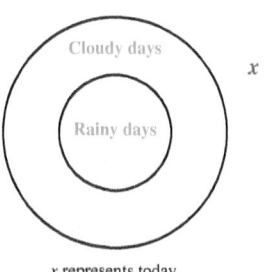

x represents today.

Figure 11

EXAMPLE 3   Using an Euler Diagram to Determine Validity

Is the following argument valid?

> All horses have hooves.
> That animal has hooves.
> That animal is a horse.

Solution

The region for "horses" goes entirely inside the region for "things that have hooves." See **Figure 12**. The x that represents "that animal" must go inside the region for "things that have hooves," but it can go either inside or outside the region for "horses." Even if the premises are true, we are not forced to accept the conclusion as true. This argument is invalid. It is a fallacy.

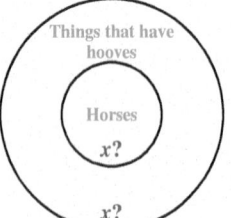

Figure 12

EXAMPLE 4   Using an Euler Diagram to Determine Validity

Is the following argument valid?

> All expensive things are desirable.
> All desirable things make you feel good.
> All things that make you feel good make you live longer.
> All expensive things make you live longer.

Solution

A diagram for the argument is given in **Figure 13**. If each premise is true, then the conclusion must be true because the region for "expensive things" lies completely within the region for "things that make you live longer." Thus, the argument is valid. (This argument shows that a *valid* argument need *not* have a true conclusion.)

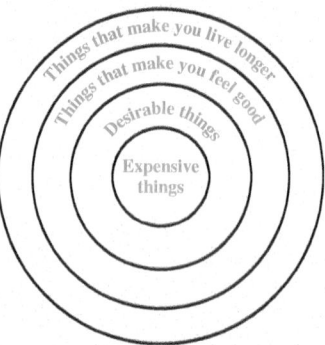

Figure 13

Arguments with Existential Quantifiers

EXAMPLE 5   Using an Euler Diagram to Determine Validity

Is the following argument valid?

> Many students drive Hondas.
> I am a student.
> I drive a Honda.

Solution

The first premise is sketched in **Figure 14,** where many (but not necessarily *all*) students drive Hondas. There are two regions in which *I* may be located, as shown in **Figure 15**. One possibility is that *I* drive a Honda. The other is that *I* don't. Since the truth of the premises does not force the conclusion to be true, the argument is invalid.

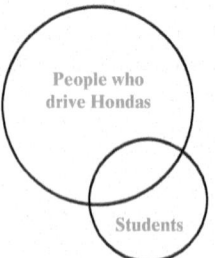

Figure 14

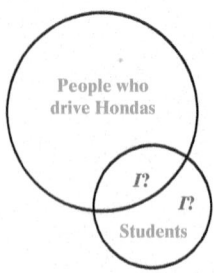

Figure 15

EXAMPLE 6   **Using an Euler Diagram to Determine Validity**

Is the following argument valid?

All fish swim.
All whales swim.
A whale is not a fish.

Solution

The premises lead to two possibilities. **Figure 16** shows the set of fish and the set of whales as intersecting, while **Figure 17** does not. Both diagrams are valid interpretations of the given premises, but only one supports the conclusion.

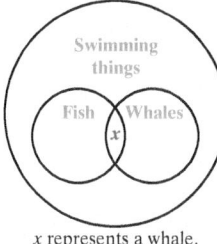

x represents a whale.	*x* represents a whale.
Figure 16	**Figure 17**

Because the truth of the premises does not force the conclusion to be true, the argument is invalid. Even though we know the conclusion to be true, this knowledge is not deduced from the premises.

3.5   EXERCISES

CONCEPT CHECK *Write* true *or* false *for each statement.*

1. In order for an argument to be valid, all of its premises must be true.

2. A valid argument may have a false conclusion.

3. A fallacy must have a false conclusion.

4. Validating an argument involves *deductive* reasoning.

Use an Euler diagram to decide whether each argument is valid *or* invalid.

5. All amusement parks have thrill rides.
   Universal Orlando is an amusement park.
   Universal Orlando has thrill rides.

6. All disc jockeys play music.
   Phlash is a disc jockey.
   Phlash plays music.

7. All celebrities have problems.
   That man has problems.
   That man is a celebrity.

8. All blogs contain writing.
   This book contains writing.
   This book is a blog.

9. All residents of Colorado know how to breathe thin air.
   Julie lives in Colorado.
   Julie knows how to breathe thin air.

10. All drivers must have a photo I.D.
   Kay has a photo I.D.
   Kay is a driver.

11. Some dinosaurs were plant eaters.
   Dino was a plant eater.
   Dino was a dinosaur.

12. Some philosophers are absent minded.
   Nicole is a philosopher.
   Nicole is absent minded.

13. Many nurses belong to unions.
   Heather is a nurse.
   Heather belongs to a union.

14. Some trucks have sound systems.
Some trucks have gun racks.

Some trucks with sound systems have gun racks.

CONCEPT CHECK *Answer each of the following.*

15. Refer to **Example 3.** If the second premise and the conclusion were interchanged, would the argument then be valid?

16. Refer to **Example 4.** Give a different conclusion from the one given there so that the argument is still valid.

Construct a valid argument based on the Euler diagram shown.

17.

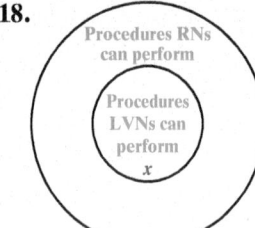

x represents Erin.

18.

Procedures RNs can perform

Procedures LVNs can perform

x

x represents vaccinations.

As mentioned in the text, an argument can have a true conclusion yet be invalid. In these exercises, each argument has a true conclusion. Identify each argument as valid *or* invalid.

19. All chickens have beaks.
All hens are chickens.

All hens have beaks.

20. All chickens have beaks.
All birds have beaks.

All chickens are birds.

21. Amarillo is northeast of Roswell.
Amarillo is northeast of El Paso.
Roswell is northeast of El Paso.

22. Beaverton is north of Salem.
Salem is north of Lebanon.

Beaverton is north of Lebanon.

23. No whole numbers are negative.
−3 is negative.

−3 is not a whole number.

24. A scalene triangle has a longest side.
A scalene triangle has a largest angle.

The largest angle in a scalene triangle is opposite the longest side.

The premises marked A, B, and C are followed by several possible conclusions. Take each conclusion in turn, and check whether the resulting argument is valid *or* invalid.

A. *All people who drive contribute to air pollution.*

B. *All people who contribute to air pollution make life a little worse.*

C. *Some people who live in a suburb make life a little worse.*

25. Some people who live in a suburb contribute to air pollution.

26. Some people who live in a suburb drive.

27. Suburban residents never drive.

28. Some people who contribute to air pollution live in a suburb.

29. Some people who make life a little worse live in a suburb.

30. All people who drive make life a little worse.

FOR FURTHER THOUGHT

Common Fallacies

Discussions in everyday conversation, politics, and advertising provide a nearly endless stream of examples of **fallacies**—arguments exhibiting illogical reasoning. There are many general forms of fallacies, and we now present descriptions and examples of some of the more common ones. (Much of this list is adapted from the document Reader Mission Critical, located on the San Jose State University Web site, www.sjsu.edu)

1. **Circular Reasoning** (also called *Begging the Question*) The person making the argument assumes to be true what he is trying to prove.

Ford owner to Chevy owner: Why do you think that Chevy trucks are the best?

Chevy owner: Because they are.

Here the Chevy owner makes a case by stating what needs to be proved.

2. **False Dilemma** (also called the *Either-Or Fallacy,* or the *Black and White Fallacy*) Presenting two options with the assumption that they are contradictions (that is, the truth of one implies the falsity of the other), when in fact they are not, is the basis of a common fallacy.

Politician: America: Love it or leave it.

This argument implies only two choices. It is possible that someone may love America and yet leave, while someone else may not love America and yet stay.

3. **Loaded Question and Complex Claims** This fallacy involves one person asking a question or making a statement that is constructed in such a way as to obtain an answer in which the responder agrees to something with which he does not actually agree.

Teenager Beth to her father: Did you enjoy embarrassing me in front of my friends?

If Beth gets the expected response "No, I didn't enjoy it," the answer allows Beth to interpret that while her father didn't enjoy it, he did indeed embarrass her.

4. **Post Hoc Reasoning** An argument that is based on the false belief that if event A preceded event B, then A must have caused B is called *post hoc reasoning*.

Johnny: I wore my Hawaiian shirt while watching all three playoff games, and my team won all three games. So I am going to wear that shirt every time I watch them.

The fact that Johnny put the same shirt on before each game has nothing to do with the outcomes of the games.

5. **Red Herring** (also called *Smoke Screen,* or *Wild Goose Chase*) This fallacy involves introducing an irrelevant topic to divert attention away from the original topic, allowing the person making the argument to seemingly prevail.

(The following script is from a political advertisement during the 2008 presidential campaign, intended to show that John McCain lacked understanding of the economy.)

> Maybe you're struggling just to pay the mortgage on your home. But recently, John McCain said, "The fundamentals of our economy are strong." Hmm. Then again, that same day, when asked how many houses he owns, McCain lost track. He couldn't remember. Well, it's seven. Seven houses. And here's one house America can't afford to let John McCain move into (showing a picture of the White House).

The advertisement shifted the focus to the number of houses McCain owned, which had nothing to do with the economy, or citizens' ability to make their mortgage payments.

6. **Shifting the Burden of Proof** In this fallacy, if a claim is difficult to support, the person making it turns the burden of proof of that claim over to someone else.

Employee: You accuse me of embezzling money? That's ridiculous.

Employer: Well, until you can prove otherwise, you will just have to accept it as true.

If money has been disappearing, it is up to the employer to prove that this employee is guilty. The burden of proof is on the employer, but he is insinuating that the employee must prove that he is not the one taking the money.

7. **Straw Man** This fallacy involves creating a false image (like a scarecrow, or straw man) of someone else's position in an argument.

During the 2016 Democratic primary, Senator Bernie Sanders responded as follows to Secretary Clinton's accusation that he had been too critical of President Barack Obama.

> . . . Last I heard we lived in a democratic society. Last I heard, a United States senator had the right to disagree with the president, including a president who has done such an extraordinary job.

The senator made Secretary Clinton's position out to be that one must agree with the president on every issue. This extreme view is much easier to attack than her actual claim that he had been overly critical of the current president.

For Group or Individual Investigation

Use the Internet to investigate the following additional logical fallacies.

Appeal to Authority	Appeal to Common	Common Practice
Two Wrongs	Belief	Wishful
Appeal to Fear	Indirect Consequences	Thinking
		Appeal to Pity
Appeal to Prejudice	Appeal to Loyalty	Appeal to Vanity
Guilt by Association	Appeal to Spite	Hasty
	Slippery Slope	Generalization

Using Truth Tables to Determine Validity

In the previous section, we used Euler diagrams to test the validity of arguments. While Euler diagrams often work well for simple arguments, difficulties can develop with more complex ones, because Euler diagrams must show every possible case. In complex arguments, it is hard to be sure that all cases have been considered.

In deciding whether to use Euler diagrams to test the validity of an argument, look for quantifiers such as "all," "some," or "no." These words often indicate arguments best tested by Euler diagrams. If these words are absent, it may be better to use truth tables to test the validity of an argument.

EXAMPLE 1   **Using a Truth Table to Determine Validity**

Determine whether the argument is *valid* or *invalid*.

> If the tank is empty, then I must fill it.
> The tank is empty.
> _____
> I must fill it.

Solution

To test the validity of this argument, we begin by assigning the letters p and q to represent these statements.

p represents "The tank is empty."

q represents "I must fill it."

Now we write the two premises and the conclusion in symbols.

Premise 1: $p \rightarrow q$
Premise 2: p
_____
Conclusion: q

To decide whether this argument is valid, we must determine whether the conjunction of both premises implies the conclusion for all possible combinations of truth values for p and q. Therefore, write the conjunction of the premises as the antecedent of a conditional statement, and write the conclusion as the consequent.

$$[(p \rightarrow q) \quad \wedge \quad p] \quad \rightarrow \quad q$$

premise   and   premise   implies   conclusion

Finally, construct the truth table for this conditional statement, as shown below.

p	q	$p \rightarrow q$	$(p \rightarrow q) \wedge p$	$[(p \rightarrow q) \wedge p] \rightarrow q$
T	T	T	T	T
T	F	F	F	T
F	T	T	F	T
F	F	T	F	T

Because the final column, shown in color, indicates that the conditional statement that represents the argument is true for all possible truth values of p and q, the statement is a tautology. Thus, the argument is valid.

Answer to the Light Bulb question on the previous page:

Label the switches 1, 2, and 3. Turn switch 1 on and leave it on for several minutes. Then turn switch 1 off, turn switch 2 on, and then immediately enter the room. If the bulb is on, then you know that switch 2 controls it. If the bulb is off, touch it to see if it is still warm. If it is, then switch 1 controls it. If the bulb is not warm, then switch 3 controls it.

The pattern of the argument in **Example 1**

$$p \to q$$
$$\frac{p}{q}$$

is called **modus ponens,** or the *law of detachment.*

To test the validity of an argument using a truth table, follow the steps in the box.

TESTING THE VALIDITY OF AN ARGUMENT WITH A TRUTH TABLE

Step 1 Assign a letter to represent each component statement in the argument.

Step 2 Express each premise and the conclusion symbolically.

Step 3 Form the symbolic statement of the entire argument by writing the *conjunction* of *all* the premises as the antecedent of a conditional statement, and the conclusion of the argument as the consequent.

Step 4 Complete the truth table for the conditional statement formed in Step 3. If it is a tautology, then the argument is valid; otherwise, it is invalid.

EXAMPLE 2 **Using a Truth Table to Determine Validity**

Determine whether the argument is *valid* or *invalid*.

> If our guests arrive before lunch, we'll eat out.
> We ate out.
> ———————————————
> Our guests arrived before lunch.

Solution

Let p represent "Our guests arrive (arrived) before lunch." Let q represent "We'll eat (We ate) out." The argument can be written as follows.

$$p \to q$$
$$\frac{q}{p}$$

To test for validity, construct a truth table for the statement $[(p \to q) \land q] \to p$.

p	q	$p \to q$	$(p \to q) \land q$	$[(p \to q) \land q] \to p$
T	T	T	T	T
T	F	F	F	T
F	T	T	T	F
F	F	T	F	T

Concluding p from q wrongly assumes $q \to p$, the *converse* of the given premise $p \to q$.

The final column of the truth table contains an F. The argument is invalid.

If a conditional and its converse were logically equivalent, then an argument of the type found in **Example 2** would be valid. Because a conditional and its converse are *not* equivalent, the argument is an example of what is sometimes called the **fallacy of the converse.**

With reasoning similar to that used to name the fallacy of the converse, the fallacy

$$p \rightarrow q$$
$$\underline{\sim p}$$
$$\sim q$$

Concluding $\sim q$ from $\sim p$ wrongly assumes $\sim p \rightarrow \sim q$, the *inverse* of the given premise $p \rightarrow q$.

is called the **fallacy of the inverse.** An example of such a fallacy is "If it rains, I get wet. It doesn't rain. Therefore, I don't get wet."

EXAMPLE 3   Using a Truth Table to Determine Validity

Determine whether the argument is *valid* or *invalid*.

> If I can avoid sweets, I can avoid the dentist.
> I can't avoid the dentist.
> _____
> I can't avoid sweets.

Solution

If p represents "I can avoid sweets" and q represents "I can avoid the dentist," the argument is written as follows.

$$p \rightarrow q$$
$$\underline{\sim q}$$
$$\sim p$$

The symbolic statement of the entire argument is as follows.

$$[(p \rightarrow q) \wedge \sim q] \rightarrow \sim p$$

The truth table for this argument indicates a tautology, and the argument is valid.

p	q	$p \rightarrow q$	$\sim q$	$(p \rightarrow q) \wedge \sim q$	$\sim p$	$[(p \rightarrow q) \wedge \sim q] \rightarrow \sim p$
T	T	T	F	F	F	T
T	F	F	T	F	F	T
F	T	T	F	F	T	T
F	F	T	T	T	T	T

The pattern of reasoning of this example is called **modus tollens,** or the *law of contraposition*, or *indirect reasoning*.

EXAMPLE 4   Using a Truth Table to Determine Validity

Determine whether the argument is *valid* or *invalid*.

> I'll buy a car or I'll take a vacation.
> I won't buy a car.
> _____
> I'll take a vacation.

Solution

If p represents "I'll buy a car" and q represents "I'll take a vacation," the argument is symbolized as follows.

$$p \vee q$$
$$\underline{\sim p}$$
$$q$$

We must set up a truth table for the statement $[(p \vee q) \wedge \sim p] \rightarrow q$.

In a scene near the beginning of the 1974 film *Monty Python and the Holy Grail,* an amazing application of **poor logic** leads to the apparent demise of a supposed witch. Some peasants have forced a young woman to wear a nose made of wood. The convoluted argument they make is this: Witches and wood are both burned, and because witches are made of wood, and wood floats, and ducks also float, if she weighs the same as a duck, then she is made of wood and, therefore, is a witch!

p	q	$p \lor q$	$\sim p$	$(p \lor q) \land \sim p$	$[(p \lor q) \land \sim p] \to q$
T	T	T	F	F	T
T	F	T	F	F	T
F	T	T	T	T	T
F	F	F	T	F	T

The statement is a tautology and the argument is valid. Any argument of this form is valid by the law of **disjunctive syllogism.**

Advertising by Transitivity   A radio ad for a sandwich shop featured this script:

"When you buy a burger for lunch instead of a sandwich with fresh ingredients, you feel lethargic. When you feel lethargic, you drink coffee. When you drink coffee, you get a stomach ache. When you get a stomach ache, you miss a meeting. When you miss a meeting, you lose your job. When you lose your job, you sell your action-figure collection to pay rent. Don't sell your action-figure collection."

The premises of the argument may be expressed symbolically as follows.

$$b \to l$$
$$l \to c$$
$$c \to s$$
$$s \to m$$
$$m \to j$$
$$j \to a$$
$$\sim a$$

What valid conclusion would the advertiser like the listener to reach? (The answer is on the next page.)

EXAMPLE 5   Using a Truth Table to Determine Validity

Determine whether the argument is *valid* or *invalid*.

If it squeaks, then I need WD-40.
If I need WD-40, then I must go to the hardware store.
If it squeaks, then I must go to the hardware store.

Solution

Let p represent "It squeaks," let q represent "I need WD-40," and let r represent "I must go to the hardware store." The argument takes on the following form.

$$p \to q$$
$$q \to r$$
$$\overline{p \to r}$$

Make a truth table for this statement, which requires eight rows.

$$[(p \to q) \land (q \to r)] \to (p \to r)$$

p	q	r	$p \to q$	$q \to r$	$p \to r$	$(p \to q) \land (q \to r)$	$[(p \to q) \land (q \to r)] \to (p \to r)$
T	T	T	T	T	T	T	T
T	T	F	T	F	F	F	T
T	F	T	F	T	T	F	T
T	F	F	F	T	F	F	T
F	T	T	T	T	T	T	T
F	T	F	T	F	T	F	T
F	F	T	T	T	T	T	T
F	F	F	T	T	T	T	T

This argument is valid because the final statement is a tautology. This pattern of argument is called **reasoning by transitivity,** or the *law of hypothetical syllogism.*

Valid and Invalid Argument Forms

A summary of the valid forms of argument presented so far follows.

VALID ARGUMENT FORMS

Modus Ponens	Modus Tollens	Disjunctive Syllogism	Reasoning by Transitivity
$p \to q$	$p \to q$	$p \lor q$	$p \to q$
p	$\sim q$	$\sim p$	$q \to r$
q	$\sim p$	q	$p \to r$

Answer to the Advertising question on the previous page:

The advertiser would like the listener to conclude

$$\sim b,$$

"Don't buy a burger for lunch."

The following is a summary of invalid forms (or fallacies).

INVALID ARGUMENT FORMS (FALLACIES)

Fallacy of the Converse	Fallacy of the Inverse
$p \rightarrow q$	$p \rightarrow q$
q	$\sim p$
p	$\sim q$

Setting the Table Correctly If an argument has the form

$$p_1$$
$$p_2$$
$$\vdots$$
$$p_n,$$
$$c$$

then **Step 3** in the testing process calls for the statement

$$(p_1 \wedge p_2 \wedge \ldots \wedge p_n) \rightarrow c.$$

EXAMPLE 6 Using a Truth Table to Determine Validity

Determine whether the argument is *valid* or *invalid*.

> If Eddie goes to town, then Mabel stays at home.
> If Mabel does not stay at home, then Rita will cook.
> Rita will not cook.
> Therefore, Eddie does not go to town.

Solution

In an argument written in this manner, the premises are given first, and the conclusion is the statement that follows the word "Therefore." Let p represent "Eddie goes to town," let q represent "Mabel stays at home," and let r represent "Rita will cook." Then the argument is symbolized as follows.

$$p \rightarrow q$$
$$\sim q \rightarrow r$$
$$\sim r$$
$$\overline{\sim p}$$

It is necessary to determine the truth values of the conjunction of *all* premises contained in the argument.

> *If at least one premise in a conjunction of several premises is false, then the entire conjunction is false.*

To test validity, set up a truth table for this statement.

$$[(p \rightarrow q) \wedge (\sim q \rightarrow r) \wedge \sim r] \rightarrow \sim p$$

p	q	r	$p \rightarrow q$	$\sim q$	$\sim q \rightarrow r$	$\sim r$	$(p \rightarrow q) \wedge (\sim q \rightarrow r) \wedge \sim r$	$\sim p$	$[(p \rightarrow q) \wedge (\sim q \rightarrow r) \wedge \sim r] \rightarrow \sim p$
T	T	T	T	F	T	F	F	F	T
T	T	F	T	F	T	T	T	F	F
T	F	T	F	T	T	F	F	F	T
T	F	F	F	T	F	T	F	F	T
F	T	T	T	F	T	F	F	T	T
F	T	F	T	F	T	T	T	T	T
F	F	T	T	T	T	F	F	T	T
F	F	F	T	T	F	T	F	T	T

Because the final column does not contain all Ts, the statement is not a tautology. The argument is invalid.

Alice's Adventures in Wonderland is the most famous work of **Charles Dodgson** (1832–1898), better known as **Lewis Carroll,** who was a mathematician and logician. He popularized recreational mathematics with this story and its sequel, *Through the Looking-Glass.* More than a century later, Raymond Smullyan continued this genre in his book *Alice in Puzzle-land* and many others.

Arguments of Lewis Carroll

Consider the following verse, which has been around for many years.

For want of a nail, the shoe was lost.
For want of a shoe, the horse was lost.
For want of a horse, the rider was lost.
For want of a rider, the battle was lost.
For want of a battle, the war was lost.
Therefore, for want of a nail, the war was lost.

Each line of the verse may be written as an *if . . . then* statement. For example, the first line may be restated as "If a nail is lost, then the shoe is lost." The conclusion, "For want of a nail, the war was lost," follows from the premises, because repeated use of the law of transitivity applies. Arguments such as the one used by Lewis Carroll in the next example often take a similar form.

EXAMPLE 7 **Supplying a Conclusion to Ensure Validity**

Supply a conclusion that yields a valid argument for the following premises.

> Babies are illogical.
> Nobody is despised who can manage a crocodile.
> Illogical persons are despised.

Solution
First, write each premise in the form *if . . . then . . . .*

> If you are a baby, then you are illogical.
> If you can manage a crocodile, then you are not despised.
> If you are illogical, then you are despised.

Let p represent "you are a baby," let q represent "you are logical," let r represent "you can manage a crocodile," and let s represent "you are despised." The statements can be written symbolically.

$$p \rightarrow \sim q$$
$$r \rightarrow \sim s$$
$$\sim q \rightarrow s$$

Begin with any letter that appears only once. Here p appears only once. Using the contrapositive of $r \rightarrow \sim s$, which is $s \rightarrow \sim r$, rearrange the statements as follows.

$$p \rightarrow \sim q$$
$$\sim q \rightarrow s$$
$$s \rightarrow \sim r$$

From the three statements, repeated use of reasoning by transitivity gives the conclusion

$$p \rightarrow \sim r, \text{ which leads to a valid argument.}$$

In words, the conclusion is "If you are a baby, then you cannot manage a crocodile," or, as Lewis Carroll would have written it, "Babies cannot manage crocodiles."

132 CHAPTER 3 Introduction to Logic

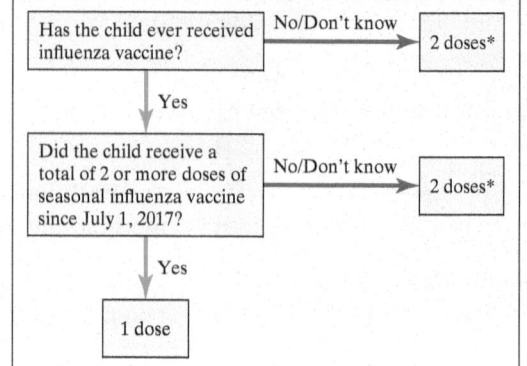

Influenza Dosing Algorithm

* Doses should be administered at least 4 weeks apart.
Source: cdc.gov

Suppose you are a pediatric nurse administering flu vaccination for a 6-year-old patient. The flowchart is a dosing algorithm provided by the CDC for children 6 months to 8 years of age.

The patient was vaccinated for the first time last year, so you place the patient on a 2-dose regimen. Use a truth table to determine the validity of your action.

We let   *p*   be "The child has received flu vaccine in the past,"

             q   be "The child has received 2 or more doses of flu
                     vaccine since July 1, 2017,"

and   *r*   be "The child needs 2 doses this season."

Then the algorithm and your treatment decision can be expressed symbolically by the following argument.

$$(p \wedge q) \leftrightarrow \sim r$$
$$\frac{\sim q}{r}$$

Make a truth table for the argument. Since there are three component statements, we require eight rows.

p	q	r	[((p	∧	q)	↔	~r)	∧	~q]	→	r
T	T	T	T	T	T	F	F	F	F	T	T
T	T	F	T	T	T	T	T	F	F	T	F
T	F	T	T	F	F	T	F	T	T	T	T
T	F	F	T	F	F	F	T	F	T	T	F
F	T	T	F	F	T	T	F	F	F	T	T
F	T	F	F	F	T	F	T	F	F	T	F
F	F	T	F	F	F	T	F	T	T	T	T
F	F	F	F	F	F	F	T	F	T	T	F
			①	②	①	④	③	⑥	⑤	⑧	⑦

The truth table shows that your argument is a tautology, so your action was valid according to the CDC.

3.6 EXERCISES

CONCEPT CHECK *Fill in the blank with the correct response.*

1. An argument consists of one or more _____ and a(n) _____.

2. When phrased as a conditional statement, an argument is valid if and only if it is a(n) _____.

3. In order for a statement to be a tautology, the final column in its truth table must contain no _____ values.

4. An argument of the form $[(p \to q) \wedge q] \to p$ is called the fallacy of the _____.

5. An argument is a conditional statement having as its antecedent the _____ of its premises.

6. An argument of the form $[(p \to q) \wedge (q \to r)] \to (p \to r)$ is called _____ _____ _____.

Each argument either is valid by one of the forms of valid arguments discussed in this section, or is a fallacy by one of the forms of invalid arguments discussed. (See the summary boxes.) Decide whether the argument is valid *or a* fallacy, *and give the form that applies.*

7. If Rascal Flatts comes to town, then I will go to the concert.
   If I go to the concert, then I'll take a vacation day.

   If Rascal Flatts comes to town, then I'll take a vacation day.

8. If you use binoculars, then you get a glimpse of the bald eagle.
   If you get a glimpse of the bald eagle, then you'll be amazed.

   If you use binoculars, then you'll be amazed.

9. If Marina works hard enough, she will get a promotion.
   Marina works hard enough.

   She will get a promotion.

10. If Josh's hip heals on time, he'll play this season.
    His hip heals on time.

    He'll play this season.

11. If he doesn't have to get up at 3:00 A.M., he's ecstatic.
    He's ecstatic.

    He doesn't have to get up at 3:00 A.M.

12. "A mathematician is a device for turning coffee into theorems." (quote from Paul Erdos)
    You turn coffee into theorems.

    You are a mathematician.

13. "If you're going through hell, keep going." (quote from Winston Churchill)
    You're not going through hell.

    Don't keep going.

14. "If you can't get rid of the skeleton in your closet, you'd best teach it to dance." (quote from George Bernard Shaw)
    You can get rid of the skeleton in your closet.

    You'd best not teach it to dance.

15. If Justin pitches, the Astros win.
    The Astros do not win.

    Justin does not pitch.

16. If Max plays, the opponent gets shut out.
    The opponent does not get shut out.

    Max does not play.

17. She uses Android pay or she pays by credit card.
    She does not pay by credit card.

    She uses Android pay.

18. Mia kicks or Drew passes.
    Drew does not pass.

    Mia kicks.

Use a truth table to determine whether the argument is valid *or* invalid.

19. $\sim p \to \sim q$
    $\dfrac{q}{p}$

20. $p \vee \sim q$
    $\dfrac{p}{\sim q}$

21. $p \to q$
    $\dfrac{q \to p}{p \wedge q}$

22. $\sim p \to q$
    $\dfrac{p}{\sim q}$

23. $p \to \sim q$
    $\dfrac{q}{\sim p}$

24. $p \to \sim q$
    $\dfrac{\sim p}{\sim q}$

25. $(\sim p \vee q) \wedge (\sim p \to q)$
    $\dfrac{p}{\sim q}$

26. $(r \wedge p) \to (r \vee q)$
    $\dfrac{q \wedge p}{r \vee p}$

27. $(p \to q) \wedge (q \to p)$
    $\dfrac{p}{p \vee q}$

28. $(p \wedge q) \vee (p \vee q)$
    $\dfrac{q}{p}$

29. $(\sim p \wedge r) \to (p \vee q)$
    $\dfrac{\sim r \to p}{q \to r}$

30. $(p \to \sim q) \vee (q \to \sim r)$
    $\dfrac{p \vee \sim r}{r \to p}$

CONCEPT CHECK   *Answer each of the following.*

31. Earlier we showed how to analyze arguments using Euler diagrams. Refer to **Example 5** in this section, restate each premise and the conclusion using a quantifier, and then draw an Euler diagram to illustrate the relationship.

32. Explain in a few sentences how to determine the statement for which a truth table will be constructed so that the arguments in the exercises that follow can be analyzed for validity.

Determine whether each argument is valid *or* invalid.

33. Joey loves to watch movies. If Terry likes to jog, then Joey does not love to watch movies. If Terry does not like to jog, then Carrie drives a school bus. Therefore, Carrie drives a school bus.

34. If Hurricane Maria hit that grove of trees, then the trees are devastated. People plant trees when disasters strike and the trees are not devastated. Therefore, if people plant trees when disasters strike, then Hurricane Maria did not hit that grove of trees.

35. The Cowboys will make the playoffs if and only if Troy comes back to play. Jerry doesn't coach the Cowboys or Troy comes back to play. Jerry does coach the Cowboys. Therefore, the Cowboys will not be in the playoffs.

36. If I've got you under my skin, then you are deep in the heart of me. If you are deep in the heart of me, then you are really a part of me. You are not deep in the heart of me or you are really a part of me. Therefore, if I've got you under my skin, then you are really a part of me.

37. If Dr. Hardy is a department chairman, then he lives in Atlanta. He lives in Atlanta and his first name is Larry. Therefore, if his first name is not Larry, then he is not a department chairman.

38. If I were your woman and you were my man, then I'd never stop loving you. I've stopped loving you. Therefore, I am not your woman or you are not my man.

39. All men are created equal. All people who are created equal are women. Therefore, all men are women.

40. All men are mortal. Socrates is a man. Therefore, Socrates is mortal.

41. A recent DirecTV commercial had the following script: "When the cable company keeps you on hold, you feel trapped. When you feel trapped, you need to feel free. When you need to feel free, you try hang-gliding. When you try hang-gliding, you crash into things. When you crash into things, the grid goes down. When the grid goes down, crime goes up, and when crime goes up, your dad gets punched over a can of soup . . ."

(a) Use reasoning by transitivity and all the component statements to draw a valid conclusion.

(b) If we added the line, "Your dad does not get punched over a can of soup," what valid conclusion could be drawn?

42. Molly made the following observation: "If I want to determine whether an argument leading to the statement

$$[(p \rightarrow q) \wedge \sim q] \rightarrow \sim p$$

is valid, I only need to consider the lines of the truth table that lead to T for the column that is headed $(p \rightarrow q) \wedge \sim q$." Molly was very perceptive. Can you explain why her observation was correct?

In the arguments used by Lewis Carroll, it is helpful to restate a premise in if . . . then *form in order to more easily identify a valid conclusion. The following premises come from Lewis Carroll. Write each premise in* if . . . then *form.*

43. All my poultry are ducks.

44. None of your sons can do logic.

45. Guinea pigs are hopelessly ignorant of music.

46. No teetotalers are pawnbrokers.

47. I have not filed any of them that I can read.

48. All of them written on blue paper are filed.

49. No teachable kitten has green eyes.

50. Opium-eaters have no self-command.

The following exercises involve premises from Lewis Carroll. Write each premise in symbols, and then, in the final part, give a conclusion that yields a valid argument.

51. Let *p* be "it is a duck," *q* be "it is my poultry," *r* be "one is an officer," and *s* be "one is willing to waltz."

(a) No ducks are willing to waltz.

(b) No officers ever decline to waltz.

(c) All my poultry are ducks.

(d) Give a conclusion that yields a valid argument.

52. Let *p* be "it will hold water," *q* be "it is cracked," *r* be "it is a jug," and *s* be "it is old."

   (a) All of the old articles in this cupboard are cracked.

   (b) No jug in this cupboard is new.

   (c) Nothing in this cupboard that is cracked will hold water.

   (d) Give a conclusion that yields a valid argument.

53. Let *p* be "one is honest," *q* be "one is a pawnbroker," *r* be "one is a promise-breaker," *s* be "one is trustworthy," *t* be "one is very communicative," and *u* be "one is a wine-drinker."

   (a) Promise-breakers are untrustworthy.

   (b) Wine-drinkers are very communicative.

   (c) A person who keeps a promise is honest.

   (d) No teetotalers are pawnbrokers. (*Hint:* Assume "teetotaler" is the opposite of "wine-drinker.")

   (e) One can always trust a very communicative person.

   (f) Give a conclusion that yields a valid argument.

54. Let *p* be "it is a guinea pig," *q* be "it is hopelessly ignorant of music," *r* be "it keeps silent while the *Moonlight Sonata* is being played," and *s* be "it appreciates Beethoven."

   (a) Nobody who really appreciates Beethoven fails to keep silent while the *Moonlight Sonata* is being played.

   (b) Guinea pigs are hopelessly ignorant of music.

   (c) No one who is hopelessly ignorant of music ever keeps silent while the *Moonlight Sonata* is being played.

   (d) Give a conclusion that yields a valid argument.

55. Let *p* be "it begins with 'Dear Sir'," *q* be "it is crossed," *r* be "it is dated," *s* be "it is filed," *t* be "it is in black ink," *u* be "it is in the third person," *v* be "I can read it," *w* be "it is on blue paper," *x* be "it is on one sheet," and *y* be "it is written by Brown."

   (a) All the dated letters are written on blue paper.

   (b) None of them are in black ink, except those that are written in the third person.

   (c) I have not filed any of them that I can read.

   (d) None of them that are written on one sheet are undated.

   (e) All of them that are not crossed are in black ink.

   (f) All of them written by Brown begin with "Dear Sir."

   (g) All of them written on blue paper are filed.

   (h) None of them written on more than one sheet are crossed.

   (i) None of them that begin with "Dear Sir" are written in the third person.

   (j) Give a conclusion that yields a valid argument.

56. Let *p* be "it is able to dance a minuet," *q* be "it is certain that it is well fitted out," *r* be "it is contemptible," *s* be "it has three rows of teeth," *t* be "it is heavy," *u* be "it is kind to children," and *v* be "it is a shark."

   (a) No shark ever doubts that it is well fitted out.

   (b) A fish that cannot dance a minuet is contemptible.

   (c) No fish is quite certain that it is well fitted out unless it has three rows of teeth.

   (d) All fish except sharks are kind to children.

   (e) No heavy fish can dance a minuet

   (f) A fish with three rows of teeth is not to be despised.

   (g) Give a conclusion that yields a valid argument.

Chapter **3** SUMMARY

KEY TERMS

3.1
symbolic logic
truth value
statement
compound statement
component statements
connectives
negation
quantifiers

3.2
conjunction
truth table
disjunction
equivalent statements

3.3
conditional statement
antecedent
consequent
tautology

3.4
converse
inverse
contrapositive
biconditional

3.5
argument
premises
conclusion
valid

fallacy
Euler diagram

3.6
modus ponens
modus tollens
disjunctive syllogism
fallacy of the converse
fallacy of the inverse
reasoning by
   transitivity

NEW SYMBOLS

∨ disjunction
∧ conjunction
~ negation

→ conditional
↔ biconditional
≡ equivalence

TEST YOUR WORD POWER

See how well you have learned the vocabulary in this chapter.

1. A **statement** is
  A. a sentence that asks a question, the answer to which may be true or false.
  B. a directive giving specific instructions.
  C. a sentence declaring something that is either true or false, but not both at the same time.
  D. a paradoxical sentence with no truth value.

2. A **disjunction** (inclusive) is
  A. a compound statement that is true only if both of its component statements are true.
  B. a compound statement that is true if one or both of its component statements is/are true.
  C. a compound statement that is false if either of its component statements is false.
  D. a compound statement that is true if exactly one of its component statements is true.

3. A **conditional** statement is
  A. a statement that may be true or false, depending on some condition.
  B. an idea that can be stated only under certain conditions.
  C. a statement using the connective *if . . . then*.
  D. a statement the antecedent of which is implied by the consequent.

4. The **inverse** of a conditional statement is
  A. the result when the antecedent and consequent are negated.
  B. the result when the antecedent and consequent are interchanged.
  C. the result when the antecedent and consequent are interchanged and negated.
  D. logically equivalent to the conditional.

5. A **fallacy** is
  A. an argument with a false conclusion.
  B. an argument whose conclusion is not supported by the premises.
  C. a valid argument.
  D. an argument containing at least one false premise.

6. **Fallacy of the inverse** is
  A. the reason the converse of a conditional is not equivalent to the conditional.
  B. an invalid argument form that assumes the converse of a premise.
  C. an invalid argument form that denies the converse of a premise.
  D. an invalid argument form that assumes the inverse of a premise.

ANSWERS
1. C **2.** B **3.** C **4.** A **5.** B **6.** D

QUICK REVIEW

Concepts

Examples

 3.1 STATEMENTS AND QUANTIFIERS

A **statement** is a declarative sentence that is either true or false (not both simultaneously).

A **compound statement** is made up of two or more **component statements** joined by **connectives** (*not, and, or, if . . . then*).

Quantifiers indicate how many members in a group being considered exhibit a particular property or characteristic. Universal quantifiers indicate *all* members, and existential quantifiers indicate *at least one* member.

The **negation** of a statement has the opposite truth value of that statement in all cases.

Consider the following statement.

  "If it rains this month, then we'll have a green spring."

This is a compound statement made up of the following two component statements joined by the connective *if, then.*

  "It rains this month"   and   "We'll have a green spring"

The statement

        "All five of those birds can fly"

contains a universal quantifier. Its negation is

        "At least one of those five birds cannot fly,"

which contains an existential quantifier.

Concepts   *Examples*

3.2 TRUTH TABLES AND EQUIVALENT STATEMENTS

Suppose two component statements p and q are given. The **conjunction** is symbolized $p \wedge q$ and is true only when both component statements are true.

Consider the truth table.

p	q	$p \wedge q$	$p \vee q$
T	T	T	T
T	F	F	T
F	T	F	T
F	F	F	F

The **disjunction,** symbolized $p \vee q$, is false only when both component statements are false.

If p represents "$7 < 10$" and q represents "$4 < 3$," then the second row of the truth table above shows that $p \wedge q$ is false and $p \vee q$ is true.

The truth value of a compound statement is found by substituting T or F for each component statement, and then working from inside parentheses out, determining truth values for larger parts of the overall statement, until the entire statement has been evaluated.

When this process is carried out for all possible combinations of truth values for the component statements, a **truth table** results.

Consider the truth table for the statement $\sim p \vee (q \wedge p)$.

p	q	$\sim p$	$\vee$	$(q$	$\wedge$	$p)$
T	T	F	T	T	T	T
T	F	F	F	F	F	T
F	T	T	T	T	F	F
F	F	T	T	F	F	F
		①	③	①	②	①

The circled numbers indicate the order in which columns were determined.

Equivalent statements have the same truth value for all combinations of truth values for the component statements. To determine whether two statements are equivalent, construct truth tables for both and see if the final truth values agree in all rows.

The statements $\sim(p \wedge q)$ and $\sim p \vee \sim q$ are equivalent, as shown in the truth table.

p	q	$\sim$	$(p$	$\wedge$	$q)$	$\sim p$	$\vee$	$\sim q$
T	T	F	T	T	T	F	F	F
T	F	T	T	F	F	F	T	T
F	T	T	F	F	T	T	T	F
F	F	T	F	F	F	T	T	T

De Morgan's laws can be used to quickly find negations of disjunctions and conjunctions.

$$\sim(p \vee q) \equiv \sim p \wedge \sim q$$
$$\sim(p \wedge q) \equiv \sim p \vee \sim q$$

To find the negation of the statement

"I love chess and I had breakfast,"

let p represent "I love chess" and let q represent "I had breakfast." Then the above statement becomes $p \wedge q$. Its negation $\sim(p \wedge q) \equiv \sim p \vee \sim q$ translates to

"I don't love chess or I didn't have breakfast."

3.3 THE CONDITIONAL AND CIRCUITS

A **conditional statement** uses the *if . . . then* connective and is symbolized $p \rightarrow q$, where p is the **antecedent** and q is the **consequent.**

If p represents "You are mighty" and q represents "I am flighty," then the conditional statement $p \rightarrow q$ is expressed as

"*If* you are mighty, *then* I am flighty."

The conditional is false if the antecedent is true and the consequent is false. Otherwise, the conditional is true. This is because q is only *required* to be true on the *condition* that p is true, but q may "voluntarily" be true even if p is false. That is, p is sufficient for q, but not necessary.

The statement $(6 < 1) \rightarrow (3 = 7)$ is true because the antecedent is false.

The statement "If you are reading this book, then it is the year 1937" is false, because the antecedent is true and the consequent is false.

Concepts	Examples

Concepts

The conditional $p \to q$ is equivalent to the disjunction $\sim p \vee q$. Its negation is $p \wedge \sim q$.

Examples

p	q	p	$\to$	q	$\sim p$	$\vee$	q	p	$\wedge$	$\sim q$
T	T	T	T	T	F	T	T	T	F	F
T	F	T	F	F	F	F	F	T	T	T
F	T	F	T	T	T	T	T	F	F	F
F	F	F	T	F	T	T	F	F	F	T

The statement "All mice love cheese" can be stated, "*If* it's a mouse, *then* it loves cheese." This is equivalent to saying, "It's not a mouse or it loves cheese." The negation of this statement is "It's a mouse and it does not love cheese."

Electrical circuits are analogous to logical statements, with *parallel* circuits corresponding to disjunctions, and *series* circuits corresponding to conjunctions. Each switch (modeled by an arrow) represents a component statement. When a switch is closed, it allows current to pass through. The circuit represents a true statement when current flows from one end of the circuit to the other.

The following circuit corresponds to the logical statement

$p \vee (\sim p \wedge q)$.

Current will flow from one end to the other if p is true *or* if p is false and q is true. This statement is equivalent to

$(p \vee \sim p) \wedge (p \vee q)$, which simplifies to $p \vee q$.

3.4 THE CONDITIONAL AND RELATED STATEMENTS

Three related statements for the **conditional statement** $p \to q$, its **converse, inverse,** and **contrapositive,** are defined as follows.

> **Converse:** $q \to p$
> **Inverse:** $\sim p \to \sim q$
> **Contrapositive:** $\sim q \to \sim p$

Common Translations of the Conditional $p \to q$

If p, then q.	p is sufficient for q.
If p, q.	q is necessary for p.
p implies q.	All p are q.
p only if q.	q if p.

For the **biconditional** statement "p if and only if q,"

$$p \leftrightarrow q \equiv (p \to q) \wedge (q \to p).$$

It is true only when p and q have the same truth value.

Consider the statement "If it's a pie, it tastes good."

Converse: "If it tastes good, it's a pie."

Inverse: "If it's not a pie, it doesn't taste good."

Contrapositive: "If it doesn't taste good, it's not a pie."

Statement	If . . . then form
You'll be sorry if I go.	If I go, then you'll be sorry.
Today is Tuesday only if yesterday was Monday.	If today is Tuesday, then yesterday was Monday.
All nurses wear comfortable shoes.	If you are a nurse, then you wear comfortable shoes.

The statement "$5 < 9$ if and only if $3 > 7$" is false because the component statements have opposite truth values.

Summary of Basic Truth Tables

1. $\sim p$, the **negation** of p, has truth value opposite that of p.

2. $p \wedge q$, the **conjunction,** is true only when both p and q are true.

3. $p \vee q$, the **disjunction,** is false only when both p and q are false.

4. $p \to q$, the **conditional,** is false only when p is true and q is false.

5. $p \leftrightarrow q$, the **biconditional,** is true only when both p and q have the same truth value.

Consider the following truth tables.

p	$\sim p$
T	F
F	T

p	q	p	$\wedge$	q	p	$\vee$	q	p	$\to$	q	p	$\leftrightarrow$	q
T	T	T	T	T	T	T	T	T	T	T	T	T	T
T	F	T	F	F	T	T	F	T	F	F	T	F	F
F	T	F	F	T	F	T	T	F	T	T	F	F	T
F	F	F	F	F	F	F	F	F	T	F	F	T	F

Concepts   *Examples*

3.5 ANALYZING ARGUMENTS WITH EULER DIAGRAMS

A logical **argument** consists of premises and a conclusion. An argument is considered **valid** if the truth of the premises forces the conclusion to be true. Otherwise, it is **invalid.**

An **Euler diagram** can be used to determine whether an argument is valid or invalid.

Drawing an Euler Diagram

Step 1   Use the first premise to draw regions. (Arguments with multiple premises may involve multiple regions.)

Step 2   Place an *x* in the diagram to represent the subject of the argument.

Consider this argument. Notice the universal quantifier "all."

All dogs are animals.

Dotty is a dog.
_____

Dotty is an animal.

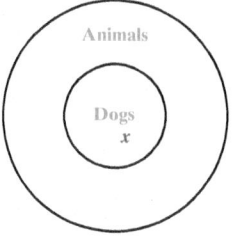

x represents Dotty.

We see from the Euler diagram that the truth of the premises forces the conclusion, that Dotty is an animal, to be true. Thus the argument is valid.

Consider this argument. Notice the existential quantifier "some."

Some animals are warmblooded.

Albie is an animal.
_____

Albie is warmblooded.

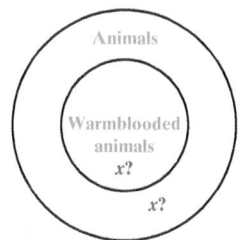

x represents Albie.

We see from the Euler diagram that the location of *x* is uncertain, so the truth of the premises does not force the conclusion to be true. Thus, the argument is invalid.

3.6 ANALYZING ARGUMENTS WITH TRUTH TABLES

An argument with premises $p_1, p_2 \ldots, p_n$ and conclusion c can be tested for validity by constructing a truth table for the statement

$$(p_1 \wedge p_2 \wedge \ldots \wedge p_n) \rightarrow c.$$

If all rows yield T for this statement (that is, if it is a tautology), then the argument is valid. Otherwise, the argument is invalid.

Valid Argument Forms

Modus Ponens	Modus Tollens	Disjunctive Syllogism	Reasoning by Transitivity
$p \rightarrow q$	$p \rightarrow q$	$p \vee q$	$p \rightarrow q$
p	$\sim q$	$\sim p$	$q \rightarrow r$
q	$\sim p$	q	$p \rightarrow r$

Consider this argument. "If I eat ice cream, I regret it later. I didn't regret it later. Therefore, I didn't eat ice cream."

Let p represent "I eat ice cream" and q represent "I regret it later." Then the argument can be expressed as the compound statement

$$[(p \rightarrow q) \wedge \sim q] \rightarrow \sim p.$$

Test its validity using a truth table, which shows that the statement is a tautology. The argument is valid by **modus tollens.**

p	q	$[(p$	$\rightarrow$	$q)$	$\wedge$	$\sim q]$	$\rightarrow$	$\sim p$
T	T	T	T	T	F	F	T	F
T	F	T	F	F	F	T	T	F
F	T	F	T	T	F	F	T	T
F	F	F	T	F	T	T	T	T

Concepts	Examples

Invalid Argument Forms

Fallacy of the Converse	Fallacy of the Inverse
$p \to q$	$p \to q$
q	$\sim p$
p	$\sim q$

Consider this argument.

> If I drink coffee, I get jittery.
> I didn't drink coffee.
> I don't get jittery.

Let p represent "I drink coffee" and q represent "I get jittery."

Test the argument, using a truth table for the statement

$$[(p \to q) \land \sim p] \to \sim q.$$

p	q	[(p	$\to$	q)	$\land$	$\sim p$]	$\to$	$\sim q$
T	T	T	T	T	F	F	T	F
T	F	T	F	F	F	F	T	T
F	T	F	T	T	T	T	F	F
F	F	F	T	F	T	T	T	T

This argument is invalid by **fallacy of the inverse.**

Chapter 3 TEST

Write a negation for each statement.

1. $6 - 3 = 3$ **2.** All roses are red.

3. Some members of the class went on the field trip.

4. If I fall in love, it will be forever.

5. She applied and did not get a student loan.

Let p represent "You will help me" and let q represent "I will help you." Write each statement in symbols.

6. If you won't help me, then I will help you.

7. I will help you if you will help me.

8. I won't help you if and only if you won't help me.

Using the same statements as above, write each of the following in words.

9. $\sim p \land q$ **10.** $\sim (p \lor \sim q)$

In each of the following, assume that p is true and that q and r are false. Find the truth value of each statement.

11. $\sim q \land \sim r$ **12.** $r \lor (p \land \sim q)$

13. $p \leftrightarrow (p \to q)$

14. $r \to (s \lor r)$ (The truth value of the statement s is unknown.)

15. Explain in your own words why, if p is a statement, the biconditional $p \leftrightarrow \sim p$ must be false.

16. State the necessary conditions for each of the following.
   (a) a conditional statement to be false
   (b) a conjunction to be true
   (c) a disjunction to be false
   (d) a biconditional to be true

Construct a truth table for each of the following.

17. $p \land (\sim p \lor \sim q)$ **18.** $\sim (p \land q) \to (\sim p \lor \sim q)$

Decide whether each statement is true *or* false.

19. Some negative integers are whole numbers.

20. All irrational numbers are real numbers.

Write each conditional statement in if . . . then form.

21. All integers are rational numbers.

22. Being a rhombus is sufficient for a polygon to be a quadrilateral.

23. Being divisible by 2 is necessary for a number to be divisible by 4.

24. She digs dinosaur bones only if she is a paleontologist.

For each statement, write (**a**) *the converse,* (**b**) *the inverse, and* (**c**) *the contrapositive.*

25. If a picture paints a thousand words, the graph will help me understand it.

26. $\sim p \rightarrow (q \wedge r)$ (Use one of De Morgan's laws as necessary.)

Solve each problem.

27. Use an Euler diagram to determine whether the argument is *valid* or *invalid.*

All members of that athletic club save money.
Don is a member of that athletic club.
_____
Don saves money.

28. Match each argument in parts (a)–(d) in the next column with the law that justifies its validity, or the fallacy of which it is an example, in choices A–F.

  A. Modus ponens

  B. Modus tollens

  C. Reasoning by transitivity

  D. Disjunctive syllogism

  E. Fallacy of the converse

  F. Fallacy of the inverse

(a) If he eats liver, then he'll eat anything.
He eats liver.
_____
He'll eat anything.

(b) If you use your seat belt, you will be safer.
You don't use your seat belt.
_____
You won't be safer.

(c) If I hear *Mr. Bojangles,* I think of her.
If I think of her, I smile.
_____
If I hear *Mr. Bojangles,* I smile.

(d) She sings or she dances.
She does not sing.
_____
She dances.

Use a truth table to determine whether each argument is valid *or* invalid.

29. If I write a check, it will bounce. If the bank guarantees it, then it does not bounce. The bank guarantees it. Therefore, I don't write a check.

30. $\sim p \rightarrow \sim q$
   $\underline{q \rightarrow p}$
   $p \vee q$

4 Numeration Systems

Web designers strive to create Web sites that will enhance the browsing experience of their clients' customers. The appearance of a Web page influences visitors either to stay or to leave, and the job of the Web designer is to draw them in, convince them (even subconsciously) to stay for a while, and thus encourage them to return for future visits.

One of the elements of Web design is the use of complementary colors to make some items stand out by contrast, or perhaps to promote others in a subtle way. A theoretical understanding of how computers generate color is foundational to an efficient system for intentional Web design. Some of the numeration systems covered in this chapter are applied in digitizing color.

4.1 HISTORICAL NUMERATION SYSTEMS

Basics of Numeration

The various ways of symbolizing and working with the counting numbers are called **numeration systems.** The symbols representing the numbers are called **numerals.**

Numeration systems have developed over many millennia of human history. Ancient documents provide insight into methods used by the early Sumerian peoples, the Egyptians, the Babylonians, the Greeks, the Romans, the Chinese, the Hindus, and the Mayan people, as well as others.

Keeping accounts by matching may have developed as humans established permanent settlements and began to grow crops and raise livestock. People might have kept track of the number of sheep in a flock by matching pebbles with the sheep, for example. The pebbles could then be kept as a record of the number of sheep.

A more efficient method is to keep a **tally stick.** With a tally stick, one notch or **tally** is made on a stick for each sheep. Although an improvement over pebbles, tally marks are still crude and inefficient compared to modern methods. For example, the numeral for the number thirteen might be written as follows.

| | | | | | | | | | | | | ← 13 tally marks

This requires the recording of 13 symbols, and later interpretation requires careful counting of symbols.

Even today, tally marks are used, especially when keeping track of things that occur one or a few at a time, over space or time. To facilitate the counting of the tally, we often use a sort of "grouping" technique as we go.

 ← Numeral (tally) for 13

History has recorded a long evolution of numeration systems progressing from tally marks to our own modern system, the **Hindu-Arabic system,** which utilizes the set of symbols

$$\{1, 2, 3, 4, 5, 6, 7, 8, 9, 0\}.$$

Ancient Egyptian Numeration

An essential feature common to all more advanced numeration systems is **grouping,** which allows for less repetition of symbols, making numerals easier to interpret. Most historical systems, including our own, have used groups of ten, reflecting the common practice of learning to count by using the fingers. The size of the groupings (again, usually ten) is called the **base** of the number system.

The ancient Egyptian system is an example of a **simple grouping system.** It utilized ten as its base, and its various symbols are shown in **Table 1** on the next page. The symbol for 1 (I) is repeated, in a tally scheme, for 2, 3, and so on up to 9. A new symbol is introduced for 10 (∩), and that symbol is repeated for 20, 30, and so on, up to 90. This pattern enabled the Egyptians to express numbers up to 9,999,999 with just the seven symbols shown in the table.

The numbers denoted by the seven Egyptian symbols are all *powers* of the base ten.

$$10^0 = 1, \quad 10^1 = 10, \quad 10^2 = 100, \quad 10^3 = 1000, \quad 10^4 = 10,000,$$

$$10^5 = 100,000, \quad 10^6 = 1,000,000$$

These expressions are called *exponential expressions.* In the expression 10^4, for example, 10 is the *base* and 4 is the *exponent.* Recall that the exponent indicates the number of repeated factors of the base to be multiplied.

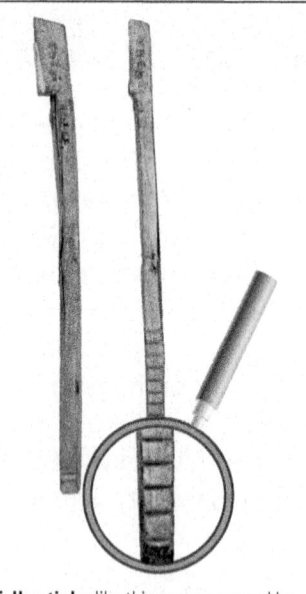

Tally sticks like this one were used by the English in about 1400 A.D. to keep track of financial transactions. Each notch stands for one pound sterling.

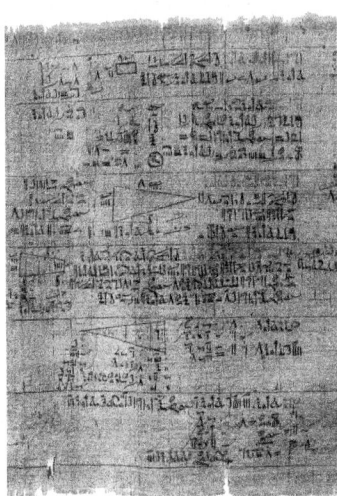

Much of our knowledge of **Egyptian mathematics** comes from the **Rhind papyrus,** from about 3800 years ago. A small portion of this papyrus, showing methods for finding the area of a triangle, is reproduced here.

Table 1 Early Egyptian Symbols

Number	Symbol	Description
1	I	Stroke
10	∩	Heel bone
100	9	Scroll
1000	⌇	Lotus flower
10,000	∅	Pointing finger
100,000	☋	Burbot fish
1,000,000	☥	Astonished person

EXAMPLE 1 Interpreting an Egyptian Numeral

Write the number below in Hindu-Arabic form.

☋☋ ⌇⌇⌇ 999 ∩∩∩∩∩III
 99 ∩∩∩IIII

Solution

Refer to **Table 1** for the values of the Egyptian symbols. Each ☋ represents 100,000. Therefore, two ☋s represent $2 \cdot 100{,}000$, or 200,000. Proceed as shown.

two	☋	$2 \cdot 100{,}000 =$	200,000
three	⌇	$3 \cdot 1000 =$	3000
five	9	$5 \cdot 100 =$	500
nine	∩	$9 \cdot 10 =$	90
seven	I	$7 \cdot 1 =$	7

203,597 ← Answer

EXAMPLE 2 Creating an Egyptian Numeral

Write 376,248 in Egyptian form.

Solution

3 7 6, 2 4 8
↓ ↓ ↓   ↓ ↓ ↓

☋☋ III ⌇⌇⌇ 99∩∩IIII
 ☋ IIII⌇⌇⌇ ∩∩IIII

Refer to Table 1 as needed.

The position or order of the symbols makes no difference in a simple grouping system. Each of the numerals 99∩∩∩IIII, IIII∩∩∩99, and II∩∩99∩II would be interpreted as 234. In **Examples 1 and 2,** like symbols are grouped together, and groups of greater-valued symbols are positioned to the left.

A simple grouping system is well suited to addition and subtraction.

An Egyptian tomb painting shows scribes tallying the count of a grain harvest. **Egyptian mathematics** was oriented more to practicality than was Greek or Babylonian mathematics, although the Egyptians did have a formula for finding the volume of a certain portion of a pyramid.

⌇⌇ 99 ∩∩∩ II

+ ⌇ 999 ∩ IIIIII

Sum: ⌇⌇⌇ 999 ∩∩ IIII
 99 ∩∩ IIII

We use a + sign for convenience and draw a line under the numbers being added, although the Egyptians did not do this.

Two Is plus six Is is equal to eight Is, and so on.

Archaeological investigation has provided much of what we know about the numeration systems of ancient peoples.

Regrouping, or "carrying," is needed when more than nine of the same symbol result.

$$\begin{array}{r} \mathscr{l} \; \mathscr{LL} \; 99 \; \overset{\cap\cap\cap\cap}{\cap\cap\cap} \; || \\ + \; \mathscr{ll} \quad\;\; 999 \; \overset{\cap\cap\cap}{\cap\cap} \; \overset{||}{|||} \\ \hline \end{array}$$

Sum: $\;\mathscr{lll}\,\mathscr{LL}\; \overset{999}{99}\, \overset{\cap\cap\cap\cap\cap\cap||||}{\cap\cap\cap\cap\cap\cap|||}$

Regrouped answer: $\;\mathscr{lll}\,\mathscr{LL}\; \overset{999}{999}\,\cap\cap\overset{|||}{||||}$

ten $\cap$ = one 9

Subtraction is done in much the same way, as shown in the next example.

EXAMPLE 3   **Subtracting Egyptian Numerals**

Work each subtraction problem.

(a)
$$\begin{array}{r} 999 \; \cap\cap \; |||| \\ 99 \; \cap\cap \; ||| \\ - \; 999 \; \cap \; |||| \\ \hline \end{array}$$

(b)
$$\begin{array}{r} 99\cap\cap\cap\cap \; || \\ - \; 9 \quad \cap\cap \quad |||| \\ \hline \end{array}$$

Solution

(a)
$$\begin{array}{r} 999 \; \cap\cap \; |||| \\ 99 \; \cap\cap \; ||| \\ - \; 999 \; \cap \; |||| \\ \hline \end{array}$$

As with addition, work from right to left and subtract.

Difference: $\quad 99 \; \cap\cap\cap \; |||$

(b) To subtract four |s from two |s, "borrow" one heel bone, which is equivalent to ten |s. Finish the problem after writing ten additional |s on the right.

Regrouped: $\quad 99 \; \cap\cap\cap \; \overset{||||||}{||||||}$ one $\cap$ = ten |s

$$\begin{array}{r} - \; 9 \; \cap\cap \; |||| \\ \hline \end{array}$$

Difference: $\quad\; 9 \;\; \cap \; ||||||||$

A procedure such as those described above is called an **algorithm:** a rule or method for working a problem. The Egyptians used an interesting algorithm for multiplication that requires only an ability to add and to double numbers, as shown in **Example 4.** For convenience, this example uses our symbols rather than theirs.

EXAMPLE 4   **Using the Egyptian Multiplication Algorithm**

A rectangular room in an archaeological excavation measures 19 cubits by 70 cubits. (A cubit, based on the length of the forearm, from the elbow to the tip of the middle finger, was approximately 18 inches.) Find the area of the room.

Solution

Multiply the width and length to find the area of a rectangle. Build two columns of numbers as shown at the top of the next page. Start the first column with 1, the second with 70. Each column is built downward by doubling the number above. Keep going until the first column contains numbers that can be added to equal 19. Then add the corresponding numbers from the second column.

$$1 + 2 + 16 = 19$$

→ 1	70 ←
→ 2	140 ←
4	280
8	560
→ 16	1120 ←

$$70 + 140 + 1120 = \mathbf{1330}$$

Thus $19 \cdot 70 = \mathbf{1330}$ and the area of the given room is 1330 square cubits.

Ancient Roman Numeration

Roman numerals are still used today, mainly for decorative purposes, on clock faces, for heading numbers in outlines, chapter numbers in books, copyright dates of movies, and so on. The base is again ten, with distinct symbols for 1, 10, 100, and 1000. The Romans, however, deviated from pure simple grouping in several ways. For the symbols and some examples, see **Tables 2 and 3,** respectively.

Clocks with Roman numerals often use IIII for 4.

Table 2 Roman Symbols

Number	Symbol
1	I
5	V
10	X
50	L
100	C
500	D
1000	M

FEATURES OF THE ROMAN SYSTEM

1. In addition to symbols for 1, 10, 100, and 1000, "extra" symbols denote 5, 50, and 500. This allows less symbol repetition within a numeral. It is like a secondary base-five grouping functioning within the base-ten simple grouping.

2. A *subtractive feature* was introduced, whereby a lesser-valued symbol, placed immediately to the left of one of greater value, meant to subtract. Thus IV = 4, while VI = 6. Only certain combinations were used in this way:
   (a) I preceded only V or X.
   (b) X preceded only L or C.
   (c) C preceded only D or M.

3. A *multiplicative feature,* rather than more symbols, allowed for larger numbers:
   (a) A bar over a numeral meant to multiply by 1000.
   (b) A double bar meant to multiply by 1000^2—that is, by 1,000,000.

Table 3

Selected Roman Numerals

Number	Numeral
6	VI
12	XII
19	XIX
30	XXX
49	XLIX
85	LXXXV
25,040	$\overline{\text{XXV}}$ XL
35,000	$\overline{\text{XXXV}}$
5,105,004	$\overline{\overline{\text{V}}}$ $\overline{\text{CV}}$ IV
7,000,000	$\overline{\overline{\text{VII}}}$

Adding and subtracting with Roman numerals is very similar to the Egyptian method, except that the subtractive feature of the Roman system sometimes makes the processes more involved. With Roman numerals we cannot add IV and VII to get the sum VVIII by simply combining like symbols. (Even XIII would be incorrect.) The safest method is to rewrite IV as IIII, then add IIII and VII, getting VIIIIII. We convert this to VVI, and then to XI by regrouping. Subtraction, which is similar, is shown in the following example.

Avoiding L

..., XLVIII, XLIX, 50, LI, LII, ...

Roman numerals have always been used in Super Bowl logos, with one exception: In 2016, the NFL opted to use the Hindu-Arabic 50 rather than a stand-alone Roman L.

EXAMPLE 5 Subtracting Roman Numerals

Janus, a Roman official, has invited 26 guests to a banquet on Saturday. If 14 of the invitees decide instead to attend the Crying Citharas concert at the Forum, how many guests will be at the Saturday banquet?

Solution

To find the answer, we subtract XIV from XXVI. Set up the problem in terms of simple grouping numerals (that is, XIV is rewritten as XIIII).

Problem: XXVI
          − XIV

Problem restated without subtractive notation: XXVI
                                                − XIIII

Regrouped: XXIIIIII
            − XIIII
              XII ← Answer

Since four **I**s cannot be subtracted from one **I**, we have "borrowed" in the top numeral, writing XXVI as XXIIIIII. The subtraction can then be carried out. Janus will have 12 guests at the banquet.

Computation, in early forms, was often aided by mechanical devices just as it is today. The Roman merchants, in particular, did their figuring on a counting board, or **counter,** on which lines or grooves represented 1s, 10s, 100s, etc., and on which the spaces between the lines represented 5s, 50s, 500s, and so on. Discs or beads (called *calculi,* the word for "pebbles") were positioned on the board to denote numbers, and *calculations* were carried out by moving the discs around and simplifying.

Roman Numeral
Abacus Addition

EXAMPLE 6 Adding on a Roman Counting Board

A Roman merchant wants to calculate the sum 934 + 286. Use counting boards to carry out the following steps.

(a) Represent the first number, 934.

(b) Represent the second number, 286, beside the first.

(c) Represent the sum, in simplified form.

Solution

(a) See **Figure 1.**   **(b)** See **Figure 2.**

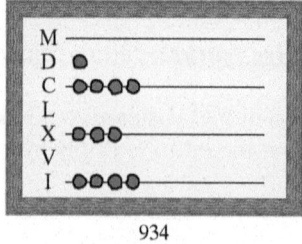

934

Figure 1

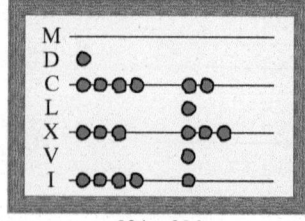

934 + 286

Figure 2

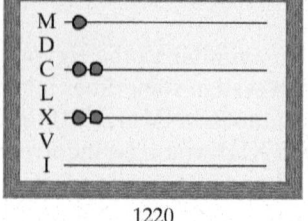

1220

Figure 3

(c) See **Figure 3.** The simplified answer is MCCXX, or 1220. In the process of simplification, five discs on the bottom line were replaced by a single disc in the V space. This made two Vs that were replaced by an additional disc on the X line. Five of those on the X line were then replaced by one in the L space, and this process continued until the disc on the M line finally appeared.

Table 4

Chinese Symbols

Number	Symbol
1	∼
2	⼆
3	⺌
4	⼝
5	⽳
6	大
7	⺁
8	入
9	ℏ
10	⼗
100	百
1000	千
0	零

Classical Chinese Numeration

The preceding examples show that simple grouping, although an improvement over tallying, still requires considerable repetition of symbols. To denote 90, for example, the ancient Egyptian system must utilize nine ∩s: ∩∩∩∩∩ ∩∩∩∩ . If an additional symbol (a "multiplier") were introduced to represent nine, say "9," then 90 could be denoted 9 ∩. All possible numbers of repetitions of powers of the base could be handled by introducing a separate multiplier symbol for each counting number less than the base.

Just such a system was developed long ago in China. One version used the symbols shown in **Table 4.** We call this type of system a **multiplicative grouping system.** In general, a numeral in such a system would contain pairs of symbols, each pair containing a multiplier (with some counting-number value less than the base) and then a power of the base. The Chinese numerals are read from top to bottom rather than from left to right.

If the Chinese system were *pure* multiplicative grouping, the number 2019 would be denoted as shown in **Figure 4.** But three special features of the system show that they had started to move beyond multiplicative grouping toward something more efficient.

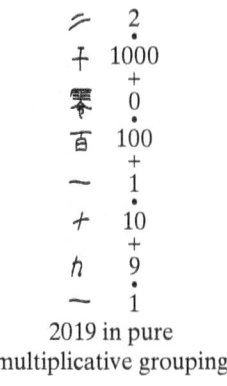

2019 in pure
multiplicative grouping

Figure 4

FEATURES OF THE CHINESE SYSTEM

1. A single symbol, rather than a pair, denotes the number of 1s. The multiplier (1, 2, 3, 4, ... , or 9) is written, but the power of the base (10^0) is omitted. See **Figure 5** below (and also **Examples 7(a), (b), and (c)** on the next page).

2. In the 10s pair, if the multiplier is 1 it is omitted. See **Figure 6** below (and **Example 8(a)** on the next page).

3. For missing powers of the base, the following rules apply.

   • When a particular power of the base is missing completely, the omission is denoted with the zero symbol. See **Figure 7** below (and **Examples 7(b) and 8(b)** on the next page).

   • If two or more consecutive powers are missing, just one zero symbol denotes the total omission. (See **Example 7(c)** on the next page.)

   • The omission of 1s and 10s and any other powers occurring consecutively at the bottom of a numeral need not be denoted at all. (See **Example 7(d)** on the next page.)

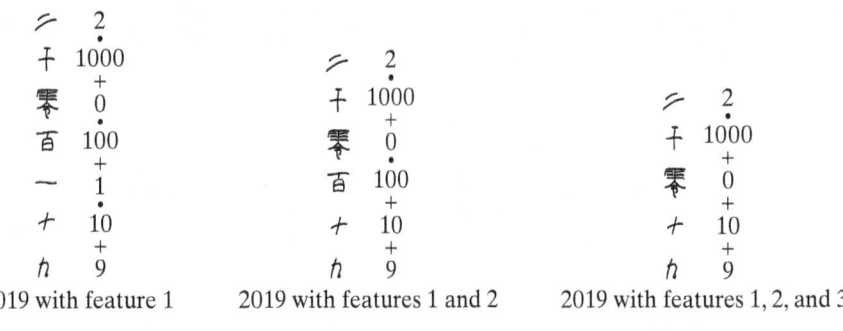

2019 with feature 1   2019 with features 1 and 2   2019 with features 1, 2, and 3

Figure 5   **Figure 6**   **Figure 7**

This illustration is of a **quipu.** In
*Ethnomathematics: A Multicultural View
of Mathematical Ideas*, Marcia Ascher
writes:

*A quipu is an assemblage of colored
knotted cotton cords. Cotton cordage and
cloth were of unparalleled importance in
Inca culture. The colors of the cords, the
way the cords are connected, the relative
placement of the cords, the spaces
between the cords, the types of knots
on the individual cords, and the relative
placement of the knots are all part of the
logical-numerical recording.*

Note that, for clarification in the following examples, we have emphasized the grouping into pairs by spacing and by colored braces. These features were *not* part of the
actual numerals in practice.

Interpreting Chinese Numerals

Interpret each Chinese numeral.

(a) **(b)** **(c)** **(d)**

Solution

(a) $3 \cdot 1000 = 3000$

$1 \cdot 100 = 100$

$6 \cdot 10 = 60$

$4(\cdot 1) = \underline{4}$

Total: 3164

(b) $7 \cdot 100 = 700$

$0(\cdot 10) = 00$

$3(\cdot 1) = \underline{3}$

Total: 703

(c) $5 \cdot 1000 = 5000$

$0(\cdot 100) = 000$

$0(\cdot 10) = 00$

$9(\cdot 1) = \underline{9}$

Total: 5009

(d) $4 \cdot 1000 = 4000$

$2 \cdot 100 = \underline{200}$

Total: 4200

Creating Chinese Numerals

Write a Chinese numeral for each number.

(a) 814   **(b)** 5090

Solution

(a) The number 814 is made up of eight 100s, one 10,
and four 1s, as depicted at the right.

$8 \cdot 100$:

$(1 \cdot)10$:

$4(\cdot 1)$:

(b) The number 5090 consists of five 1000s, no 100s,
and nine 10s (no 1s).

$5 \cdot 1000$:

$0(\cdot 100)$:

$9 \cdot 10$:

4.1 EXERCISES

CONCEPT CHECK *Choose all that apply.*

A. Ancient Egyptian   **B.** Ancient Roman
C. Classical Chinese

1. This numeration system is a *simple* grouping system.

2. In this numeration system, numerals are read from top to bottom.

3. This numeration system uses ten as its base.

4. This numeration system has a subtractive feature in which a lesser-valued symbol is placed immediately to the left of a greater-valued symbol.

5. This system is a multiplicative grouping system.

6. This system uses a multiplication algorithm that involves a doubling process.

Convert each Egyptian numeral to Hindu-Arabic form.

7. 𓂃𓏧𓏧 ∩∩∩ ||||

8. 𓏧𓏧 𓏲𓏲𓏲𓏲 ∩ ||

9. 𓏧𓏧𓏧𓏧 𓆼𓆼𓆼 𓂃 𓏲𓏲𓏲 ∩∩ |||||
𓏧𓏧𓏧 𓆼𓆼𓆼 𓏲𓏲𓏲𓏲 ∩∩ ||||

10. 𓏧𓏧𓏧 𓂃𓂃𓂃𓂃𓂃 𓏲𓏲 ∩∩∩ |

Convert each Hindu-Arabic numeral to Egyptian form.

11. 23,135   **12.** 427

13. 8,657,000   **14.** 306,090

An ancient manuscript describes a census taken of draft-eligible men from several tribes in the vicinity of Egypt about 1450 B.C. Write an Egyptian numeral for the number of available men from each tribe listed.

15. 59,300 from the tribe of Simeon

16. 35,400 from the tribe of Benjamin

17. 74,600 from the tribe of Judah

18. 62,700 from the tribe of Dan

Convert each Roman numeral to Hindu-Arabic form.

19. CLXXIII   **20.** MCDXCVII

21. $\overline{\text{XIV}}$ **22.** $\overline{\text{V}}$ CXXICD

Convert each Hindu-Arabic numeral to Roman form.

23. 2861   **24.** 649

25. 25,619   **26.** 6,402,524

Convert each Chinese numeral to Hindu-Arabic form.

27. 九   **28.** 三   **29.** 二   **30.** 四
百   百   十   十
三   四   零   七
十   十   九   百
六   五   零
                                    二

Convert each Hindu-Arabic numeral to Chinese form.

31. 965   **32.** 63

33. 7012   **34.** 2416

Though Chinese art forms began before written history, their highest development was achieved during four particular dynasties. Write traditional Chinese numerals for the beginning and ending dates of each dynasty listed.

35. T'ang (618 to 907)

36. Han (202 B.C. to A.D. 220)

37. Ming (1368 to 1644)

38. Sung (960 to 1279)

Work each addition or subtraction problem, using regrouping as necessary. Convert each answer to Hindu-Arabic form.

39.   𓏲 ∩∩ ||
       +𓏲 ∩∩∩ ||||
     _____

40.   𓏲𓏲∩∩∩ ||||
             |||
       + 𓏲 ∩∩∩ ||
           ∩∩ |||
     _____

41.   𓏲𓏲∩∩∩ |||||
       − 𓏲 ∩∩ |
     _____

42.   ∩∩∩ |||
           ∩∩ |||
       −∩∩∩ ||||
     _____

43.   𓂃𓂃𓂃 𓏧𓏧   ∩∩∩ |||
                   ∩∩
       + 𓂃𓂃 𓏧 𓏲𓏲 ∩∩ ||||
            𓏲𓏲𓏲 ∩∩ ||||
     _____

44.   𓂃 𓏧   𓏲𓏲 ||||
       − 𓏧𓏧𓏧 𓏲𓏲𓏲 |||
            𓏲𓏲𓏲 |||
     _____

45. MCDXII
    + DCIX

46. $\overline{\text{XXIII}}$CXIX
    + $\overline{\text{XIV}}$CDXII

47. MCCCXXII
    − CDXIX

48. $\overline{\text{XII}}$CCCVI
    − MMCXXXII

Use the Egyptian algorithm to find each product.

49. 32 · 47

50. 29 · 75

51. 64 · 127

52. 52 · 131

Convert all numbers to Egyptian numerals. Multiply using the Egyptian algorithm, and add using the Egyptian symbols. Give the final answer using a Hindu-Arabic numeral.

53. *Value of Ancient Treasure* An ancient text lists donations made by a king toward the construction of a temple. They included thirty golden basins, a thousand silver basins, four hundred ten silver bowls, and thirty golden bowls. Find the total value of this treasure if each golden basin was worth 3000 shekels, each silver basin was worth 500 shekels, each silver bowl was worth 50 shekels, and each golden bowl was worth 400 shekels.

54. *Total Construction Costs* Ancient scrolls record negotiations for a construction project near the coast of the Mediterranean. Find the total construction costs if the king of Tyre supplied the following: 5500 tree cutters at 2 shekels per week each, for seven weeks; 4600 sawers of wood at 3 shekels per week each, for 32 weeks; and 900 sailors at 1 shekel per week each, for 16 weeks.

Explain why each step would be an improvement in the development of numeration systems.

55. progressing from carrying groups of pebbles to making tally marks on a stick

56. progressing from tallying to simple grouping

57. utilizing a subtractive technique within simple grouping, as the Romans did

58. progressing from simple grouping to multiplicative grouping

The ancient Egyptian system described in this section was simple grouping, used a base of ten, and contained seven distinct symbols. The largest number expressible in that system is 9,999,999. Identify the largest number expressible in each of the following simple grouping systems. (In Exercises 63 and 64, d can be any counting number.)

59. base ten, five distinct symbols

60. base ten, ten distinct symbols

61. base five, five distinct symbols

62. base five, ten distinct symbols

63. base ten, d distinct symbols

64. base b, d distinct symbols (where b is any counting number 2 or greater)

65. The Chinese system presented in the text has symbols for 1 through 9 and also for 10, 100, and 1000. What is the greatest number expressible in that system?

66. The Chinese system did eventually adopt two additional symbols, for 10,000 and 100,000. What greatest number could then be expressed?

4.2 MORE HISTORICAL NUMERATION SYSTEMS

OBJECTIVES

1 Discover the basic features of positional numeration systems.
2 Use Hindu-Arabic numeration.
3 Use Babylonian numeration.
4 Use Mayan numeration.
5 Use Greek numeration.

Basics of Positional Numeration

A simple grouping system relies on repetition of symbols to denote the number of each power of the base. A multiplicative grouping system uses multipliers in place of repetition, which is more efficient. The ultimate in efficiency is attained with a **positional system** in which only multipliers are used. The various powers of the base require no separate symbols, because the power associated with each multiplier can be understood from the position that the multiplier occupies in the numeral.

If the Chinese system had evolved into a positional system, then the numeral for 7482 could have been written

rather than.

In the positional version on the left, the lowest symbol is understood to represent two 1s (10^0), the next one up denotes eight 10s (10^1), then four 100s (10^2), then seven 1000s (10^3). Each symbol in a numeral now has both a *face value*, associated with that particular symbol (the multiplier value), and a *place value* (a power of the base), associated with the place, or position, occupied by the symbol.

POSITIONAL NUMERATION

In a positional numeral, each symbol (called a **digit**) conveys two things:

1. **face value**—the inherent value of the symbol

2. **place value**—the power of the base that is associated with the position that the digit occupies in the numeral

Hindu-Arabic Numeration

The place values in a Hindu-Arabic numeral, from right to left, are 1, 10, 100, 1000, and so on. The three 4s in the number 46,424 all have the same face value but different place values. The first 4, on the left, denotes four 10,000s, the next one denotes four 100s, and the one on the right denotes four 1s. Place values (in base ten) are named as shown here.

Billions,	Hundred millions	Ten millions	Millions,	Hundred thousands	Ten thousands	Thousands,	Hundreds	Tens	Units	Decimal point
8,	3	2	1,	4	5	6,	7	9	5	.

This numeral is read as eight billion, three hundred twenty-one million, four hundred fifty-six thousand, seven hundred ninety-five.

To work successfully, a positional system must have a symbol for zero to serve as a **placeholder** in case one or more powers of the base are not needed. Because of this requirement, some early numeration systems took a long time to evolve into a positional form, or never did. Although the traditional Chinese system does utilize a zero symbol, it never did incorporate all the features of a positional system, but remained essentially a multiplicative grouping system.

The one numeration system that did achieve the maximum efficiency of positional form is our own system, the **Hindu-Arabic** system. Its symbols have been traced to the Hindus of 200 B.C. They were adopted by the Arabs and eventually transmitted to Spain, where a late tenth-century version appeared, as shown here.

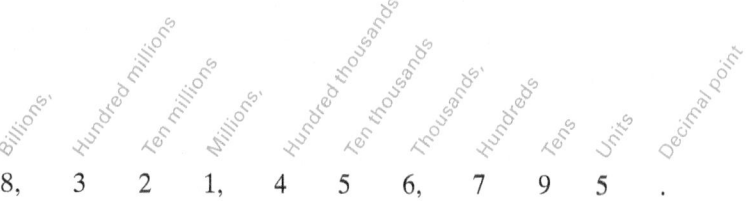

The earliest stages of the system evolved under the influence of navigational, trade, engineering, and military requirements. And in early modern times, the advance of astronomy and other sciences led to a structure well suited to fast and accurate computation.

The purely positional form that the system finally assumed was introduced to the West by Leonardo Fibonacci of Pisa (1170–1250) early in the thirteenth century, but widespread acceptance of standardized symbols and form was not achieved until the invention of printing during the fifteenth century. Since that time, no better system of numeration has been devised, and the positional base-ten Hindu-Arabic system is commonly used around the world today.

The Hindu-Arabic system and notation will be investigated further later in this chapter. The systems we consider next, the Babylonian and the Mayan, achieved the main ideas of positional numeration without fully developing them.

Babylonian Numeration

The Babylonians used a base of sixty in their system. Because of this, in theory they would then need distinct symbols for numbers from 1 through 59 (just as we have symbols for 1 through 9). However, the Babylonian method of writing on clay with wedge-shaped sticks gave rise to only *two* symbols, as shown in **Table 5.** The number 47 would be written

Table 5

Babylonian Symbols

Number	Symbol
1	▼
10	‹

$$\text{‹‹‹‹▼▼▼▼▼▼▼} \quad \text{or} \quad \text{‹‹▼▼▼▼} \\ \text{‹‹▼▼▼}$$ The number 47

Since the Babylonian system had base sixty, the "digit" on the right in a multi-digit number represented the number of 1s, with the second "digit" from the right giving the number of 60s. The third digit would give the number of 3600s $(60 \cdot 60 = 3600)$, and so on.

FEATURES OF THE BABYLONIAN SYSTEM

1. Rather than using distinct symbols for each number less than the base (sixty), the Babylonians expressed face values in base-ten simple grouping, using only the two symbols

   ‹ for 10   and   ▼ for 1.

   The system is, therefore, base-ten simple grouping *within* base-sixty positional.

2. The earliest Babylonian system lacked a placeholder symbol (zero), so missing powers of the base were difficult to express. Blank spaces within a numeral would be subject to misinterpretation.

EXAMPLE 1   Converting Babylonian Numerals to Hindu-Arabic

Convert each Babylonian numeral to Hindu-Arabic form.

(a) ‹‹‹‹‹▼▼▼   **(b)** ‹‹‹▼▼▼▼ ‹‹▼▼   **(c)** ‹‹▼▼▼▼▼‹▼‹‹‹▼▼▼▼▼▼
                          ‹‹ ▼▼▼▼ ‹‹▼

Solution

(a) Here we have five 10s and three 1s.

$$5 \cdot 10 = 50$$
$$3 \cdot 1 = \underline{3}$$
$$\ 53 \ \leftarrow \text{Answer}$$

(b) This "two-digit" Babylonian number represents fifty-eight 60s and twenty-two 1s.

$$58 \cdot 60 = 3480$$
$$22 \cdot 1 = \underline{22}$$
$$\ 3502 \ \leftarrow \text{Answer}$$

(c) Here we have a three-digit number.

$$25 \cdot 3600 = 90,000$$
$$11 \cdot 60 = 660$$
$$36 \cdot 1 = \underline{36}$$
$$\ 90,696 \ \leftarrow \text{Answer}$$

EXAMPLE 2 Converting Hindu-Arabic Numerals to Babylonian

Convert each Hindu-Arabic numeral to Babylonian form.

(a) 733   **(b)** 75,904   **(c)** 43,233

Solution

(a) To write 733 in Babylonian, we will need some 60s and some 1s. Divide 733 by 60. The quotient is 12, with a remainder of 13. Thus we need twelve 60s and thirteen 1s.

⟨TT⟨TTT ← 733

(b) For 75,904, we need some 3600s, as well as some 60s and some 1s. Divide 75,904 by 3600. The quotient is 21, with a remainder of 304. Divide 304 by 60. The quotient is 5, with a remainder of 4.

⟨⟨T TTTTT TTTT ← 75,904

(c) Divide 43,233 by 3600. The quotient is 12, with a remainder of 33. We need no 60s here. In a system such as ours we would use a 0 to show that no 60s are needed. Since the early Babylonians had no such symbol, they merely left a space.

⟨TT ⟨⟨⟨TTT ← 43,233

Example 2(c) illustrates the problem presented by the lack of a symbol for zero. In our system we know that 202 is not the same as 2002 or 20,002. The lack of a zero symbol was a major difficulty with the very early Babylonian system. A symbol for zero was introduced about 300 B.C.

Mayan Numeration

Table 6
Mayan Symbols

Number	Symbol
0	◎
1	·
5	—

The Mayan Indians of Central America and Mexico also used what is essentially a positional system. Like the Babylonians, the Mayans did not use base ten—they used base twenty, with a twist. In a true base-twenty system, the digits would represent 1s, 20s, $20 \cdot 20 = 400$s, $20 \cdot 400 = 8000$s, and so on. The Mayans used 1s, 20s, $18 \cdot 20 = 360$s, $20 \cdot 360 = 7200$s, and so on. It is possible that they multiplied 20 by 18 (instead of 20) because $18 \cdot 20$ is close to the number of days in a year, convenient for astronomy. The symbols of the Mayan system are shown in **Table 6.**

The Mayans were one of the first civilizations to invent a placeholder. They had a zero symbol many hundreds of years before it reached western Europe. Mayan numerals are written from top to bottom, just as in the classical Chinese system.

FEATURES OF THE MAYAN SYSTEM

1. Rather than using distinct symbols for each number less than the base (twenty), the Mayans expressed face values in base-five simple grouping, using only the two symbols — for 5 and · for 1. The system is, therefore, base-five simple grouping *within* base-twenty positional.

2. Place values in base-twenty would normally be

$$1, \quad 20, \quad 20^2 = 400, \quad 20^3 = 8000, \quad 20^4 = 160,000, \quad \text{and so on.}$$

However, the Mayans multiplied by 18 rather than 20 in just one case, so the place values they used are

$$1, \quad 20, \quad 20 \cdot 18 = 360, \quad 360 \cdot 20 = 7200, \quad 7200 \cdot 20 = 144,000, \quad \text{and so on.}$$

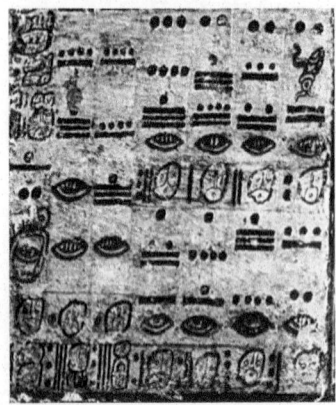

The **Dresden Codex,** written between the twelfth and fourteenth centuries, is the oldest known book from the Americas. It contains seventy-eight "pages" of hieroglyphs, including the symbols for the Mayan numeration system. This codex has been instrumental in deciphering Mayan glyphs, and it contains both astronomical and astrological tables providing insights into Mayan ceremonial practices.

   The Mayan calendar has proved to be impressively precise. Fortunately, apocalyptic predictions citing it haven't panned out. Before the end of 2012, some doomsday prognosticators claimed the coincidence of December 21, 2012 with the end of the thirteenth *baktun* (a 144,000-day period in the Mayan calendar) would mean the "end of the world."

EXAMPLE 3   Converting Mayan Numerals to Hindu-Arabic

Convert each Mayan numeral to Hindu-Arabic form.

(a)   (b)

Solution

(a) The top group of symbols represents two 360s, the middle group represents nine 20s, and the bottom group represents eleven 1s.

$$\begin{aligned} \cdots &\to 2 \cdot 360 = 720 \\ &\to 9 \cdot 20 = 180 \\ &\to 11 \cdot 1 = \underline{11} \\ & 911 \leftarrow \text{Answer} \end{aligned}$$

(b)

$$\begin{aligned} &\to 6 \cdot 360 = 2160 \\ &\to 0 \cdot 20 = 0 \\ &\to 13 \cdot 1 = \underline{13} \\ & 2173 \leftarrow \text{Answer} \end{aligned}$$

EXAMPLE 4   Converting Hindu-Arabic Numerals to Mayan

Convert each Hindu-Arabic numeral to Mayan form.

(a) 277   (b) 1238

Solution

(a) The number 277 requires thirteen 20s (divide 277 by 20) and seventeen 1s.

$$\left. \begin{aligned} 13 \cdot 20 &\to \\ 17 \cdot 1 &\to \end{aligned} \right\} 277$$

(b) Divide 1238 by 360. The quotient is 3, with remainder 158. Divide 158 by 20. The quotient is 7, with remainder 18. Thus we need three 360s, seven 20s, and eighteen 1s.

$$\left. \begin{aligned} 3 \cdot 360 &\to \\ 7 \cdot 20 &\to \\ 18 \cdot 1 &\to \end{aligned} \right\} 1238$$

Greek Numeration

The classical Greeks of Ionia assigned values to the 24 letters of their ordinary alphabet, together with three obsolete Phoenician letters (the digamma ς for 6, the koppa ϙ for 90, and the sampi λ for 900). See **Table 7** on the next page. This scheme, usually called a **ciphered system,** makes all counting numbers less than 1000 easily represented. It avoids repetitions of symbols but requires vast multiplication tables for 27 distinct symbols. Computation would be very burdensome. The base is ten, but the system is quite different from simple grouping, multiplicative grouping, or positional.

Table 7 Greek Symbols

Number	Symbol
1	α
2	β
3	γ
4	δ
5	ϵ
6	ς
7	ζ
8	η
9	θ
10	ι
20	κ
30	λ
40	μ
50	ν
60	ξ
70	o
80	π
90	q
100	ρ
200	σ
300	τ
400	υ
500	ϕ
600	χ
700	ψ
800	ω
900	λ

EXAMPLE 5 **Converting Greek Numerals to Hindu-Arabic**

Convert each Greek numeral to Hindu-Arabic form.

(a) $\lambda\alpha$ (b) $\tau\xi\epsilon$ (c) $\lambda\mathsf{q}\theta$ (d) $\chi\delta$

Solution

(a) 31 (b) 365 (c) 999 (d) 604

For numbers larger than 999, the Greeks introduced two additional techniques.

FEATURES OF THE GREEK SYSTEM

1. Multiples of 1000 (up to 9000) are indicated with a small stroke next to a units symbol. For example, 9000 would be denoted $\prime\theta$.

2. Multiples of 10,000 are indicated by the letter M (from the word *myriad*, meaning ten thousand) with the multiple (a units symbol) shown above the M. For example, the number 50,000 would be denoted $\overset{\epsilon}{\mathrm{M}}$.

EXAMPLE 6 **Converting Hindu-Arabic Numerals to Greek**

Convert each Hindu-Arabic numeral to Greek form.

(a) 3000 (b) 40,000 (c) 7692 (d) 88,888

Solution

(a) $\prime\gamma$ (b) $\overset{\delta}{\mathrm{M}}$ (c) $\prime\zeta\chi\mathsf{q}\beta$ (d) $\overset{\eta}{\mathrm{M}}\prime\eta\omega\pi\eta$

4.2 EXERCISES

CONCEPT CHECK *Choose all that apply.*

A. Hindu-Arabic   B. Babylonian
C. Mayan   D. Greek

1. This numeration system is positional.

2. This is a base-ten numeration system.

3. This numeration system features a symbol for zero.

4. Of the numeration systems covered in this section, this one utilizes the fewest symbols.

5. This system uses base-five simple grouping within a base-twenty positional structure.

6. This system is neither a grouping system nor a positional system.

Identify each numeral as Babylonian, Mayan, or Greek. Give the equivalent in the Hindu-Arabic system.

7. ≛

8. ⟨⟨▼▼

9. ⟨⟨⟨▼▼

10. ≝

11. $\sigma\lambda\delta$

12. $\omega o\beta$

13. ⟨⟨▼▼ ⟨⟨▼ ⟨⟨▼ ⟨⟨

14. ⟨⟨⟨▼▼▼ ⟨⟨▼ ⟨⟨ ▼▼ ⟨⟨⟨▼

15. $\overset{\alpha}{\mathrm{M}}\prime\epsilon\rho\mu\theta$

16. $\overset{\eta}{\mathrm{M}}\omega\eta$

17. ⠇

18. ≛

19. $\overset{\cdots}{\underset{\underset{\cdots\cdot}{\cdots\cdot}}{\cdot}}$

20. $\overset{\doubleunderline{\cdots}}{\underset{\equiv}{\underset{\equiv}{\cdots}}}$

Write each number as a Mayan numeral.

35. 12   **36.** 32   **37.** 151

38. 208   **39.** 4694   **40.** 4328

21. $\overset{\theta}{\mathrm{M}} , \theta \lambda \, \overset{?}{} \theta$ **22.** $\overset{\beta}{\mathrm{M}} \kappa$

41. ⟨ 𝖸𝖸𝖸𝖸 ⟨⟨ 𝖸𝖸𝖸𝖸 ⟨⟨ 𝖸 / 𝖸𝖸𝖸 ⟨⟨ ⟨ 𝖸𝖸𝖸𝖸 ⟨⟨ 𝖸

42. ⟨ 𝖸𝖸𝖸𝖸 𝖸𝖸𝖸 ⟨ 𝖸𝖸𝖸𝖸 / 𝖸𝖸𝖸 𝖸𝖸𝖸 ⟨ 𝖸𝖸𝖸𝖸

23. ⟨⟨ 𝖸𝖸 ⟨⟨ 𝖸 ⟨ 𝖸𝖸𝖸𝖸

43. $\tau \nu \theta$ **44.** $\tau \pi \alpha$

24. ⟨⟨ 𝖸𝖸 ⟨⟨ 𝖸𝖸𝖸𝖸 ⟨⟨ 𝖸𝖸𝖸 / ⟨⟨ 𝖸𝖸 ⟨ 𝖸𝖸𝖸𝖸 ⟨⟨ 𝖸𝖸𝖸

Write each number as a Greek numeral.

45. 39   **46.** 51

Write each number as a Babylonian numeral.

47. 412   **48.** 381

25. 21   **26.** 32   **27.** 293

49. 2769   **50.** 9814

28. 412   **29.** 5190   **30.** 43,205

51. 54,726   **52.** 80,102

31. $\overset{\cdot\cdot}{\underset{\bigcirc}{\cdots\cdot}}$ **32.** $\overset{\cdot}{\underset{\cdot}{\bigcirc}}$

53. ⟨ 𝖸 ⟨⟨ 𝖸   **54.** 𝖸 ⟨ 𝖸 ⟨ 𝖸𝖸

55. $\overset{\cdot\cdot}{\underset{\cdots}{\cdots}}$ **56.** $\overset{\cdots}{\underset{\cdots}{\overset{\cdot\cdot}{\bigcirc}}}$

33. $\chi \kappa \epsilon$ **34.** $\overset{\beta}{\mathrm{M}}$

4.3   **ARITHMETIC IN THE HINDU-ARABIC SYSTEM**

OBJECTIVES

1 Express Hindu-Arabic numerals in expanded form.

2 Explore historical calculation devices.

This Iranian stamp should remind us that counting on fingers (and toes) is an age-old practice. In fact, our word **digit,** referring to the numerals 0–9, comes from a Latin word for "finger" (or "toe"). Aristotle first noted the relationships between fingers and base ten in Greek numeration. Anthropologists go along with the notion. Some cultures, however, have used two, three, or four as number bases, for example, counting on the joints of the fingers or the spaces between them.

Expanded Form

The historical development of numeration culminated in positional systems. The most successful of these is the Hindu-Arabic system, which has base ten and, therefore, has place values that are powers of 10. Exponential expressions, or powers, are the basis of expanded form in a positional system.

EXAMPLE 1   **Evaluating Powers**

Find each power.

(a) 10^3 **(b)** 7^2 **(c)** 2^8

Solution

(a) $10^3 = 10 \cdot 10 \cdot 10 = 1000$
(10^3 is read "10 cubed," or "10 to the third power.")

(b) $7^2 = 7 \cdot 7 = 49$
(7^2 is read "7 squared," or "7 to the second power.")

(c) $2^8 = 2 \cdot 2 \cdot 2 \cdot 2 \cdot 2 \cdot 2 \cdot 2 \cdot 2 = 256$
(2^8 is read "2 to the eighth power.")

To simplify work with exponents, it is agreed that

$$a^0 = 1, \quad \text{for any nonzero number } a.$$

Thus, $7^0 = 1$, $52^0 = 1$, and so on. At the same time,

$$a^1 = a, \quad \text{for any number } a.$$

For example, $8^1 = 8$, and $25^1 = 25$. The exponent 1 is usually omitted.

By using exponents, we can write numbers in **expanded form** in which the value of the digit in each position is made clear. For example,

$$924 = 900 + 20 + 4$$
$$= (9 \cdot 100) + (2 \cdot 10) + (4 \cdot 1)$$
$$= (9 \cdot 10^2) + (2 \cdot 10^1) + (4 \cdot 10^0). \quad \text{100 = 10}^2\text{, 10 = 10}^1\text{, and 1 = 10}^0$$

EXAMPLE 2   Writing Numbers in Expanded Form

Write each number in expanded form.

(a) 1906   **(b)** 46,424

Solution

(a) $1906 = (1 \cdot 10^3) + (9 \cdot 10^2) + (0 \cdot 10^1) + (6 \cdot 10^0)$

Because $0 \cdot 10^1 = 0$, this term could be omitted, but the form is clearer with it included.

(b) $46{,}424 = (4 \cdot 10^4) + (6 \cdot 10^3) + (4 \cdot 10^2) + (2 \cdot 10^1) + (4 \cdot 10^0)$

EXAMPLE 3   Simplifying Expanded Numbers

Simplify each expansion.

(a) $(3 \cdot 10^5) + (2 \cdot 10^4) + (6 \cdot 10^3) + (8 \cdot 10^2) + (7 \cdot 10^1) + (9 \cdot 10^0)$

(b) $(2 \cdot 10^1) + (8 \cdot 10^0)$

Solution

(a) $(3 \cdot 10^5) + (2 \cdot 10^4) + (6 \cdot 10^3) + (8 \cdot 10^2) + (7 \cdot 10^1) + (9 \cdot 10^0) = 326{,}879$

(b) $(2 \cdot 10^1) + (8 \cdot 10^0) = 28$

Expanded notation and the **distributive property** can be used to see why standard algorithms for addition and subtraction really work.

DISTRIBUTIVE PROPERTY

For all real numbers a, b, and c,

$$(b \cdot a) + (c \cdot a) = (b + c) \cdot a.$$

Example:
$$(3 \cdot 10^4) + (2 \cdot 10^4) = (3 + 2) \cdot 10^4$$
$$= 5 \cdot 10^4$$

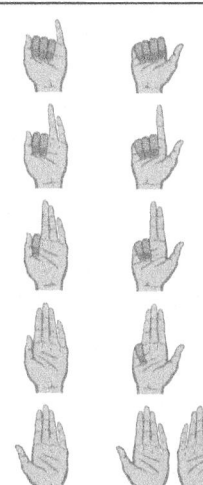

Finger Counting The first digits many people used for counting were their fingers. In Africa the Zulu used the method shown here to count to ten. They started on the left hand with palm up and fist closed. The Zulu finger positions for 1–5 are shown above on the left. The Zulu finger positions for 6–10 are shown on the right.

EXAMPLE 4   Adding Expanded Forms

Use expanded notation to add 23 and 64.

Solution

$$\begin{array}{r} 23 = (2 \cdot 10^1) + (3 \cdot 10^0) \\ + \ 64 = (6 \cdot 10^1) + (4 \cdot 10^0) \\ \hline (8 \cdot 10^1) + (7 \cdot 10^0) = 87 \quad \text{Sum} \end{array}$$

EXAMPLE 5 Subtracting Expanded Forms

Use expanded notation to subtract 254 from 695.

Solution

$$695 = (6 \cdot 10^2) + (9 \cdot 10^1) + (5 \cdot 10^0)$$
$$\underline{-254 = (2 \cdot 10^2) + (5 \cdot 10^1) + (4 \cdot 10^0)}$$
$$(4 \cdot 10^2) + (4 \cdot 10^1) + (1 \cdot 10^0) = 441 \quad \text{Difference}$$

EXAMPLE 6 Carrying in Expanded Form

Use expanded notation to add 75 and 48.

Solution

$$75 = (7 \cdot 10^1) + (5 \cdot 10^0)$$
$$\underline{+48 = (4 \cdot 10^1) + (8 \cdot 10^0)}$$
$$(11 \cdot 10^1) + (13 \cdot 10^0)$$

The units position (10^0) has room for only one digit, so we modify $13 \cdot 10^0$.

$$13 \cdot 10^0 = (10 \cdot 10^0) + (3 \cdot 10^0) \quad \text{Distributive property}$$
$$= (1 \cdot 10^1) + (3 \cdot 10^0) \quad 10 \cdot 10^0 = 1 \cdot 10^1$$

The 1 from 13 moved to the left (carried) from the units position to the tens position.

$$\overbrace{13 \cdot 10^0}$$
$$(11 \cdot 10^1) + \overbrace{(1 \cdot 10^1) + (3 \cdot 10^0)}$$

$$= (12 \cdot 10^1) + (3 \cdot 10^0) \quad \text{Distributive property}$$
$$= (10 \cdot 10^1) + (2 \cdot 10^1) + (3 \cdot 10^0) \quad \text{Modify } 12 \cdot 10^1.$$
$$= (1 \cdot 10^2) + (2 \cdot 10^1) + (3 \cdot 10^0) \quad 10 \cdot 10^1 = 1 \cdot 10^2$$
$$= 123 \quad \text{Sum}$$

EXAMPLE 7 Borrowing in Expanded Form

Use expanded notation to subtract 186 from 364.

Solution

$$364 = (3 \cdot 10^2) + (6 \cdot 10^1) + (4 \cdot 10^0)$$
$$\underline{-186 = (1 \cdot 10^2) + (8 \cdot 10^1) + (6 \cdot 10^0)}$$

We cannot subtract 6 from 4. The units position borrows from the tens position.

$$(3 \cdot 10^2) + (6 \cdot 10^1) + (4 \cdot 10^0)$$
$$= (3 \cdot 10^2) + \overbrace{(5 \cdot 10^1) + (1 \cdot 10^1)} + (4 \cdot 10^0) \quad \text{Distributive property}$$
$$= (3 \cdot 10^2) + (5 \cdot 10^1) + \underbrace{(10 \cdot 10^0) + (4 \cdot 10^0)} \quad 1 \cdot 10^1 = 10 \cdot 10^0$$
$$= (3 \cdot 10^2) + (5 \cdot 10^1) + (14 \cdot 10^0) \quad \text{Distributive property}$$

Sun	Mon	Tues	Wed
			1
5	6	7	8

December was the tenth month in an old form of the calendar. It is interesting to note that *decem* became *dix* in the French language; a ten-dollar bill, called a "dixie," was in use in New Orleans before the Civil War. "Dixie Land" was a nickname for that city before Dixie came to refer to all the Southern states, as in Daniel D. Emmett's song, written in 1859.

We cannot take 8 from 5 in the tens position, so we borrow from the hundreds.

$$(3 \cdot 10^2) + (5 \cdot 10^1) + (14 \cdot 10^0)$$

$$= \overline{(2 \cdot 10^2) + (1 \cdot 10^2)} + (5 \cdot 10^1) + (14 \cdot 10^0) \quad \text{Distributive property}$$

$$= (2 \cdot 10^2) + \underbrace{(10 \cdot 10^1) + (5 \cdot 10^1)} + (14 \cdot 10^0) \quad 1 \cdot 10^2 = 10 \cdot 10^1$$

$$= (2 \cdot 10^2) + (15 \cdot 10^1) + (14 \cdot 10^0) \quad \text{Distributive property}$$

Now we can complete the subtraction.

$$\begin{array}{r}
(2 \cdot 10^2) + (15 \cdot 10^1) + (14 \cdot 10^0) \\
- (1 \cdot 10^2) + (8 \cdot 10^1) + (6 \cdot 10^0) \\
\hline
(1 \cdot 10^2) + (7 \cdot 10^1) + (8 \cdot 10^0) = 178 \quad \text{Difference}
\end{array}$$

Examples 4–7 used expanded notation and the distributive property to clarify our usual addition and subtraction methods. In practice, our actual work for these four problems would appear as follows.

$$\begin{array}{cccc}
 & & & \overset{\text{1}}{} \quad \overset{2\ 15}{} \\
23 & 695 & 75 & 3\,{6}^{1}4 \\
+64 & -254 & +48 & -186 \\
\hline
87 & 441 & 123 & 178
\end{array}$$

These procedures seen in this section also work for positional systems with bases other than ten.

Historical Calculation Devices

Because our numeration system is based on powers of ten, it is often called the **decimal system,** from the Latin word *decem,* meaning ten. Over the years, many methods have been devised for speeding calculations in the decimal system.

One of the oldest calculation methods uses the **abacus,** a device made with a series of rods with sliding beads and a dividing bar. Reading from right to left, the rods have values of 1, 10, 100, 1000, and so on. The bead above the bar has five times the value of those below. Beads moved *toward* the bar are in the "active" position, and those toward the frame are ignored. In our illustrations of *abaci* (plural form of *abacus*), such as in **Figure 8,** the activated beads are shown in black.

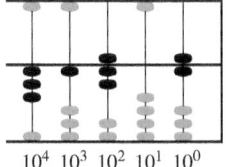

$$10^4 \quad 10^3 \quad 10^2 \quad 10^1 \quad 10^0$$

Figure 8

The **speed and accuracy of the abacus** are well known, according to http://www.ucmasusa.com. In a contest held between a Japanese **soroban** (the Japanese version of the abacus) expert and a highly skilled desk-calculator operator, the abacus won on addition, subtraction, division, and combinations of these operations. The electronic calculator won only on multiplication.

EXAMPLE 8   Reading an Abacus

What number is shown on the abacus in **Figure 8?**

Solution

Find the number as follows.

> Beads above the bar have five times the value.

$$(3 \cdot 10,000) + (1 \cdot 1000) + [(1 \cdot 500) + (2 \cdot 100)] + 0 \cdot 10 + [(1 \cdot 5) + (1 \cdot 1)]$$

$$= 30,000 + 1000 + (500 + 200) + 0 + (5 + 1)$$

$$= 31,706$$

As paper became more readily available, people switched to paper-and-pencil methods of calculation. One early scheme used in India and Persia was the **lattice method,** which arranged products of single digits into a diagonalized lattice.

John Napier's most significant mathematical contribution, developed over a period of at least 20 years, was the concept of **logarithms,** which, among other things, allow multiplication and division to be accomplished with addition and subtraction. It was a great computational advantage given the state of mathematics at the time (1614).

Napier, a supporter of John Knox and James I, published a widely read anti-Catholic work that analyzed the biblical book of Revelation. He concluded that the Pope was the Antichrist and that the Creator would end the world between 1688 and 1700. Napier is one of many who, over the years, have miscalculated the end of the world.

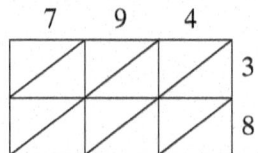 **Using the Lattice Method for Products**

Find the product 38 · 794 by the lattice method.

Solution

Step 1 Write the problem, with one number at the side and one across the top.

Step 2 Within the lattice, write the products of all pairs of digits from the top and side.

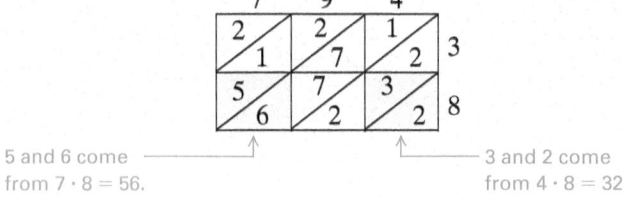

5 and 6 come from 7 · 8 = 56.   3 and 2 come from 4 · 8 = 32.

Step 3 Starting at the right of the lattice, add diagonally, carrying as necessary.

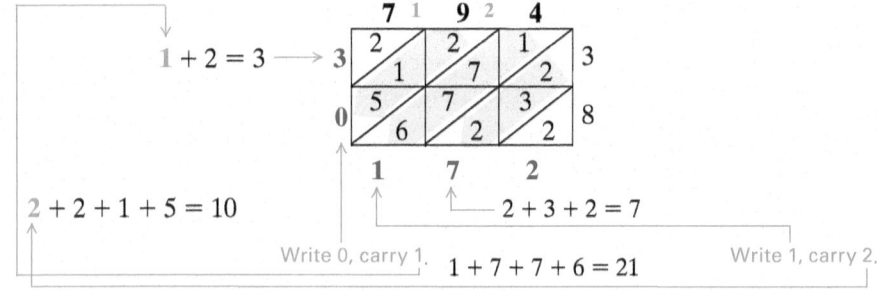

$1 + 2 = 3$

$2 + 2 + 1 + 5 = 10$ $2 + 3 + 2 = 7$

Write 0, carry 1. $1 + 7 + 7 + 6 = 21$ Write 1, carry 2.

Step 4 Read the answer around the left side and bottom: 38 · 794 = **30,172.**

The Scottish mathematician John Napier (1550–1617) introduced a significant calculating tool called **Napier's rods,** or **Napier's bones.** Napier's invention, based on the lattice method of multiplication, is widely acknowledged as a very early forerunner of modern computers. It consisted of a set of strips, several for each digit 0 through 9, on which multiples of each digit appeared in a sort of lattice column. See **Figure 9.**

An additional strip, the *index,* is laid beside any of the others to indicate the multiplier at each level. **Figure 10** shows that 7 × 2806 = **19,642**

Napier's rods were an early step toward modern computers.

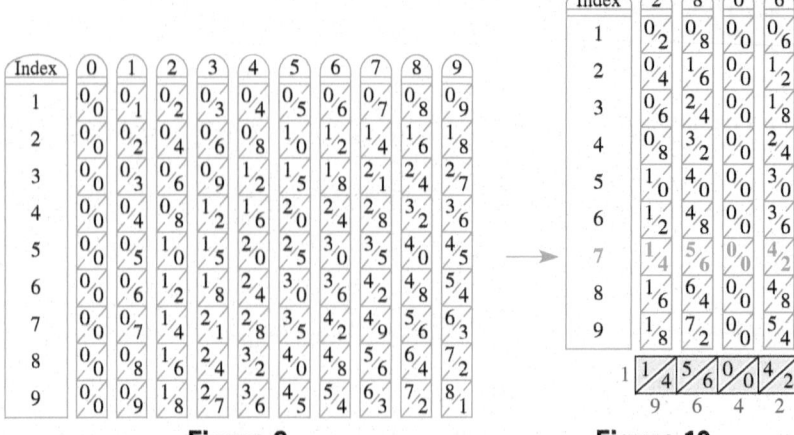

Figure 9   **Figure 10**

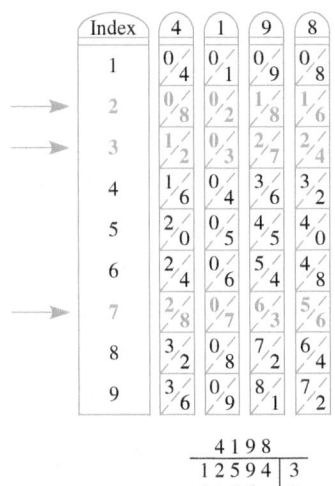

Index	4	1	9	8
1	0/4	0/1	0/9	0/8
2	0/8	0/2	1/8	1/6
3	1/2	0/3	2/7	2/4
4	1/6	0/4	3/6	3/2
5	2/0	0/5	4/5	4/0
6	2/4	0/6	5/4	4/8
7	2/8	0/7	6/3	5/6
8	3/2	0/8	7/2	6/4
9	3/6	0/9	8/1	7/2

```
      4 1 9 8
    1 2 5 9 4  3
      8 3 9 6  2
    2 9 3 8 6  7
    3 0 3 5 1 5 4
```

Figure 11

EXAMPLE 10 Multiplying with Napier's Rods

Use Napier's rods to find the product of 723 and 4198.

Solution

We line up the rods for 4, 1, 9, and 8 next to the index, as in **Figure 11.** The product $3 \cdot 4198$ is found as described in **Example 9** and written at the bottom of the figure. Then $2 \cdot 4198$ is found similarly and written below, shifted one place to the left. (Why?) Finally, the product $7 \cdot 4198$ is written shifted two places to the left.

   The final answer is found by addition.

$$723 \cdot 4198 = 3{,}035{,}154$$

Another paper-and-pencil method of multiplication is the **Russian peasant method**, which is similar to the Egyptian method of doubling. To multiply 37 and 42 by the Russian peasant method, make two columns headed by 37 and 42. Form the first column by dividing 37 by 2 again and again, ignoring any remainders. Stop when 1 is obtained. Form the second column by doubling each number down the column.

Divide by 2, ignoring remainders.	37	42	Double each number.
	18	84	
	9	168	
	4	336	
	2	672	
	1	1344	

Now add up only the second-column numbers that correspond to odd numbers in the first column. Omit those corresponding to even numbers in the first column.

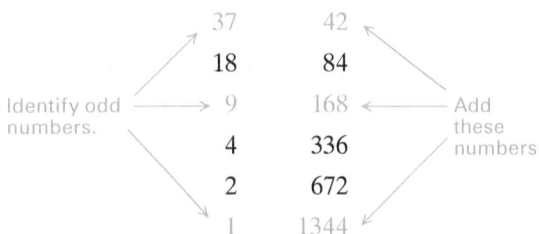

$$37 \cdot 42 = 42 + 168 + 1344 = \mathbf{1554} \leftarrow \text{Answer}$$

Most people use standard algorithms for adding and subtracting, carrying or borrowing when appropriate, as illustrated following **Example 7.** An interesting alternative is the **nines complement method** for subtracting. To use this method, we first agree that

the nines complement of a digit n is $9 - n$.

For example, the nines complement of 0 is 9, of 1 is 8, of 2 is 7, and so on, down to the nines complement of 9, which is 0.

   To carry out the nines complement method, complete the following steps:

Step 1 Align the digits as in the standard subtraction algorithm.

Step 2 Add leading zeros, if necessary, in the subtrahend so that both numbers have the same number of digits.

Step 3 Replace each digit in the subtrahend with its nines complement, and then add.

Step 4 Delete the leading digit (1), and add 1 to the remaining part of the sum.

For a way to include a little magic with your calculations, check out http://digicc.com/fido.

EXAMPLE 11   **Using the Nines Complement Method**

Use the nines complement method to subtract $2803 - 647$.

Solution

	Step 1	Step 2	Step 3	Step 4
	2803	2803	2803	2155
	$-\;\;647$	$-\;0647$	$+\;9352$	$+\;\;\;1$
			12,155	2156 Difference

4.3   EXERCISES

CONCEPT CHECK *Write* true *or* false *for each statement.*

1. In the Hindu-Arabic system, place values are all powers of 10.

2. The associative property for addition is used in adding and subtracting numbers in expanded form.

3. Each bead above the bar on the third rod (from the right) of an abacus has a value of 100.

4. In a base-ten positional system, the nines complement of 3 is $10 - 3 = 7$.

Write each number in expanded form.

5. 73   **6.** 265   **7.** 8335   **8.** 12,398

9. three thousand, six hundred twenty-four

10. sixty-two thousand, two hundred eighteen

11. fourteen million, two hundred six thousand, forty

12. two hundred twelve million, eleven thousand, nine hundred sixteen

Simplify each expansion.

13. $(7 \cdot 10^1) + (5 \cdot 10^0)$

14. $(8 \cdot 10^2) + (2 \cdot 10^1) + (0 \cdot 10^0)$

15. $(4 \cdot 10^3) + (3 \cdot 10^2) + (8 \cdot 10^1) + (0 \cdot 10^0)$

16. $(5 \cdot 10^5) + (0 \cdot 10^4) + (3 \cdot 10^3) + (5 \cdot 10^2) + (6 \cdot 10^1) + (8 \cdot 10^0)$

17. $(7 \cdot 10^7) + (4 \cdot 10^5) + (1 \cdot 10^3) + (9 \cdot 10^0)$

18. $(3 \cdot 10^8) + (8 \cdot 10^6) + (2 \cdot 10^4) + (3 \cdot 10^0)$

Perform each computation using expanded notation.

19. $37 + 42$ **20.** $732 + 417$

21. $85 - 42$ **22.** $935 - 534$

23. $64 + 45$ **24.** $663 + 272$

25. $434 + 299$ **26.** $6755 + 4827$

27. $54 - 48$ **28.** $383 - 78$

29. $855 - 649$ **30.** $816 - 335$

Perform each subtraction using the nines complement method.

31. $273 - 31$ **32.** $736 - 625$

33. $50,000 - 199$ **34.** $40,002 - 4846$

Identify the number represented on each abacus.

35. 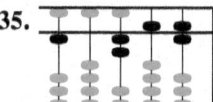   **36.**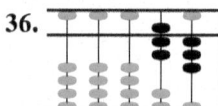

37. 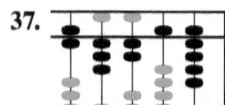   **38.**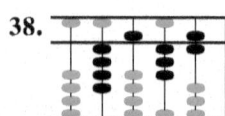

Sketch an abacus to show each number.

39. 38   **40.** 183

41. 2547   **42.** 70,163

Use the lattice method to find each product.

43. $65 \cdot 29$ **44.** $32 \cdot 741$

45. $225 \cdot 73$ **46.** $912 \cdot 483$

*Refer to **Example 10** where Napier's rods were used. Then complete each problem.*

47. Find the product of 723 and 4198 by completing the lattice process shown here.

48. Explain how Napier's rods could have been used in **Example 10** to set up one complete lattice product rather than adding three individual (shifted) lattice products. Illustrate with a sketch.

*Use Napier's rods (**Figure 9**) to find each product.*

49. 8 · 62

50. 32 · 73

51. 26 · 8354

52. 526 · 4863

Use the Russian peasant method to find each product.

53. 5 · 92

54. 41 · 53

55. 62 · 529

56. 145 · 63

The Hindu-Arabic system is positional and uses ten as the base. Describe any advantages or disadvantages that may have resulted in each case.

57. Suppose the base had been greater, say twelve or twenty.

58. Suppose the base had been less, maybe five or eight.

Write a short answer for each problem.

59. If a method similar to the nines complement method for subtraction were used in a base-five positional system, what might the method be called? Explain.

60. Suppose you are subtracting in a base-three system. Describe the steps involved in twos complement subtraction.

61. If a base-one positional system existed, how many symbols would it use? Discuss whether such a system would be useful.

62. Describe how you would construct an abacus to perform calculations in a base-eight positional system. Explain your design.

FOR FURTHER THOUGHT

Calculating on the Abacus

The abacus has been (and still is) used to perform rapid calculations. Add 526 and 362 as shown.

Start with 526 on the abacus.

To add 362, start by "activating" an additional 2 on the 1s rod.

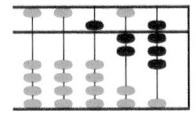

Next, activate an additional 6 on the 10s rod.

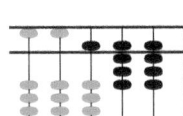

Finally, activate an additional 3 on the 100s rod.

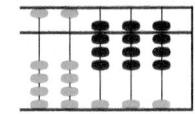

The sum, read from the abacus, is 888.
 For problems where carrying or borrowing is required, it takes a little more thought and skill.

For Group or Individual Investigation

1. Use an abacus to add: 13,728 + 61,455. Explain each step of your procedure.

2. Use an abacus to subtract: 6512 − 4816. Explain each step of your procedure.

4.4   CONVERSION BETWEEN NUMBER BASES

OBJECTIVES

1 Convert numerals from one base to another.

2 Use different bases in the context of computer mathematics.

General Base Conversions

In this section we consider bases other than ten, but we use the familiar Hindu-Arabic symbols. We indicate bases other than ten with a spelled-out subscript, as in the numeral 43_{five}. ***Whenever a number appears without a subscript, it is assumed that the intended base is ten.*** Be careful how you read (or verbalize) numerals here. The numeral 43_{five} is read "four three base five." (Do *not* read it as "forty-three," as that terminology implies base ten and names a totally different number.)

Powers of some numbers used as alternative bases are given in **Table 8**.

Table 8 Selected Powers of Some Alternative Number Bases

	Fourth Power	**Third Power**	**Second Power**	**First Power**	**Zero Power**
Base Two	16	8	4	2	1
Base Five	625	125	25	5	1
Base Seven	2401	343	49	7	1
Base Eight	4096	512	64	8	1
Base Sixteen	65,536	4096	256	16	1

For example, the base-two row of **Table 8** indicates that

$$2^4 = 16, \quad 2^3 = 8, \quad 2^2 = 4, \quad 2^1 = 2, \quad \text{and} \quad 2^0 = 1.$$

We begin with the base-five system, which requires just five distinct symbols: 0, 1, 2, 3, and 4. **Table 9** compares base-five and decimal (base-ten) numerals for the whole numbers 0 through 31. Notice that because only the symbols 0, 1, 2, 3, and 4 are used in base five, we must use two digits in base five when we get to 5_{ten}.

5_{ten} is expressed as one 5 and no 1s—that is, as 10_{five}.

6_{ten} becomes 11_{five} (one 5 and one 1).

While base five uses fewer distinct symbols than base ten (an apparent advantage because there are fewer symbols to learn), it often requires more digits than base ten to express the same number (a disadvantage because more symbols must be written).

You will find that in any base, if you denote the base "b," then the base itself will be 10_b, just as occurred in base five. For example,

$$7_{\text{ten}} = 10_{\text{seven}}, \quad 16_{\text{ten}} = 10_{\text{sixteen}}, \quad \text{and so on.}$$

Table 9

Base-Five Numerals

Base Ten	Base Five	Base Ten	Base Five
0	0	16	31
1	1	17	32
2	2	18	33
3	3	19	34
4	4	20	40
5	10	21	41
6	11	22	42
7	12	23	43
8	13	24	44
9	14	25	100
10	20	26	101
11	21	27	102
12	22	28	103
13	23	29	104
14	24	30	110
15	30	31	111

EXAMPLE 1 **Converting from Base Five to Base Ten**

Convert 1342_{five} to decimal form.

Solution

Refer to the powers of five in **Table 8**.

$$1342_{\text{five}} = (1 \cdot 125) + (3 \cdot 25) + (4 \cdot 5) + (2 \cdot 1)$$

This number has one 125, three 25s, four 5s, and two 1s.

$$= 125 + 75 + 20 + 2$$

$$= 222$$

A shortcut for converting from base five to decimal form, which is *particularly useful when you use a calculator,* can be derived as follows.

$$1342_{\text{five}} = (1 \cdot 5^3) + (3 \cdot 5^2) + (4 \cdot 5) + 2$$

$$= ((1 \cdot 5^2) + (3 \cdot 5) + 4) \cdot 5 + 2$$

Factor 5 out of the three quantities in parentheses.

$$= (((1 \cdot 5) + 3) \cdot 5 + 4) \cdot 5 + 2$$

Factor 5 out of the two "inner" quantities.

The inner parentheses around $1 \cdot 5$ are not needed because the product is done automatically before the 3 is added. Therefore, we can write

$$1342_{\text{five}} = ((1 \cdot 5 + 3) \cdot 5 + 4) \cdot 5 + 2 = 222.$$

NORMAL FLOAT AUTO REAL DEGREE MP

1*5+3
                          8
Ans*5+4
                        44
Ans*5+2
                      222
(((1*5+3)*5+4)*5+2
                      222

This TI-84 Plus C screen shows two options for the calculations of **Example 1.**

This series of products and sums is easily done as an uninterrupted sequence of operations on a calculator, with no intermediate results written down. The same method works for converting to base ten from any other base.

CALCULATOR SHORTCUT FOR BASE CONVERSION

To convert from another base to decimal form, follow these steps.

Step 1 Start with the first digit on the left and multiply by the base.

Step 2 Then add the next digit, multiply again by the base, and so on.

Step 3 Add the last digit on the right. ***Do not multiply it by the base.***

EXAMPLE 2 Using the Calculator Shortcut

Use the calculator shortcut to convert 244314_{five} to decimal form.

Solution

$$244314_{\text{five}} = ((((2 \cdot 5 + 4) \cdot 5 + 4) \cdot 5 + 3) \cdot 5 + 1) \cdot 5 + 4$$

$$= 9334 \quad \boxed{\text{Note the four left parentheses for a 6-digit numeral.}}$$

EXAMPLE 3 Converting from Base Ten to Base Five

Convert 497 from decimal form to base five.

Solution

The base-five place values, starting from the right, are

$$1, \quad 5, \quad 25, \quad 125, \quad 625, \quad \text{and so on.}$$

Because 497 is between 125 and 625, it will require no 625s, but some 125s, as well as possibly some 25s, 5s, and 1s.

- Dividing 497 by 125 yields a quotient of 3, which is the proper number of 125s.

- The remainder, 122, is divided by 25 (the next place value) to find the proper number of 25s. The quotient is 4, with remainder 22, so we need four 25s.

- Dividing 22 by 5 yields 4, with remainder 2, so we need four 5s.

- Dividing 2 by 1 yields 2 (with remainder 0), so we need two 1s.

Thus 497 consists of three 125s, four 25s, four 5s, and two 1s, so $497 = 3442_{\text{five}}$.
More concisely, this process can be written as follows.

$$497 \div 125 = 3 \qquad \text{Remainder 122}$$
$$122 \div 25 = 4 \qquad \text{Remainder 22}$$
$$22 \div 5 = 4 \qquad \text{Remainder 2}$$
$$2 \div 1 = 2 \qquad \text{Remainder 0}$$
$$497 = 3442_{\text{five}}$$

CHECK $\quad 3442_{\text{five}} = (3 \cdot 125) + (4 \cdot 25) + (4 \cdot 5) + (2 \cdot 1)$

$$= 375 + 100 + 20 + 2$$

$$= 497 \quad \checkmark$$

The symbol here is the ancient Chinese **"yin-yang,"** in which the black and the white enfold each other, each containing a part of the other. A kind of duality is conveyed between destructive (yin) and beneficial (yang) aspects.

Leibniz (1646–1716) studied Chinese ideograms in search of a universal symbolic language and promoted East–West cultural contact.

Niels Bohr (1885–1962), famous Danish Nobel laureate in physics (atomic theory), adopted the yin-yang symbol in his coat of arms to depict his principle of *complementarity*, which he believed was fundamental to reality at the deepest levels. Bohr also pushed for East–West cooperation.

In its 1992 edition, *The World Book Dictionary* first judged "yin-yang" to have been used enough to become a permanent part of our ever-changing language, assigning to it the definition "made up of opposites."

Base-Two Textiles Woven fabric is a binary (base-two) system of threads going lengthwise (warp threads—tan in the diagram above) and threads going crosswise (weft or woof). At any point in a fabric, either warp or weft is on top, and the variation creates the pattern.

Nineteenth-century looms for weaving operated using punched cards "programmed" for a pattern. The looms were set up with hooked needles, the hooks holding the warp. Where there were holes in cards, the needles moved, the warp lifted, and the weft passed under. Where no holes were, the warp did not lift, and the weft was on top. The system parallels the on–off system in calculators and computers. In fact, these looms were models in the development of modern calculating machinery.

Joseph Marie Jacquard (1752–1834) is credited with improving the mechanical loom so that mass production of fabric was feasible.

The calculator shortcut for converting from another base to decimal form involved repeated *multiplications* by the other base. (See **Example 2.**) A shortcut for converting from decimal form to another base makes use of repeated *divisions* by the other base.

Just divide the original decimal numeral, and the resulting quotients in turn, by the desired base until the quotient 0 appears.

EXAMPLE 4 Using a Shortcut to Convert from Base Ten

Repeat **Example 3** using the shortcut just described.

Solution

Remainder

$$
\begin{array}{r}
5\,\lfloor 497 \\
5\,\lfloor 99 \\
5\,\lfloor 19 \\
5\,\lfloor 3 \\
0
\end{array}
\begin{array}{l}
\\
\leftarrow\ 2 \\
\leftarrow\ 4 \\
\leftarrow\ 4 \\
\leftarrow\ 3
\end{array}
$$

Read the answer from the remainder column, reading from the bottom up.

$$497 = 3442_{\text{five}}$$

To see why this shortcut works, notice the following:

- The first division shows that four hundred ninety-seven 1s are equivalent to ninety-nine 5s and two 1s. (The two 1s are set aside and account for the last digit of the answer.)

- The second division shows that ninety-nine 5s are equivalent to nineteen 25s and four 5s. (The four 5s account for the next digit of the answer.)

- The third division shows that nineteen 25s are equivalent to three 125s and four 25s. (The four 25s account for the next digit of the answer.)

- The fourth (and final) division shows that the three 125s are equivalent to no 625s and three 125s. The remainders, as they are obtained *from top to bottom,* give the number of 1s, then 5s, then 25s, then 125s.

The methods for converting between bases ten and five, including the shortcuts, can be adapted for conversions between base ten and any other base.

EXAMPLE 5 Converting from Base Seven to Base Ten

Convert 6343_{seven} to decimal form, by expanding in powers, and by using the calculator shortcut.

Solution

$$6343_{\text{seven}} = (6 \cdot 7^3) + (3 \cdot 7^2) + (4 \cdot 7^1) + (3 \cdot 7^0)$$
$$= (6 \cdot 343) + (3 \cdot 49) + (4 \cdot 7) + (3 \cdot 1)$$
$$= 2236$$

Calculator shortcut:

$$6343_{\text{seven}} = ((6 \cdot 7 + 3) \cdot 7 + 4) \cdot 7 + 3$$
$$= 2236$$

EXAMPLE 6 Converting from Base Ten to Base Seven

Convert 7508 to base seven.

Solution

Divide 7508 by 7, then divide the resulting quotient by 7, until a quotient of 0 results.

Remainder

```
7 | 7508
7 | 1072   ←——  4
  7 | 153   ←——  1
    7 | 21   ←——  6
      7 | 3   ←——  0
          0   ←——  3
```

From the remainders, reading bottom to top, $7508 = 30614_{\text{seven}}$.

To handle conversions between arbitrary bases (where neither is ten), go from the given base to base ten and then to the desired base.

EXAMPLE 7 Converting between Two Bases Other Than Ten

Convert 3164_{seven} to base five.

Solution

Convert to decimal form.

$$3164_{\text{seven}} = (3 \cdot 7^3) + (1 \cdot 7^2) + (6 \cdot 7^1) + (4 \cdot 7^0)$$
$$= (3 \cdot 343) + (1 \cdot 49) + (6 \cdot 7) + (4 \cdot 1)$$
$$= 1029 + 49 + 42 + 4$$
$$= 1124$$

Convert this decimal result to base five.

Remainder

```
5 | 1124
  5 | 224   ←——  4
    5 | 44   ←——  4
      5 | 8   ←——  4
        5 | 1   ←——  3
            0   ←——  1
```

From the remainders, reading bottom to top, $3164_{\text{seven}} = 13444_{\text{five}}$.

Computer Mathematics

There are three alternative base systems that are most useful in computer applications—**binary** (base two), **octal** (base eight), and **hexadecimal** (base sixteen).

Computers and handheld calculators use the binary system for their internal calculations because that system consists of only two symbols, 0 and 1. All numbers can then be represented by electronic "switches," where "on" indicates 1 and "off" indicates 0. The octal and hexadecimal systems have been used extensively by programmers who work with internal computer codes and for communication between the CPU (central processing unit) and a printer or other output device.

The binary system is extreme in that it has only two available symbols (0 and 1). Thus, representing numbers in binary form requires more digits than in any other base. **Table 10** shows the whole numbers up to 31 expressed in binary form.

Table 10

Base-Two Numerals

Base Ten	Base Two	Base Ten	Base Two
0	0	16	10000
1	1	17	10001
2	10	18	10010
3	11	19	10011
4	100	20	10100
5	101	21	10101
6	110	22	10110
7	111	23	10111
8	1000	24	11000
9	1001	25	11001
10	1010	26	11010
11	1011	27	11011
12	1100	28	11100
13	1101	29	11101
14	1110	30	11110
15	1111	31	11111

EXAMPLE 8 Converting from Binary to Decimal

Convert 110101_{two} to decimal form.

Solution

$$110101_{\text{two}} = (1 \cdot 2^5) + (1 \cdot 2^4) + (0 \cdot 2^3) + (1 \cdot 2^2) + (0 \cdot 2^1) + (1 \cdot 2^0)$$

$$= (1 \cdot 32) + (1 \cdot 16) + (0 \cdot 8) + (1 \cdot 4) + (0 \cdot 2) + (1 \cdot 1)$$

$$= 32 + 16 + 0 + 4 + 0 + 1$$

$$= 53$$

Calculator shortcut: $110101_{\text{two}} = ((((1 \cdot 2 + 1) \cdot 2 + 0) \cdot 2 + 1) \cdot 2 + 0) \cdot 2 + 1$

$$= 53$$

> Note the four left parentheses for a 6-digit numeral.

EXAMPLE 9 Converting from Decimal to Octal

Convert 9583 to octal form.

Solution

Divide repeatedly by 8, writing the remainders at the side.

Remainder

```
8 | 9583
8 | 1197  ←  7
8 |  149  ←  5
8 |   18  ←  5
8 |    2  ←  2
        0  ←  2
```

From the remainders, $9583 = 22557_{\text{eight}}$.

The hexadecimal system, having base sixteen (and 16 being greater than 10), presents a new problem. Because distinct symbols are needed for all whole numbers from 0 up to one less than the base, base sixteen requires more symbols than are normally used in our decimal system. Computer programmers commonly use

the letters **A, B, C, D, E, and F** as hexadecimal digits
for **the numbers ten through fifteen, respectively.**

Trick or Tree? The octal number 31 is equal to the decimal number 25. This may be written as

$$31 \text{ OCT} = 25 \text{ DEC}$$

Does this mean that Halloween and Christmas fall on the same day of the year?

EXAMPLE 10 Converting from Hexadecimal to Decimal

Convert $FA5_{\text{sixteen}}$ to decimal form.

Solution

$$\mathbf{FA5}_{\text{sixteen}} = (\mathbf{15} \cdot 16^2) + (\mathbf{10} \cdot 16^1) + (5 \cdot 16^0)$$

F and A represent 15 and 10, respectively.

$$= 3840 + 160 + 5$$

$$= 4005$$

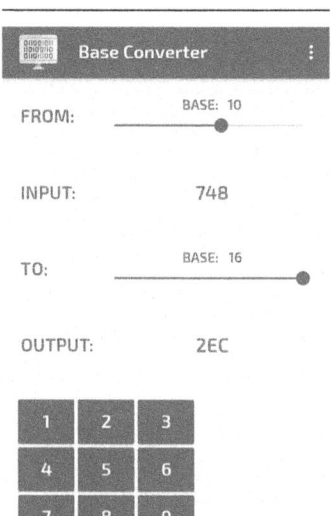

Conversion Apps There are several smartphone apps, as well as some scientific calculators, that convert among decimal, binary, octal, and hexadecimal. Instant conversions can be performed in Microsoft EXCEL or online.

EXAMPLE 11 Converting from Decimal to Hexadecimal

Convert 748 from decimal form to hexadecimal form.

Solution

Use repeated division by 16.

$$
\begin{array}{r}
\\
\text{Remainder} \quad \text{notation}\\
16\,\overline{)748}\\
16\,\overline{)46} \leftarrow \ 12 \leftarrow \ \text{C}\\
16\,\overline{)2} \leftarrow \ 14 \leftarrow \ \text{E}\\
0 \leftarrow \ 2 \leftarrow \ 2
\end{array}
$$

From the remainders at the right, $748 = 2\text{EC}_{\text{sixteen}}$.

The decimal whole numbers 0 through 17 are shown in **Table 11,** along with their equivalents in the common computer-oriented bases (two, eight, and sixteen). Because **both eight and sixteen are powers of two,** conversions among binary, octal, and hexadecimal can easily be accomplished using the shortcuts illustrated in the remaining examples.

Table 11 Some Decimal Equivalents in the Common Computer-Oriented Bases

Decimal (Base Ten)	Hexadecimal (Base Sixteen)	Octal (Base Eight)	Binary (Base Two)
0	0	0	0
1	1	1	1
2	2	2	10
3	3	3	11
4	4	4	100
5	5	5	101
6	6	6	110
7	7	7	111
8	8	10	1000
9	9	11	1001
10	A	12	1010
11	B	13	1011
12	C	14	1100
13	D	15	1101
14	E	16	1110
15	F	17	1111
16	10	20	10000
17	11	21	10001

Table 12

Octal	Binary
0	000
1	001
2	010
3	011
4	100
5	101
6	110
7	111

The binary system is the natural one for internal computer workings because of its compatibility with two-state electronic switches. It is cumbersome for human use, however, because so many digits occur even in the numerals for relatively small numbers. The octal and hexadecimal systems are the choices of computer programmers mainly because of their close relationship with the binary system.

When conversions involve one base that is a power of the other, there is a quick conversion shortcut available. For example, because $8 = 2^3$, every octal digit (0 through 7) can be expressed as a 3-digit binary numeral. See **Table 12.**

EXAMPLE 12 Converting from Octal to Binary

Convert 473_{eight} to binary form.

Solution

Replace each octal digit with its 3-digit binary equivalent. (Leading zeros can be omitted only when they occur in the leftmost group.) Combine the equivalents into a single binary numeral.

$$
\begin{array}{ccc}
4 & 7 & 3_{\text{eight}} \\
\downarrow & \downarrow & \downarrow \\
100 & 111 & 011_{\text{two}}
\end{array}
$$

$$473_{\text{eight}} = 100111011_{\text{two}}$$

EXAMPLE 13 Converting from Binary to Octal

Convert 10011110_{two} to octal form.

Solution

Start at the right and break the digits into groups of three. Then convert the groups to their octal equivalents.

Finally, $10011110_{\text{two}} = 236_{\text{eight}}.$

$$
\begin{array}{ccc}
10 & 011 & 110_{\text{two}} \\
\downarrow & \downarrow & \downarrow \\
2 & 3 & 6_{\text{eight}}
\end{array}
$$

Because $16 = 2^4$, every hexadecimal digit equates to a 4-digit binary numeral (see **Table 13** on the next page). Conversions between binary and hexadecimal forms can be done in a manner similar to that used in **Examples 12 and 13.**

EXAMPLE 14 Converting from Hexadecimal to Binary

Convert $8B4F_{\text{sixteen}}$ to binary form.

Solution

Each hexadecimal digit yields a 4-digit binary equivalent.

$$
\begin{array}{cccc}
8 & B & 4 & F_{\text{sixteen}} \\
\downarrow & \downarrow & \downarrow & \downarrow \\
1000 & 1011 & 0100 & 1111_{\text{two}}
\end{array}
$$

Combining these groups of digits, $8B4F_{\text{sixteen}} = 1000101101001111_{\text{two}}.$

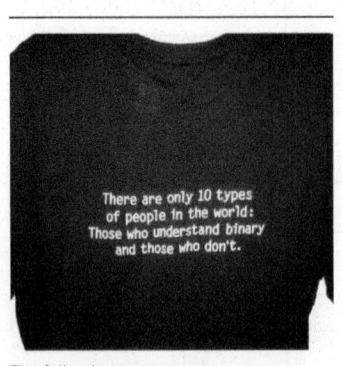

The following message was seen on the front of a T-shirt. "There are only 10 types of people in the world: Those who understand **binary** and those who don't." Do YOU understand this message?

Several games are based on the binary system. Suppose a game involves picking a "secret" number from 1 to 31. All you need is for a player to tell you the columns of **Table 14** (on the next page) that contain the number. If the columns are A, C, and E, you simply add the numbers in the top row of these columns:

The secret number is $1 + 4 + 16 = 21.$

Do you see how this game works? (See **Exercises 63–66.**)

Table 13

Hexadecimal	Binary
0	0000
1	0001
2	0010
3	0011
4	0100
5	0101
6	0110
7	0111
8	1000
9	1001
A	1010
B	1011
C	1100
D	1101
E	1110
F	1111

Table 14

A	B	C	D	E
1	2	4	8	16
3	3	5	9	17
5	6	6	10	18
7	7	7	11	19
9	10	12	12	20
11	11	13	13	21
13	14	14	14	22
15	15	15	15	23
17	18	20	24	24
19	19	21	25	25
21	22	22	26	26
23	23	23	27	27
25	26	28	28	28
27	27	29	29	29
29	30	30	30	30
31	31	31	31	31

In the world of computers, each binary digit is called a **bit,** and eight bits are called a **byte.** It is common for technology specifications to be reported in bits. For example, Web browsers may use 128-bit encryption technology, and a computer may have a 64-bit CPU and 24-bit "true color" graphics capability. Settings are often shown in blocks of decimals ranging from 0 to 255, in hexadecimal form or in binary form.

A computer with 24-bit RGB color assigns eight bits to each of the component colors red, green, and blue to generate colors in the pixels of a monitor. Thus there are

$2^8 = 256$ intensity settings (ranging from 0 to 255) for each component color.

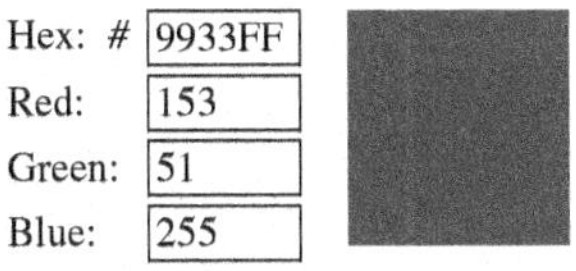

Hex: #	9933FF
Red:	153
Green:	51
Blue:	255

Figure 12

EXAMPLE 15 Converting a Color Code

Convert the hexadecimal color code 9933FF to decimal notation.

Solution

The hex code for the given color is 9933FF. Each hexadecimal digit equates to four bits, so each pair of hexadecimal digits equates to eight bits (one byte) and corresponds to a decimal numeral ranging from 0 to 255.

$$99_{\text{sixteen}} = (9 \cdot 16^1) + (9 \cdot 16^0) = 144 + 9 = 153$$

$$33_{\text{sixteen}} = (3 \cdot 16^1) + (3 \cdot 16^0) = 48 + 3 = 51$$

$$FF_{\text{sixteen}} = (15 \cdot 16^1) + (15 \cdot 16^0) = 240 + 15 = 255$$

Thus the decimal (R, G, B) notation for this color is (153, 51, 255), as indicated in **Figure 12.**

WHEN WILL I **EVER** USE THIS?

Suppose you are a Web designer, and one of your clients sells products via the Internet to a global customer base. This company has hired you to enhance their Web site in hopes of increasing Web traffic (and revenue). As part of the site enhancement, you need to choose colors that will complement each other and make products stand out.

The 256 intensity settings for each of the primary colors (red, green and blue) would normally allow for $256^3 = 16,777,216$ different color choices. However, since a significant percentage of your client's customer base tends to have older computers, you will limit your color choices to $6^3 = 216$ colors that tend to be rendered accurately on monitors with limited capability (see **Figure 13**). In this cubic model of the color palette, complementary colors are located symmetrically about the center of the color cube.

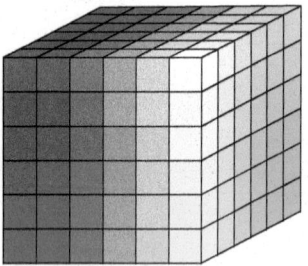

Figure 13

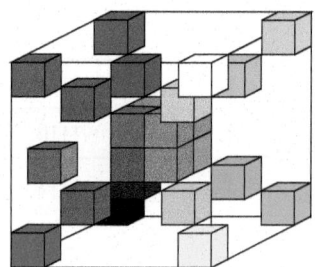

Figure 14

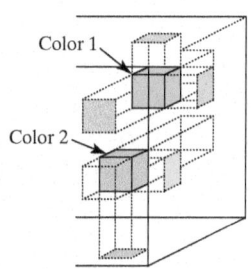

Figure 15

For example, black (RGB (0,0,0)) and white (RGB (255, 255, 255)) are complementary, as are green (RGB (0, 255, 0)) and magenta (RGB (255, 0, 255)) (see **Figure 14**).

You decide to use two main colors (labeled Color 1 (RGB (102, 204, 204)) and Color 2 (RGB (204, 204, 102)) in **Figure 15**. One way to find the complementary colors is to convert the decimal RGB notation to binary and then "switch" all the 1s to 0s and all the 0s to 1s. The result is the **ones complement**. Finally, convert back to decimal. The results of this process for Color 1 are shown in **Table 15**.

Table 15

	Color 1
Decimal (R, G, B)	(102, 204, 204)
Binary	(01100110, 11001100, 11001100)
Ones Complement	(10011001, 00110011, 00110011)
Decimal (R, G, B) for Complementary Color	(153, 51, 51)

Find the complement for Color 2.

Answer: Binary: (11001100, 11001100, 01100110); Ones complement: (00110011, 00110011, 10011001); Decimal (R, G, B) for complementary color: (51, 51, 153)

4.4 EXERCISES

CONCEPT CHECK *Choose all that apply.*

A. Binary system   **B.** Base-five system
C. Base-seven system   **D.** Octal system
E. Decimal system   **F.** Hexadecimal system

1. This system has the least number of distinct symbols.

2. 11 is a prime number in this system.

3. This numeration system is very useful in computer applications.

4. 32 is a prime number in this system.

List the first twenty counting numbers in each base.

5. seven (Only digits 0 through 6 are used in base seven.)

6. eight (Only digits 0 through 7 are used.)

7. nine (Only digits 0 through 8 are used.)

8. sixteen (The digits 0, 1, 2, … , 9, A, B, C, D, E, F are used in base sixteen.)

Write (in the same base) the counting numbers just before and just after the given number. (Do not convert to base ten.)

9. 14_{five} **10.** 555_{six}

11. $\text{B6F}_{\text{sixteen}}$ **12.** 10111_{two}

Determine, in each base, the least and greatest four-digit numbers and their decimal equivalents.

13. five   **14.** sixteen

Convert each number to decimal form by expanding in powers and by using the calculator shortcut.

15. $3\text{BC}_{\text{sixteen}}$ **16.** 34432_{five}

17. 2356_{seven} **18.** 101101110_{two}

19. 70266_{eight} **20.** $\text{ABCD}_{\text{sixteen}}$

21. 2023_{four} **22.** 6185_{nine}

23. 31544_{six} **24.** 88703_{nine}

Convert each number from decimal form to the given base.

25. 147 to base sixteen   **26.** 2730 to base sixteen

27. 36401 to base five   **28.** 70893 to base seven

29. 587 to base two   **30.** 12888 to base eight

31. 9346 to base six   **32.** 99999 to base nine

33. 8407 to base three   **34.** 11028 to base four

Make each conversion as indicated.

35. 43_{five} to base seven   **36.** 27_{sixteen} to base five

37. 6748_{nine} to base four   **38.** $\text{C02}_{\text{sixteen}}$ to base seven

Convert each number from octal form to binary form.

39. 367_{eight} **40.** 2406_{eight}

Convert each number from binary form to octal form.

41. 100110111_{two} **42.** 11010111101_{two}

Make each conversion as indicated.

43. $\text{AC}_{\text{sixteen}}$ to binary

44. $\text{F111}_{\text{sixteen}}$ to binary

45. 101101_{two} to hexadecimal

46. $101111011101011_{\text{two}}$ to hexadecimal

Identify the greatest number from each list.

47. 42_{seven}, 37_{eight}, $1\text{D}_{\text{sixteen}}$

48. 1101110_{two}, 414_{five}, $6\text{F}_{\text{sixteen}}$

*There is a theory that twelve would be a better base than ten for general use. This is mainly because twelve has more divisors (1, 2, 3, 4, 6, 12) than ten (1, 2, 5, 10), which makes fractions easier in base twelve. The base-twelve system is called the **duodecimal system.** In the decimal system we speak of a one, a ten, and a hundred (and so on); in the duodecimal system we say a one, a dozen (twelve), and a gross (twelve squared, or one hundred forty-four).*

49. Adam's clients ordered 9 gross, 10 dozen, and 11 copies of *Broad Band* during 2018. How many copies was that in base ten?

50. Which amount is greater: 3 gross, 6 dozen or 2 gross, 19 dozen?

Write a short answer for each problem.

51. Explain why the octal and hexadecimal systems are convenient for people who code for computers.

52. A four-bit string is sometimes called a "nibble." Explain why this would make sense, given the meaning of a byte.

*One common method of converting symbols into binary digits for computer processing is called **ASCII** (American Standard Code of Information Interchange). The uppercase letters A through Z are assigned the numbers 65 through 90, so A has binary code 01000001 and Z has code 01011010. Lowercase letters a through z have codes 97 through 122 (that is, 01100001 through 01111010). ASCII codes, as well as other numerical computer output, normally appear without commas.*

Write the binary code (eight bits) for each letter.

53. C   **54.** X

55. k   **56.** q

Break each code into groups of eight digits and write as letters.

57. 0100001001000001010100110101000101

58. 0101010001010111010011111

Translate each word into an ASCII string of binary digits, and then convert to hexadecimal code. (Be sure to distinguish uppercase and lowercase letters.)

59. New   **60.** Orleans

Solve each problem.

61. A base-four system is also called a *quaternary* system. There are eleven counting numbers whose octal numerals and quaternary numerals have the same number of digits. Find the least and the greatest of these numbers.

62. There are thirty-seven counting numbers whose base-eight numerals contain two digits but whose base-three numerals contain four digits. Find the least and greatest of these numbers.

*Refer to **Table 14** for Exercises 63–66.*

63. After observing the binary forms of the numbers 1–31, identify a common property of all **Table 14** numbers in each of the following columns.

   (a) Column A

   (b) Column B

   (c) Column C

   (d) Column D

   (e) Column E

64. Explain how the "game" of **Table 14** works.

65. How many columns would be needed for **Table 14** to include all integers from 1 to 63?

66. How many columns would be needed for **Table 14** to include all integers from 1 to 127?

In our decimal system, we distinguish odd and even numbers by looking at their ones (or units) digits. If the ones digit is even (0, 2, 4, 6, or 8), the number is even. If the ones digit is odd (1, 3, 5, 7, or 9), the number is odd. For Exercises 67–74, determine whether this same criterion works for numbers expressed in the given bases.

67. two   **68.** three

69. four   **70.** five

71. six   **72.** seven

73. eight   **74.** nine

Write a short answer for each problem.

75. Consider all even bases. If the above criterion works for all, explain why. If not, find a criterion that does work for all even bases.

76. Consider all odd bases. If the above criterion works for all, explain why. If not, find a criterion that does work for all odd bases.

Determine whether the given base-five numeral represents one that is divisible by five.

77. 3204_{five} **78.** 200_{five}

79. 2310_{five} **80.** 342_{five}

CONCEPT CHECK *Write a short answer for each question.*

81. If you want to multiply a decimal numeral by 10, what do you do?

82. If you want to multiply a binary numeral by 2, what do you do?

83. If a base-b numeral is a multiple of b, what will the last digit be?

84. If a binary numeral ends in 0, then it is even. How would you divide such a number in half?

Recall that conversions between binary and octal are simplified because eight is a power of 2:

$$8 = 2^3.$$

*(See **Examples 12 and 13.**) The same is true of conversions between binary and hexadecimal, because*

$$16 = 2^4.$$

*(See **Example 14.**) Direct conversion between octal and hexadecimal does not work the same way, because 16 is not a power of 8. Explain how to carry out each conversion without using base ten, and give an example.*

85. hexadecimal to octal

86. octal to hexadecimal

Devise a method (similar to the one for conversions between binary, octal, and hexadecimal) for converting between base three and base nine, and use it to carry out each conversion.

87. 6504_{nine} to base three

88. 81170_{nine} to base three

89. 212201221_{three} to base nine

90. 200121021_{three} to base nine

Arithmetic in Other Bases
Generalizing concepts can often give us deeper insights about them, helping us see not only how to use them, but also why they work. For example, when we add numbers in base ten, we carry when the expanded form contains more than nine of any power of the ten. The same principle should apply in other bases. For example, we can add $23 + 18$ *in binary as follows.*

Addition and Multiplication
in Base 5

$$\begin{array}{r} 1\ \overset{1}{0}\ \overset{1}{1}\ 1\ 1 \\ +\ 1\ 0\ 0\ 1\ 0 \\ \hline 1\ 0\ 1\ 0\ 0\ 1_{two} \end{array} \quad = 32 + 8 + 1 = 41$$

In Exercises 91–96, work the problems in the given base. Then convert your answers to decimal form to check.

91. Borrowing should work in base two just as it does in base ten. Subtract $21 - 18$ in binary.

92. The multiplication algorithm should work the same way in both systems. Multiply $15 \cdot 3$ in binary.

93. The *nines* complement method applies in the decimal system, so the *ones* complement should similarly apply to binary subtraction. Use it to subtract $28 - 13$ in base two.

94. Recall that the *decimal* point separates the whole part of a numeral from the fractional part, and place values continue to be divided by 10 as we move to the right. Similarly, the *ternary point* (in the base-three system) separates the whole part of a ternary numeral from fractional parts. Consider the place values to the right of the ternary point, and use the idea to perform the following subtraction. (Note that the minuend and subtrahend are repeating *decimals*, but using base three eliminates this problem.)

Subtract $5\frac{2}{9} - 2\frac{1}{3}$ in base three.

95. Divide 36 by 3 in base four.

96. Add $3A42 + 123$ in base sixteen.

Chapter 4 SUMMARY

KEY TERMS

 4.1

numeral
numeration system
tally stick
tally
Hindu-Arabic system
grouping
base
simple grouping system

algorithm
multiplicative
 grouping system

4.2

positional system
digit
face value
place value

placeholder
ciphered system

4.3

expanded form
distributive property
decimal system
abacus
lattice method

Napier's rods
Russian peasant method
nines complement method

4.4

binary
octal
hexadecimal
bit
byte

TEST YOUR WORD POWER

See how well you have learned the vocabulary in this chapter.

1. A **simple grouping system** is
   A. a numeration system that uses group position to determine values of numerals.
   B. a numeration system in which the order of the symbols in a numeral makes no difference.
   C. a numeration system that uses grouping symbols (such as parentheses) to simplify numerals.
   D. a system that constructs numerals by grouping different numbers together.

2. A **positional system** is
   A. a numeration system that features a different symbol for each power of the base.
   B. a numeration system that works differently depending on the context in which it is used.
   C. a numeration system that is extremely efficient because only multipliers are used.
   D. a system that uses position to determine the face values of the digits.

3. **Napier's rods (or bones)** is the name of
   A. a calculating device based on the lattice method for multiplication.
   B. a device containing sliding beads and a dividing bar used to speed calculations.
   C. a device based on the Russian peasant method for multiplication.
   D. an adding machine designed around the nines complement method of subtraction.

4. The **hexadecimal** numeration system is
   A. a numeration system needing twelve distinct symbols.
   B. a base-sixteen numeration system using symbols 0–9, A, B, C, D, E, and F.
   C. a base-six numeration system with multipliers corresponding to the vertices of a hexagon.
   D. a system having fewer symbols than the octal numeration system.

ANSWERS
1. B   2. C   3. A   4. B

QUICK REVIEW

Concepts   *Examples*

4.1 HISTORICAL NUMERATION SYSTEMS

A **numeration system** uses symbols called **numerals** to represent numbers.

Ancient Egyptian Numeration
This utilizes a **simple grouping system** that is base ten. A different symbol represents each power of 10 (from 10^0 to 10^6).

Early Egyptian Symbols

Number	Symbol
1	I
10	∩
100	9
1000	ℐ
10,000	𝓁
100,000	ᗧ
1,000,000	𝕏

Add 123,223 + 3104 using ancient Egyptian numeration.

The sum can be translated to our Hindu-Arabic system as

126,327.

Ancient Roman Numeration
This system is also base ten.

1. It has "extra" symbols for 5, 50, and 500.

2. A subtractive feature involves writing a lesser-valued symbol immediately to the left of a greater-valued symbol.

3. A multiplicative feature involves using a bar over a symbol to indicate multiplication by 1000 and a double bar to indicate multiplication by 1,000,000.

Express the difference 20,236 − 10,125 in Roman numerals.

$$\overline{XX}CCXXXVI$$
$$- \quad \overline{X}CXXV$$
$$\overline{X}CXI$$

Translating the result to Hindu-Arabic numeration gives

10,111.

Concepts	*Examples*

Classical Chinese Numeration
This is a base-ten **multiplicative grouping system,** with some modifications.

1. Symbols are written in pairs, each including a multiplier (with value less than ten) and a symbol representing a power of 10.

2. If the multiplier in the 10s pair is 1, it is omitted.

3. For missing powers of the base, the following rules apply.

   - If a single power of the base is missing, this is denoted with the zero symbol.

   - If two or more consecutive powers are missing, just one zero symbol denotes the entire omission.

   - The omission of consecutive powers including 1s at the bottom of a numeral need not be denoted at all.

Express the numeral 2019 in the classical Chinese numeration system.

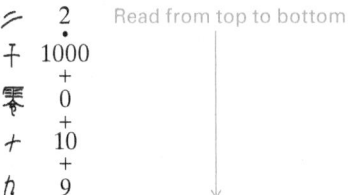

Read from top to bottom.

4.2 MORE HISTORICAL NUMERATION SYSTEMS

Positional systems rely on the position of each symbol to determine its **place value** (that is, for which power of the base it is a multiplier).

Babylonian Numeration
This system is base-ten simple grouping within base-sixty positional. It features only two symbols.

Babylonian Symbols

Number	Symbol
1	▼
10	‹

These symbols are grouped to express the numbers 1–59. The groups are then positioned to represent powers of 60.

Mayan Numeration
This system is base-five simple grouping within base-twenty positional. It uses three symbols.

Mayan Symbols

Number	Symbol
0	◯
1	.
5	—

1. Face values from 1 to 19 are expressed in base-five simple grouping.

2. Place values are 1, 20, $20 \cdot 18 = 360$, $360 \cdot 20 = 7200$, $7200 \cdot 20 = 144{,}000$, and so on.

In our Hindu-Arabic (base-ten) numeration system, the numeral 3024 means "three thousand twenty-four." The digit 3 is in the *thousands* place, the placeholder 0 is in the *hundreds* place, the digit 2 is in the *tens* place, and 4 is in the *ones* place.

The Babylonian numeral

$$\underbrace{\text{‹‹▼▼▼}}_{23 \cdot 3600} \quad \underbrace{\text{‹▼▼}}_{12 \cdot 60} \quad \underbrace{\text{▼▼▼▼}}_{4 \cdot 1}$$

represents

$23 \cdot 3600 + 12 \cdot 60 + 4 \cdot 1 = 83{,}524.$ Hindu-Arabic form

To convert the numeral 83,524 to a Babylonian numeral, divide 83,524 by 3600 (since 3600 is the greatest power of 60 that divides into 83,524) to get a quotient of 23 with remainder 724. Divide 724 by 60 to get a quotient of 12 with remainder 4. Then express these multiples of powers of 60 using the symbols as shown above.

The Mayan numeral

$$\begin{array}{l} \cdots \longleftarrow 8 \cdot 360 \\ \mathbf{◯} \longleftarrow 0 \cdot 20 \\ \equiv \longleftarrow 12 \cdot 1 \end{array}$$

represents

$8 \cdot 360 + 0 \cdot 20 + 12 \cdot 1 = 2892.$ Hindu-Arabic form

To convert the Hindu-Arabic numeral 948 to a Mayan numeral, divide by 360, and divide the remainder by 20 to see that

$$948 = 2 \cdot 360 + 11 \cdot 20 + 8.$$

In Mayan, this is written as follows.

Concepts	*Examples*

Greek Numeration

This is a **ciphered system,** using 27 distinct symbols to express numbers less than 1000 easily.

1. Multiples of 1000 (up to 9000) are indicated with a small stroke next to a units symbol.

2. Multiples of 10,000 are indicated by the letter M with the multiplier shown above the M.

Since α represents 1, ϕ represents 500, μ represents 40, and β represents 2, the Greek numeral $\overset{\alpha}{\text{M}}\phi\mu\beta$ represents

$$10{,}000 + 500 + 40 + 2 = 10{,}542.$$

The Hindu-Arabic numeral 837 is expressed in Greek numeration as $\omega\lambda\zeta$.

4.3 ARITHMETIC IN THE HINDU-ARABIC SYSTEM

The **Hindu-Arabic system** is a base-ten positional numeration system.

Writing a numeral in **expanded form** shows

• that the digits in the numeral are multipliers for powers of 10.

• the need to "carry" when the sum includes more than nine of any power of 10.

• the need to "borrow" when subtracting a greater number of a power of 10 from a lesser number.

Write the numeral 4132 in expanded form.

$$4 \cdot 10^3 + 1 \cdot 10^2 + 3 \cdot 10^1 + 2 \cdot 10^0$$

Add 67 and 56 using expanded notation.

$$67 = (6 \cdot 10^1) + (7 \cdot 10^0)$$
$$\underline{+\ 56 = (5 \cdot 10^1) + (6 \cdot 10^0)}$$
$$(11 \cdot 10^1) + (13 \cdot 10^0)$$

We must rewrite both products in this sum.

$$11 \cdot 10^1 = (10 + 1) \cdot 10^1 = 1 \cdot 10^2 + 1 \cdot 10^1$$
$$13 \cdot 10^0 = (10 + 3) \cdot 10^0 = 1 \cdot 10^1 + 3 \cdot 10^0$$

The sum $11 \cdot 10^1 + 13 \cdot 10^0$ becomes

$$(1 \cdot 10^2 + 1 \cdot 10^1) + (1 \cdot 10^1 + 3 \cdot 10^0)$$
$$= 1 \cdot 10^2 + (1 \cdot 10^1 + 1 \cdot 10^1) + 3 \cdot 10^0$$
$$= 1 \cdot 10^2 + 2 \cdot 10^1 + 3 \cdot 10^0$$
$$= 123.$$

Several calculation devices have been invented to make calculation with Hindu-Arabic numerals more efficient. Two examples are **Napier's rods** and the **abacus.**

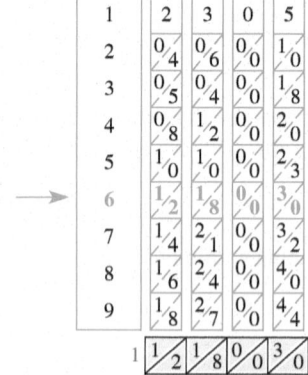

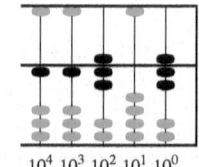

$10^4\ 10^3\ 10^2\ 10^1\ 10^0$

The abacus shows the number 11,707.

The product is 13,830.

Napier's rods show the multiplication of $6 \cdot 2305$.

Nines Complement Method

To subtract, replace the digits in the subtrahend with their nines complements, add the resulting number to the minuend, delete the leading 1, and add 1 to the result.

Calculate the difference $625 - 48$ using the nines complement method.

625	625	576
-048	$+951$	$+1$
	1576	577 ← Difference

Concepts   *Examples*

4.4 CONVERSION BETWEEN NUMBER BASES

Any base-n positional numeration system needs n distinct symbols for expressing numbers.

Converting from Another Base to Decimal Form

Step 1 Start with the first digit on the left and multiply by the base.

Step 2 Then add the next digit, multiply again by the base, and so on.

Step 3 Add the last digit on the right. ***Do not multiply it by the base.***

Converting a Decimal Numeral to Another Base n

Step 1 Divide the decimal numeral by n and record the quotient and remainder.

Step 2 Divide the quotient by n, and record the remainder.

Step 3 Repeat Step 2 until the quotient 0 appears.

Step 4 Write the remainders in reverse order.

Computers are binary machines, processing 1s and 0s. Because 8 and 16 are powers of 2, the octal and hexadecimal systems are useful in condensing data and code into shorter strings of symbols.

- Three binary digits (bits) correspond to one octal digit.

- Four bits correspond to one hexadecimal digit.

Decimal-to-binary conversion is helpful when working with computer graphics and other settings.

The hexadecimal (base-sixteen) system uses 0–9 and A, B, C, D, E, and F as digits.

Convert the base-five numeral 1214_{five} to decimal form.

$$1214_{\text{five}} = ((1 \cdot 5 + 2) \cdot 5 + 1) \cdot 5 + 4$$
$$= 184$$

Convert 395 to base four as shown.

$$
\begin{array}{r}
4\,\lfloor 395 \\
4\,\lfloor 98 \leftarrow 3 \\
4\,\lfloor 24 \leftarrow 2 \\
4\,\lfloor 6 \leftarrow 0 \\
4\,\lfloor 1 \leftarrow 2 \\
4\,\lfloor 0 \leftarrow 1
\end{array}
\quad \} \quad 395 = 12023_{\text{four}}
$$

Remainder

Convert $110100011010_{\text{two}}$ to octal and hexadecimal.

$$
\begin{array}{cccc}
6 & 4 & 3 & 2_{\text{eight}} \\
\end{array}
$$
$$110100011010_{\text{two}}$$
$$
\begin{array}{ccc}
D & 1 & A_{\text{sixteen}}
\end{array}
$$

Express color code RGB (128, 206, 155) in binary form.

(10000000, 11001110, 10011011)

Chapter **4** TEST

In each case, identify the numeration system, and give the Hindu-Arabic equivalent.

1. (numeral symbols)

2. $\overline{\text{XCDLXXIV}}$

3. (numeral symbols)

4. (cuneiform symbols)

5. (symbols)

6. $\overset{\epsilon}{\text{M}}{,}\gamma\phi\kappa\delta$

Perform each operation using the alternative algorithm specified.

7. $23 \cdot 45$ (Russian peasant or Egyptian method)

8. $246 \cdot 97$ (Lattice method)

9. $21{,}425 - 8198$ (Nines complement method)

10. $12 \cdot 123$ (Napier's rods)

Convert each number to base ten.

11. 100101_{two} **12.** 243_{five}

13. 346_{eight} **14.** $\text{BEEF}_{\text{sixteen}}$

Convert as indicated.

15. 49 to binary

16. 2930 to base five

17. 10101110_{two} to octal

18. 7215_{eight} to hexadecimal

19. 5041_{six} to decimal

20. $BAD_{sixteen}$ to binary

Write a short answer for each problem.

21. Explain the advantage of multiplicative grouping over simple grouping.

22. Explain the advantage of positional over multiplicative grouping.

23. Explain the advantage, in a positional numeration system, of a lesser base over a greater base.

24. Explain the advantage, in a positional numeration system, of a greater base over a lesser base.

25. Explain a quick method to convert a base-nine numeral to base three.

26. Illustrate your method from **Exercise 25** by converting 765_{nine} to base three.

5 Number Theory

The modern criminal justice system, like most other areas of government, the military, commerce, and personal life, is highly dependent on computer technology and the mathematical theory and algorithms underlying it. Our money, safety, and identity are protected by public key **cryptography,** which includes the processes of **encrypting** (coding) and **decrypting** (decoding) crucial information. Cryptography involves two kinds of problems:

1. Given two prime numbers, find their product.
2. Given the product, find its two prime factors.

The success of encryption depends on the fact that, given today's state of computer hardware and software, and given large enough primes, **Problem 1** above can be done, but **Problem 2** *cannot* be done, even with the most powerful computers available.

In this chapter you will learn about prime numbers, including the extremely large ones used in cryptography systems.

5.1 PRIME AND COMPOSITE NUMBERS

OBJECTIVES

1 Identify prime and composite numbers.

2 Apply divisibility tests for natural numbers.

3 Apply the fundamental theorem of arithmetic.

Primes, Composites, and Divisibility

The famous German mathematician Carl Friedrich Gauss once remarked, "Mathematics is the Queen of Science, and number theory is the Queen of Mathematics." **Number theory** is devoted to the study of the properties of the **natural numbers** (also called the **counting numbers** or the **positive integers**). The natural numbers are the most basic of the six sets of numbers introduced earlier.

$$N = \{1, 2, 3, \dots\}$$

A key concept of number theory is the idea of *divisibility*. One counting number is *divisible* by another if dividing the first by the second leaves a remainder 0.

Do not confuse $b \mid a$ **with** *b/a*. The expression $b \mid a$ denotes the **statement** "*b* divides *a*." For example, $3 \mid 12$ is a true statement, while $5 \mid 14$ is a false statement. On the other hand, *b/a* denotes the **operation** "*b* divided by *a*." For example, 28/4 yields the result 7.

DIVISIBILITY

The natural number a is **divisible** by the natural number b if there exists a natural number k such that $a = bk$. If b divides a, then we write $b \mid a.$

Notice that if b divides a, then the quotient a/b or $\frac{a}{b}$ is a natural number. For example, 4 divides 20 because there exists a natural number k such that

$$20 = 4k.$$

The value of k here is 5, because

$$20 = 4 \cdot 5.$$

The ideas of **even** and **odd natural numbers** are based on the concept of divisibility. A natural number is even if it is divisible by 2 and odd if it is not. Every even number can be written in the form $2k$ (for some natural number k), while every odd number can be written in the form $2k - 1$. Here is another way to say the same thing: 2 divides every even number but fails to divide every odd number. (If a is even, then $2 \mid a$, whereas if a is odd, then $2 \nmid a$.)

The natural number 20 is not divisible by 7, since there is no natural number k satisfying $20 = 7k$. Alternatively, "20 divided by 7 gives quotient 2 with remainder 6," and since there is a nonzero remainder, divisibility does not hold. We write $7 \nmid 20$ to indicate that 7 does *not* divide 20.

If the natural number a is divisible by the natural number b, then b is a **factor** (or **divisor**) of a, and a is a **multiple** of b. For example, 5 is a factor of 30, and 30 is a multiple of 5. Also, 6 is a factor of 30, and 30 is a multiple of 6. The number 30 equals $6 \cdot 5$. This product $6 \cdot 5$ is called a **factorization** of 30. Other factorizations of 30 include

$$3 \cdot 10, \quad 2 \cdot 15, \quad 1 \cdot 30, \quad \text{and} \quad 2 \cdot 3 \cdot 5.$$

EXAMPLE 1   Checking Divisibility

Decide whether the first number is divisible by the second.

(a) 45; 9   **(b)** 60; 7   **(c)** 19; 19   **(d)** 26; 1

Solution

(a) Is there a natural number k that satisfies $45 = 9k$? The answer is yes, because $45 = 9 \cdot 5$, and 5 is a natural number. Therefore, 9 divides 45, written $9 \mid 45$.

(b) Because the quotient $60 \div 7$ is not a natural number, 60 is not divisible by 7, written $7 \nmid 60$.

(c) The quotient $19 \div 19$ is the natural number 1, so 19 is divisible by 19. (***In fact, any natural number is divisible by itself.***)

For any natural number a, it is true that $a \mid a$ and also that $1 \mid a$.

(d) The quotient $26 \div 1$ is the natural number 26, so 26 is divisible by 1. (***In fact, any natural number is divisible by 1.***)

EXAMPLE 2   **Finding Factors**

Find all the natural number factors of each number.

(a) 36   **(b)** 50   **(c)** 11

Solution

(a) To find the factors of 36, try to divide 36 by 1, 2, 3, 4, 5, 6, and so on. The natural number factors of 36 are 1, 2, 3, 4, 6, 9, 12, 18, and 36.

(b) The factors of 50 are 1, 2, 5, 10, 25, and 50.

(c) The only natural number factors of 11 are 11 and 1.

PRIME AND COMPOSITE NUMBERS

A natural number greater than 1 that has only itself and 1 as factors is called a **prime number.** A natural number greater than 1 that is not prime is called **composite.**

Mathematicians agree that the natural number 1 is neither prime nor composite.

ALTERNATIVE DEFINITION OF A PRIME NUMBER

A **prime number** is a natural number that has *exactly* two different natural number factors (which clarifies that 1 is not a prime).

There is a systematic method for identifying prime numbers in a list of numbers: 2, 3, . . . , *n*. The method, known as the **Sieve of Eratosthenes**, is named after the Greek geographer, poet, astronomer, and mathematician (about 276–192 B.C.).

To construct such a sieve, list all the natural numbers from 2 through some given natural number *n*, such as 100. The number 2 is prime, but all other multiples of 2 (4, 6, 8, 10, and so on) are composite. Circle the prime 2, and cross out all other multiples of 2. The next number not crossed out and not circled is 3, the next prime. Circle the 3, and cross out all other multiples of 3 (6, 9, 12, 15, and so on) that are not already crossed out. Circle the next prime, 5, and cross out all other multiples of 5 not already crossed out. Continue this process for all primes less than or equal to the square root of the last number in the list. For this list, we may stop with 7, because the next prime, 11, is greater than the square root of 100, which is 10. At this stage, simply circle all remaining numbers that are not crossed out.

Table 1 shows the Sieve of Eratosthenes for 2, 3, 4, . . . , 100.

Table 1   Sieve of Eratosthenes

The 25 primes between 1 and 100 are circled.

List of Primes The following program, written by Charles W. Gantner and provided courtesy of Texas Instruments, can be used on the TI-84 Plus C calculator to list all primes less than or equal to a given natural number *N*.

```
PROGRAM: PRIMES
: Disp "INPUT N ≥ 2"
: Disp "TO GET"
: Disp "PRIMES ≤ N"
: Input N
: 2 → T
: Disp T
: 1 → A
: Lbl 1
: A + 2 → A
: 3 → B
: If A > N
: Stop
: Lbl 2
: If B ≤ √(A)
: Goto 3
: Disp A
: Pause
: Goto 1
: Lbl 3
: If A/B ≤ int (A/B)
: Goto 1
: B + 2 → B
: Goto 2
```

```
NORMAL FLOAT AUTO REAL DEGREE MP
prgmPRIMES
INPUT N ≥ 2
TO GET PRIMES ≤ N
?6
                          2
                          3
                          5
                       Done
```

The display indicates that the primes less than or equal to 6 are 2, 3, and 5.

For greater numbers of primes, the ENTER key may be pressed repeatedly to see them all.

EXAMPLE 3   **Identifying Prime and Composite Numbers**

Decide whether each number is prime or composite.

(a) 89   **(b)** 83,572   **(c)** 629

Solution

(a) Because 89 is circled in **Table 1,** it is prime. If 89 had a smaller prime factor, 89 would have been crossed out as a multiple of that factor.

(b) The number 83,572 is even, so it is divisible by 2. It is composite.

There is only one even prime, the number 2 itself.

(c) For 629 to be composite, there must be a number other than 629 and 1 that divides into it with remainder 0. Start by trying 2, and then 3. Neither works. There is no need to try 4. (If 4 divides with remainder 0 into a number, then 2 will also.) Try 5. There is no need to try 6 or any succeeding even number. (Why?) Try 7. Try 11. (Why not try 9?) Try 13. Keep trying numbers until one works, or until a number is tried whose square exceeds the given number, 629. Try 17.

$$629 \div 17 = 37$$

The number 629 is composite, since

$$629 = 17 \cdot 37.$$

An aid in determining whether a natural number is divisible by another natural number is called a **divisibility test. Table 2** shows tests for divisibility by the natural numbers 2 through 12 (except for 7 and 11, which are covered in the exercises).

Table 2 Divisibility Tests for Natural Numbers

Divisible by	Test	Example
2	Number ends in 0, 2, 4, 6, or 8. (The last digit is even.)	9,489,994 ends in 4; it is divisible by 2.
3	Sum of the digits is divisible by 3.	897,432 is divisible by 3, because $8 + 9 + 7 + 4 + 3 + 2 = 33$ is divisible by 3.
4	Last two digits form a number divisible by 4.	7,693,432 is divisible by 4, because 32 is divisible by 4.
5	Number ends in 0 or 5.	890 and 7635 are divisible by 5.
6	Number is divisible by both 2 and 3.	27,342 is divisible by 6, because it is divisible by both 2 and 3.
8	Last three digits form a number divisible by 8.	1,437,816 is divisible by 8, because 816 is divisible by 8.
9	Sum of the digits is divisible by 9.	428,376,105 is divisible by 9, because the sum of the digits is 36, which is divisible by 9.
10	The last digit is 0.	897,463,940 is divisible by 10.
12	Number is divisible by both 4 and 3.	376,984,032 is divisible by 12.

Prime Factorization This program produces the prime factorization of an entered natural number.

```
PROGRAM:FACTOR
:Disp "ENTER N ≥ 2"
:Disp "FOR THE PRIME"
:Disp "FACTORIZATION"
:Input N
:While N/2 = int(N/2)
:Disp 2
: N/2 → N
:End
: 1 → A
:Lbl 1
: A + 2 → A
: 3 → B
:If A > N
:Stop
:Lbl 2
:If B ≤ √A
:Goto 3
:While N/A = int(N/A)
: N/A → N
:Disp A
:Pause
:End
:Goto 1
:Lbl 3
:If A/B = int(A/B)
:Goto 1
: B + 2 → B
:Goto 2
```

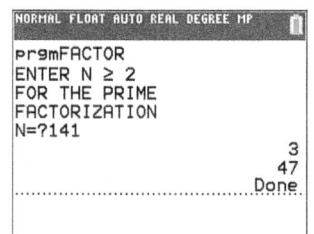

The display indicates that the prime factorization of 141 is 3 · 47.

EXAMPLE 4 Applying Divisibility Tests

In each case, decide whether the first number is divisible by the second.

(a) 2,984,094;   4   **(b)** 2,429,806,514;   9

Solution

(a) The last two digits form the number 94. Since 94 is not divisible by 4, the given number is not divisible by 4.

(b) The sum of the digits is

$$2 + 4 + 2 + 9 + 8 + 0 + 6 + 5 + 1 + 4 = 41,$$

which is not divisible by 9. The given number is, therefore, not divisible by 9.

The Fundamental Theorem of Arithmetic

A *composite* number can be thought of as "composed" of smaller factors. For example, 42 is composite since $42 = 6 \cdot 7$. If the smaller factors are all primes, then we have a *prime factorization*. For example, $42 = 2 \cdot 3 \cdot 7$.

THE FUNDAMENTAL THEOREM OF ARITHMETIC

Every natural number can be expressed in one and only one way as a product of primes (if the order of the factors is disregarded). This unique product of primes is called the **prime factorization** of the natural number.

Because a prime natural number is not composed of smaller factors, its prime factorization is simply itself. For example, $17 = 17$.

EXAMPLE 5 Finding the Unique Prime Factorization of a Composite Number

Find the prime factorization of the number 1320.

Solution

We use a "factor tree." The factor tree can start with $1320 = 2 \cdot 660$, as shown below on the left. Then $660 = 2 \cdot 330$, and so on, until every branch of the tree ends with a prime. All the resulting prime factors are shown circled in the diagram.

Alternatively, the same factorization is obtained by repeated division by primes, as shown on the right. (In general, you would divide by the primes 2, 3, 5, 7, 11, and so on, each as many times as possible, until the answer is no longer composite.)

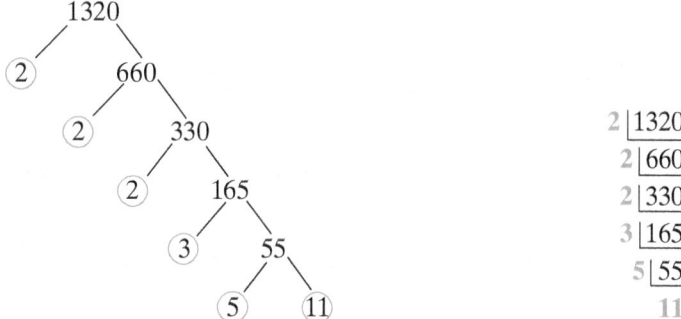

By either method, the prime factorization, in exponential form, is

$$1320 = 2^3 \cdot 3 \cdot 5 \cdot 11. \quad {\scriptstyle 2 \cdot 2 \cdot 2 = 2^3}$$

5.1   EXERCISES

CONCEPT CHECK *Write* true *or* false *for each statement.*

1. 1 is the least prime number.

2. Every natural number is both a factor and a multiple of itself.

3. Every natural number is divisible by 1.

4. There are no even prime numbers.

5. If n is a natural number and $9 \mid n$, then $3 \mid n$.

6. If n is a natural number and $5 \mid n$, then $10 \mid n$.

7. If 16 divides a natural number, then 2, 4, and 8 must also divide that natural number.

8. The prime number 53 has exactly two natural number factors.

Find all natural number factors of each number.

9. 12   10. 18

11. 28   12. 171

Use divisibility tests to decide whether the given number is divisible by each number.

    (a) 2   **(b)** 3   **(c)** 4

    (d) 5   **(e)** 6   **(f)** 8

    (g) 9   **(h)** 10   **(i)** 12

13. 321   14. 330

15. 36,360   16. 135,792,468

Solve each problem.

17. In constructing the Sieve of Eratosthenes for 2 through 100, we said that any composite in that range had to be a multiple of some prime less than or equal to 7 (since the next prime, 11, is greater than the square root of 100). Explain.

18. To extend the Sieve of Eratosthenes to 200, what is the largest prime whose multiples would have to be considered?

19. Complete this statement: In seeking prime factors of a given number, we need only consider all primes up to and including the _____ _____ of that number, since a prime factor greater than the _____ _____ can occur only if there is at least one other prime factor less than the _____ _____.

20. Complete this statement: If no prime less than or equal to $\sqrt{n}$ divides n, then n is a _____ number.

21. Continue the Sieve of Eratosthenes in **Table 1** from 101 to 200 and list the primes between 100 and 200. How many are there?

22. From your list in part (a), verify that the numbers 197 and 199 are both prime.

23. In your list for **Exercise 21,** consider the six largest primes less than 200. Which pairs of these would have products that end in the digit 7?

24. By checking your pairs of primes from **Exercise 23,** give the prime factorization of 35,657.

25. List two primes that are consecutive natural numbers. Can there be any others?

26. Can there be three primes that are consecutive natural numbers? Explain.

27. For a natural number to be divisible by both 2 and 5, what must be true about its last digit?

28. Consider the divisibility tests for 2, 4, and 8 (all powers of 2). Use inductive reasoning to predict the divisibility test for 16. Then use the test to show that 456,882,320 is divisible by 16.

29. Redraw the factor tree of **Example 5,** assuming that you first observe that

$$1320 = 12 \cdot 110, \quad \text{then} \quad 12 = 3 \cdot 4 \quad \text{and} \quad 110 = 10 \cdot 11.$$

Complete the process and give the resulting prime factorization.

30. Explain how your result in **Exercise 29** illustrates the fundamental theorem of arithmetic.

Find the prime factorization of each composite number.

31. 126   32. 306

33. 1183   34. 605

35. 570   36. 1035

Here is a divisibility test for 7.

(a) *Double the last digit of the given number, and subtract this value from the given number with the last digit omitted.*

(b) *Repeat the process of part (a) as many times as necessary until it is clear whether the number obtained is divisible by 7.*

(c) *If the final number obtained is divisible by 7, then the given number also is divisible by 7. If the final number is not divisible by 7, then neither is the given number.*

Use this divisibility test to determine whether each number is divisible by 7.

37. 496,312   **38.** 422,142

39. 226,233   **40.** 340,659

41. 232,875   **42.** 979,608

Here is a divisibility test for 11.

(a) *Starting at the left of the given number, add together every other digit.*

(b) *Add together the remaining digits.*

(c) *Subtract the smaller of the two sums from the larger. (If they are the same, the difference is* 0.)

(d) *If the final number obtained is divisible by* 11, *then the given number also is divisible by* 11. *If the final number is not divisible by* 11, *then neither is the given number.*

Use this divisibility test to determine whether each number is divisible by 11.

43. 6,524,846   **44.** 120,121,279

45. 323,911,357   **46.** 500,590,279

47 60,128,459,358   **48.** 22,896,232,942

Solve each problem.

49. Consider the divisibility test for the composite number 6, and make a conjecture for the divisibility test for the composite number 15.

50. Give two factorizations of the number 75 that are not prime factorizations.

Determine all possible digit replacements for x so that the first number is divisible by the second. For example, the number 37,58x *is divisible by* 2 *if*

$$x = 0, 2, 4, 6, or\ 8.$$

51. 398,87x; 2   **52.** 2,43x,765; 3

53. 64,537,84x; 4   **54.** 2,135,89x; 5

55. 985,23x; 6   **56.** 32,54x,290; 10

*There is a method to determine the **number of divisors** of a composite number. To do this, write the composite number in its prime factored form, using exponents. Add 1 to each exponent and multiply these numbers. Their product gives the number of divisors of the composite number. For example,*

$$24 = 2^3 \cdot 3 = 2^3 \cdot 3^1.$$

Now add 1 *to each exponent:*

$$3 + 1 = 4, 1 + 1 = 2.$$

Multiply $4 \cdot 2$ *to get* 8. *There are* 8 *divisors of* 24. *(Because* 24 *is rather small, this can be verified easily. The divisors are* 1, 2, 3, 4, 6, 8, 12, *and* 24 — *a total of eight, as predicted.)*

Find the number of divisors of each composite number.

57. 105   **58.** 234

59. $5^8 \cdot 29^2$ **60.** $2^3 \cdot 5^2 \cdot 13^3$

61. $2^3 \cdot 5^7 \cdot 23$ **62.** $7^4 \cdot 11^5 \cdot 31^3$

Leap years occur when the year number is divisible by 4. *An exception to this occurs when the year number is divisible by* 100 *(that is, it ends in two zeros). In such a case, the number must be divisible by* 400 *in order for the year to be a leap year. Determine which years are leap years.*

63. 1556   **64.** 2024

65. 2200   **66.** 2600

Solve each problem.

67. Why is the following *not* a valid divisibility test for the number 8?

   "A number is divisible by 8 if it is divisible by both 4 and 2."

   Support your answer with an appropriate example.

68. Choose any three consecutive natural numbers, multiply them together, and divide the product by 6. Repeat this several times, using different choices of three consecutive numbers. Make a conjecture concerning the result.

69. Explain why the product of three consecutive natural numbers must be divisible by 6. Include examples in your explanation.

70. Choose any 6-digit number consisting of three digits followed by the same three digits in the same order (for example, 467,467). Divide by 13. Divide by 11. Divide by 7. What do you notice? Why do you think this happens?

One of the authors has three sons who were born, from eldest to youngest, on August 30, August 31, and October 14. For most (but not all) of each year, their ages are spaced two years from the eldest to the middle and three years from the middle to the youngest. In 2011, their ages were three consecutive prime numbers for a period of exactly 44 days. This same situation had also occurred in exactly two previous years. Use this information to solve **Exercises 71–74.** *(Hint: These are challenging. Consult the Sieve of Eratosthenes for a listing of all primes less than 100.)*

71. Which were the "two previous years" referred to above?

72. What were the years of birth of the three sons?

73. In what year will the same situation next occur?

74. What will be the ages of the three sons at that time?

FOR FURTHER THOUGHT

Prime Factor Splicing

Prime factor splicing (PFS) is discussed in the article "Home Primes and Their Families," in *Mathematics Teacher*, vol. 107, no. 8, April 2014. Like many topics in number theory, PFS concerns easily stated problems but quickly leads to difficult and unanswered questions. Here is the process:

- Begin with a composite natural number, say 15.
- Obtain its prime factorization: $15 = 3 \cdot 5$
- Form a new natural number by arranging all the prime factors in ascending order: 35
- Repeat the steps until a prime number is produced, called the **home prime**. For 15, we can summarize the entire process as shown here.

$$15 = 3 \cdot 5 \rightarrow 35 = 5 \cdot 7 \rightarrow 57 = 3 \cdot 19 \rightarrow 319 = 11 \cdot 29$$

$$\rightarrow 1129 \text{ (a prime)}$$

If we call the prime 1129 the "parent," then the composite 15 is its "child." If two or more children are found to have the same parent, call them "siblings." The Sieve of Eratosthenes (**Table 1** of this section) provides plenty of composites to get started. We consider just a few arbitrarily chosen examples here. To do very much prime factor splicing, you would want to use some computer algebra system (CAS) for the factoring. In the case above, the child is 15, the parent is 1129, and the four arrows show that it took four iterations to get from child to parent.

For Group or Individual Investigation

Complete the table, and then solve each problem.

Child	Number of Iterations	Parent
4	——————	——————
6	——————	——————
9	——————	——————
10	——————	——————
12	——————	——————
22	——————	——————
25	——————	——————
33	——————	——————
46	——————	——————
55	——————	——————

1. Is it possible for a child to have more than one parent?
2. Can a parent have more than one child?
3. The child _____ shows that it is possible for PFS to require at least _____ iterations.
4. It is possible for a family to include at least _____ siblings.
5. The ten "children" in the table represent _____ groups of siblings.
6. Show that if 49 and 77 both have parents, their parents must be the same number.

5.2 LARGE PRIME NUMBERS

OBJECTIVES

1 Prove the infinitude of primes.

2 Identify prime numbers of several categories.

3 Explore several potential prime-number-generating formulas.

The Infinitude of Primes

One important basic result about prime numbers was proved by Euclid around 300 B.C., namely that there are infinitely many primes. This means that no matter how large a prime we identify, there are always others even larger. Euclid's proof remains today as one of the most elegant proofs in all of mathematics. (An *elegant* mathematical proof is one that demonstrates the desired result in a most direct, concise manner. Mathematicians strive for elegance in their proofs.) It is called a **proof by contradiction.**

A statement can be proved by contradiction as follows: Assume that the negation of the statement is true, and use that assumption to produce some sort of contradiction, or absurdity. Logically, the fact that the negation of the original statement leads to a contradiction means that the original statement must be true.

To better understand a particular part of the proof that there are infinitely many primes, first examine the following argument.

Suppose that $M = 2 \cdot 3 \cdot 5 \cdot 7 + 1 = 211$. Now M is the product of the first four prime numbers, plus 1. If we divide 211 by each of the primes 2, 3, 5, and 7, the remainder is always 1.

$$
\begin{array}{llll}
105 & 70 & 42 & 30 \\
2)\overline{211} & 3)\overline{211} & 5)\overline{211} & 7)\overline{211} \\
210 & 210 & 210 & 210 \\
\;\;1 & \;\;1 & \;\;1 & \;\;1
\end{array}
$$

All remainders are 1.

Thus 211 is not divisible by any of the primes 2, 3, 5, and 7.

Now we can present Euclid's proof that there are infinitely many primes. If *there is no largest prime number,* then there must be infinitely many primes.

The Riemann Hypothesis is an insightful conjecture stated in the mid-1800s by **Georg Friedrich Bernhard Riemann** (1826–1866). It concerns how the prime numbers are distributed on the number line and is undoubtedly the most important unproven claim in all of mathematics. Thousands of other "theorems" are built upon the assumption of its truth. If it were ever disproved, those results would fall apart. A proof, on the other hand, may provide sufficient understanding of the primes to demolish public key cryptography, upon which rests the security of all Internet commerce (among other things).

In late 2018, British-Lebanese mathematician **Sir Michael Atiyah** (nearly 90 years old at the time) claimed that he had proved the hypothesis. As of this writing, that "proof" is being evaluated by the mathematical community and may or may not be accepted as correct.

EXAMPLE 1 **Proving the Infinitude of Primes**

Prove by contradiction that there are infinitely many primes.

Solution

Suppose there is a largest prime number, called P. Form the number M such that

$$M = p_1 \cdot p_2 \cdot p_3 \cdot \cdots \cdot P + 1,$$

where $p_1, p_2, p_3, \ldots, P$ represent all the primes less than or equal to P. Now the number M must be either prime or composite.

Case 1 Suppose that M is prime.
M is obviously larger than P, so if M is prime, it is larger than the assumed largest prime P. We have reached a *contradiction*.

Case 2 Suppose that M is composite.
If M is composite, it must have a prime factor. But none of $p_1, p_2, p_3, \ldots, P$ is a factor of M, because division by each will leave a remainder of 1. (Recall the above argument.) So if M has a prime factor, it must be greater than P. But this is a *contradiction,* because P is the assumed largest prime.

In either case 1 or case 2, we reach a contradiction. The whole argument was based on the assumption that a largest prime exists, but this leads to contradictions, so there must be no largest prime, or, equivalently, ***there are infinitely many primes.***

Partitions (noun) are commonly used to partition (verb) a large meeting hall (a set) into a number of smaller rooms (subsets). The people meeting in each smaller room are associated—in a sense "equated"— so that instead of, say, hundreds of people, we can think of just a few subsets of people. (At right, we describe how the "forms" $4k$, $4k + 1$, $4k + 2$, and $4k + 3$ serve to partition the natural numbers into four disjoint subsets.)

In a mathematical "partitioning," it is essential that the union of the various subsets be equal to the original set.

We could never investigate all infinitely many primes directly. So, historically, people have observed properties of the smaller, familiar, ones and then tried to generally either "disprove" the property (usually by finding counterexamples) or prove it (usually by some deductive argument).

Here is one way to **partition** the natural numbers (divide them into disjoint subsets whose union is the entire set).

Set A: all natural numbers of the form $4k$

Set B: all natural numbers of the form $4k + 1$

Set C: all natural numbers of the form $4k + 2$

Set D: all natural numbers of the form $4k + 3$

In each case, k is some whole number, except that for set A, k cannot be 0. (Why?) Every natural number is now contained in exactly one of the sets $A, B, C,$ or D. Any number of the form $4k + 4$ would be in subset A, because

$$4k + 4 = 4(k + 1).$$

Any number of the form $4k + 5$ would be in subset B, because

$$4k + 5 = 4(k + 1) + 1, \quad \text{and so on.}$$

The subset A consists of all the multiples of 4. Each of these is divisible by 4; hence, each is composite. Now consider C. Because

$$4k + 2 = 2(2k + 1),$$

each member is divisible by 2; hence, each is composite (except for 2 itself).

Under this partitioning, all primes other than 2 must be in either B or D.

EXAMPLE 2   **Partitioning the Natural Numbers**

List the first eight members of each of the infinite sets $A, B, C,$ and D.

Solution

$A = \{4, 8, 12, 16, 20, 24, 28, 32, \dots\}$; $B = \{1, 5, 9, 13, 17, 21, 25, 29, \dots\}$

$C = \{2, 6, 10, 14, 18, 22, 26, 30, \dots\}$; $D = \{3, 7, 11, 15, 19, 23, 27, 31, \dots\}$

As mentioned earlier, sets A and C contain no primes except 2. All other primes must lie in sets B and D. (In fact, there are infinitely many primes in each.)

EXAMPLE 3   **Identifying Primes of the Forms $4k + 1$ and $4k + 3$**

Identify all primes *specifically listed* in the following infinite sets of **Example 2.**

(a) set B **(b)** set D

Solution

(a) 5, 13, 17, 29   **(b)** 3, 7, 11, 19, 23, 31

Pierre de Fermat (about 1601–1665) proved that every prime number of the form $4k + 1$ can be expressed as the sum of two squares.

EXAMPLE 4 Expressing Primes as Sums of Squares

Express each prime in the solution of **Example 3(a)** as a sum of two squares.

Solution

$5 = 1^2 + 2^2; \qquad 13 = 2^2 + 3^2; \qquad 17 = 1^2 + 4^2; \qquad 29 = 2^2 + 5^2$

The Search for Large Primes

Identifying larger and larger prime numbers and factoring large composite numbers into their prime components are of great practical importance today, because they are the basis of modern **cryptography systems**. (See the chapter opener.)

No reasonable formula has ever been found that will consistently generate prime numbers, much less "generate all primes." The most useful attempt, named to honor the French monk Marin Mersenne (1588–1648), follows.

MERSENNE NUMBERS AND MERSENNE PRIMES

For $n = 1, 2, 3, \ldots$, the **Mersenne numbers** are those generated by the formula

$$M_n = 2^n - 1.$$

(1) If n is composite, then M_n is also composite.

(2) If n is prime, then M_n may be either prime or composite.

The prime values of M_n are called the **Mersenne primes.** Large primes being verified currently are typically Mersenne primes.

Marin Mersenne (1588–1648), in his *Cogitata Physico-Mathematica* (1644), claimed that M_n was prime for $n = 2, 3, 5, 7, 13, 17, 19, 31, 67, 127,$ and 257, and composite for all other prime numbers n less than 257. Other mathematicians at the time knew that Mersenne could not have actually tested all these values, but no one else could prove or disprove them either. It was more then 300 years before all primes up to 257 were legitimately checked out, and Mersenne was finally revealed to have made five errors:

M_{61} is prime.
M_{67} is composite.
M_{89} is prime.
M_{107} is prime.
M_{257} is composite.

Long before Mersenne's time, there was general agreement on statement (1) in the box. (**Exercises 23–25** show how to find a factor of $2^n - 1$ whenever n is composite.) However, some early writers did not agree with statement (2), believing instead (incorrectly) that a prime n would always produce a prime M_n.

EXAMPLE 5 Finding Mersenne Numbers

Find each Mersenne number M_n for $n = 2, 3,$ and 5.

Solution

$$M_2 = 2^2 - 1 = 3 \qquad {\scriptstyle 2^2 = 2 \cdot 2 = 4}$$
$$M_3 = 2^3 - 1 = 7 \qquad {\scriptstyle 2^3 = 2 \cdot 2 \cdot 2 = 8}$$
$$M_5 = 2^5 - 1 = 31 \qquad {\scriptstyle 2^5 = 2 \cdot 2 \cdot 2 \cdot 2 \cdot 2 = 32}$$

Note that all three values, 3, 7, and 31, are indeed primes.

It turns out that $M_7 = 2^7 - 1 = 127$ is also a prime (see **Exercise 21** of the previous section), but it was discovered in 1536 that

$$M_{11} = 2^{11} - 1 = 2047 \quad \text{is not prime (because it is } 23 \cdot 89\text{).}$$

So prime values of n do not always produce prime M_n. Which prime values of n do produce prime Mersenne numbers (the so-called **Mersenne primes**)? No way was ever found to identify, in general, which prime values of n result in Mersenne primes. It is a matter of checking out each prime n value individually—not an easy task given that the Mersenne numbers rapidly become very large.

Prime Does Pay   Since 2008, the GIMPS program has offered participants (individuals or groups) these awards:

- $3,000 GIMPS research award for a new Mersenne prime with fewer than 100,000,000 (decimal) digits (accomplished several times)

- $50,000 for the first prime discovered with at least 100,000,000 (decimal) digits (not yet accomplished)

The Mersenne prime search yielded results slowly. By about 1600, M_n had been verified as prime for all prime n up to 19 (except for 11, as mentioned above). The next one was M_{31}, verified by Euler sometime between 1752 and 1772.

In 1876, French mathematician Edouard Lucas used a clever test he had developed to show that M_{127} (a 39-digit number) is prime. In the 1930s Lucas's method was further simplified by D. H. Lehmer, and the testing of Mersenne numbers for primality has been done ever since with the Lucas-Lehmer test. In 1952 an early computer verified that M_{521}, M_{607}, M_{1279}, M_{2203}, and M_{2281} are primes.

Since 1996, the **Great Internet Mersenne Prime Search (GIMPS)** has invited individuals and their computers (currently around 1 million CPUs worldwide) to carry out a cooperative search for Mersenne primes. Free software is provided to anyone wishing to participate. (See the margin note at below left describing "distributed computing.") Of the 50 Mersenne primes presently known, the GIMPS program has discovered the 16 largest ones. The largest known Mersenne prime is listed in **Table 5** in the next section.

During the same general period that Mersenne was thinking about prime numbers, Pierre de Fermat (about 1601–1665) conjectured that the formula

$$2^{2^n} + 1$$

would always produce a prime, for any whole number value of n. **Table 3** shows how this formula generates the first four **Fermat numbers,** which are all primes. The fifth Fermat number (from $n = 4$) is likewise prime. Fermat had verified these first five by around 1630. But the sixth Fermat number (from $n = 5$) turns out to be 4,294,967,297, which is *not* prime. (See **Exercises 13 and 14.**)

As of 2018, no more Fermat primes had been found. All F_n were known to be composite for $5 \leq n \leq 32$, but only F_0 to F_{11} had been completely factored.

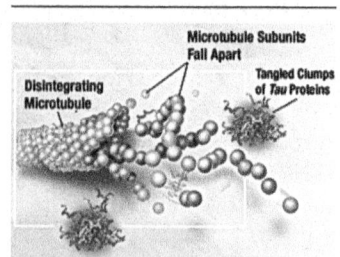

Distributed computing is a way of achieving great computer power by having lots of individual machines do separate parts of the computation. One example is the **Great Internet Mersenne Prime Search (GIMPS)**, described in this section. Another example is **SETI@ home (Search for Extraterrestrial Intelligence)**, which assigns the analysis of signal data from small patches of the "sky" to participants.

A third example, **Folding@home,** based at Stanford University, investigates the folding of proteins in living organisms into complex shapes and how they interact with other biological molecules. An example involves research into the cause of Alzheimer's disease (depicted in the image above).

Data from http://www.mersenne.org; http://setiathome.berkeley.edu; http://folding.stanford.edu

Table 3 The Generation of Fermat Numbers

n	2^n	2^{2^n}	$2^{2^n} + 1$
0	1	2	3
1	2	4	5
2	4	16	17
3	8	256	257

Of historical interest are a couple of **polynomial formulas** that produce primes. (A *polynomial* in a given variable involves adding or subtracting integer multiples of whole number powers of the variable. Discussed later in this text, polynomials are among the most basic mathematical functions.) In 1732, Leonhard Euler offered the formula

$$n^2 - n + 41, \quad \text{Euler's formula}$$

which generates primes for n up to 40 and fails at $n = 41$. In 1879, E. B. Escott produced more primes with the formula

$$n^2 - 79n + 1601, \quad \text{Escott's formula}$$

which first fails at $n = 80$.

EXAMPLE 6 Finding Prime Numbers Using Euler's and Escott's Formulas

Find the first five numbers produced by each of the polynomial formulas of Euler and Escott.

Solution

Table 4 shows the required numbers.

Table 4 A Few Polynomial-Generated Prime Numbers

n	Euler formula $n^2 - n + 41$	Escott formula $n^2 - 79n + 1601$
1	41	1523
2	43	1447
3	47	1373
4	53	1301
5	61	1231

All values found here are primes. (Use **Table 1** to verify the Euler values.)

EXAMPLE 7 Confirming the Failure of Euler's and Escott's Formulas

(a) Show that Euler's formula works for some but not all values of $n > 40$.

(b) Show that Escott's formula works for some but not all values of $n > 79$.

Solution

(a) For $n = 41$, $n^2 - n + 41 = 1681 = 41^2$ (composite).

For $n = 42$, $n^2 - n + 41 = 1763 = 41 \cdot 43$ (also composite).

For $n = 43$, $n^2 - n + 41 = 1847$ (prime).

(b) For $n = 80$, $n^2 - 79n + 1601 = 1681 = 41^2$ (composite).

For $n = 81$, $n^2 - 79n + 1601 = 1763 = 41 \cdot 43$ (also composite).

For $n = 82$, $n^2 - 79n + 1601 = 1847$ (prime).

Euler's formula first generates a composite for $n = 41$, and Escott's formula first generates a composite for $n = 80$. Beyond those values, both formulas sometimes produce composites and sometimes produce primes. (Notice that Euler's formula produces the same values for $n = 41, 42$, and 43 as Escott's formula does for $n = 80$, 81, and 82.)

Actually, it is not hard to prove that there can be no polynomial that will consistently generate primes. More complicated mathematical formulas exist for generating primes, but none produced so far can be practically applied in a reasonable amount of time, even using the fastest computers.

WHEN WILL I **EVER** USE THIS**?**

Imagine working for the U.S. Department of Justice as part of the effort to control cybercrime.

The chapter opener described the central role of prime numbers and factoring in cryptography. Another key mathematical component is **modular systems**. Partitioning the natural numbers described in this section is an example. In that case every natural number n was equated with one of just four numbers, 0, 1, 2, 3—the remainders when n is divided by 4. The divisor, any natural number greater than 1, is called the **modulus** (m), and the remainder, called the **residue**, is denoted n **(mod m)**. For example, 849,657,221 (mod 215) denotes the remainder when 849,657,221 is divided by 215.

$$\frac{849{,}657{,}221}{215} = 3951894.051 \quad \text{Use a calculator.}$$

Subtract the integer part, 3951894. (We want the remainder only, represented by the fractional part.) Now the calculator may display 0.0511628. Multiply this result by the divisor, 215, to obtain 11 (round, if necessary)—the desired remainder, or residue:

$$849{,}657{,}221 = 3{,}951{,}894 \cdot 215 + 11.$$

Now consider an even larger number. Try to find the remainder when 11^{16} is divided by 18. The number 11^{16} is so large that the calculator displays it in exponential notation:

$$11^{16} = 4.594972986\text{E}16 \quad \text{and} \quad \frac{11^{16}}{18} = 2.55276277\text{E}15.$$

The calculator cannot display enough significant digits to reveal the integer portion of the result, so we proceed as follows.

$$11^{16} \text{ (mod 18)} = 11^{8+8} \text{ (mod 18)} \qquad 16 = 8 + 8$$

$$= (11^8 \cdot 11^8) \text{ (mod 18)} \qquad \text{Rule of exponents}$$

$$= [11^8 \text{ (mod 18)}] \cdot [11^8 \text{ (mod 18)}] \qquad \text{Modular arithmetic}$$

Using a calculator,

$$11^8 = 214{,}358{,}881,$$

$$\frac{214{,}358{,}881}{18} = 11908826.7\overline{2},$$

and $\quad 0.7\overline{2} \cdot 18 = 13.$

Thus, $\quad 11^8 \text{ (mod 18)} = 13,$

$$11^{16} \text{ (mod 18)} = 13 \cdot 13 = 169$$

$$\frac{169}{18} = 9.3\overline{8}$$

$$0.3\overline{8} \cdot 18 = 7.$$

```
NORMAL FLOAT AUTO REAL DEGREE MP
11^8
                      214358881.
Ans/18
                   11908826.72
Ans-11908826
                      .722222
Ans*18
                    12.999996
```

```
NORMAL FLOAT AUTO REAL DEGREE MP
13*13
                         169.
Ans/18
                 9.388888889
Ans-9
                  .3888888889
Ans*18
                          7.
```

The desired remainder, or residue, is 7.

In practice the numbers would be larger, and high-powered computers would be used.

5.2 EXERCISES

CONCEPT CHECK *Write* true *or* false *for each statement.*

1. If n is prime, then $2^n - 1$ must be prime also.

2. There are infinitely many Fermat numbers.

3. A proof by contradiction assumes the negation of a statement and proceeds until a contradiction is encountered.

4. Marin Mersenne gave the first proof that there are infinitely many prime numbers.

5. There can be no polynomial formula that consistently generates prime numbers.

6. As of 2018, only five Fermat primes had ever been found.

Solve each problem.

7. Find the next three primes, of the form $4k + 1$, *not* listed specifically in **Example 3(a),** and express each as the sum of two squares.

8. Recall the first few perfect squares: 1, 4, 9, 16, 25. Try writing the numbers of the form $4k + 3$ listed in **Example 3(b)** as sums of two squares. Then complete this statement: The primes tested, of the form $4k + 3$, _____ be expressed as the sum of two squares.
(can / cannot)
(Fermat claimed, but did not prove, that *no* prime of the form $4k + 3$ was the sum of two squares. Euler proved it 100 years later.)

The following exercises relate to modular systems, briefly explained in the When Will I Ever Use This feature.

9. How many different remainders are possible when any positive integer is divided by 6? (*MT* calendar problem)

10. Find the remainder when 13^{20} is divided by 11. (*MT* calendar problem)

11. The International Standard Book Number (ISBN) is a unique numeric commercial book identifier. The tenth digit of ISBN-10 is a check digit. To find it, multiply the first digit by 10, the second by 9, the third by 8, and so on until the ninth digit is multiplied by 2. The tenth digit added to the sum of these nine products must result in 0 modulo 11. Find the check digit, c, for 0-9773045-6-c, (*MT* calendar problem)

12. The ISBN-13 check digit is found by multiplying each of the first twelve digits by 1 and 3 alternately. The sum of these twelve products plus the check digit must be 0 modulo 10. Find the check digit for The ISBN-13 number 978-097730458-c. (*MT* calendar problem)

13. (a) Evaluate the Fermat number $F_4 = 2^{2^n} + 1$ for $n = 4$.

(b) In seeking possible prime factors of the Fermat number of part (a), what is the largest potential prime factor that one would have to try? (As stated in the text, F_4 is in fact prime.)

14. (a) Verify the value given in the text for the "sixth" Fermat number ($2^{2^5} + 1$).

(b) Divide this Fermat number by 641 and express it in factored form. (Euler discovered this factorization in 1732, proving that the sixth Fermat number is not prime.)

Write a short answer for each problem.

15. Explain how the expressions $4k$, $4k + 1$, $4k + 2$, and $4k + 3$ serve to "partition" the natural numbers.

16. Explain why **Example 4** does not prove that every prime of the form $4k + 1$ can be expressed as a sum of two squares.

17. The 50th Mersenne prime (in order of discovery) was announced on January 3, 2018. Write a short report on when, how, and by whom it was found.

18. The margin note on Marin Mersenne cites a 1644 claim that was not totally resolved for some 300 years. Find out when, and by whom, Mersenne's five errors were demonstrated. (*Hint:* One was mentioned in a margin note in the previous section.)

19. Why do you suppose it normally takes up to a few years to discover each new Mersenne prime?

Solve each problem.

20. In Euclid's proof that there is no largest prime, we formed a number M by taking the product of primes and adding 1. Observe the pattern below.

$M = 2 + 1 = 3$	(3 is prime)
$M = 2 \cdot 3 + 1 = 7$	(7 is prime)
$M = 2 \cdot 3 \cdot 5 + 1 = 31$	(31 is prime)
$M = 2 \cdot 3 \cdot 5 \cdot 7 + 1 = 211$	(211 is prime)
$M = 2 \cdot 3 \cdot 5 \cdot 7 \cdot 11 + 1 = 2311$	(2311 is prime)

It may seem as though this pattern will always yield a prime number. Now evaluate

$$M = 2 \cdot 3 \cdot 5 \cdot 7 \cdot 11 \cdot 13 + 1.$$

21. Is the final value of M found in **Exercise 20** prime or composite? If it is composite, give its prime factorization.

22. The result of **Exercise 21** illustrates that, even though

$$p_1 \cdot p_2 \cdot p_3 \cdot \dots \cdot p_n + 1$$

is not divisible by any of the primes in the product, it may be divisible by another *prime* greater than any of them, which makes the expression _____.
(composite / prime)

The Mersenne number M_n is composite whenever n is composite. Exercises 23–25 develop one way in which you can always find a factor of such a Mersenne number.

23. For the composite number $n = 6$, find

$$M_n = 2^n - 1.$$

24. Notice that $p = 3$ is a prime factor of $n = 6$. Find $2^p - 1$ for $p = 3$. Is $2^p - 1$ a factor of $2^n - 1$?

25. Complete this statement: If p is a prime factor of n, then ____ is a factor of the Mersenne number $2^n - 1$.

26. Find $M_n = 2^n - 1$ for $n = 10$.

27. Use the statement of **Exercise 25** to find two distinct factors of M_{10}.

28. Do you think this procedure will always produce *prime* factors of M_n for composite n? (*Hint:* Consider $n = 22$ and its prime factor $p = 11$, and recall the statement following **Example 5.**) Explain.

Write a short answer for each problem.

29. Explain why large prime numbers are important in modern cryptography systems.

30. Describe the difference between Mersenne *numbers* and Mersenne *primes*.

5.3 SELECTED TOPICS FROM NUMBER THEORY

OBJECTIVES

1 Identify perfect numbers.

2 Identify deficient and abundant numbers.

3 Identify amicable (friendly) numbers.

4 State and evaluate examples of Goldbach's conjecture.

5 Identify twin primes.

6 State and evaluate Fermat's Last Theorem.

Perfect Numbers

Earlier we introduced figurate numbers, a topic investigated by the Pythagoreans, a group of Greek mathematicians and musicians who held their meetings in secret. In this section we examine some of the other special numbers that fascinated the Pythagoreans and are still studied by mathematicians today.

The number b is a "divisor" of the number a if $\frac{a}{b}$ is a natural number. The **proper divisors** of a natural number include all divisors of the number except the number itself. For example, the proper divisors of 8 are 1, 2, and 4. (8 is *not* a proper divisor of 8.)

PERFECT NUMBERS

A natural number is said to be **perfect** if it is equal to the sum of its proper divisors.

Is 8 perfect? No, because $1 + 2 + 4 = 7$, and $7 \neq 8$. The least of the perfect numbers is 6. The proper divisors of 6 are 1, 2, and 3, and

$$1 + 2 + 3 = 6. \quad \text{6 is perfect.}$$

EXAMPLE 1 Verifying a Perfect Number

Show that 28 is a perfect number.

Solution

The proper divisors of 28 are 1, 2, 4, 7, and 14. The sum of these is 28:

$$1 + 2 + 4 + 7 + 14 = 28.$$

By the definition, 28 is perfect.

The numbers 6 and 28 are the two least perfect numbers. The next two are 496 and 8128. These first four perfect numbers led early writers to conjecture that

1. The nth perfect number contains exactly n digits.

2. The even perfect numbers end in the digits 6 and 8, alternately.

} Conjectures NOT NECESSARILY TRUE

(**Exercises 43–45** will help you analyze these conjectures.)

A Dull Number? The Indian mathematician **Srinivasa Ramanujan** (1887–1920) developed many ideas in number theory. His friend and collaborator on occasion was G. H. Hardy, also a number theorist and professor at Cambridge University in England.

A story has been told about Ramanujan that illustrates his genius. Hardy once mentioned to Ramanujan that he had just taken a taxicab with a rather dull number: 1729. Ramanujan countered by saying that this number isn't dull at all; it is the least natural number that can be expressed as the sum of two cubes in two different ways:

$$1^3 + 12^3 = 1729$$

and $\quad 9^3 + 10^3 = 1729.$

Today, nearly all number theorists develop and test their ideas with the aid of powerful computers.

There still are many unanswered questions about perfect numbers. Euclid showed that the following is true.

> If $2^n - 1$ is prime, then $2^{n-1}(2^n - 1)$ is perfect, and conversely.

Because the prime values of $2^n - 1$ are the Mersenne primes (discussed in the previous section), this means that for every new Mersenne prime discovered, another perfect number is automatically revealed. (Hence, as of 2018, there were 50 known perfect numbers.) Also, it is known that the following is true.

> *All* even perfect numbers must take the form $2^{n-1}(2^n - 1).$

It is strongly suspected that no odd perfect numbers exist. (Any odd one would have at least eight different prime factors and would have at least 300 decimal digits.)

Deficient and Abundant Numbers

Earlier we saw that 8 is not perfect because it is not equal to the sum of its proper divisors ($8 \neq 7$). Next we define two alternative categories for natural numbers that are *not* perfect.

DEFICIENT AND ABUNDANT NUMBERS

A natural number is **deficient** if it is greater than the sum of its proper divisors. It is **abundant** if it is less than the sum of its proper divisors.

Based on this definition, a *deficient number* is one with proper divisors that add up to less than the number itself, while an *abundant number* is one with proper divisors that add up to more than the number itself. For example, because the proper divisors of 8 (1, 2, and 4) add up to 7, which is less than 8, the number 8 is deficient.

EXAMPLE 2 Identifying Deficient and Abundant Numbers

Decide whether each number is deficient or abundant.

(a) 12   **(b)** 10

Solution

(a) The proper divisors of 12 are 1, 2, 3, 4, and 6. The sum of these divisors is 16. Because $16 > 12$, the number 12 is abundant.

(b) The proper divisors of 10 are 1, 2, and 5. Because $1 + 2 + 5 = 8$, and $8 < 10$, the number 10 is deficient.

Amicable (Friendly) Numbers

Suppose that we add the proper divisors of 284.

$$1 + 2 + 4 + 71 + 142 = 220$$

Their sum is 220. Now, add the proper divisors of 220.

$$1 + 2 + 4 + 5 + 10 + 11 + 20 + 22 + 44 + 55 + 110 = 284$$

The sum of the proper divisors of 220 is 284, while the sum of the proper divisors of 284 is 220. Number pairs with this property are said to be *amicable,* or *friendly.*

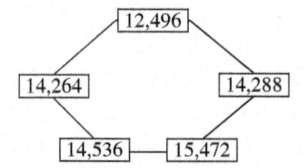

An extension of the idea of amicable numbers results in **sociable numbers.** In a chain of sociable numbers, the sum of the proper divisors of each number is the next number in the chain, and the sum of the proper divisors of the last number in the chain is the first number. A 5-link chain of sociable numbers is shown above.

   The number 14,316 starts a 28-link chain of sociable numbers.

AMICABLE, OR FRIENDLY, NUMBERS

The natural numbers a and b are **amicable,** or **friendly,** if the sum of the proper divisors of a is b, and the sum of the proper divisors of b is a.

   The smallest pair of amicable numbers, 220 and 284, was known to the Pythagoreans, but it was not until 1636 that Fermat found the next pair: 17,296 and 18,416. Many more pairs were found over the next few decades (mostly by Euler, who developed an ingenius procedure for doing so), but it took a 16-year-old Italian boy named Nicolo Paganini to discover, in the year 1866, that the pair of amicable numbers 1184 and 1210 had been overlooked for centuries!

   Today, powerful computers continually extend the lists of known amicable pairs. It still is unknown whether there are infinitely many such pairs. No one has found an amicable pair without prime factors in common, but the possibility of such a pair has not been eliminated.

Goldbach's Conjecture

The mathematician Christian Goldbach (1690–1764) stated the following conjecture (guess), which is one of the most famous unsolved problems in mathematics. Mathematicians have tried to prove the conjecture but have not succeeded. However, the conjecture has been verified (as of late 2018) for numbers up to 4×10^{18}.

GOLDBACH'S CONJECTURE (NOT PROVED)

Every even number greater than 2 can be written as the sum of two prime numbers.

   Examples: $8 = 5 + 3$
$$10 = 5 + 5 \text{ (or } 10 = 7 + 3\text{)}$$

EXAMPLE 3   Expressing Numbers as Sums of Primes

Write each even number as the sum of two primes.

(a) 18   **(b)** 60

Solution

(a) $18 = 5 + 13$. Another way of writing it is $7 + 11$. Notice that $1 + 17$ is *not* valid because by definition 1 is not a prime number.

(b) $60 = 7 + 53$. Can you find other ways? Why is $3 + 57$ not valid?

Twin Primes

Prime numbers that differ by 2 are called **twin primes.** Some twin prime pairs are

3 and 5,   5 and 7,   11 and 13,   and so on.

Like Goldbach's conjecture, the following conjecture about twin primes has never been proved. However, substantial progress in the last few years has pushed mathematicians significantly closer to proofs of the two famous conjectures. And the methods employed may also turn out to be useful in confirming the very important Riemann hypothesis (described in a margin note earlier in this chapter).

Mathematics professor Gregory Larkin, played by Jeff Bridges, woos colleague Rose Morgan (Barbra Streisand) in the 1996 film *The Mirror Has Two Faces*. Larkin's research and book focus on the **twin prime conjecture,** which he correctly states in a dinner scene. He is amazed that his nonmathematician friend actually understands what he is talking about.

TWIN PRIME CONJECTURE (NOT PROVED)

There are infinitely many pairs of twin primes.

You may wish to verify that there are eight such pairs less than 100, using the Sieve of Eratosthenes in **Table 1.** As of 2018, the greatest known twin primes were

$$2{,}996{,}863{,}034{,}895 \cdot 2^{1{,}290{,}000} \pm 1. \quad \text{Each contains 388,342 decimal digits.}$$

Recall from the previous section that Euclid's proof of the infinitude of primes used numbers of the form

$$p_1 \cdot p_2 \cdot p_3 \cdot \cdots \cdot p_n + 1,$$

that is, the product of the first n primes, plus 1. It may seem that any such number must be prime, but that is not so. (See **Exercises 20 and 21** of the previous section.) However, this form often does produce primes (as does the same form with the plus replaced by a minus).

When *all* the primes up to p_n are included, the resulting numbers, if prime, are called **primorial primes.** They are denoted

$$p\# \pm 1.$$

For example, $5\# + 1 = 2 \cdot 3 \cdot 5 + 1 = 31$ is a primorial prime. The greatest known primorial prime is given in **Table 5.**

Table 5 Greatest Known Primes of Several Types (as of 2018)

Type	Prime Number	Number of Decimal Digits
Mersenne	$M_{77{,}232{,}917} = 2^{77{,}232{,}917} - 1$	23,249,425
Factorial (see **Exercises 56–62**)	$208{,}003! - 1$	1,015,843
Primorial	$1{,}098{,}133\# - 1$	476,311
Twin	$2{,}996{,}863{,}034{,}895 \cdot 2^{1{,}290{,}000} \pm 1$	388,342
Sophie Germain (see **Exercises 52–55**)	$2{,}618{,}163{,}402{,}417 \cdot 2^{1{,}290{,}000} - 1$	388,342

Data from www.primes.utm.edu

EXAMPLE 4 Verifying Twin Primes

Verify that the primorial formula $p\# \pm 1$ produces twin prime pairs for both **(a)** $p = 3$ and **(b)** $p = 5$.

Solution

(a) $3\# \pm 1 = 2 \cdot 3 \pm 1 = 6 \pm 1 = 5$ and 7   Twin primes

> Multiply, then add and subtract.

(b) $5\# \pm 1 = 2 \cdot 3 \cdot 5 \pm 1 = 30 \pm 1 = 29$ and 31   Twin primes

The popular animated television series *The Simpsons* provides not only humor and social commentary but also lessons in mathematics. One episode depicted the equation

$$1782^{12} + 1841^{12} = 1922^{12},$$

which, according to **Fermat's Last Theorem,** cannot be true. Your calculator may indicate that the equation is true, but this is because it cannot accurately display powers of this size. Actually, 1782^{12} must be an *even* number because *an even number to any power is even.* Also, 1841^{12} must be *odd,* because *an odd number to any power is odd.* So the sum on the left must be *odd,* because

even + odd = odd.

Similarly, 1922^{12} must be *even.* So the equation states that an odd number equals an even number, which is impossible. (See www.simpsonsmath.com)

Fermat's Last Theorem

In any right triangle with shorter sides (legs) a and b, and longest side (hypotenuse) c, the equation $a^2 + b^2 = c^2$ will hold true. This is the famous Pythagorean theorem. For example,

$$a^2 + b^2 = c^2$$
$$3^2 + 4^2 = 5^2 \qquad \text{Let } a = 3, b = 4, \text{ and } c = 5.$$
$$9 + 16 = 25 \qquad \text{Apply the exponents.}$$
$$25 = 25. \qquad \text{True}$$

It is known that there are infinitely many such triples (a, b, c) of natural numbers that satisfy the equation $a^2 + b^2 = c^2$. Is something similar true of the equation

$$a^n + b^n = c^n$$

for natural numbers $n \geq 3$? Pierre de Fermat, profiled in the next section, thought that not only were there not infinitely many such triples, but that there were, in fact, none. He made the following claim in the 1600s.

FERMAT'S LAST THEOREM (PROVED IN THE 1990s)

For *any* natural number $n \geq 3$, there are *no* triples (a, b, c) that satisfy the equation

$$a^n + b^n = c^n.$$

Fermat's assertion was the object of some 350 years of attempts by mathematicians to provide a suitable proof. While it was verified for many specific cases (Fermat himself proved it for $n = 3$), a proof of the general case could not be found until the Princeton mathematician Andrew Wiles announced a proof in the spring of 1993. Although some flaws were discovered in his argument, Wiles was able, by the fall of 1994, to repair the proof.

There were probably about 100 mathematicians around the world qualified to understand the Wiles proof. ***Today Fermat's Last Theorem finally is regarded by the mathematics community as officially proved.***

EXAMPLE 5 **Applying a Theorem Proved by Fermat**

One of the theorems legitimately proved by Fermat is as follows:

Every odd prime can be expressed as the difference
of two squares in one and only one way.

Express each odd prime as the difference of two squares.

(a) 3   **(b)** 7

Solution

(a) $3 = 4 - 1 = 2^2 - 1^2$ **(b)** $7 = 16 - 9 = 4^2 - 3^2$

5.3 EXERCISES

CONCEPT CHECK *Write* true *or* false *for each statement.*

1. Given a prime number, no matter how large, there is always another prime even larger.

2. The prime numbers 2 and 3 are twin primes.

3. All prime numbers are deficient.

4. The equation $17 + 51 = 68$ verifies Goldbach's conjecture for the number 68.

5. The first and third perfect numbers both end in the digit 6, and the second and fourth perfect numbers both end in the digits 28.

6. For every Mersenne prime, there is a corresponding perfect number.

7. The number $2^5(2^6 - 1)$ is perfect.

8. Every natural number greater than 1 must be one of the following: prime, abundant, or deficient.

9. Even perfect numbers are more plentiful than Mersenne primes.

10. The twin prime conjecture was proved in 2013.

Solve each problem.

11. The proper divisors of 496 are 1, 2, 4, 8, 16, 31, 62, 124, and 248. Use this information to verify that 496 is perfect.

12. The proper divisors of 8128 are 1, 2, 4, 8, 16, 32, 64, 127, 254, 508, 1016, 2032, and 4064. Use this information to verify that 8128 is perfect.

13. As mentioned in the text, when $2^n - 1$ is prime,

$$2^{n-1}(2^n - 1)$$

is perfect. By letting $n = 2, 3, 5,$ and 7, we obtain the first four perfect numbers. Show that $2^n - 1$ is prime for $n = 13$, and then find the decimal digit representation for the fifth perfect number.

14. According to **Table 5,** the largest known Mersenne prime number in 2018 was $2^{77,232,917} - 1$. Use the formula in **Exercise 13** to write an expression for the perfect number generated by this prime number.

15. Express 85 as a sum of two squares in two different ways.

16. Express 200 as a sum of three squares.

Determine whether each number is abundant *or* deficient.

17. 32   **18.** 60   **19.** 84   **20.** 75

21. There are four abundant numbers between 1 and 25. Find them. (*Hint:* They are all even, and no prime number is abundant.)

22. A prime number p must be deficient because the sum of its proper divisors is ____, which is less than p.

23. There are no odd abundant numbers less than 945. If 945 has proper divisors 1, 3, 5, 7, 9, 15, 21, 27, 35, 45, 63, 105, 135, 189, and 315, determine whether it is abundant or deficient.

24. A margin note in this section displayed a 5-link chain of sociable numbers. Does that chain proceed clockwise or counterclockwise?

25. The proper divisors of 1184 are 1, 2, 4, 8, 16, 32, 37, 74, 148, 296, and 592. The proper divisors of 1210 are 1, 2, 5, 10, 11, 22, 55, 110, 121, 242, and 605. Verify that 16-year-old Nicolo Paganini was correct about these two numbers in 1866. See the subsection "Amicable (Friendly) Numbers."

26. An Arabian mathematician of the ninth century stated the following: "If the three numbers

$$x = 3 \cdot 2^{n-1} - 1,$$
$$y = 3 \cdot 2^n - 1,$$
and $$z = 9 \cdot 2^{2n-1} - 1$$

are all prime and $n \geq 2$, then $2^n xy$ and $2^n z$ are amicable numbers."

(a) Use $n = 2$, and show that the result is the least pair of amicable numbers, namely 220 and 284.

(b) Use $n = 4$ to obtain another pair of amicable numbers.

Write each even number as the sum of two primes. (In general there may be more than one way to do this.)

27. 12   **28.** 24   **29.** 40   **30.** 54

31. Joseph Louis Lagrange (1736–1813) conjectured that every odd natural number greater than 5 can be written as a sum $a + 2b$, where a and b are both primes.

(a) Verify this for the odd natural number 11.

(b) Verify that the odd natural number 17 can be written in this form in four different ways.

32. Another unproved conjecture in number theory states that every natural number multiple of 6 can be written as the difference of two primes. Verify this for 12 and 24.

Find one pair of twin primes between the given numbers.

33. 45, 65

34. 65, 85

In 1982, the mathematician **Albert Wilansky,** *when phoning his brother-in-law, Mr. Smith, noticed an interesting property concerning Smith's phone number (493-7775). The number 4,937,775 is composite, and its prime factorization is*

$$3 \cdot 5 \cdot 5 \cdot 65,837.$$

When the digits of the phone number are added, the result, 42, is equal to the sum of the digits in the prime factors:

$$3 + 5 + 5 + 6 + 5 + 8 + 3 + 7 = 42.$$

Wilansky termed a composite number with this property a **Smith number.**

There is one Smith number less than 10, and there are six less than 100, forty-nine less than 1000, and infinitely many altogether (proved in 1985). But there remain many unanswered questions about them. The second through the tenth are

22, 27, 58, 85, 94, 121, 166, 202, 265.

35. Identify the least (first) Smith number.

36. Identify those among the first ten Smith numbers that are palindromes (read the same forward and backward).

37. Identify a pair of Smith numbers such that one is the other with its digits reversed.

38. Confirm that 265 is a Smith number.

Pierre de Fermat provided proofs of many theorems in number theory. Exercises 39–42 investigate some of these theorems.

39. If p is prime and the natural numbers a and p have no common factor except 1, then $a^{p-1} - 1$ is divisible by p.
   (a) Verify this for $p = 5$ and $a = 3$.
   (b) Verify this for $p = 7$ and $a = 2$.

40. Every odd prime can be expressed as the difference of two squares in one and only one way.
   (a) Find this one way for the prime number 5.
   (b) Find this one way for the prime number 11.

41. There is only one solution in natural numbers for $a^2 + 2 = b^3$, and it is $a = 5, b = 3$. Verify this solution.

42. There are only two solutions in integers for $a^2 + 4 = b^3$. One solution is $a = 2, b = 2$. Find the other solution.

The first four perfect numbers were identified in the text: 6, 28, 496, and 8128. The next two are 33,550,336 and 8,589,869,056. Use this information about perfect numbers to work Exercises 43–45.

43. Verify that each of these six perfect numbers ends in either 6 or 28. (In fact, this is true of all even perfect numbers.)

44. Is conjecture (1) in the text (that the nth perfect number contains exactly n digits) true or false? Explain.

45. Is conjecture (2) in the text (that the even perfect numbers end in the digits 6 and 8, alternately) true or false? Explain.

According to the Web site www.shyamsundergupta.com/ amicable.htm, a natural number is **happy** *if the process of repeatedly summing the squares of its decimal digits finally ends in 1. For example, the least natural number (greater than 1) that is happy is 7, as shown here.*

$$7^2 = 49, \quad 4^2 + 9^2 = 97, \quad 9^2 + 7^2 = 130,$$
$$1^2 + 3^2 + 0^2 = 10, \quad 1^2 + 0^2 = 1$$

An amicable pair is a **happy amicable pair** *if and only if both members of the pair are happy numbers. (The first 5000 amicable pairs include only 111 that are happy amicable pairs.) For each amicable pair, determine whether neither, one, or both of the members are happy, and whether the pair is a happy amicable pair.*

46. 220 and 284

47. 1184 and 1210

48. 10,572,550 and 10,854,650

49. 35,361,326 and 40,117,714

50. Refer to the primorial formula.
   (a) What two numbers does the primorial formula produce for $p = 7$?
   (b) Which, if either, of these numbers is prime?

51. Choose the correct completion: The primorial formula produces twin primes
   A. never.   **B.** sometimes.   **C.** always.

Sophie Germain (1776–1831) studied at the École Polytechnique in Paris in a day when female students were not admitted. A **Sophie Germain prime** *is a prime p for which $2p + 1$ is also prime. Note that the prime 11 is a Sophie Germain prime because $2 \cdot 11 + 1 = 23$ is also a prime; but the prime 13 is not a Sophie Germain prime because $2 \cdot 13 + 1 = 27$ is not a prime. (The greatest known Sophie Germain prime is given in* **Table 5.***) Complete the table on the next page.*

	p	$2p + 1$	Is p a Sophie Germain prime?
52.	2	___	___
53.	3	___	___
54.	5	___	___
55.	7	___	___

Factorial primes *are of the form* $n! \pm 1$ *for natural numbers* n. *(n! denotes "n factorial," the product of all natural numbers up to n, not just the primes as in the primorial primes. For example,* $4! = 1 \cdot 2 \cdot 3 \cdot 4 = 24$.) *(The greatest known factorial prime is given in* **Table 5**. *Find the missing entries in the following table.*

	n	$n!$	$n! - 1$	$n! + 1$	Is $n! - 1$ prime?	Is $n! + 1$ prime?
	2	2	1	3	no	yes
56.	3	___	___	___	___	___
57.	4	___	___	___	___	___
58.	5	___	___	___	___	___

59. Explain why the factorial prime formula does not give twin primes for $n = 2$.

Based on the preceding table, complete each statement with one of the following:

A. *never,*   **B.** *sometimes, or*   **C.** *always.*

When applied to particular values of n, the factorial prime formula $n! \pm 1$ produces

60. no primes ____.   **61.** exactly one prime ____.

62. twin primes ____.

Because it does not equal the sum of its proper divisors, an abundant number is not perfect, but it is called **pseudoperfect** *if it is equal to the sum of a subset of its proper divisors.*

 For example, 12, *an abundant number, has proper divisors* 1, 2, 3, 4, *and* 6. $12 \neq 1 + 2 + 3 + 4 + 6$, *but*

$$12 = 2 + 4 + 6 \quad and \ also \quad 12 = 1 + 2 + 3 + 6.$$

(Either of these sums is sufficient to show that 12 *is pseudoperfect.)*

 In general, to confirm whether an abundant number n is pseudoperfect, carry out the following steps.

  1. List the proper divisors of n, and compute their sum s.

  2. Evaluate $a = s - n$, the amount by which s exceeds n.

  3. **(a)** If there is a subset of proper divisors with sum a, omit that subset. The remaining proper divisors will sum to n, so n is pseudoperfect by definition.

   (b) If no subset of proper divisors has sum a, then n is not pseudoperfect.

Show that each number is pseudoperfect.

63. 20 (Show the sum that confirms it.)

64. 42 (Show a sum that confirms it.)

65. 18 (Show it with two different sums.)

66. 24 (Show it with five different sums.)

Abundant numbers are so commonly pseudoperfect that when we find one that isn't, we call it **weird***. There are no weird numbers less than* 70. *Do the following exercises to investigate* 70.

67. Show that 70 is abundant.

68. Show that 70 is *not* pseudoperfect and must therefore be the smallest weird number. (Among the first 10,000 counting numbers, there are only seven weird ones: 70, 836, 4030, 5830, 7192, 7912, and 9272.)

Do the following exercises to confirm that the second number listed above, 836, *is indeed weird.*

69. Give the prime factorization of 836.

70. By observing exponents in the prime factorization, give the total number of divisors of 836.

71. Give the number of proper divisors of 836, list them, and give their sum.

72. Confirm that 836 is abundant.

73. Give the amount by which the sum of the proper divisors of 836 exceeds 836. Is there a subset of the proper divisors of 836 that sums to this amount?

74. What do you conclude?

The following number is known as **Belphegor's prime.**

One nonillion, sixty-six quadrillion, six hundred trillion, one

Belphegor, who is referred to in many literary works, is one of the seven princes of hell, known for tempting us toward particular evils. Exercises 75–78 refer to Belphegor's prime.

75. Write out the decimal form of the number. (It is a "palindrome"—that is, it reads the same backward and forward. And in keeping with its name, it actually is a prime.)

76. What are the "middle three" digits? (The book of Revelation calls this "the number of the beast." It is also a palindrome, as well as a Smith number.)

69. How many zeroes are on each side of the middle three digits?

78. With what digit does Belphegor's prime begin and end?

FOR FURTHER THOUGHT

Curious and Interesting

One of the most remarkable books on number theory is *The Penguin Dictionary of Curious and Interesting Numbers* (1997) by David Wells. This book contains fascinating numbers and their properties, including the following.

- There are only three sets of three digits that form prime numbers in all possible arrangements: {1, 1, 3}, {1, 9, 9}, {3, 3, 7}.

- Find the sum of the cubes of the digits of 136:

$$1^3 + 3^3 + 6^3 = 244.$$

Repeat the process with the digits of 244:

$$2^3 + 4^3 + 4^3 = 136.$$

We're back to where we started.

- 635,318,657 is the least number that can be expressed as the sum of two fourth powers in two ways:

$$635,318,657 = 59^4 + 158^4 = 133^4 + 134^4.$$

- The number 24,678,050 has an interesting property:

$$24,678,050 = 2^8 + 4^8 + 6^8 + 7^8 + 8^8 + 0^8$$
$$+ 5^8 + 0^8.$$

- The number 54,748 has a similar interesting property:

$$54,748 = 5^5 + 4^5 + 7^5 + 4^5 + 8^5.$$

- The number 3435 has this property:

$$3435 = 3^3 + 4^4 + 3^3 + 5^5.$$

For anyone whose curiosity is piqued by such facts, the book mentioned above is for you!

For Group or Individual Investigation

Have each student in the class choose a three-digit number that is a multiple of 3. Add the cubes of the digits. Repeat the process until the same number is obtained over and over. Then, have the students compare their results. What is curious and interesting about this process?

 5.4

GREATEST COMMON FACTOR AND LEAST COMMON MULTIPLE

OBJECTIVES

1 Find the greatest common factor by several methods.

2 Find the least common multiple by several methods.

Greatest Common Factor

The *greatest common factor* is defined as follows.

GREATEST COMMON FACTOR

The **greatest common factor (GCF)** of a group of natural numbers is the largest natural number that is a factor of all the numbers in the group.

Examples: 18 is the GCF of 36 and 54, because 18 is the largest natural number that divides both 36 and 54.

1 is the GCF of 7 and 16.

The greatest common factor is often called the *greatest common divisor*. Greatest common factors can be found by using prime factorizations. To determine the GCF of 36 and 54, first write the prime factorization of each number.

$$36 = 2^2 \cdot 3^2$$

$$54 = 2^1 \cdot 3^3$$

The GCF is the product of the primes common to the factorizations, with each prime raised to the power indicated by the *least* exponent that it has in any factorization. Here, the prime 2 has 1 as the least exponent (in $54 = 2^1 \cdot 3^3$), while the prime 3 has 2 as the least exponent (in $36 = 2^2 \cdot 3^2$).

$$\text{GCF} = 2^1 \cdot 3^2 = 2 \cdot 9 = 18$$

We summarize as follows.

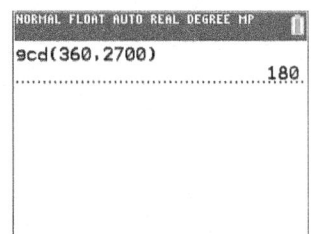

The calculator shows that the greatest common divisor (factor) of 360 and 2700 is 180. Compare with **Example 1**.

FINDING THE GREATEST COMMON FACTOR (PRIME FACTORS METHOD)

Step 1 Write the prime factorization of each number.

Step 2 Choose all primes common to *all* factorizations, with each prime raised to the *least* exponent that it has in any factorization.

Step 3 Form the product of all the numbers in Step 2. This product is the greatest common factor.

EXAMPLE 1 **Finding the Greatest Common Factor by the Prime Factors Method**

Find the greatest common factor of 360 and 2700.

Solution

Write the prime factorization of each number.

$$360 = 2^3 \cdot 3^2 \cdot 5 \qquad 2700 = 2^2 \cdot 3^3 \cdot 5^2$$

Find the primes common to both factorizations, with each prime having as its exponent the *least* exponent from either product.

Use the least exponents.

$$\text{GCF} = 2^2 \cdot 3^2 \cdot 5 = 180$$

The greatest common factor of 360 and 2700 is 180.

EXAMPLE 2 **Finding the Greatest Common Factor by the Prime Factors Method**

Find the greatest common factor of 720, 1000, and 1800.

Solution

Write the prime factorization for each number.

$$720 = 2^4 \cdot 3^2 \cdot 5$$
$$1000 = 2^3 \cdot 5^3$$
$$1800 = 2^3 \cdot 3^2 \cdot 5^2$$

Use the least exponent on each prime common to the factorizations.

$$\text{GCF} = 2^3 \cdot 5 = 40$$

(The prime 3 is not used in the greatest common factor because it does not appear in the prime factorization of 1000.)

Excursions in Number Theory, first published in 1966, is one of many popular works that have helped to grab the minds of interested readers and awaken them to a fundamental field of mathematics. (Carl Friedrich Gauss said that "mathematics is the queen of the sciences, and the theory of numbers is the queen of mathematics.")

EXAMPLE 3   **Finding the Greatest Common Factor by the Prime Factors Method**

Find the greatest common factor of 80 and 63.

Solution

$$80 = 2^4 \cdot 5 \quad \text{and} \quad 63 = 3^2 \cdot 7$$

There are no primes in common here, so the GCF is 1. The number 1 is the greatest number that will divide into both 80 and 63.

Two numbers, such as 80 and 63, with a greatest common factor of 1 are called **relatively prime numbers**—that is, they are prime *relative* to one another.

Another method of finding the greatest common factor involves dividing the numbers by common prime factors.

FINDING THE GREATEST COMMON FACTOR (DIVIDING BY PRIME FACTORS METHOD)

Step 1   Write the numbers in a row.

Step 2   Divide each of the numbers by a common prime factor. Try 2, then try 3, and so on.

Step 3   Divide the quotients by a common prime factor. Continue until no prime will divide into all the quotients.

Step 4   The product of the primes in Steps 2 and 3 is the greatest common factor.

EXAMPLE 4   **Finding the Greatest Common Factor by Dividing by Prime Factors**

Find the greatest common factor of 12, 18, and 60.

Solution

Write the numbers in a row and divide by 2.

$$
\begin{array}{r|rrr}
2 & 12 & 18 & 60 \\
\hline
 & 6 & 9 & 30
\end{array}
$$

The numbers 6, 9, and 30 are not all divisible by 2, but they are divisible by 3.

$$
\begin{array}{r|rrr}
2 & 12 & 18 & 60 \\
3 & 6 & 9 & 30 \\
\hline
 & 2 & 3 & 10
\end{array}
$$

No prime divides into 2, 3, and 10, so the greatest common factor of the numbers 12, 18, and 60 is given by the product of the primes on the left, 2 and 3.

$$
\left.
\begin{array}{r|rrr}
2 & 12 & 18 & 60 \\
3 & 6 & 9 & 30 \\
\hline
 & 2 & 3 & 10
\end{array}
\right\} \quad 2 \cdot 3 = 6
$$

The GCF of 12, 18, and 60 is 6.

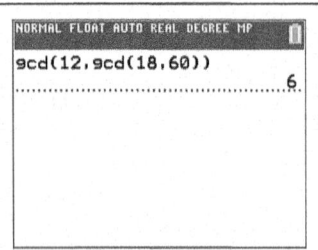

NORMAL FLOAT AUTO REAL DEGREE MP

gcd(12,gcd(18,60))
                              6.

The **greatest common divisor** is symbolized GCD or gcd. This screen uses the fact that

gcd(*a*, *b*, *c*) = gcd(*a*, gcd(*b*, *c*)).

Compare with **Example 4.**

Pierre de Fermat (about 1601–1665), a government official who did not interest himself in mathematics until he was past 30, devoted leisure time to its study. He was a worthy scholar, best known for his work in number theory. His other major contributions involved certain applications in geometry and his original work in probability.

Much of Fermat's best work survived only on loose sheets or jotted, without proof, in the margins of works that he read.

Another method of finding the greatest common factor of two numbers (but not more than two) is the **Euclidean algorithm.**

EXAMPLE 5 **Finding the Greatest Common Factor Using the Euclidean Algorithm**

Use the Euclidean algorithm to find the greatest common factor of 90 and 168.

Solution

Step 1 Begin by dividing the larger, 168, by the smaller, 90. Disregard the quotient, but note the remainder.

$$\begin{array}{r} 1 \\ 90\overline{)168} \\ 90 \\ \hline 78 \end{array}$$

Step 2 Divide the smaller of the two numbers by the remainder obtained in Step 1. Once again, note the remainder.

$$\begin{array}{r} 1 \\ 78\overline{)90} \\ 78 \\ \hline 12 \end{array}$$

Step 3 Continue dividing the successive remainders as many times as necessary to obtain a remainder of 0.

$$\begin{array}{r} 6 \\ 12\overline{)78} \\ 72 \\ \hline 6 \end{array}$$ Greatest common factor

Step 4 The *last positive remainder* in this process is the greatest common factor of 90 and 168. It can be seen that their GCF is 6.

$$\begin{array}{r} 2 \\ 6\overline{)12} \\ 12 \\ \hline 0 \end{array}$$

The Euclidean algorithm is particularly useful if the two numbers are difficult to factor into primes. We summarize the algorithm here.

FINDING THE GREATEST COMMON FACTOR (EUCLIDEAN ALGORITHM)

To find the greatest common factor of two unequal numbers, divide the larger by the smaller. Note the remainder, and divide the previous divisor by this remainder. Continue the process until a remainder of 0 is obtained. The greatest common factor is the last positive remainder obtained.

Least Common Multiple, Greatest Common Divisor

Least Common Multiple

Closely related to the idea of the greatest common factor is the concept of the *least common multiple.*

LEAST COMMON MULTIPLE

The **least common multiple (LCM)** of a group of natural numbers is the least natural number that is a multiple of all the numbers in the group.

Example: 30 is the LCM of 15 and 10 because 30 is the least number that appears in both sets of multiples.

Multiples of 15: { 15, 30, 45, 60, 75, 90, 105, . . . }

Multiples of 10: { 10, 20, 30, 40, 50, 60, 70, . . . }

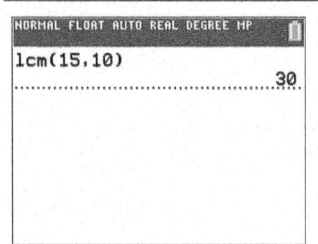

The least common multiple of 15 and 10 is 30.

The set of natural numbers that are multiples of *both* 15 and 10 form the set of *common multiples:*

$$\{30, 60, 90, 120, \ldots\}.$$

While there are infinitely many common multiples, the *least* common multiple is observed to be 30.

A method similar to the first one given for the greatest common factor may be used to find the least common multiple of a group of numbers.

FINDING THE LEAST COMMON MULTIPLE (PRIME FACTORS METHOD)

Step 1 Write the prime factorization of each number.

Step 2 Choose all primes belonging to *any* factorization, with each prime raised to the power indicated by the *greatest* exponent that it has in any factorization.

Step 3 Form the product of all the numbers in Step 2. This product is the least common multiple.

EXAMPLE 6 **Finding the Least Common Multiple by the Prime Factors Method**

Find the least common multiple of 135, 280, and 300.

Solution

Write the prime factorizations:

$$135 = 3^3 \cdot 5, \quad 280 = 2^3 \cdot 5 \cdot 7, \quad \text{and} \quad 300 = 2^2 \cdot 3 \cdot 5^2.$$

Form the product of all the primes that appear in *any* of the factorizations. Use the *greatest* exponent from any factorization.

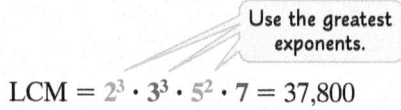

Use the greatest exponents.

$$\text{LCM} = 2^3 \cdot 3^3 \cdot 5^2 \cdot 7 = 37{,}800$$

The least natural number divisible by 135, 280, and 300 is 37,800.

The least common multiple of a group of numbers can also be found by dividing by prime factors. The process is slightly different from that for finding the GCF.

FINDING THE LEAST COMMON MULTIPLE (DIVIDING BY PRIME FACTORS METHOD)

Step 1 Write the numbers in a row.

Step 2 Divide each of the numbers by a common prime factor. Try 2, then try 3, and so on.

Step 3 Divide the quotients by a common prime factor. When no prime will divide all quotients, but a prime will divide some of them, divide where possible and bring any nondivisible quotients down. Continue until no prime will divide any two quotients.

Step 4 The product of all prime divisors from Steps 2 and 3 as well as all remaining quotients is the least common multiple.

The least common multiple of 135, 280, and 300 is 37,800. Compare with **Example 6.**

| EXAMPLE 7 | Finding the Least Common Multiple by Dividing by Prime Factors |

Find the least common multiple of 12, 18, and 60.

Solution

Proceed just as in **Example 4** to obtain the following.

$$
\begin{array}{r|rrr}
2 & 12 & 18 & 60 \\
\hline
3 & 6 & 9 & 30 \\
\hline
 & 2 & 3 & 10
\end{array}
$$

Now, even though no prime will divide 2, 3, and 10, the prime 2 will divide 2 and 10. Divide the 2 and the 10 and bring down the 3.

$$
\begin{array}{r|rrr}
2 & 12 & 18 & 60 \\
\hline
3 & 6 & 9 & 30 \\
\hline
2 & 2 & 3 & 10 \\
\hline
 & 1 & 3 & 5
\end{array} \qquad 2 \cdot 3 \cdot 2 \cdot 1 \cdot 3 \cdot 5 = 180
$$

The LCM of 12, 18, and 60 is 180.

The least common multiple of two numbers m and n can be obtained by dividing their product by their greatest common factor.

FINDING THE LEAST COMMON MULTIPLE (FORMULA)

The least common multiple of m and n can be computed as follows.

$$
\textbf{LCM} = \frac{m \cdot n}{\textbf{GCF of } m \textbf{ and } n}
$$

(This method works only for two numbers, not for more than two.)

| EXAMPLE 8 | Finding the Least Common Multiple by Formula |

Use the formula to find the least common multiple of each pair of number.

(a) 90 and 168   **(b)** 360 and 2700

Solution

(a) In **Example 5** we used the Euclidean algorithm to find that the greatest common factor of 90 and 168 is 6. Therefore, the formula gives

$$
\text{Least common multiple of 90 and 168} = \frac{90 \cdot 168}{6} = 2520.
$$

(b) With the results of Example 1, the formula gives

$$
\text{Least common multiple of 360 and 2700} = \frac{360 \cdot 2700}{180} = 5400.
$$

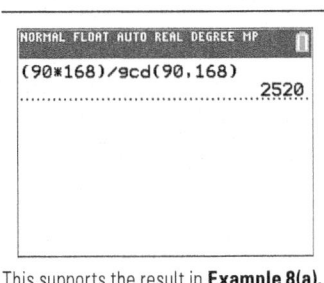

NORMAL FLOAT AUTO REAL DEGREE MP

(90*168)/gcd(90,168)

2520

This supports the result in **Example 8(a)**.

Problems that deal with questions such as "How many objects will there be in each group if each group contains the same number of objects?" and "When will two events occur at the same time?" can sometimes be solved using the ideas of greatest common factor and least common multiple.

EXAMPLE 9   Finding Common Starting Times of Movie Cycles

The King Theatre and the Star Theatre run movies continuously, and each starts its first feature at 1:00 P.M. If the movie shown at the King lasts 80 minutes and the movie shown at the Star lasts 2 hours, when will the two movies again start at the same time?

Solution

First, convert 2 hours to 120 minutes. The question can be restated as follows: "What is the least number of minutes it will take for the two movies to start at the same time again?" This is equivalent to asking,

"What is the least common multiple of 80 and 120?"

Using any of the methods described in this section, we find that the

least common multiple of 80 and 120 = 240.

Therefore, it will take

$$240 \text{ minutes, } \quad \text{or} \quad \frac{240}{60} = 4 \text{ hours,}$$

for the movies to start at the same time again. By adding 4 hours to 1:00 P.M., we find that they will start together again at 5:00 P.M.

EXAMPLE 10   Finding the Greatest Common Size of Stacks of Cards

Joshua has 450 football cards and 840 baseball cards. He wants to place them in stacks on a table so that each stack has the same number of cards, and no stack has different types of cards within it. What is the greatest number of cards that he can have in each stack?

Solution

Here, we are looking for the greatest number that will divide evenly into 450 and 840. This is, of course, the greatest common factor of 450 and 840. Using any of the methods described in this section, we find that the

greatest common factor of 450 and 840 = 30.

Therefore, the greatest number of cards he can have in each stack is 30. He will have 15 stacks of 30 football cards and 28 stacks of 30 baseball cards.

WHEN WILL I **EVER** USE THIS**?**

Addition and subtraction of fractions involve identifying the least common denominator, which is the LCM of all denominators. Suppose, for example, that a nurse needs the total displacement (volume) of the following quantities combined.

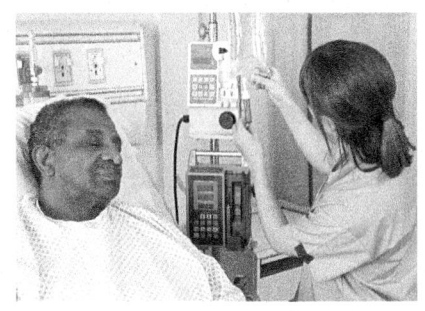

$\frac{1}{5}$ teacup, $\frac{2}{3}$ cup, 5 teaspoons, and 3 tablespoons

(In practice, the above fractions would normally first be converted to decimal equivalents in metric units, but we illustrate the math here with the common fractions and non-metric units.) The conversions necessary to obtain common units (in this case, ounces) are shown in the table.

1 teaspoon	$=$	$\frac{1}{6}$ ounce
1 tablespoon	$=$	$\frac{1}{2}$ ounce
1 teacup	$=$	6 ounces
1 cup	$=$	8 ounces

Thus, $\qquad \frac{1}{5}$ teacup $= \frac{1}{5}$ (6 ounces) $= \frac{6}{5}$ ounces,

$$\frac{2}{3} \text{ cup} = \frac{2}{3} \text{ (8 ounces)} = \frac{16}{3} \text{ ounces,}$$

$$5 \text{ teaspoons} = 5\left(\frac{1}{6} \text{ ounce}\right) = \frac{5}{6} \text{ ounce,}$$

$$3 \text{ tablespoons} = 3\left(\frac{1}{2} \text{ ounce}\right) = \frac{3}{2} \text{ ounces.}$$

We must add $\frac{6}{5} + \frac{16}{3} + \frac{5}{6} + \frac{3}{2}$. One method of finding the least common denominator is dividing by prime factors.

$$\begin{array}{c|cccc} 2 & 2 & 3 & 5 & 6 \\ \hline 3 & 1 & 3 & 5 & 3 \\ \hline & 1 & 1 & 5 & 1 \end{array} \quad \text{LCM} = 2 \cdot 3 \cdot 1 \cdot 1 \cdot 5 \cdot 1 = 30$$

$$\frac{6}{5} + \frac{16}{3} + \frac{5}{6} + \frac{3}{2} = \frac{6}{5} \cdot \frac{6}{6} + \frac{16}{3} \cdot \frac{10}{10} + \frac{5}{6} \cdot \frac{5}{5} + \frac{3}{2} \cdot \frac{15}{15} \quad \text{Use 30 as the least common denominator.}$$

$$= \frac{36}{30} + \frac{160}{30} + \frac{25}{30} + \frac{45}{30}$$

$$= \frac{36 + 160 + 25 + 45}{30}$$

$$= \frac{266}{30}, \quad \text{or} \quad 8\frac{26}{30}$$

The combined total is $8\frac{26}{30}$ ounces, or, in decimal form, about 8.867 ounces.

5.4 EXERCISES

CONCEPT CHECK *Write* true *or* false *for each statement.*

1. Any two natural numbers have at least one common factor.

2. The least common multiple of two different primes is their product.

3. If p is a prime number, then the greatest common factor of p and p^2 is p^2.

4. If p is a prime number, then the least common multiple of p and p^2 is p^3.

5. There is no prime number p such that the greatest common factor of p and 2 is 2.

6. The set of all common multiples of two given natural numbers is infinite.

7. No two even natural numbers can be relatively prime.

8. Two different prime numbers must be relatively prime.

9. No two composite numbers can be relatively prime.

10. The product of any two natural numbers is equal to the product of their least common multiple and their greatest common factor.

Use the prime factors method to find the greatest common factor of each group of numbers.

11. 84 and 140   **12.** 225 and 150

13. 275 and 132   **14.** 252 and 396

15. 68, 102, and 425   **16.** 651, 525, and 1050

Use the method of dividing by prime factors to find the greatest common factor of each group of numbers.

17. 150 and 260   **18.** 330 and 255

19. 600 and 90   **20.** 237 and 395

21. 84, 90, and 210   **22.** 585, 1680, and 990

Use the Euclidean algorithm to find the greatest common factor of each group of numbers.

23. 18 and 60   **24.** 77 and 84

25. 36 and 90   **26.** 72 and 96

27. 945 and 450   **28.** 200 and 350

Use the prime factors method to find the least common multiple of each group of numbers.

29. 48 and 60   **30.** 28 and 35

31. 81 and 45   **32.** 84 and 70

33. 20, 30, and 50   **34.** 14, 21, and 40

Use the method of dividing by prime factors to find the least common multiple of each group of numbers.

35. 27 and 36   **36.** 14 and 63

37. 63 and 99   **38.** 154, 165, and 2310

39. 48, 54, and 60   **40.** 27, 120, and 32

Use the formula given in the text and the results of Exercises 23–28 to find the least common multiple of each group of numbers.

41. 18 and 60   **42.** 77 and 84

43. 36 and 90   **44.** 72 and 96

45. 945 and 450   **46.** 200 and 350

Solve each problem.

47. Suppose that p, q, and r are different primes, and a, b, and c are natural numbers such that $a < b < c$.

   (a) What is the greatest common factor of $p^a q^c r^b$ and $p^b q^a r^c$?

   (b) What is the least common multiple of $p^b q^a$, $q^b r^c$, and $p^a r^b$?

48. Find **(a)** the greatest common factor and **(b)** the least common multiple of $2^{25} \cdot 5^{17} \cdot 7^{21}$ and $2^{28} \cdot 5^{22} \cdot 7^{13}$. Leave your answers in prime factored form.

It is possible to extend the Euclidean algorithm in order to find the greatest common factor of more than two numbers. For example, if we wish to find the greatest common factor of 150, 210, and 240, we can first use the algorithm to find the greatest common factor of two of these (say, 150 and 210). Then we find the greatest common factor of that result and the third number, 240. The final result is the greatest common factor of the original group of numbers.

Use the Euclidean algorithm as just described to find the greatest common factor of each group of numbers.

49. 90, 105, and 315   **50.** 72, 84, and 825

51. 144, 180, and 192   **52.** 375, 210, and 700

Solve each problem.

53. Suppose that the least common multiple of p and q is pq. What can we say about p and q?

54. Suppose that the least common multiple of p and q is q. What can we say about p and q?

*Refer to **Examples 9 and 10** to solve each problem.*

55. Inspecting Calculators Jameel and Fahima work on an assembly line, inspecting electronic calculators. Jameel inspects the electronics of every sixteenth calculator, while Fahima inspects the workmanship of every thirty-sixth calculator. If they both start working at the same time, which calculator will be the first that they both inspect?

56. Night Off for Security Guards Jomas and Raquel work as security guards at a factory. Jomas has every sixth night off, and Raquel has every tenth night off. If both are off on July 1, what is the next night that they will both be off together?

57. Stacking Coins Suyín has 240 pennies and 288 nickels. She wants to place them all in stacks so that each stack has the same number of coins, and each stack contains only one denomination of coin. What is the greatest number of coins that she can place in each stack?

58. Bicycle Racing Tomas and Felipe are in a bicycle race, following a circular track. If they start at the same place and travel in the same direction, and Tomas completes a revolution every 40 seconds, while Felipe takes 48 seconds to complete each revolution, how long will it take them before they reach the starting point again simultaneously?

59. Selling Books Azad sold some books at $24 each and used the money to buy some concert tickets at $50 each. He had no money left over after buying the tickets. What is the least amount of money he could have earned from selling the books? What is the least number of books he could have sold?

60. Sawing Lumber Terri has some pieces of two-by-four lumber. Some are 60 inches long, and some are 72 inches long. All of them must be sawn into shorter pieces. If all sawn pieces must be the same length, what is the longest such piece so that no lumber is left over?

61. Selling Firewood Jack runs a business sawing, splitting, delivering, and stacking (extra-cost option) firewood. His supplier brings him logs of three different lengths: 112 inches, 208 inches, and 240 inches. If Jack wants to sell only one length of firewood, and not waste any part of any log, what is the maximum length he can cut?

62. Selling Firewood Jack's supplier of logs tells him that this season he will provide only one log length. In order to continue cutting some logs into 16-inch pieces, and cut other logs into 18-inch pieces (for a new group of customers), and also keep avoiding any waste, what length logs should he order? Give the answer in feet.

5.5 THE FIBONACCI SEQUENCE AND THE GOLDEN RATIO

OBJECTIVES

1 Apply the properties of the Fibonacci sequence.

2 Calculate approximations of the Golden Ratio.

3 Discover relationships between the Fibonacci sequence and the Golden Ratio.

The Fibonacci Sequence

One of the most famous problems in elementary mathematics comes from the book *Liber Abaci,* written in 1202 by Leonardo of Pisa, a.k.a. Fibonacci. The problem is as follows:

> A man put a pair of rabbits in a cage. During the first month the rabbits produced no offspring, but each month thereafter produced one new pair of rabbits. If each new pair thus produced reproduces in the same manner, how many pairs of rabbits will there be at the end of one year?

The solution of this problem leads to a sequence of numbers known as the **Fibonacci sequence.** Here are the first fifteen terms of the Fibonacci sequence:

$$1, 1, 2, 3, 5, 8, 13, 21, 34, 55, 89, 144, 233, 377, 610.$$

After the first two terms (both 1) in the sequence, each term is obtained by adding the two previous terms. For example, the third term is obtained by adding $1 + 1$ to get 2, the fourth term is obtained by adding $1 + 2$ to get 3, and so on. This can be described by a mathematical formula known as a **recursion formula.**

If F_n represents the Fibonacci number in the nth position in the sequence, then

$$F_1 = 1$$
$$F_2 = 1$$
$$F_n = F_{n-2} + F_{n-1}, \quad \text{for } n \geq 3.$$

The solution of **Fibonacci's rabbit problem** is examined in **Chapter 1.**

To observe one of the many interesting properties of the Fibonacci sequence, do the following.

1. Choose any term after the first and square it.
2. Multiply the terms before and after the term chosen in Step 1.
3. Subtract the smaller value from the larger.
4. What is your result?

Try this procedure beginning with several different Fibonacci terms. **The result is always the same.**

Using the recursion formula $F_n = F_{n-2} + F_{n-1}$, we obtain

$$F_3 = F_1 + F_2 = 1 + 1 = 2, \quad F_4 = F_2 + F_3 = 1 + 2 = 3, \quad \text{and so on.}$$

The Fibonacci sequence exhibits many interesting patterns, and by inductive reasoning we can make many conjectures about these patterns. However, simply observing a finite number of examples does not provide a proof of a statement. Proofs of the properties of the Fibonacci sequence often involve mathematical induction (covered in college algebra texts). Here we simply observe some of the patterns and do not attempt to provide proofs.

EXAMPLE 1 **Observing a Pattern of the Fibonacci Numbers**

Find the sum of the squares of the first n Fibonacci numbers for $n = 1, 2, 3, 4, 5$, and examine the pattern. Generalize this relationship.

Solution

$$1^2 = 1 = 1 \cdot 1 = F_1 \cdot F_2$$
$$1^2 + 1^2 = 2 = 1 \cdot 2 = F_2 \cdot F_3$$
$$1^2 + 1^2 + 2^2 = 6 = 2 \cdot 3 = F_3 \cdot F_4$$
$$1^2 + 1^2 + 2^2 + 3^2 = 15 = 3 \cdot 5 = F_4 \cdot F_5$$
$$1^2 + 1^2 + 2^2 + 3^2 + 5^2 = 40 = 5 \cdot 8 = F_5 \cdot F_6$$

The sum of the squares of the first n Fibonacci numbers seems to always be the product of F_n and F_{n+1}. This has been proved to be true, in general, using mathematical induction.

The following program for the TI-84 Plus C utilizes the *Binet form* of the nth Fibonacci number (see **Exercises 43 and 44**) to determine its value.

```
PROGRAM: FIB
: ClrHome
: Disp "WHICH TERM"
: Disp "OF THE SEQUENCE"
: Disp "DO YOU WANT?"
: Input N
: (1 + √(5))/2 → A
: (1 − √(5))/2 → B
: (A ^N − B ^N)/√(5) → F
: Disp F
```

```
NORMAL FLOAT AUTO REAL DEGREE MP

WHICH TERM
OF THE SEQUENCE
DO YOU WANT?
?20
                        6765
                        Done
```

This screen indicates that the twentieth Fibonacci number is 6765.

EXAMPLE 2 **Observing the Fibonacci Sequence in a Long Division Problem**

Observe the steps of the long-division algorithm used to find the first few decimal places of the reciprocal of 89, the eleventh Fibonacci number. Locate occurrences of the terms of the Fibonacci sequence in the algorithm.

Solution

```
        .011235 ...
89) 1.000000 ...
    89
    ‾‾‾
    110
     89
    ‾‾‾
    210
    178
    ‾‾‾
    320
    267
    ‾‾‾
    530
    445
    ‾‾‾
    850 ...
```

Notice that after the 0 in the tenths place, the next five digits are the first five terms of the Fibonacci sequence. In addition, as indicated in color in the process, the digits 1, 1, 2, 3, 5, 8 appear in the division steps. Now, look at the digits next to the ones in color, beginning with the second "1"; they, too, are

$$1, 1, 2, 3, 5, \ldots.$$

If the division process is continued past the final step shown here, the pattern seems to stop, because to ten decimal places, $\frac{1}{89} \approx 0.0112359551$. (The decimal representation actually begins to repeat later in the process, because $\frac{1}{89}$ is a rational number.).

The sum below indicates how the Fibonacci numbers are actually "hidden" in this decimal.

$$0.01$$
$$0.001$$
$$0.0002$$
$$0.00003$$
$$0.000005$$
$$0.0000008$$
$$0.00000013$$
$$0.000000021$$
$$0.0000000034$$
$$0.00000000055$$
$$\underline{0.000000000089}$$

$$\frac{1}{89} = 0.0112359550\ldots\ .$$

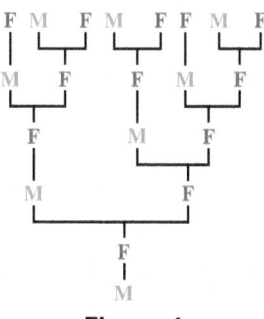

Figure 1

Fibonacci patterns have been found in numerous places in nature. For example, male honeybees (drones) hatch from eggs that have not been fertilized, so a male bee has only one parent, a female. On the other hand, female honeybees hatch from fertilized eggs, so a female has two parents, one male and one female. **Figure 1** shows several generations of ancestors for a male honeybee.

Notice that in the first generation, starting at the bottom, there is 1 bee, in the second there is 1 bee, in the third there are 2 bees, and so on. These are the terms of the Fibonacci sequence. Furthermore, beginning with the second generation, the numbers of female bees form the sequence, and beginning with the third generation, the numbers of male bees also form the sequence.

Successive terms in the Fibonacci sequence also appear in some plants. For example, the photo on the left below shows the double spiraling of a daisy head, with 21 clockwise spirals and 34 counterclockwise spirals. These numbers are successive terms in the sequence.

Most pineapples (see the photo on the right above) exhibit the Fibonacci sequence in the following way: Count the spirals formed by the "scales" of the cone, first counting from lower left to upper right. Then count the spirals from lower right to upper left. You should find that in one direction you get 8 spirals, and in the other you get 13 spirals, once again successive terms of the Fibonacci sequence. Many pinecones exhibit 5 and 8 spirals, and the cone of the giant sequoia has 3 and 5 spirals.

A fraction such as

$$1 + \cfrac{1}{1 + \cfrac{1}{1 + \cfrac{1}{1 + \cdots}}}$$

is called a **continued fraction.** Although the infinite number of steps required to evaluate this continued fraction directly could never be carried out, we can use a "trick" and some algebra to evaluate it.

Let $x = 1 + \cfrac{1}{1 + \cfrac{1}{1 + \cdots}}$

Then $x = 1 + \dfrac{1}{x}$

$$x^2 = x + 1$$
$$x^2 - x - 1 = 0.$$

By the quadratic formula from algebra,

$$x = \frac{1 \pm \sqrt{1 - 4(1)(-1)}}{2(1)}$$

$$x = \frac{1 \pm \sqrt{5}}{2}.$$

Notice that the positive solution is the **Golden Ratio.**

The Golden Ratio

If we consider the quotients of successive Fibonacci numbers, a pattern emerges. Read down the first column, then go to the next column, and so on.

$\dfrac{1}{1} = 1$	$\dfrac{5}{3} = 1.666\ldots$	$\dfrac{21}{13} \approx 1.615384615$	$\dfrac{89}{55} \approx 1.618181818$
$\dfrac{2}{1} = 2$	$\dfrac{8}{5} = 1.6$	$\dfrac{34}{21} \approx 1.619047619$	$\dfrac{144}{89} \approx 1.617977528$
$\dfrac{3}{2} = 1.5$	$\dfrac{13}{8} = 1.625$	$\dfrac{55}{34} \approx 1.617647059$	$\dfrac{233}{144} \approx 1.618055556$

These quotients seem to be approaching some "limiting value" close to 1.618. In fact, as we go farther into the sequence, these quotients approach the number

$$\frac{1 + \sqrt{5}}{2}. \quad \text{Golden Ratio}$$

This number is known as the **Golden Ratio** and often symbolized by ϕ, the Greek letter phi.

The Golden Ratio appears over and over in art, architecture, music, and nature. The ancient Greeks were aware of the ratio and believed that the most aesthetically pleasing proportions for a rectangle resulted when dividing the longer side by the shorter produced the Golden Ratio. They called such a rectangle a "Golden Rectangle."

A **Golden Rectangle** is one that can be divided into a square and another (smaller) rectangle *with the same ratio of dimensions as the original rectangle.* (See **Figure 2.**) If we let the smaller rectangle have length L and width W, as shown in the figure, then we see that the original rectangle has length $L + W$ and width L. Both rectangles (being "golden") have their lengths and widths in the Golden Ratio, ϕ, so we have the following.

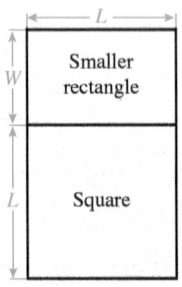

Figure 2

$$\frac{L}{W} = \frac{L + W}{L}$$

$$\frac{L}{W} = \frac{L}{L} + \frac{W}{L} \quad \text{Write the right side as the sum of two fractions.}$$

$$\phi = 1 + \frac{1}{\phi} \quad \text{Substitute } \frac{L}{W} = \phi, \frac{L}{L} = 1, \text{ and } \frac{W}{L} = \frac{1}{\phi}.$$

$$\phi^2 = \phi + 1 \quad \text{Multiply both sides by } \phi.$$

$$\phi^2 - \phi - 1 = 0 \quad \text{Write in standard quadratic form.}$$

Using the quadratic formula from algebra reveals that the positive solution of this equation is $\frac{1 + \sqrt{5}}{2} \approx 1.618033989$, the Golden Ratio.

The Parthenon (see the photo), built on the Acropolis in ancient Athens during the fifth century B.C., is an example of architecture exhibiting many distinct Golden Rectangles.

To see an interesting connection among the terms of the Fibonacci sequence, the Golden Ratio, and a phenomenon of nature, we can start with a rectangle measuring 89 by 55 units. (See **Figure 3** on the next page.) This is a very close approximation to a Golden Rectangle. Within this rectangle a square is then constructed, 55 units on a side. The remaining rectangle is also approximately a Golden Rectangle, measuring 55 units by 34 units. Each time this process is repeated, a square and an approximate Golden Rectangle are formed.

A Golden Rectangle in Art The rectangle outlining the figure in *St. Jerome* by Leonardo da Vinci is an example of a Golden Rectangle.

As indicated in **Figure 3,** vertices of the squares may be joined by a smooth curve known as a *spiral*. This spiral resembles the outline of a cross section of the shell of the chambered nautilus, as shown in the photo.

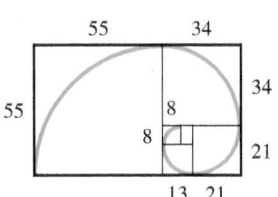

Figure 3

5.5 EXERCISES

CONCEPT CHECK *Write* true *or* false *for each statement.*

1. Although there are many counting numbers that are not Fibonacci numbers, there are as many Fibonacci numbers as there are counting numbers.

2. For all counting numbers n, it is true that $F_n \cdot F_{n+1} = F_{n+2}$.

3. The first two-digit Fibonacci number is F_6.

4. As n gets larger and larger, the quotient F_{n+1}/F_n gets closer and closer to the Golden Ratio.

5. The length of a Golden Rectangle is *exactly* 1.618 times its width.

6. The Golden Ratio ϕ satisfies the equation $2\phi = 1 + \sqrt{5}$.

Answer each question concerning the Fibonacci sequence or the Golden Ratio.

7. The fifteenth Fibonacci number is 610 and the seventeenth Fibonacci number is 1597. What is the value of the sixteenth Fibonacci number?

8. Recall that F_n represents the Fibonacci number in the nth position in the sequence. Find all values of n such that $F_n = n$.

9. Write the "parity" (o for odd, e for even) of each of the first twelve terms of the Fibonacci sequence.

10. From the pattern shown in the parity sequence, determine the least $n \geq 100$ such that F_n is even.

11. What is the exact value of the Golden Ratio?

12. What is the approximate value of the Golden Ratio to the nearest thousandth?

In each problem, a pattern is established involving terms of the Fibonacci sequence. Use inductive reasoning to make a conjecture concerning the next equation in the pattern, and verify it. You may wish to refer to the first few terms of the sequence given in the text.

13. $1 = 1$
    $1 + 2 = 3$
    $1 + 2 + 5 = 8$
    $1 + 2 + 5 + 13 = 21$

14. $1^2 + 1^2 = 2$
    $1^2 + 2^2 = 5$
    $2^2 + 3^2 = 13$
    $3^2 + 5^2 = 34$

15. $1 = 2 - 1$
    $1 + 1 = 3 - 1$
    $1 + 1 + 2 = 5 - 1$
    $1 + 1 + 2 + 3 = 8 - 1$

16. $1 = 2 - 1$
    $1 + 3 = 5 - 1$
    $1 + 3 + 8 = 13 - 1$
    $1 + 3 + 8 + 21 = 34 - 1$

17. $1 = 1^2$
    $1 - 2 = -1^2$
    $1 - 2 + 5 = 2^2$
    $1 - 2 + 5 - 13 = -3^2$

18. $1 - 1 = -1 + 1$
    $1 - 1 + 2 = 1 + 1$
    $1 - 1 + 2 - 3 = -2 + 1$
    $1 - 1 + 2 - 3 + 5 = 3 + 1$

19. $2^2 - 1^2 = 3$
$3^2 - 1^2 = 8$
$5^2 - 2^2 = 21$
$8^2 - 3^2 = 55$

20. $2^3 + 1^3 - 1^3 = 8$
$3^3 + 2^3 - 1^3 = 34$
$5^3 + 3^3 - 2^3 = 144$
$8^3 + 5^3 - 3^3 = 610$

Solve each problem.

21. Every natural number can be expressed as a sum of Fibonacci numbers, where no number is used more than once. For example,

$$25 = 21 + 3 + 1.$$

Express each of the following in this way.

(a) 39   **(b)** 59   **(c)** 99

22. For any prime number p except 2 or 5, either F_{p+1} or F_{p-1} is divisible by p. Show that this is true for the following values of p.

(a) $p = 3$ **(b)** $p = 7$ **(c)** $p = 11$

23. It has been shown that if the greatest common factor of m and n is r, then the greatest common factor of F_m and F_n is F_r. Show that this is true for the following values of m and n.

(a) $m = 10, n = 4$ **(b)** $m = 12, n = 6$
(c) $m = 14, n = 6$

24. It has been shown that if m divides n, then F_m is a factor of F_n. Show that this is true for the following values of m and n.

(a) $m = 3, n = 6$ **(b)** $m = 4, n = 12$
(c) $m = 5, n = 15$

25. In a margin note we saw that if a term of the Fibonacci sequence is squared and then the product of the terms on each side of the term is found, there will always be a difference of 1. Follow the steps below, choosing the sixth Fibonacci number, 8.

(a) Square 8. Multiply the terms of the sequence two positions away from 8 (i.e., 3 and 21). Subtract the smaller result from the larger, and record your answer.

(b) Square 8. Multiply the terms of the sequence three positions away from 8. Once again, subtract the smaller result from the larger, and record your answer.

(c) Repeat the process, moving four terms away from 8.

(d) Make a conjecture about what will happen when you repeat the process, moving five terms away. Verify your answer.

26. *A Number Trick* Here is a number trick that you can perform. Ask someone to pick any two numbers at random and to write them down. Ask the person to determine a third number by adding the first and second, a fourth number by adding the second and third, and so on, until ten numbers are determined. Then ask the person to add these ten numbers. You will be able to give the sum before the person even completes the list, because the sum will always be 11 times the seventh number in the list. Verify that this is true, by using x and y as the first two numbers arbitrarily chosen. (*Hint:* Remember the distributive property from algebra.)

*Another Fibonacci-type sequence that has been studied by mathematicians is the **Lucas sequence,** named after a French mathematician of the nineteenth century. The first nine terms of the Lucas sequence are*

$$1, 3, 4, 7, 11, 18, 29, 47, 76.$$

27. What is the tenth term of the Lucas sequence?

28. Choose any term of the Lucas sequence and square it. Then multiply the terms on either side of the one you chose. Subtract the smaller result from the larger. Repeat this for a different term of the sequence. Do you get the same result? Make a conjecture about this pattern.

29. The first term of the Lucas sequence is 1. Add the first and third terms. Record your answer. Now add the first, third, and fifth terms and record your answer. Continue this pattern, each time adding another term that is in an *odd* position in the sequence. What do you notice about all of your sums?

30. The second term of the Lucas sequence is 3. Add the second and fourth terms. Record your answer. Now add the second, fourth, and sixth terms and record your answer. Continue this pattern, each time adding another term that is in an *even* position of the sequence. What do you notice about all of your sums?

31. Many interesting patterns exist among the terms of the Fibonacci sequence and the Lucas sequence. Make a conjecture about the next equation that would appear in each of the lists, and then verify it.

(a) $1 \cdot 1 = 1$
$1 \cdot 3 = 3$
$2 \cdot 4 = 8$
$3 \cdot 7 = 21$
$5 \cdot 11 = 55$

(b) $1 + 2 = 3$
$1 + 3 = 4$
$2 + 5 = 7$
$3 + 8 = 11$
$5 + 13 = 18$

(c) $1 + 1 = 2 \cdot 1$
$1 + 3 = 2 \cdot 2$
$2 + 4 = 2 \cdot 3$
$3 + 7 = 2 \cdot 5$
$5 + 11 = 2 \cdot 8$

(d) $1 + 4 = 5 \cdot 1$
$3 + 7 = 5 \cdot 2$
$4 + 11 = 5 \cdot 3$
$7 + 18 = 5 \cdot 5$
$11 + 29 = 5 \cdot 8$

32. In the text we illustrate that the quotients of successive terms of the Fibonacci sequence approach the Golden Ratio. Make a similar observation for the terms of the Lucas sequence; that is, find the decimal approximations for the quotients

$$\frac{3}{1}, \frac{4}{3}, \frac{7}{4}, \frac{11}{7}, \frac{18}{11}, \frac{29}{18},$$

and so on, using a calculator. Then make a conjecture about what seems to be happening.

*Recall the **Pythagorean theorem** from geometry: If a right triangle has legs of lengths a and b and hypotenuse of length c, then*

$$a^2 + b^2 = c^2.$$

*Suppose that we choose any four successive terms of the Fibonacci sequence. Multiply the first and fourth. Double the product of the second and third. Add the squares of the second and third. The three results obtained form a **Pythagorean triple** (three numbers that satisfy the equation $a^2 + b^2 = c^2$). Find the Pythagorean triple obtained this way, using the four given successive terms of the Fibonacci sequence.*

33. 1, 1, 2, 3   **34.** 2, 3, 5, 8   **35.** 5, 8, 13, 21

36. Look at the values of the hypotenuse (c) in the answers to **Exercises 33–35.** What do you notice about each of them?

37. The following array of numbers is called **Pascal's triangle.**

```
            1
          1   1
        1   2   1
      1   3   3   1
    1   4   6   4   1
  1   5  10  10   5   1
1   6  15  20  15   6   1
```

This array is important in the study of counting techniques and probability (see later chapters) and appears in algebra in the binomial theorem. If the triangular array is written in a different form, as follows, and the sums along the diagonals as indicated by the colored lines are found, there is an interesting occurrence. What do you find when the numbers are added?

```
1
1   1
1   2   1
1   3   3   1
1   4   6   4   1
1   5  10  10   5   1
1   6  15  20  15   6   1
```

The following exercises require a scientific calculator.

38. Evaluate the infinite "nested" square root

$$\sqrt{1 + \sqrt{1 + \sqrt{1 + \sqrt{1 + \cdots}}}}$$

by justifying each of the following steps.

(a) $x = \sqrt{1 + \sqrt{1 + \sqrt{1 + \sqrt{1 + \cdots}}}}$

(b) $x = \sqrt{1 + x}$

(c) $x^2 = 1 + x$

(d) $x^2 - x - 1 = 0$

(e) $x = \dfrac{1 \pm \sqrt{5}}{2}$

(f) The positive solution is the value of the original nested square root. What is that value?

39. **Exercise 38** produced a positive solution (which is the Golden Ratio ϕ) and a negative solution (call it $\bar{\phi}$). Evaluate both solutions to nine decimal places. What similarity do you notice between the two decimals?

40. Evaluate, to nine decimal places, the reciprocals of ϕ and $\bar{\phi}$—that is, $\dfrac{1}{\phi}$ and $\dfrac{1}{\bar{\phi}}$. What similarity do you notice between the two decimals?

Enter the correct choice in each blank.

41. The quantity $\dfrac{1}{\phi}$ is the _____ of $\bar{\phi}$.
(reciprocal/negative)

42. The quantity $\dfrac{1}{\bar{\phi}}$ is the _____ of ϕ.
(reciprocal/negative)

A remarkable relationship existwbetween the two solutions of the equation $x^2 - x - 1 = 0$, which are

$$\phi = \frac{1 + \sqrt{5}}{2} \quad and \quad \bar{\phi} = \frac{1 - \sqrt{5}}{2},$$

and the Fibonacci numbers. To find the nth Fibonacci number without using the recursion formula, use a calculator to evaluate

$$\frac{\phi^n - \bar{\phi}^n}{\sqrt{5}}.$$

Thus, to find the thirteenth Fibonacci number, evaluate

$$\frac{\left(\dfrac{1 + \sqrt{5}}{2}\right)^{13} - \left(\dfrac{1 - \sqrt{5}}{2}\right)^{13}}{\sqrt{5}}.$$

*This form is known as the **Binet form** of the nth Fibonacci number. Use the Binet form and a calculator to find the nth Fibonacci number for each of the following values of n.*

43. $n = 16$ **44.** $n = 25$

FOR FURTHER THOUGHT

Mathematical Animation

The 1959 animated film *Donald in Mathmagic Land* is a classic. It provides a 25-minute trip with Donald Duck, led by the Spirit of Mathematics, through the world of mathematics. Several minutes of the film are devoted to the Golden Ratio (or, as it is termed there, the Golden Section).

© Disney Enterprises, Inc.

Disney provides animation to explain the Golden Ratio in a way that the printed word simply cannot do. The Golden Ratio is seen in architecture, nature, and the human body.

For Group or Individual Investigation

1. Verify the following Fibonacci pattern in the conifer family. Obtain a pineapple, and count spirals formed by the "scales" of the cone, first counting from lower left to upper right. Then count the spirals from lower right to upper left. What do you find?

2. Two popular sizes of index cards are 3" by 5" and 5" by 8". Why do you think that these are industry-standard sizes?

3. Divide your height by the height to your navel. Find a class average. What value does this come close to?

Chapter 5 SUMMARY

KEY TERMS

5.1
number theory
prime number
composite number
divisibility
divisibility tests
factor
divisor
multiple
factorization
prime factorization

Sieve of Eratosthenes
fundamental theorem
    of arithmetic
cryptography
modular systems

5.2
proof by contradiction
Mersenne number
Mersenne prime
partition

5.3
perfect number
deficient number
abundant number
amicable (friendly)
    numbers
Goldbach's conjecture
twin primes
primorial primes
Fermat's Last
    Theorem

5.4
greatest common factor (GCF)
relatively prime numbers
Euclidean algorithm
least common multiple (LCM)

5.5
Fibonacci sequence
Fibonacci numbers
Golden Ratio
Golden Rectangle

NEW SYMBOLS

$a\,|\,b$ *a* divides *b*

ϕ (phi) Golden Ratio

TEST YOUR WORD POWER

See how well you have learned the vocabulary in this chapter.

1. In mathematics, **number theory** in general deals with
   A. the properties of the rational numbers.
   B. operations on fractions.
   C. the properties of the natural numbers.
   D. only those numbers that are prime.

2. A **prime number** is
   A. divisible by 2, whereas a composite number is not.
   B. divisible by itself and 1 only.
   C. one of a finite set discovered by Marin Mersenne.
   D. a number of the form $2^p - 1$, where *p* is a prime.

3. A **partition** of a set is
   A. a rule that separates half of its elements from the other half.
   B. a subset that contains some but not all of the elements of the set.
   C. a rule by which all elements of the set can be identified.
   D. a division of all elements of the set into disjoint subsets.

4. A natural number that is less than the sum of its **proper divisors** is
   A. an **abundant** number.
   B. a **perfect** number.
   C. a **deficient** number.
   D. a **Fermat number.**

5. An **amicable** pair consists of two numbers that are
   A. either both prime or both composite.
   B. either both odd or both even.
   C. each equal to the sum of the proper divisors of the other.
   D. one Fermat number and one Mersenne number.

6. Goldbach's conjecture
   A. was formulated over 300 years before it was proved.
   B. has to do with two *n*th powers summing to a third *n*th power.
   C. has to do with expressing natural numbers as products of primes.
   D. has to do with expressing natural numbers as sums of primes.

7. The so-called **twin primes**
   A. always occur in pairs that differ by 2.
   B. always occur in pairs that differ by 1.
   C. cannot exceed 10^{100}.
   D. have finally all been identified.

8. Fermat's Last Theorem
   A. was proved by Euler about 100 years after Fermat stated it.
   B. has never been proved, but most mathematicians believe it is true.
   C. was proved by Wiles in the 1990s.
   D. says that the equation $a^n + b^n = c^n$ cannot be true for $n = 2$.

ANSWERS
1. C **2.** B **3.** D **4.** A
5. C **6.** D **7.** A **8.** C

QUICK REVIEW

Concepts

Examples

5.1 PRIME AND COMPOSITE NUMBERS

The natural number a is divisible by the natural number b, or, equivalently, "b divides a" (denoted $b \mid a$), if there exists a natural number k such that $a = bk$.

$2 \mid 14$, since $14 = 2 \cdot 7$ ($k = 7$). Here 2 is a factor of 14, and 14 is a multiple of 2.

In the equation $a = bk$, we say that b is a **factor** (or **divisor**) of a and that a is a **multiple** of b.

$2 \nmid 13$, since there is no suitable k. Here $13 = 2 \cdot 6.5$ (and $k = 6.5$ is *not* a natural number).

A **factorization** of a natural number expresses it as a product of some, or all, of its divisors.

$2 \cdot 12, \quad 3 \cdot 8, \quad 4 \cdot 6, \quad 2 \cdot 2 \cdot 6, \quad 2 \cdot 3 \cdot 4, \quad 2 \cdot 2 \cdot 2 \cdot 3$
Factorizations of 24

A **prime number** is greater than 1 and has only itself and 1 as factors.

The number 17 is prime. Its only factors are 1 and 17.

A **composite number** is greater than 1 and is *not* prime—that is, it has at least one factor other than itself and 1.

The number 10 is composite. In addition to 1 and 10, it has the factors 2 and 5.

Divisibility tests can help determine the factors of a given natural number.

54,086,226 is divisible by 3 because the sum of its digits (33) is divisible by 3.

$$5 + 4 + 0 + 8 + 6 + 2 + 2 + 6 = 33$$

258,784 is *not* divisible by 3 because the sum of its digits (34) is *not* divisible by 3.

$$2 + 5 + 8 + 7 + 8 + 4 = 34$$

A factorization with prime factors only is a number's **prime factorization.**

The prime factorization of 24 is

$$2 \cdot 2 \cdot 2 \cdot 3.$$

Concepts	*Examples*
The **fundamental theorem of arithmetic** states that the prime factorization of a number is unique (if the order of the factors is disregarded). A natural number's prime factorization can be determined by using a **factor tree** or by using **repeated division by primes.**	The number 24 has no other prime factorization except rearrangements of the factors shown above, such as $$2 \cdot 3 \cdot 2 \cdot 2 \quad \text{and} \quad 3 \cdot 2 \cdot 2 \cdot 2.$$ 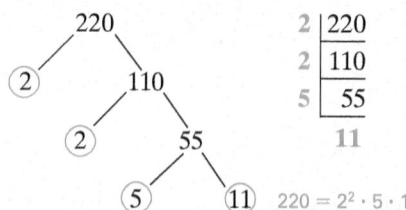 $220 = 2^2 \cdot 5 \cdot 11$

5.2   LARGE PRIME NUMBERS

Modern **cryptography systems** are based on large prime numbers. Large primes are found today mainly among the **Mersenne numbers.**	Cryptography systems, in practice, use primes so huge that the most powerful computers can multiply them but *cannot* factor their product. The largest Mersenne prime discovered, as of 2018, is $2^{77,232,917} - 1$. It has 23,249,425 decimal digits.

5.3   SELECTED TOPICS FROM NUMBER THEORY

The **proper divisors** of a natural number are all of its divisors except the number itself.	The number 42 has proper divisors 1, 2, 3, 6, 7, 14, and 21.
A natural number is **perfect** if the sum of its proper divisors equals the number itself.	The number 28 has proper divisors 1, 2, 4, 7, and 14. $$28 = 1 + 2 + 4 + 7 + 14$$
All even perfect numbers are of the form $2^{n-1}(2^n - 1)$, where $2^n - 1$ is prime.	The number $2^3 - 1 = 7$ is prime, so $$2^{3-1}(2^3 - 1) = 4(7) = 28 \quad \text{is perfect.}$$
A natural number is **abundant** if the sum of its proper divisors is greater than the number itself.	$1 + 2 + 3 + 6 + 7 + 14 + 21 = 54$ and $54 > 42 \leftarrow$ Abundant
A natural number is **deficient** if the sum of its proper divisors is less than the number itself.	$1 + 2 + 4 + 8 = 15$ and $15 < 16 \leftarrow$ Deficient
Two natural numbers are **amicable,** or **friendly,** if each equals the sum of the other's proper divisors.	The smallest amicable pair is 220 and 284. Proper divisors of 220: 1, 2, 4, 5, 10, 11, 20, 22, 44, 55, 110 Proper divisors of 284: 1, 2, 4, 71, 142 $1 + 2 + 4 + 5 + 10 + 11 + 20 + 22 + 44 + 55 + 110 = 284$ $1 + 2 + 4 + 71 + 142 = 220$
Goldbach's conjecture (which is not proved) states that every even number greater than 2 can be written as the sum of two prime numbers.	$12 = 5 + 7 \qquad 46 = 3 + 43 \quad$ Goldbach's conjecture
Twin primes are two primes that differ by 2.	17 and 19 $\qquad$ 29 and 31 $\quad$ Twin primes
Fermat's Last Theorem (proved in the 1990s) states that there are no triples of natural numbers (a, b, c) that satisfy the equation $a^n + b^n = c^n$ for any natural number $n \geq 3$.	$2^3 + 6^3 = 8 + 216 = 224$, but 224 is not a perfect cube.

Concepts   *Examples*

(5.4) GREATEST COMMON FACTOR AND LEAST COMMON MULTIPLE

The **greatest common factor (GCF)** of a group of natural numbers is the greatest natural number that is a factor of all the numbers in the group.

Find the greatest common factor of 450 and 660.
Prime factors method:

$$450 = 2 \cdot 3^2 \cdot 5^2$$
$$660 = 2^2 \cdot 3 \cdot 5 \cdot 11$$
$$\text{GCF} = 2 \cdot 3 \cdot 5 = 30$$

Dividing by prime factors method:

```
2 | 450   660
3 | 225   330
5 |  75   110
      15    22    GCF = 2 · 3 · 5 = 30
```

Euclidean algorithm:

```
           1
      450)660
          450
          210

             2
      210)450
          420
           30    GCF

            7
       30)210
          210
            0
```

The **least common multiple (LCM)** of a group of natural numbers is the least natural number that is a multiple of all the numbers in the group.

Find the least common multiple of 450 and 660.
Prime factors method:

$$450 = 2 \cdot 3^2 \cdot 5^2$$
$$660 = 2^2 \cdot 3 \cdot 5 \cdot 11$$
$$\text{LCM} = 2^2 \cdot 3^2 \cdot 5^2 \cdot 11 = 9900$$

Dividing by prime factors method:

```
2 | 450   660
2 | 225   330
3 | 225   165
3 |  75    55
5 |  25    55
       5    11    LCM = 2 · 2 · 3 · 3 · 5 · 5 · 11 = 9900
```

Formula:

$$\text{LCM} = \frac{450 \cdot 660}{\text{GCF of 450 and 660}} = \frac{297{,}000}{30} = 9900$$

(5.5) THE FIBONACCI SEQUENCE AND THE GOLDEN RATIO

The **Fibonacci sequence** begins with two 1s, and each succeeding term is the sum of the preceding two terms.

1, 1, 2, 3, 5, 8, 13, 21, 34, 55, 89, 144, 233, 377, . . .

Fibonacci sequence

Concepts	Examples
The **Golden Ratio,** $$\frac{1+\sqrt{5}}{2},$$ arises in many ways. One is as the limit of the ratio of adjacent terms in the Fibonacci sequence.	The ratios $$\frac{1}{1}=1,\quad \frac{2}{1}=2,\quad \frac{3}{2}=1.5,\quad \frac{5}{3}\approx 1.667,\quad \frac{8}{5}=1.6,\ldots$$ are approaching $\frac{1+\sqrt{5}}{2}\approx 1.618.$

Chapter 5 TEST

Write true *or* false *for each statement.*

1. For all natural numbers n, 1 is a factor of n, and n is a multiple of n.

2. If a natural number is not perfect, then it must be abundant.

3. If a natural number is divisible by 9, then it must also be divisible by 3.

4. If p and q are different primes, 1 is their greatest common factor, and pq is their least common multiple.

5. No two prime numbers differ by 1.

6. There are infinitely many prime numbers.

Solve each problem.

7. Use divisibility tests to determine whether the number

$$391,848,870$$

is divisible by each of the following.

(a) 2   (b) 3   (c) 4
(d) 5   (e) 6   (f) 8
(g) 9   (h) 10   (i) 12

8. Decide whether each number is prime, composite, or neither.

(a) 119   (b) 107   (c) 1

9. Give the prime factorization of 5460.

10. In your own words state the fundamental theorem of arithmetic.

11. Decide whether each number is perfect, deficient, or abundant.

(a) 24   (b) 34   (c) 28

12. Which of the following statements is false?

A. There are no known odd perfect numbers.
B. Every even perfect number must end in 8 or 26.
C. Goldbach's conjecture for the number 12 is illustrated by the equation $12 = 7 + 5$.

13. Give a pair of twin primes between 60 and 80.

14. Find the greatest common factor of 99, 135, and 216.

15. Find the least common multiple of 91 and 154.

16. *Day Off for Fast-food Workers* Both Katherine and Josh work at a fast-food outlet. Katherine has every sixth day off, and Josh has every fourth day off. If they are both off on Wednesday of this week, what will be the day of the week that they are next off together?

17. The twenty-first Fibonacci number is 10,946 and the twenty-second Fibonacci number is 17,711. What is the twenty-third Fibonacci number?

18. Make a conjecture about the next equation in the following list, and verify it.

$$8 - (1+1+2+3) = 1$$
$$13 - (1+2+3+5) = 2$$
$$21 - (2+3+5+8) = 3$$
$$34 - (3+5+8+13) = 5$$

19. Choose the correct completion of this statement: If p is a prime number, then $2^p - 1$ is _____ prime.

A. never   **B.** sometimes   **C.** always

20. Determine the eighth term of a Fibonacci-type sequence with first term 2 and second term 4.

21. Choose any term after the first in the sequence described in **Exercise 20.** Square it. Multiply the two terms on either side of it. Subtract the smaller result from the larger. Now repeat the process with a different term. Make a conjecture about what this process will yield for any term of the sequence.

22. Which one of the following is the *exact* value of the Golden Ratio?

A. $\frac{1+\sqrt{5}}{2}$ **B.** $\frac{1-\sqrt{5}}{2}$ **C.** 1.6   **D.** 1.618

6 The Real Numbers and Their Representations

In the development of the real numbers (the topic of this chapter), the unit number 1 is the place at which everything starts. (0 showed up much later in our number system.) The concept of percent has become a standard by which portions of one whole thing are described.

Percent means "per one hundred," and the symbol % is used to represent percent. Thus,

$$\textbf{1\% means one one-hundredth, or } \frac{1}{100}, \textbf{ or } 0.01.$$

One hundred percent, or 100%, means 1 whole thing.

One application of percent in everyday life involves tipping in a restaurant. Suppose that the bill is $50 and you decide that the service was so good that you want to tip 20% of this amount. To find

20% of 50,

you might mentally calculate using one of the following methods.

1. You think, "Well, 20% means $\frac{1}{5}$, and to find $\frac{1}{5}$ of something I divide by 5, so 50 divided by 5 is 10. The answer is 10."

2. You think, "20% is twice 10%, and to find 10% of something I move the decimal point one place to the left. So 10% of 50 is 5.0, or just 5, and I simply double 5 to get 10. The answer is 10."

For those who don't think this way, there is the tried-and-true method of multiplying 50 by 0.20.

This chapter introduces the real number system, the operations of arithmetic with real numbers, and selected applications of fractions, decimals, and percents in occupations and trades.

6.1   REAL NUMBERS, ORDER, AND ABSOLUTE VALUE

OBJECTIVES

1 Represent a number on a number line.

2 Identify a number as positive, negative, or zero.

3 Identify a number as belonging to one or more sets of numbers.

4 Given two numbers *a* and *b*, determine whether $a = b$, $a < b$, or $a > b$.

5 Given a number *a*, determine its additive inverse and absolute value.

6 Interpret signed numbers in tables of economic and occupations data.

Sets of Real Numbers

The mathematician Leopold Kronecker (1823–1891) once made the statement, "God made the integers, all the rest is the work of man." The *natural numbers* are those numbers with which we count discrete objects. By including 0 in the set, we obtain the set of *whole numbers*.

NATURAL NUMBERS

$\{1, 2, 3, 4, \dots\}$ is the set of **natural numbers.**

WHOLE NUMBERS

$\{0, 1, 2, 3, \dots\}$ is the set of **whole numbers.**

These numbers, along with many others, can be represented on **number lines** like the one pictured in **Figure 1.** We draw a number line by locating any point on the line and labeling it 0. This is the **origin.** Choose any point to the right of 0 and label it 1. The distance between 0 and 1 gives a unit of measure used to locate other points, as shown in **Figure 1.** The numbers labeled and those continuing in the same way to the right correspond to the set of whole numbers.

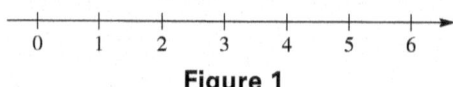

Figure 1

All the whole numbers starting with 1 are located to the right of 0 on the number line. But numbers may also be placed to the left of 0. These numbers, written −1, −2, −3, and so on, are shown in **Figure 2.** (The negative sign is used to show that the numbers are located to the *left* of 0.)

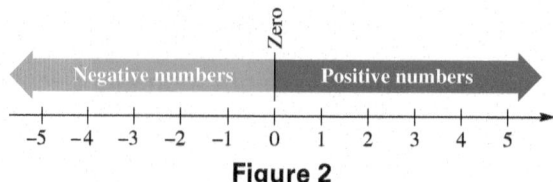

Figure 2

The numbers to the *left* of 0 are **negative numbers.** The numbers to the *right* of 0 are **positive numbers.** Positive numbers and negative numbers are called **signed numbers.** The number 0 itself is neither positive nor negative.

There are many practical applications of negative numbers.

- Temperatures may fall below zero. The lowest temperature ever recorded in meteorological records was $-128.6°F$ at Vostok, Antarctica, on July 21, 1983.

- Altitudes below sea level can be represented by negative numbers. The shore surrounding the Dead Sea is 1312 feet below sea level, written -1312 feet.

The natural numbers, their negatives, and zero make up the set of *integers*.

INTEGERS

$\{ \ldots, -3, -2, -1, 0, 1, 2, 3, \ldots \}$ is the set of **integers.**

Not all numbers are integers. For example, $\frac{1}{2}$ is a number *halfway* between the integers 0 and 1. Several numbers that are not integers are *graphed* in **Figure 3**. The **graph** of a number is a point on the number line representing that number. Think of the graph of a set of numbers as a "picture" of the set. All the numbers indicated in **Figure 3** can be written as quotients of integers and are examples of *rational numbers*. An integer, such as 2, is also a rational number. This is true because $2 = \frac{2}{1}$.

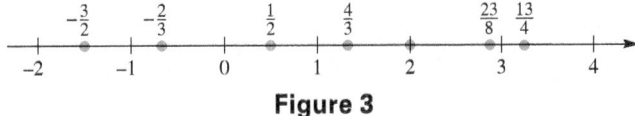

Figure 3

RATIONAL NUMBERS

$\{x \mid x$ is a quotient of two integers, with denominator not equal to $0\}$ is the set of **rational numbers.**

(Read the part in the braces as "the set of all numbers x such that x is a quotient of two integers, with denominator not equal to 0.")

The set symbolism used in the definition of rational numbers,

$$\{x \mid x \text{ has a certain property}\},$$

is called **set-builder notation.** This notation is convenient to use when it is not possible, or practical, to list all the elements of the set.

There are other numbers on the number line that are not rational. The first such number ever to be discovered was $\sqrt{2}$. The Pythagorean theorem, named for the Greek mathematician Pythagoras, says that in a right triangle, the length of the side opposite the right angle (the hypotenuse) is equal to the square root of the sum of the squares of the two perpendicular sides (the legs). See **Figure 4(a)**. This is symbolized

$$c = \sqrt{a^2 + b^2}.$$

A right triangle having both legs a and b of length 1 is placed on a number line as shown in **Figure 4(b)**. Thus, the length of the hypotenuse c is found as follows.

$$c = \sqrt{1^2 + 1^2}$$
$$c = \sqrt{1 + 1}$$
$$c = \sqrt{2}$$

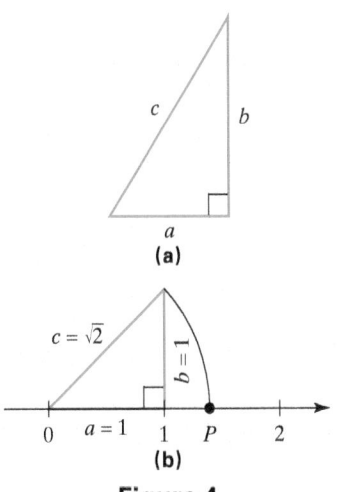

Figure 4

The Origins of Negative Numbers
Negative numbers can be traced back to the Chinese between 200 B.C. and A.D. 220. Mathematicians at first found negative numbers ugly and unpleasant, even though they kept cropping up in the solutions of problems. For example, an Indian text of about A.D. 1150 gives the solution of an equation as −5 and then makes fun of anything so useless.

Leonardo of Pisa (Fibonacci), while working on a financial problem, was forced to conclude that the solution must be a negative number (that is, a financial loss). In A.D. 1545, the rules governing operations with negative numbers were published by **Girolamo Cardano** in his *Ars Magna* (Great Art).

Figure 4(b) also shows an arc of a circle, centered at 0 with radius $\sqrt{2}$, intersecting the number line at point P. The coordinate of P is $\sqrt{2}$. The Greeks discovered that this number cannot be expressed as a quotient of integers—it is *irrational*.

IRRATIONAL NUMBERS

$\{x \mid x$ is a number on the number line that is not rational$\}$ is the set of **irrational numbers.**

Irrational numbers include $\sqrt{3}$, $\sqrt{7}$, $-\sqrt{10}$, and π, which is the ratio of the distance around a circle (its *circumference*) to the distance across it (its *diameter*). All numbers that can be represented by points on the number line (the union of the sets of rational and irrational numbers) are called *real numbers*.

REAL NUMBERS

$\{x \mid x$ is a number that can be represented by a point on a number line$\}$ is the set of **real numbers.**

Real numbers can be written as decimal numbers. Any rational number can be written as a decimal that will either come to an end (terminate) or repeat in a fixed "block" of digits. For example,

$$\frac{2}{5} = 0.4 \quad \text{and} \quad \frac{27}{100} = 0.27 \quad \text{are terminating decimals.}$$

$$\frac{1}{3} = 0.3333\ldots \quad \text{and} \quad \frac{3}{11} = 0.27272727\ldots \quad \text{are repeating decimals.}$$

The decimal representation of an irrational number neither terminates nor repeats. Decimal representations of rational and irrational numbers will be discussed further later in this chapter.

Figure 5 illustrates two ways to represent the relationships among the various sets of real numbers.

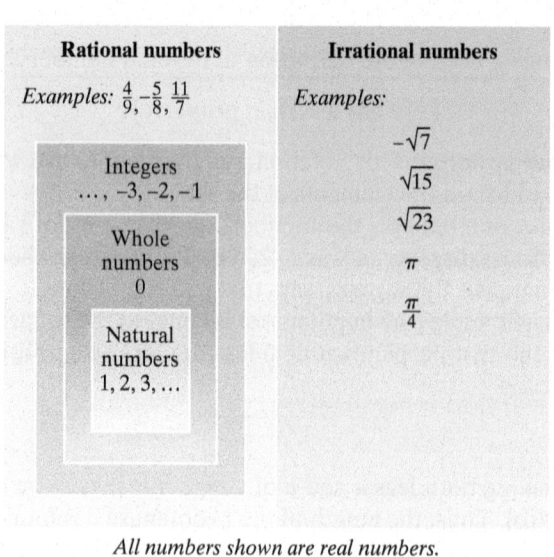

All numbers shown are real numbers.

(a)

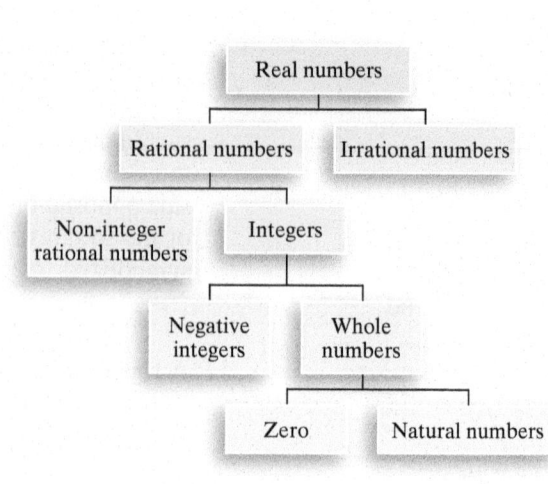

(b)

Figure 5

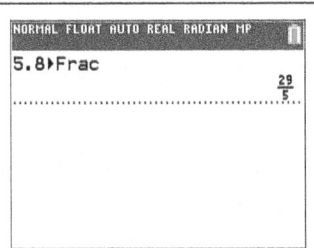

The TI-84 Plus C calculator will convert a decimal to a fraction. See **Example 1(d)**.

The symbol for equality, =, was first introduced by the Englishman **Robert Recorde** in his 1557 algebra text *The Whetstone of Witte*. He used two parallel line segments, because, he claimed, no two things can be more equal.

The symbols for order relationships, < and >, were first used by **Thomas Harriot** (1560–1621), another Englishman. These symbols were not immediately adopted by other mathematicians.

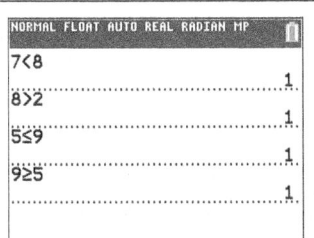

The calculator returns a 1 for each of these statements of inequality, signifying that each is true.

EXAMPLE 1   **Identifying Elements of a Set of Numbers**

List the numbers in the set that belong to each set of numbers.

$$\left\{ -5, -\frac{2}{3}, 0, \sqrt{2}, \frac{13}{4}, 5, 5.8 \right\}$$

(a) natural numbers   **(b)** whole numbers   **(c)** integers

(d) rational numbers   **(e)** irrational numbers   **(f)** real numbers

Solution

(a) The only natural number in the set is 5.

(b) The whole numbers consist of the natural numbers and 0, so the elements of the set that are whole numbers are 0 and 5.

(c) The integers in the set are $-5, 0,$ and 5.

(d) The rational numbers are $-5, -\frac{2}{3}, 0, \frac{13}{4}, 5,$ and 5.8, because each of these numbers *can* be written as the quotient of two integers. For example, $5.8 = \frac{58}{10} = \frac{29}{5}$.

(e) The only irrational number in the set is $\sqrt{2}$.

(f) All the numbers in the set are real numbers.

Order in the Real Numbers

Suppose that a and b represent two real numbers. If their graphs on a number line are the same point, then a **is equal to** b. If the graph of a lies to the left of the graph of b, then a **is less than** b, and if the graph of a lies to the right of the graph of b, then a **is greater than** b.

LAW OF TRICHOTOMY

For any real numbers a and b, one and only one of the following holds true.

$$a = b, \quad a < b, \quad a > b$$

When read from left to right, the inequality $a < b$ is read "a is less than b."

$$7 < 8 \quad \text{7 is less than 8.}$$

The inequality $a > b$ means "a is greater than b."

$$8 > 2 \quad \text{8 is greater than 2.}$$

Notice that the symbol always points to the lesser number.

$$\text{Lesser number} \longrightarrow \; 8 < 15$$

The symbol $\leq$, read from left to right, means "is less than or equal to."

$$5 \leq 9 \quad \text{5 is less than or equal to 9.}$$

This statement is true because $5 < 9$ is true. *If either the < part or the = part is true, then the inequality $\leq$ is true.* Also, $8 \leq 8$ is true because $8 = 8$ is true. But it is not true that $13 \leq 9$ because neither $13 < 9$ nor $13 = 9$ is true.

The symbol $\geq$ means "is greater than or equal to."

$$9 \geq 5 \quad \text{9 is greater than or equal to 5.}$$

This statement is true because $9 > 5$ is true.

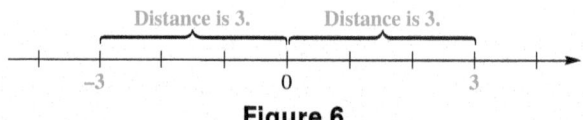

EXAMPLE 2   **Comparing Real Numbers**

Determine whether each statement is *true* or *false*.

(a) $6 \neq 6$ **(b)** $5 < 19$ **(c)** $15 \leq 20$ **(d)** $25 \geq 30$ **(e)** $12 \geq 12$

Solution

(a) The statement $6 \neq 6$ is false because 6 *is equal to* 6.

(b) Because 5 is indeed less than 19, this statement is true.

(c) The statement $15 \leq 20$ is true because $15 < 20$ is true.

(d) Both $25 > 30$ and $25 = 30$ are false, so $25 \geq 30$ is false.

(e) Because $12 = 12$, the statement $12 \geq 12$ is true.

Additive Inverses and Absolute Value

For any nonzero real number x, there is exactly one number on the number line the same distance from 0 as x but on the opposite side of 0. In **Figure 6,** the numbers 3 and -3 are the same distance from 0 but are on opposite sides of 0. Thus, 3 and -3 are called **additive inverses, negatives,** or **opposites** of each other.

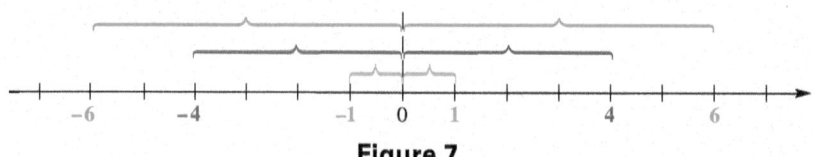

Figure 6

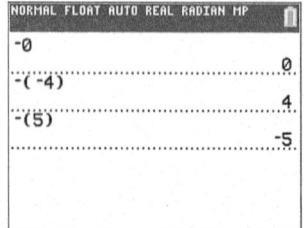

The additive inverse of the number 0 is 0 itself. This makes 0 the only real number that is its own additive inverse. Other additive inverses occur in distinct pairs. For example, 4 and -4 are additive inverses, as are 6 and -6. Several pairs of additive inverses are shown in **Figure 7.**

Figure 7

The additive inverse of a number can be indicated by writing the symbol $-$ in front of the number. With this symbol, the additive inverse of 7 is written -7. The additive inverse of -4 is written $-(-4)$ and can be read "the opposite of -4" or "the negative of -4." **Figure 7** suggests that 4 is an additive inverse of -4. A number can have only one additive inverse, so the symbols 4 and $-(-4)$ must represent the same number, which means that

$$-(-4) = 4.$$

DOUBLE NEGATIVE RULE

For any real number x, the following holds true.

$$-(-x) = x$$

Table 1

Number	Additive Inverse
−4	−(−4) or 4
0	0
19	−19
$-\frac{2}{3}$	$\frac{2}{3}$

Table 1 shows several numbers and their additive inverses. An important property of additive inverses will be studied later in this chapter:

$$a + (-a) = (-a) + a = 0, \quad \text{for all real numbers } a.$$

As mentioned above, additive inverses are numbers that are the same distance from 0 (but in opposite directions) on the number line. See **Figure 7** on the previous page. This idea can also be expressed by saying that a number and its additive inverse have the same *absolute value*. The **absolute value** of a real number can be defined as the undirected distance between 0 and the number on a number line.

The symbol for the absolute value of the number x is $|x|$, read **"the absolute value of x."** For example, the distance between 0 and 2 on a number line is 2 units, so

$$|2| = 2.$$

Because the distance between 0 and −2 on the number line is also 2 units,

$$|-2| = 2.$$

Absolute value is a measure of undirected distance, so *the absolute value of a number is never negative.* Because 0 is a distance of 0 units from 0, $|0| = 0$.

FORMAL DEFINITION OF ABSOLUTE VALUE

For any real number x, the absolute value of x is defined as follows.

$$|x| = \begin{cases} x & \text{if } x \geq 0 \\ -x & \text{if } x < 0 \end{cases}$$

If x is a positive number or 0, then its absolute value is x itself. For example, because 8 is a positive number, $|8| = 8$. By the second part of the definition, *if x is a negative number, then its absolute value is the additive inverse of x.* For example, if $x = -9$, then $|-9| = -(-9) = 9$, because the additive inverse of −9 is 9.

The formal definition of absolute value can be confusing if it is not read carefully. The "−x" in the second part of the definition *does not* represent a negative number. Since x is negative in the second part, −x represents the opposite of a negative number, which is a positive number.

NORMAL FLOAT AUTO REAL RADIAN MP

| -5|
....5
-| -14|
....-14
-|8-2|
....-6

This screen supports the results of
Example 3(b), (d), and (f).

EXAMPLE 3 Using Absolute Value

Simplify by finding the absolute value.

(a) $|5|$ (b) $|-5|$ (c) $-|5|$

(d) $-|-14|$ (e) $|8-2|$ (f) $-|8-2|$

Solution

(a) $|5| = 5$

(b) $|-5| = -(-5) = 5$

(c) $-|5| = -(5) = -5$

(d) $-|-14| = -(14) = -14$

(e) $|8-2| = |6| = 6$

(f) $-|8-2| = -|6| = -6$

Example 3(e) shows that absolute value bars also serve as grouping symbols. *Perform any operations that appear inside absolute value symbols before finding the absolute value.*

Applications of Real Numbers

EXAMPLE 4 Interpreting Signed Numbers in a Table

The consumer price index (CPI) is a measure of the average prices of goods and services purchased by urban consumers in the United States. **Table 2** shows the change in the CPI for selected categories of goods and services from 2014 to 2015 and from 2015 to 2016. Use the table to answer each question.

Table 2

Category	Change from 2014 to 2015	Change from 2015 to 2016
Transportation	−16.8	−4.2
Education	0.7	0.9
Food and beverages	4.8	0.9
Housing	4.9	5.9
Medical care	11.5	16.9

Data from U.S. Bureau of Labor Statistics.

(a) Which category in which years represents the greatest CPI decrease?

(b) Which category in which years represents the least CPI change?

Solution

(a) We must find the negative number with the greatest absolute value. The number that satisfies this condition is −16.8, so the greatest CPI decrease was shown by transportation from 2014 to 2015.

(b) We must find the number (either positive, negative, or zero) with the least absolute value. From 2014 to 2015, education showed the least change, a CPI increase of 0.7.

In the next example, we see how the need for certain occupations during the upcoming years is changing, some decreasing and some increasing.

EXAMPLE 5 Comparing Occupational Rates of Change

Table 3 shows the total rates of change projected to occur in demand for selected growing occupations and selected declining occupations from 2016 to 2026.

Table 3

Occupation (2016–2026)	Total Rate of Change (in percent)
Travel agents	−12
Computer programmers	−7
Photographers	−6
Physicians' assistants	37
Personal care aides	39
Solar photovoltaic installers	105

Data from Bureau of Labor Statistics.

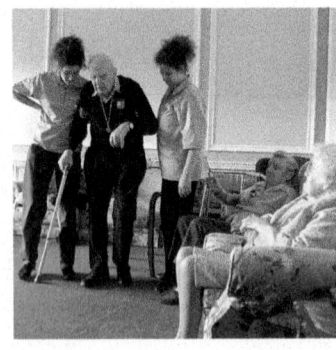

Which occupation in **Table 3** on the preceding page is expected to see

(a) the greatest change?   **(b)** the least change?

Solution

(a) We want the greatest change, without regard to whether the percent change is an increase or a decrease. Look for the number with the greatest absolute value. That number is for solar photovoltaic installers:

$$|105| = 105.$$

(b) Using similar reasoning, the least change is for photographers:

$$|-6| = 6.$$

6.1 EXERCISES

CONCEPT CHECK *Give a number that satisfies the given condition.*

1. An integer between 4.5 and 5.5

2. A rational number between 2.8 and 2.9

3. A whole number that is not positive and is less than 1

4. A whole number greater than 4.5

5. An irrational number that is between $\sqrt{17}$ and $\sqrt{19}$

6. A real number that is neither negative nor positive

CONCEPT CHECK *Write* true *or* false *for each statement.*

7. Every natural number is positive.

8. Every whole number is positive.

9. Every integer is a rational number.

10. Every rational number is a real number.

List all numbers from each set that are **(a)** *natural numbers;* **(b)** *whole numbers;* **(c)** *integers;* **(d)** *rational numbers;* **(e)** *irrational numbers;* **(f)** *real numbers.*

11. $\left\{-9, -\sqrt{7}, -1\frac{1}{4}, -\frac{3}{5}, 0, \sqrt{5}, 3, 5.9, 7\right\}$

12. $\left\{-5.3, -5, -\sqrt{3}, -1, -\frac{1}{9}, 0, 1.2, 1.8, 3, \sqrt{11}\right\}$

Use an integer or decimal to express each number in bold-face print representing a change or measurement in each of the following.

13. *U.S. Population* Between July 1, 2016, and July 1, 2017, the population of the United States increased by approximately **2,592,000.** (Data from U.S. Census Bureau.)

14. *Number of Movie Screens* Between 2013 and 2014, the number of indoor movie screens in the United States decreased by **12.** (Data from National Association of Theaters.)

15. *World Series Attendance* From 2016 to 2017, attendance at the first game of the World Series went from 38,091 to 54,253, an increase of **16,162.** (Data from Major League Baseball.)

16. *Number of U. S. Banks* In 1935, there were 15,295 banks in the United States. By 2014, the number was 6799, representing a decrease of **8496.** (Data from Federal Deposit Insurance Corporation.)

17. *Dow Jones Average* On Friday, February 2, 2018, the Dow Jones Industrial Average (DJIA) closed at 25,520.96. On the previous day, it had closed at 26,186.71. Thus, on Friday it closed down **665.75** points. (Data from *The Washington Post.*)

18. *NASDAQ* On Friday, February 23, 2018, the NASDAQ closed at 7337.39. On the previous day, it had closed at 7210.08. Thus, on Friday it closed up **127.31** points. (Data from *The Washington Post.*)

19. *Height of Mt. Arenal in Costa Rica* The height of Mt. Arenal, an active volcano in Costa Rica, is **5436** feet above sea level. (Data from *The New York Times Almanac.*)

20. *Elevation of New Orleans* The city of New Orleans, devastated by Hurricane Katrina in 2005, lies **8** feet below sea level. (Data from U.S. Geological Survey, *Elevations and Distances in the United States.*)

21. *Melting Point of Fluorine* The melting point of fluorine gas is **220°** below 0° Celsius.

22. *Boiling Point of Chlorine* The boiling point of chlorine is approximately **30°** below 0° Fahrenheit.

Depths and Heights of Seas and Mountains *The chart gives selected depths and heights of bodies of water and mountains.*

Bodies of Water	Average Depth in Feet (as a negative number)	Mountains	Altitude in Feet (as a positive number)
Pacific Ocean	−14,040	McKinley	20,237
South China Sea	−4802	Point Success	14,164
Gulf of California	−2375	Matlalcueyetl	14,636
Caribbean Sea	−8448	Rainier	14,416
Indian Ocean	−12,800	Steele	16,642

Data from *The World Almanac and Book of Facts.*

23. List the bodies of water in order, starting with the deepest and ending with the shallowest.

24. List the mountains in order, starting with the lowest and ending with the highest.

25. *True or false:* The absolute value of the depth of the Pacific Ocean is greater than the absolute value of the depth of the Indian Ocean.

26. *True or false:* The absolute value of the depth of the Gulf of California is greater than the absolute value of the depth of the Caribbean Sea.

Graph each group of numbers on a number line.

27. $-2, -6, -4, 3, 4$

28. $-5, -3, -2, 0, 4$

29. $\frac{1}{4}, 2\frac{1}{2}, -3\frac{4}{5}, -4, -1\frac{5}{8}$

30. $5\frac{1}{4}, 4\frac{5}{9}, -2\frac{1}{3}, 0, -3\frac{2}{5}$

31. CONCEPT CHECK Match each expression in Column I with its value in Column II. Some choices in Column II may not be used.

I	II
(a) $\lvert -7 \rvert$	**A.** 7
(b) $-(-7)$	**B.** −7
(c) $-\lvert -7 \rvert$	**C.** neither A nor B
(d) $-\lvert -(-7) \rvert$	**D.** both A and B

32. CONCEPT CHECK Fill in the blanks with the correct values: The opposite of −2 is ___, while the absolute value of −2 is ___. The additive inverse of −2 is ___, while the additive inverse of the absolute value of −2 is ___.

Find **(a)** *the additive inverse (or opposite) of each number and* **(b)** *the absolute value of each number.*

33. −2 **34.** −8 **35.** 6

36. 11 **37.** $7 - 4$ **38.** $8 - 3$

39. $7 - 7$ **40.** $3 - 3$

Select the lesser of the two given numbers.

41. $-12, -4$ **42.** $-9, -14$

43. $-8, -1$ **44.** $-15, -16$

45. $3, \lvert -4 \rvert$ **46.** $5, \lvert -2 \rvert$

47. $\lvert -3 \rvert, \lvert -4 \rvert$ **48.** $\lvert -8 \rvert, \lvert -9 \rvert$

49. $-\lvert -6 \rvert, -\lvert -4 \rvert$ **50.** $-\lvert -2 \rvert, -\lvert -3 \rvert$

51. $\lvert 5 - 3 \rvert, \lvert 6 - 2 \rvert$ **52.** $\lvert 7 - 2 \rvert, \lvert 8 - 1 \rvert$

Write true or false for each statement.

53. $6 > -(-2)$ **54.** $-8 > -(-2)$

55. $-4 \le -(-5)$ **56.** $-6 \le -(-3)$

57. $\lvert -6 \rvert < \lvert -9 \rvert$ **58.** $\lvert -12 \rvert < \lvert -20 \rvert$

59. $-\lvert 8 \rvert > \lvert -9 \rvert$ **60.** $-\lvert 12 \rvert > \lvert -15 \rvert$

61. $-\lvert -5 \rvert \ge -\lvert -9 \rvert$ **62.** $-\lvert -12 \rvert \le -\lvert -15 \rvert$

63. $\lvert 6 - 5 \rvert \ge \lvert 6 - 2 \rvert$ **64.** $\lvert 13 - 8 \rvert \le \lvert 7 - 4 \rvert$

65. *Population Change* The table shows the estimated percent change in population from 2010 to 2016 for selected metropolitan areas.

Metropolitan Area	Percent Change
Austin	19.93
San Francisco	8.15
Chicago	0.35
Cleveland	−2.77
Phoenix	11.72
Detroit	−5.74

Data from U. S. Census Bureau.

(a) Which metropolitan area has the greatest percent change in estimated population? What is this change? Is it an increase or a decrease?

(b) Which metropolitan area has the least estimated change in population? What is this change? Is it an increase or a decrease?

66. *Trade Balance* The table gives the net trade balance, in millions of dollars, for selected U.S. trade partners for 2018.

Country	Trade Balance (in millions of dollars)
Canada	−10,958
China	−347,016
The Netherlands	23,576
Mexico	−64,354
Japan	−68,810

Data from U.S. Census Bureau.

A negative balance means that imports to the United States exceeded exports from the United States. A positive balance means that exports exceeded imports.

(a) Which country had the greatest discrepancy between exports and imports? Explain.

(b) Which country had the least discrepancy between exports and imports? Explain.

(c) Which two countries were closest in their trade balance figures?

67. *CPI Data* Refer to **Table 2** in **Example 4.** Of the data for 2015 to 2016, which of the categories Transportation or Education shows the greater change (without regard to sign)?

68. *Change in Occupations* Refer to **Table 3** in **Example 5.** Did the demand for computer programmers or for travel agents show the lesser change (without regard to sign)?

 6.2

OPERATIONS, PROPERTIES, AND APPLICATIONS OF REAL NUMBERS

OBJECTIVES

1 Perform the operations of addition, subtraction, multiplication, and division of signed numbers.

2 Apply the rules for order of operations.

3 Identify and apply properties of addition and multiplication of real numbers.

4 Determine change in investment and meteorological data using subtraction and absolute value.

Operations on Real Numbers

The result of adding two numbers is their **sum.** The numbers being added are **addends** (or **terms**).

ADDING REAL NUMBERS

Like Signs Add two numbers with the *same* sign by adding their absolute values. The sign of the sum (either + or −) is the same as the sign of the two addends.

Unlike Signs Add two numbers with *different* signs by subtracting the lesser absolute value from the greater to find the absolute value of the sum. The sum is positive if the positive number has the greater absolute value. The sum is negative if the negative number has the greater absolute value.

For example, to add −12 and −8, first find their absolute values.

$$|-12| = 12 \quad \text{and} \quad |-8| = 8$$

−12 and −8 have the *same* sign, so add their absolute values: 12 + 8 = 20. Give the sum the sign of the two numbers. Because both numbers are negative, the sum is negative and

$$-12 + (-8) = -20.$$

Find −17 + 11 by subtracting the lesser absolute value from the greater, because these numbers have different signs.

$$|-17| = 17 \quad \text{and} \quad |11| = 11$$

$$17 - 11 = 6$$

Give the result the sign of the number with the larger absolute value.

$$-17 + 11 = -6$$

Negative because $|-17| > |11|$

EXAMPLE 1 Adding Signed Numbers

Find each sum.

(a) $-6 + (-3)$ **(b)** $-12 + (-4)$ **(c)** $4 + (-1)$

(d) $-9 + 16$ **(e)** $-16 + 12$ **(f)** $-4 + 4$

Solution

(a) $-6 + (-3) = -(6 + 3) = -9$ **(b)** $-12 + (-4) = -(12 + 4) = -16$

(c) $4 + (-1) = 3$ **(d)** $-9 + 16 = 7$

(e) $-16 + 12 = -4$ **(f)** $-4 + 4 = 0$ The sum of additive inverses is 0.

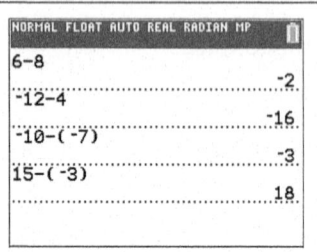

The calculator supports the results of **Example 1(a), (c), (e), and (f).**

The result of subtracting two numbers is their **difference.** In $a - b$, a is the **minuend,** and b is the **subtrahend.** Compare the two statements below.

$$7 - 5 = 2 \quad \text{and} \quad 7 + (-5) = 2$$

In a similar way,

$$9 - 3 = 9 + (-3).$$

That is, to subtract 3 from 9, add the additive inverse of 3 to 9. These examples suggest the following rule for subtraction.

DEFINITION OF SUBTRACTION

For all real numbers a and b,

$$a - b = a + (-b).$$

(Change the sign of the subtrahend and add.)

A general agreement is that the phrase "the difference of a and b" is interpreted as $a - b$.

EXAMPLE 2 Subtracting Signed Numbers

Find each difference.

(a) $6 - 8$ **(b)** $-12 - 4$ **(c)** $-10 - (-7)$ **(d)** $15 - (-3)$

Solution

Change to addition.

Change sign of the subtrahend and add.

(a) $6 - 8 = 6 + (-8) = -2$

Change to addition.

Sign is changed.

(b) $-12 - 4 = -12 + (-4) = -16$

(c) $-10 - (-7) = -10 + [-(-7)]$ This step can be omitted.

$$= -10 + 7$$

$$= -3$$

(d) $15 - (-3) = 15 + 3 = 18$

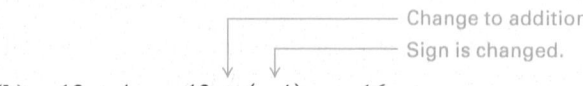

The calculator supports the results of **Example 2.**

Practical Arithmetic   From the time of Egyptian and Babylonian merchants, practical aspects of arithmetic complemented mystical (or "Pythagorean") tendencies. This was certainly true in the time of **Adam Riese** (1489–1559), a "reckon master" influential when commerce was growing in Northern Europe. He championed new methods of reckoning using Hindu-Arabic numerals and quill pens. (The Roman methods then in common use moved counters on a ruled board.) Riese thus fulfilled Fibonacci's efforts 300 years earlier to supplant Roman numerals and methods.

The result of multiplying two numbers is their **product.** The two numbers being multiplied are **factors.**

Any rules for multiplication with negative real numbers should be consistent with the usual rules for multiplication of positive real numbers and zero. Observe the pattern of products below.

$$
\begin{aligned}
4 \cdot 5 &= 20 \\
4 \cdot 4 &= 16 \\
4 \cdot 3 &= 12 \\
4 \cdot 2 &= 8 \\
4 \cdot 1 &= 4 \\
4 \cdot 0 &= 0 \\
4 \cdot (-1) &= ?
\end{aligned}
$$

Products decrease by 4 in each line.

What number must be assigned as the product $4 \cdot (-1)$ so that the pattern is maintained? The numbers just to the left of the equality symbols decrease by 1 each time, and the products to the right decrease by 4 each time. To maintain the pattern, the number to the right in the bottom equation must be 4 less than 0, which is -4, so

$$4 \cdot (-1) = -4.$$

The pattern continues with

$$
\begin{aligned}
4 \cdot (-2) &= -8 \\
4 \cdot (-3) &= -12 \\
4 \cdot (-4) &= -16,
\end{aligned}
$$

and so on. In the same way,

$$
\begin{aligned}
-4 \cdot 2 &= -8 \\
-4 \cdot 3 &= -12 \\
-4 \cdot 4 &= -16,
\end{aligned}
$$

and so on. A similar observation can be made about the product of two negative real numbers. Look at the pattern that follows.

$$
\begin{aligned}
-5 \cdot 4 &= -20 \\
-5 \cdot 3 &= -15 \\
-5 \cdot 2 &= -10 \\
-5 \cdot 1 &= -5 \\
-5 \cdot 0 &= 0 \\
-5 \cdot (-1) &= ?
\end{aligned}
$$

Products increase by 5 in each line.

The numbers just to the left of the equality symbols decrease by 1 each time. The products on the right increase by 5 each time. To maintain the pattern, the product $-5 \cdot (-1)$ must be 5 more than 0, so it seems reasonable for the following to be true.

$$-5 \cdot (-1) = 5$$

Continuing this pattern gives the following.

$$
\begin{aligned}
-5 \cdot (-2) &= 10 \\
-5 \cdot (-3) &= 15 \\
-5 \cdot (-4) &= 20 \\
&\vdots
\end{aligned}
$$

Sign Rules for Multiplication

$$(+) \cdot (+) = +$$
$$(-) \cdot (-) = +$$
$$(+) \cdot (-) = -$$
$$(-) \cdot (+) = -$$

MULTIPLYING REAL NUMBERS

Like Signs   Multiply two numbers with the *same* sign by multiplying their absolute values to find the absolute value of the product. The product is positive.

Unlike Signs   Multiply two numbers with *different* signs by multiplying their absolute values to find the absolute value of the product. The product is negative.

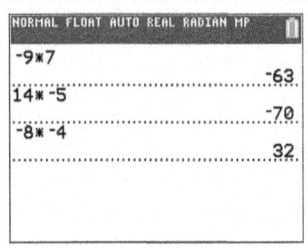

The symbol ∗ represents multiplication on this screen. The display supports the results of **Example 3.**

EXAMPLE 3   **Multiplying Signed Numbers**

Find each product.

(a) $-9 \cdot 7$ **(b)** $14 \cdot (-5)$ **(c)** $-8 \cdot (-4)$

Solution

(a) $-9 \cdot 7 = -63$ **(b)** $14 \cdot (-5) = -70$ **(c)** $-8 \cdot (-4) = 32$

The result of dividing two numbers is their **quotient.** In the quotient $a \div b$ (or $\frac{a}{b}$), where $b \neq 0$, a is the **dividend** (or **numerator**), and b is the **divisor** (or **denominator**). For real numbers a, b, and c,

$$\text{if} \quad \frac{a}{b} = c, \quad \text{then} \quad a = b \cdot c.$$

To illustrate this, consider the quotient

$$\frac{10}{-2}.$$

The value of this quotient is obtained by asking, "What number multiplied by -2 gives 10?" From our discussion of multiplication, the answer to this question must be "-5." Therefore,

$$\frac{10}{-2} = -5, \quad \text{because} \quad 10 = -2 \cdot (-5).$$

Similarly,

$$\frac{-10}{2} = -5 \quad \text{and} \quad \frac{-10}{-2} = 5.$$

These facts, along with the fact that the quotient of two positive numbers is positive, lead to the following rules for division.

Sign Rules for Division

$$(+)/(+) = +$$
$$(-)/(-) = +$$
$$(+)/(-) = -$$
$$(-)/(+) = -$$

DIVIDING REAL NUMBERS

Like Signs   Divide two numbers with the *same* sign by dividing their absolute values to find the absolute value of the quotient. The quotient is positive.

Unlike Signs   Divide two numbers with *different* signs by dividing their absolute values to find the absolute value of the quotient. The quotient is negative.

A general agreement is that the phrase "the quotient of a and b" is interpreted as

$$\frac{a}{b} \; (\text{or } a \div b).$$

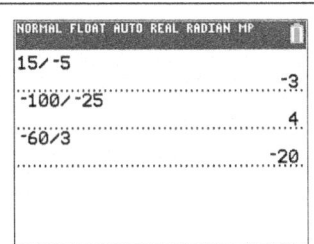

The division operation is represented by a slash (/). This screen supports the results of **Example 4.**

EXAMPLE 4   Dividing Signed Numbers

Find each quotient.

(a) $\dfrac{15}{-5}$ **(b)** $\dfrac{-100}{-25}$ **(c)** $\dfrac{-60}{3}$

Solution

(a) $\dfrac{15}{-5} = -3$ This is true because $-5 \cdot (-3) = 15.$ **(b)** $\dfrac{-100}{-25} = 4$ **(c)** $\dfrac{-60}{3} = -20$

If 0 is divided by a nonzero number, the quotient is 0.

$$\frac{0}{a} = 0, \quad \text{for } a \neq 0$$

This is true because $a \cdot 0 = 0$. However, we cannot divide by 0. There is a good reason for this. Whenever a division is performed, we want to obtain one and only one quotient. Now consider this division problem.

$$\frac{7}{0} = ?$$

We must ask ourselves, "*What number multiplied by 0 gives 7?*" There is no such number because the product of 0 and any number is zero. Now consider this quotient.

$$\frac{0}{0} = ?$$

There are infinitely many answers to the question, "What number multiplied by 0 gives 0?" Division by 0 does not yield a *unique* quotient, and thus it is not permitted.

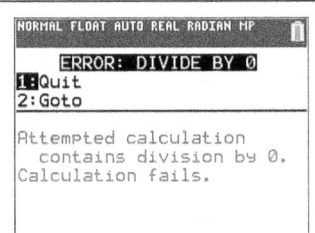

Dividing by zero leads to this message on the TI-84 Plus C.

> **DIVISION INVOLVING ZERO**
>
> $\dfrac{a}{0}$ **is undefined for all** a. $\dfrac{0}{a} = 0$ **for all nonzero** a.

Order of Operations

Given an expression such as $5 + 2 \cdot 3$, should 5 and 2 be added first, or should 2 and 3 be multiplied first? (This is answered in **Example 5(a)** on the next page.) When an expression involves more than one operation, we use the following rules for **order of operations.**

What result does the calculator give? The order of operations determines the answer. (See **Example 5(a)** on the next page.)

> **ORDER OF OPERATIONS**
>
> *If parentheses or square brackets are present:*
>
> **Step 1** Work separately above and below any **fraction bar.**
>
> **Step 2** Use the rules below within each set of **parentheses or square brackets.** Start with the innermost set and work outward.
>
> *If no parentheses or brackets are present:*
>
> **Step 1** Apply any **exponents.**
>
> **Step 2** Do any **multiplications or divisions** in the order in which they occur, working from left to right.
>
> **Step 3** Do any **additions or subtractions** in the order in which they occur, working from left to right.

The sentence **"Please excuse my dear Aunt Sally"** is often used to help us remember the rules for order of operations. The letters **P, E, M, D, A, S** are the first letters of the words of the sentence, and they stand for *parentheses, exponents, multiply, divide, add, subtract.* (*Remember also that M and D have equal priority, as do A and S. Operations with equal priority are performed in order from left to right.*)

NORMAL FLOAT AUTO REAL RADIAN MP

$(-2)^6$
$\qquad\qquad$64.
-2^6
$\qquad\qquad$-64.

Notice the difference in the two expressions and results. This supports $(-2)^6 \neq -2^6$.

When evaluating an exponential expression that involves a negative sign, be aware that $(-a)^n$ and $-a^n$ do not necessarily represent the same quantity. For example, if $a = -2$ and $n = 6$, then

$$(-2)^6 = (-2)(-2)(-2)(-2)(-2)(-2) = 64 \qquad \text{The base is } -2.$$

while $\qquad\qquad -2^6 = -(2 \cdot 2 \cdot 2 \cdot 2 \cdot 2 \cdot 2) = -64. \qquad \text{The base is 2.}$

EXAMPLE 5 Using the Order of Operations

Use the order of operations to simplify each expression.

(a) $5 + 2 \cdot 3$ **(b)** $4 \cdot 3^2 + 7 - (2 + 8)$

(c) $\dfrac{2(8 - 12) - 11(4)}{5(-2) - 3}$ **(d)** -6^4

(e) $(-6)^4$ **(f)** $(-8)(-3) - [4 - (3 - 6)]$

Solution | Be careful! Multiply first.

(a) $5 + 2 \cdot 3 = 5 + 6 \qquad$ Multiply.

$\qquad\qquad\quad = 11 \qquad$ Add.

(b) $4 \cdot 3^2 + 7 - (2 + 8) = 4 \cdot 3^2 + 7 - 10 \qquad$ Work within parentheses first.

| 3^2 means $3 \cdot 3$, not $3 \cdot 2$. $= 4 \cdot 9 + 7 - 10 \qquad$ Apply the exponent.

$\qquad\qquad\qquad\qquad\quad = 36 + 7 - 10 \qquad$ Multiply.

$\qquad\qquad\qquad\qquad\quad = 43 - 10 \qquad$ Add.

$\qquad\qquad\qquad\qquad\quad = 33 \qquad$ Subtract.

(c) $\dfrac{2(8 - 12) - 11(4)}{5(-2) - 3} = \dfrac{2(-4) - 11(4)}{5(-2) - 3} \qquad$ Work separately above and below fraction bar.

$\qquad\qquad\qquad\qquad = \dfrac{-8 - 44}{-10 - 3} \qquad$ Multiply.

$\qquad\qquad\qquad\qquad = \dfrac{-52}{-13} \qquad$ Subtract.

$\qquad\qquad\qquad\qquad = 4 \qquad$ Divide.

(d) $-6^4 = -(6 \cdot 6 \cdot 6 \cdot 6) = -1296$

| The base is 6, not -6.

(e) $(-6)^4 = (-6)(-6)(-6)(-6) = 1296$

| The base is -6 here.

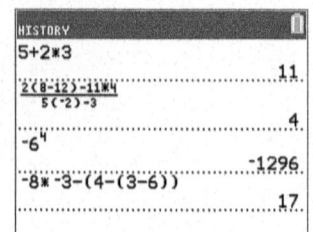

HISTORY
5+2*3
$\qquad\qquad$11.
$\frac{2(8-12)-11*4}{5(-2)-3}$
$\qquad\qquad$4.
-6^4
$\qquad\qquad$-1296.
$-8*-3-(4-(3-6))$
$\qquad\qquad$17.

The calculator supports the results in **Example 5(a), (c), (d), and (f).**

(f) $-8(-3) - [4 - (3 - 6)] = -8(-3) - [4 - (-3)] \qquad$ Work within parentheses.

| Start here. $= -8(-3) - [4 + 3] \qquad$ Definition of subtraction

$\qquad\qquad\qquad\qquad\qquad = -8(-3) - 7 \qquad$ Work within brackets.

$\qquad\qquad\qquad\qquad\qquad = 24 - 7 \qquad$ Multiply.

$\qquad\qquad\qquad\qquad\qquad = 17 \qquad$ Subtract.

Properties of Addition and Multiplication of Real Numbers

PROPERTIES OF ADDITION AND MULTIPLICATION

For real numbers a, b, and c, the following properties hold.

Closure Properties $a + b$ and $a \cdot b$ are real numbers.

Commutative Properties $a + b = b + a$ $a \cdot b = b \cdot a$

Associative Properties $(a + b) + c = a + (b + c)$

$(a \cdot b) \cdot c = a \cdot (b \cdot c)$

Identity Properties There is a real number 0 such that

$$a + 0 = a \quad \text{and} \quad 0 + a = a.$$

There is a real number 1 such that

$$a \cdot 1 = a \quad \text{and} \quad 1 \cdot a = a.$$

Inverse Properties For each real number a, there is a single real number $-a$ such that

$$a + (-a) = 0 \quad \text{and} \quad (-a) + a = 0.$$

For each nonzero real number a, there is a single real number $\frac{1}{a}$ such that

$$a \cdot \frac{1}{a} = 1 \quad \text{and} \quad \frac{1}{a} \cdot a = 1.$$

Distributive Property of Multiplication with Respect to Addition $a \cdot (b + c) = a \cdot b + a \cdot c$

$(b + c) \cdot a = b \cdot a + c \cdot a$

- The set of real numbers is said to be **closed with respect to the operations of addition and multiplication.** This means that the sum of two real numbers and the product of two real numbers are themselves real numbers.

- The **commutative properties** state that two real numbers may be added or multiplied in either order without affecting the result.

- The **associative properties** allow us to group terms or factors in any manner we wish without affecting the result.

- The number 0 is the **identity element for addition.** Adding 0 to a real number will always yield that real number.

- The number 1 is the **identity element for multiplication.** Multiplying a real number by 1 will always yield that real number.

- Each real number a has an **additive inverse,** $-a$, such that the sum of a and its additive inverse is the additive identity element 0.

- Each nonzero real number a has a **multiplicative inverse,** or **reciprocal,** $\frac{1}{a}$, such that the product of a and its multiplicative inverse is the multiplicative identity element 1.

- The **distributive property** allows us to express certain products as sums and certain sums as products.

EXAMPLE 6 **Identifying Properties of Addition and Multiplication**

Identify the property illustrated in each statement.

(a) $5 + 7$ is a real number. **(b)** $5 + (6 + 8) = (5 + 6) + 8$

(c) $8 + 0 = 8$ **(d)** $-4\left(-\dfrac{1}{4}\right) = 1$

(e) $4 + (3 + 9) = 4 + (9 + 3)$ **(f)** $5(x + y) = 5x + 5y$

Solution

(a) The statement that the sum of two real numbers is also a real number is an example of the *closure property of addition.*

(b) Because the grouping of the terms is different on the two sides of the equation, this illustrates the *associative property of addition.*

(c) Adding 0 to a number yields the number itself. This is an example of the *identity property of addition.*

(d) Multiplying a number by its reciprocal yields 1, and this illustrates the *inverse property of multiplication.*

(e) The order of the addends (terms) 3 and 9 is different, so this is justified by the *commutative property of addition.*

(f) The factor 5 is distributed to the terms x and y. This is an example of the *distributive property of multiplication with respect to addition.*

Applications of Signed Numbers

Problem-Solving Strategy

When problems deal with gains and losses, the gains may be interpreted as positive numbers and the losses as negative numbers.

- The phrases "in the **black**" (for making money) and "in the **red**" (for losing money) go back to the days when bookkeepers used black ink to represent gains and red ink to represent losses.

- Temperatures above 0° are positive, and those below 0° are negative.

- Altitudes above sea level are considered positive, and those below sea level are considered negative.

If the temperature in New Orleans was 54°F yesterday and the high today is 68°F, we interpret this as an increase of 14°. The high temperature today is *greater than* it was yesterday, so there is an *increase,* represented by a *positive* number. That number can be found mathematically.

Subtract 54 from 68 to obtain 14.

If the opposite situation had occurred, similar reasoning would say that there was a *decrease,* which would be represented by a *negative* number.

Subtract 68 from 54 to obtain −14.

In this example, both measures are positive. However, if one or both of the measures are negative, determining the amount of change may not be as easy. We can mathematically define the magnitude (or amount) of change from one to the other by taking the absolute value of their difference. ***Because opposites have the same absolute value, the order in which we subtract does not matter.***

DETERMINING CHANGE FROM *a* TO *b*

Suppose that a and b are real numbers. The amount (*without regard to sign*) of change *from a to b* is given by

$$|a - b|, \quad \text{or, equivalently,} \quad |b - a|.$$

The change is interpreted to be a **positive change** if $a < b$, and it is a **negative change** if $a > b$. If $a = b$, there is no change.

StatCrunch Interpreting Change from a Table

EXAMPLE 7   **Interpreting Change from a Graph**

Suppose that you are an employee of a fund management company and you come across the results shown in **Figure 8** in a trade publication. Amounts are rounded to the nearest percent.

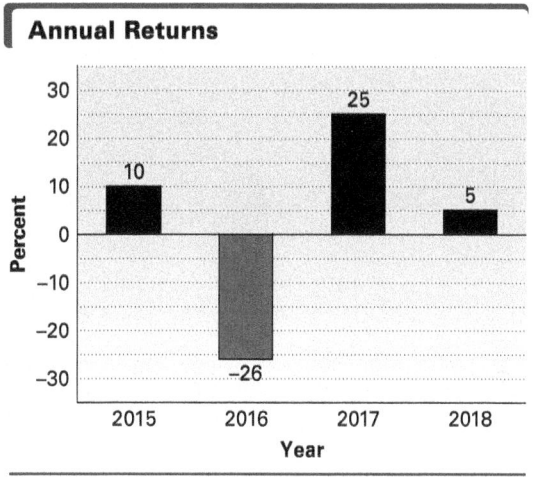

Annual Returns

Figure 8

Illustrate by subtraction and absolute value how to determine the change in percent of annual returns from **(a)** 2015 to 2016 and **(b)** 2016 to 2017.

Solution

(a) From 2015 to 2016, the annual returns went from 10 percent to -26 percent. This is a decrease, so our answer will be *negative*. The amount of the decrease is

$$|10 - (-26)| = |10 + 26| = |36| = 36.$$

(Note that we could have subtracted in the order $-26 - 10 = -36$, which has the same absolute value as 36.) The change from 2015 to 2016 is

$$-36 \text{ percent.}$$

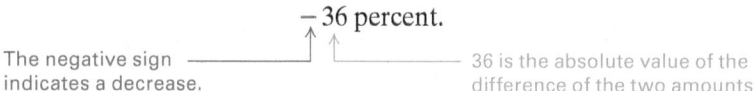

The negative sign indicates a decrease.   36 is the absolute value of the difference of the two amounts.

(b) To find the change from 2016 to 2017, which by inspection is an increase, find the absolute value of the difference of the two amounts in either order, and interpret it as a *positive* amount.

$$|-26 - 25| = |-51| = 51 \quad \text{or} \quad |25 - (-26)| = |51| = 51$$

The change from 2016 to 2017 is $+51$ percent.

The $+$ sign is understood if not specifically indicated.

EXAMPLE 8 Interpreting Temperature Change from a Map

The daily highs for selected cities during a particularly cold two-day period in Minnesota are shown in **Figure 9.** Determine the change in temperature from Monday to Tuesday for each city shown.

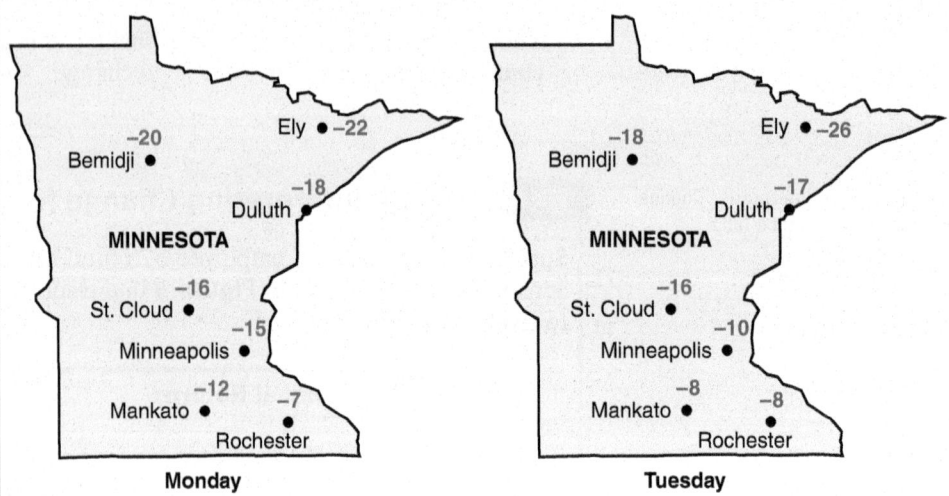

Figure 9

Solution

In each case we determine the sign of the change by observation. If the temperature on Tuesday *is greater than* that on Monday, the change is *positive*. If Tuesday's temperature *is less than* that on Monday, the change is *negative*. Equal temperatures indicate *no change*.

In Ely, the high temperature was less on Tuesday than on Monday because $-26 < -22$, so the change is negative. The amount is

$$|-26 - (-22)| = |-26 + 22| = |-4| = 4.$$

Thus the change is -4 degrees.

In Bemidji, the high temperature went from -20 to -18, which is an increase. Subtracting these temperatures in either order and taking the absolute value gives

$$|-20 - (-18)| = |-20 + 18| = |-2| = 2.$$

Thus, the change is $+2$ degrees. The other changes can be found similarly.

Duluth: +1 degree   St. Cloud: no change

Minneapolis: +5 degrees   Mankato: +4 degrees

Rochester: −1 degree

6.2 **EXERCISES**

CONCEPT CHECK *Complete each statement and give an example.*

1. The sum of two positive numbers is a _____ number.

2. The sum of two negative numbers is a _____ number.

3. The sum of a positive number and a negative number is negative if the negative number has the _____ absolute value.

4. The sum of a positive number and a negative number is positive if the positive number has the _____ absolute value.

5. The difference of two positive numbers is negative if _____.

6. The difference of two negative numbers is negative if _____.

7. The sum of a positive number and a negative number is 0 if the numbers are _____.

8. The product of two numbers with the same sign is _____.

9. The product of two numbers with different signs is _____.

10. The quotient formed by any nonzero number divided by 0 is _____, and the quotient formed by 0 divided by any nonzero number is _____.

Perform the indicated operations, using the order of operations as necessary.

11. $-12 + (-8)$

12. $-5 + (-2)$

13. $12 + (-16)$

14. $-6 + 17$

15. $-12 - (-1)$

16. $-3 - (-8)$

17. $-5 + 11 + 3$

18. $-9 + 16 + 5$

19. $12 - (-3) - (-5)$

20. $15 - (-6) - (-8)$

21. $(-12)(-2)$

22. $(-3)(-5)$

23. $9(-12)(-4)(-1)(3)$

24. $-5(-17)(2)(-2)(4)$

25. $\dfrac{-18}{-3}$

26. $\dfrac{-100}{-50}$

27. $\dfrac{36}{-6}$

28. $\dfrac{52}{-13}$

29. $\dfrac{0}{12}$

30. $\dfrac{0}{-7}$

31. $-6 + [5 - (3 + 2)]$

32. $-8[4 + (7 - 8)]$

33. $-4 - 3(-2) + 5^2$

34. $-6 - 5(-8) + 3^2$

35. $\dfrac{2(-5) + (-3)(-2^2)}{-3^2 + 9}$

36. $\dfrac{3(-4) + (-5)(-2)}{2^3 - 2 + (-6)}$

37. $\dfrac{(-8 + 6) \cdot (-5)}{-5 - 5}$

38. $\dfrac{(-10 + 4) \cdot (-3)}{-7 - 2}$

39. $-8(-2) - [(4^2) + (7 - 3)]$

40. $-7(-3) - [2^3 - (3 - 4)]$

41. $\dfrac{(-6 + 3) \cdot (-4)}{-5 - 1} - \dfrac{(-9 + 6) \cdot (-3)}{-4 + 3}$

42. $\dfrac{2(-5 + 3)}{-2^2} - \dfrac{(-3^2 + 2)(3)}{3 - (-4)}$

43. $-\dfrac{1}{4}[3(-5) + 7(-5) + 1(-2)]$

44. $\dfrac{5 - 3\left(\dfrac{-5 - 9}{-7}\right) - 6}{-9 - 11 + 3 \cdot 7}$

45. CONCEPT CHECK Which of the following expressions are undefined?

A. $\dfrac{8}{0}$ **B.** $\dfrac{9}{6 - 6}$ **C.** $\dfrac{4 - 4}{5 - 5}$ **D.** $\dfrac{0}{-1}$

46. CONCEPT CHECK If you have no money in your pocket and you divide it equally among your three siblings, how much does each get? Use this situation to explain division of zero by a positive integer.

Identify the property illustrated by each statement.

47. $6 + 9 = 9 + 6$

48. $8 \cdot 4 = 4 \cdot 8$

49. $9 + (-9) = 0$

50. $12 + 0 = 12$

51. $9 \cdot 1 = 9$

52. $\left(\dfrac{1}{-3}\right) \cdot (-3) = 1$

53. $0 + 283 = 283$

54. $(3 \cdot 5) \cdot 4 = 4 \cdot (3 \cdot 5)$

55. $6 \cdot (4 \cdot 2) = (6 \cdot 4) \cdot 2$ **56.** $0 = -8 + 8$

57. $19 \cdot 12$ is a real number.

58. $19 + 12$ is a real number.

59. $7 + (2 + 5) = (7 + 2) + 5$

60. $2 \cdot (4 + 3) = 2 \cdot 4 + 2 \cdot 3$

61. $9 \cdot 6 + 9 \cdot 8 = 9 \cdot (6 + 8)$

Write a short answer for each problem.

62. One of the authors received an email message from an old friend, Frank Capek. Frank said that his grandson had to evaluate

$$9 + 15 \div 3.$$

Frank and his wife, Barbara, said that the answer is 8, but the grandson said that the correct answer is 14. The grandson's reasoning is "There is a rule called the Order of Process so that you proceed from right to left rather than from left to right."

(a) What is the correct answer?

(b) Is his grandson's reasoning correct? Explain.

63. The following conversation actually took place between one of the authors of this text and his son, Jack, when Jack was four years old.

DADDY: "Jack, what is $3 + 0$?"

JACK: "3"

DADDY: "Jack, what is $4 + 0$?"

JACK: "4 ... and Daddy, *string* plus zero equals *string!*"

What property of addition of real numbers did Jack recognize?

64. Many everyday activities are commutative. The order in which they occur does not affect the outcome. For example, "putting on your shirt" and "putting on your pants" are commutative operations. Decide whether the given activities are commutative.

   (a) putting on your shoes; putting on your socks

   (b) getting dressed; taking a shower

   (c) combing your hair; brushing your teeth

65. Many everyday occurrences can be thought of as operations that have opposites or inverses. For example, the inverse operation for "going to sleep" is "waking up." For each of the given activities, specify its inverse activity.

   (a) cleaning up your room

   (b) earning money

   (c) increasing the volume on your telephone

66. The distributive property holds for multiplication with respect to addition. Does the distributive property hold for addition with respect to multiplication? That is, is

$$a + (b \cdot c) = (a + b) \cdot (a + c)$$

   true for all values of a, b, and c? (*Hint:* Let $a = 2$, $b = 3$, and $c = 4$.)

Each expression is equal to either 81 or −81. Decide which of these is the correct value in each case.

67. -3^4 68. $-(3^4)$ 69. $(-3)^4$

70. $-(-3^4)$ 71. $-(-3)^4$ 72. $[-(-3)]^4$

73. $-[-(-3)]^4$ 74. $-[-(-3^4)]$

Solve each problem.

75. **S&P 500 Index Fund** The graph shows annual returns (to the nearest percent) for Class A shares of the Invesco S&P 500 Index Fund from 2009 to 2013. Use subtraction and absolute value to determine the change in returns from one year to the next. Give your answer as both a signed number (in percent) and a word description.

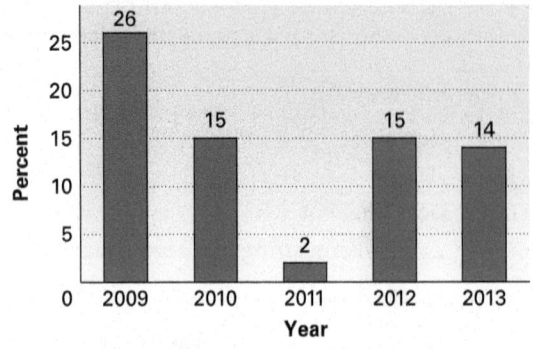

   (a) 2009 to 2010   (b) 2010 to 2011
   (c) 2011 to 2012   (d) 2012 to 2013

76. **Company Profits and Losses** The graph shows profits and losses for a private company for the years 2013 through 2017. Use subtraction and absolute value to determine the change in profit or loss from one year to the next. Give your answer as both a signed number (in thousands of dollars) and a word description.

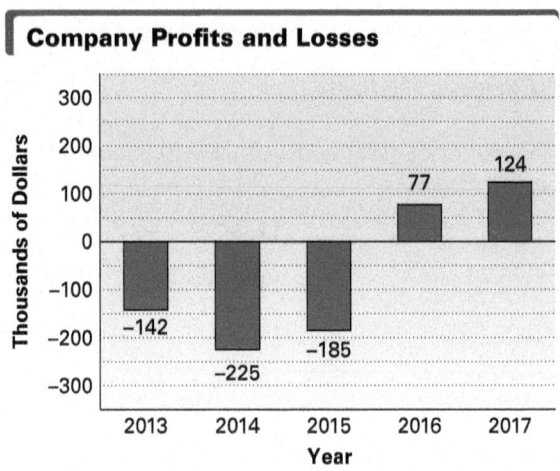

   (a) 2013 to 2014   (b) 2014 to 2015
   (c) 2015 to 2016   (d) 2016 to 2017

Heights of Mountains and Depths of Trenches   *The table shows the heights of some selected mountains and the depths of some selected ocean trenches.*

Mountain	Height (in feet)	Trench	Depth (in feet, as a negative number)
Foraker	17,400	Philippine	−32,995
Wilson	14,246	Cayman	−24,721
Pikes Peak	14,110	Java	−23,376

Data from *The World Almanac and Book of Facts.*

Use the information given to answer each question.

77. What is the difference between the height of Mt. Foraker and the depth of the Philippine Trench?

78. What is the difference between the height of Pikes Peak and the depth of the Java Trench?

79. How much deeper is the Cayman Trench than the Java Trench?

80. How much deeper is the Philippine Trench than the Cayman Trench?

81. How much higher is Mt. Wilson than Pikes Peak?

82. If Mt. Wilson and Pikes Peak were stacked one on top of the other, how much higher would they be than Mt. Foraker?

Golf Scores *The table gives scores above or below par (that is, above or below the score "standard") for selected golfers during the 2017 PGA Tour Championship. Write a signed number that represents the total score for the four rounds (above or below par) for each golfer.*

	Golfer	Round 1	Round 2	Round 3	Round 4
83.	Xander Schauffele	−1	−4	−5	−2
84.	Justin Thomas	−3	−4	0	−4
85.	Sergio Garcia	+3*	−4	−2	−3
86.	Rickie Fowler	+3	+4	0	−1

*Golf scoring commonly includes a + sign with a score over par.
Data from PGA.

Temperature Change *During a cold two-day period in North Dakota, the daily highs for selected cities were as shown in the figure. Determine the change in temperature from Thursday to Friday for each city listed.* **All temperatures in Exercises 87–96 are in degrees Fahrenheit.**

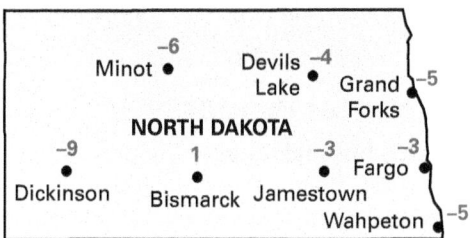

Thursday

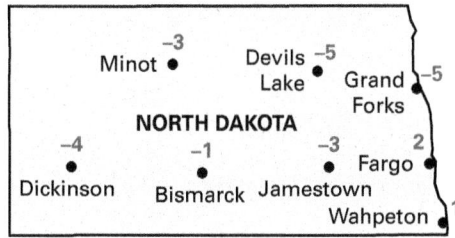

Friday

87. Dickinson

88. Minot

89. Bismarck

90. Devils Lake

91. Jamestown

92. Grand Forks

93. Fargo

94. Wahpeton

95. *24-Hour Temperature Change* Television weather forecasters often show maps indicating how the current temperature compares to that of 24 hours earlier. Refer to the map for Thursday used in **Exercises 87–94,** and imagine that the temperatures listed indicated the change in temperature compared to 24 hours earlier. Use the table to determine the temperature in each city 24 hours earlier. (When no sign is shown on the map, + is understood.)

City	Current Temperature	Temperature 24 Hours Earlier
Dickinson	34	
Minot	28	
Bismarck	30	
Devils Lake	32	
Jamestown	35	
Grand Forks	31	
Fargo	34	
Wahpeton	36	

96. *24-Hour Temperature Change* Repeat **Exercise 95** for the Friday map used in **Exercises 87–94.**

Solve each problem.

97. *Breaching of Humpback Whales* Humpback whale researchers Mark and Debbie noticed that one of their favorite whales, "Pineapple," breached 15 feet above the surface of the ocean while her mate cruised 12 feet below the surface. What is the difference between these two levels?

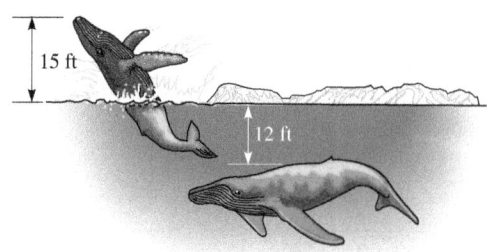

98. *Altitude of Hikers* The surface, or rim, of a canyon is at altitude 0. On a hike down into the canyon, a party of hikers stop for a rest at 130 meters below the surface. They then descend another 54 meters. What is their new altitude? (Write the altitude as a signed number.)

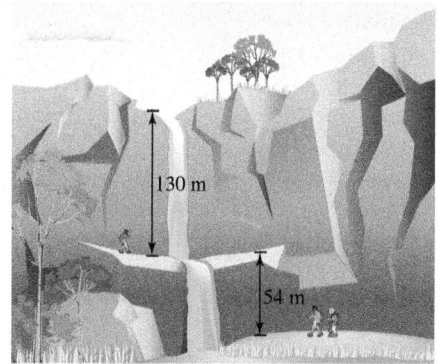

99. *Pythagoras and Euclid* The Greek mathematicians Pythagoras and Euclid died in 495 B.C. and 285 B.C., respectively. Interpret these years as negative numbers, and write a subtraction problem indicating that these two mathematicians died 210 years apart.

100. *Archimedes and Eratosthenes* The Greek mathematicians Archimedes and Eratosthenes died in 212 B.C. and 194 B.C., respectively. Interpret these years as negative numbers, and write a subtraction problem indicating that these two mathematicians died 18 years apart.

101. *Birth Date of a Greek Mathematician* A certain Greek mathematician was born in 428 B.C. Her father was born 41 years earlier. In what year was her father born?

102. *Birth Date of a Roman Philosopher* A certain Roman philosopher was born in 325 B.C. Her mother was born 35 years earlier. In what year was her mother born?

103. *Record Low Temperatures* The lowest temperature ever recorded in Illinois was −36°F. The lowest temperature ever recorded in North Dakota was 24°F lower than Illinois's record low. What is the record low temperature in North Dakota? (Data from National Climatic Data Center.)

104. *Record Low Temperatures* The lowest temperature ever recorded in South Carolina was −19°F. The lowest temperature ever recorded in Wyoming was 47°F lower than South Carolina's record low. What is the record low temperature in Wyoming? (Data from National Climatic Data Center.)

105. *Temperature Extremes* The lowest temperature ever recorded in Indiana was −36°F. The highest temperature ever recorded there was 152°F more than the lowest. What is this highest temperature? (Data from National Climatic Data Center.)

106. *Drastic Temperature Change* On January 23, 1943, the temperature rose 49°F in two minutes in Spearfish, South Dakota. If the starting temperature was −4°F, what was the temperature two minutes later? (Data from National Climatic Data Center.)

6.3 RATIONAL NUMBERS AND DECIMAL REPRESENTATION

OBJECTIVES

1 Define and identify rational numbers.

2 Write a rational number in lowest terms.

3 Add, subtract, multiply, and divide rational numbers in fraction form.

4 Solve a carpentry problem using operations with fractions.

5 Apply the density property and find arithmetic mean.

6 Convert a rational number in fraction form to a decimal number.

7 Convert a terminating or repeating decimal to a rational number in fraction form.

Definition and the Fundamental Property

Recall that quotients of integers are called **rational numbers.** Think of the rational numbers as being made up of all the fractions (quotients of integers with denominator not equal to zero) and all the integers. Any integer can be written as a quotient of two integers. For example, the integer 9 can be written as the quotient

$$\frac{9}{1}, \quad \text{or} \quad \frac{18}{2}, \quad \text{or} \quad \frac{27}{3}, \quad \text{and so on.}$$

Also, −5 can be expressed as a quotient of integers as

$$\frac{-5}{1}, \quad \text{or} \quad \frac{-10}{2}, \quad \text{and so on.}$$

(How can the integer 0 be written as a quotient of integers?)

RATIONAL NUMBERS

Rational numbers = $\{x \mid x$ is a quotient of two integers, with denominator not $0\}$

A rational number is in **lowest terms** if the greatest common factor of the numerator (top number) and the denominator (bottom number) is 1. Rational numbers are written in lowest terms using the following property.

FUNDAMENTAL PROPERTY OF RATIONAL NUMBERS

If a, b, and k are integers with $b \neq 0$ and $k \neq 0$, then the following holds true.

$$\frac{a \cdot k}{b \cdot k} = \frac{a}{b}$$

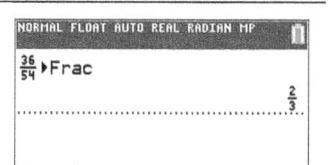

The calculator gives $\frac{36}{54}$ in lowest terms, as illustrated in **Example 1**.

EXAMPLE 1 Writing a Fraction in Lowest Terms

Write $\frac{36}{54}$ in lowest terms.

Solution

The greatest common factor of 36 and 54 is 18.

$$\frac{36}{54} = \frac{2 \cdot 18}{3 \cdot 18} = \frac{2}{3} \qquad \text{Use the fundamental property with } k = 18.$$

In **Example 1**, we see that $\frac{36}{54} = \frac{2}{3}$. If we multiply the numerator of the fraction on the left by the denominator of the fraction on the right, we obtain $36 \cdot 3 = 108$. If we multiply the denominator of the fraction on the left by the numerator of the fraction on the right, we obtain $54 \cdot 2 = 108$. The result is the same in both cases.

One way of determining whether two fractions are equal is to perform this test. If the product of the **"extremes"** (36 and 3 in this case) equals the product of the **"means"** (54 and 2), the fractions are equal. This is the **cross-product test.**

CROSS-PRODUCT TEST FOR EQUALITY OF RATIONAL NUMBERS

For rational numbers $\frac{a}{b}$ and $\frac{c}{d}$, where $b \neq 0$, $d \neq 0$, the following holds true.

$$\frac{a}{b} = \frac{c}{d} \quad \text{if and only if} \quad a \cdot d = b \cdot c$$

Operations with Rational Numbers

The operation of addition of rational numbers can be illustrated by the sketches in **Figure 10.** The rectangle at the top left is divided into three equal portions, with one of the portions in color. The rectangle at the top right is divided into five equal parts, with two of them in color.

The total of the areas in color is represented by the sum

$$\frac{1}{3} + \frac{2}{5}.$$

To evaluate this sum, the areas in color must be redrawn in terms of a common unit. Since the least common multiple of 3 and 5 is 15, redraw both rectangles with 15 parts. See **Figure 11.** In the figure, 11 of the small rectangles are in color, so

$$\frac{1}{3} + \frac{2}{5} = \frac{5}{15} + \frac{6}{15} = \frac{11}{15}.$$

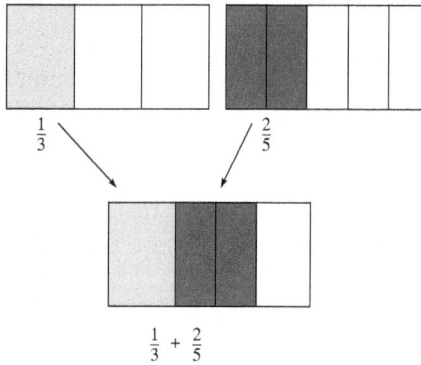

Figure 10

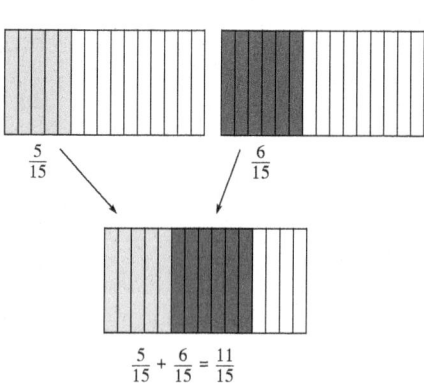

Figure 11

ADDING AND SUBTRACTING RATIONAL NUMBERS

If $\frac{a}{b}$ and $\frac{c}{d}$ are rational numbers, then the following hold true.

$$\frac{a}{b} + \frac{c}{d} = \frac{ad + bc}{bd} \quad \text{and} \quad \frac{a}{b} - \frac{c}{d} = \frac{ad - bc}{bd}$$

This formal definition is seldom used in practice. We usually first rewrite the fractions with the least common multiple of their denominators, called the **least common denominator (LCD).**

EXAMPLE 2 **Adding and Subtracting Rational Numbers**

Perform each operation.

(a) $\dfrac{2}{15} + \dfrac{1}{10}$ **(b)** $\dfrac{173}{180} - \dfrac{69}{1200}$

Solution

(a) Because $30 \div 15 = 2$, $\dfrac{2}{15} = \dfrac{2 \cdot 2}{15 \cdot 2} = \dfrac{4}{30}$

and because $30 \div 10 = 3$, $\dfrac{1}{10} = \dfrac{1 \cdot 3}{10 \cdot 3} = \dfrac{3}{30}.$

> The LCD is 30.

Thus, $\dfrac{2}{15} + \dfrac{1}{10} = \dfrac{4}{30} + \dfrac{3}{30} = \dfrac{7}{30}.$

(b) The least common multiple of 180 and 1200 is 3600.

$$\frac{173}{180} - \frac{69}{1200} = \frac{3460}{3600} - \frac{207}{3600} = \frac{3460 - 207}{3600} = \frac{3253}{3600}$$

Fractions that are greater than 1 may be expressed in **mixed number** form.

The mixed number form of $\dfrac{3}{2}$ is $1\dfrac{1}{2}.$

It is understood that the whole number part and the fraction part of a mixed number are added (even though no addition symbol is shown).

EXAMPLE 3 **Conversions Involving Mixed Numbers**

Perform each conversion.

(a) $2\dfrac{5}{8}$ to a fraction   **(b)** $\dfrac{86}{5}$ to a mixed number

Solution

(a) $2\dfrac{5}{8} = 2 + \dfrac{5}{8}$ The addition is understood.

$= \dfrac{16}{8} + \dfrac{5}{8}$ Write 2 as a fraction with denominator 8.

$= \dfrac{21}{8}$ Add the fractions.

Animation
→

Least Common Multiple, Greatest Common Divisor

NORMAL FLOAT AUTO REAL RADIAN MP

$\frac{2}{15} + \frac{1}{10}$

$\frac{7}{30}$

$\frac{173}{180} - \frac{69}{1200}$

$\frac{3253}{3600}$

The results of **Example 2** are illustrated in this screen.

A shortcut for writing $2\frac{5}{8}$ as a fraction leads to the same result. We multiply the denominator (8) by the whole number part (2) and add the numerator (5) to find the numerator of the mixed number.

$$(8 \times 2) + 5 = 21$$

Then use the original denominator of the fraction part (8) as the denominator.

$$2\frac{5}{8} = \frac{21}{8}$$

(b) To convert $\frac{86}{5}$ to a mixed number, we use long division. The quotient is the whole number part of the mixed number. The remainder is the numerator of the fraction part, while the divisor is the denominator.

$$
\begin{array}{r}
17 \leftarrow \text{Whole number part} \\
\text{Denominator of the fraction} \longrightarrow 5\overline{)86} \\
\underline{5} \\
36 \\
\underline{35} \\
1 \leftarrow \text{Numerator of the fraction}
\end{array}
$$

Thus $\dfrac{86}{5} = 17\dfrac{1}{5}$.

MULTIPLYING RATIONAL NUMBERS

If $\frac{a}{b}$ and $\frac{c}{d}$ are rational numbers, then the following holds true.

$$\frac{a}{b} \cdot \frac{c}{d} = \frac{ac}{bd}$$

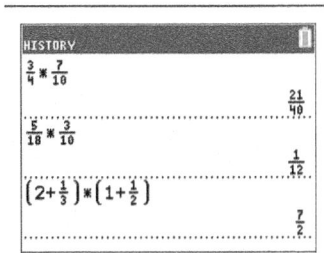

This screen supports the results of
Example 4.

EXAMPLE 4   **Multiplying Rational Numbers**

Find each product. If it is greater than 1, write as a mixed number.

(a) $\dfrac{3}{4} \cdot \dfrac{7}{10}$ **(b)** $\dfrac{5}{18} \cdot \dfrac{3}{10}$ **(c)** $2\dfrac{1}{3} \cdot 1\dfrac{1}{2}$

Solution

(a) $\dfrac{3}{4} \cdot \dfrac{7}{10} = \dfrac{3 \cdot 7}{4 \cdot 10} = \dfrac{21}{40}$

(b) $\dfrac{5}{18} \cdot \dfrac{3}{10} = \dfrac{5 \cdot 3}{18 \cdot 10} = \dfrac{15}{180} = \dfrac{1 \cdot 15}{12 \cdot 15} = \dfrac{1}{12}$

In practice, a multiplication problem such as this is often solved by using slash marks to indicate that common factors have been divided out of the numerator and denominator.

$$\frac{\overset{1}{5}}{\underset{6}{18}} \cdot \frac{\overset{1}{3}}{\underset{2}{10}} = \frac{1}{6} \cdot \frac{1}{2} \qquad \begin{array}{l} \text{3 is divided out of 3 and 18.} \\ \text{5 is divided out of 5 and 10.} \end{array}$$

$$= \frac{1}{12}$$

(c) $2\frac{1}{3} \cdot 1\frac{1}{2} = \frac{7}{3} \cdot \frac{3}{2}$ Convert mixed numbers to fractions.

$= \frac{7}{3} \cdot \frac{\overset{1}{3}}{2}$ Divide out the common factor 3.

$= \frac{7}{2}$ Multiply the fractions.

$= 3\frac{1}{2}$ Write as a mixed number.

A fraction bar indicates the operation of division. The multiplicative inverse of the nonzero number b is $\frac{1}{b}$. We define division using the multiplicative inverse.

DEFINITION OF DIVISION

If a and b are real numbers, where $b \neq 0$, then the following holds true.

$$\frac{a}{b} = a \cdot \frac{1}{b}$$

Early U.S. cents and **half cents** used fractions to denote their denominations. The half cent used $\frac{1}{200}$ and the cent used $\frac{1}{100}$. (See **Exercise 28** for a photo of an interesting error coin.)

The coins shown here were part of the collection of Louis E. Eliasberg, Sr. **Louis Eliasberg** was the only person ever to assemble a complete collection of United States coins.

You probably have heard the rule "To divide fractions, invert the divisor and multiply." To illustrate this rule, suppose that you have $\frac{7}{8}$ of a gallon of milk and you wish to find how many quarts you have. A quart is $\frac{1}{4}$ of a gallon, so you must ask yourself, "*How many $\frac{1}{4}$s are there in $\frac{7}{8}$?*" This would be interpreted as

$$\frac{7}{8} \div \frac{1}{4}, \quad \text{or} \quad \frac{\frac{7}{8}}{\frac{1}{4}}.$$

The fundamental property of rational numbers can be extended to rational number values of a, b, and k.

$$\frac{a}{b} = \frac{a \cdot k}{b \cdot k} = \frac{\frac{7}{8} \cdot 4}{\frac{1}{4} \cdot 4} = \frac{\frac{7}{8} \cdot 4}{1} = \frac{7}{8} \cdot \frac{4}{1}$$ Let $a = \frac{7}{8}$, $b = \frac{1}{4}$, and $k = 4$ (the reciprocal of $b = \frac{1}{4}$).

To divide $\frac{7}{8}$ by $\frac{1}{4}$ is equivalent to multiplying $\frac{7}{8}$ by the reciprocal, $\frac{4}{1}$.

$$\frac{7}{8} \cdot \frac{4}{1} = \frac{28}{8} = \frac{7}{2}$$ Multiply and reduce to lowest terms.

Thus there are $\frac{7}{2}$, or $3\frac{1}{2}$, quarts in $\frac{7}{8}$ gallon.

DIVIDING RATIONAL NUMBERS

If $\frac{a}{b}$ and $\frac{c}{d}$ are rational numbers, where $\frac{c}{d} \neq 0$, then the following holds true.

$$\frac{a}{b} \div \frac{c}{d} = \frac{a}{b} \cdot \frac{d}{c} = \frac{ad}{bc}$$

EXAMPLE 5 Dividing Rational Numbers

Find each quotient.

(a) $\dfrac{3}{5} \div \dfrac{7}{15}$ (b) $\dfrac{-4}{7} \div \dfrac{3}{14}$ (c) $\dfrac{2}{9} \div 4$ (d) $-9 \div \dfrac{3}{5}$

Solution

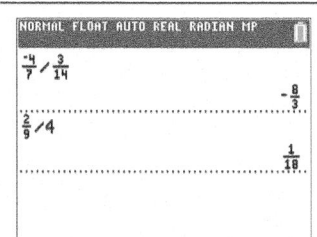

This screen supports the results in
Example 5(b) and (c).

(a) $\dfrac{3}{5} \div \dfrac{7}{15} = \dfrac{3}{5} \cdot \dfrac{15}{7} = \dfrac{45}{35} = \dfrac{9 \cdot 5}{7 \cdot 5} = \dfrac{9}{7}$

(b) $\dfrac{-4}{7} \div \dfrac{3}{14} = \dfrac{-4}{7} \cdot \dfrac{14}{3} = \dfrac{-56}{21} = \dfrac{-8 \cdot 7}{3 \cdot 7} = \dfrac{-8}{3} = -\dfrac{8}{3}$ $\dfrac{-a}{b}, \dfrac{a}{-b}$, and $-\dfrac{a}{b}$ are all equal.

(c) $\dfrac{2}{9} \div 4 = \dfrac{2}{9} \div \dfrac{4}{1} = \dfrac{2}{9} \cdot \dfrac{1}{4} = \dfrac{\overset{1}{2}}{9} \cdot \dfrac{1}{\underset{2}{4}} = \dfrac{1}{18}$

(d) $-9 \div \dfrac{3}{5} = \dfrac{\overset{-3}{\cancel{-9}}}{1} \cdot \dfrac{5}{\underset{1}{\cancel{3}}} = -15$

EXAMPLE 6 Applying Operations with Fractions to Carpentry

A carpenter has a board 72 inches long that she must cut into 10 pieces of equal length. This will require 9 cuts. See **Figure 12.**

(a) If each cut causes a waste of $\dfrac{3}{16}$ inch, how many inches of actual board will remain after the cuts?

(b) What will be the length of each of the resulting pieces?

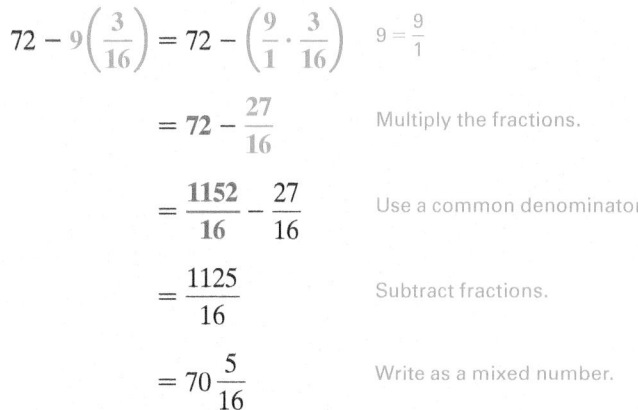

Figure 12

Solution

(a) We start with 72 inches and must subtract $\dfrac{3}{16}$ nine times. Rather than writing out each of these, we represent the amount subtracted by $9\left(\dfrac{3}{16}\right)$.

$$72 - 9\left(\dfrac{3}{16}\right) = 72 - \left(\dfrac{9}{1} \cdot \dfrac{3}{16}\right) \qquad 9 = \dfrac{9}{1}$$

$$= 72 - \dfrac{27}{16} \qquad \text{Multiply the fractions.}$$

$$= \dfrac{1152}{16} - \dfrac{27}{16} \qquad \text{Use a common denominator.}$$

$$= \dfrac{1125}{16} \qquad \text{Subtract fractions.}$$

$$= 70\dfrac{5}{16} \qquad \text{Write as a mixed number.}$$

Thus $70\dfrac{5}{16}$ inches of actual board will remain.

Benjamin Banneker (1731–1806) spent the first half of his life tending a farm in Maryland. He gained a reputation locally for his mechanical skills and his ability in mathematical problem solving. In 1772 he acquired astronomy books from a neighbor and devoted himself to learning astronomy, observing the skies, and making calculations. In 1789 Banneker joined the team that surveyed what is now the District of Columbia.

Banneker published almanacs yearly from 1792 to 1802. He sent a copy of his first almanac to **Thomas Jefferson,** along with an impassioned letter against slavery. Jefferson subsequently championed the cause of this early African-American mathematician.

(b) The $70\frac{5}{16}$, or $\frac{1125}{16}$, inches of board from part (a) will be divided into 10 pieces of equal length.

$$\frac{1125}{16} \div 10 = \frac{1125}{16} \cdot \frac{1}{10} \qquad \text{Definition of division}$$

$$= \frac{\overset{225}{\cancel{1125}}}{16} \cdot \frac{1}{\underset{2}{\cancel{10}}} \qquad \text{Divide out the common factor 5.}$$

$$= \frac{225}{32} \qquad \text{Multiply.}$$

$$= 7\frac{1}{32} \qquad \text{Convert to a mixed number.}$$

Each piece will measure $7\frac{1}{32}$ inches.

Density and the Arithmetic Mean

There is no integer between two consecutive integers, such as 3 and 4. However, a rational number can always be found between any two distinct rational numbers. For this reason, the set of rational numbers is said to be *dense.*

DENSITY PROPERTY OF THE RATIONAL NUMBERS

If r and t are distinct rational numbers, with $r < t$, then there exists a rational number s such that

$$r < s < t.$$

Repeated applications of the density property lead to the following conclusion.

There are infinitely many rational numbers between two distinct rational numbers.

One example of a rational number that is between two distinct rational numbers is the *arithmetic mean* of the two numbers. To find the **arithmetic mean,** or **average,** of n numbers, we add the numbers and then divide the sum by n. For two numbers, the number that lies halfway between them on a number line is their average.

Table 4

Academic Year	Amount of Tuition and Fees
2013–14	$31,570
2014–15	32,140
2015–16	33,180
2016–17	34,100
2017–18	34,740

Data from trends.collegeboard.org

EXAMPLE 7 Finding an Arithmetic Mean (Average)

Table 4 shows the average amount in 2017 dollars of tuition and fees at private 4-year institutions for five academic years. What is the average cost for this period?

Solution

To find this average, divide the sum of the amounts by the number of amounts, 5.

$$\frac{31,570 + 32,140 + 33,180 + 34,100 + 34,740}{5} = \frac{165,730}{5}$$

$$= 33,146$$

The average amount for the period is $33,146.

Simon Stevin (1548–1620) worked as a bookkeeper in Belgium and became an engineer in the Netherlands army. He is usually given credit for the development of **decimals**.

EXAMPLE 8 Finding a Measurement in a Socket Wrench Set

A carpenter owns a socket wrench set that has measurements in the English system. As is the custom in carpentry, measurements are given in half-, quarter-, eighth-, and sixteenth-inches. He finds that his $\frac{3}{8}$-inch socket is just a bit too small for his job, while his $\frac{1}{2}$-inch socket is just a bit too large. He suspects that he will need to use the socket with the measure that is halfway between these (their arithmetic mean, or average). What measure socket must he use?

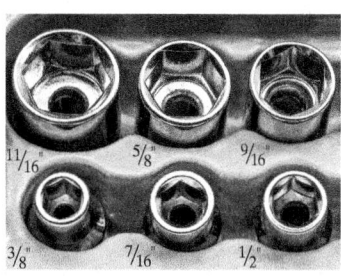

Solution

The carpenter needs to find the average of $\frac{3}{8}$ and $\frac{1}{2}$. To do this, find their sum and divide by 2.

$$\frac{3}{8} + \frac{1}{2} = \frac{3}{8} + \frac{4}{8} \qquad \text{Find a common denominator.}$$

$$= \frac{7}{8} \qquad \text{Add the numerators, and keep the same denominator.}$$

Now divide $\frac{7}{8}$ by 2.

$$\frac{7}{8} \div 2 = \frac{7}{8} \cdot \frac{1}{2} \qquad \text{Definition of division}$$

$$= \frac{7}{16} \qquad \text{Multiply the numerators and multiply the denominators.}$$

He must use the $\frac{7}{16}$-inch socket.

Decimal Form of Rational Numbers

Rational numbers can be expressed as decimals. Decimal numerals have place values that are powers of 10. The place values are as shown here.

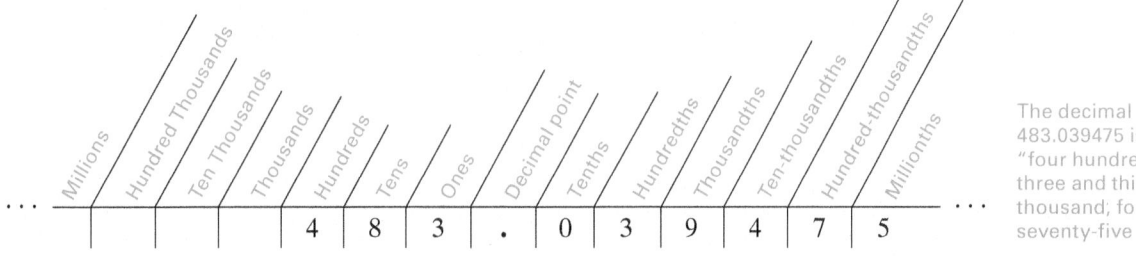

The decimal numeral 483.039475 is read "four hundred eighty-three and thirty-nine thousand; four hundred seventy-five millionths."

A rational number in the form $\frac{a}{b}$ can be expressed as a decimal most easily by entering it into a calculator. For example, to write $\frac{3}{8}$ as a decimal, enter 3, then enter the operation of division, then enter 8. Press the equals (or enter) key to find the following equivalence.

$$\frac{3}{8} = 0.375$$

```
  0.375       0.3636...
8)3.000     11)4.00000...
  24          33
  ──          ──
  60          70
  56          66
  ──          ──
  40          40
  40          33
  ──          ──
   0          70
              66
              ──
              40
               ⋰
```

This same result may be obtained by long division, as shown in the margin. By this result, the rational number $\frac{3}{8}$ is the same as the decimal 0.375. A decimal such as 0.375, which stops, is called a **terminating decimal.**

$$\frac{1}{4} = 0.25, \quad \frac{7}{10} = 0.7, \quad \text{and} \quad \frac{89}{1000} = 0.089 \quad \text{Examples of terminating decimals}$$

Not all rational numbers can be represented by terminating decimals. For example, convert $\frac{4}{11}$ into a decimal by dividing 11 into 4 using a calculator. The display shows

$$0.3636363636, \quad \text{or perhaps} \quad 0.363636364.$$

However, the long division process shown in the margin indicates that we will actually get 0.3636..., with the digits 36 repeating over and over indefinitely. To indicate this, we write a bar (called a *vinculum*) over the "block" of digits that repeats.

$$\frac{4}{11} = 0.\overline{36} \quad \text{0.} \overline{36} \text{ means 0.3636....}$$

A decimal such as $0.\overline{36}$, which continues indefinitely, is called a **repeating decimal.**

$$\frac{5}{11} = 0.\overline{45}, \quad \frac{1}{3} = 0.\overline{3}, \quad \text{and} \quad \frac{5}{6} = 0.8\overline{3} \quad \text{Examples of repeating decimals}$$

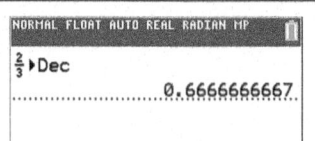

Although $\frac{2}{3}$ has a repeating decimal representation ($\frac{2}{3} = 0.\overline{6}$), the calculator rounds off in the final decimal place displayed.

While we distinguish between *terminating* and *repeating* decimals in this text, some mathematicians prefer to consider all rational numbers as repeating decimals. This can be justified by thinking this way: If the division process leads to a remainder of 0, then zeros repeat without end in the decimal form. For example, we can consider the decimal form of $\frac{3}{4}$ as follows.

$$\frac{3}{4} = 0.75\overline{0}$$

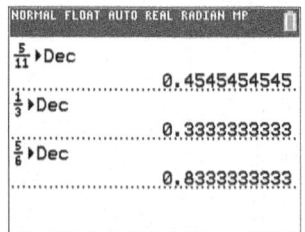

Although only ten decimal digits are shown, all three fractions have decimals that repeat endlessly.

By considering the possible remainders that may be obtained when converting a quotient of integers to a decimal, we can draw an important conclusion about the decimal form of rational numbers. If the remainder is never zero, the division will produce a repeating decimal. This happens because each step of the division process must produce a remainder that is less than the divisor. Since the number of different possible remainders is less than the divisor, the remainders must eventually begin to repeat. This makes the digits of the quotient repeat, producing a repeating decimal.

DECIMAL REPRESENTATION OF RATIONAL NUMBERS

Any rational number can be expressed as either a terminating decimal or a repeating decimal.

To determine whether the decimal form of a quotient of integers will terminate or repeat, we use the following rule. Justification of this rule is based on the fact that the prime factors of 10 are 2 and 5, and the decimal system uses ten as its base.

CRITERIA FOR TERMINATING AND REPEATING DECIMALS

A rational number $\frac{a}{b}$ *in lowest terms* results in a **terminating decimal** if the only prime factor of the denominator is 2 or 5 (or both).

A rational number $\frac{a}{b}$ *in lowest terms* results in a **repeating decimal** if a prime other than 2 or 5 appears in the prime factorization of the denominator.

To find a baseball player's batting average, we divide the number of hits by the number of at-bats. A surprising paradox exists concerning averages. It is possible for Player A to have a higher batting average than Player B in each of two successive years, yet for the two-year period, Player B can have a higher total batting average.

Year	Casey	Boomer
2016	$\dfrac{191}{589} = .324$	$\dfrac{41}{116} = .353$
2017	$\dfrac{115}{419} = .274$	$\dfrac{155}{554} = .280$
Two-year total	$\dfrac{306}{1008} = .304$	$\dfrac{196}{670} = .293$

The statistics shown actually occurred in two successive years in Major League Baseball, but the names and dates have been changed.

In both individual years, Boomer had a higher average, but for the two-year period, Casey had the higher average. This is an example of **Simpson's paradox** from statistics. (The authors thank Carol Merrigan for her input.)

```
NORMAL FLOAT AUTO REAL RADIAN MP

.437▶Frac
                            437
                           ————
                           1000
8.2▶Frac
                            41
                           ——
                            5
```

The results of **Example 10(a) and (b)** are supported in this screen.

EXAMPLE 9 **Determining Whether a Decimal Terminates or Repeats**

Determine whether the decimal form terminates or repeats.

(a) $\dfrac{7}{8}$ **(b)** $\dfrac{13}{150}$ **(c)** $\dfrac{6}{75}$

Solution

(a) The rational number $\dfrac{7}{8}$ is in lowest terms. Its denominator is 8, and since 8 factors as 2^3, the decimal form will terminate. No primes other than 2 or 5 divide the denominator.

(b) The rational number $\dfrac{13}{150}$ is in lowest terms with denominator

$$150 = 2 \cdot 3 \cdot 5^2.$$

Because 3 appears as a prime factor of the denominator, the decimal form will repeat.

(c) First write the rational number $\dfrac{6}{75}$ in lowest terms.

$$\dfrac{6}{75} = \dfrac{2}{25} \quad \text{Denominator is 25.}$$

Because $25 = 5^2$, the decimal form will terminate.

We have seen that a rational number will be represented by either a terminating or a repeating decimal. *Must a terminating decimal or a repeating decimal represent a rational number?* The answer is *yes*. For example, the terminating decimal 0.6 represents a rational number.

$$0.6 = \dfrac{6}{10} = \dfrac{3}{5}$$

EXAMPLE 10 **Writing Decimals as Quotients of Integers**

Write each decimal as a quotient of integers.

(a) 0.437 **(b)** 8.2 **(c)** $0.\overline{85}$

Solution

(a) $0.437 = \dfrac{437}{1000}$ Read as "four hundred thirty-seven thousandths" and then write as a fraction.

(b) $8.2 = 8 + \dfrac{2}{10} = \dfrac{82}{10} = \dfrac{41}{5}$ Read as a decimal, write as a sum, and then add.

(c) To convert a repeating decimal to a quotient of integers, we use some simple algebra.

Step 1 Let $x = 0.\overline{85}$, so $x = 0.858585\dots$.

Step 2 Multiply both sides of the equation $x = 0.858585\dots$ by 100. (Use 100 since there are **two** digits in the part that repeats, and $100 = 10^2$.)

$$x = 0.858585\dots$$
$$100x = 100(0.858585\dots)$$
$$100x = 85.858585\dots$$

$1 = 0.99999999999^{9^{9^{9^{9^{9}}}}}$

Terminating or Repeating? One of the most baffling truths of elementary mathematics is the following:

$$1 = 0.9999\ldots.$$

Most people believe that $0.\overline{9}$ has to be less than 1, but this is not the case. The following argument shows why. Let $x = 0.9999\ldots$ Then

$$10x = 9.9999\ldots$$
$$\underline{x = 0.9999\ldots}$$
$$9x = 9 \qquad \text{Subtract.}$$
$$x = 1. \qquad \text{Divide.}$$

Therefore, $1 = 0.9999\ldots$ Similarly, it can be shown that any terminating decimal can be represented as a repeating decimal with an endless string of 9s. For example, $0.5 = 0.49999\ldots$ and $2.6 = 2.59999\ldots$ This is a way of justifying that any rational number may be represented as a repeating decimal.

Step 3 Subtract the expressions in Step 1 from the final expressions in Step 2.

$$100x = 85.858585\ldots \qquad \text{(Recall that } x = 1x \text{ and } 100x - x = 99x.\text{)}$$
$$\underline{x = 0.858585\ldots}$$
$$99x = 85 \qquad \text{Subtract.}$$

Step 4 Solve the equation $99x = 85$ by dividing both sides by 99.

$$99x = 85$$
$$\frac{99x}{99} = \frac{85}{99} \qquad \text{Divide by 99.}$$
$$x = \frac{85}{99} \qquad \frac{99x}{99} = x$$
$$0.\overline{85} = \frac{85}{99} \qquad x = 0.\overline{85}$$

When checking with a calculator, remember that the calculator will show only a finite number of decimal places and may round off in the final decimal place shown.

6.3 EXERCISES

CONCEPT CHECK *Complete each of the following, based on the discussion in this section.*

1. A number that can be represented as a quotient of integers with denominator not 0 is called a(n) _____.

2. In the fraction $\frac{7}{13}$, _____ is the numerator and _____ is the denominator.

3. A number of the type in **Exercise 1** is in lowest terms if its numerator and denominator have _____ as their greatest common factor.

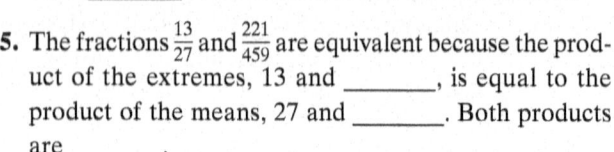

4. The fundamental property of rational numbers allows us to write $\frac{18}{27}$ as the equivalent $\frac{2}{3}$ because there is a common factor of _____ in the numerator and the denominator.

5. The fractions $\frac{13}{27}$ and $\frac{221}{459}$ are equivalent because the product of the extremes, 13 and _____, is equal to the product of the means, 27 and _____. Both products are _____.

6. In the mixed number $3\frac{1}{7}$ there is a(n) _____ sign understood between the whole number part and the fraction part.

7. To divide $\frac{7}{12}$ by $\frac{3}{4}$, we multiply _____ by the reciprocal of the divisor. That reciprocal is _____.

8. The rational numbers exhibit the density property, which says that between any two distinct rational numbers there exists _____.

9. The integers do not exhibit the density property. For example, between 5 and _____ there is no other integer.

10. A rational number will have a decimal representation that will show one of two patterns: It will _____ or it will _____.

Choose the expression(s) that is (are) equivalent to the given rational number.

11. $\frac{4}{8}$

 A. $\frac{1}{2}$ **B.** $\frac{8}{4}$ **C.** 0.5   **D.** $0.5\overline{0}$ **E.** $0.\overline{55}$

12. $\frac{2}{3}$

    A. 0.67 **B.** $0.\overline{6}$ **C.** $\frac{20}{30}$ **D.** 0.6 **E.** 0.666...

13. $\frac{5}{9}$

    A. 0.56 **B.** 0.55 **C.** $0.\overline{5}$ **D.** $\frac{9}{5}$ **E.** $1\frac{4}{5}$

14. $\frac{1}{4}$

    A. 0.25 **B.** $0.24\overline{9}$ **C.** $\frac{25}{100}$ **D.** 4 **E.** $\frac{10}{400}$

Write each fraction in lowest terms.

15. $\frac{16}{48}$ **16.** $\frac{21}{28}$ **17.** $-\frac{15}{35}$ **18.** $-\frac{8}{48}$

Write each fraction in three other ways.

19. $\frac{3}{8}$ **20.** $\frac{9}{10}$

21. $-\frac{5}{7}$ **22.** $-\frac{7}{12}$

23. *Fractional Parts* Write a fraction in lowest terms that represents the portion of each figure that is in color.

(a) 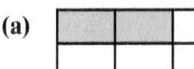   **(b)**

(c) 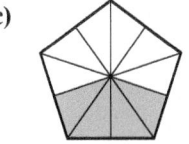   **(d)**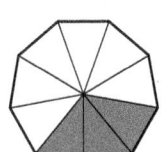

24. *Fractional Parts* Write a fraction in lowest terms that represents the region described in parts (a)–(d).

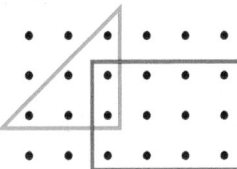

(a) the dots in the rectangle as a part of the dots in the entire figure

(b) the dots in the triangle as a part of the dots in the entire figure

(c) the dots in the rectangle as a part of the dots in the union of the triangle and the rectangle

(d) the dots in the intersection of the triangle and the rectangle as a part of the dots in the union of the triangle and the rectangle

25. *Fractional Parts* Refer to the figure for **Exercise 24,** and write a description of the region that is represented by the fraction $\frac{1}{12}$.

26. *Batting Averages* In a softball league, Paula got 8 hits in 20 at-bats, and Josh got 12 hits in 30 at-bats. Which player (if either) had the higher batting average?

27. *Batting Averages* After ten games, the following statistics were obtained.

Player	At-bats	Hits	Home Runs
Terry	40	9	2
Callie	36	12	3
Jessica	11	5	1
Johnny	16	8	0
Rocky	20	10	2

Answer using estimation skills as necessary.

(a) Which player got a hit in exactly $\frac{1}{3}$ of his or her at-bats?

(b) Which player got a hit in just less than $\frac{1}{2}$ of his or her at-bats?

(c) Which player got a home run in just less than $\frac{1}{10}$ of his or her at-bats?

(d) Which player got a hit in just less than $\frac{1}{4}$ of his or her at-bats?

(e) Which two players got hits in exactly the same fractional parts of their at-bats? What was the fractional part, reduced to lowest terms?

28. *Error Coin* Refer to the margin note discussing the use of common fractions on early U.S. copper coinage. The photo here shows an error near the bottom that occurred on an 1802 large cent. Discuss the error and how it represents a mathematical impossibility.

Perform the indicated operations and express answers in lowest terms. Use the order of operations as necessary.

29. $\dfrac{3}{8} + \dfrac{1}{8}$ **30.** $\dfrac{7}{9} + \dfrac{1}{9}$ **31.** $\dfrac{5}{16} + \dfrac{7}{12}$

32. $\dfrac{1}{15} + \dfrac{7}{18}$ **33.** $\dfrac{2}{3} - \dfrac{7}{8}$ **34.** $\dfrac{13}{20} - \dfrac{5}{12}$

35. $\dfrac{5}{8} - \dfrac{3}{14}$ **36.** $\dfrac{19}{15} - \dfrac{7}{12}$ **37.** $\dfrac{3}{4} \cdot \dfrac{9}{5}$

38. $\dfrac{3}{8} \cdot \dfrac{2}{7}$ **39.** $-\dfrac{2}{3} \cdot \left(-\dfrac{5}{8}\right)$ **40.** $-\dfrac{2}{4} \cdot \dfrac{3}{9}$

41. $\dfrac{5}{12} \div \dfrac{15}{4}$ **42.** $\dfrac{15}{16} \div \dfrac{30}{8}$

43. $-\dfrac{9}{16} \div \left(-\dfrac{3}{8}\right)$ **44.** $-\dfrac{3}{8} \div \left(-\dfrac{5}{4}\right)$

45. $\left(\dfrac{1}{3} \div \dfrac{1}{2}\right) + \dfrac{5}{6}$ **46.** $\dfrac{2}{5} \div \left(-\dfrac{4}{5} \div \dfrac{3}{10}\right)$

Convert each mixed number to a fraction, and convert each fraction to a mixed number.

47. $4\dfrac{1}{3}$ **48.** $3\dfrac{7}{8}$ **49.** $2\dfrac{9}{10}$

50. $\dfrac{18}{5}$ **51.** $\dfrac{27}{4}$ **52.** $\dfrac{19}{3}$

Perform each operation and express your answer as a mixed number.

53. $3\dfrac{1}{4} + 2\dfrac{7}{8}$ **54.** $6\dfrac{1}{5} - 2\dfrac{7}{15}$

55. $-4\dfrac{7}{8} \cdot 3\dfrac{2}{3}$ **56.** $-4\dfrac{1}{6} \div 1\dfrac{2}{3}$

Carpenter Calculations *Carpentry applications often require calculations involving fractions and mixed numbers with denominators of 2, 4, 8, and 16. Perform each of the following, and express each answer as a mixed number.*

57. $\dfrac{9}{16} + 2\left(\dfrac{5}{8}\right) - \dfrac{1}{2}$ **58.** $\dfrac{5}{16} + 3\left(\dfrac{3}{8}\right) - \dfrac{1}{4}$

59. $9\left(2\dfrac{1}{2}\right) + 3\left(5\dfrac{1}{8}\right)$ **60.** $8\left(4\dfrac{1}{4}\right) - 3\left(7\dfrac{1}{2}\right)$

61. $5\left(\dfrac{1}{16}\right) + 3\left(2\dfrac{1}{2}\right) - \dfrac{7}{16}$ **62.** $8\left(\dfrac{3}{16}\right) + 2\left(6\dfrac{1}{4}\right) - \dfrac{13}{16}$

63. $4\left[3\left(\dfrac{1}{8}\right) + 2\left(\dfrac{7}{16}\right)\right] - 2\left(\dfrac{1}{2} - \dfrac{1}{16}\right)$

64. $6\left[2\left(\dfrac{5}{8}\right) + 3\left(\dfrac{3}{16}\right)\right] - 3\left(\dfrac{3}{8} - \dfrac{3}{16}\right)$

Solve each problem.

65. Socket Wrench Measurements A hardware store sells a 22-piece socket wrench set. The measure of the largest socket is $\dfrac{3}{4}$ in., while the measure of the smallest socket is $\dfrac{3}{16}$ in. What is the difference between these measures?

66. TV Guide First published in 1953, the digest-sized *TV Guide* has changed to a "full-sized" magazine, as shown in the figure. What is the difference in their heights? (Data from *TV Guide*.)

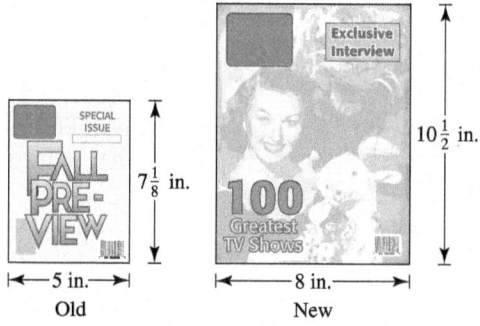

Old   New

67. Desk Height The diagram appears in a woodworker's book. Find the height of the desk to the top of the writing surface.

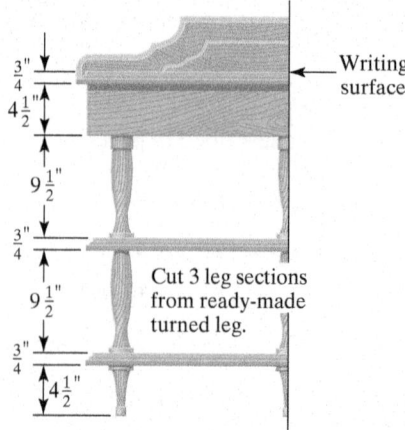

68. Legs of a Desk The desk in **Exercise 67** has four legs, each of which consists of three individual pieces. What is the total length of these twelve pieces?

69. Golf Tees The Pride Golf Tee Company, the only U.S. manufacturer of wooden golf tees, has created the Professional Tee system shown in the figure. (Data from *The Gazette*.)

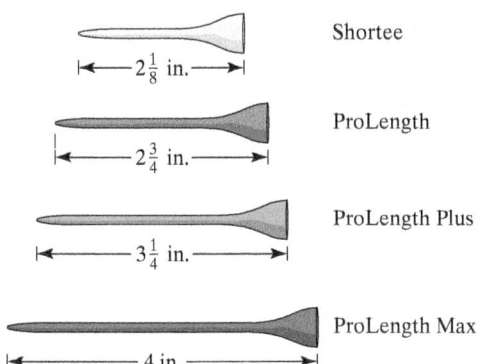

  Shortee $2\frac{1}{8}$ in.

  ProLength $2\frac{3}{4}$ in.

  ProLength Plus $3\frac{1}{4}$ in.

  ProLength Max   4 in.

 (a) Find the difference between the lengths of the ProLength Plus and the once-standard Shortee.

 (b) The ProLength Max tee is the longest tee allowed by the U.S. Golf Association's Rules of Golf. How much longer is the ProLength Max than the Shortee?

70. Recipe for Grits The following chart appears on a package of Quaker® Quick Grits.

Microwave		Stove Top		
Servings	**1**	**1**	**4**	**6**
Water	$\frac{3}{4}$ cup	1 cup	3 cups	4 cups
Grits	3 Tbsp	3 Tbsp	$\frac{3}{4}$ cup	1 cup
Salt (optional)	dash	dash	$\frac{1}{4}$ tsp	$\frac{1}{2}$ tsp

 (a) How many cups of water would be needed for 6 microwave servings?

 (b) How many cups of grits would be needed for 5 stove-top servings? (*Hint:* 5 is halfway between 4 and 6.)

71. Cuts on a Board A board is 48 in. long. It must be divided into four pieces of equal length, and each cut will cause a waste of $\frac{3}{16}$ in. How long will each piece be after the cuts are made?

72. Cuts on a Board A board is 72 in. long. It must be divided into six pieces of equal length, and each cut will cause a waste of $\frac{1}{8}$ in. How long will each piece be after the cuts are made?

73. Cake Recipe A cake recipe calls for $1\frac{3}{4}$ cups of sugar. A caterer has $15\frac{1}{2}$ cups of sugar on hand. How many cakes can he make?

74. Fabric It takes $2\frac{1}{4}$ yd of fabric to cover a chair of a particular design. How many chairs can be covered with $23\frac{2}{3}$ yd of fabric?

75. Fabric It takes $2\frac{3}{8}$ yd of fabric to make a costume for a school play. How much fabric would be needed for seven costumes?

76. Cookie Recipe A cookie recipe calls for $2\frac{2}{3}$ cups of sugar. How much sugar would be needed to make four batches of cookies?

77. Public 4-Year Institution Costs The table below shows the average amount in 2017 dollars of tuition and fees at public 4-year institutions for in-state students for five academic years. What is the average cost for this period (to the nearest dollar)? (Data from trends.collegeboard.org)

Academic Year	Amount of Tuition and Fees
2013–14	$9310
2014–15	9400
2015–16	9670
2016–17	9840
2017–18	9970

78. Public 2-Year Institution Costs The table below shows the average amount in 2017 dollars of tuition and fees at public 2-year institutions for in-state students for five academic years. What is the average cost for this period (to the nearest dollar)? (Data from trends.collegeboard.org)

Academic Year	Amount of Tuition and Fees
2013–14	$3400
2014–15	3430
2015–16	3490
2016–17	3530
2017–18	3570

Find the rational number halfway between the two given rational numbers.

79. $\frac{1}{2}, \frac{3}{4}$ **80.** $\frac{1}{3}, \frac{5}{12}$

81. $\frac{3}{5}, \frac{2}{3}$ **82.** $\frac{7}{12}, \frac{5}{8}$

83. $-\frac{2}{3}, -\frac{5}{6}$ **84.** $-\frac{5}{16}, -\frac{5}{2}$

*Use the method of **Example 9** to decide whether each rational number would yield a repeating or a terminating decimal. (Hint: Write in lowest terms before trying to decide.)*

85. $\frac{8}{15}$ **86.** $\frac{8}{35}$ **87.** $\frac{13}{125}$

88. $\frac{3}{24}$ **89.** $\frac{22}{55}$ **90.** $\frac{24}{75}$

Convert each rational number into either a repeating or a terminating decimal. Use a calculator if your instructor so allows.

91. $\dfrac{3}{4}$

92. $\dfrac{7}{8}$

93. $\dfrac{3}{16}$

94. $\dfrac{9}{32}$

95. $\dfrac{3}{11}$

96. $\dfrac{9}{11}$

97. $\dfrac{2}{7}$

98. $\dfrac{11}{15}$

Convert each decimal into a quotient of integers, written in lowest terms.

99. 0.4

100. 0.9

101. 0.85

102. 0.105

103. 0.934

104. 0.7984

105. $0.\overline{67}$

106. $0.\overline{53}$

107. $0.0\overline{42}$

108. $0.0\overline{86}$

109. $0.0\overline{1}$

110. $0.1\overline{2}$

111. *Hard to Believe?* Follow through on all parts of this exercise in order.

(a) Find the decimal for $\frac{1}{3}$.

(b) Find the decimal for $\frac{2}{3}$.

(c) By adding the decimal expressions obtained in parts (a) and (b), obtain a decimal expression for $\frac{1}{3} + \frac{2}{3} = \frac{3}{3} = 1$.

(d) State your result. Read the margin note on terminating and repeating decimals in this section, which refers to this idea.

112. *Hard to Believe?* It is a fact that $\frac{1}{3} = 0.333\ldots$. Multiply both sides of this equation by 3. Does your answer bother you? See the margin note on terminating and repeating decimals in this section.

FOR FURTHER THOUGHT

The Influence of Spanish Coinage on Stock Prices

Until August 28, 2000, when decimalization of the U.S. stock market began, market prices were reported with fractions having denominators that were powers of 2, such as

$$17\frac{3}{4} \quad {\scriptstyle 4 = 2^2} \quad \text{and} \quad 112\frac{5}{8}. \quad {\scriptstyle 8 = 2^3}$$

Did you ever wonder why this was done?

During the early years of the United States, prior to the minting of its own coinage, the Spanish eight-reales coin, also known as the Spanish milled dollar, circulated freely in the states. Its fractional parts, the four reales, two reales, and one real, were known as **pieces of eight** and were described as such in pirate and treasure lore. When the New York Stock Exchange was founded in 1792, it chose to use the Spanish milled dollar as its price basis, rather than the decimal base proposed by Thomas Jefferson that same year. All prices on the U.S. stock markets are now reported in decimals.

(*Source:* "Stock price tables go to decimal listings," *The Times Picayune,* June 27, 2000.)

For Group or Individual Investigation

Consider this: Have you ever heard this old cheer?

"Two bits, four bits, six bits, a dollar. All for the (home team), stand up and holler."

The term **two bits** refers to 25 cents.

Discuss how this cheer is based on the Spanish eight-reales coin.

6.4 **IRRATIONAL NUMBERS AND DECIMAL REPRESENTATION**

OBJECTIVES

1 Illustrate how irrational numbers differ from rational numbers in their decimal representations.

2 Follow the proof that $\sqrt{2}$ is an irrational number.

3 Use a calculator to find square roots.

4 Apply the product and quotient rules for square roots.

5 Rationalize a denominator.

6 Explain the relevance of the irrational numbers π, ϕ, and e.

Definition and Basic Concepts

Every rational number has a decimal form that terminates or repeats, and every repeating or terminating decimal represents a rational number. However,

$$0.10200100020000100000 2 \ldots$$

does not terminate and does not repeat. (It is true that there is a pattern in this decimal, but no single block of digits ever repeats indefinitely.)*

IRRATIONAL NUMBERS

Irrational numbers $= \{x \,|\, x$ is a number represented by a nonrepeating, nonterminating decimal$\}$

As the name implies, an irrational number cannot be represented as a quotient of integers.

The decimal number mentioned at the top of this page is an irrational number. Other irrational numbers include $\sqrt{2}$, $\frac{1 + \sqrt{5}}{2}$ (ϕ, the Golden Ratio), π (the ratio of the circumference of a circle to its diameter), and e (a constant *approximately equal to* 2.71828). There are infinitely many irrational numbers.

Irrationality of $\sqrt{2}$ and Proof by Contradiction

Figure 13 illustrates how a point with coordinate $\sqrt{2}$ can be located on a number line. (This was mentioned in the first section of this chapter.)

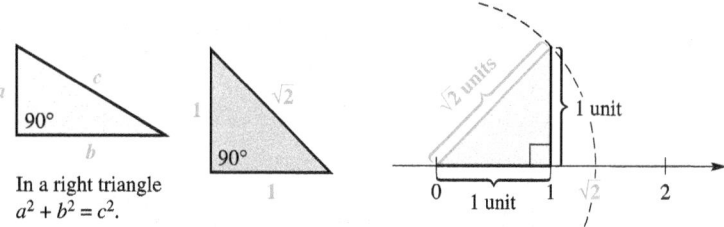

Figure 13

The irrational number $\sqrt{2}$ was discovered by the Pythagoreans in about 500 B.C. This discovery was a great setback to their philosophy that everything is based on the whole numbers. The Pythagoreans kept their findings secret, and legend has it that members of the group who divulged this discovery were sent out to sea and, according to Proclus (410–485), "perished in a shipwreck, to a man."

The proof that $\sqrt{2}$ is irrational is a classic example of a **proof by contradiction.** We begin by assuming that $\sqrt{2}$ is rational, which leads to a contradiction, or absurdity. The method is also called **reductio ad absurdum** (Latin for "reduce to the absurd"). In order to understand the proof, we consider three preliminary facts:

1. When a rational number is written in lowest terms, the greatest common factor of the numerator and denominator is 1.

2. If an integer is even, then it has 2 as a factor and may be written in the form $2k$, where k is an integer.

3. If a perfect square is even, then its square root is even.

---

*In this section, we will assume that the digits of a number such as this continue indefinitely in the pattern established. The next few digits would be 000000100000002, and so on.

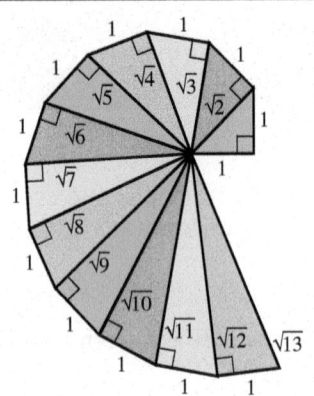

An interesting way to represent the lengths corresponding to

$$\sqrt{2},\ \sqrt{3},\ \sqrt{4},\ \sqrt{5},$$

and so on, is shown in the figure. Use the **Pythagorean theorem** to verify the lengths in the figure.

THEOREM

Statement: $\sqrt{2}$ is an irrational number.

Proof: Assume that $\sqrt{2}$ is a rational number. Then by definition,

$$\sqrt{2} = \frac{p}{q}, \quad \text{for some integers } p \text{ and } q.$$

Furthermore, assume that $\frac{p}{q}$ is the form of $\sqrt{2}$ that is written in lowest terms, so the greatest common factor of p and q is 1.

$$2 = \frac{p^2}{q^2} \qquad \text{Square both sides of the equation.}$$

$$2q^2 = p^2 \qquad \text{Multiply by } q^2.$$

The last equation, $2q^2 = p^2$, indicates that 2 is a factor of p^2. So p^2 is even, and thus p is even. Since p is even, it may be written in the form $2k$, where k is an integer.

Now, substitute $2k$ for p in the last equation and simplify.

$$2q^2 = (2k)^2 \qquad \text{Let } p = 2k.$$

$$2q^2 = 4k^2 \qquad (2k)^2 = 2k \cdot 2k = 4k^2$$

$$q^2 = 2k^2 \qquad \text{Divide by 2.}$$

Since q^2 has 2 as a factor, q^2 must be even, and thus q must be even. This leads to a contradiction: p and q cannot both be even because they would then have a common factor of 2. It was assumed that their greatest common factor is 1. The original assumption that $\sqrt{2}$ is rational has led to a contradiction, so $\sqrt{2}$ is irrational.

Operations with Square Roots

In everyday mathematical work, nearly all of our calculations deal with rational numbers, usually in decimal form. However, we must sometimes perform operations with irrational numbers. ***Recall that $\sqrt{a}$, for $a \geq 0$, is the nonnegative number whose square is a. That is, $(\sqrt{a})^2 = a$.***

$$\sqrt{2}, \quad \sqrt{3}, \quad \text{and} \quad \sqrt{13} \qquad \text{Examples of square roots that are irrational}$$

$$\sqrt{4} = 2, \quad \sqrt{36} = 6, \quad \text{and} \quad \sqrt{100} = 10 \qquad \text{Examples of square roots that are rational}$$

If n is a positive integer that is not the square of an integer, then $\sqrt{n}$ is an irrational number.

A calculator with a square root key can give approximations of square roots of numbers that are not perfect squares. ***We use the $\approx$ symbol to indicate "is approximately equal to."*** Sometimes, for convenience, the $=$ symbol is used even if the statement is actually one of approximation, such as $\pi = 3.14$.

EXAMPLE 1   Using a Calculator to Approximate Square Roots

Use a calculator to verify the following approximations.

(a) $\sqrt{2} \approx 1.414213562$ **(b)** $\sqrt{6} \approx 2.449489743$ **(c)** $\sqrt{1949} \approx 44.14748011$

Solution

Use a calculator to verify these approximations. See **Figure 14**. Depending on the model, fewer or more digits may be displayed, and because of different rounding procedures, final digits may differ slightly.

```
NORMAL FLOAT AUTO REAL RADIAN MP
√2
                      1.414213562
√6
                      2.449489743
√1949
                      44.14748011
```

Figure 14

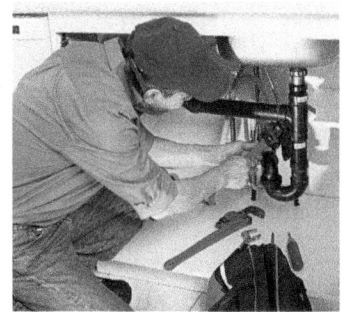

This screen supports the result of **Example 2,** using decimal equivalents of the values of *d* and *D*.

EXAMPLE 2 **Applying a Formula for Pipe Flow**

The following formula gives the number *N* of smaller pipes of diameter *d* that are necessary to supply the same total flow as one larger pipe of diameter *D*.

$$N = \sqrt{\left(\frac{D}{d}\right)^5}$$

This formula takes into account the extra friction caused by the smaller pipes. Use this formula to determine the number of $\frac{1}{2}$-inch pipes that will provide the same flow as one $1\frac{1}{2}$-inch pipe. (*Source:* Saunders, Hal M., and Robert A. Carman. *Mathematics for the Trades—A Guided Approach.* Pearson, 2015.)

Solution

Let $d = \frac{1}{2}$ and $D = 1\frac{1}{2} = \frac{3}{2}$ in the formula.

$$N = \sqrt{\left(\frac{\frac{3}{2}}{\frac{1}{2}}\right)^5} \qquad d = \frac{1}{2}, D = \frac{3}{2}$$

$$= \sqrt{3^5} \qquad \frac{3}{2} \div \frac{1}{2} = \frac{3}{2} \cdot \frac{2}{1} = 3$$

$$= \sqrt{243} \qquad 3^5 = 243$$

$$= 15.59 \qquad \text{Approximate with a calculator.}$$

In this case the answer must be a whole number, so we round 15.59 up to 16. A total of 16 pipes measuring $\frac{1}{2}$ inch will be needed.

We will now look at some simple operations with square roots. Notice that

$$\sqrt{4} \cdot \sqrt{9} = 2 \cdot 3 = 6$$

and

$$\sqrt{4 \cdot 9} = \sqrt{36} = 6.$$

Thus, $\sqrt{4} \cdot \sqrt{9} = \sqrt{4 \cdot 9}$. This is a particular case of the following rule.

PRODUCT RULE FOR SQUARE ROOTS

For nonnegative real numbers *a* and *b*, the following holds true.

$$\boldsymbol{\sqrt{a} \cdot \sqrt{b} = \sqrt{a \cdot b}}$$

Just as every rational number $\frac{a}{b}$ can be written in *lowest terms* (by using the fundamental property of rational numbers), every square root radical has a *simplified form*.

CONDITIONS FOR A SIMPLIFIED SQUARE ROOT RADICAL

A square root radical is in **simplified form** if the following three conditions are met.

1. The number under the radical (**radicand**) has no factor (except 1) that is a perfect square.

2. The radicand has no fractions.

3. No denominator contains a radical.

Tsu Ch'ung-chih (about A.D. 500), the Chinese mathematician honored on the above stamp, investigated the digits of π. **Aryabhata,** his Indian contemporary, gave 3.1416 as the value.

EXAMPLE 3   Simplifying a Square Root Radical (Product Rule)

Simplify $\sqrt{27}$.

Solution

Since 9 is a factor of 27, and 9 is a perfect square, $\sqrt{27}$ is not in simplified form. The first condition for simplified form is not met. We simplify as follows.

$$\sqrt{27} = \sqrt{9 \cdot 3}$$

Simplified form → $= \sqrt{9} \cdot \sqrt{3}$ Use the product rule.

$= 3\sqrt{3}$ $\sqrt{9} = 3$ because $3^2 = 9$.

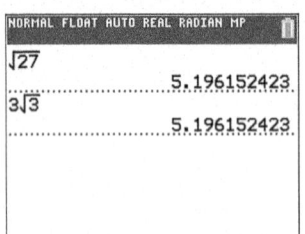

Figure 15

Expressions such as $\sqrt{27}$ and $3\sqrt{3}$ represent the *exact value* of the square root of 27. If we use the square root key of a calculator, we find

$$\sqrt{27} \approx 5.196152423.$$

If we find $\sqrt{3}$ and then multiply the result by 3, we obtain

$$3\sqrt{3} \approx 3(1.732050808) \approx 5.196152423.$$

These approximations are the same, as we would expect. See **Figure 15.** The work in **Example 3** provides the mathematical justification that they are indeed equal.

QUOTIENT RULE FOR SQUARE ROOTS

For nonnegative real numbers a and positive real numbers b, the following holds true.

$$\frac{\sqrt{a}}{\sqrt{b}} = \sqrt{\frac{a}{b}}$$

EXAMPLE 4   Simplifying Square Root Radicals (Quotient Rule)

Simplify each radical.

(a) $\sqrt{\dfrac{25}{9}}$ **(b)** $\sqrt{\dfrac{3}{4}}$ **(c)** $\sqrt{\dfrac{1}{2}}$

Solution

(a) The radicand contains a fraction, so the radical expression is not simplified.

$$\sqrt{\frac{25}{9}} = \frac{\sqrt{25}}{\sqrt{9}} = \frac{5}{3}$$ Use the quotient rule.

(b) $\sqrt{\dfrac{3}{4}} = \dfrac{\sqrt{3}}{\sqrt{4}} = \dfrac{\sqrt{3}}{2}$ Use the quotient rule.

(c) $\sqrt{\dfrac{1}{2}} = \dfrac{\sqrt{1}}{\sqrt{2}} = \dfrac{1}{\sqrt{2}}$ ◁ This is not yet simplified.

Condition 3 for simplified form is not met. To find an equivalent expression with no radical in the denominator, we **rationalize the denominator.**

Carl Gauss Square roots of negative numbers were called **imaginary numbers** by early mathematicians. Eventually the symbol i came to represent the **imaginary unit** $\sqrt{-1}$, and numbers of the form $a + bi$, where a and b are real numbers, were named **complex numbers.**

In about 1831, **Carl Gauss** was able to show that numbers of the form $a + bi$ can be represented as points on the plane just as real numbers are. He shared this contribution with **Robert Argand,** a bookkeeper in Paris, who wrote an essay on the geometry of the complex numbers in 1806. This went unnoticed at the time.

The complex number i has the property that its whole number powers repeat in a cycle, with four values.

$$i^0 = 1 \quad i^1 = i \quad i^2 = -1 \quad i^3 = -i$$

The pattern continues on and on this way.

To rationalize the denominator here, we multiply $\frac{1}{\sqrt{2}}$ by $\frac{\sqrt{2}}{\sqrt{2}}$, which is a form of 1, the identity element for multiplication.

$$\frac{1}{\sqrt{2}} = \frac{1}{\sqrt{2}} \cdot \frac{\sqrt{2}}{\sqrt{2}} = \frac{\sqrt{2}}{2} \quad {\scriptstyle \sqrt{2}\, \cdot\, \sqrt{2}\, =\, 2}$$

> This is the simplified form of $\sqrt{\frac{1}{2}}$.

Is $\sqrt{4} + \sqrt{9} = \sqrt{4 + 9}$ *a true statement?* The answer is *no*, because

$$\sqrt{4} + \sqrt{9} = 2 + 3 = 5, \quad \text{while} \quad \sqrt{4 + 9} = \sqrt{13}, \quad \text{and} \quad 5 \neq \sqrt{13}.$$

Square root radicals may be combined, however, if they have the same radicand. Such radicals are called **like radicals.** We add (and subtract) like radicals using the distributive property.

EXAMPLE 5 **Adding and Subtracting Square Root Radicals**

Add or subtract as indicated.

(a) $3\sqrt{6} + 4\sqrt{6}$ **(b)** $\sqrt{18} - \sqrt{32}$

Solution

(a) Since both terms contain $\sqrt{6}$, they are like radicals, and may be combined.

$$3\sqrt{6} + 4\sqrt{6} = (3 + 4)\sqrt{6} \quad \text{Distributive property}$$

$$= 7\sqrt{6} \quad \text{Add.}$$

(b) If we simplify $\sqrt{18}$ and $\sqrt{32}$, then this operation can be performed.

$$\sqrt{18} - \sqrt{32} = \sqrt{9 \cdot 2} - \sqrt{16 \cdot 2} \quad \text{Factor so that perfect squares are in the radicands.}$$

$$= \sqrt{9} \cdot \sqrt{2} - \sqrt{16} \cdot \sqrt{2} \quad \text{Product rule}$$

$$= 3\sqrt{2} - 4\sqrt{2} \quad \text{Take square roots.}$$

$$= (3 - 4)\sqrt{2} \quad \text{Distributive property}$$

$$= -1\sqrt{2} \quad \text{Subtract.}$$

$$= -\sqrt{2} \quad {\scriptstyle -1\, \cdot\, a\, =\, -a}$$

Like radicals may be added or subtracted by adding or subtracting their coefficients (the numbers by which they are multiplied) and keeping the same radical.

Examples: $9\sqrt{7} + 8\sqrt{7} = 17\sqrt{7}$ (because $9 + 8 = 17$)

$4\sqrt{3} - 12\sqrt{3} = -8\sqrt{3}$ (because $4 - 12 = -8$)

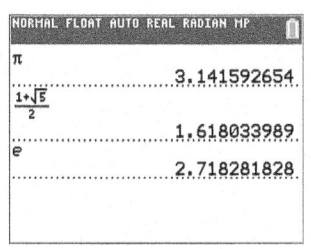

Figure 16

The Irrational Numbers π, ϕ, and e

Figure 16 shows approximations for three important irrational numbers. The first, π, represents the ratio of the circumference of a circle to its diameter. The second, ϕ, is the Golden Ratio. Its exact value is $\frac{1 + \sqrt{5}}{2}$. The third, e, is a fundamental number in our universe. It is the base of the *natural exponential* and *natural logarithmic* functions. The letter e was chosen to honor Leonhard Euler, who published extensive research on the number in 1746.

This poem, dedicated to **Archimedes** ("the immortal Syracusan"), allows us to learn the first 31 digits of the decimal representation of π. By replacing each word with the number of letters it contains, with a decimal point following the initial 3, the decimal is found. The poem was written by A. C. Orr, and it appeared in the *Literary Digest* in 1906.

Now I, even I, would celebrate
In rhymes unapt, the great
Immortal Syracusan, rivaled nevermore,
Who in his wondrous lore
Passed on before,
Left men his guidance
How to circles mensurate.

Pi (π)

$$\pi \approx 3.1415926535897932384626433383279$$

The computation of the digits of π has fascinated mathematicians since ancient times. Archimedes was the first to explore it extensively. As of November 11, 2016, its value had been computed to over 22 trillion decimal places. Some of today's foremost researchers of the digits of π are Yasumasa Kanada, Gregory and David Chudnovsky, and the team of Alexander J. Yee and Shigeru Kondo.

A History of π, by Petr Beckmann, is a classic that is now in its third edition. Numerous Web sites, including the following, are devoted to the history and methods of computation of pi.

- www.joyofpi.com/
- www.math.utah.edu/~alfeld/Archimedes/Archimedes.html
- www.super-computing.org/
- www.pbs.org/wgbh/nova/physics/approximating-pi.html
- www.numberworld.org/misc_runs/pi-5t/details.html

One of the methods of computing pi involves the topic of *infinite series*.

EXAMPLE 6   **Computing the Digits of Pi Using an Infinite Series**

It is shown in higher mathematics that the *infinite series*

$$1 - \frac{1}{3} + \frac{1}{5} - \frac{1}{7} + \frac{1}{9} + \ldots \text{ "converges" to } \frac{\pi}{4}.$$

That is, as more and more terms are considered, its value becomes closer and closer to $\frac{\pi}{4}$. With a calculator, approximate the value of pi using twenty-one terms of this series.

Solution

Figure 17 shows the necessary calculation on the TI-84 Plus C calculator through the term $+\frac{1}{41}$. The sum of the first twenty-one terms is multiplied by 4, to obtain the approximation

$$3.189184782.$$

This series converges slowly. Although this is correct only to the first decimal place, better approximations are obtained using more terms of the series.

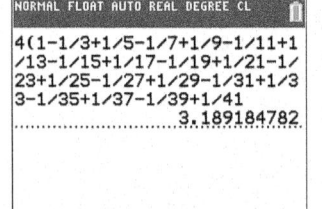

NORMAL FLOAT AUTO REAL DEGREE CL

4(1–1/3+1/5–1/7+1/9–1/11+1
/13–1/15+1/17–1/19+1/21–1/
23+1/25–1/27+1/29–1/31+1/3
3–1/35+1/37–1/39+1/41
............................3.189184782

Figure 17

A rectangle that satisfies the condition that the ratio of its length to its width is equal to the ratio of the sum of its length and width to its length is called a **Golden Rectangle.** See **Figure 18.** This ratio is called the **Golden Ratio.** Its exact value is the irrational number $\frac{1 + \sqrt{5}}{2}$. It is represented by the Greek letter ϕ (phi).

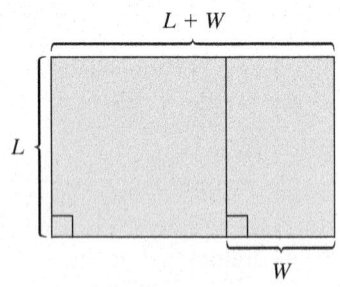

$L + W$

L

W

In a Golden Rectangle,

$$\frac{L + W}{L} = \frac{L}{W}.$$

Figure 18

Phi (ϕ)

$$\phi = \frac{1 + \sqrt{5}}{2} \approx 1.6180339887498948482045868343656$$

Two books on phi are *The Divine Proportion, A Study in Mathematical Beauty,* by H. E. Huntley, and the more recent *The Golden Ratio,* by Mario Livio. Web sites devoted to this irrational number include the following.

- www.mcs.surrey.ac.uk/hosted-sites/R.Knott/Fibonacci/fib.html
- www.goldennumber.net/
- www.mathforum.org/dr.math/faq/faq.golden.ratio.html
- www.geom.uiuc.edu/~demo5337/s97b/art.htm

EXAMPLE 7   **Computing the Digits of Phi Using the Fibonacci Sequence**

The first twelve terms of the Fibonacci sequence are

$$1, 1, 2, 3, 5, 8, 13, 21, 34, 55, 89, 144.$$

Each term after the first two terms is obtained by adding the two previous terms. Thus, the thirteenth term is $89 + 144 = 233$. As one goes farther out in the sequence, the ratio of a term to its predecessor gets closer to ϕ. How far out must one go in order to approximate ϕ so that the first five decimal places agree?

Solution

After 144, the next three Fibonacci numbers are 233, 377, and 610. **Figure 19** shows that $\frac{610}{377} \approx 1.618037135$, which agrees with ϕ to the fifth decimal place.

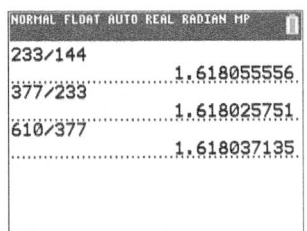

Figure 19

The irrational number e is a fundamental constant in mathematics.

e

$$e \approx 2.718281828459045235360287471353$$

As x takes on larger and larger values, the expression $(1 + \frac{1}{x})^x$ approaches e. Because of its nature, e is less understood by the layperson than π (or even ϕ, for that matter). The 1994 book *e: The Story of a Number,* by Eli Maor, attempted to rectify this situation. The following Web sites also give information on e.

- www.mathforum.org/dr.math/faq/faq.e.html
- www-groups.dcs.st-and.ac.uk/~history/HistTopics/e.html
- http://antwrp.gsfc.nasa.gov/htmltest/gifcity/e.1mil
- www.math.toronto.edu/mathnet/answers/ereal.html

EXAMPLE 8   **Computing the Digits of e Using an Infinite Series**

The infinite series

$$2 + \frac{1}{1 \cdot 2} + \frac{1}{1 \cdot 2 \cdot 3} + \frac{1}{1 \cdot 2 \cdot 3 \cdot 4} + \dots \text{ converges to } e.$$

Use a calculator to approximate e using the first seven terms of this series.

Solution

Figure 20 shows the sum of the first seven terms. (The denominators have all been multiplied out.) The sum is 2.718253968, which agrees with e to four decimal places. This series converges more rapidly than the series for π in **Example 6.**

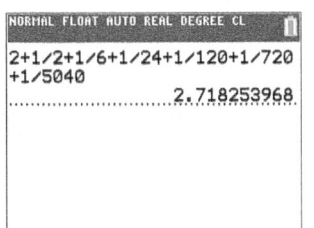

Figure 20

6.4   EXERCISES

CONCEPT CHECK *Write* true *or* false *for each statement.*

1. A real number must be either rational or irrational and cannot be both.

2. The square root of 36 is an irrational number.

3. The irrational number π represents the ratio of the circumference of a circle to its diameter.

4. The irrational number ϕ represents the ratio of the length of a Golden Rectangle to its width.

5. The irrational number e was named after Albert Einstein.

6. The sum of two irrational numbers must be an irrational number.

Identify each number as rational *or* irrational.

7. $\dfrac{4}{9}$ 8. $\dfrac{7}{8}$ 9. $\sqrt{10}$ 10. $\sqrt{14}$

11. 1.618   12. 2.718   13. $0.\overline{41}$ 14. $0.\overline{32}$

15. π 16. e 17. 3.14159   18. $\dfrac{22}{7}$

19. 0.8787787777877778 …   20. $\dfrac{1 + \sqrt{5}}{2}$

In Exercises 21 and 22, work parts (a) and (b) in order.

21. **(a)** Find the sum.   0.272772777277772 …
                       +0.616116111611116 …

    (b) Based on the result of part (a), we can conclude that the sum of two _____ numbers may be a(n) _____ number.

22. **(a)** Find the sum.   0.010110111011110 …
                       +0.252552555255552 …

    (b) Based on the result of part (a), we can conclude that the sum of two _____ numbers may be a(n) _____ number.

Use a calculator to find a rational decimal approximation for each irrational number.

23. $\sqrt{39}$ 24. $\sqrt{44}$ 25. $\sqrt{15.1}$ 26. $\sqrt{33.6}$

27. $\sqrt{884}$ 28. $\sqrt{643}$ 29. $\sqrt{\dfrac{9}{8}}$ 30. $\sqrt{\dfrac{6}{5}}$

Solve each problem. Use a calculator as necessary, and give approximations to the nearest tenth unless specified otherwise.

31. *Plumbing*   Use the formula from **Example 2,**

$$N = \sqrt{\left(\dfrac{D}{d}\right)^5},$$

to find the number of $\frac{3}{4}$-inch pipes that would be necessary to provide the same total flow as a single $1\frac{1}{2}$-inch pipe.

32. *Plumbing*   Use the formula from **Example 2** to find the number of $\frac{5}{8}$-inch pipes that would be necessary to provide the same total flow as a single 2-inch pipe.

33. *Allied Health*   Body surface area (BSA) is used in the allied health field to calculate proper dosages of drugs based on a patient's height and weight. The formula

$$\text{BSA} = \sqrt{\dfrac{\text{weight in kg} \times \text{height in cm}}{3600}}$$

will give the patient's BSA in square meters. (*Source: Saunders, Hal M., and Robert A. Carman. Mathematics for the Trades—A Guided Approach, Tenth Edition.* Pearson, 2015.) What is the BSA for a person who weighs 100 kilograms and is 200 centimeters tall?

34. *Allied Health*   Use the formula from **Exercise 33** to find the BSA of a person who weighs 94 kilograms and is 190 centimeters tall.

35. *Diagonal of a Box*   The length of the diagonal of a box is given by

$$D = \sqrt{L^2 + W^2 + H^2},$$

where L, W, and H are the length, width, and height of the box. Find the length of the diagonal, D, of a box that is 4 feet long, 3 feet wide, and 2 feet high.

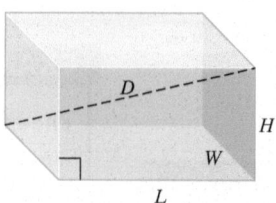

36. *Distance to the Horizon*   A meteorologist provided the answer at the top of the next page to one of the readers of his newspaper column, a 6-foot man who lives in an apartment 150 feet above the ground.

To find the distance to the horizon in miles, take the square root of the height of your view in feet and multiply that result by 1.224. That will give you the number of miles to the horizon.

Assuming the viewer's eyes are 6 feet above his floor, the total height from the ground is $150 + 6 = 156$ feet. How far can he see to the horizon?

37. **Period of a Pendulum** The period of a pendulum, in seconds, depends on its length, L, in feet, and is given by the formula

$$P = 2\pi\sqrt{\frac{L}{32}}.$$

If a pendulum is 5.1 feet long, what is its period? Use 3.14 for π.

38. **Radius of an Aluminum Can** The radius of the circular top or bottom of an aluminum can with surface area S in square cm and height h in cm is given by

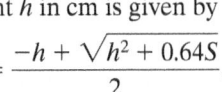

$$r = \frac{-h + \sqrt{h^2 + 0.64S}}{2}.$$

What radius should be used to make a can with height 12 cm and surface area 400 square cm?

39. **Electronics Formula** The formula

$$I = \sqrt{\frac{2P}{L}}$$

relates the coefficient of self-induction L (in henrys), the energy P stored in an electronic circuit (in joules), and the current I (in amps). Find the value of I if $P = 120$ joules and $L = 80$ henrys.

40. **Law of Tensions** In the study of sound, one version of the law of tensions is

$$f_1 = f_2\sqrt{\frac{F_1}{F_2}}.$$

Find f_1 to the nearest unit if $F_1 = 300$, $F_2 = 60$, and $f_2 = 260$.

Accident Reconstruction *Police sometimes use the following procedure to estimate the speed at which a car was traveling at the time of an accident. A police officer drives the car involved in the accident under conditions similar to those during which the accident took place and then skids to a stop. If the car is driven at 30 miles per hour, then the speed at the time of the accident is given by*

$$s = 30\sqrt{\frac{a}{p}}.$$

Here, a is the length of the skid marks left at the time of the accident, and p is the length of the skid marks in the police test. Find s, to the nearest unit, for the following values of a and p.

41. $a = 862$ feet; $p = 156$ feet

42. $a = 382$ feet; $p = 96$ feet

43. $a = 84$ feet; $p = 26$ feet

44. $a = 90$ feet; $p = 35$ feet

45. **Area of the Bermuda Triangle** **Heron's formula** gives a method of finding the area of a triangle if the lengths of its sides are known. Suppose that a, b, and c are the lengths of the sides. Let s denote one-half of the perimeter of the triangle (called the **semiperimeter**); that is,

$$s = \frac{1}{2}(a + b + c).$$

Then the area $\mathcal{A}$ of the triangle is given by

$$\mathcal{A} = \sqrt{s(s-a)(s-b)(s-c)}.$$

Find the area of the Bermuda Triangle, if the "sides" of this triangle measure approximately 850 miles, 925 miles, and 1300 miles. Give your answer to the nearest thousand square miles.

46. **Area Enclosed by the Vietnam Veterans' Memorial** The Vietnam Veterans' Memorial in Washington, D.C., is in the shape of an unenclosed isosceles triangle with equal sides of length 246.75 feet. If the triangle were enclosed, the third side would have length 438.14 feet. Use Heron's formula from the previous exercise to find the area of this enclosure to the nearest hundred square feet. (*Source:* Information pamphlet obtained at the Vietnam Veterans' Memorial.)

47. **Perfect Triangles** A **perfect triangle** is a triangle whose sides have whole number lengths and whose area is numerically equal to its perimeter. Use Heron's formula to show that the triangle with sides of length 9, 10, and 17 is perfect.

48. **Heron Triangles** A **Heron triangle** is a triangle having integer sides and area. Use Heron's formula to show that each of the following is a Heron triangle.
(a) $a = 11, b = 13, c = 20$
(b) $a = 13, b = 14, c = 15$
(c) $a = 7, b = 15, c = 20$

*Use the methods of **Examples 3 and 4** to simplify each expression. Then, use a calculator to approximate both the given expression and the simplified expression. (Both should be the same.)*

49. $\sqrt{50}$ 50. $\sqrt{32}$

51. $\sqrt{75}$ 52. $\sqrt{150}$

53. $\sqrt{288}$ 54. $\sqrt{200}$

55. $\dfrac{5}{\sqrt{6}}$ 56. $\dfrac{3}{\sqrt{2}}$

57. $\sqrt{\dfrac{7}{4}}$ **58.** $\sqrt{\dfrac{8}{9}}$

59. $\sqrt{\dfrac{7}{3}}$ **60.** $\sqrt{\dfrac{14}{5}}$

*Use the method of **Example 5** to perform the indicated operations.*

61. $\sqrt{17} + 2\sqrt{17}$ **62.** $3\sqrt{19} + \sqrt{19}$

63. $5\sqrt{7} - \sqrt{7}$ **64.** $3\sqrt{27} - \sqrt{27}$

65. $3\sqrt{18} + \sqrt{2}$ **66.** $2\sqrt{48} - \sqrt{3}$

67. $-\sqrt{12} + \sqrt{75}$ **68.** $2\sqrt{27} - \sqrt{300}$

Irrational Investigations *The following exercises deal with the irrational numbers π, ϕ, e, and $\sqrt{3}$. Use a calculator or computer as necessary.*

69. Move one matchstick to make the equation approximately true. (*Source:* http://www.joyofpi.com)

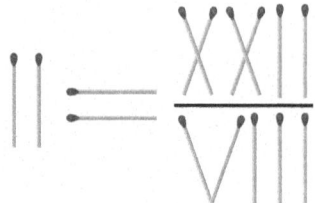

70. Find the square root of $\dfrac{2143}{22}$ using a calculator. Then find the square root of that result. Compare your result to the decimal given for π in the text. What do you notice?

71. Use a calculator to find the first eight digits in the decimal for $\dfrac{355}{113}$. Compare the result to the decimal for π given in the text. What do you notice?

72. You may have seen the statements

"Use $\dfrac{22}{7}$ for π" and "Use 3.14 for π."

Since $\dfrac{22}{7}$ is the quotient of two integers, and 3.14 is a terminating decimal, do these statements suggest that π is rational?

73. An ancient Jewish manuscript describes a circular pool as "ten cubits across" and "thirty cubits round about." What value of π does this imply?

74. The ancient Egyptians used a method for finding the area of a circle that is equivalent to a value of 3.1605 for π. Write this decimal as a mixed number.

75. The computation of π has fascinated mathematicians and others for centuries. In the nineteenth century, the British mathematician William Shanks spent many years of his life calculating π to 707 decimal places. It turned out that only the first 527 were correct. Use an Internet search to find the 528th decimal digit of π (following the whole number part 3).

76. A **mnemonic device** is a scheme whereby one is able to recall facts by memorizing something completely unrelated to the facts. One way of learning the first few digits of the decimal for π is to memorize a sentence (or several sentences) and count the letters in each word of the sentence. For example,

"See, I know a digit."

will give the first 5 digits of π: "See" has 3 letters, "I" has 1 letter, "know" has 4 letters, "a" has 1 letter, and "digit" has 5 letters. So the first five digits are 3.1415.

Verify that the following mnemonic devices work.

(a) "May I have a large container of coffee?"

(b) "See, I have a rhyme assisting my feeble brain, its tasks ofttimes resisting."

(c) "How I want a drink, alcoholic of course, after the heavy lectures involving quantum mechanics."

77. In the second season of the original *Star Trek* series, the episode "Wolf in the Fold" told the story of an alien entity that had taken over the computer of the starship *Enterprise*. To drive the entity out of the computer, Mr. Spock gave the alien the compulsory directive to compute π to the last digit. Explain why this strategy proved successful.

78. *Northern Exposure* ran between 1990 and 1995 on the CBS network. In the episode "Nothing's Perfect," the local disc jockey Chris Stevens meets and develops a relationship with a mathematician after accidentally running over her dog. Her area of research is computation of the decimal digits of pi. She mentions that a string of eight 8s appears in the decimal relatively early in the expansion. Use an Internet search to determine the position at which this string appears.

79. Use a calculator to find the decimal approximations for

$$\phi = \frac{1 + \sqrt{5}}{2} \quad \text{and its } \textbf{conjugate,} \quad \frac{1 - \sqrt{5}}{2}.$$

Comment on the similarities and differences in the two decimals.

80. In some literature, the Golden Ratio is defined to be the reciprocal of

$$\frac{1 + \sqrt{5}}{2}, \quad \text{which is} \quad \frac{2}{1 + \sqrt{5}}.$$

Use a calculator to find a decimal approximation for $\dfrac{2}{1 + \sqrt{5}}$, and compare it to ϕ as defined in this text. What do you observe?

81. Near the end of the 2008 movie *Harold & Kumar Escape from Guantanamo Bay,* Kumar (Kal Penn) recites a poem dealing with the square root of 3, written by the late David Feinberg. The text of the poem can be found on the Internet by searching

"David Feinberg Square Root of 3."

There are references to irrational numbers, integers, an approximation for the square root of 3, and the product rule for square root radicals. Why would Kumar prefer the 3 to be a 9?

82. See **Exercise 81.** What is the decimal approximation for $\sqrt{3}$ (to four decimal places) given in Feinberg's poem?

83. An approximation for *e* is

2.718281828.

A student noticed that there seems to be a repetition of four digits in this number (1, 8, 2, 8) and concluded that *e* is rational, because repeating decimals represent rational numbers. Was the student correct? Why or why not?

84. Use a calculator with an exponential key to find values for the following:

$$(1.1)^{10}, \quad (1.01)^{100}, \quad (1.001)^{1000},$$
$$(1.0001)^{10,000}, \quad \text{and} \quad (1.00001)^{100,000}.$$

Compare your results to the approximation given for *e* in this section. What do you find?

Roots Other Than Square Roots *The concept of square (second) root can be extended to* **cube (third) root, fourth root,** *and so on. For example,*

$$\sqrt[3]{8} = 2 \quad because \quad 2^3 = 8,$$
$$\sqrt[3]{1000} = 10 \quad because \quad 10^3 = 1000,$$
$$\sqrt[4]{81} = 3 \quad because \quad 3^4 = 81, \quad and\ so\ on.$$

If $n \geq 2$ and a is a nonnegative number, then $\sqrt[n]{a}$ represents the nonnegative number whose nth power is a.

Find each root.

85. $\sqrt[3]{64}$ **86.** $\sqrt[3]{125}$ **87.** $\sqrt[3]{343}$ **88.** $\sqrt[3]{729}$

89. $\sqrt[3]{216}$ **90.** $\sqrt[3]{512}$ **91.** $\sqrt[4]{1}$ **92.** $\sqrt[4]{16}$

93. $\sqrt[4]{256}$ **94.** $\sqrt[4]{625}$ **95.** $\sqrt[4]{4096}$ **96.** $\sqrt[4]{2401}$

Use a calculator to approximate each root. (Hint: To find the fourth root, find the square root of the square root.)

97. $\sqrt[3]{43}$ **98.** $\sqrt[3]{87}$

99. $\sqrt[3]{198}$ **100.** $\sqrt[4]{2107}$

101. $\sqrt[4]{10,265.2}$ **102.** $\sqrt[4]{863.5}$

103. $\sqrt[4]{968.1}$ **104.** $\sqrt[4]{12,966.4}$

6.5   APPLICATIONS OF DECIMALS AND PERCENTS

OBJECTIVES

1 Perform operations of arithmetic with decimal numbers.

2 Round whole numbers and decimals to a given place value.

3 Perform computations using percent.

4 Convert among forms of fractions, decimals, and percents.

5 Find percent increase and percent decrease.

6 Apply formulas involving fractions and decimals from the allied health industry.

Operations with Decimals

Because calculators have, for the most part, replaced paper-and-pencil methods for operations with decimals and percent, we will only briefly mention these latter methods. ***We strongly suggest that the work in this section be done with a calculator at hand.***

ADDITION AND SUBTRACTION OF DECIMALS

To add or subtract decimal numbers, line up the decimal points in a column and perform the operation.

EXAMPLE 1 Adding and Subtracting Decimal Numbers

Find each of the following.

(a) $0.46 + 3.9 + 12.58$ **(b)** $12.1 - 8.723$

Solution

(a)
```
   0.46      Line up decimal points.
   3.90      Attach a zero as a placeholder.
 +12.58
 ─────
  16.94   ← Sum
```

(b)
```
  12.100    Attach zeros.
 − 8.723
 ──────
   3.377  ← Difference
```

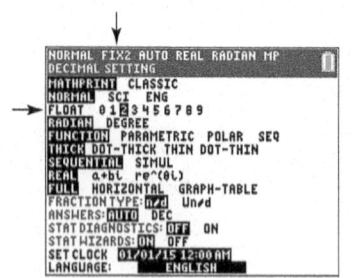

This screen supports the results in **Example 1.**

When two numbers are multiplied, the numbers are **factors** and the answer is the **product.** When two numbers are divided, the number being divided is the **dividend,** the number doing the dividing is the **divisor,** and the answer is the **quotient.**

MULTIPLICATION AND DIVISION OF DECIMALS

Multiplication   To multiply decimals, multiply in the same manner as integers are multiplied. The number of decimal places to the right of the decimal point in the product is the *sum* of the numbers of places to the right of the decimal points in the factors.

Division   To divide decimals, move the decimal point to the right the same number of places in the divisor and the dividend so as to obtain a whole number in the divisor. Divide in the same manner as integers are divided. The number of decimal places to the right of the decimal point in the quotient is the same as the number of places to the right of the decimal in the dividend.

EXAMPLE 2   **Multiplying and Dividing Decimal Numbers**

Find each of the following.

(a) 4.613×2.52 **(b)** $65.175 \div 8.25$

Solution

(a)
$$
\begin{array}{r}
4.613 \leftarrow \text{3 decimal places} \\
\times \quad 2.52 \leftarrow \text{2 decimal places} \\
\hline
9226 \\
23065 \\
9226 \\
\hline
11.62476 \leftarrow 3 + 2 = \text{5 decimal places}
\end{array}
$$

(b) $8.25)\overline{65.175} \rightarrow$
$$
\begin{array}{r}
7.9 \\
825)\overline{6517.5} \\
5775 \\
\hline
7425 \\
7425 \\
\hline
0
\end{array}
$$

Bring the decimal point straight up in the answer.

Here the calculator is set to round the answer to two decimal places.

Rounding Methods

To round, or approximate, a whole number or a decimal number to a given place value, use the following procedure. (You may wish to refer to the diagram in the third section of this chapter that describes place values determined by powers of 10.)

RULES FOR ROUNDING WHOLE NUMBERS

Step 1   Locate the **place** to which the whole number is being rounded.

Step 2   Look at the next **digit to the right** of the place to which the number is being rounded.

Step 3A   If the digit to the right is **less than 5,** replace it and all digits following it with zeros. Do not change the digit in the place to which the number is being rounded.

Step 3B   If the digit to the right is **5 or greater,** replace it and all digits following it with zeros, but also add 1 to the digit in the place to which the number is being rounded. (If adding 1 to a 9, replace the 9 with 0 and add 1 to the next digit to the left. If that next digit is also 9, repeat this procedure; and so on, to the left.)

"**Technology** pervades the world outside school. There is no question that students will be expected to use calculators in other settings; this technology is now part of our culture ... students no longer have the same need to perform these [paper-and-pencil] procedures with large numbers of lengthy expressions that they might have had in the past without ready access to technology."

From *Computation, Calculators, and Common Sense (A Position Paper of the National Council of Teachers of Mathematics).*

RULES FOR ROUNDING DECIMAL NUMBERS

Steps 1 and 2 are the same as for whole numbers.

Step 3A If the digit to the right is **less than 5,** drop all digits to the right of the place to which the number is being rounded. Do not change the digit in the place to which it is being rounded.

Step 3B If the digit to the right is **5 or greater,** drop it and all digits following it, but also add 1 to the digit in the place to which the number is being rounded. (When adding 1 to a 9, use the same procedure as for whole numbers.)

Note: Some disciplines have guidelines specifying that in certain cases, a *downward* roundoff is made. We will not investigate these.

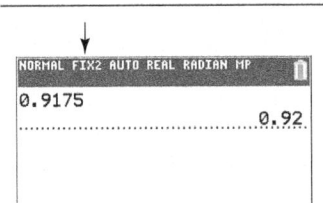

The calculator rounds 0.9175 to the nearest hundredth.

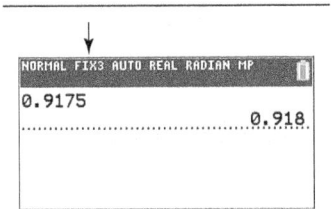

The calculator rounds 0.9175 *up* to 0.918.

EXAMPLE 3 Rounding Numbers

Round to the place values indicated.

(a) 0.9175 to the nearest tenth, hundredth, and thousandth

(b) 1358 to the nearest thousand, hundred, and ten

Solution

(a) Round 0.9175 as follows: 0.9 to the nearest tenth,
0.92 to the nearest hundredth,
0.918 to the nearest thousandth.

(b) Round 1358 as follows: 1000 to the nearest thousand,
1400 to the nearest hundred,
1360 to the nearest ten.

Percent

The word **percent** means **"per hundred."** The symbol % represents "percent."

PERCENT

$$1\% = \frac{1}{100} = 0.01$$

EXAMPLE 4 Converting Percents to Decimals

Convert each percent to a decimal.

(a) 98%   **(b)** 3.4%   **(c)** 0.2%   **(d)** 150%

Solution

(a) $98\% = 98(1\%) = 98(0.01) = 0.98$

(b) $3.4\% = 3.4(1\%) = 3.4(0.01) = 0.034$

(c) $0.2\% = 0.2(1\%) = 0.2(0.01) = 0.002$

(d) $150\% = 150(1\%) = 150(0.01) = 1.5$

The percent symbol, %, probably evolved from a symbol introduced in an Italian manuscript of 1425. Instead of "per 100," "P 100," or "P cento," which were common at that time, the author used "P^R." By about 1650 the ^{R} had become $\frac{0}{0}$, so "per $\frac{0}{0}$" was often used. Finally the "per" was dropped, leaving $\frac{0}{0}$ or %.

(*Source: Historical Topics for the Mathematics Classroom,* the Thirty-first Yearbook of the National Council of Teachers of Mathematics, 1969.)

EXAMPLE 5   **Converting Decimals to Percents**

Convert each decimal to a percent.

(a) 0.13   **(b)** 0.532   **(c)** 2.3   **(d)** 0.07

Solution

Reverse the procedure used in **Example 4.**

(a) $0.13 = 13(0.01) = 13(1\%) = 13\%$

(b) $0.532 = 53.2(0.01) = 53.2(1\%) = 53.2\%$

(c) $2.3 = 230(0.01) = 230(1\%) = 230\%$

(d) $0.07 = 7(0.01) = 7(1\%) = 7\%$

Examples 4 and 5 suggest the following shortcut methods for converting.

CONVERTING BETWEEN DECIMALS AND PERCENTS

To convert a percent to a decimal, drop the percent symbol (%) and move the decimal point two places to the left, inserting zeros as placeholders if necessary.

To convert a decimal to a percent, move the decimal point two places to the right, inserting zeros as placeholders if necessary, and attach the percent symbol (%).

EXAMPLE 6   **Converting Fractions to Percents**

Convert each fraction to a percent.

(a) $\frac{3}{5}$ **(b)** $\frac{14}{25}$ **(c)** $2\frac{7}{10}$

Solution

(a) First write $\frac{3}{5}$ as a decimal. Dividing 3 by 5 gives $\frac{3}{5} = 0.6 = 60\%$.

(b) $\frac{14}{25} = \frac{14 \cdot 4}{25 \cdot 4} = \frac{56}{100} = 0.56 = 56\%$ **(c)** $2\frac{7}{10} = 2.7 = 270\%$

CONVERTING A FRACTION TO A PERCENT

To convert a fraction to a percent, convert the fraction to a decimal, and then convert the decimal to a percent.

In the following examples involving percents, three methods are shown. They illustrate some basic ideas of solving equations. The second method in each case involves using cross products. The third method involves the percent key of a basic calculator. (Keystrokes may vary among models.)

EXAMPLE 7   **Finding a Percent of a Number**

A publisher requires that an author reduce the page count in her next edition by 18%. The current edition has 250 pages. How many pages must the author cut from her book?

Solution

This problem can be reworded as follows: Find 18% of 250.

Method 1   The key word "of" translates as "times."

$$18\%(250) = 0.18(250) = 45$$

Method 2   Think "18 is to 100 as what (x) is to 250?" This translates as follows.

$$\frac{18}{100} = \frac{x}{250}$$

$$100x = 18 \cdot 250 \qquad \text{\small $\frac{a}{b} = \frac{c}{d}$ leads to } ad = bc.$$

$$x = \frac{18 \cdot 250}{100} \qquad \text{\small Divide by 100.}$$

$$x = 45 \qquad \text{\small Simplify.}$$

Method 3   Use the percent key on a calculator.

   **45** ◁— Final display

• Thus, 18% of 250 is 45. The author must reduce her page count by 45 pages.

EXAMPLE 8   Finding What Percent One Number Is of Another

A kindergarten teacher has submitted 6 of the required 50 class preparations to the principal of her school. What percent of her requirement has she completed?

Solution

This problem can be reworded as follows: What percent of 50 is 6?

Method 1   Let the phrase "what percent" be represented by $x \cdot 1\%$ or $0.01x$. Again the word "of" translates as "times," and "is" translates as "equals."

$$0.01x \cdot 50 = 6$$

$$0.50x = 6 \qquad \text{\small Multiply on the left side.}$$

$$50x = 600 \qquad \text{\small Multiply by 100 to clear decimals.}$$

$$x = 12 \qquad \text{\small Divide by 50.}$$

Method 2   Think "What (x) is to 100 as 6 is to 50?"

$$\frac{x}{100} = \frac{6}{50}$$

$$50x = 600 \qquad \text{\small Cross products}$$

$$x = 12 \qquad \text{\small Divide by 50.}$$

Method 3   Use a calculator.

6 ÷ 5 0 %   **12** ◁— Final display

• Thus, 6 is 12% of 50. She has completed 12% of her requirement.

EXAMPLE 9   Finding a Number That Is a Given Percent of a Given Number

A government employee working for the county judicial system chooses 5% of a jury pool to question for possible service. This amounts to 38 people. What is the size of the entire jury pool?

Solution

This problem can be reworded as follows: 38 is 5% of what number?

Method 1
$$38 = 0.05x \qquad \text{\small Let x represent the number.}$$

$$x = \frac{38}{0.05} \qquad \text{\small Divide by 0.05.}$$

$$x = 760 \qquad \text{\small Simplify.}$$

Method 2 Think "38 is to what number (x) as 5 is to 100?"

$$\frac{38}{x} = \frac{5}{100}$$

$$5x = 3800 \qquad \text{Cross products}$$

$$x = 760 \qquad \text{Divide by 5.}$$

Method 3 Use a calculator.

$$\boxed{3}\ \boxed{8}\ \boxed{\div}\ \boxed{5}\ \boxed{\%} \qquad\qquad \textbf{760} \quad \longleftarrow \boxed{\text{Final display}}$$

Each method shows us that 38 is 5% of 760. There are 760 in the jury pool.

Consumers often encounter figures in the media that involve **percent change,** which includes **percent increase** and **percent decrease.** The following guidelines are helpful in understanding these concepts.

FINDING PERCENT INCREASE OR DECREASE

1. To find the **percent increase from *a* to *b*,** where $b > a$, subtract a from b, and divide this result by a. Convert to a percent.

   Example: The *percent increase* from 4 to 7 is $\frac{7-4}{4} = \frac{3}{4} = 75\%$.

2. To find the **percent decrease from *a* to *b*,** where $b < a$, subtract b from a, and divide this result by a. Convert to a percent.

   Example: The *percent decrease* from 8 to 6 is $\frac{8-6}{8} = \frac{2}{8} = \frac{1}{4} = 25\%$.

EXAMPLE 10 **Solving Problems about Percent Change**

Solve each problem involving percent change.

(a) An electronics store marked up a laptop computer from the store's cost of $1200 to a selling price of $1464. What was the percent markup?

(b) The enrollment in a community college declined from 12,750 during one school year to 11,350 the following year. Find the percent decrease to the nearest tenth of a percent.

Solution

(a) "Markup" is a name for an increase. Let $x =$ the percent increase (as a decimal).

$$\text{percent increase} = \frac{\text{amount of increase}}{\text{original amount}}$$

Subtract to find the amount of increase.

$$x = \frac{1464 - 1200}{1200} \qquad \text{Substitute the given values.}$$

Use the original cost.

$$x = \frac{264}{1200}$$

$$x = 0.22 \qquad \text{Use a calculator.}$$

The computer was marked up 22%.

(b) Let $x =$ the percent decrease (as a decimal).

$$\text{percent decrease} = \frac{\text{amount of decrease}}{\text{original amount}}$$

Subtract to find the *amount* of decrease.

$$x = \frac{12{,}750 - 11{,}350}{12{,}750} \qquad \text{Substitute the given values.}$$

Use the original enrollment.

$$x = \frac{1400}{12{,}750}$$

$$x \approx 0.110 \qquad \text{Use a calculator.}$$

The college enrollment decreased by about 11.0%.

When calculating a percent increase or a percent decrease, use the original number (before the increase or decrease) as the base. A common error is to use the final number (*after* the increase or decrease) in the denominator of the fraction.

Applications of decimal numbers sometimes involve formulas.

EXAMPLE 11 Determining a Child's Body Surface Area

If a child weighs k kilograms, then the child's body surface area S in square meters (m^2) is determined by the following formula.

$$S = \frac{4k + 7}{k + 90}$$

What is the body surface area of a child who weighs 40 pounds? (*Source:* Hegstad, Lorrie N., and Wilma Hayek. *Essential Drug Dosage Calculations, 4th ed.* Prentice Hall, 2001.)

Solution

Tables and Web sites for conversion between the metric and English systems are readily available. Using www.metric-conversions.org, we find that 40 pounds is approximately 18.144 kilograms. Use this value for k in the formula.

$$S = \frac{4(18.144) + 7}{18.144 + 90} \qquad \text{Let } k = 18.144 \text{ in the formula.}$$

$$S = 0.74 \qquad \text{Use a calculator. Round to the nearest hundredth.}$$

The body surface area is approximately 0.74 m^2.

EXAMPLE 12 Determining a Child's Dose of a Drug

If D represents the usual adult dose of a drug, the corresponding child's dose C is calculated by the following formula.

$$C = \frac{\text{body surface area in square meters}}{1.7} \times D$$

Determine the appropriate dose for a child weighing 40 pounds if the usual adult dose is 50 milligrams. (*Source:* Hegstad, Lorrie N., and Wilma Hayek. *Essential Drug Dosage Calculations, 4th ed.* Prentice Hall, 2001.)

Solution

From **Example 11,** the body surface area of a child weighing 40 pounds is 0.74 m².

$$C = \frac{\text{body surface area}}{1.7} \times D$$

$$C = \frac{0.74}{1.7} \times 50 \qquad \text{Body surface area} = 0.74, D = 50.$$

$$C = 22 \qquad\qquad \text{Use a calculator. Round to the nearest unit.}$$

The child's dose is 22 milligrams.

WHEN WILL I **EVER** USE THIS?

Cynthia (not her real name) had worked for eight years in an upscale women's clothing store in New Orleans. She decided to go back to school with the goal of becoming a veterinarian. Her curriculum required a course in college algebra. Her instructor (one of the authors) made an offhand comment one day about *percent increase.*

"If you want to mark up an item by a given percent, such as 15%, all you need to do is multiply the original price by 1.15. That's your final amount."

Cynthia realized that when she marked up items for retail sale, as her first step she would find the amount of markup. Then, in a separate step she would add it to the wholesale cost of the item.

For example, if the cost was $25.80, she would calculate **15% of $25.80** on her calculator to obtain **$3.87.** Then she would add.

$3.87 + **$25.80** = **$29.67.** ← Selling price

She now realized that the selling price could be obtained in a single step.

$25.80 · 1.15 = **$29.67.** ← Selling price

This method works because of the identity property of multiplication and the distributive property. Suppose that C represents the wholesale cost of an item. Then $0.15C$ represents 15% of C, the amount of the markup. The selling price S is found by adding wholesale cost to the amount of markup.

$S = C + 0.15C$ Selling price = wholesale cost + markup

$S = 1C + 0.15C$ Identity property of multiplication

$S = (1 + 0.15)C$ Distributive property

$S = 1.15C$ $1 + 0.15 = 1.15$

A similar situation holds if a discount is to be applied. For example, to mark an item down 25%, just subtract from 100% to get 75%, and then multiply the original amount by 0.75. The result is the discounted price.

Cynthia went on to get her doctor's degree in veterinary medicine and open her own practice in the New Orleans area.

6.5 EXERCISES

CONCEPT CHECK *Write* true *or* false *for each statement.*

1. 300% of 12 is 36.

2. 25% of a quantity is the same as $\frac{1}{4}$ of that quantity.

3. To find 50% of a quantity, we may simply divide the quantity by 2.

4. A soccer team that has won 12 games and lost 8 games has a winning percentage of 60%.

5. If 70% is the lowest passing grade on a quiz that has 50 items of equal value, then answering at least 35 items correctly will assure you of a passing grade.

6. 30 is more than 40% of 120.

7. .99¢ = 99 cents

8. If an item usually costs $70.00 and it is discounted 10%, then the discount price is $7.00.

Calculate each of the following using either a calculator or paper-and-pencil methods, as directed by your instructor.

9. 8.53 + 2.785

10. 9.358 + 7.2137

11. 8.74 − 12.955

12. 2.41 − 3.997

13. 25.7 × 0.032

14. 45.1 × 8.344

15. 1019.825 ÷ 21.47

16. −262.563 ÷ 125.03

17. $\dfrac{118.5}{1.45 + 2.3}$

18. 2.45(1.2 + 3.4 − 5.6)

Personal Finance *Solve each problem.*

19. Martin has $48.35 in his checking account. He uses his debit card to make purchases of $35.99 and $20.00, which overdraws his account. His bank charges his account an overdraft fee of $28.50. He then deposits his paycheck for $66.27 from his part-time job at Arby's. What is the balance in his account?

20. Kellen has $37.60 in her checking account. She uses her debit card to make purchases of $25.99 and $19.34, which overdraws her account. Her bank charges her account an overdraft fee of $25.00. She then deposits her paycheck for $58.66 from her part-time job at Subway. What is the balance in her account?

21. Ahmad owes $382.45 on his Visa account. He returns two items costing $25.10 and $34.50 for credit. Then he makes purchases of $45.00 and $98.17.

(a) How much should his payment be if he wants to pay off the balance on the account?

(b) Instead of paying off the balance, he makes a payment of $300 and then incurs a finance charge of $24.66. What is the balance on his account?

22. Marin owes $237.59 on her MasterCard account. She returns one item costing $47.25 for credit and then makes two purchases of $12.39 and $20.00.

(a) How much should her payment be if she wants to pay off the balance on the account?

(b) Instead of paying off the balance, she makes a payment of $75.00 and incurs a finance charge of $32.06. What is the balance on her account?

23. *Bank Account Balance* In August, Erin began with a bank account balance of $904.89. Her withdrawals and deposits for August are given below:

Withdrawals	Deposits
$35.84	$85.00
$26.14	$120.76
$3.12	$205.00
$21.46	

Assuming no other transactions, what was her account balance at the end of August?

24. *Bank Account Balance* In September, José began with a bank account balance of $904.89. His withdrawals and deposits for September are given below:

Withdrawals	Deposits
$41.29	$80.59
$13.66	$276.13
$84.40	$550.00
$93.00	

Assuming no other transactions, what was his account balance at the end of September?

Rounding Round each number to the place value indicated. *For example, in part (a) of Exercise 25, round 54,793 to the nearest ten thousand. (Hint: Always round from the original number.)*

25. 54,793

    (a) ten thousand   **(b)** thousand

    (c) hundred   **(d)** ten

26. 453,258

    (a) hundred thousand   **(b)** thousand

    (c) hundred   **(d)** ten

27. 0.892451

    (a) hundred-thousandth

    (b) ten-thousandth

    (c) thousandth

    (d) hundredth

    (e) tenth

    (f) one or unit

28. 22.483956

    (a) hundred-thousandth

    (b) ten-thousandth

    (c) thousandth

    (d) hundredth

    (e) tenth

    (f) ten

Percent Concepts Use the concept of percent in each exercise.

29. Match the percents in Group I with their equivalent fractions in Group II.

I		II	
(a) 25%	**(b)** 10%	**A.** $\frac{1}{3}$	**B.** $\frac{1}{50}$
(c) 2%	**(d)** 20%	**C.** $\frac{3}{4}$	**D.** $\frac{1}{10}$
(e) 75%	**(f)** $33\frac{1}{3}$%	**E.** $\frac{1}{4}$	**F.** $\frac{1}{5}$

30. Fill in each blank with the correct numerical response.

    (a) 5% means _____ in every 100.

    (b) 25% means 6 in every _____.

    (c) 200% means _____ for every 4.

    (d) 0.5% means _____ in every 100.

    (e) _____ % means 12 for every 2.

31. The figures in **Exercise 23** of the third section of this chapter are reproduced here. Express the fractional parts represented by the shaded areas as percents.

  (a) 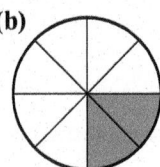   **(b)**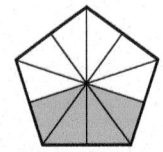

  (c) 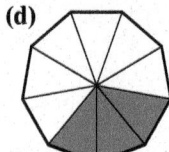   **(d)**

32. The Venn diagram shows the numbers of elements in the four regions formed.

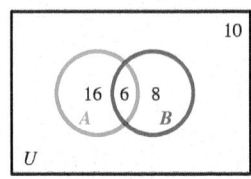

    (a) What percent of the elements in the universe are in $A \cap B$?

    (b) What percent of the elements in the universe are in A but not in B?

    (c) What percent of the elements in $A \cup B$ are in $A \cap B$?

    (d) What percent of the elements in the universe are in neither A nor B?

Convert each decimal to a percent.

33. 0.42	**34.** 0.87	**35.** 0.365
36. 0.792	**37.** 0.008	**38.** 0.0093
39. 2.1	**40.** 8.9	

Convert each percent to a decimal.

41. 96%	**42.** 23%	**43.** 5.46%
44. 2.99%	**45.** 0.3%	**46.** 0.6%
47. 400%	**48.** 260%	**49.** $\frac{1}{2}$%
50. $\frac{2}{5}$%	**51.** $3\frac{1}{2}$%	**52.** $8\frac{3}{4}$%

Convert each fraction to a percent.

53. $\frac{1}{5}$ **54.** $\frac{2}{5}$ **55.** $\frac{1}{100}$ **56.** $\frac{1}{50}$

57. $\frac{3}{8}$ **58.** $\frac{5}{6}$ **59.** $\frac{3}{2}$ **60.** $\frac{7}{4}$

61. Explain the difference between $\frac{1}{2}$ of a quantity and $\frac{1}{2}\%$ of the quantity.

62. Explain the difference between 9% and .9%.

Solve each problem involving percent.

63. What is 26% of 480?

64. What is 38% of 12?

65. What is 10.5% of 28?

66. What is 48.6% of 19?

67. What percent of 30 is 45?

68. What percent of 48 is 20?

69. 25% of what number is 150?

70. 12% of what number is 3600?

71. 0.392 is what percent of 28?

72. 78.84 is what percent of 292?

Solve each problem involving percent increase or decrease.

73. *Percent Increase* After one year on the job, Grady got a raise from $10.50 per hour to $11.34 per hour. What was the percent increase in his hourly wage?

74. *Percent Discount* Clayton bought a ticket to a rock concert at a discount. The regular price of the ticket was $70.00, but he paid only $59.50. What was the percent discount?

75. *Percent Decrease* Between 2010 and 2018, the estimated population of Baltimore, Maryland, declined from 620,961 to 601,188. What was the percent decrease to the nearest tenth? (Data from U.S. Census Bureau.)

76. *Percent Increase* Between 2010 and 2018, the estimated population of Oklahoma City, Oklahoma, grew from 597,999 to 653,865. What was the percent increase to the nearest tenth? (Data from U.S. Census Bureau.)

77. *Percent Discount* The DVD release of Season 8 of *Modern Family* was available at www.amazon.com for $14.97. The list price of this DVD was $19.98. To the nearest tenth, what was the percent discount? (Data from www.amazon.com)

78. *Percent Discount* The Blu-ray of *Hidden Figures* had a list price of $19.99 and was for sale at www.amazon.com for $12.84. To the nearest tenth, what was the percent discount? (Data from www.amazon.com)

79. *Value of 1916-D Mercury Dime* The 1916 Mercury dime minted in Denver is quite rare. In 1979 its value in Extremely Fine condition was $625. The 2017 value had increased to $5250. What was the percent increase in the value of this coin from 1979 to 2017? (Data from *A Guide Book of United States Coins; Coin World Coin Values.*)

80. *Value of 1903-O Morgan Dollar* In 1963, the value of a 1903 Morgan dollar minted in New Orleans in typical Uncirculated condition was $1500. Due to a discovery of a large hoard of these dollars late that year, the value plummeted. Its value in 2017 was $625. What was the percent decrease in its value from 1963 to 2017? (Data from *A Guide Book of United States Coins; Coin World Coin Values.*)

Crude Oil Prices *The line graph shows the average price, adjusted for inflation, of domestic crude oil for selected years between 1956 and 2016. Use this information to answer each question.*

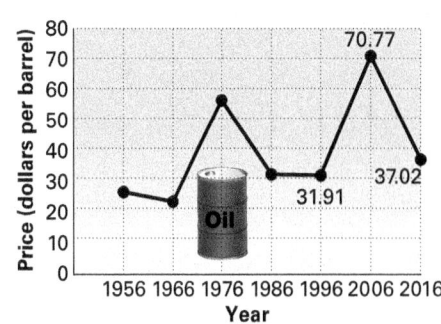

Data from www.inflationdata.com

81. By what percent did prices increase from 1996 to 2006?

82. By what percent did prices decrease from 2006 to 2016?

Business Earnings Report *An article in the business section of* The Gazette *covered various facets of the earnings report of a major corporation. Answer each of the following questions about a particular phase of the report. Round to the nearest tenth if applicable.*

83. In one year, profit fell from $1.56 billion to $1.23 billion. What was the percent decrease in profit?

84. Sales rose 2.3% to $8.7 billion. What were the sales prior to this rise?

85. Profit in one division dropped 24% to $680 million. What was the profit prior to this drop?

86. Shares of the company closed on one recent day at $17.98 per share, down $1.19. What percent decrease did this represent?

Use mental techniques to answer each question. Try to avoid using paper and pencil or a calculator.

87. Allowance Increase Carly's allowance was raised from $4.00 per week to $5.00 per week. What was the percent of the increase?

  A. 25%   **B.** 20%   **C.** 50%   **D.** 30%

88. Boat Purchase and Sale Susan bought a boat five years ago for $5000 and sold it this year for $2000. What percent of her original purchase did she lose on the sale?

  A. 40%   **B.** 50%   **C.** 20%   **D.** 60%

89. Population of Alabama A 2016 report indicated that the population of Alabama was 4,863,000, with 26.0% represented by African Americans. What is the best estimate of the African American population in Alabama? (Data from U.S. Census Bureau.)

  A. 500,000   **B.** 1,500,000   **C.** 1,250,000   **D.** 750,000

90. Population of Hawaii A 2016 report indicated that the population of Hawaii was 1,429,000, with 19.4% of the population being of two or more races. What is the best estimate of this racial demographic population of Hawaii? (Data from U.S. Census Bureau.)

  A. 280,000   **B.** 300,000   **C.** 21,400   **D.** 24,000

91. Discount and Markup Suppose that an item regularly costs $100.00 and is discounted 20%. If it is then marked up 20%, is the resulting price $100.00? If not, what is it?

92. Computing a Tip Suppose that you have decided that you will always tip 25% when dining in restaurants. By what whole number should you divide the bill to find the amount of the tip?

Body Surface Area *Use the formula from **Example 11** for determining a child's body surface area*

$$S = \frac{4k + 7}{k + 90}$$

to solve each problem. Round to the nearest hundredth.

93. Find the body surface area S in square meters of a child who weighs 20 kilograms.

94. Find the body surface area S in square meters of a child who weighs 26 kilograms.

95. Find the body surface area S in square meters of a child who weighs 20 pounds. (Use $k = 9.07$.)

96. Find the body surface area S in square meters of a child who weighs 26 pounds. (Use $k = 11.79$.)

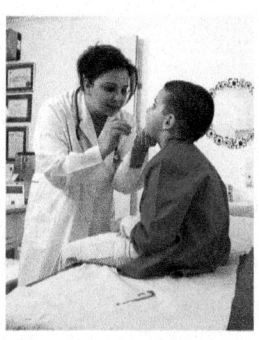

Child's Medication Dosage *Use the formula from **Example 12** for determining a child's dose of a drug*

$$C = \frac{\text{body surface area in square meters}}{1.7} \times D$$

to solve each exercise. Round to the nearest unit.

97. If the usual adult dose D of a drug is 250 mg, what is the child's dose for a child weighing 20 kilograms? (*Hint:* Use the result of **Exercise 93**.)

98. If the usual adult dose D of a drug is 250 mg, what is the child's dose for a child weighing 26 kilograms? (*Hint:* Use the result of **Exercise 94**.)

99. If the usual adult dose D of a drug is 500 mg, what is the child's dose for a child weighing 20 pounds? (*Hint:* Use the result of **Exercise 95**.)

100. If the usual adult dose D of a drug is 500 mg, what is the child's dose for a child weighing 26 pounds? (*Hint:* Use the result of **Exercise 96**.)

Metabolic Units *One way to measure a person's cardio fitness is to calculate how many METs, or metabolic units, he or she can reach at peak exertion. One MET is the amount of energy used when sitting quietly. To calculate ideal METs, we can use one of the following formulas.*

$$\text{MET} = 14.7 - \text{age} \cdot 0.13$$
$$\text{MET} = 14.7 - \text{age} \cdot 0.11$$

(*Source: New England Journal of Medicine.*)

101. A 40-year-old woman wishes to find her ideal MET.

(a) Write the expression using her age.

(b) Calculate her ideal MET.

(c) Researchers recommend that people reach about 85% of their MET when exercising. Calculate 85% of the ideal MET from part (b). Then refer to the following table. What activity can the woman do that is approximately this value?

Activity	METs	Activity	METs
Golf (with cart)	2.5	Skiing (water or downhill)	6.8
Walking (3 mph)	3.3	Swimming	7.0
Mowing lawn (power mower)	4.5	Walking (5 mph)	8.0
Ballroom or square dancing	5.5	Jogging	10.2

Data from Harvard School of Public Health.

102. Repeat **Exercise 101** for a 55-year-old man.

These exercises are based on formulas found in Auto Math Handbook: Mathematical Calculations, Theory, and Formulas for Automotive Enthusiasts, *by John Lawlor (1991, HP Books).*

103. *Blood Alcohol Concentration* The blood alcohol concentration (BAC) of a person who has been drinking is given by the formula

$$BAC = \frac{(\text{ounces} \times \text{percent alcohol} \times 0.075)}{\text{body weight in lb}}$$
$$- (\text{hours of drinking} \times 0.015).$$

Suppose a policeman stops a 190-pound man who, in two hours, has ingested four 12-ounce beers, each having a 3.2% alcohol content. The formula would then read

$$BAC = \frac{[(4 \times 12) \times 3.2 \times 0.075]}{190} - (2 \times 0.015).$$

(a) Find this BAC.

(b) Find the BAC for a 135-pound woman who, in three hours, has drunk three 12-ounce beers, each having a 4.0% alcohol content.

104. *Approximate Automobile Speed* The approximate speed of an automobile in miles per hour (mph) can be found in terms of the engine's revolutions per minute (rpm), the tire diameter in inches, and the overall gear ratio by the following formula.

$$mph = \frac{rpm \times \text{tire diameter}}{\text{gear ratio} \times 336}$$

If a certain automobile has an rpm of 5600, a tire diameter of 26 inches, and a gear ratio of 3.12, what is its approximate speed (mph)?

105. *Engine Horsepower* Horsepower can be found from mean effective pressure (mep) in pounds per square inch, engine displacement in cubic inches, and revolutions per minute (rpm) using the following formula.

$$\text{Horsepower} = \frac{\text{mep} \times \text{displacement} \times \text{rpm}}{792{,}000}$$

An engine has displacement of 302 cubic inches and indicated mep of 195 pounds per square inch at 4000 rpm. What is its approximate horsepower?

106. *Torque Approximation* To determine the torque at a given value of rpm, the following formula applies.

$$\text{Torque} = \frac{5252 \times \text{horsepower}}{\text{rpm}}$$

If the horsepower of a certain vehicle is 400 at 4500 rpm, what is the approximate torque?

Win-Loss Record These exercises deal with winning percentage in the standings of sports teams.

107. At the end of the regular 2017 Major League Baseball season, the standings of the East Division of the American League were as shown. Winning percentage is commonly expressed as a decimal rounded to the nearest thousandth. To find the winning percentage of a team, divide the number of wins (W) by the total number of games played (W + L). Find the winning percentage of each team.

(a) Boston   (b) Baltimore

(c) New York Yankees   (d) Toronto

Team	W	L	Pct.
Boston	93	69	
New York Yankees	91	71	
Tampa Bay	80	82	.494
Toronto	76	86	
Baltimore	75	87	

Data from The World Almanac and Book of Facts.

108. Repeat **Exercise 107** for the following standings for the East Division of the National League.

(a) Atlanta   (b) New York Mets

(c) Philadelphia   (d) Miami

Team	W	L	Pct.
Washington	97	65	.599
Miami	77	85	
Atlanta	72	90	
New York Mets	70	92	
Philadelphia	66	96	

Data from *The World Almanac and Book of Facts.*

Tipping Procedure *It is customary in our society to "tip" the wait staff when dining in restaurants. One common rate for tipping is 15%. A quick way of figuring a tip that will give a close approximation of 15% follows.*

Step 1 *Round off the bill to the nearest dollar.*

Step 2 *Find 10% of this amount by moving the decimal point one place to the left.*

Step 3 *Take half of the amount obtained in Step 2 and add it to the result of Step 2.*

These steps will give approximately 15% of the bill. The amount obtained in Step 3 is 5%, and

$$10\% + 5\% = 15\%.$$

Use the method above to find an approximation of 15% for each restaurant bill.

109. $29.57   **110.** $38.32

111. $5.15   **112.** $7.89

For Excellent Service *Suppose that you get extremely good service and decide to tip 20%. You can use the first two steps listed earlier and then, in Step 3, double the amount you obtained in Step 2. Use this method to find an approximation of 20% for each restaurant bill.*

113. $59.96   **114.** $40.24

115. $180.43   **116.** $199.86

Postage Stamp Pricing *Refer to the margin note on decimal point abuse. At one time, the United States Postal Service sold rolls of 33-cent stamps that featured fruit berries. One such stamp is suggested on the left. On the right is a photo of the pricing information found on the cellophane wrapper of such a roll.*

100 STAMPS PSA
.33¢ ea. TOTAL $33.00
FRUIT BERRIES
ITEM 7757
BCA

117. Look at the second line of the pricing information. According to the price listed *per stamp*, how many stamps should you be able to purchase for one cent?

118. The total price listed is the amount the Postal Service actually charges. If you were to multiply the listed price *per stamp* by the number of stamps, what should the total price be?

Pricing of Pie and Coffee *The photos here were taken at a flea market near Natchez, Mississippi. The handwritten signs indicate that a piece of pie costs .10¢ and a cup of coffee ("ffee") costs .5¢. Assuming these are the actual prices, answer the questions.*

119. How much will 10 pieces of pie and 10 cups of coffee cost?

120. How much will 20 pieces of pie and 10 cups of coffee cost?

121. How many pieces of pie can you get for $1.00?

122. How many cups of coffee can you get for $1.00?

Percents in Movies *Answer each of these by watching a movie.*

123. In the 1967 movie *The Producers*, Leo Bloom (Gene Wilder) and Max Bialystock (Zero Mostel) scheme to make a fortune by overfinancing what they think will be a Broadway flop. After enumerating the percent of profits all of Max's little old ladies have been offered in the production, reality sets in. Watch the movie to see what percent of the profits was sold.

124. There are several appearances of percent in the 1971 movie *Willy Wonka and the Chocolate Factory*. In one of them, upon preparing a mixture in his laboratory, Willy Wonka (Gene Wilder) states an impossible percent analysis as he drinks his latest concoction. Watch the movie to see what percent the ingredients total up to be.

Chapter 6 SUMMARY

KEY TERMS

6.1

natural numbers
whole numbers
number line
origin
negative numbers
positive numbers
signed numbers
integers
graph
rational numbers
irrational numbers
real numbers
additive inverses
  (negatives, opposites)
absolute value

6.2

sum
addends (terms)
difference
minuend
subtrahend
product
factors
quotient
dividend (numerator)
divisor (denominator)
positive change
negative change

6.3

extremes
means
cross-product test (for
  equality of fractions)
least common
  denominator (LCD)
mixed number
arithmetic mean
  (average)
terminating decimal
repeating decimal

6.4

irrational number
radicand
like radicals
Golden Rectangle
Golden Ratio

6.5

factors
product
dividend
divisor
quotient
percent

NEW SYMBOLS

$\{x \mid x \text{ has a certain property}\}$ set-builder notation

$a = b$ a is equal to b

$a < b$ a is less than b

$a \leq b$ a is less than or equal to b

$a > b$ a is greater than b

$a \geq b$ a is greater than or equal to b

$|x|$ absolute value of x

π irrational number $\pi \approx 3.14159$

ϕ irrational number $\phi \approx 1.618$

e irrational number $e \approx 2.71828$

%   percent

¢   cents

TEST YOUR WORD POWER

See how well you have learned the vocabulary in this chapter.

1. The **absolute value** of a real number is
   A. the same as the opposite of the number in all cases.
   B. the same as the number itself in all cases.
   C. never zero.
   D. its undirected distance from zero on the number line.

2. The **sum** of two numbers is the result obtained by
   A. addition.
   B. subtraction.
   C. multiplication.
   D. division.

3. The **identity element for multiplication** is
   A. 0.   **B.** 1.   **C.** −1.   **D.** −a, for the real number a.

4. An example of an **irrational number** is
   A. ϕ.   **B.** e.   **C.** π.   **D.** all of these.

5. An example of a rational number that has a **terminating decimal** representation is
   A. $\frac{1}{3}$.   **B.** $\frac{1}{7}$.   **C.** $\frac{1}{2}$.   **D.** $\frac{1}{9}$.

6. The irrational number π represents the
   A. circumference of a circle divided by its radius.
   B. radius of a circle divided by its diameter.
   C. diameter of a circle divided by its radius.
   D. circumference of a circle divided by its diameter.

ANSWERS
1. D **2.** A **3.** B **4.** D **5.** C **6.** D

QUICK REVIEW

Concepts

Examples

 6.1 REAL NUMBERS, ORDER, AND ABSOLUTE VALUE

Sets of Numbers

Natural Numbers　$\{1, 2, 3, 4, \dots\}$

Whole Numbers　$\{0, 1, 2, 3, 4, \dots\}$

Integers　$\{\dots, -2, -1, 0, 1, 2, \dots\}$

Rational Numbers
$\{x \mid x$ is a quotient of two integers, with denominator not equal to $0\}$

Irrational Numbers
$\{x \mid x$ is a number on the number line that is not rational$\}$

Real Numbers
$\{x \mid x$ is a number that can be represented by a point on a number line$\}$

$10, 25, 143$	Natural numbers
$0, 8, 47$	Whole numbers
$-22, -7, 0, 4, 9$	Integers
$-\dfrac{2}{3}, -0.14, 0, \dfrac{15}{8},$ $6, 0.33333\dots, \sqrt{4}$	Rational numbers
$-\sqrt{22}, \sqrt{3}, \pi$	Irrational numbers
$-3, -\dfrac{2}{7}, 0.7, \pi, \sqrt{11}$	Real numbers

Order in the Real Numbers
Suppose that a and b are real numbers.

$a = b$　if a and b are represented by the same point on a number line.

$a < b$　if the graph of a lies to the left of the graph of b.

$a > b$　if the graph of a lies to the right of the graph of b.

$$2 = \tfrac{4}{2} \qquad 0 < 1 \qquad 2 > -1$$

Law of Trichotomy
For any real numbers a and b, one and only one of the following holds true.

$$a = b, \qquad a < b, \qquad a > b$$

Additive Inverse
The additive inverse of a is $-a$. For all a,

$$a + (-a) = (-a) + a = 0.$$

$$-(-5) = 5 \qquad -\left(-\dfrac{2}{3}\right) = \dfrac{2}{3}$$

$$5 + (-5) = 0$$

Absolute Value

$$|x| = \begin{cases} x & \text{if } x \geq 0 \\ -x & \text{if } x < 0 \end{cases}$$

$$|-1| = 1 \qquad |0| = 0 \qquad |2| = 2$$

6.2 OPERATIONS, PROPERTIES, AND APPLICATIONS OF REAL NUMBERS

Adding Real Numbers

Like Signs　Add the absolute values. The sum has the same sign as the given numbers.

Unlike Signs　Find the absolute values of the numbers, and subtract the lesser absolute value from the greater. The sum has the same sign as the number with the greater absolute value.

Add.
$$9 + 4 = 13$$
$$-8 + (-5) = -13$$
$$7 + (-12) = -5$$
$$-5 + 13 = 8$$

Subtracting Real Numbers
For all real numbers a and b,

$$a - b = a + (-b).$$

Subtract.
$$-3 - 4 = -3 + (-4) = -7$$
$$-2 - (-6) = -2 + 6 = 4$$
$$13 - (-8) = 13 + 8 = 21$$

Concepts	Examples

Multiplying and Dividing Real Numbers
Like Signs   The product or quotient of two numbers with like signs is positive.

Multiply or divide.

$6 \cdot 5 = 30$ $-7(-8) = 56$ $\dfrac{20}{4} = 5$ $\dfrac{-24}{-6} = 4$

Unlike Signs   The product or quotient of two numbers with different signs is negative.

$-6(5) = -30$ $6(-5) = -30$ $\dfrac{-18}{9} = -2$ $\dfrac{49}{-7} = -7$

Division Involving 0

$\dfrac{a}{0}$ is undefined for all a. $\dfrac{0}{a} = 0$ for all nonzero a.

$\dfrac{5}{0}$ is undefined. $\dfrac{0}{5} = 0$

Order of Operations
If parentheses or square brackets are present:

Step 1   Work separately above and below any fraction bar.

Step 2   Use the rules below within each set of parentheses or square brackets. Start with the innermost set and work outward.

If no parentheses or brackets are present:

Step 1   Apply any exponents.

Step 2   Do any multiplications or divisions in the order in which they occur, working from left to right.

Step 3   Do any additions or subtractions in the order in which they occur, working from left to right.

Simplify.

$(-6)[2^2 - (3 + 4)] + 3$ Work inside the innermost parentheses.

$= (-6)[2^2 - 7] + 3$

$= (-6)[4 - 7] + 3$ Work inside the brackets.

$= (-6)[-3] + 3$

$= 18 + 3$ Multiply.

$= 21$ Add.

Properties of Real Numbers
For real numbers a, b, and c, the following properties hold.

Closure Properties

$a + b$ and $a \cdot b$ are real numbers.

$3 + 4$ and $3 \cdot 4$ are real numbers.

Commutative Properties

$a + b = b + a$
$a \cdot b = b \cdot a$

$7 + (-1) = -1 + 7$
$5(-3) = (-3)5$

Associative Properties

$(a + b) + c = a + (b + c)$
$(a \cdot b) \cdot c = a \cdot (b \cdot c)$

$(3 + 4) + 8 = 3 + (4 + 8)$
$[-2(6)]4 = -2[(6)4]$

Identity Properties

$a + 0 = a$ $0 + a = a$
$a \cdot 1 = a$ $1 \cdot a = a$

$-7 + 0 = -7$ $0 + (-7) = -7$
$9 \cdot 1 = 9$ $1 \cdot 9 = 9$

Inverse Properties

$a + (-a) = 0$ $-a + a = 0$
$a \cdot \dfrac{1}{a} = 1$ $\dfrac{1}{a} \cdot a = 1$ $(a \ne 0)$

$7 + (-7) = 0$ $-7 + 7 = 0$
$-2\left(-\dfrac{1}{2}\right) = 1$ $-\dfrac{1}{2}(-2) = 1$

Distributive Property

$a \cdot (b + c) = a \cdot b + a \cdot c$
$(b + c) \cdot a = b \cdot a + c \cdot a$

$5(4 + 2) = 5(4) + 5(2)$
$(4 + 2)5 = 4(5) + 2(5)$

Concepts

Examples

6.3 RATIONAL NUMBERS AND DECIMAL REPRESENTATION

Fundamental Property of Rational Numbers

$$\frac{a \cdot k}{b \cdot k} = \frac{a}{b} \quad (b \neq 0, k \neq 0)$$

Write $\frac{8}{12}$ in lowest terms.

$$\frac{8}{12} = \frac{2 \cdot 4}{3 \cdot 4} = \frac{2}{3}$$

Cross-Product Test for Equality of Rational Numbers

$$\frac{a}{b} = \frac{c}{d} \quad \text{if and only if} \quad a \cdot d = b \cdot c \quad (b \neq 0, d \neq 0)$$

Is $\frac{25}{36} = \frac{5}{6}$ a true statement?

$$25 \cdot 6 = 150 \quad \text{and} \quad 36 \cdot 5 = 180$$

Because $150 \neq 180$, the statement is false.

Adding and Subtracting Rational Numbers

$$\frac{a}{b} + \frac{c}{d} = \frac{ad + bc}{bd} \quad \text{and} \quad \frac{a}{b} - \frac{c}{d} = \frac{ad - bc}{bd}$$

In practice, we usually find the least common denominator to add and subtract fractions.

Add or subtract.

$$\frac{2}{5} + \frac{7}{5} = \frac{2 + 7}{5} = \frac{9}{5}, \quad \text{or} \quad 1\frac{4}{5}$$

$$\frac{2}{3} - \frac{1}{2} = \frac{4}{6} - \frac{3}{6} = \frac{1}{6} \quad \text{6 is the LCD.}$$

Multiplying and Dividing Rational Numbers

$$\frac{a}{b} \cdot \frac{c}{d} = \frac{ac}{bd} \quad (b \neq 0, d \neq 0)$$

$$\frac{a}{b} \div \frac{c}{d} = \frac{a}{b} \cdot \frac{d}{c} = \frac{ad}{bc} \quad (b \neq 0, c \neq 0, d \neq 0)$$

Multiply or divide.

$$\frac{4}{3} \cdot \frac{5}{6} = \frac{4 \cdot 5}{3 \cdot 6} = \frac{20}{18} = \frac{10}{9}, \quad \text{or} \quad 1\frac{1}{9}$$

$$\frac{6}{5} \div \frac{1}{4} = \frac{6}{5} \cdot \frac{4}{1} = \frac{24}{5}, \quad \text{or} \quad 4\frac{4}{5}$$

Density Property of the Rational Numbers
If r and t are distinct rational numbers, with $r < t$, then there exists a rational number s such that

$$r < s < t.$$

Find the arithmetic mean (average) of $\frac{2}{3}$ and $\frac{3}{4}$.

$$\frac{2}{3} + \frac{3}{4} = \frac{8}{12} + \frac{9}{12} = \frac{17}{12} \quad \text{Add.}$$

$$\frac{1}{2} \cdot \frac{17}{12} = \frac{17}{24} \quad \text{Find half of the sum.} \\ \leftarrow \text{Average of } \frac{2}{3} \text{ and } \frac{3}{4}$$

The decimal representation of a rational number will either terminate or will repeat indefinitely in a "block" of digits.

$$\frac{3}{16} = 0.1875 \quad \text{Terminating decimal}$$

$$\frac{2}{3} = 0.\overline{6}, \text{ or } 0.666\ldots \quad \text{Repeating decimal}$$

6.4 IRRATIONAL NUMBERS AND DECIMAL REPRESENTATION

Decimal Representation
The decimal for an irrational number neither terminates nor repeats.

$$\sqrt{2}, \quad \sqrt{10}, \quad 0.10110111011110\ldots$$

Approximations
Approximations for irrational numbers (for example, square roots of whole numbers that are not perfect squares) can be found with a calculator.

$$\sqrt{2} \approx 1.414213562 \qquad \sqrt{10} \approx 3.16227766$$

Product and Quotient Rules for Square Roots

$$\sqrt{a} \cdot \sqrt{b} = \sqrt{a \cdot b} \quad (a \geq 0, b \geq 0)$$

$$\frac{\sqrt{a}}{\sqrt{b}} = \sqrt{\frac{a}{b}} \quad (a \geq 0, b > 0)$$

Simplify.

$$\sqrt{54} = \sqrt{9 \cdot 6} = \sqrt{9} \cdot \sqrt{6} = 3\sqrt{6}$$

$$\frac{\sqrt{36}}{\sqrt{4}} = \sqrt{\frac{36}{4}} = \sqrt{9} = 3$$

Concepts	Examples

Conditions for a Simplified Square Root Radical

1. The number under the radical (radicand) has no factor (except 1) that is a perfect square.

2. The radicand has no fractions.

3. No denominator contains a radical.

Simplify.

$$\sqrt{\frac{5}{6}} = \frac{\sqrt{5}}{\sqrt{6}} = \frac{\sqrt{5}\cdot\sqrt{6}}{\sqrt{6}\cdot\sqrt{6}} = \frac{\sqrt{30}}{6}$$

Three Important Irrational Numbers

$\pi \approx 3.14159265358979323846264338327\mathbf{9}$

$\phi = \dfrac{1 + \sqrt{5}}{2} \approx 1.61803398874989484820458683436\mathbf{5}$

$e \approx 2.71828182845904523536028747135\mathbf{3}$

The rational number $\frac{355}{113}$ gives an excellent approximation for the irrational number π.

The Golden Ratio ϕ appears throughout nature.

The irrational number e is important in higher mathematics.

6.5 APPLICATIONS OF DECIMALS AND PERCENTS

Adding and Subtracting Decimals
To add or subtract decimal numbers, line up the decimal points in a column and perform the operation.

Add or subtract.

$1.2 + 36.158 + 9.26$

$$\begin{array}{r} 1.200 \\ 36.158 \\ \underline{9.260} \\ 46.618 \end{array}$$

$93.86 - 42.9142$

$$\begin{array}{r} 93.8600 \\ -42.9142 \\ \hline 50.9458 \end{array}$$ Attach zeros as place holders.

Multiplying and Dividing Decimals

Multiplication Multiply in the same manner as integers are multiplied. The number of decimal places to the right of the decimal point in the product is the *sum* of the numbers of places to the right of the decimal points in the factors.

Division Move the decimal point to the right the same number of places in the divisor and the dividend so as to obtain a whole number in the divisor. Divide in the same manner as integers are divided. The number of decimal places to the right of the decimal point in the quotient is the same as the number of places to the right of the decimal point in the dividend.

Multiply or divide.

51.6×2.3

$$\begin{array}{r} 51.6 \\ \underline{2.3} \\ 1548 \\ \underline{1032} \\ 118.68 \end{array}$$

$35.38 \div 6.1$

$$\begin{array}{r} 5.8 \\ 61.\overline{)353.8} \\ \underline{305} \\ 488 \\ \underline{488} \\ 0 \end{array}$$

Rules for Rounding
See pages 276–277.

Round 745.2935 to the given place value.

hundreds: 700   tenths: 745.3

ones or units: 745   thousandths: 745.294

Percent
The word **percent** means **"per hundred."** The symbol % represents "percent."

$$1\% = \frac{1}{100} = 0.01$$

Converting between Decimals and Percents
To convert a percent to a decimal, drop the percent symbol and move the decimal point two places to the left, inserting zeros as placeholders if necessary.

Convert 0.8% to a decimal.

$$0.8\% = 0.008$$

To convert a decimal to a percent, move the decimal point two places to the right, inserting zeros as placeholders if necessary, and attach the percent symbol.

Convert 0.8 to a percent.

$$0.8 = 0.80 = 80\%$$

Concepts	*Examples*
Converting a Fraction to a Percent To convert a fraction to a percent, convert the fraction to a decimal, and then convert the decimal to a percent.	Convert $\frac{11}{4}$ to a percent. $$\frac{11}{4} = 2\frac{3}{4} = 2.75 = 275\%$$
Finding Percent Increase or Decrease To find the percent increase from a to b, where $b > a$, subtract a from b, and divide this result by a. Convert to a percent. To find the percent decrease from a to b, where $b < a$, subtract b from a, and divide this result by a. Convert to a percent.	The sales of a textbook decreased from \$3,500,000 in its third edition to \$2,975,000 in its fourth edition. What was the percent decrease? $$\text{Percent decrease} = \frac{\$3,500,000 - \$2,975,000}{\$3,500,000}$$ $$= \frac{\$525,000}{\$3,500,000}$$ $$= 0.15$$ The percent decrease was 15%.

Chapter 6 TEST

1. Consider $\{-4, -\sqrt{5}, -\frac{3}{2}, -0.5, 0, \sqrt{3}, 4.1, 12\}$. List the elements of the set that belong to each of the following.

   (a) natural numbers

   (b) whole numbers

   (c) integers

   (d) rational numbers

   (e) irrational numbers

   (f) real numbers

2. Match each set in (a)–(d) with the correct set-builder notation description in A–D.

   (a) $\{\ldots, -4, -3, -2, -1\}$

   (b) $\{3, 4, 5, 6, \ldots\}$

   (c) $\{1, 2, 3, 4, \ldots\}$

   (d) $\{-12, \ldots, -2, -1, 0, 1, 2, \ldots, 12\}$

   A. $\{x \mid x$ is an integer with absolute value less than or equal to 12$\}$

   B. $\{x \mid x$ is an integer greater than 2.5$\}$

   C. $\{x \mid x$ is a negative integer$\}$

   D. $\{x \mid x$ is a positive integer$\}$

3. Write *true* or *false* for each statement.

   (a) The absolute value of a number must be positive.

   (b) $|-7| = -(-7)$

   (c) $\frac{2}{5}$ is an example of a real number that is not an integer.

   (d) Every real number is either positive or negative.

Perform the indicated operations. Use the order of operations as necessary.

4. $6^2 - 4(9 - 1)$

5. $\dfrac{(-8 + 3) - (5 + 10)}{7 - 9}$

6. $(-3)(-2) - [5 + (8 - 10)]$

7. $5(-6) - (3 - 8)^2 + 12$

8. **Temperature Extremes** The record high temperature in the United States was 134° Fahrenheit, recorded at Death Valley, California, in 1913. The record low was $-79.8°$F, at Prospect Creek, Alaska, in 1971. How much greater was the highest temperature than the lowest temperature? (Data from *The World Almanac and Book of Facts.*)

9. **Altitude of a Plane** The surface of the Dead Sea has altitude 1299 ft below sea level. A pilot is flying 80 ft above that surface. How much altitude must she gain to clear a 3852-ft pass by 225 ft? (Data from *The World Almanac and Book of Facts.*)

10. Match each statement in (a)–(f) with the property that justifies it in A–F.

(a) $7 \cdot (8 \cdot 5) = (7 \cdot 8) \cdot 5$

(b) $3x + 3y = 3(x + y)$

(c) $8 \cdot 1 = 1 \cdot 8 = 8$

(d) $7 + (6 + 9) = (6 + 9) + 7$

(e) $9 + (-9) = -9 + 9 = 0$

(f) $5 \cdot 8$ is a real number.

A. Distributive property

B. Identity property

C. Closure property

D. Commutative property

E. Associative property

F. Inverse property

11. *Basketball Shot Statistics*   Six players on a local high school basketball team had shooting statistics as shown in the table below. Answer each question, using estimation skills as necessary.

Player	Field Goal Attempts	Field Goals Made
Priya	40	13
Jackie	10	4
Mabel	20	8
Mary	6	4
Charlene	7	2
Janie	7	6

(a) Which players made more than half of her attempts?

(b) Which players made just less than $\frac{1}{3}$ of her attempts?

(c) Which player made exactly $\frac{2}{3}$ of her attempts?

(d) Which two players made the same fractional parts of her attempts? What was the fractional part, reduced to lowest terms?

(e) Which player made the greatest fractional part of her attempts?

Perform each operation. Write your answer in lowest terms.

12. $\dfrac{3}{16} + \dfrac{1}{2}$

13. $\dfrac{9}{20} - \dfrac{3}{32}$

14. $\dfrac{3}{8} \cdot \left(-\dfrac{16}{15}\right)$

15. $\dfrac{7}{9} \div 1\dfrac{3}{4}$

16. Convert each rational number into a repeating or terminating decimal. Use a calculator if your instructor so allows.

(a) $\dfrac{9}{20}$

(b) $\dfrac{5}{12}$

17. Convert each decimal into a quotient of integers, reduced to lowest terms.

(a) 0.72

(b) $0.\overline{58}$

18. Identify each number as rational or irrational.

(a) $\sqrt{10}$

(b) $\sqrt{16}$

(c) 0.01

(d) $0.\overline{01}$

(e) $0.0101101110\ldots$

(f) π

For each of the following, (a) use a calculator to find a decimal approximation and (b) simplify the radical according to the guidelines in this chapter.

19. $\sqrt{150}$

20. $\dfrac{13}{\sqrt{7}}$

21. $2\sqrt{32} - 5\sqrt{128}$

22. *Rate of Return on an Investment*   If an investment of P dollars grows to A dollars in two years, the annual rate of return on the investment is given by

$$r = \frac{\sqrt{A} - \sqrt{P}}{\sqrt{P}}.$$

What is the rate if an investment of \$50,000 grows to \$58,320?

23. Work each of the following using either a calculator or paper-and-pencil methods, as directed by your instructor.

(a) $4.6 + 9.21$

(b) $12 - 3.725 - 8.59$

(c) $86(0.45)$

(d) $236.439 \div (-9.73)$

24. Round 346.0449 to the given place values.

(a) tens

(b) hundredths

(c) thousandths

25. *Sale Price*   A dress that originally sold for \$100.00 is discounted 40% for a sale. Then the owner decides to offer an additional 10% off of the sale price. What is the final price of the dress?

26. *Sales of Books*   Use estimation techniques to answer the following: In 2017, Carol sold \$300,000 worth of books. In 2018, she sold \$900,000. Her 2018 sales were _____ of her 2017 sales.

A. 30%   B. $33\dfrac{1}{3}$%   C. 3%   D. 300%

27. *Foreign-Born Population* Approximately 40,107,000 people living in the United States in 2013 were born in other countries. The graph below gives the percent for each region of birth for these people.

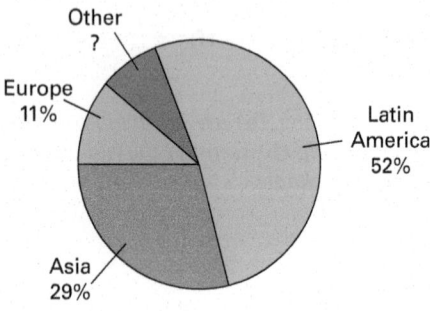

U.S. Foreign-Born Population by Region of Birth

Other ?
Europe 11%
Latin America 52%
Asia 29%

Data from U.S. Census Bureau.

(a) What percent of the U.S. foreign-born population was from a region other than the ones specified?

(b) What percent of the foreign-born population was from Latin America or Asia?

(c) About how many of the foreign-born people were born in Europe?

28. Consider the figure.

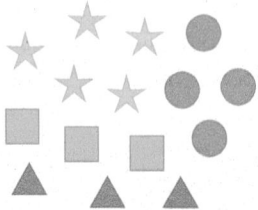

(a) What percent of the total number of shapes are circles?

(b) What percent of the total number of shapes are not stars?

29. *Creature Comforts* From a list of "everyday items" often taken for granted, adults were asked to indicate those items they wouldn't want to live without. Complete the results shown in the table if 2400 adults were surveyed.

Item	Percent That Wouldn't Want to Live Without	Number That Wouldn't Want to Live Without
Toilet paper	69%	
Zipper	42%	
Frozen food		384
Self-stick note pads		144

(Other items included tape, hairspray, pantyhose, paper clips, and Velcro.)

Data from Market Facts for Kleenex Cottonelle.

30. *Child's Drug Dosage* If D represents the usual adult dose of a drug, the corresponding child's dose C is calculated by the following formula.

$$C = \frac{\text{body surface area in square meters}}{1.7} \times D$$

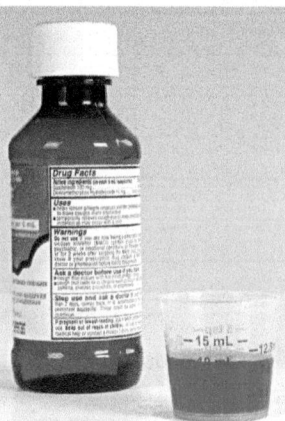

Determine the appropriate dose for a child with body surface area 0.85 m² if the usual adult dose is 150 mg.

7 The Basic Concepts of Algebra

The word "algebra" has its roots in a ninth-century work by **Mohammad ibn Musa al-Khowarizmi** titled *Hisab al-jabr w'al muqualbalah,* which translates as

"the science of transposition and cancellation."

From Latin versions of this text, *al-jabr* became the broad term covering the art of equation solving. The word "algorithm" is derived from the name of the author.

The concepts of algebra date back to the Babylonians of 2000 B.C. The Egyptians also worked problems in algebra, but the problems were not as complex as those of the Babylonians. Many Hindu and Greek works on mathematics were preserved only because Muslim scholars from about 750 to 1250 translated them. The Hindus developed methods for solving problems involving interest, discounts, and partnerships.

In this chapter we present the basic algebraic concepts applied in solving some of the simple equations of algebra.

7.1 **LINEAR EQUATIONS**

OBJECTIVES

1 Solve a linear equation.

2 Identify a linear equation as a conditional equation, an identity, or a contradiction.

3 Solve a literal equation or a formula for a specified variable.

4 Use a linear model in an application.

Solving Linear Equations

An **algebraic expression** involves only the basic operations of addition, subtraction, multiplication, division (except by 0), raising to powers, and taking roots on any collection of variables and numbers.

$$5, \quad 8x + 9, \quad \sqrt{y} + 4, \quad \text{and} \quad \frac{x^3 y^8}{z} \quad \text{Examples of algebraic expressions}$$

An **equation** is a statement that two algebraic expressions are equal. A *linear equation in one variable* involves only real numbers and one variable.

$$x + 1 = -2, \quad x - 3 = 5, \quad \text{and} \quad 2x + 5 = 10 \quad \text{Examples of linear equations}$$

LINEAR EQUATION IN ONE VARIABLE

An equation in the variable x is a **linear equation** if it can be written in the form

$$Ax + B = C,$$

where A, B, and C are real numbers, with $A \neq 0$.

A linear equation in one variable is also called a **first-degree equation,** because the greatest power on the variable is one.

If the variable in an equation is replaced by a real number that makes the statement true, then that number is a **solution** of the equation. For example, 8 is a solution of the equation

$$x - 3 = 5,$$

because replacing x with 8 gives a true statement. An equation is solved by finding its **solution set,** the set of all solutions. The solution set of the equation $x - 3 = 5$ is $\{8\}$.

Equivalent equations are equations with the same solution set. Equations generally are solved by starting with a given equation and producing a sequence of simpler equivalent equations. For example,

$$8x + 1 = 17, \quad 8x = 16, \quad \text{and} \quad x = 2 \quad \text{Equivalent equations}$$

are equivalent equations because each has the same solution set, $\{2\}$. We use the addition and multiplication properties of equality to produce equivalent equations.

ADDITION PROPERTY OF EQUALITY

For all real numbers A, B, and C, the equations

$$A = B \quad \text{and} \quad A + C = B + C$$

are equivalent. (The same number may be added to both sides of an equation without changing the solution set.)

algebrista, algebra   In Spain under Muslim rule, the word **algebrista** referred to the person who restored (reset) broken bones. Signs outside barber shops read *Algebrista y Sangrador* (bonesetter and bloodletter). Such services were part of the barber's trade. The traditional red-and-white striped barber pole symbolizes blood and bandages.

MULTIPLICATION PROPERTY OF EQUALITY

For all real numbers A, B, and C, where $C \neq 0$, the equations

$$A = B \quad \text{and} \quad AC = BC$$

are equivalent. (Both sides of an equation may be multiplied by the same non-zero number without changing the solution set.)

Because subtraction and division are defined in terms of addition and multiplication, respectively, we are able to subtract the same number from both sides of the equation and divide both sides by the same nonzero number without affecting the solution set.

The following steps can be used to solve a linear equation in one variable.

SOLVING A LINEAR EQUATION IN ONE VARIABLE

Step 1   **Clear fractions.**   Eliminate any fractions by multiplying both sides of the equation by a common denominator.

Step 2   **Simplify each side separately.**   Use the distributive property to clear parentheses, and combine like terms as needed.

Step 3   **Isolate the variable terms on one side.**   Use the addition property of equality to transform the equation so that all terms with variables are on one side and all terms without variables are on the other.

Step 4   **Transform so that the coefficient of the variable is 1.**   Use the multiplication property of equality to obtain an equation with only the variable (with coefficient 1) on one side.

Step 5   **Check.**   Substitute the solution into the original equation.

EXAMPLE 1   **Using the Distributive Property to Solve a Linear Equation**

Solve $2(x - 5) + 3x = x + 6$.

Solution

Step 1   Because there are no fractions in this equation, Step 1 does not apply.

Step 2   Use the distributive property to simplify and combine terms on the left.

$$2(x - 5) + 3x = x + 6 \quad \text{Original equation}$$
$$2x - 10 + 3x = x + 6 \quad \text{Distributive property}$$
$$5x - 10 = x + 6 \quad \text{Combine like terms.}$$

Step 3   Next, use the addition property of equality.

$$5x - 10 + 10 = x + 6 + 10 \quad \text{Add 10.}$$
$$5x = x + 16 \quad \text{Combine like terms.}$$
$$5x - x = x + 16 - x \quad \text{Subtract } x.$$
$$4x = 16 \quad \text{Combine like terms.}$$

The problem-solving strategy of guessing and checking, discussed earlier, was actually used by the early Egyptians in equation solving. This method, called the **Rule of False Position,** involved making an initial guess at the solution of an equation, and then following up with an adjustment in the likely event that the guess was incorrect. For example (using our modern notation), if the equation

$$6x + 2x = 32$$

was to be solved, an initial guess might have been $x = 3$. Substituting 3 for x gives

$$6(3) + 2(3) \overset{?}{=} 32$$
$$18 + 6 \overset{?}{=} 32$$
$$24 = 32. \quad \text{False}$$

The guess, 3, gives a value (24) that is smaller than the desired value (32). Since 24 is $\frac{3}{4}$ of 32, the guess, 3, is $\frac{3}{4}$ of the actual solution. The actual solution, therefore, must be 4, since 3 is $\frac{3}{4}$ of 4.

Use the methods explained in this section to verify this result.

Step 4 Use the multiplication property of equality to isolate x on the left.

$$\frac{4x}{4} = \frac{16}{4} \quad \text{Divide by 4.}$$

$$x = 4 \quad \text{Simplify.}$$

Step 5 *Check* by substituting 4 for x in the original equation.

$$2(x - 5) + 3x = x + 6 \quad \text{Original equation}$$
$$2(4 - 5) + 3(4) \overset{?}{=} 4 + 6 \quad \text{Let } x = 4.$$
$$2(-1) + 3(4) \overset{?}{=} 10 \quad \text{Work inside the parentheses on the left, and add on the right.}$$
$$-2 + 12 \overset{?}{=} 10 \quad \text{Multiply.}$$
$$10 = 10 \quad \checkmark \quad \text{True}$$

A true statement indicates that $\{4\}$ is the solution set.

EXAMPLE 2 **Solving a Linear Equation with Fractions**

Solve $\dfrac{x + 7}{6} + \dfrac{2x - 8}{2} = -4$.

Solution

Step 1
$$6\left(\frac{x + 7}{6} + \frac{2x - 8}{2}\right) = 6(-4) \quad \text{Multiply each side by the least common denominator (LCD), 6, to eliminate the fractions.}$$

Step 2
$$6\left(\frac{x + 7}{6}\right) + 6\left(\frac{2x - 8}{2}\right) = 6(-4) \quad \text{Distributive property}$$

Multiply each term by 6.

$$x + 7 + 3(2x - 8) = -24 \quad \text{Multiply.}$$
$$x + 7 + 6x - 24 = -24 \quad \text{Distributive property}$$
$$7x - 17 = -24 \quad \text{Combine like terms.}$$

Step 3
$$7x - 17 + 17 = -24 + 17 \quad \text{Add 17.}$$
$$7x = -7 \quad \text{Combine like terms.}$$

Step 4
$$\frac{7x}{7} = \frac{-7}{7} \quad \text{Divide by 7.}$$
$$x = -1 \quad \text{Simplify.}$$

Step 5 *Check:*
$$\frac{x + 7}{6} + \frac{2x - 8}{2} = -4 \quad \text{Original equation}$$
$$\frac{-1 + 7}{6} + \frac{2(-1) - 8}{2} \overset{?}{=} -4 \quad \text{Let } x = -1.$$
$$\frac{6}{6} + \frac{-10}{2} \overset{?}{=} -4 \quad \text{Simplify each fraction.}$$
$$1 - 5 \overset{?}{=} -4 \quad \text{Simplify terms.}$$
$$-4 = -4 \quad \checkmark \quad \text{True}$$

The solution -1 checks, so the solution set is $\{-1\}$.

Sofia Kovalevskaya (1850–1891) was the most widely known Russian mathematician in the late nineteenth century. She did most of her work in the theory of **differential equations**— equations invaluable for expressing rates of change. For example, in biology, the rate of growth of a population, say of microbes, can be precisely stated by using differential equations.

Kovalevskaya studied privately because public lectures were not open to women. She eventually received a degree (1874) from the University of Göttingen, Germany. In 1884 she became a lecturer at the University of Stockholm and later was appointed professor of higher mathematics.

EXAMPLE 3 Solving a Linear Equation with Decimals

Solve $0.06x + 0.09(15 - x) = 0.07(15)$.

Solution

Because each decimal number is in hundredths, multiply both sides of the equation by 100. This is done by moving the decimal points two places to the right.

$$0.06x + 0.09(15 - x) = 0.07(15) \qquad \text{Original equation}$$

$$0.06x + 0.09(15 - x) = 0.07(15) \qquad \text{Multiply each term by 100.}$$

$$6x + 9(15 - x) = 7(15)$$

$$6x + 9(15) - 9x = 105 \qquad \text{Distributive property; multiply.}$$

$$6x + 135 - 9x = 105 \qquad \text{Multiply.}$$

$$-3x + 135 = 105 \qquad \text{Combine like terms.}$$

$$-3x + 135 - 135 = 105 - 135 \qquad \text{Subtract 135.}$$

$$-3x = -30 \qquad \text{Combine like terms.}$$

$$\frac{-3x}{-3} = \frac{-30}{-3} \qquad \text{Divide by } -3.$$

$$x = 10 \qquad \text{Simplify.}$$

Check to verify that the solution set is $\{10\}$.

Special Kinds of Linear Equations

The preceding equations had solution sets containing one element. For example,

$$2(x - 5) + 3x = x + 6 \quad \text{has solution set} \quad \{4\}.$$

Some equations have no solutions. Others have an infinite number of solutions.

EXAMPLE 4 Recognizing Conditional Equations, Identities, and Contradictions

Solve each equation. Decide whether it is a *conditional equation*, an *identity*, or a *contradiction*.

(a) $5x - 9 = 4(x - 3)$ **(b)** $5x - 15 = 5(x - 3)$ **(c)** $5x - 15 = 5(x - 4)$

Solution

(a)

$$5x - 9 = 4(x - 3) \qquad \text{Original equation}$$

$$5x - 9 = 4x - 12 \qquad \text{Distributive property}$$

$$5x - 9 - 4x = 4x - 12 - 4x \qquad \text{Subtract } 4x.$$

$$x - 9 = -12 \qquad \text{Combine like terms.}$$

$$x - 9 + 9 = -12 + 9 \qquad \text{Add 9.}$$

$$x = -3 \qquad \text{Solution set } \{-3\}$$

The solution set has one element, so $5x - 9 = 4(x - 3)$ is a **conditional equation.**

François Viète (1540–1603) was a lawyer at the court of Henry IV of France and studied equations. Viète simplified the notation of algebra and was among the first to use letters to represent numbers.

(b) $5x - 15 = 5(x - 3)$ Original equation

$5x - 15 = 5x - 15$ Distributive property

$0 = 0$ Subtract 5x and add 15.

The final line, $0 = 0$, indicates that the solution set is {all real numbers}, and the equation $5x - 15 = 5(x - 3)$ is an **identity.** (*Note:* The first step yielded $5x - 15 = 5x - 15$, which is true for all values of x, implying an identity there.)

(c) $5x - 15 = 5(x - 4)$ Original equation

$5x - 15 = 5x - 20$ Distributive property

$5x - 15 - 5x = 5x - 20 - 5x$ Subtract 5x.

$-15 = -20$ False

Because the result, $-15 = -20$, is *false*, the equation has no solution. The solution set is $\varnothing$, and the equation is a **contradiction.**

Table 1 summarizes this discussion.

Table 1 Types of Linear Equations

Type of Equation	Number of Solutions	Indication When Solving
Conditional	One	Final line is x = a number. (See **Example 4(a).**)
Identity	Infinite; solution set {all real numbers}	Final line is true, such as $0 = 0$. (See **Example 4(b).**)
Contradiction	None; solution set $\varnothing$	Final line is false, such as $-15 = -20$. (See **Example 4(c).**)

Literal Equations and Formulas

An equation involving *variables* (or letters), such as $cx + d = e$, is called a **literal equation.** The most useful examples of literal equations are *formulas.* The solution of a problem in algebra often depends on the use of a mathematical statement or **formula** in which more than one letter is used to express a relationship.

$$d = rt, \quad I = prt, \quad \text{and} \quad S = \frac{\mathscr{A}}{5T} + \frac{d}{20} \quad \text{Examples of formulas}$$

In some cases, a formula must be solved for one of its variables. This process is called **solving for a specified variable.** The steps used are similar to those used in solving linear equations.

When you are solving for a specified variable, the key is to treat that variable as if it were the only one. Treat all other variables like numbers (constants).

SOLVING FOR A SPECIFIED VARIABLE

Step 1   If the equation contains fractions, multiply both sides by the LCD to clear the fractions.

Step 2   Transform so that all terms with the specified variable are on one side and all terms without that variable are on the other side.

Step 3   Divide each side by the factor that is the coefficient of the specified variable.

EXAMPLE 5   Solving for a Specified Variable

To determine how much welded parts shrink in groove joints, the formula

$$S = \frac{\mathcal{A}}{5T} + \frac{d}{20}$$

is used. Here, S represents transverse shrinkage, $\mathcal{A}$ represents the cross-sectional area of the weld in square millimeters, T is the thickness of the plates in millimeters, and d is the root opening in millimeters. (*Source:* Saunders, Hal M., and Robert A. Carman. *Mathematics for the Trades—A Guided Approach*, 10th ed. Pearson, 2015.) Solve this welding formula for d.

Solution

$$S = \frac{\mathcal{A}}{5T} + \frac{d}{20} \qquad \text{Given formula}$$

$$20T \cdot S = 20T \left(\frac{\mathcal{A}}{5T} + \frac{d}{20} \right) \qquad \text{Multiply by the LCD of the fractions, } 20T.$$

$$20TS = 20T \left(\frac{\mathcal{A}}{5T} \right) + 20T \left(\frac{d}{20} \right) \qquad \text{Distributive property}$$

$$20TS = 4\mathcal{A} + Td \qquad \text{Multiply and simplify.}$$

$$20TS - 4\mathcal{A} = Td \qquad \text{Subtract } 4\mathcal{A}.$$

$$\frac{20TS - 4\mathcal{A}}{T} = d \qquad \text{Divide by } T.$$

Models

An equation or an inequality that expresses a relationship among various quantities is an example of a **mathematical model.** The relationship between the Fahrenheit and Celsius temperature scales is an example of a *linear model.*

EXAMPLE 6   Using the Formulas for Fahrenheit and Celsius

In the Fahrenheit temperature scale, water freezes at 32 degrees and boils at 212 degrees. The equivalent measures for the Celsius scale are 0 degrees and 100 degrees. Using elementary algebra, we can justify the formula

$$F = \frac{9}{5} C + 32$$

for converting Celsius C to Fahrenheit F.

(a) Find the Fahrenheit temperature that corresponds to 30 degrees Celsius.

(b) Solve the formula for C.

Solution

(a) Let $C = 30$ in the formula.

$$F = \frac{9}{5} (30) + 32 \qquad \text{Let } C = 30.$$

$$F = 54 + 32 \qquad \text{Multiply.}$$

$$F = 86 \qquad \text{Add.}$$

The Celsius temperature 30 degrees corresponds to 86 degrees Fahrenheit.

(b)

$$F = \frac{9}{5}C + 32 \qquad \text{Given formula}$$

$$F - 32 = \frac{9}{5}C \qquad \text{Subtract 32.}$$

$$\frac{5}{9}(F - 32) = \frac{5}{9}\left(\frac{9}{5}C\right) \qquad \text{Multiply by } \frac{5}{9}.$$

$$\frac{5}{9}(F - 32) = C \qquad \text{Simplify on the right.}$$

$$C = \frac{5}{9}(F - 32) \qquad \text{Rewrite.}$$

This final line is the form of the equation to convert Fahrenheit to Celsius.

7.1 EXERCISES

CONCEPT CHECK *Answer each question.*

1. Which equation is linear in x?

   A. $2x + 3 = 9x - 6$ **B.** $\dfrac{1}{x} = 5$

   C. $x^2 = 4$ **D.** $3x - \dfrac{4}{x} = 9$

2. Which choice is a solution of $-3x = 12$?

   A. 4   **B.** -4

   C. $\dfrac{1}{4}$ **D.** $-\dfrac{1}{4}$

3. What does the term "equivalent equations" mean?

4. What is the only solution of $6x = -6x$?

5. To solve the linear equation
$$\frac{1}{3}x + \frac{1}{2}x = \frac{1}{5},$$
we can begin by multiplying both sides by the least common denominator of all the fractional coefficients. What is this LCD?

6. To solve the linear equation
$$0.05x + 0.12(x + 5000) = 940,$$
we can multiply both sides by a power of 10 so that all coefficients are integers. What is the least power of 10 that will accomplish this?

Solve each equation.

7. $7x + 8 = 1$ **8.** $5x - 4 = 21$

9. $8 - 8x = -16$ **10.** $9 - 2x = 15$

11. $7x - 5x + 15 = x + 8$

12. $2x + 4 - x = 4x - 5$

13. $12x + 15x - 9 + 5 = -3x - 4$

14. $-4x + 5x - 8 + 4 = 6x - 4$

15. $2(x + 3) = -4(x + 1)$

16. $4(x - 9) = 8(x + 3)$

17. $3(2x + 1) - 2(x - 2) = 5$

18. $4(x - 2) + 2(x + 3) = 6$

19. $2x + 3(x - 4) = 2(x - 3)$

20. $6x - 3(5x + 2) = 4(1 - x)$

21. $6x - 4(3 - 2x) = 5(x - 4) - 10$

22. $-2x - 3(4 - 2x) = 2(x - 3) + 2$

23. $-[2x - (5x + 2)] = 2 + (2x + 7)$

24. $-[6x - (4x + 8)] = 9 + (6x + 3)$

25. $-3x + 6 - 5(x - 1) = -(2x - 4) - 5x + 5$

26. $4(x + 2) - 8x - 5 = -3x + 9 - 2(x + 6)$

27. $\dfrac{3x}{4} + \dfrac{5x}{2} = 13$

28. $\dfrac{8x}{3} - \dfrac{2x}{4} = -13$

29. $\dfrac{x - 8}{5} + \dfrac{8}{5} = -\dfrac{x}{3}$

30. $\dfrac{2x-3}{7} + \dfrac{3}{7} = -\dfrac{x}{3}$

31. $\dfrac{3x+7}{6} + \dfrac{x+7}{6} = \dfrac{x+6}{4}$

32. $\dfrac{2x+5}{5} = \dfrac{3x+1}{2} + \dfrac{-x+7}{2}$

33. $0.05x + 0.12(x+5000) = 940$

34. $0.09x + 0.13(x+300) = 61$

35. $0.02(50) + 0.08x = 0.04(50+x)$

36. $0.20(14{,}000) + 0.14x = 0.18(14{,}000+x)$

37. $0.05x + 0.10(200-x) = 0.45x$

38. $0.08x + 0.12(260-x) = 0.48x$

Decide whether each equation is conditional, an identity, or a contradiction. Give the solution set.

39. $-2x + 5x - 9 = 3(x-4) - 5$

40. $-6x + 2x - 11 = -2(2x-3) + 4$

41. $6x + 2(x-2) = 9x + 4$

42. $4x + 6(x-8) = 15x - 18$

43. $-11x + 4(x-3) + 6x = 4x - 12$

44. $3x - 5(x+4) + 9 = -11 + 15x$

45. $7[2 - (3+4x)] - 2x = -9 + 2(1-15x)$

46. $4[6 - (1+2x)] + 10x = 2(10-3x) + 8x$

Formulas from Trades and Occupations *The formulas here are found in various trades and occupations. Solve each formula for the specified variable. (Sources: Saunders, Hal M., and Robert A. Carman.* Mathematics for the Trades—A Guided Approach, *10th ed. Pearson, 2015; Timmons, Daniel L., and Catherine W. Johnson.* Math Skills for Allied Health Careers. *Pearson, 2008.)*

47. $p = \dfrac{wA}{150}$; for A (allied health)

48. $p = \dfrac{mA}{150}$; for m (allied health)

49. $\text{ACH} = \dfrac{60\ell}{v}$; for ℓ (energy efficiency)

50. $\text{HP} = \dfrac{tr}{5252}$; for r (automotive)

51. $\ell = a + b + \dfrac{1}{2}t$; for t (sheet metal)

52. $r = c + h + \dfrac{1}{4}s$; for s (forestry)

53. $r_1 x = r_2(\ell + x)$; for ℓ (electrical)

54. $v = (21.78 - 0.101a)h$; for a (allied health)

55. $\ell = -h^2 + 2rh$; for r (sheet metal)

56. $r = \dfrac{d^2 n}{2.5}$; for n (automotive)

57. $S = \dfrac{A}{5T} + \dfrac{d}{20}$; for A (welding)

58. $A = \dfrac{T(100S - 5d)}{20}$; for S (welding)

Mathematical Formulas *Solve each formula for the specified variable.*

59. $d = rt$; for t (distance)

60. $I = prt$; for r (simple interest)

61. $A = bh$; for b (area of a parallelogram)

62. $P = 2L + 2W$; for L (perimeter of a rectangle)

63. $P = a + b + c$; for a (perimeter of a triangle)

64. $V = LWH$; for W (volume of a rectangular solid)

65. $A = \dfrac{1}{2}bh$; for b (area of a triangle)

66. $C = 2\pi r$; for r (circumference of a circle)

67. $S = 2\pi rh + 2\pi r^2$; for h
(surface area of a right circular cylinder)

68. $A = \dfrac{1}{2}h(B+b)$; for B (area of a trapezoid)

69. $V = \dfrac{1}{3}\pi r^2 h$; for h (volume of a cone)

70. $V = \dfrac{1}{3}Bh$; for h (volume of a right pyramid)

Use the formulas in **Example 6** *to make the following degree conversions.*

71. 15 degrees Celsius to Fahrenheit

72. 25 degrees Celsius to Fahrenheit

73. -20 degrees Celsius to Fahrenheit

74. −40 degrees Celsius to Fahrenheit

75. 41 degrees Fahrenheit to Celsius

76. 59 degrees Fahrenheit to Celsius

77. −13 degrees Fahrenheit to Celsius

78. −40 degrees Fahrenheit to Celsius

FOR FURTHER THOUGHT

The Axioms of Equality

When we solve an equation, we must make sure that it remains "balanced"—that is, any operation that is performed on one side of the equation must also be performed on the other side in order to ensure that the set of solutions remains the same.

Underlying the rules for solving equations are the four axioms of equality listed below. For all real numbers a, b, and c, the following are true.

1. Reflexive axiom

$a = a$

2. Symmetric axiom

If $a = b$, then $b = a$.

3. Transitive axiom

If $a = b$ and $b = c$, then $a = c$.

4. Substitution axiom

If $a = b$, then a may replace b in any statement without affecting the truth or falsity of the statement.

A relation, such as equality, that satisfies the first three of these axioms (reflexive, symmetric, and transitive), is called an *equivalence relation*.

For Group or Individual Investigation

1. Give an example of an everyday relation that does not satisfy the symmetric axiom.

2. Does the transitive axiom hold in sports competition with the relation "defeats"?

3. Give an example of a relation that does not satisfy the transitive axiom.

7.2 APPLICATIONS OF LINEAR EQUATIONS

OBJECTIVES

1 Translate verbal expressions involving the operations of arithmetic into symbolic mathematical expressions.

2 Solve applications to find unknown numbers.

3 Solve applications involving mixture and simple interest.

4 Solve applications involving different denominations of money.

5 Solve applications involving motion.

Translating Words into Symbols

There are key words and phrases that translate into mathematical expressions involving addition, subtraction, multiplication, and division. The ability to make these translations is important when solving applications. The chart on the following page illustrates such translations.

The equality symbol, =, is often indicated by the word *is*. In fact, because equal mathematical expressions represent different names for the same number, words that indicate the idea of "sameness" translate as =. For example,

If the product of a number and 12 is decreased by 7, the result is 105

translates into the mathematical equation

$$12x - 7 = 105,$$

where x represents the unknown number. Note that

$$7 - 12x = 105$$

is incorrect because subtraction is not commutative.

TRANSLATING FROM WORDS TO MATHEMATICAL EXPRESSIONS

Verbal Expression	Mathematical Expression (where x and y are numbers)
Addition	
The **sum** of a number and 7	$x + 7$
6 **more than** a number	$x + 6$
3 **plus** a number	$3 + x$
24 **added to** a number	$x + 24$
A number **increased by** 5	$x + 5$
The **sum** of two numbers	$x + y$
Subtraction	
2 **less than** a number	$x - 2$
12 **minus** a number	$12 - x$
A number **decreased by** 12	$x - 12$
The **difference between** two numbers	$x - y$
A number **subtracted from** 10	$10 - x$
Multiplication	
16 **times** a number	$16x$
A number **multiplied by** 6	$6x$
$\frac{2}{3}$ **of** a number (as applied to fractions and percent)	$\frac{2}{3}x$
Twice (2 times) a number	$2x$
Triple (3 times) a number	$3x$
The **product** of two numbers	xy
Division	
The **quotient** of 8 and a number	$\frac{8}{x} \;\; (x \neq 0)$
A number **divided by** 13	$\frac{x}{13}$
The **ratio** of two numbers or the **quotient** of two numbers	$\frac{x}{y} \;\; (y \neq 0)$

Guidelines for Applications

To solve applied problems, the following six steps are helpful.

George Polya's problem-solving **procedure** can be adapted to applications of algebra as seen in the steps in the box.
Steps 1 and 2 make up the first stage of Polya's procedure (*Understand the Problem*),
Step 3 forms the second stage (*Devise a Plan*),
Step 4 comprises the third stage (*Carry Out the Plan*).
Steps 5 and 6 form the last stage (*Look Back*).

SOLVING AN APPLIED PROBLEM

Step 1 **Read** the problem carefully until you understand what is given and what is to be found.

Step 2 **Assign a variable** to represent the unknown value, using diagrams or tables as needed. Write down what the variable represents. If necessary, express any other unknown values in terms of the variable.

Step 3 **Write an equation** using the variable expression(s).

Step 4 **Solve** the equation.

Step 5 **State the answer.** Does it seem reasonable?

Step 6 **Check** the answer in the words of the *original* problem.

Finding Unknown Quantities

> **Problem-Solving Strategy**
>
> A common type of problem involves finding two quantities when the sum of the quantities is known. Choose a variable to represent one of the unknowns, and then represent the other quantity in terms of the same variable, using information from the problem. Then write an equation based on the words of the problem.

EXAMPLE 1 **Finding Numbers of Olympic Medals**

In the 2016 Summer Olympics in Rio de Janeiro, Brazil, the United States won 51 more medals than China. The two countries won a total of 191 medals. How many medals did each country win? (Data from *The World Almanac and Book of Facts.*)

Solution

Step 1 **Read** the problem. We are asked to find the number of medals each country won.

Step 2 **Assign a variable** to represent the number of medals for one of the countries.

$$\text{Let } x = \text{the number of medals China won.}$$

We must also find the number of medals for the United States. Because the United States had 51 more medals than China,

$$x + 51 = \text{the number of medals for the United States.}$$

Step 3 **Write an equation.** The sum of the numbers of medals is 191. Translate this to write the appropriate equation.

China's medals	+	U.S. medals	=	Total
↓		↓		↓
x	$+$	$(x + 51)$	$=$	191

Step 4 **Solve** the equation.

$$2x + 51 = 191 \qquad \text{Combine like terms.}$$
$$2x + 51 - 51 = 191 - 51 \qquad \text{Subtract 51.}$$
$$2x = 140 \qquad \text{Combine like terms.}$$
$$\frac{2x}{2} = \frac{140}{2} \qquad \text{Divide by 2.}$$

Don't stop here.

$$x = 70$$

Step 5 **State the answer.** The variable x represents the number of medals for China, so China had 70. Then the number of U.S. medals is

$$x + 51 = 70 + 51 = 121.$$

Be sure to find the second answer.

Step 6 **Check.** 121 is 51 more than 70, and the sum of 70 and 121 is 191.

Here is an application of linear equations, taken from the **Greek Anthology** (about A.D. 500), a group of 46 number problems.

Demochares has lived a fourth of his life as a boy, a fifth as a youth, and a third as a man, and he has spent 13 years in his dotage. How old is he?

(Answer: 60 years old)

EXAMPLE 2 Finding Lengths of Pieces of Wood

A carpenter is working on a project that calls for three pieces of wood. The longest piece must be twice the length of the middle-sized piece, and the shortest piece must be 10 inches shorter than the middle-sized piece. If the three pieces are to totally utilize a board 70 inches long, how long must each piece be?

Solution

Step 1 **Read** the problem. Three lengths must be found.

Step 2 **Assign a variable.** Because the middle-sized piece appears in both comparisons, let x represent the length, in inches, of the middle-sized piece.

$$x = \text{the length of the middle-sized piece,}$$

$$2x = \text{the length of the longest piece, and}$$

$$x - 10 = \text{the length of the shortest piece. See \textbf{Figure 1}.}$$

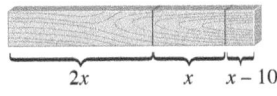

Figure 1

Step 3 **Write an equation.**

Longest plus middle-sized plus shortest is total length.

$$2x + x + (x - 10) = 70$$

Step 4 **Solve.**

$$4x - 10 = 70 \qquad \text{Combine like terms.}$$

$$4x - 10 + 10 = 70 + 10 \qquad \text{Add 10.}$$

$$4x = 80 \qquad \text{Combine like terms.}$$

$$\frac{4x}{4} = \frac{80}{4} \qquad \text{Divide by 4.}$$

$$x = 20$$

Step 5 **State the answer.** The middle-sized piece is 20 inches long, the longest piece is $2(20) = 40$ inches long, and the shortest piece is $20 - 10 = 10$ inches long.

Step 6 **Check.** The sum of the lengths is 70 inches. All conditions of the problem are satisfied.

Mixture and Interest Problems

Problems involving age have been around since antiquity. The *Greek Anthology* gives the only information known about the life of the mathematician **Diophantus:**

Diophantus passed $\frac{1}{6}$ of his life in childhood, $\frac{1}{12}$ in youth, and $\frac{1}{7}$ more as a bachelor. Five years after his marriage was born a son who died 4 years before his father, at $\frac{1}{2}$ his father's final age.

Try to write an equation and solve it to show that Diophantus was 84 years old when he died.

Problem-Solving Strategy

Percents often are used in problems involving mixing different concentrations of a substance or different interest rates. In each case, to obtain the amount of pure substance or the interest, we multiply.

MIXTURE PROBLEMS	**INTEREST PROBLEMS (ANNUAL)**
base × rate (%) = percentage	principal × rate (%) = interest
$b \quad \times \quad r \quad = \quad p$	$P \quad \times \quad r \quad = \quad I$

In an equation, the percent should be written as a decimal.

EXAMPLE 3 Using Percents in Applications

Use percents in each of the following.

(a) A nurse has 40 liters of a 35% alcohol solution. How much pure alcohol does that solution contain?

(b) If $1300 is invested for one year at 2% simple interest, how much interest is earned in one year?

Solution

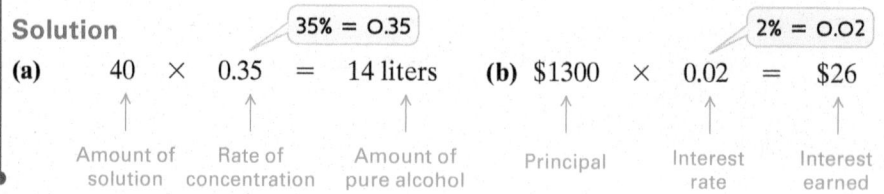

| 35% = 0.35 | | | 2% = 0.02 | | |

(a) $\quad 40 \quad \times \quad 0.35 \quad = \quad 14 \text{ liters}$ **(b)** $\$1300 \quad \times \quad 0.02 \quad = \quad \$26$

| Amount of solution | Rate of concentration | Amount of pure alcohol | Principal | Interest rate | Interest earned |

Problem-Solving Strategy

Creating a table, as seen in the next example, makes it easier to set up an equation for a problem. This is usually the most difficult step.

EXAMPLE 4 Solving a Mixture Problem

A pharmacist must combine 8 liters of a 40% acid solution with some 70% solution to obtain a 50% solution. How much of the 70% solution should be used?

Solution

Step 1 **Read** the problem. We must find the amount of 70% solution to be used.

Step 2 **Assign a variable.** Let x = the number of liters of 70% solution to be used. The information in the problem is illustrated in **Figure 2**.

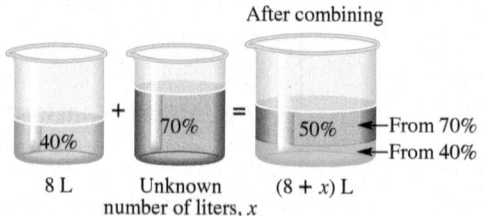

Figure 2

The rhyme below is derived from an old application found in the **Rhind papyrus,** an Egyptian manuscript that dates back to about 1650 B.C.

> As I was going to St. Ives,
> I met a man with seven wives.
> Every wife had seven sacks.
> Every sack had seven cats.
> Every cat had seven kits.
> Kits, cats, sacks, and wives,
> How many were going to St. Ives?

The "trick" answer is 1. The only person going *to* St. Ives was the person making the rhyme.

Use the given information to complete the table.

Percent (as a decimal)	Number of Liters	Liters of Pure Acid
40% = 0.40	8	0.40(8) = 3.2
70% = 0.70	x	0.70x
50% = 0.50	8 + x	0.50(8 + x)

Sum must equal

The numbers in the right column were found by multiplying the strengths and the numbers of liters. The number of liters of pure acid in the 40% solution plus the number of liters of pure acid in the 70% solution must equal the number of liters of pure acid in the 50% solution.

Step 3 **Write an equation.**

$$3.2 + 0.70x = 0.50(8 + x)$$

Step 4 **Solve.**

$$3.2 + 0.70x = 4 + 0.50x \qquad \text{Distributive property}$$

$$32 + 7x = 40 + 5x \qquad \text{Multiply by 10 to clear decimals.}$$

$$2x = 8 \qquad \text{Subtract 32 and 5x.}$$

$$x = 4 \qquad \text{Divide by 2.}$$

Step 5 **State the answer.** The pharmacist should use 4 liters of the 70% solution.

Step 6 **Check.** 8 liters of 40% solution plus 4 liters of 70% solution is

$$8(0.40) + 4(0.70) = 6 \text{ liters}$$

of acid. Similarly, $8 + 4$ or 12 liters of 50% solution has

$$12(0.50) = 6 \text{ liters}$$

of acid in the mixture. The total amount of pure acid is 6 liters both before and after mixing, so the answer checks.

EXAMPLE 5 **Solving an Investment Problem**

After receiving an inheritance, Theo has $40,000 to invest. He will put part of the money in an account paying 2% interest and the remainder into stocks paying 3% interest. His accountant tells him that the total annual income from these investments should be $1020. How much should he invest at each rate?

Solution

Step 1 **Read** the problem again. We must find the two amounts.

Step 2 **Assign a variable.**

Let $x =$ the amount to invest at 2%.

Then $40,000 - x =$ the amount to invest at 3%.

The formula for interest is $I = prt$. Here the time, t, is 1 year.

Rate (as a decimal)	Principal	Interest
2% = 0.02	x	$0.02x$
3% = 0.03	$40,000 - x$	$0.03(40,000 - x)$
	40,000	1020

← Totals

Step 3 **Write an equation.** The last column of the table gives the equation.

Interest at 2% plus interest at 3% is total interest.

$$0.02x + 0.03(40,000 - x) = 1020$$

Step 4 **Solve.**

$$0.02x + 0.03(40,000) - 0.03x = 1020 \qquad \text{Distributive property}$$

$$2x + 3(40,000) - 3x = 102,000 \qquad \text{Multiply by 100.}$$

$$-x + 120,000 = 102,000 \qquad \text{Combine like terms.}$$

$$-x = -18,000 \qquad \text{Subtract 120,000.}$$

$$x = 18,000 \qquad \text{Multiply by } -1.$$

Step 5 **State the answer.** Theo should invest $18,000 at 2%. At 3%, he should invest $40,000 − $18,000 = $22,000.

Step 6 **Check** by finding the annual interest at each rate and taking the sum.

$$0.02(\$18,000) = \$360 \quad \text{and} \quad 0.03(\$22,000) = \$660$$

$$\$360 + \$660 = \$1020, \quad \text{as required.}$$

Monetary Denomination Problems

Problem-Solving Strategy

Problems that involve money are similar to mixture and investment problems.

MONEY DENOMINATION PROBLEMS
Number × Value of one = Total value

For example, if a jar contains 37 quarters, the monetary value of the coins is

$$37 \quad \times \quad \$0.25 \quad = \quad \$9.25.$$

Number of coins   Denomination   Monetary value

EXAMPLE 6   **Solving a Monetary Denomination Problem**

A jar of 25 coins consisting of nickels and quarters has a value of $5.65. How many of each denomination does the jar contain?

Solution

Step 1 **Read** the problem. We must find the number of each denomination.

Step 2 **Assign a variable.**

Let x = the number of nickels.

Then $25 − x$ = the number of quarters.

Denomination	Number of Coins	Value
$0.05	x	$0.05x$
$0.25	$25 − x$	$0.25(25 − x)$
	25	5.65

Sum must equal

Step 3 **Write an equation.** The last column of the table gives the following.

$$0.05x + 0.25(25 − x) = 5.65$$

Step 4 **Solve.**

$$5x + 25(25 − x) = 565 \qquad \text{Multiply by 100.}$$

$$5x + 625 − 25x = 565 \qquad \text{Distributive property}$$

$$−20x = −60 \qquad \text{Subtract 625. Combine like terms.}$$

$$x = 3 \qquad \text{Divide by −20.}$$

Step 5 **State the answer.** The group has 3 nickels and 25 − 3 = 22 quarters.

Step 6 **Check.** There are 3 + 22 = 25 coins, and the value of the coins is

$$\$0.05(3) + \$0.25(22) = \$5.65, \quad \text{as required.}$$

Motion Problems

If an automobile travels at an average rate of 50 miles per hour for 2 hours, then it travels $50 \times 2 = 100$ miles. This is an example of the basic relationship

distance = rate × time.

This is given by the formula $d = rt$. By solving, in turn, for r and t, we obtain two other equivalent forms of the formula. The three forms are given below.

DISTANCE, RATE, TIME RELATIONSHIP

$$d = rt \qquad r = \frac{d}{t} \qquad t = \frac{d}{r}$$

| EXAMPLE 7 | Using the Distance, Rate, Time Relationship |

Use a formula relating distance, rate, and time in each of the following.

(a) The speed of sound is 1088 feet per second at sea level at 32°F. Under these conditions, how far does sound travel in 5 seconds?

(b) The winner of the first Indianapolis 500 race (in 1911) was Ray Harroun, driving a Marmon Wasp at an average speed of 74.59 miles per hour. How long did it take him to complete the 500-mile course? (Data from *The Universal Almanac*.)

(c) At the 2016 Olympic Games, United States swimmer Lilly King won the women's 100-m breast stroke, swimming the event in 64.93 seconds. What was her rate? (Data from www.olympic.org)

Solution

(a)

$$\underset{\text{Rate}}{1088} \quad \underset{\times}{\times} \quad \underset{\text{Time}}{5} \quad \underset{=}{=} \quad \underset{\text{Distance}}{5440 \text{ feet}}$$

(b) To complete the 500 miles, it took Harroun

$$\underset{\text{Rate}}{\underset{\longrightarrow}{\overset{\text{Distance} \longrightarrow}{\frac{500}{74.59}}}} = 6.70 \text{ hours} \quad (\text{rounded}). \; \longleftarrow \text{Time}$$

Here, we found time given rate and distance, using $t = \frac{d}{r}$. To convert 0.70 hour to minutes, multiply by 60 to get $0.70(60) = 42$ minutes. The race took him 6 hours, 42 minutes to complete.

(c)

$$\text{Rate} = \underset{\text{Time}}{\overset{\text{Distance} \longrightarrow}{\underset{\longrightarrow}{}}} \frac{100}{64.93} = 1.54 \text{ meters per second} \; (\text{rounded})$$

Problem-Solving Strategy

Motion problems use the distance formula,

$d = rt$.

In this formula, **when rate (or speed) is given in miles per hour, time must be given in hours.** To solve such problems, **draw a sketch** to illustrate what is happening in the problem, and **make a table** to summarize the given information.

Can we average averages? A car travels from *A* to *B* at 40 miles per hour and returns at 60 miles per hour. What is its rate for the entire trip?

The correct answer is not 50 miles per hour, as you might expect. Remembering the distance, rate, time relationship and letting *x* = the distance between *A* and *B*, we can simplify a complex fraction to find the correct answer.

$$\frac{\text{Average rate for}}{\text{entire trip}} = \frac{\text{Total distance}}{\text{Total time}}$$

$$= \frac{x + x}{\frac{x}{40} + \frac{x}{60}}$$

$$= \frac{2x}{\frac{3x}{120} + \frac{2x}{120}}$$

$$= \frac{2x}{\frac{5x}{120}}$$

$$= 2x \cdot \frac{120}{5x}$$

$$= 48$$

The average rate for the entire trip is 48 miles per hour.

Step 2 in our problem-solving steps involves assigning a variable. In most cases, we assign the variable to the unknown quantity that we are directed to find. However, in some cases, like the one in **Example 8,** it is easier to assign the variable to a different quantity and then, later in the process, use its value to find the final answer. Knowing when to do this comes from experience in problem solving, and the way to gain that experience is to solve lots of problems.

EXAMPLE 8 **Solving a Motion Problem**

Greg can bike from home to work in $\frac{3}{4}$ hour. By bus, the trip takes $\frac{1}{4}$ hour. If the bus travels 20 mph faster than Greg rides his bike, how far is it to his workplace?

Solution

Step 1 **Read** the problem. We must find the distance between Greg's home and his workplace.

Step 2 **Assign a variable.** Although the problem asks for a distance, it is easier here to let *x* be his speed when he rides his bike to work. Then the speed of the bus is *x* + 20.

$$d = rt = x \cdot \frac{3}{4} = \frac{3}{4}x, \quad \text{Distance of trip by bike}$$

and

$$d = rt = (x + 20) \cdot \frac{1}{4} = \frac{1}{4}(x + 20) \quad \text{Distance of trip by bus}$$

We summarize this information in a table.

	Rate	Time	Distance	
Bike	x	$\frac{3}{4}$	$\frac{3}{4}x$	←
Bus	$x + 20$	$\frac{1}{4}$	$\frac{1}{4}(x + 20)$	←

Same distance

Step 3 **Write an equation.** The key to setting up the correct equation is to recognize that the distance is the *same* in both cases. See **Figure 3.**

Home   Workplace

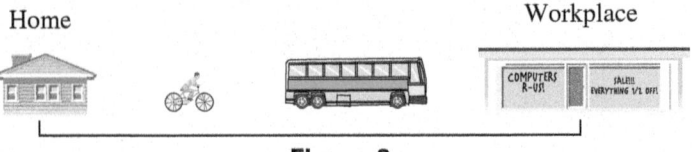

Figure 3

$$\frac{3}{4}x = \frac{1}{4}(x + 20) \quad \text{The distance is the same.}$$

Step 4 **Solve.**

$$4\left(\frac{3}{4}x\right) = 4\left(\frac{1}{4}\right)(x + 20) \quad \text{Multiply by 4.}$$

$$3x = x + 20 \quad \text{Multiply.}$$

$$2x = 20 \quad \text{Subtract } x.$$

$$x = 10 \quad \text{Divide by 2.}$$

Step 5 **State the answer.** The required distance is given by $d = rt$.

$$d = \frac{3}{4}x = \frac{3}{4}(10) = \frac{30}{4} = 7.5 \text{ miles} \quad \text{Distance by bike}$$

Step 6 **Check** by finding the distance by bus.

$$d = \frac{1}{4}(x + 20) = \frac{1}{4}(10 + 20) = \frac{30}{4} = 7.5 \text{ miles} \quad \text{Distance by bus}$$

This yields the same result. It is 7.5 miles to Greg's workplace.

7.2 EXERCISES

CONCEPT CHECK *Which choice would **not** be a reasonable answer? Justify your response.*

A problem requires finding

1. the number of cars on a dealer's lot.

   A. 0   **B.** 38   **C.** 1   **D.** $18\frac{1}{2}$

2. the number of hours a light bulb is on during one day.
   A. 0   **B.** 4.5   **C.** 18   **D.** 26

3. the distance traveled by a bus (in miles).

   A. −10   **B.** 1.7   **C.** $13\frac{1}{2}$ **D.** 50

4. the time (in minutes) elapsed since departing.

   A. $\frac{1}{60}$ **B.** 11.5   **C.** −5   **D.** 90

Decide whether each of the following translates into an expression or an equation.

5. the product of a number and 6

6. 39% of a number

7. $\frac{2}{3}$ of a number is 36.

8. 9 is 5 more than a number.

9. the ratio of a number and 24

10. 48 divided by a number is 12.

Translate each verbal phrase into a mathematical expression. Use x to represent the unknown number.

11. a number decreased by 12

12. 7 more than a number

13. the product of 6 less than a number and 4 more than the number

14. the quotient of a number and 9

15. the ratio of 25 and a nonzero number

16. $\frac{6}{7}$ of a number

Unknown Numbers *Let x represent the number, write an equation for the sentence, and then solve.*

17. If 2 is added to five times a number, the result is equal to 5 more than four times the number. Find the number.

18. If four times a number is added to 8, the result is three times the number added to 5. Find the number.

19. If 2 is subtracted from a number and this difference is tripled, the result is 6 more than the number. Find the number.

20. If 3 is added to a number and this sum is doubled, the result is 2 more than the number. Find the number.

21. The sum of three times a number and 7 more than the number is the same as the difference between −11 and twice the number. What is the number?

22. If 4 is added to twice a number and this sum is multiplied by 2, the result is the same as if the number is multiplied by 3 and 4 is added to the product. What is the number?

*Use the methods of **Examples 1 and 2** or your own method to solve each problem.*

23. *Concert Revenues* Beyoncé and Guns N' Roses had the two top-grossing North American concert tours for 2016, together generating $300 million in ticket sales. If Guns N' Roses took in $38 million less than Beyoncé, how much did each tour generate? (Data from Pollstar.)

24. *Automobile Sales* The Toyota Camry and the Honda Civic were the top-selling passenger cars in the United States in 2016. Honda Civic sales were 22 thousand less than Toyota Camry sales, and 756 thousand of these two cars were sold. How many of each model of car were sold? (Data from www.businessinsider.com)

25. NBA Record In the 2016–2017 NBA regular season, the Golden State Warriors won 7 more than four times as many games as they lost. The Warriors played 82 games. How many wins and losses did the team have? (Data from www.NBA.com)

26. MLB Record In the 2017 Major League Baseball season, the Cleveland Indians won 18 fewer than twice as many games as they lost. They played 162 regular season games. How many wins and losses did the team have? (Data from www.mlb.com)

27. U.S. Senate During the 115th Congress, the U.S. Senate had a total of 98 Democrats and Republicans. There were 6 fewer Democrats than Republicans. How many Democrats and Republicans were there in the Senate? (Data from *The World Almanac and Book of Facts.*)

28. U.S. House of Representatives The total number of Democrats and Republicans in the U.S. House of Representatives during the 115th Congress was 434. There were 46 fewer Democrats than Republicans. How many members of each party were there? (Data from *The World Almanac and Book of Facts.*)

29. Submarine Sandwich Nagaraj has a party-length sandwich that is 59 inches long and is to be cut into three pieces. The middle piece will be 5 inches longer than the shortest piece, and the shortest piece will be 9 inches shorter than the longest piece. How long will the pieces be?

30. Office Manager Duties In one week, an office manager booked 55 tickets, divided among three airlines. He booked 7 more tickets on American Airlines than on United Airlines. On Southwest Airlines, he booked 4 more than twice as many tickets as on United. How many tickets did he book on each airline?

31. Olympic Medals China earned a total of 70 medals at the 2016 Olympics. The number of gold medals was the same as the number of bronze medals. The number of silver medals was 8 less than the number of bronze medals. How many of each kind of medal did China earn? (Data from *The World Almanac and Book of Facts.*)

32. Olympic Medals The United States earned 121 medals at the 2016 Summer Olympics. The number of silver medals earned was 1 less than the number of bronze medals. The number of gold medals earned was 8 more than the number of bronze medals. How many of each kind of medal did the United States earn? (Data from *The World Almanac and Book of Facts.*)

Use basic formulas, as in **Example 3,** *to solve each problem.*

33. Acid Mixture How much pure acid is in 500 milliliters of a 14% acid solution?

34. Alcohol Mixture How much pure alcohol is in 300 liters of a 30% alcohol solution?

35. Interest Earned If $10,000 is invested for 1 year at 2.5% simple interest, how much interest is earned?

36. Interest Earned If $50,000 is invested at 3% simple interest for 2 years, how much interest is earned?

37. Monetary Value of Coins What is the monetary amount of 497 nickels?

38. Monetary Value of Coins What is the monetary amount of 89 half-dollars?

Use the method of **Example 4** *or your own method to solve each problem.*

39. Alcohol Mixture In a chemistry class, 12 liters of a 12% alcohol solution must be mixed with a 20% solution to get a 14% solution. How many liters of the 20% solution are needed?

Strength	Liters of Solution	Liters of Alcohol
12%	12	
20%		
14%		

40. Alcohol Mixture How many liters of a 10% alcohol solution must be mixed with 40 liters of a 50% solution to obtain a 40% solution?

41. Antifreeze Mixture A car radiator needs a 40% antifreeze solution. The radiator now holds 20 liters of a 20% solution. How many liters of this should be drained and replaced with 100% antifreeze to obtain the desired strength?

42. Chemical Mixture A tank holds 80 liters of a chemical solution. Currently, the solution has a strength of 30%. How much of this should be drained and replaced with a 70% solution to obtain a final strength of 40%?

43. Insecticide Mixture How much water must be added to 3 gallons of a 4% insecticide solution to reduce the concentration to 3%? (*Hint:* Water is 0% insecticide.)

44. Alcohol Mixture in First Aid Spray A medicated first aid spray on the market is 78% alcohol by volume. If the manufacturer has 50 liters of the spray containing 70% alcohol, how much pure alcohol should be added so that the final mixture is the required 78% alcohol? (*Hint:* Pure alcohol is 100% alcohol.)

*Use the method of **Example 5** or your own method to solve each problem. Assume all rates and amounts are annual.*

45. Investments at Different Rates Milos earned $12,000 last year by giving tennis lessons. He invested part at 1.5% simple interest and the rest at 2%. He earned a total of $220 in interest. How much did he invest at each rate?

Rate (as a Decimal)	Principal	Interest in One Year
0.015		
0.02		
	12,000	220

46. Investments at Different Rates Kim won $60,000 on a slot machine in Las Vegas. She invested part at 2% simple interest and the rest at 3%. She earned a total of $1600 in interest. How much was invested at each rate?

47. Investments at Different Rates Derrick invested some money at 2.25% simple interest and $1000 less than twice this amount at 1.5%. His total income from the interest was $510. How much was invested at each rate?

48. Investments at Different Rates Laquenia invested some money at 3.5% simple interest, and $5000 more than 3 times this amount at 2%. She earned $860 in interest. How much did she invest at each rate?

*Use the method of **Example 6** or your own method to solve each problem.*

49. Coin Mixture Carlos has a box of coins that he uses when playing poker with his friends. The box currently contains 44 coins, consisting of pennies, dimes, and quarters. The number of pennies is equal to the number of dimes, and the total value is $4.37. How many of each denomination of coin does he have in the box?

Denomination	Number of Coins	Value	
0.01	x	0.01x	
	x		
0.25			
	44	4.37	Totals

50. Coin Mixture Kathy found some coins while looking under her sofa pillows. There were equal numbers of nickels and quarters, and twice as many half-dollars as quarters. If she found $2.60 in all, how many of each denomination of coin did she find?

51. Attendance at a School Play For opening night of a school production of *Our Town*, 410 tickets were sold. Students paid $6 each, while nonstudents paid $14 each. If a total of $3300 was collected, how many students and how many nonstudents attended?

52. Attendance at a Concert A total of 1100 people attended a yacht rock concert. Floor tickets cost $40 each, while balcony tickets cost $28 each. If a total of $41,600 was collected, how many of each type of ticket were sold?

53. Attendance at a Sporting Event At the local minor league hockey arena home games, Row 1 seats cost $35 each, and Row 2 seats cost $30 each. The 105 seats in these rows were sold out for the season. The total receipts for them were $3420. How many of each type of seat were sold?

54. Coin Mixture In the nineteenth century, the United States minted two-cent and three-cent pieces. Frances has three times as many three-cent pieces as two-cent pieces, and the face value of these coins is $1.76. How many of each denomination does she have?

Use the formula d = rt in Exercises 55–58.

55. Distance between Cities A small plane traveled from Warsaw to Rome, averaging 164 miles per hour. The trip took two hours. What is the distance from Warsaw to Rome?

56. Distance between Cities A driver averaged 53 miles per hour and took 10 hours to travel from Memphis to Chicago. What is the distance between Memphis and Chicago?

57. Distance Traveled Suppose that an automobile averages 55 miles per hour, and travels for 30 minutes. Is the distance traveled

$$55 \cdot 30 = 1650 \text{ miles?}$$

If not, give the correct distance.

58. Average Speed Which of the following choices is the best *estimate* for the average speed of a trip of 350 miles that lasted 6.8 hours?

A. 50 miles per hour

B. 30 miles per hour

C. 60 miles per hour

D. 40 miles per hour

Automobile Racing   In Exercises 59–62, find the time. Use a calculator and round your answers to the nearest thousandth. (Data from The World Almanac and Book of Facts.)

	Event and Year	Participant	Distance	Rate
59.	Indianapolis 500, 2017	Takuma Sato (Dallara-Honda)	500 miles	155.395 mph
60.	Daytona 500, 2017	Kurt Busch (Ford)	500 miles	143.187 mph
61.	Indianapolis 500, 1980	Johnny Rutherford (Hy-Gain McLaren/ Goodyear)	255 miles*	148.725 mph
62.	Indianapolis 500, 1975	Bobby Unser (Jorgensen Eagle)	435 miles*	149.213 mph

*rain-shortened

Olympic Results   In Exercises 63–66, find the rate. Use a calculator and round answers to the nearest hundredth. All events were at the 2016 Olympics. (Data from The World Almanac and Book of Facts.)

	Event	Participant	Distance	Time
63.	200-m run, Women	Elaine Thompson, Jamaica	200 meters	21.78 seconds
64.	400-m run, Women	Shaunae Miller, The Bahamas	400 meters	49.44 seconds
65.	200-m run, Men	Usain Bolt, Jamaica	200 meters	19.78 seconds
66.	110-m hurdles, Men	Omar McLeod, Jamaica	110 meters	13.05 seconds

*Use the method of **Example 8** or your own method to solve each problem.*

67. Distance Traveled to Work   When Glen drives his car to work, the trip takes 30 minutes. When he rides the bus, it takes 45 minutes. The average speed of the bus is 12 miles per hour less than his speed when driving. Find the distance he travels to work.

68. Distance Traveled to School   Theresa can get to school in 15 minutes if she rides her bike. It takes her 45 minutes if she walks. Her speed when walking is 10 miles per hour slower than her speed when riding. How far does she travel to school?

69. Travel Times of Trains   A train leaves Little Rock, Arkansas, and travels north at 85 kilometers per hour. Another train leaves at the same time and travels south at 95 kilometers per hour. How long will it take before they are 315 kilometers apart?

	Rate	Time	Distance
First train	85	t	
Second train			

70. Travel Times of Steamers   Two steamers leave a port on a river at the same time, traveling in opposite directions. Each is traveling 22 miles per hour. How long will it take for them to be 110 miles apart?

	Rate	Time	Distance
First steamer	22	t	
Second steamer			

71. Travel Times of Commuters   Nancy and Mark commute to work, traveling in opposite directions. Nancy leaves the house at 9:00 A.M. and averages 35 miles per hour. Mark leaves at 9:15 A.M. and averages 40 miles per hour. At what time will they be 140 miles apart?

72. Travel Times of Bicyclers   Jeff leaves his house on his bicycle at 7:30 A.M. and averages 5 miles per hour. His wife, Joan, leaves at 8:00 A.M., following the same path and averaging 8 miles per hour. At what time will Joan catch up with Jeff?

73. *Time Traveled by a Pleasure Boat* A pleasure boat on the Mississippi River traveled from New Roads, Louisiana, to New Orleans with a stop at White Castle. On the first part of the trip, the boat traveled at an average speed of 10 miles per hour. From White Castle to New Orleans the average speed was 15 miles per hour. The entire trip covered 100 miles. How long did the entire trip take if the two parts each took the same number of hours?

74. *Time Traveled on a Visit* Steve leaves Nashville to visit his cousin David in Napa, 80 miles away. He travels at an average speed of 50 miles per hour. One-half hour later David leaves to visit Steve, traveling at an average speed of 60 miles per hour. How long after David leaves will they meet?

7.3 RATIO, PROPORTION, AND VARIATION

OBJECTIVES

1 Write a comparison of two quantities as a ratio.
2 Find unit price.
3 Solve proportions using equation-solving methods.
4 Write proportions in applications from nursing and biological field studies and solve them.
5 Determine the constant of variation in direct, inverse, joint, and combined variation, and solve variation applications.

Writing Ratios

One of the most frequently used mathematical concepts in everyday life is *ratio*. A baseball player's batting average is actually a ratio. The slope, or pitch, of a roof on a building may be expressed as a ratio. A percent is a ratio (a comparison to 100). Ratios provide a way of comparing two numbers or quantities.

RATIO

A **ratio** is a comparison of two quantities expressed as a quotient. The ratio of the number a to the number b is written in any of the following ways.

$$a \text{ to } b, \qquad \frac{a}{b}, \qquad a:b$$

When ratios are used in comparing units of measure, the units should be the same.

EXAMPLE 1 Writing Ratios

Write a ratio for each word phrase.

(a) 5 hours to 3 hours **(b)** 6 inches to 14 inches **(c)** 6 hours to 3 days

Solution

(a) The ratio of 5 hr to 3 hr is

$$\frac{5 \text{ hr}}{3 \text{ hr}} = \frac{5}{3}. \quad \text{The ratio is 5 to 3.}$$

(b) The ratio of 6 in. to 14 in. is

$$\frac{6 \text{ in.}}{14 \text{ in.}} = \frac{6}{14} = \frac{3}{7}. \quad \text{Write in lowest terms.}$$

(c) To find the ratio of 6 hr to 3 days, first convert 3 days to hours.

$$3 \text{ days} = 3 \text{ days} \cdot \frac{24 \text{ hr}}{1 \text{ day}} = 72 \text{ hr}$$

The ratio of 6 hr to 3 days is found as follows.

$$\frac{6 \text{ hr}}{3 \text{ days}} = \frac{6 \text{ hr}}{72 \text{ hr}} = \frac{6}{72} = \frac{1}{12} \quad \text{The ratio is 1 to 12.}$$

Unit Pricing

Ratios can be applied in unit pricing, to see which size of an item offered in different sizes has the best price per unit.

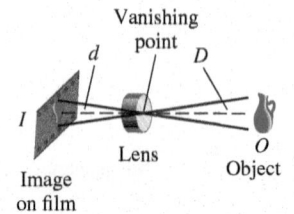

Vanishing point

d *D*

I

Lens *O*
Object

Image on film

When you look a long way down a straight road or railroad track, it seems to narrow as it vanishes in the distance. The point where the sides seem to touch is called the **vanishing point.**

The same thing occurs in the lens of a camera, as shown in the figure. Suppose *I* represents the length of the image, *O* the length of the object, *d* the distance from the lens to the film, and *D* the distance from the lens to the object. These four quantities are related by the following proportion.

$$\frac{\text{Image length}}{\text{Object length}} = \frac{\text{Image distance}}{\text{Object distance}}$$

or

$$\frac{I}{O} = \frac{d}{D}$$

Given the length of the image on the film and its distance from the lens, the length of the object determines the distance the lens must be from the object.

EXAMPLE 2 Finding Price per Unit

A supermarket charges the following for a jar of extra crunchy peanut butter.

Peanut Butter

Size	Price
18-oz	$3.49
28-oz	$4.99
40-oz	$6.79

Which size is the best buy? That is, which size has the lowest unit price?

Solution

Write ratios comparing the price for each size jar to the number of units (ounces) per jar. Then divide to obtain the price per unit (ounce). See **Table 2**.

Table 2

Size	Unit Cost (dollars per ounce)
18-oz	$\frac{\$3.49}{18} = \$0.194$
28-oz	$\frac{\$4.99}{28} = \$0.178$
40-oz	$\frac{\$6.79}{40} = \$0.170$ ⟵ The best buy

The results are rounded to the nearest thousandth.

The 40-oz size produces the lowest unit cost, so it is the best buy. Buying the largest size does not always provide the best buy, although it often does, as in this case.

Solving Proportions

> **PROPORTION**
>
> A **proportion** is a statement that two ratios are equal.

For example, $\dfrac{3}{4} = \dfrac{15}{20}$ Proportion

is a proportion that says the ratios $\frac{3}{4}$ and $\frac{15}{20}$ are equal. In the proportion

$$\frac{a}{b} = \frac{c}{d} \quad (b, d \neq 0),$$

a, *b*, *c*, and *d* are the **terms** of the proportion. The *a* and *d* terms are called the **extremes,** and the *b* and *c* terms are called the **means.** We read the proportion $\frac{a}{b} = \frac{c}{d}$ as "*a* is to *b* as *c* is to *d*."

Multiply each side of this proportion by the common denominator, bd.

$$bd \cdot \frac{a}{b} = bd \cdot \frac{c}{d} \qquad \text{Multiply each side by } bd.$$

$$\frac{b}{b}(d \cdot a) = \frac{d}{d}(b \cdot c) \qquad \text{Associative and commutative properties}$$

$$ad = bc \qquad \text{Commutative and identity properties}$$

We can also find the products ad and bc by multiplying diagonally.

$$\frac{a}{b} \overset{bc}{\underset{ad}{=}} \frac{c}{d}$$

For this reason, ad and bc are called **cross products of the proportion.**

CROSS PRODUCTS

If $\dfrac{a}{b} = \dfrac{c}{d},$ then the cross products of the proportion, ad and bc, are equal.

Also, if $ad = bc,$ then $\dfrac{a}{b} = \dfrac{c}{d}$ (where $b \neq 0, d \neq 0$).

For a proportion to be true, the product of the extremes must equal the product of the means.

$$\text{If } \frac{a}{c} = \frac{b}{d}, \text{then} \quad ad = cb, \quad \text{or} \quad ad = bc.$$

This means that these two corresponding proportions are equivalent.

$$\frac{a}{b} = \frac{c}{d} \quad \textit{can also be written as} \quad \frac{a}{c} = \frac{b}{d}.$$

The proportions in **Examples 3–7** are examples of linear equations.

EXAMPLE 3   Solving a Proportion

Solve the proportion $\dfrac{x}{63} = \dfrac{5}{9}.$

Solution

$$\frac{x}{63} = \frac{5}{9} \qquad \text{Start with the given proportion.}$$

$$9x = 63 \cdot 5 \qquad \text{Set the cross products equal.}$$

$$9x = 315 \qquad \text{Multiply.}$$

$$x = 35 \qquad \text{Divide by 9.}$$

The solution set is $\{35\}$.

EXAMPLE 4 **Solving an Equation Using Cross Products**

Solve $\dfrac{x-2}{5} = \dfrac{x+1}{3}$.

Solution

Find the cross products, and set them equal to each other.

$$3(x-2) = 5(x+1) \qquad \text{Cross products are equal.}$$

> Be sure to use parentheses.

$$3x - 6 = 5x + 5 \qquad \text{Distributive property}$$

$$3x = 5x + 11 \qquad \text{Add 6.}$$

$$-2x = 11 \qquad \text{Subtract } 5x.$$

$$x = -\frac{11}{2} \qquad \text{Divide by } -2.$$

The solution set is $\left\{-\frac{11}{2}\right\}$.

We now consider a proportion problem that occurs in the allied health field.

EXAMPLE 5 **Using a Proportion to Determine Proper Dosage of a Drug**

Nurses use proportions to determine the amount of a drug to administer when the dose is measured in milligrams (mg) but the drug is packaged in a diluted form in milliliters (mL). For example, to find the number of milliliters of fluid needed to administer 300 mg of a drug that comes packaged as 120 mg in 2 mL of fluid, a nurse sets up the proportion

$$\frac{120 \text{ mg}}{2 \text{ mL}} = \frac{300 \text{ mg}}{x \text{ mL}},$$

where x represents the amount to administer in milliliters. (Data from Hoyles, Celia, Richard Noss, and Stefano Pozzi. "Proportional Reasoning in Nursing Practice." *Journal for Research in Mathematics Education*, 32, 1, 4–27.)

(a) Solve the proportion above.

(b) Determine the correct dose if 120 mg of Amikacin is packaged as 100 mg in 2-mL vials.

Solution

(a) We must solve the following proportion.

$$\frac{120}{2} = \frac{300}{x}$$

$$120x = 2 \cdot 300 \qquad \text{Cross products are equal.}$$

$$120x = 600 \qquad \text{Multiply.}$$

$$x = 5 \qquad \text{Divide by 120.}$$

Thus, 5 milliliters of fluid are needed.

(b) Let x represent the correct dose. Set up the proportion as in part (a).

$$\frac{100 \text{ mg}}{2 \text{ mL}} = \frac{120 \text{ mg}}{x \text{ mL}}$$ Set up the ratios consistently.

Now solve for x.

$$\frac{100}{2} = \frac{120}{x}$$ Multiply by mL; divide by mg.

$$100x = 240$$ Cross products are equal.

$$x = 2.4$$ Divide by 100.

Thus, 2.4 milliliters are needed.

Researchers use sampling to make predictions about populations. In the biological sciences, populations of birds, fish, and other wildlife species are made using proportions. The mathematics of how reliable these predictions are (confidence intervals) is a topic for statistics courses, and the Internet provides information about this topic. In the next example, we show how a proportion is used in sampling.

EXAMPLE 6   **Using a Proportion to Approximate the Number of Fish in a Lake**

To estimate a population, biologists obtain a sample of fish in a lake and mark each specimen with a harmless tag. Later they return, get a similar sample of fish from the same areas of the lake, and determine the proportion of previously tagged fish in the new sample. The total fish population is estimated by assuming that the proportion of tagged fish in the new sample is the same as the proportion of tagged fish in the entire lake.

Suppose biologists tag 300 fish on May 1. When they return on June 1 and take a new sample of 400 fish, they discover that 5 of the 400 were previously tagged. Estimate the number of fish in the lake.

Solution

Let x represent the number of fish in the lake. Set up and solve a proportion.

Tagged fish on May 1 $\longrightarrow$ $\dfrac{300}{x} = \dfrac{5}{400}$ $\longleftarrow$ Tagged fish in the June 1 sample
Total fish in the lake $\longrightarrow$ $\longleftarrow$ Total number in the June 1 sample

$$5x = 120{,}000$$ Cross products are equal.

$$x = 24{,}000$$ Divide by 5.

Based on this sampling procedure, there are about 24,000 fish in the lake.

Direct Variation

Suppose that the cost of gasoline is $3.50 per gallon, and you decide to buy 5 gallons. **Table 3** shows the relationship between the number of gallons you pump and the price shown on the gas pump for that number of gallons. There is a *direct* relationship between the number of gallons and the total cost.

Table 3

Number of Gallons	Cost of This Number of Gallons
1	$3.50
2	$7.00
3	$10.50
4	$14.00
5	$17.50

THIS SALE $

GALLONS

PRICE PER GALLON $

ALL TAXES INCLUDED

Suppose that we let x represent the number of gallons pumped and let y represent the price in dollars for that number of gallons. Then the relationship between x and y is given by the equation

$$y = 3.50x,$$

or

$$\frac{y}{x} = 3.50.$$

This relationship is an example of **direct variation.** Here the quotient $\frac{y}{x}$ is always 3.50, and it is the **constant of variation.**

DIRECT VARIATION

y varies directly as x, or **y is directly proportional to x,** if there exists a nonzero constant k such that

$$y = kx, \quad \text{or, equivalently,} \quad \frac{y}{x} = k.$$

The constant k is the **constant of variation.**

In direct variation where $k > 0$, as x increases, y increases, and similarly as x decreases, y decreases.

SOLVING A VARIATION PROBLEM

Step 1 Write the variation equation.

Step 2 Substitute the initial values and solve for k.

Step 3 Rewrite the variation equation with the value of k from Step 2.

Step 4 Substitute the remaining values, solve for the unknown, and find the required answer.

EXAMPLE 7 **Solving a Direct Variation Problem**

Hooke's law for an elastic spring states that the distance a spring stretches is directly proportional to the force applied. If a force of 150 pounds stretches a certain spring 8 centimeters, how much will a force of 400 pounds stretch the spring?

Solution

Step 1 See **Figure 4.** If d is the distance the spring stretches, and f is the force applied, then $d = kf$ for some constant k.

$$d = kf \qquad \text{Variation equation}$$

Step 2 Substitute $d = 8$ and $f = 150$, and solve for k.

$$8 = k \cdot 150 \qquad \text{Let } d = 8 \text{ and } f = 150.$$

$$k = \frac{8}{150} \qquad \text{Divide by 150 and interchange sides.}$$

$$k = \frac{4}{75} \qquad \text{Lowest terms}$$

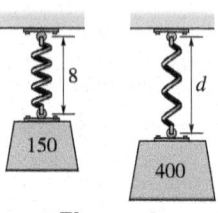

Figure 4

Step 3 Thus $d = \frac{4}{75}f$.

Step 4 Now let $f = 400$.

$$d = \frac{4}{75}(400) \quad \text{Let } f = 400.$$

$$d = \frac{64}{3} \quad \text{Multiply and simplify.}$$

The spring will stretch $21\frac{1}{3}$ centimeters if a force of 400 pounds is applied.

In some cases, one quantity will vary directly as a power of another.

DIRECT VARIATION AS A POWER

y varies directly as the nth power of x if there exists a nonzero real number k such that

$$y = kx^n.$$

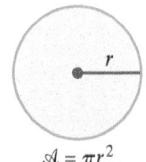

$\mathcal{A} = \pi r^2$

Figure 5

An example of direct variation as a power involves the area of a circle. See **Figure 5.** The formula for the area of a circle is $\mathcal{A} = \pi r^2$. Here, π is the constant of variation, and the area $\mathcal{A}$ varies directly as the square of the radius r.

EXAMPLE 8 Solving a Direct Variation Problem

The distance a body falls from rest varies directly as the square of the time it falls (here we disregard air resistance). If a skydiver falls 64 feet in 2 seconds, how far will she fall in 8 seconds?

Solution

Step 1 If d represents the distance the skydiver falls, and t represents the time it takes to fall, then for some constant k, $d = kt^2$.

Step 2 To find the value of k, use the fact that the skydiver falls 64 feet in 2 seconds.

$$d = kt^2 \quad \text{Formula}$$

$$64 = k(2)^2 \quad \text{Let } d = 64 \text{ and } t = 2.$$

$$k = 16 \quad \text{Evaluate } k.$$

Step 3 With this result, the variation equation can be written.

$$d = 16t^2$$

Step 4 Now let $t = 8$ to find the number of feet the skydiver will fall in 8 seconds.

$$d = 16t^2$$

$$d = 16(8)^2$$

$$d = 1024 \quad \boxed{8^2 = 8 \cdot 8 = 64}$$

The skydiver will fall 1024 feet in 8 seconds.

Inverse Variation

Another type of variation is *inverse variation*. **With inverse variation, for k > 0, as one variable increases, the other variable decreases.**

For example, in a closed space, volume decreases as pressure increases, which can be illustrated by a trash compactor. See **Figure 6.** As the compactor presses down, the pressure on the trash increases, and in turn, the trash occupies a smaller space.

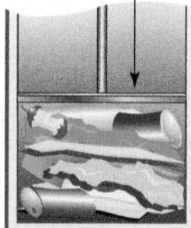

As pressure on trash increases, volume of trash decreases.

Figure 6

INVERSE VARIATION

y varies inversely as x if there exists a nonzero real number k such that

$$y = \frac{k}{x}, \quad \text{or, equivalently,} \quad xy = k.$$

Also, **y varies inversely as the nth power of x** if there exists a nonzero real number k such that

$$y = \frac{k}{x^n}.$$

EXAMPLE 9   Solving an Inverse Variation Problem

The weight of an object above Earth varies inversely as the square of its distance from the center of Earth. A space vehicle in an elliptical orbit has a maximum distance from the center of Earth (apogee) of 6700 miles. Its minimum distance from the center of Earth (perigee) is 4090 miles. See **Figure 7** (not to scale). If an astronaut in the vehicle weighs 57 pounds at its apogee, what does the astronaut weigh at the perigee?

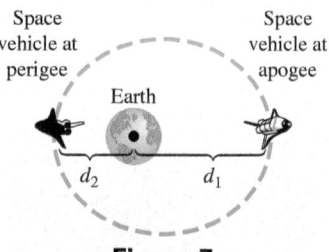

Figure 7

Solution

If w is the weight and d is the distance from the center of Earth, then the variation equation is

$$w = \frac{k}{d^2}, \quad \text{for some constant } k.$$

At the apogee the astronaut weighs 57 pounds and the distance from the center of Earth is 6700 miles. Use these values to find k.

$$57 = \frac{k}{(6700)^2} \qquad \text{Let } w = 57 \text{ and } d = 6700.$$

$$k = 57(6700)^2 \qquad \text{Multiply by } (6700)^2. \text{ Rewrite.}$$

Then the weight at the perigee with $d = 4090$ miles is

$$w = \frac{57(6700)^2}{(4090)^2} \approx 153 \text{ pounds.} \qquad \text{Use a calculator.}$$

Joint and Combined Variation

If one variable varies as the product of several other variables (perhaps raised to powers), the first variable is said to be in **joint variation** as the others.

EXAMPLE 10   Solving a Joint Variation Problem

The strength of a rectangular beam varies jointly as its width and the square of its depth. If the strength of a beam 2 inches wide by 10 inches deep is 1000 pounds per square inch, what is the strength of a beam 4 inches wide and 8 inches deep?

Solution

If S represents the strength, w the width, and d the depth, then for some constant k, the variation equation is $S = kwd^2$.

$$S = kwd^2 \qquad \boxed{10^2 = 10 \cdot 10 = 100}$$

$$1000 = k \cdot 2 \cdot 10^2 \qquad \text{Let } S = 1000,\ w = 2,\ \text{and } d = 10.$$

$$1000 = 200k \qquad \text{Apply the exponent. Multiply.}$$

$$k = 5 \qquad \text{Divide by 200. Rewrite.}$$

Thus, $S = 5wd^2$. Find S for $w = 4$ and $d = 8$ by substitution.

$$S = 5 \cdot 4 \cdot 8^2 \qquad \text{Let } w = 4 \text{ and } d = 8.$$

$$S = 1280$$

The strength of the beam is 1280 pounds per square inch.

Combined variation involves combinations of direct and inverse variation.

EXAMPLE 11   Solving a Combined Variation Problem

Body mass index, or BMI, is used by physicians to assess a person's body fat measurement. BMI varies directly as an individual's weight in pounds and inversely as the square of the individual's height in inches. A person who weighs 118 lb and is 64 in. tall has a BMI of 20.25. (BMI is usually rounded to the nearest whole number.) (Data from www.cdc.gov)

Grady weighs 200 pounds and is 6 feet, 2 inches tall. Find his BMI.

Solution

Let B represent the BMI, w the weight, and h the height. Use the given information to determine k.

$$B = \frac{kw}{h^2} \qquad \begin{array}{l} \text{BMI varies directly as the weight.} \\ \text{BMI varies inversely as the square of the height.} \end{array}$$

$$20.25 = \frac{k(118)}{64^2} \qquad \text{Let } B = 20.25,\ W = 118,\ \text{and } h = 64.$$

$$k = \frac{20.25(64^2)}{118} \qquad \text{Multiply by } 64^2. \text{ Divide by 118.}$$

$$k \approx 703 \qquad \text{Use a calculator.}$$

Use $k = 703$, $w = 200$, and $h = (6 \times 12) + 2 = 74$ to find B, Grady's BMI.

$$B = \frac{703(200)}{74^2} \approx 25.68 \approx 26 \qquad \text{Nearest whole number}$$

WHEN WILL I **EVER** USE THIS**?**

"I've never had to use algebra in real life."

Most algebra teachers have heard something like this at one time or another. Those who have uttered it probably do use algebra but don't realize it.

Consider the little word *per,* which is interpreted "as compared to." The concept of percent refers to a comparison to 100. When we shop for a new car, we are often interested in its miles-per-gallon rating, which can be symbolized mi/gal or mpg. The slash indicates division, as does the "p" in mpg. Both indicate a *ratio* of miles to gallons, also called a *rate.*

Suppose that a manufacturer says that a hybrid car has a rating of 39 mpg—that is, 39 miles compared to 1 gallon of gas.

Can this car travel 468 *miles from your home to your lake cabin and back on one tank of gas?*

One way to answer this question is to use a *proportion.* Suppose that we let x represent the number of gallons that the tank holds. We can now write an *equation.*

$$\frac{39 \text{ miles}}{1 \text{ gallon}} = \frac{468 \text{ miles}}{x \text{ gallons}}$$

We solve this equation by applying some basic rules of algebra, one being that for this statement to be true, the product 39 times x must equal the product 1 times 468.

$$\frac{39}{1} = \frac{468}{x}$$

$$39 \cdot x = 1 \cdot 468$$

$$39x = 468$$

$$\frac{39x}{39} = \frac{468}{39}$$

$$x = 12$$

We also need to consider the size of the tank of the car we will be driving. If the tank holds 12 gallons (or more) when full, we won't need to fill up along the way if we leave with a full tank.

You might say,

"I could have done that problem a lot more easily by just dividing 468 by 39 at the beginning."

In this case, you have a good understanding of rates, or ratios. You are also probably one of those people who has a knack for computing percents. *And you just used algebra even if you didn't realize it.*

7.3   EXERCISES

CONCEPT CHECK *Solve each problem.*

1. Ratios are used to _____ two numbers or quantities. Which of the following indicate the ratio of *a* to *b*?

   A. $\dfrac{a}{b}$ **B.** $\dfrac{b}{a}$ **C.** $a \cdot b$ **D.** $a:b$

2. A proportion says that two _____ are equal. The equation

$$\frac{a}{b} = \frac{c}{d} \quad (\text{where } b, d \neq 0)$$

is a _____, where *ad* and *bc* are the _____ _____.

3. Match each ratio in Column I with the ratio equivalent to it in Column II.

I	II
(a) 75 to 100	**A.** 80 to 100
(b) 5 to 4	**B.** 50 to 100
(c) $\dfrac{1}{2}$	**C.** 3 to 4
(d) 4 to 5	**D.** 15 to 12

4. Which of the following represent a ratio of 15 to 100?

   A. 15%   **B.** $\dfrac{100}{15}$

   C. $\dfrac{3}{20}$ **D.** 30 to 200

   E. $\dfrac{15}{100}$ **F.** $\dfrac{100}{15}$%

Determine the ratio and write it in lowest terms.

5. 50 feet to 80 feet   **6.** 12 miles to 36 miles

7. 17 dollars to 68 dollars   **8.** 600 people to 500 people

9. 288 inches to 12 feet   **10.** 60 inches to 2 yards

11. 5 days to 40 hours   **12.** 75 minutes to 4 hours

Merchandise Pricing   *A supermarket was surveyed to find the prices charged for items in various sizes. Find the best buy (based on price per unit) for each item.*

13.

Granulated Sugar	
Size	**Price**
4-lb	$1.79
10-lb	$4.29

14.

Ground Coffee	
Size	**Price**
15-oz	$3.43
34.5-oz	$6.98

15.

Salad Dressing	
Size	**Price**
16-oz	$2.44
32-oz	$2.98
48-oz	$4.95

16.

Black Pepper	
Size	**Price**
2-oz	$2.23
4-oz	$2.49
8-oz	$6.59

17.

Vegetable Oil	
Size	**Price**
16-oz	$1.66
32-oz	$2.59
64-oz	$4.29
128-oz	$6.49

18.

Mouthwash	
Size	**Price**
8.5-oz	$0.99
16.9-oz	$1.87
33.8-oz	$2.49
50.7-oz	$2.99

19.

Tomato Ketchup	
Size	**Price**
14-oz	$1.39
24-oz	$1.55
36-oz	$1.78
64-oz	$3.99

20.

Grape Jelly	
Size	**Price**
12-oz	$1.05
18-oz	$1.73
32-oz	$1.84
48-oz	$2.88

Solve each equation.

21. $\dfrac{x}{4} = \dfrac{175}{20}$ **22.** $\dfrac{x}{8} = \dfrac{49}{56}$

23. $\dfrac{3x-2}{5} = \dfrac{6x-5}{11}$ **24.** $\dfrac{5+x}{3} = \dfrac{x+7}{5}$

25. $\dfrac{3x+1}{7} = \dfrac{2x-3}{6}$ **26.** $\dfrac{2x+7}{3} = \dfrac{x-1}{4}$

Drug Dosage Calculations   *Solve the following problems involving drug dosages.*

27. A physician orders 37.5 milligrams of Cylert (permoline). It is available in 18.75-milligram tablets. How many tablets should the nurse administer? (Data from Hegstad, Lorrie N., and Wilma Hayek. *Essential Drug Dosage Calculations,* 4th ed. Prentice Hall, 2001.)

28. K-Tab (potassium chloride) is available in 750-milligram (0.75-gram) tablets. A physician orders that 1.5 grams of K-Tab should be given with meals. How many tablets should the nurse administer with each meal? (Data from Hegstad, Lorrie N., and Wilma Hayek. *Essential Drug Dosage Calculations,* 4th ed. Prentice Hall, 2001.)

29. A patient's order is 40 milligrams of codeine phosphate. Available in stock is 50 milligrams in 1 milliliter of liquid. How many milliliters should be administered? (Data from Timmons, Daniel L., and Catherine W. Johnson. *Math Skills for Allied Health Careers.* Pearson Prentice Hall, 2008.)

30. A nurse is asked to administer 200 milligrams of fluconazole to a patient. There is a stock solution that provides 40 milligrams per milliliter. How much of the solution should be given to the patient? (Data from Timmons, Daniel L., and Catherine W. Johnson. *Math Skills for Allied Health Careers.* Pearson Prentice Hall, 2008.)

31. A doctor prescribed ampicillin for a child who weighs 9.1 kilograms. The label on the box recommends a daily dosage of 100 milligrams per kilogram of weight. How much ampicillin should the child receive per day? (Data from Timmons, Daniel L., and Catherine W. Johnson. *Math Skills for Allied Health Careers.* Pearson Prentice Hall, 2008.)

32. An order is for 250 milligrams of a drug given orally. You have scored tablets in 100-milligram dosages. How many tablets should you give? (Data from Lesmeister, Michele Benjamin. *Math Basics for the Health Care Professional,* 4th ed. Pearson, 2014.)

33. A drug label notes that the client's medicine has 250 milligrams in 5 milliliters of syrup. How many milliliters would deliver 375 milligrams of medicine? (Data from Lesmeister, Michele Benjamin. *Math Basics for the Health Care Professional,* 4th ed. Pearson, 2014.)

34. A physician orders 50 milligrams of hydroxyzine pamoate by mouth every 12 hours. Each teaspoonful contains 25 milligrams. How many teaspoons should be administered every 12 hours? (Data from Lesmeister, Michele Benjamin. *Math Basics for the Health Care Professional,* 4th ed. Pearson, 2014.)

35. A label for lorazepam indicates that each milliliter contains 4 milligrams of the drug. How many milligrams are in 3 milliliters? (Data from Lesmeister, Michele Benjamin. *Math Basics for the Health Care Professional,* 4th ed. Pearson, 2014.)

36. Flow rate in milliliters per hour *F* is found by dividing volume in millilters *V* by time in hours *t*. A doctor has prescribed 750 milligrams of ampicillin in 125 milliliters to infuse (to be given) over 45 minutes. What is the infusion rate in milliliters per hour? (Data from Lesmeister, Michele Benjamin. *Math Basics for the Health Care Professional,* 4th ed. Pearson, 2014.)

Solve each problem involving proportions.

37. *Price of Gasoline* If 6 gallons of premium unleaded gasoline cost $17.82, how much would it cost to completely fill a 15-gallon tank?

38. *Sales Tax* If the sales tax on a $16.00 DVD is $1.40, how much would the sales tax be on a $120.00 Blu-ray disc player?

39. *Tagging Fish for a Population Estimate* Louisiana biologists tagged 250 fish in the oxbow lake False River on October 5. On a later date they found 7 tagged fish in a sample of 350. Estimate the total number of fish in False River to the nearest hundred.

40. *Tagging Fish for a Population Estimate* On May 13 researchers at Spirit Lake tagged 420 fish. When they returned a few weeks later, their sample of 500 fish contained 9 that were tagged. Give an approximation of the fish population in Spirit Lake to the nearest hundred.

41. *Distance between Cities* The distance between Singapore and Tokyo is 3300 miles. On a certain wall map, this distance is represented by 11 inches. The actual distance between Mexico City and Cairo is 7700 miles. How far apart are they on the same map?

42. *Distance between Cities* A wall map of the United States has a distance of 8.5 inches between Memphis and Denver, two cities that are actually 1040 miles apart. The actual distance between St. Louis and Des Moines is 333 miles. How far apart are St. Louis and Des Moines on this map?

43. *Distance between Cities* On a world globe, the distance between Capetown and Bangkok, two cities that are actually 10,080 kilometers apart, is 12.4 inches. The actual distance between Moscow and Berlin is 1610 kilometers. How far apart are Moscow and Berlin on this globe?

44. *Distance between Cities* On a world globe, the distance between Rio de Janeiro and Hong Kong, two cities that are actually 17,615 kilometers apart, is 21.5 inches. The actual distance between Paris and Stockholm is 1605 kilometers. How far apart are Paris and Stockholm on this globe?

45. *Cleaning Mixture* According to the directions on a bottle of Armstrong® Concentrated Floor Cleaner, for routine cleaning, $\frac{1}{4}$ cup of cleaner should be mixed with 1 gallon of warm water. How much cleaner should be mixed with $10\frac{1}{2}$ gallons of water?

46. *Cleaning Mixture* See **Exercise 45.** The directions also specify that for extra-strength cleaning, $\frac{1}{2}$ cup of cleaner should be used for each gallon of water. How much cleaner should be mixed with $15\frac{1}{2}$ gallons of water?

47. Suppose you are the one hundred twentieth person in line at airport security.

(a) In 75 seconds, 5 people are able to pass through security. Assuming this rate stays the same, how long (in minutes) will you be waiting in the security line?

(b) It takes 15 minutes to walk from security to the gate for your flight. If the flight stops boarding in 60 minutes, will you arrive at the gate in time?

48. Suppose you are the thirtieth person in line to renew your driver's license at the Department of Motor Vehicles.

(a) In 150 seconds, 2 people are helped. Assuming this rate stays the same, how long (in minutes) will it take to reach the service counter?

(b) It takes 15 minutes to drive back to work. If your lunch hours is 45 minutes long, will you arrive back at work on time?

Solve each problem involving variation.

49. *Force Required to Compress a Spring* The force required to compress a spring varies directly as the change in the length of the spring. If a force of 12 pounds is required to compress a certain spring 3 inches, how much force is required to compress the spring 5 inches?

50. *Interest on an Investment* The interest earned on an investment varies directly as the rate of interest. If the interest is $48 when the interest rate is 2%, find the interest when the rate is 1.5%.

51. *Weight of a Moose* The weight of an object on the moon varies directly as the weight of the object on Earth. According to *Guinness World Records,* "Shad," a goat, weighs 352 pounds. Shad would weigh about 59 pounds on the moon. A bull moose weighing 1800 pounds was discovered in Canada. How much would the moose have weighed on the moon?

52. *Voyage in a Paddleboat* According to *Guinness World Records,* the longest recorded voyage in a paddle boat is 2226 miles in 103 days by the foot power of two boaters down the Mississippi River. Assuming a constant rate, how far would they have gone if they had traveled 120 days? (Distance varies directly as time.)

53. *Exchange Rate (Dollars and Euros)* The euro is the common currency used by most European countries, including Italy. On April 11, 2018, the exchange rate between euros and U.S. dollars was 0.74 euro to 1 dollar. Ashley went to Rome and exchanged her U.S. currency for euros, receiving 333 euros. How much in U.S. dollars did she exchange?

54. *Exchange Rate (U.S. and Mexico)* If 8 U.S. dollars can be exchanged for 145.45 Mexican pesos, how many pesos, to the nearest hundredth, can be obtained for $65?

55. *Pressure Exerted by a Liquid* The pressure exerted by a certain liquid at a given point varies directly as the depth of the point beneath the surface of the liquid. The pressure at a depth of 10 feet is 50 pounds per square inch. What is the pressure at a depth of 20 feet?

56. *Pressure of a Gas in a Container* If the volume is constant, the pressure of a gas in a container varies directly as the temperature. (Temperature must be measured in *kelvins* (K), a unit of measurement used in physics.) If the pressure is 5 pounds per square inch at a temperature of 200 K, what is the pressure at a temperature of 300 K?

57. *Falling Body* For a body falling freely from rest (disregarding air resistance), the distance the body falls varies directly as the square of the time. If an object is dropped from the top of a tower 400 feet high and hits the ground in 5 seconds, how far did it fall in the first 3 seconds?

58. *Illumination from a Light Source* The illumination produced by a light source varies inversely as the square of the distance from the source. If the illumination produced 4 feet from a light source is 75 foot-candles, find the illumination produced 9 feet from the same source.

59. *Skidding Car* The force needed to keep a car from skidding on a curve varies inversely as the radius of the curve and jointly as the weight of the car and the square of the rate (speed). If 242 pounds of force keep a 2000-pound car from skidding on a curve of radius 500 feet at 30 miles per hour, what force would keep the same car from skidding on a curve of radius 750 feet at 50 miles per hour?

60. *Load Supported by a Column* The maximum load that a cylindrical column with a circular cross section can hold varies directly as the fourth power of the diameter of the cross section and inversely as the square of the height. A 9-meter column 1 meter in diameter will support 8 metric tons. How many metric tons can be supported by a column 12 meters high and $\frac{2}{3}$ meter in diameter?

BMI Formula *Use the formula*

$$B = \frac{kw}{h^2}$$

*from **Example 11** to find the BMI of a man with the given weight and height. Round to the nearest whole number.*

61. weight: 180 pounds
height: 72 inches

62. weight: 195 pounds
height: 68 inches

63. weight: 260 pounds
height: 66 inches

64. weight: 140 pounds
height: 64 inches

Estimating Fish Weight *Girth is the distance around the body of a fish.*

65. The weight of a trout varies jointly as its length and the square of its girth. One angler caught a trout that weighed 10.5 pounds and measured 26 inches long with an 18-inch girth. Find the weight of a trout that is 22 inches long with a 15-inch girth.

66. The weight of a bass varies jointly as its girth and the square of its length. A prize-winning bass weighed in at 22.7 pounds and measured 36 inches long with a 21-inch girth. How much would a bass 28 inches long with an 18-inch girth weigh?

Consumer Price Index *The consumer price index (CPI), issued by the U.S. Bureau of Labor Statistics, provides a means of determining the purchasing power of the U.S. dollar from one year to the next. Using the period from 1982 to 1984 as a measure of* 100.0, *the CPI figures for selected years from 2002 to 2016 are shown here.*

Year	Consumer Price Index
2002	179.9
2004	188.9
2006	201.6
2008	215.3
2010	218.1
2012	229.6
2014	236.7
2016	240.0

Data from Bureau of Labor Statistics.

To use the CPI to predict a price in a particular year, we can set up a proportion and compare it with a known price in another year, as follows.

$$\frac{\text{Price in year } A}{\text{Index in year } A} = \frac{\text{Price in year } B}{\text{Index in year } B}$$

Use the CPI figures in the table to find the amount that would be charged for using the same amount of electricity that cost $225 *in 2002. Give your answer to the nearest dollar.*

67. in 2006   **68.** in 2008   **69.** in 2010   **70.** in 2014

Solve each problem from the world of television and film.

71. *The Andy Griffith Show* In an episode of *The Andy Griffith Show,* Andy told Opie (Ron Howard) that there were 400 needy boys in their county, which calculated to be $1\frac{1}{2}$ boys per square mile. Find the area of their county.

72. *Mean Girls* In the movie *Mean Girls,* Regina George (Rachel McAdams) tells Cady (Lindsay Lohan) that her candy bar has 120 calories and 48 of them are from fat. If her candy bar had 180 calories, how many would be from fat?

7.4 LINEAR INEQUALITIES

OBJECTIVES

1 Graph an interval on a number line.

2 Solve a linear inequality using the addition and multiplication properties of inequality.

3 Solve an application that leads to a linear inequality.

Number Lines and Interval Notation

Inequalities are algebraic expressions related by any of these symbols.

$<$ "is less than" $\leq$ "is less than or equal to"

$>$ "is greater than" $\geq$ "is greater than or equal to"

Unless otherwise specified, we solve an inequality by finding all real number solutions for it. For example, the solution set of $x \leq 2$ includes *all real numbers* that are less than or equal to 2, not just the *integers* less than or equal to 2.

Archimedes, one of the greatest mathematicians of antiquity, was born in the Greek city of Syracuse about 287 B.C.

A colorful story about Archimedes relates his reaction to one of his discoveries. While taking a bath, he noticed that an immersed object, if heavier than a fluid, "will, if placed in it, descend to the bottom of the fluid, and the solid will, when weighed in the fluid, be lighter than its true weight by the weight of the fluid displaced." This discovery so excited him that he ran through the streets shouting "Eureka!" ("I have found it!") without bothering to clothe himself.

Archimedes met his death at age 75 during the pillage of Syracuse. He was using a sand tray to draw geometric figures when a Roman soldier came upon him. He ordered the soldier to move clear of his "circles," and the soldier obliged by killing him.

We show the solution set of this inequality by graphing the real numbers satisfying $x \leq 2$. We do this by placing a square bracket at 2 on a number line and drawing an arrow from the bracket to the left (to show that all numbers less than 2 are also part of the graph).* See **Figure 8.**

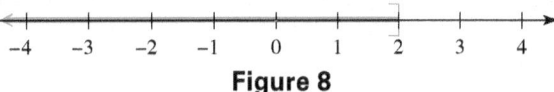

Figure 8

The set of numbers less than or equal to 2 is an example of an **interval** on the number line. To write intervals, we use **interval notation.** For example, using this notation, the interval of all numbers less than or equal to 2 is written as $(-\infty, 2]$. Interval notation often uses the **infinity symbol, ∞,** The **negative infinity symbol, $-\infty$,** does not indicate a number. It is used to show that the interval includes all real numbers less than 2. As on the number line, the square bracket indicates that 2 is part of the solution. *A parenthesis is always used next to the infinity symbol.*

The set of real numbers is written in interval notation as $(-\infty, \infty)$. Examples of other sets written in interval notation are shown in **Table 4.** In these intervals, assume that $a < b$.

Table 4

Type of Interval	Set-Builder Notation	Interval Notation	Graph
Open interval	$\{x \mid a < x < b\}$	(a, b)	
Closed interval	$\{x \mid a \leq x \leq b\}$	$[a, b]$	
Half-open (or half-closed) interval	$\{x \mid a \leq x < b\}$	$[a, b)$	
	$\{x \mid a < x \leq b\}$	$(a, b]$	
Disjoint interval	$\{x \mid x < a \text{ or } x > b\}$	$(-\infty, a) \cup (b, \infty)$	
Infinite interval	$\{x \mid x > a\}$	(a, ∞)	
	$\{x \mid x \geq a\}$	$[a, \infty)$	
	$\{x \mid x < a\}$	$(-\infty, a)$	
	$\{x \mid x \leq a\}$	$(-\infty, a]$	
	$\{x \mid x \text{ is a real number}\}$	$(-\infty, \infty)$	

*Some texts use solid circles rather than square brackets to indicate that the end point is included on a number line graph. Open circles are also used to indicate noninclusion, in place of the parentheses whose use is described in **Example 1(a).**

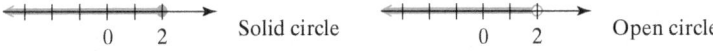

Bernard Bolzano (1781–1848) was an early exponent of rigor and precision in mathematics. Many early results in such areas as calculus were produced by the masters in the field. These masters knew what they were doing and produced accurate results. However, their sloppy arguments caused trouble in the hands of the less gifted. The work of Bolzano and others helped put mathematics on a strong footing.

EXAMPLE 1 Graphing Intervals Written in Interval Notation on a Number Line

Write each inequality in interval notation, and graph the interval.

(a) $x > -5$ **(b)** $-1 \leq x < 3$

Solution

(a) The statement $x > -5$ says that x can be any number greater than -5 but cannot be -5. The interval is written $(-5, \infty)$. Place a parenthesis at -5 and draw an arrow to the right, as shown in **Figure 9.** The parenthesis indicates that -5 is not part of the graph.

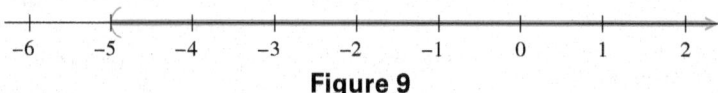

Figure 9

(b) The statement $-1 \leq x < 3$ is read "-1 is less than or equal to x *and* x is less than 3." Thus, we want the set of numbers that are *between* -1 and 3, with -1 included and 3 excluded. In interval notation, we write $[-1, 3)$, using a square bracket at -1 because -1 is part of the graph, and a parenthesis at 3 because 3 is not part of the graph. The graph is shown in **Figure 10.**

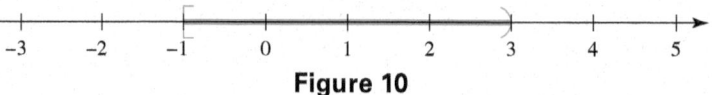

Figure 10

Addition Property of Inequality

LINEAR INEQUALITY IN ONE VARIABLE

A **linear inequality in one variable** can be written in the form

$$Ax + B < C,$$

where $A, B,$ and C are real numbers, with $A \neq 0$. (The symbol $<$ may be replaced by $>$, $\leq$, or $\geq$.)

Just as we can add the same quantity to both sides of an *equation,* we can do the same for an *inequality.* This is the **addition property of inequality.**

ADDITION PROPERTY OF INEQUALITY

For any real number expressions $A, B,$ and C, the inequalities

$$A < B \quad \text{and} \quad A + C < B + C$$

have the same solutions. (The same number may be added to each side of an inequality without changing the solutions.)

The same number may also be *subtracted* from each side of an inequality.

Amalie ("Emmy") Noether (1882–1935) was an outstanding mathematician in the field of **abstract algebra**. She studied and worked in Germany at a time when it was very difficult for a woman to do so. At the University of Erlangen in 1900, Noether was one of only two women. Although she could attend classes, professors could and did deny her the right to take the exams for their courses. Not until 1904 was Noether allowed to officially register. She completed her doctorate four years later.

In 1916 Emmy Noether went to Göttingen to work with David Hilbert on the general theory of relativity. But even with Hilbert's backing and prestige, it was three years before the faculty voted to make Noether a *Privatdozent*, the lowest rank in the faculty. In 1922 Noether was made an unofficial professor (or assistant). She received no pay for this post, although she was given a small stipend to lecture in algebra.

EXAMPLE 2   Using the Addition Property of Inequality

Solve $7 + 3x > 2x - 5$.

Solution

Follow the same general procedure as in solving linear equations.

$$7 + 3x > 2x - 5$$
$$7 + 3x - 2x > 2x - 5 - 2x \qquad \text{Subtract } 2x.$$
$$7 + x > -5 \qquad \text{Combine like terms.}$$
$$7 + x - 7 > -5 - 7 \qquad \text{Subtract 7.}$$
$$x > -12 \qquad \text{Combine like terms.}$$

The solution set is $(-12, \infty)$. Its graph is shown in **Figure 11.**

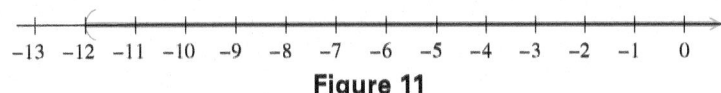

Figure 11

Multiplication Property of Inequality

The addition property of inequality cannot be used to solve inequalities such as

$$4x \geq 28.$$

These inequalities require the *multiplication property of inequality*. To see how this property works, we look at some examples.

Multiply each side of the true inequality $3 < 7$ by the positive number 2. The result is $6 < 14$, which is a true statement. However, if we multiply by -2, we obtain $-6 < -14$, which is *false*.

In solving inequalities, when multiplying both sides by a negative number, we must reverse the direction of the inequality symbol to maintain the truth value of the statement. In summary, the **multiplication property of inequality** has two parts.

MULTIPLICATION PROPERTY OF INEQUALITY

Let A, B, and C be real number expressions, with $C \neq 0$.

1. If C is *positive,* then the inequalities

$$A < B \quad \text{and} \quad AC < BC \qquad \text{Here } C \text{ is positive.}$$

have the same solutions.

2. If C is *negative,* then the inequalities

$$A < B \quad \text{and} \quad AC > BC \qquad \text{Here } C \text{ is negative.}$$

have the same solutions.

(Each side of an inequality may be multiplied by the same positive number without changing the solutions. *If the multiplier is negative, we must reverse the direction of the inequality symbol.*)

The multiplication property of inequality also permits *division* of each side of an inequality by the same nonzero number. In **Example 3** that follows, notice how the inequality symbol is reversed *only* for part (b), where both sides of the inequality are divided by a negative number.

EXAMPLE 3 Using the Multiplication Property of Inequality

Solve each inequality and graph the solution set.

(a) $3x < -18$ **(b)** $-4x \geq 8$

Solution

(a)
$$3x < -18$$

> 3 is a positive number, so the inequality symbol does not change.

$$\frac{3x}{3} < \frac{-18}{3} \quad \text{Divide by 3, a } \textit{positive} \text{ number.}$$

$$x < -6$$

The solution set is $(-\infty, -6)$. The graph is shown in **Figure 12.**

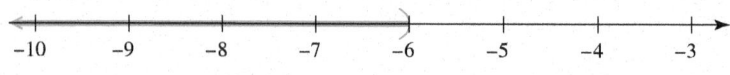

Figure 12

(b)
$$-4x \geq 8$$

$$\frac{-4x}{-4} \leq \frac{8}{-4} \quad \begin{array}{l}\text{Divide by } -4, \text{ a } \textit{negative} \text{ number.} \\ \text{Reverse the inequality symbol.}\end{array}$$

> Reverse the inequality when multiplying or dividing by a negative number.

$$x \leq -2$$

The solution set $(-\infty, -2]$ is graphed in **Figure 13.**

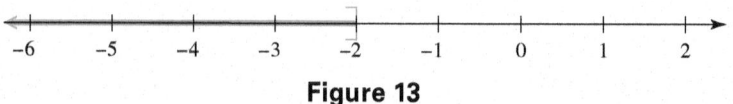

Figure 13

Solving Linear Inequalities

The general guidelines for solving linear inequalities are similar to those for solving linear equations.

SOLVING A LINEAR INEQUALITY IN ONE VARIABLE

Step 1 **Simplify each side separately.** Use the distributive property to clear parentheses and combine like terms on each side as needed.

Step 2 **Isolate the variable terms on one side.** Use the addition property of inequality to get all terms with variables on one side of the inequality and all numbers on the other side.

Step 3 **Isolate the variable.** Use the multiplication property of inequality to change the inequality in the variable x to the form

$$x < k, \quad x > k, \quad x \leq k, \quad \text{or} \quad x \geq k, \quad \text{where } k \text{ is a number.}$$

Remember: Reverse the direction of the inequality symbol only when multiplying or dividing each side of an inequality by a negative number.

EXAMPLE 4   Solving a Linear Inequality

Solve $5(x - 3) - 7x \geq 4(x - 3) + 9$. Give the solution set in interval form, and then graph.

Solution

Step 1
$$5(x - 3) - 7x \geq 4(x - 3) + 9$$

$$5x - 15 - 7x \geq 4x - 12 + 9 \qquad \text{Distributive property}$$

$$-2x - 15 \geq 4x - 3 \qquad \text{Combine like terms.}$$

Step 2
$$-2x - 15 - 4x \geq 4x - 3 - 4x \qquad \text{Use the addition property of inequality and subtract } 4x.$$

$$-6x - 15 \geq -3$$

$$-6x - 15 + 15 \geq -3 + 15 \qquad \text{Add 15.}$$

$$-6x \geq 12 \qquad \text{Combine like terms.}$$

Step 3
$$\frac{-6x}{-6} \leq \frac{12}{-6} \qquad \text{Use the multiplication property of inequality. Divide by } -6, \text{ a } \textit{negative} \text{ number. Reverse the symbol.}$$

$$x \leq -2$$

The solution set is $(-\infty, -2]$. Its graph is shown in **Figure 14.**

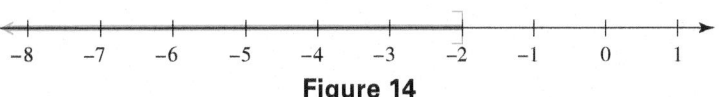

Figure 14

In Step 3 of **Example 4** we used the property that says the following.

> ***Multiplying or dividing an inequality by a negative number
> requires reversing the direction of the inequality symbol.***

In that step, we divided both sides of $-6x \geq 12$ by -6 and changed $\geq$ to $\leq$. Note that if it had been

$$6x \geq -12,$$

we would have divided by 6 and the symbol would NOT have been reversed. The fact that the sign of -12 is negative is irrelevant when this rule for solving inequalities is applied.

Applications

Problem-Solving Strategy

The table gives some of the more common phrases that suggest inequality, along with examples and translations.

Phrase	Example	Inequality
Is greater than	A number *is greater than* 4.	$x > 4$
Is less than	A number *is less than* -12.	$x < -12$
Is at least	A number *is at least* 6.	$x \geq 6$
Is at most	A number *is at most* 8.	$x \leq 8$

EXAMPLE 5 Comparing Rate Plans for a Rental Truck

Newlyweds Bryce and Lauren need to rent a truck for one day to move their belongings to their new home. They can rent a truck of the size they need from We're Outta Here for $59.95 per day with a mileage charge of $0.28 per mile, or from Moving On Out for $64.95 per day plus $0.25 per mile. They are not sure how many miles they will need to drive during their move. After how many miles will the Moving On Out rental be a better deal than the one from We're Outta Here?

Solution

For the Moving On Out rental to be a better deal, its price must be less than that of We're Outta Here. Let x represent the number of miles traveled. Then

$$59.95 + 0.28x \quad \text{represents the price for We're Outta Here,}$$

and

$$64.95 + 0.25x \quad \text{represents the price for Moving On Out.}$$

We must solve the following inequality.

$$\overbrace{64.95 + 0.25x}^{\text{Price for Moving On Out}} < \overbrace{59.95 + 0.28x}^{\text{Price for We're Outta Here}}$$

$6495 + 25x$	$<$	$5995 + 28x$	Multiply by 100.
$25x - 28x$	$<$	$5995 - 6495$	Subtract $28x$ and 6495.
$-3x$	$<$	-500	Subtract.
x	$>$	$166.\overline{6}$	Divide by -3. Reverse the inequality symbol.

The price for Moving On Out *will be less than* that of We're Outta Here when the number of miles traveled *is greater than* about 167 miles. To make their decision, Bryce and Lauren need to determine the number of miles they plan to drive and compare it to 167 miles.

Compound Inequalities

Inequalities that say that one number is *between* two other numbers are **compound inequalities.** For example,

$$-3 < 5 < 7 \quad \text{says that 5 is between } -3 \text{ and } 7.$$

Notice that the symbols must point in the same direction and toward the lesser number.

For some applications, it is necessary to work with an inequality such as

$$3 < x + 2 < 8, \quad \text{where } x + 2 \text{ is between 3 and 8.}$$

To solve this inequality, we subtract 2 from each of the three parts.

$$3 - 2 < x + 2 - 2 < 8 - 2 \quad \text{Subtract 2 from each part.}$$

$$1 < \quad x \quad < 6 \quad \text{Simplify.}$$

The idea is to use "is less than" to obtain the inequality in the form

$$\textbf{a number} < x < \textbf{another number.}$$

EXAMPLE 6   Solving a Compound Inequality

Solve $4 \leq 3x - 5 < 6$. Give the solution set in interval form, and then graph.

Solution

$$4 \leq \quad 3x - 5 \quad < 6$$

$$4 + 5 \leq 3x - 5 + 5 < 6 + 5 \qquad \text{Add 5 to each part.}$$

$$9 \leq \quad 3x \quad < 11 \qquad \text{Simplify.}$$

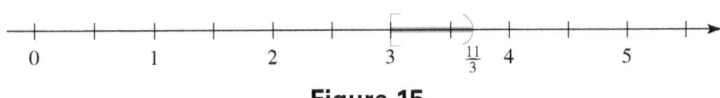 Remember to divide all three parts by 3.

$$\frac{9}{3} \leq \quad \frac{3x}{3} \quad < \frac{11}{3} \qquad \text{Divide each part by 3.}$$

$$3 \leq \quad x \quad < \frac{11}{3} \qquad \text{Simplify.}$$

The solution set is $\left[3, \frac{11}{3}\right)$. Its graph is shown in **Figure 15.**

Figure 15

EXAMPLE 7   Interpreting Tolerance Using an Inequality

Suppose a bottle indicates that it contains 32 ounces of a sports drink. It may contain slightly more or less, depending on the **tolerance** allowed by the packager. If the agreed-upon tolerance is 0.25 ounce, then the bottle may contain 0.25 ounce more or 0.25 ounce less and still satisfy the requirement.

(a) Write a compound inequality to illustrate this last statement. (Assume that equality also holds.) Let x represent the actual amount in the bottle.

(b) Solve the inequality and interpret the solution.

Solution

(a) The difference between x and 32 must be greater than or equal to -0.25 and less than or equal to 0.25. This is written as follows.

$$-0.25 \leq \quad x - 32 \quad \leq 0.25$$

(b)

$$-0.25 + 32 \leq x - 32 + 32 \leq 0.25 + 32 \qquad \text{Add 32.}$$

$$31.75 \leq \quad x \quad \leq 32.25 \qquad \text{Add.}$$

This final inequality indicates that the actual amount in the bottle must be between and inclusive of 31.75 ounces and 32.25 ounces.

WHEN WILL I **EVER** USE THIS?

Have you ever wondered what grade you will need on a test in order to keep a certain average in a course? For example, suppose that you are about to take your fourth test. Your previous scores (each out of 100) are 86, 88, and 78, and you want to keep an average of at least 80.

Step 1 **Read** the problem again.

Step 2 **Assign a variable.**

Let x = the score on your fourth test.

Step 3 **Write an inequality.** To find your average after 4 tests, add the test scores and divide by 4.

$$\underset{\text{Average}}{\underbrace{\frac{86 + 88 + 78 + x}{4}}} \underset{\substack{\text{is at} \\ \text{least 80.}}}{\geq} 80$$

Step 4 **Solve.**

$$\frac{252 + x}{4} \geq 80 \qquad \text{Add the known scores.}$$

$$4\left(\frac{252 + x}{4}\right) \geq 4(80) \qquad \text{Multiply by 4 to clear the fraction.}$$

$$252 + x \geq 320$$

$$252 + x - 252 \geq 320 - 252 \qquad \text{Subtract 252.}$$

$$x \geq 68 \qquad \text{Combine like terms.}$$

Step 5 **State the answer.** You must score 68 or more on the fourth test to have an average of *at least* 80.

Step 6 **Check.** Confirm that this minimum score of 68 gives an average of 80.

$$\frac{86 + 88 + 78 + 68}{4} = \frac{320}{4} = 80 \quad \checkmark$$

7.4 EXERCISES

CONCEPT CHECK *Solve each problem.*

1. When graphing an inequality, we use a parenthesis if the inequality symbol is _____ or _____. We use a square bracket if the inequality symbol is _____ or _____.

2. *True or false?* In interval notation, a square bracket is sometimes used next to an infinity symbol.

3. In interval notation, the set $\{x \mid x > 0\}$ is written _____.

4. In interval notation, the set of all real numbers is written _____.

Match each set in Group I with the correct graph or interval notation in A–F in Group II.

I

5. $\{x \mid x \le 3\}$

6. $\{x \mid x > 3\}$

7. $\{x \mid x < 3\}$

8. $\{x \mid x \ge 3\}$

9. $\{x \mid -3 \le x \le 3\}$

10. $\{x \mid -3 < x < 3\}$

II

A.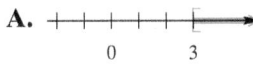

B.

C. $(3, \infty)$

D. $(-\infty, 3]$

E. $(-3, 3)$

F. $[-3, 3]$

Solve each inequality. Give the solution set in both interval and graph forms.

11. $10x \ge 70$

12. $12x \le 72$

13. $4x + 1 \ge 21$

14. $5x + 2 \ge 52$

15. $\dfrac{3x - 1}{4} > 5$

16. $\dfrac{5x - 6}{8} < 8$

17. $-4x < 16$

18. $-2x > 10$

19. $-\dfrac{3}{4}x \ge 30$

20. $-1.5x \le -\dfrac{9}{2}$

21. $-1.3x \ge -5.2$

22. $-2.5x \le -1.25$

23. $\dfrac{2x - 5}{-4} > 5$

24. $\dfrac{3x - 2}{-5} < 6$

25. $x + 4(2x - 1) \ge x$

26. $x - 2(x - 4) \le 3x$

27. $-(4 + x) + 2 - 3x < -14$

28. $-(9 + x) - 5 + 4x \ge 4$

29. $-3(x - 6) > 2x - 2$

30. $-2(x + 4) \le 6x + 16$

31. $\dfrac{2}{3}(3x - 1) \ge \dfrac{3}{2}(2x - 3)$

32. $\dfrac{7}{5}(10x - 1) < \dfrac{2}{3}(6x + 5)$

Solve each compound inequality. Give the solution set in both interval and graph forms.

33. $-4 < x - 5 < 6$

34. $-1 < x + 1 < 8$

35. $-9 \le x + 5 \le 15$

36. $-4 \le x + 3 \le 10$

37. $-6 \le 2x + 4 \le 16$

38. $-15 < 3x + 6 < -12$

39. $-19 \le 3x - 5 \le 1$

40. $-16 < 3x + 2 < -10$

41. $-1 \le \dfrac{2x - 5}{6} \le 5$

42. $-3 \le \dfrac{3x + 1}{4} \le 3$

43. $4 \le 5 - 9x < 8$

44. $4 \le 3 - 2x < 8$

Solve each problem.

45. **Truck Rental** John and Suzanne needed to rent a truck for a day, so they priced what they needed at two rental firms. Agency A wanted $28 for one day, plus a fee of $0.25 per mile, while Agency B wanted $32 for one day, plus a fee of $0.15 per mile. After how many miles will the deal that Agency B offers be better than that of Agency A?

46. **Car Rental** Cody and Margaret needed to rent a car for a day, so they checked out two rental firms. Agency X wanted $47 for one day, plus a fee of $0.20 per mile, while Agency Y wanted $38 for one day, plus a fee of $0.35 per mile. After how many miles will the deal that Agency X offers be better than that of Agency Y?

47. *Taxicab Fare* In some situations, taxicabs charge $3.00 for the first $\frac{1}{5}$ mile and $0.50 for each additional $\frac{1}{5}$ mile. Frank has only $7.50. What is the maximum distance he can travel (not including a tip for the cabbie)?

48. *Taxicab Fare* Fifteen years ago taxicab fares in the city in **Exercise 47** were $0.90 for the first $\frac{1}{7}$ mile and $0.10 for each additional $\frac{1}{7}$ mile. Based on the information given there and the answer you found, how much farther could Frank have traveled at that time?

49. *Grade Average* Mohammed earned scores of 90 and 82 on his first two tests in English Literature. What score must he make on his third test to keep an average of 84 or greater?

50. *Grade Average* Letitia scored 92 and 96 on her first two tests in Methods in Photography. What score must she make on her third test to keep an average of 90 or greater?

51. *Grade Average* Pedro has scores of 68, 84, 82, and 76 after four tests. What score must he make on his fifth test to have an average of 80 or greater?

52. *Grade Average* Laquenia has scores of 96, 88, 92, and 90 after four tests. What score must she make on her fifth test to have an average of 90 or greater?

53. *Grade Average* Katrina has scores of 88, 92, and 78 after three tests. Her final exam counts as two test scores. What score must she make on her final to have an average of 80 or greater?

54. *Grade Average* Selena has scores of 90, 84, and 96 after three tests. Her final exam counts as two test scores. What score must she make on her final to have an average of 80 or greater?

Tolerance In Exercises 55–58 at the top of the next column, a container capacity and a tolerance are given. Let x represent the amount in a particular container. Give a compound inequality that describes the conditions that must be fulfilled to stay within the given tolerance. Then solve the inequality to determine the acceptable limits of the contents of the container.

55. capacity: 18 ounces;   tolerance: 0.15 ounces

56. capacity: 22 ounces;   tolerance: 0.20 ounces

57. capacity: 52.5 milliliters;   tolerance: 0.75 milliliters

58. capacity: 125.8 milliliters;   tolerance: 0.50 milliliters

Personal Health Solve each problem.

59. A BMI (body mass index) between 19 and 25 is generally considered healthy. Use the formula

$$\text{BMI} = \frac{703 \times (\text{weight in pounds})}{(\text{height in inches})^2}$$

to find the weight range w, to the nearest pound, that gives a healthy BMI (body mass index) for each of the following heights. (Data from www.bmi-calculator.net)

(a) 72 inches

(b) Your height in inches

60. To achieve the maximum benefit from exercising, the heart rate in beats per minute should be in the target heart rate (THR) zone. For a person aged A, the formula is

$$0.7(220 - A) \le \text{THR} \le 0.85(220 - A).$$

(Data from Hockey, Robert V. *Physical Fitness: The Pathway to Healthful Living.* Times Mirror/Mosby College Publishing.)

Find THR to the nearest whole number for each age.

(a) 35

(b) Your age

Profit/Cost Analysis A product yields a profit only when the revenue R from selling the product exceeds the cost C of producing it (R and C in dollars). Find the least whole number of units x that must be sold for the business to show a profit for the item described.

61. Peripheral Visions, Inc. finds that the cost to produce x studio-quality DVDs is

$$C = 20x + 100,$$

and the revenue produced from them is $R = 24x$.

62. Speedy Delivery finds that the cost to make x customer deliveries is

$$C = 3x + 2300,$$

and the revenue produced from them is $R = 5.50x$.

7.5 **PROPERTIES OF EXPONENTS AND SCIENTIFIC NOTATION**

OBJECTIVES

1 Identify the base and the exponent in an exponential expression, and evaluate exponential expressions with natural number exponents.

2 Use zero and negative integer exponents.

3 Apply the product, quotient, power, and special rules for exponents.

4 Convert a number in standard notation to scientific notation and a number in scientific notation to standard notation.

5 Compute with numbers in scientific notation.

Exponents and Exponential Expressions

Exponents are used to write products of repeated factors.

$$\underbrace{3 \cdot 3 \cdot 3 \cdot 3}_{4 \text{ factors of } 3} = 3^{\underset{\uparrow}{4}}. \longleftarrow \text{Exponent}$$
Base

The number 4 shows that 3 appears as a factor four times. The number 4 is the **exponent** and 3 is the **base.** The quantity 3^4 is called an **exponential expression.** Read 3^4 as "3 to the fourth power" or "3 to the fourth." Multiplying out the four 3s gives 81.

$$3^4 = 3 \cdot 3 \cdot 3 \cdot 3 = 81$$

EXPONENTIAL EXPRESSION

If a is a real number and n is a natural number, then the exponential expression a^n is defined as follows.

$$a^n = \underbrace{a \cdot a \cdot a \cdot \ldots \cdot a}_{n \text{ factors of } a}$$

The number a is the *base* and n is the *exponent*.

EXAMPLE 1 **Evaluating Exponential Expressions**

Evaluate each exponential expression.

(a) 10^2 **(b)** 10^3 **(c)** $(-2)^4$ **(d)** $(-2)^5$ **(e)** 10^1

Solution | $10^2 = 10 \cdot 10$, *not* $10 \cdot 2$.

(a) $10^2 = 10 \cdot 10 = 100$ Read 10^2 as "10 squared."

(b) $10^3 = 10 \cdot 10 \cdot 10 = 1000$ Read 10^3 as "10 cubed."

(c) $(-2)^4 = (-2)(-2)(-2)(-2) = 16$

(d) $(-2)^5 = (-2)(-2)(-2)(-2)(-2) = -32$

(e) $10^1 = 10$

In the exponential expression $3x^7$, the base of the exponent 7 is x, *not* $3x$.

$$3x^7 = 3 \cdot x \cdot x \cdot x \cdot x \cdot x \cdot x \cdot x \qquad \text{Base is } x.$$

$$(3x)^7 = (3x)(3x)(3x)(3x)(3x)(3x)(3x) \qquad \text{Base is } 3x.$$

For $(-2)^6$, the parentheses around -2 indicate that the base is -2.

$$(-2)^6 = (-2)(-2)(-2)(-2)(-2)(-2) = 64 \qquad \text{Base is } -2.$$

In the expression -2^6, the base is 2, *not* -2. The $-$ sign tells us to find the negative, or additive inverse, of 2^6. It acts as a symbol for the factor -1.

$$-2^6 = -(2 \cdot 2 \cdot 2 \cdot 2 \cdot 2 \cdot 2) = -64 \qquad \text{Base is } 2.$$

Therefore, because $64 \neq -64$, $(-2)^6 \neq -2^6$.

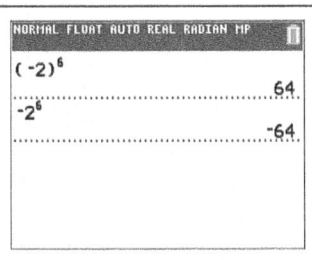

NORMAL FLOAT AUTO REAL RADIAN MP

$(-2)^6$

64.

-2^6

-64.

This screen supports the discussion in the text.

EXAMPLE 2   Evaluating Exponential Expressions

Evaluate each exponential expression.

(a) -10^2 **(b)** -10^4 **(c)** $(-10)^4$

Solution

(a) $-10^2 = -(10 \cdot 10) = -100$ **(b)** $-10^4 = -(10 \cdot 10 \cdot 10 \cdot 10) = -10{,}000$

(c) $(-10)^4 = (-10)(-10)(-10)(-10) = 10{,}000$

The Product Rule

Consider the product $2^5 \cdot 2^3$, which can be simplified as follows.

$$\overset{\displaystyle \longleftarrow 5+3=8 \longrightarrow}{2^5 \cdot 2^3 = (2 \cdot 2 \cdot 2 \cdot 2 \cdot 2)(2 \cdot 2 \cdot 2) = 2^8}$$

This result—products of exponential expressions with the same base are found by adding exponents—is generalized as the **product rule for exponents.**

PRODUCT RULE FOR EXPONENTS

If m and n are natural numbers and a is any real number, then

$$a^m \cdot a^n = a^{m+n}.$$

EXAMPLE 3   Applying the Product Rule

Apply the product rule for exponents in each case.

(a) $10^4 \cdot 10^7$ **(b)** $10^3 \cdot 10$ **(c)** $10^3 \cdot 10^8 \cdot 10^2$

Solution

(a) $10^4 \cdot 10^7 = 10^{4+7} = 10^{11}$ **(b)** $10^3 \cdot 10 = 10^3 \cdot 10^1 = 10^{3+1} = 10^4$

> Do not make the error of writing $10 \cdot 10 = 100$ as the base.

(c) $10^3 \cdot 10^8 \cdot 10^2 = 10^{3+8+2} = 10^{13}$ Product rule extended to three powers

Zero and Negative Exponents

We now consider 0 as an exponent. How can we define an expression such as 4^0 so that it is consistent with the product rule? By the product rule, we should have

$$4^2 \cdot 4^0 = 4^{2+0} = 4^2.$$

For the product rule to hold true, 4^0 must equal 1. This leads to the definition of a^0 for any nonzero real number a.

ZERO EXPONENT

If a is any nonzero real number, then $a^0 = 1.$

The expression 0^0 **is undefined.***

---

*In advanced studies, 0^0 is called an *indeterminate form.*

The term **googol,** meaning 10^{100}, was coined by Professor Edward Kasner of Columbia University. A googol is made up of a 1 with one hundred zeros following it. This number exceeds the estimated number of electrons in the universe, which is 10^{79}.

The Web search engine Google is named after a googol. Sergey Brin, president and cofounder of Google, Inc., was a mathematics major. He chose the name Google to describe the vast reach of this search engine. The term "googling" is now part of the English language.

If a googol isn't big enough for you, try a **googolplex:**

googolplex = 10^{googol}.

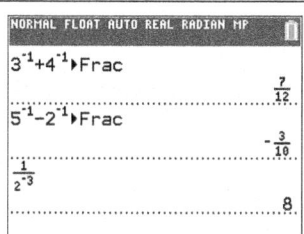

This screen supports the results in parts (b), (c), and (d) of **Example 4.**

EXAMPLE 4 Applying the Definition of Zero Exponent

Evaluate each expression.

(a) 10^0 **(b)** $(-10)^0$ **(c)** -10^0 **(d)** $5^0 + 12^0$

Solution

(a) $10^0 = 1$ **(b)** $(-10)^0 = 1$ The base is -10.

(c) $-10^0 = -(10^0) = -1$ The base is 10.   **(d)** $5^0 + 12^0 = 1 + 1 = 2$

We now define a negative exponent. Using the product rule again, we have
$$8^2 \cdot 8^{-2} = 8^{2+(-2)} = 8^0 = 1.$$
This indicates that 8^{-2} is the reciprocal of 8^2. But $\frac{1}{8^2}$ is the reciprocal of 8^2, and a number can have only one reciprocal. Therefore, we conclude that $8^{-2} = \frac{1}{8^2}$.

NEGATIVE EXPONENT

For any natural number n and any nonzero real number a,
$$a^{-n} = \frac{1}{a^n}.$$

With this definition and the ones given earlier, the expression a^n is meaningful for any integer exponent n and any nonzero real number a.

EXAMPLE 5 Applying the Definition of Negative Exponents

Write the following expressions with only positive exponents and simplify.

(a) 2^{-3} **(b)** 3^{-2} **(c)** 10^{-1} **(d)** 10^{-3}

Solution

(a) $2^{-3} = \frac{1}{2^3} = \frac{1}{8}$ **(b)** $3^{-2} = \frac{1}{3^2} = \frac{1}{9}$

(c) $10^{-1} = \frac{1}{10^1} = \frac{1}{10}$ **(d)** $10^{-3} = \frac{1}{10^3} = \frac{1}{1000}$

EXAMPLE 6 Evaluating Exponential Expressions

Evaluate each expression.

(a) $3^{-1} + 4^{-1}$ **(b)** $5^{-1} - 2^{-1}$ **(c)** $\dfrac{1}{2^{-3}}$ **(d)** $\dfrac{10^{-3}}{10^{-2}}$

Solution

(a) $3^{-1} + 4^{-1} = \frac{1}{3} + \frac{1}{4} = \frac{4}{12} + \frac{3}{12} = \frac{7}{12}$ $3^{-1} = \frac{1}{3}; 4^{-1} = \frac{1}{4}$

(b) $5^{-1} - 2^{-1} = \frac{1}{5} - \frac{1}{2} = \frac{2}{10} - \frac{5}{10} = -\frac{3}{10}$

This screen supports the results in parts (a), (b), and (c) of **Example 6.**

(c) $\dfrac{1}{2^{-3}} = \dfrac{1}{\frac{1}{2^3}} = 1 \div \dfrac{1}{2^3} = 1 \cdot \dfrac{2^3}{1} = 2^3 = 8$

> To divide, multiply by the reciprocal of the divisor.

(d) $\dfrac{10^{-3}}{10^{-2}} = \dfrac{\frac{1}{10^3}}{\frac{1}{10^2}} = \dfrac{1}{10^3} \div \dfrac{1}{10^2} = \dfrac{1}{10^3} \cdot \dfrac{10^2}{1} = \dfrac{10^2}{10^3} = \dfrac{100}{1000} = \dfrac{1}{10^1}$, or $\dfrac{1}{10}$

The following special rules for negative exponents can be verified using the definition of a negative exponent.

SPECIAL RULES FOR NEGATIVE EXPONENTS

If $a \neq 0$ and $b \neq 0$, then $\dfrac{1}{a^{-n}} = a^n$ and $\dfrac{a^{-n}}{b^{-m}} = \dfrac{b^m}{a^n}$.

The Quotient Rule

A quotient, such as $\dfrac{a^8}{a^3}$, can be simplified in much the same way as a product. (Assume that the denominator is not 0.) Using the definition of an exponent,

$$\dfrac{a^8}{a^3} = \dfrac{a \cdot a \cdot a \cdot a \cdot a \cdot a \cdot a \cdot a}{a \cdot a \cdot a} = a \cdot a \cdot a \cdot a \cdot a = a^5.$$

Notice that $8 - 3 = 5$. In the same way,

$$\dfrac{a^3}{a^8} = \dfrac{a \cdot a \cdot a}{a \cdot a \cdot a \cdot a \cdot a \cdot a \cdot a \cdot a} = \dfrac{1}{a^5} = a^{-5}.$$

Here, $3 - 8 = -5$. These examples suggest the **quotient rule for exponents.**

QUOTIENT RULE FOR EXPONENTS

If a is any nonzero real number and m and n are integers, then

$$\dfrac{a^m}{a^n} = a^{m-n}.$$

EXAMPLE 7 Applying the Quotient Rule

Apply the quotient rule for exponents in each case.

(a) $\dfrac{10^7}{10^2}$ (b) $\dfrac{2^{10}}{2^9}$ (c) $\dfrac{10^4}{10^6}$

Solution

Numerator exponent
Denominator exponent

(a) $\dfrac{10^7}{10^2} = 10^{7-2} = 10^5$

Subtract.

> Use the definition of negative exponent.

(b) $\dfrac{2^{10}}{2^9} = 2^{10-9} = 2^1 = 2$ (c) $\dfrac{10^4}{10^6} = 10^{4-6} = 10^{-2} = \dfrac{1}{10^2}$

EXAMPLE 8 Applying the Quotient Rule

Write each quotient using only positive exponents.

(a) $\dfrac{2^7}{2^{-3}}$ (b) $\dfrac{10^{-2}}{10^5}$ (c) $\dfrac{10^{-5}}{10^{-2}}$ (d) $\dfrac{4}{4^{-1}}$

Solution

> Be careful when subtracting a negative number.

(a) $\dfrac{2^7}{2^{-3}} = 2^{7-(-3)} = 2^{10}$

(b) $\dfrac{10^{-2}}{10^5} = 10^{-2-5} = 10^{-7} = \dfrac{1}{10^7}$

(c) $\dfrac{10^{-5}}{10^{-2}} = 10^{-5-(-2)} = 10^{-3} = \dfrac{1}{10^3}$

(d) $\dfrac{4}{4^{-1}} = \dfrac{4^1}{4^{-1}} = 4^{1-(-1)} = 4^2$

The Power Rules

The expression $(3^4)^2$ can be simplified as

$$(3^4)^2 = 3^4 \cdot 3^4 = 3^{4+4} = 3^8, \quad \text{where } 4 \cdot 2 = 8.$$

This example suggests the first of the **power rules for exponents.** The other two parts can be demonstrated with similar examples.

POWER RULES FOR EXPONENTS

If a and b are real numbers, and m and n are integers, then

$$(a^m)^n = a^{mn}, \quad (ab)^m = a^m b^m, \quad \text{and} \quad \left(\dfrac{a}{b}\right)^m = \dfrac{a^m}{b^m} \quad (b \neq 0).$$

In the statements of rules for exponents, we always assume that zero never appears to a negative power or to the power zero.

EXAMPLE 9 Applying the Power Rules

Use one or more power rules in each case.

(a) $(10^8)^3$ (b) $\left(\dfrac{2}{3}\right)^4$ (c) $(3 \cdot 10)^4$ (d) $(6 \cdot 10^3)^2$

Solution

(a) $(10^8)^3 = 10^{8 \cdot 3} = 10^{24}$

(b) $\left(\dfrac{2}{3}\right)^4 = \dfrac{2^4}{3^4} = \dfrac{16}{81}$

(c) $(3 \cdot 10)^4 = 3^4 \cdot 10^4 = 81 \cdot 10{,}000 = 810{,}000$

(d) $(6 \cdot 10^3)^2 = 6^2 \cdot 10^6 = 36 \cdot 1{,}000{,}000 = 36{,}000{,}000$

Notice that

$$6^{-3} = \left(\dfrac{1}{6}\right)^3 = \dfrac{1}{216} \quad \text{and} \quad \left(\dfrac{2}{3}\right)^{-2} = \left(\dfrac{3}{2}\right)^2 = \dfrac{9}{4}.$$

These are examples of two special rules for negative exponents.

SPECIAL RULES FOR NEGATIVE EXPONENTS

If $a \neq 0$ and $b \neq 0$ and n is an integer, then

$$a^{-n} = \left(\frac{1}{a}\right)^n \quad \text{and} \quad \left(\frac{a}{b}\right)^{-n} = \left(\frac{b}{a}\right)^n.$$

EXAMPLE 10 Applying Special Rules for Negative Exponents

Write each expression with only positive exponents, and then evaluate.

(a) 10^{-2} **(b)** $\left(\frac{3}{7}\right)^{-2}$ **(c)** $\left(\frac{4}{5}\right)^{-3}$

Solution

(a) $10^{-2} = \left(\frac{1}{10}\right)^2 = \frac{1}{100}$ **(b)** $\left(\frac{3}{7}\right)^{-2} = \left(\frac{7}{3}\right)^2 = \frac{49}{9}$

(c) $\left(\frac{4}{5}\right)^{-3} = \left(\frac{5}{4}\right)^3 = \frac{125}{64}$

NORMAL FLOAT AUTO REAL RADIAN MP

$\frac{3}{7}^{-2}$ ▸Frac
$\frac{49}{9}$

$\frac{4}{5}^{-3}$ ▸Frac
$\frac{125}{64}$

This screen supports the results in parts (b) and (c) of **Example 10.**

EXAMPLE 11 Writing Expressions with No Negative Exponents

Simplify each expression so that no negative exponents appear in the final result.

(a) $3^2 \cdot 3^{-5}$ **(b)** $(4^{-2})^{-5}$ **(c)** $(10^{-4})^6$

Solution

(a) $3^2 \cdot 3^{-5} = 3^{2+(-5)} = 3^{-3} = \frac{1}{3^3}$, or $\frac{1}{27}$

(b) $(4^{-2})^{-5} = 4^{-2(-5)} = 4^{10}$ **(c)** $(10^{-4})^6 = 10^{(-4)6} = 10^{-24} = \frac{1}{10^{24}}$

Scientific Notation

Many of the numbers that occur in science are very large or very small. Writing these numbers is simplified by using *scientific notation*.

SCIENTIFIC NOTATION

A number is written in **scientific notation** when it is expressed in the form

$$a \times 10^n, \quad \text{where } 1 \leq |a| < 10, \text{ and } n \text{ is an integer.}$$

Scientific notation requires that the number be written as a product of a number between 1 and 10 (or −1 and −10) and some integer power of 10. (1 and −1 are allowed as values of a, but 10 and −10 are not.) For example,

$$8000 = 8 \cdot 1000 = 8 \cdot 10^3, \quad \text{or} \quad 8 \times 10^3. \quad \longleftarrow \text{Scientific notation}$$

In scientific notation, it is customary to use × instead of a multiplication dot.

The steps involved in writing a number in scientific notation follow. (If the number is negative, ignore the negative sign, go through these steps, and then attach a negative sign to the result.)

CONVERTING A POSITIVE NUMBER TO SCIENTIFIC NOTATION

Step 1 **Position the decimal point.** Place a caret, ∧, to the right of the first nonzero digit, where the decimal point will be placed.

Step 2 **Determine the numeral for the exponent.** Count the number of digits from the decimal point to the caret. This number gives the absolute value of the exponent on 10.

Step 3 **Determine the sign for the exponent.** Decide whether multiplying by 10^n should make the result of Step 1 larger or smaller. The exponent should be positive to make the result larger. It should be negative to make the result smaller.

It is helpful to remember that for $n \geq 1$, $\quad 10^{-n} < 1 \quad$ and $\quad 10^n \geq 10$.

EXAMPLE 12 Converting to Scientific Notation

Convert each number from standard notation to scientific notation.

(a) 8,200,000   **(b)** 0.000072

Solution

(a) Place a caret to the right of the 8 (the first nonzero digit) to mark the new location of the decimal point.

$$8_\wedge 200,000$$

Count from the decimal point, which is understood to be after the last 0, to the caret.

$$8_\wedge 200,000. \longleftarrow \text{Decimal point}$$
$$\text{Count 6 places.}$$

Because the number 8.2 is to be made larger, the exponent on 10 is positive.

$$8,200,000 = 8.2 \times 10^6$$

(b)
$$0.00007_\wedge 2 \quad \text{Count from left to right.}$$
$$\text{5 places}$$

The number 7.2 is to be made smaller, so the exponent on 10 is negative.

$$0.000072 = 7.2 \times 10^{-5}$$

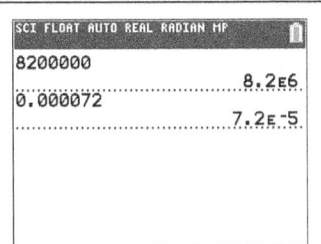

Standard Form and Scientific Notation

If a graphing calculator is set in scientific notation mode, it will give results as shown here. E6 means "times 10^6" and E−5 means "times 10^{-5}". Compare to the results of **Example 12.**

CONVERTING A POSITIVE NUMBER FROM SCIENTIFIC NOTATION TO STANDARD NOTATION

- Multiplying a positive number by a positive power of 10 makes the number larger, so move the decimal point to the right if n is positive in 10^n.

- Multiplying a positive number by a negative power of 10 makes the number smaller, so move the decimal point to the left if n is negative.

- If n is zero, do not move the decimal point.

Standard Form and Scientific Notation

EXAMPLE 13 Converting from Scientific Notation

Convert each number from scientific notation to standard notation.

(a) 6.93×10^5 **(b)** 4.7×10^{-6} **(c)** -1.083×10^0

Solution

(a) $6.93 \times 10^5 = 6.93\underbrace{000}_{\text{5 places}}$ Attach 0s as necessary.

$6.93 \times 10^5 = 693,000$ The decimal point was moved 5 places *to the right*.

(b) $4.7 \times 10^{-6} = \underbrace{000004}_{\text{6 places}}.7$ Attach 0s as necessary.

$4.7 \times 10^{-6} = 0.0000047$ The decimal point was moved 6 places *to the left*.

(c) $-1.083 \times 10^0 = -1.083$

EXAMPLE 14 Using Scientific Notation in Computation

Evaluate $\dfrac{1,920,000 \times 0.0015}{0.000032 \times 45,000}$ using scientific notation.

Solution

$$\frac{1,920,000 \times 0.0015}{0.000032 \times 45,000} = \frac{1.92 \times 10^6 \times 1.5 \times 10^{-3}}{3.2 \times 10^{-5} \times 4.5 \times 10^4}$$ Express all numbers in scientific notation.

$$= \frac{1.92 \times 1.5 \times 10^6 \times 10^{-3}}{3.2 \times 4.5 \times 10^{-5} \times 10^4}$$ Commutative and associative properties

$$= \frac{1.92 \times 1.5}{3.2 \times 4.5} \times 10^4$$ Product and quotient rules

Don't stop here. $= 0.2 \times 10^4$ Simplify.

$$= (2 \times 10^{-1}) \times 10^4$$ Write 0.2 using scientific notation.

$$= 2 \times 10^3, \quad \text{or} \quad 2000$$ Associative property; product rule; multiply.

EXAMPLE 15 Using Scientific Notation in an Application

As of November 13, 2017, the gross federal debt was about $\$2.050 \times 10^{13}$ (which is more than \$20 trillion). The U.S. population was approximately 326 million that year. About how much would each person have had to contribute in order to pay off the federal debt? (Data from www.usgovernmentdebt.us; www.census.gov)

Solution

Divide the debt by the population to obtain the per-person contribution.

$$\frac{2.050 \times 10^{13}}{326,000,000} = \frac{2.050 \times 10^{13}}{3.26 \times 10^8}$$ Write 326 million in scientific notation.

$$= \frac{2.050}{3.26} \times 10^5$$ Quotient rule

$$\approx 0.62883 \times 10^5$$ Divide.

$$\approx 62,883$$ Write in standard notation.

Each person would have had to pay nearly \$63,000.

7.5 EXERCISES

CONCEPT CHECK *Match each number written in scientific notation in Column I with the correct choice from Column II. Not all choices in Column II will be used.*

I	II
1. (a) 4.6×10^{-4}	**A.** 46,000
(b) 4.6×10^{4}	**B.** 460,000
(c) 4.6×10^{5}	**C.** 0.00046
(d) 4.6×10^{-5}	**D.** 0.000046
	E. 4600

I	II
2. (a) 1×10^{9}	**A.** 1 billion
(b) 1×10^{6}	**B.** 100 million
(c) 1×10^{8}	**C.** 1 million
(d) 1×10^{10}	**D.** 10 billion
	E. 100 billion

3. CONCEPT CHECK Write each number in scientific notation.

(a) 63,000   The first nonzero digit is _____. The decimal point should be moved _____ places.

$$63{,}000 = \underline{\quad} \times 10^{\overline{\quad}}$$

(b) 0.0571   The first nonzero digit is _____. The decimal point should be moved _____ places.

$$0.0571 = \underline{\quad} \times 10^{\overline{\quad}}$$

4. CONCEPT CHECK Write each number in standard notation.

(a) 4.2×10^{3}
Move the decimal point _____ places to the _____.

$$4.2 \times 10^{3} = \underline{\quad}$$

(b) 6.42×10^{-3}
Move the decimal point _____ places to the _____.

$$6.42 \times 10^{-3} = \underline{\quad}$$

Evaluate each exponential expression.

5. 2^4 **6.** 3^5

7. 5^4 **8.** 10^3

9. $(-2)^5$ **10.** $(-5)^4$

11. -2^3 **12.** -3^2

13. $-(-3)^4$ **14.** $-(-5)^2$

15. 8^{-1} **16.** 9^{-1}

17. 7^{-2} **18.** 4^{-2}

19. -7^{-2} **20.** -16^{-1}

21. $\dfrac{5^{-1}}{4^{-2}}$ **22.** $\dfrac{2^{-3}}{3^{-2}}$

23. $\left(\dfrac{1}{5}\right)^{-3}$ **24.** $\left(\dfrac{1}{2}\right)^{-3}$

25. $\left(\dfrac{4}{5}\right)^{-2}$ **26.** $\left(\dfrac{2}{3}\right)^{-2}$

27. $4^{-1} + 5^{-1}$ **28.** $3^{-1} + 2^{-1}$

29. 12^0 **30.** 8^0

31. $(-4)^0$ **32.** $(-23)^0$

33. $3^0 - 4^0$ **34.** $10^0 - 8^0$

Use the product, quotient, and power rules to simplify each expression. Write answers with only positive exponents.

35. $10^{12} \cdot 10^4$ **36.** $10^6 \cdot 10^8$

37. $10^{-6} \cdot 10^5$ **38.** $10^{-9} \cdot 10^8$

39. $\dfrac{5^{16}}{5^{17}}$ **40.** $\dfrac{3^{12}}{3^{13}}$

41. $\dfrac{10^5}{10^3}$ **42.** $\dfrac{10^8}{10^5}$

43. $\dfrac{10^{-2}}{10^{-5}}$ **44.** $\dfrac{10^{-2}}{10^{-6}}$

45. $10^{-8} \cdot 10^6 \cdot 10^3$ **46.** $10^{-12} \cdot 10^{10} \cdot 10^3$

47. $(5^{-2})^{-1}$ **48.** $(6^{-1})^{-2}$

49. $\dfrac{(10^{-3})^{-2}}{10^{-5}}$ **50.** $\dfrac{(10^{-1})^{-5}}{10^{-3}}$

51. $\dfrac{10^8 \cdot 10^4}{10^6 \cdot 10^{12}}$ **52.** $\dfrac{10^{10} \cdot 10^5}{10^8 \cdot 10^9}$

53. $\dfrac{(6 \times 10^{-2})^2}{(3 \times 10^5)^2}$ **54.** $\dfrac{(4 \times 10^{-2})^3}{(2 \times 10^3)^3}$

Convert each number from standard notation to scientific notation.

55. 230   **56.** 46,500

57. 8,100,000,000   **58.** 7,250,000

59. 1,400,000,000,000   **60.** 1,300,000,000,000,000

61. 0.02   **62.** 0.0051

63. 0.5623   **64.** 0.1234

65. 0.0000044   **66.** 0.00000099

Convert each number from scientific notation to standard notation.

67. 6.5×10^3

68. 2.317×10^5

69. 2.8×10^7

70. 9.1×10^8

71. 1.5×10^{10}

72. 4.4×10^{11}

73. 4.2×10^{-1}

74. 5.3×10^{-1}

75. 1.52×10^{-2}

76. 1.63×10^{-4}

77. 8.33×10^{-6}

78. 7.51×10^{-6}

Use scientific notation to perform each of the following computations. Leave the answers in scientific notation.

79. $\dfrac{0.002 \times 3900}{0.000013}$

80. $\dfrac{0.009 \times 600}{0.02}$

81. $\dfrac{0.0004 \times 56{,}000}{0.000112}$

82. $\dfrac{0.018 \times 20{,}000}{300 \times 0.0004}$

83. $\dfrac{840{,}000 \times 0.03}{0.00021 \times 600}$

84. $\dfrac{28 \times 0.0045}{140 \times 1500}$

Solve each problem.

85. National Debt In 2014, the national debt of the United States was about \$17.27 trillion. The U.S. population at that time was approximately 317.4 million. (Data from U. S. Census Bureau.)

(a) Write the population using scientific notation.

(b) Write the amount of the national debt using scientific notation.

(c) Using the answers for parts (a) and (b), calculate how much debt this is per citizen. Write this amount in standard notation to the nearest ten dollars.

86. U.S. Population In 2017, the population of the United States was approximately 326.4 million. (Data from U.S. Census Bureau.)

(a) Write the population using scientific notation.

(b) Write \$1 trillion (\$1,000,000,000,000) using scientific notation.

(c) Using the answers from parts (a) and (b), calculate how much each person in the United States would have to contribute in order to make someone a trillionaire. Write this amount in standard notation to the nearest dollar.

87. Broadway Tickets During the 2016–2017 season, Broadway shows grossed a total of $\$1.45 \times 10^9$. Total attendance for the season was 1.33×10^7. What was the average ticket price (to the nearest cent) for a Broadway show? (Data from The Broadway League.)

88. Motion Pictures In 2016, $\$1.14 \times 10^{10}$ was spent to attend motion pictures in the United States and Canada. The total number of tickets sold was 1.32 billion. What was the average ticket price (to the nearest cent) for a movie? (Data from Motion Picture Association of America.)

89. Distance of Uranus from the Sun A parsec, a unit of length used in astronomy, is 1.9×10^{13} miles. The mean distance of Uranus from the sun is 1.8×10^7 miles. How many parsecs is Uranus from the sun?

90. Number of Inches in a Mile An inch is approximately 1.57828×10^{-5} mile. Find the reciprocal of this number to determine the number of inches in a mile.

91. Speed of Light The speed of light is approximately 3×10^{10} centimeters per second. How long will it take light to travel 9×10^{12} centimeters?

92. Probe from Earth to the Sun On August 12, 2018, NASA's Parker Solar Probe was launched on a Delta IV Heavy rocket. It arrived at the sun to make 24 sweeps through the sun's atmosphere in November of 2018. Assuming that it took exactly three months (90 days) to make the trip, and the average distance from Earth to the sun is 9.3×10^7 miles, what is the average speed of the probe in miles per hour? (Round to the nearest thousand.)

93. Miles in a Light-Year A *light-year* is the distance that light travels in one year. Find the number of miles in a light-year if light travels 1.86×10^5 miles per second.

94. Time for Light to Travel Use the information given in the previous two exercises to find the number of minutes necessary for light from the sun to reach Earth.

95. Population of Luxembourg In 2017, the population of Luxembourg was approximately 5.94130×10^5, and the population density was about 595 people per square mile. (Data from *The World Almanac and Book of Facts*.)

(a) Write the population density in scientific notation.

(b) With this information, calculate the area of Luxembourg to the nearest square mile.

96. *Population of Costa Rica* In 2017, the population of Costa Rica was approximately 4.930258×10^6. The population density was about 96.6 people per square kilometer. (Data from *The World Almanac and Book of Facts*.)

(a) Write the population density in scientific notation.

(b) With this information, calculate the area of Costa Rica to the nearest hundred square kilometers.

97. *Star Trek Mathematics* In a television episode of *Star Trek,* Captain Kirk once said that a booster would increase the sound on the bridge of the ship by a factor of "one to the fourth power." Explain his error.

98. *Star Trek Mathematics* Captain Kirk probably meant to use scientific notation in his statement in **Exercise 97.** What single-word change would accomplish this?

7.6 POLYNOMIALS AND FACTORING

OBJECTIVES

1 Identify polynomials using the basic terminology of polynomials.

2 Add and subtract polynomials.

3 Multiply polynomials.

4 Multiply two binomials using the FOIL method.

5 Find special products of binomials.

6 Factor the greatest common factor from a polynomial.

7 Factor a trinomial using the FOIL method in reverse.

8 Factor a perfect square trinomial and a difference of squares.

Basic Terminology

A **term,** or **monomial,** is defined to be a number, a variable, or a product of numbers and variables. A **polynomial** is a term, or a finite sum or difference of terms, with only nonnegative integer exponents permitted on the variables. If the terms of a polynomial contain only the variable x, then the polynomial is called a **polynomial in x.** (Polynomials in other variables are defined similarly.)

$$5x^3 - 8x^2 + 7x - 4, \quad 9x^5 - 3, \quad 8x^2, \quad \text{and} \quad 6 \quad \text{Examples of polynomials in } x$$

The expression $9x^2 - 4x - \frac{6}{x}$ is not a polynomial because of the presence of $-\frac{6}{x}$. The terms of a polynomial cannot have variables in a denominator.

The greatest exponent in a polynomial in one variable is the **degree** of the polynomial. A nonzero constant has degree 0. (The polynomial 0 has no degree.)

$$3x^6 - 5x^2 + 2x + 3 \quad \text{is a polynomial of degree 6.}$$

A polynomial containing exactly three terms is called a **trinomial,** and one containing exactly two terms is a **binomial. Table 5** gives examples.

Table 5

Polynomial	Degree	Type
$9x^4 - 4x^3 + 8x^2$	4	Trinomial
$29x^3 + 8x^2$	3	Binomial
$-10x$	1	Monomial

Addition and Subtraction

Because the variables used in polynomials represent real numbers, a polynomial represents a real number. This means that all the properties of the real numbers hold for polynomials. Here is an application of the distributive property.

$$3x^5 - 7x^5 = (3 - 7)\, x^5 = -4x^5 \quad \text{Combine like terms.}$$

Like terms are terms that have the exact same variable factors. Thus, polynomials are added by adding coefficients of like terms and are subtracted by subtracting coefficients of like terms.

EXAMPLE 1 Adding and Subtracting Polynomials

Add or subtract, as indicated.

(a) $(2x^4 - 3x^2 + x) + (4x^4 + 7x^2 + 6x)$

(b) $(-3x^3 - 8x^2 + 4) - (x^3 + 7x^2 - 3)$

(c) $4(x^2 - 3x + 7) - 5(2x^2 - 8x - 4)$

Solution

(a) $(2x^4 - 3x^2 + x) + (4x^4 + 7x^2 + 6x)$

$\qquad = (2 + 4)x^4 + (-3 + 7)x^2 + (1 + 6)x \quad \text{Combine like terms.}$

$\qquad = 6x^4 + 4x^2 + 7x$

(b) $(-3x^3 - 8x^2 + 4) - (x^3 + 7x^2 - 3)$

$\qquad = (-3 - 1)x^3 + (-8 - 7)x^2 + [4 - (-3)]$

$\qquad = -4x^3 - 15x^2 + 7$

(c) $4(x^2 - 3x + 7) - 5(2x^2 - 8x - 4)$

$\qquad = 4x^2 - 4(3x) + 4(7) - 5(2x^2) - 5(-8x) - 5(-4) \quad \text{Distributive property}$

$\qquad = 4x^2 - 12x + 28 - 10x^2 + 40x + 20 \quad\quad\quad\quad \text{Associative property}$

$\qquad = -6x^2 + 28x + 48 \quad\quad\quad\quad\quad\quad\quad\quad\quad \text{Combine like terms.}$

As shown in **Example 1,** polynomials are often written with their terms arranged in **descending powers.** The term of greatest degree is first, the one with the next greatest degree is second, and so on.

Multiplication

The associative and distributive properties, together with the properties of exponents, can also be used to find the product of two polynomials.

$$(3x - 4)(2x^2 - 3x + 5)$$

$$= (3x - 4)(2x^2) - (3x - 4)(3x) + (3x - 4)(5)$$

$$= 3x(2x^2) - 4(2x^2) - 3x(3x) - (-4)(3x) + 3x(5) - 4(5)$$

$$= 6x^3 - 8x^2 - 9x^2 + 12x + 15x - 20$$

$$= 6x^3 - 17x^2 + 27x - 20$$

Such a product can be written vertically, as shown in **Example 2.**

EXAMPLE 2   Multiplying Polynomials Vertically

Multiply $(3x - 4)(2x^2 - 3x + 5)$.

Solution

$$
\begin{array}{r}
2x^2 - 3x + 5 \\
3x - 4 \\
\hline
-8x^2 + 12x - 20 \\
6x^3 - 9x^2 + 15x \\
\hline
6x^3 - 17x^2 + 27x - 20
\end{array}
$$

Multiply each term of the first polynomial by each term of the second.

Be sure to place like terms in columns.

$\longleftarrow -4(2x^2 - 3x + 5)$

$\longleftarrow 3x(2x^2 - 3x + 5)$

Add in columns.

The memory aid "FOIL" is a convenient way to find the product of two binomials. FOIL (for First, Outside, Inside, Last) gives the pairs of terms to be multiplied to get the product, as shown in the next examples.

EXAMPLE 3   Using the FOIL Method

Find each product.

(a) $(6x + 1)(4x - 3)$ **(b)** $(2x + 7)(2x - 7)$ **(c)** $(x^2 - 4)(x^2 + 4)$

Solution

(a) $(6x + 1)(4x - 3)$

$$
\; \overset{\text{F}}{} \quad \overset{\text{O}}{} \quad \overset{\text{I}}{} \quad \overset{\text{L}}{}
$$

$$= (6x)(4x) + (6x)(-3) + 1(4x) + 1(-3)$$

$$= 24x^2 - 18x + 4x - 3$$

$$= 24x^2 - 14x - 3 \qquad \text{Combine like terms.}$$

(b) $(2x + 7)(2x - 7)$

$$= 4x^2 - 14x + 14x - 49 \qquad \text{FOIL method}$$

$$= 4x^2 - 49 \qquad \text{Combine like terms.}$$

(c) $(x^2 - 4)(x^2 + 4)$

$$= x^4 + 4x^2 - 4x^2 - 16 \qquad \text{FOIL method}$$

$$= x^4 - 16 \qquad \text{Combine like terms.}$$

The **special product**

$$(x + y)(x - y) = x^2 - y^2$$

can be used to solve some multiplication problems.

$$51 \times 49 = (50 + 1)(50 - 1)$$
$$= 50^2 - 1^2$$
$$= 2500 - 1$$
$$= 2499$$

$$102 \times 98 = (100 + 2)(100 - 2)$$
$$= 100^2 - 2^2$$
$$= 10{,}000 - 4$$
$$= 9996$$

Once these patterns are recognized, multiplications of this type can be done mentally.

Special Products

As shown in **Examples 3(b) and (c)**, *the product of two binomials of the forms x + y and x − y is always a binomial.* Check that the following is true.

PRODUCT OF THE SUM AND DIFFERENCE OF TWO TERMS

$$(x + y)(x - y) = x^2 - y^2$$

The product $x^2 - y^2$ is called a **difference of two squares.**

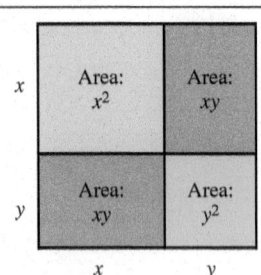

The **special product**

$$(x + y)^2 = x^2 + 2xy + y^2$$

can be illustrated geometrically using the diagram shown here. Each side of the largest square has length $x + y$, so the area of the square is

$$(x + y)^2.$$

The largest square is made up of two smaller squares and two congruent rectangles. The sum of the areas of these figures is

$$x^2 + 2xy + y^2.$$

Since these expressions represent the same quantity, they must be equal, thus giving us the pattern for squaring a binomial.

EXAMPLE 4 Multiplying $(x + y)(x - y)$

Find each product.

(a) $(3x + 1)(3x - 1)$ (b) $(5x^3 - 3)(5x^3 + 3)$

(c) $(9x - 11)(9x + 11)$

Solution

(a) Using the pattern discussed earlier, replace x with $3x$ and replace y with 1.

$$(3x + 1)(3x - 1)$$
$$= (3x)^2 - 1^2$$
$$= 9x^2 - 1$$

(b) $(5x^3 - 3)(5x^3 + 3)$ (c) $(9x - 11)(9x + 11)$
$$= (5x^3)^2 - 3^2$$ $$= (9x)^2 - (11)^2$$
$$= 25x^6 - 9$$ $$= 81x^2 - 121$$

Squares of binomials are also special products. These are verified by first writing as two factors and then using the FOIL method.

SQUARES OF BINOMIALS

$$(x + y)^2 = x^2 + 2xy + y^2$$
$$(x - y)^2 = x^2 - 2xy + y^2$$

EXAMPLE 5 Squaring Binomials

Find each product.

(a) $(x + 6)^2$ (b) $(2x + 5)^2$ (c) $(3x - 7)^2$

Solution

(a) $(x + 6)^2 = x^2 + 2(x)(6) + 6^2$ *The square of a binomial has three terms. $(x + y)^2 \neq x^2 + y^2$*
$$= x^2 + 12x + 36$$

(b) $(2x + 5)^2$ (c) $(3x - 7)^2$
$$= (2x)^2 + 2(2x)(5) + (5)^2$$ $$= (3x)^2 - 2(3x)(7) + (7)^2$$
$$= 4x^2 + 20x + 25$$ $$= 9x^2 - 42x + 49$$

Factoring

The process of finding polynomials whose product equals a given polynomial is called **factoring.** For example, since

$$4x + 12 = 4(x + 3),$$

both 4 and $x + 3$ are called **factors** of $4x + 12$. Also, $4(x + 3)$ is called a **factored form** of $4x + 12$. A polynomial that cannot be written as a product of two polynomials with integer coefficients is a **prime polynomial.** A polynomial is **factored completely** when it is written as a product of prime polynomials with integer coefficients.

Factoring is the inverse of multiplying, so a check requires that the product of the factored form yields the original polynomial.

Factoring Out the Greatest Common Factor

When factoring a polynomial, we first look for a monomial that is the greatest common factor (GCF) of all the terms of the polynomial.

EXAMPLE 6 Factoring Out the Greatest Common Factor

Factor out the greatest common factor from each polynomial.

(a) $9x^5 + x^2$ **(b)** $5x^4 - 10x^3 + 20x^2$ **(c)** $6x^4 + 9x^3 + 18x^2 + 15x$

Solution

(a) $9x^5 + x^2 = x^2 \cdot 9x^3 + x^2 \cdot 1$ The greatest common factor is x^2.

$\qquad\qquad = x^2(9x^3 + 1)$

(b) $5x^4 - 10x^3 + 20x^2$

$\qquad\qquad = 5x^2(x^2 - 2x + 4)$ GCF $= 5x^2$

(c) $6x^4 + 9x^3 + 18x^2 + 15x$

$\qquad\qquad = 3x(2x^3) + 3x(3x^2) + 3x(6x) + 3x(5)$ GCF $= 3x$

$\qquad\qquad = 3x(2x^3 + 3x^2 + 6x + 5)$

Factoring can always be checked by multiplying out the factored form.

Factoring Trinomials

Factoring trinomials requires using the FOIL method in reverse.

EXAMPLE 7 Factoring Trinomials

Factor each trinomial.

(a) $4x^2 - 11x + 6$ **(b)** $6x^2 - 7x - 5$

Solution

(a) To factor this polynomial, we must find integers a, b, c, and d such that

$$4x^2 - 11x + 6 = (ax + b)(cx + d).$$

By using FOIL, we see that $ac = 4$ and $bd = 6$. The positive factors of 4 are 4 and 1 or 2 and 2. Since the middle term is negative, we consider only negative factors of 6. The possibilities are -2 and -3 or -1 and -6. Try various arrangements.

$(2x - 1)(2x - 6) = 4x^2 - 14x + 6$ Incorrect

$(2x - 2)(2x - 3) = 4x^2 - 10x + 6$ Incorrect

$(x - 2)(4x - 3) = 4x^2 - 11x + 6$ Correct

The last trial gives the correct factorization.

(b) Again, we try various possibilities. The positive factors of 6 could be 2 and 3 or 1 and 6. As factors of -5, we have only -1 and 5 or -5 and 1.

$(2x - 5)(3x + 1) = 6x^2 - 13x - 5$ Incorrect

$(3x - 5)(2x + 1) = 6x^2 - 7x - 5$ Correct

Thus, $6x^2 - 7x - 5$ factors as $(3x - 5)(2x + 1)$.

PERFECT SQUARE TRINOMIALS

$$x^2 + 2xy + y^2 = (x + y)^2$$
$$x^2 - 2xy + y^2 = (x - y)^2$$

EXAMPLE 8 Factoring Perfect Square Trinomials

Factor each polynomial.

(a) $16x^2 - 40x + 25$ **(b)** $169x^2 + 104x + 16$

Solution

(a) First verify that the middle term of the trinomial being factored, $-40x$ here, is twice the product of the two terms in the binomial $4x - 5$.

$$-40x = 2(4x)(-5)$$

Because $16x^2 = (4x)^2$ and $25 = 5^2$, use the second pattern shown above.

$$16x^2 - 40x + 25 = (4x)^2 - 2(4x)(5) + (5)^2$$
$$= (4x - 5)^2$$

(b) $169x^2 + 104x + 16 = (13x + 4)^2$.

The trinomial is a perfect square because $2(13x)(4) = 104x$.

Factoring Special Binomials

DIFFERENCE OF SQUARES

$$x^2 - y^2 = (x + y)(x - y)$$

EXAMPLE 9 Factoring Differences of Squares

Factor each polynomial.

(a) $4x^2 - 9$ **(b)** $256x^4 - 625$ **(c)** $16x^4 - 1$

Solution

(a) $4x^2 - 9$

$= (2x)^2 - 3^2$

$= (2x + 3)(2x - 3)$ Difference of squares

(b) $256x^4 - 625$

$= (16x^2)^2 - 25^2$

$= (16x^2 + 25)(16x^2 - 25)$ Difference of squares

$= (16x^2 + 25)(4x + 5)(4x - 5)$ Difference of squares

(c) $16x^4 - 1$

$= (4x^2 + 1)(4x^2 - 1)$ Difference of squares

$= (4x^2 + 1)(2x + 1)(2x - 1)$ Difference of squares

In parts (b) and (c), the sum of squares cannot be factored further.

7.6 EXERCISES

CONCEPT CHECK *Match each polynomial in Column I with the correct description in Column II.*

I	II
1. $2x^2 - 3x + 7$	**A.** a monomial
2. $x^2 - 9$	**B.** a trinomial of degree 2
3. $9x^3 - 2x^2 + 3$	**C.** a polynomial of degree 3
4. $-11x^5$	**D.** a binomial

Find each sum or difference.

5. $(3x^2 - 4x + 5) + (-2x^2 + 3x - 2)$

6. $(4x^3 - 3x^2 + 5) + (-3x^3 - x^2 + 5)$

7. $(12x^2 - 8x + 6) - (3x^2 - 4x + 2)$

8. $(8x^2 - 5x) - (3x^2 - 2x + 4)$

9. $(6x^4 - 3x^2 + x) - (2x^3 + 5x^2 + 4x) + (x^2 - x)$

10. $-(8x^3 + x - 3) + (2x^3 + x^2) - (4x^2 + 3x - 1)$

11. $5(2x^2 - 3x + 7) - 2(6x^2 - x + 12)$

12. $6(3x^2 - 4x + 8) - 3(2x^2 - 3x + 1)$

Find each product.

13. $(x + 3)(x - 8)$ **14.** $(x - 3)(x - 9)$

15. $(4x - 1)(7x + 2)$ **16.** $(5x - 6)(3x + 4)$

17. $4x^2(3x^3 - 5x + 1)$ **18.** $2x^3(x^2 - 4x + 3)$

19. $(2x + 3)(2x - 3)$ **20.** $(8x - 3)(8x + 3)$

21. $(x + 7)^2$ **22.** $(x + 9)^2$

23. $(5x - 3)^2$ **24.** $(6x - 5)^2$

25. $(x + 3)(10x^3 + 4x^2 - 5x + 2)$

26. $(x + 2)(12x^3 - 3x^2 + x + 1)$

Factor the greatest common factor from each polynomial.

27. $12x + 36$ **28.** $15x + 75$

29. $8x^3 - 4x^2 + 6x$ **30.** $9x^3 + 12x^2 - 36x$

31. $8x^4 + 6x^3 - 12x^2$ **32.** $2x^5 - 10x^4 + 16x^3$

Factor each trinomial.

33. $x^2 - 2x - 15$ **34.** $x^2 + 8x + 12$

35. $x^2 + 2x - 35$ **36.** $x^2 - 7x + 6$

37. $6x^2 - 48x - 120$ **38.** $8x^2 - 24x - 320$

39. $3x^3 + 12x^2 + 9x$ **40.** $9x^4 - 54x^3 + 45x^2$

41. $6x^2 + 5x - 6$ **42.** $14x^2 + 11x - 15$

43. $5x^2 - 7x - 6$ **44.** $12x^2 + 11x - 5$

45. $21x^2 - x - 2$ **46.** $30x^2 + x - 1$

47. $24x^4 + 10x^3 - 4x^2$ **48.** $18x^5 + 15x^4 - 75x^3$

Factor each perfect square trinomial. It may be necessary to factor out a common factor first.

49. $x^2 - 8x + 16$ **50.** $x^2 - 12x + 36$

51. $x^2 + 14x + 49$ **52.** $x^2 + 18x + 81$

53. $9x^2 - 12x + 4$ **54.** $16x^2 - 40x + 25$

55. $32x^2 - 48x + 18$ **56.** $20x^2 - 100x + 125$

57. $4x^2 + 28x + 49$ **58.** $9x^2 - 12x + 4$

Factor each difference of squares.

59. $x^2 - 36$ **60.** $x^2 - 64$

61. $100 - x^2$ **62.** $25 - x^2$

63. $9x^2 - 16$ **64.** $16x^2 - 25$

65. $25x^4 - 9$ **66.** $36x^4 - 25$

67. $x^4 - 625$ **68.** $x^4 - 81$

Decide on a factoring method, and then factor the polynomial completely.

69. $13x^3 + 39x^2 + 52x$ **70.** $8x^2 - 10x - 3$

71. $12x^2 + 16x - 35$ **72.** $36x^2 + 60x + 25$

73. $4x^2 + 28x + 49$ **74.** $6x^4 + 7x^2 - 3$

75. $x^8 - 81$ **76.** $100x^4 - 100$

7.7 **QUADRATIC EQUATIONS AND APPLICATIONS**

OBJECTIVES

1 Recognize a quadratic equation in standard form.

2 Use the zero-factor property to solve a quadratic equation (when applicable).

3 Use the square root property to solve a quadratic equation (when applicable).

4 Use the quadratic formula to solve a quadratic equation.

5 Solve an application that leads to quadratic equations.

Quadratic Equations and the Zero-Factor Property

We have considered linear equations and inequalities in earlier sections. Now we investigate *second-degree*, or *quadratic*, equations.

QUADRATIC EQUATION

An equation that can be written in the form

$$ax^2 + bx + c = 0 \quad \text{Standard form}$$

where a, b, and c are real numbers, with $a \neq 0$, is a **quadratic equation.** The form of the equation given above is called **standard form.**

One method of solving a quadratic equation depends on the following property.

ZERO-FACTOR PROPERTY

If $ab = 0$, then $a = 0$ or $b = 0$ or both.

Solving a quadratic equation by the zero-factor property requires that the equation be in standard form before factoring.

EXAMPLE 1 **Using the Zero-Factor Property**

Solve $6x^2 + 7x = 3$.

Solution

$6x^2 + 7x = 3$		Given equation
$6x^2 + 7x - 3 = 0$		Write in standard form.
$(3x - 1)(2x + 3) = 0$		Factor.
$3x - 1 = 0$ or $2x + 3 = 0$		Zero-factor property
$3x = 1$ or $2x = -3$		Solve each equation.
$x = \dfrac{1}{3}$ or $x = -\dfrac{3}{2}$		Divide.

Check. Substitute $\frac{1}{3}$ and $-\frac{3}{2}$ in the original equation. The solution set is $\left\{\frac{1}{3}, -\frac{3}{2}\right\}$.

The Square Root Property

A quadratic equation of the form $x^2 = k$, $k \geq 0$, can be solved as follows.

$x^2 = k$		
$x^2 - k = 0$		Subtract k.
$(x + \sqrt{k})(x - \sqrt{k}) = 0$		Factor, using radicals.
$x + \sqrt{k} = 0$ or $x - \sqrt{k} = 0$		Zero-factor property
$x = -\sqrt{k}$ or $x = \sqrt{k}$		Solve each equation.

Historical Controversy For centuries, mathematicians wrestled with finding a formula that could solve **cubic (third-degree) equations.** In sixteenth-century Italy, **Niccolo Tartaglia** had developed a method of solving a cubic equation of the form

$$x^3 + mx = n.$$

Girolamo Cardano begged to know Tartaglia's method, and after he was told, he was sworn to secrecy. Nonetheless, Cardano published Tartaglia's method in his 1545 work *Ars Magna* (although he did give Tartaglia credit).

The formula for finding one real solution of the equation $x^3 + mx = n$ is

$$x = \sqrt[3]{\frac{n}{2} + \sqrt{\left(\frac{n}{2}\right)^2 \left(\frac{m}{3}\right)^3}}$$
$$- \sqrt[3]{-\left(\frac{n}{2}\right) + \sqrt{\left(\frac{n}{2}\right)^2 + \left(\frac{m}{3}\right)^3}}.$$

Show that 2 is a solution of $x^3 + 9x = 26$ using this formula, with $m = 9$ and $n = 26$.

This leads to the **square root property** for solving equations.

> **SQUARE ROOT PROPERTY**
>
> If $k \geq 0$, then the solutions of $x^2 = k$ are $\pm\sqrt{k}$.

If $k > 0$, the equation $x^2 = k$ has two real solutions. If $k = 0$, there is only one solution, 0. If $k < 0$, there are no real solutions. (In this last case, there *are complex* solutions. Complex numbers are covered in more extensive algebra texts.)

EXAMPLE 2 **Using the Square Root Property**

Use the square root property to solve each quadratic equation for real solutions.

(a) $x^2 = 25$ **(b)** $x^2 = 18$ **(c)** $x^2 = -3$ **(d)** $(x - 4)^2 = 12$

Solution

(a) Since $\sqrt{25} = 5$, the solution set of the equation $x^2 = 25$ is

$$\{5, -5\}, \quad \text{which may be abbreviated} \quad \{\pm 5\}.$$

(b)
$$x^2 = 18$$
$$x = \pm\sqrt{18} \qquad \text{Square root property}$$
$$x = \pm\sqrt{9 \cdot 2} \qquad \text{Factor 18 as 9 \cdot 2.}$$
$$x = \pm\sqrt{9} \cdot \sqrt{2} \qquad \text{Product rule for square roots}$$
$$x = \pm 3\sqrt{2} \qquad \sqrt{9} = 3$$

The solution set is $\left\{\pm 3\sqrt{2}\right\}$.

(c) Because $-3 < 0$, the equation $x^2 = -3$ has no real solutions. The solution set is $\varnothing$.

(d) To solve this equation, we extend the square root property to a binomial base with exponent 2.

$$(x - 4)^2 = 12$$
$$x - 4 = \pm\sqrt{12} \qquad \text{Square root property extended}$$
$$x = 4 \pm\sqrt{12} \qquad \text{Add 4.}$$
$$x = 4 \pm\sqrt{4 \cdot 3} \qquad \text{Factor 12 as 4 \cdot 3.}$$
$$x = 4 \pm 2\sqrt{3} \qquad \sqrt{4 \cdot 3} = 2\sqrt{3}$$

The solution set is $\left\{4 \pm 2\sqrt{3}\right\}$.

Completing the square, a technique used in deriving the quadratic formula, has important applications in algebra. To transform the expression $x^2 + kx$ into the square of a binomial, we add to it the square of half the coefficient of x.

$$\left[\left(\frac{1}{2}\right)k\right]^2 = \frac{k^2}{4}$$

We then obtain

$$x^2 + kx + \frac{k^2}{4} = \left(x + \frac{k}{2}\right)^2.$$

For example, to make $x^2 + 6x$ the square of a binomial, we add 9, because $9 = \left[\frac{1}{2}(6)\right]^2$. This results in the trinomial $x^2 + 6x + 9$, which is equal to $(x + 3)^2$.

The Greeks had a method of completing the square geometrically. For example, to complete the square for $x^2 + 6x$, begin with a square of side x. Add three rectangles of width 1 and length x to the right side and the bottom. Each rectangle has area $1x$, or x, so the total area of the figure is now $x^2 + 6x$. To fill in the corner (that is, "complete the square"), we add 9 1-by-1 squares as shown. The new completed square has sides of length $x + 3$ and area

$$(x + 3)^2 = x^2 + 6x + 9.$$

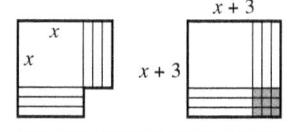

The Quadratic Formula

By *completing the square* (see the margin note at left), we can derive one of the most important formulas in algebra, the *quadratic formula*.

A Radical Departure from the Other Methods of Evaluating the Golden Ratio Recall that the Golden Ratio is found in numerous places in mathematics, art, and nature. In an earlier margin note we showed that the *"continued" fraction*

$$1 + \cfrac{1}{1 + \cfrac{1}{1 + \cfrac{1}{1 + \cdots}}}$$

is equal to the Golden Ratio, $\frac{1 + \sqrt{5}}{2}$. Now consider this *"nested" radical:*

$$\sqrt{1 + \sqrt{1 + \sqrt{1 + \cdots}}}$$

Let *x* represent this radical expression. Because it appears "within itself," we can write

$$x = \sqrt{1 + x}$$
$$x^2 = 1 + x$$
$$x^2 - x - 1 = 0.$$

Using the quadratic formula, with $a = 1$, $b = -1$, and $c = -1$, it can be shown that the positive solution of this equation, and thus the value of the nested radical, is (you guessed it!) the Golden Ratio.

$$ax^2 + bx + c = 0 \qquad \text{Standard quadratic equation } (a > 0)$$

$$x^2 + \frac{b}{a}x + \frac{c}{a} = 0 \qquad \text{Divide by } a.$$

$$x^2 + \frac{b}{a}x = -\frac{c}{a} \qquad \text{Add } -\frac{c}{a}.$$

$$x^2 + \frac{b}{a}x + \frac{b^2}{4a^2} = \frac{b^2}{4a^2} - \frac{c}{a} \qquad \text{Add } \frac{b^2}{4a^2}.$$

$$\left(x + \frac{b}{2a}\right)^2 = \frac{b^2 - 4ac}{4a^2} \qquad \begin{array}{l}\text{Factor on the left.}\\ \text{Combine terms on the right.}\end{array}$$

$$x + \frac{b}{2a} = \pm \sqrt{\frac{b^2 - 4ac}{4a^2}} \qquad \text{Square root property}$$

$$x + \frac{b}{2a} = \pm \frac{\sqrt{b^2 - 4ac}}{\sqrt{4a^2}} \qquad \text{Quotient rule for square roots}$$

$$x = -\frac{b}{2a} \pm \frac{\sqrt{b^2 - 4ac}}{2a} \qquad \text{Subtract } \frac{b}{2a}.$$

Be careful; $-b \pm \sqrt{b^2 - 4ac}$ is all written over 2*a*.

$$x = \frac{-b \pm \sqrt{b^2 - 4ac}}{2a} \qquad \begin{array}{l}\text{Combine like terms.}\\ \text{This is also valid for } a < 0.\end{array}$$

QUADRATIC FORMULA

The solutions of $ax^2 + bx + c = 0$, $a \neq 0$, are given by the **quadratic formula.**

$$x = \frac{-b \pm \sqrt{b^2 - 4ac}}{2a}$$

EXAMPLE 3 **Using the Quadratic Formula**

Solve $x^2 - 4x + 2 = 0$.

Solution

$$x = \frac{-b \pm \sqrt{b^2 - 4ac}}{2a} \qquad \text{Quadratic formula}$$

$$x = \frac{-(-4) \pm \sqrt{(-4)^2 - 4(1)(2)}}{2(1)} \qquad a = 1, b = -4, c = 2$$

$$x = \frac{4 \pm \sqrt{16 - 8}}{2} \qquad \text{Start to simplify.}$$

$$x = \frac{4 \pm 2\sqrt{2}}{2} \qquad \sqrt{16 - 8} = \sqrt{8} = 2\sqrt{2}$$

$$x = \frac{2(2 \pm \sqrt{2})}{2} \qquad \text{Factor out 2 in the numerator.}$$

Factor, and then divide out the common factor.

$$x = 2 \pm \sqrt{2} \qquad \text{Divide out the common factor.}$$

The solution set is $\left\{2 + \sqrt{2}, 2 - \sqrt{2}\right\}$, which is abbreviated $\left\{2 \pm \sqrt{2}\right\}$.

EXAMPLE 4 Using the Quadratic Formula

Solve $2x^2 = x + 4$.

Solution

First write the equation in standard form as $2x^2 - x - 4 = 0$.

$$x = \frac{-(-1) \pm \sqrt{(-1)^2 - 4(2)(-4)}}{2(2)}$$ Quadratic formula with $a = 2$, $b = -1$, $c = -4$

$$x = \frac{1 \pm \sqrt{1 + 32}}{4}$$ Simplify the radicand.

$$x = \frac{1 \pm \sqrt{33}}{4}$$ Add.

The solution set is $\left\{ \frac{1 \pm \sqrt{33}}{4} \right\}$.

Applications

EXAMPLE 5 Applying a Quadratic Equation

Two cars left an intersection at the same time, one heading due north, and the other due west. When they were exactly 100 miles apart, the car headed north had gone 20 miles farther than the car headed west. How far had each car traveled?

Solution

Step 1 **Read** the problem carefully.

Step 2 **Assign a variable.** See **Figure 16**.

    Let $x =$ the distance traveled by the car headed west.

    Then $(x + 20) =$ the distance traveled by the car headed north.

    The cars are 100 miles apart. The hypotenuse of the right triangle equals 100.

Step 3 **Write an equation.**

$$c^2 = a^2 + b^2$$ Pythagorean theorem

$$100^2 = x^2 + (x + 20)^2$$ Substitute.

Step 4 **Solve.** $10,000 = x^2 + x^2 + 40x + 400$ Square the binomial.

$$2x^2 + 40x - 9600 = 0$$ Standard form

$$x^2 + 20x - 4800 = 0$$ Divide both sides by 2.

Use the quadratic formula to find x.

$$x = \frac{-20 \pm \sqrt{20^2 - 4(1)(-4800)}}{2(1)}$$ $a = 1$, $b = 20$, $c = -4800$

$$x = \frac{-20 \pm \sqrt{19,600}}{2}$$ Simplify under the radical.

$$x = 60 \quad \text{or} \quad x = -80$$ Use a calculator.

Step 5 **State the answer.** Distance cannot be negative here, so discard the negative solution. The required distances are 60 miles and $60 + 20 = 80$ miles.

Step 6 **Check.** Because $60^2 + 80^2 = 100^2$, the answer is correct.

Évariste Galois (1811–1832), as a young Frenchman, agreed to fight a duel. He had been engaged for some time in mathematical research that centered on solving equations by using **group theory**. Now, anticipating the possibility of his death, he summarized the essentials of his discoveries in a letter to a friend. The next day Galois was killed. He was not yet 21 years old when he died.

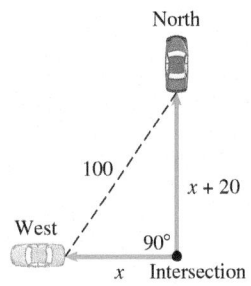

Figure 16

Niels Henrik Abel (1802–1829) of Norway was identified in childhood as a mathematical genius but never received in his lifetime the professional recognition his work deserved. He proved that a general formula for solving **fifth-degree equations** does not exist. The quadratic formula (for equations of degree 2) is well known, and formulas do exist for solving third- and fourth-degree equations. Abel's accomplishment ended a search that had lasted for years.

EXAMPLE 6 **Applying a Quadratic Equation**

If an object on Earth is projected upward from the top of a 144-foot building with an initial velocity of 112 feet per second, its position (in feet above the ground) is given by

$$s = -16t^2 + 112t + 144,$$

where t is time in seconds after it was projected and air resistance is neglected. How long does it take for the object to hit the ground?

Solution

When the object hits the ground, its distance above the ground is 0. Find t when s is 0.

$0 = -16t^2 + 112t + 144$ Let $s = 0$.

$0 = t^2 - 7t - 9$ Divide both sides by -16.

$t = \dfrac{7 \pm \sqrt{49 + 36}}{2}$ Quadratic formula

$t = \dfrac{7 \pm \sqrt{85}}{2}$ Add.   This must be rejected.

$t \approx 8.1$ or $t \approx -1.1$ Use a calculator.

Because it cannot hit the ground before it is projected, discard the negative solution. It will hit the ground about 8.1 seconds after it is projected.

7.7 EXERCISES

CONCEPT CHECK *Complete each statement.*

1. In the form $(x + 2)(x - 3) = 6$, the zero-factor property _____ be used to solve the equation.
(can/cannot)

2. To solve the quadratic equation $4x^2 + 3x - 7 = 0$ by the quadratic formula, use $a = $ ___, $b = $ ___, and $c = $ ___.

3. The equation $x^2 = 11$ has _____ real solutions.
(how many?)

4. The equation $(2x + 1)^2 = -4$ has _____ real solutions.
(how many?)

Solve each equation by the zero-factor property.

5. $(x + 3)(x - 9) = 0$

6. $(x + 6)(x + 4) = 0$

7. $(2x - 7)(5x + 1) = 0$

8. $(7x - 3)(6x + 4) = 0$

9. $x(x + 7) = 0$

10. $x(x - 8) = 0$

11. $x^2 - x - 12 = 0$

12. $x^2 + 4x - 5 = 0$

13. $x^2 + 9x + 14 = 0$

14. $x^2 + 3x - 4 = 0$

15. $2x^2 + 3x - 5 = 0$

16. $3x^2 - 5x - 28 = 0$

17. $12x^2 + 4x = 1$

18. $15x^2 + 7x = 2$

19. $(x + 4)(x - 6) = -16$

20. $(x - 1)(3x + 2) = 4x$

Solve each equation for real solutions by using the square root property.

21. $x^2 = 64$

22. $x^2 = 16$

23. $x^2 = 24$

24. $x^2 = 48$

25. $x^2 = -5$

26. $x^2 = -10$

27. $(x - 4)^2 = 9$

28. $(x + 3)^2 = 25$

29. $(4 - x)^2 = 3$

30. $(3 + x)^2 = 11$

31. $(2x - 5)^2 = 13$

32. $(4x + 1)^2 = 19$

Solve each equation for real solutions by the quadratic formula.

33. $x^2 + 3x - 7 = 0$

34. $x^2 + 2x - 5 = 0$

35. $2x^2 = 2x + 1$ **36.** $9x^2 = 6x + 1$

37. $x^2 - 1 = x$ **38.** $2x^2 - 5 = 4x$

39. $4x(x + 1) = 1$ **40.** $4x(x - 1) = 19$

41. $(x + 2)(x - 3) = 1$ **42.** $(x - 5)(x + 2) = 6$

43. $x^2 - 6x = -14$ **44.** $x^2 - 2x = -2$

Solve each problem. Use a calculator as necessary, and round approximations to the nearest tenth.

45. ***Distances Traveled by Ships***   Two ships leave port at the same time, one heading due south and the other heading due east. Several hours later, they are 170 miles apart. If the ship traveling south travels 70 miles farther than the other, how many miles does each travel?

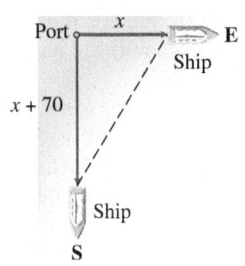

46. ***Distances Traveled by Ships***   Two ships leave port at the same time, one heading due north and the other heading due east. Several hours later, they are 20 miles apart. If the ship traveling north travels 4 miles farther than the other, how many miles does each travel?

47. ***Height of a Kite***   Paulette is flying a kite that is 30 feet farther above her hand than its horizontal distance from her. The string from her hand to the kite is 150 feet long. How far is the kite above her hand?

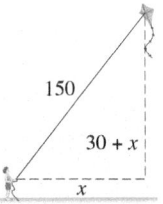

48. ***Leaning Ladder***   A 13-foot ladder is leaning against a house. The distance from the bottom of the ladder to the house is 7 feet less than the distance from the top of the ladder to the ground. How far is the bottom of the ladder from the house?

49. ***Length of a Wire***   Suppose that a wire is attached to a top of a building 400 feet high and is pulled tight. It is attached to the ground 100 feet from the base of the building, as shown in the figure. How long is the wire?

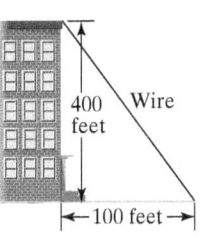

50. ***Length of a Wire***   Suppose that a wire is attached to the top of a building 407 feet high and is pulled tight. The length of the wire is twice the distance between the base of the building and the point on the ground where the wire is attached. How long is the wire?

51. ***Side Lengths of a Triangle***   Find the lengths of the sides of the right triangle.

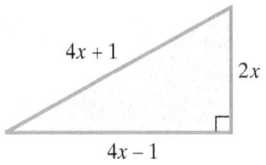

52. ***Side Lengths of a Triangle***   Find the lengths of the sides of the right triangle.

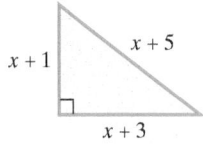

53. ***Size of a Toy Piece***   A manufacturer needs a piece of plastic in the shape of a right triangle with the longer leg 2 centimeters more than twice as long as the shorter leg, and the hypotenuse 1 centimeter more than the longer leg. How long should the three sides of the piece be?

54. ***Size of a Developer's Property***   A developer owns a piece of land enclosed on three sides by streets, giving it the shape of a right triangle. The hypotenuse is 8 meters longer than the longer leg, and the shorter leg is 9 meters shorter than the hypotenuse. Find the lengths of the three sides of the property.

55. ***Height of a Projectile***   An object is projected directly upward from the ground. After t seconds its distance in feet above the ground is $s = 144t - 16t^2$.

(a) After how many seconds will the object be 128 feet above the ground? (*Hint:* Look for a common factor before solving the equation.)

(b) When does the object strike the ground?

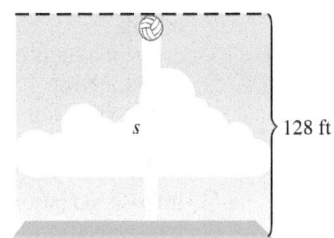

Ground level

56. ***Height of a Projectile***   A building is 400 feet high. Suppose that a ball is projected upward from the top, and its position s in feet above the ground is given by the equation

$$s = -16t^2 + 45t + 400,$$

where t is the number of seconds elapsed. How long will it take for the ball to reach a height of 200 feet above the ground?

57. *Height of a Projectile* A high-rise condominium building is 407 feet high. Suppose that a ball is projected upward from the top and its position s in feet above the ground is given by the equation $s = -16t^2 + 75t + 407$, where t is the number of seconds elapsed. How long will it take for the ball to reach a height of 450 feet above the ground?

58. *Height of a Projectile* Refer to the equations in **Exercises 56 and 57.** Suppose that the first sentence in each problem did not give the height. How could you use the equation to determine the height?

59. *Dimensions of a Strip of Flooring around a Rug* Jason and Marin want to buy a rug for a room that is 15 feet by 20 feet. They want to leave an even strip of flooring uncovered around the edges of the room. How wide a strip will they have if they buy a rug with an area of 234 square feet?

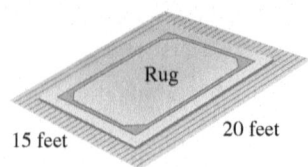

60. *Dimensions of a Border around a Pool* A club swimming pool is 30 feet wide and 40 feet long. The club members want an exposed aggregate border in a strip of uniform width around the pool. They have enough material for 296 square feet. How wide can the strip be?

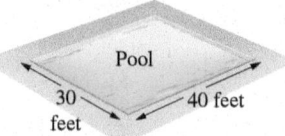

61. *Dimensions of a Garden* Arif's backyard is 20 meters by 30 meters. He wants to put a flower garden in the middle of the backyard, leaving a strip of grass of uniform width around the flower garden. To be happy, he must have 184 square meters of grass. Under these conditions, what will the length and width of the garden be?

62. *Dimensions of a Piece of Sheet Metal* A rectangular piece of sheet metal has a length that is 4 inches less than twice the width. A square piece 2 inches on a side is cut from each corner. The sides are then turned up to form an uncovered box of volume 256 cubic inches. Find the length and width of the original piece of metal.

Chapter **7** SUMMARY

KEY TERMS

7.1

algebraic expression
equation
linear equation
  (first-degree equation)
solution
solution set
equivalent equations
conditional equation
identity
contradiction
literal equation
formula
solving for a specified
  variable
mathematical model

7.3

ratio
proportion
terms (of a
  proportion)
extremes
means
cross products of a
  proportion
direct variation (direct
  proportion)
constant of variation
inverse variation
joint variation
combined variation

7.4

interval
interval notation
linear inequality
compound inequality
tolerance

7.5

exponent
base
exponential expression
scientific notation

7.6

term (monomial)
polynomial

degree (of a polynomial)
like terms
descending powers
difference of two squares
factoring
factor
factored form
prime polynomial
factored completely

7.7

quadratic equation
standard form
square root property
quadratic formula

NEW SYMBOLS

a to b, $\dfrac{a}{b}$, $a:b$ ratio of a to b

∞, $-\infty$ infinity, negative infinity

$(-\infty, 2], (-3, \infty), (-3, 2]$
    examples of interval notation

a^n a to the nth power

TEST YOUR WORD POWER

See how well you have learned the vocabulary in this chapter.

1. A **solution set** is the set of numbers that make
   A. an expression undefined.
   B. an equation false.
   C. an equation true.
   D. an expression equal to 0.

2. An **inequality** is
   A. a statement that two algebraic expressions are equal.
   B. a point on a number line.
   C. an equation with no solutions.
   D. a statement with algebraic expressions related by $<, \leq, >,$ or $\geq$.

3. A **ratio**
   A. compares two quantities using a quotient.
   B. says that two quotients are equal.
   C. is a product of two quantities.
   D. is a difference between two quantities.

4. A **proportion**
   A. compares two quantities using a quotient.
   B. says that two ratios are equal.
   C. is a product of two quantities.
   D. is a difference between two quantities.

5. A **monomial** is a polynomial with
   A. only one term.
   B. exactly two terms.
   C. exactly three terms.
   D. more than three terms.

6. A **binomial** is a polynomial with
   A. only one term.
   B. exactly two terms.
   C. exactly three terms.
   D. more than three terms.

7. A **trinomial** is a polynomial with
   A. only one term.
   B. exactly two terms.
   C. exactly three terms.
   D. more than three terms.

8. The **FOIL** method is used to
   A. add two binomials.
   B. add two trinomials.
   C. multiply two binomials.
   D. multiply two trinomials.

9. **Factoring** is
   A. a method of multiplying polynomials.
   B. the process of writing a polynomial as a product.
   C. the answer in a multiplication problem.
   D. a way to add the terms of a polynomial.

10. A **quadratic equation** is an equation that can be written in the form
   A. $Ax + B = C.$ **B.** $ax^2 + bx + c = 0$ $(a \neq 0).$
   C. $x = k.$ **D.** $x = k^2.$

ANSWERS

1. C   **2.** D   **3.** A   **4.** B   **5.** A
6. B   **7.** C   **8.** C   **9.** B   **10.** B

QUICK REVIEW

Concepts	*Examples*

7.1 LINEAR EQUATIONS

Addition and Multiplication Properties of Equality The same number may be added to (or subtracted from) each side of an equation to obtain an equivalent equation. Similarly, the same nonzero number may be multiplied by (or divided into) each side of an equation to obtain an equivalent equation.	Solve the equation.

$$4(8 - 3x) = 32 - 8(x + 2)$$

$32 - 12x = 32 - 8x - 16$ Distributive property

$32 - 12x = 16 - 8x$ Combine like terms.

Solving a Linear Equation in One Variable

Step 1 Clear fractions.

Step 2 Simplify each side separately.

Step 3 Isolate the variable terms on one side.

Step 4 Transform the equation so that the coefficient of the variable is 1.

Step 5 Check.

$32 - 12x + 12x = 16 - 8x + 12x$ Add 12x.

$32 = 16 + 4x$ Combine like terms.

$32 - 16 = 16 + 4x - 16$ Subtract 16.

$16 = 4x$ Combine like terms.

$\dfrac{16}{4} = \dfrac{4x}{4}$ Divide by 4.

$x = 4$ Interchange sides.

To check, substitute 4 for x in the original equation. A true statement results, and the solution set is $\{4\}$.

Concepts	Examples

Solving for a Specified Variable

Step 1 If the equation contains fractions, multiply both sides by the LCD to clear the fractions.

Step 2 Transform the equation so that all terms with the specified variable are on one side and all terms without that variable are on the other side.

Step 3 Divide each side by the factor that is the coefficient of the specified variable.

Solve the formula for h.

$$\mathcal{A} = \frac{1}{2}bh \qquad \text{Area of a triangle}$$

$$2\mathcal{A} = 2\left(\frac{1}{2}bh\right) \qquad \text{Multiply by 2.}$$

$$2\mathcal{A} = bh \qquad 2 \cdot \frac{1}{2} = 1$$

$$\frac{2\mathcal{A}}{b} = h, \quad \text{or} \quad h = \frac{2\mathcal{A}}{b} \qquad \begin{array}{l}\text{Divide by } b.\\ \text{Interchange sides.}\end{array}$$

7.2 APPLICATIONS OF LINEAR EQUATIONS

Solving an Applied Problem

Step 1 Read the problem.

Step 2 Assign a variable.

Step 3 Write an equation.

Step 4 Solve the equation.

Step 5 State the answer.

Step 6 Check.

How many liters of 30% alcohol solution and 80% alcohol solution must be mixed to obtain 100 L of 50% alcohol solution?

Let $\quad x$ = number of liters of 30% solution.

Then $100 - x$ = number of liters of 80% solution.

Percent (as a decimal)	Liters of Solution	Liters of Pure Alcohol
0.30	x	$0.30x$
0.80	$100 - x$	$0.80(100 - x)$
0.50	100	$0.50(100)$

From the last column of the table, write an equation.

$$0.30x + 0.80(100 - x) = 0.50(100)$$

$$0.30x + 80 - 0.80x = 50 \qquad \text{Distributive property}$$

$$3x + 800 - 8x = 500 \qquad \begin{array}{l}\text{Multiply by 10 to clear}\\ \text{decimals.}\end{array}$$

$$-5x + 800 = 500 \qquad \text{Combine like terms.}$$

$$-5x = -300 \qquad \text{Subtract 800.}$$

$$x = 60 \qquad \text{Divide by } -5.$$

60 liters of 30% solution and

$$100 - 60 = 40 \text{ liters}$$

of 80% solution are required.

7.3 RATIO, PROPORTION, AND VARIATION

Ratio

To write a ratio, express quantities in the same units and write as a quotient.

Write a ratio for the word phrase 4 feet to 8 inches.

$$\frac{4 \text{ feet}}{8 \text{ inches}} = \frac{48 \text{ inches}}{8 \text{ inches}} = \frac{48}{8} = \frac{6}{1}$$

Concepts   *Examples*

Proportion

To solve a proportion, use the method of cross products.

Solve the proportion.

$$\frac{x}{12} = \frac{35}{60}$$

$60x = 12 \cdot 35$ Set cross products equal.

$60x = 420$ Multiply.

$x = 7$ Divide by 60.

The solution set is $\{7\}$.

Solving a Variation Problem

Step 1   Write the variation equation.

$$y = kx \quad \text{or} \quad y = kx^n \quad \text{Direct variation}$$

$$y = \frac{k}{x} \quad \text{or} \quad y = \frac{k}{x^n} \quad \text{Inverse variation}$$

Step 2   Find k by substituting the appropriate given values of x and y into the equation.

Step 3   Rewrite the variation equation with the value of k from Step 2.

Step 4   Substitute the remaining values, and solve for the unknown.

The weight y of an object on Earth is directly proportional to the weight x of that same object on the moon. If a 200-pound astronaut weighs 32 pounds on the moon, how much would a dog weighing 8 pounds on the moon weigh on Earth?

$$y = kx \qquad \text{Variation equation (Step 1)}$$

$$200 = k \cdot 32 \quad \text{Let } y = 200 \text{ and } x = 32.$$

$$\frac{200}{32} = k \qquad \text{Solve for } k. \text{ (Step 2)}$$

$$k = \frac{25}{4} \qquad \text{$\frac{200}{32}$ in lowest terms is $\frac{25}{4}$.}$$

Rewrite the variation equation $y = kx$ with the value of k.

$$y = \frac{25}{4}x \qquad \text{Let } k = \frac{25}{4}. \text{ (Step 3)}$$

$$y = \frac{25}{4}(8) \qquad \text{Let } x = 8. \text{ (Step 4)}$$

$$y = 50 \qquad \text{Solve for } y.$$

The dog would weigh 50 pounds on Earth.

7.4 LINEAR INEQUALITIES

Solving Linear Inequalities in One Variable

Step 1   Simplify each side of the inequality by clearing parentheses and combining like terms.

Step 2   Use the addition property of inequality to get all terms with variables on one side and all terms without variables on the other side.

Step 3   Use the multiplication property of inequality to write the inequality in the form $x < k$ or $x > k$.

When an inequality is multiplied or divided by a negative number, the direction of the inequality symbol must be reversed.

Solve the inequality.

$$3(x + 2) - 5x \leq 12$$

$$3x + 6 - 5x \leq 12 \qquad \text{Distributive property}$$

$$-2x + 6 \leq 12 \qquad \text{Combine like terms.}$$

$$-2x + 6 - 6 \leq 12 - 6 \quad \text{Subtract 6.}$$

$$-2x \leq 6 \qquad \text{Combine like terms.}$$

$$\frac{-2x}{-2} \geq \frac{6}{-2} \qquad \begin{array}{l}\text{Divide by -2.}\\ \text{Change $\leq$ to $\geq$.}\end{array}$$

$$x \geq -3$$

The solution set $[-3, \infty)$ is graphed below.

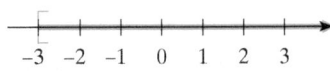

Concepts	Examples
To solve a compound inequality, work with all three parts at the same time.	Solve the compound inequality. $$-4 < \quad 2x + 3 \quad \leq 7$$ $$-4 - 3 < 2x + 3 - 3 \leq 7 - 3 \quad \text{Subtract 3.}$$ $$-7 < \quad 2x \quad \leq 4$$ $$\frac{-7}{2} < \quad \frac{2x}{2} \quad \leq \frac{4}{2} \quad \text{Divide by 2.}$$ $$-\frac{7}{2} < \quad x \quad \leq 2$$ The solution set $\left(-\frac{7}{2}, 2\right]$ is graphed below. 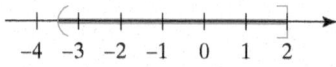

7.5 PROPERTIES OF EXPONENTS AND SCIENTIFIC NOTATION

Definitions and Rules for Exponents For all integers m and n and all real numbers a and b, the following rules apply.	Apply the rules for exponents.
Product Rule $\quad a^m \cdot a^n = a^{m+n}$	$3^4 \cdot 3^2 = 3^6$
Quotient Rule $\quad \dfrac{a^m}{a^n} = a^{m-n} \quad (a \neq 0)$	$\dfrac{2^5}{2^3} = 2^2$
Zero Exponent $\quad a^0 = 1 \quad (a \neq 0)$	$27^0 = 1 \qquad (-5)^0 = 1$
Negative Exponent $\quad a^{-n} = \dfrac{1}{a^n} \quad (a \neq 0)$	$5^{-2} = \dfrac{1}{5^2}$
Power Rules (a) $(a^m)^n = a^{mn}$	$(6^3)^4 = 6^{12}$
(b) $(ab)^m = a^m b^m$	$(5 \cdot 2)^4 = 5^4 \cdot 2^4$
(c) $\left(\dfrac{a}{b}\right)^n = \dfrac{a^n}{b^n} \quad (b \neq 0)$	$\left(\dfrac{2}{3}\right)^5 = \dfrac{2^5}{3^5}$
Special Rules for Negative Exponents $\dfrac{1}{a^{-n}} = a^n \quad (a \neq 0) \qquad \dfrac{a^{-n}}{b^{-m}} = \dfrac{b^m}{a^n} \quad (a, b \neq 0)$ $a^{-n} = \left(\dfrac{1}{a}\right)^n \quad (a \neq 0) \qquad \left(\dfrac{a}{b}\right)^{-n} = \left(\dfrac{b}{a}\right)^n \quad (a, b \neq 0)$	Apply the rules for exponents. $\dfrac{1}{10^{-3}} = 10^3 \qquad \dfrac{10^{-3}}{10^{-4}} = \dfrac{10^4}{10^3}$ $4^{-3} = \left(\dfrac{1}{4}\right)^3 \qquad \left(\dfrac{4}{7}\right)^{-2} = \left(\dfrac{7}{4}\right)^2 = \dfrac{7^2}{4^2}$
Scientific Notation A number is in scientific notation when it is expressed in the form $a \times 10^n$, where $1 \leq |a| < 10$, and n is an integer.	Write 23,500,000,000 in scientific notation. $$23{,}500{,}000{,}000 = 2.35 \times 10^{10}$$ Write 4.3×10^{-6} in standard notation. $$4.3 \times 10^{-6} = 0.0000043$$

Concepts   *Examples*

7.6 POLYNOMIALS AND FACTORING

Adding and Subtracting Polynomials
Add or subtract polynomials by combining like terms.

Perform the indicated operations.
$$(5x^4 + 3x^2) - (7x^4 + x^2 - x)$$
$$= 5x^4 + 3x^2 - 7x^4 - x^2 + x$$
$$= -2x^4 + 2x^2 + x$$

Multiplying Polynomials
To multiply two polynomials, multiply each term of one by each term of the other.

$$(x^3 + 3x)(4x^2 - 5x + 2)$$
$$= 4x^5 - 5x^4 + 2x^3 + 12x^3 - 15x^2 + 6x$$
$$= 4x^5 - 5x^4 + 14x^3 - 15x^2 + 6x$$

To multiply two binomials, use the **FOIL** method. Multiply the **First** terms, the **Outer** terms, the **Inner** terms, and the **Last** terms. Then add these products.

$$(2x + 3)(x - 7)$$
$$= 2x(x) + 2x(-7) + 3x + 3(-7) \quad \text{FOIL method}$$
$$\qquad\qquad\qquad\qquad\qquad\qquad \text{Multiply.}$$
$$= 2x^2 - 11x - 21 \quad \text{Combine like terms.}$$

Special Products
$$(x + y)(x - y) = x^2 - y^2$$
$$(x + y)^2 = x^2 + 2xy + y^2$$
$$(x - y)^2 = x^2 - 2xy + y^2$$

$$(3x + 8)(3x - 8) = 9x^2 - 64$$
$$(5x + 3)^2 = 25x^2 + 30x + 9$$
$$(2x - 1)^2 = 4x^2 - 4x + 1$$

Factoring Out the Greatest Common Factor
Use the distributive property to write the given polynomial as a product of two factors, one of which is the greatest common factor of the terms of the polynomial.

Factor each polynomial.
$$4x^5 + 12x^3 - 8x^2 \quad \text{GCF} = 4x^2$$
$$= 4x^2(x^3 + 3x - 2) \quad \text{Check by multiplying.}$$

Factoring Trinomials
To factor a trinomial, choose factors of the first term and factors of the last term. Then, place them in a pair of parentheses of this form.

$$(\qquad)(\qquad)$$

Try various combinations of the factors until the correct product yields the given trinomial.

$$15x^2 + 14x - 8$$

The factors of 15 are 5 and 3, and 15 and 1.

The factors of -8 are

$$-4 \text{ and } 2, \quad 4 \text{ and } -2, \quad -1 \text{ and } 8, \quad \text{and } 1 \text{ and } -8.$$

Various combinations of these factors lead to the following.

$$15x^2 + 14x - 8$$
$$= (5x - 2)(3x + 4) \quad \text{Check by multiplying.}$$

Special Factoring
Use the special products above in reverse to factor a difference of squares or a perfect square trinomial.

$$4x^2 - 25$$
$$= (2x)^2 - (5)^2$$
$$= (2x + 5)(2x - 5)$$

$$9x^2 + 6x + 1 \qquad\qquad 16x^2 - 56x + 49$$
$$= (3x + 1)^2 \qquad\qquad = (4x - 7)^2$$

Concepts   *Examples*

7.7 QUADRATIC EQUATIONS AND APPLICATIONS

In some cases, quadratic equations can be solved using the zero-factor property or the square root property. The quadratic formula can always be used.

Zero-Factor Property
If $ab = 0$, then $a = 0$ or $b = 0$ or both.

Square Root Property
If $k \geq 0$, then the solutions of $x^2 = k$ are $\pm\sqrt{k}$.

Quadratic Formula
The solutions of $ax^2 + bx + c = 0$, where $a \neq 0$, are given by the following formula.

$$x = \frac{-b \pm \sqrt{b^2 - 4ac}}{2a}$$

Solve $3x^2 - 10x - 8 = 0$.

Using the zero-factor property:

$3x^2 - 10x - 8 = 0$

$(3x + 2)(x - 4) = 0$ Factor.

$3x + 2 = 0$ or $x - 4 = 0$ Set each factor equal to 0.

$x = -\dfrac{2}{3}$ or $x = 4$ Solve.

Using the quadratic formula:

$a = 3, \quad b = -10, \quad c = -8$

$x = \dfrac{-(-10) \pm \sqrt{(-10)^2 - 4(3)(-8)}}{2(3)}$ Substitute.

$x = \dfrac{10 \pm \sqrt{100 + 96}}{6}$ Perform the operations.

$x = \dfrac{10 \pm \sqrt{196}}{6}$ Add under the radical.

$x = \dfrac{10 \pm 14}{6}$ $\sqrt{196} = 14$

$x = 4$ or $x = -\dfrac{2}{3}$ Simplify to find two solutions.

The solution set is $\left\{-\frac{2}{3}, 4\right\}$.

Chapter 7 TEST

Solve each equation.

1. $5x - 3 + 2x = 3(x - 2) + 11$

2. $\dfrac{2x - 1}{3} + \dfrac{x + 1}{4} = \dfrac{43}{12}$

3. Decide whether the equation

$$3x - (2 - x) + 4x = 7x - 2 - (-x)$$

is conditional, an identity, or a contradiction. Give its solution set.

4. Solve for v: $S = vt - 16t^2$.

Solve each application.

5. Areas of Hawaiian Islands Three islands in the Hawaiian island chain are Hawaii (the Big Island), Maui, and Kauai. Together, their areas total 5300 square miles. The island of Hawaii is  3293 square miles larger than the island of Maui, and Maui is 177 square miles larger than Kauai. What is the area of each island?

6. Chemical Mixture How many liters of a 20% solution of a chemical should Roxanne mix with 10 liters of a 50% solution to obtain a mixture that is 40% chemical?

7. Speeds of Trains A passenger train and a freight train leave a town at the same time and travel in opposite directions. Their speeds are 60 mph and 75 mph, respectively. How long will it take for the trains to be 297 miles apart?

8. Merchandise Pricing Which is the better buy for processed cheese slices: 16 slices for $4.38 or 12 slices for $3.30?

9. Drug Dosage A physician orders 7.5 milligrams daily for Zyprexa (olanzipine). How many 2.5-milligram tablets of this antipsychotic drug should be administered? (Data from Giangrasso, Anthony Patrick, and Delores M. Shrimpton. *Ratio & Proportion Dosage Calculations*, 2nd ed. Pearson, 2014.)

10. Distance between Cities The distance between Boston and Milwaukee is 1050 miles. On a certain map this distance is represented by 21 inches. On the same map Seattle and Cincinnati are 46 inches apart. What is the actual distance between Seattle and Cincinnati?

11. Current in a Circuit The current in a simple electrical circuit is inversely proportional to the resistance. If the current is 80 amps when the resistance is 30 ohms, find the current when the resistance is 12 ohms.

Solve each inequality. Give the solution set in both interval and graph forms.

12. $-4x + 2(x - 3) \geq 4x - (3 + 5x) - 7$

13. $-10 < 3x - 4 \leq 14$

14. Grade Average Quinvenia has scores of 83, 76, and 79 on her first three tests in African American Studies. If she wants an average of at least 80 after her fourth test, what are the possible scores she can make on her fourth test?

Evaluate each exponential expression.

15. $\left(\dfrac{4}{3}\right)^2$

16. $-(-2)^6$

17. $\left(\dfrac{3}{4}\right)^{-3}$

18. $-5^0 + (-5)^0$

Use the properties of exponents to simplify each expression. Write answers with positive exponents only.

19. $\dfrac{(10^{-2})^{-3}}{10^4 \cdot 10^2}$

20. $\dfrac{10^{-4} \cdot 10^{12}}{10^7}$

21. Write each number in standard notation.

(a) 6.93×10^8 (b) 1.25×10^{-7}

22. Use scientific notation to evaluate the following. Leave the answer in scientific notation.
$$\frac{(2,500,000)(0.00003)}{(0.05)(5,000,000)}$$

23. Time Traveled for a Radio Signal The mean distance to Earth from Pluto is 4.58×10^9 kilometers. The first U.S. space probe to Pluto transmitted radio signals from Pluto to Earth at the speed of light, 3.00×10^5 kilometers per second. How long (in minutes) did it take for the signals to reach Earth?

Perform the indicated operations.

24. $(3x^3 - 5x^2 + 8x - 2) - (3x^3 - 9x^2 + 2x - 12)$

25. $(5x + 2)(3x - 4)$

26. $(4x^2 - 3)(4x^2 + 3)$

27. $(x + 6)^2$

Factor each polynomial completely.

28. $x^2 + 3x - 28$

29. $2x^2 + 19x + 45$

30. $25x^2 - 49$

31. $x^2 - 18x + 81$

32. $2x^3 - 18x^2 + 40x$

Solve each quadratic equation.

33. $6x^2 + 7x - 3 = 0$

34. $x^2 - 13 = 0$

35. $x^2 - x = 7$

36. Time an Object Has Descended The equation
$$s = 16t^2 + 15t$$
gives the distance s in feet an object dropped off a building has descended in t seconds. Find the time t when the object has descended 25 feet. Use a calculator and round the answer to the nearest hundredth.

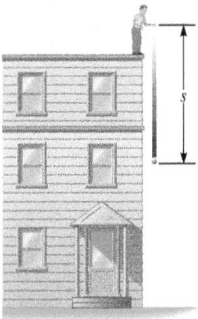

8 Graphs, Functions, and Systems of Equations and Inequalities

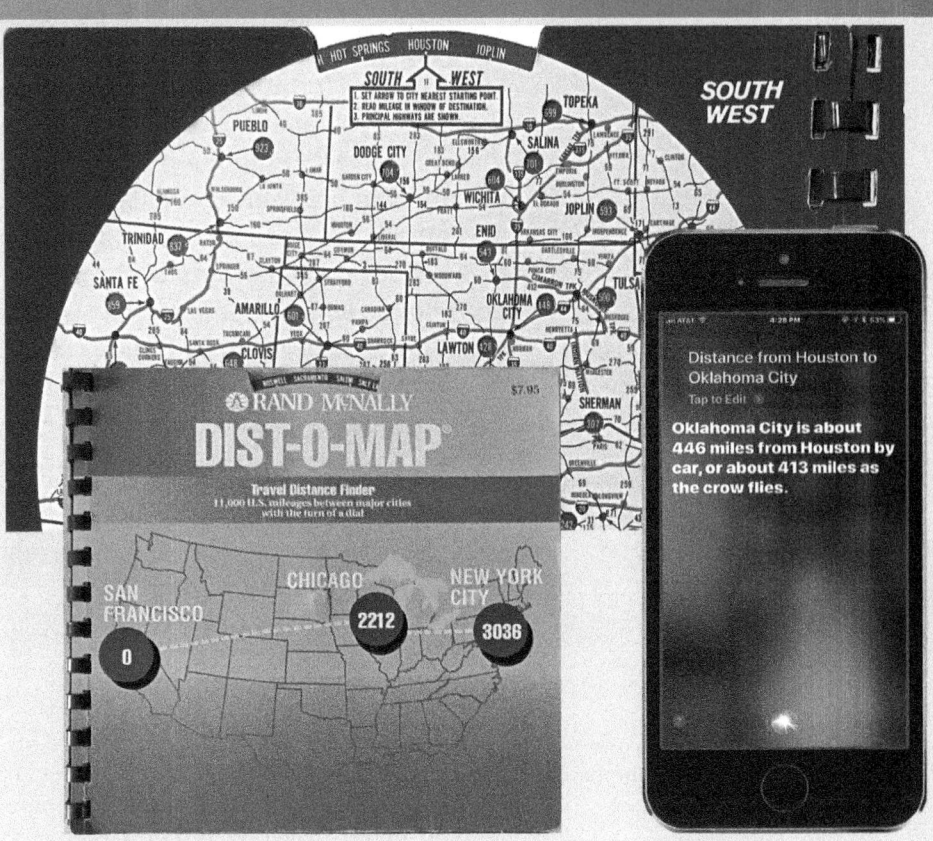

If you drive 100 miles in 2 hours, what is your rate?

This is a straightforward question, and most drivers can probably answer it by just dividing 100 by 2 to obtain the answer, 50 miles per hour. The statement "Distance equals rate times time" is fundamental in the study of algebra.

The concepts of rate and ratio are a common simple application of mathematics found in everyday life. The ratio of vertical change to horizontal change along a line is called the slope of the line. The idea of how one quantity (for example, distance) changes in relation to another (time) was crucial to the development of differential calculus in the seventeenth century.

The correspondence between elements of one set of numbers with those of another can be described by a relation. Relations, and special relations called functions, can be represented in several ways, one of which is graphing in a rectangular coordinate plane.

When Euclid presented his *Elements* around 300 B.C., the concepts of point, line, and plane were not associated with numbers. But the development of analytic geometry by René Descartes and Pierre de Fermat (also in the seventeenth century) changed this, and the result is reflected in many of the depictions of mathematics that we see on a daily basis. Graphs in mass media, meteorological charts, and GPS coordinates are just a few of these.

In a rectangular coordinate plane, points are represented by ordered pairs of numbers. Given any two points, the distance between them can be found using a formula that is justified by the Pythagorean theorem. A circle is the set of all points in a plane that are equidistant from a given point. In today's world, finding the distance between two points, such as the number of miles between two cities, can easily be accomplished by a variety of methods. You can just ask a smart phone, for example, and the answer given is based on a program justified by logic and mathematics. But if you don't have a smart phone, you can use the Dist-O-Map, seen on the previous page. It was developed by Rand-McNally, and it uses rotating circular dials to indicate the distances between major cities of the United States.

These ideas, among others, are introduced in this chapter.

8.1 THE RECTANGULAR COORDINATE SYSTEM AND CIRCLES

OBJECTIVES

1 Plot ordered pairs in a rectangular coordinate system.

2 Find the distance between two points using the distance formula.

3 Find the midpoint of a segment using the midpoint formula.

4 Find the equation of a circle given the coordinates of the center and the radius.

5 Find the center and radius of a circle given its equation.

6 Apply the midpoint formula.

7 Apply the definition of a circle to locating a geographical point.

Rectangular Coordinates

Each pair of numbers $(1, 2)$, $(-1, 5)$, and $(3, 7)$ is an example of an **ordered pair**—a pair of numbers written within parentheses in which the positions of the numbers are relevant. The two numbers are the **components** of the ordered pair. Two ordered pairs are **equal** if and only if their first components are equal and their second components are equal.

An ordered pair is graphed using two number lines that intersect at right angles at the zero points, as shown in **Figure 1.** The common zero point is called the **origin.** The horizontal line, the ***x*-axis,** represents the first number in an ordered pair, and the vertical line, the ***y*-axis,** represents the second. The *x*-axis and the *y*-axis make up a **rectangular** (or **Cartesian**) **coordinate system.** The axes form four **quadrants,** numbered I, II, III, and IV as shown in **Figure 2.** (A point on an axis is not considered to be in any of the four quadrants.)

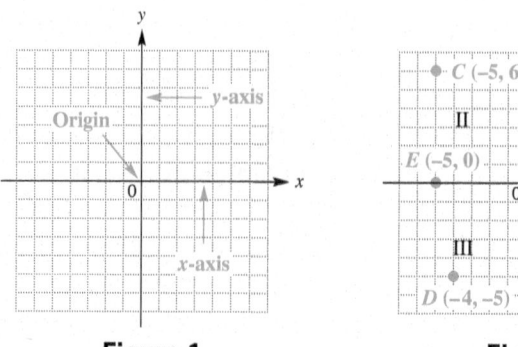

Figure 1 **Figure 2**

We locate, or plot, the point on the graph that corresponds to the ordered pair $(3, 1)$ by moving three units from zero to the right along the x-axis, and then one unit up parallel to the y-axis; this point is labeled A in **Figure 2.** The phrase "the point corresponding to the ordered pair $(3, 1)$" is abbreviated "the point $(3, 1)$." The numbers in an ordered pair are called the **coordinates** of the corresponding point.

The symbolism for an ordered pair may also represent an open interval (introduced earlier in this text). The context of the discussion tells us whether we are discussing points in the Cartesian coordinate system or an open interval on a number line.

Distance Formula

To find the distance between two points, say $(3, -4)$ and $(-5, 3)$, we use the Pythagorean theorem. In **Figure 3,** we see that the vertical line through $(-5, 3)$ and the horizontal line through $(3, -4)$ intersect at the point $(-5, -4)$. Thus, the point $(-5, -4)$ becomes the vertex of the right angle in a right triangle. By the Pythagorean theorem, the square of the length of the hypotenuse, c, in **Figure 3** is equal to the sum of the squares of the lengths of the two legs a and b.

$$c^2 = a^2 + b^2 \quad \text{Equation of the Pythagorean theorem}$$

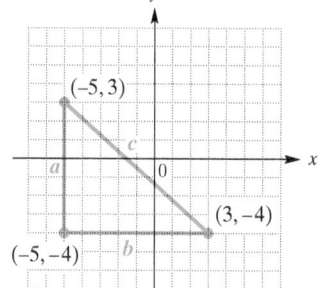

Figure 3

The length a is the distance between the endpoints of that leg. Because the x-coordinate of both points is -5, the side is vertical, and we can find a by finding the difference between the y-coordinates. Subtract -4 from 3 to get a positive value of a. Similarly, find b by subtracting -5 from 3.

$$a = 3 - (-4) = 7$$
$$b = 3 - (-5) = 8$$

Use parentheses when subtracting a negative number.

Substitute these values into the equation.

$$c^2 = a^2 + b^2$$
$$c^2 = 7^2 + 8^2 \quad \text{Let } a = 7 \text{ and } b = 8.$$
$$c^2 = 49 + 64 \quad \text{Apply the exponents.}$$
$$c^2 = 113 \quad \text{Add.}$$
$$c = \sqrt{113} \quad \text{Square root property, } c > 0$$

Therefore, the distance between $(3, -4)$ and $(-5, 3)$ is $\sqrt{113}$.

This result can be generalized. **Figure 4** shows the two different points (x_1, y_1) and (x_2, y_2). We can find a formula for the distance c between these two points.

Double Descartes After the French postal service issued the above stamp in honor of **René Descartes,** sharp eyes noticed that the title of Descartes's most famous book was wrong. Thus a second stamp (see the next page) was issued with the correct title. The book in question, *Discourse on Method,* appeared in 1637. His method was *analysis,* going from self-evident truths step-by-step to more distant and more general truths. One of these truths is his famous statement, "I think, therefore I am."

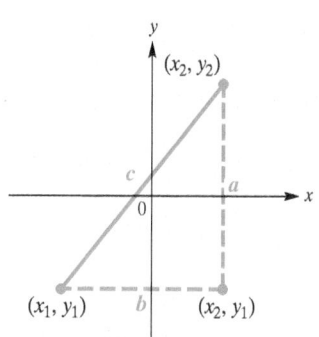

Figure 4

In **Figure 4,** the distance between (x_2, y_2) and (x_2, y_1) is given by $a = y_2 - y_1$, and the distance between (x_1, y_1) and (x_2, y_1) is given by $b = x_2 - x_1$. Thus,

$$c^2 = (x_2 - x_1)^2 + (y_2 - y_1)^2. \quad \text{Pythagorean theorem}$$

From this, we obtain the **distance formula.** (Here we use d to represent distance.)

DISTANCE FORMULA

The distance d between the points (x_1, y_1) and (x_2, y_2) is given by the following formula.

$$d = \sqrt{(x_2 - x_1)^2 + (y_2 - y_1)^2}$$

The small numbers 1 and 2 in the ordered pairs (x_1, y_1) and (x_2, y_2) are called **subscripts.** We read x_1 as "*x* sub 1." Subscripts are used to distinguish between different values of a variable that have a common property.

EXAMPLE 1 Finding the Distance between Two Points

Find the distance between the points $(-3, 5)$ and $(6, 4)$.

Solution

Designating the points as (x_1, y_1) and (x_2, y_2) is arbitrary. Let us choose $(x_1, y_1) = (-3, 5)$ and $(x_2, y_2) = (6, 4)$.

$$d = \sqrt{(x_2 - x_1)^2 + (y_2 - y_1)^2} \quad \text{Distance formula}$$

$$= \sqrt{(6 - (-3))^2 + (4 - 5)^2} \quad x_2 = 6,\ y_2 = 4,\ x_1 = -3,\ y_1 = 5$$

> Begin with the *x*- and *y*-values of the same point.

$$= \sqrt{9^2 + (-1)^2} \quad \text{Subtract twice.}$$

$$= \sqrt{82} \quad \text{Square and add.}$$

Midpoint Formula

The **midpoint** of a line segment is the point on the segment that is equidistant from both endpoints. Given the coordinates of the two endpoints of a line segment, we can find the coordinates of the midpoint of the segment.

Descartes wrote his *Geometry* as an application of his method. It was published as an appendix to the *Discourse.* His attempts to unify algebra and geometry influenced the creation of what became coordinate geometry and influenced the development of calculus by Newton and Leibniz in the next generation.

In 1649 Descartes went to Sweden to tutor Queen Christina. She preferred working in the unheated castle in the early morning. Descartes was used to staying in bed until noon. The rigors of the Swedish winter proved too much for him, and he died less than a year later.

MIDPOINT FORMULA

The coordinates of the midpoint M of the line segment with endpoints (x_1, y_1) and (x_2, y_2) are given by the following formula.

$$M = \left(\frac{x_1 + x_2}{2}, \frac{y_1 + y_2}{2} \right)$$

Thus, the coordinates of the midpoint of a line segment are found by calculating the averages of the x- and y-coordinates of the endpoints.

EXAMPLE 2 Finding the Midpoint of a Line Segment

Find the coordinates of the midpoint M of the line segment with endpoints $(8, -4)$ and $(-9, 6)$.

Solution

Use the midpoint formula to find the coordinates of M.

$$M = \left(\frac{8 + (-9)}{2}, \frac{-4 + 6}{2} \right) = \left(-\frac{1}{2}, 1 \right) \qquad \begin{array}{l} x_1 = 8, \ x_2 = -9, \\ y_1 = -4, \ y_2 = 6 \end{array}$$

An Application of the Midpoint Formula

EXAMPLE 3 Applying the Midpoint Formula to Data

Figure 5 shows a line graph depicting personal spending (in billions of dollars) on medical care in the United States from 2005 through 2015. Use the midpoint formula and the two given points to estimate how much was spent on medical care in 2010, and compare it to the actual figure of $2196 billion.

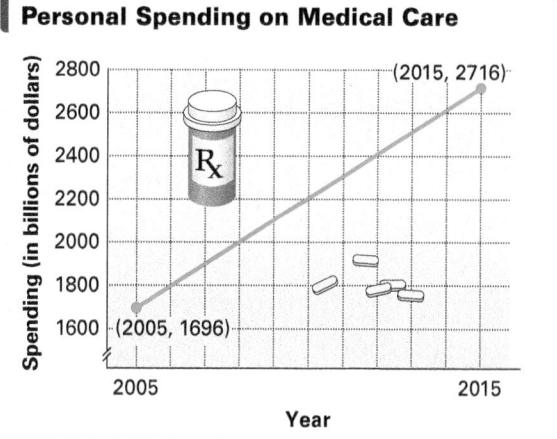

Personal Spending on Medical Care

Data from Centers for Medicare and Medicaid Services.

Figure 5

Solution

The year 2010 lies halfway between 2005 and 2015, so we must find the coordinates of the midpoint M of the segment that has endpoints

$$(2005, 1696) \quad \text{and} \quad (2015, 2716).$$

(Here, the second component is in billions of dollars.)

$$M = \left(\frac{2005 + 2015}{2}, \frac{1696 + 2716}{2} \right) = (2010, 2206)$$

Thus our estimate is $2206 billion, which is slightly more than the actual figure of $2196 billion.

Circles

An application of the distance formula leads to one of the most familiar shapes in geometry, the circle. A **circle** is the set of all points in a plane that lie a fixed distance from a fixed point. The fixed point is the **center** and the fixed distance is the **radius.**

EXAMPLE 4 Finding an Equation of a Circle

Find an equation of the circle with radius 4 and center at $(0, 0)$, and graph the circle.

Solution

If the point (x, y) is on the circle, the distance from (x, y) to the center $(0, 0)$ is 4, as shown in **Figure 6.**

$$\sqrt{(x_2 - x_1)^2 + (y_2 - y_1)^2} = d \qquad \text{Distance formula}$$

$$\sqrt{(x - 0)^2 + (y - 0)^2} = 4 \qquad x_1 = 0,\ y_1 = 0,\ x_2 = x,\ y_2 = y,\ d = 4$$

$$x^2 + y^2 = 16 \qquad \text{Square both sides.}$$

An equation of this circle is $x^2 + y^2 = 16$. It can be graphed by locating all points four units from the origin.

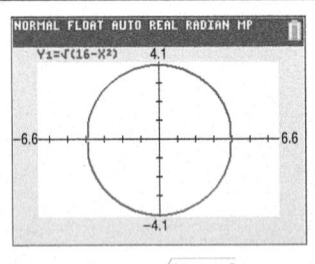

Example 4 can be generalized to obtain the equation of a circle with radius r and center at (h, k). If (x, y) is a point on the circle, the distance from the center (h, k) to the point (x, y) is r. Then, by the distance formula,

$$\sqrt{(x - h)^2 + (y - k)^2} = r.$$

Squaring both sides gives the **center-radius form** of the equation of a circle.

EQUATION OF A CIRCLE

The equation of a circle of radius r with center at (h, k) is

$$(x - h)^2 + (y - k)^2 = r^2.$$

In particular, a circle of radius r with center at the origin has equation

$$x^2 + y^2 = r^2.$$

To graph $x^2 + y^2 = 16$, in **Example 4,** we solve for y to obtain

$$y_1 = \sqrt{16 - x^2} \text{ and } y_2 = -\sqrt{16 - x^2}.$$

Then we graph both in a square window.

In these equations, r^2 will be greater than or equal to 0. If the constant on the right is 0, the graph is a single point. If it is negative, there is no graph.

EXAMPLE 5 Finding an Equation of a Circle and Graphing

Find an equation of the circle that has its center at $(4, -3)$ and radius 5, and graph the circle.

Solution

Use the equation of a circle with $h = 4$, $k = -3$, and $r = 5$.

$$(x - h)^2 + (y - k)^2 = r^2 \qquad \text{Center-radius form}$$

$$(x - 4)^2 + [y - (-3)]^2 = 5^2 \qquad \text{Substitute.}$$

$$(x - 4)^2 + (y + 3)^2 = 25$$

> Be careful with signs.

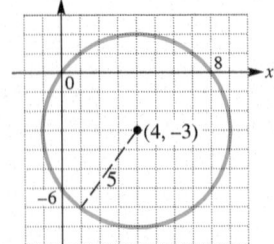

$(x - 4)^2 + (y + 3)^2 = 25$

Figure 7

The graph of this circle is shown in **Figure 7.**

In the equation found in **Example 5,** multiplying out $(x - 4)^2$ and $(y + 3)^2$ and then combining like terms give the following.

$$(x - 4)^2 + (y + 3)^2 = 25$$

$$x^2 - 8x + 16 + y^2 + 6y + 9 = 25$$

$$x^2 + y^2 - 8x + 6y = 0$$

This result suggests that an equation that has both x^2- and y^2-terms with equal coefficients may represent a circle. The next example shows how to determine this, using a method from algebra called **completing the square.**

EXAMPLE 6 Completing the Square and Graphing a Circle

Graph $x^2 + y^2 + 2x + 6y - 15 = 0$.

Solution

Because the equation has x^2- and y^2-terms with equal coefficients, its graph could be that of a circle. To find the center and radius, complete the squares in x and y as follows.

$$x^2 + y^2 + 2x + 6y = 15 \qquad \text{Add 15 to both sides.}$$

$$(x^2 + 2x \quad) + (y^2 + 6y \quad) = 15 \qquad \text{Rewrite in anticipation of completing the square.}$$

$$(x^2 + 2x + 1) + (y^2 + 6y + 9) = 15 + 1 + 9 \qquad \text{Complete the squares in both } x \text{ and } y.$$

$$(x + 1)^2 + (y + 3)^2 = 25 \qquad \text{Factor on the left and add on the right.}$$

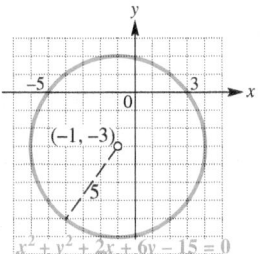

Figure 8

The final equation shows that the graph is a circle with center at $(-1, -3)$ and radius 5. The graph is shown in **Figure 8.**

An Application of Circles

EXAMPLE 7 Locating the Epicenter of an Earthquake

Seismologists can locate the epicenter of an earthquake by determining the intersection of three circles. The radii of these circles represent the distances from the epicenter to each of three receiving stations. The centers of the circles represent the receiving stations.

Suppose receiving stations A, B, and C are located on a coordinate plane at the points $(1, 4)$, $(-3, -1)$, and $(5, 2)$, respectively. Let the distances from the earthquake epicenter to the stations be 2 units, 5 units, and 4 units, respectively. See **Figure 9.** Where on the coordinate plane is the epicenter located?

Solution

In **Figure 9,** the epicenter appears to be located at $(1, 2)$. To check this algebraically, determine the equation for each circle, and substitute $x = 1$ and $y = 2$.

Station A:	Station B:	Station C:
$(x - 1)^2 + (y - 4)^2 = 2^2$	$(x + 3)^2 + (y + 1)^2 = 5^2$	$(x - 5)^2 + (y - 2)^2 = 4^2$
$(1 - 1)^2 + (2 - 4)^2 = 4$	$(1 + 3)^2 + (2 + 1)^2 = 25$	$(1 - 5)^2 + (2 - 2)^2 = 16$
$0 + 4 = 4$	$16 + 9 = 25$	$16 + 0 = 16$
$4 = 4$	$25 = 25$	$16 = 16$

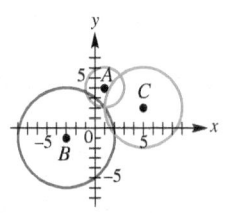

Figure 9

The point $(1, 2)$ does lie on all three graphs. Thus, it must be the epicenter.

8.1 **EXERCISES**

CONCEPT CHECK *Fill in each blank with the correct response.*

1. If a point lies on the *x*-axis, its _____ -coordinate is 0.

2. If a point lies on the *y*-axis, its _____ -coordinate is 0.

3. The point on the *y*-axis 4 units below the origin has coordinates _____.

4. The point on the *x*-axis 5 units to the left of the origin has coordinates _____.

5. The distance from the origin to the point $(3, 4)$ is _____.

6. The coordinates of the midpoint of the segment having endpoints $(-2, 6)$ and $(4, 10)$ are _____.

7. The circle with equation $x^2 + y^2 = 49$ has center with coordinates _____ and radius equal to _____.

8. The circle with center $(3, 6)$ and radius 4 has equation _____.

9. The graph of $(x - 4)^2 + (y + 7)^2 = 9$ has center with coordinates _____ and radius equal to _____.

10. The points $(0, 6)$ and $(0, -6)$ lie on the circle with equation $x^2 + y^2 =$ _____.

Name the quadrant, if any, in which each point is located.

11. (a) $(1, 6)$
(b) $(-4, -2)$
(c) $(-3, 6)$
(d) $(7, -5)$
(e) $(-3, 0)$

12. (a) $(-2, -10)$
(b) $(4, 8)$
(c) $(-9, 12)$
(d) $(3, -9)$
(e) $(0, -8)$

Locate the following points on the rectangular coordinate system, using a graph similar to **Figure 2**.

13. $(2, 3)$
14. $(-1, 2)$
15. $(-3, -2)$
16. $(1, -4)$
17. $(0, 5)$
18. $(0, -4)$
19. $(-2, 0)$
20. $(3, 0)$
21. $(-2, 4)$
22. $(3, -3)$

Find each of the following.

(a) *the distance between the two points in each pair*

(b) *the coordinates of the midpoint of the segment having the points as endpoints*

23. $(3, 4)$ and $(-2, 1)$

24. $(-2, 1)$ and $(3, -2)$

25. $(-2, 4)$ and $(3, -2)$

26. $(1, -5)$ and $(6, 3)$

27. $(-3, 7)$ and $(2, -4)$

28. $(0, 5)$ and $(-3, 12)$

Find each of the following.

(a) *the distance between P and Q*

(b) *the coordinates of the midpoint of the segment joining P and Q*

29.

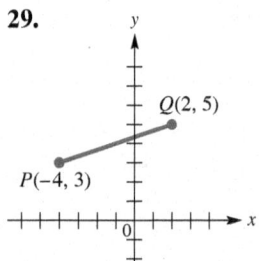

30.

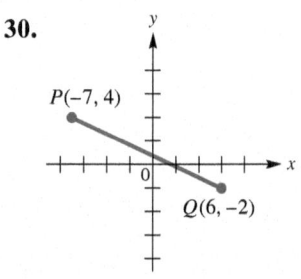

Match each center-radius form of the equation of a circle with the correct graph from choices A–D.

31. $(x - 3)^2 + (y - 2)^2 = 25$

32. $(x - 3)^2 + (y + 2)^2 = 25$

33. $(x + 3)^2 + (y - 2)^2 = 25$

34. $(x + 3)^2 + (y + 2)^2 = 25$

A.

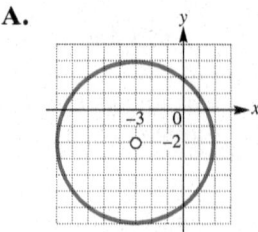

B.

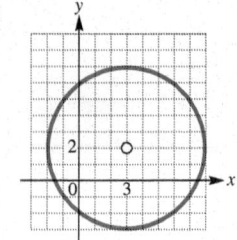

C.

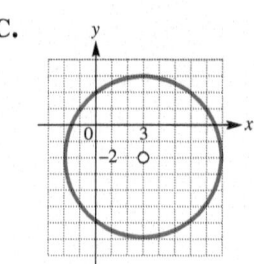

D.

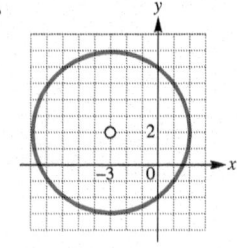

Write an equation of the circle with the given center and radius.

35. $(0, 0); r = 6$

36. $(0, 0); r = 5$

37. $(-1, 3); r = 4$

38. $(2, -2); r = 3$

39. $(0, 4); r = \sqrt{3}$

40. $(-2, 0); r = \sqrt{5}$

Find the center and the radius of each circle.

41. $x^2 + y^2 + 4x + 6y + 9 = 0$

42. $x^2 + y^2 - 8x - 12y + 3 = 0$

43. $x^2 + y^2 + 10x - 14y - 7 = 0$

44. $x^2 + y^2 - 2x + 4y - 4 = 0$

45. $x^2 + y^2 - 4x - 8y + 4 = 0$

46. $x^2 + y^2 + 10x + 8y + 5 = 0$

Graph each circle.

47. $x^2 + y^2 = 36$

48. $x^2 + y^2 = 81$

49. $(x - 2)^2 + y^2 = 36$

50. $x^2 + (y + 3)^2 = 49$

51. $(x + 2)^2 + (y - 5)^2 = 16$

52. $(x - 4)^2 + (y - 3)^2 = 25$

53. $(x + 3)^2 + (y + 2)^2 = 36$

54. $(x - 5)^2 + (y + 4)^2 = 49$

Solve each problem.

55. *Bachelor's Degree Attainment* The graph shows a straight line segment that approximates the percentage of Americans 25 years old or older who earned bachelor's degrees or higher during the years 1990–2012.

Percent Earning Bachelor's Degrees or Higher

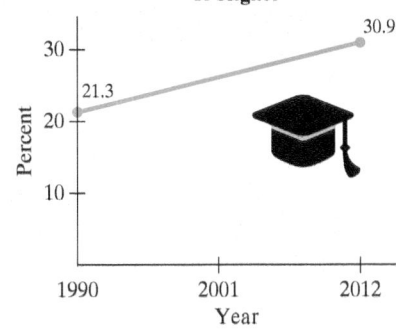

Data from U.S. Census Bureau.

(a) Use the midpoint formula and the two given points to estimate the percent in 2001.

(b) Compare your answer with the actual figure, 26.2%.

56. *Motor Vehicle Sales* The graph shows a straight line segment that approximates all vehicle sales in the United States during the years 2012–2016.

U.S. Vehicle Sales

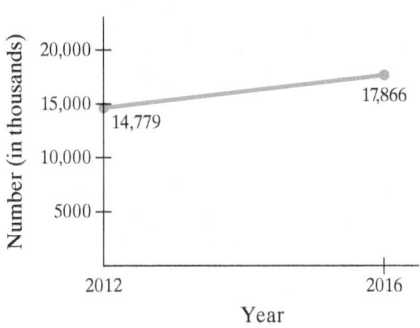

Data from Wards Auto.

(a) Use the midpoint formula and the two given points to estimate the number of sales in 2014.

(b) Compare your answer with the actual number of sales, 16,860 thousand.

57. *Poverty-Level Income Cutoffs* The table shows how poverty-level income cutoffs (in dollars) for a family of four have changed over time. Use the midpoint formula to approximate the poverty-level cutoff in 2013, to the nearest dollar.

Year	Income (in dollars)
1980	8414
1990	13,359
2000	17,604
2010	22,315
2016	24,563

Data from U.S. Census Bureau.

58. *Public College Enrollment* Enrollments in public colleges for recent years are shown in the table. Use the midpoint formula to estimate the enrollments for each year to the nearest whole number of thousands.

(a) 2003

(b) 2009

Year	Enrollment (in thousands)
2000	11,753
2006	13,180
2012	14,880

Data from U.S. Census Bureau.

59. *Drawing a Circle* A circle can be drawn on a piece of posterboard by fastening one end of a string, pulling the string taut with a pencil, and tracing a curve as shown in the figure. Explain why this method works.

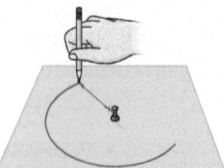

60. *Crawfish Racing* This figure shows how the crawfish race is held at the Crawfish Festival in Breaux Bridge, Louisiana. Explain why a circular "race-track" is appropriate for such a race.

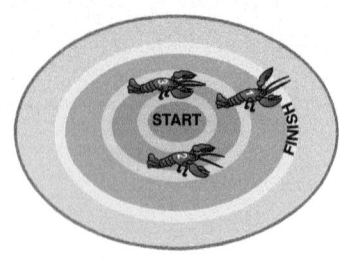

61. *Epicenter of an Earthquake* Three receiving stations record the occurrence of an earthquake. The locations of the receiving stations and the distances to the epicenter are contained in the following three equations:

$$(x - 2)^2 + (y - 1)^2 = 25,$$

$$(x + 2)^2 + (y - 2)^2 = 16,$$

and $$(x - 1)^2 + (y + 2)^2 = 9.$$

Graph the circles and determine the location of the earthquake epicenter.

62. *Epicenter of an Earthquake* Show algebraically that if three receiving stations at

$$(1, 4), \quad (-6, 0), \quad \text{and} \quad (5, -2)$$

record distances to an earthquake epicenter of 4 units, 5 units, and 10 units, respectively, the epicenter lies at $(-3, 4)$.

FOR FURTHER THOUGHT

Locating Nothing

Phlash Phelps is the morning radio personality on Sirius XM Satellite Radio's *Sixties on Six* Decades channel. Phlash is an expert on U.S. geography. He loves traveling around the country to strange, out-of-the-way locations. The photo shows Phlash visiting a small Arizona settlement called *Nothing.*

(Nothing is so small that it's not named on current maps.) The sign indicates that Nothing is 50 mi from Wickenburg, AZ, 75 mi from Kingman, AZ, 105 mi from Phoenix, AZ, and 180 mi from Las Vegas, NV.

For Group or Individual Investigation

Discuss how the concepts of **Example 7** can be used to locate Nothing, AZ, on a map of the area.

8.2 LINES, SLOPE, AND AVERAGE RATE OF CHANGE

OBJECTIVES

1 Determine ordered pairs and graph a linear equation in two variables.

2 Find the x- and y-intercepts of the graph of a linear equation.

Linear Equations in Two Variables

Earlier we studied linear equations in a single variable. A solution of such an equation is a real number.

A linear equation in *two* variables has solutions written as ordered pairs. Equations with two variables will, in general, have an infinite number of solutions. To find ordered pairs that satisfy the equation, select any number for one of the variables, substitute it into the equation for that variable, and then solve for the other variable.

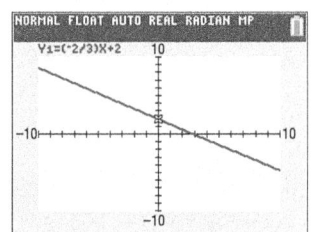

This is a calculator graph of the line shown in **Figure 10(b)**. We must first write $2x + 3y = 6$ as $y = -\frac{2}{3}x + 2$ to input the equation.

For example, suppose $x = 0$ in the equation $2x + 3y = 6$.

$$2x + 3y = 6 \quad \text{Given equation}$$
$$2(0) + 3y = 6 \quad \text{Let } x = 0.$$
$$0 + 3y = 6 \quad \text{Multiply.}$$
$$3y = 6 \quad \text{Add.}$$
$$y = 2 \quad \text{Divide by 3.}$$

This gives the ordered pair $(0, 2)$. Other ordered pairs satisfying $2x + 3y = 6$ include

$$(6, -2), \quad (3, 0), \quad (-3, 4), \quad \text{and} \quad (9, -4).$$

The equation $2x + 3y = 6$ is graphed by plotting the ordered pairs mentioned above. These are shown in **Figure 10(a)**. The resulting points appear to lie on a straight line. If all the ordered pairs that satisfy the equation $2x + 3y = 6$ were graphed, they would form the straight line shown in **Figure 10(b)**. **Figure 10(c)** shows a **table of values** for selected points on the graph.

The graph of any first-degree equation in two variables is a straight line.

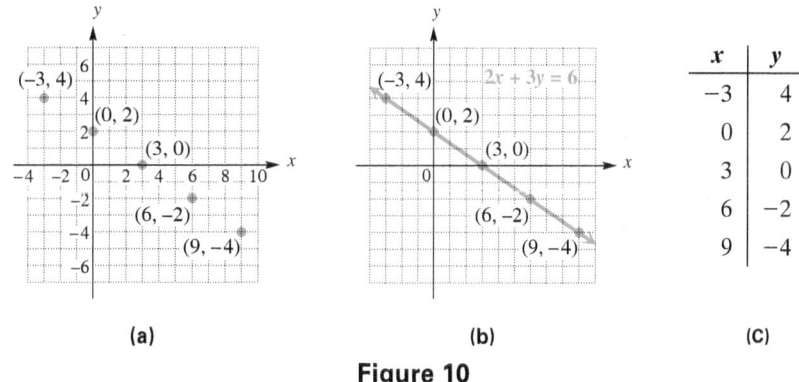

(a)   **(b)**   **(C)**

Figure 10

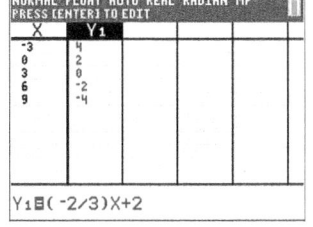

This is a table of values for the equation $y = -\frac{2}{3}x + 2$.

LINEAR EQUATION IN TWO VARIABLES

An equation that can be written in the form

$$\boldsymbol{Ax + By = C} \quad \text{(where } A \text{ and } B \text{ are not both 0)}$$

is a **linear equation in two variables.** This form is called **standard form.**

The equation $2x + 3y = 6$ is in standard form. When we provide answers in standard form, we do so with $A > 0$ (or, if $A = 0$, then with $B > 0$) and with integers A, B, and C having greatest common factor 1.

Intercepts

Two points that are useful for graphing lines are the x- and y-intercepts. The **x-intercept** is the point (if any) where the line crosses the x-axis, and the **y-intercept** is the point (if any) where the line crosses the y-axis. (*Note:* In many texts, the intercepts are defined as numbers and not as points. However, in this text we will refer to intercepts as points.)

INTERCEPTS FOR GRAPHS OF LINEAR EQUATIONS

- To find the x-intercept, let $y = 0$ and solve for x.
- To find the y-intercept, let $x = 0$ and solve for y.

EXAMPLE 1 Graphing an Equation Using Intercepts

Find the x- and y-intercepts of $4x - y = -3$, and graph the equation.

Solution

To find the x-intercept, let $y = 0$.

$$4x - 0 = -3 \quad \text{Let } y = 0.$$

$$4x = -3$$

$$x = -\frac{3}{4} \quad x\text{-intercept is } \left(-\frac{3}{4}, 0\right).$$

To find the y-intercept, let $x = 0$.

$$4(0) - y = -3 \quad \text{Let } x = 0.$$

$$-y = -3$$

$$y = 3 \quad y\text{-intercept is } (0, 3).$$

The intercepts are the two points $\left(-\frac{3}{4}, 0\right)$ and $(0, 3)$. Use these two points to draw the graph, as shown in **Figure 11**.

x	y
$-\frac{3}{4}$	0
0	3

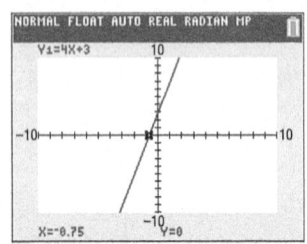

The display at the bottom of the screen supports the fact that $\left(-\frac{3}{4}, 0\right)$ is the x-intercept of the line in **Figure 11**. We could locate the y-intercept similarly.

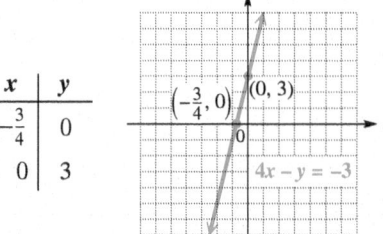

Figure 11

EXAMPLE 2 Graphing Lines with a Single Intercept

Graph each line.

(a) $y = 2$ **(b)** $x = -1$

Solution

(a) Writing the equation $y = 2$ as

$$0x + 1y = 2$$

shows that any value of x, including $x = 0$, gives $y = 2$, making the y-intercept $(0, 2)$. Because y is always 2, there is no value of x corresponding to $y = 0$, so the graph has no x-intercept. The graph, shown in **Figure 12(a),** is a horizontal line.

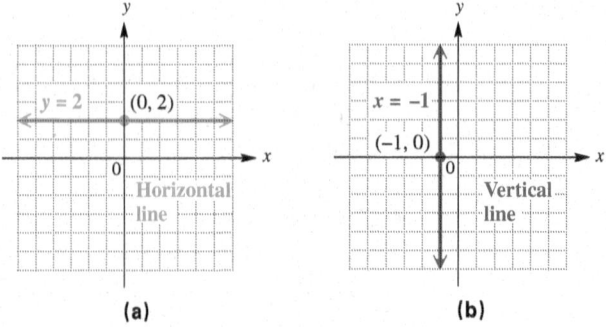

(a)   (b)

Figure 12

(b) In this equation, $x = -1$ for all y. No value of y makes $x = 0$. The graph has no y-intercept and thus must be vertical, as shown in **Figure 12(b)**.

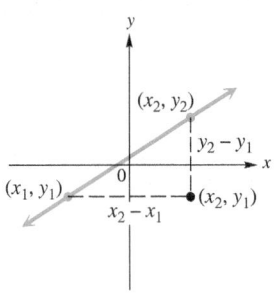

Figure 13

Slope

Two distinct points determine a unique line. A line also can be determined by a point on the line and some measure of the "steepness" of the line. One way to measure the steepness of a line is to compare the vertical change in the line (the *rise*) to the horizontal change (the *run*) while moving along the line from one fixed point to another.

Suppose that (x_1, y_1) and (x_2, y_2) are two different points on a line. Then, moving along the line from (x_1, y_1) to (x_2, y_2), the y-value changes from y_1 to y_2, an amount equal to $y_2 - y_1$. As y changes from y_1 to y_2, the value of x changes from x_1 to x_2 by the amount $x_2 - x_1$. See **Figure 13**. The ratio of the change in y to the change in x is called the **slope** of the line. The letter m is used to denote the slope.

SLOPE FORMULA

If $x_1 \neq x_2$, the slope m of the line through the distinct points (x_1, y_1) and (x_2, y_2) is given by the formula

$$m = \frac{\text{rise}}{\text{run}} = \frac{\text{change in } y}{\text{change in } x} = \frac{y_2 - y_1}{x_2 - x_1}.$$

StatCrunch Using the Slope Formula

EXAMPLE 3   **Using the Slope Formula**

Find the slope of the line that passes through the points $(2, -1)$ and $(-5, 3)$.

Solution

Let $(2, -1) = (x_1, y_1)$ and $(-5, 3) = (x_2, y_2)$, and use the slope formula.

$$m = \frac{y_2 - y_1}{x_2 - x_1} = \frac{3 - (-1)}{-5 - 2} = \frac{4}{-7} = -\frac{4}{7}$$

> Start with the x- and y-values of the same point.

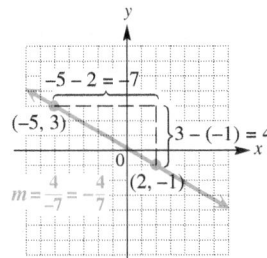

Figure 14

See **Figure 14**. Now let $(2, -1) = (x_2, y_2)$ and $(-5, 3) = (x_1, y_1)$.

$$m = \frac{-1 - 3}{2 - (-5)} = \frac{-4}{7} = -\frac{4}{7} \quad \text{The same answer}$$

This example suggests that the slope is the same no matter which point is considered first. Using similar triangles from geometry, we can show that the slope is the same for *any* two different points chosen on the line.

EXAMPLE 4   **Finding Slopes of Vertical and Horizontal Lines**

Find the slope, if possible, of each of the following lines.

(a) $x = -3$ **(b)** $y = 5$

Solution

(a) By inspection, $(-3, 5)$ and $(-3, -4)$ are two points that satisfy the equation $x = -3$. Use these two points to find the slope.

$$m = \frac{-4 - 5}{-3 - (-3)} = \frac{-9}{0} \quad \text{Undefined slope}$$

Division by zero is undefined, so the slope is undefined. (This is why the definition of slope includes the restriction that $x_1 \neq x_2$.)

Visitors to Epcot Center at Walt Disney World in Florida love the ride **Test Track**. This sign at the beginning of the ride indicates an upward **slope** of about $\frac{1}{4}$.

(b) Find the slope by selecting two different points on the line, such as $(3, 5)$ and $(-1, 5)$, and by using the definition of slope.

$$m = \frac{5 - 5}{3 - (-1)} = \frac{0}{4} = 0 \quad \text{Slope 0}$$

In **Example 2,** $x = -1$ has a graph that is a vertical line, and $y = 2$ has a graph that is a horizontal line. Generalizing from those results and the results of **Example 4,** we can make the following statements about vertical and horizontal lines.

VERTICAL AND HORIZONTAL LINES

- A **vertical line** has an equation of the form $x = a$, where a is a real number, and its **slope is undefined.**

- A **horizontal line** has an equation of the form $y = b$, where b is a real number, and its **slope is 0.**

EXAMPLE 5   **Graphing a Line Using Slope and a Point**

Graph the line that has slope $\frac{2}{3}$ and passes through the point $(-1, 4)$.

Solution

First locate the point $(-1, 4)$ on a graph as shown in **Figure 15**. Then,

$$m = \frac{\text{change in } y}{\text{change in } x} = \frac{2}{3}. \quad \text{Definition of slope}$$

Recall that the geometric interpretation of slope is "rise over run." Move *up* 2 units in the *y*-direction and then *right* 3 units in the *x*-direction to locate another point on the graph (labeled *P*). The line through $(-1, 4)$ and *P* is the required graph.

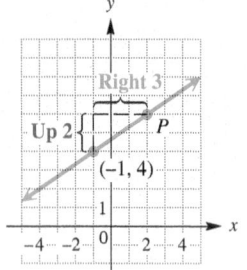

Figure 15

The line graphed in **Figure 14** has a negative slope, $-\frac{4}{7}$, and the line *falls* from left to right. In contrast, the line graphed in **Figure 15** has a positive slope, $\frac{2}{3}$, and it *rises* from left to right. These ideas can be generalized. (**Figure 16** shows lines of positive, negative, zero, and undefined slopes.)

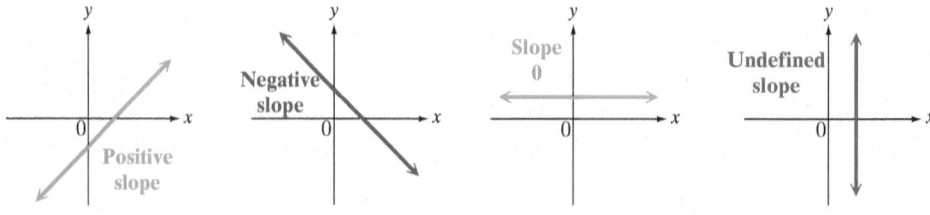

Figure 16

POSITIVE AND NEGATIVE SLOPES

- A line with a positive slope rises from left to right.

- A line with a negative slope falls from left to right.

Parallel and Perpendicular Lines

The slope of a line is a measure of the steepness of the line. **Parallel** lines have equal steepness, and thus their slopes (if defined) also must be equal. Also, lines with the same slope are parallel.

SLOPES OF PARALLEL LINES

- Two nonvertical lines with the same slope are parallel.
- Two nonvertical parallel lines have the same slope.
- Furthermore, any two vertical lines are parallel.

Perpendicular lines are lines that meet at right angles. It can be shown that the slopes of perpendicular lines have a product of -1, provided that neither line is vertical. For example, if the slope of a line is $\frac{3}{4}$, then any line perpendicular to it has slope $-\frac{4}{3}$, because $\frac{3}{4}\left(-\frac{4}{3}\right) = -1$.

SLOPES OF PERPENDICULAR LINES

- If neither is vertical, two perpendicular lines have slopes that are negative reciprocals—that is, their product is -1.
- Two lines with slopes that are negative reciprocals are perpendicular.
- Every vertical line is perpendicular to every horizontal line.

EXAMPLE 6   Determining Whether Two Lines Are Parallel or Perpendicular

Consider the four lines described below.

$$L_1 \text{ passes through } (-2, 1) \text{ and } (4, 5).$$
$$L_2 \text{ passes through } (3, 0) \text{ and } (0, -2).$$
$$L_3 \text{ passes through } (0, -3) \text{ and } (2, 0).$$
$$L_4 \text{ passes through } (-3, 0) \text{ and } (0, -2).$$

Find the slopes of these lines, and determine which two lines are parallel and which two are perpendicular.

Solution

We apply the slope formula in each case.

$$L_1\colon \quad m_1 = \frac{5 - 1}{4 - (-2)} = \frac{4}{6} = \frac{2}{3} \qquad\qquad L_2\colon \quad m_2 = \frac{-2 - 0}{0 - 3} = \frac{-2}{-3} = \frac{2}{3}$$

$$L_3\colon \quad m_3 = \frac{0 - (-3)}{2 - 0} = \frac{3}{2} \qquad\qquad L_4\colon \quad m_4 = \frac{-2 - 0}{0 - (-3)} = -\frac{2}{3}$$

Lines L_1 and L_2 have slope $\frac{2}{3}$, so they are parallel. Lines L_3 and L_4 have slopes that are negative reciprocals $\left(\frac{3}{2}\left(-\frac{2}{3}\right) = -1\right)$, so they are perpendicular.

Average Rate of Change

The slope formula applied to any two points on a line gives the **average rate of change** in y per unit change in x, where the value of y depends on the value of x.

For example, suppose the height of a boy increased from 60 to 68 inches between the ages of 12 and 16, as shown in **Figure 17.** We can find the boy's average growth rate (or average change in height per year).

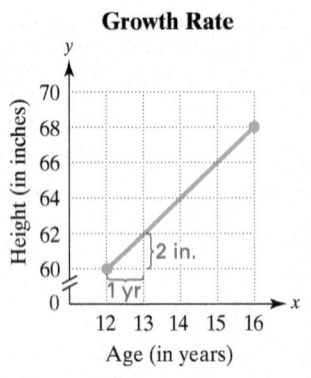

Growth Rate

Figure 17

$$\begin{array}{l} \text{Change in height } y \longrightarrow \\ \text{Change in age } x \longrightarrow \end{array} \quad \frac{68 - 60}{16 - 12} = \frac{8}{4} = 2 \text{ inches per year}$$

The boy may actually have grown more than 2 inches during some years and less than 2 inches during other years. If we plotted ordered pairs of the form

(age, height)

for those years and drew a line connecting any two of those points, the average rate of change we would find would probably be slightly different from that found above. However, using the data for ages 12 and 16, we find that the boy's *average* change in height was 2 inches per year over these years.

EXAMPLE 7   Interpreting Slope as Average Rate of Change

The graph in **Figure 18** shows the number of high-speed Internet subscribers in the United States from 2012 to 2017. Find the average rate of change in number of subscribers per year.

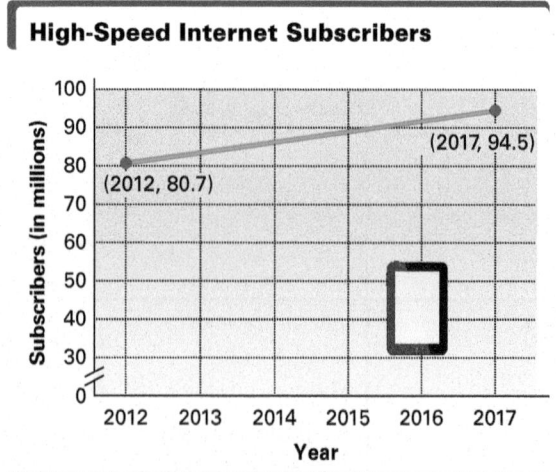

High-Speed Internet Subscribers

Data from Leichtman Research Group.

Figure 18

Solution

To find the average rate of change, we use the two pairs of data (2012, 80.7) and (2017, 94.5). Apply the slope formula.

$$\text{average rate of change} = \frac{94.5 - 80.7}{2017 - 2012} = \frac{13.8}{5} = 2.76 \quad \boxed{\begin{array}{l} \text{A positive slope} \\ \text{indicates an increase.} \end{array}}$$

The number of high-speed Internet subscribers *increased* by an average of 2.76 million per year between 2012 and 2017.

EXAMPLE 8 Interpreting Slope as Average Rate of Change

The graph in **Figure 19** shows the number of U.S. households with cable television from 2012 to 2017. Find the average rate of change in this number per year.

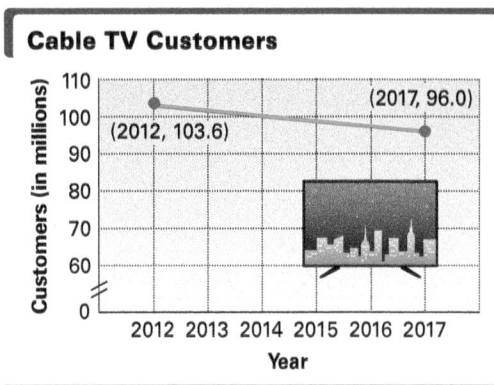

Data from Nielson Media Research.

Figure 19

Solution

As earlier, we use the two pairs of data shown.

$$\text{average rate of change} = \frac{103.6 - 96.0}{2012 - 2017} = \frac{7.6}{-5} = -1.52$$

> A negative slope indicates a decrease.

The number of households in the United States that had cable television *decreased* by an average of 1.52 million per year from 2012 to 2017.

8.2   EXERCISES

CONCEPT CHECK *Answer each of the following.*

1. What is the slope (or pitch) of this roof?

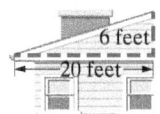

2. What is the slope (or grade) of this hill?

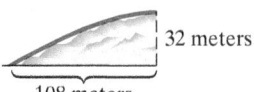

CONCEPT CHECK *Use the indicated points to find the slope of each line. (Use the "rise over run" interpretation.)*

3.

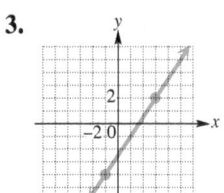

4.

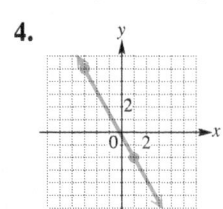

5. CONCEPT CHECK Determine whether the slope of the given line in (a)–(d) is *positive, negative, zero,* or *undefined.*

(a)

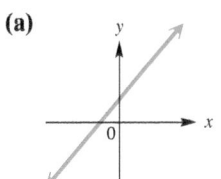

(b)

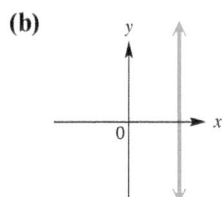

(c)

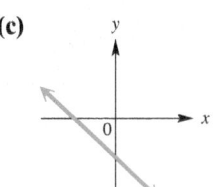

(d)

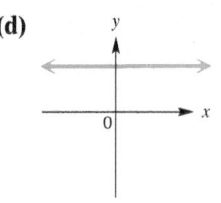

6. Plot any two points on the line $3x - y = 4$, and then use the "rise over run" interpretation to find the slope of the line.

Complete the given ordered pairs or table for each equation. Then graph the equation.

7. $x - y = 4$; $(0, \_\_)$, $(\_\_, 0)$,
    $(2, \_\_)$, $(\_\_, -1)$

8. $x + y = 5$; $(0, \_\_)$, $(\_\_, 0)$,
    $(2, \_\_)$, $(4, \_\_)$

9. $2x + y = 5$; $(0, \_\_)$, $(\_\_, 0)$,
    $(1, \_\_)$, $(\_\_, 1)$

10. $x + 3y = 12$; $(0, \_\_)$, $(\_\_, 0)$,
    $(3, \_\_)$, $(\_\_, 6)$

11. $4x + 5y = 20$; $(0, \_\_)$, $(\_\_, 0)$
    $(3, \_\_)$, $(\_\_, 2)$

12. $3x - 4y = 24$; $(0, \_\_)$, $(\_\_, 0)$,
    $(6, \_\_)$, $(\_\_, -3)$

13. $3x + 2y = 8$

x	y
0	—
—	0
2	—
—	-2

14. $5x + y = 12$

x	y
0	—
—	0
—	-3
2	—

15. $x + 3y = 9$

x	y
0	—
3	—
—	0
—	-1

16. $x - 2y = 8$

x	y
0	—
2	—
—	0
—	-2

For each equation, give the x-intercept and the y-intercept. Then graph the equation.

17. $3x + 2y = 12$

18. $2x + 5y = 10$

19. $5x + 6y = 10$

20. $x + 3y = 6$

21. $2x - y = 5$

22. $3x - 2y = 4$

23. $x - 3y = 2$

24. $4x - y = -3$

25. $x + y = 0$

26. $2x - y = 0$

27. $3x - y = 0$

28. $x + 4y = 0$

29. $x = 2$

30. $x = -2$

31. $y = 4$

32. $y = -3$

In Exercises 33–40, match the equation with the figure in choices A–D below that most closely resembles its graph.

A.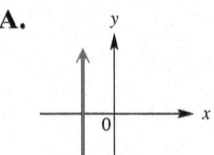

B.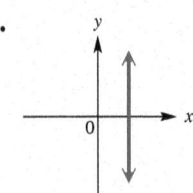

C.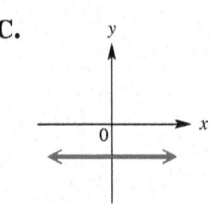

D.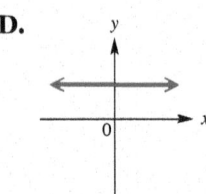

33. $y + 2 = 0$

34. $y + 4 = 0$

35. $x + 3 = 0$

36. $x + 7 = 0$

37. $y - 2 = 0$

38. $y - 4 = 0$

39. $x - 3 = 0$

40. $x - 7 = 0$

Find the slope of the line through each pair of points by using the slope formula.

41. $(-2, -3)$ and $(-1, 5)$

42. $(-4, 3)$ and $(-3, 4)$

43. $(8, 1)$ and $(2, 6)$

44. $(13, -3)$ and $(5, 6)$

45. $(2, 4)$ and $(-4, 4)$

46. $(-6, 3)$ and $(2, 3)$

47. $(1, 6)$ and $(1, 8)$

48. $(-3, 8)$ and $(-3, 2)$

Graph each line with the given slope and passing through the given point.

49. $m = \frac{1}{2}$; $(-3, 2)$

50. $m = \frac{2}{3}$; $(0, 1)$

51. $m = -\frac{5}{4}$; $(-2, -1)$

52. $m = -\frac{3}{2}$; $(-1, -2)$

53. $m = -2$; $(-1, -4)$

54. $m = 3$; $(1, 2)$

55. $m = 0$; $(2, -5)$

56. $m = 0$; $(2, 4)$

57. undefined slope; $(4, 2)$

58. undefined slope; $(-3, 1)$

Determine whether the lines described are parallel, perpendicular, *or* neither parallel nor perpendicular.

59. L_1 through $(4, 6)$ and $(-8, 7)$, and L_2 through $(7, 4)$ and $(-5, 5)$

60. L_1 through $(9, 15)$ and $(-7, 12)$, and L_2 through $(-4, 8)$ and $(-20, 5)$

61. L_1 through $(2, 0)$ and $(5, 4)$, and L_2 through $(6, 1)$ and $(2, 4)$

62. L_1 through $(1, 2)$ and $(-7, -2)$, and L_2 through $(1, -1)$ and $(5, -9)$

63. L_1 through $(0, 1)$ and $(2, -3)$, and L_2 through $(10, 8)$ and $(5, 3)$

64. L_1 through $(0, -7)$ and $(2, 3)$, and L_2 through $(0, -3)$ and $(1, -2)$

Use the concept of slope to solve each problem.

65. *Steepness of an Upper Deck* The upper deck at Guaranteed Rate Field in Chicago has produced, among other complaints, displeasure with its steepness. It has been compared to a ski jump. It is 160 ft from home plate to the front of the upper deck and 250 ft from home plate to the back. The top of the upper deck is 63 ft above the bottom. What is its slope?

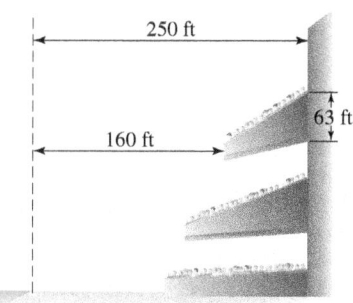

66. *Grade (Slope) of a Ramp* When designing the TD Banknorth Garden arena in Boston to replace the old Boston Garden, architects were careful to design the ramps leading up to the entrances so that elephants would be able to walk up the ramps. The maximum grade (or slope) that an elephant will walk on is 13%. Suppose that such a ramp were constructed with a horizontal run of 150 ft. What would be the maximum vertical rise the architects could use?

Rate of Change *Find and interpret the average rate of change illustrated in each graph.*

67.

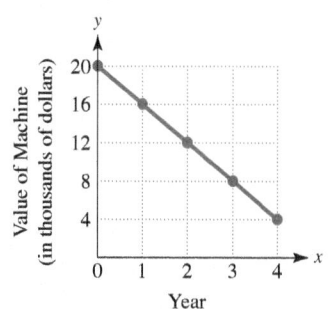

68.

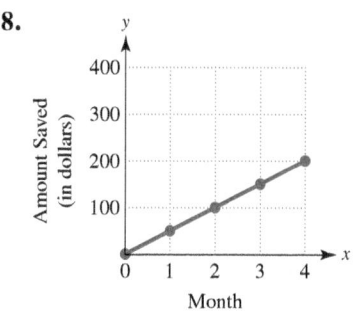

Solve each problem.

69. *Wireless Subscriber Connections* The graph shows the number of wireless subscriber connections (that is, active devices, including smartphones, feature phones, tablets, etc.) in millions in the United States for the years 2011 to 2016.

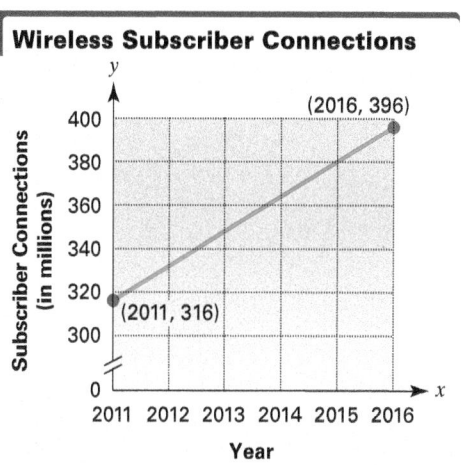

Data from CTIA.

(a) In the context of this graph, what does the ordered pair (2016, 396) mean?

(b) Use the given ordered pairs to find the slope of the line.

(c) Interpret the slope in the context of this problem.

70. *Wireless-Only Households* The graph shows the percent of households in the United States that were wireless-only for the years 2011 to 2016.

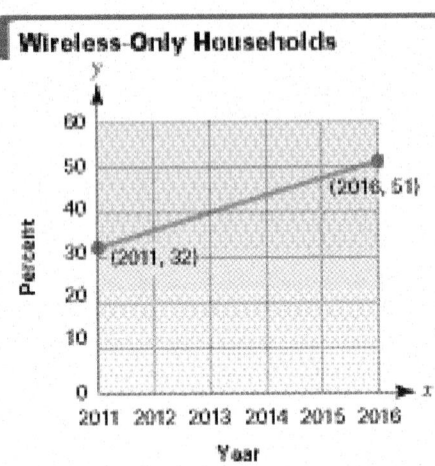

Data from CTIA.

(a) In the context of this graph, what does the ordered pair (2016, 51) mean?

(b) Use the given ordered pairs to find the slope of the line.

(c) Interpret the slope in the context of this problem.

71. *Drive-In Theaters* The graph shows the number of drive-in movie theaters in the United States from 2010 through 2017.

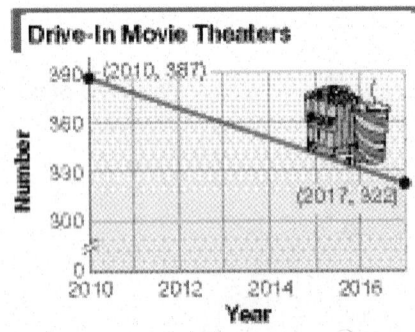

Data from www.drive-ins.com

(a) Use the given ordered pairs to find the average rate of change in the number of drive-in theaters per year during this period. Round the answer to the nearest whole number.

(b) Explain how a negative slope is interpreted in this situation.

72. *U.S. Travelers to Canada* The graph shows the number of U.S. travelers to Canada (in thousands) from 2000 through 2016.

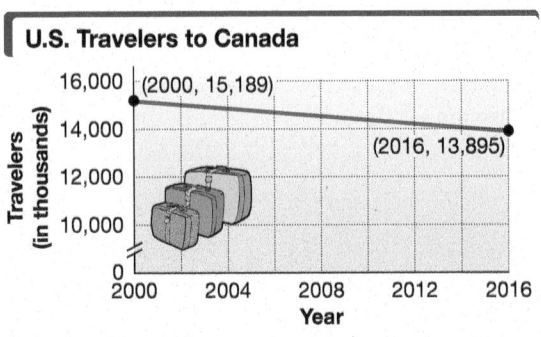

Data from U.S. Department of Commerce.

(a) Use the given ordered pairs to find the average rate of change in the number of U.S. travelers to Canada per year during this period. Round the answer to the nearest thousand.

(b) Explain how a negative slope is interpreted in this situation.

73. *Gasoline Prices* The average price of a gallon of gasoline in 1996 was $1.25. In 2016, the average price was $2.25. Find and interpret the average rate of change in the price of a gallon of gasoline per year to the nearest cent. (Data from Energy Information Administration.)

74. *Movie Ticket Prices* The average price of a movie ticket in 2000 was $5.39. In 2016, the average price was $8.65. Find and interpret the average rate of change in the price of a movie ticket per year to the nearest cent. (Data from Motion Picture Association of America.)

75. *Digital Cameras* In 2010, the number of digital cameras shipped worldwide totaled 122 million. There were 24 million shipped in 2016. Find and interpret the average rate of change in the number of digital cameras shipped per year to the nearest million. (Data from CIPA.)

76. *Desktop Computers* In 2010, worldwide shipments of desktop computers totaled 157 million units. In 2016, worldwide shipments were 103 million units. Find and interpret the average rate of change in worldwide shipments of desktop computers per year to the nearest million units. (Data from IDC.)

8.3 EQUATIONS OF LINES

Point-Slope Form

If the slope of a line and a particular point on a line are known, it is possible to find an equation of the line. Suppose that the slope of a line is m and that (x_1, y_1) is a particular point on the line. Let (x, y) be any other point on the line. Then, by the definition of slope, we have the following.

$$m = \frac{y - y_1}{x - x_1} \qquad \text{Slope formula}$$

$$m(x - x_1) = \frac{y - y_1}{x - x_1}(x - x_1) \qquad \text{Multiply by } x - x_1.$$

$$m(x - x_1) = y - y_1 \qquad \text{Simplify the right side.}$$

Interchanging the sides of this last equation gives the *point-slope form* of the equation of a line.

POINT-SLOPE FORM

The equation of a line through (x_1, y_1) with slope m is written in **point-slope form** as follows.

$$y - y_1 = m(x - x_1)$$

EXAMPLE 1 Finding an Equation Given the Slope and a Point

Find the standard form of an equation of the line with slope $\frac{1}{3}$, passing through the point $(-2, 5)$.

Solution

$$y - y_1 = m(x - x_1) \qquad \text{Point-slope form}$$

$$y - 5 = \frac{1}{3}[x - (-2)] \qquad \text{Let } (x_1, y_1) = (-2, 5) \text{ and } m = \frac{1}{3}.$$

Substitute carefully.

$$3(y - 5) = 3\left[\frac{1}{3}(x + 2)\right] \qquad \text{Multiply by 3.}$$

$$3(y - 5) = x + 2 \qquad 3 \cdot \frac{1}{3} = 1$$

$$3y - 15 = x + 2 \qquad \text{Distributive property.}$$

$$x - 3y = -17 \qquad \text{Standard form}$$

If two points on a line are known, it is possible to find an equation of the line. One method is to first find the slope using the slope formula, and then use the slope with one of the given points in the point-slope form.

EXAMPLE 2   **Finding an Equation Given Two Points**

Find the standard form of an equation of the line passing through the points $(-4, 3)$ and $(6, -2)$.

Solution

First find the slope, using the definition.

$$m = \frac{-2 - 3}{6 - (-4)} = \frac{-5}{10} = -\frac{1}{2} \qquad \begin{array}{l} x_1 = -4, \; y_1 = 3, \\ x_2 = 6, \; y_2 = -2 \end{array}$$

The slope of the line passing through the given points is $-\frac{1}{2}$. Either $(-4, 3)$ or $(6, -2)$ may be used as (x_1, y_1) in the point-slope form of the equation of the line. If $(-4, 3)$ is used, then $x_1 = -4$ and $y_1 = 3$.

$$y - y_1 = m(x - x_1) \qquad \text{Point-slope form}$$

$$y - 3 = -\frac{1}{2}\left[x - (-4)\right] \qquad \text{Let } (x_1, \, y_1) = (-4, 3) \text{ and } m = -\frac{1}{2}.$$

$$y - 3 = -\frac{1}{2}(x + 4) \qquad \text{Definition of subtraction}$$

$$2(y - 3) = 2\left[-\frac{1}{2}(x + 4)\right] \qquad \text{Multiply by 2.}$$

$$2(y - 3) = -(x + 4) \qquad 2\left(-\frac{1}{2}\right) = -1$$

$$2y - 6 = -x - 4 \qquad \text{Distributive property}$$

$$x + 2y = 2 \qquad \text{Standard form}$$

Slope-Intercept Form

Suppose that the slope m of a line is known, and the y-intercept of the line has coordinates $(0, b)$. Then we have the following.

$$y - y_1 = m(x - x_1) \qquad \text{Point-slope form}$$

$$y - b = m(x - 0) \qquad \text{Let } (x_1, \, y_1) = (0, b).$$

$$y - b = mx \qquad \text{Subtract. Multiply.}$$

$$y = mx + b \qquad \text{Add } b \text{ to both sides.}$$

This result is known as the *slope-intercept form* of the equation of the line.

Maria Gaetana Agnesi (1719–1799) did much of her mathematical work in coordinate geometry. She grew up in a scholarly atmosphere. Her father was a mathematician on the faculty at the University of Bologna. In a larger sense she was an heir to the long tradition of Italian mathematicians.

Maria was fluent in several languages by age 13, but she chose mathematics over literature. The curve shown below, called the **witch of Agnesi,** is studied in analytic geometry courses.

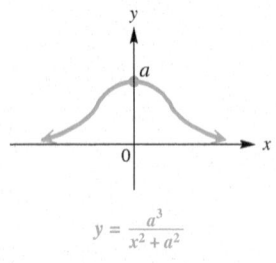

$$y = \frac{a^3}{x^2 + a^2}$$

SLOPE-INTERCEPT FORM

The equation of a line with slope m and y-intercept $(0, b)$ is written in **slope-intercept form** as follows.

$$y = mx + b$$

Slope y-intercept is $(0, b)$.

EXAMPLE 3 Writing an Equation in Slope-Intercept Form

Write each of the following equations in slope-intercept form.

(a) The line $x - 3y = -17$ described in **Example 1**

(b) The line $x + 2y = 2$ described in **Example 2**

Solution

(a) Solve for y to obtain the slope-intercept form.

$$x - 3y = -17$$

$$-3y = -x - 17 \quad \text{Subtract } x.$$

$$y = \frac{1}{3}x + \frac{17}{3} \quad \text{Divide by } -3.$$

The slope is $\frac{1}{3}$ and the y-intercept is $\left(0, \frac{17}{3}\right)$.

(b)
$$x + 2y = 2$$

$$2y = -x + 2 \quad \text{Subtract } x.$$

$$y = -\frac{1}{2}x + 1 \quad \text{Multiply by } \frac{1}{2}.$$

The slope is $-\frac{1}{2}$ and the y-intercept is $(0, 1)$.

If the slope-intercept form of the equation of a line is known, the method of graphing described in the following example can be used to graph the line.

EXAMPLE 4 Graphing a Line Using Slope and y-Intercept

Graph the line with the equation $y = -\frac{2}{3}x + 3$.

Solution

Because the equation is given in slope-intercept form, we can see that the slope is $-\frac{2}{3}$ and the y-intercept is $(0, 3)$. We will interpret $-\frac{2}{3}$ as $\frac{-2}{3}$. Plot the point $(0, 3)$, and then, using the "rise over run" interpretation of slope, move *down* 2 units (because of the -2 in the numerator of the slope) and to the *right* 3 units (because of the 3 in the denominator). We arrive at the point $(3, 1)$. Plot the point $(3, 1)$, and join the two points with a line, as shown in **Figure 20.**

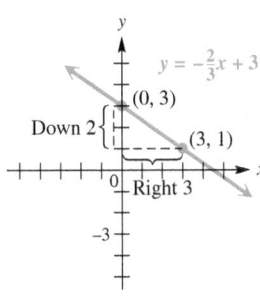

Figure 20

We could interpret $-\frac{2}{3}$ as $\frac{2}{-3}$ and obtain a different second point. The line would be the same.

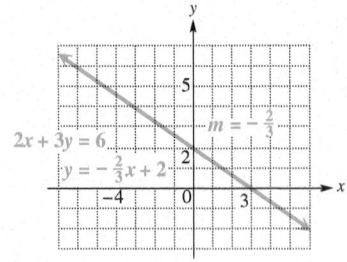

Figure 21

EXAMPLE 5 Using a Slope Relationship (Parallel Lines)

Find the slope-intercept form of the equation of the line parallel to the graph of $2x + 3y = 6$, passing through the point $(-2, -1)$. Graph both lines.

Solution

The slope of the graph of $2x + 3y = 6$ in **Figure 21** can be found by solving for y.

$$2x + 3y = 6$$

$$3y = -2x + 6 \qquad \text{Subtract } 2x.$$

$$y = -\frac{2}{3}x + 2 \qquad \text{Divide by 3.}$$

$$\underset{\textstyle\uparrow}{} \;\; \text{Slope}$$

The slope is given by the coefficient of x, so $m = -\frac{2}{3}$. The line through $(-2, -1)$ and parallel to $2x + 3y = 6$ must have the same slope. To find the required equation, use the point-slope form with $(x_1, y_1) = (-2, -1)$ and $m = -\frac{2}{3}$.

$$y - (-1) = -\frac{2}{3}[x - (-2)] \qquad y_1 = -1,\; m = -\frac{2}{3},\; x_1 = -2$$

$$y + 1 = -\frac{2}{3}(x + 2) \qquad \text{Definition of subtraction}$$

$$y + 1 = -\frac{2}{3}x - \frac{4}{3} \qquad \text{Distributive property}$$

$$y = -\frac{2}{3}x - \frac{4}{3} - 1 \qquad \text{Subtract 1.}$$

$$y = -\frac{2}{3}x - \frac{7}{3} \qquad \text{Subtract.}$$

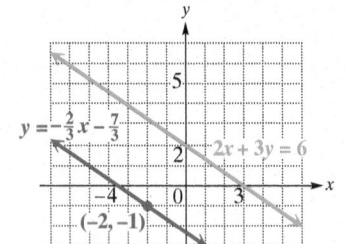

Figure 22

Both lines are shown in **Figure 22.**

EXAMPLE 6 Using a Slope Relationship (Perpendicular Lines)

Find the slope-intercept form of the equation of the line perpendicular to the graph of $2x + 3y = 6$, passing through the point $(-4, 5)$. Graph both lines.

Solution

In **Example 5** we found that the slope of the line $2x + 3y = 6$ is $-\frac{2}{3}$. A line perpendicular to this line must have a slope that is the negative reciprocal of $-\frac{2}{3}$, which is $\frac{3}{2}$. Use the point $(-4, 5)$ and slope $\frac{3}{2}$ in the point-slope form to obtain the equation of the perpendicular line shown in **Figure 23.**

$$y - 5 = \frac{3}{2}[x - (-4)] \qquad y_1 = 5,\; m = \frac{3}{2},\; x_1 = -4$$

$$y - 5 = \frac{3}{2}(x + 4) \qquad \text{Definition of subtraction}$$

$$y - 5 = \frac{3}{2}x + 6 \qquad \text{Distributive property}$$

$$y = \frac{3}{2}x + 11 \qquad \text{Add 5.}$$

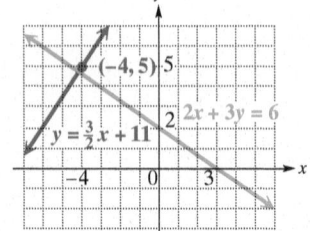

Figure 23

An Application of Linear Equations

One familiar formula that provides exact results gives the relationship between the Celsius and Fahrenheit temperature scales. Earlier we examined the formula for converting Celsius to Fahrenheit. The following example shows how that formula is derived.

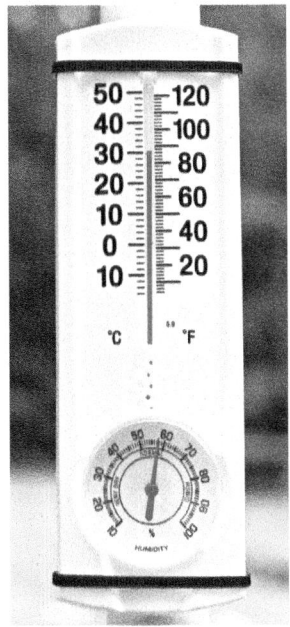

EXAMPLE 7 **Deriving the Formula for Converting Celsius to Fahrenheit**

In the Fahrenheit scale, water freezes at 32 degrees and boils at 212 degrees. In the Celsius scale, water freezes at 0 degrees and boils at 100 degrees. There is an exact linear relationship between these two scales. Use this information to derive the formula for converting Celsius to Fahrenheit.

Solution

For now, we will let x represent the Celsius degree measure and let y represent the Fahrenheit measure. The two ordered pairs

$$(0, 32) \quad \text{and} \quad (100, 212)$$

illustrate a relationship between the two scales. We now find the equation of the line containing these two points. First, find the slope of the line.

$$m = \frac{212 - 32}{100 - 0} \qquad \text{Slope formula with } x_1 = 0, \; y_1 = 32, \, x_2 = 100, \, y_2 = 212$$

$$m = \frac{180}{100} \qquad \text{Subtract.}$$

$$m = \frac{9}{5} \qquad \text{Lowest terms}$$

Now, because $(0, 32)$ is the y-intercept, we can substitute into the slope-intercept form to find the equation of the line.

$$y = mx + b \qquad \text{Slope-intercept form}$$

$$y = \frac{9}{5}x + 32 \qquad m = \frac{9}{5}, \, b = 32$$

Finally, replace x with C and y with F to represent the two scales.

$$F = \frac{9}{5}C + 32 \qquad \text{The familiar formula}$$

8.3 EXERCISES

CONCEPT CHECK *Fill in each blank with the correct response.*

1. The slope of the graph of $y = -\frac{1}{3}x + 6$ is _____.

2. The y-intercept of the graph of $y = 4x - 8$ is _____.

3. The slope of the graph of any line parallel to the graph of $4x + y = 6$ is _____.

4. The slope of the graph of any line perpendicular to the graph of $-3x + 8y = 4$ is _____.

Match each equation in Group I with the correct description given in Group II.

I

5. $y = 4x$

6. $y = \frac{1}{4}x$

7. $y = -2x + 1$

8. $y - 1 = -2(x - 4)$

II

A. slope $= -2$, through the point $(4, 1)$

B. slope $= -2$, y-intercept $(0, 1)$

C. passing through the points $(0, 0)$ and $(4, 1)$

D. passing through the points $(0, 0)$ and $(1, 4)$

In Exercises 9–16, match each equation with the graph that it most closely resembles in Choices A–H. (Hint: Determining the signs of m and b will help you make your decision.)

9. $y = 2x + 3$

10. $y = -2x + 3$

11. $y = -2x - 3$

12. $y = 2x - 3$

13. $y = 2x$

14. $y = -2x$

15. $y = 3$

16. $y = -3$

A.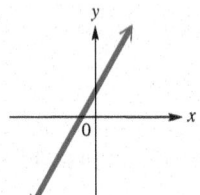

B.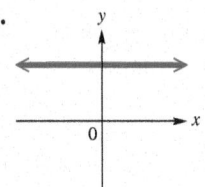

C.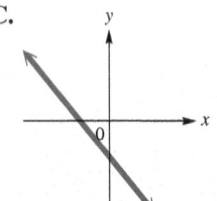

D.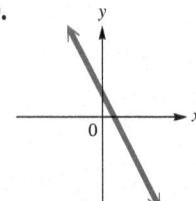

E.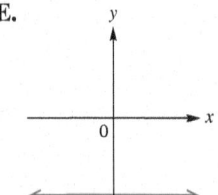

F.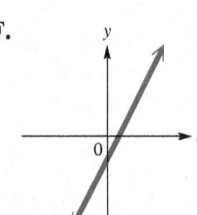

G.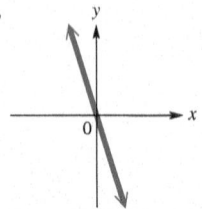

H.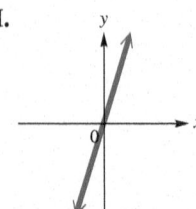

Use the geometric interpretation of slope (rise divided by run) to find the slope of each line. Then, by identifying the y-intercept from the graph, write the slope-intercept form of the equation of the line.

17.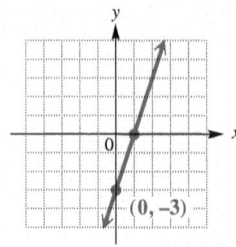

18.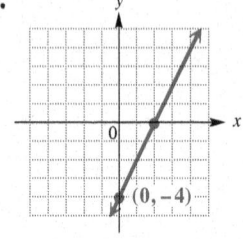

19.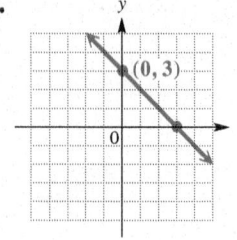

20.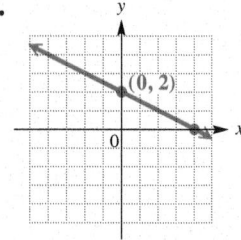

Write the slope-intercept form of the equation of the line satisfying the given conditions.

21. Through $(5, 8)$; slope -2

22. Through $(1, 4)$; slope -5

23. Through $(-2, 4)$; slope $-\frac{3}{4}$

24. Through $(-1, 6)$; slope $-\frac{5}{6}$

25. Through $(-5, 4)$; slope $\frac{1}{2}$

26. Through $(7, -2)$; slope $\frac{1}{4}$

27. x-intercept $(3, 0)$; slope 4

28. x-intercept $(-2, 0)$; slope -5

Write an equation for a line that satisfies the given conditions.

29. Through $(9, 5)$; slope 0

30. Through $(-4, -2)$; slope 0

31. Through $(9, 10)$; undefined slope

32. Through $(-2, 8)$; undefined slope

33. Through $(0.5, 0.2)$; vertical

34. Through $\left(\frac{5}{8}, \frac{2}{9}\right)$; vertical

35. Through $(-7, 8)$; horizontal

36. Through $(2, 7)$; horizontal

Write the equation, in slope-intercept form if possible, of the line passing through the two points.

37. $(3, 4)$ and $(5, 8)$

38. $(5, -2)$ and $(-3, 14)$

39. $(6, 1)$ and $(-2, 5)$

40. $(-3, 6)$ and $(6, 12)$

41. $(2, 5)$ and $(1, 5)$

42. $(-2, 2)$ and $(4, 2)$

43. $(7, 6)$ and $(7, -8)$

44. $(13, 5)$ and $(13, -1)$

45. $(1, -3)$ and $(-1, -3)$

46. $(-4, 6)$ and $(5, 6)$

Find the equation, in slope-intercept form, of the line satisfying the given conditions.

47. $m = 5$; $b = 15$

48. $m = -2$; $b = 12$

49. $m = -\frac{2}{3}$; $b = \frac{4}{5}$

50. $m = -\frac{5}{8}$; $b = -\frac{1}{3}$

51. Slope $\frac{2}{5}$; y-intercept $(0, 5)$

52. Slope $-\frac{3}{4}$; y-intercept $(0, 7)$

For each equation **(a)** *write in slope-intercept form,* **(b)** *give the slope of the line, and* **(c)** *give the y-intercept.*

53. $x + y = 12$ **54.** $x - y = 14$

55. $5x + 2y = 20$ **56.** $6x + 5y = 40$

57. $2x - 3y = 10$ **58.** $4x - 3y = 10$

Write the equation, in slope-intercept form, of the line satisfying the given conditions.

59. Through $(7, 2)$; parallel to $3x - y = 8$

60. Through $(4, 1)$; parallel to $2x + 5y = 10$

61. Through $(-2, -2)$; parallel to $-x + 2y = 10$

62. Through $(-1, 3)$; parallel to $-x + 3y = 12$

63. Through $(8, 5)$; perpendicular to $2x - y = 7$

64. Through $(2, -7)$; perpendicular to $5x + 2y = 18$

65. Through $(-2, 7)$; perpendicular to $x = 9$

66. Through $(8, 4)$; perpendicular to $x = -3$

Celsius and Fahrenheit *Refer to* **Example 7** *to answer Exercises 67–70.*

67. Determine the formula for converting Fahrenheit to Celsius by applying the method of **Example 7** to the points $(32, 0)$ and $(212, 100)$. Give the answer using C and F.

68. Refer to the formula derived in **Example 7** and solve algebraically for C. Verify that the result is the same as determined in **Exercise 67.**

69. Use either formula to find the temperature at which Celsius and Fahrenheit are equal. (*Hint:* Replace C with F and solve for the single variable.)

70. At what Celsius temperature is the corresponding Fahrenheit temperature numerically double the Celsius temperature?

8.4 LINEAR FUNCTIONS, GRAPHS, AND MODELS

OBJECTIVES

1 Interpret the terms *relation, function, domain, range, independent variable,* and *dependent variable* in mathematical exposition.

2 Use function notation.

3 Identify a linear function and graph it.

4 Perform cost analysis using linear functions.

5 Construct a linear model from a set of data points.

Relations and Functions

Mathematics provides a means of describing how values of one quantity correspond with values of another quantity. We use *relations* and *functions* to express these descriptions.

> **RELATION**
>
> A **relation** is a set of ordered pairs.

Suppose that it is time to fill up your car's gas tank. At your local station, 89-octane gas is selling for $3.10 per gallon. The final price you pay is determined by the number of gallons you buy multiplied by the price per gallon (in this case, $3.10). **Table 1** on the next page gives ordered pairs to illustrate this situation.

If we let x denote the number of gallons pumped, then the price y in dollars can be found by the linear equation

$$y = 3.10x.$$

Theoretically, there are infinitely many ordered pairs (x, y) that satisfy this equation. We are limited to nonnegative values for x here, since we cannot have a negative number of gallons. There also is a practical maximum value for x in this situation, which varies from one car to another—that value is the capacity of the tank.

Table 1

Number of Gallons Pumped	Price for This Number of Gallons
0	$0.00 = 0($3.10)$
1	$3.10 = 1($3.10)$
2	$6.20 = 2($3.10)$
3	$9.30 = 3($3.10)$
4	$12.40 = 4($3.10)$

In this example, the total price depends on the amount of gasoline pumped. For this reason, price is called the *dependent variable,* and the number of gallons is called the *independent variable.* Generalizing, if the value of the variable y depends on the value of the variable x, then y is the **dependent variable** and x the **independent variable.**

Independent variable ⌐ ⌐Dependent variable
$$(x, y)$$

The set of all values taken on by the independent variable is the **domain** of the relation. The set of all dependent variable values that result is the **range** of the relation. As mentioned earlier, a special kind of relation, called a *function,* is important in mathematics and its applications. Three variations of the definition of function are now given.

VARIATIONS OF THE DEFINITION OF FUNCTION

1. A **function** is a relation in which, for each value of the first component of the ordered pairs, there is *exactly one* value of the second component.

2. A **function** is a set of distinct ordered pairs in which no first component is repeated.

3. A **function** is defined by a rule or correspondence that assigns exactly one range value to each distinct domain value.

In a function, there is exactly one value of the dependent variable (the second component) for each value of the independent variable (the first component).

Function Notation

When a function f is defined with a rule or an equation using x and y for the independent and dependent variables, we say "y is a function of x" to emphasize that y depends on x. For a function f, we use the notation

$$y = f(x),$$

called **function notation,** to express this and read $f(x)$ as **"f of x."** (In this notation the parentheses do not indicate multiplication.) For example, in function f, if $y = 2x - 7$, we write

Do not read $f(x)$ as "f times x." ⟶ $f(x) = 2x - 7$.

Note that $f(x)$ is just another name for the dependent variable y. For example, if

$$y = f(x) = 3x - 5,$$

and $x = 2$, then we find y, or $f(x)$, by replacing x with 2.

$$y = f(2)$$
$$y = 3 \cdot 2 - 5 \quad {\scriptstyle f(x) = 3x - 5}$$
$$y = 1 \quad\quad {\scriptstyle \text{Multiply, and then subtract.}}$$

The fact that $(2, 1)$ belongs to the function f is written with function notation as

$$f(2) = 1.$$

We read $f(2)$ as "f of 2" or "f at 2."

The Gasoline Pumping Function For every number of gallons, there is one and only one sale price. The price charged *depends* on the number of gallons pumped. When there were 0 gallons pumped, the price was 0 dollars. When there were 25.825 gallons pumped, the price was $79.00. If we graph a line segment joining (0, 0) and (25.825, 79.00), what is its slope?

Now look at the prices for Supreme, Plus, Regular, and Diesel. Which fuel was purchased? (The answer is on **page 404.**)

These ideas and the symbols used to represent them can be explained as follows.

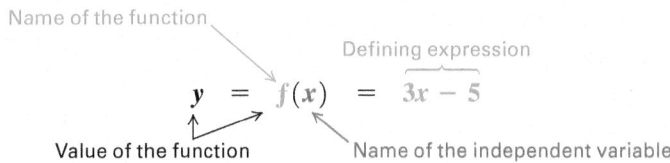

Name of the function

Defining expression

$$y = f(x) = 3x - 5$$

Value of the function

Name of the independent variable

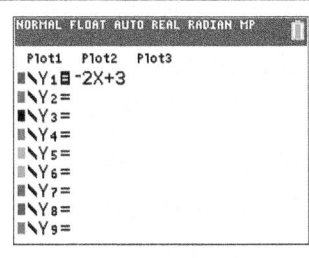

The TI-84 Plus C features function notation as shown here. Compare to the discussion in **Example 1.**

EXAMPLE 1 **Using Function Notation**

Find the following for the function

$$f(x) = -2x + 3.$$

(a) $f(0)$ **(b)** $f(1)$ **(c)** $f(3)$ **(d)** the value of x for which $f(x) = 7$

(e) Use these results to give four ordered pairs that belong to f.

Solution

(a)
$$f(x) = -2x + 3 \quad \text{Given function}$$
$$f(0) = -2(0) + 3 \quad \text{Replace } x \text{ with 0.}$$
$$= 0 + 3 \quad \text{Multiply.}$$
$$= 3 \quad \text{Add.}$$

(b) $f(1) = -2(1) + 3$ **(c)** $f(3) = -2(3) + 3$
$$= -2 + 3 \qquad\qquad = -6 + 3$$
$$= 1 \qquad\qquad\qquad = -3$$

(d) Replace $f(x)$ with 7 and solve for x.

$$f(x) = -2x + 3 \quad \text{Given function}$$
$$7 = -2x + 3 \quad \text{Replace } f(x) \text{ with 7.}$$
$$4 = -2x \quad \text{Subtract 3.}$$
$$-2 = x \quad \text{Divide by } -2.$$

Thus, when $x = -2$, $f(x) = f(-2) = 7$.

(e) The results in parts (a)–(d) indicate that these ordered pairs all belong to f.

$$(0, 3), \quad (1, 1), \quad (3, -3), \quad \text{and} \quad (-2, 7)$$

Linear Functions

In our earlier discussion, we saw that if x represents the number of gallons pumped and 3.10 is the price per gallon, then

$$y = 3.10x$$

is the total price paid. If we decide to also get a car wash for 4 dollars, then a fixed amount is added to the variable amount. In this case,

$$y = 3.10x + 4$$

represents the total cost. This is an example of a linear equation in two variables, where 3.10 is the slope m of the line and $(0, 4)$ is the y-intercept. In the context of functions, it is called a *linear function*.

LINEAR FUNCTION

A function that can be defined by a rule written in the form

$$f(x) = ax + b,$$

for real numbers a and b, is a **linear function.**

The form $f(x) = ax + b$ defining a linear function is the same as that of the slope-intercept form of the equation of a line with a representing the slope. We know that the graph of $f(x) = ax + b$ will be a line with slope a and y-intercept $(0, b)$.

EXAMPLE 2 Graphing Linear Functions

Graph each linear function. Assume that x can be any real number. Give the domain and the range.

(a) $f(x) = -2x + 3$ **(b)** $f(x) = 3$

Solution

(a) One way to graph this function is to locate the y-intercept, $(0, 3)$. From this point, use the slope $-2 = \frac{-2}{1}$ to move *down* 2 and *right* 1. This second point is used to obtain the graph in **Figure 24(a).**

Alternatively, we can use the ordered pairs found in **Example 1** to plot the graph. Those ordered pairs are also shown in the table accompanying **Figure 24(a).**

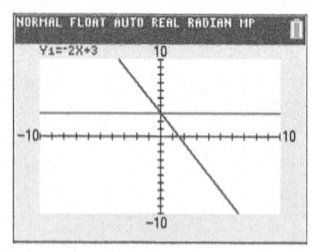

$Y_2 = 3$
The two lines in **Example 2** are shown here.

x	y
-2	7
0	3
1	1
3	-3

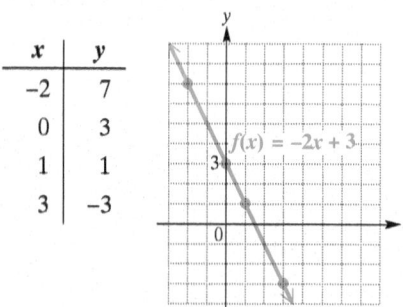

x	y
-2	3
0	3
2	3

Linear function
(a)

Constant function
(b)

Figure 24

The domain and the range are both $(-\infty, \infty)$.

(b) From the previous section, we know that the graph of $y = 3$ is a horizontal line. Therefore, the graph of $f(x) = 3$ is a horizontal line ($m = 0$) with y-intercept $(0, 3)$, as shown in **Figure 24(b).** The domain is $(-\infty, \infty)$ and the range is $\{3\}$.

The function defined in **Example 2(b)** and graphed in **Figure 24(b)** is an example of a constant function. A **constant function** is a linear function of the form

$$f(x) = b,$$

Answer to the Gasoline Pumping question on page 402 Divide the sale amount by the number of gallons to find that the price per gallon is that of diesel fuel.

where b is a real number.

In applications of linear functions, domain values are often restricted. In the case of the gas-pumping example, x can take on only non-negative values. The size of the tank would also put a restriction on the maximum value of x.

Linear Models

If an entrepreneur opens a business that specializes in selling a particular product, the goal is to earn a profit. Linear functions can model cost, revenue, and profit.

EXAMPLE 3   Analyzing Cost, Revenue, and Profit

Peripheral Visions, Inc., produces studio-quality DVDs of live concerts. The company places an ad online. The cost of the ad is $100. Each DVD costs $20 to produce, and the company charges $24 per disk.

(a) Express the cost C as a function of x, the number of DVDs produced and sold.

(b) Express the revenue R as a function of x, the number of DVDs sold.

(c) When will the company break even? That is, for what value of x does revenue equal cost?

(d) Graph the cost and revenue functions on the same coordinate system, and interpret the graph.

Solution

(a) The **fixed cost** is $100, and for each DVD produced, the **variable cost** is $20. The cost C can be expressed as a function of x, the number of DVDs produced.

$$C(x) = 20x + 100 \quad (C \text{ in dollars})$$

(b) Each DVD sells for $24, so the revenue R is given by

$$R(x) = 24x \quad (R \text{ in dollars}).$$

(c) The company will just break even (no profit and no loss) if revenue just equals cost, or $R(x) = C(x)$. Solve this equation for x.

$$R(x) = C(x)$$
$$24x = 20x + 100 \quad \text{Substitute for } R(x) \text{ and } C(x).$$
$$4x = 100 \quad \text{Subtract } 20x.$$
$$x = 25 \quad \text{Divide by 4.}$$

If 25 DVDs are produced and sold, the company will break even.

(d) **Figure 25** shows the graphs of the two functions. (For convenience, we use lines rather than discrete points with whole number y-values.) At the break-even point, we see that when 25 DVDs are produced and sold, both the cost and the revenue are $600. If fewer than 25 DVDs are produced and sold (that is, when $x < 25$), the company loses money. When more than 25 DVDs are produced and sold (that is, when $x > 25$), there is a profit.

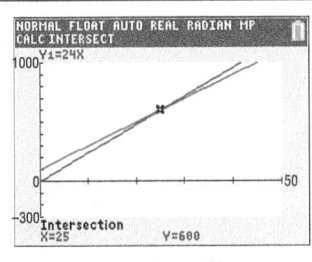

$Y_2 = 20X + 100$

The break-even point is (25, 600), as indicated at the bottom of the screen. The calculator can find the point of intersection of the graphs. Compare with **Figure 25.**

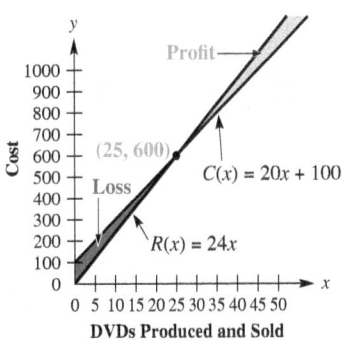

Figure 25

Given data points that lie approximately on a straight line, we can develop a linear function, or **linear model,** to describe the data.

Table 2

Year	Cost (in dollars)
2013 ($x = 0$)	3264
2014 ($x = 1$)	3347
2015 ($x = 2$)	3435
2016 ($x = 3$)	3520
2017 ($x = 4$)	3570

Data from U.S. Department of Education.

StatCrunch Modeling Tuition Costs

EXAMPLE 4   Modeling Tuition Costs

Average annual tuition and fees for in-state students at public two-year colleges are shown in **Table 2.**

(a) Graph the data. Let $x = 0$ correspond to 2013, $x = 1$ correspond to 2014, and so on. What type of equation might model the data?

(b) Find a linear function f that models the data, using the information for the years 2013 and 2017.

(c) Use the equation from part (b) to approximate the cost of tuition and fees at public two-year colleges in 2018.

Solution

(a) The points are plotted in **Figure 26,** where x corresponds to the year as described, and y represents the cost in dollars. It appears that a linear equation will model the data.

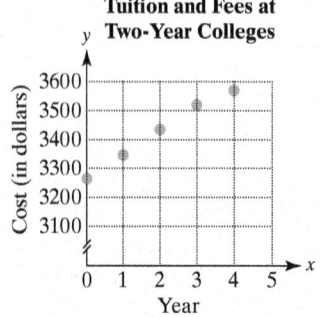

Tuition and Fees at Two-Year Colleges

Figure 26

(b) Since the points in **Figure 26** lie approximately on a straight line, we can write a linear equation that models the relationship between year x and cost y. We use the two data points for 2013 and 2017 to find the slope of the line.

$$m = \frac{3570 - 3264}{4 - 0} = \frac{306}{4} = 76.5$$

Start with the x- and y-values of the same point.

Use (0, 3264) and (4, 3570).

The slope 76.5 indicates that the cost of tuition and fees increased by about $76.50 per year from 2013 to 2017. We use this slope, the y-intercept (0, **3264**), and the slope-intercept form to write an equation of the line.

$$f(x) = 76.5x + 3264$$

(c) The value $x = 5$ corresponds to the year 2018.

$$f(x) = 76.5x + 3264 \qquad \text{Equation from part (b)}$$

$$f(5) = 76.5(5) + 3264 \qquad \text{Substitute 5 for x.}$$

$$f(5) \approx 3647 \qquad \text{Multiply and then add.}$$

According to this model, average tuition and fees for in-state students at public two-year colleges in 2018 were about $3647.

In **Example 4,** if we had chosen a different pair of data points, we would have found a slightly different equation. However, all such equations should yield similar results, because the data points are approximately linear.

WHEN WILL I **EVER** USE THIS?

The term *Eureka moment* comes from a story involving the Greek mathematician Archimedes. Legend has it that when he was in his bathtub, a law of displacement suddenly dawned on him, and he shouted, *"Eureka"*. [I have found it!]

Suppose that a teacher of composition asks her students to write a short essay on a personal experience that led to such a moment. One possible response follows.

The Day My Algebra Class Finally Made Sense

by Christopher Eric Rogers

I am a first-year college student with aspirations to become a writer while working my way through school delivering pizzas. One of my classes is algebra. I've never been a bad math student, but it has never been my favorite subject, and it's always seemed to be totally removed from any everyday experience (except maybe balancing my checkbook). Until today.

*My teacher was talking about **ordered pairs, graphs, functions, lines, slope, intercepts,** and other things that I remember hearing about in high school. I thought I would never have to hear about them again. And then the proverbial light bulb went on in my head when I least expected it.*

My gas tank had been nearly empty for two days and needed filling. I put my debit card in the slot, selected my grade, and started to pump my gas. The display showed two numbers spinning by:

the sale price and the number of gallons.

Below them was the rate: $3.289, which is basically $3.29. And then it occurred to me that price divided by number of gallons is the price per gallon. These two spinning numbers can be written as those ordered pairs she talked about:

(number of gallons, total price).

I pulled out my camera phone and took some pictures.

   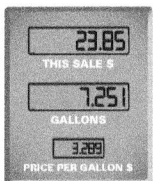

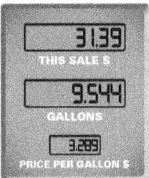 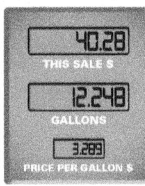 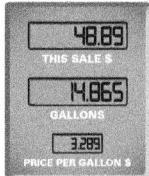

I remember my teacher saying that pairs of numbers can be represented in a table, so when I got home after work I did this:

Number of Gallons	Sale Price
0.000	0.00
2.402	7.90
5.003	16.45
7.251	23.85
9.544	31.39
12.248	40.28
14.865	48.89

She also said that points representing pairs of numbers can be graphed:

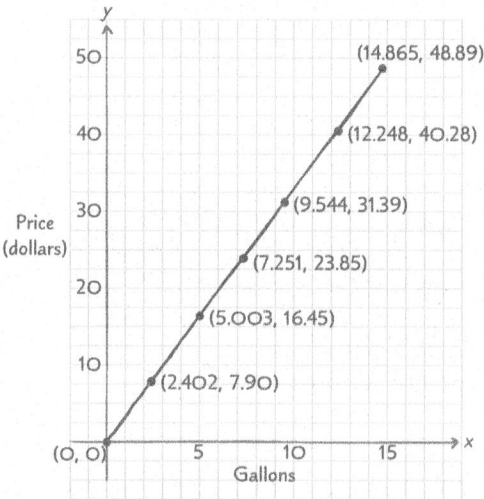

These points lie in a straight line. They can be connected to represent other pairs not shown in this table. And I had to stop when the number of gallons was 14.865, because that's all my tank would hold.

What's this? Dividing the price by the number of gallons in every picture gives the price per gallon, which is the slope of this line. (If the price had been almost $4.00 per gallon like it was last week, the line would have been steeper. Whoa! Is this what the term steeper *means when referring to increased price?)*

The teacher talked about **domain** *. . . and that's what the numbers of gallons represent. The corresponding prices belong to the* **range**. *There is a smallest domain value, 0, which showed up when I started, and a largest value, 14.865, because my tank couldn't hold any more. I also had the option of a $6.00 car wash, which I declined. It would have moved all the points upward 6 units and changed the starting point to the ordered pair (0, 6). And that's the y-intercept she talked about.*

My dad says he learns something new every day. As I drove around town during dinner time, I realized that I had learned something new in algebra today. Eureka!

8.4 EXERCISES

CONCEPT CHECK *Choose from the following list of responses for Exercises 1–6. Each choice is used once and only once.*

function	dependent	domain	range
line	horizontal	constant	linear
independent	set of ordered pairs		

1. A relation is a _____.

2. The set of all first components in a relation is the _____ of the relation. Each one is a value of the _____ variable.

3. The set of all second components in a relation is the _____ of the relation. Each one is a value of the _____ variable.

4. A relation in which each first component is paired with one and only one second component is a(n) _____.

5. $f(x) = 3x + 6$ is an example of a(n) _____ function. Its graph is a(n) _____.

6. $f(x) = 6$ is an example of a(n) _____ function. Its graph is a(n) _____ line.

Let $f(x) = 2x + 3$ and $g(x) = x^2 - 2$. Find each function value.

7. $f(1)$ 8. $f(4)$ 9. $g(2)$

10. $g(0)$ 11. $g(-1)$ 12. $g(-3)$

13. $f(-8)$ 14. $f(-5)$ 15. $f(0)$

16. $g(0)$ 17. $f\left(-\dfrac{3}{2}\right)$ 18. $g\left(\sqrt{2}\right)$

Sketch the graph of each linear function. Give the domain and range.

19. $f(x) = -2x + 5$ 20. $f(x) = 4x - 1$

21. $f(x) = \dfrac{1}{2}x + 2$ 22. $f(x) = -\dfrac{1}{4}x + 1$

23. $f(x) = 2x$ 24. $f(x) = -3x$

25. $f(x) = 5$ 26. $f(x) = -4$

Find the value of x for which $f(x) = 10$.

27. $f(x) = 3x - 2$ 28. $f(x) = 4x + 6$

29. $f(x) = \dfrac{1}{2}x + 2$ 30. $f(x) = \dfrac{1}{4}x - 8$

Cost and Revenue Models *In Exercises 31–34, do the following.*

(a) *Express the cost C as a function of x, where x represents the quantity of items as given.*

(b) *Express the revenue R as a function of x.*

(c) *Determine the value of x for which revenue equals cost.*

(d) *Graph the equations*
$$y = C(x) \quad and \quad y = R(x)$$
on the same axes, and interpret the graph.

31. Perian stuffs envelopes for extra income during her spare time. Her initial cost to obtain the necessary information for the job was $200.00. Each envelope costs $0.02 and she gets paid $0.04 per envelope stuffed. Let x represent the number of envelopes stuffed.

32. Bart runs a copying service in his home. He paid $3500 for the copier and a lifetime service contract. Each sheet of paper he uses costs $0.01, and he gets paid $0.05 per copy he makes. Let x represent the number of copies he makes.

33. Frank operates a delivery service in a southern city. His start-up costs amounted to $2300. He estimates that it costs him (in terms of gasoline, wear and tear on his car, etc.) $3.00 per delivery. He charges $5.50 per delivery. Let x represent the number of deliveries he makes.

34. Jane bakes cakes and sells them at Louisiana parish fairs. Her initial cost for the Pointe Coupee Parish fair this year was $40.00. She figures that each cake costs $2.50 to make, and she charges $6.50 per cake. Let x represent the number of cakes sold. (Assume that there were no cakes left over.)

Solve each problem.

35. **Tuition Costs** Average annual tuition and fees for full-time students at public four-year colleges are shown in the table for selected years and graphed as ordered pairs of points, where $x = 0$ represents 2010, $x = 1$ represents 2011, and so on, and $f(x)$ represents the cost in dollars.

Tuition and Fees

Year	Cost (in dollars)
2010	7132
2011	7713
2012	8070
2013	8312
2014	8543
2015	8778

Data from U.S. Department of Education.

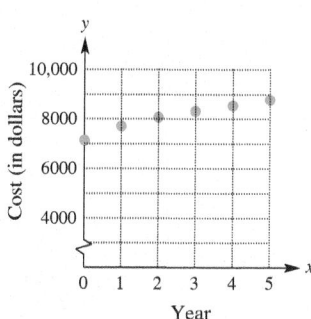

(a) Use the points from the years 2010 and 2015 to find a linear function f that models the data.

(b) Use the function from part (a) to estimate the cost, to the nearest dollar, in 2016.

36. *Tuition Costs* Average annual tuition and fees for full-time students at private two-year colleges are shown in the table for selected years and graphed as ordered pairs of points, where $x = 0$ represents 2010, $x = 1$ represents 2011, and so on, and $f(x)$ represents the cost in dollars.

Tuition and Fees

Year	Cost (in dollars)
2010	13,687
2011	13,961
2012	14,149
2013	14,170
2014	14,254
2015	14,524

Data from U.S. Department of Education.

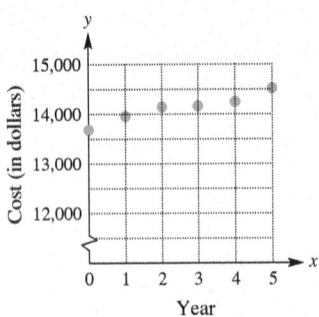

(a) Use the points from the years 2010 and 2015 to find a linear function f that models the data.

(b) Use the function from part (a) to estimate the cost, to the nearest dollar, in 2016.

37. *High School Graduates* The percentage of the U.S. population 25 years and older with at least a high school diploma is shown in the table for selected years.

High School Graduates

Year	Percent
1950	34.3
1960	41.1
1970	52.3
1980	66.5
1990	77.6
2000	84.1

Data from U.S. Census Bureau.

Let $x = 0$ represent 1950, $x = 10$ represent 1960, and so on.

(a) Use the data for 1950 and 2000 to find a linear function f that models the data.

(b) Use the function from part (a) to approximate the percentage, to the nearest tenth, of the U.S. population 25 years and older who were at least high school graduates in 2005.

38. *Tuition Costs* Average annual tuition and fees for private four-year colleges are shown in the table.

Tuition and Fees

Year	Cost (in dollars)
2010	22,677
2011	23,464
2012	24,523
2013	25,707
2014	26,740
2015	27,951

Data from U.S. Department of Education.

Let $x = 0$ represent 2010, $x = 1$ represent 2011, and so on, and let $f(x)$ represent the cost in dollars.

(a) Use the information for the years 2010 and 2015 to find a linear function f that models the data.

(b) Use the function from part (a) to estimate the cost in 2016, to the nearest dollar.

39. *Distances and Velocities of Galaxies* The table lists the distances (in megaparsecs) and velocities (in kilometers per second) of four galaxies moving away from Earth. (1 megaparsec = 3.26 million light years)

Galaxy	Distance	Velocity
Virgo	15	1600
Ursa Minor	200	15,000
Corona Borealis	290	24,000
Bootes	520	40,000

Data from Acker, A., and C. Jaschek, *Astronomical Methods and Calculations,* John Wiley and Sons. Karttunen, H. (editor), *Fundamental Astronomy,* Springer-Verlag.

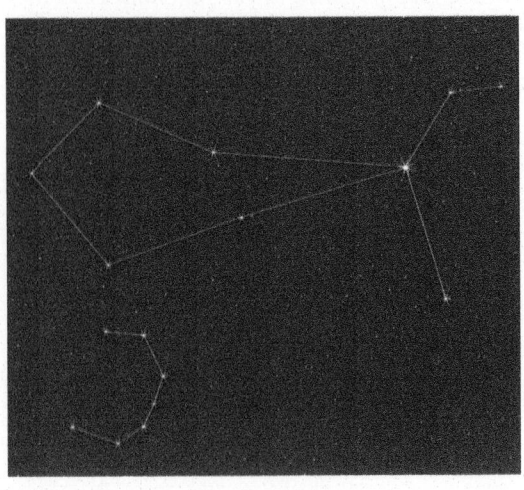

(a) Find a linear function of the form

$$f(x) = ax$$

that models these data using $(520, 40{,}000)$ and $(0, 0)$. Round a to the nearest tenth.

(b) Use the function from part (a) to predict the velocity of a galaxy 400 megaparsecs from Earth.

40. *Nuclear Waste*   The table gives the heavy metal nuclear waste (in thousands of metric tons) from spent reactor fuel stored temporarily at reactor sites, awaiting permanent storage. (Data from "Burial of Radioactive Nuclear Waste under the Seabed," *Scientific American*, January 1998, p. 62.)

Heavy Metal Nuclear Waste

Year x	Waste y
1995	32
2000	42
2010	61
2020*	76

*Estimate by U.S. Department of Energy.

Let $x = 0$ represent 1995, $x = 5$ represent 2000, and so on.

(a) Use $(0, 32)$ and $(25, 76)$ to find the equation of a line that approximates the other ordered pairs. Use the form $f(x) = ax + b$.

(b) Use the equation from part (a) to estimate the amount of nuclear waste in 2018.

41. *Limo Fares*   A limo driver charges \$2.50 per mile.

(a) Complete the table with the correct response for the price $f(x)$ she charges for a trip of x miles.

x	$f(x)$
0	
1	
2	
3	

(b) The linear function that gives a rule for the amount charged is $f(x) =$ _____.

(c) Graph this function for the domain $\{0, 1, 2, 3\}$.

42. *Forensic Studies*   Forensic scientists use the lengths of the tibia (t), a bone from the ankle to the knee, and the femur (r), the bone from the knee to the hip socket, to calculate the height of a person. A person's height (h) is determined from the lengths of these bones using functions defined by the following formulas. All measurements are in centimeters.

For men:	*For women:*
$h(r) = 69.09 + 2.24r$	$h(r) = 61.41 + 2.32r$
or $h(t) = 81.69 + 2.39t$	or $h(t) = 72.57 + 2.53t$

(a) Find the height of a man with a femur measuring 56 centimeters.

(b) Find the height of a man with a tibia measuring 40 centimeters.

(c) Find the height of a woman with a femur measuring 50 centimeters.

(d) Find the height of a woman with a tibia measuring 36 centimeters.

43. *Speeding Fines*   Speeding fines are determined by

$$f(x) = 10(x - 65) + 50, \quad x > 65,$$

where $f(x)$ is the cost, in dollars, of the fine if a person is caught driving x miles per hour.

(a) Radar clocked a driver at 76 mph. How much was the fine?

(b) While balancing his checkbook, Johnny ran across a \$100 check that his wife Gwen had written to the Department of Motor Vehicles for a speeding fine. How fast was Gwen driving?

(c) At what whole number speed are tickets first given?

(d) For what speeds is the fine greater than \$200?

44. *Cost to Mail a Package*   A package weighing x pounds costs $f(x)$ dollars to mail to a given location, where

$$f(x) = 2.75x.$$

(a) Evaluate $f(3)$.

(b) Describe what 3 and the value $f(3)$ mean in part (a), making use of the terms *independent variable* and *dependent variable*.

(c) How much would it cost to mail a 5-lb package? Write the answer using function notation.

45. *Expansion and Contraction of Gases*   In 1787, Jacques Charles noticed that gases expand when heated and contract when cooled. Suppose that a particular gas follows the model

$$f(x) = \frac{5}{3}x + 455,$$

where x is temperature in degrees Celsius and $f(x)$ is volume in cubic centimeters. (*Source:* Bushaw, D., et al., *A Sourcebook of Applications of School Mathematics*, MAA, 1980. Reprinted with permission.)

(a) Find the volume when the temperature is 27°C.

(b) What is the temperature when the volume of gas is 605 cubic centimeters?

8.5   QUADRATIC FUNCTIONS, GRAPHS, AND MODELS

OBJECTIVES

1 Interpret the terms *quadratic function, parabola, axis of symmetry (axis)*, and *vertex* in mathematical exposition.

2 Graph a quadratic function of the form $f(x) = a(x - h)^2 + k$.

3 Find the coordinates of the vertex of the graph of a quadratic function.

4 Find a maximum or minimum in an application of a quadratic function.

A parabolic dish

Quadratic Functions and Parabolas

We now look at *quadratic functions*, those defined by second-degree polynomials.

> **QUADRATIC FUNCTION**
>
> A function f is a **quadratic function** if it can be defined as follows.
>
> $$f(x) = ax^2 + bx + c,$$
>
> where a, b, and c are real numbers, with $a \neq 0$.

The simplest quadratic function f is $f(x) = x^2$. Plotting the points in the table below and drawing a smooth curve through them give the graph in **Figure 27**. This graph is called a **parabola.**

Parabolas are symmetric with respect to a line (the y-axis in **Figure 27**). The line of symmetry for a parabola is the **axis of symmetry** (or simply **axis**) of the parabola. The point where the axis intersects the parabola is the **vertex** of the parabola. The vertex is the lowest (or highest) point of a vertical parabola.

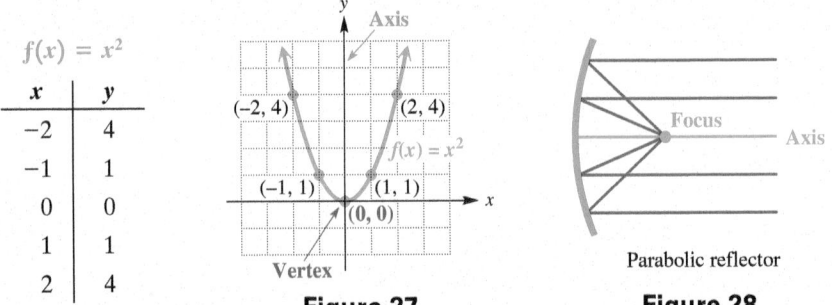

Figure 27   **Figure 28**

The *focus* of a parabola is a point on its axis that determines the curvature. See **Figure 28.** When the parabolic reflector of a solar oven is aimed at the sun, the light rays bounce off the reflector and collect at the focus, creating intense heat at that point.

Graphs of Quadratic Functions

The first example shows how the constant a affects the graph of a function of the form $g(x) = ax^2$.

EXAMPLE 1   Graphing Quadratic Functions ($g(x) = ax^2$)

Graph each function. Compare with the graph of $f(x) = x^2$.

(a) $g(x) = -x^2$ **(b)** $h(x) = \dfrac{1}{2}x^2$

Solution

(a) For a given value of x, the corresponding value of $g(x) = -x^2$ will be the negative of what it was for $f(x) = x^2$. See the table of values on the next page.

The graph of $g(x) = -x^2$ is the same shape as that of $f(x) = x^2$ but opens downward, as in **Figure 29.**

In general, the graph of $f(x) = ax^2 + bx + c$ opens downward when $a < 0$ and upward when $a > 0$.

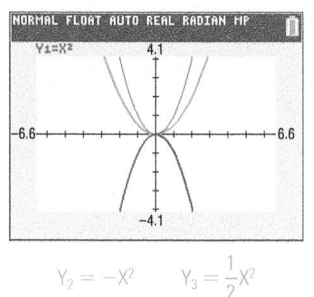

$Y_2 = -X^2$ $Y_3 = \frac{1}{2}X^2$

The screen illustrates the three graphs considered in **Example 1** and **Figures 29** and **30.**

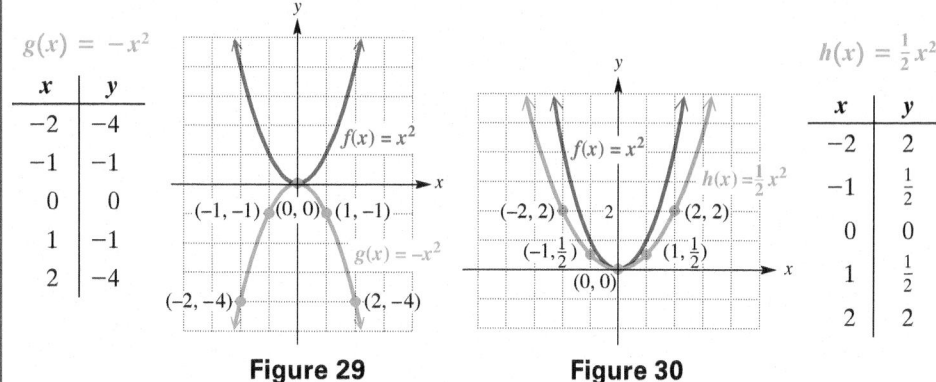

$g(x) = -x^2$

x	y
−2	−4
−1	−1
0	0
1	−1
2	−4

Figure 29

$h(x) = \frac{1}{2}x^2$

x	y
−2	2
−1	$\frac{1}{2}$
0	0
1	$\frac{1}{2}$
2	2

Figure 30

(b) Choose a value of x, and then find $h(x)$. The coefficient $\frac{1}{2}$ will cause the resulting value of $h(x)$ to be less than that of $f(x) = x^2$, making the parabola wider than the graph of $f(x) = x^2$. See **Figure 30.**

In both parabolas of this example, the axis is the vertical line $x = 0$, and the vertex is the origin $(0, 0)$.

Vertical and horizontal shifts of the graph of $f(x) = x^2$ are called **translations.**

EXAMPLE 2   Graphing a Quadratic Function (Vertical Shift)

Graph $g(x) = x^2 - 4$. Compare with the graph of $f(x) = x^2$.

Solution

By comparing the tables of values for $g(x) = x^2 - 4$ and $f(x) = x^2$ shown with **Figure 31,** we can see that for corresponding x-values, the y-values of g are each 4 less than those for f. This leads to a *vertical shift*. Thus, the graph of

$$g(x) = x^2 - 4$$

is the same as that of $f(x) = x^2$, but translated 4 units down. See **Figure 31.** The vertex of this parabola (here the lowest point) is at $(0, -4)$. The axis of the parabola is the vertical line $x = 0$.

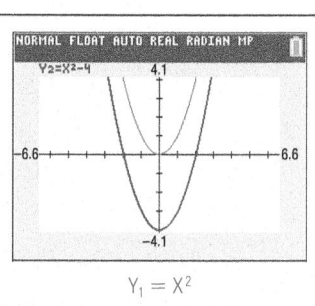

$Y_1 = X^2$

Compare with **Figure 31.**

$f(x) = x^2$

x	y
−2	4
−1	1
0	0
1	1
2	4

$g(x) = x^2 - 4$

x	y
−2	0
−1	−3
0	−4
1	−3
2	0

Figure 31

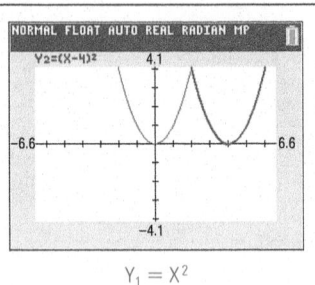

NORMAL FLOAT AUTO REAL RADIAN MP

Y₂=(X-4)²

$Y_1 = X^2$

Compare with **Figure 32.**

EXAMPLE 3 Graphing a Quadratic Function (Horizontal Shift)

Graph $g(x) = (x - 4)^2$. Compare with the graph of $f(x) = x^2$.

Solution

Comparing the tables of values shown with **Figure 32** reveals that the graph of $g(x) = (x - 4)^2$ is the same as that of $f(x) = x^2$, but translated 4 units to the right. This is a *horizontal shift*. The vertex is at $(4, 0)$. As shown in **Figure 32,** the axis of this parabola is the vertical line $x = 4$.

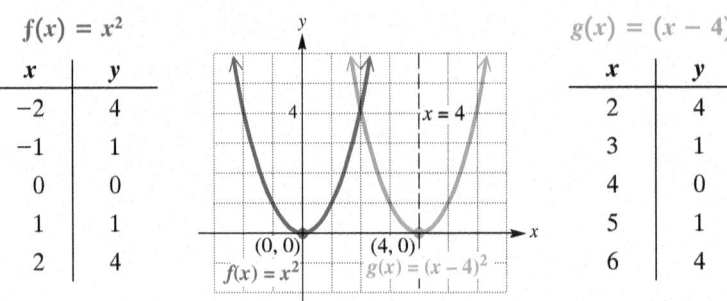

$f(x) = x^2$	
x	y
-2	4
-1	1
0	0
1	1
2	4

$g(x) = (x - 4)^2$	
x	y
2	4
3	1
4	0
5	1
6	4

Figure 32

Errors frequently occur when horizontal shifts are involved. To determine the direction and magnitude of horizontal shifts, find the value of x that would cause the expression $x - h$ to equal 0. For example, in comparison to the graph of $y = x^2$, the graph of

$$f(x) = (x - 5)^2 \text{ would be shifted 5 units to the } right,$$

because $x = +5$ would cause $x - 5$ to equal 0. On the other hand, the graph of

$$f(x) = (x + 4)^2 \text{ would be shifted 4 units to the } left,$$

because $x = -4$ would cause $x + 4$ to equal 0.

GENERAL PRINCIPLES FOR GRAPHS OF QUADRATIC FUNCTIONS

1. The graph of the quadratic function

$$f(x) = a(x - h)^2 + k, \quad a \neq 0,$$

   is a parabola with vertex (h, k) and with the vertical line $x = h$ as axis.

2. The graph opens upward if a is positive and downward if a is negative.

3. The graph is wider than that of $y = x^2$ if $0 < |a| < 1$.
   The graph is narrower than that of $y = x^2$ if $|a| > 1$.

The trajectory of a shell fired from a cannon is a **parabola.** It is shown in calculus that to reach the maximum range with a cannon, the muzzle must be set at 45°. If the muzzle is elevated above 45°, the shell goes too high and falls too soon. If the muzzle is set below 45°, the shell is rapidly pulled to Earth by gravity.

EXAMPLE 4 Graphing a Quadratic Function Using General Principles

Graph $f(x) = -2(x + 3)^2 + 4$.

Solution

The parabola opens downward (because $a < 0$), and it is narrower than the graph of $f(x) = x^2$, since $a = -2$, and $|-2| > 1$. It has vertex at $(-3, 4)$, as shown in **Figure 33** on the next page. To complete the graph, we plotted the additional ordered pairs $(-5, -4)$, $(-4, 2)$, $(-2, 2)$, and $(-1, -4)$.

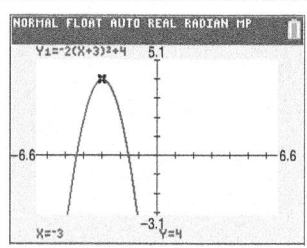

Compare with **Figure 33**. The vertex is $(-3, 4)$.

$$f(x) = -2(x + 3)^2 + 4$$

x	y
-5	-4
-4	2
-3	4
-2	2
-1	-4

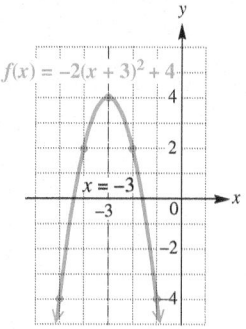

Figure 33

Vertex of a Parabola

When the equation of a parabola is given in the form $f(x) = ax^2 + bx + c$, we can determine the vertex in two ways.

1. By completing the square (**Example 5**)

2. By using a formula that can be derived by completing the square (**Example 6**)

EXAMPLE 5 Finding the Vertex by Completing the Square

Find the vertex of the graph of $f(x) = x^2 - 4x + 5$.

Solution

To find the vertex, we need to express $x^2 - 4x + 5$ in the form $(x - h)^2 + k$. This is done by completing the square, which is covered earlier in this text.

$$y = x^2 - 4x + 5 \qquad \text{Replace } f(x) \text{ with } y.$$

$$y - 5 = x^2 - 4x \qquad \text{Transform so that the constant term is on the left.}$$

$$y - 5 + 4 = x^2 - 4x + 4 \qquad \text{Half of } -4 \text{ is } -2; (-2)^2 = 4. \text{ Add 4 to both sides.}$$

$$y - 1 = (x - 2)^2 \qquad \text{Combine terms on the left, and factor on the right.}$$

$$y = (x - 2)^2 + 1 \qquad \text{Add 1 to both sides.}$$

Now write the original equation as $f(x) = (x - 2)^2 + 1$. As shown earlier, the vertex of this parabola is $(2, 1)$.

A formula for the vertex of the graph of a quadratic function can be found by completing the square for the general form of the equation $y = ax^2 + bx + c$. Using this process, we arrive at the following.

$$y = a\left[x - \left(\underbrace{-\frac{b}{2a}}_{h}\right)\right]^2 + \underbrace{\frac{4ac - b^2}{4a}}_{k} \qquad \text{Write addition as subtraction of the opposite in the brackets.}$$

The final equation shows that the vertex (h, k) can be expressed in terms of a, b, and c. It is not necessary to memorize the expression for k, because it can be obtained by replacing x with $-\frac{b}{2a}$.

VERTEX FORMULA

The vertex of the graph of $f(x) = ax^2 + bx + c$ $(a \neq 0)$ has these coordinates.

$$\left(-\frac{b}{2a}, \ f\left(-\frac{b}{2a}\right)\right) \qquad \text{Vertex}$$

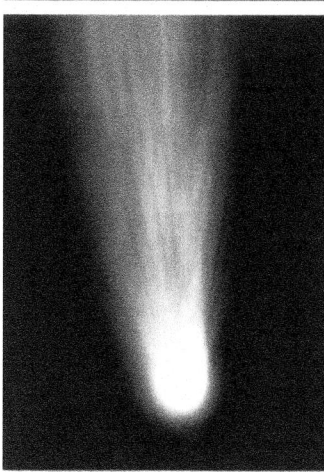

Johann Kepler (1571–1630) established the importance of a curve called an **ellipse** in 1609, when he discovered that the orbits of the planets around the sun were elliptical, not circular. The orbit of Halley's comet, shown here, also is elliptical.

See **For Further Thought** following the exercise set for more on ellipses.

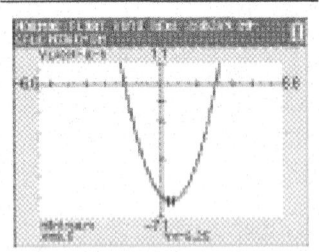

The vertex of the graph of the function in **Example 6** is a minimum point.

EXAMPLE 6 Finding the Vertex Using the Formula

Use the vertex formula to find the vertex of the graph of the function

$$f(x) = x^2 - x - 6.$$

Solution

For $f(x) = x^2 - x - 6$, or $1x^2 - 1x - 6$, the values are $a = 1$, $b = -1$, and $c = -6$. The x-coordinate of the vertex of the parabola is given by

$$-\frac{b}{2a} = -\frac{(-1)}{2(1)} = \frac{1}{2}.$$

The y-coordinate is $f\left(-\frac{b}{2a}\right) = f\left(\frac{1}{2}\right).$

$$f\left(\frac{1}{2}\right) = \left(\frac{1}{2}\right)^2 - \frac{1}{2} - 6 = \frac{1}{4} - \frac{1}{2} - 6 = -\frac{25}{4}$$

Finally, the vertex is $\left(\frac{1}{2}, -\frac{25}{4}\right).$

General Graphing Guidelines

We now examine a general approach to graphing quadratic functions using intercepts and the vertex.

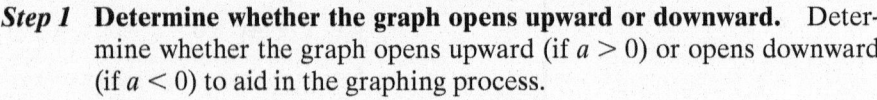

GRAPHING A QUADRATIC FUNCTION $f(x) = ax^2 + bx + c$

Step 1 **Determine whether the graph opens upward or downward.** Determine whether the graph opens upward (if $a > 0$) or opens downward (if $a < 0$) to aid in the graphing process.

Step 2 **Find the vertex.** Find the vertex either by using the formula or by completing the square.

Step 3 **Find the y-intercept.** Find the y-intercept by evaluating $f(0)$.

Step 4 **Find the x-intercepts.** Find the x-intercepts, if any, by solving $f(x) = 0$.

Step 5 **Complete the graph.** Find and plot additional points as needed, using the symmetry with respect to the axis.

Galileo Galilei (1564–1642) died in the year Newton was born. His work was important in Newton's development of calculus. The idea of **function** is implicit in Galileo's analysis of the parabolic path of a projectile, where height and range are functions (in our terms) of the angle of elevation and the initial velocity.

According to legend, Galileo dropped objects of different weights from the tower of Pisa to disprove the Aristotelian view that heavier objects fall faster than lighter objects. He developed a formula for freely falling objects that is described by

$$d = 16t^2,$$

where d is the distance in feet that a given object falls (discounting air resistance) in a given time t, in seconds, regardless of weight.

EXAMPLE 7 Graphing a Quadratic Function Using General Guidelines

Graph the quadratic function $f(x) = x^2 - x - 6.$

Solution

Step 1 From the equation, $a = 1 > 0$, so the graph of the function opens upward.

Step 2 The vertex, $\left(\frac{1}{2}, -\frac{25}{4}\right)$, was found in **Example 6** using the vertex formula.

Step 3 To find the y-intercept, evaluate $f(0)$.

$$f(x) = x^2 - x - 6$$
$$f(0) = 0^2 - 0 - 6 \quad \text{Let } x = 0.$$
$$f(0) = -6 \quad \text{Simplify.}$$

The y-intercept is $(0, -6)$.

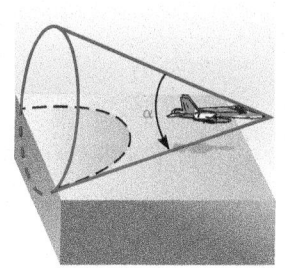

A **sonic boom** is a loud explosive sound caused by the shock wave that accompanies an aircraft traveling at supersonic speed. The sonic boom shock wave has the shape of a cone, and it intersects the ground in one branch of a curve known as a **hyperbola.** Everyone located along the hyperbolic curve on the ground hears the sound at the same time.

See **For Further Thought** following the exercise set for more on hyperbolas.

Step 4 Find any *x*-intercepts. Because the vertex $\left(\frac{1}{2}, -\frac{25}{4}\right)$ is in quadrant IV and the graph opens upward, there will be two *x*-intercepts. To find them, let $f(x) = 0$ and solve the quadratic equation.

$$f(x) = x^2 - x - 6$$
$$0 = x^2 - x - 6 \qquad \text{Let } f(x) = 0.$$
$$0 = (x - 3)(x + 2) \qquad \text{Factor.}$$
$$x - 3 = 0 \quad \text{or} \quad x + 2 = 0 \qquad \text{Zero-factor property}$$
$$x = 3 \quad \text{or} \qquad x = -2 \qquad \text{Solve for } x.$$

The *x*-intercepts are $(3, 0)$ and $(-2, 0)$.

Step 5 Plot the points found so far, and plot any additional points as needed. The symmetry of the graph is helpful here. The graph is shown in **Figure 34.**

$$f(x) = x^2 - x - 6$$

x	y
-2	0
-1	-4
0	-6
$\frac{1}{2}$	$-\frac{25}{4}$
2	-4
3	0

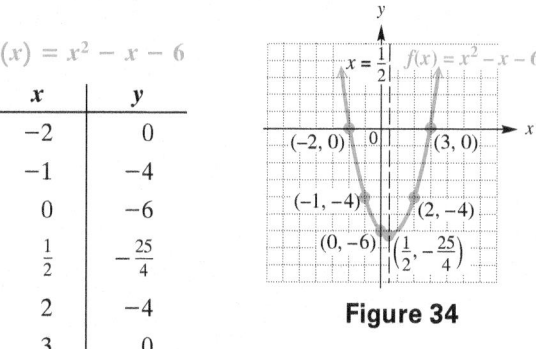

Figure 34

A Model for Optimization

As we have seen, the vertex of a vertical parabola is either the highest or the lowest point of the parabola. The *y*-value of the vertex gives the maximum or minimum value of *y*, and the *x*-value tells where that maximum or minimum occurs. Often a model can be constructed so that *y* can be *optimized.*

Problem-Solving Strategy

We often want to know the least or greatest value of some quantity. When that quantity can be expressed using a quadratic function $f(x) = ax^2 + bx + c$, as in the next example, the vertex can be used to find the desired value.

EXAMPLE 8 **Finding a Maximum Area**

A farmer has 120 feet of fencing. He wants to put a fence around three sides of a rectangular plot of land, with the side of a barn forming the fourth side. Find the maximum area he can enclose. What dimensions give this area?

Solution

Figure 35 shows the enclosed area. Let *x* represent the width. Then, because there are 120 feet of fencing, we can represent the length in terms of *x* as follows.

$$x + x + \text{length} = 120 \qquad \text{Sum of the three fenced sides is 120 feet.}$$
$$2x + \text{length} = 120 \qquad \text{Combine terms.}$$
$$\text{length} = 120 - 2x \qquad \text{Subtract } 2x.$$

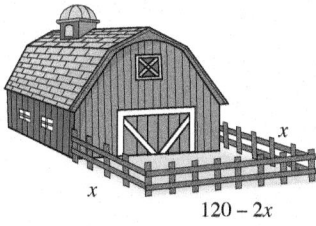

Figure 35

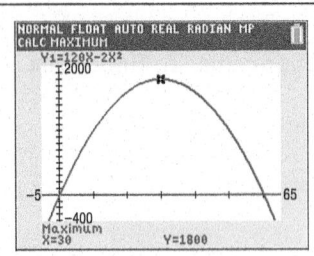

The vertex is (30, 1800), supporting the analytic result in **Example 8.**

The area $\mathcal{A}(x)$ is modeled by the product of the length and width.

$$\mathcal{A}(x) = (120 - 2x)x$$
$$= 120x - 2x^2$$

To make $120x - 2x^2$ (and thus the area) as large as possible, first find the vertex of the graph of the function $\mathcal{A}(x) = 120x - 2x^2$.

$$\mathcal{A}(x) = -2x^2 + 120x \quad \text{Standard form}$$

Here we have $a = -2$ and $b = 120$. The x-coordinate of the vertex is

$$-\frac{b}{2a} = -\frac{120}{2(-2)} = 30.$$

The vertex of the parabola is $(30, 1800)$. The vertex is a maximum point (since $a = -2 < 0$), so the maximum area that the farmer can enclose is

$$\mathcal{A}(30) = -2(30)^2 + 120(30) = 1800 \quad \text{square feet.}$$

The farmer can enclose a maximum area of **1800** square feet, when the width of the plot is 30 feet and the length is $120 - 2(30) = 60$ feet.

> ***Be careful when interpreting the meanings of the coordinates of the vertex in problems involving maximum or minimum values.*** Read the problem carefully to determine whether you are asked to find the value of the independent variable, the value of the dependent variable (that is, the function value), or both.

8.5   EXERCISES

CONCEPT CHECK *Answer each question.*

1. What is the y-intercept of the parabola $y = -3x^2 + x + 4$?

2. What are the x-intercepts of the parabola $y = 2x^2 + 5x - 3$?

3. What is the equation of the axis of symmetry of the parabola $y = (x + 3)^2 - 5$?

4. Which of the parabolas in **Exercises 1–3** open upward?

5. Is the graph of $g(x) = 4x^2$ *wider* or *narrower* than that of $f(x) = x^2$?

6. If the graph of $f(x) = ax^2 + bx + c$ is a parabola that opens downward, what must be true of the value of a?

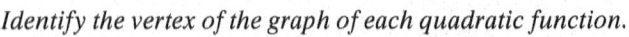

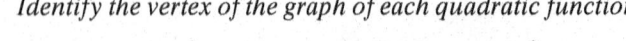

In Exercises 7–12, match each equation with the figure in A–F that most closely resembles its graph.

7. $f(x) = x^2 - 5$

8. $f(x) = -x^2 + 4$

9. $f(x) = (x - 1)^2$

10. $f(x) = (x + 1)^2$

11. $f(x) = (x - 1)^2 + 1$

12. $f(x) = (x + 1)^2 + 1$

Identify the vertex of the graph of each quadratic function.

13. $f(x) = -3x^2$

14. $f(x) = -0.5x^2$

15. $f(x) = x^2 + 4$

16. $f(x) = x^2 - 4$

17. $f(x) = (x - 1)^2$

18. $f(x) = (x + 3)^2$

19. $f(x) = (x + 3)^2 - 4$

20. $f(x) = (x - 5)^2 - 8$

For each quadratic function, tell whether the graph opens upward or downward, and tell whether the graph is wider than, narrower than, or the same as the graph of $f(x) = x^2$.

21. $f(x) = -2x^2$

22. $f(x) = -3x^2 + 1$

23. $f(x) = 0.5x^2$

24. $f(x) = \frac{2}{3}x^2 - 4$

Sketch the graph of each quadratic function using the methods described in this section. Indicate the vertex and two additional points on each graph.

25. $f(x) = 3x^2$

26. $f(x) = -2x^2$

27. $f(x) = -\frac{1}{4}x^2$

28. $f(x) = \frac{1}{3}x^2$

29. $f(x) = x^2 - 1$

30. $f(x) = x^2 + 3$

31. $f(x) = -x^2 + 2$

32. $f(x) = -x^2 - 4$

33. $f(x) = 2x^2 - 2$

34. $f(x) = -3x^2 + 1$

35. $f(x) = (x - 4)^2$

36. $f(x) = (x - 3)^2$

37. $f(x) = 3(x + 1)^2$

38. $f(x) = -2(x + 1)^2$

39. $f(x) = (x + 1)^2 - 2$

40. $f(x) = (x - 2)^2 + 3$

Sketch the graph of each quadratic function. Indicate the coordinates of the vertex of the graph.

41. $f(x) = x^2 + 8x + 14$

42. $f(x) = x^2 + 10x + 23$

43. $f(x) = x^2 + 2x - 4$

44. $f(x) = 3x^2 - 9x + 8$

45. $f(x) = -2x^2 + 4x + 5$

46. $f(x) = -5x^2 - 10x + 2$

Solve each problem.

47. *Dimensions of an Exercise Run* Steve has 100 meters of fencing material to enclose a rectangular exercise run for his dog. What width will give the enclosure the maximum area? What is the enclosed area?

48. *Dimensions of a Parking Lot* Morgan's Department Store wants to construct a rectangular parking lot on land bordered on one side by a highway. It has 280 feet of fencing that is to be used to fence off the other three sides. What should be the dimensions of the lot if the enclosed area is to be a maximum? What is the maximum area?

49. *Height of a Projected Object* If an object on Earth is projected upward with an initial velocity of 32 feet per second, then its height after x seconds is given by

$$f(x) = -16x^2 + 32x.$$

Find the maximum height attained by the object and the number of seconds it takes to hit the ground.

50. *Height of a Projected Object* A projectile on Earth is fired straight upward so that its distance (in feet) above the ground x seconds after firing is given by

$$f(x) = -16x^2 + 400x.$$

Find the maximum height it reaches and the number of seconds it takes to reach that height.

51. *Height of a Projected Object* If air resistance is neglected, a projectile on Earth shot straight upward with an initial velocity of 40 meters per second will be at a height s (in meters) given by the function

$$f(x) = -4.9x^2 + 40x,$$

where x is the number of seconds elapsed after projection. After how many seconds will it reach its maximum height, and what is this maximum height? Round your answers to the nearest tenth.

52. *Height of a Projected Object* A space robot is projected from the moon with its height (in feet) given by

$$f(x) = -0.0013x^2 + 1.727x,$$

where x is time in seconds. Find the maximum height the robot can reach and the time it takes to get there. Round answers to the nearest tenth.

53. *Automobile Stopping Distance* Selected values of the stopping distance y (in feet) of a car traveling x mph are given in the table.

Speed x (in mph)	Stopping Distance y (in feet)
20	46
30	87
40	140
50	240
60	282
70	371

Data from *National Safety Institute Student Workbook.*

The quadratic function

$$y = f(x) = 0.056057x^2 + 1.06657x$$

is one model of the data. Find and interpret $f(45)$.

54. *Path of an Object on a Planet* When an object moves under the influence of constant force (without air resistance), its path is parabolic. This is the path of a ball thrown near the surface of a planet or other celestial object. Suppose two balls are simultaneously thrown upward at a 45° angle on two different planets. If their initial velocities are both 30 mph, then their rectangular coordinates (in feet) can be determined by the function

$$f(x) = x - \frac{g}{1922}x^2,$$

where g is the acceleration due to gravity. The value of g will vary with the mass and size of the planet. (*Source:* Zeilik, M., S. Gregory, and E. Smith, *Introductory Astronomy and Astrophysics*, Saunders College Publishers.)

On Earth, $g = 32.2$, and on Mars, $g = 12.6$. Find the two equations, and determine the difference in the horizontal distances traveled by the two balls.

FOR FURTHER THOUGHT

The Conic Sections

The circle, introduced in the first section of this chapter, the parabola, the ellipse, and the hyperbola are known as **conic sections**. As seen in **Figure 36,** each of these geometric shapes can be obtained when a plane and an infinite cone (made up of two *nappes*) intersect.

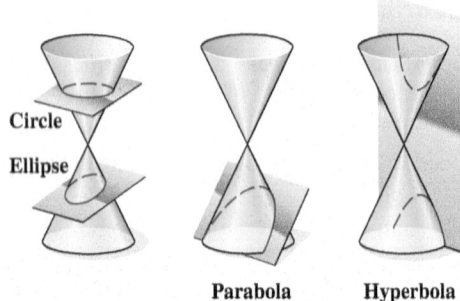

Figure 36

The Greek geometer Apollonius (c. 225 B.C.) was also an astronomer, and his classic work *Conic Sections* thoroughly investigated these figures. Apollonius is responsible for the names "ellipse," "parabola," and "hyperbola." The margin notes in this section show some ways in which these figures appear in the world around us.

For Group or Individual Investigation

1. The terms "ellipse," "parabola," and "hyperbola" are similar to the terms "ellipsis," "parable," and "hyperbole." What do these latter three terms mean? You might want to investigate the similarities between the mathematical terminology and these language-related terms.

2. Identify some places in the world around you where conic sections are encountered.

3. The accompanying photo shows how an ellipse can be drawn using tacks and string. Have a class member volunteer to go to the board and, using string and chalk, modify the method to draw a circle. Then have two class members work together to draw an ellipse. (*Hint:* Press hard!)

8.6 EXPONENTIAL AND LOGARITHMIC FUNCTIONS, GRAPHS, AND MODELS

OBJECTIVES

1 Interpret the terms *exponential function,* *horizontal asymptote,* *logarithmic function,* and *vertical asymptote* in mathematical exposition.

2 Find powers of the number *e* using a calculator.

3 Use the compound interest formulas.

4 Find the natural logarithm of a number using a calculator.

5 Use doubling time and half-life in applications of exponential growth and decay.

Exponential Functions and Applications

In this section we introduce two new types of functions: *exponential* and *logarithmic.*

EXPONENTIAL FUNCTION

An **exponential function with base *b*,** where $b > 0$ and $b \neq 1$, is a function f of the following form.

$$f(x) = b^x, \qquad \text{where } x \text{ is any real number}$$

Notice that in the definition of exponential function, the base b is restricted to positive numbers, with $b \neq 1$.

Thus far, we have defined only integer exponents. In the definition of exponential function, we allow x to take on any real number value. By using methods not discussed in this text, expressions such as

$$2^{9/7}, \qquad \left(\frac{1}{2}\right)^{1.5}, \qquad \text{and} \qquad 10^{\sqrt{3}}$$

can be approximated. A scientific or graphing calculator is capable of determining approximations for these numbers. (See the calculator screen in the margin.)

The graphs of $f(x) = 2^x$, $g(x) = \left(\frac{1}{2}\right)^x$, and $h(x) = 10^x$ are shown in **Figure 37**. In each case, a table of selected points is given. The points are joined with a smooth curve. For each graph, the curve approaches but does not intersect the x-axis. For this reason, the x-axis is called the **horizontal asymptote** of the graph.

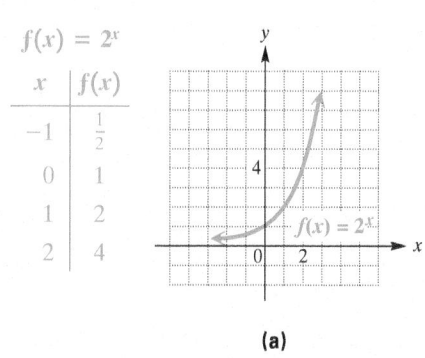

(a)

(b)

(c)

Figure 37

CHARACTERISTICS OF THE GRAPH OF $f(x) = b^x$

1. The graph always will contain the point $(0, 1)$, because $b^0 = 1$.

2. When $b > 1$, the graph *rises* from left to right (as in **Figures 37(a) and (c)** above, with $b = 2$ and $b = 10$).

   When $0 < b < 1$, the graph *falls* from left to right (as in **Figure 37(b)**, with $b = \frac{1}{2}$).

3. The x-axis is the horizontal asymptote.

4. The domain is $(-\infty, \infty)$ and the range is $(0, \infty)$.

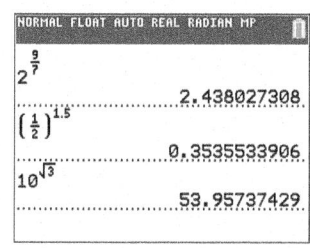

Compare with the discussion in the text.

The **natural exponential function** has the base e. The irrational number e is named after Leonhard Euler (1707–1783) and is approximately 2.718281828. As n gets larger without bound,

$$\left(1 + \frac{1}{n}\right)^n \text{ approaches } e.$$

This is expressed as follows.

$$\text{As } n \to \infty, \quad \left(1 + \frac{1}{n}\right)^n \to e.$$

See **Table 3**.

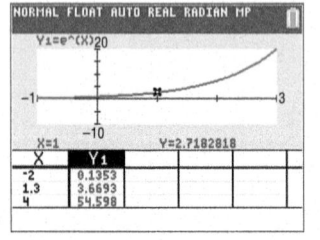

This table shows selected values for

$$Y_1 = \left(1 + \frac{1}{x}\right)^x. \text{ Compare with } \textbf{Table 3}.$$

Table 3

n	Approximate Value of $\left(1 + \dfrac{1}{n}\right)^n$
1	2
100	2.70481
1000	2.71692
10,000	2.71815
1,000,000	2.71828

e

$$e \approx 2.718281828459045235360287471353$$

Powers of e can be approximated on a scientific or graphing calculator. Some powers of e obtained on a calculator are

$$e^{-2} \approx 0.1353352832, \qquad e^{1.3} \approx 3.669296668, \qquad e^4 \approx 54.59815003.$$

The graph of the function $f(x) = e^x$ is shown in **Figure 38**.

The values for e, e^{-2}, $e^{1.3}$, and e^4 are approximated in this split-screen graph of $Y_1 = e^X$.

x	y
-2	≈ 0.14
-1	≈ 0.37
0	1
1	≈ 2.72
2	≈ 7.39

Figure 38

A real-life application of exponential functions occurs in the computation of **compound interest**.

COMPOUND INTEREST FORMULA

Suppose that a principal of P dollars is invested at an annual interest rate r (in percent, expressed as a decimal), compounded n times per year. Then the amount A accumulated after t years is given by the following formula.

$$A = P\left(1 + \frac{r}{n}\right)^{nt}$$

EXAMPLE 1   Applying the Compound Interest Formula

Suppose that $1000 is invested at an annual rate of 2%, compounded quarterly (four times per year). Find the total amount in the account after ten years if no withdrawals are made.

Solution

$$A = P\left(1 + \frac{r}{n}\right)^{nt}$$ Compound interest formula

$$A = 1000\left(1 + \frac{0.02}{4}\right)^{4 \cdot 10}$$ $P = 1000$, $r = 0.02$, $n = 4$, $t = 10$

$$A = 1000(1.005)^{40}$$ Simplify.

$$A \approx 1000(1.22079)$$ Evaluate 1.005^{40} with a calculator.

$$A = 1220.79$$ To the nearest cent

$\approx$ means "is approximately equal to."

There would be $1220.79 in the account at the end of ten years.

The compounding formula given earlier applies if the financial institution compounds interest for a finite number of compounding periods annually. Theoretically, the number of compounding periods per year can get larger and larger (quarterly, monthly, daily, etc.), and if n is allowed to approach infinity, we say that interest is compounded *continuously*. The formula for **continuous compounding** involves the number e. It depends on the fact, mentioned earlier, that as $n \rightarrow \infty$, $\left(1 + \frac{1}{n}\right)^n \rightarrow e$.

CONTINUOUS COMPOUND INTEREST FORMULA

Suppose that a principal of P dollars is invested at an annual interest rate r (in percent, expressed as a decimal), compounded continuously. Then the amount A accumulated after t years is given by the following formula.

$$A = Pe^{rt}$$

EXAMPLE 2   Applying the Continuous Compound Interest Formula

Suppose that $5000 is invested at an annual rate of 2.5%, compounded continuously. Find the total amount in the account after four years if no withdrawals are made.

Solution

$$A = Pe^{rt}$$ Continuous compound interest formula

$$A = 5000e^{0.025(4)}$$ $P = 5000$, $r = 0.025$, $t = 4$

$$A = 5000\,e^{0.1}$$ Simplify.

$$A \approx 5000(1.10517)$$ Use the e^x key on a calculator.

$$A = 5525.85$$ To the nearest cent

There will be $5525.85 in the account after four years.

Animation

Linear, Quadratic, and Exponential Growth

NORMAL FIX2 AUTO REAL RADIAN MP

$1000\left(1+\frac{0.02}{4}\right)^{4*10}$

1220.79

The computation in **Example 1** is shown here.

NORMAL FIX2 AUTO REAL RADIAN MP

$5000e^{0.025*4}$

5525.85

The computation in **Example 2** is shown here.

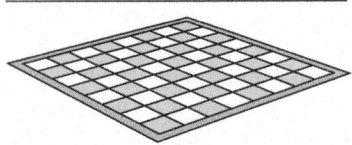

A variation of **exponential growth** is found in the legend of a Persian king, who wanted to please his executive officer, the Grand Vizier, with a gift of his choice. The Grand Vizier explained that he would like to be able to use his chessboard to accumulate wheat. A single grain of wheat would be received for the first square on the board, two grains would be received for the second square, four grains for the third, and so on, doubling the number of grains for each of the 64 squares on the board. As unlikely as it may seem, the number of grains would total 18.5 quintillion!

The continuous compound interest formula is an example of an **exponential growth function.** In situations involving growth or decay of a quantity, the amount or number present at time t can often be approximated by a function of the form

$$A(t) = A_0 e^{kt},$$
If $k > 0$, there is exponential *growth*.
If $k < 0$, there is exponential *decay*.

where A_0 represents the amount or number present at time $t = 0$, and k is a constant.

Logarithmic Functions and Applications

Consider the statement

$$2^3 = 8.$$

Here 3 is the exponent (or power) to which the base two must be raised in order to obtain 8. The exponent 3 is called the base-two **logarithm** of 8, and this is written

$$3 = \log_2 8.$$

DEFINITION OF $\log_b x$

For $b > 0, b \neq 1$:

$$\text{If} \quad b^y = x, \quad \text{then} \quad y = \log_b x.$$

$$\log_b x$$

is the exponent to which b must be raised in order to obtain x.

Table 4 illustrates equivalent exponential equations and logarithmic equations.

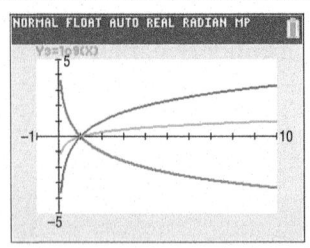

$Y_1 = \log_2 X \qquad Y_2 = \log_{1/2} X$

Compare with **Figure 39** on the next page. Graphs of logarithmic functions with bases other than 10 and e can be obtained with the use of the **change-of-base rule:**

$$\log_a x = \frac{\log x}{\log a} = \frac{\ln x}{\ln a}.$$

Some recent calculators allow direct graphing as well.

Note that log x denotes $\log_{10} x$ and that ln x denotes $\log_e x$.

Table 4

Exponential Equation	Logarithmic Equation
$3^4 = 81$	$4 = \log_3 81$
$10^5 = 100{,}000$	$5 = \log_{10} 100{,}000$
$\left(\frac{1}{2}\right)^{-4} = 16$	$-4 = \log_{1/2} 16$
$10^0 = 1$	$0 = \log_{10} 1$
$4^{-3} = \frac{1}{64}$	$-3 = \log_4 \frac{1}{64}$

A logarithm is an exponent.

The concept of inverse functions (studied in more advanced algebra courses) leads us to the definition of a second new function.

LOGARITHMIC FUNCTION

A **logarithmic function with base b,** where $b > 0$ and $b \neq 1$, is a function g of the following form.

$$g(x) = \log_b x, \quad \text{where } x > 0$$

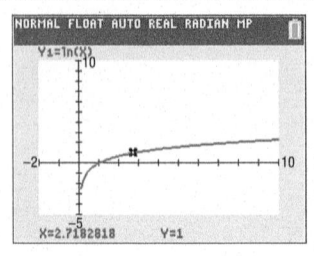

Notice that
when X $= e \approx 2.7182818$, Y $= \ln e = 1$.

The graph of the logarithmic function $g(x) = \log_b x$ can be found by interchanging the roles of x and y in the graph of the exponential function $f(x) = b^x$. Geometrically, this is accomplished by reflecting the graph of $f(x) = b^x$ across the line $y = x$.

The graphs of

$$F(x) = \log_2 x, \quad G(x) = \log_{1/2} x, \quad \text{and} \quad H(x) = \log_{10} x$$

are shown in **Figure 39.** The points in each table were obtained by interchanging the roles of x and y in the tables of points given in **Figure 37.** The points are joined with a smooth curve, typical of the graphs of logarithmic functions. For each graph, the curve approaches but does not intersect the y-axis. Thus, the y-axis is called the **vertical asymptote** of the graph.

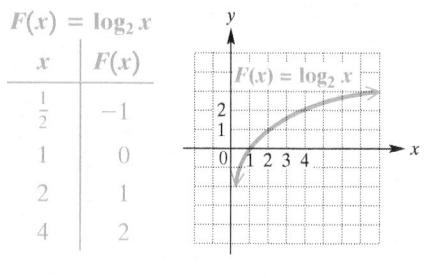

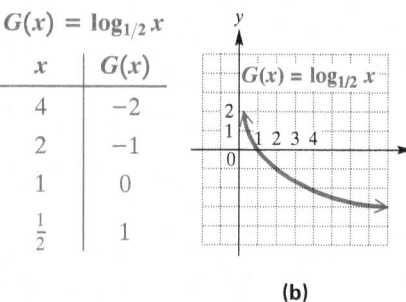

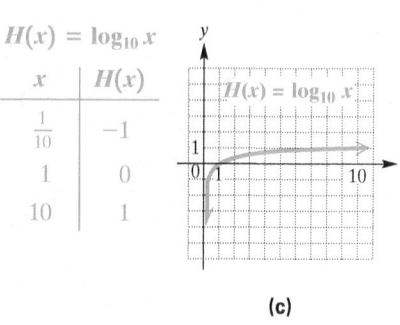

(a)   (b)   (c)

Figure 39

CHARACTERISTICS OF THE GRAPH OF $g(x) = \log_b x$

1. The graph will always contain the point $(1, 0)$, because $\log_b 1 = 0$.

2. When $b > 1$, the graph will *rise* from left to right, from the fourth quadrant to the first (as in **Figure 39(a) and (c),** with $b = 2$ and $b = 10$).

   When $0 < b < 1$, the graph will *fall* from left to right, from the first quadrant to the fourth (as in **Figure 39(b),** with $b = \frac{1}{2}$).

3. The y-axis is the vertical asymptote.

4. The domain is $(0, \infty)$ and the range is $(-\infty, \infty)$.

An important logarithmic function is the function with base e. If we interchange the roles of x and y in the graph of $f(x) = e^x$ **(Figure 38),** we obtain the graph of $g(x) = \log_e x$. The special symbol for the base-e logarithm of x is $\ln x$. That is,

$$\ln x = \log_e x.$$

Figure 40 shows the graph of $g(x) = \ln x$, which is the **natural logarithmic function.**

x	y
≈ 0.14	-2
≈ 0.37	-1
1	0
≈ 2.72	1
≈ 7.39	2

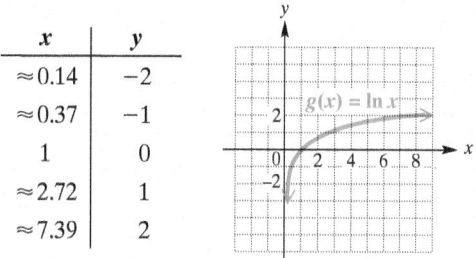

Figure 40

The expression $\ln e^k$ is the unique exponent to which the base e must be raised in order to obtain e^k. Thus, for all real numbers k,

$$\ln e^k = k.$$

> This will be used in applications to follow.

The number **e** is named in honor of **Leonhard Euler** (1707–1783), the prolific Swiss mathematician. The value of e can be expressed as an **infinite series.**

$$e = 1 + \frac{1}{1} + \frac{1}{1 \cdot 2}$$
$$+ \frac{1}{1 \cdot 2 \cdot 3}$$
$$+ \frac{1}{1 \cdot 2 \cdot 3 \cdot 4} + \cdots$$

It can also be expressed using a **continued fraction.**

$$e = 2 + \cfrac{1}{1 + \cfrac{1}{2 + \cfrac{2}{3 + \cfrac{3}{4 + \cfrac{4}{\ddots}}}}}$$

The expression **log x,** which does not indicate a base, is understood to be the base-ten logarithm, or **common logarithm,** of x. Common logarithms were devised hundreds of years ago to assist scientists in performing involved calculations. They are still used in some modern applications, such as measurement of earthquakes on the Richter scale, and the measure of acidity or alkalinity of substances in chemistry (pH).

Exponential Models

If an amount of money is left on deposit and not touched, the amount of time it takes for the money to double at a particular interest rate is the **doubling time.**

EXAMPLE 3 **Finding the Time for an Amount to Double**

Suppose that a certain amount P is invested at an annual rate of 1.5%, compounded continuously. How long will it take for the amount to double?

Solution

We must find the value of t in the continuous compound interest formula that will make A equal to $2P$ (because we want the initial investment, P, to double).

$$A = Pe^{rt}$$ Continuous compound interest formula

$$2P = Pe^{0.015t}$$ Substitute $2P$ for A and 0.015 for r.

$$2 = e^{0.015t}$$ Divide both sides by P.

$$\ln 2 = \ln e^{0.015t}$$ Take the natural logarithm of both sides.

$$\ln 2 = 0.015t$$ Use the fact that $\ln e^k = k$.

$$t = \frac{\ln 2}{0.015}$$ Divide both sides by 0.015.

The computation in **Example 3** is shown here.

A calculator shows that $\ln 2 \approx 0.6931471806$. Dividing this by 0.015 gives

$$t \approx 46.21$$

to the nearest hundredth. Therefore, it will take about 46.21 years for any initial investment P to double under the given conditions.

Further Applications

EXAMPLE 4 **Modeling Population Growth**

According to data from the United Nations, Department of Economics and Social Affairs, Population Division, the World Population Forecast for 2020 through 2050 can be projected by the model

$$f(x) = 2049.73\, e^{0.007506x}$$

where x is the year.

(a) Based on this model, what will the world population be in 2030?

(b) Based on this model, in what year will the world population reach 9 billion?

See www.worldometers.com for a dynamic representation of the current world population.

Solution

(a) We find $f(x)$ when x is 2030.

$$f(x) = 2049.73e^{0.007506x}$$ Given function

$$f(2030) = 2049.73e^{0.007506(2030)}$$ Let $x = 2030$.

$$f(2030) \approx 8,494,000,000$$ Use a calculator.

According to this model, the population will be about 8.494 billion in 2030.

(b)

$$f(x) = 2049.73e^{0.007506x} \quad \text{Given function}$$

$$9 \times 10^9 = 2049.73\,e^{0.007506x} \quad \text{Let } f(x) = 9 \times 10^9 \text{ (9 billion).}$$

$$\frac{9 \times 10^9}{2049.73} = e^{0.007506x} \quad \text{Divide by 2049.73.}$$

$$\ln\left(\frac{9 \times 10^9}{2049.73}\right) = 0.007506x \quad \text{In } e^x = x$$

$$x = \frac{\ln\left(\dfrac{9 \times 10^9}{2049.73}\right)}{0.007506} \quad \text{Divide by 0.007506. Rewrite.}$$

$$x \approx 2038 \quad \text{Use a calculator.}$$

According to this model, the world population will reach 9 billion in 2038.

The computations in **Example 4** are shown here.

Radioactive materials disintegrate according to exponential decay functions. The **half-life** of a substance that decays exponentially is the amount of time that it takes for the amount to decay to half its initial value.

EXAMPLE 5   **Carbon-14 Dating**

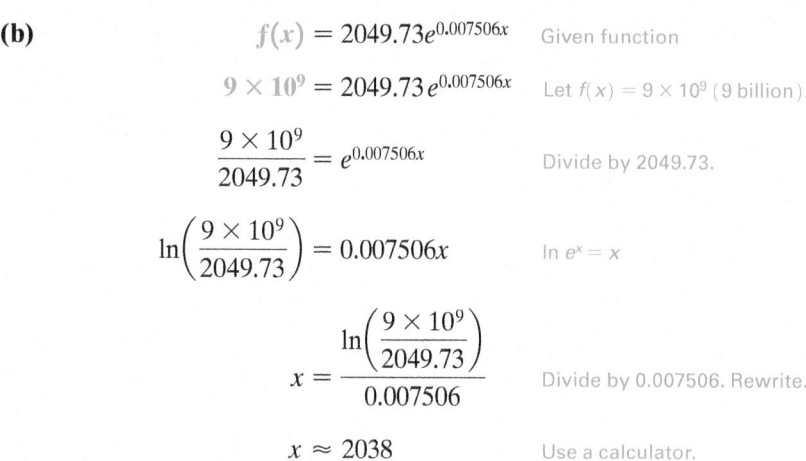

Carbon-14 is a radioactive form of carbon that is found in all living plants and animals. After a plant or animal dies, the radiocarbon disintegrates. Scientists determine the age of the remains by comparing the amount of carbon-14 present with the amount found in living plants and animals. The amount of carbon-14 present after x years is modeled by the exponential function

$$f(x) = A_0 e^{-0.0001216x},$$

where A_0 represents the initial amount.

(a) What is the half-life of carbon-14?

(b) If an initial sample contains 1 gram of carbon-14, how much will be left after 10,000 years?

Solution

(a) To find the half-life, let $f(x) = \frac{1}{2}A_0$ in the equation.

$$\frac{1}{2}A_0 = A_0 e^{-0.0001216x} \quad \text{Let } f(x) = \frac{1}{2}A_0.$$

$$\frac{1}{2} = e^{-0.0001216x} \quad \text{Divide by } A_0.$$

$$\ln\left(\frac{1}{2}\right) = -0.0001216x \quad \text{Take natural logarithms on both sides.}$$

$$x \approx 5700 \quad \text{Use a calculator.}$$

The half-life of carbon-14 is about 5700 years.

(b) Evaluate $f(x)$ for $x = 10{,}000$ and $A_0 = 1$.

$$f(10{,}000) = 1e^{-0.0001216(10{,}000)} \approx 0.30 \quad \text{Use a calculator.}$$

There will be about 0.30 gram remaining.

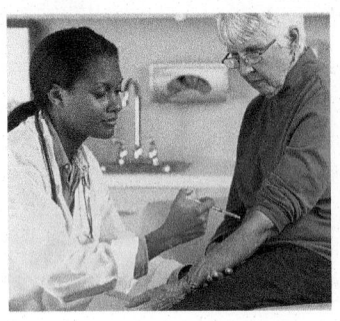

EXAMPLE 6 Determining a Time Interval

The concentration of a drug injected into the bloodstream decreases with time. The intervals of time T, in hours, at which the drug should be administered are given by

$$T = \frac{1}{k} \ln \frac{C_2}{C_1},$$

where k is a constant determined by the drug in use, C_2 is the concentration at which the drug is harmful, and C_1 is the concentration below which the drug is ineffective. (*Source:* Horelick, Brindell, and Sinan Koont, "Applications of Calculus to Medicine: Prescribing Safe and Effective Dosage, *UMAP Module 202.*)

For a certain drug, $k = \frac{1}{3}$, $C_2 = 5$, and $C_1 = 2$. How often should the drug be administered? Round down to the nearest hour.

Solution

$$T = \frac{1}{k} \ln \frac{C_2}{C_1} = \frac{1}{\frac{1}{3}} \ln \frac{5}{2} = 3 \ln 2.5 = 2.75$$

Rounding down to the nearest hour, the drug should be administered every 2 hours.

WHEN WILL I **EVER** USE THIS?

The **Richter scale** for measuring earthquake intensity was devised by Charles F. Richter in 1935. It is a base-ten logarithmic scale—each unit increase in the measure represents a 10-fold increase in intensity. For example, an earthquake measuring 6.0 on this scale releases ten times as much energy as one that measures 5.0.

Suppose that an earthquake measuring 5.7 occurred in southern California, followed by an aftershock that measured 5.4.

How do these earthquakes compare?

Because these measures are exponents on the base ten, we evaluate

$$\frac{10^{5.7}}{10^{5.4}}.$$

We subtract the exponents to obtain $10^{0.3}$. Use a calculator and raise 10 to the power 0.3 to obtain approximately 1.995, which is very close to 2. The earthquake was about twice as powerful as its aftershock.

Verify that the 9.1 earthquake that caused the tsunami that devastated Japan in 2010 was about 250 times as powerful as the 6.7 Northridge earthquake that struck the Los Angeles area in 1994.

8.6   EXERCISES

To the student: The number of digits shown for calculator approximations in the exercises will vary depending on the model of calculator used.

CONCEPT CHECK *Fill in each blank with the correct response.*

1. For an exponential function
$$f(x) = a^x,$$
if $a > 1$, the graph _____ from left to right.
(rises/falls)
If $0 < a < 1$, the graph _____ from left to right.
(rises/falls)

2. The *y*-intercept of the graph of
$$f(x) = a^x$$
is _____.

3. The graph of the exponential function
$$f(x) = a^x$$
_____ have an *x*-intercept.
(does/does not)

4. The point $(2, \_\_\_)$ is on the graph of $f(x) = 4^x$.

5. For a logarithmic function
$$g(x) = \log_a x,$$
if $a > 1$, the graph _____ from left to right.
(rises/falls)
If $0 < a < 1$, the graph _____ from left to right.
(rises/falls)

6. The *x*-intercept of the graph of $y = \log_a x$ is ____.

7. The graph of the logarithmic function
$$g(x) = \log_a x$$
_____ have a *y*-intercept.
(does/does not)

8. The point $(100, \_\_\_)$ lies on the graph of $g(x) = \log_{10} x$.

Use a calculator to find an approximation for each number. Give as many digits as the calculator displays.

9. $9^{3/7}$

10. $14^{2/7}$

11. $(0.83)^{-1.2}$

12. $(0.97)^{3.4}$

13. $\left(\sqrt{6}\right)^{\sqrt{5}}$

14. $\left(\sqrt{7}\right)^{\sqrt{3}}$

15. $\left(\dfrac{1}{3}\right)^{9.8}$

16. $\left(\dfrac{2}{5}\right)^{8.1}$

Sketch the graph of each function.

17. $f(x) = 3^x$

18. $f(x) = 5^x$

19. $f(x) = \left(\dfrac{1}{4}\right)^x$

20. $f(x) = \left(\dfrac{1}{3}\right)^x$

Use a calculator to approximate each number. Give as many digits as the calculator displays.

21. e^3

22. e^4

23. e^{-4}

24. e^{-3}

In Exercises 25–28, rewrite the exponential equation as a logarithmic equation. In Exercises 29–32, rewrite the logarithmic equation as an exponential equation.

25. $4^2 = 16$

26. $5^3 = 125$

27. $\left(\dfrac{2}{3}\right)^{-3} = \dfrac{27}{8}$

28. $\left(\dfrac{1}{10}\right)^{-4} = 10{,}000$

29. $5 = \log_2 32$

30. $3 = \log_4 64$

31. $1 = \log_3 3$

32. $0 = \log_{12} 1$

Use a calculator to approximate each number. Give as many digits as the calculator displays.

33. $\ln 4$

34. $\ln 6$

35. $\ln 0.35$

36. $\ln 2.45$

Sketch the graph of each function. (Hint: Use the graphs of the exponential functions in Exercises 17–20 to help.)

37. $g(x) = \log_3 x$

38. $g(x) = \log_5 x$

39. $g(x) = \log_{1/4} x$

40. $g(x) = \log_{1/3} x$

Investment Determine the amount of money that will be accumulated in an account that pays compound interest, given the initial principal in each of the following.

41. $20,000 invested at 1% annual interest for 4 years compounded **(a)** annually;   **(b)** semiannually.

42. $35,000 invested at 1.2% annual interest for 3 years compounded **(a)** annually;   **(b)** quarterly.

43. $27,500 invested at 0.5% annual interest for 5 years compounded **(a)** daily $(n = 365)$;   **(b)** continuously.

44. $15,800 invested at 0.4% annual interest for 6.5 years compounded **(a)** quarterly;   **(b)** continuously.

Comparing Investment Plans   *In Exercises 45 and 46, decide which of the two plans will provide a better yield. (Interest rates stated are annual rates.)*

45. *Plan A:* $40,000 invested for 3 years at 1.5%, compounded quarterly

Plan B: $40,000 invested for 3 years at 1.4%, compounded continuously

46. *Plan A:* $50,000 invested for 10 years at 1.75%, compounded daily ($n = 365$)

Plan B: $50,000 invested for 10 years at 1.7%, compounded continuously

47. *Comparing Investments*   Roop wants to invest $60,000 in a pension plan. One investment offers 2% compounded quarterly. Another offers 1.75% compounded continuously.

   (a) Which investment will earn more interest in 5 years?

   (b) How much more will the better plan earn?

48. *Growth of an Account*   If Roop (see **Exercise 47**) chooses the plan with continuous compounding, how long will it take for his $60,000 to grow to $80,000?

Solve each problem.

49. *Doubling Time*   Find the doubling time of an investment earning 2.5% annual interest if interest is compounded continuously.

50. *Tripling Time*   Find the time it would take the investment in **Exercise 49** to triple.

51. *Growth of an Account*   How long will it take an investment of $1000 to grow to $1500, if interest is compounded continuously at 0.75% annually?

52. *Growth of an Account*   What annual interest rate is required for an investment of $1000 to grow to $1200 in 10 years if interest is compounded continuously?

53. *World Population Growth*   As seen in **Example 4,** the world population closely fits the exponential function

$$f(x) = 2049.73e^{0.007506x},$$

where x is the year.

   (a) Predict the population in 2025.

   (b) Predict when the population will reach 9.6 billion.

54. *Atmospheric Pressure*   The atmospheric pressure (in millibars) at a given altitude x (in meters) is approximated by the function

$$f(x) = 1013e^{-0.0001341x}.$$

Give answers to the nearest unit.

   (a) Predict the atmospheric pressure at 1500 meters.

   (b) Predict the atmospheric pressure at 11,000 meters.

55. *Carbon-14 Dating*   Suppose an Egyptian mummy is discovered in which the amount of carbon-14 present is only about one-third the amount found in living human beings. About how long ago did the Egyptian die?

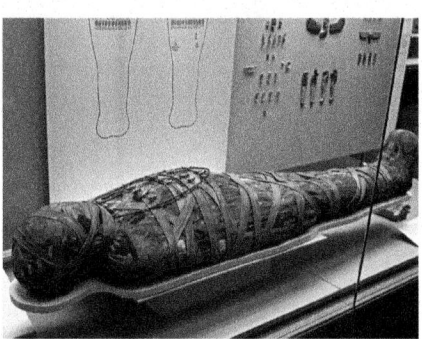

56. *Carbon-14 Dating*   A sample from a refuse deposit near the Strait of Magellan had 60% of the carbon-14 of a contemporary living sample. How old was the sample?

57. *Decay of Lead*   A sample of 500 grams of radioactive lead-210 decays to polonium-210 according to the function

$$f(x) = 500e^{-0.032x},$$

where x is time in years. Find the amount of the sample remaining after

   (a) 4 years   **(b)** 8 years

   (c) 20 years   **(d)** Find the half-life.

58. *Decay of Plutonium*   Repeat **Exercise 57** for 500 grams of plutonium-241, which decays according to this function, where x is time in years.

$$f(x) = A_0e^{-0.053x}$$

59. *Decay of Radium*   Find the half-life of radium-226, which decays according to this function, where x is time in years.

$$f(x) = A_0e^{-0.00043x}$$

60. *Decay of Iodine* How long will it take any quantity of iodine-131 to decay to 25% of its initial amount, given that it decays according to this function, where x is time in days?

$$f(x) = A_0 e^{-0.087x}$$

Time Intervals for Drugs *Use the formula given in **Example 6** to determine how often the drug with the given values should be administered. Round down to the nearest hour.*

61. $k = \dfrac{1}{4}, C_2 = 6, C_1 = 2$ **62.** $k = \dfrac{1}{2}, C_2 = 4, C_1 = 2$

Climate Change *In 1990 a major scientific periodical published an article dealing with the problem of climate change. The article was accompanied by a graph that illustrated two possible scenarios.*

(a) *The change might be modeled by an exponential function of the form*

$$f(x) = (1.046 \times 10^{-38})(1.0444^x).$$

(b) *The change might be modeled by a linear function of the form*

$$g(x) = 0.009x - 17.67.$$

In both cases, x represents the year, and the function value represents the increase in degrees Celsius. Use these functions to approximate the increase in temperature for each of the following years.

63. 2000   **64.** 2010

65. 2020   **66.** 2040

8.7 SYSTEMS OF LINEAR EQUATIONS

OBJECTIVES

1 Interpret the use of *system of linear equations* in mathematical exposition.

2 Identify the solution of a system of linear equations from a graph, and classify the system.

3 Solve a system of two linear equations in two variables using the elimination method.

4 Solve a system of two linear equations in two variables using the substitution method.

Linear Systems

In recent years, the number of Americans who subscribe to Netflix has increased, while the number who subscribe to cable TV has decreased. These trends can be seen in the graph in **Figure 41.** The two straight-line graphs intersect at the time when Netflix and cable TV had the *same* number of subscribers.

Cable vs. Netflix

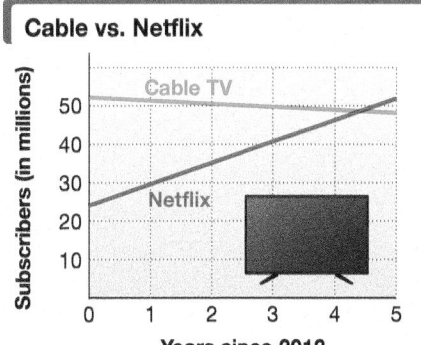

Data from Netflix, Leichtman Research Group.

Figure 41

$-5.61x + y = 23.9$ Linear system
$0.77x + y = 52.1$ of equations

(Here, $x = 0$ represents 2012, $x = 1$ represents 2013, and so on. y represents million of subscribers.)

As shown beside **Figure 41,** we can use a linear equation to model the graph of the number of Netflix subscribers (red graph) and another linear equation to model the graph of the number of cable TV subscribers (blue graph).

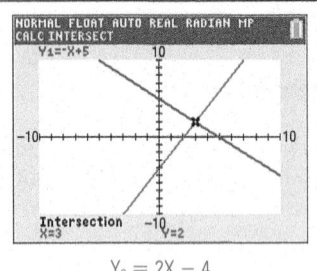

A graphing calculator supports our statement that (3, 2) is the solution of the system

$$x + y = 5$$
$$2x - y = 4.$$

Such a set of equations is a **system of linear equations.** The point where the graphs in **Figure 41** intersect is a solution of each of the individual equations. It is also the solution of the system of linear equations.

In **Figure 42,** the two linear equations $x + y = 5$ and $2x - y = 4$ are graphed in the same coordinate system. Notice that they intersect at the point $(3, 2)$. Because $(3, 2)$ is the only ordered pair that satisfies both equations, we say that $\{(3, 2)\}$ is the solution set of the system

$$x + y = 5$$

$$2x - y = 4.$$

Because the graph of a linear equation in two variables is a line, there are three possibilities for the number of solutions in the solution set of a linear system.

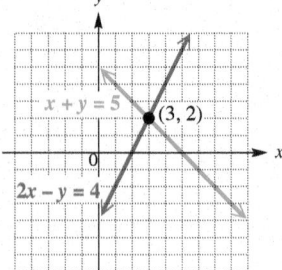

Figure 42

> **GRAPHS OF A LINEAR SYSTEM IN TWO VARIABLES (THE THREE POSSIBILITIES)**

1. The two graphs intersect in a single point. The coordinates of this point give the only solution of the system. In this case, the system is **consistent** and the equations are **independent.** This is the most common case. See **Figure 43(a).**

2. The graphs are parallel lines. In this case, the system is **inconsistent** and the equations are **independent.** That is, there is no solution common to both equations of the system, and the solution set is $\varnothing$. See **Figure 43(b).**

3. The graphs are the same line. In this case, the system is **consistent** and the equations are **dependent,** because any solution of one equation of the system is also a solution of the other. The solution set is an infinite set of ordered pairs representing the points on the line. See **Figure 43(c).**

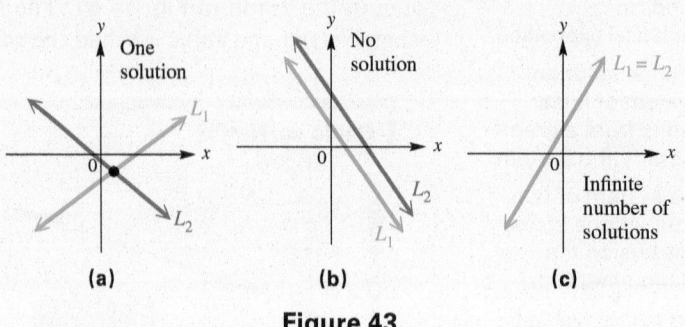

Figure 43

Elimination Method

In most cases, we cannot rely on graphing to solve systems, so we use algebraic methods. The **elimination method** involves combining the two equations of the system in such a way that one variable is eliminated. This is done using the following rule.

If $a = b$ **and** $c = d$, **then** $a + c = b + d$.

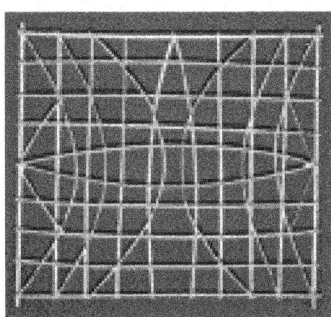

The solution of **systems of equations** of graphs more complicated than straight lines is the principle behind the **mattang,** a stick chart used by the people of the Marshall Islands in the Pacific. A mattang is made of roots tied together with coconut fibers, and it shows the wave patterns found when approaching an island.

SOLVING LINEAR SYSTEMS BY ELIMINATION

Step 1   **Write both equations in standard form** $Ax + By = C.$

Step 2   **Make the coefficients of one pair of variable terms opposites.** Multiply one or both equations by appropriate numbers so that the sum of the coefficients of either x or y is zero.

Step 3   **Add** the new equations to eliminate a variable. The sum should be an equation with just one variable.

Step 4   **Solve** the equation from Step 3.

Step 5   **Find the other value.**   Substitute the result of Step 4 into either of the given equations, and solve for the other variable.

Step 6   **Find the solution set.**   Check the solution in both of the given equations. Then write the solution set.

EXAMPLE 1   Solving a System by Elimination

Solve the system.

$$5x - 2y = 4 \quad {\scriptstyle (1)}$$
$$2x + 3y = 13 \quad {\scriptstyle (2)}$$

Solution

Step 1   Both equations are already in standard form.

Step 2   Our goal is to add the two equations so that one of the variables is eliminated. To eliminate the variable x, we must first transform one or both equations so that the coefficients of x are opposites. Then, when we combine the equations, the term with x will have a coefficient of 0, and we will be able to solve for y. We begin by multiplying equation (1) by 2 and equation (2) by -5.

$$10x - 4y = 8 \quad {\scriptstyle \text{2 times each side of equation (1)}}$$
$$-10x - 15y = -65 \quad {\scriptstyle -5 \text{ times each side of equation (2)}}$$

Step 3   Now add the two equations to eliminate x.

$$\begin{array}{r} 10x - 4y = 8 \\ \underline{-10x - 15y = -65} \\ -19y = -57 \quad {\scriptstyle \text{Add.}} \end{array}$$

Step 4   Solve the equation from Step 3 to get $y = 3$. (Divide by -19.)

Step 5   To find x, we substitute 3 for y in either of the original equations.

$$2x + 3y = 13 \quad {\scriptstyle \text{We use equation (2).}}$$
$$2x + 3(3) = 13 \quad {\scriptstyle \text{Let } y = 3.}$$
$$2x + 9 = 13 \quad {\scriptstyle \text{Multiply.}}$$
$$2x = 4 \quad {\scriptstyle \text{Subtract 9.}}$$
$$x = 2 \quad {\scriptstyle \text{Divide by 2.}}$$

Step 6   The solution appears to be $(2, 3)$. To check, substitute 2 for x and 3 for y in both of the original equations.

$Y_2 = (13 - 2X)/3$

The solution in **Example 1** is supported by a graphing calculator. We solve each equation for y, graph them both, and find the point of intersection of the two lines: $(2, 3)$.

Check:

$5x - 2y = 4$ (1)	$2x + 3y = 13$ (2)
$5(2) - 2(3) \stackrel{?}{=} 4$	$2(2) + 3(3) \stackrel{?}{=} 13$
$10 - 6 \stackrel{?}{=} 4$	$4 + 9 \stackrel{?}{=} 13$
$4 = 4$ ✓ True	$13 = 13$ ✓ True

The solution set is $\{(2, 3)\}$.

Substitution Method

Linear systems can also be solved by the **substitution method.**

SOLVING LINEAR SYSTEMS BY SUBSTITUTION

Step 1 **Solve for one variable in terms of the other.** Solve one of the equations for either variable. (If one of the variables has coefficient 1 or -1, choose it, since the substitution method is usually easier this way.)

Step 2 **Substitute** for that variable in the other equation. The result should be an equation with just one variable.

Step 3 **Solve** the equation from Step 2.

Step 4 **Find the other value.** Substitute the result from Step 3 into the equation from Step 1 to find the value of the other variable.

Step 5 **Find the solution set.** Check the solution in both of the given equations. Then write the solution set.

EXAMPLE 2 Solving a System by Substitution

Solve the system.

$$3x + 2y = 13 \quad (1)$$
$$4x - y = -1 \quad (2)$$

Solution

Step 1 First solve one of the equations for either x or y. Because the coefficient of y in equation (2) is -1, we choose to solve for y in equation (2).

$$-y = -1 - 4x \quad \text{Equation (2) rearranged}$$
$$y = 1 + 4x \quad \text{Multiply by } -1.$$

Step 2 Substitute $1 + 4x$ for y in equation (1) to obtain an equation in x.

$$3x + 2y = 13 \quad (1)$$
$$3x + 2(1 + 4x) = 13 \quad \text{Let } y = 1 + 4x.$$

Step 3 Solve for x in the equation just obtained.

$$3x + 2 + 8x = 13 \quad \text{Distributive property}$$
$$11x = 11 \quad \text{Combine terms; subtract 2.}$$
$$x = 1 \quad \text{Divide by 11.}$$

Step 4 Now solve for y. Because $y = 1 + 4x$, we find that $y = 1 + 4(1) = 5$.

Step 5 Check to see that the ordered pair $(1, 5)$ satisfies both of the original equations. The solution set is $\{(1, 5)\}$.

EXAMPLE 3   **Solving Systems of Special Cases**

Solve each system.

(a) $3x - 2y = 4$ (1)   **(b)** $-4x + y = 2$ (3)

$\quad\quad -6x + 4y = 7$ (2) $\quad\quad 8x - 2y = -4$ (4)

Solution

(a) Eliminate x by multiplying both sides of equation (1) by 2 and then adding.

$$
\begin{array}{rl}
6x - 4y = 8 & \text{2 times equation (1)} \\
\underline{-6x + 4y = 7} & \text{(2)} \\
0 = 15 & \text{False}
\end{array}
$$

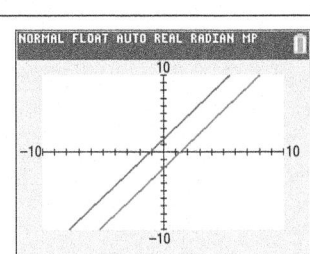

$Y_1 = (7 + 6X)/4$ $Y_2 = (3X - 4)/2$

The graphs of the equations in **Example 3(a)** are parallel. There are no solutions.

Both variables were eliminated here, leaving the false statement $0 = 15$, which indicates that these two equations have no solutions in common. The system is inconsistent, with the empty set $\varnothing$ as the solution set.

(b) Eliminate x by multiplying both sides of equation (3) by 2 and then adding the result to equation (4).

$$
\begin{array}{rl}
-8x + 2y = 4 & \text{2 times equation (3)} \\
\underline{8x - 2y = -4} & \text{(4)} \\
0 = 0 & \text{True}
\end{array}
$$

This true statement, $0 = 0$, indicates that a solution of one equation is also a solution of the other, so the solution set is an infinite set of ordered pairs. The two equations are dependent.

We write the solution set of a system of dependent equations as a set of ordered pairs by expressing x in terms of y as follows. Choose either equation and solve for x. We arbitrarily choose equation (3).

$$-4x + y = 2 \quad\quad (3)$$

$$x = \frac{2 - y}{-4}$$

$$x = \frac{y - 2}{4} \quad \text{Multiply by } \frac{-1}{-1}.$$

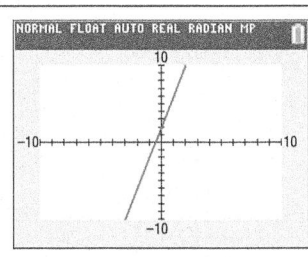

$Y_1 = (8X + 4)/2$ $Y_2 = 4X + 2$

The graphs of the equations in **Example 3(b)** coincide. We see only one line. There are infinitely many solutions.

The solution set can be written as

$$\left\{ \left(\frac{y - 2}{4}, y \right) \right\}.$$

By selecting values for y and calculating the corresponding values for x, we can find individual ordered pairs of the solution set. For example,

$$\text{if}\quad y = -2, \quad \text{then}\quad x = \frac{-2 - 2}{4} = -1,$$

and the ordered pair $(-1, -2)$ is a solution.

8.7   **EXERCISES**

CONCEPT CHECK *Fill in each blank with the correct response.*

1. If $(3, -6)$ is a solution of a linear system in two variables, then substituting _____ for x and _____ for y leads to true statements in *both* equations.

2. A solution of a system of independent linear equations in two variables is a(n) _____ _____ of numbers.

3. Which ordered pair could possibly be a solution of the graphed system of equations? Why?

A. $(3,0)$
B. $(-3,3)$
C. $(-3,-3)$
D. $(3,-3)$

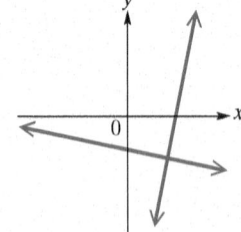

4. Which ordered pair could possibly be a solution of the graphed system of equations? Why?

A. $(3,0)$
B. $(-3,0)$
C. $(0,3)$
D. $(0,-3)$

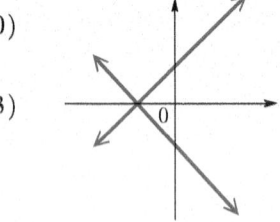

Decide whether the ordered pair is a solution of the given system.

5. $x + y = 6;$ $(5,1)$
$\quad x - y = 4$

6. $x - y = 17;$ $(8,-9)$
$\quad x + y = -1$

7. $2x - y = 8;$ $(5,2)$
$\quad 3x + 2y = 20$

8. $3x - 5y = -12;$ $(-1,2)$
$\quad x - y = 1$

Match each system in Exercises 9–12 with the correct graph in A–D below.

9. $x + y = 6$
$\quad x - y = 0$

10. $x + y = -6$
$\quad x - y = 0$

11. $x + y = 0$
$\quad x - y = -6$

12. $x + y = 0$
$\quad x - y = 6$

A.
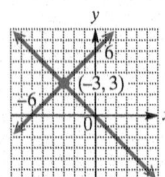

B.
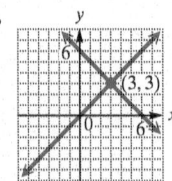

C.
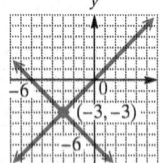

D.
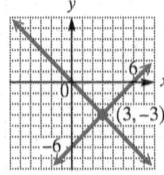

Solve each system by graphing.

13. $x + y = 4$
$\quad 2x - y = 2$

14. $x + y = -5$
$\quad -2x + y = 1$

Solve each system by elimination.

15. $2x - 5y = 11$
$\quad 3x + y = 8$

16. $-2x + 3y = 1$
$\quad -4x + y = -3$

17. $3x + 4y = -6$
$\quad 5x + 3y = 1$

18. $4x + 3y = 1$
$\quad 3x + 2y = 2$

19. $3x + 3y = 0$
$\quad 4x + 2y = 3$

20. $8x + 4y = 0$
$\quad 4x - 2y = 2$

21. $\quad 7x + 2y = 6$
$\quad -14x - 4y = -12$

22. $\quad x - 4y = 2$
$\quad 4x - 16y = 8$

23. $5x - 5y = 3$
$\quad x - y = 12$

24. $\quad 2x - 3y = 7$
$\quad -4x + 6y = 14$

25. $\dfrac{x}{2} + \dfrac{y}{3} = -\dfrac{1}{3}$
$\quad \dfrac{x}{2} + 2y = -7$

26. $\dfrac{x}{5} + y = -\dfrac{12}{5}$
$\quad \dfrac{x}{10} + \dfrac{y}{3} = -\dfrac{11}{30}$

Solve each system by substitution.

27. $4x + y = 6$
$\quad y = 2x$

28. $2x - y = 6$
$\quad y = 5x$

29. $\quad 3x - 4y = -22$
$\quad -3x + y = 0$

30. $-3x + y = -5$
$\quad x + 2y = 0$

31. $-x - 4y = -14$
$\quad 2x = y + 1$

32. $-3x - 5y = -17$
$\quad 4x = y - 8$

33. $5x - 4y = 9$
$\quad 3 - 2y = -x$

34. $6x - y = -9$
$\quad 4 + 7x = -y$

35. $4x + 4y = 1$
$\quad 2x + 2y = 8$

36. $-3x + 5y = 6$
$\quad 3x - 5y = 10$

37. $6x + 2y = 4$
$\quad y = 2 - 3x$

38. $8x + 4y = 12$
$\quad y = 3 - 2x$

39. $x = 3y + 5$
$\quad y = \dfrac{2}{3}x$

40. $x = 6y - 2$
$\quad y = \dfrac{4}{3}x$

41. $\dfrac{1}{2}x + \dfrac{1}{3}y = 3$
$\quad y = 3x$

42. $\dfrac{1}{4}x - \dfrac{1}{5}y = 9$
$\quad y = 5x$

Media Delivery *Use the graph given at the beginning of this section (repeated here) to work Exercises 43–46.*

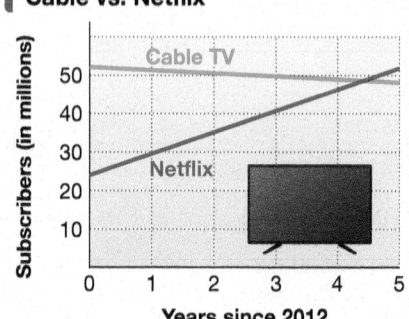

Data from Netflix, Leichtman Research Group.

43. Which type of subscription was more popular in 2017?

44. Between which two years did the two services have the same number of subscribers? About how many subscribers were there at that point?

45. If $x = 0$ represents 2012 and $x = 5$ represents 2017, the number of subscribers y (in millions) to the two services can be modeled by the linear equations in the following system.

$$-5.61x + y = 23.9 \quad \text{Netflix}$$
$$0.77x + y = 52.1 \quad \text{Cable TV}$$

Solve this system. Express values as decimals rounded to the nearest tenth. Write the solution set with an ordered pair of the form (year, number of subscribers).

46. Interpret the answer for **Exercise 45,** rounding down for the year. How does it compare to the estimate from **Exercise 44?**

AIDS and Race *The graph in the next column shows a comparison of the number of African Americans and the number of Whites living with AIDS in the United States during 1993–2000. Use this graph to work Exercises 47–50.*

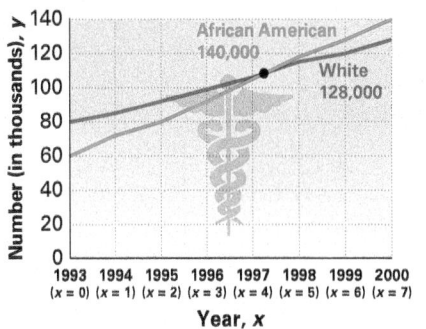

People Living with AIDS

Data from U.S. Centers for Disease Control.

47. Use the points $(0, 60)$ and $(7, 140)$ to find an equation for a line approximating the data for the graph in blue. Write it in standard form.

48. Repeat **Exercise 47,** using the points $(0, 80)$ and $(7, 128)$ for the graph in red.

49. Solve the system formed by the two equations in **Exercises 47 and 48.**

50. Describe how the solution is related to the graph.

8.8 APPLICATIONS OF LINEAR SYSTEMS

OBJECTIVES

1. Solve an application involving a rectangular area using a system of linear equations.

2. Solve an application involving different ticket prices using a system of linear equations.

3. Solve a mixture problem involving percent concentrations using a system of linear equations.

4. Solve a motion problem using a system of linear equations.

Introduction

Problems like the one in the margin on the next page, which comes from a Hindu work that dates back to about A.D. 850, have been around for thousands of years. It can be solved by using a system of equations. In this section we illustrate some strategies for solving applications using systems. The following steps, based on the six-step problem-solving method introduced earlier, give a strategy for solving problems using more than one variable.

SOLVING AN APPLIED PROBLEM BY WRITING A SYSTEM OF EQUATIONS

Step 1 **Read** the problem carefully until you understand what is given and what is to be found.

Step 2 **Assign variables** to represent the unknown values, using diagrams or tables as needed. *Write down* what each variable represents.

Step 3 **Write a system of equations** that relates the unknowns.

Step 4 **Solve** the system of equations.

Step 5 **State the answer** to the problem. Does it seem reasonable?

Step 6 **Check** the answer in the words of the original problem.

A problem from Hindu Mathematics, A.D. 850   *The mixed price of 9 citrons [a lemonlike fruit shown in the photo] and 7 fragrant wood apples is 107; again, the mixed price of 7 citrons and 9 fragrant wood apples is 101. O you arithmetician, tell me quickly the price of a citron and the price of a wood apple here, having distinctly separated those prices well.*

The answer is on **page 441.**

Applications

EXAMPLE 1 **Solving a Perimeter Problem**

A rectangular soccer field may have a width between 50 and 100 yards and a length between 50 and 100 yards. Suppose that one particular field has a perimeter of 320 yards. Its length measures 40 yards more than its width. What are the dimensions of this field?

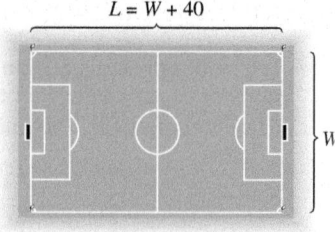

Figure 44

Solution

Step 1 **Read** the problem again. We are asked to find the dimensions of the field.

Step 2 **Assign variables.**

$$\text{Let } L = \text{the length} \quad \text{and} \quad W = \text{the width.}$$

Step 3 **Write a system of equations.** Because the perimeter is 320 yards, we find one equation by using the perimeter formula.

$$2L + 2W = 320$$

Because the length is 40 yards more than the width, we have

$$L = W + 40.$$

See **Figure 44.** We now have a system to solve.

$$2L + 2W = 320 \quad \text{(1)}$$
$$L = W + 40 \quad \text{(2)}$$

Step 4 **Solve** the system of equations. Since equation (2) is solved for L, we can substitute $W + 40$ for L in equation (1) and solve for W.

$$2L + 2W = 320 \quad \text{(1)}$$
$$2(W + 40) + 2W = 320 \quad \text{Let } L = W + 40.$$
$$2W + 80 + 2W = 320 \quad \text{Distributive property}$$
$$4W + 80 = 320 \quad \text{Combine like terms.}$$
$$4W = 240 \quad \text{Subtract 80.}$$
$$W = 60 \quad \text{Divide by 4.}$$

Let $W = 60$ in the equation $L = W + 40$ to find L.

$$L = 60 + 40 = 100$$

Step 5 **State the answer.** The length is 100 yards, and the width is 60 yards. Both dimensions are within the ranges given in the problem.

Step 6 **Check.** The answer is correct, because the perimeter of this soccer field is

$$2(100) + 2(60) = 320 \text{ yards,}$$

and the length, 100 yards, is indeed 40 yards more than the width, because

$$100 - 40 = 60.$$

EXAMPLE 2   Solving a Problem about Ticket Prices

For the 2015–2016 National Football League and National Basketball Association seasons, two football tickets and one basketball ticket purchased at their average prices would have cost $241.84. One football ticket and two basketball tickets would have cost $204.74. What were the average ticket prices for the two sports? (Data from Team Marketing Report, Chicago.)

Solution

Step 1   **Read** the problem again. There are two unknowns.

Step 2   **Assign variables.**

Let f = the average price for a football ticket

and b = the average price for a basketball ticket.

Step 3   **Write a system of equations.** We write one equation using the fact that two football tickets and one basketball ticket cost $241.84.

$$2f + b = 241.84$$

By similar reasoning, we can write a second equation.

$$f + 2b = 204.74$$

Therefore, the system is as follows.

$$2f + b = 241.84 \quad \text{(1)}$$
$$f + 2b = 204.74 \quad \text{(2)}$$

Step 4   **Solve** the system of equations. We eliminate f.

$$2f + b = 241.84 \quad \text{(1)}$$
$$\underline{-2f - 4b = -409.48} \quad \text{Multiply each side of (2) by } -2.$$
$$-3b = -167.64 \quad \text{Add.}$$
$$b = 55.88 \quad \text{Divide by } -3.$$

To find the value of f, let $b = 55.88$ in equation (2).

$$f + 2b = 204.74 \quad \text{(2)}$$
$$f + 2(55.88) = 204.74 \quad \text{Let } b = 55.88.$$
$$f + 111.76 = 204.74 \quad \text{Multiply.}$$
$$f = 92.98 \quad \text{Subtract } 111.76.$$

Step 5   **State the answer.** The average price for one basketball ticket was $55.88. For one football ticket, the average price was $92.98.

Step 6   **Check** that these values satisfy the conditions stated in the problem.

$$2(\$92.98) + \$55.88 = \$241.84, \quad \text{as required.}$$
$$\$92.98 + 2(\$55.88) = \$204.74, \quad \text{as required.}$$

We solved mixture problems earlier using one variable. Another approach is to use two variables and a system of equations.

EXAMPLE 3 Solving a Mixture Problem

How many ounces each of 2% hydrochloric acid and 8% hydrochloric acid must be combined to obtain 12 oz of solution that is 4% hydrochloric acid?

Solution

Step 1 **Read** the problem. Two solutions of different strengths are being mixed together to get a specific amount of a solution with an "in-between" strength.

Step 2 **Assign variables.** Let x represent the number of ounces of 2% solution, and let y represent the number of ounces of 8% solution.

Percent (as a decimal)	Ounces of Solution	Ounces of Pure Acid
2% = 0.02	x	$0.02x$
8% = 0.08	y	$0.08y$
4% = 0.04	12	$0.04(12)$

Summarize the information from the problem in a table.

Figure 45 also illustrates what is happening in the problem.

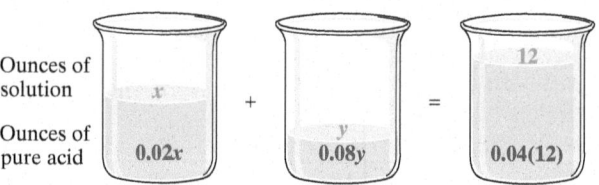

Ounces of solution
Ounces of pure acid

x
$0.02x$
+
y
$0.08y$
=
12
$0.04(12)$

Figure 45

Step 3 **Write a system of equations.** When x ounces of 2% solution and y ounces of 8% solution are combined, the total number of ounces is 12.

$$x + y = 12 \quad \text{(1)}$$

The ounces of pure acid in the 2% solution $(0.02x)$ plus the ounces of pure acid in the 8% solution $(0.08y)$ should equal the total ounces of pure acid in the mixture, which is $0.04(12)$, or 0.48.

$$0.02x + 0.08y = 0.48 \quad \text{(2)}$$

Notice that these equations can be quickly determined by reading down in the table or using the labels in **Figure 45.**

Step 4 **Solve** the system of equations (1) and (2). We eliminate x.

$$
\begin{array}{rl}
2x + 8y = 48 & \text{Multiply each side of (2) by 100.} \\
\underline{-2x - 2y = -24} & \text{Multiply each side of (1) by } -2. \\
6y = 24 & \text{Add.} \\
y = 4 & \text{Divide by 6.}
\end{array}
$$

Because $y = 4$ and $x + y = 12$, it follows that $x = 8$.

Step 5 **State the answer.** The desired mixture will require 8 ounces of the 2% solution and 4 ounces of the 8% solution.

Step 6 **Check** that these values satisfy both equations of the system.

François Viète, a mathematician of sixteenth-century France, contributed significantly to the evolution of **mathematical symbols.** Before his time, different symbols were often used for different powers of a quantity. Viète used the same letter with a description of the power and the coefficient. According to Howard Eves in *An Introduction to the History of Mathematics*, Viète would have written

$$5BA^2 - 2CA + A^3 = D$$

as

*B*5 in *A* quad − *C* plano 2 in

A + *A* cub aequatur *D* solido.

Solution to the Citron/Wood Apple problem Let *c* represent the price of a single citron, and let *w* represent the price of a wood apple. The system to solve is

$$9c + 7w = 107$$
$$7c + 9w = 101.$$

We find that the prices are 8 for a citron and 5 for a wood apple.

EXAMPLE 4 Solving a Motion Problem

Two executives in cities 400 miles apart leave at the same time to drive to a business meeting at a location on the line between their cities. They meet after 4 hours. Find the rate of each car if one car travels 20 miles per hour faster than the other.

Solution

Step 1 **Read** the problem carefully.

Step 2 **Assign variables.**

Let x = the speed of the faster car,

and y = the speed of the slower car.

We use the formula $d = rt$. Each car travels for 4 hours. See the table.

	r	t	d	
Faster car	x	4	$4x$	Find *d* from $d = rt$.
Slower car	y	4	$4y$	

Sketch what is happening in the problem. See **Figure 46.**

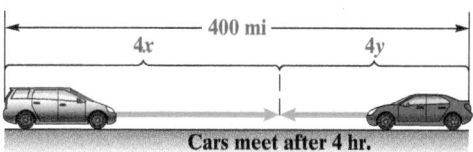

Figure 46

Step 3 **Write two equations.** The total distance traveled by both cars is 400 miles.

$$4x + 4y = 400 \quad \text{(1)}$$

The faster car goes 20 miles per hour faster than the slower car.

$$x = 20 + y \quad \text{(2)}$$

Step 4 **Solve** this system of equations by substitution.

$$4x + 4y = 400 \quad \text{(1)}$$
$$x = 20 + y \quad \text{(2)}$$

Replace x with $20 + y$ in equation (1) and solve for y.

$$4(20 + y) + 4y = 400 \quad \text{Let } x = 20 + y.$$
$$80 + 4y + 4y = 400 \quad \text{Distributive property}$$
$$80 + 8y = 400 \quad \text{Combine like terms.}$$
$$8y = 320 \quad \text{Subtract 80.}$$
$$y = 40 \quad \text{Divide by 8.}$$

Because $x = 20 + y$ and $y = 40$, $x = 20 + 40 = 60$.

Step 5 **State the answer.** The rates are 40 miles per hour and 60 miles per hour.

Step 6 **Check.** Because each car travels for 4 hours, the total distance traveled is

$$4(60) + 4(40) = 240 + 160 = 400 \text{ miles,} \quad \text{as required.}$$

CONCEPT CHECK *Answer the following using these formulas.*

$p = br$ (percentage = base × rate),

$I = prt$ (simple interest = principal × rate × time),

and

$d = rt$ (distance = rate × time)

1. If a container of liquid contains 120 ounces of solution, what is the number of ounces of pure acid if the given solution contains the following acid concentrations?

   (a) 10% (b) 25%
   (c) 40% (d) 50%

2. If $50,000 is invested in an account paying simple annual interest, how much interest will be earned during the first year at the following rates?

   (a) 0.5% (b) 1%
   (c) 1.25% (d) 2.5%

3. If a pound of ham costs $4.29, give an expression for the cost of x pounds.

4. If a ticket to the movie *Wonder Woman* costs $12, and y tickets are sold, give an expression for the amount collected.

5. If the speed of a boat in still water is 10 mph, and the speed of the current of a river is x mph, what is the speed of the boat in each situation?

   (a) going upstream (that is, against the current, which slows the boat down)

   (b) going downstream (that is, with the current, which speeds the boat up)

6. If the speed of a killer whale is 25 mph and the whale swims for y hours, give an expression for the number of miles the whale travels.

7. What expression represents the monetary value of x dimes and y quarters (in dollars)?

8. What expression represents the amount of pure alcohol in p ounces of 20% alcohol and q ounces of 35% alcohol (in ounces)?

Solve each problem by using a system of equations.

9. **Win–Loss Record** During the 2017 Major League Baseball regular season, the Cleveland Indians played 162 games and won the Central Division of the American League. They won 42 more games than they lost. What was their win–loss record that year?

2017 MLB Final Regular Season Standings, American League Central

Team	W	L
Cleveland	—	—
Minnesota	85	77
Kansas City	80	82
Chicago White Sox	67	95
Detroit	64	98

Data from www.mlb.com

10. **Win–Loss Record** During the 2017 Major League Baseball season, the Houston Astros played 162 games and won the Western Division of the American League. They won 40 more games than they lost. What was their win–loss record that year?

2017 MLB Final Regular Season Standings, American League West

Team	W	L
Houston	—	—
L.A. Angels	80	82
Seattle	78	84
Texas	78	84
Oakland	75	87

Data from www.mlb.com

11. **Dimensions of a Basketball Court** LeBron and José found that the width of their basketball court was 44 feet less than the length. If the perimeter was 288 feet, what were the length and the width of their court?

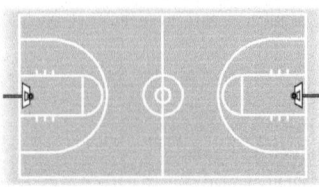

12. **Dimensions of a Tennis Court** Venus and Serena measured a tennis court and found that it was 42 feet longer than it was wide. It had a perimeter of 228 feet. What were the length and the width of the tennis court?

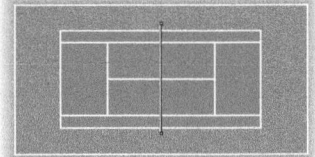

13. *Cost of Clay* For his art class, Bryce bought 2 kilograms of dark clay and 3 kilograms of light clay, paying $22 for the clay. He later needed 1 kilogram of dark clay and 2 kilograms of light clay, which cost $13 altogether. What was the cost per kilogram for each type of clay?

14. *Cost of Art Supplies* For an art project, Nicole bought 8 sheets of colored paper and 3 marker pens for $6.50. She later needed 2 sheets of colored paper and 2 marker pens. These items cost $3.00. Find the cost of 1 marker pen and the cost of 1 sheet of colored paper.

15. *Dimensions of a Square and a Triangle* The side of a square is 4 centimeters longer than the side of an equilateral triangle. The perimeter of the square is 24 centimeters more than the perimeter of the triangle. Find the length of a side of the square and the length of a side of the triangle.

16. *Dimensions of a Rectangle* The length of a rectangle is 7 feet more than the width. If the length were decreased by 3 feet and the width were increased by 2 feet, the perimeter would be 32 feet. Find the length and width of the original rectangle.

17. *Coffee Prices* At a business meeting at Panera Bread, the bill (without tax) for two cappuccinos and three caffe lattes was $19.75. At another table, the bill for one cappuccino and two caffe lattes was $11.97. How much did each type of beverage cost? (Data from Panera Bread.)

18. *Ticket Prices* Tickets to a concert at Lake Sumter Community College cost $5 for general admission or $4 with a student ID. If 184 people paid to see the concert and $812 was collected, how many of each type of ticket were sold?

19. *Meal Prices* Brianna is a waitress at Bonefish Grill. On one particular day she sold 15 ribeye steak dinners and 20 grilled salmon dinners, totaling $886.50. Another day she sold 25 ribeye steak dinners and 10 grilled salmon dinners, totaling $973.50. How much did each type of dinner cost? (Data from Bonefish Grill.)

20. *Ticket Prices* For the 2016 Major League Baseball season, based on average ticket prices, three tickets to a Cleveland Indians game and two tickets to a Chicago Cubs game cost $179.49. Two tickets to an Indians game and one ticket to a Cubs game cost $102.55. What were the average ticket prices for the two teams? (Data from Team Marketing Report.)

21. *Travel Costs* On the basis of total cost per day for business travel to New York City and Washington, DC (including a hotel room, car rental, and three meals), 2 days in New York and 3 days in Washington cost $2484. Four days in New York and 2 days in Washington cost $3120. What was the average cost per day in each city? (Data from *Business Travel News*.)

22. *Theme Park Admissions* Two days at Busch Gardens (Tampa Bay) and 3 days at Universal Studios Florida (Orlando) cost $509.98, while 4 days at Busch Gardens and 2 days at Universal Studios cost $579.96. (Prices are based on single-day admissions.) What was the cost per day for each park? (Data from Busch Gardens and Universal Studios.)

23. *Telecommunication Giants* In 2016, the two American telecommunication companies with the greatest revenues were AT&T and Verizon. The two companies had combined revenues of $288.9 billion. AT&T's revenue was $38.7 billion more than that of Verizon. What was the revenue for each company? (Data from Verizon and AT&T's Annual Reports.)

24. *Imports and Exports* In 2017, U.S. exports to Canada were $39.4 billion more than exports to Mexico. Together, exports to these two countries totaled $525.4 billion. How much were exports to each country? (Data from U.S. Census Bureau.)

Fan Cost Index The Fan Cost Index (FCI) represents the cost of four average-price tickets, four small soft drinks, two small beers, four hot dogs, parking for one car, two game programs, and two souvenir caps to a sporting event. (Data from Team Marketing Report.)

Use the concept of FCI in Exercises 25 and 26.

25. For the 2016 Major League Baseball season, the FCI prices for the Cleveland Indians and the Boston Red Sox totaled $540.10. The Boston FCI was $181.22 more than that of Cleveland. What were the FCIs for these teams?

26. In 2016, the FCI prices for Major League Baseball and the National Football League totaled $722.37. The football FCI was $283.31 more than that of baseball. What were the FCIs for these sports?

27. *Acid Mixture* How many liters each of 15% acid and 33% acid should be mixed to obtain 40 liters of 21% acid?

Kind of Solution	Liters of Solution	Amount of Pure Acid
0.15	x	
0.33	y	
0.21	40	

28. *Alcohol Mixture* How many gallons each of 25% alcohol and 35% alcohol should be mixed to obtain 20 gallons of 32% alcohol?

29. *Antifreeze Mixture* A truck radiator holds 18 liters of fluid. How much pure antifreeze must be added to a mixture that is 4% antifreeze in order to fill the radiator with a mixture that is 20% antifreeze?

30. *Acid Mixture* Pure acid is to be added to a 10% acid solution to obtain 27 liters of a 20% acid solution. What amounts of each should be used?

31. *Fruit Drink Mixture* A popular fruit drink is made by mixing fruit juices. Such a mixture with 50% juice is to be mixed with another mixture that is 30% juice to get 200 liters of a mixture that is 45% juice. How much of each should be used?

Kind of Juice	Number of Liters	Amount of Pure Juice
0.50	x	$0.50x$
0.30	y	
0.45		

32. *Candy Mixture* Madeline plans to mix pecan clusters that sell for $3.60 per pound with chocolate truffles that sell for $7.20 per pound to get a mixture that she can sell in Valentine boxes for $4.95 per pound. How much of the $3.60 clusters and how much of the $7.20 truffles should she use to create 80 pounds of the mix?

33. *Candy Mixture* A grocer plans to mix candy that sells for $1.20 per pound with candy that sells for $2.40 per pound to get a mixture that he plans to sell for $1.65 per pound. How much of the $1.20 candy and how much of the $2.40 candy should he use if he wants 160 pounds of the mix?

34. *Candy Mixture* A confectioner plans to mix candy that sells for $3.00 per pound with candy that sells for $6.00 per pound to obtain 9 pounds of a mixture that sells for $3.84 per pound. How much of the $3.00 candy and how much of the $6.00 candy should she use?

35. *Investment Mix* Julio must invest a total of $15,000 in two accounts, one paying 4% simple annual interest, and the other paying 3%. If he wants to earn $550 annual interest, how much should he invest at each rate?

Principal	Rate	Interest
x	0.04	
y	0.03	
15,000		

36. *Investment Mix* A total of $3000 is invested, part at 2% simple interest and part at 4%. If the total annual return from the two investments is $100, how much is invested at each rate?

37. *Investment Mix* A total of $3000 is invested, part at 1% simple interest and part at 2%. If the total annual return from the two investments is $50, how much is invested at each rate?

38. *Investment Mix* An investor will invest a total of $15,000 in two accounts, one paying 2% annual simple interest and the other 1.5%. He determines that he will earn $275 annual interest. How much will he invest at each rate?

39. *Speeds of Trains* A train travels 150 miles in the same time that a plane covers 400 miles. If the speed of the plane is 20 miles per hour less than 3 times the speed of the train, find both speeds.

40. *Speeds of Trains* A freight train and an express train leave towns 390 miles apart at the same time, traveling toward one another. The freight train travels 30 mph slower than the express train. They pass one another 3 hours later. What are their speeds?

41. *Speeds of Boat and Current* In his motorboat, Tran travels upstream at top speed to his favorite fishing spot, a distance of 36 miles, in 2 hours. Returning, he finds that the trip downstream, still at top speed, takes only 1.5 hours. Find the speed of Tran's boat and the speed of the current.

42. *Speeds of Snow Speeder and Wind* Braving blizzard conditions on the planet Hoth, Luke Skywalker sets out at top speed in his snow speeder for a rebel base 2400 km away. He travels into a steady headwind and makes the trip in 2.4 hours. Returning, he finds that the trip back, still at top speed but now with a tailwind, takes only 2 hours. Find the top speed of Luke's snow speeder and the speed of the wind.

8.9 LINEAR INEQUALITIES, SYSTEMS, AND LINEAR PROGRAMMING

OBJECTIVES

1 Graph a linear inequality in two variables.

2 Graph a system of linear inequalities in two variables.

3 Apply linear programming to maximize or minimize a value.

Linear Inequalities in Two Variables

Linear inequalities with *one variable* were graphed on number lines earlier in this text. Here we graph linear inequalities in *two variables* in a rectangular coordinate system.

LINEAR INEQUALITY IN TWO VARIABLES

An inequality that can be written as

$$Ax + By < C \qquad \text{or} \qquad Ax + By > C,$$

where A, B, and C are real numbers and A and B are not both 0, is a **linear inequality in two variables.** The symbols $\leq$ and $\geq$ may replace $<$ and $>$ in this definition.

A line divides the plane into three regions: the line itself and the two half-planes on either side of the line. Recall that the graphs of linear inequalities in one variable are *intervals* on the number line that may include an endpoint. The graphs of linear inequalities in two variables are *regions* in the real number plane and may include a *boundary line*. The **boundary line** for the inequality $Ax + By < C$ or $Ax + By > C$ is the graph of the *equation $Ax + By = C$.*

To graph a linear inequality, we follow these steps.

GRAPHING A LINEAR INEQUALITY

Step 1 **Draw the boundary.**   Draw the graph of the straight line that is the boundary. Make the line solid if the inequality involves $\leq$ or $\geq$. Make the line dashed if the inequality involves $<$ or $>$.

Step 2 **Choose a test point.**   Choose any point not on the line as a test point.

Step 3 **Shade the appropriate region.**   Shade the region that includes the test point if it satisfies the original inequality. Otherwise, shade the region on the other side of the boundary line.

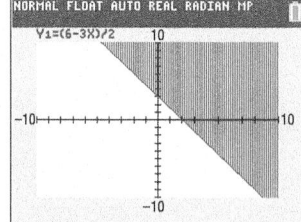

The TI-84 Plus C allows us to shade the appropriate region for an inequality. Compare with **Figure 47.**

EXAMPLE 1   Graphing a Linear Inequality

Graph $3x + 2y \geq 6$.

Solution

First graph the straight line $3x + 2y = 6$. The graph of this line, the boundary of the graph of the inequality, is shown in **Figure 47.** The graph of the inequality $3x + 2y \geq 6$ includes the points of the line $3x + 2y = 6$, and either the points *above* the line $3x + 2y = 6$ or the points *below* that line.

To decide which, select any point not on the line $3x + 2y = 6$ as a test point. The origin, $(0, 0)$, often is a good choice. Substitute the values from $(0, 0)$ for x and y in the linear inequality $3x + 2y \geq 6$.

$$3(0) + 2(0) \overset{?}{\geq} 6$$

$$0 \geq 6 \quad \text{False}$$

Because the result is false, $(0, 0)$ does not satisfy the inequality, and so the solution set includes all points on the *other* side of the line, shaded in **Figure 47.**

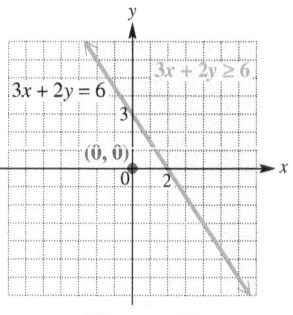

Figure 47

EXAMPLE 2 Graphing a Linear Inequality

Graph $x - 3y > 4$.

Solution

First graph the boundary line, $x - 3y = 4$. The line is shown in **Figure 48.** The points of the boundary line do not belong to the inequality $x - 3y > 4$ (because the inequality symbol does not involve equality). For this reason, the line is dashed.

To decide which side of the line is the graph of the solution set, choose any point that is not on the line—say $(2, -2)$. Substitute 2 for x and -2 for y in the original inequality.

$$2 - 3(-2) \overset{?}{>} 4$$

$$8 > 4 \quad \text{True}$$

Because the result is true, the solution set lies on the side of the boundary line that contains the test point $(2, -2)$. The solution set, graphed in **Figure 48,** includes only those points in the shaded region (not those on the line).

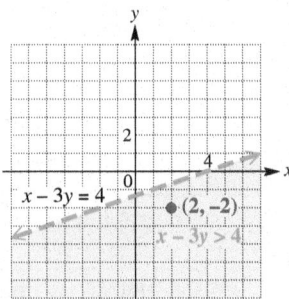

Figure 48

Systems of Inequalities

Systems of inequalities with two variables may be solved by graphing. A system of linear inequalities consists of two or more such inequalities, and the solution set of such a system consists of all points that make all the inequalities true.

GRAPHING A SYSTEM OF LINEAR INEQUALITIES

Step 1 **Graph each inequality in the same coordinate system.** Graph each inequality in the system, using the method described in **Examples 1 and 2.**

Step 2 **Find the intersection of the regions of solutions.** Indicate the intersection of the regions of solutions of the individual inequalities. This is the solution set of the system.

EXAMPLE 3 Graphing a System of Inequalities

Graph the solution set of the linear system.

$$3x + 2y \leq 6$$

$$2x - 5y \geq 10$$

Solution

To graph $3x + 2y \leq 6$, graph $3x + 2y = 6$ as a solid line. Because $(0, 0)$ makes the inequality *true,* shade the region containing $(0, 0)$. See **Figure 49** on the next page.

Now graph $2x - 5y \geq 10$. The solid-line boundary is the graph of $2x - 5y = 10$. Because $(0, 0)$ makes the inequality *false*, shade the region that does not contain $(0, 0)$, as shown in **Figure 50.**

The solution set of the system is given by the intersection (overlap) of the regions of the graphs in **Figures 49 and 50.** The solution set is the shaded region in **Figure 51,** and includes portions of the two boundary lines.

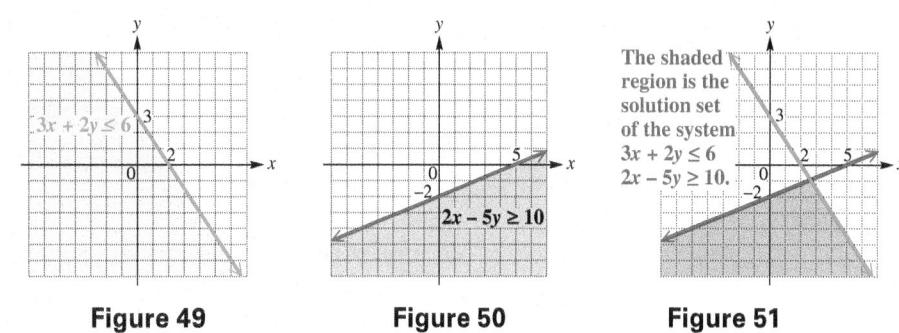

Figure 49   Figure 50   Figure 51

In practice, we usually do all the work in one coordinate system.

EXAMPLE 4   **Graphing a System of Inequalities**

Graph the solution set of the linear system.

$$2x + 3y \geq 12$$
$$7x + 4y \geq 28$$
$$y \leq 6$$
$$x \leq 5$$

Solution

Graph the four inequalities in one coordinate system, and shade the region common to all four as shown in **Figure 52.** As shown, the boundary lines are all solid.

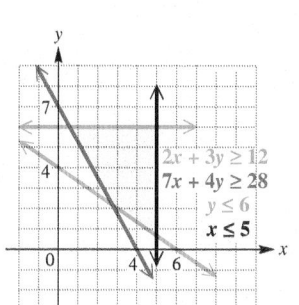

Figure 52

Linear Programming

One important application of mathematics to business and social science is **linear programming.** Linear programming is used to find an optimum value—for example, minimum cost or maximum profit. Procedures for solving linear programming problems were developed in 1947 by George Dantzig, while he was working on a problem of allocating supplies for the Air Force in a way that minimized total cost.

To solve a linear programming problem in general, use the following steps. (The italicized terms are defined in **Example 5.**)

George B. Dantzig (1914–2005) of Stanford University was one of the key people behind **operations research** (OR). As a management science, OR is not a single discipline. It draws from mathematics, probability theory, statistics, and economics. The name given to this "multiplex" shows its historical origins in World War II, when operations of a military nature called forth the efforts of many scientists to research their fields for applications to the war effort and to solve tactical problems.

SOLVING A LINEAR PROGRAMMING PROBLEM

Step 1   Write all necessary *constraints* and the *objective function*.

Step 2   Graph the *region of feasible solutions*.

Step 3   Identify all *vertices*.

Step 4   Find the value of the *objective function* at each *vertex*.

Step 5   The solution is given by the *vertex* producing the optimum value of the *objective function*.

EXAMPLE 5   Maximizing Rescue Efforts

Earthquake victims in China need medical supplies and bottled water. Each medical kit measures 1 cubic foot and weighs 10 pounds. Each container of water is also 1 cubic foot and weighs 20 pounds. The plane can carry only 80,000 pounds with a total volume of 6000 cubic feet. Each medical kit will aid 6 people, and each container of water will serve 10 people. How many of each should be sent in order to maximize the number of people aided?

Solution

Step 1 We translate the statements of the problem into symbols as follows.

> Let x = the number of medical kits to be sent,
>
> and y = the number of containers of water to be sent.

Because negative values of x and y are not valid for this problem, these two inequalities must be satisfied.

$$x \geq 0$$

$$y \geq 0$$

Each medical kit and each container of water will occupy 1 cubic foot of space, and there is a maximum of 6000 cubic feet available.

$$1x + 1y \leq 6000$$

or $x + y \leq 6000$

Each medical kit weighs 10 pounds, and each water container weighs 20 pounds. The total weight cannot exceed 80,000 pounds.

$$10x + 20y \leq 80{,}000$$

or, equivalently, $x + 2y \leq 8000$ Divide by 10.

The four inequalities in color form a system of linear inequalities.

$$x \geq 0$$
$$y \geq 0$$
$$x + y \leq 6000$$
$$x + 2y \leq 8000$$

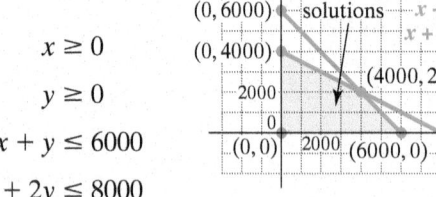

Figure 53

These are the **constraints** on the variables in this application.

Because each medical kit will aid 6 victims and each container of water will serve 10 victims, the total number of victims served is represented by the following **objective function.**

Number of victims served = $6x + 10y$ Multiply the number of items by the number of victims served, and add.

Step 2 The maximum number of victims served, subject to these constraints, is found by sketching the graph of the solution set of the system. See **Figure 53.** The only feasible values of x and y are those that satisfy all constraints. These values correspond to points that lie on the boundary or in the shaded region, which is called the **region of feasible solutions.**

Step 3 The problem may now be stated as follows: Find values of x and y in the region of feasible solutions as shown in **Figure 53** that will produce the maximum possible value of $6x + 10y$. It can be shown that any optimum value (maximum or minimum) will always occur at a **vertex** (or **corner point**) of the region of feasible solutions.

Step 4 Locate the point (x, y) that gives the maximum value by checking the coordinates of the vertices, shown in **Figure 53** and in **Table 5**. Find the number of victims served that corresponds to each coordinate pair.

Table 5

Points are from Figure 53.

Point	Number of Victims Served $= 6x + 10y$
$(0, 0)$	$6(0) + 10(0) = 0$
$(0, 4000)$	$6(0) + 10(4000) = 40{,}000$
$(4000, 2000)$	$6(4000) + 10(2000) = 44{,}000$
$(6000, 0)$	$6(6000) + 10(0) = 36{,}000$

44,000 is the maximum number.

Step 5 Choose the vertex that gives the maximum number. The maximum number of victims served is 44,000, when 4000 medical kits and 2000 containers of water are sent.

8.9 EXERCISES

CONCEPT CHECK *Fill in each blank with the correct response.*

1. The boundary line of the graph of the inequality $-2x + 5y \leq 10$ is a _____ line.
(dashed/solid)

2. The ordered pair $(0, 0)$ makes the inequality $x - 3y \leq 2$ a _____ statement.
(true/false)

3. The ordered pair $(4, -2)$ _____ be used as a test point in graphing $-x + 2y \leq -8$.
(can/cannot)

4. For the objective function $3x + 8y$, the value is _____ for the ordered pair $(100, 200)$.

Graph each linear inequality.

5. $x + y \leq 2$

6. $x - y \geq -3$

7. $4x - y \leq 5$

8. $3x + y \geq 6$

9. $x + 3y \geq -2$

10. $x + 2y \leq -5$

11. $4x + 6y \leq -3$

12. $2x - 4y \leq 3$

13. $4x - 3y < 12$

14. $5x + 3y > 15$

15. $y > -x$

16. $y < x$

In Exercises 17–20, match each system of inequalities with the correct graph from choices A–D.

17. $x \geq 5$
  $y \leq -3$

18. $x \leq 5$
  $y \geq -3$

19. $x > 5$
  $y < -3$

20. $x < 5$
  $y > -3$

A.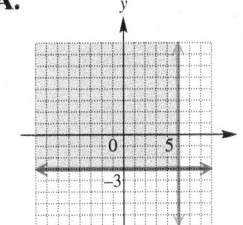

B.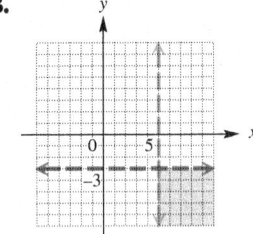

C.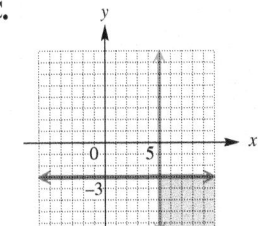

D.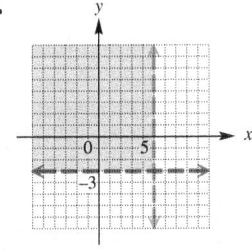

Graph each system of inequalities.

21. $x + y \leq 1$
  $x \geq 0$

22. $3x - 4y \leq 6$
  $y \geq 1$

23. $2x - y \geq 1$
  $3x + 2y \geq 6$

24. $x + 3y \geq 6$
  $3x - 4y \leq 12$

25. $x + y > -5$
  $x - y \leq 3$

26. $6x - 4y < 8$
  $x + 2y \geq 4$

Exercises 27 and 28 show regions of feasible solutions. Find the maximum and minimum values of the given objective functions.

27. $3x + 5y$

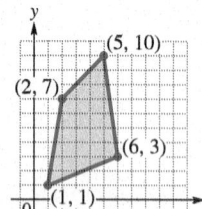

28. $40x + 75y$

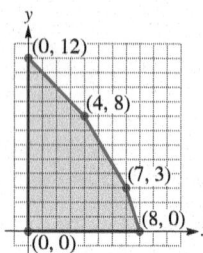

Use graphical methods to find values of x and y that satisfy the given conditions. (It may be necessary to solve a system of equations in order to find vertices.) Find the value of the maximum or minimum of the objective function, as directed.

29.
$$x \geq 0$$
$$y \geq 0$$
$$2x + 3y \leq 5$$
$$4x + y \leq 5$$
Maximize $5x + 2y$.

30.
$$x \geq 0$$
$$y \geq 0$$
$$x + y \leq 6$$
$$5x + 2y \leq 24$$
Minimize $x + 3y$.

31.
$$x \geq 2$$
$$y \geq 5$$
$$3x - y \geq 12$$
$$x + y \leq 15$$
Minimize $2x + y$.

32.
$$x \geq 10$$
$$y \geq 20$$
$$2x + 3y \leq 100$$
$$5x + 4y \leq 200$$
Maximize $x + 3y$.

Solve each linear programming problem.

33. **Refrigerator Shipping Costs** A manufacturer of refrigerators must ship at least 100 refrigerators to its two West Coast warehouses. Each warehouse holds a maximum of 100 refrigerators. Warehouse A holds 25 refrigerators already, while warehouse B has 20 on hand. It costs $12 to ship a refrigerator to warehouse A and $10 to ship one to warehouse B. How many refrigerators should be shipped to each warehouse to minimize cost? What is the minimum cost?

34. **Gasoline and Fuel Oil Costs** A manufacturing process requires that oil refineries manufacture at least 2 gallons of gasoline for each gallon of fuel oil. To meet the winter demand for fuel oil, at least 3 million gallons a day must be produced. The demand for gasoline is no more than 6.4 million gallons per day. If the price of gasoline is $2.90 per gallon and the price of fuel oil is $3.50 per gallon, how much of each should be produced to maximize revenue?

35. **Vitamin Pill Costs** Robin takes vitamin pills. Each day, she must have at least 16 units of vitamin A, at least 5 units of vitamin B_1, and at least 20 units of vitamin C. She can choose between red pills costing 10¢ each that contain 8 units of A, 1 of B_1, and 2 of C; and blue pills that cost 20¢ each and contain 2 units of A, 1 of B_1, and 7 of C. How many of each pill should she take in order to minimize her cost and yet fulfill her daily requirements?

36. **Food Supplement Costs** Renée requires two food supplements, I and II. She can get these supplements from two different products, A and B. Product A provides 3 grams per serving of supplement I, and 2 grams per serving of supplement II. Product B provides 2 grams per serving of supplement I, and 4 grams per serving of supplement II. Her dietician has recommended that she include at least 15 grams of each supplement in her daily diet. If product A costs $0.25 per serving and product B costs $0.40 per serving, how can she satisfy her requirements most economically?

37. **Cake and Cookie Production** A bakery makes both cakes and cookies. Each batch of cakes requires two hours in the oven and three hours in the decorating room. Each batch of cookies needs one and a half hours in the oven and two-thirds of an hour in the decorating room. The oven is available no more than 15 hours a day, and the decorating room can be used no more than 13 hours a day. How many batches of cakes and cookies should the bakery make in order to maximize profits if cookies produce a profit of $20 per batch and cakes produce a profit of $30 per batch?

38. **Profitable Production** The Smartski Company makes two products: gadgets and gizmos. Each gadget gives a profit of $3, and each gizmo gives a profit of $7. The company must manufacture at least 1 gadget per day to satisfy one of its customers, but no more than 5 because of production problems. Also, the number of gizmos produced cannot exceed 6 per day. As a further requirement, the number of gadgets cannot exceed the number of gizmos. How many of each should the company manufacture per day in order to obtain the maximum profit?

Chapter 8 SUMMARY

KEY TERMS

8.1
ordered pair
components
equal ordered pairs
origin
x-axis
y-axis
rectangular (Cartesian)
   coordinate system
quadrants
coordinates
midpoint
circle
center
radius

8.2
table of values
linear equation in two
   variables

standard form
x-intercept
y-intercept
slope
parallel
perpendicular
average rate of change

8.3
point-slope form
slope-intercept form

8.4
relation
dependent variable
independent variable
function
function notation
linear function
constant function

8.5
quadratic function
parabola
axis of symmetry (axis)
vertex

8.6
exponential function
horizontal asymptote
natural exponential function
compound interest
continuous compounding
exponential growth function
logarithm
logarithmic function
vertical asymptote
natural logarithmic function
common logarithm
doubling time
half-life

8.7
system of linear equations
consistent system
independent equations
inconsistent system
dependent equations
elimination method
substitution method

8.9
linear inequality in two
   variables
boundary line
linear programming
constraints
objective function
region of feasible solutions
vertex (corner point)

NEW SYMBOLS

(a, b) ordered pair
x_1 a specific value of the variable x
      (read "x-sub-one")
m slope
$f(x)$ function f of x (read "f of x" or "f at x")

$\log_a x$ logarithm of x with base a
e a constant, approximately 2.718281828
$\ln x$ natural (base e) logarithm of x
$\log x$ common (base ten) logarithm of x

TEST YOUR WORD POWER

See how well you have learned the vocabulary in this chapter.

1. An **ordered pair** is a pair of numbers written
   A. in numerical order between brackets.
   B. between parentheses or brackets.
   C. between parentheses in which order is important.
   D. between parentheses in which order does not matter.

2. A **linear equation in two variables** is an equation that can be written in the form
   A. $Ax + By < C$.   B. $ax = b$.
   C. $y = x^2$.   D. $Ax + By = C$.

3. An **intercept** is
   A. the point where the *x*-axis and the *y*-axis intersect.
   B. a pair of numbers written between parentheses in which order matters.
   C. one of the four regions determined by a rectangular coordinate system.
   D. a point where a graph intersects the *x*-axis or the *y*-axis.

4. The **slope** of a line is
   A. the measure of the run over the rise of the line.
   B. the distance between two points on the line.
   C. the ratio of the change in *y* to the change in *x* along the line.
   D. the horizontal change compared to the vertical change of two points on the line.

5. A **relation** is
   A. a set of ordered pairs.
   B. the ratio of the change in *y* to the change in *x* along a line.
   C. the set of all possible values of the independent variable.
   D. all the second components of a set of ordered pairs.

6. A **function** is
- **A.** the numbers in an ordered pair.
- **B.** a set of ordered pairs in which each distinct x-value corresponds to exactly one y-value.
- **C.** a pair of numbers written between parentheses in which order matters.
- **D.** the set of all ordered pairs that satisfy an equation.

7. The **domain** of a function is
- **A.** the set of all possible values of the dependent variable y.
- **B.** a set of ordered pairs.
- **C.** the difference between the x-values.
- **D.** the set of all possible values of the independent variable x.

8. The **range** of a function is
- **A.** the set of all possible values of the dependent variable y.
- **B.** a set of ordered pairs.
- **C.** the difference between the y-values.
- **D.** the set of all possible values of the independent variable x.

9. The **vertex** of a parabola is
- **A.** the point where the graph intersects the y-axis.
- **B.** the point where the graph intersects the x-axis.
- **C.** the lowest point on a parabola that opens up or the highest point on a parabola that opens down.
- **D.** the origin.

10. A **system of linear equations** consists of
- **A.** linear equations with second-degree variables.
- **B.** linear equations that have an infinite number of solutions.
- **C.** linear equations that are to be solved at the same time.
- **D.** linear inequalities that are to be solved at the same time.

ANSWERS
1. C **2.** D **3.** D **4.** C **5.** A
6. B **7.** D **8.** A **9.** C **10.** C

QUICK REVIEW

Concepts   *Examples*

8.1   THE RECTANGULAR COORDINATE SYSTEM AND CIRCLES

Ordered Pairs
Plot the ordered pair $(-3, 4)$ by starting at the origin, moving 3 units to the left, and then moving 4 units up.

Plot the point $(-3, 4)$.

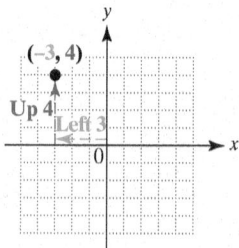

Distance Formula
The distance d between the points (x_1, y_1) and (x_2, y_2) is given by the following formula.

$$d = \sqrt{(x_2 - x_1)^2 + (y_2 - y_1)^2}$$

Find the distance d between $(3, -2)$ and $(-1, 1)$.

$$d = \sqrt{(-1 - 3)^2 + [1 - (-2)]^2}$$
$$= \sqrt{(-4)^2 + 3^2}$$
$$= \sqrt{16 + 9}$$
$$= \sqrt{25}$$
$$= 5$$

Midpoint Formula
The coordinates of the midpoint M of the line segment with endpoints (x_1, y_1) and (x_2, y_2) are given by the following formula.

$$M = \left(\frac{x_1 + x_2}{2}, \frac{y_1 + y_2}{2} \right)$$

Find the midpoint M of the segment with endpoints $(4, -7)$ and $(-10, -13)$.

$$M = \left(\frac{4 + (-10)}{2}, \frac{-7 + (-13)}{2} \right) = (-3, -10)$$

Concepts	Examples

Circle

A circle with radius r and center at (h, k) has an equation of the following form.

$$(x - h)^2 + (y - k)^2 = r^2$$

A circle of radius r and center at the origin has an equation of the following form.

$$x^2 + y^2 = r^2$$

Graph $(x + 2)^2 + (y - 3)^2 = 25$.

This equation can be written $[x - (-2)]^2 + (y - 3)^2 = 5^2$. It is a circle with center at $(-2, 3)$ and radius 5.

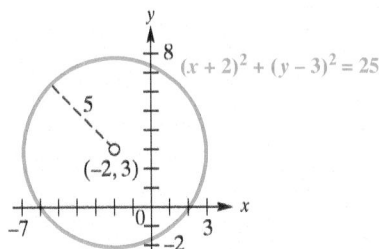

8.2 LINES, SLOPE, AND AVERAGE RATE OF CHANGE

Linear Equation in Two Variables

An equation that can be written in the form

$$Ax + By = C \quad \text{(where A and B are not both 0)}$$

is a linear equation in two variables.

Graph the linear equation $x - 2y = 4$ using the intercepts.

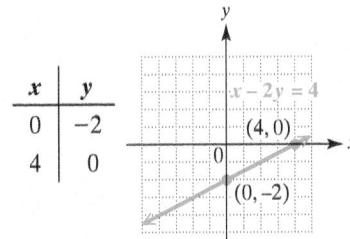

x	y
0	-2
4	0

Slope

If $x_1 \neq x_2$, then the slope m of the line through (x_1, y_1) and (x_2, y_2) is given by the following formula.

$$\text{slope } m = \frac{\text{rise}}{\text{run}} = \frac{\text{change in } y}{\text{change in } x} = \frac{y_2 - y_1}{x_2 - x_1}$$

Find the slope of the graph of $2x + 3y = 12$.

Use the intercepts $(6, 0)$ and $(0, 4)$ and the slope formula.

$$m = \frac{4 - 0}{0 - 6} = \frac{4}{-6} = -\frac{2}{3} \quad \begin{array}{l} (x_1, y_1) = (6, 0) \\ (x_2, y_2) = (0, 4) \end{array}$$

Special Cases of Slope

A horizontal line has slope 0.

A vertical line has undefined slope.

The graph of the line $y = -5$ has slope $m = 0$.

The graph of the line $x = 3$ has undefined slope.

Slopes of Parallel and Perpendicular Lines

Parallel lines have equal slopes.

The lines $y = 2x + 3$ and $4x - 2y = 6$ are parallel. Both have slope 2.

$$
\begin{array}{l|l}
y = 2x + 3 & 4x - 2y = 6 \\
 & -2y = -4x + 6 \\
 & y = 2x - 3
\end{array}
$$

Perpendicular lines, neither of which is vertical, have slopes that are negative reciprocals (with a product of -1).

The lines $y = 3x - 1$ and $x + 3y = 4$ are perpendicular. The slopes are negative reciprocals, because $3\left(-\frac{1}{3}\right) = -1$.

$$
\begin{array}{l|l}
y = 3x - 1 & x + 3y = 4 \\
 & 3y = -x + 4 \\
 & y = -\frac{1}{3}x + \frac{4}{3}
\end{array}
$$

Concepts	Examples

Average Rate of Change

To find the average rate of change from (x_1, y_1) to (x_2, y_2), find the slope of the line containing them.

Find the average rate of change from year 0 to year 3.

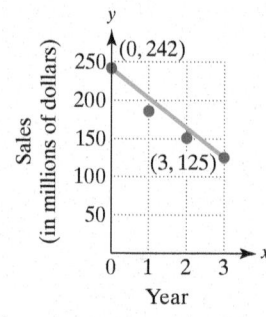

$$m = \frac{125 - 242}{3 - 0} = \frac{-117}{3} = -39$$

From year 0 to year 3, the sales *decreased* by an average of 39 million dollars per year.

8.3 EQUATIONS OF LINES

Point-Slope Form

The equation of a line through (x_1, y_1) with slope m is written in point-slope form as follows.

$$y - y_1 = m(x - x_1)$$

Find the standard form of the equation of the line with slope 3, passing through $(4, 1)$.

$$y - y_1 = m(x - x_1) \quad \text{Point-slope form}$$

$$y - 1 = 3(x - 4) \quad m = 3,\ x_1 = 4,\ y_1 = 1$$

$$y - 1 = 3x - 12 \quad \text{Distributive property}$$

$$11 = 3x - y \quad \text{Subtract } y \text{ and add 12.}$$

$$3x - y = 11 \quad \text{Standard form}$$

Slope-Intercept Form

The equation of a line with slope m and y-intercept $(0, b)$ is written in slope-intercept form as follows.

$$y = mx + b$$

Find an equation in slope-intercept form of the line passing through $(0, 5)$ and $(2, 9)$.

The y-intercept is $(0, 5)$, so $b = 5$. The slope is

$$m = \frac{9 - 5}{2 - 0} = \frac{4}{2} = 2.$$

The equation in slope-intercept form is $y = 2x + 5$.

8.4 LINEAR FUNCTIONS, GRAPHS, AND MODELS

Relations and Functions

A relation is a set of ordered pairs.

A function is a relation in which, for each value of the first component of the ordered pairs, there is *exactly* one value of the second component.

Represent the relation $\{(0, 1), (1, 2), (2, 1), (3, 0)\}$ as a table of values and as a graph.

x	y
0	1
1	2
2	1
3	0

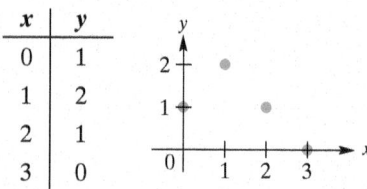

This is a function because each independent variable value (from the domain) is paired with one and only one dependent variable value (from the range).

Concepts	*Examples*

Function Notation

For a function f, the symbol $f(x)$ is used to represent the dependent variable associated with the independent variable x.

Find $f(-2)$ if $f(x) = -3x + 6$.

$$f(x) = -3x + 6$$
$$f(-2) = -3(-2) + 6$$
$$= 12$$

Linear Function

A function f that can be defined by a rule written in the form

$$f(x) = ax + b$$

for real numbers a and b (where $a \neq 0$) is a linear function.

Graph $f(x) = 2x - 5$.

x	$f(x)$
0	-5
3	1
-1	-7

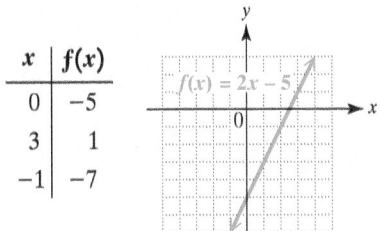

8.5 QUADRATIC FUNCTIONS, GRAPHS, AND MODELS

Graph of a Quadratic Function

The graph of the quadratic function

$$f(x) = a(x - h)^2 + k \quad \text{(where } a \neq 0\text{)}$$

is a parabola with vertex at (h, k) and with the vertical line $x = h$ as axis of symmetry.

- The graph opens upward if a is positive and downward if a is negative.
- The graph is wider than the graph of $y = x^2$ if $0 < |a| < 1$ and narrower if $|a| > 1$.

Graph $f(x) = -(x + 3)^2 + 1$.

The graph opens down because $a < 0$.

Vertex: $(-3, 1)$

Axis of symmetry: $x = -3$

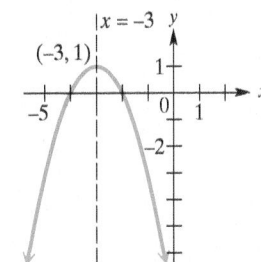

Vertex Formula

The vertex of the graph of

$$f(x) = ax^2 + bx + c \text{ (where } a \neq 0\text{)}$$

has coordinates as follows.

$$\left(-\frac{b}{2a}, f\left(-\frac{b}{2a}\right)\right)$$

Graphing a Quadratic Function

Step 1 Determine whether the graph opens up or down.

Step 2 Find the vertex.

Step 3 Find the y-intercept.

Step 4 Find the x-intercept(s), if any.

Step 5 Find and plot additional points as needed.

Graph $f(x) = x^2 + 4x + 3$.

The graph opens up because $a > 0$.

Vertex: $(-2, -1)$

The solutions of the equation $x^2 + 4x + 3 = 0$ are -1 and -3, so the x-intercepts are $(-1, 0)$ and $(-3, 0)$.

$f(0) = 3$, so the y-intercept is $(0, 3)$.

Axis of symmetry: $x = -2$

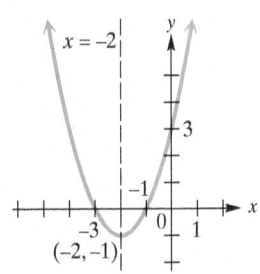

Concepts	*Examples*

8.6   EXPONENTIAL AND LOGARITHMIC FUNCTIONS, GRAPHS, AND MODELS

Exponential Function For $b > 0$, $b \neq 1$, $f(x) = b^x$ is the exponential function with base b.	$f(x) = 3^x$ is the exponential function with base 3.

Graph of $f(x) = b^x$

- The graph contains the point $(0, 1)$.
- When $b > 1$, the graph rises from left to right.
When $0 < b < 1$, the graph falls from left to right.
- The x-axis is an asymptote.
- The domain is $(-\infty, \infty)$ and the range is $(0, \infty)$.

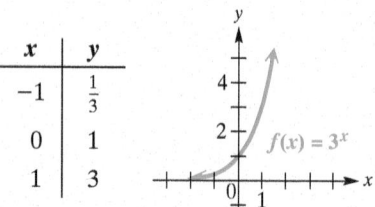

x	y
-1	$\frac{1}{3}$
0	1
1	3

Logarithmic Function

$y = \log_b x$ means $x = b^y$.

For $b > 0$, $b \neq 1$, $\log_b b = 1$ and $\log_b 1 = 0$.

For $b > 0$, $b \neq 1$, $x > 0$, $g(x) = \log_b x$ is the logarithmic function with base b.

$y = \log_2 x$ means $x = 2^y$.

$\log_3 3 = 1$ $\log_5 1 = 0$

$g(x) = \log_3 x$ is the logarithmic function with base 3.

Graph of $g(x) = \log_b x$

- The graph contains the point $(1, 0)$.
- When $b > 1$, the graph rises from left to right.
When $0 < b < 1$, the graph falls from left to right.
- The y-axis is an asymptote.
- The domain is $(0, \infty)$ and the range is $(-\infty, \infty)$.

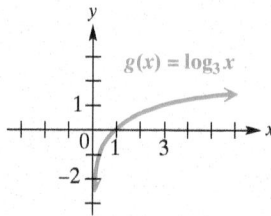

8.7   SYSTEMS OF LINEAR EQUATIONS

Solving a Linear System by Elimination

Step 1   Write both equations in standard form.

Step 2   Make the coefficients of one pair of variable terms opposites.

Step 3   Add the new equations. The sum should be an equation with just one variable.

Step 4   Solve the equation from Step 3.

Step 5   Find the value of the other variable by substituting the result from Step 4 into either of the original equations.

Step 6   Check the ordered-pair solution in *both* of the *original* equations. Then write the solution set.

Solve by elimination.

$$5x + y = 2 \quad \text{(1)}$$
$$2x - 3y = 11 \quad \text{(2)}$$

To eliminate y, multiply equation (1) by 3, and add the result to equation (2).

$$\begin{array}{rl} 15x + 3y = 6 & \text{3 times equation (1)} \\ \underline{2x - 3y = 11} & \text{(2)} \\ 17x \phantom{{}- 3y} = 17 & \text{Add.} \\ x = 1 & \text{Divide by 17.} \end{array}$$

Let $x = 1$ in equation (1), and solve for y.

$$5(1) + y = 2$$
$$y = -3$$

Check to verify that $\{(1, -3)\}$ is the solution set.

Concepts	Examples
Solving a Linear System by Substitution *Step 1* Solve one of the equations for either variable. *Step 2* Substitute for that variable in the other equation. The result should be an equation with just one variable. *Step 3* Solve the equation from Step 2. *Step 4* Find the value of the other variable by substituting the result from Step 3 into the equation from Step 1. *Step 5* Check the ordered-pair solution in *both* of the *original* equations. Then write the solution set.	Solve by substitution. $$4x - y = 7 \quad \text{(1)}$$ $$3x + 2y = 30 \quad \text{(2)}$$ Solve for y in equation (1). $$y = 4x - 7$$ Substitute $4x - 7$ for y in equation (2), and solve for x. $$3x + 2y = 30 \quad \text{(2)}$$ $$3x + 2(4x - 7) = 30 \quad \text{Let } y = 4x - 7.$$ $$3x + 8x - 14 = 30 \quad \text{Distributive property}$$ $$11x - 14 = 30 \quad \text{Combine like terms.}$$ $$11x = 44 \quad \text{Add 14.}$$ $$x = 4 \quad \text{Divide by 11.}$$ Substitute 4 for x in the equation $y = 4x - 7$ to find y. $$y = 4(4) - 7$$ $$y = 9$$ Check to verify that $\{(4, 9)\}$ is the solution set.

8.8 APPLICATIONS OF LINEAR SYSTEMS

Problem-Solving Method *Step 1* Read the problem carefully. *Step 2* Assign variables. *Step 3* Write a system of equations that relates the unknowns. *Step 4* Solve the system. *Step 5* State the answer. *Step 6* Check.	The perimeter of a rectangle is 18 ft. The length is 3 ft more than twice the width. What are the dimensions of the rectangle? Let x represent the length and y represent the width. From the perimeter formula, one equation is $2x + 2y = 18$. From the problem, another equation is $x = 2y + 3$. We must solve the following system. $$2x + 2y = 18 \quad \text{(1)}$$ $$x = 2y + 3 \quad \text{(2)}$$ 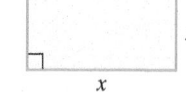 Substitute $2y + 3$ for x in equation (1). $$2x + 2y = 18 \quad \text{(1)}$$ $$2(2y + 3) + 2y = 18 \quad \text{Let } x = 2y + 3.$$ $$4y + 6 + 2y = 18 \quad \text{Distributive property}$$ $$6y = 12 \quad \text{Combine like terms. Subtract 6.}$$ $$y = 2 \quad \text{Divide by 6.}$$ Because $y = 2, \quad x = 2(2) + 3 = 7.$ Check that the length is 7 ft and the width is 2 ft.

Concepts   *Examples*

8.9 LINEAR INEQUALITIES, SYSTEMS, AND LINEAR PROGRAMMING

Graphing a Linear Inequality

Step 1   Graph the line that is the boundary of the region. Make it solid if the inequality is $\leq$ or $\geq$. Make it dashed if the inequality is $<$ or $>$.

Step 2   Use any point not on the line as a test point. Substitute for x and y in the inequality.

Step 3   If the result is true, shade the region of the line containing the test point. If the result is false, shade the other region.

Graph $2x + y \leq 5$.

Graph the boundary line $2x + y = 5$. Make it solid because of the equality portion of the symbol $\leq$.

Use $(0, 0)$ as a test point.

$$2x + y < 5$$
$$2(0) + 0 \overset{?}{<} 5$$
$$0 < 5 \quad \text{True}$$

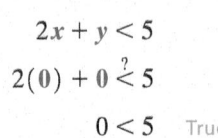

Shade the side of the line containing $(0, 0)$.

Solving a Linear Programming Problem

Step 1   Write all necessary *constraints* and the *objective function*.

Step 2   Graph the *region of feasible solutions*.

Step 3   Identify all *vertices*.

Step 4   Find the value of the *objective function* at each *vertex*.

Step 5   The solution is given by the *vertex* producing the optimum value of the *objective function*.

Find the maximum value of the objective function $8x + 12y$ for the given constraints.

$$x \geq 0, \quad y \geq 0, \quad x + 2y \leq 14, \quad 3x + 4y \leq 36$$

The region of feasible solutions is given here.

Vertex Point (x, y)	Value of $8x + 12y$
$(0, 0)$	0
$(0, 7)$	84
$(12, 0)$	96
$(8, 3)$	100 ← Maximum

The objective function is maximized at 100 for

$$x = 8 \quad \text{and} \quad y = 3.$$

Chapter 8 TEST

1. Find the distance between the points $(-3, 5)$ and $(2, 1)$.

2. Find an equation of the circle whose center has coordinates $(-1, 2)$, with radius 3. Sketch its graph.

3. **Wireless Subscribers** The graph shows the number of wireless subscriber connections (that is, active devices, including smartphones, feature phones, tablets, and so on) in millions in the United States for the years 2012 through 2016.

   Use the midpoint formula to estimate the number of wireless subscriber connections in 2014.

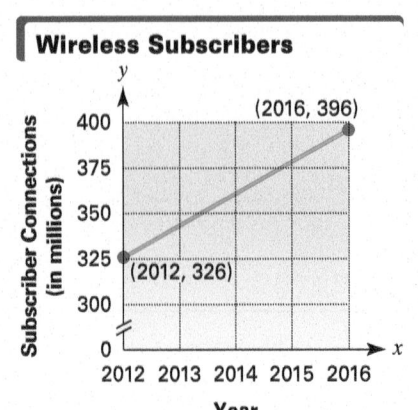

4. Find the x- and y-intercepts of the graph of

$$3x - 2y = 8,$$

and graph the equation.

5. Find the slope of the line passing through the points $(6, 4)$ and $(-1, 2)$.

6. Find the slope-intercept form of the equation of the line described.

(a) passing through the point $(-1, 3)$, with slope $-\frac{2}{5}$

(b) passing through the point $(-7, 2)$ and perpendicular to $y = 2x$

(c) the line through $(-2, 3)$ and $(6, -1)$

7. Which one of the following has a positive slope and a negative y-coordinate for its y-intercept?

A. 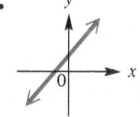 **B.**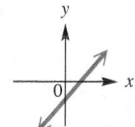

C. 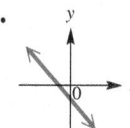 **D.**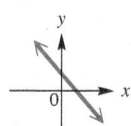

8. *Hospital Expenditures* The table shows hospital expenditures in the United States for the years 2011–2016.

Year	Expenditures (in billions of dollars)
2011	852
2012	903
2013	938
2014	978
2015	1033
2016	1082

Data from Centers for Medicare & Medicaid Services.

(a) Use $x = 0$ to represent 2011 and $x = 5$ to represent 2016 to find a linear function f that models expenditures (in billions) in year x.

(b) Use the function in part (a) to estimate expenditures in 2017.

9. *Internet Subscribers* In 2015, there were approximately 90 million high-speed Internet subscribers in the United States. In 2017, the number had increased to 95 million. What was the average rate of change in subscribers per year during this period? (Data from Leichtman Research Group.)

10. *Library Fines* It costs a borrower $0.05 per day for an overdue book, plus a flat $0.50 charge for all books borrowed. Let x represent the number of days the book is overdue, and let y represent the total fine to the tardy user.

(a) Write an equation in the form $y = mx + b$ for this situation.

(b) Give three ordered pairs with x-values of 1, 5, and 10 that satisfy the equation.

11. Write the slope-intercept form of the equation of the line by observing the graph.

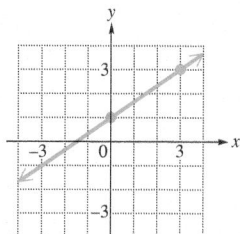

12. Find each of the following for $f(x) = -4x + 7$.

(a) $f(3)$ (b) the value of x for which $f(x) = -1$

13. *Calculator Production* If the cost to produce x units of calculators is

$$C(x) = 50x + 5000 \text{ dollars,}$$

and the revenue is

$$R(x) = 60x \text{ dollars,}$$

find the number of units of calculators that the manufacturer must produce in order to break even. What is the revenue at the break-even point?

14. Graph the following quadratic function.

$$f(x) = -(x + 3)^2 + 4$$

Give the equation of its axis of symmetry and the coordinates of its vertex.

15. *Dimensions of a Parking Lot* Kirkwood Community College wants to construct a rectangular parking lot on land bordered on one side by a highway. It has 320 ft of fencing with which to fence off the other three sides. What should be the dimensions of the lot if the enclosed area is to be a maximum? What is this maximum area?

16. Use a calculator to find an approximation of each of the following. The number of digits shown will vary depending on the model of calculator used.

    (a) $5.1^{4.7}$ (b) $e^{-1.85}$ (c) $\ln 23.56$

17. **Investment** Suppose that $12,000 is invested in an account that pays 2% annual interest, and it is left untouched for 3 years. How much will be in the account in each of the following cases?

    (a) Interest is compounded quarterly (four times per year).

    (b) Interest is compounded continuously.

18. **Decay of Plutonium-241** Suppose that the amount, in grams, of plutonium-241 present in a given sample is determined by the function

$$f(x) = 2.00e^{-0.053x},$$

    where x is measured in years. Find the amount present in the sample after the given number of years.

    (a) 4   (b) 10   (c) 20

    (d) What was the initial amount present?

Solve each system by using elimination, substitution, or a combination of the two methods.

19. $2x + 3y = 2$
    $3x - 4y = 20$

20. $x + 3y = 7$
    $-2x - 6y = -7$

Solve each problem by using a system of equations.

21. **Ticket Prices** In 2017, on the secondary market, the average prices for Major League Baseball tickets indicated that 2 Arizona Diamondbacks tickets and 5 Baltimore Orioles tickets would cost $493. Three Arizona tickets and 4 Baltimore tickets would cost $477. What was the average ticket price for each team? (Data from www.barrystickets.com)

22. **Candy Mixture** Sweet's Candy Store is offering a special mix for Valentine's Day. Ms. Sweet will mix some $6-per-lb nuts with some $3-per-lb chocolate candy to get 100 lb of mix, which she will sell at $3.90 per lb. How many pounds of each should she use?

23. Graph the solution set of the system of inequalities.

$$x + y \le 6$$
$$2x - y \ge 3$$

24. **Profit from Farm Animals** Jan raises only pigs and geese. She wants to raise no more than 16 animals, with no more than 12 geese. She spends $50 to raise a pig and $20 to raise a goose. She has $500 available for this purpose. Find the maximum profit she can make if she makes a profit of $80 per goose and $40 per pig, and determine how many pigs and geese she should raise to achieve this maximum.

9 Geometry

Video game developers and programmers of flight simulators, architectural design tools, and other virtual reality applications use software programs to render objects in their virtual worlds in three dimensions. The software can track the location of objects in a "world" and then show only the ones in the part of the world where the user is looking.

Some of the concepts from this chapter are foundational to the rendering process involved in such applications.

9.1   POINTS, LINES, PLANES, AND ANGLES

OBJECTIVES

1 Identify relationships among points, lines, and planes.

2 Classify and calculate angles.

The Geometry of Euclid

Let no one unversed in geometry enter here.
—Motto over the door of Plato's Academy

To the ancient Greeks, mathematics meant geometry above all—a rigid kind of geometry from a modern-day point of view. The Greeks studied the properties of figures identical in shape and size (congruent figures) as well as figures identical in shape but not necessarily in size (similar figures). They absorbed ideas about area and volume from the Egyptians and Babylonians and established general formulas. The Greeks were the first to insist that statements in geometry be given rigorous proof.

The most basic ideas of geometry are **point, line,** and **plane.** In fact, it is not really possible to define them with other words. Euclid defined a point as "that which has no part" and a line as "that which has breadthless length." These definitions are vague. Based on our experience, however, we know what Euclid meant. The drawings that we use for points are dots. Lines have properties of no thickness and no width, and they extend indefinitely in two directions.

Euclid's definition of a plane, "a surface which lies evenly with the straight lines on itself," is represented by a flat surface, such as a tabletop or a page in a book.

Points, Lines, and Planes

Certain universally accepted conventions and symbols are used to represent points, lines, and planes.

- A point is usually denoted by a single capital letter.

- A line may be named by two capital letters representing points that lie on the line, or by a single lowercase letter (sometimes subscripted to distinguish distinct lines).

- A plane may be named by three capital letters representing three points that lie in the plane, or by a single Greek letter, such as α (alpha), β (beta), or γ (gamma).

Figure 1 depicts a plane that may be represented either as α or as plane ADE. Contained in the plane are lines AD and DE (or, equivalently, lines DA and ED, respectively), which are also labeled ℓ_1 and ℓ_2.

Selecting any point on a line divides the line into three parts: the point itself, and two **half-lines,** one on each side of the point. For example, point A divides the line shown in **Figure 2** into three parts: A itself and two half-lines. Point A belongs to neither half-line. As the figure suggests, each half-line extends indefinitely in the direction opposite the other half-line.

Euclid's *Elements* as translated by Billingsley appeared in 1570 and was the first English-language translation of the most influential geometry text ever written.

Unfortunately, no copy of *Elements* exists that dates back to the time of Euclid (circa 300 B.C.), and most current translations are based on a revision of the work prepared by Theon of Alexandria.

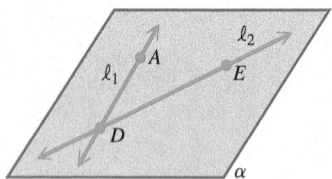

Figure 1

Figure 2

Figure 3

Including an initial point with a half-line gives a **ray.** A ray is named with two letters: one for the initial point of the ray, and one for another point contained in the half-line. In **Figure 3** ray AB has initial point A and extends in the direction of B. On the other hand, ray BA has B as its initial point and extends in the direction of A.

Given any three points that are not
in a straight line, a plane can be passed
through the points. That is why **camera
tripods** have three legs—no matter how
irregular the surface, the tips of the three
legs determine a plane. A camera support
with four legs would wobble unless all
four legs were carefully extended just the
right amount.

A **line segment** includes both endpoints and is named by its endpoints. **Figure 3** shows line segment AB, which may also be designated as line segment BA.

Table 1 shows these figures along with the symbols used to represent them.

Table 1

Name	Figure	Symbol
Line AB or line BA		$\overleftrightarrow{AB}$ or $\overleftrightarrow{BA}$
Half-line AB		$\overset{\circ}{\overrightarrow{AB}}$
Half-line BA		$\overset{\circ}{\overleftarrow{BA}}$
Ray AB		$\overset{\bullet}{\overrightarrow{AB}}$
Ray BA		$\overset{\bullet}{\overleftarrow{BA}}$
Segment AB or segment BA		$\overset{\bullet\,\bullet}{AB}$ or $\overset{\bullet\,\bullet}{BA}$

For a line, the symbol above the two letters shows two arrowheads, indicating that the line extends indefinitely in both directions. For half-lines and rays, only one arrowhead is used because these extend in only one direction. An open circle is used for a half-line to show that the endpoint is not included, while a solid circle is used for a ray to indicate the inclusion of the endpoint. Since a segment includes both endpoints and does not extend in either direction, solid circles are used to indicate endpoints of line segments.

The geometric definitions of "parallel" and "intersecting" apply to two or more lines or planes. (See **Figure 4.**) **Parallel lines** lie in the same plane and never meet. However, **intersecting lines** do meet.

If two distinct lines intersect, they intersect in one and only one point.

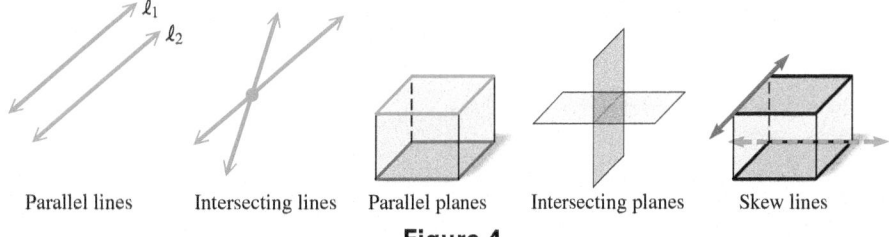

Parallel lines   Intersecting lines   Parallel planes   Intersecting planes   Skew lines

Figure 4

We use the symbol $\parallel$ to denote parallelism. If ℓ_1 and ℓ_2 are parallel lines, as in **Figure 4,** then this may be indicated as

$$\ell_1 \parallel \ell_2.$$

Parallel planes also never meet. Two distinct **intersecting planes** form a straight line, the one and only line they have in common. **Skew lines** do not lie in a common plane, so they are neither parallel nor intersecting.

Angles are the key to the study of **geodesy,** the measurement of distances on the earth's surface.

Angles

An **angle** is the union of two rays that have a common endpoint. The angle is formed by points on the rays themselves, and no other points. In **Figure 5,** point X is *not* a point on the angle. (It is said to be in the *interior* of the angle.)

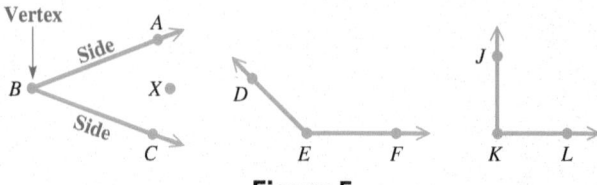

Figure 5

The rays forming an angle are its **sides.** The common endpoint of the rays is the **vertex** of the angle. There are two standard ways of naming angles using letters. If no confusion will result, an angle can be named with the letter marking its vertex. Using this method, the angles in **Figure 5** can be named, respectively, angle B, angle E, and angle K.

Angles also can be named with three letters: the first letter names a point on one side of the angle; the middle letter names the vertex; the third names a point on the other side of the angle. In this system, the angles in the figure can be named angle ABC, angle DEF, and angle JKL. The symbol for representing an angle is ∡. Rather than writing "angle ABC," we may write "∡ABC."

An angle can be associated with an amount of rotation. For example, in **Figure 6(a),** we let $\overrightarrow{BA}$ first coincide with $\overrightarrow{BC}$—as though they were the same ray. We then rotate $\overrightarrow{BA}$ (the endpoint remains fixed) in a counterclockwise direction to form ∡ABC.

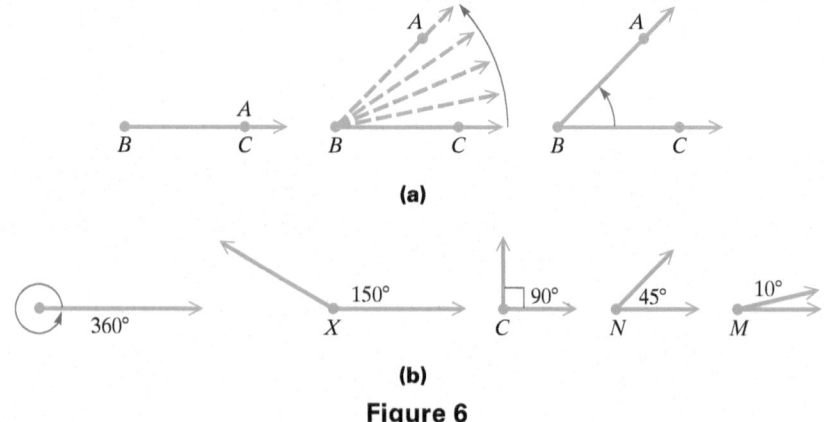

(a)

(b)

Figure 6

360°

Why 360? The use of the number 360 goes back to the Babylonian culture. There are several theories regarding why 360 was chosen for the number of degrees in a complete rotation around a circle. One says that 360 was chosen because it is close to the number of days in a year and because it is conveniently divisible by 2, 3, 4, 5, 6, 8, 9, 10, 12, and other numbers.

Angles are measured by the amount of rotation, using a system that dates back to the Babylonians. Babylonian astronomers chose the number 360 to represent a complete rotation of a ray back onto itself. **One degree,** written 1°, is defined to be $\frac{1}{360}$ of a complete rotation.

Angles are classified and named with reference to their degree measures. Refer to **Figure 6(b).**

- An angle whose measure is between 0° and 90° is an **acute angle.** (See ∡M and ∡N.)

- An angle that measures 90° is a **right angle.** The squared symbol ⌐ at the vertex denotes a right angle. (See ∡C.)

- An angle that measures more than 90° but less than 180° is an **obtuse angle.** (See ∡X.)

- An angle that measures 180° is a **straight angle.** Its sides from a straight line.

A **protractor** is a tool used to measure angles. **Figure 7** shows a protractor measuring an angle whose measure is 135°.

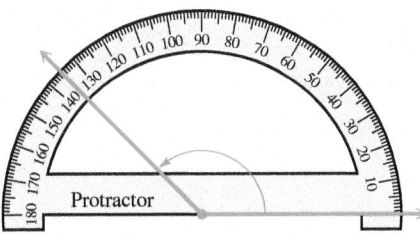

To use a protractor, position the hole (or dot) of the protractor on the vertex of the angle. With the 0-degree measure on the protractor placed on one side of the angle, the other side will show the degree measure of the angle.

Figure 7

When two lines intersect to form right angles, they are **perpendicular lines.** In **Figure 8,** the sides of ∡*NMP* have been extended to form another angle, ∡*RMQ*. The pair ∡*NMP* and ∡*RMQ* are called **vertical angles.** Another pair of vertical angles have been formed at the same time. They are ∡*NMQ* and ∡*PMR*.

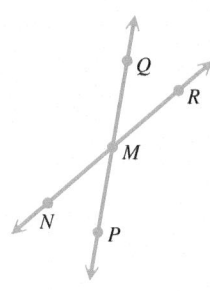

Figure 8

PROPERTY OF VERTICAL ANGLES

Vertical angles have equal measures.

EXAMPLE 1 Finding Angle Measures

Find the measure of each marked angle in the given figure.

(a) Figure 9   **(b) Figure 10**

Solution

(a) Because the marked angles are vertical angles, they have equal measures.

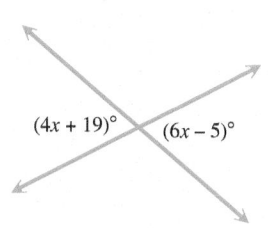

Figure 9

$$4x + 19 = 6x - 5 \qquad \text{Set measures equal.}$$
$$4x + 19 - 4x = 6x - 5 - 4x \qquad \text{Subtract } 4x.$$
$$19 = 2x - 5 \qquad \text{Combine like terms.}$$
$$19 + 5 = 2x - 5 + 5 \qquad \text{Add 5.}$$
$$24 = 2x \qquad \text{Add.}$$

Don't stop here. → $12 = x$ Divide by 2.

Since $x = 12$, one angle has measure $4(12) + 19 = 67$ degrees. The other has the same measure, because $6(12) - 5 = 67$ as well. Each angle measures 67°.

(b) The marked angles form a straight angle, so their measures must add to 180°.

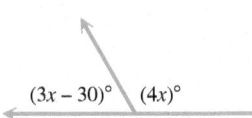

Figure 10

$$(3x - 30) + 4x = 180 \qquad \text{The angle sum is 180.}$$
$$7x - 30 = 180 \qquad \text{Combine like terms.}$$
$$7x - 30 + 30 = 180 + 30 \qquad \text{Add 30.}$$
$$7x = 210 \qquad \text{Add.}$$

Don't stop here. → $x = 30$ Divide by 7.

To find the measures of the angles, replace x with 30 in the two expressions.

$$3x - 30 = 3(30) - 30 = 90 - 30 = 60$$
$$4x = 4(30) = 120$$

The two angle measures are 60° and 120°.

If the sum of the measures of two acute angles is 90°, the angles are said to be **complementary,** and each is the *complement* of the other. For example, angles measuring 40° and 50° are complementary angles, because

$$40° + 50° = 90°.$$

If two angles have a sum of 180°, they are **supplementary.** The *supplement* of an angle whose measure is 40° is an angle whose measure is 140°, because

$$40° + 140° = 180°.$$

> *If x represents the degree measure of an angle, 90 − x represents the measure of its complement, and 180 − x represents the measure of its supplement.*

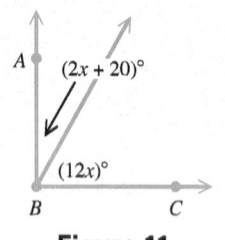

Figure 11

EXAMPLE 2   Finding Angle Measures

Find the measures of the angles in **Figure 11,** given that $\angle ABC$ is a right angle.

Solution

The sum of the measures of the two acute angles is 90° (that is, they are complementary), because they form a right angle.

$$(2x + 20) + 12x = 90 \qquad \text{The angle measures sum to 90°.}$$
$$14x + 20 = 90 \qquad \text{Combine like terms.}$$
$$14x = 70 \qquad \text{Subtract 20.}$$
$$x = 5 \qquad \text{Divide by 14.}$$

The value of x is 5. Therefore, replace x with 5 in the two expressions.

$$2x + 20 = 2(5) + 20 = 30$$
$$12x = 12(5) = 60$$

The measures of the two angles are 30° and 60°.

EXAMPLE 3   Using Complementary and Supplementary Angles

The supplement of an angle measures 10° more than three times its complement. Find the measure of the angle.

Solution

Let $x =$ the degree measure of the angle.

Then $180 - x =$ the degree measure of its supplement,

and $90 - x =$ the degree measure of its complement.

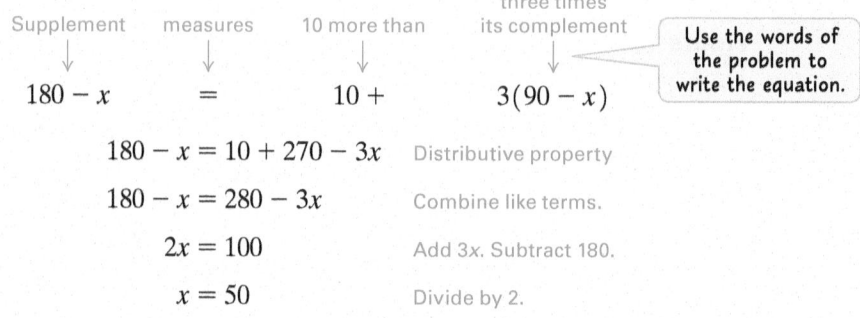

$$180 - x = 10 + 270 - 3x \qquad \text{Distributive property}$$
$$180 - x = 280 - 3x \qquad \text{Combine like terms.}$$
$$2x = 100 \qquad \text{Add 3x. Subtract 180.}$$
$$x = 50 \qquad \text{Divide by 2.}$$

The angle measures 50°. Because its supplement (130°) is 10° more than three times its complement (40°) (that is, $130 = 10 + 3(40)$ is true), the answer checks.

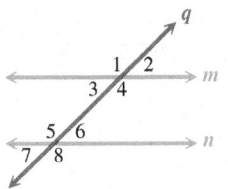

Figure 12

A set of parallel lines with equidistant spacing intersects an identical set, but at a small angle. The result is a **moiré pattern,** named after the fabric *moiré* ("watered") *silk.* Moiré patterns are related to **periodic functions,** which describe regular recurring phenomena (wave patterns such as heartbeats or business cycles). Moirés thus apply to the study of electromagnetic, sound, and water waves, to crystal structure, and to other wave phenomena.

Figure 12 shows parallel lines m and n. When a line q intersects two parallel lines, q is a **transversal.** In **Figure 12,** the transversal intersecting the parallel lines forms eight angles, indicated by numbers. Angles 1 through 8 in the figure possess some special properties regarding their degree measures, as shown in **Table 2.**

Table 2

Name	Figure	Rule
Alternate interior angles	5 4 *q m n* (also 3 and 6)	Angle measures are equal.
Alternate exterior angles	1 *q m n* 8 (also 2 and 7)	Angle measures are equal.
Interior angles on same side of transversal	*q m* 4 6 *n* (also 3 and 5)	Angle measures add to 180°.
Corresponding angles	*q* 2 *m* 6 *n* (also 1 and 5, 3 and 7, 4 and 8)	Angle measures are equal.

The converses of the above also are true. That is, if alternate interior angles are equal, then the lines are parallel. Similar results are valid for alternate exterior angles, interior angles on the same side of a transversal, and corresponding angles.

EXAMPLE 4   **Finding Angle Measures**

Find the measure of each marked angle in **Figure 13,** given that lines m and n are parallel.

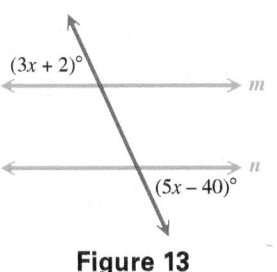

Figure 13

Solution

The marked angles are alternate exterior angles, which are equal.

$$3x + 2 = 5x - 40 \qquad \text{Set angle measures equal.}$$
$$42 = 2x \qquad \text{Subtract 3x. Add 40.}$$

Don't stop here.

$$21 = x \qquad \text{Divide by 2.}$$

Thus, $3x + 2 = 3 \cdot 21 + 2 = 65$ and $5x - 40 = 5 \cdot 21 - 40 = 65$.

So both angles measure 65°.

9.1 EXERCISES

CONCEPT CHECK *Fill in each blank with the correct response.*

1. The sum of the measures of two complementary angles is _____ degrees.

2. The sum of the measures of two supplementary angles is _____ degrees.

3. The measures of two vertical angles are _____. (equal/not equal)

4. The measures of _____ right angles add up to the measure of a straight angle.

CONCEPT CHECK *Write* true *or* false *for each statement.*

5. A line segment has two endpoints.

6. A ray has one endpoint.

7. If A and B are distinct points on a line, then ray AB and ray BA represent the same set of points.

8. If two lines intersect, they lie in the same plane.

9. If two lines are parallel, they lie in the same plane.

10. If two lines are neither parallel nor intersecting, they cannot lie in the same plane.

11. If two lines do not intersect, they must be parallel.

12. There is no angle that is its own complement.

13. Segment AB and segment BA represent the same set of points.

14. There is no angle whose complement and supplement are the same.

15. There is no angle that is its own supplement.

16. The use of the degree as a unit of measure of an angle goes back to the Egyptians.

Exercises 17–24 name portions of the line shown. For each exercise, **(a)** *give the symbol that represents the portion of the line named, and* **(b)** *draw a figure showing just the portion named, including all labeled points.*

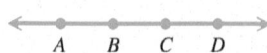

17. line segment AB

18. line segment AD

19. ray CB

20. ray BC

21. half-line BC

22. half-line AD

23. ray BA

24. line segment CA

Lines, rays, half-lines, and segments may be considered sets of points. Match each symbol in Group I with the symbol in Group II that names the same set of points, based on the figure.

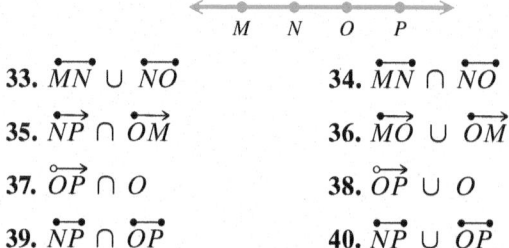

I		II	
25. $\overrightarrow{PQ}$	26. $\overrightarrow{QR}$	**A.** $\overrightarrow{QS}$	**B.** $\overrightarrow{RQ}$
27. $\overleftrightarrow{QR}$	28. $\overleftrightarrow{PQ}$	**C.** $\overleftrightarrow{SR}$	**D.** $\overrightarrow{QS}$
29. $\overleftrightarrow{RP}$	30. $\overleftrightarrow{SQ}$	**E.** $\overrightarrow{SP}$	**F.** $\overrightarrow{QP}$
31. $\overrightarrow{PS}$	32. $\overleftrightarrow{PS}$	**G.** $\overleftrightarrow{RS}$	**H.** none of these

The **intersection** *(symbolized ∩) of two sets is composed of all elements common to both sets, and the* **union** *(symbolized ∪) of two sets is composed of all elements found in at least one of the two sets.*

Based on the figure below, specify each of the sets given in Exercises 33–40 in a simpler way.

33. $\overleftrightarrow{MN} \cup \overrightarrow{NO}$

34. $\overleftrightarrow{MN} \cap \overrightarrow{NO}$

35. $\overrightarrow{NP} \cap \overrightarrow{OM}$

36. $\overrightarrow{MO} \cup \overrightarrow{OM}$

37. $\overrightarrow{OP} \cap O$

38. $\overrightarrow{OP} \cup O$

39. $\overrightarrow{NP} \cap \overrightarrow{OP}$

40. $\overleftrightarrow{NP} \cup \overrightarrow{OP}$

Give the measure of the complement of each angle.

41. $38°$ 42. $22°$ 43. $89°$ 44. $45°$

45. $x°$ 46. $(90 - x)°$ 47. $(x + 80)°$ 48. $(70 - x)°$

Give the measure of the supplement of each angle.

49. $132°$ 50. $105°$ 51. $26°$

52. $90°$ 53. $y°$ 54. $(180 - y)°$

55. $(90 - x)°$ 56. $(170 - y)°$

Complementary and Supplementary Angles *Solve each problem.*

57. If the supplement of an angle is 40° more than twice its complement, what is the measure of the angle?

58. If an angle measures 15° less than twice its complement, what is the measure of the angle?

59. Half the supplement of an angle is 12° less than twice the complement of the angle. Find the measure of the angle.

60. The supplement of an angle measures 25° more than twice its complement. Find the measure of the angle.

Name all pairs of vertical angles in each figure.

61.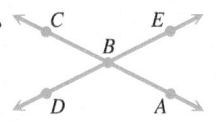

62.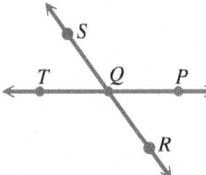

63. In **Exercise 61**, if ∡*ABE* has a measure of 52°, find the measures of the following angles.

(a) ∡*CBD*   (b) ∡*CBE*

64. In **Exercise 62**, if ∡*SQP* has a measure of 126°, find the measures of the following angles.

(a) ∡*TQR*   (b) ∡*PQR*

Find the measure of each marked angle.

65.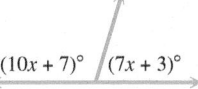

66.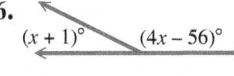

67.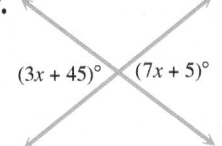

68.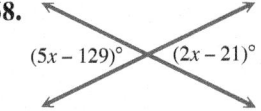

69.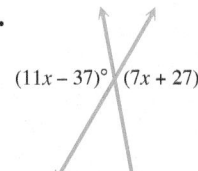

70.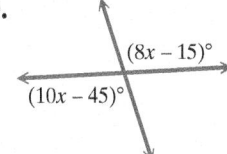

71.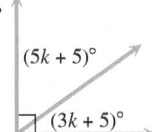

72.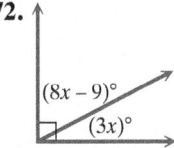

In Exercises 73–76, assume that lines m and n are parallel, and find the measure of each marked angle.

73.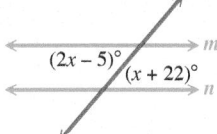

74.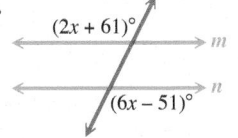

75.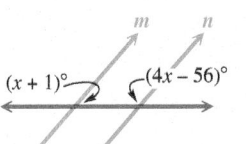

76.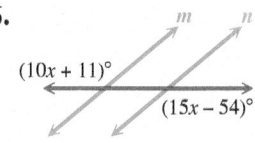

77. The sketch shows parallel lines m and n cut by a transversal q. Complete the steps to prove that alternate exterior angles have the same measure.

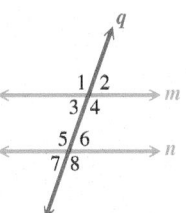

(a) Measure of ∡2 = measure of ∡ _____, since they are vertical angles.

(b) Measure of ∡3 = measure of ∡ _____, since they are alternate interior angles.

(c) Measure of ∡6 = measure of ∡ _____, since they are vertical angles.

(d) By the results of parts (a)–(c), the measure of ∡2 must equal the measure of ∡ _____, showing that alternate _____ angles have equal measures.

78. Use the sketch to find the measure of each numbered angle. Assume that $m \parallel n$.

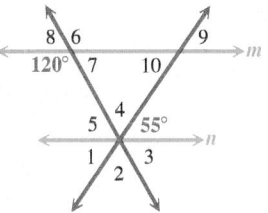

79. Complete these steps in the proof that vertical angles have equal measures. In this exercise, m(∡*x*) means "the measure of the angle *x*." Use the figure below.

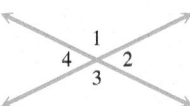

(a) m(∡1) + m(∡2) = _____ °

(b) m(∡2) + m(∡3) = _____ °

(c) Subtract the equation in part (b) from the equation in part (a) to get [m(∡1) + m(∡2)] − [m(∡2) + m(∡3)] = _____ ° − _____ °.

(d) m(∡1) + m(∡2) − m(∡2) − m(∡3) = _____ °

(e) m(∡1) − m(∡3) = _____ °

(f) m(∡1) = m(∡ _____)

9.2 CURVES, POLYGONS, CIRCLES, AND GEOMETRIC CONSTRUCTIONS

Curves

The term *curve* is used for describing figures in the plane.

SIMPLE CURVE AND CLOSED CURVE

A **simple curve** can be drawn without lifting the pencil from the paper and without passing through any point twice.

A **closed curve** has its starting and ending points the same and is drawn without lifting the pencil from the paper.

Some examples are shown in **Figure 14.**

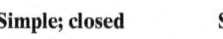

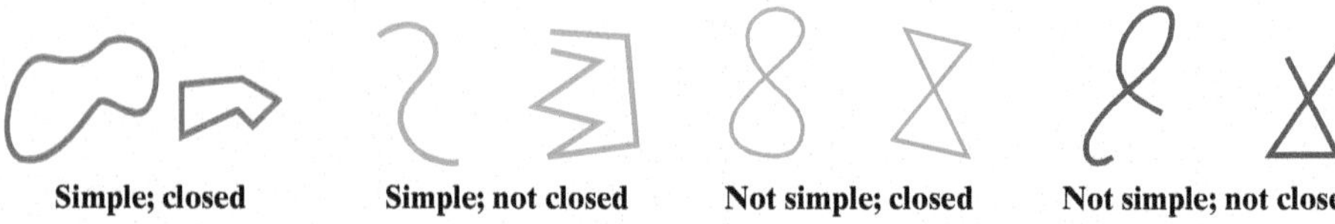

Simple; closed   Simple; not closed   Not simple; closed   Not simple; not closed

Figure 14

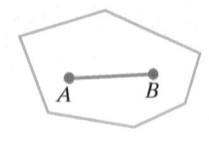

Convex

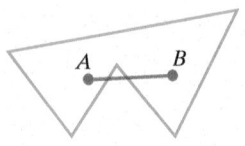

Not convex

Figure 15

A simple closed figure is said to be **convex** if, for any two points A and B inside the figure, the line segment AB (that is, $\overrightarrow{AB}$) is always completely inside the figure. **Figure 15** shows a convex figure and one that is not convex.

Among the most common types of curves in mathematics are those that are both simple and closed, and perhaps the most important of these are *polygons*. A **polygon** is a simple closed curve made up only of straight line segments. The line segments are the *sides*, and the points at which the sides meet are the *vertices* (singular: *vertex*).

Polygons are classified according to the number of line segments used as sides. **Table 3** gives the special names. *In general, if a polygon has* **n** *sides, and no particular value of* **n** *is specified, it is called an* **n-gon**.

Some examples of polygons are shown in **Figure 16**. A polygon may or may not be convex. Polygons with all sides equal and all angles equal are **regular polygons.**

Table 3 Classification of Polygons According to Number of Sides

Number of Sides	Name
3	triangle
4	quadrilateral
5	pentagon
6	hexagon
7	heptagon
8	octagon
9	nonagon
10	decagon

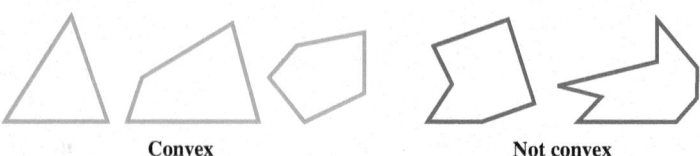

Convex   Not convex

Polygons are simple closed curves made up of straight line segments.

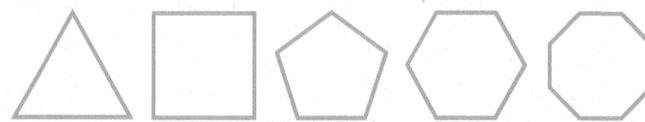

Regular polygons have equal sides and equal angles.

Figure 16

The puzzle game above comes from China, where it has been a popular amusement for centuries. Each figure above is a **tangram.** Any tangram is composed of the same set of seven tans (the pieces making up the figure).

Mathematicians have described various properties of tangrams. Each tan is convex, but only 13 convex tangrams are possible. All others, like the figure on the left, are not convex.

Triangles and Quadrilaterals

Triangles (three-sided polygons) are classified by measures of angles as well as by number of equal sides, as shown in the following box. (Notice that tick marks are used in the bottom three figures to show how side lengths are related.)

TYPES OF TRIANGLES

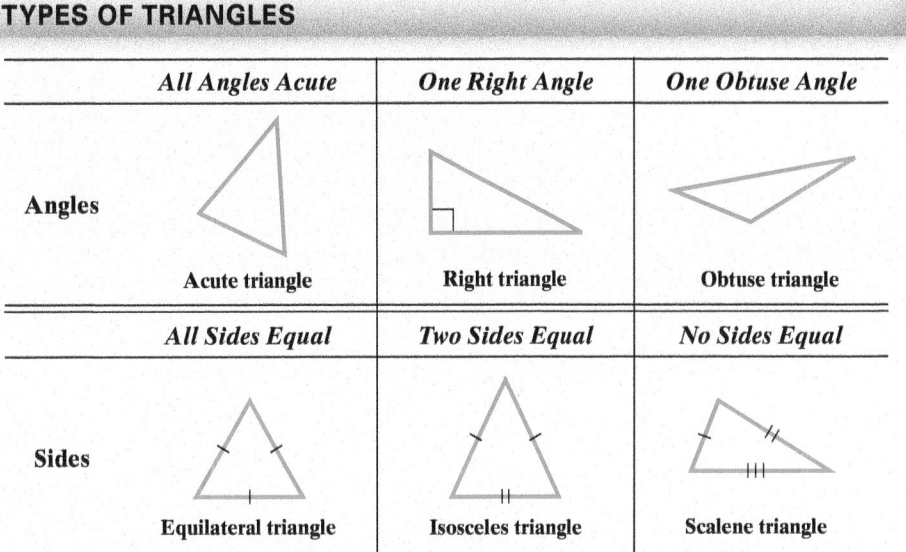

	All Angles Acute	*One Right Angle*	*One Obtuse Angle*
Angles	Acute triangle	Right triangle	Obtuse triangle
	All Sides Equal	*Two Sides Equal*	*No Sides Equal*
Sides	Equilateral triangle	Isosceles triangle	Scalene triangle

   Quadrilaterals (four-sided polygons) are classified by sides and angles. An important distinction involving quadrilaterals is whether one or more pairs of sides are parallel.

TYPES OF QUADRILATERALS

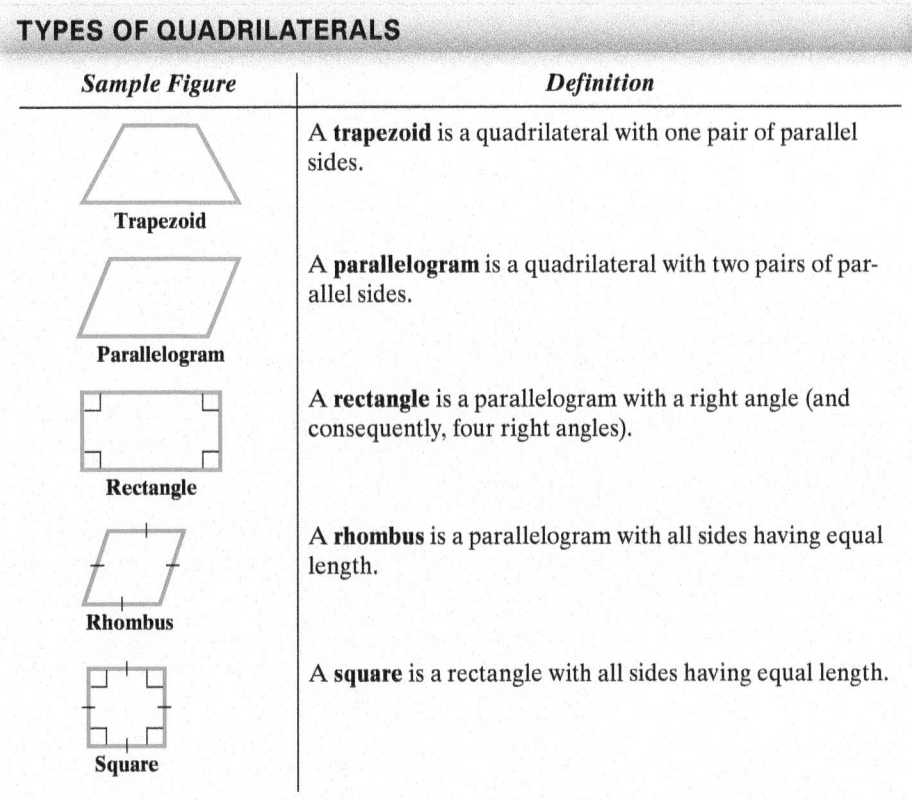

Sample Figure	*Definition*
Trapezoid	A **trapezoid** is a quadrilateral with one pair of parallel sides.
Parallelogram	A **parallelogram** is a quadrilateral with two pairs of parallel sides.
Rectangle	A **rectangle** is a parallelogram with a right angle (and consequently, four right angles).
Rhombus	A **rhombus** is a parallelogram with all sides having equal length.
Square	A **square** is a rectangle with all sides having equal length.

An important property of triangles that was first proved by the Greek geometers deals with the sum of the measures of the angles of any triangle.

ANGLE SUM OF A TRIANGLE

The sum of the measures of the interior angles of any triangle is 180°.

Although it is not an actual proof, a rather convincing argument for the truth of this statement can be given using any size triangle cut from a piece of paper. Tear each corner from the triangle, as suggested in **Figure 17(a)**. You should be able to rearrange the pieces so that the three angles form a straight angle, as shown in **Figure 17(b)**.

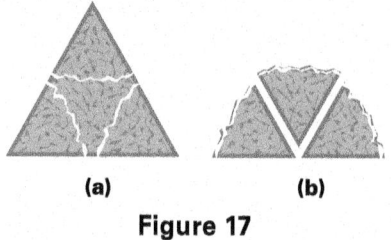

(a)   (b)

Figure 17

EXAMPLE 1   **Finding Angle Measures in a Triangle**

Find the measure of each angle in the triangle of **Figure 18**.

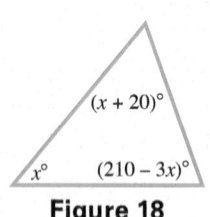

Figure 18

Solution

By the angle sum relationship, the three angle measures must add up to 180°.

$$x + (x + 20) + (210 - 3x) = 180 \qquad \text{Sum is 180.}$$

$$-x + 230 = 180 \qquad \text{Combine like terms.}$$

$$-x = -50 \qquad \text{Subtract 230.}$$

There are two more values to find. $\qquad x = 50 \qquad$ Divide by −1.

Because $x = 50$,

$$x + 20 = 50 + 20 = 70 \quad \text{and} \quad 210 - 3x = 210 - 3(50) = 60.$$

The three angles measure 50°, 70°, and 60°. Because $50° + 70° + 60° = 180°$, the answers satisfy the angle sum relationship.

In the triangle shown in **Figure 19**, angles 1, 2, and 3 are **interior angles**, while angles 4, 5, and 6 are **exterior angles** of the triangle.

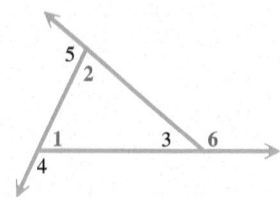

Figure 19

Using the fact that the sum of the interior angle measures of any triangle is 180°, and the fact that a straight angle also measures 180°, the following property may be deduced.

EXTERIOR ANGLE MEASURE

The measure of an exterior angle of a triangle is equal to the sum of the measures of the two opposite interior angles.

In **Figure 19,** the measure of angle 6 is equal to the sum of the measures of angles 1 and 2. Two other such statements can be made.

EXAMPLE 2   Finding Interior and Exterior Angle Measures

Find the measures of interior angles A, B, and C of the triangle in **Figure 20,** and the measure of exterior angle BCD.

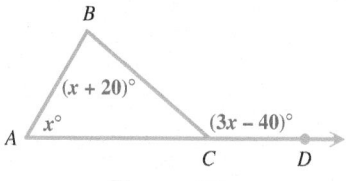

Figure 20

Solution

By the property concerning exterior angles, the sum of the measures of interior angles A and B must equal the measure of angle BCD.

$$x + (x + 20) = 3x - 40$$

$$2x + 20 = 3x - 40 \quad \text{Combine like terms.}$$

$$-x = -60 \quad \text{Subtract } 3x. \text{ Subtract } 20.$$

$$x = 60 \quad \text{Divide by } -1.$$

Because the value of x is 60, we have the following.

> m(angle A) denotes the measure of angle A.

$$\text{m}(\text{Interior angle } A) = 60°$$

$$\text{m}(\text{Interior angle } B) = (60 + 20)° = 80°$$

$$\text{m}(\text{Interior angle } C) = 180° - (60° + 80°) = 40°$$

$$\text{m}(\text{Exterior angle } BCD) = [3(60) - 40]° = 140°$$

Circles

A circle is a simple closed curve defined as follows.

CIRCLE

A **circle** is a set of points in a plane, each of which is the same distance from a fixed point.

A circle may be physically constructed with a compass, where the spike leg remains fixed and the other leg swings around to construct the circle. A string may also be used to draw a circle. For example, loop a piece of chalk on one end of a piece of string. Hold the other end in a fixed position on a chalkboard, and pull the string taut. Then swing the chalk end around to draw a circle.

KEY TERMS INVOLVING CIRCLES

Sample Figure	Definition
Center: O Radius: $\overrightarrow{OP}$ Chord: $\overleftrightarrow{PM}$ Diameter: $\overleftrightarrow{MN}$ Arc: $\overparen{NP}$ Semicircle: $\overparen{MN}$ Minor arc: $\overparen{PM}$ Major arc: $\overparen{MNP}$	The **center** of a circle is the point from which all points on the circle are an equal distance. A **radius** (plural *radii*) of a circle is a segment having the center as one endpoint and a point on the circle as its other endpoint. A **chord** is a segment with both endpoints lying on a circle. A **diameter** is a chord passing through the center. An **arc** of a circle consists of two endpoints on a circle and all points of the circle between the endpoints. A **semicircle** is an arc sharing endpoints with a diameter of a circle. A **minor arc** is an arc less than a semicircle. A **major arc** is more than a semicircle. (*Note:* An arc named with two points is understood to be a minor arc or semicircle. A third point is used to indicate a major arc or to specify a semicircle.)
Tangent line: $\overleftrightarrow{RN}$ Secant line: $\overleftrightarrow{MP}$ Inscribed angle: $\angle PQM$ Intercepted arc: $\overparen{PM}$	A line **tangent** to a circle touches (intersects) the circle in only one point, called the point of tangency. A **secant** line is a line that intersects a circle in two points. An **inscribed angle** is formed by two chords of a circle with a common endpoint (the vertex of the angle). An **intercepted arc** is an arc between (and including) the points where the sides of an inscribed angle intersect the circle.

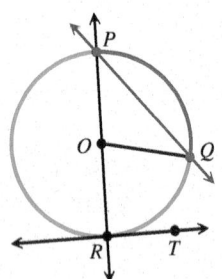

Figure 21

EXAMPLE 3 **Identifying Features of a Circle**

Identify all examples of each feature appearing in **Figure 21.**

(a) radius   **(b)** chord   **(c)** arc

(d) tangent line   **(e)** secant line   **(f)** inscribed angle

Solution

(a) The radii in **Figure 21** are $\overrightarrow{OP}$, $\overrightarrow{OQ}$, and $\overrightarrow{OR}$.

(b) The chords appearing in the figure are $\overleftrightarrow{PQ}$ and $\overleftrightarrow{PR}$. ($\overleftrightarrow{PR}$ is also a diameter.)

(c) The arcs appearing in the figure are $\overparen{PQ}$, $\overparen{PRQ}$, $\overparen{PR}$ (two semicircular arcs), $\overparen{QR}$, and $\overparen{QPR}$.

(d) $\overleftrightarrow{RT}$ is tangent to the circle at R.

(e) $\overleftrightarrow{PQ}$ and $\overleftrightarrow{PR}$ are secant lines.

(f) $\angle RPQ$ is inscribed in the circle.

Thales is credited with the first proof of the **inscribed angle theorem.** Legend records that he studied for a time in Egypt and then introduced geometry to Greece, where he attempted to apply the principles of Greek logic to his newly learned subject.

The Greeks were the first to insist that all propositions, or **theorems,** about geometry be given rigorous proofs before being accepted. One of the theorems receiving such a proof was this one.

INSCRIBED ANGLE

Any angle inscribed in a circle has degree measure half of that of its intercepted arc.

Figure 22(a) states this theorem symbolically. For example, if *arc PR* measures $80°$, then inscribed angle PQR measures $\left(\frac{80}{2}\right)° = 40°$. A special case of this theorem is indicated in **Figure 22(b).** *Any angle inscribed in a semicircle is a right angle.*

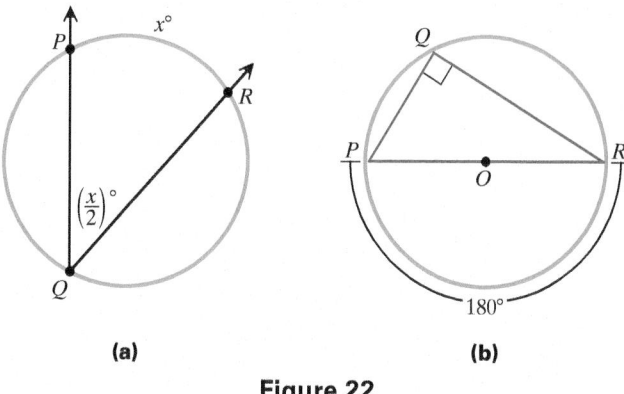

(a)   **(b)**

Figure 22

Geometric Constructions

To the Greeks, geometry was the highest expression of mathematics. It is no surprise then that they required **geometric constructions** to have abstract beauty. This was achieved by allowing only a minimal set of tools: a compass (also called compasses, or a pair of compasses) for drawing circles and arcs of circles, and an unmarked straightedge for drawing straight line segments.

Here are three basic constructions. Their justifications are based on the *congruence properties* of the next section.

Perpendicular Bisector **Construct the perpendicular bisector of a given line segment.**

Let the segment have endpoints A and B. Adjust the compass for any radius greater than half the length of AB. Place the point of the compass at A and draw an arc; then draw another arc of the same size at B. The line drawn through the points of intersection of these two arcs is the desired perpendicular bisector. See **Figure 23.**

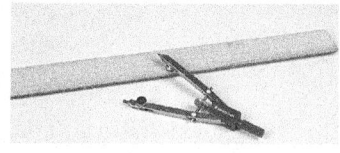

A **compass** may be used to:
1. Swing a circular arc with given center and radius.
2. Reproduce a given length. (The Greeks used "collapsing" compasses, which would not allow distances to be copied directly. But Proposition II from Book I of Euclid's *Elements* provides a construction for transferring distances using a collapsing compass. Thus, for our purposes, we will use a "fixed" compass.)

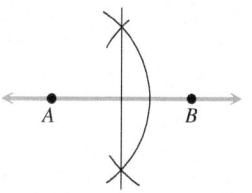

Figure 23

In his first effort as a director, Mel Gibson starred in the 1993 movie *The Man Without a Face*. As disfigured former teacher Justin McLeod, he tutors teenager Chuck Norstadt (portrayed by Nick Stahl). McLeod explains to Norstadt how to find the center of a circle using any three points on the circle as he sketches the diagram on a windowpane. His explanation is based on the fact that the **perpendicular bisector** of any chord of a circle passes through the center of the circle. See the figure.

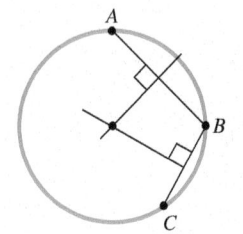

Perpendicular to a Line **Construct a perpendicular from a point off a line to the line.**

1. Let *A* be the point, *r* the line. Place the point of the compass at *A* and draw an arc, cutting *r* in two points.

2. Swing arcs of equal radius from each of the two points on *r* that were constructed in (1). The line drawn through the intersection of the two arcs and point *A* is perpendicular to *r*. See **Figure 24**.

Figure 24

Copied Angle **Copy an angle.**

1. In order to copy an angle *ABC* on line *r*, place the point of the compass at *B* and draw an arc. Then place the point of the compass on *r'* at some point *P* and draw the same arc, as in **Figure 25**.

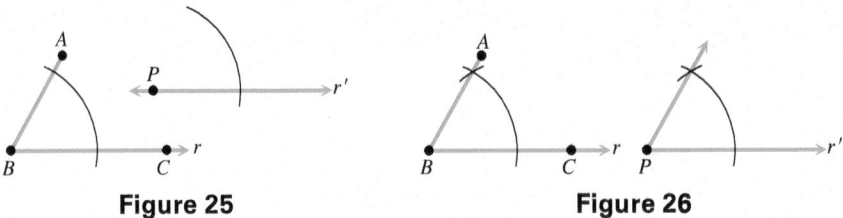

Figure 25   **Figure 26**

2. Measure, with your compass, the distance between the points where the arc intersects the angle, and transfer this distance, as shown in **Figure 26**. Use a straightedge to join *P* to the point of intersection. The angle is now copied.

There are other basic constructions that can be found in texts on plane geometry.

 | 9.2 | **EXERCISES**

CONCEPT CHECK *Fill in each blank with the correct response.*

1. A segment joining two points on a circle is called a(n) _____.

2. A segment joining the center of a circle and a point on the circle is called a(n) _____.

3. A regular triangle is called a(n) _____ triangle.

4. A chord that contains the center of a circle is called a(n) _____.

CONCEPT CHECK *Write* true *or* false *for each statement.*

5. A rhombus is an example of a regular polygon.

6. If a triangle is isosceles, then it is not scalene.

7. A triangle can have more than one obtuse angle.

8. A square is both a rectangle and a parallelogram.

9. A square must be a rhombus.

10. A rhombus must be a square.

Write a short answer for each problem.

11. In your own words, explain the distinction between a chord and a secant.

12. What common traffic sign in the United States is in the shape of an octagon?

Identify each curve as simple, closed, both, *or* neither.

13.    14.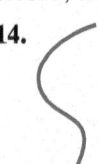

15.    16.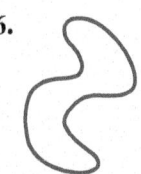

17.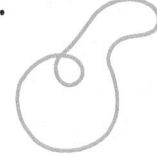

18.

19.

20.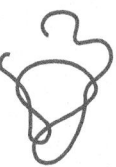

Decide whether each figure is convex *or* not convex.

21.

22.

23.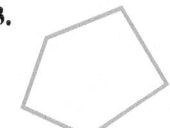

24.

25.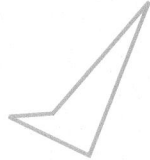

26.

Classify each triangle as acute, right, *or* obtuse. *Also classify each as* equilateral, isosceles, *or* scalene.

27.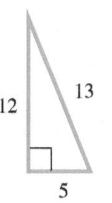

28.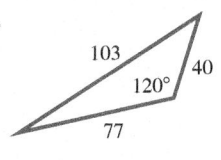

29.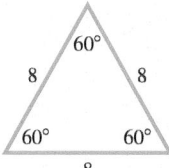

30.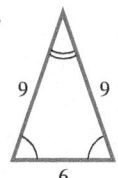

31.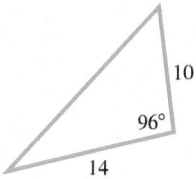

32.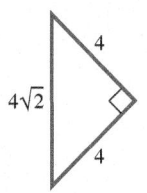

33.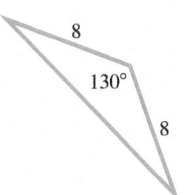

34.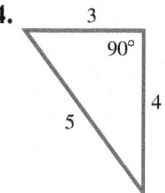

35.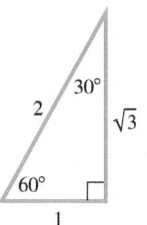

36.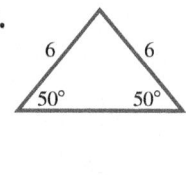

Write a short answer for each problem.

37. Give a definition of *isosceles right triangle*.

38. Can a triangle be both right and obtuse?

Find the measure of each angle in triangle ABC.

39.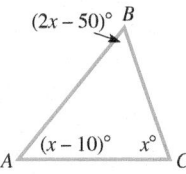

40.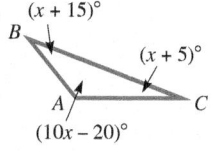

41.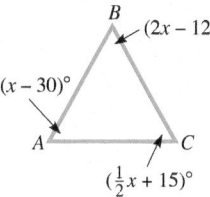

42.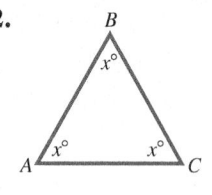

43. *Angle Measures*   In triangle ABC, angles A and B have the same measure, while the measure of angle C is 15 degrees larger than the measure of each of A and B. What are the measures of the three angles?

44. *Angle Measures*   In triangle ABC, the measure of angle A is 3 degrees more than the measure of angle B. The measure of angle B is the same as the measure of angle C. Find the measure of each angle.

In each triangle, find the measure of exterior angle BCD.

45.

46.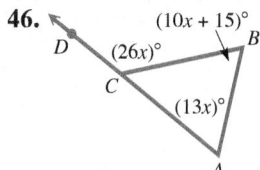

*Exercises 47–50 require ingenuity, but all can be solved using the concepts presented so far in this chapter.**

47. Find the sum, in degrees, of the six labeled angles. (*MT* calendar problem)

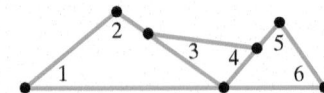

48. Triangle *ABC* is an isosceles triangle with *AB = AC*. Points *P* and *R* are on side *AC*, and point *Q* is on side *AB*. Moreover, *BC = BP = QP = AR*. Find the measure of ∡*BAC*. (*MT* calendar problem)

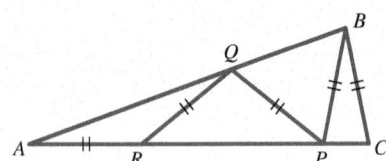

49. At the end of each segment making up a triangle, perpendiculars are drawn, creating an angle at each vertex. Find the sum of the angle measures α + β + γ. (*MT* calendar problem)

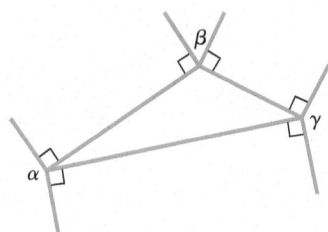

50. Find the measure of the angle θ in the following diagram, given that the two horizontal segments are parallel. (*MT* calendar problem)

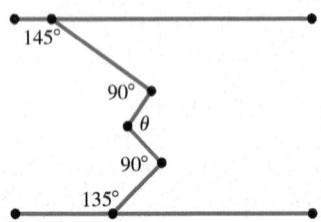

51. Using the points, segments, and lines in the figure, list all parts of the circle.

(a) center

(b) radii

(c) diameters

(d) chords

(e) secants

(f) tangents

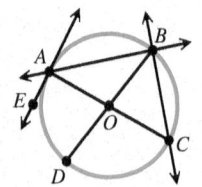

These and other, similar problems in this chapter have been adapted from Mathematics Teacher *calendar problems.*

52. In the classic 1939 movie *The Wizard of Oz*, the Scarecrow, upon getting a brain, says the following:

"The sum of the square roots of any two sides of an isosceles triangle is equal to the square root of the remaining side."

Give an example to show that his statement is incorrect.

In Exercises 53 and 54, construct the perpendicular bisector of segment PQ.

53.

54.

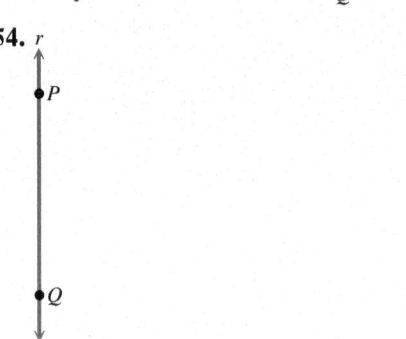

In Exercises 55 and 56, construct a perpendicular from P to the line r.

55. **56.**

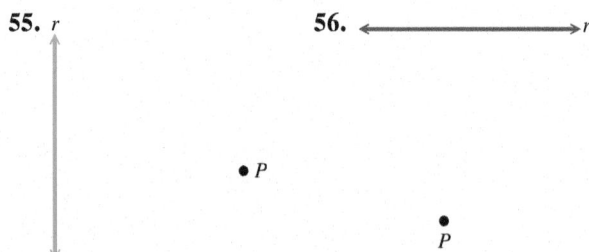

In Exercises 57 and 58, construct a perpendicular through the line r at P.

57.

58. *r*

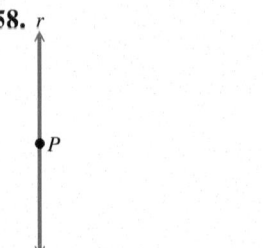

In Exercises 59 and 60, copy the given angle.

59. **60.**

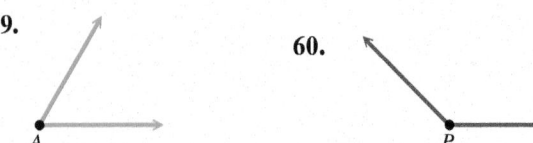

Write a short answer for each problem.

61. Investigate Proposition II of Book I of Euclid's *Elements*, and explain why the construction it provides enables us to use a "fixed" compass without guilt.

62. Use a software program to duplicate any of the constructions in **Exercises 53–60.** Print your results and show your instructor. Some excellent free programs (www.geogebra.org, for example) are available online.

9.3 THE GEOMETRY OF TRIANGLES: CONGRUENCE, SIMILARITY, AND THE PYTHAGOREAN THEOREM

OBJECTIVES

1 Recognize congruent triangles.

2 Determine similar triangles.

3 Apply the Pythagorean theorem.

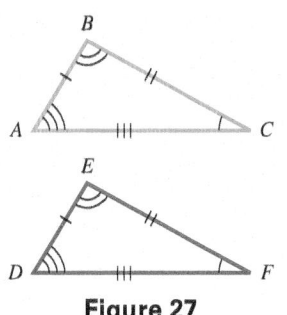

Figure 27

Congruent Triangles

Triangles that are both the same size and the same shape are **congruent triangles.** Informally speaking, if two triangles are congruent, then it is possible to pick up one of them and place it on top of the other so that they coincide exactly. As an everyday example of congruent triangles, consider the trusses used in constructing the roofs on homes (see the margin note), which are carefully produced with exactly the same dimensions each time.

We use the symbol $\triangle$ to designate triangles. **Figure 27** illustrates two congruent triangles, $\triangle ABC$ and $\triangle DEF$. The symbol $\cong$ denotes congruence, so

$$\triangle ABC \cong \triangle DEF.$$

Notice how the angles and sides are marked to indicate which angles are congruent and which sides are congruent. (Using precise terminology, we refer to angles or sides as being *congruent,* while the *measures* of congruent angles or congruent sides are *equal.* We will often use the terms "equal angles" and "equal sides" to describe angles of equal measure and sides of equal measure.)

In geometry the following properties are used to prove that two triangles are congruent.

CONGRUENCE PROPERTIES

Side-Angle-Side (SAS)   If two sides and the included angle of one triangle are equal, respectively, to two sides and the included angle of a second triangle, then the triangles are congruent.

Angle-Side-Angle (ASA)   If two angles and the included side of one triangle are equal, respectively, to two angles and the included side of a second triangle, then the triangles are congruent.

Side-Side-Side (SSS)   If three sides of one triangle are equal, respectively, to three sides of a second triangle, then the triangles are congruent.

Construction Congruence   Triangles have long been used in construction to provide rigid structure and symmetry. This photo shows congruent triangular sections of roof trusses ready for use.

What would happen if these triangles were not identical in both size and shape?

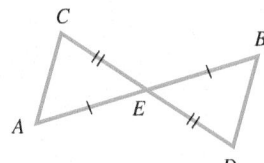

Figure 28

EXAMPLE 1   **Proving Congruence**

Refer to **Figure 28.**

Given: $CE = ED$

   $AE = EB$

Prove: $\triangle ACE \cong \triangle BDE$

PROOF

	STATEMENTS	REASONS
	1. $CE = ED$	**1.** Given
	2. $AE = EB$	**2.** Given
	3. $\angle CEA = \angle DEB$	**3.** Vertical angles are equal.
	4. $\triangle ACE \cong \triangle BDE$	**4.** SAS congruence property

Congruence and Symmetry Images and structures exhibiting symmetry involve congruent figures. The iron dome in the photo above contains several red regular pentagons. If the "spokes" bisect the interior angles of the pentagon, what congruence properties may be used to prove the congruence of the five red triangles sharing a side with one of these regular pentagons?

EXAMPLE 2 Proving Congruence

Refer to **Figure 29**.

Given: $\angle ADB = \angle CBD$

$\angle ABD = \angle CDB$

Prove: $\triangle ADB \cong \triangle CBD$

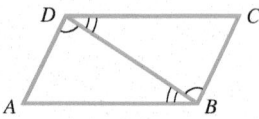

Figure 29

PROOF	STATEMENTS	REASONS
	1. $\angle ADB = \angle CBD$	1. Given
	2. $\angle ABD = \angle CDB$	2. Given
	3. $DB = DB$	3. Reflexive property (a quantity is equal to itself)
	4. $\triangle ADB \cong \triangle CBD$	4. ASA congruence property

EXAMPLE 3 Proving Congruence

Refer to **Figure 30**.

Given: $AD = CD$

$AB = CB$

Prove: $\triangle ABD \cong \triangle CBD$

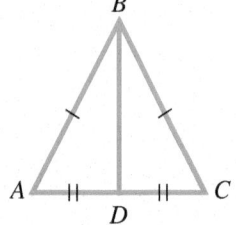

Figure 30

PROOF	STATEMENTS	REASONS
	1. $AD = CD$	1. Given
	2. $AB = CB$	2. Given
	3. $BD = BD$	3. Reflexive property
	4. $\triangle ABD \cong \triangle CBD$	4. SSS congruence property

Isosceles triangles (**Example 3** and **Figure 31**) have several important features.

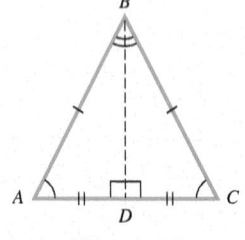

Figure 31

IMPORTANT STATEMENTS ABOUT ISOSCELES TRIANGLES

If $\triangle ABC$ is an isosceles triangle with $AB = CB$, and if D is the midpoint of the base AC (that is, $AD = CD$), then the following properties hold.

1. The base angles A and C are equal.

2. Angles ABD and CBD are equal.

3. Angles ADB and CDB are both right angles.

Similar Triangles

Similar triangles are pairs of triangles that are exactly the same shape but not necessarily the same size. **Figure 32** shows three pairs of similar triangles. (*Note:* The triangles do not need to be oriented in the same fashion in order to be similar.)

Figure 32

Suppose a correspondence between two triangles ABC and DEF is established.

$\angle A$ corresponds to $\angle D$ side AB corresponds to side DE

$\angle B$ corresponds to $\angle E$ side BC corresponds to side EF

$\angle C$ corresponds to $\angle F$ side AC corresponds to side DF

For triangle ABC to be similar to triangle DEF, these conditions must hold.

  1. Corresponding angles must have the same measure.
  2. The ratios of the corresponding sides must be constant. That is, the corresponding sides are proportional.

By showing that either of these conditions holds in a pair of triangles, we may conclude that the triangles are similar.

EXAMPLE 4   Verifying Similarity

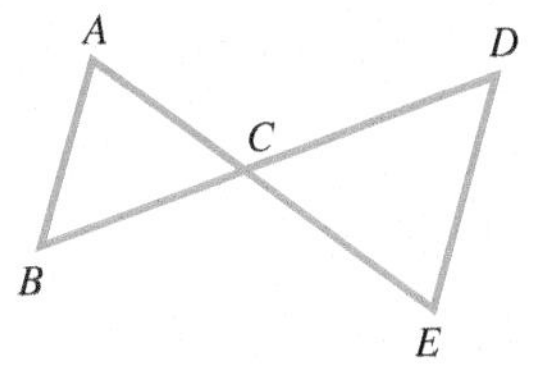

In **Figure 33**, $\overleftrightarrow{AB}$ is parallel to $\overleftrightarrow{ED}$. Verify that $\triangle ABC$ is similar to $\triangle EDC$.

Solution

Because $\overleftrightarrow{AB}$ is parallel to $\overleftrightarrow{ED}$, the transversal $\overleftrightarrow{BD}$ forms equal alternate interior angles ABC and EDC. Also, transversal $\overleftrightarrow{AE}$ forms equal alternate interior angles BAC and DEC. We know that $\angle ACB = \angle ECD$, because they are vertical angles. Because the corresponding angles have the same measures in triangles ABC and EDC, the triangles are similar.

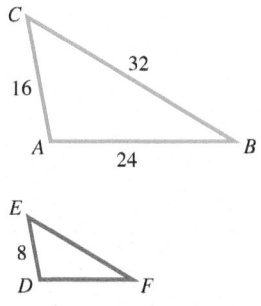

Figure 33

Once we have shown that two angles of one triangle are equal to the two corresponding angles of a second triangle, it is not necessary to show the same for the third angle. In any triangle the sum of the angles equals $180°$, so the measures of the remaining angles *must* be equal. This leads to the following property.

ANGLE-ANGLE (AA) SIMILARITY PROPERTY

If the measures of two angles of one triangle are equal to those of two corresponding angles of a second triangle, then the two triangles are similar.

EXAMPLE 5   Finding Side Lengths in Similar Triangles

In **Figure 34**, $\triangle EDF$ is similar to $\triangle CAB$. Find the unknown side lengths in $\triangle EDF$.

Solution

Side DF of the small triangle corresponds to side AB of the larger one, and sides DE and AC correspond. Similar triangles have corresponding sides in proportion.

$$\frac{8}{16} = \frac{DF}{24}$$

Using algebra, set the two cross products equal. (As an alternative method of solution, multiply both sides by 48, the least common multiple of 16 and 24.)

$8(24) = 16DF$ Cross products are equal.

$192 = 16DF$ Multiply.

$12 = DF$ Divide by 16.

Side DF has length 12.

Figure 34

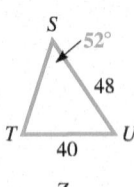

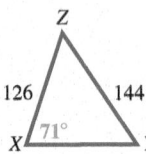

Figure 35

Side *EF* corresponds to side *CB*. This leads to another proportion.

$$\frac{8}{16} = \frac{EF}{32}$$

$$\frac{1}{2} = \frac{EF}{32} \qquad \frac{8}{16} = \frac{1}{2}$$

$$EF = 16 \qquad \text{Multiply by 32.}$$

● Side *EF* has length 16.

EXAMPLE 6   **Finding Unknown Measures in Similar Triangles**

Find the measures of the unknown parts of the similar triangles *STU* and *ZXY* in **Figure 35.**

Solution

Here angles *X* and *T* correspond, as do angles *Y* and *U*, and angles *Z* and *S*. Since angles *Z* and *S* correspond and since angle *S* is 52°, angle *Z* also must be 52°. The sum of the angles of any triangle is 180°. In the larger triangle, *X* = 71° and *Z* = 52°.

$$X + Y + Z = 180 \qquad \text{The angle sum is 180°.}$$

$$71 + Y + 52 = 180 \qquad \text{Substitute, and solve for } Y.$$

$$123 + Y = 180 \qquad \text{Add.}$$

$$Y = 57 \qquad \text{Subtract 123.}$$

Angle *Y* measures 57°. Because angles *Y* and *U* correspond, *U* = 57° also.

Now find the unknown sides. Sides *SU* and *ZY* correspond, as do *TS* and *XZ*, and *TU* and *XY*, leading to the following proportions.

$\dfrac{SU}{ZY} = \dfrac{TS}{XZ}$	$\dfrac{XY}{TU} = \dfrac{ZY}{SU}$	Write the proportions.
$\dfrac{48}{144} = \dfrac{TS}{126}$	$\dfrac{XY}{40} = \dfrac{144}{48}$	Substitute.
$\dfrac{1}{3} = \dfrac{TS}{126}$	$\dfrac{XY}{40} = \dfrac{3}{1}$	Lowest terms
$3TS = 126$	$XY = 120$	Cross products are equal. Solve.
$TS = 42$		

● Side *TS* has length 42, and side *XY* has length 120.

The Pythagorean Theorem

In a right triangle, the side opposite the right angle (the longest side) is the **hypotenuse.** The other two sides, which are perpendicular, are the **legs.**

PYTHAGOREAN THEOREM

If the two legs of a right triangle have lengths *a* and *b*, and the hypotenuse has length *c*, then

$$a^2 + b^2 = c^2.$$

That is, the sum of the squares of the lengths of the legs is equal to the square of the hypotenuse.

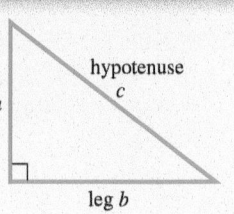

Our knowledge of the mathematics of the Babylonians of Mesopotamia is based largely on archaeological discoveries of thousands of clay tablets. On the tablet labeled **Plimpton 322,** there are several columns of inscriptions that represent numbers. In one portion of the tablet (not shown here), the far right column is simply one that serves to number the lines, but two other columns contain numbers that some scholars have interpreted to be hypotenuses and legs of right triangles.

   This interpretation of the tablet would suggest that what we call the Pythagorean theorem was known more than 1000 years before the time of Pythagoras. Other interpretations exist, however, which cast doubt on ancient Babylonian knowledge of the theorem. (For another interpretation of Plimpton 322, see "Words and Pictures: New Light on Plimpton 322" in *The American Mathematical Monthly*, vol. 109, 2002, pp. 105–120.)

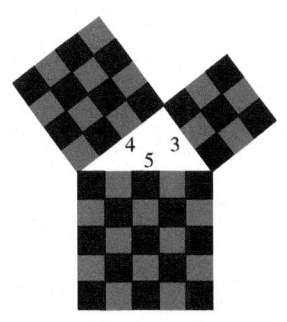

Figure 36

$$a^2 + b^2 = c^2$$
$$3^2 + 4^2 = 5^2$$
$$9 + 16 = 25$$
$$25 = 25 \checkmark$$

Figure 36 illustrates the theorem. The side of the square along the hypotenuse measures 5 units. Those along the legs measure 3 and 4 units. If $a = 3$, $b = 4$, and $c = 5$, the equation of the Pythagorean theorem is satisfied. As a result, the natural numbers 3, 4, and 5 form the **Pythagorean triple** $(3, 4, 5)$. There are infinitely many such triples.

EXAMPLE 7   Using the Pythagorean Theorem

Find the length a in the right triangle shown in **Figure 37**.

Solution

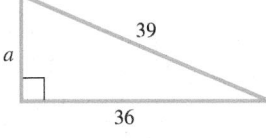

$$a^2 + b^2 = c^2 \qquad \text{Pythagorean theorem}$$
$$a^2 + 36^2 = 39^2 \qquad b = 36,\ c = 39$$
$$a^2 + 1296 = 1521 \qquad \text{Square.}$$
$$a^2 = 225 \qquad \text{Subtract 1296 from both sides.}$$
$$a = 15 \qquad \text{Choose the positive square root, because } a > 0.$$

Figure 37

Verify that $(15, 36, 39)$ is a Pythagorean triple as a check.

The statement of the Pythagorean theorem is an *if . . . then* statement. If the antecedent (the statement following the word "if") and the consequent (the statement following the word "then") are interchanged, the new statement is the *converse* of the original. The *converse* of the Pythagorean theorem *is* also a true statement.

CONVERSE OF THE PYTHAGOREAN THEOREM

If a triangle has sides of lengths a, b, and c, where c is the length of the longest side, and if $a^2 + b^2 = c^2$, then the triangle is a right triangle.

EXAMPLE 8   Using the Converse of the Pythagorean Theorem

Efren has been contracted to complete an unfinished 8-foot-by-12-foot laundry room in an existing house. He finds that the previous contractor built the floor so that the length of its diagonal is 14 feet, 8 inches. Is the floor "squared off" properly?

Solution

Because 14 feet, 8 inches $= 14\frac{2}{3}$ feet, he must check to see whether the following statement is true.

$$8^2 + 12^2 \stackrel{?}{=} \left(14\frac{2}{3}\right)^2 \qquad a^2 + b^2 = c^2$$

$$8^2 + 12^2 \stackrel{?}{=} \left(\frac{44}{3}\right)^2 \qquad 14\frac{2}{3} = \frac{44}{3}$$

$$208 \stackrel{?}{=} \frac{1936}{9} \qquad \text{Simplify.}$$

$$208 \neq 215\frac{1}{9} \qquad \text{The two values are not equal.}$$

Following **Hurricane Katrina** in August 2005, the pine trees of southeastern Louisiana provided thousands of examples of **right triangles.** See the photo.

Suppose the vertical distance from the base of a broken tree to the point of the break is 55 inches. The length of the broken part is 144 inches.

How far along the ground is it from the base of the tree to the point where the broken part touches the ground?

Efren needs to fix the problem, since the diagonal, which measures 14 feet, 8 inches, should actually measure $\sqrt{208} \approx 14.4 \approx 14$ feet, 5 inches.

WHEN WILL I **EVER** USE THIS**?**

Suppose you work for the Department of Forestry and have been called in to cut down a diseased tree at a summer camp. The tree is quite tall and is in an area surrounded by cabins. You measure the distance from the base of the tree to the nearby cabins, and you find that the cabins will be safe if the tree falls within 140 feet of its base.

Is it safe to cut the tree at ground level?
If not, how high up the trunk must you cut
to guarantee that the tip of the tree will hit
the ground within the safe zone?

Assume that the tree stays "hinged" at the location of the cut.

In early afternoon, you find that a 72-inch fence post casts an 18-inch shadow at the same time the diseased tree casts a 50-foot shadow. See **Figure 38**.

Let *y* be the height of the tree. Solve the proportion.

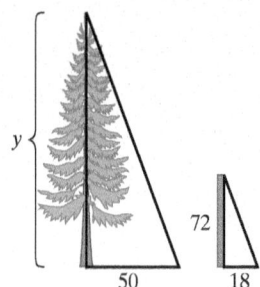

$$\frac{y}{50} = \frac{72}{18}$$

$$\frac{y}{50} = 4 \qquad \text{Reduce: } \frac{72}{18} = \frac{4}{1}.$$

$$y = 200 \qquad \text{Multiply by 50.}$$

Figure 38

Since the tree is 200 feet tall, you cannot cut it at ground level. Refer to **Figure 39**.

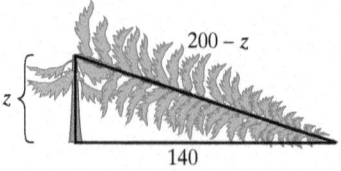

Figure 39

Let *z* be the height above the ground at which to cut. We can use the Pythagorean theorem to find *z*.

$$z^2 + 140^2 = (200 - z)^2$$

$$z^2 + 19{,}600 = 40{,}000 - 400z + z^2 \qquad \text{FOIL method}$$

$$19{,}600 = 40{,}000 - 400z \qquad \text{Subtract } z^2.$$

$$400z = 20{,}400 \qquad \text{Add } 400z \text{ and subtract } 19{,}600.$$

$$z = 51 \qquad \text{Divide by 400.}$$

You need to cut the tree 51 feet above the ground.

9.3   **EXERCISES**

CONCEPT CHECK *Write* true *or* false *for each statement.*

1. Similar triangles are the same shape and the same size.

2. If two sides and an angle (not included) of one triangle are equal to two sides and an angle (not included) of a second triangle, then the triangles are congruent.

3. If two angles of one triangle are equal to the two corresponding angles of a second triangle, then the two triangles are similar.

4. In a right triangle, the sum of the lengths of the legs is equal to the length of the hypotenuse.

Provide a STATEMENTS/REASONS proof similar to the ones in **Examples 1–3.**

5. Given: $AC = BD$; $AD = BC$

   Prove: $\triangle ABD \cong \triangle BAC$

   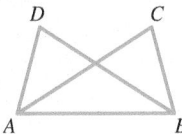

6. Given: $AC = BC$; D is the midpoint of AB.

   Prove: $\triangle ADC \cong \triangle BDC$

   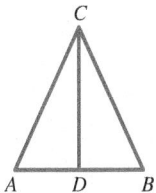

7. Given: $\angle BAC = \angle DAC$; $\angle BCA = \angle DCA$

   Prove: $\triangle ABC \cong \triangle ADC$.

   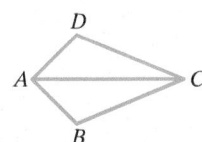

8. Given: $BC = BA$; $\angle 1 = \angle 2$

   Prove: $\triangle DBC \cong \triangle DBA$

   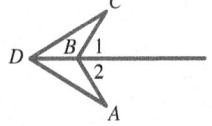

9. Given: $\overleftrightarrow{DB}$ is perpendicular to $\overleftrightarrow{AC}$; $AB = BC$

   Prove: $\triangle ABD \cong \triangle CBD$

   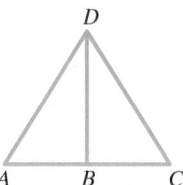

10. Given: $BO = OE$; $\overleftrightarrow{OB}$ is perpendicular to $\overleftrightarrow{AC}$; $\overleftrightarrow{OE}$ is perpendicular to $\overleftrightarrow{DF}$

    Prove: $\triangle AOB \cong \triangle FOE$.

    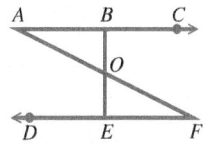

Exercises 11–14 refer to the given figure, which includes an isosceles triangle with $AB = BC$.

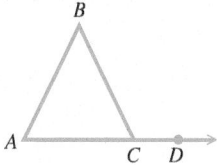

11. What is the measure of $\angle BCD$ if $\angle B$ measures 36°?

12. If $\angle BCA$ measures 57°, what is the measure of $\angle B$?

13. If $\angle B$ measures 46°, then $\angle A$ measures _____ and $\angle BCA$ measures _____.

14. What is the measure of $\angle B$ if $\angle BCD$ measures 100°?

Write a short answer for each problem.

15. Explain why all equilateral triangles must be similar.

16. Explain how this figure demonstrates why the Congruence Properties just before **Example 1** did *not* include a Side-Side-Angle (SSA) property.

    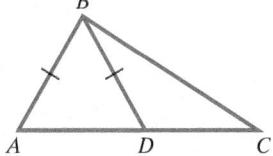

Name the corresponding angles and the corresponding sides for each of the following pairs of similar triangles.

17.
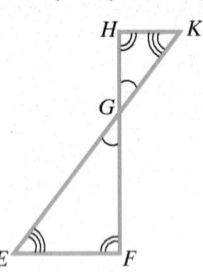

18.
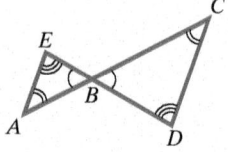

19.
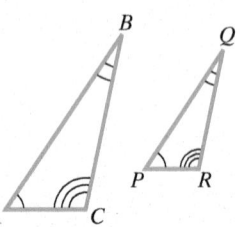

20.
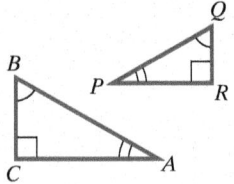

Find all unknown angle measures in each pair of similar triangles.

21.
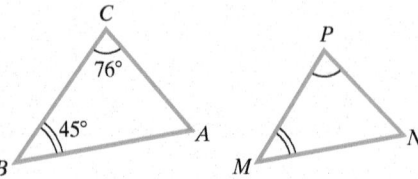

22.
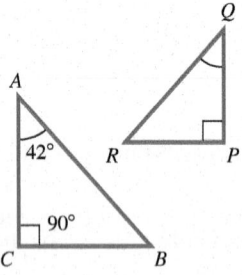

23.
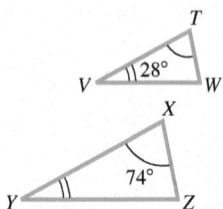

24.
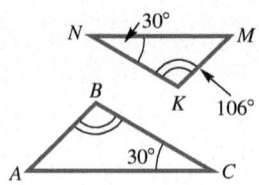

25.
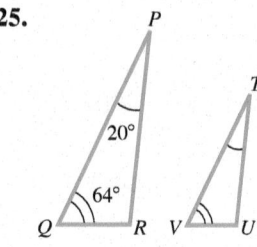

26.
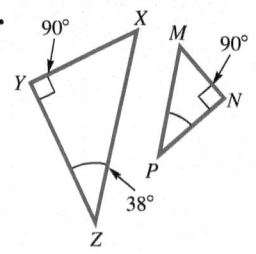

Find the unknown side lengths in each pair of similar triangles.

27.
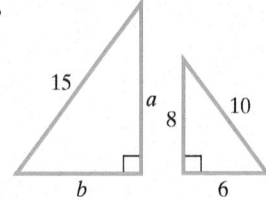

28.
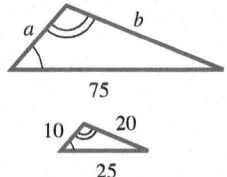

29.
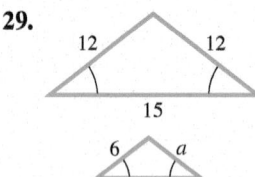

30.
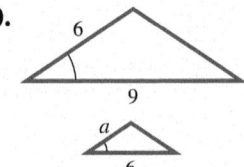

31.
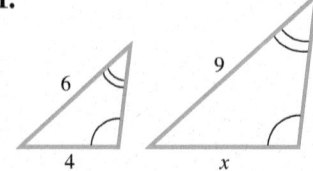

32.
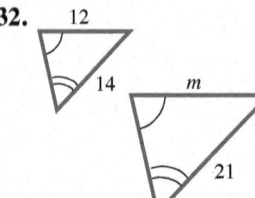

In each diagram, there are two similar triangles. Find the unknown measurement in each. (Hint: In the figure for Exercise 33, the side of length 150 in the smaller triangle corresponds to a side of length

$$150 + 180 = 330$$

in the larger triangle.)

33.

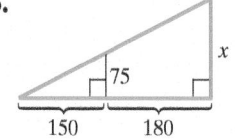

34.

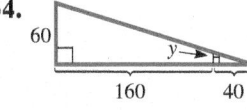

35.

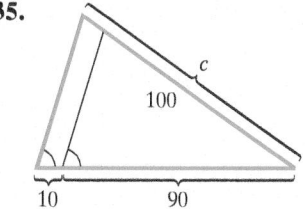

36.

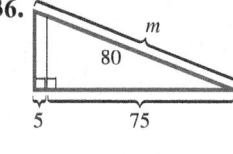

37.

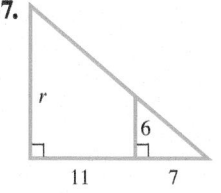

38.

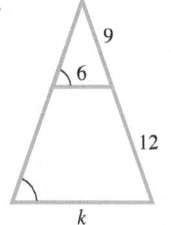

Solve each problem.

39. *Height of a Tree* A tree casts a shadow 45 m long. At the same time, the shadow cast by a vertical 4-m stick is 3 m long. Find the height of the tree.

40. *Height of a Tower* A forest fire lookout tower casts a shadow 80 ft long at the same time that the shadow of a 9-ft truck is 5 ft long. Find the height of the tower.

41. *Lengths of Sides of a Photograph* On a photograph of a triangular piece of land, the lengths of the three sides are 4 cm, 5 cm, and 7 cm, respectively. The shortest side of the actual piece of land is 200 m long. Find the lengths of the other two sides.

42. *Height of a Lighthouse Keeper* The Santa Cruz lighthouse is 14 m tall and casts a shadow 28 m long at 7 P.M. At the same time, the shadow of the lighthouse keeper is 3.5 m long. How tall is she?

43. *Height of a Building* A house is 15 ft tall. Its shadow is 40 ft long at the same time the shadow of a nearby building is 300 ft long. Find the height of the building.

44. *Height of the World's Tallest Human* Robert Wadlow was the tallest human being ever recorded. When a 6-ft stick cast a shadow 24 in., Robert would cast a shadow 35.7 in. How tall was he?

When proportional relationships are expressed graphically, similar triangles result. In Exercises 45 and 46, use similar triangles to answer the question.

45. *Feeding an Army* A waffle recipe that serves 6 calls for $1\frac{1}{2}$ cups of flour. How much flour will be needed if the chef must serve 40 people?

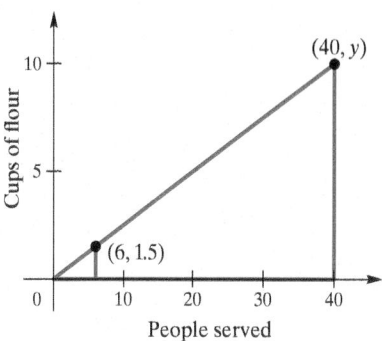

46. *How Much Herbicide?* A landscape maintenance service is spraying for weed control. Eight ounces of the chemical being used treats 500 square feet of lawn. How much of the chemical will be needed for treating 3250 square feet of lawn?

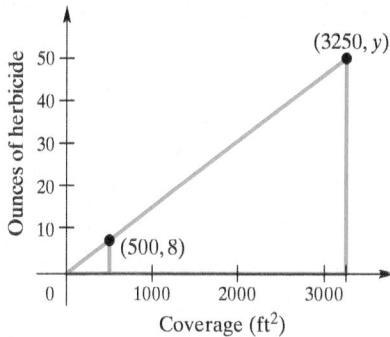

In Exercises 47–54, a and b represent the two legs of a right triangle, and c represents the hypotenuse. Find the lengths of the unknown sides.

47.

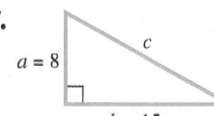

48.

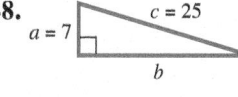

49.

50.

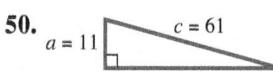

51. $a = 14$ m; $b = 48$ m

52. $a = 28$ km; $c = 100$ km

53. $b = 21$ in.; $c = 29$ in.

54. $b = 120$ ft; $c = 169$ ft

Write a short answer for each problem.

55. Refer to **Exercise 52** in the previous section. Correct the Scarecrow's statement, using language similar to his.

56. Show that if $a^2 + b^2 = c^2$, then it is not necessarily true that $a + b = c$.

There are various formulas that will generate Pythagorean triples. For example, if we choose positive integers r and s, with r > s, then the set of equations

$$a = r^2 - s^2, \quad b = 2rs, \quad c = r^2 + s^2$$

generates a Pythagorean triple (a, b, c). Use the values of r and s given in each of Exercises 57–60 to generate a Pythagorean triple using this method.

57. $r = 2, s = 1$ **58.** $r = 3, s = 2$

59. $r = 4, s = 3$ **60.** $r = 4, s = 2$

61. Show that the formula given for **Exercises 57–60** actually satisfies $a^2 + b^2 = c^2$.

62. It can be shown that if $(x, x + 1, y)$ is a Pythagorean triple, then so is

$$(3x + 2y + 1, \quad 3x + 2y + 2, \quad 4x + 3y + 2).$$

Use this idea to find three more Pythagorean triples, starting with $(3, 4, 5)$. *(Hint: Here, $x = 3$ and $y = 5$.)*

If m is an odd positive integer greater than 1, then

$$\left(m, \frac{m^2 - 1}{2}, \frac{m^2 + 1}{2} \right)$$

is a Pythagorean triple. Use this to find the Pythagorean triple generated by each value of m in Exercises 63–66.

63. $m = 3$ **64.** $m = 5$

65. $m = 7$ **66.** $m = 9$

67. Show that the expressions in the directions for **Exercises 63–66** actually satisfy $a^2 + b^2 = c^2$.

68. Show why $(6, 8, 10)$ is the only Pythagorean triple consisting of consecutive even numbers.

For any integer n greater than 1,

$$(2n, \quad n^2 - 1, \quad n^2 + 1)$$

is a Pythagorean triple. Use this pattern to find the Pythagorean triple generated by each value of n in Exercises 69–72.

69. $n = 2$ **70.** $n = 3$

71. $n = 6$ **72.** $n = 7$

73. Show that the expressions in the directions for **Exercises 69–72** actually satisfy

$$a^2 + b^2 = c^2.$$

74. Can an isosceles right triangle have sides with integer lengths? Why or why not?

Solve each problem. (You may wish to review quadratic equations from algebra.)

75. *Side Length of a Triangle*   The hypotenuse of a right triangle is 1 m more than the longer leg, and the shorter leg is 7 m. Find the length of the longer leg.

76. *Side Lengths of a Triangle*   The hypotenuse of a right triangle is 3 cm more than twice the shorter leg, and the longer leg is 3 cm less than three times the shorter leg. Find the lengths of the three sides of the triangle.

77. *Height of a Tree*   At a point on the ground 24 ft from the base of a tree, the distance to the top of the tree is 6 ft less than twice the height of the tree. Find the height of the tree.

78. *Dimensions of a Rectangle*   The length of a rectangle is 2 in. less than twice the width. The diagonal is 5 in. Find the length and width of the rectangle.

79. *Height of a Break in Bamboo*   (Problem of the broken bamboo, from the Chinese work *Arithmetic in Nine Sections*, 1261) There is a bamboo 10 ft high, the upper end of which, being broken, reaches the ground 3 ft from the stem. Find the height of the break.

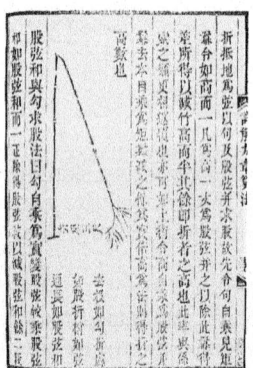

80. *Depth of a Pond*   (Adapted from *Arithmetic in Nine Sections*) There grows in the middle of a circular pond 10 ft in diameter a reed that projects 1 ft out of the water. When it is drawn down, it just reaches the edge of the pond. How deep is the water?

Squaring Off a Floor under Construction   *Imagine that you are a carpenter building the floor of a rectangular room. What must the diagonal of the room measure if your floor is to be squared off properly, given the dimensions in Exercises 81–84? Give your answer to the nearest inch.*

81. 12 ft by 15 ft

82. 14 ft by 20 ft

83. 16 ft by 24 ft

84. 20 ft by 32 ft

Exercises 85–92 require ingenuity, but all can be solved using the concepts presented so far in this chapter.

85. Unknown Length Points A, B, and C lie along the circumference of a circle of radius 4, as shown. The measure of $\angle ABC$ is 45°. Find the length of chord AC. (*MT* calendar problem)

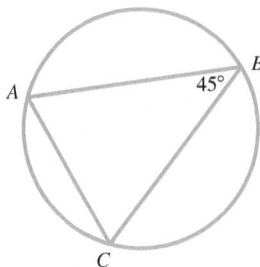

86. Unknown Angle Measure $ABCD$ is a square and ABE is an equilateral triangle. What is the measure of $\angle AED$? (*MT* calendar problem)

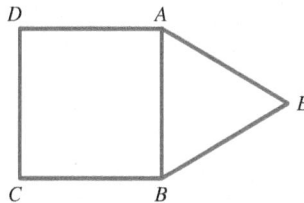

87. Unknown Radius The ratio of the radii of two concentric circles is 1:3. AC is a diameter of the larger circle, BC is a chord of the larger circle that is tangent to the smaller circle, and $AB = 12$. Find the radius of the larger circle. (*MT* calendar problem)

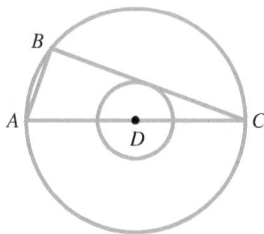

88. Unknown Length In $\triangle ABC$, $AB = 8$, $BC = 7$, $CA = 6$, and BC is extended to a point P, so that $\triangle PAB$ is similar to $\triangle PCA$. Find the length of PC. (*MT* calendar problem)

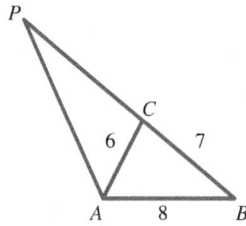

89. Unknown Angle Measure In the figure, $\triangle ABC$ has a right angle at B and $\angle A = 20°$. If CD is the bisector of $\angle ACB$, find $\angle CDB$. (*MT* calendar problem)

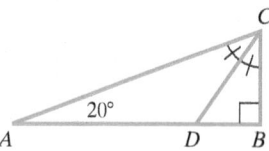

90. Unknown Length Triangle CDE is equilateral with $DE = 60$ units. A is the foot of the perpendicular from D to $\overleftrightarrow{CE}$, and B is the midpoint of $\overleftrightarrow{DA}$. Find the length CB. (*MT* calendar problem)

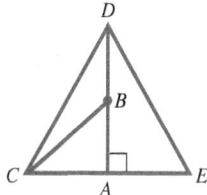

91. Unknown Side Ratio Given that $DB = DC$, $BA = BC$, and $\angle D = 30°$, find the exact value of DB/BC. (*MT* calendar problem)

92. Unknown Length In the right triangle ACD with right angle C,

$$AB + AD = BC + DC.$$

If $BC = 8$ and $DC = 10$, find the length AB. (*MT* calendar problem)

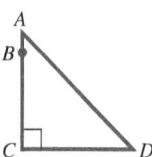

In Exercises 93–95, verify that the following constructions from the previous section are valid. Use a STATEMENTS/REASONS proof.

93. Perpendicular Bisector

94. Perpendicular to a Line

95. Copied Angle

96. Given a line m and a point A on m, construct a line perpendicular to m through A. Prove that your construction is valid.

FOR FURTHER THOUGHT

Proving the Pythagorean Theorem

The Pythagorean theorem has probably been proved in more different ways than any theorem in mathematics. *The Pythagorean Proposition*, by Elisha Scott Loomis, first published in 1927, contains more than 250 different proofs of the theorem.

One of the most popular proofs of the theorem follows. This proof requires two formulas for area. The area $\mathcal{A}$ of a square is given by

$$\mathcal{A} = s^2,$$

where s is the length of a side of the square. The area $\mathcal{A}$ of a triangle is given by

$$\mathcal{A} = \frac{1}{2}bh,$$

where b is the base of the triangle and h is the height corresponding to that base.

For Group or Individual Investigation

1. ***A Classic Proof*** In the figure below, the area of the large square must always be the same. It is made up of four right triangles and a smaller square.

   (a) The length of a side of the large square is _____, so its area is (_____)^2, or _____.

   (b) The area of the large square can also be found by adding the areas of the four right triangles and the smaller square.

The area of each right triangle is _____, so the sum of the areas of the four right triangles is _____. The area of the smaller square is _____.

   (c) The sum of the areas of the four right triangles and the smaller square is _____.

   (d) Since the areas in (a) and (c) represent the area of the same figure, the expressions there must be equal. Setting them equal to each other, we obtain

   _____ = _____.

   (e) Subtract $2ab$ from each side of the equation in (d) to obtain the desired result:

   _____ = _____.

2. ***Proof by Similar Triangles*** In the figure, right triangles *ABC*, *CBD*, and *ACD* are similar. To prove the Pythagorean theorem, fill in the blanks with the correct responses.

   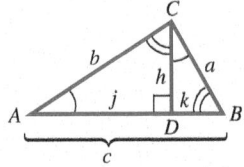

   (a) Using proportions, we have $\frac{c}{b} = \dfrac{\underline{\quad}}{j}$.

   (b) We also have $\frac{c}{a} = \dfrac{a}{\underline{\quad}}$.

   (c) From part (a), $b^2 = $ _____.

   (d) From part (b), $a^2 = $ _____.

   (e) Using the results of parts (c) and (d) and factoring, $a^2 + b^2 = c(\underline{\quad})$. Since _____ $= c$, it follows that _____.

3. ***Animated Proof*** Refer to

   www.davis-inc.com/pythagor/proof2.html

   to view an animation that provides a proof of the Pythagorean theorem.

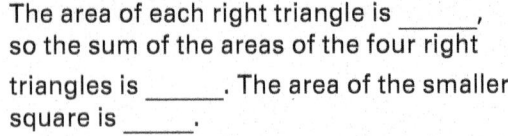

9.4   PERIMETER, AREA, AND CIRCUMFERENCE

OBJECTIVES

1 Determine the perimeter of a polygon.

2 Determine the areas of polygons and circles.

3 Determine the circumference of a circle.

Perimeter of a Polygon

We are sometimes required to find the "distance around", or *perimeter*, of a polygon.

> **PERIMETER**
>
> The **perimeter** of any polygon is the sum of the measures of the line segments that form its sides. Perimeter is measured in *linear units*.

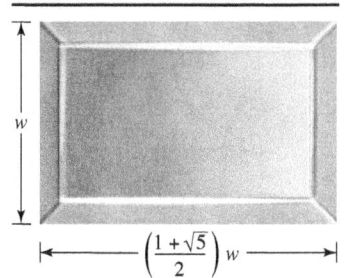

A **Golden Rectangle** is one having length ℓ and width w such that

$$\frac{\ell}{w} = \frac{\ell + w}{\ell} = \frac{1 + \sqrt{5}}{2}.$$

(This was shown in an earlier chapter.) Because the perimeter of a rectangle is

$$P = 2(\ell + w),$$

we see that the ratio of perimeter to length is

$$\frac{P}{\ell} = \frac{2(\ell + w)}{\ell} = 1 + \sqrt{5}.$$

The simplest polygon is a triangle. If a triangle has sides of lengths a, b, and c, then to find its perimeter we simply find the sum of a, b, and c, as shown below.

PERIMETER OF A TRIANGLE

The perimeter P of a triangle with sides of lengths a, b, and c is given by the following formula.

$$P = a + b + c$$

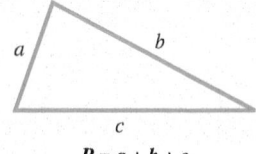

$$P = a + b + c$$

Because a rectangle is made up of two pairs of sides with the two sides in each pair equal in length, the formula for the perimeter of a rectangle may be stated as follows.

PERIMETER OF A RECTANGLE

The perimeter P of a rectangle with length ℓ and width w is given by the following formula.

$$P = 2\ell + 2w, \quad \text{or, equivalently,} \quad P = 2(\ell + w)$$

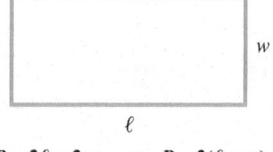

$$P = 2\ell + 2w, \quad \text{or} \quad P = 2(\ell + w)$$

EXAMPLE 1   **Using Perimeter to Determine Amount of Fencing**

A plot of land is in the shape of a rectangle. If it has length 50 feet and width 26 feet, how much fencing would be needed to completely enclose the plot?

Solution

We must find the distance around the plot of land.

$$P = 2\ell + 2w \qquad \text{Perimeter formula}$$
$$P = 2(50) + 2(26) \qquad \ell = 50,\ w = 26$$
$$P = 100 + 52 \qquad \text{Multiply.}$$
$$P = 152 \qquad \text{Add.}$$

The perimeter is 152 feet, so 152 feet of fencing is required.

A square is a rectangle with four sides of equal length. The formula for the perimeter of a square is a special case of the formula for the perimeter of a rectangle.

PERIMETER OF A SQUARE

The perimeter P of a square with all sides of length s is given by the following formula.

$$P = 4s$$

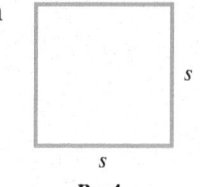

$$P = 4s$$

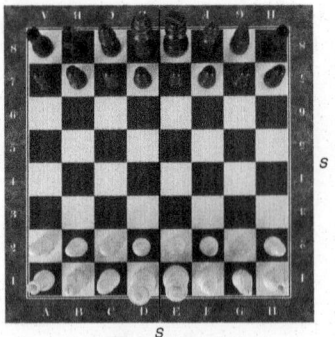

s

Figure 40

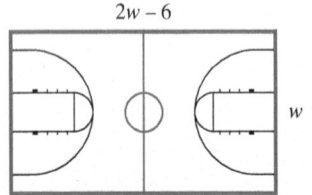

2*w* − 6

w

Figure 41

EXAMPLE 2 **Using the Formula for Perimeter of a Square**

A square chess board has perimeter 34 inches. See **Figure 40.** What is the length of each side?

Solution

$$P = 4s \quad \text{Perimeter formula}$$

$$34 = 4s \quad P = 34$$

$$s = 8.5 \quad \text{Divide by 4.}$$

Each side has a measure of 8.5 inches.

EXAMPLE 3 **Finding Length and Width of a Rectangle**

The length of a basketball court is 6 feet less than twice the width. The perimeter is 288 feet. Find the length and width.

Solution

Step 1 **Read the problem.** We must find the length and width.

Step 2 **Assign a variable.** Let *w* represent the width. Then 2*w* − 6 can represent the length, because the length is 6 less than twice the width. **Figure 41** shows a diagram of the basketball court.

Step 3 **Write an equation.** In the formula $P = 2\ell + 2w$, replace ℓ with 2*w* − 6, and replace *P* with 288, because the perimeter is 288 feet.

$$288 = 2(2w - 6) + 2w$$

Step 4 **Solve the equation.**

$$288 = 2(2w - 6) + 2w$$

$$288 = 4w - 12 + 2w \qquad \text{Distributive property}$$

$$288 = 6w - 12 \qquad \text{Combine like terms.}$$

$$300 = 6w \qquad \text{Add 12.}$$

$$50 = w \qquad \text{Divide by 6.}$$

Step 5 **State the answer.** Because $w = 50$, the width is 50 feet and the length is 2*w* − 6 = 2(50) − 6 = 94 feet.

Step 6 **Check.** Because 94 is 6 less than twice 50, and because the perimeter is 2(94) + 2(50) = 288, the answer is correct.

Problem-Solving Strategy

The six-step method of solving an applied problem from an earlier chapter can be used to solve problems involving geometric figures, as shown in **Example 3.**

Area of a Polygon

AREA

The amount of plane surface covered by a polygon is its **area.** Area is measured in *square units*.

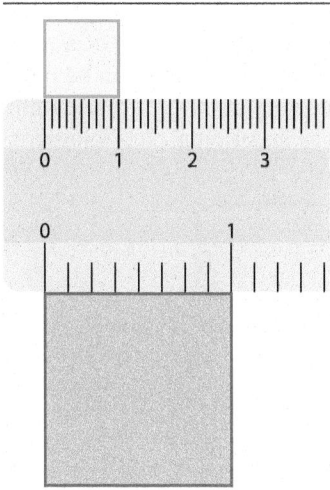

Converting units requires a ratio called a **conversion factor.** The ruler (shown above) provides such a ratio and shows that each inch is equal to 2.54 cm. To see how many square centimeters "fit" into one square inch, we use this factor.

$$1 \text{ in.}^2 \cdot \left(\frac{2.54 \text{ cm}}{1 \text{ in.}}\right)^2 \approx 6.45 \text{ cm}^2$$

Metric units will be used extensively in this chapter. Further help with the **metric system,** including unit conversion, is available in MyLab Math or at www.pearsonhighered.com/mathstatsresources.

Determining the **area** of a figure requires a basic *unit of area.* One that is commonly used is the *square centimeter,* abbreviated cm². One square centimeter, or 1 cm², is the area of a square one centimeter on a side. In place of 1 cm², the basic unit of area could be 1 in.², 1 ft², 1 m², or any appropriate unit.

As an example, we calculate the area of the rectangle shown in **Figure 42(a)**. Using the basic 1-cm² unit, **Figure 42(b)** shows that six squares, each 1 cm on a side, can be laid off horizontally, and four such squares can be laid off vertically. A total of 4 · 6 = 24 of the small squares are needed to cover the large rectangle. Thus, the area of the large rectangle is 24 cm².

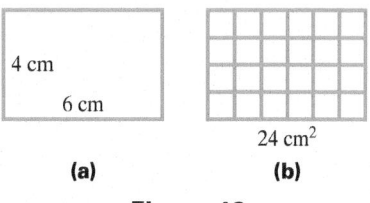

(a)   **(b)**

Figure 42

AREA OF A RECTANGLE

The area $\mathscr{A}$ of a rectangle with length ℓ and width w is given by the following formula.

$$\mathscr{A} = \ell w$$

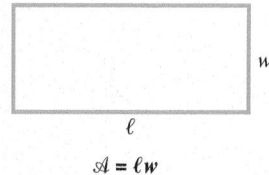

$$\mathscr{A} = \ell w$$

The formula for the area of a rectangle $\mathscr{A} = \ell w$ can be used to find formulas for the areas of other figures.

AREA OF A SQUARE

The area $\mathscr{A}$ of a square with all sides of length s is given by the following formula.

$$\mathscr{A} = s^2$$

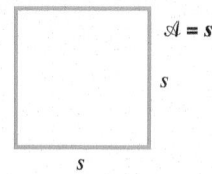

$$\mathscr{A} = s^2$$

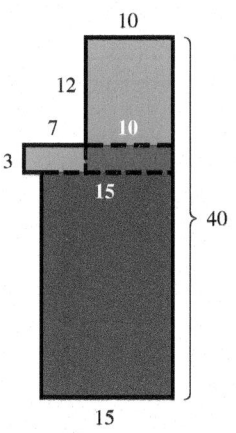

Figure 43

EXAMPLE 4   **Using Area to Determine Amount of Carpet**

Figure 43 shows the floor plan of a building, made up of various rectangles. Each length given is in meters. How many square meters of carpet would be required to carpet the building?

Solution

The dashed lines in the figure break up the floor area into rectangles. The areas of the various rectangles that result are as follows.

$$10 \cdot 12 = 120, \qquad 3 \cdot 10 = 30,$$
$$3 \cdot 7 = 21, \qquad 15 \text{m} \cdot 25 = 375$$

40 − 12 − 3 = 25

$$(120 + 30 + 21 + 375) = 546$$

The amount of carpet needed is 546 m².

A **parallelogram** is a four-sided figure with both pairs of opposite sides parallel. Because a parallelogram need not be a rectangle, the formula for the area of a rectangle cannot be used directly for a parallelogram. However, this formula can be used indirectly, as shown in **Figure 44.** Cut off the triangle in color, and slide it to the right end. The resulting figure is a rectangle with the same area as the original parallelogram.

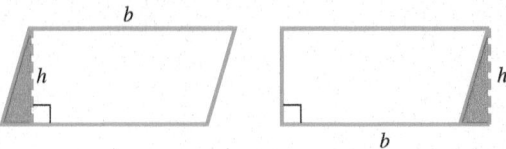

Figure 44

The *height* of the parallelogram is the perpendicular distance between the top and bottom and is denoted by h in the figure. The width of the rectangle equals the height of the parallelogram, and the length of the rectangle is the base b, so

$$\mathcal{A} = \text{length} \cdot \text{width} \quad \text{becomes} \quad \mathcal{A} = \text{base} \cdot \text{height}.$$

AREA OF A PARALLELOGRAM

The area $\mathcal{A}$ of a parallelogram with height h and base b is given by the following formula.

$$\mathcal{A} = bh$$

(*Note:* The height h must be perpendicular to the base b. If the parallelogram is not a rectangle, then h is not the length of a side.)

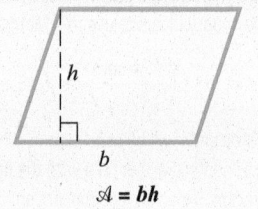

EXAMPLE 5   **Using the Formula for Area of a Parallelogram**

Find the area of the parallelogram in **Figure 45.**

Solution

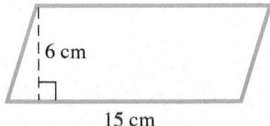

Figure 45

$$\mathcal{A} = bh \quad \text{Area formula}$$
$$\mathcal{A} = 15 \cdot 6 \quad b = 15, h = 6$$
$$\mathcal{A} = 90 \quad \text{Multiply.}$$

The area of the parallelogram is 90 cm².

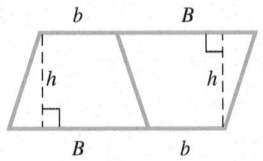

Figure 46

Figure 46 shows a parallelogram made up of two trapezoids, each having height h, shorter base b, and longer base B. The area of the parallelogram is found by multiplying the height h by the base of the parallelogram, $b + B$—that is, $h(b + B)$. Because the area of the parallelogram is *twice* the area of each trapezoid, the area of each trapezoid is *half* the area of the parallelogram.

AREA OF A TRAPEZOID

The area $\mathcal{A}$ of a trapezoid with parallel bases b and B and height h is given by the following formula.

$$\mathcal{A} = \frac{1}{2}h(b + B)$$

(*Note:* h must be perpendicular to both bases.)

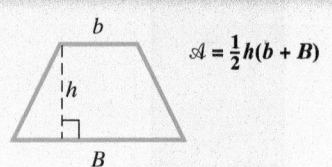

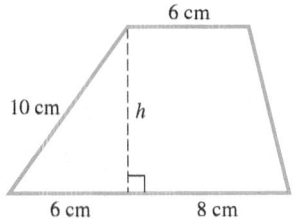

Figure 47

EXAMPLE 6   **Using the Formula for Area of a Trapezoid**

Find the area of the trapezoid in **Figure 47.**

Solution

First, use the Pythagorean theorem to find the height h.

$$h^2 + 6^2 = 10^2 \qquad \text{Pythagorean theorem}$$

$$h^2 = 10^2 - 6^2 = 100 - 36 = 64 \qquad \text{Subtract and simplify}$$

$$h = 8 \qquad \sqrt{64} = 8$$

The height is 8 cm. Now use the area formula.

$$\mathscr{A} = \frac{1}{2}h(B + b) \qquad \text{Area formula}$$

$$\mathscr{A} = \frac{1}{2}(8)(14 + 6) \qquad h = 8,\, B = 14,\, b = 6$$

$$\mathscr{A} = \frac{1}{2}(8)(20) \qquad \text{Add.}$$

$$\mathscr{A} = 80 \qquad \text{Multiply.}$$

The area of the trapezoid is 80 cm².

The formula for the area of a triangle can be found from the formula for the area of a parallelogram. In **Figure 48,** the triangle with vertices A, B, and C has been combined with another copy of itself, rotated 180° about the midpoint of $\overset{\bullet\,\bullet}{BC}$, to form a parallelogram.

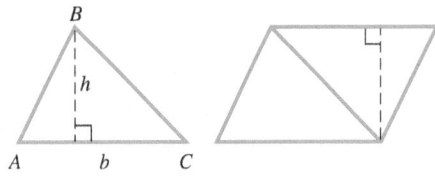

Figure 48

The area of this parallelogram is

$$\mathscr{A} = \text{base} \cdot \text{height}, \quad \text{or} \quad \mathscr{A} = bh.$$

However, the parallelogram has *twice* the area of the triangle, so the area of the triangle is *half* the area of the parallelogram.

AREA OF A TRIANGLE

The area $\mathscr{A}$ of a triangle with height h and base b is given by the following formula.

$$\mathscr{A} = \frac{1}{2}bh$$

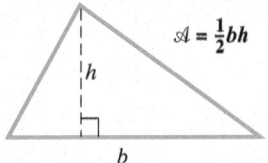

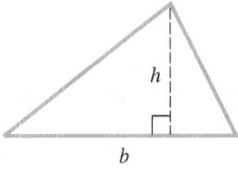

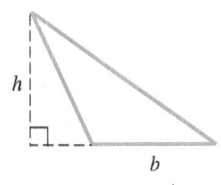

In each case, $\mathscr{A} = \frac{1}{2}bh$.

Figure 49

When applying the formula for the area of a triangle, remember that the height is the perpendicular distance between a vertex and the opposite side (or the extension of that side). See **Figure 49.**

Heron's formula gives the area of a triangle without using its height. Heron of Alexandria lived during the first century A.D., and although the formula is named after him, there is evidence that it was known to Archimedes several centuries earlier.

Given a triangle with side lengths a, b and c, let

$$s = \frac{1}{2}(a + b + c)$$

represent the **semiperimeter.** Then the area of the triangle is given by the formula

$$\mathscr{A} = \sqrt{s(s-a)(s-b)(s-c)}.$$

The Vietnam Veterans Memorial in Washington, D.C., is in the shape of an unenclosed isosceles triangle. Its walls measure 246.75 feet. The distance between the ends of the wall is 438.14 feet. Use Heron's formula to show that the area enclosed by the triangular shape is approximately 24,900 ft².

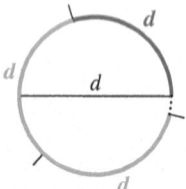

Figure 51

EXAMPLE 7 Finding the Height of a Triangular Sail

The area of a triangular sail of a sailboat is 126 ft². The base of the sail is 12 ft. Find the height of the sail.

Solution

Step 1 **Read the problem.** We must find the height of the triangular sail.

Step 2 **Assign a variable.** Let $h =$ the height of the sail in feet. See **Figure 50.**

Step 3 **Write an equation.** Using the information given in the problem, we substitute 126 for $\mathscr{A}$ and 12 for b in the formula for the area of a triangle.

$$\mathscr{A} = \frac{1}{2}bh \qquad \text{Area formula}$$

$$126 = \frac{1}{2}(12)h \qquad \mathscr{A} = 126,\ b = 12$$

Step 4 **Solve.** $126 = 6h$ Multiply.

$21 = h$ Divide by 6.

Step 5 **State the answer.** The height of the sail is 21 feet.

Step 6 **Check** to see that the values $\mathscr{A} = 126$, $b = 12$, and $h = 21$ satisfy the formula for the area of a triangle.

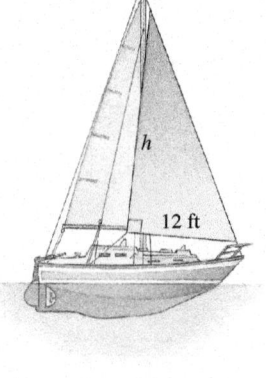

Figure 50

Circumference of a Circle

The distance around a circle is its **circumference** (rather than its perimeter). To understand the formula for the circumference of a circle, use a piece of string to measure the distance around a circular object. Then find the object's diameter and divide the circumference by the diameter. The quotient is the same, no matter what the size of the circular object is, and it will be an approximation for the number π.

$$\pi = \frac{\text{circumference}}{\text{diameter}} = \frac{C}{d}, \quad \text{or, alternatively,} \quad C = \pi d$$

Recall that π is not a rational number. **Figure 51** shows that it takes slightly more than three (about 3.14) diameters to make the circumference. In this chapter we will use 3.14 as an approximation for π when one is required.

CIRCUMFERENCE OF A CIRCLE

The circumference C of a circle of diameter d is given by the following formula.

$$C = \pi d$$

Also, since $d = 2r$, the circumference C of a circle of radius r is given by the following formula.

$$C = 2\pi r$$

$C = \pi d$

$C = 2\pi r$

EXAMPLE 8 **Finding the Circumference of a Circle**

Find the circumference of each circle described. Use $\pi \approx 3.14$.

(a) A circle with diameter 12.6 centimeters

(b) A circle with radius 1.70 meters

Solution

(a)

$$C = \pi d \qquad \text{Circumference formula}$$

$$C \approx (3.14)(12.6) \qquad \pi \approx 3.14, d = 12.6$$

$$C = 39.6 \qquad \text{Multiply.}$$

The circumference is about 39.6 centimeters, rounded to the nearest tenth.

(b)

$$C = 2\pi r \qquad \text{Circumference formula}$$

$$C \approx 2(3.14)(1.70) \qquad \pi \approx 3.14, r = 1.70$$

$$C = 10.7 \qquad \text{Multiply.}$$

The circumference is approximately 10.7 meters.

Area of a Circle

Start with a circle as shown in **Figure 52(a),** divided into many equal wedge-shaped pieces **(sectors).** Rearrange the pieces into an approximate rectangle as shown in **Figure 52(b).** The circle has circumference $2\pi r$, so the "length" of the approximate rectangle is one-half of the circumference, or $\frac{1}{2}(2\pi r) = \pi r$, while its "width" is r. The area of the approximate rectangle is length times width, or $(\pi r)r = \pi r^2$. As we choose smaller and smaller sectors, the figure becomes closer and closer to a rectangle, so its area becomes closer and closer to πr^2.

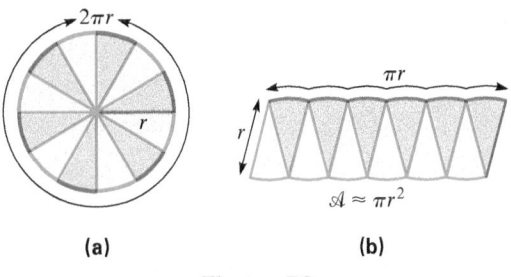

(a)   (b)

Figure 52

On March 8, 2014, Malaysia Airlines Flight 370 (a Boeing 777) disappeared while en route from Malaysia to China. Officials estimated that the plane had about a half-hour of fuel left when satellites received last contact over the Indian Ocean.

If a Boeing 777 can fly 950 km/hr at top speed, and if the plane's flight direction was unknown after its last known location, approximately how large an area should be searched? The answer can be found on the next page.

AREA OF A CIRCLE

The area $\mathcal{A}$ of a circle with radius r is given by the following formula.

$$\mathcal{A} = \pi r^2$$

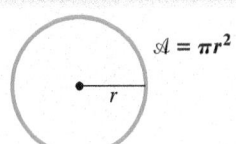

Problem-Solving Strategy

The formula for the area of a circle can be used to determine the best value for your money the next time you purchase a pizza. The next example uses the idea of unit pricing.

Solution to Margin Note problem If the plane travels at full speed of 950 km/hr for half an hour in an unknown direction, then the search area would be a circle centered at the last known location with a radius of

$$r = \frac{950}{2} = 475 \text{ km.}$$

The area of this circle would be

$$\mathcal{A} = \pi (475 \text{ km})^2$$

$$\mathcal{A} \approx 710{,}000 \text{ km}^2.$$

EXAMPLE 9   Using Area to Determine Better Value for Pizza

Paw-Paw Johnny's delivers pizza. The price of an 8-inch-diameter pizza is $8.99, and the price of a 16-inch-diameter pizza is $19.98. Which is the better buy?

Solution

To determine which pizza is the better value for the money, we must find the area of each, and then divide the price by the area to determine the price per square inch.

8-inch-diameter pizza area $= \pi (4 \text{ in.})^2 \approx 50.2 \text{ in.}^2$ Radius is $\frac{1}{2}(8 \text{ in.}) = 4 \text{ in.}$

16-inch-diameter pizza area $= \pi (8 \text{ in.})^2 \approx 201 \text{ in.}^2$ Radius is $\frac{1}{2}(16 \text{ in.}) = 8 \text{ in.}$

The price per square inch for the 8-inch pizza is

$$\frac{\$8.99}{50.2} \approx 17.9\,\cancel{c}.$$

The price per square inch for the 16-inch pizza is

$$\frac{\$19.98}{201} \approx 9.9\,\cancel{c}.$$

The 16-inch pizza is the better buy. It costs less per square inch.

WHEN WILL I **EVER** USE THIS**?**

Suppose you are a pest control specialist and need to treat the perimeter of a home to prevent termite infestation. The insecticide concentrate that you use calls for you to dig a narrow trench around the perimeter of the home, to add $\frac{3}{4}$ cup of insecticide to every gallon of water you use, and to apply 4 gallons of water for every 10 feet of trench.

How much of the concentrate will you need?

Your measurements around the perimeter of the home, including a semicircular wood deck, are shown in **Figure 53** (with dimensions given in feet). Starting at the lower left corner of the figure, the perimeter is calculated as follows.

$$P = 35 + 40 + 10 + 15 + 45 + 25 + \frac{1}{2}(2 \cdot \pi \cdot 10) + 10$$

$\frac{1}{2}(55 - 35)$

Circumference of deck

$P = 180 + 10\pi$ Simplify.

$P \approx 180 + 10(3.14)$ Use 3.14 for π.

$P = 211.4 \text{ ft}$ Multiply, then add.

Once the perimeter is determined, you can calculate the amount of concentrate needed.

$$211.4 \text{ ft}\left(\frac{3 \text{ cups}}{4 \text{ gal}}\right)\left(\frac{4 \text{ gal}}{10 \text{ ft}}\right) = 63.42 \text{ cups}\left(\frac{1 \text{ gal}}{16 \text{ cups}}\right) \approx 4 \text{ gallons}$$

1 gal = 16 cups

You will need about 4 gallons of concentrate to effectively protect the home from termite infestation.

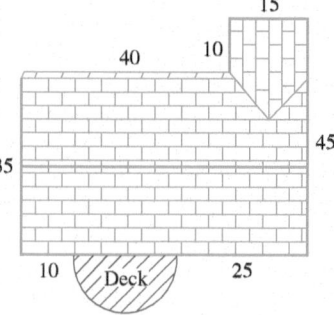

Figure 53

9.4 EXERCISES

CONCEPT CHECK *Fill in the blank with the correct response.*

1. The perimeter of an equilateral triangle with side length equal to_____inches is the same as the perimeter of a rectangle with length 15 inches and width 12 inches.

2. A square with area 64 cm² has perimeter _____ cm.

3. If the area of a certain triangle is 48 square inches, and the base measures 12 inches, then the height must measure _____ inches.

4. If the radius of a circle is tripled, then its area is multiplied by a factor of _____.

5. Circumference is to a circle as_____is to a polygon.

6. *Perimeter or Area?* *Decide whether perimeter or area would be used to solve a problem concerning the measure of the quantity.*

   (a) Sod for a lawn

   (b) Carpeting for a bedroom

   (c) Baseboards for a living room

   (d) Fencing for a yard

   (e) Fertilizer for a garden

   (f) Tile for a bathroom

   (g) Determining the cost of planting rye grass in a lawn for the winter

   (h) Determining the cost of replacing a linoleum floor with a wood floor

*Use the formulas of this section to find **(a)** the perimeter, and **(b)** the area, of each figure.*

7.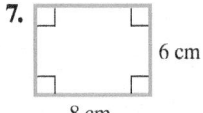
6 cm
8 cm

8.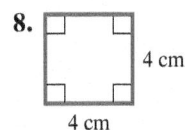
4 cm
4 cm

9.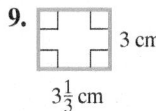
3 cm
$3\frac{1}{3}$ cm

10.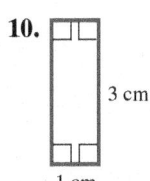
3 cm
1 cm

11.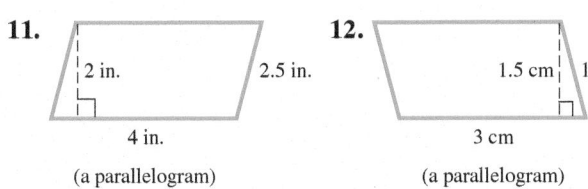
2 in.   2.5 in.
4 in.
(a parallelogram)

12.
1.5 cm   1.6 cm
3 cm
(a parallelogram)

13.
38 mm   39.6 mm
22 mm

14.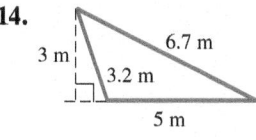
3 m   6.7 m
3.2 m
5 m

15.
b = 3 cm
h = 2 cm   2.8 cm
B = 5 cm
(a trapezoid)

16.
b = 5 cm
5.1 cm   h = 5 cm   5.1 cm
B = 7 cm
(a trapezoid)

*Use the formulas from this section to find **(a)** the circumference and **(b)** the area. Use 3.14 as an approximation for π, and round to the nearest hundredth.*

17.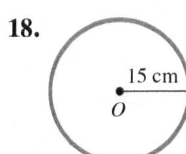
1 cm
O

18.
15 cm
O

19.
36 m
O

20.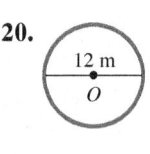
12 m
O

Solve each problem.

21. *Window Side Length* A stained-glass window in a church is in the shape of a rhombus. The perimeter of the rhombus is 7 times the length of a side in meters, decreased by 12. Find the length of a side of the window.

22. *Dimensions of a Rectangle* A video rental machine has a rectangular display beside it advertising several movies inside. The display's length is 18 in. more than the width, and the perimeter is 180 in. What are the dimensions of the display?

23. *Dimensions of a Lot* A lot is in the shape of a triangle. One side is 100 ft longer than the shortest side. The third side is 200 ft longer than the shortest side. The perimeter of the lot is 1200 ft. Find the lengths of the sides.

24. *Pennant Side Lengths* A wall pennant is in the shape of an isosceles triangle. Each of the two equal sides measures 18 in. more than the third side, and the perimeter of the triangle is 54 in. What are the lengths of the sides of the pennant?

25. *Radius of a Circular Foundation* A hotel is in the shape of a cylinder, with a circular foundation. The circumference of the foundation is 6 times the radius, increased by 14 ft. Find the radius of the circular foundation. (Use 3.14 as an approximation for π.)

26. *Radius of a Circle* If the radius of a certain circle is tripled, with 8.2 cm then added, the result is the circumference of the circle. Find the radius of the circle. (Use 3.14 as an approximation for π.)

27. *Area of Two Lots* The survey plat in the figure below shows two lots that form a trapezoid. The measures of the parallel sides are 115.80 ft and 171.00 ft. The height of the trapezoid is 165.97 ft. Find the combined area of the two lots. Round your answer to the nearest hundredth of a square foot.

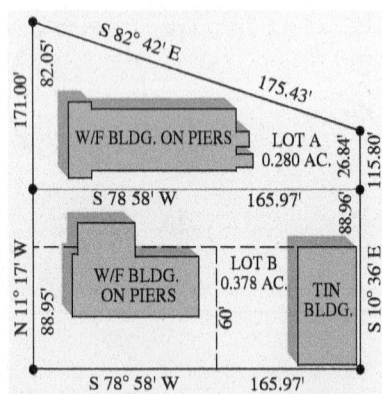

28. *Area of a Lot* Lot A in the figure is in the shape of a trapezoid. The parallel sides measure 26.84 ft and 82.05 ft. The height of the trapezoid is 165.97 ft. Find the area of Lot A. Round your answer to the nearest hundredth of a square foot.

29. *Search Area* A search plane carries radar equipment that can detect metal objects (like submarine periscopes or plane wreckage) on the ocean surface up to 15.5 miles away. If the plane completes a circular flight pattern of 471 miles in circumference, how much area will it search? (Use 3.14 as an approximation for π, and round to the nearest 100 mi^2.)

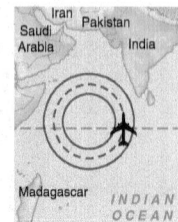

30. *Flight Path* If the search plane from **Exercise 29** needs to search a circular swath with an area of 20,000 square miles, how far will the plane need to fly (in a circle)? (Round to the nearest mile.)

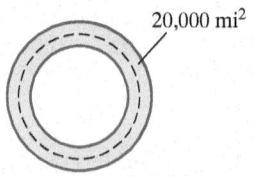

20,000 mi^2

In the chart below, the value of r (radius), d (diameter), C (circumference), or A (area) is given for a particular circle. Find the remaining three values. Leave π in your answers.

	r	d	C	$\mathscr{A}$
31.	6 in.			
32.	9 in.			
33.		10 ft		
34.		40 ft		
35.			12π cm	
36.			18π cm	
37.				100π in.^{2}
38.				256π in.^{2}
39.				$\frac{400}{\pi}$ yd^2
40.			60 m	

Each figure has the perimeter indicated. (Figures are not necessarily to scale.) Find the value of x.

41. $P = 68$

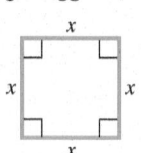

42. $P = 54$

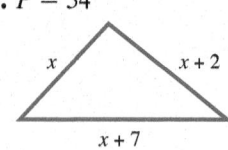

43. $P = 38$

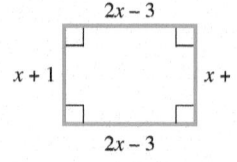

44. $P = 278$

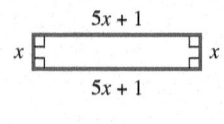

Each figure has the area indicated. Find the value of x.

45. $\mathscr{A} = 32.49$

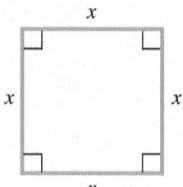

46. $\mathscr{A} = 28$

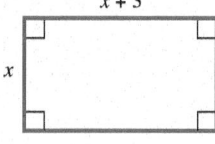

47. $\mathcal{A} = 21$

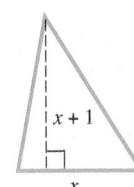

48. $\mathcal{A} = 30$

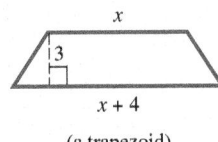

(a trapezoid)

Each circle has the circumference or area indicated. Find the value of x. Use 3.14 as an approximation for π.

49. $C = 37.68$

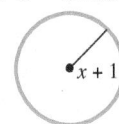

50. $C = 54.95$

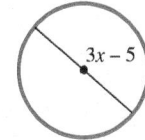

51. $\mathcal{A} = 18.0864$

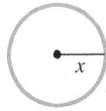

52. $\mathcal{A} = 28.26$

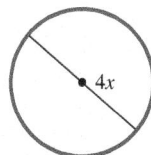

53. Work the parts of this exercise in order, and make a generalization concerning areas of rectangles.

(a) Find the area of a rectangle 4 cm by 5 cm.

(b) Find the area of a rectangle 8 cm by 10 cm.

(c) Find the area of a rectangle 12 cm by 15 cm.

(d) Find the area of a rectangle 16 cm by 20 cm.

(e) The rectangle in part (b) had sides twice as long as the sides of the rectangle in part (a). Divide the larger area by the smaller. Doubling the sides made the area increase _____ times.

(f) To get the rectangle in part (c), each side of the rectangle in part (a) was multiplied by _____. This made the larger area _____ times the size of the smaller area.

(g) To get the rectangle of part (d), each side of the rectangle of part (a) was multiplied by _____. This made the area increase to _____ times what it was originally.

(h) In general, if the length of each side of a rectangle is multiplied by n, the area is multiplied by _____.

*Use the results of **Exercise 53** to complete the following.*

54. If the height of a triangle is multiplied by n and the base length remains the same, then the area of the triangle is multiplied by _____.

55. If the radius of a circle is multiplied by n, then the area of the circle is multiplied by _____.

*Use the results of **Exercise 53** to solve each problem.*

56. *Cost of Paint* A ceiling measuring 9 ft by 15 ft can be painted for $60. How much would it cost to paint a ceiling 18 ft by 30 ft?

57. *Cost of Carpet* Suppose carpet for a room 10 ft by 12 ft costs $200. Find the cost of carpet for a room 20 ft by 24 ft.

58. *Amount of Shampoo* A carpet cleaner uses 8 oz of shampoo to clean an area 31 ft by 31 ft. How much shampoo would be needed for an area 93 ft by 93 ft?

Total Area as the Sum of Areas *By considering total area as the sum of the areas of all of its parts, we can determine the area of a figure such as those in Exercises 59–62. Find the total area of each figure. Use 3.14 as an approximation for π in Exercises 61 and 62, and round to the nearest hundredth.*

59.

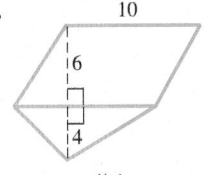

(a parallelogram and a triangle)

60.

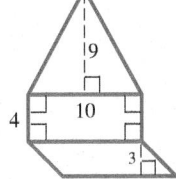

(a triangle, a rectangle, and a parallelogram)

61.

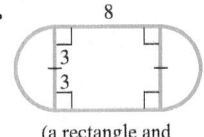

(a rectangle and two semicircles)

62.

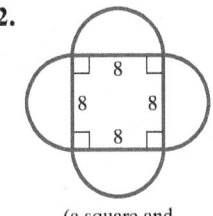

(a square and four semicircles)

Area of a Shaded Portion of a Plane Figure *The shaded areas of the figures in Exercises 63–68 may be found by subtracting the area of the unshaded portion from the total area of the figure. Use this approach to find the area of the shaded portion. Use 3.14 as an approximation for π in Exercises 66–68, and round to the nearest hundredth.*

63.

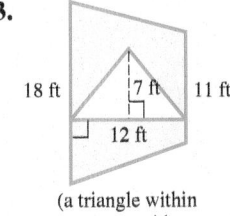

(a triangle within a trapezoid)

64.

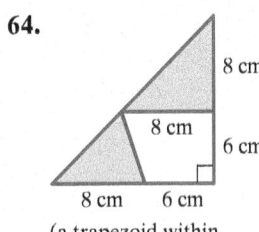

(a trapezoid within a triangle)

65.

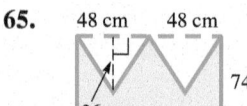

48 cm   48 cm
74 cm
36 cm

(two congruent triangles
within a rectangle)

66.

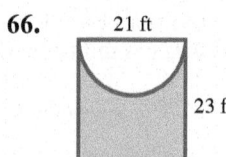

21 ft
23 ft

(a semicircle within
a rectangle)

67.

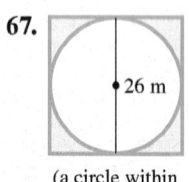

• 26 m

(a circle within
a square)

68.

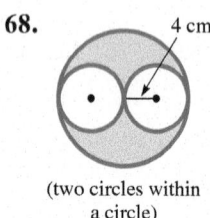

4 cm

(two circles within
a circle)

Pizza Pricing *The following exercises show prices actually charged by Old Town Pizza, a local pizzeria. The diameters of the pizzas are given. Find the best buy.*

Menu Item	Prices			
	10-in.	12-in.	14-in.	16-in.
69. Cheese pizza with one topping	$11	$14	$18	$21
70. Cheese pizza with two toppings	$12.50	$15.75	$20	$23.25
71. Choo Choo Chicken	$15	$19	$24	$28
72. Steam Engine plus two toppings	$18	$22.50	$28	$32.50

James Garfield's Proof of the Pythagorean Theorem *James A. Garfield, the twentieth president of the United States, provided a proof of the Pythagorean theorem using the figure in the next column. Supply the required information in each of Exercises 73–76, in order, to follow his proof.*

73. Find the area of the trapezoid $WXYZ$ using the formula for the area of a trapezoid.

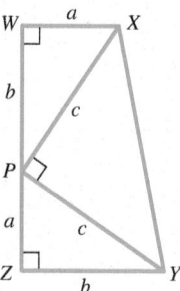

74. Find the area of each of the right triangles PWX, PZY, and PXY.

75. Because the sum of the areas of the three right triangles must equal the area of the trapezoid, set the expression from **Exercise 73** equal to the sum of the three expressions from **Exercise 74.**

76. Simplify the terms of the equation from **Exercise 75** as much as possible. What is the result?

A polygon can be inscribed within a circle or circumscribed about a circle. In the figure, triangle ABC is inscribed within the circle, while square WXYZ is circumscribed about it. These ideas will be used in some of the remaining exercises in this section and later in this chapter.

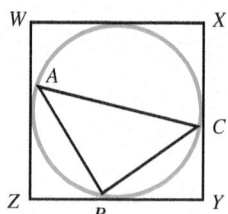

Exercises 77–88 require some ingenuity, but all may be solved using the concepts presented so far in this chapter.

77. Shaded Area C is the center of the accompanying circle, and F is a point on the circle such that $BCDF$ is a 3 cm × 4 cm rectangle. What is the area of the shaded region? (*MT* calendar problem)

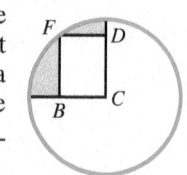

78. Perimeter of a Rectangle If the equilateral triangle and the rectangle in the figure have the same area and the triangle has a perimeter of 3, what is the perimeter of the rectangle? (*MT* calendar problem)

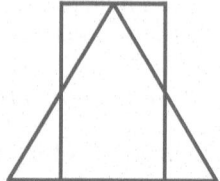

79. *Area of a Rectangle* In a circle of radius r, two parallel chords of length r form opposite sides of a rectangle. What is the area of the inscribed rectangle in terms of r? (*MT* calendar problem)

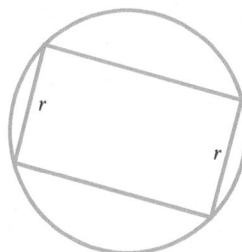

80. *Area of a Quadrilateral* The rectangle $ABCD$ has length twice the width. If P, Q, R, and S are the midpoints of the sides, and the perimeter of $ABCD$ is 96 in., what is the area of quadrilateral $PQRS$?

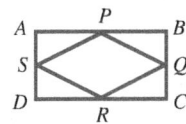

81. *Area of a Quadrilateral* In the figure, E and F are the midpoints of two sides of quadrilateral $ABCD$. If the area of quadrilateral $FAEC$ is 13 in.^{2}, what is the area of quadrilateral $ABCD$? (*MT* calendar problem)

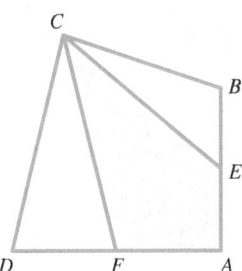

82. *Area of Trapezoid* The three squares have the dimensions indicated in the diagram. What is the area of the shaded trapezoid? (*MT* calendar problem)

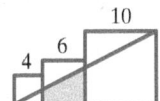

83. *Area of Shaded Region* The radius of each circle is r. Find the area of the shaded region. (*MT* calendar problem)

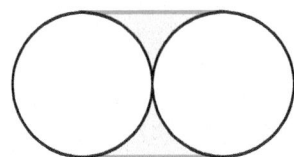

84. *Area of a Pentagon* In the figure, pentagon $PQRST$ is formed by a square and an equilateral triangle such that $PQ = QR = RS = ST = PT$. The perimeter of the pentagon is 80. Find the area of the pentagon.

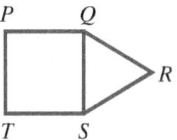

85. *Area of a Quadrilateral* Square $ABCD$ has side length 8. Find the area of quadrilateral $PQRS$ if $AP = 1$, $DQ = 2$, $CR = 3$, and $BS = 4$. (*MT* calendar problem)

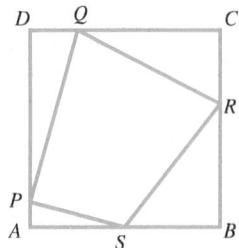

86. *Base Measure of an Isosceles Triangle* An isosceles triangle has a base of 24 and two sides of 13. What other base measure can an isosceles triangle with equal sides of 13 have and still have the same area as the given triangle?

Exercises 87 and 88 refer to the given figure. The center of the circle is O.

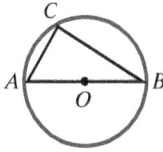

87. *Radius of a Circle* If $\overline{AC}$ measures 6 in. and $\overline{BC}$ measures 8 in., what is the radius of the circle?

88. *Lengths of Chords of a Circle* If $\overline{AB}$ measures 13 cm, and the length of $\overline{BC}$ is 7 cm more than the length of $\overline{AC}$, what are the lengths of $\overline{BC}$ and $\overline{AC}$?

9.5 VOLUME AND SURFACE AREA

Space Figures

Thus far, this chapter has discussed only **plane figures**—figures that can be drawn completely in the plane of a sheet of paper. However, it takes the three dimensions of space to represent the solid world around us. For example, **Figure 54** shows a "box" (a **rectangular parallelepiped**). The *faces* of a box are rectangles. The faces meet at *edges;* the "corners" are *vertices* (plural of *vertex*—the same word that is used for the "corner" of an angle).

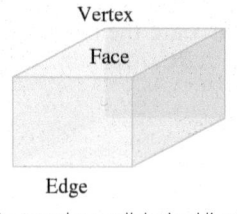

Rectangular parallelepiped (box)

Figure 54

Boxes are one kind of **space figure** belonging to an important group called **polyhedra,** the faces of which are made only of polygons. Perhaps the most interesting polyhedra are the *regular polyhedra*. Recall that a *regular polygon* is a polygon with all sides equal and all angles equal. A regular polyhedron is a space figure, the faces of which are only one kind of regular polygon. It turns out that there are only five different regular polyhedra. They are shown in **Figure 55**. A **tetrahedron** is composed of four equilateral triangles, each three of which meet in a point. Use the figure to verify that there are four faces, four vertices, and six edges.

Polyhedral dice such as the ones shown here are often used in today's role-playing games. (Can you identify one that is not regular?)

The five regular polyhedra are also known as **Platonic solids,** named for the Greek philosopher Plato. He considered them as "building blocks" of nature and assigned fire to the tetrahedron, earth to the cube, air to the octahedron, and water to the icosahedron. Because the dodecahedron is different from the others due to its pentagonal faces, he assigned to it the cosmos (stars and planets). (*Source:* http://platonicrealms.com/encyclopedia/Platonic-solid) An animated view of the Platonic solids can be found at www.3quarks.com/en/PlatonicSolids.

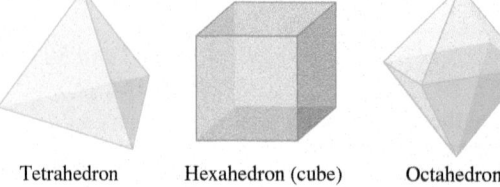 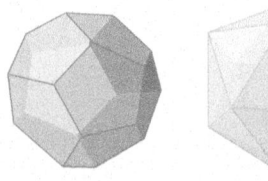

Tetrahedron Hexahedron (cube) Octahedron Dodecahedron Icosahedron

Figure 55

The four remaining regular polyhedra are the **hexahedron,** the **octahedron,** the **dodecahedron,** and the **icosahedron.** The hexahedron, or cube, is composed of six squares, each three of which meet at a point. The octahedron is composed of groups of four regular (i.e., equilateral) triangles meeting at a point. The dodecahedron is formed by groups of three regular pentagons, and the icosahedron is made up of groups of five regular triangles.

Two other types of polyhedra are familiar space figures: pyramids and prisms. **Pyramids** are made of triangular sides and a polygonal base. **Prisms** have two faces in parallel planes; these faces are congruent polygons. The remaining faces of a prism are all parallelograms. (See **Figures 56(a)** and **(b)** on the next page.) By this definition, a box is also a prism.

Figure 56(c) shows space figures made up in part of circles, including *right circular cones* and *right circular cylinders.* It also shows how a circle can generate a *torus,* a doughnut-shaped solid that has interesting topological properties (covered later in this chapter).

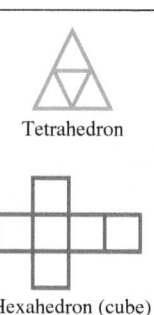

Tetrahedron

Hexahedron (cube)

Octahedron

Dodecahedron

Icosahedron

Patterns such as these may be used to construct three-dimensional models of the **regular polyhedra.** See

www.korthalsaltes.com

for some examples.

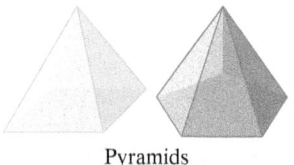
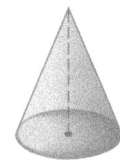

Pyramids

(a)

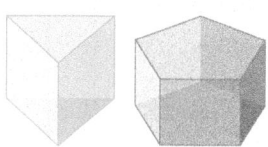
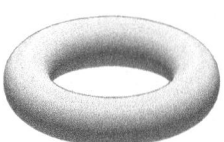

Prisms

(b)

Right circular cone   Right circular cylinder   A rotating circle generates a torus.

(c)

Figure 56

Volume and Surface Area of Space Figures

While area is a measure of surface covered by a plane figure, **volume** is a measure of capacity of a space figure. Volume is measured in *cubic* units. For example, a cube with edge measuring 1 cm has volume 1 cubic centimeter, which is also written as 1 cm³, or 1 cc. The **surface area** is the total area that would be covered if the space figure were "peeled" and the peel laid flat, as shown in the margin. Surface area is measured in *square* units.

VOLUME AND SURFACE AREA OF A BOX

Suppose that a box has length ℓ, width w, and height h. Then the volume V and the surface area S are given by the following formulas.

$$V = \ell wh \qquad \text{and} \qquad S = 2\ell w + 2\ell h + 2hw$$

If the box is a cube with edge of length s, the formulas are as follows.

$$V = s^3 \qquad \text{and} \qquad S = 6s^2$$

$V = \ell wh$

$S = 2\ell w + 2\ell h + 2hw$

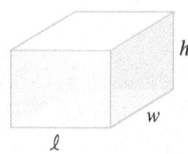
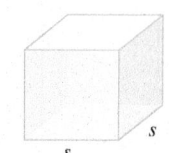

$V = s^3$

$S = 6s^2$

EXAMPLE 1   Using the Formulas for a Box

Find the volume V and the surface area S of the box shown in **Figure 57.**

Solution

$$V = \ell wh \qquad \text{Volume formula}$$
$$V = 14 \cdot 7 \cdot 5 \qquad \text{Substitute.}$$
$$V = 490 \qquad \text{Multiply.}$$

Volume is measured in cubic units, so the volume of the box is 490 cubic centimeters, or 490 cm³.

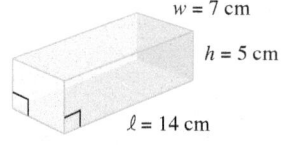

$w = 7$ cm

$h = 5$ cm

$\ell = 14$ cm

Figure 57

Polyhedral Origami Constructed entirely from folded strips of paper, this dodecahedral 11-hole torus, created by **David Honda,** was awarded Best textile, sculpture, or other medium at the 2018 Mathematical Art Exhibition. (For this and other examples of polyhedral paper structures, see http://snaporigami.weebly.com.)

$$S = 2\ell w + 2\ell h + 2hw \qquad \text{Surface area formula}$$

$$S = 2(14)(7) + 2(14)(5) + 2(5)(7) \qquad \text{Substitute.}$$

$$S = 196 + 140 + 70 \qquad \text{Multiply.}$$

$$S = 406 \qquad \text{Add.}$$

Surface areas of space figures are measured in square units, so the surface area of the box is 406 square centimeters, or 406 cm^2.

A typical tin can is an example of a **right circular cylinder.**

VOLUME AND SURFACE AREA OF A RIGHT CIRCULAR CYLINDER

If a right circular cylinder has height h and the radius of its base is equal to r, then the volume V and the surface area S are given by the following formulas.

$$V = \pi r^2 h$$

and $\qquad S = 2\pi rh + 2\pi r^2$

$$V = \pi r^2 h$$
$$S = 2\pi rh + 2\pi r^2$$

(In the formula for S, the areas of the top and bottom are included.)

EXAMPLE 2 **Using the Formulas for a Right Circular Cylinder**

In **Figure 58,** the volume of medication in the syringe is 10 mL (which is equivalent to 10 cm^3). Find each measure. Use 3.14 as an approximation for π.

(a) The radius of the cylindrical syringe (round to the nearest 0.1 cm)

(b) The surface area of the medication (round to the nearest 0.1 cm^2)

Solution

(a)

$$V = \pi r^2 h \qquad \text{Volume formula}$$

$$10 = \pi r^2 (5.0) \qquad V = 10, \ h = 5.0$$

$$\frac{2}{\pi} = r^2 \qquad \text{Divide by } 5.0\pi.$$

$$r = \sqrt{\frac{2}{\pi}} \approx 0.8 \qquad \text{Take the square root and approximate.}$$

The radius is approximately 0.8 cm.

(b)

$$S = 2\pi rh + 2\pi r^2 \qquad \text{Surface area formula}$$

$$S = 2\pi(0.8)(5.0) + 2\pi(0.8)^2 \qquad r = 0.8, \ h = 5.0$$

$$S = 9.28\pi \qquad \text{Multiply; add.}$$

$$S \approx 29.1 \qquad \text{Approximate, using 3.14 for } \pi.$$

The surface area is approximately 29.1 cm^2.

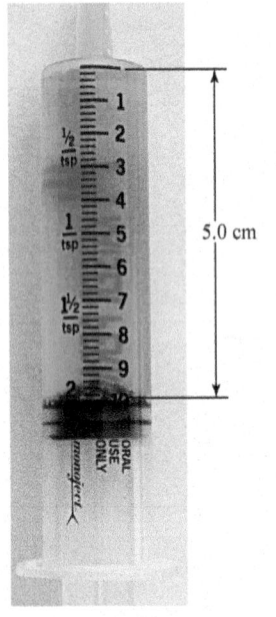

5.0 cm

Figure 58

The three-dimensional analogue of a circle is a **sphere.** It is defined by replacing the word "plane" with "space" in the definition of a circle.

VOLUME AND SURFACE AREA OF A SPHERE

If a sphere has radius r, then the volume V and the surface area S are given by the following formulas.

$$V = \frac{4}{3}\pi r^3 \quad \text{and} \quad S = 4\pi r^2$$

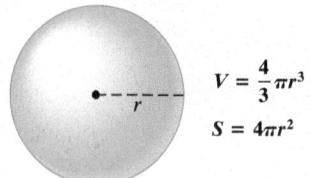

$$V = \frac{4}{3}\pi r^3$$
$$S = 4\pi r^2$$

A cone with circular base having its apex (highest point) directly above the center of its base is a **right circular cone.**

VOLUME AND SURFACE AREA OF A RIGHT CIRCULAR CONE

If a right circular cone has height h and the radius of its circular base is r, then the volume V and the surface area S are given by the following formulas.

$$V = \frac{1}{3}\pi r^2 h$$

and

$$S = \pi r \sqrt{r^2 + h^2} + \pi r^2$$

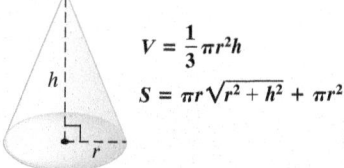

$$V = \frac{1}{3}\pi r^2 h$$
$$S = \pi r \sqrt{r^2 + h^2} + \pi r^2$$

(In the formula for S, the area of the circular base is included.)

EXAMPLE 3   **Comparing Volumes Using Ratios**

Figure 59 shows a right circular cone inscribed in a semi-sphere of radius r. What is the ratio of the volume of the cone to the volume of the hemisphere?

Solution

First, use the formula for the volume of a cone. Note that because the cone is inscribed in the hemisphere, its height is equal to its radius.

$$V_1 = \text{Volume of the cone} = \frac{1}{3}\pi r^2 h = \frac{1}{3}\pi r^3 \quad \text{Use } h = r.$$

The hemisphere will have half the volume of a sphere of radius r.

$$V_2 = \text{Volume of the hemisphere} = \frac{1}{2} \cdot \frac{4}{3}\pi r^3 = \frac{2}{3}\pi r^3$$

Now find the ratio of the first volume to the second.

$$\frac{V_1}{V_2} = \frac{\frac{1}{3}\pi r^3}{\frac{2}{3}\pi r^3} = \frac{1}{2} \qquad \text{The ratio is } \frac{1}{2}.$$

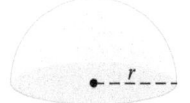

Figure 59

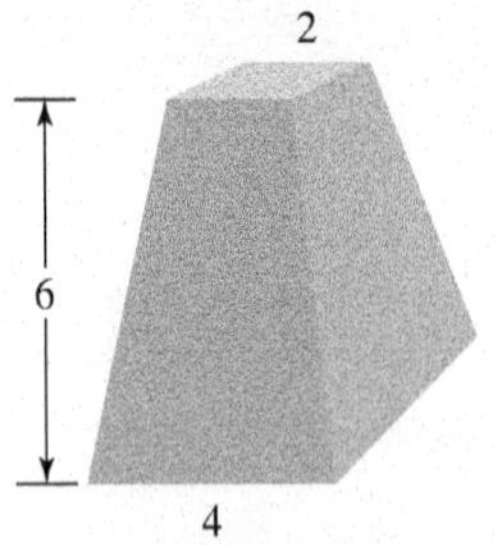

2

6

4

A problem concerning the **frustum of a pyramid** like the one shown above is included in the **Moscow papyrus**, which dates back to about 1850 B.C. Problem 14 in the document reads:

> You are given a truncated pyramid of 6 for the vertical height by 4 on the base by 2 on the top. You are to square this 4, result 16. You are to double 4, result 8. You are to square 2, result 4. You are to add the 16, the 8, and the 4, result 28. You are to take one-third of 6, result 2. You are to take 28 twice, result 56. See, it is 56. You will find it right.

The formula for finding the volume of the frustum of a pyramid with square bases is

$$V = \frac{1}{3}h(B^2 + Bb + b^2),$$

where B is the side length of the lower base, b is the side length of the upper base, and h is the height (or altitude).

A **pyramid** is a space figure having a polygonal base and triangular sides.

VOLUME OF A PYRAMID

If B represents the area of the base of a pyramid, and h represents the height (that is, the perpendicular distance from the top, or apex, to the base), then the volume V is given by the following formula.

$$V = \frac{1}{3}Bh$$

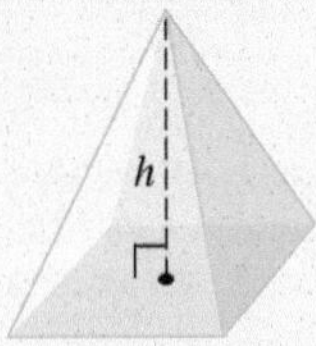

$V = \frac{1}{3}Bh,$ where B is the area of the base

EXAMPLE 4 Using the Volume Formula for a Pyramid

The Great Pyramid at Giza has a square base. When originally constructed, its base measured *about* 230 meters on a side, and it was *about* 147 meters high. Find its volume to the nearest ten thousand cubic meters.

Solution

Use the formula for the volume of a pyramid. Since the base is square, the area of the base is the square of the side length.

$V = \frac{1}{3}Bh$ Volume formula

$V \approx \frac{1}{3}(230)^2(147)$ $B \approx (230)^2, h \approx 147$

$V \approx 2{,}590{,}000 \text{ m}^3$ Use a calculator, round to the nearest ten thousand.

The volume was approximately 2,590,000 m³.

EXAMPLE 5 Using a Combination of Volume Formulas

Find the volume of the snow cone in **Figure 60** to the nearest cubic centimeter. Use 3.14 as an approximation for π.

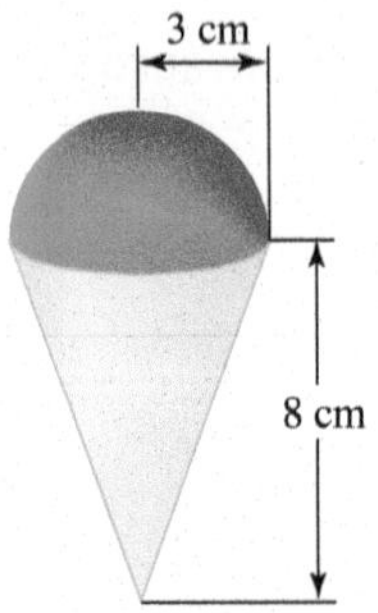

Solution

Use the formula for the volume of a cone and the formula for the volume of a hemisphere.

3 cm

8 cm

Figure 60

$V = \frac{1}{3}\pi r^2 h + \frac{1}{2}\cdot\frac{4}{3}\pi r^3$ Volume formulas for a cone and a hemisphere

$V = \frac{1}{3}\pi(3)^2(8) + \frac{1}{2}\cdot\frac{4}{3}\pi(3)^3$ $r = 3, h = 4$

$V = 42\pi$ Multiply; add.

$V \approx 132$ Approximate, using 3.14 for π

The volume is approximately 132 cm³.

WHEN WILL I **EVER** USE THIS?

As a video game designer, you have created a "virtual world" where users can navigate and see objects within a "volume of visibility" called the *view frustum*. It is the frustum of a pyramid with its apex at the vantage point of the player (see **Figure 61**). Its *near plane* (the screen) and *far plane* (at the limit of the visible range) are shown in green in **Figures 61 and 62**.

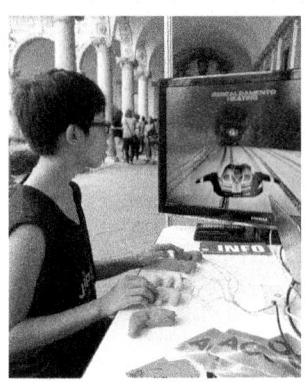

The software you are using for rendering three-dimensional objects allows for *frustum culling,* which instructs the graphics hardware to render only objects in the view frustum, rather than *all* of the objects in the virtual world. This allows for a much more efficient use of memory resources, leading to shorter loading times and a better player experience.

Suppose the screen is square, measuring about $\frac{1}{3}$ meter on a side, and is about 1 meter away from the user's eye. The range of visibility is 300 meters. *What is the volume of the view frustum? If the entire virtual world can be represented by a circular cylinder of radius 1000 meters and height 200 meters, what percentage of the world is contained within a single view frustum?*

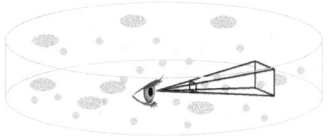

Figure 61	Figure 62

The view frustum is shown in **Figure 62**. It has square bases, so its volume is given by

$$V = \frac{1}{3}h(B^2 + Bb + b^2).$$ *B* is the side length of the larger base; *b* is the side length of the smaller base.

To find the value of *B*, we use similar triangles.

$$\frac{300}{1} = \frac{B}{\frac{1}{3}}$$ Ratios of corresponding sides are equal.

$$100 = B$$ Multiply using cross products.

Then

$$V = \frac{1}{3}h(B^2 + Bb + b^2)$$ Volume formula

$$V = \frac{1}{3}(299)\left(100^2 + 100\left(\frac{1}{3}\right) + \left(\frac{1}{3}\right)^2\right)$$ $h = 299, B = 100, b = \frac{1}{3}$

$$V \approx 1{,}000{,}000.$$ Simplify.

The volume of the view frustum is about 1,000,000 m³.

The volume of the entire virtual world is found as follows.

$$V = \pi r^2 h$$ Volume of a right circular cylinder

$$V = \pi(1000)^2(200)$$ $r = 1000, h = 200$

$$V \approx 3.14(1{,}000{,}000)(200)$$ $\pi \approx 3.14$; Multiply.

$$V = 628{,}000{,}000$$ Multiply.

A single view frustum contains about $\frac{1}{628} \approx 0.16\%$ of the entire virtual world.

9.5 **EXERCISES**

CONCEPT CHECK *Write* true *or false* *for each statement.*

1. A cube with volume 125 cubic inches has surface area 150 square inches.

2. A tetrahedron has the same number of faces as vertices.

3. A sphere with a 1-unit radius has three times as many units of surface area as it has units of volume.

4. Each face of an octahedron is an octagon.

5. If you double the length of the edge of a cube, the new cube will have a volume that is four times the volume of the original cube.

6. A dodecahedron can be used as a model for a calendar for a given year, where each face of the dodecahedron contains a calendar for a single month, and there are no faces left over.

Find **(a)** *the volume and* **(b)** *the surface area of each space figure. When necessary, use* 3.14 *as an approximation for* π, *and round answers to the nearest hundredth.*

7.

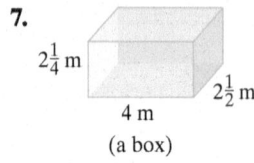

$2\frac{1}{4}$ m

4 m

$2\frac{1}{2}$ m

(a box)

8.

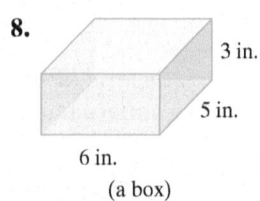

3 in.

5 in.

6 in.

(a box)

9.

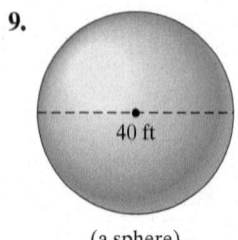

40 ft

(a sphere)

10.

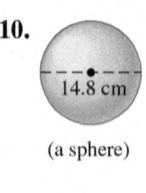

14.8 cm

(a sphere)

11.

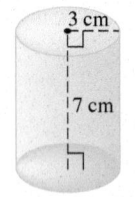

3 cm

7 cm

(a right circular cylinder)

12.

12 m

4 m

(a right circular cylinder)

13.

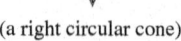

3 m

7 m

(a right circular cone)

14.

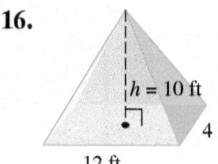

6 cm

4 cm

(a right circular cone)

Find the volume of each pyramid. In each case, the base is a rectangle.

15.

$h = 7$ in.

9 in.

8 in.

16.

$h = 10$ ft

4 ft

12 ft

Volumes of Common Objects *Find each volume. Use* 3.14 *as an approximation for* π *when necessary.*

17. a coffee can, radius 6.3 cm and height 15.8 cm

18. a soup can, radius 3.2 cm and height 9.5 cm

19. a pork-and-beans can, diameter 7.2 cm and height 10.5 cm

20. a cardboard mailing tube, diameter 2 in. and height 40 in.

21. a basketball, diameter 9 in.

22. a bottle of glue, diameter 3 cm and height 4.3 cm

23. the Red Pyramid at Dahshur, near Cairo—its base is a square 220 m on a side, and its height is 105 m

24. a grain silo in the shape of a right circular cylinder with a base radius of 7 m and a height of 25 m

25. a road construction marker, a cone with height 2 m and base radius $\frac{1}{2}$ m

26. the conical portion of a wizard's hat for a Halloween costume, with height 12 in. and base radius 4 in.

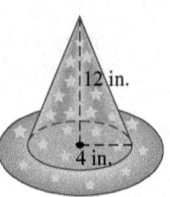

12 in.

4 in.

In the chart below, the value of r (radius), V (volume), or S (surface area) is given for a particular sphere. Find the remaining two values. Leave π in your answers.

	r	V	S
27.	6 in.		
28.	9 in.		
29.		$\frac{32}{3}\pi$ cm^3	
30.		$\frac{256}{3}\pi$ cm^3	
31.			4π m^2
32.			144π m^2

Solve each problem.

33. *Volume or Surface Area?* In order to determine the amount of liquid a spherical tank will hold, would you need to use volume or surface area?

34. *Volume or Surface Area?* In order to determine the amount of leather it would take to manufacture a basketball, would you need to use volume or surface area?

35. *Irrigation Tank* An irrigation tank is formed of a concrete "box" that is 5 ft wide, 5 ft deep, and 10 ft long, along with two cylindrical sleeves, each 2 ft high and 2.5 ft in diameter. What is the total volume of the tank? (Use 3.14 as an approximation for π, and round to the nearest cubic foot.)

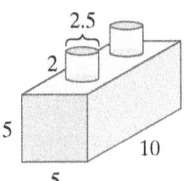

36. *Underground Bunker* An underground bunker is to be made of concrete 2 ft thick. It will be semi-spherical with outer radius of 20 feet. How many cubic feet of concrete will be needed to construct the bunker? (Use 3.14 as an approximation for π, and round to the nearest cubic foot.)

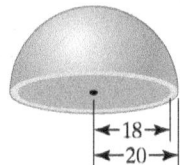

37. *Side Length of a Cube* One of the three famous construction problems of Greek mathematics required the construction of an edge of a cube with twice the volume of a given cube. If the length of each side of the given cube is x, what would be the length of each side of a cube with twice the original volume?

38. Work through the parts of this exercise in order, and use them to make a generalization concerning volumes of spheres. Leave answers in terms of π.

(a) Find the volume of a sphere having radius of 1 m.

(b) Suppose the radius is doubled to 2 m. What is the volume?

(c) When the radius was doubled, by how many times did the volume increase? (To find out, divide the answer for part (b) by the answer for part (a).)

(d) Suppose the radius of the sphere from part (a) is tripled to 3 m. What is the volume?

(e) When the radius was tripled, by how many times did the volume increase?

(f) In general, if the radius of a sphere is multiplied by n, the volume is multiplied by ____.

Cost to Fill a Spherical Tank *If a spherical tank 2 m in diameter can be filled with a liquid for \$300, find the cost to fill tanks of each diameter.*

39. 6 m   **40.** 8 m   **41.** 10 m

42. Use the logic of **Exercise 38** to answer the following: If the radius of a sphere is multiplied by n, then the surface area of the sphere is multiplied by ____.

43. *Volume Decrease* The radius of a sphere is decreased by 30%. By what percent does the volume decrease? Round to the nearest 0.1%.

44. *Surface Area Decrease* The length of each edge of a cube is decreased by 40%. By what percent does the surface area decrease?

Each of the following figures has the volume indicated. Find the value of x.

45. $V = 60$

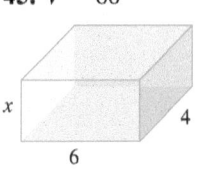

(a box)

46. $V = 450$

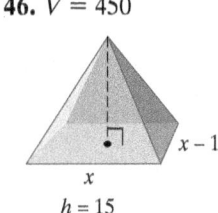

$h = 15$
Base is a rectangle.
(a pyramid)

47. $V = 36\pi$

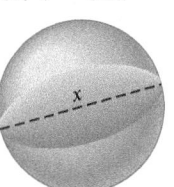

(a sphere)

48. $V = 245\pi$

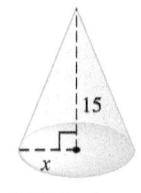

(a right circular cone)

Exercises 49–58 require some ingenuity, but all can be solved using the concepts presented so far in this chapter.

49. *Side Length of a Cube* The total surface area of a cube, expressed in square centimeters, has the same value as the cube's volume, expressed in cubic centimeters. Compute the length, in centimeters, of a side of the cube. (*MT* calendar problem)

50. *Ratio of Volumes* Three tennis balls are stacked in a cylindrical container that touches the stack on all sides, on the top, and on the bottom. What is the ratio of the volume filled with tennis balls to the volume of empty space in the container?

51. *Equal Area and Volume* The inhabitants of Planet Volarea have a unit of distance called a *volar*. The number of square volars in the planet's surface area is the same as the number of cubic volars in the planet's volume. If the diameter of Volarea is 1800 miles, how many miles are in a volar?

52. *Volume and Surface Area* What are the volume V and the surface area S of the right cylinder capped by two hemispheres pictured below? (*MT* calendar problem)

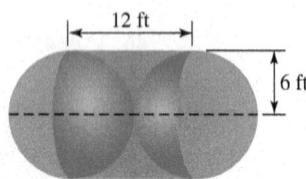

12 ft
6 ft

53. *Sum of Reciprocals* A cylinder with radius r and height h has volume 1 and total surface area 12. Compute $\frac{1}{r} + \frac{1}{h}$. (*MT* calendar problem)

54. *Increasing Sales* A company sells peanut butter in cylindrical jars. Marketing research suggests that using wider jars will increase sales. If the diameter of the jars is increased by 25% without altering the volume, by what percentage must the height be decreased? (*MT* calendar problem)

55. *Surface Area and Volume* Find the surface area S and volume V of a triangular prism, all of whose edges are $\sqrt{3}$. (*MT* calendar problem)

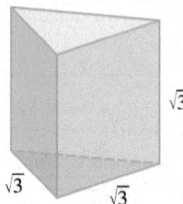

$\sqrt{3}$
$\sqrt{3}$
$\sqrt{3}$

56. *Ratio of Volumes* A cube is inscribed in a sphere. Find the ratio of the volume of the sphere to the volume of the cube.

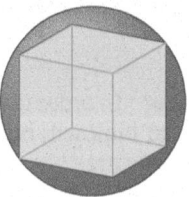

57. *Frustum of a Pyramid* The feature "When Will I Ever Use This?" at the end of this section shows an application of the *frustum* of a pyramid. Use the figure below to help verify the formula for the volume of the frustum of a pyramid with square base. (Note that in this formula, B and b are the lengths of the sides of the bases, *not* the areas of the bases.)

$$V = \frac{1}{3} h \left(B^2 + Bb + b^2 \right)$$

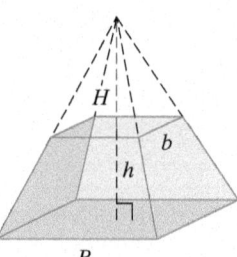

H
b
h
B

58. *Ratio of Surface Area to Base Area* The figure shows a pyramid with square base and with height equal to the side length of the base. Find the ratio of the entire surface area to the area of the base.

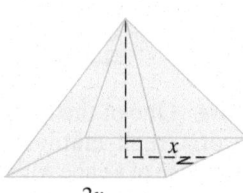

x
$2x$

Euler's Formula *Many crystals and biological structures are constructed in the shapes of regular polyhedra.*

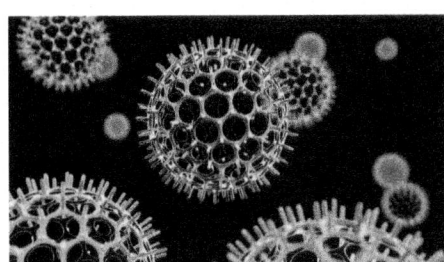

Radiolara

Leonhard Euler investigated a remarkable relationship among the numbers of faces (F), vertices (V), and edges (E) for the five regular polyhedra. Complete the chart in Exercises 59–63, and then draw a conclusion in Exercise 64.

Polyhedron	Faces (F)	Vertices (V)	Edges (E)	Value of F + V − E
59. Tetrahedron				
60. Hexahedron (cube)				
61. Octahedron				
62. Dodecahedron				
63. Icosahedron				

64. Euler's formula is $F + V - E =$ _____ .

9.6 TRANSFORMATIONAL GEOMETRY

OBJECTIVES

1 Transform geometric figures by reflection.

2 Transform geometric figures by translation.

3 Transform geometric figures by rotation.

4 Perform size transformations on geometric figures.

Exploring Reflection and Rotation Symmetry

In this chapter we have studied concepts of Euclidean geometry. Another branch of geometry, known as **transformational geometry,** investigates how one geometric figure can be transformed into another. In transformational geometry we reflect, rotate, and change the size of figures using concepts that we now discuss.

Reflections

One way to transform one geometric figure into another is by **reflection.** In **Figure 63,** line m is perpendicular to the line segment AA' and bisects this line segment. We call point A' the **reflection image** of point A about line m. Line m is the **line of reflection** for points A and A'. In the figure, we use a dashed line segment to connect points A and A' to show that these two points are images of each other under this transformation.

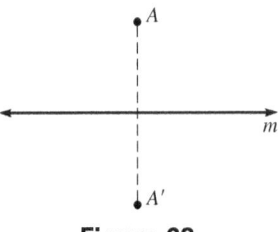

Figure 63

Point A' is the reflection image of point A only for line m. If a different line were used, A would have a different reflection image. Think of the reflection image of a point A about a line m as follows:

Place a drop of ink at point A, and fold the paper along line m. The spot made by the ink on the other side of m is the reflection image of A. If A' is the image of A about line m, then A is the image of A' about the same line m.

To find the reflection image of a figure, find the reflection image of each point of the figure. The set of all reflection images of the points of the original figure is the reflection image of the figure. **Figure 64** on the next page shows several figures (in black) and their reflection images (in color) about the lines shown.

An example of a **reflection.**

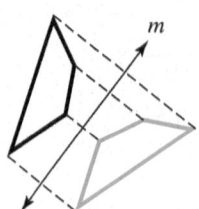 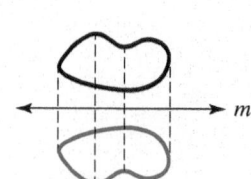 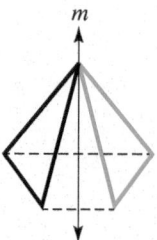

Figure 64

Each point in a plane has exactly one reflection image point with respect to a given line of reflection. Also, each reflection image point has exactly one original point. Thus, two distinct points cannot have the same reflection image. This means there is a *1-to-1 correspondence* between the set of points of the plane and the image points with respect to a given line of reflection. Any operation, such as reflection, in which there is a 1-to-1 correspondence between the points of the plane and their image points is a **transformation.** We call reflection about a line the **reflection transformation.**

If a point A and its image, A', under a certain transformation are the same point, then point A is an **invariant point** of the transformation. The only invariant points of the reflection transformation are the points of the line of reflection.

Three points that lie on the same straight line are **collinear.** In **Figure 65,**

points A, B, and C are collinear,

and it can be shown that the reflection images A', B', and C' are also collinear. Thus, the reflection image of a line is also a line. We express this by saying that **reflection preserves collinearity.**

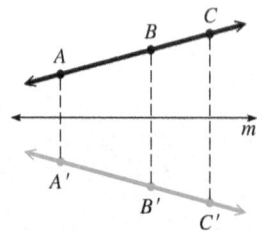

Figure 65

Distance is also preserved by the reflection transformation. Thus, in **Figure 66,** the distance between points A and B, written $|AB|$, is equal to the distance between the reflection images A' and B', or

$$|AB| = |A'B'|.$$

To prove this, we can use the definition of reflection image to verify that

$$|AM| = |MA'|, \quad \text{and} \quad |BN| = |NB'|.$$

Construct segments CB and $C'B'$, each perpendicular to BB'. Note that $CBB'C'$ is a rectangle. Because the opposite sides of a rectangle are equal and parallel, we have

$$|CB| = |C'B'|. \tag{Side (1)}$$

Because $CBB'C'$ is a rectangle, we can also say

$$m \angle ACB = m \angle A'C'B' = 90°, \tag{Angle (2)}$$

where we use $m \angle ACB$ to represent the measure of angle ACB.

We know $|AM| = |MA'|$ and can show $|CM| = |MC'|$, so that

$$|AC| = |A'C'|. \tag{Side (3)}$$

From statements (1), (2), and (3) above, we conclude that in triangles ABC and $A'B'C'$, two sides and the included angle of one are equal in measure to the corresponding two sides and included angle of the other and, thus, are congruent by SAS. Corresponding sides of congruent triangles are equal in length, so

$$|AB| = |A'B'|,$$

which is what we wanted to show. Hence, the distance between two points equals the distance between their reflection images, and thus, reflection preserves distance.

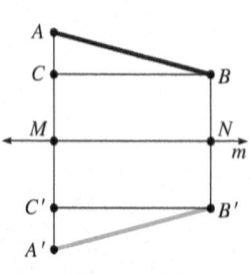

Figure 66

How many lines of **symmetry** do you see here?

(The proof we have given is not really complete, because we have tacitly assumed that AB is not parallel to $A'B'$ and that A and B are on the same side of the line of reflection. Some modification would have to be made in the proof above to include these other cases.)

The figures shown in **Figure 67** are their own reflection images about the lines of reflection shown. In this case, the line of reflection is a **line of symmetry** for the figure. **Figure 67(a)** has three lines of symmetry. A circle has every line through its center as a line of symmetry.

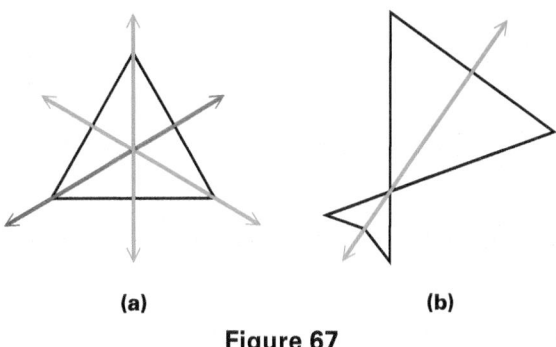

(a)   (b)

Figure 67

Translations and Rotations

We shall use the symbol r_m to represent a reflection about line m, and let us use $r_n \cdot r_m$ to represent a reflection about line m followed by a reflection about line n. We call $r_n \cdot r_m$ the **composition,** or **product,** of the two reflections r_n and r_m. **Figure 68** shows an example of the composition of two reflections about parallel lines m and n.

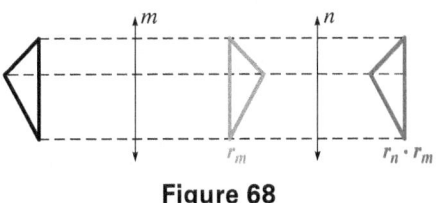

Figure 68

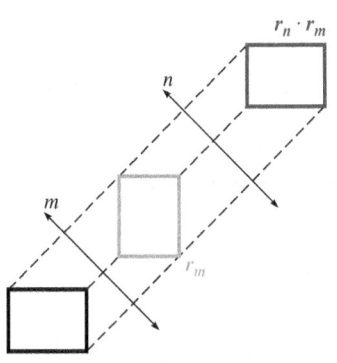

Figure 69

In **Figure 68,** both the original figure and its image under the composition $r_n \cdot r_m$ appear to be oriented the same way and to have the same "tilt." In fact, it appears that the original figure could be slid along the dashed lines with no rotation, so as to cover the image. This composite transformation is a **translation.**

Thus far, our discussion of **Figure 68** has illustrated that a composition of reflections about parallel lines constitutes a translation. Conversely, **Figure 69** shows that the image of a translation can be obtained as a composition of reflections about two parallel lines. Check that the distance between a point and its image under a translation is twice the distance between the two parallel lines. The distance between a point and its image under a translation is the **magnitude** of the translation.

A translation of magnitude zero leaves every point of the plane unchanged and, thus, is the **identity translation.** A translation of magnitude k, followed by a similar translation of magnitude k but of opposite direction, returns a point to its original position, so two such translations are **inverses** of each other. Check that there are no invariant points in a translation of magnitude $k > 0$.

A translation preserves collinearity (three points on the same line have image points that also lie on a line) and distance (the distance between two points is the same as the distance between the images of the points).

Exploring Reflection and
Rotation Symmetry

M. C. Escher (1898 –1972) was a Dutch
graphic artist, most recognized for spatial
illusions, impossible buildings, repeating
geometric patterns (tessellations), and his
incredible techniques in woodcutting and
lithography. He was a humble man who
considered himself neither an artist nor a
mathematician.

The tessellation pictured above is
reminiscent of Escher's work, which may
be viewed online at www.escher.com.

Figure 70 shows a composition of two reflections about nonparallel lines m and n. Because the original figure could be rotated so as to cover the image, we call the composition of two reflections about nonparallel lines a **rotation.** The point of intersection of the two lines is the **center of rotation.** The black triangle of **Figure 70** was reflected about line m, and then the blue image of reflection r_m was reflected about line n, resulting in the image of $r_n \cdot r_m$, which is also the image of a rotation with center at B. The dashed arcs in color represent the paths of the vertices of the triangle under rotation. It can be shown that $m \sphericalangle ABA'$ is twice as large as $m \sphericalangle MBN$. The measure of angle ABA' is the **magnitude** of the rotation.

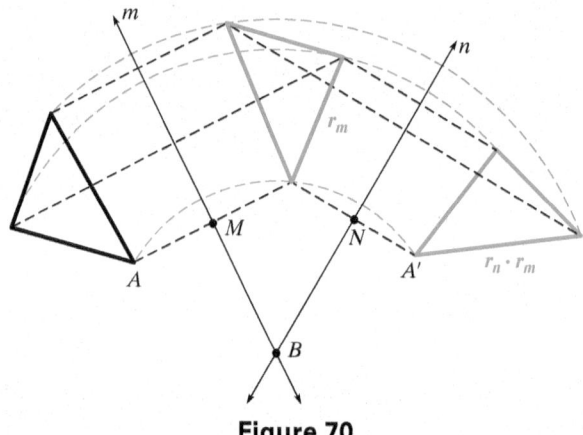

Figure 70

Rotations also preserve collinearity and distance. The **identity rotation** is a rotation of 0° or 360°, and rotations of, say, 240° and 120° (or, in general, $x°$ and $360° - x°$, $0 \leq x \leq 360°$) are inverses of each other. The center of rotation is the only invariant point of any rotation except the identity rotation.

We have defined rotations as the composition of two reflections about nonparallel lines of reflection. We can also define a rotation by specifying its center, the angle of rotation, and a direction of rotation, as shown by the following example.

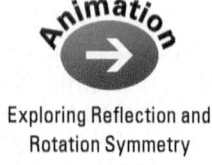

Exploring Reflection and
Rotation Symmetry

EXAMPLE 1 **Finding an Image under a Rotation**

Find the image of a point P under a rotation transformation having center at a point Q and magnitude 135° in a clockwise direction.

Solution

To find P', the image of P, first draw angle PQM having measure 135°. Then draw an arc of a circle with center at Q and radius $|PQ|$. The point where this arc intersects side QM is P'. See **Figure 71.**

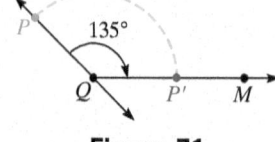

Figure 71

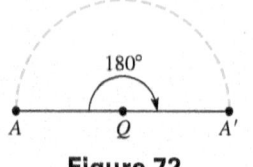

Figure 72

Figure 72 shows a rotation transformation having center Q and magnitude 180° in a clockwise direction. Point Q bisects the line segment from a point A to its image A', and for this reason this rotation is sometimes called a **point reflection.**

Exploring Reflection and
Rotation Symmetry

EXAMPLE 2 Finding Point Reflection Images

Find the point reflection images about point Q for each of the following figures.

(a)

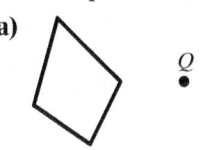

(b)

Solution

(a)

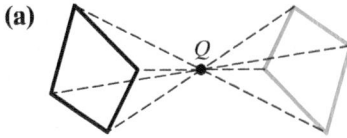

(b)

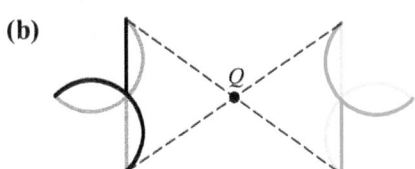

The point reflection images are shown in color.

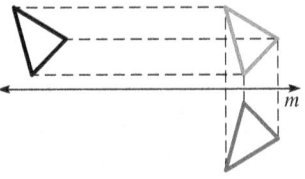

Figure 73

Let r_m be a reflection about line m, and let T be a translation having nonzero magnitude and a direction parallel to m. Then the composition of T and r_m is a **glide reflection,** as seen in **Figure 73**. Here a reflection followed by a translation is the same as a translation followed by a reflection, so in this case,

$$T \cdot r_m = r_m \cdot T.$$

Because a translation is the composition of *two* reflections, a glide reflection is the composition of *three* reflections. Because a glide reflection involves a translation of nonzero magnitude, there is no identity glide reflection.

All the transformations of this section discussed so far are **isometries,** or transformations in which the image of a figure has the same size and shape as the original figure. Any isometry is either a reflection or the composition of two or more reflections.

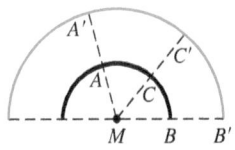

Figure 74

Size Transformations

Figure 74 shows a semicircle in black, a point M, and an image semicircle in color. Distance $A'M$ is twice the distance AM, and distance $B'M$ is twice the distance BM. In fact, every point of the image semicircle, such as C', was obtained by drawing a line through M and C, and then locating C' such that $|MC'| = 2\,|MC|$.

Such a transformation is a **size transformation** with center M and magnitude 2. We shall assume that a size transformation can have any positive real number k as magnitude. A size transformation having magnitude $k > 1$ is a **dilation,** or **stretch.** A size transformation having magnitude $k < 1$ is a **contraction,** or **shrink.**

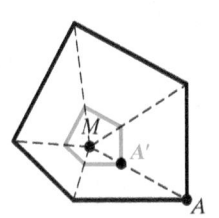

EXAMPLE 3 Applying Size Transformations

Apply a size transformation with center M and magnitude $\frac{1}{3}$ to the two pentagons shown in black in **Figure 75**.

Solution

To find the images of these pentagons, we can find the image points of some sample points. For example, if we select point A on each of the original pentagons, we can find the image points by drawing a line through A and M, and locating a point A' such that $|MA'| = \frac{1}{3}|MA|$. By doing this for all points of each of the black pentagons, we get the images shown in color in **Figure 75**.

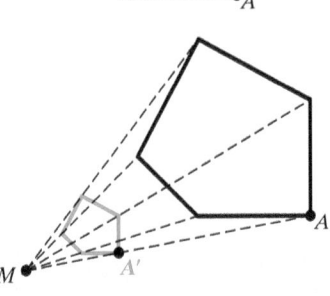

Figure 75

The identity transformation is a size transformation of magnitude 1, while size transformations of magnitude k and $\frac{1}{k}$, having the same center, are inverses of each other. The only invariant point of a size transformation of magnitude $k \neq 1$ is the center of the transformation.

EXAMPLE 4   Investigating Size Transformations

Does a size transformation preserve **(a)** collinearity? **(b)** distance?

Solution

(a) **Figure 76** shows three collinear points, A, B, and C, and their images under two different size transformations with center at M: one of magnitude 3 and one of magnitude $\frac{1}{3}$. In each case the image points appear to be collinear, and it can be proved that they are, using similar triangles. In fact, the image of a line not through the center of the transformation is a line parallel to the original line.

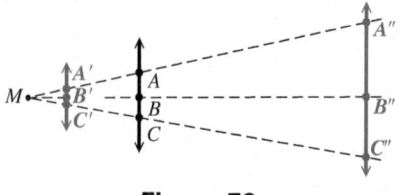

Figure 76

(b) As shown in **Figure 76**, $|AB| \neq |A'B'|$. Thus, a size transformation of magnitude $k \neq 1$ does not preserve distance and is not an isometry.

Transformations are summarized in Table 4.

Table 4   Summary of Transformations

	Reflection	Translation	Rotation	Glide Reflection	Size Transformation
Example					
Preserve collinearity?	Yes	Yes	Yes	Yes	Yes
Preserve distance?	Yes	Yes	Yes	Yes	No
Identity transformation?	None	Magnitude 0	Magnitude 360°	None	Magnitude 1
Inverse transformation?	None	Same magnitude; opposite direction	Same center; magnitude 360° − x°	None	Same center; magnitude $\frac{1}{k}$
Composition of n reflections?	$n = 1$	$n = 2$, parallel	$n = 2$, nonparallel	$n = 3$	No
Isometry?	Yes	Yes	Yes	Yes	No
Invariant points?	Line of reflection	None	Center of rotation	None	Center of transformation

9.6 EXERCISES

CONCEPT CHECK *Write* true *or* false *for each statement.*

1. Any transformation has at least one invariant point.

2. There are some translations that are not equivalent to a composition of reflections.

3. A clockwise rotation of 20° about a point *P* and a clockwise rotation of 340° about *P* are inverses of each other.

4. When a point undergoes a transformation, the result is called an image point.

5. A rotation of 180° may also be called a point reflection.

6. A dilation having magnitude $k > 1$ may be accomplished by a composition of reflections.

Find the reflection images of the given figures about the given lines.

7.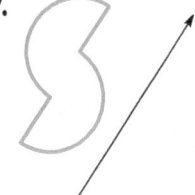

8.

9.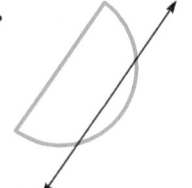

10.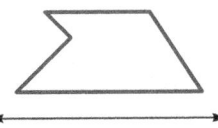

11.

12.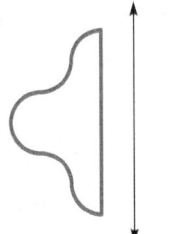

13.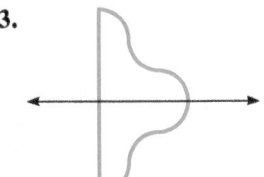

14.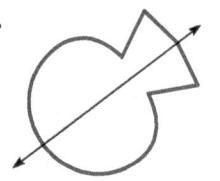

Find any lines of symmetry of the given figures.

15.

16.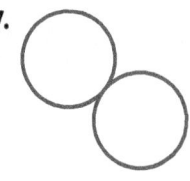

17.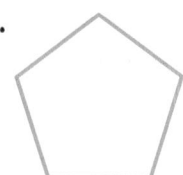

18.

First reflect the given figure about line m. Then reflect about line n.

19.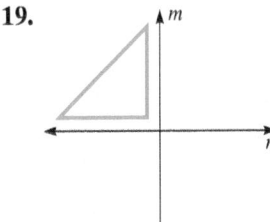

20.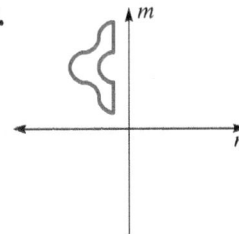

21.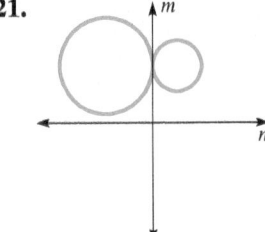

22.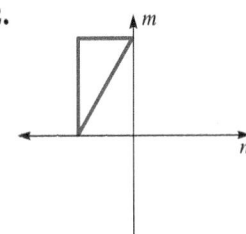

23.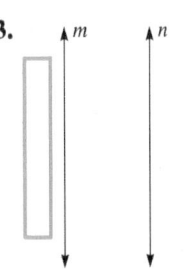

24.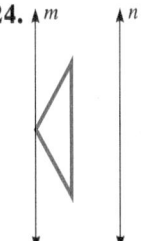

25.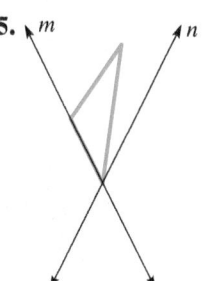

26.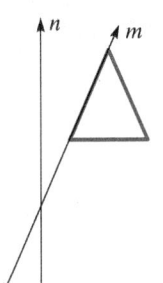

In Exercises 27–40, let T be a translation having magnitude 1.5 cm to the right in a direction parallel to the bottom edge of the page. Let r_m be a reflection about line m, and let R_P be a rotation about point P having magnitude 60° clockwise. In each of Exercises 27–38, perform the given transformations on point A of the figure below to obtain final image point A'.

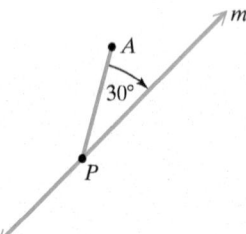

27. r_m **28.** R_P **29.** T

30. $r_m \cdot r_m$ **31.** $T \cdot T$ **32.** $R_P \cdot R_P$

33. $R_P \cdot r_m$ **34.** $r_m \cdot R_P$ **35.** $R_P \cdot T$

36. $T \cdot R_P$ **37.** $T \cdot r_m$ **38.** $r_m \cdot T$

39. Is $T \cdot r_m$ a glide reflection here?

40. Is $T \cdot r_m = r_m \cdot T$ true?

41. Suppose a rotation is given by $r_m \cdot r_n$, as shown in the figure below. Find the images of A, B, and C.

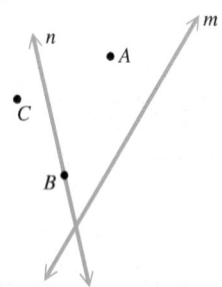

42. Does a glide reflection preserve

   (a) collinearity?   **(b)** distance?

Find the point reflection images of each of the following figures with the given point Q as center.

43.

44.

45.
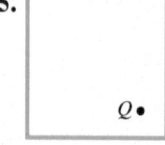

46.
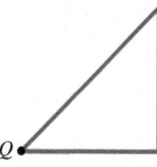

Perform the indicated size transformation.

47. magnitude 2; center M

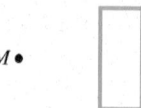

48. magnitude $\frac{1}{2}$; center M

49. magnitude $\frac{1}{3}$; center M

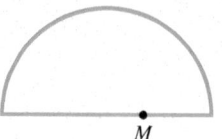

50. magnitude $\frac{1}{3}$; center M

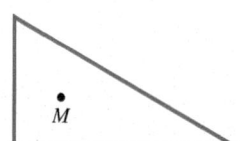

51. magnitude $\frac{1}{2}$; center M

52. magnitude 2; center M

Exercises 53–56 refer to the figure, which consists of a sequence of quadrilaterals spiraling toward a single point P. Each successive quadrilateral is obtained by a clockwise 60° rotation about P, followed by a contraction of magnitude $\frac{4}{5}$ with center P. We will call the rotation R_P and the contraction C_P.

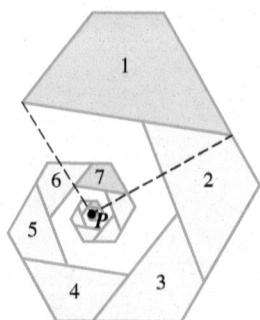

53. A composition of ____ clockwise rotations of 60° results in an identically oriented quadrilateral.

54. Quadrilateral 1 may be transformed into Quadrilateral 7 by the composition

$$C_P \cdot R_P \cdot C_P \cdot R_P \cdot C_P \cdot R_P \cdot C_P \cdot R_P \cdot C_P \cdot R_P \cdot C_P \cdot R_P.$$

Use your conclusion from **Exercise 53** to simplify this to a single contraction. Express the magnitude as a fraction.

55. Quadrilateral 1 may be transformed into Quadrilateral $6n + 1$ by repeating C_P $6n$ times. This can be expressed concisely as $C_P{}^{6n}$. Find a similar concise expression for the transformation of Quadrilateral 1 into Quadrilateral $6n + 3$.

56. What transformation will transform Quadrilateral n into Quadrilateral $n + 3$?

Solve each problem.

57. *Discovery by Transformation* Consider $\triangle ABC$ below. M is the midpoint of side BC, and c is the length of side AB.

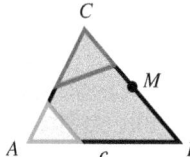

(a) Perform a translation of the blue triangle of magnitude c units to the right.

(b) Rotate the red triangle $180°$ about M. (You may instead perform a point reflection about M.)

(c) What characteristic of triangles is illustrated by the resulting figure? (Refer to the discussion surrounding **Figure 17** in this chapter to confirm your conclusion.)

58. *Transformation to Gold* In square $ABCD$, point M is the midpoint of side BC, and E is the image of the counterclockwise rotation of point D about M through the angle necessary for E to be located on the extension of side BC, as shown. Let x be the length of segment MC.

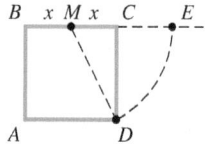

(a) Let F be the image of a translation of magnitude $2x$ of point E in a downward direction. Draw the resulting rectangle $ABEF$.

(b) Verify that rectangle $ABEF$ is a **Golden Rectangle.**

(*Hint:* Recall that the **Golden Ratio** is $\frac{1 + \sqrt{5}}{2}$.)

59. *Formula by Transformation* In trapezoid $PQRS$, M is the midpoint of side RS, b_1 is the larger base, b_2 is the smaller base, and h is the height. Rotate this trapezoid $180°$ about M, and use the resulting figure to find the area of the trapezoid in terms of b_1, b_2, and h. (*Hint:* Recall that the area of a parallelogram is equal to the product of its base and height.)

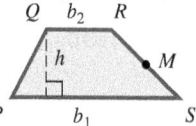

60. *Reflections on a Translation* In the figure below, $\triangle ABC$ is translated, and the resulting image is $\triangle A'B'C'$.

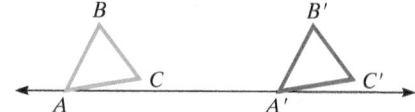

(a) Draw segment BB'.

(b) Construct the perpendicular bisector of segment BB' to locate the midpoint of segment BB'. Label the midpoint M.

(c) Choose any point D on $\overleftrightarrow{AA'}$ and copy the length of segment BM onto $\overleftrightarrow{AA'}$ stretching from D to a new point E. (Note: the location of point D on $\overleftrightarrow{AA'}$ is arbitrary. For the purposes of this exercise, it is recommended that you choose a point between points A and A'.)

(d) Construct lines m and n perpendicular to $\overleftrightarrow{AA'}$ through D and E, respectively.

(e) Reflect $\triangle ABC$ about line m, and reflect the resulting image about line n.

(f) What do you notice about the image resulting from the composition of these two reflections about m and n?

(g) Since the choice of point D was arbitrary, what can you conclude about the relationship between the magnitude of a translation and the distance between the two lines of reflection?

FOR FURTHER THOUGHT

Tessellations

The authors wish to thank Suzanne Alejandre for permission to reprint this article on tessellations, which first appeared at www.mathforum.org/sum95/suzanne/whattess.html.

tessellate (verb), **tessellation** (noun): from Latin *tessera* "a square tablet" or "a die used for gambling." Latin *tessera* may have been borrowed from Greek *tessares*, meaning "four," since a square tile has four sides. The diminutive of *tessera* was *tessella*, a small, square piece of stone or a cubical tile used in mosaics. Since a mosaic extends over a given area without leaving any region uncovered, the geometric meaning of the word *tessellate* is "to cover the plane with a pattern in such a way as to leave no region uncovered." By extension, space or hyperspace may also be tessellated.

Definition

A dictionary will tell you that the word *tessellate* means "to form or arrange small squares in a checkered or mosaic pattern". The word *tessellate* is derived from the Ionic version of the Greek word *tesseres*, which in English means "four." The first tilings were made from square tiles.

A regular polygon has 3 or 4 or 5 or more sides and angles, all equal. A **regular tessellation** means a tessellation made up of congruent regular polygons. [Remember: *Regular* means that the sides of the polygon are all the same length. *Congruent* means that the polygons that you put together are all the same size and shape.]

Only three regular polygons tessellate in the Euclidean plane: triangles, squares, or hexagons. We can't show the entire plane, but imagine that these are pieces taken from planes that have been tiled. Here are examples of

a tessellation of triangles

a tessellation of squares

a tessellation of hexagons

When you look at these three samples you can easily notice that the squares are lined up with each other, but the triangles and hexagons are not.

Also, if you look at six triangles at a time, they form a hexagon, so the tiling of triangles and the tiling of hexagons are similar and cannot be formed by directly lining shapes up under each other—a slide (or a glide!) is involved.

You can work out the interior measure of the angles for each of these polygons:

Shape	Angle Measure in Degrees
triangle	60
square	90
pentagon	108
hexagon	120
more than six sides	more than 120 degrees

Since the regular polygons in a tessellation must fill the plane at each vertex, the interior angle must be an exact divisor of 360 degrees. This works for the triangle, square, and hexagon, and you can show working tessellations for these figures. For all the others, the interior angles are not exact divisors of 360 degrees, and therefore, those figures cannot tile the plane.

Naming Conventions

A tessellation of squares is named "4.4.4.4." Here's how: Choose a vertex, and then look at one of the polygons that touches that vertex. How many sides does it have?

Since it's a square, it has four sides, and that's where the first "4" comes from. Now keep going around the vertex in either direction, finding the number of sides of the polygons until you get back to the polygon you started with. How many polygons did you count?

There are four polygons, and each has four sides.

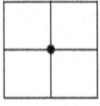

4.4.4.4

For a tessellation of regular congruent hexagons, if you choose a vertex and count the sides of the polygons that touch it, you'll see that there are three polygons and each has six sides, so this tessellation is called "6.6.6":

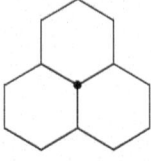

6.6.6

A tessellation of triangles has six polygons surrounding a vertex, and each of them has three sides: "3.3.3.3.3.3."

3.3.3.3.3.3

Semi-regular Tessellations

You can also use a variety of regular polygons to make **semi-regular tessellations**.

A semi-regular tessellation has two properties:

1. It is formed by regular polygons.

2. The arrangement of polygons at every vertex point is identical.

Here are the eight semi-regular tessellations:

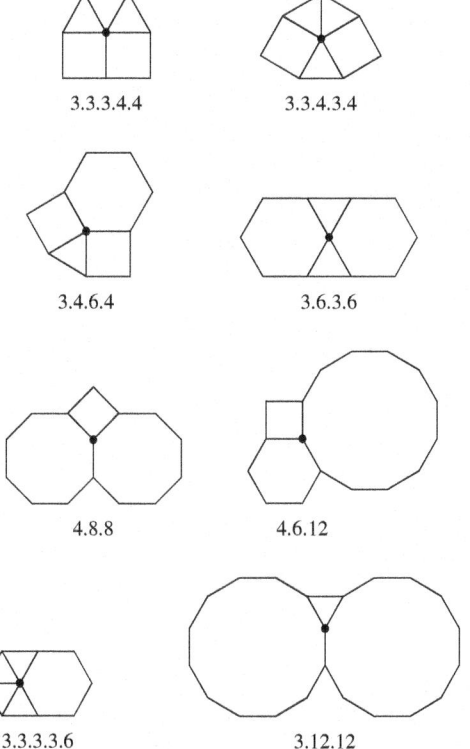

3.3.3.4.4   3.3.4.3.4

3.4.6.4   3.6.3.6

4.8.8   4.6.12

3.3.3.3.6   3.12.12

Interestingly, there are other combinations that seem like they should tile the plane because the arrangements of the regular polygons fill the space around a point. For example:

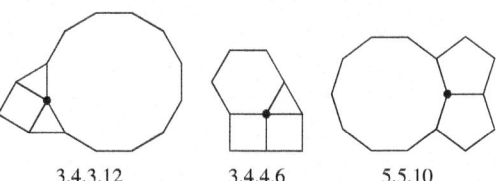

3.4.3.12   3.4.4.6   5.5.10

But If you try tiling the plane with these units of tessellation, you will find that they cannot be extended infinitely.

There is an infinite number of tessellations that can be made of patterns that do not have the same combination of angles at every vertex point. There are also tessellations made of polygons that do not share common edges and vertices.

For Group or Individual Investigation

1. Use the naming conventions to name each of these semi-regular tessellations.

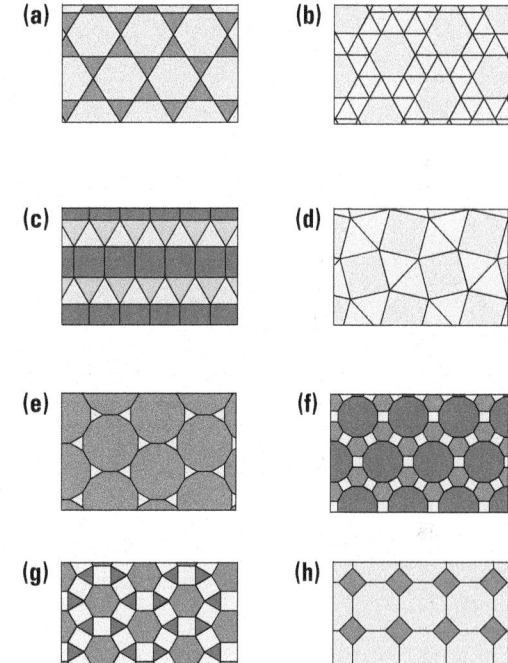

(a)   (b)

(c)   (d)

(e)   (f)

(g)   (h)

2. Why isn't this a semi-regular tessellation?

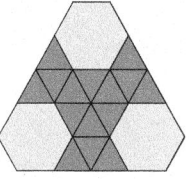

Escher-like Tessellations:
Parallelogram Translations

9.7 NON-EUCLIDEAN GEOMETRY AND TOPOLOGY

Euclid's Postulates and Axioms

The *Elements* of Euclid is quite possibly the most influential mathematics text ever written. It begins with definitions of basic ideas such as

point, line, and plane.

Euclid then gives five postulates providing the foundation of all that follows.

Next, Euclid lists five axioms that he views as general truths and not just facts about geometry. See **Table 5.** (To some of the Greek writers, postulates were truths about a particular field, while axioms were general truths. Today, "axiom" is used in either case.) Using only these ten statements and the basic rules of logic, Euclid was able to prove a large number of "propositions" about geometric figures.

Table 5

Euclid's Postulates	Euclid's Axioms
1. Two points determine one and only one straight line.	6. Things equal to the same thing are equal to each other.
2. A straight line extends indefinitely far in either direction.	7. If equals are added to equals, the sums are equal.
3. A circle may be drawn with any given center and any given radius.	8. If equals are subtracted from equals, the remainders are equal.
4. All right angles are equal.	9. Figures that can be made to coincide are equal.
5. Given a line k and a point P not on the line, there exists one and only one line m through P that is parallel to k.	10. The whole is greater than any of its parts.

John Playfair (1748–1819) wrote his *Elements of Geometry* in 1795. Playfair's Axiom is: Given a line k and a point P not on the line, there exists one and only one line m through P that is parallel to k. This is equivalent to Euclid's Postulate 5.

The statement for Postulate 5 given above is actually known as Playfair's axiom on parallel lines, which is equivalent to Euclid's fifth postulate. To understand why this postulate caused trouble for so many mathematicians for so long, we must examine the original formulation.

The Parallel Postulate (Euclid's Fifth Postulate)

In its original form, Euclid's fifth postulate states the following:

*If two lines (k and m in **Figure 77**) are such that a third line, n, intersects them so that the sum of the two interior angles (A and B) on one side of line n is less than (the sum of) two right angles, then the two lines, if extended far enough, will meet on the same side of n that has the sum of the interior angles less than (the sum of) two right angles.*

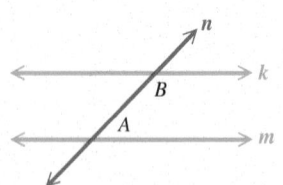

Figure 77

Euclid's parallel postulate is quite different from the other nine postulates and axioms. It is long and wordy, and difficult to understand without a sketch. It was commonly believed that this was not a postulate at all but a theorem to be proved. For more than 2000 years, mathematicians tried repeatedly to prove it.

The most dedicated attempt came from an Italian Jesuit, Girolamo Saccheri (1667–1733). He attempted to prove the parallel postulate in an indirect way, by so-called "reduction to absurdity." This strategy of proof involves making an

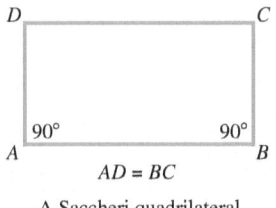

A Saccheri quadrilateral

Figure 78

assumption and showing that it leads to an absurdity (a contradiction of a known truth), thus showing the assumption to be wrong. Saccheri would assume the postulate to be false, and then show that this leads to a contradiction. He began with a quadrilateral (as in **Figure 78**) with right angles A and B and equal sides AD and BC. His plan was as follows:

1. Assume that angles C and D are obtuse angles, and show that this leads to a contradiction (and the conclusion that angles C and D cannot be obtuse).

2. Assume that angles C and D are acute angles, and show that this leads to a contradiction (and the conclusion that angles C and D cannot be acute).

3. Use parts 1 and 2 to conclude that angles C and D must be right angles.

4. If angles C and D are right angles, then the fifth postulate can be proved true and is therefore a theorem rather than a postulate.

Saccheri had no trouble with part 1. However, he did not actually reach a contradiction in the second part but produced some theorems so "repugnant" that he convinced himself he had vindicated Euclid. In fact, he published a book called in English *Euclid Freed of Every Flaw*. However, today we know that the fifth postulate is indeed an axiom and not a theorem. It is *consistent* with Euclid's other axioms.

The ten axioms of Euclid describe the world around us with remarkable accuracy. We now realize that the fifth postulate is necessary in Euclidean geometry to establish *flatness*. That is, the axioms of Euclid describe the geometry of *plane surfaces*. By changing the fifth postulate, we can describe the geometry of other surfaces. Thus, other geometric systems exist as much as Euclidean geometry exists, and they can even be demonstrated in our world. A system of geometry in which the fifth postulate is changed is a **non-Euclidean geometry.**

The Origins of Non-Euclidean Geometry

One non-Euclidean system was developed by three people working separately at about the same time. Early in the nineteenth century, Carl Friedrich Gauss worked out a consistent geometry replacing Euclid's fifth postulate. He never published his work, however, because he feared the ridicule of people who were committed to the prevailing thought of the time.

Nikolai Ivanovich Lobachevsky (1792–1856) published a similar system in 1830 in the Russian language. At the same time, Janos Bolyai (1802–1860), a Hungarian army officer, worked out a similar system, which he published in 1832, not knowing about Lobachevsky's work. Bolyai never recovered from the disappointment of not being the first and did no further work in mathematics.

Lobachevsky replaced Euclid's fifth postulate with the following.

Angles C and D in the quadrilateral of Saccheri are acute angles.

Lobachevsky's replacement

This postulate of Lobachevsky can be rephrased as follows.

*Through a point P off a line k (**Figure 79**), at least two different lines can be drawn parallel to k.*

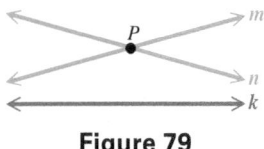

Figure 79

Compare this form of Lobachevsky's postulate to the geometry of Euclid, where only one line can be drawn through P and parallel to k. At first glance, the postulate of Lobachevsky does not agree with our "Euclidean" thinking about the world around us.

A song titled simply **Lobachevsky** appears on the CD *Tom Lehrer Revisited.* The songwriter described this song as an account of "one way to get ahead in mathematics (which happens to be the author's own academic specialty) or any other academic field." Lehrer's music is available for download. Listen to *Lobachevsky,* and see what Lehrer suggests!

Further consideration of this replacement postulate confirms that it leads to some surprising results. For example, the sum of the measures of the angles in any triangle is *less* than 180°. Also, triangles having different side lengths cannot have equal angles, so similar triangles do not exist.

The non-Euclidean geometry of Lobachevsky can be represented as a curved surface called a **pseudosphere.** This surface is formed by revolving a curve called a **tractrix** about the line *AB* in **Figure 80.**

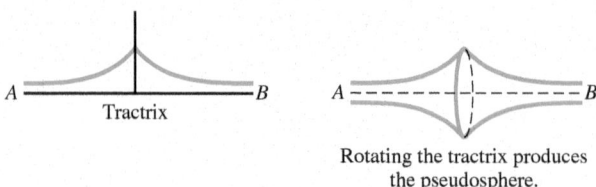

Rotating the tractrix produces
the pseudosphere.

Figure 80

A second non-Euclidean system was developed by Georg Riemann (1826–1866). He pointed out the difference between a line that continues indefinitely and a line having infinite length. For example, a circle on the surface of a sphere continues indefinitely but does not have infinite length. Riemann developed the idea of geometry on a sphere and replaced Euclid's fifth postulate with the following.

Angles C and D of the Saccheri quadrilateral are obtuse angles.

Riemann's replacement

In terms of parallel lines, Riemann's postulate is stated this way:

Through a point P off a line k, no line can be drawn that is parallel to k.

Riemannian geometry is important in navigation. "Lines" in this geometry are really *great circles,* or circles whose centers are at the center of the sphere. The shortest distance between two points on a sphere lies along an arc of a great circle. Great circle routes on a globe don't look at all like the shortest distance when the globe is flattened out to form a map, but this is part of the distortion that occurs when the earth is represented as a flat surface. See **Figure 81.** The sides of a triangle drawn on a sphere would be arcs of great circles. And, in Riemannian geometry, the sum of the measures of the angles in any triangle is *more* than 180°.

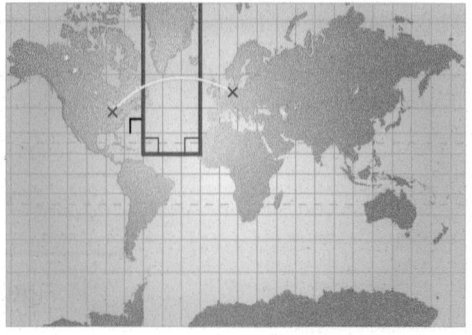
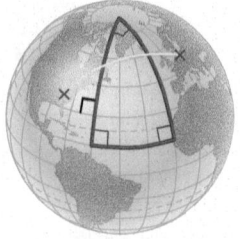

Figure 81

Many of the theorems of Euclidean geometry are valid in the geometries of Lobachevsky and Riemann, but several are not. These differences arise as a result of the respective replacements of Euclid's fifth postulate. A summary comparison of Euclidean, Lobachevskian, and Riemannian geometries is given in **Table 6.**

Table 6 Three Geometries

EUCLIDEAN	NON-EUCLIDEAN	
Dates back to about 300 B.C.	**Lobachevskian (about 1830)**	**Riemannian (about 1850)**
Lines have *infinite* length.		Lines have *finite* length.
Geometry on a plane	Geometry on a surface like a pseudosphere	Geometry on a sphere
Angles C and D of a Saccheri quadrilateral are *right* angles.	Angles C and D are *acute* angles.	Angles C and D are *obtuse* angles.
Given point P off line k, exactly *one* line can be drawn through P and parallel to k.	*More than one* line can be drawn through P and parallel to k.	*No* line can be drawn through P and parallel to k.
Typical triangle ABC	Typical triangle ABC	Typical triangle ABC
Two triangles with the same size angles can have different size sides (similarity as well as congruence).	Two triangles with the same size angles must have the same size sides (congruence only).	
Circles of radius r have circumference $2\pi r$.	Circles of radius r have circumference greater than $2\pi r$.	Circles of radius r have circumference less than $2\pi r$.

How should artists paint a realistic view of railroad tracks vanishing into the horizon? In reality, the tracks are always a constant distance apart, but they cannot be drawn that way except from overhead. The artist must make the tracks converge at a "vanishing point" to show how things look from the perspective of an observer.

Projective Geometry

Beginning in the fifteenth century, artists led by Leone Battista Alberti (1404–1472), Leonardo da Vinci (1452–1519), and Albrecht Dürer (1471–1528) began to study the problems of representing three dimensions in two. What artists initiated, mathematicians developed into another non-Euclidean geometry, that of **projective geometry.**

Gerard Desargues (1591–1661), a French architect and engineer, published in 1636 and 1639 a treatise and proposals about **perspective,** thus inventing projective geometry. However, these innovations were hidden for nearly 200 years until a manuscript by Desargues was discovered in 1840, about 30 years after projective geometry had been revived by Jean-Victor Poncelet (1788–1867).

There are many projective geometries. An example of one that projects three-dimensional space onto a **projective sphere** is depicted in **Figure 82.** The projected image of a point P is the pair of points of intersection of the sphere with the line passing through P and the center of the sphere. These two points (labeled P' and P'' in the figure) are **antipodal points** and are considered undistinguishable (that is, they are treated as a single point). The eye of the observer is thought to be at the center of the sphere, and the images of objects in space are projected onto the spherical "screen." Notice that parallel lines m and n in Euclidean space correspond to two great circles in the sphere that intersect on the "equator" of the sphere. In this configuration, the equator corresponds to infinity, which is just another great circle (line) in this geometry.

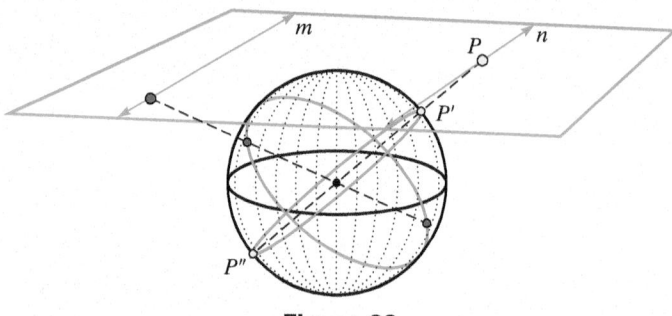

Figure 82

Figure 83 shows that intersecting lines in Euclidean space are also mapped to great circles that intersect at a single point (antipodal pair).

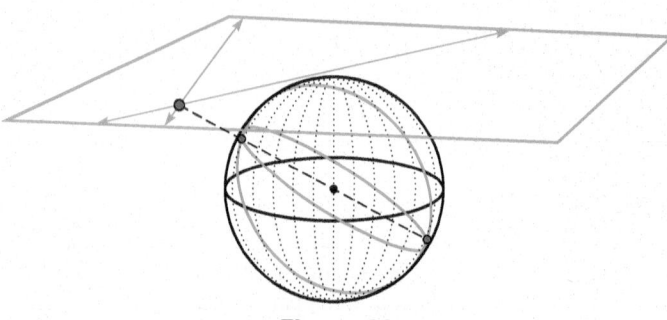

Figure 83

Projective geometries preserve points and intersections, but not distances. **Figure 83** demonstrates that a point of intersection in Euclidean space will be projected to a point of intersection in the projective plane. But **Figure 82** illustrates that distance (such as that between parallel lines) in Euclidean space does not survive the projection process.

EXAMPLE 1 Determining the Projective Image of a Plane Figure

Given a triangle in Euclidean space, draw a picture showing how the triangle would be projected onto the projective sphere.

Solution

Since antipodal points are not distinguished, we use a hemisphere for simplicity. We begin with a triangle contained in a plane above the hemisphere, and draw segments from the vertices of the triangle to the center of the sphere. We then do our best to estimate the three points where those segments intersect the sphere. Remembering that lines in the projective geometry are great circles, we connect the three points with arcs of great circles. The finished drawing is shown in **Figure 84.**

Figure 84

Consider the following statements.

Any two distinct points are contained in a unique line.

Any two distinct lines contain a unique point.

The first statement is true in Euclidean plane geometry, but the second is not. Projective geometry removes the obstacle of non-intersecting parallel lines. As shown in **Figures 82 and 83,** every pair of lines (great circles) has a point of intersection. Surfaces for which both of the above statements are true are **projective planes.** These statements are identical, except that the instances of "point" and "line" are interchanged, and the containment relationship is exchanged for its inverse.

The preceding pair of statements is one example of the **duality** that is characteristic of projective geometry. Every true statement about a relationship between points and lines has its dual—a statement formed by interchanging "point" with "line" and replacing the relationship with its inverse.

EXAMPLE 2 Expressing a Dual

Express the dual of the following statement.

There are four distinct points, no three of which are contained in the same line.

Solution

We interchange "point" and "line" in the statement, and exchange the relationship "contained in" with its inverse, "contains." These changes give this statement.

There are four distinct lines, no three of which contain the same point.

Topology

The plane and space figures studied in the Euclidean system are distinguished by differences in size, shape, angularity, and so on. For a given figure such properties are permanent. Thus we can ask sensible questions about congruence and similarity. Suppose we studied "figures" made of rubber bands that could be distorted without tearing or scattering. **Topology** does just that.

Topological questions concern the basic structure of objects rather than size or arrangement. A typical topological question has to do with the number of holes in an object, a basic structural property that does not change during deformation. Consider these examples.

- We cannot deform a rubber ball to get a rubber band without tearing it—making a hole in it. Thus the two objects are not topologically equivalent.

- On the other hand, a doughnut and a coffee cup are topologically equivalent, because one could be stretched to form the other without changing the basic structural property.

Topology and geometry software, including games for users age 10 and up, can be found at www.geometrygames.org, a site developed by Jeff Weeks. Included are Torus Games, Kali, KaleidoTile, and investigations into Curved Spaces.

EXAMPLE 3 Determining Topological Equivalence

Decide whether the figures in each pair are topologically equivalent.

(a) football and a cereal box

(b) a doughnut and an unzipped coat

Solution

(a) If we assume that a football is made of a perfectly elastic substance such as rubber or dough, it could be twisted or kneaded into the same shape as a cereal box. Thus, the two figures are topologically equivalent.

(b) A doughnut has one hole, while the coat has two (the sleeve openings). Thus, a doughnut could not be stretched and twisted into the shape of the coat without tearing another hole in it. Because of this, the doughnut and the coat are not topologically equivalent.

In topology, figures are classified according to their **genus**—that is, the number of cuts that can be made without cutting the figures into two pieces. The genus of an object is the number of holes in it. See **Figure 85.**

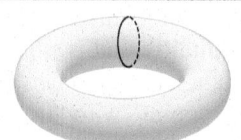

Torus One of the most useful figures in topology is the torus, a doughnut-like surface. Its properties are different from those of a sphere, for example. Imagine a sphere covered with hair. You cannot comb the hairs in a completely smooth way. One fixed point remains, as you can find on your own head. In the same way, on the surface of the earth the winds are not a smooth system. There is a calm point somewhere. However, the hair on a torus could be combed smooth.

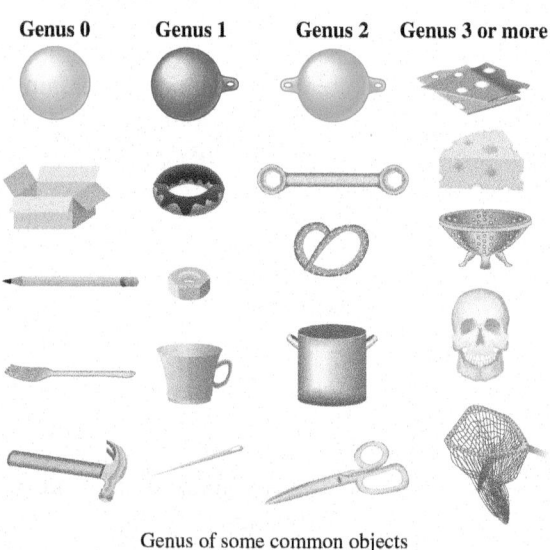

Genus of some common objects

Figure 85

9.7 EXERCISES

CONCEPT CHECK *Fill in each blank with the correct response. If necessary, refer to* **Table 6** *in the text.*

1. In _____ geometry, the sum of the measures of the angles of a triangle is equal to 180°.

2. In _____ geometry, the sum of the measures of the angles of a triangle is greater than 180°.

3. In _____ geometry, the sum of the measures of the angles of a triangle is less than 180°.

4. In a quadrilateral $ABCD$ in Lobachevskian geometry, the sum of the measures of the angles must be _____ 360°.
(less than / greater than)

5. In a quadrilateral $ABCD$ in Riemannian geometry, the sum of the measures of the angles must be _____ 360°.
(less than / greater than)

6. Suppose m and n represent lines through P that are both parallel to k. This is possible in _____ geometry.

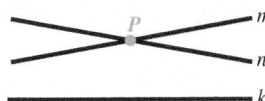

7. Suppose m and n below *must* meet at a point. This is possible in _____ geometry.

8. A globe representing the earth is a model for a surface in _____ geometry.

9. In _____ geometry, the following statement is possible. "Triangle ABC and triangle DEF are such that $\angle A = \angle D, \angle B = \angle E$, and $\angle C = \angle F$, and they have different perimeters."

10. In _____ geometry, the circumference of a circle having radius r is less than $2\pi r$.

In Riemannian geometry, the distance between two points is the length of the shortest *path from one point to the other along the surface of the sphere. In Exercises 11–14, consider a sphere of radius 1, and the great circle forming the equator of the sphere, as shown.*

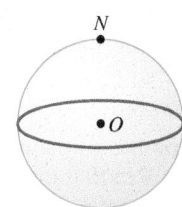

11. In Euclidean geometry, the center of a circle, though not part of the circle, is still located in the plane. The same is true in Riemannian geometry (although the plane is a curved sphere). Which point in the Riemannian geometry is the center of the red circle?

12. What is the radius of the red circle? [*Hint:* The radius is the *distance* from the center of a circle to the circle, and the path used to determine a distance must lie along the surface of the sphere.]

13. Let r represent the radius found in **Exercise 12.** Find the circumference of the equator and compare it to $2\pi r$.

14. To draw a circle of radius r, a Euclidean geometer may spread the points of a compass r units apart and swing an arc, or may use a piece of string that has a length of r units when stretched tight. Which of these two tools would be the better choice for a Riemannian geometer? Explain.

In Exercises 15 and 16, project the given shape onto the hemisphere.

15.

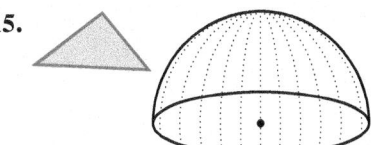

16.

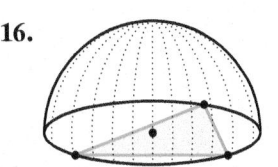

*Exercises 17–22 refer to the **Fano Plane,** shown here. It is a finite projective geometry containing seven points and seven lines.*

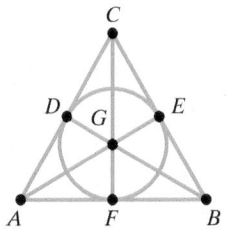

Points: A, B, C, D, E, F, and G

Lines: ADC, AFB, AGE, BEC, BGD, CGF, and DEF

(Note: There is no requirement that a line be "straight" as in Euclidean geometry.)

Verify that the Fano Plane satisfies each of the following statements, and state the dual in each case.

17. Any two distinct points have at least one line in common.

18. No two distinct points have more than one line in common.

19. Any two lines in a plane have at least one point of the plane in common.

20. There is at least one line on a plane.

21. Every line contains at least three points of the plane.

22. Not all of the points are contained in the same line.

Topological Equivalence *Someone once described a topologist as "a mathematician who doesn't know the difference between a doughnut and a coffee cup." This is because both objects are of genus 1—they are topologically equivalent! Based on this interpretation, would a topologist know the difference between each pair of objects in Exercises 23–26?*

23. a spoon and a fork

24. a mixing bowl and a colander

25. a slice of American cheese and a slice of Swiss cheese

26. a compact disc and a phonograph record

In Exercises 27–34 each figure may be topologically equivalent to none or some of the objects labeled A–E. List all topological equivalences (by letter) for each figure.

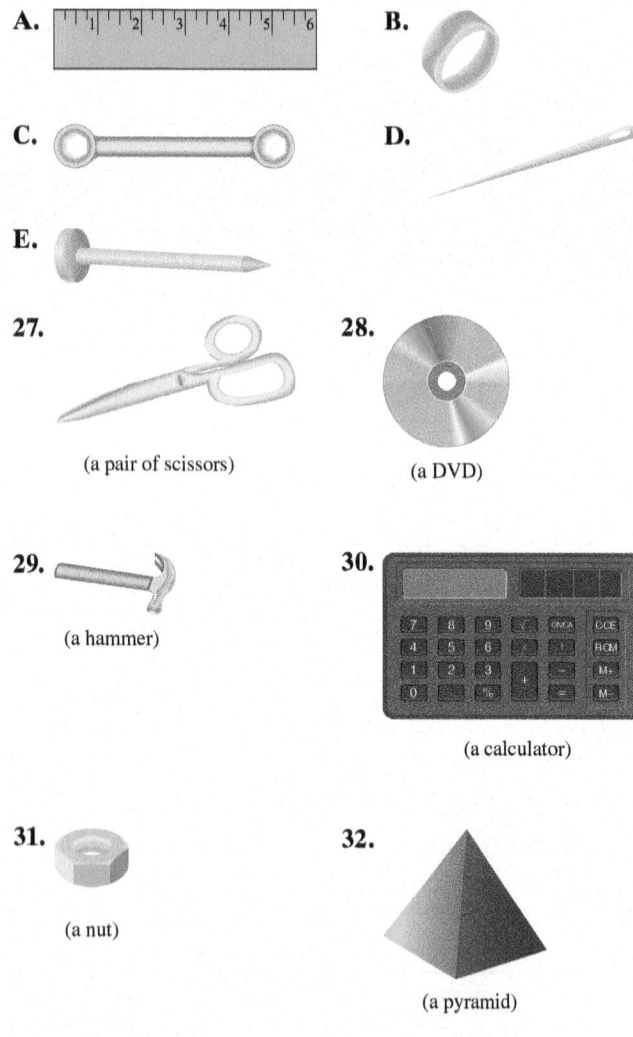

A.

B.

C.

D.

E.

27.

(a pair of scissors)

28.

(a DVD)

29.

(a hammer)

30.

(a calculator)

31.

(a nut)

32.

(a pyramid)

33.

(a coin)

34.

(a helmet)

Give the genus of each object.

35. a compact disc

36. a phonograph record

37. a sheet of loose-leaf paper made for a three-ring binder

38. a sheet of loose-leaf paper made for a two-ring binder

39. a wedding band

40. a postage stamp

41. ***Theorem of Pappus for Hexagons*** Pappus, a Greek mathematician in Alexandria about A.D. 320, wrote a commentary on the geometry of the times. We will work out a theorem of his about a hexagon inscribed in two intersecting lines.

First we define an old word in a new way:

> A **hexagon** consists of any six lines in a plane, no three of which meet in the same point.

In the figure, the vertices of several hexagons are labeled with numbers. Thus 1–2 represents a line segment joining vertices 1 and 2. Segments 1–2 and 4–5 are opposite sides of a hexagon, as are segments 2–3 and 5–6, and segments 3–4 and 6–1.

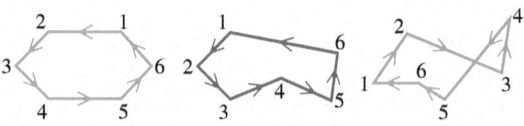

(a) Draw an angle less than 180°.

(b) Choose three points on one side of the angle. Label them 1, 5, 3 in that order, beginning with the point nearest the vertex.

(c) Choose three points on the other side of the angle. Label them 6, 2, 4 in that order, beginning with the point nearest the vertex.

(d) Draw line segments 1–6 and 3–4. Draw lines through the segments so that they extend to meet in a point. Call it *N*.

(e) Let lines through 1–2 and 4–5 meet in point *M*.

(f) Let lines through 2–3 and 5–6 meet in point *P*.

(g) Draw a straight line through points *M*, *N*, and *P*.

(h) Write in your own words a theorem generalizing your result.

42. The following theorem comes from projective geometry:

Theorem of Desargues in a Plane In a plane, if two triangles are placed so that lines joining corresponding vertices meet in a point, then corresponding sides, when extended, will meet in three collinear points. (*Collinear* points are points lying on the same line.)

Draw a figure that illustrates Desargues' theorem.

FOR FURTHER THOUGHT

Two Interesting Topological Surfaces

Two examples of topological surfaces are the **Möbius strip** and the **Klein bottle.** The Möbius strip is a single-sided surface named after August Ferdinand Möbius (1790–1868), a pupil of Gauss.

    To construct a Möbius strip, cut out a rectangular strip of paper, perhaps 3 cm by 25 cm. Paste together the two 3-cm ends after giving the paper a half-twist. To see how the strip now has only one side, mark an *x* on the strip and then mark another *x* on what appears to be the other "side." Begin at one of the *x*'s you have drawn, and trace a path along the strip. You will eventually come to the other *x* without crossing the edge of the strip.

    A branch of chemistry called chemical topology studies the structures of chemical configurations. A recent advance in this area was the synthesis of the first molecular Möbius strip, which was formed by joining the ends of a double-stranded strip of carbon and oxygen atoms.

A mathematician confided

That a Möbius strip is one-sided.

And you'll get quite a laugh

If you cut one in half,

For it stays in one piece when divided.

Möbius strip

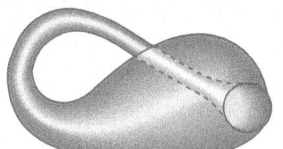

Klein bottle

    Whereas a Möbius strip results from giving a paper *strip* a half-twist and then connecting it to itself, if we could do the same thing with a paper *tube* we would obtain a Klein bottle, named after Felix Klein (1849–1925). Klein produced important results in several areas, including non-Euclidean geometry and the early beginnings of group theory.

A mathematician named Klein

Thought the Möbius strip was divine.

Said he, "If you glue

The edges of two

You'll get a weird bottle like mine."

For Group or Individual Investigation

1. The Möbius strip has other interesting properties. With a pair of scissors, cut the strip lengthwise. Do you get two strips? Repeat the process with what you have obtained from the first cut. What happens?

2. Now construct another Möbius strip, and start cutting lengthwise about $\frac{1}{3}$ of the way from one edge. What happens?

3. What would be the advantage of a conveyor belt with the configuration of a Möbius strip?

9.8   CHAOS AND FRACTAL GEOMETRY

OBJECTIVES

1 Investigate the origins and meaning of chaos.

2 Find attractors in sequences.

3 Explore the basics of fractal geometry.

Chaos

Does Chaos Rule the Cosmos?

    —One of the ten great unanswered questions of science, as found in the November 1992 issue of *Discover*

Ironically, this question about the degree of disorder and unpredictability in the universe couldn't be suitably addressed until the age of the highly predictable and algorithmic computer. In the early 1960s, Edward Lorenz of MIT discovered (accidentally) that even tiny changes in conditions could result in drastic and unpredictable outcomes in his models for predicting the weather.

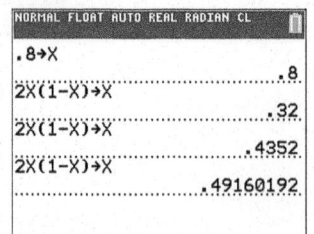

These two screens show how the TI-84 Plus C calculator can produce the sequence described.

The theory that arose from this discovery was dubbed **chaos** in a paper by the mathematician James A. Yorke, of the University of Maryland at College Park. In the ensuing decades, chaos theory has been applied in countless fields. As we begin our investigation of chaos, we will see that mathematical models are useful in demonstrating how a nearly imperceptible change in an initial condition can determine whether a system is stable or completely unpredictable.

Attractors

Consider the equation

$$y = kx(1 - x).$$

Choosing $k = 2$ gives the equation

$$y = 2x(1 - x),$$

which can be "iterated" by starting with an arbitrary x-value between 0 and 1, calculating the resulting y-value, substituting that y-value back in as x, calculating the resulting y-value, substituting that y-value back in as x, calculating another y-value, and so on. For example, a starting value of $x = 0.8$ produces the following sequence (which you can verify with a calculator):

$$0.8, \ 0.32, \ 0.435, \ 0.492, \ 0.500, \ 0.500, \ 0.500, \quad \text{and so on.}$$

The sequence seems to begin randomly but quickly stabilizes at the value 0.500. A different initial x-value would produce another sequence that would also "converge" to 0.500. The value 0.500 can be called an *attractor* for the sequence generated by the equation $y = 2x(1 - x)$. The values of the sequence are "attracted" toward 0.500.

EXAMPLE 1   Finding Attractors

For the equation $y = kx(1 - x)$ with $k = 3$, begin with $x = 0.7$ and iterate with a calculator. What pattern emerges? How many attractors are there?

Solution

Using a TI-84 Plus C calculator, we find that the seventeenth through twentieth iterations give this sequence of terms.

$$0.6354387337, \quad 0.6949690482, \quad 0.6359612107, \quad 0.6945436475$$

The sequence apparently converges in a manner different from the initial discussion, alternating between values near 0.636 and 0.695. Therefore, for $k = 3$, the sequence tends alternately toward *two* distinct attractors.

It happens that the equation in **Example 1** exhibits the same behavior for any initial value of x between 0 and 1.

These screens support the results of **Example 1.**

EXAMPLE 2   Finding Attractors

In the equation of **Example 1,** change the multiplier k to 3.5, and find the forty-fourth through fifty-first terms. What pattern emerges? How many attractors are there?

Solution

Again, using a TI-84 Plus C calculator and rounding to three decimal places, we get

$$0.383, \ 0.827, \ 0.501, \ 0.875, \ 0.383, \ 0.827, \ 0.501, \ 0.875.$$

This sequence seems to stabilize around *four* alternating attractors, approximately 0.383, 0.827, 0.501, and 0.875.

NORMAL FLOAT AUTO REAL RADIAN MP
PRESS ENTER TO EDIT

n	$u(n)$			
0	.7			
1	.735			
2	.68171			
3	.75943			
4	.63943			
5	.80695			
6	.54523			
7	.86784			
8	.40142			
9	.84099			
10	.46804			

$u(n)▤3.5u(n-1)(1-u(n-1))$

NORMAL FLOAT AUTO REAL RADIAN MP
PRESS ◆ TO EDIT FUNCTION

n	$u(n)$			
44	.38282			
45	.82694			
46	.50088			
47	.875			
48	.38282			
49	.82694			
50	.50088			
51	.875			

$u(n)=.8749972636$

These screens support the discussion in **Example 2.**

John Nash (1928–2015), a notable modern American mathematician, first came to the attention of the general public through his biography *A Beautiful Mind* (and the movie of the same name). In 1958 Nash narrowly lost out to René Thom (pictured on the next page), topologist and inventor of catastrophe theory, for the Fields Medal. This is the mathematical equivalent of the Nobel prize.

Although his brilliant career was sadly interrupted by mental illness for a period of about thirty years, in 1994 Nash was awarded "the Central Bank of Sweden Prize in Economic Science in Memory of Alfred Nobel," generally regarded as equivalent to the Nobel prize. This award was for Nash's equilibrium theorem, published in his doctoral thesis in 1950. It turned out that Nash's work established a significant new way of analyzing rational conflict and cooperation in economics and other social sciences.

Notice that in our initial discussion, for $k = 2$, the sequence converged to *one* attractor. In **Example 1,** for $k = 3$, it converged to *two* attractors, and in **Example 2,** for $k = 3.5$, it converged to *four* attractors.

It turns out that as k is increased further, the number of attractors doubles over and over again, more and more often. In fact, this doubling has occurred infinitely many times before k even gets as large as 4. When we look closely at groups of these doublings, we find that they are always similar to earlier groups but on a smaller scale. This is called *self-similarity,* or *scaling,* an idea that is not new but has taken on new significance in recent years. Somewhere before k reaches 4, the resulting sequence becomes apparently totally random, with no attractors and no stability. This type of condition is one instance of chaos.

The equation $y = kx(1 - x)$ does not look all that complicated, but the intricate behavior exhibited by it and similar equations has occupied some of the brightest minds (not to mention computers) in various fields—ecology, biology, physics, genetics, economics, mathematics—since about 1960. Such an equation might represent, for example, the population of some animal species, where the value of k is determined by factors (such as food supply or predators that prey on the species) that affect the increase or decrease of the population. Under certain conditions there is a long-run steady-state population (a single attractor). Under other conditions the population will eventually fluctuate between two alternating levels (two attractors), or four, or eight, and so on. But after a certain value of k, the long-term population becomes totally chaotic and unpredictable.

As long as k is small enough, there will be some number of attractors, and the long-term behavior of the sequence (or population) is the same regardless of the initial x-value. But once k is large enough to cause chaos, the long-term behavior of the system will change drastically when the initial x-value is changed only slightly. For example, consider the following two sequences, both generated from $y = 4x(1 - x)$.

0.600, 0.960, 0.154, 0.520, 0.998, 0.006, 0.025, . . . Starting with $x = 0.600$

0.610, 0.952, 0.184, 0.601, 0.959, 0.157, 0.529, . . . Starting with $x = 0.610$

The fact that the two sequences wander apart from one another is partly due to roundoff errors along the way. But Yorke and others have shown that even "exact" calculations of the iterates would quickly produce divergent sequences just because of the slightly different initial values.

Patterns like those in the sequences above are more than just numerical oddities. Similar patterns apply to a great many phenomena in the physical, biological, and social sciences, many of them seemingly common natural systems that have been studied for hundreds of years. The measurement of a coastline; the description of the patterns in a branching tree, or a mountain range, or a cloud formation, or intergalactic cosmic dust; the prediction of weather patterns; the turbulent behavior of fluids of all kinds; the circulatory and neurological systems of the human body; fluctuations in populations and economic systems—these and many other phenomena remain mysteries, concealing their true nature somewhere beyond the reach of even our brightest minds and our fastest computers.

Continuous phenomena are easily dealt with. A change in one quantity produces a predictable change in another. (For example, a little more pressure on the gas pedal produces a little more speed.) Mathematical functions that represent continuous events can be graphed by unbroken lines or curves, or perhaps smooth, gradually changing surfaces. The governing equations for such phenomena are "linear," and extensive mathematical methods of solving them have been developed. On the other hand, erratic events associated with certain other equations are harder to describe or predict. The science of chaos, made possible by modern computers, continues to open up new ways to deal with such events.

René Thom
(1923–2002)

One early attempt to deal with discontinuous processes in a new way, generally acknowledged as a forerunner of chaos theory, was that of the French mathematician René Thom, who, in the 1960s, applied the methods of topology. To emphasize the feature of sudden change, Thom referred to events such as a heartbeat, a buckling beam, a stock market crash, a riot, or a tornado as *catastrophes*. He proved that all catastrophic events (in our four-dimensional space-time) are combinations of seven elementary catastrophes. (In higher dimensions the number quickly approaches infinity.)

Each of the seven elementary catastrophes has a characteristic topological shape. Two examples are shown in **Figure 86.** The figure on the left is called a *cusp*. The figure on the right is an *elliptic umbilicus* (a belly button with an oval cross-section). Thom's work became known as **catastrophe theory.**

Figure 86

Computer graphics have been indispensable in the study of chaotic processes. The plotting of large numbers of points has revealed patterns that would otherwise not have been observed. (The underlying reasons for many of these patterns, however, have still not been explained.) The first two images in **Figure 87** were created using chaotic processes. Interestingly, chaos theory has been used to explain the Great Red Spot of Jupiter.

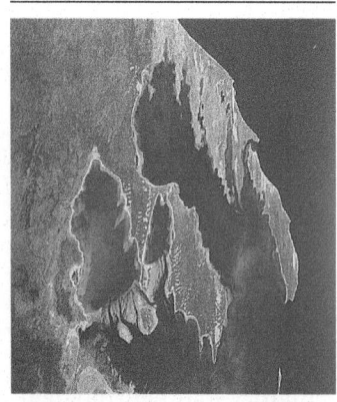

The surface of the earth, consisting of continents, mountains, oceans, valleys, and so on, has **fractal dimension** 2.2.

Aside from providing a geometric structure for chaotic processes in nature, fractal geometry is viewed by many as a significant art form. (To appreciate why, see the 1986 publication *The Beauty of Fractals,* by H. O. Peitgen and P. H. Richter, which contains 184 figures, many in color.) Peitgen and others have also published *Fractals for the Classroom: Strategic Activities Volume One* (Springer-Verlag, 1991).

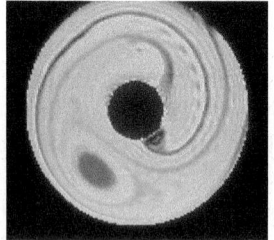 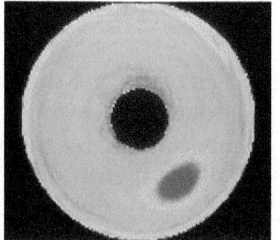 

Figure 87

Fractals

If there is one structure that has provided a key for the new study of nonlinear processes, it is **fractal geometry,** developed over a period of years mainly by the IBM mathematician Benoit Mandelbrot (1924–2010). For his work in this field, and at the recommendation of the National Science Foundation, Columbia University awarded him the 1985 Bernard Medal for Meritorious Service to Science.

Lines have a single dimension. Plane figures have two dimensions, and we live in a three-dimensional spatial world. In a paper published in 1967, Mandelbrot investigated the idea of measuring the length of a coastline. He concluded that such a shape defies conventional Euclidean geometry and that, rather than having a natural number dimension, it has a "fractional dimension." A coastline is an example of a *self-similar shape*—a shape that repeats itself over and over on different scales. From a distance, the bays and inlets cannot be individually observed, but as one moves closer they become more apparent. The branching of a tree, from twig to limb to trunk, also exhibits a shape that repeats itself.

Mandelbrot named such self-similar objects with fractional dimension **fractals.** One of the most famous fractals is generated by a set of numbers called the Mandelbrot set in his honor. **Figure 88** shows images of the fractal. The image on the right is an enlarged version of the region bounded by the red rectangle and displays the scaled self-similarity characteristic of fractals.

 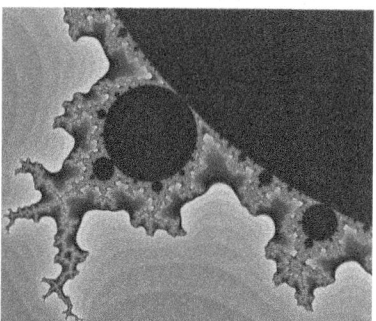

Figure 88

The theory of fractals today finds application in many areas of science and technology. It has been used to analyze the turbulence of liquids, the branching of rivers, and price variation in economics—and even to identify certain types of cancer cells. Hollywood has used fractals in the special effects found in some blockbuster movies.

An interesting account of the science of chaos is found in the popular 1987 book *Chaos,* by James Gleick. Mandelbrot published two books on fractals. They are *Fractals: Form, Chance, and Dimension* (1975) and *The Fractal Geometry of Nature* (1982). He is also featured in several videos about fractals, including *Clouds Are Not Spheres.*

9.8 EXERCISES

CONCEPT CHECK *Write* true *or* false *for each statement.*

1. A sequence of numbers generated by the iteration of an equation can have at most one attractor.

2. When a system is completely unpredictable and unstable, it is an example of chaos.

3. It is possible for a system to have five attractors.

4. Fractal geometry deals with shapes having fractional dimension.

5. Chaos theory predated catastrophe theory.

Use a calculator to determine the pattern of attractors for the equation

$$y = kx(1 - x)$$

for the given value of k and the given initial value of x.

6. $k = 3.25$
$x = 0.7$

7. $k = 3.4$
$x = 0.8$

8. $k = 3.55$
$x = 0.7$

Exercises 9–33 are taken from an issue of Student Math Notes, published by the National Council of Teachers of Mathematics. They were written by Dr. Tami S. Martin, Mathematics Department, Illinois State University, and the authors wish to thank N.C.T.M. and Tami Martin for permission to reproduce this activity. Because the exercises should be done in numerical order, answers to all exercises (both even- and odd-numbered) appear in the answer section of the student edition of this text.

Most of the mathematical objects you have studied have dimensions that are whole numbers. For example, such solids as cubes and icosahedrons have dimension three. Squares, triangles, and many other planar figures are two-dimensional. Lines are one-dimensional, and points have dimension zero.

Consider a square with side of length 1. Gather several of these squares by cutting out or using patterning blocks.

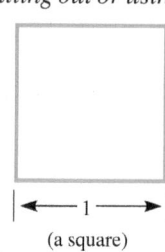
(a square)

The size of a figure is calculated by counting the number of replicas (small pieces) that make it up. Here, a replica is the original square with edges of length 1.

9. What is the least number of these squares that can be put together edge to edge to form a larger square?

The original square is made up of one small square, so its size is 1.

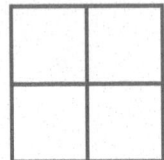

10. What is the size of the new square?

11. What is the length of each edge of the new square?

*Similar figures have the same shape but are not necessarily the same size. The **scale factor** between two similar figures can be found by calculating the ratio of corresponding edges:*

$$\frac{\text{new length}}{\text{old length}}.$$

12. What is the scale factor between the large square and the small square?

13. Find the ratio $\frac{\text{new size}}{\text{old size}}$ for the two squares.

14. Form an even larger square that is three units long on each edge. Compare this square to the small square. What is the scale factor between the two squares? What is the ratio of the new size to the old size?

15. Form an even larger square that is four units long on each edge. Compare this square to the small square. What is the scale factor between the two squares? What is the ratio of the new size to the old size?

16. Complete the table for squares.

Scale factor	2	3	4	5	6	10
Ratio of new size to old size						

17. How are the two rows in the table related?

Consider an equilateral triangle. The length of an edge of the triangle is one unit. The size of this triangle is 1.

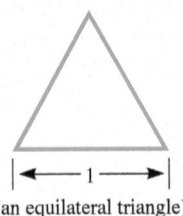

(an equilateral triangle)

18. What is the least number of equilateral triangles that can be put together edge to edge to form a similar larger triangle?

19. Complete the table for triangles.

Scale factor	2	3	4	5	6	10
Ratio of new size to old size						

20. How does the relationship between the two rows in this table compare with the one you found in the table for squares?

One way to define the dimension, d, of a figure relates the scale factor, the new size, and the old size:

$$(\text{scale factor})^d = \frac{\text{new size}}{\text{old size}}.$$

Using a scale factor of two for squares or equilateral triangles, we can see that $2^d = \frac{4}{1}$; that is, $2^d = 4$. Because $2^2 = 4$, the dimension, d, must be two. This definition of dimension confirms what we already know—that squares and equilateral triangles are two-dimensional figures.

21. Use this definition and your completed tables to confirm that the square and the equilateral triangle are two-dimensional figures for scale factors other than two.

Consider a cube, with edges of length 1. Let the size of the cube be 1.

22. What is the least number of these cubes that can be put together face to face to form a larger cube?

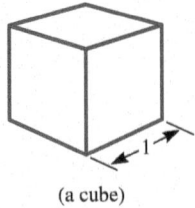

(a cube)

23. What is the scale factor between these two cubes? What is the ratio of the new size to the old size for the two cubes?

24. Complete the table for cubes.

Scale factor	2	3	4	5	6	10
Ratio of new size to old size						

25. How are the two rows in the table related?

26. Use the definition of dimension and a scale factor of two to verify that a cube is a three-dimensional object.

We have explored scale factors and sizes associated with two- and three-dimensional figures. Is it possible for mathematical objects to have fractional dimensions? Consider each figure formed by replacing the middle third of a line segment of length 1 by one upside-down V, each of whose two sides is equal in length to the segment removed. The first four stages in the development of this figure are shown.

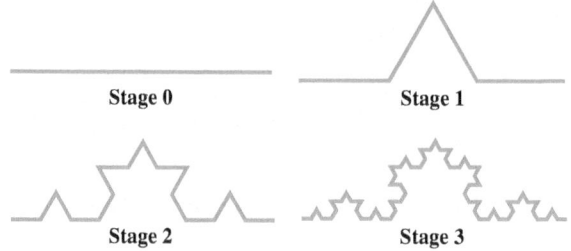

Finding the scale factor for this sequence of figures is difficult, because the overall length of a representative portion of the figure remains the same while the number of pieces increases. To simplify the procedure, follow these steps.

Step 1 *Start with any stage (e.g., Stage 1).*

Step 2 *Draw the next stage (e.g., Stage 2) of the sequence and "blow it up" so that it contains an exact copy of the preceding stage (in this example, Stage 1).*

Notice that Stage 2 contains four copies, or replicas, of Stage 1 and is three times as long as Stage 1.

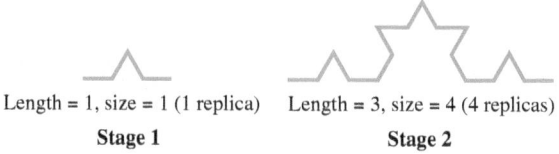

27. The scale factor is equal to the ratio $\frac{\text{new length}}{\text{old length}}$ between any two consecutive stages. The scale factor between Stage 1 and Stage 2 is _____.

28. The size can be determined by counting the number of replicas of Stage 1 found in Stage 2. Old size $= 1$, new size $=$ _____.

Use the definition of dimension to compute the dimension, d, of the figure formed by this process: $3^d = \frac{4}{1}$; that is, $3^d = 4$. Since $3^1 = 3$ and $3^2 = 9$, for $3^d = 4$ the dimension of the figure must be greater than one but less than two:

$$1 < d < 2.$$

29. Use your calculator to estimate d. Remember that d is the exponent that makes 3^d equal 4. For example, because d must be between 1 and 2, try $d = 1.5$. But $3^{1.5} = 5.196\ldots$, which is greater than 4; thus, d must be smaller than 1.5. Continue until you approximate d to three decimal places. (Use logarithms for maximum accuracy.)

*The original figure was a one-dimensional line segment. By iteratively adding to the line segment, an object of dimension greater than one but less than two was generated. Objects with fractional dimension are known as **fractals**. Fractals are infinitely self-similar objects formed by repeated additions to, or removals from, a figure. The object attained at the limit of the repeated procedure is the fractal.*

Next consider a two-dimensional object with sections removed iteratively. In each stage of the fractal's development, a triangle is removed from the center of each triangular region.

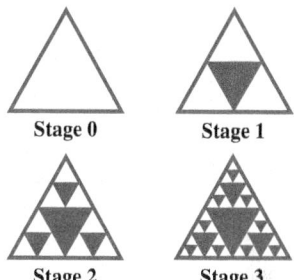

Use the process from the last example to help answer the following questions.

30. What is the scale factor of the fractal?

31. Old size $= 1$, new size $=$ _____.

32. The dimension of the fractal is between what two whole number values?

33. Use the definition of dimension and your calculator to approximate the dimension of this fractal to three decimal places.

34. The **Koch snowflake** begins with an equilateral triangle with a side length of 1 (Stage 0). Each side is trisected, and a smaller equilateral triangle is attached to the center of each side to obtain the Stage-1 snowflake, as shown in the figure. The process is repeated to obtain the Stage-2 snowflake. Find the perimeter P and the area $\mathcal{A}$ of the Stage-2 snowflake. (*MT* calendar problem)

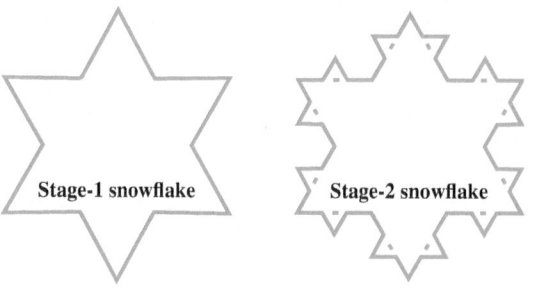

35. Find the dimension of the Koch snowflake. [*Hint:* Consider your work in **Exercises 27–29**.]

Chapter **9** SUMMARY

KEY TERMS

9.1

point
line
plane
half-line
ray
line segment
parallel lines
intersecting lines
parallel planes
intersecting planes
skew lines
angle
sides
vertex
one degree
acute angle
right angle
obtuse angle
straight angle
protractor
perpendicular lines
vertical angles
complementary
supplementary
transversal
alternate interior angles
alternate exterior angles
interior angles on same
  side of transversal
corresponding angles

9.2

simple curve
closed curve

convex
polygon
regular polygon
triangle
acute triangle
right triangle
obtuse triangle
equilateral triangle
isosceles triangle
scalene triangle
quadrilateral
trapezoid
parallelogram
rectangle
square
rhombus
interior angle
exterior angle
circle
center
radius
chord
diameter
semicircle
tangent
secant
arc
minor arc
major arc
inscribed
geometric
  construction
compass
perpendicular bisector

9.3

congruent triangles
similar triangles
hypotenuse
legs
Pythagorean triple

9.4

perimeter
area
circumference

9.5

plane figure
rectangular
  parallelopiped
space figure
polyhedra
tetrahedron
hexahedron
octahedron
dodecahedron
icosahedron
pyramid
prism
volume
surface area
right circular cylinder
sphere
right circular cone

9.6

transformational
  geometry

reflection
reflection image
line of reflection
transformation
invariant point
collinear
line of symmetry
composition
translation
magnitude
identity translation
inverses
rotation
center of rotation
identity rotation
point reflection
glide reflection
isometry
size transformation
dilation
contraction

9.7

pseudosphere
tractrix
perspective
projective sphere
antipodal points
projective plane
duality
genus

9.8

chaos
catastrophe theory
fractal

NEW SYMBOLS

$\longleftrightarrow$ line
$\circ\!\!\longrightarrow$ half-line
$\bullet\!\!\longrightarrow$ ray
$\bullet\!\!\longrightarrow\!\!\bullet$ segment

$\parallel$ parallel
$\measuredangle$ angle
$\urcorner$ right angle
$\frown$ arc

$\triangle$ triangle
$\cong$ congruence
A' image of A

TEST YOUR WORD POWER

See how well you have learned the vocabulary in this chapter.

1. **Vertical angles** are
   A. angles whose measures have a sum of 180 degrees.
   B. formed by intersecting lines and have equal measure.
   C. formed by intersecting lines and do not have equal measure.
   D. angles with one side pointing vertically upward.

2. A **scalene triangle** is
   A. a triangle having at least two equal sides.
   B. a triangle that cannot possibly be a right triangle.
   C. a triangle having no two sides equal.
   D. a regular polygon having three sides.

3. A **Pythagorean triple** is
   A. a special triangle with side length ratio 1:2:3.
   B. an ordered triple (a, b, c) for which $a^2 + b^2 = c^2$.
   C. a triangle with largest angle having measure three times that of the smallest angle.
   D. a triangle with base equaling triple the height.

4. A **tetrahedron** is
   A. a regular polyhedron having four sides.
   B. a pyramid with a square base.
   C. a space figure having two faces in parallel planes.
   D. a space figure with parallelogram faces.

5. An **invariant point** of a transformation is a point
   A. that is its own image under the transformation.
   B. whose image is the same distance from the line of reflection as the point itself.
   C. contained in both the image and the original figure.
   D. that cannot be changed by any transformation.

6. In topology, the **genus** of a figure is
   A. the area of its cross section.
   B. the shape of its image projected onto the unit sphere.
   C. the number of distinct surfaces it has.
   D. the number of cuts that can be made without cutting it into two pieces.

ANSWERS
1. B   2. C   3. B   4. A   5. A   6. D

QUICK REVIEW

Concepts | *Examples*

9.1 POINTS, LINES, PLANES, AND ANGLES

A **point** is represented by a dot. A **line** extends indefinitely in two directions, has no width, and is determined by two distinct points. A **plane** is represented by a flat surface and is determined by three points not all on the same line.

A **ray** has a starting point and extends indefinitely in one direction, and a **half-line** is a ray without a starting point.

A **line segment** includes a starting point, an ending point, and all points between them.

Two lines are **parallel** if they are in the same plane but never intersect. Lines not in the same plane are **skew.**

An **angle** is made up of two rays with a common endpoint (**vertex**). The measure of an angle is given in **degrees,** where 1 degree (1°) is $\frac{1}{360}$ of a full rotation.

- An angle measuring between 0° and 90° is an **acute angle.**
- An angle measuring between 90° and 180° is an **obtuse angle.**
- An angle measuring exactly 90° is a **right angle.**
- An angle measuring 180° is a **straight angle.**
- Two angles whose measures have a sum of 90° are **complementary angles.**
- Two angles whose measures have a sum of 180° are **supplementary angles.**

Lines intersecting at 90° angles are **perpendicular lines.**

A line intersecting parallel lines is a **transversal.**

Intersecting lines create two pairs of **vertical angles.**

Vertical angles have equal measure.

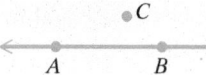

The figure shows a line passing through points A and B, and a plane determined by points A, B, and C.

- Ray AB ($\overrightarrow{AB}$) consists of point A and all points on the line to the right of point A.
- Half-line $\overset{\circ}{\longrightarrow}AB$ consists only of the points to the right of point A.
- Line segment $\overline{AB}$ contains the two endpoints and all points in between.

In the figure below, lines m and n are parallel. The transversal q, along with lines m and n, form eight angles.

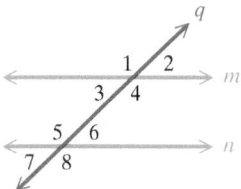

∡1 is an **obtuse angle,** and ∡2 is an **acute angle.**

∡1 and ∡2 are **supplementary angles,** and together they form a **straight angle.**

∡1 and ∡4 are **vertical angles.**

∡1 and ∡5 are **corresponding angles** and have equal measure.

∡3 and ∡6 are **alternate interior angles** and have equal measure.

∡2 and ∡7 are **alternate exterior angles** and have equal measure.

∡3 and ∡5 are **interior angles on the same side of the transversal** and are supplementary.

Concepts   *Examples*

9.2 CURVES, POLYGONS, CIRCLES, AND GEOMETRIC CONSTRUCTIONS

Simple curves can be drawn without lifting the pencil from the paper and without passing through the same point twice. A **closed curve** has the same starting and ending point, and it can be drawn without lifting the pencil from the paper.

Simple;   Closed;   Neither
closed   not simple

A **polygon** is a simple closed curve made up of straight line segments; it is **regular** if all sides are equal and all angles are equal.

A **triangle** is a three-sided polygon classified by the number of equal sides and by the measures of its angles relative to 90°.

A **quadrilateral** has four sides and is classified by characteristics of its sides and angles.

The sum of the interior angles of any triangle is 180°.

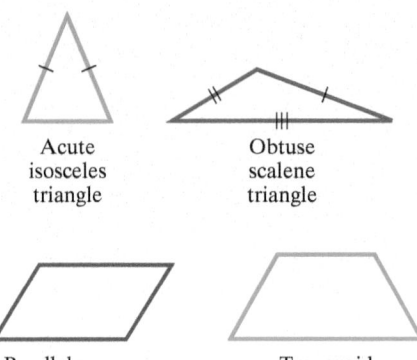

Acute   Obtuse
isosceles   scalene
triangle   triangle

Parallelogram   Trapezoid

Any exterior angle has measure equal to the sum of the two opposite interior angles.

Solve for x.

$$3x + 2x - 5 = 3x + 41$$

$$5x - 5 = 3x + 41 \quad \text{Combine like terms.}$$

$$2x = 46 \quad \text{Subtract } 3x. \text{ Add 5.}$$

$$x = 23 \quad \text{Divide by 2.}$$

A **circle** is a set of points in the plane that are equidistant from a fixed point called the **center.**

- A **chord** is a line segment having endpoints on the circle.
- An **arc** is formed by the endpoints of a chord and the part of the circle between those endpoints.
- A chord passing through the center is a **diameter.**
- An angle with its vertex on the circle and its sides intercepting two (other) points on the circle is **inscribed** in the circle.

A straightedge and compass may be used to complete **geometric constructions.**

The figure below shows the inscribed angle PQR whose sides are chords PQ and QR. Chord PQ contains center O and is therefore a diameter.

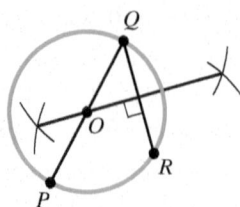

The figure also shows the construction of the perpendicular bisector of chord QR, which intersects diameter PQ at center O.

Concepts   *Examples*

9.3 THE GEOMETRY OF TRIANGLES: CONGRUENCE, SIMILARITY, AND THE PYTHAGOREAN THEOREM

Congruent triangles have exactly the same shape and size. This means that corresponding sides have equal length and corresponding angles have equal measure.

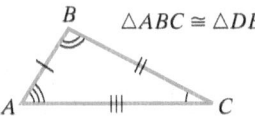 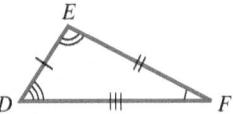

Congruence is often verified in a two-column proof.

Given: $AC = BC; \angle ACD = \angle BCD$
Prove: $\triangle ADC \cong \triangle BDC$

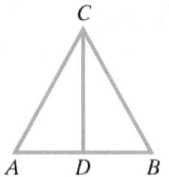

PROOF

Statements	Reasons
1. $AC = BC$	1. Given
2. $\angle ACD = \angle BCD$	2. Given
3. $CD = CD$	3. Reflexive property
4. $\triangle ADC \cong \triangle BDC$	4. SAS congruence property

Similar triangles have the same shape (corresponding angles are equal) but not necessarily the same size. Corresponding sides are proportional.

In the figure shown, $\overleftrightarrow{AB}$ is parallel to $\overleftrightarrow{DE}$. Verify that $\triangle ABC$ is similar to $\triangle EDC$.

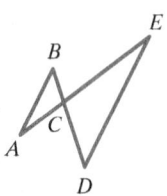

$\overleftrightarrow{AB}$ is parallel to $\overleftrightarrow{DE}$, so $\overleftrightarrow{AE}$ is a transversal, and thus $\angle BAC = \angle DEC$ because they are alternate interior angles. Also, $\angle BCA = \angle DCE$, since they are vertical angles.

Therefore, $\triangle ABC$ is similar to $\triangle EDC$ by the Angle-Angle (AA) similarity property.

Properties of similar triangles can be used to solve for unknown parts of triangles, and this has many applications.

A flagpole casts a 99-ft shadow at the same time that a lamp post 10 ft high casts an 18-ft shadow. Find the height of the flagpole.

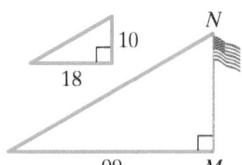

$$\frac{MN}{10} = \frac{99}{18} \quad \text{Write a proportion using similar triangles.}$$

$$MN \cdot 18 = 10 \cdot 99 \quad \text{Cross products}$$

$$MN = 55 \quad \text{Solve for } MN.$$

The flagpole is 55 ft tall.

Pythagorean Theorem
If the two legs of a right triangle have lengths a and b, and the hypotenuse has length c, then

$$a^2 + b^2 = c^2.$$

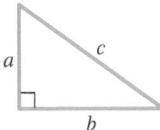

Find the length of the hypotenuse of the right triangle with the flagpole above. Let c be the length of the hypotenuse.

$$99^2 + (MN)^2 = c^2 \quad \text{Pythagorean theorem}$$

$$99^2 + 55^2 = c^2 \quad \text{Let } MN = 55.$$

$$9801 + 3025 = c^2 \quad \text{Apply the exponents.}$$

$$12{,}826 = c^2 \quad \text{Add.}$$

$$c = \sqrt{12{,}826} \quad \text{Take the square root.}$$

$$c \approx 113.25 \quad \text{Approximate.}$$

The length of the hypotenuse is approximately 113.25 ft.

Concepts   *Examples*

9.4 PERIMETER, AREA, AND CIRCUMFERENCE

The **perimeter** of a polygon is the sum of the lengths of its sides.

The **area** of a polygon is the amount of plane surface it covers.

Polygon	Perimeter	Area
Triangle 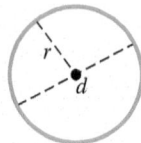	$P = a + b + c$	$\mathcal{A} = \dfrac{1}{2}bh$
Rectangle	$P = 2\ell + 2w$	$\mathcal{A} = \ell w$
Square	$P = 4s$	$\mathcal{A} = s^2$
Parallelogram		$\mathcal{A} = bh$
Trapezoid		$\mathcal{A} = \dfrac{1}{2}h(B + b)$

The distance around a circle is its **circumference.**

$C = \pi d = 2\pi r$

$\mathcal{A} = \pi r^2$

Find the perimeter of the entire figure and the area of the shaded region.

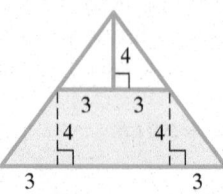

The hypotenuse of each triangle with legs of length 3 and 4 is

$$\sqrt{3^2 + 4^2} = \sqrt{25} = 5.$$

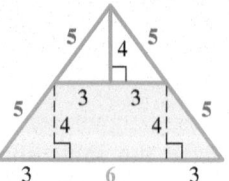

The *entire figure* is an isosceles triangle with side lengths of 12, 10, and 10.

$$P = 12 + 10 + 10 = 32$$

The shaded region is a trapezoid, with $B = 12$, $b = 6$, and $h = 4$.

$$\mathcal{A} = \frac{1}{2}h(B + b)$$

$$\mathcal{A} = \frac{1}{2}(4)(12 + 6)$$

$$\mathcal{A} = 36$$

Find the perimeter and area of the following figure.

The figure is made up of a rectangle and a semicircle.

$$P = (4 + 8 + 4) + \left(\frac{1}{2} \cdot \pi \cdot 8\right)$$

$$P = 4\pi + 16$$

$$\mathcal{A} = (8 \cdot 4) + \left(\frac{1}{2} \cdot \pi \cdot 4^2\right)$$

$$\mathcal{A} = 8\pi + 32$$

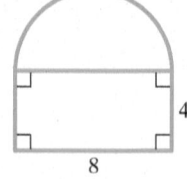

Concepts

Examples

9.5 VOLUME AND SURFACE AREA

Space figures occupy three-dimensional space and have both **volume** and **surface area.** A **polyhedron** is a space figure, the faces of which are polygons. A *regular* polyhedron is one for which all faces are a copy of a single regular polygon. There are only five regular polyhedra.

The **volume** of a space figure is a measure of its capacity.

The **surface area** is the total area that would be covered if the space figure were "peeled" and the peel laid flat.

Space Figure	Volume and Surface Area
Box h, w, ℓ	$V = \ell wh$ $S = 2\ell w + 2\ell h + 2hw$
Cube s, s, s	$V = s^3$ $S = 6s^2$
Right Circular Cylinder h, r	$V = \pi r^2 h$ $S = 2\pi rh + 2\pi r^2$
Sphere r	$V = \dfrac{4}{3}\pi r^3$ $S = 4\pi r^2$
Right Circular Cone h, r	$V = \dfrac{1}{3}\pi r^2 h$ $S = \pi r\sqrt{r^2 + h^2} + \pi r^2$
Pyramid h	$V = \dfrac{1}{3}Bh$ (B is the area of the base.)

Regular Polyhedra

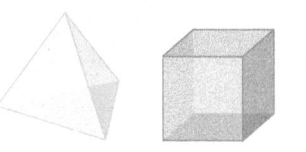
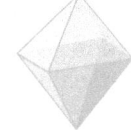

Tetrahedron   Hexahedron   Octahedron
(cube)

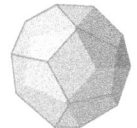

Dodecahedron   Icosahedron

Find the volume and surface area of the box.

$$V = \ell wh$$
$$V = (8)(6)(4)$$
$$V = 192\,\text{cm}^3$$

$$S = 2\ell w + 2\ell h + 2hw$$
$$S = 2(8)(6) + 2(8)(4) + 2(4)(6)$$
$$S = 96 + 64 + 48$$
$$S = 208\ \text{cm}^2$$

The grain silo consists of a half-sphere on top of a right circular cylinder. Find its volume and surface area (using $\pi \approx 3.14$).

$$V = \frac{1}{2} \cdot \frac{4}{3}\pi(4)^3 + \pi(4)^2(8)$$

$$V \approx 535.89 \text{ cubic units}$$

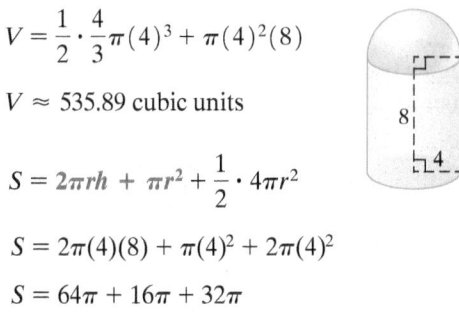

$$S = 2\pi rh + \pi r^2 + \frac{1}{2} \cdot 4\pi r^2$$

$$S = 2\pi(4)(8) + \pi(4)^2 + 2\pi(4)^2$$

$$S = 64\pi + 16\pi + 32\pi$$

$$S = 112\pi \text{ square units}$$

$$S \approx 351.68 \text{ square units}$$

Note that in the portion of surface area for the cylinder (shown in red), the formula included πr^2 instead of $2\pi r^2$, because the top of the cylinder is not an exposed surface in this case.

Concepts   *Examples*

9.6 TRANSFORMATIONAL GEOMETRY

One geometric figure can be transformed into another by one or more geometric **transformations.**

• A **reflection** gives the mirror image of a figure about a **line of reflection.**

• A **translation** "slides" a copy of a figure to a new location. It is equivalent to two reflections about parallel lines.

• A **rotation** revolves a figure about a point called the **center of rotation** through an angle (the **magnitude of rotation**).

• A **glide reflection** is the composition of a reflection and a translation along the line of reflection.

• A **size transformation** stretches (or shrinks) a figure away from (or toward) a point (the center).

The figure shows $\triangle ABC$ **reflected** about line m. The image $\triangle A'B'C'$ is then **translated** parallel to line m. The image of this translation, $\triangle A''B''C''$, is then **rotated** clockwise around point H through an angle of 45°, resulting in the image $\triangle A'''B'''C'''$. Finally, this triangle is **stretched** (or **dilated**) by a factor of 2 using M as the center of the size transformation, resulting in the image $\triangle JKL$.

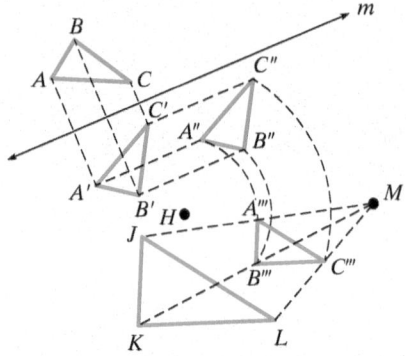

9.7 NON-EUCLIDEAN GEOMETRY AND TOPOLOGY

Euclidean geometry is based on five postulates and five axioms. The fifth postulate is equivalent to "Given a line k and a point P not on the line, there exists one and only one line m through P that is parallel to k." When this postulate is changed, a **non-Euclidean geometry** results.

Lobachevskian geometry replaces Euclid's fifth postulate with the statement "Through a point P off a line k, at least two different lines can be drawn parallel to k."

Riemannian geometry uses as its parallel postulate "Through a point P off a line k, no line can be drawn that is parallel to k."

Projective geometries exhibit **duality,** in that for each statement that holds true about points and lines in a projective plane, another true statement can be formed by interchanging "point" and "line" in the original statement.

Topology deals with properties of figures that are preserved under distortions that do not involve separating (puncturing or tearing) the figures. The topological **genus** of a figure is equal to the number of holes in the figure.

The figure shows the projection of a Euclidean triangle and line onto the sphere. The image is a Riemannian triangle with interior angles summing to more than 180°, and half of a great circle, which is a line in this projective geometry.

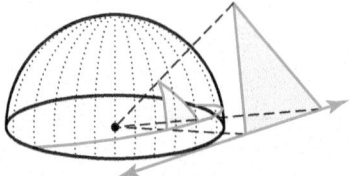

The *dual* of the statement "Every pair of distinct points determine a line" is the statement

"Every pair of distinct lines determine a point."

A sphere and a cone both have genus 0, and a torus has genus 1.

Concepts   *Examples*

9.8 CHAOS AND FRACTAL GEOMETRY

Chaos theory deals with systems that may become unstable with very small changes in governing parameter values.

Fractals are infinitely self-similar objects formed by repeated additions to, or removals from, a figure. The object attained at the limit of the procedure is the fractal.

The sequence generated by the equation

$$y = 2.5x(1 - x),$$

with intial value of $x = 0.5$, has one attractor at 0.6.

NORMAL FLOAT AUTO REAL RADIAN MP			
n	u(n)		
0	.5		
1	.625		
2	.58594		
3	.60654		
4	.59662		
5	.60166		
6	.59916		
7	.60042		
8	.59979		
9	.6001		
10	.59995		

u(n)⬛2.5u(n-1)(1-u(n-1))

Chapter **9** TEST

1. Consider a 42° angle. Answer each of the following.

   (a) What is the measure of its complement?

   (b) What is the measure of its supplement?

   (c) Classify it as acute, obtuse, right, or straight.

Find the measure of each marked angle.

2.

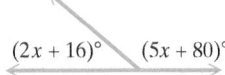

$(2x + 16)°$ $(5x + 80)°$

3.

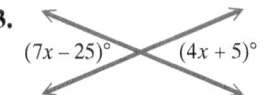

$(7x - 25)°$ $(4x + 5)°$

4.

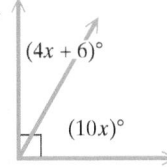

$(4x + 6)°$

$(10x)°$

5.

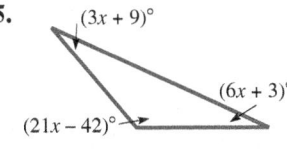

$(3x + 9)°$

$(6x + 3)°$

$(21x - 42)°$

In Exercises 6 and 7, assume that lines m and n are parallel, and find the measure of each marked angle.

6.

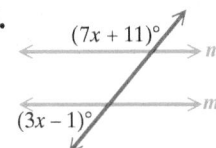

$(7x + 11)°$

$(3x - 1)°$

7.

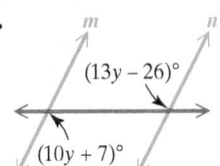

$(13y - 26)°$

$(10y + 7)°$

8. Which one of the statements A–D is false?

   A. A square is a rhombus.

   B. The acute angles of a right triangle are complementary.

   C. A triangle may have both a right angle and an obtuse angle.

   D. A trapezoid may have nonparallel sides of the same length.

9. Explain why a rhombus must be a quadrilateral, but a quadrilateral might not be a rhombus.

Identify each of the following curves as simple, closed, both, *or* neither.

10.

11.

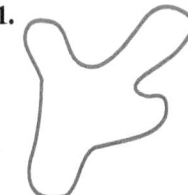

Find the area of each of the following figures.

12.

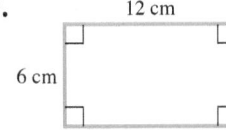

12 cm

6 cm

13.

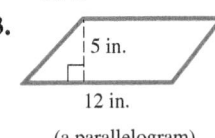

5 in.

12 in.

(a parallelogram)

14.

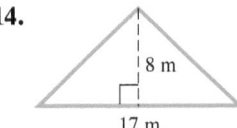

8 m

17 m

15.

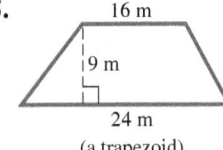

16 m

9 m

24 m

(a trapezoid)

16. *Area of a Shaded Figure* What is the area (to the nearest square unit) of the colored portion of the figure? Use 3.14 as an approximation for π.

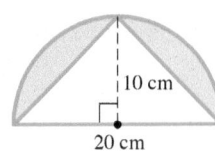

10 cm

20 cm

(a triangle within a semicircle)

17. *Circumference of a Circle* If a circle has area 144π square inches, what is its circumference?

18. *Circumference of a Dome* The Rogers Centre in Toronto, Canada, is the first stadium with a hard-shell, retractable roof. The steel dome is 630 feet in diameter. To the nearest foot, what is the circumference of this dome?

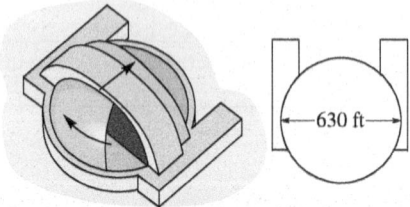

—630 ft—

19. Given: $\angle CAB = \angle DBA$; $DB = CA$
Prove: $\triangle ABD \cong \triangle BAC$

20. *Height of a Building* If an 8-ft stop sign casts a 5-ft shadow at the same time a building casts a 40-ft shadow, how tall is the building?

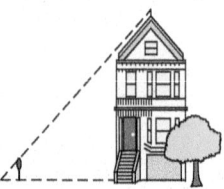

21. *Diagonal of a Rectangle* What is the measure of a diagonal of a rectangle that has width 20 m and length 21 m?

22. Reflect the given figure first about line n and then about line m.

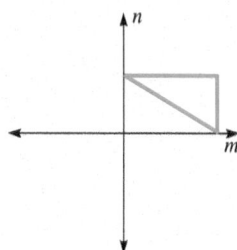

23. Find the point reflection image of the given figure with the given point as center.

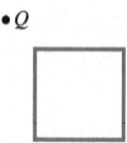

•Q

Find **(a)** *the volume and* **(b)** *the surface area of each of the following space figures. When necessary, use 3.14 as an approximation for π. In Exercises 24 and 26, round to the nearest hundredth.*

24.

6 in.

(a sphere)

25.

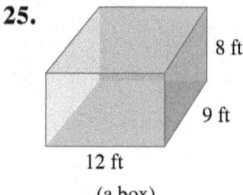

8 ft

9 ft

12 ft

(a box)

26.

6 m

14 m

(a right circular cylinder)

27. List several main distinctions between Euclidean geometry and non-Euclidean geometry.

28. *Topological Equivalence* Are the following pairs of objects topologically equivalent?

(a) A page of a book and the cover of the same book

(b) A pair of glasses with the lenses removed, and the Mona Lisa

29. *Projection Image* A line segment of length 7 cm is projected onto a sphere with radius 1 cm. Will the image also have a length of 7 cm?

30. *Attractors* Use a calculator to determine the attractors for the sequence generated by the equation

$$y = 2.1x(1 - x),$$

with initial value of $x = 0.6$.

10 Counting Methods

Number of Extras	Number of Ways to Order
0	1
1	8
2	28
3	56
4	70
5	56
6	28
7	8
8	1
Total:	**256**

Up and Down Hamburgers offers a very limited menu to ensure the highest quality in the items offered. Customers have 8 choices of extras to add to the basic bun and hamburger patty:

a second patty, cheese, mustard, mayonnaise, lettuce, tomato, onion, pickle.

A customer could choose none, some but not others, or all of these extras. In how many ways can a customer order a hamburger? It is easy to count the special cases:

1 way of ordering *no extras,*

8 ways of ordering only *1 extra,*

and   1 way of ordering *all 8 extras.*

Using concepts from this chapter, we can determine the *other* numbers of ways. See the table in the margin. A burger from Up and Down Hamburgers can be chosen in 256 different ways.

10.1 COUNTING BY SYSTEMATIC LISTING

OBJECTIVES

1 Use a systematic approach to perform a one-part counting task.

2 Use a product table to perform a two-part counting task.

3 Use a tree diagram to perform a multiple-part counting task.

4 Use systematic listing in a counting task that involves a geometric figure.

Counting

In this chapter, **counting** means finding the number of objects, of some certain type, that exist. The methods of counting presented in this section involve listing the possible results for a given task. This approach is practical only for fairly short lists. When listing possible results, it is extremely important to use a *systematic* approach, so that no possibilities are missed.

One-Part Tasks

The results for simple *one-part tasks* can often be listed easily. An example of a **one-part task** is tossing a single fair coin. The list here is *heads, tails,* with two possible results. If the task is to roll a single fair die (a cube with faces numbered 1 through 6), the different results are

1, 2, 3, 4, 5, 6, a total of six possibilities.

EXAMPLE 1 Selecting a Club President

Consider a club N with five members:

$N = \{$Alan, Bill, Cathy, David, Evelyn$\}$, abbreviated as $N = \{A, B, C, D, E\}$.

In how many ways can this group select a president? (All members are eligible.)

Solution

The task in this case is to select one of the five members as president. There are five possible results.

A, B, C, D, and E

Product Tables for Two-Part Tasks

We now consider **two-part tasks.**

EXAMPLE 2 Building Numbers from a Set of Digits

Determine the number of two-digit numbers that can be written using only the digits 1, 2, and 3.

Solution

This task consists of two parts.

Part 1 Choose a first digit.

Part 2 Choose a second digit.

The results for a two-part task can be pictured in a **product table** such as **Table 1.** From the table we obtain our list of possible results.

11, 12, 13, 21, 22, 23, 31, 32, 33

There are nine possibilities.

Table 1

	Second Digit		
First Digit	1	2	3
1	11	12	13
2	21	22	23
3	31	32	33

StatCrunch Dice Rolling

EXAMPLE 3 Rolling a Pair of Dice

Determine the number of different possible results when two ordinary dice are rolled.

Solution

Assume the dice are easily distinguishable. Perhaps one is red and the other green. Then the task consists of two parts.

Part 1 Roll the red die.

Part 2 Roll the green die.

The product table in **Table 2** shows that there are thirty-six possible results.

Table 2 Rolling Two Fair Dice

		Green Die					
		1	2	3	4	5	6
Red Die	1	(1, 1)	(1, 2)	(1, 3)	(1, 4)	(1, 5)	(1, 6)
	2	(2, 1)	(2, 2)	(2, 3)	(2, 4)	(2, 5)	(2, 6)
	3	(3, 1)	(3, 2)	(3, 3)	(3, 4)	(3, 5)	(3, 6)
	4	(4, 1)	(4, 2)	(4, 3)	(4, 4)	(4, 5)	(4, 6)
	5	(5, 1)	(5, 2)	(5, 3)	(5, 4)	(5, 5)	(5, 6)
	6	(6, 1)	(6, 2)	(6, 3)	(6, 4)	(6, 5)	(6, 6)

36 possible results

You will want to refer to **Table 2** *when various dice-rolling problems occur in the remainder of this chapter and the next.*

EXAMPLE 4 Electing Two Club Officers

In **Example 1,** we considered club N consisting of five members.

$$N = \{A, B, C, D, E\}$$

Find the number of ways that the club can elect both a president and a secretary. Assume that all members are eligible, but that no one can hold both offices.

Solution

Again, the required task has two parts.

Part 1 Determine the president.

Part 2 Determine the secretary.

Constructing **Table 3** gives us the possibilities (where, for example, AB denotes president A and secretary B, while BA denotes president B and secretary A).

Table 3 Electing Two Officers

		Secretary				
		A	B	C	D	E
President	A		AB	AC	AD	AE
	B	BA		BC	BD	BE
	C	CA	CB		CD	CE
	D	DA	DB	DC		DE
	E	EA	EB	EC	ED	

20 possible results

Polyhedral dice, based on the Platonic solids, are used in role-playing games. In addition to the common *cube* (6 faces), dice can also take the form of a *tetrahedron* (4 faces), an *octahedron* (8 faces), a *dodecahedron* (12 faces), or an *icosahedron* (20 faces).

Notice that entries down the main diagonal, from upper left to lower right, are omitted from the table because the cases AA, BB, and so on would imply one person holding both offices. Altogether, there are twenty possibilities.

Bone dice were unearthed in the remains of a Roman garrison, Vindolanda, near the border between England and Scotland. Life on the Roman frontier was occupied with gaming as well as fighting. Some of the Roman dice were loaded in favor of 6 and 1.

Life on the American frontier was reflected in cattle brands that were devised to keep alive the memories of hardships, feuds, and romances. A rancher named Ellis from Paradise Valley in Arizona designed his cattle brand in the shape of a pair of dice. You can guess that the pips were 6 and 1.

Fundamental Counting Principle:
Tree Diagrams

EXAMPLE 5 Selecting Committees for a Club

Find the number of ways that club *N* from **Example 4** can appoint a committee of two members to represent them at an association conference.

Solution

The required task again has two parts. In fact, we can refer to **Table 3** again, but this time, the order of the two letters (people) in a given pair really makes no difference. For example, *BD* and *DB* are the same committee. In **Example 4,** *BD* and *DB* were different results since the two people would be holding different offices.

In the case of committees, we can eliminate not only the main diagonal entries but also all entries below the main diagonal. The resulting list contains ten possibilities.

$$AB, \quad AC, \quad AD, \quad AE, \quad BC, \quad BD, \quad BE, \quad CD, \quad CE, \quad DE$$

Tree Diagrams for Multiple-Part Tasks

EXAMPLE 6 Building Numbers from a Set of Digits

Find the number of three-digit numbers that can be written using only the digits 1, 2, and 3, assuming that

(a) repeated digits are allowed   **(b)** repeated digits are not allowed.

Solution

(a) The task of constructing such a number has three parts.

Part 1 Select the first digit.

Part 2 Select the second digit.

Part 3 Select the third digit.

As we move from left to right through the **tree diagram** in **Figure 1,** the tree branches at the first stage to all possibilities for the first digit. Then each first-stage branch again branches, or splits, at the second stage, to all possibilities for the second digit. Finally, the third-stage branching shows the third-digit possibilities. The list of twenty-seven possible results is shown in the right-hand column.

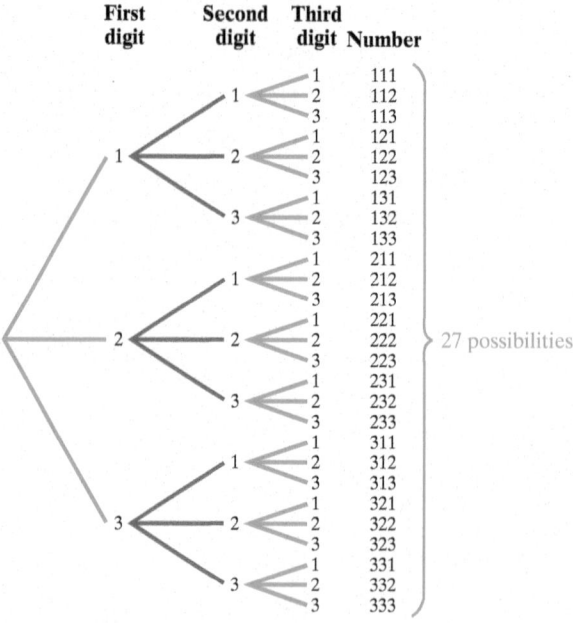

Tree diagram for three-digit numbers using digits 1, 2, and 3

Figure 1

(b) For the case of nonrepeating digits, we could construct a whole new tree diagram, as in **Figure 2,** or we could simply go down the list of numbers from the first tree diagram and strike out any that contain repeated digits. In either case we obtain only six possibilities.

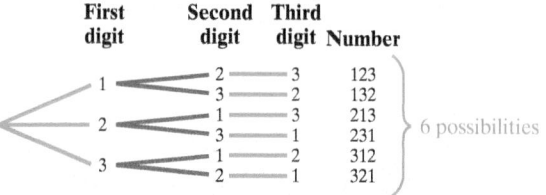

Tree diagram for nonrepeating three-digit numbers using digits 1, 2, and 3

Figure 2

Notice the distinction between parts (a) and (b) of **Example 6.** There are twenty-seven possibilities when "repetitions (of digits) are allowed," but only six possibilities when "repetitions are not allowed."

Here is another way to phrase the problem of **Example 6.**

A three-digit number is to be determined by placing three slips of paper (marked 1, 2, and 3) into a hat and drawing out three slips in succession. Find the number of possible results if the drawing is done (a) with replacement and (b) without replacement.

Drawing "with replacement" means drawing a slip, recording its digit, and replacing the slip into the hat so that it is again available for subsequent draws.

Drawing "with replacement" has the effect of "allowing repetitions," while drawing "without replacement" has the effect of "not allowing repetitions."

The words "repetitions" and "replacement" are important in the statement of a problem. In **Example 2,** no restrictions were stated, so we assumed that *repetitions* (of digits) *were allowed,* or, equivalently, that digits were to be selected *with replacement.*

Animation
→
Fundamental Counting Principle:
Tree Diagrams

EXAMPLE 7 Selecting Switch Settings on a Printer

Pamela's computer printer allows for optional settings with a panel of four on-off switches in a row. How many different settings can she select if no two adjacent switches can both be off?

Solution

This situation is typical of user-selectable options on various devices, including computer equipment, garage door openers, and other appliances. In **Figure 3,** we denote "on" and "off" with 1 and 0, respectively. The number of possible settings is eight.

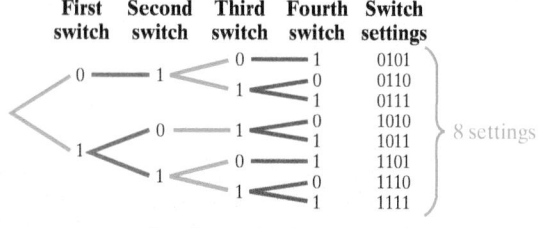

Tree diagram for printer settings

Figure 3

Notice that each time a switch is indicated as off (0), the next switch can only be on (1). This is to satisfy the restriction that no two adjacent switches can both be off.

Fundamental Counting Principle:
Tree Diagrams

EXAMPLE 8 Seating Attendees at a Concert

Arne, Bobbette, Chuck, and Deirdre have tickets for four reserved seats in a row at a concert. In how many different ways can they seat themselves so that Arne and Bobbette will sit next to each other?

Solution

Here we have a four-part task:

Assign people to the first, second, third, and fourth seats.

Let *A*, *B*, *C*, and *D* represent the four people. The tree diagram in **Figure 4** avoids repetitions, because no person can occupy more than one seat. Also, once *A* or *B* appears in the tree, the other one *must* occur at the next stage. No splitting occurs from stage three to stage four because by that time there is only one person left unassigned. The right column in the figure shows the twelve possible seating arrangements.

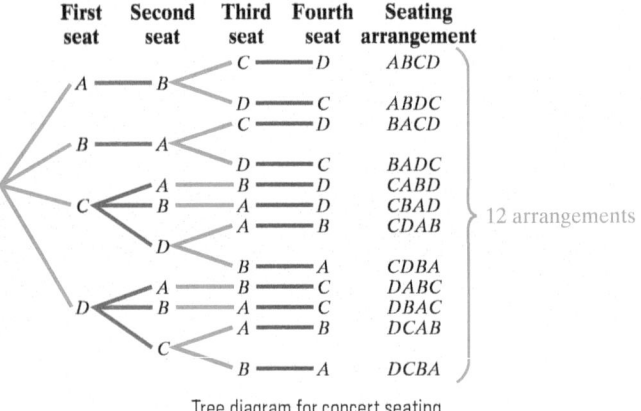

Tree diagram for concert seating

Figure 4

Although we have applied tree diagrams only to tasks with three or more parts, they can also be used for two-part or even simple, one-part tasks. Product tables, on the other hand, are practical only for two-part tasks.

Other Systematic Listing Methods

There are additional systematic ways, besides product tables and tree diagrams, to produce complete listings of possible results.

In **Example 4,** where we used a product table (**Table 3**) to list all possible president-secretary pairs for the club

$$N = \{A, B, C, D, E\},$$

we could have systematically constructed the same list using a sort of alphabetical or left-to-right approach.

First, consider the results where *A* is president. Any of the remaining members (*B*, *C*, *D*, or *E*) could then be secretary. That gives us the pairs *AB*, *AC*, *AD*, and *AE*. Next, assume *B* is president. The secretary could then be *A*, *C*, *D*, or *E*. We get the pairs *BA*, *BC*, *BD*, and *BE*. Continuing in order, we get the complete list just as in **Example 4.**

AB, AC, AD, AE, BA, BC, BD, BE, CA, CB,

CD, CE, DA, DB, DC, DE, EA, EB, EC, ED

Counting methods can be used to find the number of moves required to solve a **Rubik's Cube.** The scrambled cube must be modified so that each face is a solid color. Rubik's royalties from sales of the cube in Western countries made him Hungary's richest man.

Although the craze over the cube of the early 1980s has waned, certain groups have remained intensely interested in not only solving the scrambled cube, but also doing so as quickly as possible. And the 30-year search for an exact number of moves (called face turns) that is guaranteed to suffice in all cases, while no smaller number will suffice, finally ended in July 2010. That number is now known to be 20.

EXAMPLE 9 Counting Triangles in a Figure

How many different triangles (of any size) can be traced in **Figure 5?**

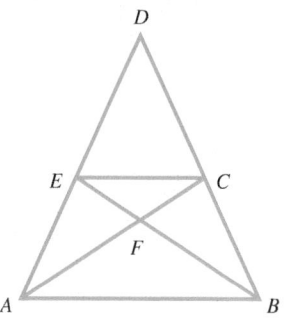

Figure 5

Solution

One systematic approach is to label points as shown, begin with A, and proceed in alphabetical order to write all three-letter combinations, and then cross out the ones that are not triangles in the figure.

$ABC, \quad ABD, \quad ABE, \quad ABF, \quad ACD, \quad ACE, \quad ACF, \quad ADE, \quad ADF, \quad AEF,$

$BCD, \quad BCE, \quad BCF, \quad BDE, \quad BDF, \quad BEF, \quad CDE, \quad CDF, \quad CEF, \quad DEF$

Finally, there are twelve different triangles in the figure. ACB and CBF (and many others) are not included in the list, because they have already been considered.

Another method might be first to identify the triangles consisting of a single region each:

$$DEC, \quad ECF, \quad AEF, \quad BCF, \quad ABF.$$

Then list those consisting of two regions each:

$$AEC, \quad BEC, \quad ABE, \quad ABC.$$

Then list those with three regions each:

$$ACD, \quad BED.$$

There are no triangles with four regions, but there is one with five:

$$ABD.$$

The total is again twelve. *Can you think of other systematic ways of getting the same list?*

Problem-Solving Strategy

Notice that in the first method shown in **Example 9,** the labeled points were considered in alphabetical order. In the second method, the single-region triangles were listed by using first a top-to-bottom order and then a left-to-right order. Using a definite system helps to ensure that we get a complete list.

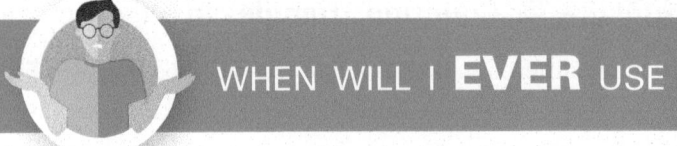

WHEN WILL I **EVER** USE THIS ?

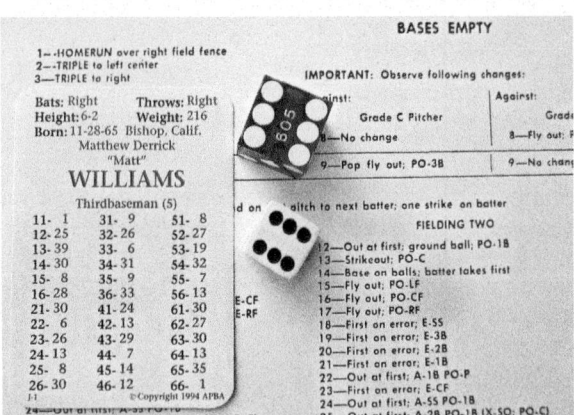

One of today's most popular sports-related pastimes is that of fantasy leagues. Members of the league draft a team and compete against other "owners." Before the advent of fantasy leagues, a similar hobby provided hours of enjoyment for rabid sports fans.

In 1951, a young entrepreneur named Richard Seitz transformed his love for baseball and its statistics into a successful game company called APBA. His idea spawned numerous competitors that feature simulation of athletes performing according to their actual statistics.

To initiate the action in a random fashion, the classic Parker Brothers game Monopoly uses the sum of a roll of two dice. But certain sums occur more than others. To ensure that all results are equally likely, in APBA Major League Baseball, the action is begun by rolling two dice of different sizes (and colors) and reading the results in such a way that every result has an equal chance to occur. Specifically, the number that will activate play is found by reading the face of the large die followed by that of the small die. (See **Table 2** in **Example 3**)

Example: See the photo of a play board and the batter card for Matt Williams, a third baseman for the San Francisco Giants in the 1990s.

- A larger red-die result of 3 followed by a small-die result of 5 is read 35. This number is located on the player card. It corresponds to a different number, which in this case is 9.

- That latter number 9 is then found on the applicable game board with a description of the play. Against a grade D pitcher, Matt gets a single over shortstop.

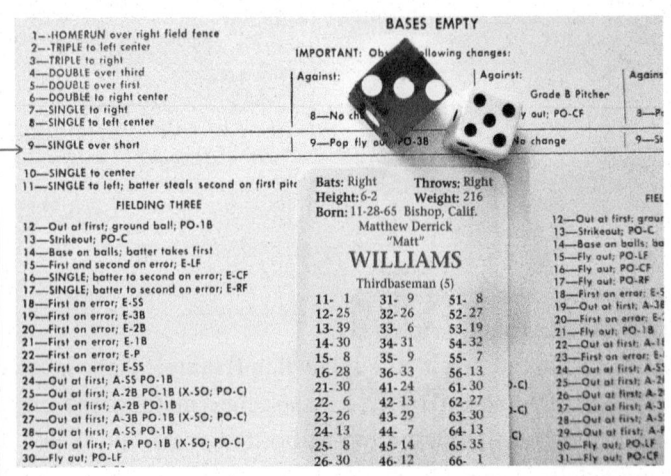

When their teams are at bat, APBA managers love to roll 66, because that number always leads to the best possible outcome for the batter. In Matt's case, a roll of 66 would lead to 1, which would be a home run over the right field fence.

The well-established formula devised years ago by the company for constructing batters' cards ensures that in the long run, their results in the game will accurately reflect their actual performances.

10.1 EXERCISES

CONCEPT CHECK *Answer each question.*

1. If there are 6 ways to perform task *A* and 7 ways to perform task *B*, then how many ways are there to perform task *A* and then task *B*?

2. If there are 9 ways to perform task *A* and 54 ways to perform task *A* and then task *B*, how many ways are there to perform task *B* alone?

3. How many squares (of any size) are there in the figure shown here?

4. How many triangles (of any size) are there in the figure shown here?

Electing Officers of a Club *Refer to* ***Examples 1 and 4,*** *involving the club*

$$N = \{ \text{Alan, Bill, Cathy, David, Evelyn} \}.$$

Assuming that all members are eligible, but no one can hold more than one office, list and count the different ways the club could elect each group of officers. (Cathy and Evelyn are women, and the others are men.)

5. a president and a treasurer

6. a president and a treasurer if the president must be a woman

7. a president and a treasurer if the two officers must be the same sex

8. a president, a secretary, and a treasurer, if the president and treasurer must be women

9. a president, a secretary, and a treasurer, if the president must be a man and the other two must be women

10. a president, a secretary, and a treasurer, if all three officers must be men

Appointing Committees *List and count the ways club N from above could appoint a committee of three members under each condition.*

11. There are no restrictions.

12. The committee must include more men than women.

13. The committee must include more women than men.

14. The committee must include no men.

Refer to **Table 2** *(the product table for rolling two dice). Of the 36 possible outcomes, determine the number for which the sum (for both dice) is the following.*

15. 2   **16.** 3   **17.** 4

18. 5   **19.** 6   **20.** 7

21. 8   **22.** 9   **23.** 10

24. 11   **25.** 12   **26.** 13

27. odd   **28.** even   **29.** prime

30. composite   **31.** greater than 4   **32.** less than 5

33. from 5 through 10 inclusive

34. from 6 through 8 inclusive

35. between 6 and 10

36. between 7 and 11

Solve each problem.

37. Construct a product table showing all possible two-digit numbers using digits from the set

$$\{2, 4, 6\}.$$

38. Construct a product table showing all possible two-digit numbers using digits from the set

$$\{2, 3, 5, 7\}.$$

Of the sixteen numbers in the product table for ***Exercise 38,*** *list the ones that belong to each category.*

39. numbers with repeating digits

40. even numbers

41. prime numbers

42. multiples of 3

43. Construct a tree diagram showing all possible results when three fair coins are tossed. Then list the ways of getting each result.

(a) at least two heads

(b) more than two heads

(c) no more than two heads

(d) fewer than two heads

44. Extend the tree diagram of **Exercise 43** for four fair coins. Then list the ways of getting each result.

(a) more than three tails

(b) fewer than three tails

(c) at least three tails

(d) no more than three tails

Counting Triangles *Determine the number of triangles (of any size) in each figure.*

45.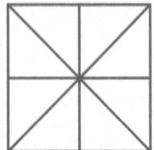

46.

47.

48.

Counting Squares *Determine the number of squares (of any size) in each figure.*

49.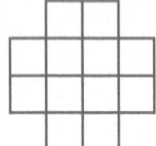

50.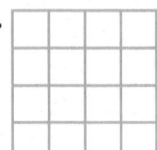

51.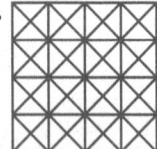

52.

Counting Cubes *Consider only the smallest individual cubes and assume solid stacks (no gaps). Determine the number of cubes in each stack that are not visible from the perspective shown.*

53.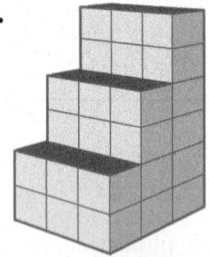

54.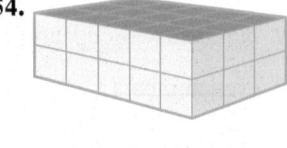

55.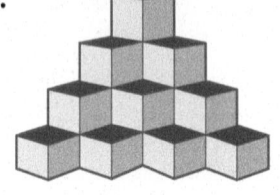

56.

Solve each problem.

57. *Pathfinding* Find the number of paths from A to B in the figure illustrated here if the directions on various segments are restricted as shown.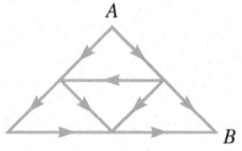

58. *Pathfinding* In the plane figure illustrated here, only movement that tends downward is allowed. Find the total number of paths from A to B.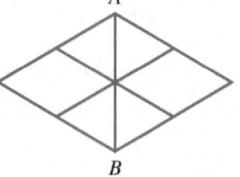

59. *Number of Triangles* How many triangles are contained in the figure? (*MT* calendar problem)

60. *Sums of Integers* How many ways are there to write 18 as the sum of three distinct positive integers? (Assume, for example, that $4 + 6 + 8$ is the same as $6 + 8 + 4$.) (*MT* calendar problem)

61. *Rolling Unusual Dice* An unusual die has the numbers

$$2, \quad 2, \quad 3, \quad 3, \quad 5, \quad \text{and} \quad 8$$

on its six faces. Two of these dice are rolled, and the two numbers on the top faces are added. How many different sums are possible? (*MT* calendar problem)

62. *Filling an Order* A customer ordered fifteen Zingers. Zingers are placed in packages of four, three, or one. In how many different ways can this order be filled? (*MT* calendar problem)

63. *Counting Three-digit Numbers* If only the digits 0, 1, 2, 3, and 4 may be used, find the number of possible three-digit numbers that can be formed.

64. *Shaking Hands in a Group* A group of six strangers sat in a circle, and each one got acquainted only with the person to the left and the person to the right. Then all six people stood up and each one shook hands (once) with each of the others who was still a stranger. How many handshakes occurred?

65. *Number of Games in a Chess Tournament* Fifty people enter a single-elimination chess tournament. If you lose one game, you're out. Assuming no ties occur, what is the number of games required to determine the tournament champion?

66. *Sums of Digits* How many positive integers less than 100 have the sum of their digits equal to a perfect square?

67. *Sums of Digits* How many three-digit numbers have the sum of their digits equal to 22?

68. *Integers Containing the Digit 2* How many integers between 100 and 400 contain the digit 2?

69. *Counting Radio Call Letters* How many 4-letter radio station call letters can be made if the first letter must be W or K and no letter may be repeated?

70. *Selecting Dinner Items* Michael and several friends are dining this evening at the Clam Shell Restaurant, where a complete dinner consists of three items.

    Choice 1 soup (clam chowder or minestrone) or salad (fresh spinach or shrimp),

    Choice 2 sourdough rolls or bran muffin, and

    Choice 3 entree (lasagna, lobster, or roast turkey).

Michael selects his meal subject to the following restrictions. He cannot stomach more than one kind of seafood at a sitting. Also, whenever he tastes minestrone, he cannot resist having lasagna as well. Use a tree diagram to determine the number of different choices Michael has.

Setting Options on a Computer Printer *For Exercises 71–74, refer to* **Example 7.** *How many different settings could Pamela choose in each case?*

71. No restrictions apply to adjacent switches.

72. No two adjacent switches can both be off *and* no two adjacent switches can both be on.

73. There are five switches rather than four, and no two adjacent switches can both be on.

74. There are six switches rather than four, and no two adjacent switches can both be off.

Solve each problem.

75. *Lattice Points on a Line Segment* A line segment joins the points

$$(8, 12) \quad \text{and} \quad (53, 234)$$

in the Cartesian plane. Including its endpoints, how many lattice points does this line segment contain? (A *lattice point* is a point with integer coordinates.)

76. *Building Numbers from Sets of Digits* Determine the number of odd, nonrepeating three-digit numbers that can be written using only the digits 0, 1, 2, and 3.

77. *Patterns in Floor Tiling* A square floor is to be tiled with square tiles as shown. There are blue tiles on the main diagonals and red tiles everywhere else. In all cases, both blue and red tiles must be used, and the two diagonals must have a common blue tile at the center of the floor.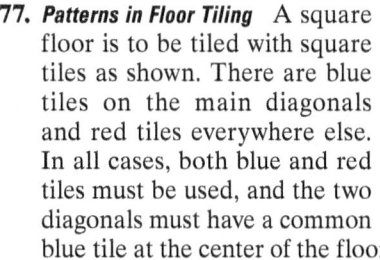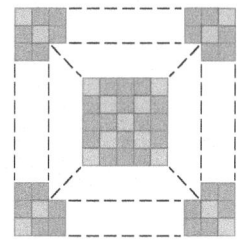

  (a) If 81 blue tiles will be used, how many red tiles will be needed?

  (b) For what numbers in place of 81 would this problem still be solvable?

  (c) Find an expression in k giving the number of red tiles required in general.

78. *Lengths of Segments Joining Lattice Points* In the pattern shown, dots are one unit apart horizontally and vertically. If a segment can join any two dots, how many segments can be drawn with each length?

  (a) 1   **(b)** 2   **(c)** 3   **(d)** 4   **(e)** 5

79. *Shaking Hands in a Group* Chris and his son were among four father-and-son pairs who gathered to trade baseball cards. As each person arrived, he shook hands with anyone he had not known previously. Each person ended up making a different number of new acquaintances (0–6), except Chris and his son, who each met the same number of people. How many hands did Chris shake?

80. *Counting Matchsticks in a Grid* Uniform-length matchsticks are used to build a rectangular grid as shown here. If the grid is 12 matchsticks high and 25 matchsticks wide, how many matchsticks are used?

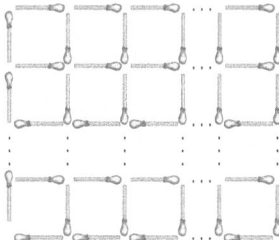

81. *Sum of Primes* Determine the number of different ways the given composite number can be written as the sum of two prime numbers. (*Hint:* Consider a sum such as $2 + 3$ the same as $3 + 2$.)

  (a) 30   **(b)** 40   **(c)** 95

82. *Sum of Composites* Determine the different number of ways the given prime number can be written as the sum of two composite numbers. (*Hint:* Consider a sum such as $6 + 4$ the same as $4 + 6$. Remember that 1 is neither prime nor composite.)

  (a) 23   **(b)** 29   **(c)** 37

10.2 USING THE FUNDAMENTAL COUNTING PRINCIPLE

Fundamental Counting Principle: Tree Diagrams

Uniformity and the Fundamental Counting Principle

In the previous section we obtained *complete lists* of all possible results for various tasks. However, if the *total number* of possibilities is all we need to know, then an actual listing usually is unnecessary and often is difficult or tedious to obtain, especially when the list is long.

Figure 6 shows all possible nonrepeating three-digit numbers using only the digits 1, 2, and 3.

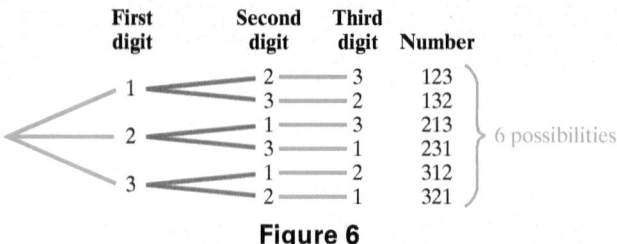

Figure 6

The tree diagram in **Figure 6** is "uniform" in the sense that a given part of the task can be done in the same number of ways, no matter which choices were selected for previous parts. For example, there are always two choices for the second digit. (If the first digit is 1, the second can be 2 or 3. If the first is 2, the second can be 1 or 3. If the first is 3, the second can be 1 or 2.)

Example 6(a) of the previous section addressed the same basic situation:

> *Find the number of three-digit numbers that can be written using the digits 1, 2, and 3.*

In that case repetitions were allowed. With repetitions allowed, there were many more possibilities (27 rather than 6—see **Figure 1**). But the uniformity criterion mentioned above still applied. No matter what the first digit is, there are three choices for the second (1, 2, 3). And no matter what the first and second digits are, there are three choices for the third. This uniformity criterion can be stated in general as follows.

UNIFORMITY CRITERION FOR MULTIPLE-PART TASKS

A multiple-part task is said to satisfy the **uniformity criterion** if the number of choices for any particular part is the same *no matter which choices were selected for previous parts.*

The uniformity criterion is not always satisfied. Refer to **Example 7** and **Figure 3** of the previous section. After the first switch (two possibilities), other switches had either one or two possible settings, depending on how previous switches were set. This "nonuniformity" arose, in that case, from the requirement that no two adjacent switches could both be off.

In the many cases where uniformity does hold, we can avoid having to construct a tree diagram by using the **fundamental counting principle,** stated as follows.

Richard Dedekind (1831–1916) studied at the University of Göttingen, where he was Gauss's last student. His work was not recognized during his lifetime, but his treatment of the infinite and of what constitutes a real number are influential even today.

While on vacation in Switzerland, Dedekind met Georg Cantor. Dedekind was interested in Cantor's work on infinite sets. Perhaps because both were working in new and unusual fields of mathematics, such as number theory, and because neither received the professional attention he deserved during his lifetime, the two struck up a lasting friendship.

FUNDAMENTAL COUNTING PRINCIPLE

When a task consists of k separate parts and satisfies the uniformity criterion, if the first part can be done in n_1 ways, the second part can then be done in n_2 ways, and so on through the kth part, which can be done in n_k ways, then the total number of ways to complete the task is given by the following product.

$$n_1 \cdot n_2 \cdot n_3 \cdot \cdots \cdot n_k$$

Problem-Solving Strategy

A problem-solving strategy suggested earlier in the text was "*If a formula applies, use it.*" The fundamental counting principle provides a formula that applies to a variety of problems.

A helpful technique when applying the fundamental counting principle is to write out all the separate parts of the task, with a blank for each one. Reason out how many ways each part can be done, and enter these numbers in the blanks. Finally, multiply these numbers together.

EXAMPLE 1   Counting the Two-Digit Numbers

How many two-digit numbers are there in our (base-ten) system of counting numbers? (*Hint:* 40 is a two-digit number, but 04 is not.)

Solution

Our "task" here is to select, or construct, a two-digit number. Work as follows.

Part of task	Select first digit	Select second digit
Number of ways	_____	_____

There are nine choices for the first digit (1 through 9). Since there were no stated or implied restrictions, we assume that repetition of digits is allowed. Therefore, no matter which nonzero digit is used as the first digit, all nine choices are available for the second digit. Also, unlike the first digit, the second digit may be zero, so we have ten choices for the second digit. We can now fill in the blanks and multiply.

Part of task	Select first digit	Select second digit	
Number of ways	9	10	= 90

There are 90 two-digit numbers. (As a check, notice that they are the numbers from 10 through 99, a total of $99 - 10 + 1 = 90$.)

Fundamental Counting Principle: Tree Diagrams

EXAMPLE 2   Building Two-Digit Numbers with Restrictions

Find the number of two-digit numbers that do not contain repeated digits.

Solution

The basic task is again to select a two-digit number, and there are two parts.

Part 1 Select the first digit.   ***Part 2*** Select the second digit.

But a new restriction applies—no repetition of digits. There are nine choices for the first digit (1 through 9). Then nine choices remain for the second digit, since one nonzero digit has been used and cannot be repeated, but zero is now available. The total number is $9 \cdot 9 = 81$.

EXAMPLE 3   **Electing Club Officers with Restrictions**

In how many ways can Club *N*, of **Examples 1** and **4** in the previous section, elect a president and a secretary if no one may hold more than one office and the secretary must be a man?

Solution

Recall that $N = \{A, B, C, D, E\} = \{$ Alan, Bill, Cathy, David, Evelyn $\}$. Considering president first, there are five choices (no restrictions). But now we have a problem with finding the number of choices for secretary. If a woman was selected president (*C* or *E*), there are three choices for secretary (*A*, *B*, and *D*). If a man was selected president, only two choices (the other two men) remain for secretary. *In other words, the uniformity criterion is not met, and our attempt to apply the fundamental counting principle has failed.*

    All is not lost, however. To find the total number of ways, we can consider secretary first. There are three choices: *A*, *B*, and *D*. Now, no matter which man was chosen secretary, both of the other men and both women are available for president, with four choices in every case. In this order, we satisfy the uniformity criterion and can use the fundamental counting principle. The total number of ways to elect a president and a secretary is $3 \cdot 4 = 12$.

Problem-Solving Strategy

Example 3 suggests a useful problem-solving strategy: Whenever one or more parts of a task have special restrictions, try considering that part (or those parts) before other parts.

EXAMPLE 4   **Counting Three-Digit Numbers with Restrictions**

How many nonrepeating odd three-digit counting numbers are there?

Solution

The most restricted digit is the third, since it must be odd. There are five choices (1, 3, 5, 7, and 9). Next, consider the first digit. It can be any nonzero digit except the one already chosen as the third digit. There are eight choices. Finally, the second digit can be any digit (including 0) except for the two nonzero digits already used. There are eight choices.

Part of task	Select third digit		Select first digit		Select second digit	
Number of ways	5	$\cdot$	8	$\cdot$	8	$= 320$

There are 320 nonrepeating odd three-digit counting numbers.

EXAMPLE 5   **Counting License Plates**

In some states, auto license plates have contained three letters of the English alphabet followed by three digits. How many such licenses are possible?

Solution

The basic task is to design a license plate with three letters followed by three digits. There are six component parts to this task. There are no restrictions on letters or digits, so the fundamental counting principle gives

$$26 \cdot 26 \cdot 26 \cdot 10 \cdot 10 \cdot 10 = 26^3 \cdot 10^3$$

$$= 17,576,000 \text{ possible licenses.}$$

EXAMPLE 6 **Building Numbers with Specified Digits**

A four-digit number is to be constructed using only the digits 1, 2, and 3.

(a) How many such numbers are possible?

(b) How many of these numbers are odd and less than 2000?

Solution

(a) To construct such a number, we must select four digits, in succession, from the given set of three digits, where the selection is done with replacement (since repetition of digits is apparently allowed). The number of possibilities is

$$3 \cdot 3 \cdot 3 \cdot 3 = 3^4 = 81. \quad \text{\small Fundamental counting principle}$$

(b) The number is less than 2000 only if the first digit is 1 (just one choice) and is odd only if the fourth digit is 1 or 3 (two choices). The second and third digits are unrestricted (three choices for each). The answer is

$$1 \cdot 3 \cdot 3 \cdot 2 = 18.$$

As a check, you may want to list the eighteen possibilities.

Problem-Solving Strategy

Two of the problem-solving strategies given earlier were "*First solve a similar simpler problem*" and "*Look for a pattern.*" In fact, a counting problem may sometimes prove to be essentially the same, or at least to fit the same pattern, as another problem already solved.

EXAMPLE 7 **Distributing Golf Clubs**

Vern has four antique golf clubs that he wants to give to his three sons, Mark, Chris, and Scott.

(a) How many ways can the clubs be distributed?

(b) How many choices are there if the power driver must go to Mark and the number 3 wood must go to either Chris or Scott?

Solution

(a) The task is to distribute four clubs among three sons. Consider the clubs in succession, and, for each one, ask how many sons could receive it. In effect, we must select four sons, in succession, from the list Mark, Chris, Scott, selecting with replacement. Compare this with **Example 6(a),** in which we selected four digits, in succession, from the digits 1, 2, and 3, selecting with replacement. In this case, we are selecting sons rather than digits, but the pattern is the same and the numbers are the same. Again our answer is

$$3^4 = 81 \text{ ways.}$$

(b) Just as in **Example 6(b),** one part of the task is now restricted to a single choice, and another part is restricted to two choices. As in that example, the number of possibilities is

$$1 \cdot 3 \cdot 3 \cdot 2 = 18.$$

EXAMPLE 8 Seating Attendees at a Concert

This is a repeat of **Example 8** in the previous section.

Arne, Bobbette, Chuck, and Deirdre have tickets for four reserved seats in a row at a concert. In how many different ways can they seat themselves so that Arne and Bobbette will sit next to each other?

Solution

Arne, Bobbette, Chuck, and Deirdre (A, B, C, and D) are to seat themselves in four adjacent seats (say 1, 2, 3, and 4) so that A and B are side-by-side. One approach to accomplish this task is to make three successive decisions as follows.

1	2	3	4
X	X	—	—
—	X	X	—
—	—	X	X

Seats available to *A* and *B*

1. *Which pair of seats should A and B occupy?* There are *three* choices: 1 and 2, 2 and 3, 3 and 4, as illustrated in the margin.

2. *Which order should A and B take?* There are *two* choices: A left of B, or B left of A.

3. *Which order should C and D take?* There are *two* choices: C left of D, or D left of C, not necessarily right next to each other.

The fundamental counting principle now gives the total number of choices.

$$3 \cdot 2 \cdot 2 = 12 \quad \text{Same result as in the previous section}$$

Factorials

This section began with a discussion of nonrepeating three-digit numbers using digits 1, 2, and 3. The number of possibilities was

$$3 \cdot 2 \cdot 1 = 6. \quad \text{Fundamental counting principle}$$

That product can also be thought of as the total number of distinct *arrangements* of the three digits 1, 2, and 3. Similarly, the number of distinct arrangements of four objects (say A, B, C, and D) is

$$4 \cdot 3 \cdot 2 \cdot 1 = 24. \quad \text{Fundamental counting principle}$$

Because this type of product occurs so commonly in applications, we give it a special name and symbol. For any counting number n, the product of *all* counting numbers from n down through 1 is called *n* **factorial,** and is denoted *n*!.

n!

The exclamation point symbol for **factorial notation** sometimes seems strange to those who are not aware of its meaning. There is a "nerdy" statement that sometimes appears on bumper stickers, Facebook posts, and the like:

Only mathematicians know that 2 + 2 is not equal to 4!

An 1891 algebra text by G. A. Wentworth indicates that a different symbol was used at that time for the factorial, as illustrated by this statement, which refers to the number of permutations of *n* different things:

". . . . the whole number arrangements is the continued product of all these numbers,

$$n(n-1)(n-2)(n-3)\cdots 3 \times 2 \times 1.$$

For the sake of brevity, this product is written ⌊n and is read **factorial n.**"

FACTORIAL FORMULA

For any counting number *n*, the quantity *n* **factorial** is defined as follows.

$$n! = n \cdot (n - 1) \cdot (n - 2) \cdot \cdots \cdot 2 \cdot 1$$

So that factorials will be defined for all whole numbers, including zero, we define 0! as follows.

DEFINITION OF ZERO FACTORIAL

$$0! = 1$$

(We will see later that this special definition makes other results easier to state.)

The first few factorial values are easily found by simple multiplication, but they rapidly become very large. The use of a calculator is advised in most cases.

Short Table of Factorials Factorial values increase rapidly. The value of 100! is a number with 158 digits.

$$0! = 1$$
$$1! = 1$$
$$2! = 2$$
$$3! = 6$$
$$4! = 24$$
$$5! = 120$$
$$6! = 720$$
$$7! = 5040$$
$$8! = 40,320$$
$$9! = 362,880$$
$$10! = 3,628,800$$

Problem-Solving Strategy

Sometimes expressions involving factorials can be evaluated easily by observing that, in general, $n! = n \cdot (n-1)!$, $n! = n \cdot (n-1) \cdot (n-2)!$, and so on. For example,

$$8! = 8 \cdot 7!, \quad 12! = 12 \cdot 11 \cdot 10 \cdot 9!, \quad \text{and so on.}$$

This pattern is especially helpful in evaluating quotients of factorials.

$$\frac{10!}{8!} = \frac{10 \cdot 9 \cdot 8!}{8!} = 10 \cdot 9 = 90$$

EXAMPLE 9 **Evaluating Expressions Containing Factorials**

Evaluate each expression.

(a) $3!$ (b) $6!$ (c) $(6-3)!$ (d) $6! - 3!$ (e) $\dfrac{6!}{3!}$

(f) $\left(\dfrac{6}{3}\right)!$ (g) $(4-4)!$ (h) $15!$ (i) $100!$

Solution

(a) $3! = 3 \cdot 2 \cdot 1 = 6$ (b) $6! = 6 \cdot 5 \cdot 4 \cdot 3 \cdot 2 \cdot 1 = 720$

(c) $(6-3)! = 3! = 6$ (d) $6! - 3! = 720 - 6 = 714$

(e) $\dfrac{6!}{3!} = \dfrac{6 \cdot 5 \cdot 4 \cdot 3!}{3!} = 6 \cdot 5 \cdot 4 = 120$ Note application of the Problem-Solving Strategy.

(f) $\left(\dfrac{6}{3}\right)! = 2! = 2 \cdot 1 = 2$ (g) $(4-4)! = 0! = 1$

(h) $15! = 1.307674368000 \times 10^{12}$ ← Done on a calculator

(i) $100! = 9.332621544 \times 10^{157}$ ← Too large for most calculators

NORMAL FLOAT AUTO REAL RADIAN MP
```
3!
                          6.
6!
                        720.
(4-4)!
                          1.
15!
             1.307674368ᴇ12
```

The results of **Example 9(a), (b), (g), and (h)** are illustrated in this calculator screen.

Notice the distinction between parts (c) and (d) and between parts (e) and (f).

Arrangements of Objects

When finding the total number of ways to *arrange* a given number of distinct objects, we can use a factorial. The fundamental counting principle would do, but factorials provide a shortcut.

ARRANGEMENTS OF *n* DISTINCT OBJECTS

The total number of different ways to arrange *n* distinct objects is

$$n!.$$

EXAMPLE 10 **Arranging Essays**

Michelle has seven essays to include in her English 1A folder. In how many different orders can she arrange them?

Solution

Seven distinct objects can be arranged in

$$7! = 5040 \text{ different ways.}$$

NORMAL FLOAT AUTO REAL RADIAN MP

7!
                              5040
13!
                        6227020800

The answers in **Examples 10 and 11** are verified here.

D₁AD₂

D₂AD₁

D₁D₂A

D₂D₁A

AD₁D₂

AD₂D₁

EXAMPLE 11 **Arranging Preschoolers**

Tricia is taking thirteen preschoolers to the park. How many ways can the children line up, in single file, to board the van?

Solution

Thirteen children can be arranged in

$$13! = 6{,}227{,}020{,}800 \text{ different ways.}$$

Distinguishable Arrangements

In counting arrangements of objects that contain look-alikes, the normal factorial formula must be modified to find the number of truly different arrangements. For example, the number of distinguishable arrangements of the letters of the word DAD is not $3! = 6$ but rather $\frac{3!}{2!} = 3$. The listing in the margin shows how the six total arrangements consist of just three groups of two, where the two in a given group look alike.

ARRANGEMENTS OF *n* OBJECTS CONTAINING LOOK-ALIKES

The number of **distinguishable arrangements** of n objects, where one or more subsets consist of look-alikes (say n_1 are of one kind, n_2 are of another kind, ..., and n_k are of yet another kind), is given by

$$\frac{n!}{n_1! \cdot n_2! \cdot \cdots \cdot n_k!}.$$

EXAMPLE 12 **Counting Distinguishable Arrangements**

Determine the number of distinguishable arrangements of the letters in each word.

(a) ATTRACT   **(b)** NIGGLING

Solution

(a) For the letters of ATTRACT, the number of distinguishable arrangements is

$$\begin{array}{l} \text{7 letters total} \longrightarrow \\ \text{3 } T\text{'s, 2 } A\text{'s} \longrightarrow \end{array} \frac{7!}{3! \cdot 2!} = 420.$$

(b) For the letters of NIGGLING, the number of distinguishable arrangements is

$$\begin{array}{l} \text{8 letters total} \longrightarrow \\ \text{2 } N\text{'s, 2 } I\text{'s, 3 } G\text{'s} \longrightarrow \end{array} \frac{8!}{2! \cdot 2! \cdot 3!} = 1680.$$

10.2 **EXERCISES**

CONCEPT CHECK *Complete each statement with the correct numerical response.*

1. The value of 6! is _____.

2. The value of 0! is _____.

3. There are _____ ways to arrange the letters A, B, C, D.

4. There are _____ ways to arrange the three people Anna, Betty, and Charmaine in a line.

5. There are _____ ways to arrange the three people Anna, Betty, and Charmaine in a line, given that Charmaine must be first in line.

6. There are _____ two-digit numerals in our numeration system.

Evaluate each expression without using a calculator.

7. $4!$ **8.** $7!$ **9.** $\dfrac{9!}{7!}$

10. $\dfrac{16!}{14!}$ **11.** $\dfrac{5!}{(5-2)!}$ **12.** $\dfrac{6!}{(6-3)!}$

13. $\dfrac{50!}{48! \cdot 2!}$ **14.** $\dfrac{37!}{35! \cdot 2!}$ **15.** $\dfrac{8!}{6!(8-6)!}$

16. $\dfrac{9!}{7!(9-7)!}$ **17.** $\dfrac{10!}{9!(10-9)!}$ **18.** $\dfrac{100!}{99!(100-99)!}$

For the given values of n and r, evaluate each expression.

(a) $\dfrac{n!}{(n-r)!}$ **(b)** $\dfrac{n!}{r!(n-r)!}$

19. $n = 10, r = 3$ **20.** $n = 11, r = 2$

21. $n = 18, r = 2$ **22.** $n = 20, r = 3$

Use a calculator to evaluate each expression.

23. $10!$ **24.** $14!$ **25.** $\dfrac{12!}{5!}$

26. $\dfrac{14!}{8!}$ **27.** $\dfrac{16!}{9! \cdot 7!}$ **28.** $\dfrac{13!}{6! \cdot 7!}$

29. $\dfrac{n!}{(n-r)!}$, where $n = 17$ and $r = 8$

30. $\dfrac{n!}{(n-r)!}$, where $n = 20$ and $r = 4$

31. $\dfrac{n!}{r!(n-r)!}$, where $n = 24$ and $r = 18$

32. $\dfrac{n!}{r!(n-r)!}$, where $n = 26$ and $r = 20$

Arranging Letters *Find the number of distinguishable arrangements of the letters of each word or phrase.*

33. GOOGOL **34.** BANANA

35. HEEBIE-JEEBIES **36.** VICE VERSA

Settings on a Switch Panel *A panel containing three on-off switches in a row is to be set.*

37. Assuming no restrictions on individual switches, use the fundamental counting principle to find the total number of possible panel settings.

38. Assuming no restrictions, construct a tree diagram to list all the possible panel settings of **Exercise 37.**

39. Now assume that no two adjacent switches can both be off. Explain why the fundamental counting principle does not apply.

40. Construct a tree diagram to list all possible panel settings under the restriction of **Exercise 39.**

Matching Club Members with Tasks *Recall the club*

$$N = \{\text{Alan, Bill, Cathy, David, Evelyn}\}.$$

In how many ways could they do each of the following?

41. line up all five members for a photograph

42. schedule one member to work in the office on each of five different days, assuming members may work more than one day

43. select a man and a woman to decorate for a party

44. select two members, one to open their next meeting and one to close it, given that Bill will not be present

Building Numbers with Specified Digits *In Exercises 45–48, counting numbers are to be formed using only the digits 3, 4, and 5. Determine the number of different possibilities for each type of number described.*

45. two-digit numbers

46. odd three-digit numbers

47. four-digit numbers with one pair of adjacent 4s and no other repeated digits

   (*Hint:* You may want to split the task of designing such a number into three parts, such as ***Part 1*** position the pair of 4s, ***Part 2*** position the 3, and ***Part 3*** position the 5.)

48. five-digit numbers beginning and ending with 3 and with unlimited repetitions allowed

Selecting Dinner Items *The Gourmet de Coeur Restaurant offers*

   five choices in the soup and salad category (two soups and three salads),

   two choices in the bread category,

and   *four choices in the entrée category.*

Find the number of dinners available in each case.

49. One item should be included from each of the three categories.

50. Only salad and entrée are to be included.

51. One soup, one salad, and one entrée are to be included.

52. One soup, one salad, and one bread are to be included.

Selecting Answers on a Test *Determine the number of possible ways to mark your answer sheet (with an answer for each question) for each test.*

53. a six-question true-or-false test

54. a fifteen-question true-or-false test

55. a ten-question multiple-choice test with five answer choices for each question

56. an eight-question multiple-choice test with four answer choices for each question

Selecting a College Class Schedule *Jessica's class schedule for next semester must consist of exactly one class from each of the four categories shown in the table.*

For each situation in Exercises 57–62, use the table to determine the number of different sets of classes Jessica can take.

Category	Choices	Number of Choices
Economics	Free Markets	2
	Controlled Markets	
Mathematics	History of Mathematics	3
	College Algebra	
	Finite Mathematics	
Education	Classroom Technology	4
	Group Dynamics	
	Language Supervision	
	Parent/Teacher Relations	
Sociology	Social Problems	5
	Sociology of the Middle East	
	Aging in America	
	Minorities in America	
	Women in American Culture	

57. All classes shown are available.

58. She is not eligible for Free Markets or for Group Dynamics.

59. All sections of Minorities in America and Women in American Culture already are filled.

60. She does not have the prerequisites for Controlled Markets, College Algebra, or Language Supervision.

61. Funding has been withdrawn for three of the Education courses and for two of the Sociology courses.

62. She must complete Finite Mathematics and Social Problems to fulfill her degree requirements.

Solve each problem.

63. ***Rolling Dice*** **Table 2** in the previous section shows that there are 36 possible outcomes when two fair dice are rolled. How many outcomes would there be if three fair dice were rolled?

64. ***Bowling*** After the rolling of the first ball of a frame in a game of 10-pin bowling, how many different pin configurations can remain (assuming all configurations are physically possible)? (*MT* calendar problem)

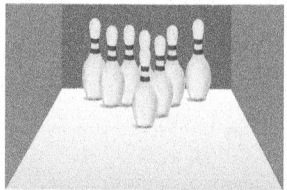

65. ***Counting Five-Digit Numbers*** How many five-digit numbers are there in our system of counting numbers?

66. ***Counting Three-Digit Numbers*** If only the digits 0, 1, 2, 3, 4, 5, and 6 may be used, find the number of even three-digit numbers.

67. ***Selecting Clothing*** Don has two pairs of shoes, four pairs of pants, and six shirts. If all items are compatible, how many different outfits can he wear?

68. ***Selecting Music Equipment*** A music equipment outlet stocks ten different guitars, three guitar cases, six amplifiers, and five special effects processors, with all items mutually compatible and all suitable for beginners. How many different complete setups could Lionel choose to start his musical career?

69. ***Counting ZIP Codes*** Tonya's ZIP code is 85726. How many ZIP codes altogether could be formed, each one using those same five digits?

70. *Listing Phone Numbers* Raj keeps the phone numbers for his seven closest friends (three men and four women) in his digital phone memory. (Refer to **Example 8.**) How many ways can he list them for the following conditions?

(a) men are listed before women

(b) men are all listed together

(c) no two men are listed next to each other?

71. *Counting Telephone Area Codes* Until 1995, the rules for three-digit area codes in the United States were as follows:

- The first digit could not be 0 or 1.
- The second digit had to be 0 or 1.
- The third digit had no such restrictions.

In 1995, the restriction on the second digit of area codes was removed. How many area codes are possible? (*MT* calendar problem)

72. *Repeated Digits* Repeat **Example 4,** but this time allow repeated digits. Does the order in which digits are considered matter in this case?

*Seating Arrangements at a Theater   In Exercises 73–76, Arne, Bobbette, Chuck, Deirdre, Ed, and Fran have reserved six seats in a row at the theater, starting at an aisle seat. (Refer to **Example 8**.)*

73. In how many ways can they arrange themselves? (*Hint:* Divide the task into the series of six parts shown below, performed in order.)

(a) If *A* is seated first, how many seats are available for him?

(b) Now, how many are available for *B*?

(c) Now, how many for *C*?

(d) Now, how many for *D*?

(e) Now, how many for *E*?

(f) Now, how many for *F*?

Now multiply your six answers above.

74. In how many ways can they arrange themselves so that Arne and Bobbette will be next to each other?

Seats available to *A* and *B*

(*Hint:* First answer the following six questions in order.)

(a) How many pairs of adjacent seats can *A* and *B* occupy?

(b) Now, given the two seats for *A* and *B*, in how many orders can they be seated?

(c) Now, how many seats are available for *C*?

(d) Now, how many for *D*?

(e) Now, how many for *E*?

(f) Now, how many for *F*?

Now multiply your six answers above.

75. In how many ways can they arrange themselves if the men and women are to alternate seats and a man must sit on the aisle? Arne, Chuck, and Ed are men, and the others are women. (*Hint:* First answer the following six questions in order.)

(a) How many choices are there for the person to occupy the first seat, next to the aisle? (It must be a man.)

(b) How many choices of people may occupy the second seat from the aisle? (It must be a woman.)

(c) Now, how many for the third seat? (one of the remaining men)

(d) Now, how many for the fourth seat? (a woman)

(e) Now, how many for the fifth seat? (a man)

(f) Now, how many for the sixth seat? (a woman)

Now multiply your six answers above.

76. In how many ways can they arrange themselves if the men and women are to alternate with either a man or a woman on the aisle? Arne, Chuck, and Ed are men, and the others are women. (*Hint:* First answer the following six questions in order.)

(a) How many choices of people are there for the aisle seat?

(b) Now, how many are there for the second seat? (This person must not be of the same gender as the person on the aisle.)

(c) Now, how many choices are there for the third seat?

(d) Now, how many for the fourth seat?

(e) Now, how many for the fifth seat?

(f) Now, how many for the sixth seat?

Now multiply your six answers above.

Solve each problem.

77. *Six-Digit Numbers* If all the six-digit numbers formed by using the digits

1, 2, 3, 4, 5, and 6,

without repetition, are listed from least to greatest, which number will be 500th in the list? (*MT* calendar problem)

78. *Arranging a Melody* The tune "Twinkle, Twinkle Little Star" has seven notes in it first line: C, C, G, G, A, A, G.

Assume that all seven notes are held for the same length of time. If the notes are rearranged at random, how many different melodies could be composed? (*MT* calendar problem)

79. *Palindromes* RACECAR is a palindromic word. (A palindrome reads the same forward and backward.) If you add the total number of distinguishable possible arrangements of these 7 letters to the number of resulting palindromic arrangements (including RACECAR), the result is a palindrome. Find that number. (*MT* calendar problem)

80. *Palindromes* How many of the anagrams (arrangements of the letters) of

INDIANA

are palindromes—that is, arrangements that read the same both forward and backward? *Hint:* One such palindrome is

INADANI.

(*MT* calendar problem)

81. *Divisibility* The number $2^7 \cdot 3^4 \cdot 5 \cdot 7^2 \cdot 11^3$ is divisible by many perfect squares. How many? (*MT* calendar problem)

82. *Distinguishability* How many distinguishable rearrangements of the letters in the word

CONTEST

start with the two vowels? (*MT* calendar problem)

FOR FURTHER THOUGHT

Stirling's Approximation for $n!$

Although all factorial values are counting numbers, they can be approximated using **Stirling's formula,**

$$n! \approx \sqrt{2\pi n} \cdot n^n \cdot e^{-n},$$

which involves two important irrational numbers, π and e. For example, while the exact value of 5! is $5 \cdot 4 \cdot 3 \cdot 2 \cdot 1 = 120$, the corresponding approximation, using Stirling's formula, is

$$5! \approx \sqrt{2\pi \cdot 5} \cdot 5^5 \cdot e^{-5} \approx 118.019168,$$

which is off by less than 2, an error of only 1.65%.

For Group or Individual Investigation

1. Use a calculator to fill in the table on the right. The column values are defined as follows.

$C = n!$ (exact value, by calculator)

$S \approx n!$ (Stirling's approximation, by calculator)

$D =$ Difference $(C - S)$

$P =$ Percentage difference $\left(\dfrac{D}{C} \cdot 100\%\right)$

n	C	S	D	P
10				
15				
20				
25				
30				

Try to obtain percentage differences accurate to two decimal places.

2. In general, is Stirling's approximation too low or too high?

3. Observe the values in the table as n grows larger.

   (a) Do the differences (D) get larger or smaller?

   (b) Do the percentage differences (P) get larger or smaller?

   (c) Does Stirling's formula become more accurate or less accurate as n increases?

4. An even better approximation is given by

$$n! \approx \sqrt{\left(2n + \tfrac{1}{3}\right)\pi} \cdot n^n \cdot e^{-n}.$$

Show that this formula gives a closer approximation to 5! than Stirling's formula.

10.3   USING PERMUTATIONS AND COMBINATIONS

OBJECTIVES

1 Solve counting problems involving permutations and the fundamental counting principle.

2 Solve counting problems involving combinations and the fundamental counting principle.

3 Solve counting problems that require deciding whether to use permutations or combinations.

Combining Counting Methods: Forming a Governing Committee

Permutations

Again recall the club

$$N = \{\text{Alan, Bill, Cathy, David, Evelyn}\} = \{A, B, C, D, E\}$$

and consider two questions.

1. How many ways can all the club members arrange themselves in a row for a photograph?

2. How many ways can the club elect a president, a secretary, and a treasurer if no one can hold more than one office?

From the previous section, the answer to the first question above is

$$5! = 5 \cdot 4 \cdot 3 \cdot 2 \cdot 1 = 120 \text{ ways,}$$

the number of possible arrangements of 5 distinct objects. We answered questions like the second one using a tree diagram or the fundamental counting principle.

$$5 \cdot 4 \cdot 3 = 60 \text{ ways}$$

A good way to think of the second question is as follows.

How many arrangements are there of five things taken three at a time?

The factors begin with 5 and proceed downward, just as in a factorial product, but do not go all the way to 1. Here the product stops when there are three factors.

In the context of counting problems, arrangements are called **permutations.** The number of permutations of n distinct things taken r at a time is denoted $_nP_r$.* The number of objects being arranged cannot exceed the total number available, so we assume that $r \leq n$. Applying the fundamental counting principle gives

$$_nP_r = n(n-1)(n-2)\cdots[n-(r-1)].$$

The first factor is $n-0$, the second is $n-1$, the third is $n-2$, and so on. The rth factor, the last one in the product, will be the one with $r-1$ subtracted from n, as shown above. We can express permutations, in general, in terms of factorials, to obtain a formula as follows.

$$_nP_r = n(n-1)(n-2)\cdots[n-(r-1)]$$

$$= n(n-1)(n-2)\cdots(n-r+1) \qquad \text{Simplify the last factor.}$$

$$= \frac{n(n-1)(n-2)\cdots(n-r+1)(n-r)(n-r-1)\cdots2\cdot1}{(n-r)(n-r-1)\cdots2\cdot1} \qquad \text{Multiply and divide by } (n-r)(n-r-1)\cdots2\cdot1.$$

$$_nP_r = \frac{n!}{(n-r)!} \qquad \text{Definition of factorial}$$

---

*Alternative notations are $P(n, r)$ and P_r^n.

FACTORIAL FORMULA FOR PERMUTATIONS

The number of **permutations,** or *arrangements,* of n distinct things taken r at a time, where $r \leq n$, can be calculated as follows.

$$_nP_r = \frac{n!}{(n-r)!}$$

Although we sometimes refer to a symbol such as

$$_4P_2$$

as "a permutation," the symbol actually represents "the number of permutations of 4 distinct things taken 2 at a time."

EXAMPLE 1   Using the Factorial Formula for Permutations

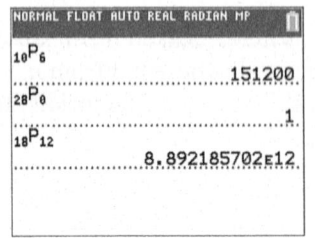

This screen uses factorials to support the results of **Example 1.**

Evaluate each permutation.

(a) $_4P_2$ **(b)** $_8P_5$ **(c)** $_5P_5$

Solution

(a) $_4P_2 = \dfrac{4!}{(4-2)!} = \dfrac{4!}{2!} = \dfrac{24}{2} = 12$

(b) $_8P_5 = \dfrac{8!}{(8-5)!} = \dfrac{8!}{3!} = \dfrac{40{,}320}{6} = 6720$

(c) $_5P_5 = \dfrac{5!}{(5-5)!} = \dfrac{5!}{0!} = \dfrac{120}{1} = 120$

Notice that $_5P_5$ is equal to 5!. The following is true for all whole numbers n.

$$_nP_n = n!$$

This is the number of arrangements of n distinct objects taken all n at a time.
   Many calculators allow direct calculation of permutations.

EXAMPLE 2   Calculating Permutations Directly

Evaluate each permutation.

(a) $_{10}P_6$ **(b)** $_{28}P_0$ **(c)** $_{18}P_{12}$

Solution

(a) $_{10}P_6 = 151{,}200$ **(b)** $_{28}P_0 = 1$ **(c)** $_{18}P_{12} = 8{,}892{,}185{,}702{,}400$

This screen uses the permutations feature to support the results of **Example 2.**

Concerning part (c), many calculators will not display this many digits, so you may obtain an answer such as 8.8921857×10^{12}.

Problem-Solving Strategy

Permutations can be used any time we need to know the number of arrangements of r objects that can be selected from a collection of n objects. *The word "arrangement" implies an ordering, so we use permutations only in cases that satisfy these conditions:*

   1. Repetitions are not allowed.   **2. Order is important.**

Change ringing, the English way of ringing church bells, combines mathematics and music. Bells are rung first in sequence, 1, 2, 3, ... Then the sequence is permuted ("changed"). On six bells, 720 different "changes" (different permutations of tone) can be rung: $_6P_6 = 6!$.

The church bells are swung by means of ropes attached to the wheels beside them. One ringer swings each bell, listening intently and watching the other ringers closely. If one ringer gets lost and stays lost, the rhythm of the ringing cannot be maintained; all the ringers have to stop.

A ringer can spend weeks just learning to keep a bell going and months learning to make the bell ring in exactly the right place. Errors of $\frac{1}{4}$ second mean that two bells are ringing at the same time. Even errors of $\frac{1}{10}$ second can be noticed.

Combining Counting Methods: Forming a Governing Committee

EXAMPLE 3 Building Numbers from a Set of Digits

How many nonrepeating three-digit numbers can be written using only the digits 3, 4, 5, 6, 7, and 8?

Solution

Repetitions are not allowed since the numbers are to be "nonrepeating." For example, 448 is not acceptable. Also, order is important. For example, 476 and 746 are *distinct* cases. So we use permutations to determine the total number. There are

$$_6P_3 = 6 \cdot 5 \cdot 4 = 120 \text{ ways.}$$

EXAMPLE 4 Designing Account Numbers

Suppose certain account numbers are to consist of two letters followed by four digits and then three more letters. Repetitions of letters or digits are not allowed *within* any of the three groups, but the last group of letters may contain one or both of those used in the first group. How many such accounts are possible?

Solution

The task of designing such a number consists of three parts.

Part 1 Determine the first set of two letters.   (There are 26 letters.)
Part 2 Determine the set of four digits.   (There are 10 digits.)
Part 3 Determine the final set of three letters.   (There are 26 letters.)

Each part requires an arrangement without repetitions, which is a permutation.

$$_{26}P_2 \cdot {}_{10}P_4 \cdot {}_{26}P_3 = \underbrace{650}_{\text{Part 1}} \cdot \underbrace{5040}_{\text{Part 2}} \cdot \underbrace{15{,}600}_{\text{Part 3}} \quad \text{Fundamental counting principle}$$

$$= 51{,}105{,}600{,}000 \text{ accounts}$$

Combinations

Permutations involve the number of arrangements of n things taken r at a time, where repetitions are not allowed. Order is important. Recall that club

$$N = \{\text{Alan, Bill, Cathy, David, Evelyn}\}$$

could elect three officers in $_5P_3 = 60$ different ways. With three-member committees, on the other hand, order is not important. The committees B, D, E and E, B, D are not different. The possible number of committees is not the number of arrangements of size 3. Rather, it is the number of *subsets* of size 3.

Recall that in the study of sets, a **set** is a collection or group of things (called **elements** or **members** of the set) commonly designated using a list within braces. This is how we have been designating the club

$$N = \{A, B, C, D, E\}.$$

The order of listing of the members (of any set) is unimportant. For example,

$$\{D, B, A, E, C\} \text{ is the same club.}$$

A **subset** of a set is a collection of some of the members. It may be all members of the original set, or even none of them, or anywhere in between.

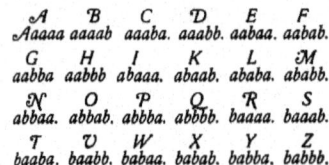

"**Bilateral cipher**" (above) was invented by **Francis Bacon** early in the seventeenth century to code political secrets. This binary code, *a* and *b* in combinations of five, has 32 permutations. Bacon's "biformed alphabet" (bottom four rows) uses two type fonts to conceal a message in some straight text. The decoder deciphers a string of *a*'s and *b*'s, groups them by fives, then deciphers letters and words. This code was applied to Shakespeare's plays in efforts to prove Bacon the rightful author.

Such subsets are **combinations**. The number of combinations of *n* things taken *r* at a time—that is, the number of size-*r* subsets, given a set of size *n*—is written $_nC_r$.*

> *There are n things available and we are choosing r of them, so we can read $_nC_r$ as "n choose r."*

In choosing subsets of a given set, repetitions are not allowed and the order of the listing of elements is irrelevant. For example, in the set *N*, $\{E, E, B\}$ is not a valid three-member subset, just as *E*, *E*, *B* is not a valid three-member arrangement.

EXAMPLE 5 Listing Subsets of a Given Size

List the subsets of $N = \{A, B, C, D, E\}$ that contain 3 members (elements). How many are there?

Solution

Using ideas from systematic listing, we find that there are 10 three-member subsets of *N*:

$$\{A, B, C\}, \quad \{A, B, D\}, \quad \{A, B, E\}, \quad \{A, C, D\}, \quad \{A, C, E\},$$
$$\{A, D, E\}, \quad \{B, C, D\}, \quad \{B, C, E\}, \quad \{B, D, E\}, \quad \{C, D, E\}.$$

To see how to find the number of such subsets without listing them all in **Example 5**, notice that each size-3 subset (combination) gives rise to six size-3 arrangements (permutations). For example, the single combination *ADE* yields these six permutations.

$$A, D, E \quad A, E, D \quad D, A, E \quad D, E, A \quad E, A, D \quad E, D, A$$

There must be six times as many size-3 permutations as there are size-3 combinations, or, equivalently, one-sixth as many combinations as permutations.

$$_5C_3 = \frac{_5P_3}{6} = \frac{60}{6} = 10$$

The 6 appears in the denominator because there are six different ways to arrange a set of three things, and $3! = 3 \cdot 2 \cdot 1 = 6$. Generalizing, we obtain the following.

$$_nC_r = \frac{_nP_r}{r!} \qquad \text{\scriptsize r things can be arranged in r! ways.}$$

$$= \frac{\dfrac{n!}{(n-r)!}}{r!} \qquad \text{\scriptsize Substitute the factorial formula for } _nP_r.$$

$$_nC_r = \frac{n!}{r!(n-r)!} \qquad \text{\scriptsize Simplify algebraically.}$$

FACTORIAL FORMULA FOR COMBINATIONS

The number of **combinations**, or *subsets*, of *n* distinct things taken *r* at a time, where $r \le n$, can be calculated as follows.

$$_nC_r = \frac{_nP_r}{r!} = \frac{n!}{r!(n-r)!}$$

In **Examples 6 and 7**, we refer to $_nC_r$ as "a combination" even though it actually represents the number of combinations of *n* distinct things taken *r* at a time.

NORMAL FLOAT AUTO REAL RADIAN MP

$\frac{9!}{7!*2!}$
                                36
$\frac{24!}{18!*6!}$
                            134596

This screen uses factorials to support the results of **Example 6**.

---

*Alternative notations are $C(n, r)$,, C_r^n, and $\binom{n}{r}$.

EXAMPLE 6 Using the Factorial Formula for Combinations

Evaluate each combination.

(a) $_9C_7$ **(b)** $_{24}C_{18}$

Solution

(a) $_9C_7 = \dfrac{9!}{7!(9-7)!} = \dfrac{9!}{7! \cdot 2!} = \dfrac{9 \cdot 8 \cdot 7 \cdot 6 \cdot 5 \cdot 4 \cdot 3 \cdot 2 \cdot 1}{7 \cdot 6 \cdot 5 \cdot 4 \cdot 3 \cdot 2 \cdot 1 \cdot 2 \cdot 1} = \dfrac{9 \cdot 8}{2 \cdot 1} = 36$

(b) $_{24}C_{18} = \dfrac{24!}{18!(24-18)!} = \dfrac{24!}{18! \cdot 6!} = 134{,}596$

NORMAL FLOAT AUTO REAL RADIAN MP

$_{14}C_6$

3003

$_{21}C_{15}$

54264

This screen uses the combinations feature to support the results of **Example 7**.

EXAMPLE 7 Calculating Combinations Directly

Evaluate each combination.

(a) $_{14}C_6$ **(b)** $_{21}C_{15}$

Solution

(a) $_{14}C_6 = 3003$ **(b)** $_{21}C_{15} = 54{,}264$ Use a calculator in each case.

Problem-Solving Strategy

Combinations share an important feature with permutations in that repetitions are not allowed, and they differ from permutations in one key distinction, which is that order is *not* important with combinations. Combinations are applied only in these situations:

1. *Repetitions are not allowed.*   2. *Order is not important.*

EXAMPLE 8 Finding the Number of Possible Poker Hands

A common form of poker involves "hands" (sets) of five cards each, dealt from a standard deck consisting of 52 different cards. How many different 5-card hands are possible?

Solution

A 5-card hand must contain five distinct cards, so repetitions are not allowed. Also, the order is not important because a given hand depends only on the cards it contains, and not on the order in which they were dealt or in which they are displayed or played. Order does not matter, so we use combinations and a calculator.

$$_{52}C_5 = \frac{52!}{5!(52-5)!} = \frac{52!}{5! \cdot 47!} = 2{,}598{,}960 \text{ hands}$$

The set of 52 playing cards in the **standard deck** has four suits.

♠ spades   ♦ diamonds
♥ hearts   ♣ clubs

Ace is the unit card. Jacks, queens, and kings are "face cards." Each suit contains thirteen denominations: ace, 2, 3, . . . , 10, jack, queen, king. In some games, ace rates above king, instead of counting as 1.

EXAMPLE 9 Finding the Number of Subsets of Paintings

Keri would like to buy ten different paintings, but she can afford only four of them. In how many ways can she make her selections?

Solution

The four paintings selected must be distinct. Repetitions are not allowed, and the order of the four chosen has no bearing in this case, so we use combinations.

$$_{10}C_4 = \frac{10!}{4!(10-4)!} = \frac{10!}{4! \cdot 6!} = 210 \text{ ways}$$

Combining Counting Methods:
Forming a Governing Committee

Notice that, according to our formula for combinations,

$$_{10}C_6 = \frac{10!}{6!(10-6)!} = \frac{10!}{6! \cdot 4!} = 210,$$

which is the same as $_{10}C_4$. In fact, **Exercise 84** asks you to prove the following fact, in general, for all whole numbers n and r, with $r \le n$.

$$_nC_r = {_nC_{n-r}}$$

Guidelines on Which Method to Use

The following table summarizes the similarities and differences between permutations and combinations, and appropriate formulas for calculating their values.

Permutations	Combinations
Number of ways of selecting r items out of n items	
Repetitions are not allowed.	
Order is important.	Order is not important.
Arrangements of n items taken r at a time	Subsets of n items taken r at a time
$_nP_r = \dfrac{n!}{(n-r)!}$	$_nC_r = \dfrac{n!}{r!(n-r)!}$
Clue words: arrangement, schedule, order	Clue words: set, group, sample, selection

In cases where r items are to be selected from n items and repetitions are allowed, it is usually best to make direct use of the fundamental counting principle.

The exercises in this section will call for permutations and/or combinations. In the case of multiple-part tasks, the fundamental counting principle may also be required. *In all cases, decide carefully whether order is important, because that determines whether to use permutations or combinations.*

Problem-Solving Strategy

The particular conditions of a counting problem will determine which specific problem-solving technique to use.

1. **If selected items can be repeated, use the fundamental counting principle.**

   Example: How many four-digit numbers are there?

   $$9 \cdot 10^3 = 9000 \text{ numbers}$$

2. **If selected items cannot be repeated, and order is important, use permutations.**

   Example: How many ways can three of eight people line up at a counter?

   $$_8P_3 = \frac{8!}{(8-3)!} = 336 \text{ ways}$$

3. **If selected items cannot be repeated, and order is *not* important, use combinations.**

   Example: How many ways can a committee of three people be selected from a group of twelve people?

   $$_{12}C_3 = \frac{12!}{3!(12-3)!} = 220 \text{ ways}$$

Combining Counting Methods:
Forming a Governing Committee

EXAMPLE 10   Distributing Toys to Children

In how many ways can a mother distribute three different toys among her seven children if a child may receive anywhere from none to all three toys?

Solution

Because a given child can be a repeat recipient, repetitions are allowed here, so we use the fundamental counting principle. Each of the three toys can go to any of the seven children. The number of possible distributions is

$$7 \cdot 7 \cdot 7 = 343.$$

Combining Counting Methods:
Forming a Governing Committee

EXAMPLE 11   Selecting Committees

How many different three-member committees could club

$$N = \{\text{Alan, Bill, Cathy, David, Evelyn}\}$$

appoint so that exactly one woman is on the committee?

Solution

Two members are women and three are men. Although the question mentioned only that the committee must include exactly one woman, to complete the committee two men must be selected as well. The task of selecting the committee members consists of two parts.

Part 1   Choose one woman.   ***Part 2***   Choose two men.

Because order is not important for committees, use combinations for the two parts. One woman can be chosen in $_2C_1 = \frac{2!}{1! \cdot 1!} = 2$ ways, and two men can be chosen in $_3C_2 = \frac{3!}{2! \cdot 1!} = 3$ ways. Finally, use the fundamental counting principle to obtain $2 \cdot 3 = 6$ different committees. This small number can be checked by listing.

$$\{C, A, B\}, \quad \{C, A, D\}, \quad \{C, B, D\}, \quad \{E, A, B\}, \quad \{E, A, D\}, \quad \{E, B, D\}$$

Combining Counting Methods:
Forming a Governing Committee

EXAMPLE 12   Selecting Attendees for an Event

Every member of the Alpha Beta Gamma fraternity would like to attend a special event this weekend, but only ten members will be allowed to attend. How many ways can the lucky ten be selected if there are a total of forty-eight members?

Solution

In this case, ten distinct men are required, so repetitions are not allowed, and the order of selection makes no difference. We use combinations.

$$_{48}C_{10} = \frac{48!}{10! \cdot 38!} = 6{,}540{,}715{,}896 \text{ ways}$$ Use a calculator.

EXAMPLE 13 Selecting Escorts

When the ten fraternity men of **Example 12** arrive at the event, four of them are selected to pair with the four homecoming queen candidates to escort them. In how many ways can this selection be made?

Solution

Of the ten, four distinct men are required, and order is important here because different orders will pair the men with different women. Use permutations.

$$_{10}P_4 = \frac{10!}{6!} = 5040 \text{ possible selections}$$

EXAMPLE 14 Dividing into Groups

In how many ways can the 9 members of a baseball lineup divide into groups of 4, 3, and 2 players?

Solution

Order is not important within the groups. The players within a group are interchangeable in their order of listing. Use combinations.

- First, 4 can be chosen from 9 in $_9C_4 = 126$ ways.
- Then, 3 can be chosen from the remaining 5 in $_5C_3 = 10$ ways.
- Then, 2 can be chosen from the remaining 2 in $_2C_2 = 1$ way.

The three groups also are not interchangeable. They all have different sizes. Apply the fundamental counting principle to find the number of ways.

$$_9C_4 \cdot {_5C_3} \cdot {_2C_2} = 126 \cdot 10 \cdot 1 = 1260 \text{ possible ways}$$

Combining Counting Methods: Forming a Governing Committee

EXAMPLE 15 Dividing into Groups

In how many ways can the 9 players of **Example 14** divide into three groups of 3?

Solution

After the pattern of **Example 14,** the answer may *seem* to be

$$_9C_3 \cdot {_6C_3} \cdot {_3C_3} = 84 \cdot 20 \cdot 1 = 1680. \quad \text{INCORRECT}$$

This would impose an *unwanted order among* the groups. Ordering the three group selections was appropriate in **Example 14,** because those three groups were distinguishable. They were all different sizes. But here, all groups are size-3. If the players are denoted $A, B, C, D, E, F, G, H,$ and $I,$ then the list

(1) BIG, HEF, CAD **(2)** BIG, CAD, HEF **(3)** HEF, BIG, CAD

(4) HEF, CAD, BIG **(5)** CAD, BIG, HEF **(6)** CAD, HEF, BIG

contains six orderings of the same three groups. Since the product 1680 calculated earlier duplicates every set of three groups in this way, we must adjust that value by dividing by $3! = 6$ to obtain the true number of *unordered* sets of three groups. The idea is the same as when we adjust the number of arrangements—orderings—of n things taken r at a time to obtain the number of unordered sets of n things taken r at a time.

The number of ways in which 9 players can divide into three groups of 3 is

$$\frac{_9C_3 \cdot {_6C_3} \cdot {_3C_3}}{3!} = \frac{1680}{6} = 280. \quad \text{Use the formula } _nC_r = \frac{_nP_r}{r!}.$$

StatCrunch Poker Hands

WHEN WILL I **EVER** USE THIS**?**

If you want to play video poker or have a poker night with your friends, you may find the following information useful.

In 5-card poker, played with a standard 52-card deck, 2,598,960 different hands are possible. See **Example 8**. The desirability of the various hands depends on their relative chance of occurrence, which, in turn, depends on the number of different ways they can occur, as shown in **Table 4**. Note that an ace can generally be positioned either below 2 (as a 1) or above king (as a 14). This is important in counting straight flush hands and straight hands.

Royal Flush

Four of a Kind

Full House

In **Exercises 85–90,** we guide you in the process of determining the values in the three blanks in the far right column.

Table 4 Categories of Hands in 5-Card Poker

Event E	Description of Event E	Number of Outcomes Favorable to E
Royal flush	Ace, king, queen, jack, and 10, all of the same suit	4
Straight flush	5 cards of consecutive denominations, all in the same suit (excluding royal flush)	36
Four of a kind	4 cards of the same denomination, plus 1 additional card	_____
Full house	3 cards of one denomination, plus 2 cards of a second denomination	3744
Flush	Any 5 cards all of the same suit (excluding royal flush and straight flush)	_____
Straight	5 cards of consecutive denominations (not all the same suit)	10,200
Three of a kind	3 cards of one denomination, plus 2 cards of two additional denominations	54,912
Two pairs	2 cards of one denomination, plus 2 cards of a second denomination, plus 1 card of a third denomination	_____
One pair	2 cards of one denomination, plus 3 additional cards of three different denominations	1,098,240
No pair	No two cards of the same denomination (and excluding any sort of flush or straight)	1,302,540
Total		**2,598,960**

10.3 EXERCISES

CONCEPT CHECK *Complete each statement with the correct numerical response.*

1. There are ____ 1-member subsets of the set $\{a, b, c\}$.

2. There are ____ permutations of the elements of the set $\{P, Q, R\}$.

3. The value of "4 things taken 2 at a time" is ____.

4. There are ____ subsets of the elements of the set $\{a, b, c, d\}$ taken 3 at a time.

5. The calculator returns a value of ____ for the screen shown here.

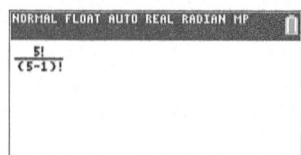

6. The calculator returns a value of ____ for the screen shown here.

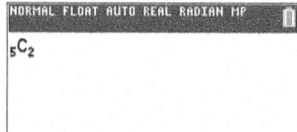

Evaluate each expression.

7. $_9P_3$	**8.** $_{10}P_6$	**9.** $_{12}P_5$
10. $_{11}P_3$	**11.** $_{11}C_7$	**12.** $_{14}C_6$
13. $_{10}C_8$	**14.** $_{12}C_9$	**15.** $_{12}C_3$

16. Find the number of different subsets of size 2 in the set $\{a, b, c, d\}$. List them.

Determine the number of permutations (arrangements) of each of the following.

17. 20 things taken 4 at a time

18. 15 things taken 5 at a time

19. 16 things taken 4 at a time

20. 13 things taken 7 at a time

Determine the number of combinations (subsets) of each of the following.

21. 9 things taken 4 at a time

22. 13 things taken 6 at a time

23. 12 things taken 5 at a time

24. 11 things taken 8 at a time

Use a calculator to evaluate each expression.

25. $_{22}P_9$

26. $_{23}P_{11}$

27. $_{32}C_{12}$

28. $_{31}C_{17}$

Decide whether each object is a permutation *or a* combination.

29. a telephone number

30. a Social Security number

31. a hand of cards in poker

32. a committee of politicians

33. the "combination" on a student gym locker combination lock

34. a lottery choice of six numbers where the order does not matter

35. an automobile license plate number

36. a smartphone passcode

Exercises 37–44 can be solved using permutations even though the problem statements will not always include a form of the word "permutation" or "arrangement" or "ordering."

37. **Arranging New Home Models** Tyler, a contractor, builds homes of eight different models and presently has five lots to build on. In how many different ways can he arrange homes on these lots? Assume five different models will be built.

38. **ATM PIN Numbers** An automated teller machine (ATM) requires a four-digit personal identification number (PIN), using the digits 0–9. (The first digit may be 0.) How many such PINs have no repeated digits?

39. *Electing Officers of a Club* How many ways can a president and a vice president be determined in a club with twelve members?

40. *Counting Prize Winners* First, second, and third prizes are to be awarded to three different people. If there are ten eligible candidates, how many outcomes are possible?

41. *Counting Prize Winners* How many ways can a teacher give five different prizes to five of her 25 students?

42. *Scheduling Security Team Visits* A security team visits 12 offices each night. How many different ways can the team order its visits?

43. *Placing in a Race* How many different ways could first-, second-, and third-place finishers occur in a race with six runners competing?

44. *Sums of Digits* How many counting numbers have four distinct nonzero digits such that the sum of the four digits is

(a) 10? **(b)** 11?

Exercises 45–52 can be solved using combinations even though the problem statements will not always include the word "combination" or "subset."

45. *Arranging New Home Models* Tyler (the contractor) is to build six homes on a block in a new subdivision, using two different models: standard and deluxe. (All standard model homes are the same, and all deluxe model homes are the same.)

(a) How many different choices does Tyler have in positioning the six houses if he decides to build three standard and three deluxe models?

(b) If Tyler builds two deluxes and four standards, how many different positionings can he use?

46. *Sampling Cell Phones* How many ways can a sample of five cell phones be selected from a shipment of twenty-four cell phones?

47. *Detecting Defective Cell Phones* If the shipment of **Exercise 46** contains six defective phones, how many of the size-five samples would not include any of the defective ones?

48. *Committees of U.S. Senators* How many different five-member committees could be formed from the 100 U.S. senators?

49. *Selecting Hands of Cards* Refer to the standard 52-card deck pictured in the margin near **Example 8,** and notice that the deck contains four aces, twelve face cards, thirteen hearts (all red), thirteen diamonds (all red), thirteen spades (all black), and thirteen clubs (all black). Of the 2,598,960 different five-card hands possible, decide how many would consist of the following cards.

(a) all diamonds

(b) all black cards

(c) all aces

50. *Selecting Lottery Entries* In a 7/39 lottery, you select seven distinct numbers from the set 1 through 39, where order makes no difference. How many different ways can you make your selection?

51. *Number of Paths from Point to Point* In a certain city, there are seven streets going north–south and four streets going east–west. How many street paths start at the southwest corner of the city, end at the northeast corner of the city, and have the shortest possible length? (*MT* calendar problem)

52. *Choosing a Monogram* Sonya Johnson wants to name her new baby so that his monogram (first, middle, and last initials) will be distinct letters in alphabetical order and he will share her last name. How many different monograms could she select?

In Exercises 53–82, use permutations, combinations, the fundamental counting principle, or other counting methods, as appropriate.

53. *Class Groupings* In how many ways can a class of twenty students be divided into three sets so that three students are in the first set, five are in the second, and twelve are in the third? (*MT* calendar problem)

54. *Calendar Pages* Calendar pages are available without the names of the months. All they have on them are the days of the week and the dates. The user inserts the name of the month. A page exists for every arrangement of dates in a month. How many distinct calendar pages are necessary? (*MT* calendar problem)

55. *Identification Numbers in Research* Subject identification numbers in a certain scientific research project consist of three letters followed by three digits and then three more letters. Assume repetitions are not allowed within any of the three groups, but letters in the first group of three may occur also in the last group of three. How many distinct identification numbers are possible?

56. *Radio Station Call Letters* Radio stations in the United States have call letters that begin with K or W. Some have three call letters, such as WBZ in Boston, WLS in Chicago, and KGO in San Francisco. Assuming no repetition of letters, how many three-letter sets of call letters are possible? (Count all possibilities even though, practically, some may be inappropriate.)

57. *Radio Station Call Letters* Most stations that were licensed after 1927 have four call letters starting with K or W, such as WXYZ in Detroit and KRLD in Dallas. Assuming no repetitions, how many four-letter sets are possible? (Count all possibilities even though, practically, some may be inappropriate.)

58. *Selecting Lottery Entries* In SuperLotto Plus, a California state lottery game, you select five distinct numbers from 1 to 47, and one MEGA number from 1 to 27, hoping that your selection will match a random list selected by lottery officials.

(a) How many different sets of six numbers can you select?

(b) Paul always includes his age and his wife's age as two of the first five numbers in his SuperLotto Plus selections. How many ways can he complete his list of six numbers?

59. *Scheduling Batting Orders in Baseball* The Coyotes, a youth league baseball team, have seven pitchers, who only pitch, and twelve other players, all of whom can play any position other than pitcher. For Saturday's game, the coach has not yet determined which nine players to use or what the batting order will be, except that the pitcher will bat last. How many different batting orders may occur?

60. *Scheduling Games in a Basketball League* Each team in an eight-team basketball league is scheduled to play each other team three times. How many games will be played altogether?

61. *Arranging a Wedding Reception Line* At a wedding reception, the bride, the groom, and four attendants will form a reception line. How many ways can they be arranged in each of the following cases?

(a) Any order will do.

(b) The bride and groom must be the last two in line.

(c) The groom must be last in line with the bride next to him.

62. *Ordering Performers in a Music Recital* A music class of five girls and four boys is having a recital. If each member is to perform once, how many ways can the program be arranged in each of the following cases?

(a) All girls must perform first.

(b) A girl must perform first, and a boy must perform last.

(c) Elisa (a girl) and Doug (a boy) will perform first and last, respectively.

(d) The entire program will alternate between girls and boys.

(e) The first, fifth, and ninth performers must be girls.

63. *Dividing People into Groups* In how many ways could fifteen people be divided into five groups containing, respectively, one, two, three, four, and five people?

64. *Dividing People into Groups* In how many ways could fifteen people be divided into five groups of three people?

65. *Dividing People into Groups* In how many ways could eight people be divided into two groups of three people and a group of two people?

66. *Scheduling Daily Reading* Carole begins each day by reading from one of seven inspirational books. How many ways can she arrange her reading for one week if the selection is done

(a) with replacement?

(b) without replacement?

67. *Counting Card Hands* How many of the possible 5-card hands from a standard 52-card deck would consist of the following cards?

(a) four clubs and one non-club

(b) two face cards and three non-face cards

(c) two red cards, two clubs, and a spade

68. *Drawing Cards* How many cards must be drawn (without replacement) from a standard deck of 52 to guarantee the following?

(a) Two of the cards will be of the same suit.

(b) Three of the cards will be of the same suit.

69. *Flush Hands in Poker* How many different 5-card poker hands would contain only cards of a single suit?

70. *Screening Computer Processors* A computer company will screen a shipment of 30 processors by testing a random sample of five of them. How many different samples are possible?

71. *Selecting Drivers and Passengers for a Trip* Natalie, her husband, her son, and four additional friends are driving in two vehicles to the seashore.

(a) If all seven people can drive, how many ways can the two drivers be selected? (Everyone wants to drive the sports car, so it is important which driver gets which car.)

(b) If the sports car must be driven by Natalie, her husband, or their son, how many ways can the drivers now be determined?

(c) If the sports car will accommodate only two people, and there are no other restrictions, how many ways can both drivers and passengers be assigned to both cars?

72. *Points and Lines in a Plane* If any two points determine a line, how many lines are determined by seven points in a plane, no three of which are collinear?

73. *Points and Triangles in a Plane* How many triangles are determined by twenty points in a plane, no three of which are collinear?

74. *Counting Possibilities on a Combination Lock* How many different three-number "combinations" are possible on a combination lock having 40 numbers on its dial? (*Hint:* "Combination" is a misleading name for these locks since repetitions are allowed and order makes a difference.)

75. *Winning the Trifecta in Horse Racing* Many race tracks offer a "trifecta" race. You win by selecting the correct first-, second-, and third-place finishers. If eight horses are entered, how many tickets must you purchase to guarantee that one of them will be a trifecta winner?

76. *Winning the Daily Double in Horse Racing* At a horse race, you win the "daily double" by purchasing a ticket and selecting the winners of two specific races. If there are six and eight horses running in those races, respectively, how many tickets must you buy to guarantee a win?

77. *Selecting Committees* Nine people are to be distributed among three committees of two, three, and four members, and a chairperson is to be selected for each committee. How many ways can this be done? (*Hint:* Break the task into the following sequence of parts.)

Part 1 Select the members of the two-person committee.

Part 2 Select the members of the three-person committee.

Part 3 Select the chair of the two-person committee.

Part 4 Select the chair of the three-person committee.

Part 5 Select the chair of the four-person committee.

78. *Selecting Committee Members* Repeat **Exercise 77** in the case where the three committees are to have three members each. (*Hint:* Use the same general sequence of task parts, but remember to adjust for *unwanted ordering* of the three committees.)

79. *Arranging New Home Models* (See **Exercise 45.**) Because of his good work, Tyler gets a contract to build homes on three additional blocks in the subdivision, with six homes on each block. He decides to build nine deluxe homes on these three blocks: two on the first block, three on the second, and four on the third. The remaining nine homes will be standard.

(a) Altogether on the three-block stretch, how many different choices does Tyler have for positioning the eighteen homes? (*Hint:* Consider the three blocks separately, and use the fundamental counting principle.)

(b) How many choices would he have if he built 2, 3, and 4 deluxe models on the three different blocks as before, but not necessarily on the first, second, and third blocks in that order?

80. *Sums of Digits* How many counting numbers consist of four distinct nonzero digits such that the sum of the four digits is the given number?

(a) 12   **(b)** 13

81. *Building Numbers from Sets of Digits* Recall that the counting numbers are

$$1, 2, 3, 4, \ldots .$$

(a) How many six-digit counting numbers use all six digits 4, 5, 6, 7, 8, and 9?

(b) Suppose all these numbers were arranged in increasing order:

$$456{,}789; \quad 456{,}798; \quad \text{and so on.}$$

Which number would be 364th in the list?

82. *Arranging Five-letter Words* The 120 permutations of AHSME are arranged in dictionary order, as if each were an ordinary five-letter word. Find the last letter of the 86th word in the list. (*MT* calendar problem)

83. Verify that $_{12}C_9 = {}_{12}C_3$.

84. Use the factorial formula for combinations to prove that in general,

$$_nC_r = {}_nC_{n-r}.$$

Poker Hands *Refer to the discussion in "When Will I Ever Use This?" to answer Exercises 85–90. As* **Table 4** *shows, a full house is a relatively rare occurrence. (Only four of a kind, straight flush, and royal flush are less likely.) To verify that there are 3744 different full house hands possible, carry out the following steps.*

85. How many different ways are there to select three aces from the deck?

86. How many different ways are there to select two 8s from the deck?

87. "Aces and 8s" is one form of a full house. Other forms are "8s and Aces," "Kings and 6s," and so forth. How many different forms of a full house are there altogether?

88. Use the results of **Exercises 85–87** to determine the total number of full house hands possible.

89. Use similar reasoning to find the three missing values in the right column of **Table 4.**

90. Confirm that the total number of possible hands is 2,598,960.

10.4 USING PASCAL'S TRIANGLE

OBJECTIVES

1 Construct Pascal's triangle and recognize the relationship between its entries and values of $_nC_r$.

2 Use Pascal's triangle to solve applications involving combinations.

3 Compute binomial coefficients and apply the binomial theorem.

Pascal's Triangle

The triangular array in **Figure 7** represents what we can call "random walks" that begin at START and proceed downward according to the following rule:

At each circle (branch point), a coin is tossed. If it lands heads, we go downward to the left. If it lands tails, we go downward to the right. At each point, left and right are equally likely.

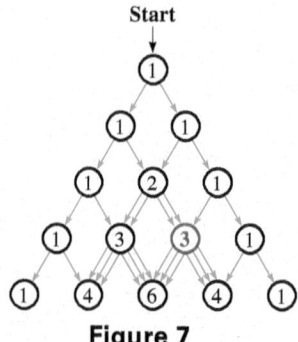

Figure 7

In each circle we have recorded the number of different routes that could bring us to that point. For example, the colored 3 can be reached as the result of three different coin-tossing sequences.

htt, tht, and tth   Ways to get to ③

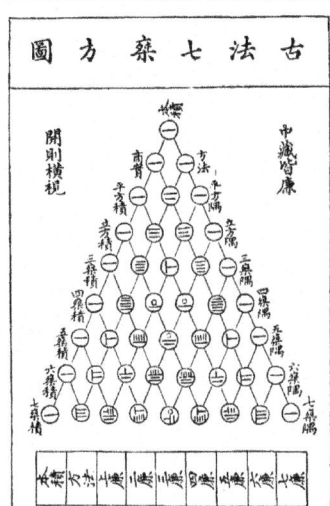

"Pascal's" triangle is shown in the 1303 text **Szu-yuen Yu-chien** (*The Precious Mirror of the Four Elements*) by the Chinese mathematician Chu Shih-chieh.

Another way to generate the same pattern of numbers is to begin with 1s down both diagonals and then fill in the interior entries by adding the two numbers just above a given position (to the left and right). For example, the green 28 in **Table 5** is the result of adding 7 and 21 in the row above it.

By continuing to add pairs of numbers, we extend the array indefinitely downward, always beginning and ending each row with 1s. The table shows just rows 0 through 10. This unending triangular array of numbers is called **Pascal's triangle,** because Blaise Pascal wrote a treatise about it in 1653. There is evidence, though, that it was known as early as around 1100 and may have been studied in China or India still earlier.

Table 5 Pascal's Triangle

Row Number												Row Sum
0						1						1
1					1		1					2
2				1		2		1				4
3			1		3		3		1			8
4		1		4		6		4		1		16
5	1		5		10		10		5		1	32
6	1	6		15		20		15		6	1	64
7	1	7	21		35		35		21	7	1	128
8	1	8	28	56		70		56	28	8	1	256
9	1	9	36	84	126		126	84	36	9	1	512
10	1	10	45	120	210	252	210	120	45	10	1	1024

At any rate, the triangle possesses many interesting properties. In counting applications, its most useful property is that, in general, entry number r in row number n is equal to $_nC_r$—the number of *combinations* of n things taken r at a time. This correspondence is shown through row 8 in **Table 6.**

The Pascal Identity We know that each interior entry in Pascal's triangle can be obtained by adding the two numbers just above it (to the left and right). This fact, known as the "Pascal identity," can be written as

$$_nC_r = {}_{n-1}C_{r-1} + {}_{n-1}C_r.$$

The factorial formula for combinations (along with some algebra) can be used to prove the Pascal identity.

Table 6 Combination Values in Pascal's Triangle

Row Number									
0					$_0C_0$				
1				$_1C_0$		$_1C_1$			
2			$_2C_0$		$_2C_1$		$_2C_2$		
3		$_3C_0$		$_3C_1$		$_3C_2$		$_3C_3$	
4	$_4C_0$		$_4C_1$		$_4C_2$		$_4C_3$		$_4C_4$
5	$_5C_0$	$_5C_1$		$_5C_2$		$_5C_3$		$_5C_4$	$_5C_5$
6	$_6C_0$	$_6C_1$	$_6C_2$		$_6C_3$		$_6C_4$	$_6C_5$	$_6C_6$
7	$_7C_0$	$_7C_1$	$_7C_2$	$_7C_3$		$_7C_4$	$_7C_5$	$_7C_6$	$_7C_7$
8	$_8C_0$	$_8C_1$	$_8C_2$	$_8C_3$	$_8C_4$	$_8C_5$	$_8C_6$	$_8C_7$	$_8C_8$

The entries in color correspond to those examined earlier in **Table 5.** Having a copy of Pascal's triangle handy gives us another option for evaluating combinations. Any time we need to know the number of combinations of n things taken r at a time, we can simply read entry number r of row number n. ***Keep in mind that the first row shown is row number 0.*** Also, the first entry of each row can be called entry number 0. This entry gives the number of subsets of size 0, which is always 1 because there is only one empty set.

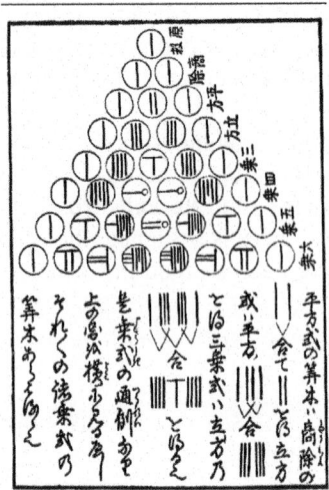

This **Japanese version** of the triangle dates from the eighteenth century. The "stick numerals" evolved from bamboo counting pieces used on a ruled board. Omar Khayyam, twelfth-century Persian mathematician and poet, may also have divined its patterns in pursuit of algebraic solutions. (The triangle lists the coefficients of the binomial expansion.)

Applications

EXAMPLE 1 Applying Pascal's Triangle to Counting People

A group of ten people includes six women and four men. If five of these people are randomly selected, how many different samples of five people are possible?

Solution

This is simply a matter of selecting a subset of five from a set of ten, which is the number of combinations of ten things taken five at a time.

$$_{10}C_5 = 252 \text{ different samples}$$ See row 10 of Pascal's triangle in **Table 5**.

EXAMPLE 2 Applying Pascal's Triangle to Counting People

Among the 252 possible samples of five people in **Example 1,** how many of them would consist of exactly two women and three men?

Solution

Two women can be selected from six women in $_6C_2$ different ways, and three men can be selected from four men in $_4C_3$ different ways. These values can be read from Pascal's triangle. Because the task of obtaining two women and three men requires both individual parts, the fundamental counting principle can be applied.

$$_6C_2 \cdot {}_4C_3 = 15 \cdot 4 = 60 \text{ samples}$$ Rows 6 and 4 of Pascal's triangle

EXAMPLE 3 Applying Pascal's Triangle to Coin Tossing

If five fair coins are tossed, in how many different ways could exactly three heads be obtained?

Solution

There are various "ways" of obtaining exactly three heads because the three heads can occur on different subsets of the coins. For example, hhtht and thhth are just two of many possibilities. When such a possibility is written down, exactly three positions are occupied by an h, the other two by a t. Each distinct way of choosing three positions from a set of five positions gives a different possibility. (Once the three positions for h are determined, each of the other two positions automatically receives a t.)

The answer is just the number of size-3 subsets of a size-5 set—that is, the number of combinations of five things taken three at a time.

$$_5C_3 = 10 \text{ different ways}$$ Row 5 of Pascal's triangle

StatCrunch Coin Flipping

Notice that row 5 of Pascal's triangle also provides answers to several other questions about tossing five fair coins. They are summarized in **Table 7.**

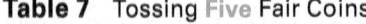

Table 7 Tossing Five Fair Coins

Number of Heads n	Ways of Obtaining Exactly n Heads	Listing
0	$_5C_0 = 1$	ttttt
1	$_5C_1 = 5$	htttt, thttt, tthtt, tttht, tttth
2	$_5C_2 = 10$	hhttt, hthtt, htthт, htttн, thhtt, ththt, thtth, tthht, tthth, ttthh
3	$_5C_3 = 10$	hhhtt, hhtht, hhtth, hthht, hthth, htthh, thhht, thhth, ththh, tthhh
4	$_5C_4 = 5$	hhhht, hhhth, hhthh, hthhh, thhhh
5	$_5C_5 = 1$	hhhhh

To analyze the tossing of a different number of fair coins, we can simply take the pertinent numbers from a different row of Pascal's triangle. Repeated coin tossing is an example of a "binomial" experiment because each toss has *two* possible outcomes:

$$\text{heads} \quad \text{and} \quad \text{tails.}$$

The Binomial Theorem

The combination values that make up Pascal's triangle also arise in a totally different mathematical context. In algebra, **binomial** refers to a two-term expression such as

$$x + y, \quad \text{or} \quad a + 2b, \quad \text{or} \quad w^3 - 4.$$

The first few powers of the binomial $x + y$ are shown here.

$$(x + y)^0 = 1$$
$$(x + y)^1 = x + y$$
$$(x + y)^2 = x^2 + 2xy + y^2$$
$$(x + y)^3 = x^3 + 3x^2y + 3xy^2 + y^3$$
$$(x + y)^4 = x^4 + 4x^3y + 6x^2y^2 + 4xy^3 + y^4$$
$$(x + y)^5 = x^5 + 5x^4y + 10x^3y^2 + 10x^2y^3 + 5xy^4 + y^5$$

The numerical coefficients of these expansions form rows 0 through 5 of Pascal's triangle. In our study of counting, we have called these numbers combinations, but in the study of algebra, they are called **binomial coefficients** and are usually denoted

$$\binom{n}{r} \quad \text{rather than} \quad {}_nC_r.$$

Generalizing the pattern of the powers yields the important result known as the **binomial theorem.**

BINOMIAL THEOREM

For any whole number n,

$$(x + y)^n$$
$$= \binom{n}{0} \cdot x^n + \binom{n}{1} \cdot x^{n-1}y$$
$$+ \binom{n}{2} \cdot x^{n-2}y^2 + \binom{n}{3} \cdot x^{n-3}y^3 +$$
$$\cdots + \binom{n}{n-1} \cdot xy^{n-1} + \binom{n}{n} \cdot y^n.$$

In the statement above, each binomial coefficient can be calculated by the formula

$$\binom{n}{r} = \frac{n!}{r!(n-r)!}.$$

Observation of the expansion of $(x + y)^n$ leads to these conclusions.

1. There are $n + 1$ terms in the expansion.
2. The first term is x^n, and the last term is y^n.
3. In each succeeding term, the exponent on x decreases by 1 and the exponent on y increases by 1.
4. The sum of the exponents on x and y in any term is n.
5. The coefficient of the term with $x^r y^{n-r}$ or $x^{n-r} y^r$ is $\binom{n}{r}$.

Furthermore, the expansion of a binomial of the form $(a - b)$ will lead to alternating $+$ and $-$ signs in the result.

EXAMPLE 4 **Applying the Binomial Theorem**

Write out the binomial expansion for $(2a + 5)^4$.

Solution

We take the initial coefficients from row 4 of Pascal's triangle and then simplify algebraically.

$$(2a + 5)^4 = \binom{4}{0} \cdot (2a)^4 + \binom{4}{1} \cdot (2a)^3 \cdot 5 + \binom{4}{2} \cdot (2a)^2 \cdot 5^2$$

$$+ \binom{4}{3} \cdot (2a) \cdot 5^3 + \binom{4}{4} \cdot 5^4$$

$$= 1 \cdot 2^4 \cdot a^4 + 4 \cdot 2^3 \cdot a^3 \cdot 5 + 6 \cdot 2^2 \cdot a^2 \cdot 5^2 + 4 \cdot 2 \cdot a \cdot 5^3 + 1 \cdot 5^4$$

$$= 16a^4 + 160a^3 + 600a^2 + 1000a + 625$$

10.4   EXERCISES

CONCEPT CHECK *Match the items in Column I with the appropriate response from Column II.*

I	**II**
1. $_3C_1$	**A.** 3
2. The third element in row 6 of Pascal's triangle (See **Table 6**)	**B.** 6
3. $_{14}C_1$	**C.** $_6C_2$
4. The last term in the expansion of $(x + 1)^7$	**D.** $\binom{7}{6}$
5. $\binom{7}{1}$	**E.** 1
6. $\binom{4}{2}$	**F.** 14

Read each combination value directly from Pascal's triangle.

7. $_4C_2$ **8.** $_5C_3$ **9.** $_6C_3$ **10.** $_7C_5$

11. $_8C_5$ **12.** $_9C_6$ **13.** $_9C_2$ **14.** $_{10}C_7$

Selecting Committees of Congressmembers *A committee of four Congressmembers will be selected from a group of seven Democrats and three Republicans. Find the number of ways of obtaining each result.*

15. exactly one Democrat

16. exactly two Democrats

17. exactly three Democrats

18. exactly four Democrats

Tossing Coins   *Suppose eight fair coins are tossed. Find the number of ways of obtaining each result.*

19. exactly three heads

20. exactly four heads

21. exactly five heads

22. exactly six heads

Selecting Classrooms   *Diana is searching for her ecology class and knows that it must be in one of nine classrooms. Since the professor does not allow people to enter after the class has begun, and there is very little time left, she decides to try just four of the rooms at random.*

23. How many different selections of four rooms are possible?

24. How many of the selections of **Exercise 23** will fail to locate the class?

25. How many of the selections of **Exercise 23** will succeed in locating the class?

26. What fraction of the possible selections will lead to "success"? (Give three decimal places.)

Write out the binomial expansion for each of the following.

27. $(x + 1)^6$

28. $(x + 1)^7$

29. $(y - 4)^4$

30. $(y - 3)^5$

31. $(2x + 5)^4$

32. $(2x - 5)^6$

Subsets   *For a set of five objects, find the number of different subsets of each size. (Use row 5 of Pascal's triangle to find the answers.)*

33. 0   **34.** 1   **35.** 2

36. 3   **37.** 4   **38.** 5

Pascal's Triangle   *Solve each problem.*

39. How many subsets (of any size) are there for a set of five elements?

40. What is the least four-digit number in Pascal's triangle? (*MT* calendar problem)

41. Which rows of Pascal's triangle have a single greatest entry?

42. For a given row in Pascal's triangle, let *n* be the row number and let *s* be the row sum.

   (a) Write an equation relating *s* and *n*.

   (b) Explain the relationship in part (a).

Patterns in Pascal's Triangle   *Over the years, many interesting patterns have been discovered in Pascal's triangle. We explore a few of them in Exercises 43–50.*

43. Refer to **Table 5.**

   (a) Choose a row whose row number is prime. Except for the 1s in this row, what is true of all the other entries?

   (b) Choose a second prime row number and see if the same pattern holds.

   (c) Use the usual method to construct row 11 in **Table 5.** Verify that the same pattern holds.

44. Name the next five numbers of the diagonal sequence in the figure. What are these numbers called?

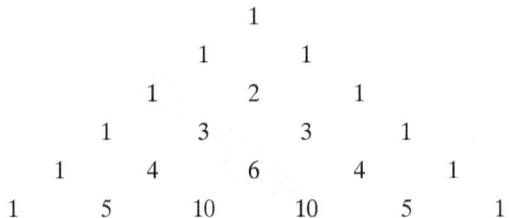

45. Complete the sequence of sums on the diagonals shown in the figure. What pattern do these sums make? What is the name of this important sequence of numbers? The presence of this sequence in the triangle apparently was not recognized by Pascal.

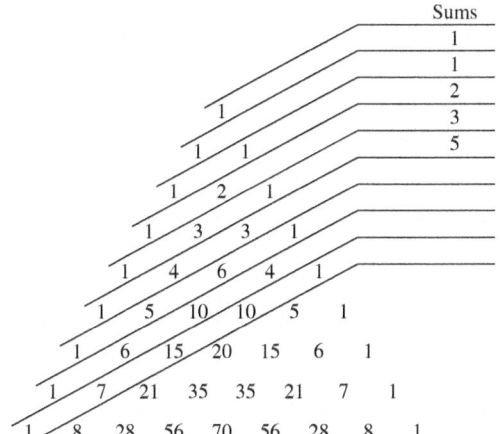

46. Construct another "triangle" by replacing every number in Pascal's triangle (rows **0** through **5**) by the remainder when it is divided by 2. What special property is shared by rows **2** and **4** of this new triangle?

47. What is the next row that would have the same property as rows **2** and **4** in **Exercise 46?**

48. (Work **Exercises 46 and 47** first.) How many even numbers are there in row **256** of Pascal's triangle?

49. The figure shows a portion of Pascal's triangle with several inverted triangular regions outlined. For any one of these regions, what can be said of the sum of the squares of the entries across its top row?

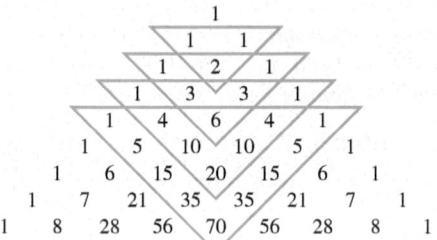

50. Show that the rule

$$\binom{n}{k} = \binom{n-1}{k-1} + \binom{n-1}{k}$$

holds for $n = 5$ and $k = 2$. Where does this particular application of the rule appear in Pascal's triangle?

Tartaglia's Rectangle *More than a century before Pascal's treatise on the triangle appeared, a work by the Italian mathematician Niccolo Tartaglia (1506–1559) came out and included the table of numbers shown here.*

1	1	1	1	1	1
1	2	3	4	5	6
1	3	6	10	15	21
1	4	10	20	35	56
1	5	15	35	70	126
1	6	21	56	126	252
1	7	28	84	210	462
1	8	36	120	330	792

The triangle that Pascal studied and published in his treatise was actually more like a truncated corner of Tartaglia's rectangle, as shown here.

1	1	1	1	1	1	1	1	1	1
1	2	3	4	5	6	7	8	9	
1	3	6	10	15	21	28	36		
1	4	10	20	35	56	84			
1	5	15	35	70	126				
1	6	21	56	126					
1	7	28	84						
1	8	36							
1	9								
1									

Each number in the truncated corner of Tartaglia's rectangle can be calculated in various ways. In each of Exercises 51–54, consider the number N to be located anywhere in the array. By checking several locations in the given array, determine how N is related to the sum of all entries in the shaded cells. Describe the relationship in words.

51.

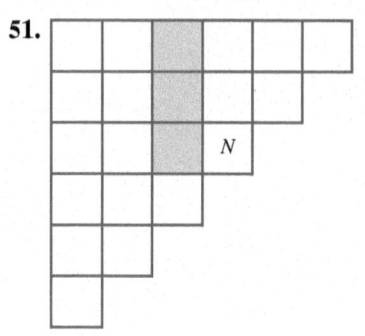

52.

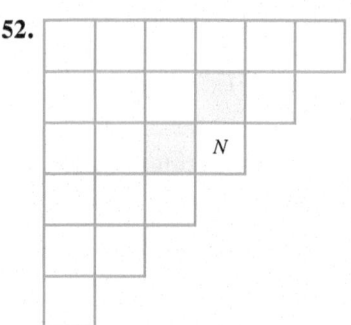

53.

54.

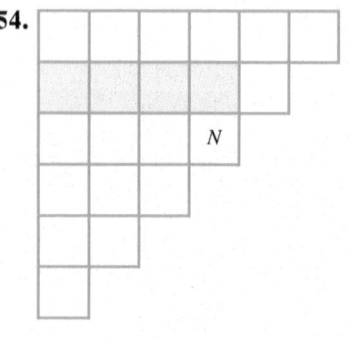

10.5 COUNTING PROBLEMS INVOLVING "NOT" AND "OR"

OBJECTIVES

1 Apply the complements principle to a counting problem by considering the problem in terms of outcomes that do not occur (a "not" statement).

2 Apply the additive principle to a counting problem by considering the problem in terms of the union of two outcomes (an "or" statement).

Set Theory/Logic/Arithmetic Correspondences

The counting techniques in this section, which can be thought of as *indirect techniques*, are based on some useful correspondences that we have noted among set theory, logic, and arithmetic, as shown in **Table 8**.

Table 8 Correspondences

	Set Theory	**Logic**	**Arithmetic**
Operation or Connective (Symbol)	Complement ($'$)	Not ($\sim$)	Subtraction ($-$)
Operation or Connective (Symbol)	Union ($\cup$)	Or ($\vee$)	Addition ($+$)

Problems Involving "Not"

Suppose U is the set of all possible results of some type. The "universal set U" is made up of all possibilities. Let A be the set of all those results that satisfy a given condition. For any set S, its cardinal number is written $n(S)$, and its complement is written S'. **Figure 8** suggests that

$$n(A) + n(A') = n(U).$$

Also, $n(A) = n(U) - n(A')$ and $n(A') = n(U) - n(A).$

We focus here on the form that expresses the following indirect counting principle based on the complement/not/subtraction correspondence from **Table 8**.

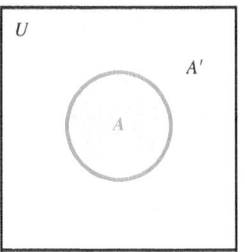

The complement of a set

Figure 8

COMPLEMENTS PRINCIPLE OF COUNTING

The number of ways in which a certain condition can be satisfied is the total number of possible results minus the number of ways the condition would **not** be satisfied. Symbolically, if A is any set within the universal set U, then the following formula holds.

$$n(A) = n(U) - n(A')$$

A **proper subset** of a set S is a subset of S that is not S itself. For example, the proper subsets of $\{a, b\}$ are $\{a\}$, $\{b\}$, and $\varnothing$. This idea is used in the following example.

EXAMPLE 1 Counting the Proper Subsets of a Set

Find the number of proper subsets of $S = \{a, b, c, d, e, f\}$.

Solution

A proper subset of S is any subset with fewer than all six elements. Subsets of several different sizes would satisfy this condition. Consider the one subset that is not proper, namely S itself. From set theory, we know that set S has a total of $2^6 = 64$ subsets. Thus, from the complements principle, the number of proper subsets is

$$64 - 1 = 63.$$

In words, the number of subsets that *are* proper is the total number of subsets minus the number of subsets that are *not* proper.

In **set theory,** we establish that a set with n elements has 2^n subsets.

Consider the tossing of three fair coins. Since each coin will land either heads (h) or tails (t), the possible results can be listed as follows.

hhh, hht, hth, thh, htt, tht, tth, ttt   Possible results of tossing three fair coins

Even without the listing, we could have concluded that there would be eight possibilities. There are two possible outcomes for each coin, so the fundamental counting principle gives $2 \cdot 2 \cdot 2 = 2^3 = 8$.

Three such possibilities

Suppose we wanted the number of ways of obtaining *at least* one head. In this case, "at least one" means one or two or three. Rather than dealing with all three cases, we can note that "at least one" is the opposite (or complement) of "fewer than one", which is zero. Because there is only one way to get zero heads (ttt), and there are a total of eight possibilities, the complements principle gives the number of ways of getting at least one head.

$$8 - 1 = 7 \text{ ways to obtain at least one head}$$

Indirect counting methods can often be applied to problems involving "at least," or "at most," or "less than," or "more than."

EXAMPLE 2 **Counting Coin-Tossing Results**

StatCrunch Coin Flipping

If four fair coins are tossed, in how many ways can at least one tail be obtained?

Solution

By the fundamental counting principle, $2^4 = 16$ different results are possible. Exactly one of these fails to satisfy the condition of "at least one tail," and that is hhhh. So the answer from the complements principle is

$$16 - 1 = 15 \text{ ways.}$$

EXAMPLE 3 **Counting Selections of Airliner Seats**

Carol and three friends are boarding an airliner. There are only ten seats left, three of which are aisle seats. How many ways can the four friends arrange themselves in available seats so that at least one of them sits on the aisle?

Solution

The word "arrange" implies that order is important, so we use permutations. "At least one aisle seat" is the opposite (complement) of "no aisle seats." The total number of ways to arrange four people among ten seats is

$$_{10}P_4 = 5040.$$

The number of ways to arrange four people among seven (non-aisle) seats is

$$_7P_4 = 840.$$

Therefore, by the complements principle, the number of arrangements with at least one aisle seat is found by subtracting.

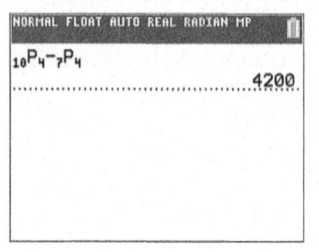

The result in **Example 3** is supported in this screen.

$$\underset{\displaystyle\downarrow}{_{10}P_4} \quad \underset{\displaystyle\downarrow}{_7P_4}$$

$$5040 - 840 = 4200$$

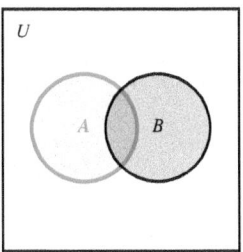

Nondisjoint sets
Figure 9

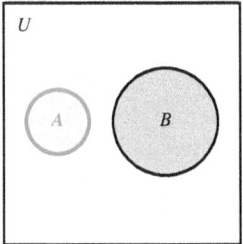

Disjoint sets
Figure 10

Problems Involving "Or"

The complements principle is one way of counting indirectly. Another technique is to count the elements of a set by breaking that set into simpler component parts. If

$$S = A \cup B,$$

the cardinal number formula from set theory says to find the number of elements in S by adding the number in A to the number in B. We must then subtract the number in the intersection $A \cap B$ if A and B are not disjoint, as in **Figure 9.** But if A and B are disjoint, as in **Figure 10,** the subtraction is not necessary.

The following principle reflects the union/or/addition correspondence from **Table 8.** (Recall that $n(A)$ means the number of elements in A.)

ADDITIVE PRINCIPLE OF COUNTING

The number of ways that one **or** the other of two conditions could be satisfied is the number of ways one of them could be satisfied plus the number of ways the other could be satisfied, minus the number of ways they could both be satisfied together.

Symbolically, if A and B are any two sets, then

$$n(A \cup B) = n(A) + n(B) - n(A \cap B).$$

If sets A and B are disjoint, then

$$n(A \cup B) = n(A) + n(B).$$

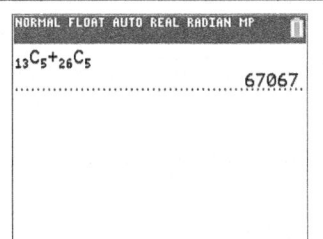

NORMAL FLOAT AUTO REAL RADIAN MP

$_{13}C_5 + _{26}C_5$

67067

The result in **Example 4** is supported in this screen.

StatCrunch Poker Hands

EXAMPLE 4 Counting Card Hands

How many five-card poker hands consist of either all clubs or all red cards?

Solution

No hand that satisfies one of these conditions could also satisfy the other, so the two sets of possibilities (all **clubs,** all **red** cards) are disjoint. Therefore, the second formula of the additive principle applies.

$n(\text{all } \mathbf{clubs} \text{ or all } \mathbf{red} \text{ cards}) = n(\text{all } \mathbf{clubs}) + n(\text{all } \mathbf{red} \text{ cards})$ Additive counting principle

$\qquad\qquad\qquad = _{13}C_5 + _{26}C_5$ 13 clubs, 26 red cards

$\qquad\qquad\qquad = \mathbf{1287} + \mathbf{65{,}780}$ Substitute values.

$\qquad\qquad\qquad = 67{,}067$ Add.

EXAMPLE 5 Counting Selections from a Diplomatic Delegation

Table 9 categorizes a diplomatic delegation of 18 congressional members as to political party and gender. If one of the members is chosen randomly to be spokesperson for the group, in how many ways could that person be a Democrat (D) or a woman (W)?

Table 9

	Men (M)	Women (W)	Totals
Republican (R)	5	3	8
Democrat (D)	4	6	10
Totals	9	9	18

Solution

D and W are not disjoint because 6 delegates are both Democrats and women. The first formula of the additive principle is required.

$$n(D \text{ or } W) = n(D \cup W) \qquad \text{Union/or correspondence}$$

$$= n(D) + n(W) - n(D \cap W) \qquad \text{Additive principle}$$

$$= 10 + 9 - 6 \qquad \text{Substitute values.}$$

$$= 13 \qquad \text{Add and subtract.}$$

EXAMPLE 6 **Counting Course Selections for a Degree Program**

Chrissy needs to take twelve more specific courses for a bachelor's degree, including four in math, three in physics, three in computer science, and two in business. If five courses are randomly chosen from these twelve for next semester's program, how many of the possible selections would include at least two math courses?

Solution

Of all the information given here, what is important is that there are four math courses and eight other courses to choose from, and that five of them are being selected for next semester. If T denotes the set of selections that include at least two math courses, then we can write

$$T = A \cup B \cup C,$$

where $A =$ the set of selections with exactly two math courses

         $B =$ the set of selections with exactly three math courses

         $C =$ the set of selections with exactly four math courses.

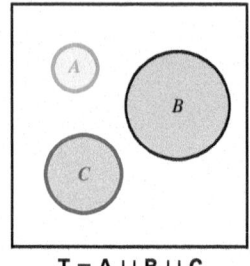

$T = A \cup B \cup C$

Figure 11

In this case, *at least two* means exactly two **or** exactly three **or** exactly four. The situation is illustrated in **Figure 11**. By previous methods, we know that

Two math courses $\longrightarrow n(A) = {}_4C_2 \cdot {}_8C_3 = 6 \cdot 56 = 336$ In each case, $8 = 12 - 4$.

Three math courses $\longrightarrow n(B) = {}_4C_3 \cdot {}_8C_2 = 4 \cdot 28 = 112$ In each case, the sum of the numbers in red is 5.

Four math courses $\longrightarrow n(C) = {}_4C_4 \cdot {}_8C_1 = 1 \cdot 8 = 8.$

By the additive principle,

$$n(T) = 336 + 112 + 8 = 456 \text{ selections.}$$

EXAMPLE 7 **Counting Three-Digit Numbers with Conditions**

How many three-digit counting numbers are multiples of 2 or multiples of 5?

Solution

- A multiple of 2 must end in an even digit $(0, 2, 4, 6, \text{ or } 8)$, so there are $9 \cdot 10 \cdot 5 = 450$ three-digit multiples of 2.

- A multiple of 5 must end in 0 or 5, so there are $9 \cdot 10 \cdot 2 = 180$ of those.

- A multiple of both 2 and 5 is a multiple of 10 and must end in 0. There are $9 \cdot 10 \cdot 1 = 90$ of those.

By the additive principle, there are

$$450 + 180 - 90 = 540 \qquad \text{\textit{Subtract} 90 because those have been counted twice.}$$

possible three-digit numbers that are multiples of 2 or multiples of 5.

StatCrunch Draw Cards

EXAMPLE 8   **Counting Card-Drawing Results**

A single card is drawn from a standard 52-card deck.

(a) In how many ways could it be a heart or a king?

(b) In how many ways could it be a club or a face card?

Solution

(a) A single card can be both a heart and a king (the king of hearts), so use the first additive formula. There are thirteen hearts, four kings, and one card that is both a heart and a king.

$$13 + 4 - 1 = 16 \text{ ways} \quad \text{Additive principle}$$

(b) There are 13 clubs, 12 face cards, and 3 cards that are both clubs and face cards.

$$13 + 12 - 3 = 22 \text{ ways} \quad \text{Additive principle}$$

EXAMPLE 9   **Counting Subsets of a Set with Conditions**

How many subsets of a 25-element set have more than three elements?

Solution

It would be a real job to count directly all subsets of size $4, 5, 6, \ldots, 25$. It is much easier to count those with three or fewer elements and apply the complements principle.

- There is $_{25}C_0 = 1$ size-0 subset.
- There are $_{25}C_1 = 25$ size-1 subsets.
- There are $_{25}C_2 = 300$ size-2 subsets.
- There are $_{25}C_3 = 2300$ size-3 subsets.

Use a calculator to find that the total number of subsets of all sizes, 0 through 25, is $2^{25} = 33{,}554{,}432$. So the number of subsets with more than three elements must be

$$33{,}554{,}432 - \underbrace{(1 + 25 + 300 + 2300)}_{\text{Additive principle}} = 33{,}554{,}432 - 2626 \quad \text{Complements principle}$$
$$= 33{,}551{,}806. \quad \text{Subtract.}$$

10.5   **EXERCISES**

CONCEPT CHECK *Complete each statement with the correct numerical response.*

1. A set with 7 elements has _____ subsets and, consequently, _____ proper subsets.

2. In a standard deck of 52 cards, there are 13 hearts and thus _____ cards that are not hearts.

3. In a standard deck of 52 cards, there are 13 hearts and 4 kings, and thus there are _____ cards that are neither a heart nor a king.

4. When 4 fair coins are tossed, there are _____ ways in which not all heads can result.

Proper Subsets *How many proper subsets are there of each set?*

5. $\{A, B, C, D\}$

6. $\{u, v, w, x, y, z\}$

7. $\{1, 2, 3, 4, 5, 6, 7, 8\}$

8. $\{a, b, c, \ldots, j\}$

Tossing Coins *If you toss seven fair coins, in how many ways can you obtain each result?*

9. at least one head
("At least one" is the complement of "none.")

10. at least two heads
("At least two" is the complement of "zero or one.")

11. at least two tails

12. at least one of each (a head and a tail)

Rolling Dice *If you roll two fair dice (say red and green), in how many ways can you obtain each result? (Refer to* **Table 2** *earlier in this chapter.)*

13. at least 2 on the green die

14. a sum of at least 3

15. a 4 on at least one of the dice

16. a different number on each die

Drawing Cards *If you draw a single card from a standard 52-card deck, in how many ways can you obtain each result?*

17. a card other than the ace of spades

18. a nonface card

19. a card other than a ten

20. a card other than a jack or queen

Identifying Properties of Counting Numbers *How many two-digit counting numbers meet each requirement?*

21. not a multiple of 10

22. not a multiple of 20

23. greater than 70 or a multiple of 10

24. less than 30 or a multiple of 20

Solve each problem.

25. *Choosing Country Music Albums* Jeanne's collection of ten country music albums includes *Everywhere,* by Tim McGraw. She will choose three of her albums to play on a drive to Nashville. (Assume order is not important.)

  (a) How many different sets of three albums could she choose?

  (b) How many of these sets would not include *Everywhere*?

  (c) How many of them would include *Everywhere*?

26. *Choosing Broadway Hits* The ten longest Broadway runs include *The Phantom of the Opera* and *Les Misérables.* Four of the ten are chosen randomly. (Assume order is not important.)

  (a) How many ways can the four be chosen?

  (b) How many of those groups of four would include neither of the two productions mentioned?

  (c) How many of them would include at least one of the two productions mentioned?

27. *Choosing Days of the Week* How many different ways could three distinct days of the week be chosen so that at least one of them begins with the letter S? (Assume order of selection is not important.)

28. *Choosing School Assignments for Completion* Diona has nine major assignments to complete for school this week. Two of them involve writing essays. Diona decides to work on two of the nine assignments tonight. How many different choices of two would include at least one essay assignment? (Assume order is not important.)

29. *Selecting Restaurants* Jason wants to dine at four different restaurants during a summer getaway. If three of eight available restaurants serve seafood, find the number of ways that at least one of the selected restaurants will serve seafood given the following conditions.

  (a) The order of selection is important.

  (b) The order of selection is not important.

30. *Groups of Lawyers* Three lawyers are to be selected from a group of 25 to work on a special project. In how many ways can the group be selected if one particular lawyer has been preselected to work on the project

31. *Seating Arrangements on an Airliner* Refer to **Example 3.** If one of the group decided at the last minute not to fly, then how many ways could the remaining three arrange themselves among the ten available seats so that at least one of them would sit on the aisle?

32. *Identifying Properties of Counting Numbers* Find the number of four-digit counting numbers containing at least one zero, under each of the following conditions.

  (a) Repeated digits are allowed.

  (b) Repeated digits are not allowed.

33. *Counting Radio Call Letters* Radio stations in the United States have call letters that begin with either K or W. Some have a total of three letters, and others have four letters. How many different call letter combinations are possible? Count all possibilities even though, practically, some may be inappropriate. (*MT* calendar problem) (*Hint:* Do *not* apply combinations.)

34. *Selecting Faculty Committees* A committee of four faculty members will be selected from a department of twenty-five, which includes professors Fontana and Spradley. In how many ways could the committee include at least one of these two professors?

35. *Selecting Search-and-Rescue Teams* A Civil Air Patrol unit of twelve members includes four officers. In how many ways can four members be selected for a search-and-rescue mission such that at least one officer is included?

36. *Choosing Team Members* Three students from a class of 12 will form a math contest team that must include at least 1 boy and at least 1 girl. If 160 different teams can be formed from the 12 students, which of the following can be the difference between the number of boys and the number of girls in the class?

A. 0   **B.** 2   **C.** 4   **D.** 6   **E.** 8

(*MT* calendar problem)

Drawing Cards *If a single card is drawn from a standard 52-card deck, in how many ways could it be the following? (Use the additive principle.)*

37. a club or a jack

38. a heart or a ten

39. a face card or a black card

40. an ace or a red card

Choosing Senators *The table categorizes 20 senators as to political party and gender. One member is chosen at random. In how many ways can the chosen person be as described?*

	Men (*M*)	Women (*W*)	Totals
Democrat (*D*)	8	4	12
Republican (*R*)	3	5	8
Totals	11	9	20

41. a woman or a Republican

42. a man or a Democrat

43. a man or a Republican woman

44. a woman or a Democrat man

Counting Card Hands *Among the 2,598,960 possible 5-card poker hands from a standard 52-card deck, how many contain the following cards?*

45. at least one card that is not a heart (complement of "all hearts")

46. cards of more than one suit (complement of "all the same suit")

47. at least one face card (complement of "no face cards")

48. at least one club, but not all clubs (complement of "no clubs or all clubs")

The Size of Subsets of a Set *If a set has ten elements, how many of its subsets have the given numbers of elements?*

49. at most two elements

50. at least eight elements

51. more than two elements

52. from three through seven elements

53. *Selecting Doughnuts* A doughnut shop has a special on its Mix-n-Match selection, which allows customers to select three doughnuts from among the following varieties: plain, maple, frosted, chocolate, glazed, and jelly. How many different Mix-n-Match selections are possible? (*MT* calendar problem)

54. *Rolling Three Dice* Three fair, standard six-faced dice of different colors are rolled. In how many ways can the dice be rolled such that the sum of the numbers rolled is 10? (*MT* calendar problem)

55. *Counting License Numbers* If license numbers consist of two letters followed by three digits, how many different licenses could be created having at least one letter or digit repeated? (*Hint:* Use the complements principle of counting.)

56. *Counting License Numbers* Transylvanian license plates consist of exactly three letters. Two license plates are considered identical if and only if they contain the same three letters in the same order. How many Transylvanian license plates are possible if the letter Q must be followed directly by the letter U? (*MT* calendar problem)

57. *Drawing Cards* If two cards are drawn from a 52-card deck without replacement (that is, the first card is not replaced in the deck before the second card is drawn), in how many different ways is it possible to obtain a king on the first draw and a heart on the second? (*Hint:* Split this event into the two disjoint components "king of hearts and then another heart" and "non-heart king and then heart." Use the fundamental counting principle on each component and then apply the additive principle.)

58. How many of the counting numbers 1 through 300 are *not* divisible by 2, 3, or 5?

Selecting National Monuments and Parks to Visit Marge, Terry, Gary, and Johnny are planning a driving tour. Although they are interested in seeing the twelve national monuments and parks listed here, they will have to settle for seeing just three of them.

New Mexico	Arizona	California
Gila Cliff Dwellings	Canyon de Chelly	Devils Postpile
Petroglyph	Organ Pipe Cactus	Joshua Tree
White Sands	Grand Canyon	Lava Beds
Aztec Ruins		Sequoia
		Yosemite

In how many ways could the three sites chosen include the following? (Assume that order of selection is not important.)

59. sites in only one state

60. at least one site not in California

61. sites in fewer than all three states

62. sites in exactly two of the three states

Counting Categories of Poker Hands **Table 4** *in this chapter described the various kinds of hands in 5-card poker. Verify each statement in Exercises 63–66. (Explain all steps of your argument.)*

63. There are four ways to get a royal flush.

64. There are 36 ways to get a straight flush.

65. There are 10,200 ways to get a straight.

66. There are 54,912 ways to get three of a kind.

Chapter **10** SUMMARY

KEY TERMS

10.1
counting
one-part task
two-part task
product table

10.2
uniformity criterion
fundamental counting
   principle
n factorial
distinguishable arrangements

10.3
permutation
combination
set
elements (members)
subset

10.4
Pascal's triangle
binomial coefficients
binomial theorem

10.5
proper subset

NEW SYMBOLS

$n!$ *n* factorial

$_nP_r$, $P(n, r)$, P_r^n permutations of *n* things taken *r* at a time

$_nC_r$, $C(n, r)$, C_r^n combinations of *n* things taken *r* at a time

$\binom{n}{r}$ binomial coefficient

TEST YOUR WORD POWER

See how well you have learned the vocabulary of this chapter.

1. In counting, a **product table** can be used to analyze
  A. a one-part task.
  B. a two-part task.
  C. a problem involving distinguishable arrangements.
  D. a problem involving Pascal's triangle.

2. The **factorial** of a natural number *n* is found by
  A. multiplying all the natural numbers less than or equal to *n*.
  B. adding all the natural numbers less than or equal to *n*.
  C. multiplying *n* by *n* + 1.
  D. multiplying *n* by *n* − 1.

3. A **permutation** of a group of objects
  A. is a subset of those objects where order is not important.
  B. is a proper subset of those objects.
  C. is an arrangement of those objects where order is important.
  D. is exemplified by a hand of cards in poker.

4. A **combination** of a group of objects
  A. is a subset of those objects.
  B. is exactly the same as a permutation of those objects.

  C. is an arrangement of those objects where order is important.
  D. is exemplified by the process of opening a combination lock.

ANSWERS
1. B **2.** A **3.** C **4.** A

QUICK REVIEW

Concepts   *Examples*

10.1 COUNTING BY SYSTEMATIC LISTING

Systematic Listing Methods
Two methods for listing the possible results of a two-part task are

(1) constructing a product table, and

(2) constructing a tree diagram.

Tree diagrams can be extended to tasks with three or more parts as well.

Construct a tree diagram showing all possible results when a die is rolled and a coin is tossed. How many possibilities are there?

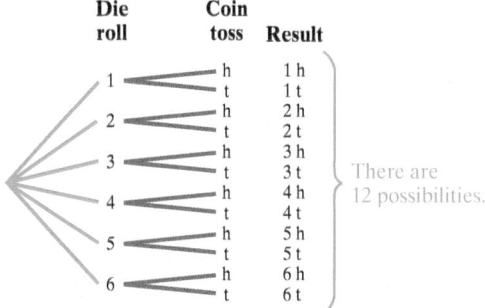

List the ways in which the die roll is even and the coin toss is heads.

2h, 4h, 6h

10.2 USING THE FUNDAMENTAL COUNTING PRINCIPLE

Uniformity Criterion
A multiple-part task is said to satisfy the uniformity criterion if the number of choices for any particular part is the same *no matter which choices were selected for previous parts.*

If each of two adjacent switches are to be set on or off, uniformity applies if there are no restrictions. However, it does not apply if, say, they must not both be off.

Fundamental Counting Principle
When a task consists of k separate parts and satisfies the uniformity criterion, if the first part can be done in n_1 ways, the second part can then be done in n_2 ways, and so on through the kth part, which can be done in n_k ways, then the total number of ways to complete the task is given by the following product.

$$n_1 \cdot n_2 \cdot n_3 \cdots \cdots n_k$$

How many ways are there to choose a meal from 3 salads, 6 entrees, and 4 desserts?

$$3 \cdot 6 \cdot 4 = 72 \text{ ways}$$

Factorial Formula
For any counting number n, the quantity n **factorial** is defined as follows.

$$n! = n \cdot (n-1) \cdot (n-2) \cdots \cdots 2 \cdot 1$$

By definition, **0! = 1.**

Evaluate each factorial.

$$5! = 5 \cdot 4 \cdot 3 \cdot 2 \cdot 1 = 120$$

$$3! \cdot 4 = (3 \cdot 2 \cdot 1) \cdot 4 = 24$$

$$\frac{8!}{6!} = \frac{8 \cdot 7 \cdot 6 \cdot 5 \cdot 4 \cdot 3 \cdot 2 \cdot 1}{6 \cdot 5 \cdot 4 \cdot 3 \cdot 2 \cdot 1} = 8 \cdot 7 = 56$$

Concepts	*Examples*
Arrangements of *n* Distinct Objects The total number of different ways to arrange *n* distinct objects is $$n!.$$	How many ways are there to arrange the letters of ROBIN? There are 5 distinct letters, so there are $$5! = 120 \text{ ways.}$$
Arrangements of *n* Objects Containing Look-Alikes The number of distinguishable arrangements of *n* objects, where one or more subsets consist of look-alikes (say n_1 are of one kind, n_2 are of another kind, ..., and n_k are of yet another kind), is given by $$\frac{n!}{n_1! \cdot n_2! \cdots \cdot n_k!}.$$	How many distinguishable arrangements are there of the letters of MAMMOGRAM? Of the 9 letters total, 4 are M and 2 are A. So there are $$\frac{9!}{4! \cdot 2!} = 7560 \text{ distinguishable arrangements.}$$

10.3 USING PERMUTATIONS AND COMBINATIONS

Factorial Formula for Permutations The number of permutations, or arrangements, of *n* distinct things taken *r* at a time, where $r \le n$, can be calculated as follows. $$_nP_r = \frac{n!}{(n-r)!}$$	How many arrangements of the letters $$W, H, I, T, E, S, U, G, A, R$$ are there if they are taken 4 at a time? There are 10 different letters. We use the permutations formula because order matters. $$_{10}P_4 = \frac{10!}{(10-4)!} = \frac{10!}{6!} = 5040 \text{ arrangements}$$
Factorial Formula for Combinations The number of combinations, or subsets, of *n* distinct things taken *r* at a time, where $r \le n$, can be calculated as follows. $$_nC_r = \frac{_nP_r}{r!} = \frac{n!}{r!(n-r)!}$$	How many committees of 3 politicians can be formed from a pool of 11 politicians? We use the combinations formula because order does not matter. $$_{11}C_3 = \frac{11!}{3!(11-3)!} = \frac{11!}{3! \cdot 8!} = 165 \text{ committees}$$

10.4 USING PASCAL'S TRIANGLE

Pascal's Triangle

```
           1
         1   1
       1   2   1
     1   3   3   1
   1   4   6   4   1
 1   5  10  10   5   1     (6 + 4 = 10)
1   6  15  20  15   6   1
```

Binomial Coefficient

$$\binom{n}{r} = \frac{n!}{r!(n-r)!}$$

If six fair coins are tossed, in how many different ways can exactly two heads be obtained?
The entry in green in the triangle represents

$$_6C_2.$$

There are 15 ways.

(The triangle entries can be expressed in terms of $_nC_r$. See **Table 6**.)

$$\binom{6}{4} = \frac{6!}{4!(6-4)!} = 15 \text{ ways}$$

Concepts	Examples
Binomial Theorem	Expand $(3x + 2)^3$.
For any whole number n,	

$$(x + y)^n$$
$$= \binom{n}{0} \cdot x^n + \binom{n}{1} \cdot x^{n-1}y$$
$$+ \binom{n}{2} \cdot x^{n-2}y^2 + \binom{n}{3} \cdot x^{n-3}y^3 +$$
$$\cdots + \binom{n}{n-1} \cdot xy^{n-1} + \binom{n}{n} \cdot y^n.$$

$$(3x + 2)^3 = 1(3x)^3 + 3(3x)^2(2) + 3(3x)(2)^2 + 1(2)^3$$
$$= 27x^3 + 54x^2 + 36x + 8$$

10.5 COUNTING PROBLEMS INVOLVING "NOT" AND "OR"

Complements Principle of Counting
The number of ways a certain condition can be satisfied is the total number of possible results minus the number of ways the condition would **not** be satisfied.

Symbolically, if A is any set within the universal set U, then

$$n(A) = n(U) - n(A').$$

If six fair coins are tossed, in how many ways can at least one head be obtained?
There are $2^6 = 64$ total ways, and only one of them (t t t t t t) does not contain a head. So there are

$$64 - 1 = 63 \text{ ways}$$

that at least one head can be obtained.

Additive Principle of Counting
The number of ways that one **or** the other of two conditions could be satisfied is the number of ways one of them could be satisfied plus the number of ways the other could be satisfied minus the number of ways they could both be satisfied together.

Symbolically, if A and B are any two sets, then

$$n(A \cup B) = n(A) + n(B) - n(A \cap B).$$

If sets A and B are disjoint, then

$$n(A \cup B) = n(A) + n(B).$$

A single card is drawn from a standard 52-card deck. In how many ways could it be black or a face card?
There are 26 black cards, 12 face cards, and 6 black face cards. So there are

$$26 + 12 - 6 = 32 \text{ ways.}$$

Chapter **10** TEST

Counting Three-digit Numbers *If only the digits 0, 1, 2, 3, 4, 5, and 6 may be used, find the number of possibilities in each category.*

1. three-digit numbers

2. odd three-digit numbers

3. three-digit numbers without repeated digits

4. three-digit multiples of five without repeated digits

Evaluate each expression.

5. $6!$ **6.** $\dfrac{8!}{6!}$ **7.** $_{12}P_3$ **8.** $_8C_5$

Solve each problem.

9. Counting Triangles in a Figure Determine the number of triangles (of any size) in the figure shown here.

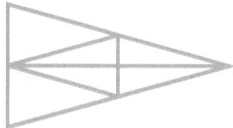

10. Tossing Coins Construct a tree diagram showing all possible results when a fair coin is tossed four times, if no two consecutive tosses can both be heads.

11. **Sums of Digits** How many nonrepeating four-digit numbers have the sum of their digits equal to 30?

12. **Arranging Children** Yeo has invited her sister Hae and her four best friends to her birthday party. In how many ways can the six children be arranged around a rectangular table with one child at each end and two on each side if Yeo must sit at an end seat and Hae must not sit next to Yeo?

13. **Building Words from Sets of Letters** How many five-letter "words" without repeated letters are possible using the English alphabet? (Assume that any five letters make a "word.")

14. **Building Words from Sets of Letters** Using the Russian alphabet (which has 32 letters), and allowing repeated letters, how many five-letter "words" are possible?

Scheduling Assignments *Eileen has seven homework assignments to complete. She wants to do two of them on Thursday and the other five on Saturday.*

15. In how many ways can she order Thursday's work?

16. Assuming she finishes Thursday's work successfully, in how many ways can she order Saturday's work?

Selecting Groups of Basketball Players *If there are ten players on a basketball team, find the number of choices the coach has in selecting each of the following.*

17. four players to carry the team equipment

18. two players for guard positions and two for forward positions

19. five starters and five subs

20. two groups of four

21. a group of three or more of the players

Choosing Switch Settings *Determine the number of possible settings for a row of five on–off switches under each condition.*

22. There are no restrictions.

23. The first and fifth switches must be on.

24. The first and fifth switches must be set the same.

25. No two adjacent switches can both be off.

26. No two adjacent switches can be set the same.

27. At least two switches must be on.

Choosing Subsets of Letters *In Exercises 28–31, three distinct letters are to be chosen from the set*

$$\{A, B, C, D, E, F, G\}.$$

Determine the number of ways to obtain a subset that includes each of the following.

28. the letter B

29. both A and E

30. either A or E, but not both

31. more consonants than vowels

32. Expand the binomial: $(x + 2)^4$

Solve each problem.

33. **Arranging Letters** Find the number of distinguishable arrangements of the letters of the word PIPPIN.

34. **Assigning Student Grades** A professor teaches a class of 60 students and another class of 40 students. Five percent of the students in each class are to receive a grade of A. How many different ways can the A grades be distributed?

35. **Number of Paths from Point to Point** A transit bus can travel in only two directions, north and east. From its starting point on the map shown, determine how many paths exist to reach the garage. (*MT* calendar problem)

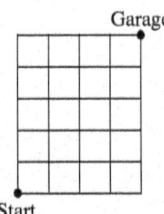

Garage

Start

36. **Pascal's Triangle** Write down the second entry of each row of Pascal's triangle (starting with row 1). What sequence of numbers do you obtain?

11 Probability

The mathematics of *probability* occurs in various occupations and experiences in daily life. An important feature distinguishes probability from other topics in this text. For example, an equation such as $4x + 6 = 14$ has a specific solution (in this case, 2) that we can state without reservation. Results found in probability are often theoretical—that is, we cannot say for sure what will actually happen in a certain case. We can only say what trends will occur "in the long run."

The many places in everyday life where probabilities or odds occur include meteorology and gaming. Meteorologists state the chances of rain based on data collected on days where the atmospheric conditions have been the same in the past. If the chance of rain is 90%, it's a good bet that rain will occur, but there is still a 10% chance that no rain will fall. Games of chance are designed so that in the long run, "the house wins." Once in a while, the player wins, but over a lifetime, the average gambler will likely lose.

11.1   BASIC CONCEPTS

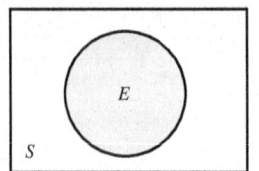

Every event *E* is a subset of the sample space *S*.

Figure 1

The Language of Probability

If you go to a supermarket and select five pounds of peaches at $2.29 per pound, you can easily predict the amount you will be charged at the checkout counter.

$$5 \cdot \$2.29 = \$11.45$$

This is an example of a **deterministic phenomenon.** It can be predicted exactly on the basis of obtainable information—in this case, number of pounds and cost per pound.

On the other hand, consider the problem faced by the produce manager of the market, who must order peaches to have on hand each day without knowing exactly how many pounds customers will buy during the day. Customer demand is an example of a **random phenomenon.** It fluctuates in such a way that its value on a given day cannot be predicted exactly with obtainable information.

The study of probability is concerned with such random phenomena. Even though we cannot be certain whether a given result will occur, we often can obtain a good measure of its *likelihood,* or **probability.** This chapter discusses various ways of finding and using probabilities.

Any observation, or measurement, of a random phenomenon is an **experiment.** The possible results of the experiment are **outcomes,** and the set of all possible outcomes is the **sample space.**

Usually we are interested in some particular collection of the possible outcomes. Any such subset of the sample space is an **event.** See the Venn Diagram in **Figure 1.** Outcomes that belong to the event are *favorable outcomes,* or *successes.* Any time a success is observed, we say that the event has *occurred.* The probability of an event, being a numerical measure of the event's likelihood, is determined in one of two ways: either *theoretically* (mathematically) or *empirically* (experimentally). We use the notation

$$P(E) \text{ to represent the } \textbf{probability of event } E.$$

StatCrunch   Simulation: Coin Flipping

Examples in Probability

EXAMPLE 1   Finding Probability When Tossing a Coin

If a single coin is tossed, find the probability that it will land heads up.

Solution

There is no apparent reason for one side of a coin to land up any more often than the other (in the long run), so we assume that heads and tails are equally likely. The experiment here is the tossing of a single fair coin, the sample space is

$$S = \{h, t\},$$

and the event whose probability we seek is $E = \{h\}$.

Because one of the two equally likely outcomes is a head, the probability of heads is the quotient of 1 and 2.

$$\text{Probability (heads)} = \frac{1}{2}, \quad \text{written} \quad P(h) = \frac{1}{2} \quad \text{or} \quad P(E) = \frac{1}{2}.$$

EXAMPLE 2 Finding Probability When Tossing a Cup

If a plastic cup is tossed, find the probability that it will land on its top.

Solution

Intuitively, it seems that such a cup will land on its side much more often than on its top or its bottom. But just how much more often is not clear. To get an idea, we performed the experiment of tossing such a cup 50 times. It landed on its side 44 times, on its top 5 times, and on its bottom just 1 time. By the frequency of "success" in this experiment, we concluded that for the cup we used,

$$P(\text{top}) \approx \frac{5}{50} = \frac{1}{10}. \quad \text{— Write in lowest terms.}$$

In **Example 1** involving the tossing of a fair coin, the number of possible outcomes was obviously two, both were equally likely, and one of the outcomes was a head. No actual experiment was required. The desired probability was obtained *theoretically.* Theoretical probabilities apply to dice rolling, card games, roulette, lotteries, and so on, and apparently to many phenomena in nature.

Laplace, in his famous *Analytic Theory of Probability,* published in 1812, gave a formula that applies to any such theoretical probability, as long as the sample space *S* is finite and all outcomes are equally likely. This is the **classical definition of probability.**

THEORETICAL PROBABILITY FORMULA

If all outcomes in a sample space *S* are equally likely, and *E* is an event within that sample space, then the **theoretical probability** of event *E* is given by the following formula.

$$P(E) = \frac{\text{number of favorable outcomes}}{\text{total number of outcomes}} = \frac{n(E)}{n(S)}$$

Example 2 involved the tossing of a cup where the likelihoods of the various outcomes were not intuitively clear. It took an actual experiment to arrive at a probability value of $\frac{1}{10}$, and that value, based on a portion of all possible tosses of the cup, should be regarded as an approximation of the true theoretical probability. The value was found according to the **experimental,** or **empirical, probability** formula.

EMPIRICAL PROBABILITY FORMULA

If *E* is an event that may happen when an experiment is performed, then an **empirical probability** of event *E* is given by the following formula.

$$P(E) = \frac{\text{number of times event } E \text{ occurred}}{\text{number of times the experiment was performed}}$$

Usually it is clear in applications which probability formula should be used.

EXAMPLE 3 Finding the Probability of Having Daughters

Kathy wants to have exactly two daughters. Assuming that boy and girl babies are equally likely, find her probability of success for the following cases.

(a) She has a total of two children.

(b) She has a total of three children.

Solution

(a) The equal-likelihood assumption allows the use of theoretical probability. We can determine the number of favorable outcomes and the total number of possible outcomes by using a tree diagram to enumerate the equally likely possibilities, as shown in **Figure 2**. From the outcome column we obtain the sample space $S = \{gg, gb, bg, bb\}$. Only one outcome, marked with an arrow, is favorable to the event of exactly two daughters: $E = \{gg\}$.

$$P(E) = \frac{n(E)}{n(S)} = \frac{1}{4} \quad \text{Theoretical probability formula}$$

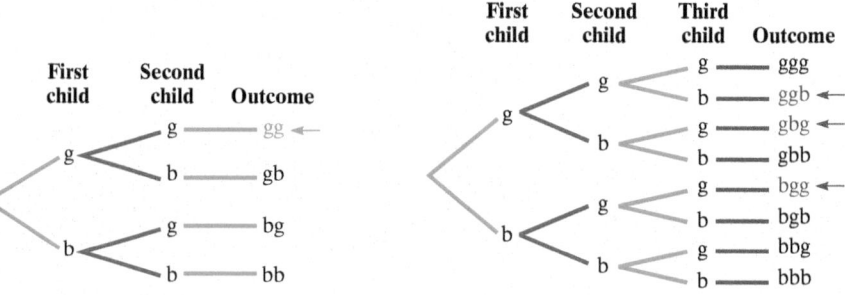

Exactly two girls among two children

Figure 2

Exactly two girls among three children

Figure 3

(b) For three children altogether, we construct another tree diagram, as shown in **Figure 3**. In this case, we see that

$$S = \{ggg, ggb, gbg, gbb, bgg, bgb, bbg, bbb\} \quad \text{and} \quad E = \{ggb, gbg, bgg\},$$

so
$$P(E) = \frac{3}{8}$$

because S has 8 elements and E has 3 elements.

When dealing or drawing cards, as in the next example, the dealing is generally done "without replacement." Once dealt, a card is *not* replaced in the deck. So all cards in a hand are distinct. Repetitions are *not* allowed. In many cases, such as building three-digit numbers, repetition of digits *is* allowed. For example, 255 is a legitimate three-digit number. So digit selection is done "with replacement."

EXAMPLE 4 Finding Probability When Dealing Cards

Find the probability of being dealt each of the following hands in five-card poker. Use a calculator to obtain answers to eight decimal places.

(a) a full house (three of one denomination and two of another)

(b) a royal flush (the five highest cards—ace, king, queen, jack, ten—of a single suit)

Table 1 Number of Poker Hands in 5-Card Poker; Nothing Wild

Event *E*	Number of Outcomes Favorable to *E*
Royal flush	4
Straight flush	36
Four of a kind	624
Full house	3744
Flush	5108
Straight	10,200
Three of a kind	54,912
Two pairs	123,552
One pair	1,098,240
No pair	1,302,540
Total	2,598,960

Solution

(a) **Table 1** summarizes the various possible kinds of five-card hands. Because the 2,598,960 possible individual hands all are equally likely, we can enter the appropriate numbers from the table into the theoretical probability formula.

$$P(\text{full house}) = \frac{3744}{2,598,960} = \frac{6}{4165} \approx 0.00144058$$

(b) The table shows that there are four royal flush hands, one for each suit.

$$P(\text{royal flush}) = \frac{4}{2,598,960} = \frac{1}{649,740} \approx 0.00000154$$

Examples 3 and 4 both utilized the theoretical probability formula because we were able to enumerate all possible outcomes, and all were equally likely. In **Example 3,** however, the equal likelihood of girl and boy babies was *assumed*. While male births typically occur a little more frequently, there usually are more females living at any given time, due to higher infant mortality rates among males and longer female life expectancy in general. **Example 5** shows a way of incorporating such empirical information.

EXAMPLE 5   **Finding the Probability of the Gender of a Resident**

According to the United States Census figures, on July 14, 2018, there were an estimated 162.4 million males and 166.9 million females. If a person were selected randomly from the population in that year, what is the probability that the person would be a male? (Data from www.census.gov)

Solution

In this case, we calculate the empirical probability from the given experimental data.

$$P(\text{male}) = \frac{\text{number of males}}{\text{total number of persons}}$$

$$= \frac{162.4 \text{ million}}{162.4 \text{ million} + 166.9 \text{ million}}$$

$$\approx 0.493$$

The Law of Large Numbers (or Law of Averages)

Recall the cup of **Example 2.** If we tossed it 50 more times, we would have 100 total tosses on which to base an empirical probability of the cup landing on its top. The new value would likely be slightly different from what we obtained before. It would still be an empirical probability, but it would be "better" in the sense that it was based on a larger set of outcomes.

If, as we increase the number of tosses, the resulting empirical probability values approach some particular number, that number can be defined as the theoretical probability of that particular cup landing on its top. We could determine this "limiting" value only as the actual number of observed tosses approached the total number of possible tosses of the cup. There are potentially an infinite number of possible tosses, so we could never actually find the theoretical probability. But we can still assume such a number exists. And as the number of actual observed tosses increases, the resulting empirical probabilities should tend ever closer to the theoretical value.

This principle is known as the **law of large numbers** or the **law of averages.**

The **law of large numbers** (or **law of averages**) also can be stated as follows.

A theoretical probability really says nothing about one, or even a few, repetitions of an experiment, but only about the proportion of successes we would expect over the long run.

Gregor Johann Mendel (1822–1884) came from a peasant family who managed to send him to school. By 1847 he had been ordained and was teaching at the Abbey of St. Thomas. He finished his education at the University of Vienna and returned to the abbey to teach mathematics and natural science.

Mendel began to carry out experiments on plants in the abbey garden, notably pea plants, whose distinct traits (unit characters) he had puzzled over. In 1865 he published his results. His work was not appreciated at the time, even though he had laid the foundation of **classical genetics.**

LAW OF LARGE NUMBERS (LAW OF AVERAGES)

As an experiment is repeated more and more times, the proportion of outcomes favorable to any particular event will tend to come closer and closer to the theoretical probability of that event.

EXAMPLE 6 Graphing a Sequence of Proportions

A fair coin was tossed 35 times, producing the following sequence of outcomes.

tthhh, ttthh, hthtt, hhthh, ttthh, thttt, hhthh

Calculate the ratio of heads to total tosses after the first toss, after the second toss, and so on through all 35 tosses, and plot these ratios on a graph.

Solution

After the first toss, we have 0 heads out of 1 toss, for a ratio of $\frac{0}{1} = 0.00$. After two tosses, we have $\frac{0}{2} = 0.00$. After three tosses, we have $\frac{1}{3} \approx 0.33$. Verify that the first six ratios are

0.00, 0.00, 0.33, 0.50, 0.60, 0.50.

The 35 ratios are plotted in **Figure 4.** The fluctuations away from 0.50 become smaller as the number of tosses increases, and the ratios appear to approach 0.50 toward the right side of the graph, in keeping with the law of large numbers.

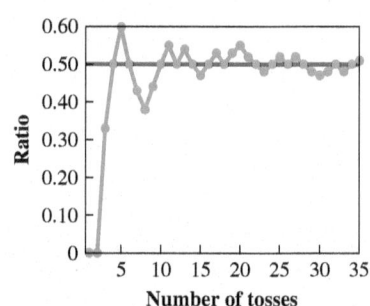

Ratio of heads to total tosses
Figure 4

COMPARING EMPIRICAL AND THEORETICAL PROBABILITIES

A series of repeated experiments provides an **empirical probability** for an event, which, by *inductive reasoning,* is an *estimate* of the event's *theoretical probability.* Increasing the number of repetitions increases the reliability of the estimate.

Likewise, an established **theoretical probability** for an event enables us, by *deductive reasoning,* to *predict* the proportion of times the event will occur in a series of repeated experiments. The prediction should be more accurate for larger numbers of repetitions.

Probability in Genetics

Probabilities, both empirical and theoretical, have been valuable tools in many areas of science. An important early example was the work of the Austrian monk Gregor Mendel, who used the idea of randomness to help establish the study of genetics.

In an effort to understand the mechanism of character transmittal from one generation to the next in plants, Mendel counted the number of occurrences of various characteristics. He found that the flower color in certain pea plants obeyed this scheme:

Pure red crossed with pure white produces red.

Mendel theorized that red is "dominant," symbolized in this explanation with the capital letter R, while white is "recessive," symbolized with the lowercase letter r. The pure red parent carried only genes for red (R), and the pure white parent carried only genes for white (r).

The offspring would receive one gene from each parent, hence one of the four combinations shown in the body of **Table 2**. Because every offspring receives one gene for red, that characteristic dominates, and each offspring exhibits the color red.

Table 2 First to Second Generation

		Second Parent	
		r	r
First Parent	R	Rr	Rr
	R	Rr	Rr

Table 3 Second to Third Generation

		Second Parent	
		R	r
First Parent	R	RR	Rr
	r	rR	rr

Now each of these second-generation offspring, though exhibiting the color red, still carries one of each gene. So when two of them are crossed, each third-generation offspring will receive one of the gene combinations shown in **Table 3.** Mendel theorized that each of the four possibilities would be equally likely, and he produced experimental counts that were close enough to support this hypothesis. (There has been some skepticism about his results.)

Super Bowl Coin Tosses During the pregame coin flip of Super Bowl XLIV in 2010, the New Orleans Saints of the National Football Conference (NFC) called "heads" and won the toss. The probability of this, of course, was $\frac{1}{2}$. For the twelve previous Super Bowls, the NFC representative in the Super Bowl had won the toss over the American Football Conference (AFC). The Saints continued this streak, and thus the NFC had won the toss for thirteen straight years.

The television announcer indicated that the odds against such a run of thirteen straight wins were

"about 8100 to 1."

Someone had done the mathematics ahead of time and evaluated

$\left(\frac{1}{2}\right)^{13}$ to obtain $\frac{1}{8192}$.

The actual theoretical odds are 8191 to 1 against a run of thirteen straight wins by one conference.

EXAMPLE 7 Finding Probabilities of Flower Colors

Referring to **Table 3,** determine the probability that a third-generation offspring will exhibit each flower color. Base the probabilities on the sample space of equally likely outcomes.

$$S = \{RR, Rr, rR, rr\}$$

(a) red   **(b)** white

Solution

(a) Since red dominates white, any combination with at least one gene for red (R) will result in red flowers. Since three of the four possibilities meet this criterion,

$$P(\text{red}) = \frac{3}{4}.$$

(b) Only the combination rr has no gene for red, so

$$P(\text{white}) = \frac{1}{4}.$$

Odds

Whereas probability compares the number of favorable outcomes to the total number of outcomes, **odds** compare the number of favorable outcomes to the number of unfavorable outcomes. Odds are commonly quoted, rather than probabilities, in horse racing, lotteries, and most other gambling situations. And the odds quoted normally are odds "against" rather than odds "in favor."

ODDS

If all outcomes in a sample space are equally likely, and if a of them are favorable to the event E, and the remaining b outcomes are unfavorable to E, then the

odds in favor of E are a to b,   and   the **odds against E** are b to a.

EXAMPLE 8 Finding the Odds of Getting an Intern Position

Theresa has been promised one of six jobs, three of which would be intern positions at the state capitol. If she has equal chances for all six jobs, find the odds *in favor of* her getting one of the intern positions.

Solution

Since three possibilities are favorable and three are not, the odds of *becoming an intern* at the capitol are 3 to 3 (or 1 to 1 in reduced terms). Odds of 1 to 1 are often termed

"even odds," or a "50–50 chance."

EXAMPLE 9 Finding the Odds of Winning a Raffle

Bob has purchased 12 tickets for an office raffle in which the winner will receive an iPad. If 104 tickets were sold altogether and each has an equal chance of winning, what are the odds *against* Bob's winning the iPad?

Solution

Bob has 12 chances to win and $104 - 12 = 92$ chances to lose, so the odds *against* winning are

92 to 12.

In practice, we often use the fact that each number can be divided by their greatest common factor, to express the odds in lower terms. Dividing each of 92 and 12 by 4 gives the equivalent odds of

23 to 3.   Lowest terms

CONVERTING BETWEEN PROBABILITY AND ODDS

Let E be an event.

- If $P(E) = \frac{a}{b}$, then the odds in favor of E are a **to** $(b - a)$.
- If the odds in favor of E are a to b, then $P(E) = \frac{a}{a + b}$.

EXAMPLE 10 Converting from Probability to Odds

There is a 30% chance of rain tomorrow. What are the odds *in favor of* rain? What are the odds *against* rain?

Solution

$$P(\text{rain}) = 0.30 = \frac{30}{100} = \frac{3}{10}$$

> Convert the decimal fraction to a quotient of integers and reduce.

By the first conversion formula, if $P(E) = \frac{a}{b}$, then the odds *in favor of* rain tomorrow are a to $(b - a)$—that is,

3 to $(10 - 3)$

or   3 to 7.   Odds in favor of rain

We can also say that the odds *against* rain tomorrow are

7 to $(10 - 7)$

or   7 to 3.   Odds against rain

EXAMPLE 11 Converting from Odds to Probability

In a certain sweepstakes, your odds of winning are 1 to 99,999. What is the probability that you will win?

Solution

Use the second conversion formula $P(E) = \frac{a}{a+b}$.

$$P(\text{win}) = \frac{1}{1 + 99{,}999} = \frac{1}{100{,}000} = 0.00001$$

WHEN WILL I **EVER** USE THIS**?**

The traditional coin flip is a tried-and-true method of making a decision when only two people (players) are involved. Each outcome of heads and tails has a probability of $\frac{1}{2}$. We say that "the chances are 50–50," meaning that each player has a 50% chance of winning. Suppose that we must determine a winner among *three* players.

Can we use coin flipping in such a way that each of the three players has an equal probability of winning?

The answer is yes. Using a method called "odd man wins," each player flips a coin, and then the results are examined. If all three players show heads or all show tails, the experiment is repeated. (This eliminates the outcomes hhh and ttt in our sample space.) There are 6 other possible outcomes, and each one shows 1 tail or 1 head. If a player has this "odd" result, that player is the winner. The tree diagram here shows that each player has a $\frac{2}{6}$, or $\frac{1}{3}$, probability of winning.

Player A result	Player B result	Player C result	Outcome	Winner
h	h	h	hhh	—
		t	hht	C
	t	h	hth	B
		t	htt	A
t	h	h	thh	A
		t	tht	B
	t	h	tth	C
		t	ttt	—

Of the 6 equally likely outcomes considered, each player wins 2 times.

Probabilities of Independent Events: Tree Diagram

11.1   **EXERCISES**

You can use StatCrunch *(available in MyLab Math) to solve exercises marked with the* StatCrunch.

CONCEPT CHECK *Give the correct numerical response.*

1. The probability *as a common fraction* that the result of a single fair die toss is a 4 is _____.

2. The probability *as a common fraction* that the king of spades is drawn from a fair standard deck of cards is _____.

3. The probability *as a common fraction* that a toss of a single fair coin results in tails is _____.

4. See **Exercise 1.** The odds *in favor of* obtaining a 4 when a single fair die is tossed are _____.

5. See **Exercise 1.** The odds *against* obtaining a 4 when a single fair die is tossed are _____.

6. See **Exercise 3.** The probability *as a percent* that a toss of a single fair coin results in tails is _____.

Give the probability that the spinner shown would land on **(a)** *red,* **(b)** *yellow, and* **(c)** *blue.*

7.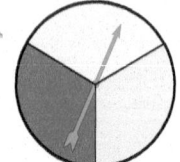

8.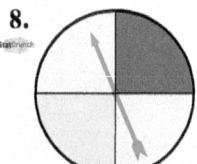

9.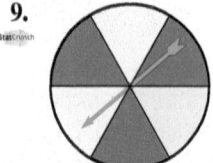

10.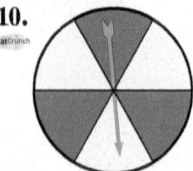

Solve each probability problem.

11. **Using Spinners to Generate Numbers** Suppose the spinner shown here is spun once, to determine a single-digit number, and we are interested in the event *E* that the resulting number is odd. Give each of the following.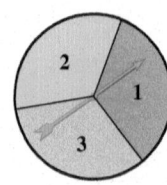

(a) the sample space

(b) the number of favorable outcomes

(c) the number of unfavorable outcomes

(d) the total number of possible outcomes

(e) the probability of an odd number

(f) the odds in favor of an odd number

12. **Lining Up Preschool Children** Kim's group of preschool children includes nine girls and seven boys. If she randomly selects one child to be first in line, with *E* being the event that the one selected is a girl, give each of the following.

(a) the total number of possible outcomes

(b) the number of favorable outcomes

(c) the number of unfavorable outcomes

(d) the probability of event *E*

(e) the odds in favor of event *E*

13. **Using Spinners to Generate Numbers** The spinner of **Exercise 11** is spun twice in succession to determine a two-digit number. Give each of the following.

(a) the sample space

(b) the probability of an odd number

(c) the probability of a number with repeated digits

(d) the probability of a number greater than 30

(e) the probability of a prime number

14. **Probabilities in Coin Tossing** Two fair coins are tossed (say a dime and a quarter). Give each of the following.

(a) the sample space

(b) the probability of heads on the dime

(c) the probability of heads on the quarter

(d) the probability of getting both heads

(e) the probability of getting the same outcome on both coins

15. **Random Selection of Fifties Music** Butch has fifty vinyl records from the fifties, including exactly one by Smiley Lewis, two by The Drifters, three by Bobby Darin, four by The Coasters, and five by Fats Domino. If he randomly selects one from his collection of fifty, find the probability that it will be by each of the following.

(a) Smiley Lewis

(b) The Drifters

(c) Bobby Darin

(d) The Coasters

(e) Fats Domino

16. **Probabilities in Coin Tossing** Three fair coins are tossed.

(a) Write out the sample space.

Determine the probability of each event.

(b) no heads

(c) exactly one head

(d) exactly two heads

(e) three heads

17. *Number Sums for Rolling Two Dice* Construct a sample space for the rolling of two dice, where the result is the sum of the values. (For example, rolling 1 and 2 results in 3.) Then find the probability of rolling each sum.

(a) 2   (b) 3   (c) 4   (d) 5

(e) 6   (f) 7   (g) 8   (h) 9

(i) 10   (j) 11   (k) 12

18. *Probabilities of Two Daughters among Four Children* In **Example 3,** what would be Kathy's probability of having exactly two daughters if she were to have four children altogether? (You may want to use a tree diagram to construct the sample space.)

Probabilities of Poker Hands *In 5-card poker, find the probability of being dealt each of the following. Give each answer to eight decimal places. (Refer to* **Table 1.***)*

19. a straight flush

20. two pairs

21. four of a kind

22. four queens

23. a hearts flush (*not* a royal flush or a straight flush)

24. no pair

In Exercises 25 and 26, give answers to three decimal places.

25. *Probability of Seed Germination* In a hybrid corn research project, 200 seeds were planted, and 175 of them germinated. Find the empirical probability that any particular seed of this type will germinate.

26. *Probability of Forest Land in California* According to *The World Almanac and Book of Facts,* California has 155,959 square miles of land area, 51,250 square miles of which are forested. Find the probability that a randomly selected location in California will be forested.

27. *Probabilities in Olympic Curling* In Olympic curling, the scoring area (shown here) consists of four concentric circles on the ice with radii of 6 inches, 2 feet, 4 feet, and 6 feet. If a team member lands a (43-pound) stone *randomly* within the scoring area, find the probability that it ends up centered on the given color.

(a) red   (b) white   (c) blue

28. *Probabilities in Dart Throwing* If a dart hits the square target shown here at random, what is the probability that it will hit in a colored region? (*Hint:* Compare the area of the colored regions to the total area of the target.)

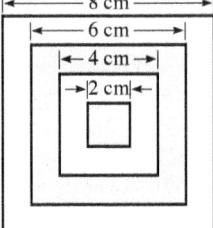

Genetics in Snapdragons *Mendel found no dominance in snapdragons (in contrast to peas) with respect to red and white flower color. When pure red and pure white parents are crossed (see* **Table 2***), the resulting* Rr *combination (one of each*  *gene) produces second-generation offspring with pink flowers. These second-generation pinks, however, still carry one red and one white gene, so when they are crossed, the third generation is still governed by* **Table 3.**

Find each probability for third-generation snapdragons.

29. *P*(red)   **30.** *P*(pink)   **31.** *P*(white)

Genetics in Pea Plants *Mendel also investigated various characteristics besides flower color. For example, round peas are dominant over recessive wrinkled peas. First, second, and third generations can again be analyzed using* **Tables 2 and 3,** *where* R *represents round and* r *represents wrinkled.*

32. Explain why crossing pure round and pure wrinkled first-generation parents will always produce round peas in the second-generation offspring.

33. When second-generation round pea plants (each of which carries both R and r genes) are crossed, find the probability that a third-generation offspring will have the following.

(a) round peas   (b) wrinkled peas

Genetics of Cystic Fibrosis *Cystic fibrosis is one of the most common inherited diseases in North America (including the United States), occurring in about 1 of every 2000 Caucasian births and about 1 of every 250,000 non-Caucasian births. Even with modern treatment, victims usually die from lung damage by their early twenties.*

If we denote a cystic fibrosis gene with a c *and a disease-free gene with a* C *(since the disease is recessive), then only a* cc *person will actually have the disease. Such persons would ordinarily die before parenting children, but a child can also inherit the disease from two* Cc *parents (who themselves are healthy—that is, have no symptoms but are "carriers" of the disease). This is like a pea plant inheriting white flowers from two red-flowered parents that both carry genes for white.*

34. Find the empirical probability (to four decimal places) that cystic fibrosis will occur in a randomly selected infant birth among U.S. Caucasians.

35. Find the empirical probability (to six decimal places) that cystic fibrosis will occur in a randomly selected infant birth among U.S. non-Caucasians.

36. Among 150,000 North American Caucasian births, about how many occurrences of cystic fibrosis would you expect?

Suppose that both partners in a marriage are cystic fibrosis carriers (a rare occurrence). Construct a chart similar to **Table 3** *and determine the probability of each of the following events.*

37. Their first child will have the disease.

38. Their first child will be a carrier.

39. Their first child will neither have nor carry the disease.

Suppose a child is born to one cystic fibrosis carrier parent and one non-carrier parent. Find the probability of each of the following events.

40. The child will have cystic fibrosis.

41. The child will be a healthy cystic fibrosis carrier.

42. The child will neither have nor carry the disease.

Genetics of Sickle-Cell Anemia   *Sickle-cell anemia occurs in about 1 of every 500 black baby births and about 1 of every 160,000 non-black baby births. It is ordinarily fatal in early childhood. There is a test to identify carriers. Unlike cystic fibrosis, which is recessive, sickle-cell anemia is* **codominant.** *This means that inheriting two sickle-cell genes causes the disease, while inheriting just one sickle-cell gene causes a mild (non-fatal) version (which is called* **sickle-cell trait**). *This is similar to a snapdragon plant manifesting pink flowers by inheriting one red gene and one white gene.*

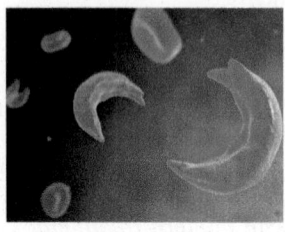

In Exercises 43 and 44, find the empirical probabilities of the given events.

43. A randomly selected black baby will have sickle-cell anemia. (Give your answer to three decimal places.)

44. A randomly selected non-black baby will have sickle-cell anemia. (Give your answer to six decimal places.)

45. Among 80,000 births of black babies, about how many occurrences of sickle-cell anemia would you expect?

Find the theoretical probability of each condition in a child both of whose parents have sickle-cell trait.

46. The child will have sickle-cell anemia.

47. The child will have sickle-cell trait.

48. The child will be healthy.

Drawing Balls from an Urn   *Anne randomly chooses a single ball from the urn shown here. Find the odds against each event.*

49. red   **50.** yellow

51. blue   **52.** red or yellow

53. yellow or blue   **54.** red or blue

Random Selection of Club Officers   *Five people (Alan, Bill, Cathy, David, and Evelyn) form a club:* $N = \{A, B, C, D, E\}$. *Cathy and Evelyn are women, and the others are men. If they choose a president randomly, find the odds against each of the following becoming president.*

55. Cathy   **56.** a woman

57. a man   **58.** a person whose name begins with a consonant

In Exercises 59 and 60, assume that the probability of an event E is

$$P(E) = 0.37.$$

Find each of the following.

59. the odds in favor of E **60.** the odds against E

In Exercises 61 and 62, make the requested conversions.

61. If the odds in favor of event E are 12 to 19, find $P(E)$.

62. If the odds against event E are 10 to 3, find $P(E)$.

63. Women's 100-Meter Run   In the history of track and field, no woman has broken the 10-second barrier in the 100-meter run.

   (a) From the statement above, find the empirical probability that a woman runner will break the 10-second barrier next year.

   (b) Can you find the theoretical probability for the event of part (a)?

   (c) Is it possible that the event of part (a) will occur?

64. On page 27 of their book *Descartes' Dream*, Philip Davis and Reuben Hersh ask, "Is probability real or is it just a cover-up for ignorance?" What do you think?

The remaining exercises require careful thought to determine n(E) and n(S). (In some cases, you may want to employ counting methods from earlier work, such as the fundamental counting principle, permutations, or combinations.)

Probabilities of Seating Arrangements *Six people (three het-erosexual married couples) arrange themselves randomly in six consecutive seats in a row.*

(a) *Determine the number of ways in which the described event can occur, and*

(b) *Determine the probability of the event.*

(Hint: In each case the denominator of the probability fraction will be 6! = 720, *the total number of ways to arrange six items.)*

65. Each husband will sit immediately to the left of his wife.

66. Each man will sit immediately to the left of a woman.

67. The women will be in three adjacent seats.

68. The women will be in three adjacent seats, as will the men.

Selecting Class Reports *Assuming that Ben, Jill, and Pam are three of the 26 members of the class, and that three of the class members will be chosen randomly to deliver their reports during the next class meeting, find the probability (to six decimal places) of each event.*

69. Ben, Jill, and Pam are selected, in that order.

70. Ben, Jill, and Pam are selected, in any order.

Solve each problem.

71. ***Altered Dice*** A six-sided die has been altered so that the side that had been a single dot is now a blank face. Another die has a blank face instead of the face with four dots. What is the probability that a sum of 7 is rolled when the two dice are thrown? (*MT* calendar problem)

72. ***Location in a Tunnel*** Mr. Davis is driving through a tunnel that is eight miles long. At this instant, what is the probability that he is at least six miles from one end of the tunnel? (*MT* calendar problem)

73. ***Slopes*** Two lines, neither of which is vertical, are perpendicular if and only if their slopes are negative reciprocals (as are $\frac{2}{3}$ and $-\frac{3}{2}$). If two distinct numbers are chosen randomly from the set

$$\left\{ -2, -\frac{4}{3}, -\frac{1}{2}, 0, \frac{1}{2}, \frac{3}{4}, 3 \right\},$$

find the probability that they will be the slopes of two perpendicular lines.

74. ***Drawing Cards*** When drawing cards without replacement from a standard 52-card deck, find the maximum number of cards you could possibly draw and still get the following.

(a) fewer than three black cards

(b) fewer than six spades

(c) fewer than four face cards

(d) fewer than two kings

75. ***Student Course Schedules*** A student plans to take three courses next term. If he selects them randomly from a list of twelve courses, five of which are science courses, what is the probability that all three courses selected will be science courses?

76. ***Symphony Performances*** Rhonda randomly selects three symphony performances to attend this season from a schedule of ten performances, three of which feature works by Beethoven. Find the probability that she will select all of the Beethoven programs.

77. ***Racing Bets*** A "trifecta" is a particular horse race in which you win by picking the "win," "place," and "show" horses (the first-, second-, and third-place winners) in their proper order. If five horses of equal ability are entered in a trifecta race, and Tracy selects an entry, what is the probability that she will be a winner?

78. ***Random Selection of Prime Numbers*** If two distinct prime numbers are randomly selected from among the first eight prime numbers, what is the probability that their sum will be 24?

79. ***Random Sums*** Two integers are randomly selected from the set {1, 2, 3, 4, 5, 6, 7, 8, 9} and are added together. Find the probability that their sum is 11 if they are selected as described.

(a) with replacement

(b) without replacement

80. ***Numbers from Sets of Digits*** The digits 1, 2, 3, 4, and 5 are randomly arranged to form a five-digit number. Find the probability of each event.

(a) The number is even.

(b) The first and last digits of the number both are even.

81. ***Divisibility of Random Products*** When a fair six-sided die is tossed on a tabletop, the bottom face cannot be seen. What is the probability that the product of the numbers on the five faces that can be seen is divisible by 6? (*MT* calendar problem)

82. ***Random Sums and Products*** Tamika selects two different numbers at random from the set {8, 9, 10} and adds them. Carlos takes two different numbers at random from the set {3, 5, 6} and multiplies them. What is the probability that Tamika's result is greater than Carlos's result? (*MT* calendar problem)

83. *Classroom Demographics* A high school class consists of 6 seniors, 10 juniors, 12 sophomores, and 4 freshmen. Exactly 2 of the juniors are female, and exactly 2 of the sophomores are male. If two students are randomly selected from this class, what is the probability that the pair consists of a male junior and a female sophomore? (*MT* calendar problem)

84. *Fractions from Dice Rolls* Lisa has one red die and one green die, which she rolls to make up fractions. The green die is the numerator, and the red die is the denominator. Some of the fractions have terminating decimal representations. How many different terminating decimal results can these two dice represent? What is the probability of rolling a fraction with a terminating decimal representation? (*MT* calendar problem)

Palindromic Numbers *Numbers that are **palindromes** read the same forward and backward.*

30203 *is a five-digit palindrome.*

If a single number is chosen randomly from each set, find the probability that it will be palindromic.

85. the set of all two-digit numbers

86. the set of all three-digit numbers

Six people, call them

A, B, C, D, E, and F,

are randomly divided into three groups of two. Find the probability of each event. (Do not impose unwanted ordering among groups.)

87. *A* and *B* are in the same group, as are *C* and *D*.

88. *E* and *F* are in the same group.

11.2   EVENTS INVOLVING "NOT" AND "OR"

OBJECTIVES

1 Apply the fact that the probability of an event is a real number between 0 and 1, inclusive of both, and know the meanings of the terms *impossible event* and *certain event*.

2 Identify the correspondences among set theory, logic, and arithmetic.

3 Determine the probability of "not *E*" given the probability of *E*.

4 Determine the probability of "*A* or *B*" given the probabilities of *A*, *B*, and "*A* and *B*."

Properties of Probability

Recall that an empirical probability, based upon experimental observation, may be the best value available but still is only an approximation to the ("true") theoretical probability. For example, no human has ever been known to jump higher than 8.5 feet vertically, so the empirical probability of such an event is zero. Observing the rate at which high-jump records have been broken, we suspect that the event is, in fact, possible and may one day occur. Hence it must have some nonzero theoretical probability, even though we have no way of assessing its exact value.

Recall also that the theoretical probability formula,

$$P(E) = \frac{n(E)}{n(S)},$$

is valid only when all outcomes in the sample space S are equally likely. For the experiment of tossing two fair coins, we can write $S = \{\text{hh, ht, th, tt}\}$ and compute

$$P(\text{both heads}) = \frac{1}{4}. \quad \text{Correct}$$

However, if we define the sample space with non-equally likely outcomes as $S = \{\text{both heads}, \text{both tails}, \text{one of each}\}$, we are led to

$$P(\text{both heads}) = \frac{1}{3}. \quad \text{Incorrect}$$

To convince yourself that $\frac{1}{4}$ is a better value than $\frac{1}{3}$, toss two fair coins 100 times or so, to see what the empirical fraction seems to approach.

More on the Birth of Probability
The modern mathematical theory of probability came mainly from the Russian scholars **P. L. Chebyshev** (1821–1922), **A. A. Markov** (1856–1922), and **Andrei Nikolaevich Kolmogorov** (1903–1987). The Dutch mathematician and scientist **Christiaan Huygens** (1629–1695) wrote a formal treatise on probability. It appeared in 1657 and was based on the Pascal–Fermat correspondence. One of the first to apply probability to matters other than gambling was the French mathematician **Pierre Simon de Laplace** (1749–1827), who is usually credited with being the "father" of probability theory.

For any event E within a sample space S, we know that $0 \le n(E) \le n(S)$. Dividing all members of this inequality by $n(S)$ gives

$$\frac{0}{n(S)} \le \frac{n(E)}{n(S)} \le \frac{n(S)}{n(S)}, \quad \text{or} \quad 0 \le P(E) \le 1.$$

In words, the probability of any event is a number from 0 through 1, inclusive.

If event E is *impossible* (cannot happen), then $n(E)$ must be 0 (E is the empty set), so

$$P(E) = 0. \quad \text{\scriptsize E is impossible.}$$

If event E is *certain* (cannot help but happen), then $n(E) = n(S)$, so

$$P(E) = \frac{n(E)}{n(S)} = \frac{n(S)}{n(S)} = 1. \quad \text{\scriptsize E is certain.}$$

PROPERTIES OF PROBABILITY

Let E be an event within the sample space S. That is, E is a subset of S. Then the following properties hold.

1. $0 \le P(E) \le 1$ The probability of an event is a number from 0 through 1, inclusive.

2. $P(\varnothing) = 0$ The probability of an impossible event is 0.

3. $P(S) = 1$ The probability of a certain event is 1.

StatCrunch   Simulation: Dice Rolling

Probabilities are often expressed in terms of percents in the media. We will, however, give them in terms of fractions or decimals.

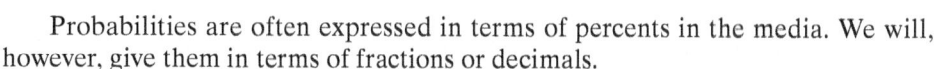

EXAMPLE 1   **Finding Probability When Rolling a Die**

When a single fair die is rolled, find the probability of each event.

(a) The number 2 is rolled.

(b) A number other than 2 is rolled.

(c) The number 7 is rolled.

(d) A number less than 7 is rolled.

Solution

(a) Since one of the six possibilities is a 2, $P(2) = \frac{1}{6}$.

(b) There are five such numbers, 1, 3, 4, 5, and 6, so $P(\text{a number other than 2}) = \frac{5}{6}$.

(c) None of the possible outcomes is 7. Thus, $P(7) = \frac{0}{6} = 0$.   Impossible event

(d) Since all six of the possible outcomes are less than 7,

$$P(\text{a number less than 7}) = \frac{6}{6} = 1. \quad \text{\scriptsize Certain event}$$

Pierre Simon de Laplace (1749–1827) began in 1773 to solve the problem of why Jupiter's orbit seems to shrink and Saturn's orbit seems to expand. Eventually Laplace worked out a complete theory of the solar system. His *Celestial Mechanics* resulted from almost a lifetime of work. In five volumes, it was published between 1799 and 1825 and gained for Laplace the reputation "Newton of France."

Laplace's work on probability was actually an adjunct to his celestial mechanics. He needed to demonstrate that probability is useful in interpreting scientific data.

Refer to the box preceding **Example 1.** No probability in that example was less than 0 or greater than 1, which illustrates Property 1. The "impossible" event of part (c) had probability 0, illustrating Property 2. The "certain" event of part (d) had probability 1, illustrating Property 3.

Events Involving "Not"

The correspondences shown in **Table 4** are the basis for the probability rules stated and applied in this section. For example, the probability of an event *not* happening involves the *complement* and *subtraction,* according to row 1 of the table.

Table 4 Set Theory/Logic/Arithmetic Correspondences

	Set Theory	**Logic**	**Arithmetic**
1. Operation or Connective (Symbol)	Complement ($'$)	Not ($\sim$)	Subtraction ($-$)
2. Operation or Connective (Symbol)	Union ($\cup$)	Or ($\vee$)	Addition ($+$)
3. Operation or Connective (Symbol)	Intersection ($\cap$)	And ($\wedge$)	Multiplication ($\cdot$)

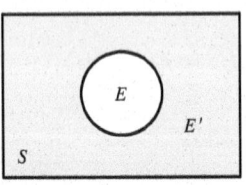

The logical connective "not" corresponds to "complement" in set theory.

$$P(\text{not } E) = P(S) - P(E)$$
$$= 1 - P(E)$$

Figure 5

The rule for the probability of a complement follows and is illustrated in **Figure 5.**

PROBABILITY OF A COMPLEMENT (FOR THE EVENT "NOT E")

The probability that an event E will *not* occur is equal to 1 minus the probability that it will occur.

$$P(\text{not } E) = 1 - P(E)$$

Notice that the events of **Examples 1(a) and (b),** namely "2" and "not 2," are complements of one another, and that their probabilities add up to 1. This illustrates the above probability rule. The equation

$$P(E) + P(E') = 1$$

is a rearrangement of the formula for the probability of a complement. Another form of the equation that is also useful at times follows.

$$P(E) = 1 - P(E')$$

EXAMPLE 2 Finding the Probability of a Complement

When a single card is drawn from a standard 52-card deck, what is the probability that it will not be a king?

Solution

$$P(\text{not a king}) = 1 - P(\text{king}) = 1 - \frac{4}{52} = \frac{48}{52} = \frac{12}{13}$$

Remember to write in lowest terms.

Mary Somerville (1780–1872) is associated with Laplace because of her brilliant exposition of his *Celestial Mechanics.*

Somerville studied Euclid thoroughly and perfected her Latin so she could read Newton's *Principia.* In about 1816 she went to London and soon became part of its literary and scientific circles.

Somerville's book on Laplace's theories came out in 1831 to great acclaim. Then followed a panoramic book, *Connection of the Physical Sciences* (1834). A statement in one of its editions suggested that irregularities in the orbit of Uranus might indicate that a more remote planet, not yet seen, existed. This caught the eye of the scientists who worked out the calculations for Neptune's orbit.

EXAMPLE 3 **Finding the Probability of a Complement**

If five fair coins are tossed, find the probability of obtaining at least two heads.

Solution

A tree diagram will show that there are $2^5 = \mathbf{32}$ possible outcomes for the experiment of tossing five fair coins. Most include at least two heads. In fact, only the outcomes

$$ttttt, \quad htttt, \quad thttt, \quad tthtt, \quad tttht, \quad \text{and} \quad tttth \quad \text{6 of these}$$

do *not* include at least two heads. If E denotes the event "at least two heads," then E' is the event "not at least two heads,"

$$P(E) = 1 - P(E') = 1 - \frac{6}{32} = \frac{26}{32} = \frac{13}{16}$$

Events Involving "Or"

Examples 2 and 3 showed how the probability of an event can be approached *indirectly,* by first considering the complement of the event. Another indirect approach is to break the event into simpler component events. Row 2 of **Table 4** indicates that the probability of one event *or* another should involve the *union* and *addition.*

EXAMPLE 4 **Selecting from a Set of Numbers**

If one number is selected randomly from the set $\{1, 2, 3, 4, 5, 6, 7, 8, 9, 10\}$, find the probability of each of the following events.

(a) The number is odd or a multiple of 4.   **(b)** The number is odd or a multiple of 3.

Solution

Define the following events.

$$S = \{1, 2, 3, 4, 5, 6, 7, 8, 9, 10\} \quad \text{Sample space}$$

$$A = \{1, 3, 5, 7, 9\} \qquad\qquad \text{Odd outcomes; } P(A) = \tfrac{5}{10}$$

$$B = \{4, 8\} \qquad\qquad\qquad \text{Multiples of 4; } P(B) = \tfrac{2}{10}$$

$$C = \{3, 6, 9\} \qquad\qquad\quad \text{Multiples of 3; } P(C) = \tfrac{3}{10}$$

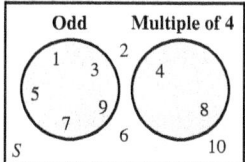

Figure 6

(a) **Figure 6** shows the positioning of the 10 integers within the sample space and within the pertinent sets A and B. The composite event "A or B" corresponds to the set $A \cup B = \{1, 3, 4, 5, 7, 8, 9\}$. $A \cup B$ has **seven** elements. By the theoretical probability formula,

$$P(A \text{ or } B) = \frac{7}{10}. \quad \text{Of 10 total outcomes, seven are favorable.}$$

(b) **Figure 7** shows the situation.

$$P(A \text{ or } C) = \frac{6}{10} = \frac{3}{5} \quad \text{Of 10 total outcomes, six are favorable.}$$

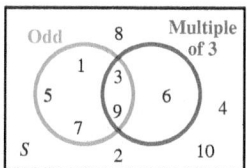

Figure 7

Would an addition formula have worked in **Example 4?** Let's check.

Part (a): $P(A \text{ or } B) = P(A) + P(B) = \dfrac{5}{10} + \dfrac{2}{10} = \dfrac{7}{10}$ Correct

Part (b): $P(A \text{ or } C) = P(A) + P(C) = \dfrac{5}{10} + \dfrac{3}{10} = \dfrac{8}{10} = \dfrac{4}{5}$ Incorrect

The trouble in part (b) is that A and C are not disjoint sets. They have outcomes in common. Just as with the additive counting principle in the previous chapter, an adjustment must be made here to compensate for counting the common outcomes twice.

$$P(A \text{ or } C) = P(A) + P(C) - P(A \text{ and } C)$$

$$= \frac{5}{10} + \frac{3}{10} - \frac{2}{10} = \frac{6}{10} = \frac{3}{5} \quad \text{Correct}$$

In probability theory, events that are disjoint sets are *mutually exclusive events*.

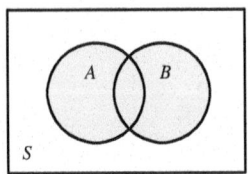

The logical connective "or" corresponds to "union" in set theory.

$P(A \text{ or } B)$
$= P(A) + P(B) - P(A \text{ and } B)$

Figure 8

MUTUALLY EXCLUSIVE EVENTS

Two events A and B are **mutually exclusive events** if they have no outcomes in common. Mutually exclusive events cannot occur simultaneously.

The results observed in **Example 4** are generalized as follows. The two possibilities are illustrated in **Figures 8 and 9**.

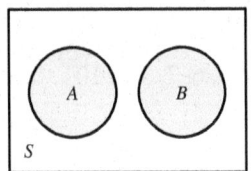

When *A* and *B* are mutually exclusive,

$P(A \text{ or } B) = P(A) + P(B)$.

Figure 9

ADDITION RULE OF PROBABILITY (FOR THE EVENT "A OR B")

If A and B are any two events, then the following holds.

$$P(A \text{ or } B) = P(A) + P(B) - P(A \text{ and } B)$$

If A and B are mutually exclusive, then the following holds.

$$P(A \text{ or } B) = P(A) + P(B)$$

Actually, the first formula in the addition rule applies in all cases. The third term on the right drops out when A and B are mutually exclusive, because $P(A \text{ and } B) = 0$. Still, it is good to remember the second formula in the preceding box for the many cases where the component events are mutually exclusive. In this section, we consider only cases where the event "A and B" is simple. We deal with more complicated composite events involving "and" in the next section.

EXAMPLE 5 Finding the Probability of an Event Involving "Or"

If a single card is drawn from a standard 52-card deck, what is the probability that it will be a spade or a red card?

Solution

First note that "spade" and "red" cannot both occur, because there are no red spades. All spades are black. Therefore, we can use the formula for mutually exclusive events. There are 13 spades and 26 red cards in the deck.

Two examples of "success" in Example 5

$$P(\text{spade or red}) = P(\text{spade}) + P(\text{red}) = \frac{13}{52} + \frac{26}{52} = \frac{39}{52} = \frac{3}{4}$$

We often need to consider composites of more than two events. When each event involved is mutually exclusive of all the others, we extend the addition rule to the appropriate number of components.

Table 5

x	P(x)
1	0.05
2	0.10
3	0.20
4	0.40
5	0.10
6	0.15

EXAMPLE 6   Treating Unions of Several Components

Amy plans to spend from 1 to 6 hours on her homework. If x represents the number of hours to be spent on a given night, then the probabilities of the various values of x, rounded to the nearest hour, are as shown in **Table 5**. Find the probabilities that Amy will spend each of the following amounts of time.

(a) fewer than 3 hours

(b) more than 2 hours

(c) more than 1 hour but no more than 5 hours

(d) fewer than 5 hours

Solution

Because the time periods in **Table 5** are mutually exclusive of one another, we can simply add the appropriate component probabilities.

(a) $P(\text{fewer than } 3) = P(1 \text{ or } 2)$ Fewer than 3 means 1 or 2.

$\qquad\qquad\qquad\quad = P(1) + P(2)$ Addition rule

$\qquad\qquad\qquad\quad = 0.05 + 0.10$ Substitute values from **Table 5**.

$\qquad\qquad\qquad\quad = 0.15$ Add.

(b) $P(\text{more than } 2) = P(3 \text{ or } 4 \text{ or } 5 \text{ or } 6)$ "More than 2" means 3, 4, 5, or 6.

$\qquad\qquad\qquad\quad = P(3) + P(4) + P(5) + P(6)$ Addition rule

$\qquad\qquad\qquad\quad = 0.20 + 0.40 + 0.10 + 0.15$ Substitute values from **Table 5**.

$\qquad\qquad\qquad\quad = 0.85$ Add.

(c) $P(\text{more than } 1 \text{ but no more than } 5)$

$\qquad\qquad = P(2 \text{ or } 3 \text{ or } 4 \text{ or } 5)$ 2, 3, 4, and 5 are more than 1 and no more than 5.

$\qquad\qquad = P(2) + P(3) + P(4) + P(5)$ Addition rule

$\qquad\qquad = 0.10 + 0.20 + 0.40 + 0.10$ Substitute values from **Table 5**.

$\qquad\qquad = 0.80$ Add.

(d) Although we could take a direct approach here, as in parts (a), (b), and (c), we will combine the complement rule with the addition rule.

$\qquad P(\text{fewer than } 5) = 1 - P(\text{not fewer than } 5)$ Complement rule

$\qquad\qquad\qquad\qquad = 1 - P(5 \text{ or more})$ 5 or more is equivalent to not fewer than 5.

$\qquad\qquad\qquad\qquad = 1 - P(5 \text{ or } 6)$ 5 or more means 5 or 6.

$\qquad\qquad\qquad\qquad = 1 - [P(5) + \text{P}(6)]$ Addition rule

$\qquad\qquad\qquad\qquad = 1 - (0.10 + 0.15)$ Substitute values from **Table 5**.

$\qquad\qquad\qquad\qquad = 1 - 0.25$ Add Inside the parentheses first.

$\qquad\qquad\qquad\qquad = 0.75$

Table 5 in **Example 6** lists all possible time intervals so the corresponding probabilities add up to 1, a necessary condition for the way part (d) was done. The time spent on homework here is an example of a **random variable.** It is "random" because we cannot predict which of its possible values will occur.

A listing like **Table 5,** which shows all possible values of a random variable, along with the probabilities that those values will occur, is a **probability distribution** for that random variable. *All* possible values are listed, so they make up the entire sample space, and thus the listed probabilities must add up to 1 (by probability Property 3). Probability distributions will be discussed in later sections.

EXAMPLE 7 **Finding the Probability of an Event Involving "Or"**

Find the probability that a single card drawn from a standard 52-card deck will be a diamond or a face card.

Solution

The component events "diamond" and "face card" can occur simultaneously. (The jack, queen, and king of diamonds belong to both events.) So we must use the first formula of the addition rule. We let D denote "diamond" and F denote "face card."

One example of a card that is both a diamond and a face card

$$P(D \text{ or } F) = P(D) + P(F) - P(D \text{ and } F) \qquad \text{Addition rule}$$

$$= \frac{13}{52} + \frac{12}{52} - \frac{3}{52} \qquad \begin{array}{l}\text{There are 13 diamonds,}\\ \text{12 face cards, and 3 that are both.}\end{array}$$

$$= \frac{22}{52} \qquad \text{Add and subtract.}$$

$$= \frac{11}{26} \qquad \text{Write in lowest terms.}$$

EXAMPLE 8 **Finding the Probability of an Event Involving "Or"**

Of 20 elective courses, Emily plans to enroll in one, which she will choose by throwing a dart at the schedule of courses. If 8 of the courses are recreational, 9 are interesting, and 3 are both recreational and interesting, find the probability that the course she chooses will have at least one of these two attributes.

Solution

If R denotes "recreational" and I denotes "interesting," then

$$P(R) = \frac{8}{20}, \quad P(I) = \frac{9}{20}, \quad \text{and} \quad P(R \text{ and } I) = \frac{3}{20}.$$

R and I are not mutually exclusive.

$$P(R \text{ or } I) = \frac{8}{20} + \frac{9}{20} - \frac{3}{20}$$

$$= \frac{14}{20} \qquad \text{Addition rule}$$

$$= \frac{7}{10} \qquad \text{Write in lowest terms.}$$

(11.2) EXERCISES

CONCEPT CHECK *Give the correct numerical response.*

1. If the probability of an event E is $\frac{2}{3}$, then the probability that E will not occur is _____.

2. If A and B are mutually exclusive events, with $P(A) = \frac{1}{3}$ and $P(B) = \frac{1}{6}$, then $P(A \text{ or } B)$ is _____.

3. If A and B are events, with $P(A) = \frac{1}{5}$, $P(B) = \frac{2}{3}$, and $P(A \text{ and } B) = \frac{1}{3}$, then $P(A \text{ or } B)$ is _____.

4. If $S = \{1, 2, 3, 4, 5, 6\}$, A is the set of elements of S that are less than 5, and B is the set of elements of S that are even, then the probability that a single element chosen from S belongs to A or B is _____.

Probabilities for Rolling a Die *For the experiment of rolling a single fair die, find the probability of each event. (Hint: Recall that 1 is neither prime nor composite.)*

5. not prime

6. not less than 2

7. even or prime

8. odd or less than 5

9. less than 3 or greater than 4

10. odd or even

Probability and Odds for Drawing a Card *For the experiment of drawing a single card from a standard 52-card deck, find* **(a)** *the probability of each event, and* **(b)** *the odds in favor of each event.*

11. king or queen

12. not an ace

13. spade or face card

14. club or heart

15. neither a heart nor a 7

16. not a heart, or a 7

Number Sums for Rolling a Pair of Dice *For the experiment of rolling an ordinary pair of dice, find the probability that the sum will be each of the following. (You may want to use a table showing the sum for each of the 36 equally likely outcomes.)*

17. even or a multiple of 3

18. 11 or 12

19. less than 3 or greater than 9

20. odd or greater than 9

Prime Results *Find the probability of getting a prime number in each case.*

21. A number is chosen randomly from the set of numbers $\{1, 2, 3, 4, \ldots, 12\}$.

22. Two dice are rolled and the sum is observed.

23. A single digit is randomly chosen from the number 578.

24. A single digit is randomly chosen from the number 1234.

25. ***Determining Whether Events Are Mutually Exclusive*** Amanda has three office assistants. If A is the event that at least two of them are men and B is the event that at least two of them are women, are A and B mutually exclusive?

26. ***Determining Whether Events Are Mutually Exclusive*** Jeanne earned her college degree several years ago. Consider the following four events.

Her alma mater is in the East.

Her alma mater is a private college.

Her alma mater is in the Northwest.

Her alma mater is in the South.

Are these events all mutually exclusive of one another?

Probabilities in Golf Scoring *The table gives Josh's probabilities of scoring in various ranges on a par-70 course. In a given round, find the probability of each event.*

x	$P(x)$
Below 60	0.04
60–64	0.06
65–69	0.14
70–74	0.30
75–79	0.23
80–84	0.09
85–89	0.06
90–94	0.04
95–99	0.03
100 or above	0.01

27. par or above

28. in the 80s

29. less than 90

30. not in the 70s, 80s, or 90s

31. 95 or higher

32. What are the odds of Josh's scoring below par?

Probabilities of Poker Hands   *If you are dealt a 5-card hand (this implies without replacement) from a standard 52-card deck, find the probability of getting each of the following. Refer to* **Table 1** *in the previous section, and give answers to six decimal places.*

33. a full house or a straight

34. a black flush or two pairs

35. nothing any better than two pairs

36. a flush or three of a kind

Probability Distribution   *Solve each problem.*

37. Let x denote the sum of two distinct numbers selected randomly from the set of numbers $\{1, 2, 3, 4, 5\}$. Construct the probability distribution for the random variable x.

38. Anne randomly chooses a single ball from the urn shown here, and x represents the color of the ball chosen. Construct a complete probability distribution for the random variable x.

Comparing Empirical and Theoretical Probabilities for Rolling Dice
Roll a pair of dice 50 *times, keeping track of the number of times the sum is "less than 3 or greater than 9" (that is 2, 10, 11, or 12).*

39. From your results, calculate an empirical probability for the event "less than 3 or greater than 9."

40. By how much does your answer differ from the *theoretical* probability of **Exercise 19?**

41. Explain the difference between the two formulas in the addition rule of probability, illustrating each one with an appropriate example.

42. *Sum of Probabilities*   Suppose, for a given experiment, that A, B, C, and D are events, all mutually exclusive of one another, such that

$$A \cup B \cup C \cup D = S \text{ (the sample space)}.$$

By extending the addition rule of probability to this case and utilizing probability Property 3, what statement can you make concerning the sum of the probabilities of A, B, C, and D?

Complementary Probability   *For Exercises 43–45, let A be an event within the sample space S, and let*

$$n(A) = a \quad \text{and} \quad n(S) = s.$$

43. Use the complements principle of counting to find an expression for $n(A')$.

44. Use the theoretical probability formula to express $P(A)$ and $P(A')$.

45. Evaluate and simplify $P(A) + P(A')$.

46. What rule have you proved?

The remaining exercises require careful thought for the determination of n(E) and n(S). (In some cases, you may want to employ counting methods from earlier in the text, such as the fundamental counting principle, permutations, or combinations.)

Building Numbers from Sets of Digits   *Suppose we want to form three-digit numbers using the set of digits*

$$\{0, 1, 2, 3, 4, 5\}.$$

For example, 501 *and* 224 *are such numbers, but* 035 *is not.*

47. How many such numbers are possible?

48. How many of these numbers are multiples of 10?

49. How many of these numbers are multiples of 5?

50. If one three-digit number is chosen at random from all those that can be made from the above set of digits, find the probability that the one chosen is not a multiple of 5.

51. *Drawing Colored Marbles from Boxes*   A bag contains fifty blue and fifty green marbles. Two marbles at a time are randomly selected. If both are green, they are placed in box A; if both are blue, in box B; if one is green and the other is blue, in box C. After all marbles are drawn, what is the probability that the numbers of marbles in box A and box B are the same? (*MT* calendar problem)

52. *Multiplying Numbers Generated by Spinners*   An experiment consists of spinning both spinners shown here and multiplying the resulting numbers together. Find the probability that the resulting product will be even.

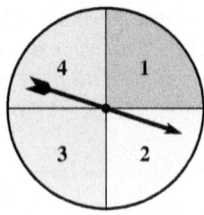 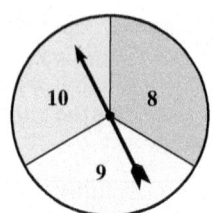

53. *Coin Tossing*   Luka and Janie are playing a coin toss game. If the coin lands heads up, Luka earns a point; otherwise, Janie earns a point. The first player to reach 20 points wins the game. If 19 of the first 37 tosses have been heads, what is the probability that Janie wins the game? (*MT* calendar problem)

54. *Random Births on the Same Day of the Week*   What is the probability that, of three people selected at random, at least two were born on the same day of the week? (*MT* calendar problem)

11.3 **CONDITIONAL PROBABILITY AND EVENTS INVOLVING "AND"**

OBJECTIVES

1 Apply the conditional probability formula.

2 Determine whether two events are independent.

3 Apply the multiplication rule for the event "*A* and *B*."

Conditional Probability

Sometimes the probability of an event must be computed using the knowledge that some other event has happened (or is happening, or will happen—the timing is not important). This type of probability is called *conditional probability*.

CONDITIONAL PROBABILITY

The probability of event *B*, computed on the assumption that event *A* has happened, is the **conditional probability of *B* given *A*** and is denoted

$$P(B|A).$$

EXAMPLE 1 Selecting from a Set of Numbers

From the sample space

$$S = \{1, 2, 3, 4, 5, 6, 7, 8, 9, 10\},$$

a single number is to be selected randomly. Find each probability given the events

A: The selected number is odd,   and B: The selected number is a multiple of 3.

(a) $P(B)$ **(b)** $P(A \text{ and } B)$ **(c)** $P(B|A)$

Solution

(a) $B = \{3, 6, 9\}$, so

$$P(B) = \frac{n(B)}{n(S)} = \frac{3}{10}.$$

(b) *A* and *B* is the set $A \cap B = \{1, 3, 5, 7, 9\} \cap \{3, 6, 9\} = \{3, 9\}$.

$$P(A \text{ and } B) = \frac{n(A \cap B)}{n(S)} = \frac{2}{10} = \frac{1}{5} \qquad \boxed{\text{Two elements}}$$

(c) The given condition, that *A* occurs, effectively reduces the sample space from *S* to *A*, and the elements of the new sample space *A* that are also in *B* are the elements of $A \cap B$.

$$P(B|A) = \frac{n(A \cap B)}{n(A)} = \frac{2}{5}$$

Example 1 illustrates some important points. First, because

$$\frac{n(A \cap B)}{n(A)} = \frac{\frac{n(A \cap B)}{n(S)}}{\frac{n(A)}{n(S)}} \qquad \text{Multiply numerator and denominator by } \frac{1}{n(S)}.$$

$$= \frac{P(A \cap B)}{P(A)}, \qquad \text{Theoretical probability formula}$$

the final line of the example gives the convenient formula on the next page.

Probability and statistics are implemented in actuarial science, which is devoted to analyzing data to provide information to insurance companies that then make financial decisions based on those data. What is the probability that you will die due to a **cosmic impact** (the collision of a meteor, comet, or asteroid with the Earth)? Such an event could be as catastrophic as full-scale nuclear war, killing a billion or more people.

The Spaceguard Survey has discovered more than half of the estimated number of near-Earth asteroids (NEAs) 1 kilometer or more in diameter and hopes to locate 90% of them in the next decade. Although the risk of finding one on a collision course with the Earth is slight, it is anticipated that, if we did, we would be able to deflect it before impact.

The photo above shows a crater in Arizona, 4000 feet in diameter and 570 feet deep. It is thought to have been formed 20,000 to 50,000 years ago by a meteorite about 50 meters across hitting the ground at several kilometers per second. (See http://en.wikipedia.org/wiki/Meteor_Crater.)

Conditional Probabilities:
Tree Diagram

CONDITIONAL PROBABILITY FORMULA

The **conditional probability of B given A** is calculated as follows.

$$P(B|A) = \frac{P(A \cap B)}{P(A)} = \frac{P(A \text{ and } B)}{P(A)}$$

A second observation from **Example 1** is that the conditional probability of B given A was $\frac{2}{5}$, whereas the "unconditional" probability of B (with no condition given) was $\frac{3}{10}$, so the condition *did* make a difference.

EXAMPLE 2   Finding Probabilities of Boys and Girls in a Family

Given a family with two children, find the probability of each event.

(a) Both are girls, given that at least one is a girl.

(b) Both are girls, given that the older child is a girl.

(Assume boys and girls are equally likely.)

Solution

We define the following events. (The older child's gender appears first.)

$S = \{gg, gb, bg, bb\}$ Sample space S; $n(S) = 4$

$A = \{gg\}$ Both are girls. $P(A) = \frac{1}{4}$

$B = \{gg, gb, bg\}$ At least one is a girl. $P(B) = \frac{3}{4}$

$C = \{gg, gb\}$ The older one is a girl. $P(C) = \frac{2}{4}$

Note that $A \cap B = \{gg\}$, and $A \cap C = \{gg\}$ as well. Thus

$$P(A \cap B) = P(A \cap C) = \frac{1}{4}.$$

(a) $P(A|B) = \dfrac{P(A \text{ and } B)}{P(B)} = \dfrac{\frac{1}{4}}{\frac{3}{4}} = \frac{1}{4} \div \frac{3}{4} = \frac{1}{4} \cdot \frac{4}{3} = \frac{1}{3}$

(b) $P(A|C) = \dfrac{P(A \text{ and } C)}{P(C)} = \dfrac{\frac{1}{4}}{\frac{2}{4}} = \frac{1}{4} \div \frac{2}{4} = \frac{1}{4} \cdot \frac{4}{2} = \frac{1}{2}$

Independent Events

Sometimes a conditional probability is no different from the corresponding unconditional probability, in which case the two events are *independent*.

INDEPENDENT EVENTS

Two events A and B are **independent events** if knowledge about the occurrence of one of them has no effect on the probability of the other one occurring. A and B are independent if

$$P(B|A) = P(B), \quad \text{or, equivalently,} \quad P(A|B) = P(A).$$

One example of a card that is both a face card and black

EXAMPLE 3 **Checking Events for Independence**

A single card is to be drawn from a standard 52-card deck. (The sample space S has 52 elements.) Find each of the following, given the events

   A: The selected card is a face card,   and B: The selected card is black.

(a) $P(B)$ **(b)** $P(B|A)$

(c) Use the results of parts (a) and (b) to determine whether events A and B are independent.

Solution

(a) There are 26 black cards in the 52-card deck.

$$P(B) = \frac{26}{52} = \frac{1}{2} \quad \text{Theoretical probability formula}$$

(b) $P(B|A) = \dfrac{P(B \text{ and } A)}{P(A)}$ Conditional probability formula

$$= \frac{\frac{6}{52}}{\frac{12}{52}} \quad \text{Of 52 cards, 12 are face cards and 6 are black face cards.}$$

$$= \frac{6}{52} \cdot \frac{52}{12} \quad \text{To divide, multiply by the reciprocal of the divisor.}$$

$$= \frac{1}{2} \quad \text{Calculate and write in lowest terms.}$$

(c) Because $P(B|A) = P(B)$, events A and B are independent.

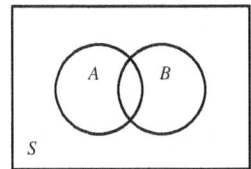

The logical connective "and" corresponds to "intersection" in set theory.

$$P(A \text{ and } B) = P(A) \cdot P(B|A)$$

Figure 10

Events Involving "And"

If we multiply both sides of the conditional probability formula by $P(A)$, we obtain an expression for $P(A \cap B)$, which applies to events of the form "A and B." The resulting formula is related to the fundamental counting principle from an earlier chapter. It is illustrated in **Figure 10**.

   Just as the calculation of $P(A \text{ or } B)$ is simpler when A and B are mutually exclusive, the calculation of $P(A \text{ and } B)$ is simpler when A and B are independent.

MULTIPLICATION RULE OF PROBABILITY (FOR THE EVENT "A AND B")

If A and B are *any two events*, then

$$P(A \text{ and } B) = P(A) \cdot P(B|A).$$

If A and B are *independent*, then

$$P(A \text{ and } B) = P(A) \cdot P(B).$$

   The first formula in the multiplication rule actually applies in all cases, because $P(B|A) = P(B)$ when A and B are independent. The independence of the component events is clear in many cases, so it is good to remember the second formula.

When Mattel Toys marketed a new **talking Barbie doll** a few years ago, some of the Barbies were programmed to say,

"Math class is tough."

The National Council of Teachers of Mathematics (NCTM), the American Association of University Women (AAUW), and numerous consumers voiced complaints about the damage such a message could do to children's attitudes toward school and mathematics. Mattel subsequently agreed to erase the phrase from the microchip to be used in future doll production.

Originally, each Barbie was programmed to say four different statements, randomly selected from a pool of 270 prerecorded statements. Therefore, the probability of getting a Barbie that said "Math class is tough" was only

$$\frac{1 \cdot {}_{269}C_3}{{}_{270}C_4} \approx 0.015 = 1.5\%.$$

Other messages included in the pool were "I love school, don't you?," "I'm studying to be a doctor," and "Let's study for the quiz."

EXAMPLE 4   **Selecting from a Set of Books**

Each year, Jacqui adds to her book collection a number of new publications that she believes will be of lasting value and interest. She has categorized each of her twenty acquisitions for 2019 as hardcover or paperback and as fiction or nonfiction. The numbers of books in the various categories are shown in **Table 6.**

Table 6   Year 2019 Books

	Fiction (F)	Nonfiction (N)	Totals
Hardcover (H)	3	5	8
Paperback (P)	8	4	12
Totals	11	9	20

Reduced sample space in part (b)

If Jacqui randomly chooses one of these 20 books, find the probability that it will be each of the following.

(a) hardcover   **(b)** fiction, given it is hardcover   **(c)** hardcover and fiction

Solution

(a) Eight of the 20 books are hardcover, so $P(H) = \frac{8}{20} = \frac{2}{5}$.

(b) The given condition that the book is hardcover reduces the sample space to eight books. Of those eight, just three are fiction, so $P(F \mid H) = \frac{3}{8}$.

(c) $P(H \text{ and } F) = P(H) \cdot P(F \mid H) = \frac{2}{5} \cdot \frac{3}{8} = \frac{3}{20}$ Multiplication rule

It is easier here if we simply notice, directly from **Table 6,** that 3 of the 20 books are "hardcover and fiction." This confirms that the general multiplication rule of probability did give us the correct answer.

EXAMPLE 5   **Selecting from a Set of Planets**

Table 7 lists the eight planets of our solar system, together with their mean distances from the sun, in millions of kilometers. Carrie must choose two distinct planets to cover in her astronomy report. If she selects randomly, find the probability that the first one selected is closer to the sun than is Mars and the second is closer to the sun than is Saturn. (Data from *The World Almanac and Book of Facts*.)

Solution

We define the following events.

   A: The first is closer than Mars. B: The second is closer than Saturn.

Then $P(A) = \frac{3}{8}$ because three of the original eight choices are favorable to event A. If the planet selected first is closer than Mars, it is also closer than Saturn, and since that planet is no longer available, four of the remaining seven are favorable to event B. Thus $P(B \mid A) = \frac{4}{7}$. Now apply the multiplication rule.

$$P(A \text{ and } B) = P(A) \cdot P(B \mid A) = \frac{3}{8} \cdot \frac{4}{7} = \frac{3}{14} \approx 0.214$$ Multiplication rule

Table 7   Mean Distance of Planets from the Sun

Planet	Distance (in millions of kilometers)
Mercury	58
Venus	108
Earth	150
Mars	228
Jupiter	778
Saturn	1430
Uranus	2870
Neptune	4500

In **Example 5,** the condition that A had occurred changed the probability of B, since the selection was done, in effect, without replacement. (Repetitions were not allowed.) Events A and B were not independent. On the other hand, in the next example the same events, A and B, are independent.

EXAMPLE 6 Selecting from a Set of Planets

Carrie must again select two planets, but this time one is for an oral report, the other is for a written report, and they need not be distinct. Here, the same planet may be selected for both reports. Again, find the probability that if she selects randomly, the first is closer than Mars to the sun and the second is closer than Saturn.

Solution

Defining events A and B as in **Example 5,** we have $P(A) = \frac{3}{8}$, just as before. But the selection is now done *with* replacement because repetitions *are* allowed. Event B is independent of event A, so we can use the second form of the multiplication rule.

$$P(A \text{ and } B) = P(A) \cdot P(B) = \frac{3}{8} \cdot \frac{5}{8} = \frac{15}{64} \approx 0.234 \quad \text{Answer is different than in \textbf{Example 5.}}$$

EXAMPLE 7 Selecting from a Deck of Cards

A single card is drawn from a standard 52-card deck. Let B denote the event that the card is black, and let D denote the event that it is a diamond. Are events B and D

(a) independent?   **(b)** mutually exclusive?

One example of a card that is black and one example of a card that is a diamond

(No single card is both black and a diamond.)

Solution

(a) For the unconditional probability of D, we get $P(D) = \frac{13}{52} = \frac{1}{4}$ because 13 of the 52 cards are diamonds. But for the conditional probability of D given B, we have $P(D \mid B) = \frac{0}{26} = 0$. None of the 26 black cards are diamonds. Since the conditional probability $P(D \mid B)$ is different from the unconditional probability $P(D)$, B and D are not independent.

(b) Mutually exclusive events are events that cannot both occur for a given performance of an experiment. Since no card in the deck is both black and a diamond, B and D are mutually exclusive.

EXAMPLE 8 Selecting from an Urn of Balls

Anne is still drawing balls from the same urn, as shown at the side. This time she draws three balls, without replacement. Find the probability that she gets red, yellow, and blue balls, in that order.

Solution

Using appropriate letters to denote the colors, and subscripts to indicate first, second, and third draws, the event can be symbolized "R_1 and Y_2 and B_3."

$$P(R_1 \text{ and } Y_2 \text{ and } B_3) = P(R_1) \cdot P(Y_2 \mid R_1) \cdot P(B_3 \mid R_1 \text{ and } Y_2)$$

$$= \frac{4}{11} \cdot \frac{5}{10} \cdot \frac{2}{9} = \frac{4}{99} \approx 0.0404$$

One example of five cards that are all hearts

The **search for extraterrestrial intelligence (SETI)** may have begun in earnest as early as 1961, when Dr. Frank Drake presented an equation for estimating the number of possible civilizations in the Milky Way galaxy whose communications we might detect. Over the years, the effort has been advanced by many scientists, including the late astronomer and exobiologist Carl Sagan, who popularized the issue in TV appearances and in his book *The Cosmic Connection: An Extraterrestrial Perspective* (Dell Paperback). "There must be other starfolk," said Sagan. In fact, some astronomers have estimated the odds against life on Earth being the only life in the universe at one hundred billion billion to one.

Other experts disagree. Freeman Dyson, a noted mathematical physicist and astronomer, says (in his book *Disturbing the Universe*) that after considering the same evidence and arguments, he believes it is just as likely as not (even odds) that there never was any other intelligent life out there.

EXAMPLE 9 Selecting from a Deck of Cards

If five cards are drawn without replacement from a standard 52-card deck, find the probability that they all are hearts.

Solution

Each time a heart is drawn, the number of available cards decreases by one and the number of hearts decreases by one.

$$P(\text{all hearts}) = \frac{13}{52} \cdot \frac{12}{51} \cdot \frac{11}{50} \cdot \frac{10}{49} \cdot \frac{9}{48} = \frac{33}{66,640} \approx 0.000495$$

We saw in the previous chapter that the problem of **Example 9** can also be solved by using the theoretical probability formula and combinations. The total possible number of 5-card hands, drawn without replacement, is $_{52}C_5$, and the number of those containing only hearts is $_{13}C_5$.

$$P(\text{all hearts}) = \frac{_{13}C_5}{_{52}C_5} = \frac{\dfrac{13!}{5! \cdot 8!}}{\dfrac{52!}{5! \cdot 47!}} \approx 0.000495 \quad \text{Use a calculator.}$$

EXAMPLE 10 Using Both Addition and Multiplication Rules

The local garage employs two mechanics, Ray and Tom. Your consumer club has found that Ray does twice as many jobs as Tom, Ray does a good job three out of four times, and Tom does a good job only two out of five times. If you take your car in for repairs and you have no say as to the mechanic selected, find the probability that a good job will be done.

Solution

We define the following events.

R: work done by Ray T: work done by Tom G: good job done

Since Ray does twice as many jobs as Tom, the unconditional probabilities of events R and T are, respectively, $\frac{2}{3}$ and $\frac{1}{3}$. Because Ray does a good job three out of four times, the probability of a good job, given that Ray did the work, is $\frac{3}{4}$. And because Tom does well two out of five times, the probability of a good job, given that Tom did the work, is $\frac{2}{5}$. (These last two probabilities are conditional.) These four values can be summarized.

$$P(R) = \frac{2}{3}, \quad P(T) = \frac{1}{3}, \quad P(G \mid R) = \frac{3}{4}, \quad \text{and} \quad P(G \mid T) = \frac{2}{5}$$

Event G can occur in two mutually exclusive ways: Ray could do the work and do a good job $(R \cap G)$, or Tom could do the work and do a good job $(T \cap G)$.

$$P(G) = P(R \cap G) + P(T \cap G) \quad \text{Addition rule}$$

$$= P(R) \cdot P(G \mid R) + P(T) \cdot P(G \mid T) \quad \text{Multiplication rule}$$

Multiply first, then add.

$$= \frac{2}{3} \cdot \frac{3}{4} + \frac{1}{3} \cdot \frac{2}{5} \quad \text{Substitute the values.}$$

$$= \frac{1}{2} + \frac{2}{15} = \frac{19}{30} \approx 0.633 \quad \leftarrow \text{Answer}$$

Conditional Probabilities:
Tree Diagram

Alternative method: The tree diagram in **Figure 11** shows another approach. Use the given information to draw the tree diagram, and then find the probability of a good job by adding the probabilities from the two indicated branches of the tree.

$$\text{Sum} = \frac{19}{30} \approx 0.633$$

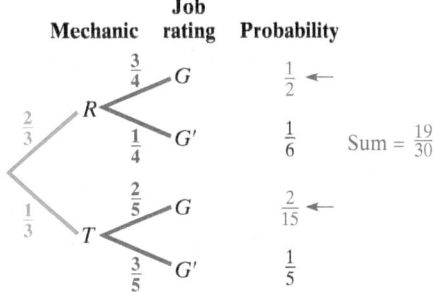

Garage mechanics experiment

Figure 11

EXAMPLE 11 Selecting Door Prizes

Rob is among five door prize winners at a holiday party. The five winners are asked to choose, without looking, from a bag that, they are told, contains five tokens, four of them redeemable for candy canes and one specific token redeemable for a $100 gift certificate. Can Rob improve his chance of getting the gift certificate by drawing first among the five people?

Solution

We denote candy cane by C, gift certificate by G, and first draw, second draw, and so on by subscripts $1, 2, \ldots$. If Rob draws first, his probability of getting the gift certificate is

$$P(G_1) = \frac{1}{5}.$$

If he draws second, his probability of getting the gift certificate is

$$
\begin{aligned}
P(G_2) &= P(C_1 \text{ and } G_2) \\
&= P(C_1) \cdot P(G_2 \mid C_1) \\
&= \frac{4}{5} \cdot \frac{1}{4} = \frac{1}{5}. \quad \text{Same result as above}
\end{aligned}
$$

For the third draw,

$$
\begin{aligned}
P(G_3) &= P(C_1 \text{ and } C_2 \text{ and } G_3) \\
&= P(C_1) \cdot P(C_2 \mid C_1) \cdot P(G_3 \mid C_1 \text{ and } C_2) \\
&= \frac{4}{5} \cdot \frac{3}{4} \cdot \frac{1}{3} = \frac{1}{5}. \quad \text{Same result as above}
\end{aligned}
$$

The probability of getting the gift certificate is also $\frac{1}{5}$ when drawing fourth or fifth. The order in which the five winners draw does not affect Rob's chances.

The **search for extraterrestrial intelligence (SETI)** has been mainly accomplished in recent years through **SETI@HOME,** a very large distributed computing program. Most of the data are collected by the world's largest radio telescope, built into a 20-acre natural bowl in Aricebo, Puerto Rico (pictured above), and processed by millions of personal computers around the world.

To learn more, or for a chance to be the first to "contact" an extraterrestrial civilization, check out
www.setiathome.ssl.berkeley.edu.

WHEN WILL I **EVER** USE THIS**?**

Monty Hall (1921–2017)

The late game show host Monty Hall has a rather famous probability problem named after him. Knowing the solution to the problem might help a contestant win a car instead of a goat.

> *Suppose that there are three doors, and we are asked to choose one of them. One of the doors hides a new car, and behind the two other doors are goats. We pick a door, say Door A. Then the host, who knows what's behind all three doors, opens another door, say Door C, and a goat appears. The host asks us whether we want to change our choice to Door B. Is it to our advantage to do so?*

One way to look at the problem, given that the car is *not* behind Door C, is that Doors A and B are now equally likely to contain the car. Thus, switching doors will neither help nor hurt our chances of winning the car.

However, there is another way to look at the problem. When we picked Door A, the probability was $\frac{1}{3}$ that it contained the car. Being shown the goat behind Door C doesn't give us any new information; after all, we knew that there was a goat behind at least one of the other doors. The probability that Door A has the car remains $\frac{1}{3}$. But because Door C has been ruled out, the probability that Door B has the car is now $\frac{2}{3}$. Thus, we should switch.

Analysis of this problem depends on the psychology of the host. If we suppose that the host must *always* show us a losing door and then give us an option to switch, then we should switch. This was not specifically stated in the problem but has been pointed out by many mathematicians.

There are several Web sites that simulate the Monty Hall problem. One of them is http://stayorswitch.com/

(The authors wish to thank David Berman of the University of New Orleans for his assistance with this explanation.)

StatCrunch Games: Let's Make a Deal

11.3 **EXERCISES**

CONCEPT CHECK *The chips shown are placed in a bag and drawn at random, one by one, without replacement. Answer each question.*

1. What is the probability that the first chip drawn is red?

2. What is the probability that the first chip drawn is blue and the second chip drawn is yellow?

3. What is the probability that the first two chips drawn are both yellow?

4. What is the probability that the first four chips drawn are all blue?

From the sample space $S = \{1, 2, 3, 4, \ldots, 15\}$, a single number is to be selected at random. Given the events

   A: The selected number is even,

   B: The selected number is a multiple of 4,

   C: The selected number is a prime number,

find each probability.

5. $P(A)$ 6. $P(B)$

7. $P(C)$ 8. $P(A \text{ and } B)$

9. $P(A \text{ and } C)$ 10. $P(B \text{ and } C)$

11. $P(A|B)$ 12. $P(B|A)$

13. $P(C|A)$ 14. $P(A|C)$

Given a family with three children, find the probability of each event in exercises 15–24. Assume that the probabilities of boy birth and girl birth are both $\frac{1}{2}$.

15. All are girls.   16. All are boys.

17. The oldest two are boys, given that there are at least two boys.

18. The youngest is a girl, given that there are at least two girls.

19. The oldest is a boy and the youngest is a girl, given that there are at least one boy and at least one girl.

20. The youngest two are girls, given that there are at least two girls.

21. The oldest two are boys, given that the oldest is a boy.

22. The oldest is a boy, given that there is at least one boy.

23. The youngest is a boy, given that the birth order alternates between girls and boys.

24. The oldest is a girl, given that the middle child is a girl.

For each experiment, determine whether the two given events are independent.

25. *Coin Tosses* A fair coin is tossed twice. The events are "head on the first" and "head on the second."

26. *Dice Rolls* A pair of dice are rolled. The events are "even on the first" and "odd on the second."

27. *Planets' Mean Distances from the Sun* Two planets are selected, without replacement, from the list in **Table 7**. The events are "the first selected planet is closer than Jupiter" and "the second selected planet is farther than Mars."

28. *Mean Distances from the Sun* Two planets are selected, with replacement, from the list in **Table 7**. The events are "the first selected planet is closer than Earth" and "the second selected planet is farther than Uranus."

29. *Answers on a Multiple-choice Test* The answers are all guessed on a twenty-question multiple-choice test. The events are "the first answer is correct" and "the last answer is correct."

30. *Committees of U.S. Senators* A committee of five is randomly selected from the 100 U.S. senators. The events are "the first member selected is a Republican" and "the second member selected is a Republican." (Assume that there are both Republicans and non-Republicans in the Senate.)

Gender and Career Motivation of College Students *One hundred college seniors attending a career fair at a university were categorized according to gender and according to primary career motivation. See the table for the results.*

	Primary Career Motivation			
	Money	Allowed to be Creative	Sense of Giving to Society	Total
Female	19	15	14	48
Male	12	23	17	52
Total	31	38	31	100

If one student is to be selected at random, find the probability that the student selected will satisfy each condition.

31. male

32. motivated primarily by creativity

33. not motivated primarily by money

34. female and motivated primarily by money

35. female, given that primary motivation is a sense of giving to society

36. motivated primarily by money or creativity, given that the student is male

Pet Selection *A no-kill shelter has seven puppies, including four huskies, two terriers, and one retriever. If Rebecka and Aaron, in that order, each select one puppy at random, with replacement (they may both select the same one), find the probability of each event.*

37. Both select a husky.

38. Rebecka selects a retriever; Aaron selects a terrier.

39. Rebecka selects a terrier; Aaron selects a retriever.

40. Both select a retriever.

Pet Selection *Two puppies are selected as earlier, but this time without replacement (Rebecka and Aaron cannot both select the same puppy). Find the probability of each event.*

41. Both select a husky.

42. Aaron selects a terrier, given that Rebecka selects a husky.

43. Aaron selects a retriever, given that Rebecka selects a husky.

44. Rebecka selects a retriever.

45. Aaron selects a retriever, given that Rebecka selects a retriever.

46. Both select a retriever.

Card Dealing *Let two cards be dealt successively, without replacement, from a standard 52-card deck. Find the probability of each event.*

47. spade dealt second, given a spade dealt first

48. club dealt second, given a diamond dealt first

49. two face cards   **50.** no face cards

51. The first card is a jack and the second is a face card.

52. The first card is the ace of hearts and the second is black.

53. The first card is black and the second is red.

54. **Proof of Formulas** Given events A and B within the sample space S, the following sequence of steps establishes formulas that can be used to compute conditional probabilities. Justify each statement.

(a) $P(A \text{ and } B) = P(A) \cdot P(B|A)$

(b) Therefore, $P(B|A) = \dfrac{P(A \text{ and } B)}{P(A)}$.

(c) Therefore, $P(B|A) = \dfrac{n(A \text{ and } B)/n(S)}{n(A)/n(S)}$.

(d) Therefore, $P(B|A) = \dfrac{n(A \text{ and } B)}{n(A)}$.

Conditions in Card Drawing *Use the results of* **Exercise 54** *to find each probability when a single card is drawn from a standard 52-card deck.*

55. $P(\text{queen} \mid \text{face card})$ **56.** $P(\text{face card} \mid \text{queen})$

57. $P(\text{red} \mid \text{diamond})$ **58.** $P(\text{diamond} \mid \text{red})$

Property of P(A and B) *Complete Exercises 59 and 60 to discover a general property of the probability of an event of the form "A and B."*

59. If one number is chosen randomly from the integers 1 through 10, the probability of getting a number that is *odd and prime,* by the multiplication rule, is

$$P(\text{odd}) \cdot P(\text{prime} \mid \text{odd}) = \frac{5}{10} \cdot \frac{3}{5} = \frac{3}{10}.$$

Compute the product $P(\text{prime}) \cdot P(\text{odd} \mid \text{prime})$, and compare to the product above.

60. What does **Exercise 59** imply, in general, about the probability of an event of the form "A and B"?

61. **Gender in Sequences of Babies** Two authors of this book each have three sons and no daughters. Assuming boy and girl babies are equally likely, what is the probability of this event?

62. *Dice Rolls* Three dice are rolled. What is the probability that the numbers shown will all be different? (*MT* calendar problem)

The remaining exercises, and groups of exercises, may require concepts from earlier sections, such as the complements principle of counting and addition rules, as well as the multiplication rule of this section.

Warehouse Grocery Shopping *Therese manages a grocery warehouse that encourages volume shopping on the part of its customers. Therese has discovered that on any given weekday, 70% of the customer sales amount to*  more than $100. *That is, any given sale on such a day has a probability of 0.70 of being for more than $100. (Actually, the conditional probabilities throughout the day would change slightly, depending on earlier sales, but this effect would be negligible for the first several sales of the day, so we can treat them as independent.)*

Find the probability of each event. (Give answers to three decimal places.)

63. The first two sales on Wednesday are both for more than $100.

64. The first three sales on Wednesday are all for more than $100.

65. None of the first three sales on Wednesday is for more than $100.

66. Exactly one of the first three sales on Wednesday is for more than $100.

Pollution from the Space Shuttle Launch Site *One problem encountered by developers of the space shuttle program was air pollution in the area surrounding the launch site. A certain direction from the launch site was considered critical in terms of hydrogen chloride pollution from the exhaust cloud. It has been determined that weather conditions would cause emission*  cloud movement in the critical direction only 5% of the time.

Find each probability. Assume that probabilities for a particular launch in no way depend on the probabilities for other launches. (Give answers to two decimal places.)

67. A given launch will not result in cloud movement in the critical direction.

68. No cloud movement in the critical direction will occur during any of 5 launches.

69. Any 5 launches will result in at least one cloud movement in the critical direction.

70. Any 10 launches will result in at least one cloud movement in the critical direction.

Job Interviews *Three men and three women are waiting to be interviewed for jobs. If they are all selected in random order, find the probability of each event.*

71. All the women will be interviewed first.

72. The first interviewee will be a man, and the remaining ones will alternate gender.

73. The first three chosen will all be the same gender.

74. No man will be interviewed until at least two women have been interviewed.

Garage Door Opener *Kevin installed a certain brand of automatic garage door opener that utilizes a transmitter control with six independent switches, each one set on or off. The receiver (wired to the door) must be set with the same pattern as the transmitter. (Exercises 75–78 are based on ideas similar to those of the "birthday problem" in the **For Further Thought** feature at the end of these exercises.)*

75. How many different ways can Kevin set the switches?

76. If one of Kevin's neighbors also has this same brand of opener, and both of them set the switches randomly, what is the probability, to four decimal places, that they will use the same settings?

77. If five neighbors with the same type of opener set their switches independently, what is the probability of at least one pair of neighbors using the same settings? (Give your answer to four decimal places.)

78. What is the minimum number of neighbors who must use this brand of opener before the probability of at least one duplication of settings is greater than $\frac{1}{2}$?

Weather Conditions on Successive Days *In November, the rain in a certain valley tends to fall in storms of several days' duration. The unconditional probability of rain on any given day of the month is 0.500. But the probability of rain on a day that follows a rainy day is 0.800, and the probability of rain on a day following a nonrainy day is 0.300. Find the probability of each event. Give answers to three decimal places.*

79. rain on two randomly selected consecutive days in November

80. rain on three randomly selected consecutive days in November

81. rain on November 1 and 2, but not on November 3 (The October 31 weather is unknown.)

82. rain on the first four days of November, given that October 31 was clear all day

Engine Failures in a Vintage Aircraft *In a certain four-engine vintage aircraft, now quite unreliable, each engine has a 10% chance of failure on any flight, as long as it is carrying its one-fourth share of the load. But if one engine fails, then the chance of failure increases to 20% for each of the other three engines. And if a second engine fails, each of the remaining two has a 30% chance of failure.*

Assuming that no two engines ever fail simultaneously, and that the aircraft can continue flying with as few as two operating engines, find each probability for a given flight of this aircraft. (Give answers to four decimal places.)

83. no engine failures

84. exactly one engine failure (any one of four engines)

85. exactly two engine failures (any two of four engines)

86. a failed flight

Fair Decisions from Biased Coins *Many everyday decisions, like who will drive to lunch or who will pay for the coffee, are made by the toss of a (presumably fair) coin and using the criterion "heads, you will; tails, I will." This criterion is not quite fair, however, if the coin is biased (perhaps due to slightly irregular construction or wear). John von Neumann suggested a way to make perfectly fair decisions, even with a possibly biased coin. If a coin, biased so that*

$$P(\text{h}) = 0.5200 \quad \text{and} \quad P(\text{t}) = 0.4800,$$

is tossed twice, find each probability. Give answers to four decimal places.

87. $P(\text{hh})$ **88.** $P(\text{ht})$

89. $P(\text{th})$ **90.** $P(\text{tt})$

One-and-one Free Throw Shooting in Basketball *In basketball, "one-and-one" free throw shooting (commonly called foul shooting) is done as follows: If the player makes the first shot (1 point), she is given a second shot. If she misses the first shot, she is not given a second shot (see the tree diagram).*

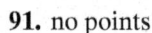

First shot	Second shot	Total points
Point	Point	2
	No point	1
No point		0

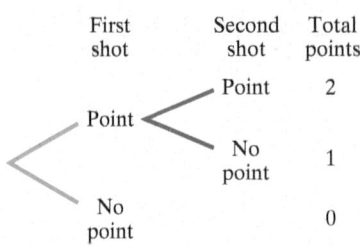

Christine, a basketball player, has a 70% free throw record. (She makes 70% of her free throws.) Find the probability that, on a given one-and-one free throw shooting opportunity, Christine will score each number of points.

91. no points   **92.** one point

93. two points   **94.** three points

Use the methods discussed so far in this chapter to solve each problem.

95. ***Card Choice*** What is the probability that a single card drawn from a standard 52-card deck will be neither a club nor a jack? (*MT* calendar problem)

96. ***Number Choice*** Two different whole numbers are randomly selected from the set $\{1, 2, 3, \ldots, 9\}$. What is the probability that their product is an even number? (*MT* calendar problem)

97. ***Number Choice*** There are 24 four-digit numbers that can be formed using each of the digits 1, 2, 5, and 7 exactly once in each number. If one of those 24 numbers is randomly selected, what is the probability that it is a prime number? (*MT* calendar problem)

98. ***Dice Total*** If three fair dice are rolled and the numbers of dots on the top of each of the dice are added together, what is the probability that the sum will be odd? (*MT* calendar problem)

99. ***Ping-Pong Ball Choice*** Fifty Ping-Pong balls, each marked with a different counting number from 1 to 50, are placed in a bag. One by one, fifteen students remove a ball from the bag, announce its number, and then return the ball to the bag, at which time the next student takes his or her turn. What is the probability that at least two students will pick the same ball? (*MT* calendar problem) (*Hint*: Refer to the method used in solving the "birthday problem" that follows this exercise set.)

100. ***Birthday Day of the Week*** What is the probability that, of three people selected at random, at least two of them were born on the same day of the week? (*MT* calendar problem)

101. ***Dice Total*** What is the probability of a sum of 6 when three fair dice are tossed? (*MT* calendar problem)

102. ***Triangle Side Lengths*** The numbers in the set $\{1, 2, 3, 4, 5\}$ are segment lengths in inches. If three numbers are randomly selected from the set with replacement, what is the probability that the three values can be the side lengths of a triangle? (*MT* calendar problem)

103. *Letter Choice* Suppose a bag contains the three letters of the word *MOM*. If you take one letter out at a time and line the letters up from left to right as you take each one out, what is the probability that the letters will spell the word *MOM*? (*MT* calendar problem)

104. *Coin Choice* Five counterfeit coins are mixed with nine authentic coins. If two coins are drawn at random, find the probability that one coin is authentic and one is counterfeit. (*MT* calendar problem)

105. *Card Choice* A deck of Italian playing cards contains 40 cards. The Italian deck contains four suits (coins, swords, cups, and clubs) and ten values (1 through 7, soldier, horse, and king). When drawing 2 cards without replacement, how much higher is the probability of drawing a pair of kings from a deck of Italian playing cards than from a standard deck of American playing cards? (*MT* calendar problem)

106. *Seating Choice* You, your friend, and three other people are randomly assigned seats in a row of five chairs. What is the probability that you and your friend will be seated next to each other? (*MT* calendar problem)

107. *Sock Choice* Seven white socks and four black socks are in a bag. If two socks are drawn at random, what is the probability that both socks are the same color? (*MT* calendar problem)

108. *Tetrahedron Die Roll* A tetrahedron has one blue, one red, one green, and one yellow face. If you toss the tetrahedron twice, what is the probability that it lands on the same color both times? (*MT* calendar problem)

109. *Chip Choice* A box contains exactly five chips, three of which are red and two of which are blue. Chips are randomly removed one at a time without replacement until all the red chips are drawn or all the blue chips are drawn. What is the probability that the last chip drawn is blue? (*MT* calendar problem)

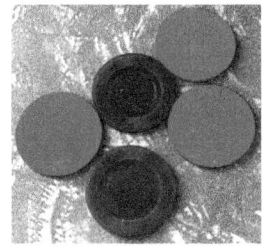

110. *Prime Choice* An integer that is greater than 5 and less than 15 is chosen at random and denoted by x. What is the probability that $x^2 - 29$ is prime? (*MT* calendar problem)

111. *Marble Choice* A pouch contains 6 red marbles and 2 blue marbles. A second pouch contains 3 red marbles and 4 blue marbles. A marble will be drawn at random from the first pouch and placed in the second pouch without its color being noted. Then a marble will be drawn from the second pouch. If the second marble is red, what is the probability that the first marble drawn is blue? (*MT* calendar problem)

112. *Card Choice* Ten people each have a standard deck of cards, which they thoroughly shuffle. If each person draws a card randomly from his or her own deck, what is the probability that one pair of cards will be the same? (*MT* calendar problem)

113. *Race Outcome* Six schools are competing in a track meet. If each school enters three runners in the mile run, and if each runner has an equal chance of winning (or of coming in second or third), what is the probability that the top three finishers will come from the same school? (*MT* calendar problem)

114. *Dice Roll* If three fair standard dice are tossed, what is the probability that the results will be three consecutive integers? (*MT* calendar problem)

115. *Letter Choice* What is the probability that when the letters in the word *mathematics* are arranged at random, they will spell *mathematics*? (*MT* calendar problem)

116. *Dice Product Outcome* A pair of 8-sided dice have sides numbered 1 through 8. Each side has the same probability of landing face up. What is the probability that the product of the two numbers on the sides that land face up exceeds 36?

A. $\dfrac{5}{32}$ **B.** $\dfrac{11}{64}$ **C.** $\dfrac{3}{16}$ **D.** $\dfrac{1}{4}$ **E.** $\dfrac{1}{2}$

(*MT* calendar problem)

117. *Dice Sum Outcome* You roll two fair, six-sided dice with dots from 1 to 6. You calculate the sum of the squares instead of the sum. What is the probability that your score is a prime number? (*MT* calendar problem)

118. *A Two-Headed Coin?* A gambler has two coins in his pocket—one fair coin and one two-headed coin. He selects a coin at random and flips it twice. If he gets two heads, what is the probability that he selected the fair coin? (*MT* calendar problem)

119. *Cube Cuts* A $4'' \times 4'' \times 4''$ cube is painted and then cut into sixty-four $1'' \times 1'' \times 1''$ cubes. A unit cube is then randomly selected and rolled. What is the probability that the top face of the rolled cube is painted? Express your answer as a common fraction. (*MT* calendar problem)

120. *Card Choice* There are three cards, one that is green on both sides, one that is red on both sides, and one that is green on one side and red on the other. One of the three cards is selected randomly and laid on the table. If it happens that the card on the table has a red side up, what is the probability that it is also red on the other side?

FOR FURTHER THOUGHT

The Birthday Problem

A classic problem (with a surprising result) involves the probability that a given group of people will include at least one pair of people with the same birthday (the same day of the year, not necessarily the same year). This problem can be analyzed using the "probability of a complement" formula and the multiplication rule of probability from this section. Suppose there are three people in the group.

P(at least one duplication of birthdays)

$= 1 - P(\text{no duplications})$ Complement formula

$= 1 - P(\text{2nd is different from 1st, and 3rd is different from 1st and 2nd})$

$= 1 - \dfrac{364}{365} \cdot \dfrac{363}{365}$ Multiplication rule

$\approx 1 - 0.992$

$= 0.008$

(To simplify the calculations, we have assumed 365 possible birth dates, ignoring February 29.)

By doing more calculations like the foregoing, we find that the smaller the group, the smaller the probability of a duplication. The larger the group, the larger the probability of a duplication. The table below shows the probability of at least one duplication for numbers of people from 2 through 52.

For Group or Individual Investigation

1. Based on the data shown in the table, what are the odds in favor of a duplication in a group of 30 people?

2. Estimate from the table the least number of people for which the probability of duplication is at least $\frac{1}{2}$.

3. How small a group is required for the probability of a duplication to be *exactly* 0?

4. How large a group is required for the probability of a duplication to be *exactly* 1?

Number of People	Probability of at Least One Duplication	Number of People	Probability of at Least One Duplication	Number of People	Probability of at Least One Duplication
2	0.003	19	0.379	36	0.832
3	0.008	20	0.411	37	0.849
4	0.016	21	0.444	38	0.864
5	0.027	22	0.476	39	0.878
6	0.040	23	0.507	40	0.891
7	0.056	24	0.538	41	0.903
8	0.074	25	0.569	42	0.914
9	0.095	26	0.598	43	0.924
10	0.117	27	0.627	44	0.933
11	0.141	28	0.654	45	0.941
12	0.167	29	0.681	46	0.948
13	0.194	30	0.706	47	0.955
14	0.223	31	0.730	48	0.961
15	0.253	32	0.753	49	0.966
16	0.284	33	0.775	50	0.970
17	0.315	34	0.795	51	0.974
18	0.347	35	0.814	52	0.978

StatCrunch   Simulation: Birthday Problem

11.4 **BINOMIAL PROBABILITY**

OBJECTIVES

1 Construct a simple binomial probability distribution.

2 Apply the binomial probability formula for an experiment involving Bernoulli trials.

Binomial Probability Distribution

Suppose the spinner in **Figure 12** is spun twice. We are interested in the number of times a 2 is obtained. (Assume that 1, 2, and 3 all are equally likely on a given spin.) We can think of the outcome 2 as a "success," while a 1 or a 3 would be a "failure."

When the outcomes of an experiment are divided into just two categories, success and failure, the associated probabilities are called "binomial." (The prefix *bi* means *two*.) Repeated performances of such an experiment, where the probability of success remains constant throughout all repetitions, are also known as repeated **Bernoulli trials** (after James Bernoulli). If we use an ordered pair to represent the result of each pair of spins, then the sample space for this experiment is

$$S = \{(1,1),(1,2),(1,3),(2,1),(2,2),(2,3),(3,1),(3,2),(3,3)\}.$$

The nine outcomes in S are all equally likely. This follows from the numbers 1, 2, and 3 being equally likely on a particular spin.

If x denotes the number of 2s occurring on each pair of spins, then x is an example of a *random variable*. Although we cannot predict the result of any particular pair of spins, we can find the probabilities of various events from the sample space listing. In S, the number of 2s is 0 in four cases, 1 in four cases, and 2 in one case, as reflected in **Table 8.** Because the table includes all possible values of x, together with their probabilities, it is an example of a *probability distribution*. In this case, we have a **binomial probability distribution.** The probability column in **Table 8** has a sum of 1, in agreement with probability Property 3 earlier in this chapter.

In order to develop a general formula for binomial probabilities, we can consider another way to obtain the probability values in **Table 8.** The various spins of the spinner are independent of one another, and on each spin the probability of success (S) is $\frac{1}{3}$ and the probability of failure (F) is $\frac{2}{3}$. We will denote success on the first spin by S_1, failure on the second by F_2, and so on.

$$
\begin{aligned}
P(x = 0) &= P(F_1 \text{ and } F_2) \\
&= P(F_1) \cdot P(F_2) \quad \text{Multiplication rule} \\
&= \frac{2}{3} \cdot \frac{2}{3} \quad \text{Substitute values.} \\
&= \frac{4}{9} \quad \text{Multiply.}
\end{aligned}
$$

$$
\begin{aligned}
P(x = 1) &= P[(S_1 \text{ and } F_2) \text{ or } (F_1 \text{ and } S_2)] \quad \text{2 ways to get } x = 1 \\
&= P(S_1 \text{ and } F_2) + P(F_1 \text{ and } S_2) \quad \text{Addition rule} \\
&= P(S_1) \cdot P(F_2) + P(F_1) \cdot P(S_2) \quad \text{Multiplication rule} \\
&= \frac{1}{3} \cdot \frac{2}{3} + \frac{2}{3} \cdot \frac{1}{3} \quad \text{Substitute values.} \\
&= \frac{2}{9} + \frac{2}{9} \quad \text{Multiply.} \\
&= \frac{4}{9} \quad \text{Add.}
\end{aligned}
$$

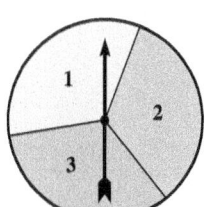

Figure 12

Table 8 Probability Distribution for the Number of 2s in Two Spins

x	$p(x)$
0	$\frac{4}{9}$
1	$\frac{4}{9}$
2	$\frac{1}{9}$
	Sum: $\frac{9}{9} = 1$

$$P(x = 2) = P(S_1 \text{ and } S_2)$$

$$= P(S_1) \cdot P(S_2) \quad \text{Multiplication rule}$$

$$= \frac{1}{3} \cdot \frac{1}{3} \quad \text{Substitute values.}$$

$$= \frac{1}{9} \quad \text{Multiply.}$$

There is only one way to get $x = 0$ (namely, F_1 and F_2). And there is only one way to get $x = 2$ (namely, S_1 and S_2). But there are two ways to get $x = 1$. One way is S_1 and F_2; the other is F_1 and S_2. There are two ways because the one success required can occur on the first spin or on the second spin. How many ways can exactly one success occur in two repeated trials? This question is equivalent to

How many size-one subsets are there of the set of two trials?

The answer is $_2C_1 = 2$. The expression $_2C_1$ denotes "combinations of 2 things taken 1 at a time." Each of the two ways to get exactly one success has a probability equal to $\frac{1}{3} \cdot \frac{2}{3}$, the probability of success times the probability of failure.

EXAMPLE 1 **Constructing a Probability Distribution**

Use a tree diagram to construct a probability distribution for the number of 2s in three spins of the spinner in **Figure 12**.

Solution

If the spinner is spun three times rather than two, then x, the number of successes, could have values of 0, 1, 2, or 3. Then the number of ways to get exactly 1 success is $_3C_1 = 3$. They are

$$S_1 \text{ and } F_2 \text{ and } F_3, \quad F_1 \text{ and } S_2 \text{ and } F_3, \quad F_1 \text{ and } F_2 \text{ and } S_3.$$

The probability of each of these three ways is $\frac{1}{3} \cdot \frac{2}{3} \cdot \frac{2}{3} = \frac{4}{27}$.

$$P(x = 1) = 3 \cdot \frac{4}{27} = \frac{12}{27} = \frac{4}{9}$$

Figure 13 shows all possibilities for three spins, and **Table 9** gives the associated probability distribution. In the tree diagram, the number of ways of obtaining two successes in three trials is 3, in agreement with the fact that $_3C_2 = 3$. Also, the sum of the $P(x)$ column in **Table 9** is again 1.

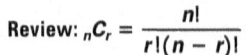

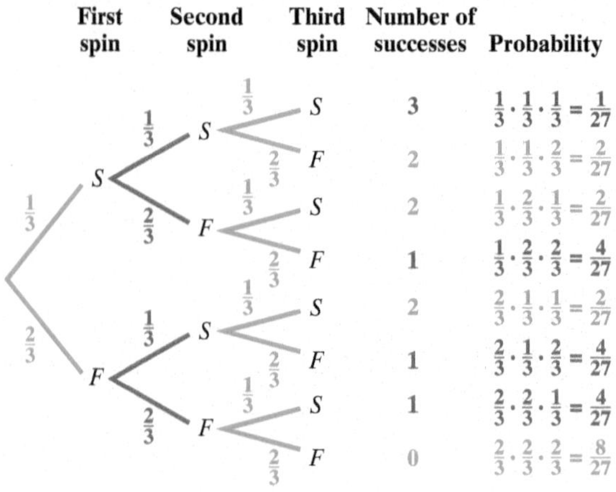

Review: $_nC_r = \dfrac{n!}{r!(n-r)!}$

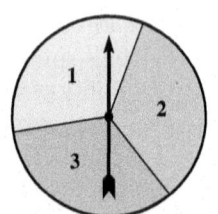

Figure 12 (repeated)

StatCrunch Spinner

Table 9 Probability Distribution for the Number of 2s in Three Spins

x	$P(x)$
0	$\frac{8}{27}$
1	$\frac{12}{27}$
2	$\frac{6}{27}$
3	$\frac{1}{27}$
Sum:	$\frac{27}{27} = 1$

	First spin	Second spin	Third spin	Number of successes	Probability
			S	3	$\frac{1}{3} \cdot \frac{1}{3} \cdot \frac{1}{3} = \frac{1}{27}$
		S	F	2	$\frac{1}{3} \cdot \frac{1}{3} \cdot \frac{2}{3} = \frac{2}{27}$
	S		S	2	$\frac{1}{3} \cdot \frac{2}{3} \cdot \frac{1}{3} = \frac{2}{27}$
		F	F	1	$\frac{1}{3} \cdot \frac{2}{3} \cdot \frac{2}{3} = \frac{4}{27}$
		S	S	2	$\frac{2}{3} \cdot \frac{1}{3} \cdot \frac{1}{3} = \frac{2}{27}$
	F		F	1	$\frac{2}{3} \cdot \frac{1}{3} \cdot \frac{2}{3} = \frac{4}{27}$
		F	S	1	$\frac{2}{3} \cdot \frac{2}{3} \cdot \frac{1}{3} = \frac{4}{27}$
			F	0	$\frac{2}{3} \cdot \frac{2}{3} \cdot \frac{2}{3} = \frac{8}{27}$

Tree diagram for three spins

Figure 13

James Bernoulli (1654–1705) is also known as Jacob or Jacques. He was charmed away from theology by the writings of Leibniz, became his pupil, and later headed the mathematics faculty at the University of Basel. His results in probability are contained in the *Art of Conjecture,* which was published in 1713, after his death, and which also included a reprint of the earlier Huygens paper. Bernoulli also made many contributions to calculus and analytic geometry.

StatCrunch Experiment: Flip Coin

From the
DISTR menu

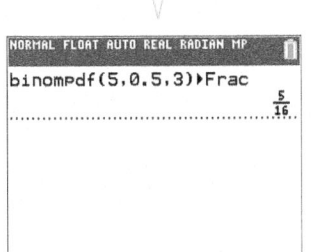

The TI-83/84 Plus C calculator will find the probability discussed in **Example 2.**

Problem-Solving Strategy

One of the various problem-solving strategies from earlier in this text was "Look for a pattern." Having constructed complete probability distributions for binomial experiments with 2 and 3 repeated trials (and probability of success $\frac{1}{3}$), we can now generalize the observed pattern to any binomial experiment, as shown next.

Binomial Probability Formula

Define the following quantities.

n = the number of repeated trials

p = the probability of success on any given trial

$q = 1 - p$ = the probability of failure on any given trial

x = the number of successes that occur

Note that p remains fixed throughout all n trials. This means that all trials are independent of one another. The random variable x representing the number of successes can have any integer value from 0 through n.

In general, x successes can be assigned among n repeated trials in $_nC_x$ different ways, because this is the number of different subsets of x positions among a set of n positions. Also, regardless of which x of the trials result in successes, there will always be x successes and $n - x$ failures, so we multiply x factors of p and $n - x$ factors of q together.

BINOMIAL PROBABILITY FORMULA

When n independent repeated trials occur, where

p = probability of success   and q = probability of failure,

with p and q (where $q = 1 - p$) remaining constant throughout all n trials, the probability of exactly x successes is calculated as follows.

$$P(x) = \,_nC_x p^x q^{n-x} = \frac{n!}{x!(n-x)!} p^x q^{n-x}$$

Binomial probabilities for particular values of n, p, and x can be found directly using tables, statistical software, and some handheld calculators. In the following examples, we use the formula derived above.

EXAMPLE 2   Finding Probability in Coin Tossing

Find the probability of obtaining exactly three heads in five tosses of a fair coin.

Solution

Let heads be "success." Then this is a binomial experiment with

$$n = 5, \quad p = \frac{1}{2}, \quad q = \frac{1}{2}, \quad \text{and} \quad x = 3.$$

$$P(x = 3) = \,_5C_3 \left(\frac{1}{2}\right)^3 \left(\frac{1}{2}\right)^{5-3} = 10 \cdot \frac{1}{8} \cdot \frac{1}{4} = \frac{5}{16} \quad \text{Binomial probability formula}$$

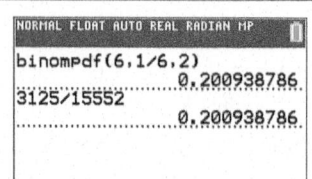

This screen supports the answer in
Example 3.

StatCrunch Experiment: Roll Dice

EXAMPLE 3 Finding Probability in Dice Rolling

Find the probability of obtaining exactly two 5s in six rolls of a fair die.

Solution

Let 5 be "success." Then $n = 6$, $p = \frac{1}{6}$, $q = \frac{5}{6}$, and $x = 2$.

$$P(x = 2) = {}_6C_2\left(\frac{1}{6}\right)^2\left(\frac{5}{6}\right)^4 = 15 \cdot \frac{1}{36} \cdot \frac{625}{1296} = \frac{3125}{15{,}552} \approx 0.201$$

In the case of repeated independent trials, when an event involves more than one specific number of successes, we can employ the binomial probability formula along with the complement and/or addition rules.

EXAMPLE 4 Finding the Probability of Female Children

A couple plans to have 5 children. Find the probability that they will have more than 3 girls. (Assume girl and boy babies are equally likely.)

Solution

Let a girl be "success." Then $n = 5$, $p = q = \frac{1}{2}$, and $x > 3$.

$$P(x > 3) = P(x = 4 \text{ or } 5) \qquad \text{More than 3 means 4 or 5.}$$

$$= P(4) + P(5) \qquad \text{Addition rule}$$

$$= {}_5C_4\left(\frac{1}{2}\right)^4\left(\frac{1}{2}\right)^1 + {}_5C_5\left(\frac{1}{2}\right)^5\left(\frac{1}{2}\right)^0 \qquad \text{Binomial probability formula}$$

$$= 5 \cdot \frac{1}{16} \cdot \frac{1}{2} + 1 \cdot \frac{1}{32} \cdot 1 \qquad \text{Simplify.}$$

$$= \frac{5}{32} + \frac{1}{32} = \frac{6}{32} = \frac{3}{16} = 0.1875$$

This screen supports the answer in
Example 4.

This screen supports the answer in
Example 5.

EXAMPLE 5 Finding the Probability of Hits in Baseball

Gary Bell, a pitcher who is also an excellent hitter, has a well-established career batting average of .300. Suppose that he bats ten times. Find the probability that he will get more than two hits in the ten at-bats.

Solution

This "experiment" involves $n = 10$ repeated Bernoulli trials, with probability of success (a hit) given by $p = 0.3$ (which implies $q = 1 - 0.3 = 0.7$). Since, in this case, "more than 2" means

"3 or 4 or 5 or 6 or 7 or 8 or 9 or 10"   (eight different possibilities),

it will be less work to apply the complement rule.

$$P(x > 2) = 1 - P(x \le 2) \qquad \text{Complement rule}$$

$$= 1 - P(x = 0 \text{ or } 1 \text{ or } 2) \qquad \text{Only three different possibilities}$$

$$= 1 - [P(0) + P(1) + P(2)] \qquad \text{Addition rule}$$

$$= 1 - [{}_{10}C_0(0.3)^0(0.7)^{10} \qquad \text{Binomial probability formula}$$

$$\qquad + {}_{10}C_1(0.3)^1(0.7)^9 + {}_{10}C_2(0.3)^2(0.7)^8]$$

$$\approx 1 - [0.0282 + 0.1211 + 0.2335] \qquad \text{Simplify.}$$

$$= 1 - 0.3828 = 0.6172$$

11.4 EXERCISES

CONCEPT CHECK *Complete each statement.*

1. In repeated Bernoulli trials, the probability of success remains _____ throughout he experiment.

2. If the probability of success on any given trial is 75%, then the probability of failure on any given trial is _____.

3. The value of

$$_6C_2 \left(\frac{1}{2}\right)^2 \left(\frac{1}{2}\right)^{6-2}$$

is _____.

4. Based on the result of **Example 4,** the probability that the couple will have 0, 1, 2, or 3 girls is 1 − _____, or _____.

For Exercises 5–26, give all numerical answers as common fractions reduced to lowest terms. For Exercises 27–59, give all numerical answers to three decimal places.

Coin Tossing *If three fair coins are tossed, find the probability of each number of heads.*

5. 0　　　　　　**6.** 1

7. 2　　　　　　**8.** 3

9. 1 or 2　　　　**10.** at least 1

11. no more than 1　　**12.** fewer than 3

Pascal's Triangle *was discussed in an earlier chapter, and a small portion is reproduced here.*

```
          1
        1   1
      1   2   1
    1   3   3   1
  1   4   6   4   1
```

13. *Relating Pascal's Triangle to Coin Tossing* Explain how the probabilities in **Exercises 5–8** here relate to row 3 of the "triangle." (We will refer to the topmost row of the triangle as "row number 0" and to the leftmost entry of each row as "entry number 0.")

14. Generalize the pattern in **Exercise 13** to complete the following statement. If *n* fair coins are tossed, the probability of exactly *x* heads is the fraction whose numerator is entry number _____ of row number _____ in Pascal's triangle, and whose denominator is the sum of the entries in row number _____.

Binomial Probability Applied to Tossing Coins *Use the pattern noted in* **Exercises 13 and 14** *to find the probabilities of each number of heads when seven fair coins are tossed.*

15. 0　　**16.** 1　　**17.** 2　　**18.** 3

19. 4　　**20.** 5　　**21.** 6　　**22.** 7

Binomial Probability Applied to Rolling Dice *A fair die is rolled three times. A 4 is considered "success," while all other outcomes are "failures." Find the probability of each number of successes.*

23. 0　　**24.** 1　　**25.** 2　　**26.** 3

For n repeated independent trials, with constant probability of success p for all trials, find the probability of exactly x successes in Exercises 27–30.

27. $n = 5, \quad p = \frac{1}{3}, \quad x = 4$

28. $n = 10, \quad p = \frac{7}{10}, \quad x = 5$

29. $n = 20, \quad p = 0.125, \quad x = 2$

30. $n = 30, \quad p = 0.6, \quad x = 22$

For the following exercises, refer to **Example 5.**

31. *Batting Averages in Baseball* If Gary's batting average is exactly .300 going into the series based on exactly 1200 career hits out of 4000 previous times at bat, what is the greatest his average could possibly be when he goes up to bat the tenth time of the series?

32. *Batting Averages in Baseball* Refer to **Exercise 31.** What is the least his average could possibly be when he goes up to bat the tenth time of the series?

33. *Batting Averages in Baseball* Does Gary's probability of a hit really remain constant at exactly 0.300 through all ten times at bat? Explain your reasoning.

34. Do you think the use of the binomial probability formula was justified in **Example 5,** even though *p* is not strictly constant? Explain your reasoning.

Random Selection of Answers on a Multiple-choice Test *Beth is taking a ten-question multiple-choice test for which each question has three answer choices, only one of which is correct. Beth decides on answers by rolling a fair die and marking the first answer choice if the die shows 1 or 2, the second if it shows 3 or 4, and the third if it shows 5 or 6. Find the probability of each event in Exercises 35–38.*

35. exactly four correct answers

36. exactly seven correct answers

37. fewer than three correct answers

38. at least seven correct answers

Side Effects of Prescription Drugs *It is known that a certain prescription drug produces undesirable side effects in 35% of all patients who use it. Among a random sample of eight patients using the drug, find the probability of each event.*

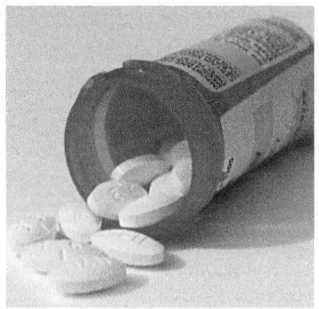

39. None has undesirable side effects.

40. Exactly one has undesirable side effects.

41. Exactly two have undesirable side effects.

42. More than two have undesirable side effects.

Likelihood of Having Brown Eyes *It has been shown that 60% of a certain group of people have brown eyes. Find the probability that, among nine of these people, each number will have brown eyes.*

43. exactly 4

44. from 4 through 6

45. all 9

46. at least 3

Solve each problem.

47. ***Ball Choices*** A bag contains only white balls and black balls. Let p be the probability that a ball selected at random is black. Each time a ball is selected, it is placed back in the bag before the next ball is selected. Four balls are selected at random. What is the probability that two of the four balls are black and two are white? (*MT* calendar problem) (*Hint:* Use the binomial probability formula to express the probability in terms of p.)

48. ***Ball Choices*** Evaluate the probability of **Exercise 47** in the case where the bag actually contains 15 black balls and 25 white balls.

49. ***Frost Survival among Orange Trees*** If it is known that 65% of all orange trees will survive a hard frost, then what is the probability that at least half of a group of six trees will survive such a frost?

50. ***Rate of Favorable Media Coverage of an Incumbent President*** During a presidential campaign, 64% of the political columns in a certain group of major newspapers were favorable to the incumbent president. If a sample of fifteen of these columns is selected at random, what is the probability that exactly ten of them will be favorable?

51. ***Student Ownership of Personal Computers*** At a large midwestern university, 90% of all students have their own personal computers. If five students at that university are selected at random, find the probability that exactly three of them have their own computers.

Taking a Random Walk *Abby is parked at a mile marker on an east-west country road. She decides to toss a fair coin 10 times, each time driving 1 mile east if it lands heads up and 1 mile west if it lands tails up. The term "random walk" applies to this process, even though she drives rather than walks. It is a simplified model of Brownian motion, which is mentioned in a margin note earlier in the chapter.*

In each of Exercises 52–59, find the probability that Abby's "walk" will end as described.

52. 6 miles east of the start

53. 10 miles east of the start

54. 5 miles west of the start

55. 6 miles west of the start

56. at least 2 miles east of the start

57. 2 miles east of the start

58. exactly at the start

59. at least 2 miles from the start

60. ***Random Walk Simulation*** Do an online search for "Random walk simulator" that will provide the estimated probability of ending within a bordered area after a given number of steps. Write a short report on how it works.

11.5 # EXPECTED VALUE AND SIMULATION

OBJECTIVES

1 Determine expected value of a random variable and expected net winnings in a game of chance.

2 Determine whether a game of chance is a fair game.

3 Use expected value to make business and insurance decisions.

4 Use simulation in genetic processes such as determination of flower color and birth gender.

Expected Value

We repeat the beginning of **Example 6** in the second section of this chapter.

Amy plans to spend from 1 to 6 hours on her homework. If x represents the number of hours to be spent on a given night, then the probabilities of the various values of x, rounded to the nearest hour, are as shown in **Table 10.**

If Amy's friend Tara asks her how many hours her studies will take, what would be her best guess? Six different time values are possible, with some more likely than others. One thing Amy could do is calculate a "weighted average" by multiplying each possible time value by its probability and then adding the six products.

$$1(0.05) + 2(0.10) + 3(0.20) + 4(0.40) + 5(0.10) + 6(0.15)$$

$$= 0.05 + 0.20 + 0.60 + 1.60 + 0.50 + 0.90$$

$$= 3.85$$

Thus 3.85 hours is the **expected value** (or the **mathematical expectation**) of the quantity of time to be spent.

Table 10

x	$P(x)$
1	0.05
2	0.10
3	0.20
4	0.40
5	0.10
6	0.15

EXPECTED VALUE

If a random variable x can have any of the values $x_1, x_2, x_3, \ldots, x_n$, and the corresponding probabilities of these values occurring are $P(x_1)$, $P(x_2)$, $P(x_3), \ldots, P(x_n)$, then $E(x)$, the **expected value of x,** is calculated as follows.

$$E(x) = x_1 \cdot P(x_1) + x_2 \cdot P(x_2) + x_3 \cdot P(x_3) + \cdots + x_n \cdot P(x_n)$$

EXAMPLE 1 Finding the Expected Number of Boys

Find the expected number of boys for a three-child family—that is, the expected value of the number of boys. Assume girls and boys are equally likely.

Solution

The sample space for this experiment is

$$S = \{ggg, ggb, gbg, bgg, gbb, bgb, bbg, bbb\}.$$

The probability distribution is shown in **Table 11.**

Table 11

Number of Boys x	Probability $P(x)$	Product $x \cdot P(x)$
0	$\frac{1}{8}$	0
1	$\frac{3}{8}$	$\frac{3}{8}$
2	$\frac{3}{8}$	$\frac{6}{8}$
3	$\frac{1}{8}$	$\frac{3}{8}$

Expected value: $E(x) = \frac{12}{8} = \frac{3}{2}$

The expected number of boys is $\frac{3}{2}$, or 1.5. This result seems reasonable. Boys and girls are equally likely, so "half" the children are expected to be boys.

The expected value for the number of boys in the family could never actually occur. It is only a kind of long-run average of the various values that *could* occur. If we record the number of boys in many different three-child families, then by the law of large numbers, or law of averages, as the number of observed families increases, the observed average number of boys should approach the expected value.

Games and Gambling

EXAMPLE 2 Finding Expected Winnings

A player pays $3 to play the following game: He tosses three fair coins and receives back "payoffs" of $1 if he tosses no heads, $2 for one head, $3 for two heads, and $4 for three heads. Find the player's expected net winnings for this game.

Solution

See **Table 12.** For each possible event, "net winnings" are "gross winnings" (payoff) minus cost to play. Probabilities are derived from the sample space.

$$S = \{\text{ttt, htt, tht, tth, hht, hth, thh, hhh}\}$$

Dollar Coins (all heads)

Table 12

Number of Heads	Payoff	Net Winnings x	Probability $P(x)$	Product $x \cdot P(x)$
0	$1	−$2	$\frac{1}{8}$	−$$\frac{2}{8}$
1	2	−1	$\frac{3}{8}$	−$\frac{3}{8}$
2	3	0	$\frac{3}{8}$	0
3	4	1	$\frac{1}{8}$	$\frac{1}{8}$

Expected value: $E(x) = -\$\frac{1}{2} = -\$0.50$

The expected net loss of 50 cents is a long-run average only. On any particular play of this game, the player would lose $2 or lose $1 or break even or win $1. Over a long series of plays, say 100, there would be some wins and some losses, but the total net result would likely be around a $100 \cdot (\$0.50) = \$50$ *loss.*

A game in which the expected net winnings are zero is called a **fair game.** The game in **Example 2** has negative expected net winnings, so it is unfair against the player. A game with positive expected net winnings is unfair in favor of the player.

EXAMPLE 3 Finding the Fair Cost to Play a Game

The $3 cost to play the game of **Example 2** makes the game unfair against the player (since the player's expected net winnings are negative). What cost would make this a fair game?

Solution

We already computed, in **Example 2,** that the $3 cost to play resulted in an expected net loss of $0.50. Therefore, we can conclude that the $3 cost was 50 cents too high. A fair cost to play the game would then be $3 − $0.50 = $2.50.

Roulette ("little wheel") was invented in France in the seventeenth or early eighteenth century. The disk is divided into red and black alternating compartments numbered 1 to 36 (but not in that order). There is a compartment also for 0 (and for 00 in the United States). The wheel is set in motion, and an ivory ball is thrown into the bowl opposite to the direction of the wheel. When the wheel stops, the ball comes to rest in one of the compartments—the number and color determine who wins.

The players bet against the banker (person in charge of the pool of money) by placing money or equivalent chips in spaces on the roulette table corresponding to the wheel's colors or numbers. Bets can be made on one number or several, on odd or even, on red or black, or on combinations. The banker pays off according to the odds against the particular bet(s). For example, the classic payoff for a winning single number is $36 for each $1 bet.

The result in **Example 3** can be verified. Disregard the cost to play, and find the expected *gross* winnings by summing the products of payoff times probability.

$$E(\text{gross winnings}) = \$1 \cdot \frac{1}{8} + \$2 \cdot \frac{3}{8} + \$3 \cdot \frac{3}{8} + \$4 \cdot \frac{1}{8} = \frac{\$20}{8} = \$2.50$$

Expected gross winnings (payoff) are $2.50, so this amount is a fair cost to play.

EXAMPLE 4   Finding the Fair Cost to Play a Game

In a certain state lottery, a player chooses three digits, in a specific order. Leading digits may be 0, so numbers such as 028 and 003 are legitimate entries. The lottery operators randomly select a three-digit sequence, and any player matching that selection receives a payoff of $600. What is a fair cost to play this game?

Solution

In this case, no cost has been proposed, so we have no choice but to compute expected *gross* winnings. The probability of selecting all three digits correctly is $\frac{1}{10} \cdot \frac{1}{10} \cdot \frac{1}{10} = \frac{1}{1000}$, and the probability of not selecting all three correctly is $1 - \frac{1}{1000} = \frac{999}{1000}$. The expected gross winnings are

$$E(\text{gross winnings}) = \$600 \cdot \frac{1}{1000} + \$0 \cdot \frac{999}{1000} = \$0.60.$$

Thus the fair cost to play this game is 60 cents. (In fact, the lottery charges $1 to play, so players should expect to lose 40 cents per play *on the average*.)

State lotteries must be unfair against players because they are designed to help fund benefits (such as the state's school system), as well as to cover administrative costs and certain other expenses. Among people's reasons for playing may be a willingness to support such causes, but most people undoubtedly play for the chance to "beat the odds" and be one of the few net winners.

Gaming casinos are major business enterprises, by no means designed to break even. The games they offer are always unfair in favor of the house. The bias does not need to be great, however, because even relatively small average losses per player, multiplied by large numbers of players, can result in huge profits for the house.

EXAMPLE 5   Finding Expected Winnings in Roulette

One simple type of *roulette* is played with an ivory ball and a wheel set in motion. The wheel contains thirty-eight compartments. Eighteen of the compartments are black, eighteen are red, one is labeled "zero," and one is labeled "double zero." (These last two are neither black nor red.) In this case, assume the player places $1 on either red or black. If the player picks the correct color of the compartment in which the ball finally lands, the payoff is $2. Otherwise, the payoff is zero. Find the expected net winnings.

Solution

By the expected value formula, expected net winnings are

$$E(\text{net winnings}) = (\$1)\frac{18}{38} + (-\$1)\frac{20}{38} = -\$\frac{1}{19}.$$

The expected net *loss* here is $\$\frac{1}{19}$, or about 5.3¢, per play.

Investments

EXAMPLE 6 Finding Expected Investment Profits

Nick has $5000 to invest and will commit the whole amount, for six months, to one of three technology stocks. A number of uncertainties could affect the prices of these stocks, but Nick is confident, based on his research, that one of only several possible profit scenarios will prove true of each one at the end of the six-month period. His complete analysis is shown in **Table 13.** For example, stock ABC could lose $400, gain $800, or gain $1500. Find the expected profit (or loss) for each of the three stocks, and select Nick's optimum choice based on these calculations.

Table 13

Company ABC		Company RST		Company XYZ	
Profit or Loss x	Probability $P(x)$	Profit or Loss x	Probability $P(x)$	Profit or Loss x	Probability $P(x)$
−$400	0.2	$500	0.8	$0	0.4
800	0.5	1000	0.2	700	0.3
1500	0.3			1200	0.1
				2000	0.2

Solution

Apply the expected value formula.

ABC: $-\$400 \cdot (0.2) + \$800 \cdot (0.5) + \$1500 \cdot (0.3) = \$770$ ← Largest

RST: $\$500 \cdot (0.8) + \$1000 \cdot (0.2) = \$600$

XYZ: $\$0 \cdot (0.4) + \$700 \cdot (0.3) + \$1200 \cdot (0.1) + \$2000 \cdot (0.2) = \$730$

The largest expected profit is $770. Nick should invest in stock ABC.

Of course, by investing in stock ABC, Nick may in fact *lose* $400 over the six months. The "expected" return of $770 is only a long-run average over many identical situations. Because this particular investment situation may never occur again, you may argue that using expected values is not the best approach for Nick to use.

An optimist would ignore most possibilities and focus on the *best* that each investment could do, while a pessimist would focus on the *worst* possibility.

EXAMPLE 7 Choosing Stock Investments

Decide which stock of **Example 6** Nick would pick in each case.

(a) He is an optimist. **(b)** He is a pessimist.

Solution

(a) Disregarding the probabilities, he would focus on the best case for each stock. Since ABC could return as much as $1500, RST as much as $1000, and XYZ as much as $2000, the optimum is $2000. He would buy stock XYZ (the best of the three *best* cases).

(b) In this situation, he would focus on the worst possible cases. Since ABC might return as little as −$400 (a $400 loss), RST as little as $500, and XYZ as little as $0, he would buy stock RST (the best of the three *worst* cases).

The first **Silver Dollar Slot Machine** was fashioned in 1929 by the Fey Manufacturing Company, San Francisco, inventors of the 3-reel, automatic payout machine (1895).

Business and Insurance

EXAMPLE 8 Finding Expected Lumber Revenue

Mike, a lumber wholesaler, is considering the purchase of a truckload of varied dimensional lumber. He calculates that the probabilities of reselling the load for $10,000, for $9000, and for $8000 are 0.22, 0.33, and 0.45, respectively. In order to ensure an *expected* profit of at least $3000, how much can Mike pay for the load?

Solution

The expected revenue (or income) from resales can be found as in **Table 14.**

Table 14 Expected Lumber Revenue

Income x	Probability $P(x)$	Product $x \cdot P(x)$
$10,000	0.22	$2200
9000	0.33	2970
8000	0.45	3600

Expected revenue: $8770

In general, we have the relationship

$$\text{profit} = \text{revenue} - \text{cost}.$$

Therefore, in terms of expectations, we have the following.

$$\text{expected profit} = \text{expected revenue} - \text{cost}$$

$$\$3000 = \$8770 - \text{cost}$$

$$\text{cost} = \$8770 - \$3000 \quad \text{Add cost and subtract \$3000.}$$

$$\text{cost} = \$5770 \quad \text{Subtract.}$$

Mike can pay up to $5770 and still maintain an expected profit of at least $3000.

EXAMPLE 9 Analyzing an Insurance Decision

Jeff, a farmer, will realize a profit of $150,000 on his wheat crop, unless there is rain before harvest, in which case he will realize only $40,000. The long-term weather forecast assigns rain a probability of 0.16. (The probability of no rain is $1 - 0.16 = 0.84$.) An insurance company offers crop insurance of $150,000 against rain for a premium of $20,000. Should he buy the insurance?

Solution

In order to make a wise decision, Jeff computes his expected profit under both options: to insure and not to insure. The complete calculations are summarized in the two "expectation" **Tables 15 and 16** on the next page.

For example, if insurance is purchased and it rains, Jeff's net profit is

$$\begin{bmatrix} \text{Insurance} \\ \text{proceeds} \end{bmatrix} + \begin{bmatrix} \text{Reduced} \\ \text{crop profit} \end{bmatrix} - \begin{bmatrix} \text{Insurance} \\ \text{premium} \end{bmatrix} \quad \text{Net profit}$$

$$\$150,000 \quad + \quad \$40,000 \quad - \quad \$20,000 \quad = \quad \$170,000.$$

Pilots, astronauts, race car drivers, and others train in **simulators.** Some of these devices, which may be viewed as very technical, high-cost versions of video games, imitate conditions to be encountered later in the "real world." A simulator session allows estimation of the likelihood, or probability, of different responses that the learner would display under actual conditions. Repeated sessions help the learner to develop more successful responses before actual lives and equipment are put at risk.

StatCrunch Games: Fair Dice?

Table 15 Expectation When Insuring

	Net Profit x	Probability $P(x)$	Product $x \cdot P(x)$
Rain	$170,000	0.16	$27,200
No rain	130,000	0.84	109,200

Expected profit: $136,400

Table 16 Expectation When Not Insuring

	Net Profit x	Probability $P(x)$	Product $x \cdot P(x)$
Rain	$40,000	0.16	$6400
No rain	150,000	0.84	126,000

Expected profit: $132,400

By comparing expected profits (136,400 > 132,400), we conclude that Jeff is better off buying the insurance.

Simulation

An important area within probability theory is the process called **simulation.** It is possible to study a complicated, or unclear, phenomenon by *simulating,* or imitating, it with a simpler phenomenon involving the same basic probabilities. **Simulation methods** (also called **Monte Carlo methods**) require huge numbers of random digits, so computers are used to produce them. A computer, however, cannot toss coins. It must use an algorithmic process, programmed into the computer, that is called a *random number generator.* It is very difficult to avoid all nonrandom patterns in the results, so the digits produced are called "pseudorandom" numbers. They must pass a battery of tests of randomness before being "approved for use."

Computer scientists and physicists have been encountering unexpected difficulties with even the most sophisticated random number generators. Therefore, these must be carefully checked along with each new simulation application proposed.

On July 19, 2018, an **online simulator for the Monty Hall problem** reported that players who chose STAY had won 202,595 cars out of 603,228 games, for a winning percentage of 34%. Those who chose SWITCH had won 214,244 cars out of 322,860 games, for a winning percentage of 66%.

StatCrunch Games: Let's Make a Deal

Simulating Genetic Traits

Recall Mendel's discovery that when two Rr pea plants (red- flowered but carrying genes for both red and white flowers) are crossed, the offspring will have red flowers if an R gene is received from either parent or from both. This is because red is dominant and white is recessive. **Table 3,** reproduced here in the margin, shows that three of the four equally likely possibilities result in red-flowered offspring.

Now suppose we want to estimate the probability that three offspring in a row will have red flowers. It is much easier (and quicker) to toss coins than to cross pea plants. And the equally likely outcomes, heads and tails, can be used to simulate the transfer of the equally likely genes, R and r. If we toss two coins (say a nickel and a penny), then we can interpret the results as follows.

hh $\Rightarrow$ RR $\Rightarrow$ red gene from first parent and red gene from second parent
$\Rightarrow$ red flowers

ht $\Rightarrow$ Rr $\Rightarrow$ red gene from first parent and white gene from second parent
$\Rightarrow$ red flowers

		Second Parent	
		R	r
First Parent	R	RR	Rr
	r	rR	rr

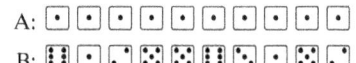

Two **possible sequences** of outcomes, when a fair die is rolled ten times in succession, are

A: 1–1–1–1–1–1–1–1–1–1
and B: 6–1–2–5–5–6–3–1–5–2.

Which of the following options would you choose?

(a) Result A is more likely.
(b) Result B is more likely.
(c) They are equally likely.

Even though the specific sequences A and B are, in fact, equally likely (choice (c)), 67% (rounded) of the students in a class of college freshman business majors, not previously trained in probability theory, initially selected choice (b), as shown here.

Answer Choice	Pre-activity Survey	Post-activity Survey
(a)	4%	0%
(b)	67%	42%
(c)	29%	58%

Note that even after a classroom session of hands-on simulation activities, 42% of the students persisted in their faulty reasoning.

(Data from: *Mathematics Teacher*, September 2014, page 126.)

th $\Rightarrow$ rR $\Rightarrow$ white gene from first parent and red gene from second parent
$\Rightarrow$ red flowers

tt $\Rightarrow$ rr $\Rightarrow$ white gene from first parent and white gene from second parent
$\Rightarrow$ white flowers

Although nothing is certain for a few tosses, the law of large numbers indicates that larger and larger numbers of tosses should become better and better indicators of general trends in the genetic process.

EXAMPLE 10 Simulating Genetic Processes

Toss two coins 50 times and use the results to approximate the probability that the crossing of Rr pea plants will produce three successive red-flowered offspring.

Solution

We actually tossed two coins 50 times and got the following sequence.

th, hh, th, tt, th, hh, ht, th, ht, th, hh, hh, tt, th, hh,
ht, ht, ht, ht, th, hh, hh, hh, tt, ht, tt, hh, ht, ht, hh, tt,
tt, tt, th, tt, tt, hh, ht, ht, ht, hh, tt, th, hh, tt, hh, ht,
tt, tt, tt

By the color interpretation described above, this gives the following sequence of flower colors in the offspring. (Only "both tails" gives white.)

red – red – red – white – red – red – red – red – red – red – red – red – white –
red – red – red – red – red – red – red – red – red – red – white – red – white –
red – red – red – red – white – white – white – red – white – white – red – red –
red – red – red – white – red – red – white – red – red – white – white – white

We now have an experimental list of 48 sets of three successive plants, the 1st, 2nd, and 3rd entries, then the 2nd, 3rd, and 4th entries, and so on. Do you see why there are 48 in all?

Now we just count up the number of these sets of three that are "red-red-red." Since there are 20 of those, our empirical probability of three successive red off-spring, obtained through simulation, is $\frac{20}{48} = \frac{5}{12}$, or about 0.417. By applying the multiplication rule of probability (with all outcomes independent of one another), we find that the theoretical value is

$$\left(\frac{3}{4}\right)^3 = \frac{27}{64}, \quad \text{or about } 0.422,$$

so our approximation obtained by simulation is very close.

Table 17

→ 51592
77876
36500
40571
04822

→ 53033
92080
01587
36006
63698

→ 17297
22841

→ 91979
96480
74949

76896
47588
45521
02472
55184

40177
84861
86937
20931
22454

→ 73219
→ 55707
48007
→ 65191
06772

94928
→ 15709
39922
96365
14655

65587
76905
12369
54219
89329

90060
06975
05050
69774
→ 78351

11464
84086
→ 51497
12307
68009

Simulating Other Phenomena

In human births, boys and girls are (essentially) equally likely. Therefore, an individual birth can be simulated by tossing a fair coin, letting a head correspond to a girl and a tail to a boy.

EXAMPLE 11 Simulating Births with Coin Tossing

A sequence of 40 actual coin tosses produced the results below.

bbggb, gbbbg, gbgbb, bggbg, bbbbg, gbbgg, gbbgg, bgbbg

(For every head we have written g for girl, and for every tail, b for boy.)

(a) How many pairs of two successive births are represented by the sequence?

(b) How many of those pairs consist of both boys?

(c) Find the empirical probability, based on this simulation, that two successive births both will be boys. Give your answer to three decimal places.

Solution

(a) Beginning with the 1st–2nd pair and ending with the 39th–40th pair, there are 39 pairs.

(b) Observing the above sequence, we count 11 pairs of two consecutive boys.

(c) Utilizing parts (a) and (b), we have $\frac{11}{39} \approx 0.282$.

Another way to simulate births, and other phenomena, is with random numbers. The spinner in **Figure 14** can be used to obtain a table of random digits, as in **Table 17**. The 250 random digits generated have been grouped conveniently so that we can easily follow down a column or across a row to carry out a simulation.

The Web site www.random.org furnishes many options for the user to generate random numbers.

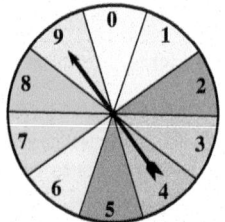

Figure 14

StatCrunch   Spinner

EXAMPLE 12 Simulating Births with Random Numbers

A couple plans to have five children. Use the random number simulation in **Table 17** to estimate the probability that they will have more than three boys.

Solution

Let each sequence of five digits represent a family with five children, and (arbitrarily) associate odd digits with boys, even digits with girls. (Recall that 0 is even.) Verify that, of the fifty families simulated, only the ten marked with arrows have more than 3 boys (4 boys or 5 boys).

$$P(\text{more than 3 boys}) = \frac{10}{50} = 0.20 \quad \text{Estimated (empirical) probability}$$

The theoretical value for the probability in **Example 12** above would be the same as that obtained in **Example 4** of the previous section. It was 0.1875. Our estimate above was fairly close. In light of the law of large numbers, a larger sampling of random digits (more than 50 simulated families) would likely yield a closer approximation.

EXAMPLE 13 Simulating Card Drawing with Random Numbers

Use the random number simulation in **Table 17** to estimate the probability that two cards drawn from a standard deck, with replacement, both will be of the same suit.

Solution

Use this correspondence: 0 and 1 mean clubs, 2 and 3 mean diamonds, 4 and 5 mean hearts, 6 and 7 mean spades, and 8 and 9 are disregarded. If we (arbitrarily) use the first digit of each five-digit group, omitting 8s and 9s, we obtain the sequence

5–7–3–4–0–5–0–3–6–1–2–7–7–4–4–0–5–4–2–2–

7–5–4–6–0–1–3–1–6–7–1–5–0–0–6–7–1–5–1–6.

First digits of all groups

> 5 gives hearts, 7 gives spades, 3 gives diamonds, and so on.

This 40-digit sequence of digits yields the sequence of suits shown next.

hearts – spades – diamonds – hearts – clubs – hearts – clubs – diamonds – spades –

clubs – diamonds – spades – spades – hearts – hearts – clubs – hearts – hearts –

diamonds – diamonds – spades – hearts – hearts – spades – clubs – clubs –

diamonds – clubs – spades – spades – clubs – hearts – clubs – clubs – spades –

spades – clubs – hearts – clubs – spades

Verify that of the 39 successive pairs of suits (hearts–spades, spades–diamonds, diamonds–hearts, etc.), 9 of them are pairs of the same suit. This makes the estimated probability $\frac{9}{39} \approx 0.23$. (For comparison, the theoretical value is 0.25.)

11.5 EXERCISES

CONCEPT CHECK *Suppose that we wish to find the expected number of girls for a four-child family. Complete the following probability distribution, assuming that girls and boys are equally likely.*

	Number of Girls, x	Probability P(x)	Product x · P(x)
	0	$\frac{1}{16}$	0
1.	1	$\frac{4}{16} = \frac{1}{4}$	____
2.	2	$\frac{6}{16} = \frac{3}{8}$	____
3.	3	____	____
4.	4	____	____

5. The expected value $E(x)$ of the number of girls in a four-child family is the sum of the products in the far right column above, which is ____.

6. Does the answer in **Exercise 5** agree with your intuition? Explain.

Expected Winnings in a Die-rolling Game A game consists of rolling a single fair die and pays off as follows: $3 for a 6, $2 for a 5, $1 for a 4, and no payoff otherwise.

7. Find the expected winnings for this game.

8. What is a fair price to pay to play this game?

Expected Winnings in a Die-rolling Game For Exercises 9 and 10, consider a game consisting of rolling a single fair die, with payoffs as follows. If an even number of spots turns up, you receive as many dollars as there are spots up. But if an odd number of spots turns up, you must pay as many dollars as there are spots up.

9. Find the expected net winnings of this game.

10. Is this game fair, or unfair against the player, or unfair in favor of the player?

Solve each problem.

11. *Expected Winnings in a Coin-tossing Game* A certain game involves tossing 3 fair coins, and it pays 10¢ for 3 heads, 5¢ for 2 heads, and 3¢ for 1 head. Is 5¢ a fair price to pay to play this game? That is, does the 5¢ cost to play make the game fair?

12. *Expected Winnings in Roulette* In a form of roulette slightly different from that in **Example 5,** a more generous management supplies a wheel having only thirty-seven compartments, with eighteen red, eighteen black, and one zero. Find the expected net winnings if you bet $1 on red in this game.

13. *Expected Number of Absences in a Math Class* In a certain mathematics class, the probabilities have been empirically determined for various numbers of absentees on any given day. These values are shown in the table below. Find the expected number of absentees on a given day. Give the answer to two decimal places.

Number absent	0	1	2	3	4
Probability	0.18	0.26	0.29	0.23	0.04

14. *Expected Profit of an Insurance Company* An insurance company will insure a $200,000 home for its total value for an annual premium of $650. If the company spends $25 per year to service such a policy, the probability of total loss for such a home in a given year is 0.002, and you assume that either total loss or no loss will occur, what is the company's expected annual gain (or profit) on each such policy?

Profits from a College Foundation Raffle A college foundation raises funds by selling raffle tickets for a new car worth $36,000.

15. If 600 tickets are sold for $120 each, determine each of the following.

(a) The expected *net* winnings of a person buying one of the tickets

(b) The total profit for the foundation, assuming that it had to purchase the car

(c) The total profit for the foundation, assuming that the car was donated

16. For the raffle described in **Exercise 15,** if 720 tickets are sold for $120 each, determine each of the following.

(a) The expected *net* winnings of a person buying one of the tickets

(b) The total profit for the foundation, assuming that it had to purchase the car

(c) The total profit for the foundation, assuming that the car was donated

Winnings and Profits of a Raffle Five thousand raffle tickets are sold. One first prize of $1000, two second prizes of $500 each, and three third prizes of $100 each will be awarded, with all winners selected randomly.

17. If you purchased one ticket, what are your expected gross winnings?

18. If you purchased ten tickets, what are your expected gross winnings?

19. If the tickets were sold for $1 each, how much profit goes to the raffle sponsor?

Solve each problem.

20. *Expected Sales at a Theater Snack Bar* A children's theater found in a random survey that 58 customers bought one snack bar item, 49 bought two items, 31 bought three items, 4 bought four items, and 8 avoided the snack bar altogether. Use this information to find the expected number of snack bar items per customer. Round your answer to the nearest tenth.

21. *Expected Number of Children to Attend an Amusement Park* An amusement park, considering adding some new attractions, conducted a study over several typical days and found that, of 10,000 families entering the park, 1020 brought just one child (defined as younger than age twelve), 3370 brought two children, 3510 brought three children, 1340 brought four children, 510 brought five children, 80 brought six children, and 170 brought no children at all. Find the expected number of children per family attending this park. Round your answer to the nearest tenth.

22. *Expected Sums of Randomly Selected Numbers* Four cards are numbered 1 through 4. Two of these cards are chosen randomly without replacement, and the numbers on them are added. Find the expected value of this sum.

23. *Prospects for Electronics Jobs in a City* In a certain California city, projections for the next year are that there is a 20% chance that electronics jobs will increase by 200, a 50% chance that they will increase by 300, and a 30% chance that they will decrease by 800. What is the expected change in the number of electronics jobs in that city in the next year?

24. *Expected Winnings in Keno* In one version of the game *keno*, the house has a pot containing 80 balls, numbered 1 through 80. A player buys a ticket for $1 and marks on it one number from 1 to 80. The house then selects 20 of the 80 numbers at random. If the number selected by the player is among the 20 selected by the management, the player is paid $3.20. Find the expected net winnings for this game.

Contractor Decisions Based on Expected Profits   Lori, a commercial building contractor, will commit her company to one of three projects, depending on her analysis of potential profits or losses as shown here.

Project A		Project B		Project C	
Profit or Loss x	Probability $P(x)$	Profit or Loss x	Probability $P(x)$	Profit or Loss x	Probability $P(x)$
$60,000	0.10	$0	0.20	$40,000	0.65
180,000	0.60	210,000	0.35	340,000	0.35
250,000	0.30	290,000	0.45		

In Exercises 25–27, determine which project Lori should choose according to each approach.

25. expected values

26. the optimist viewpoint

27. the pessimist viewpoint

28. Refer to **Examples 6 and 7.** Considering the three different approaches (expected values, optimist, and pessimist), which one seems most reasonable to you, and why?

Expected Winnings in a Game Show   A game show contestant is offered the option of receiving a computer system worth $2300 or accepting a chance to win either a luxury vacation worth $5000 or a boat worth $8000. If the second option is chosen the contestant's probabilities of winning the vacation and of winning the boat are 0.20 and 0.15, respectively.

29. If the contestant were to turn down the computer system and go for one of the other prizes, what would be the expected winnings?

30. Purely in terms of monetary value, what is the contestant's wiser choice?

Insurance Purchase   David, the promoter of an outdoor concert, expects a gate profit of $100,000 unless it rains, which would reduce the gate profit to $30,000. The probability of rain is 0.20. For a premium of $25,000 David can purchase insurance coverage that would pay him $100,000 in case of rain.

31. Find the expected net profit when the insurance is purchased.

32. Find the expected net profit when the insurance is not purchased.

33. Based on expected values, which is David's wiser choice in this situation?

34. If you were the promoter, would you base your decision on expected values? Explain your reasoning.

Expected Values in Book Sales   Jessica, an educational publisher representative, presently has five accounts, and her manager is considering assigning her three additional accounts. The new accounts would bring potential volume to her business, and some of her present accounts have potential for growth as well. See the accompanying table.

1	2	3	4	5	6
Account Number	Existing Volume	Potential Additional Volume	Probability of Getting Additional Volume	Expected Value of Additional Volume	Existing Volume plus Expected Value of Additional Volume
1	$10,000	$10,000	0.40	$4000	$14,000
2	30,000	0	—	—	30,000
3	25,000	15,000	0.20	3000	
4	35,000	0	—	—	
5	15,000	5,000	0.30		
6	0	30,000	0.10		
7	0	25,000	0.70		
8	0	45,000	0.60		

35. Compute the four missing expected values in column 5.

36. Compute the six missing amounts in column 6.

37. What is Jessica's total "expected" additional volume?

38. If Jessica achieved her expected additional volume in all accounts, what would be the total volume of all her accounts?

39. If Jessica achieved her expected additional volume in all accounts, by what percentage (to the nearest tenth of a percent) would she increase her total volume?

Solve each problem.

40. *Expected Winnings in Keno* Recall that in the game keno of **Exercise 24,** the house randomly selects 20 numbers from the counting numbers 1–80. In the variation called 6-spot keno, the player pays 60¢ for his ticket and marks 6 numbers of his choice. If the 20 numbers selected by the house contain at least 3 of those chosen by the player, he gets a payoff according to this scheme.

3 of the player's numbers among the 20	$0.35
4 of the player's numbers among the 20	2.00
5 of the player's numbers among the 20	60.00
6 of the player's numbers among the 20	1250.00

Find the player's expected net winnings in this game. (*Hint:* The four probabilities required here can be found using combinations, the fundamental counting principle, and the theoretical probability formula.)

41. *Pea Plant Reproduction Simulation with Coin Tossing* Use the sequence of flower colors of **Example 10** to approximate the probability that *four* successive offspring all will have red flowers.

42. *Pea Plant Reproduction Simulation with Coin Tossing* Explain why, in **Example 10,** fifty tosses of the coins produced only 48 sets of three successive offspring.

43. *Boy and Girl Simulation with Random Numbers* Use **Table 17** to simulate fifty families with three children. Let 0–4 correspond to boys and 5–9 to girls, and use the middle three digits of the 5-digit groupings (159, 787, 650, and so on). Estimate the probability of exactly two boys in a family of three children. Compare with the theoretical probability, which is $\frac{3}{8} = 0.375$.

44. *Empirical Probability of Successive Girls* Simulate 40 births by tossing coins yourself, and obtain an empirical probability for two successive girls.

45. *Likelihoods of Girl and Boy Births* Should the probability of two successive girl births be any different from that of two successive boy births?

One-and-One Free Throw Shooting In Exercises 91–93 of the third section of this chapter, Christine, who had a 70% foul-shooting record, had probabilities of scoring 0, 1, or 2 points of 0.30, 0.21, and 0.49, respectively.

Use **Table 17** *(with digits 0–6 representing hit and 7–9 representing miss) to simulate 50 one-and-one shooting opportunities for Christine. Begin at the top left (5, 7, 3, etc., to the bottom), then move to the second column (1, 7, 6, etc.), going until 50 one-and-one opportunities are obtained. (Some "opportunities" involve one shot and one random digit, while others involve two shots and two random digits.) Keep a tally of the numbers of times 0, 1, and 2 points are scored. (See the top of the next column.)*

Number of Points	Tally
0	
1	
2	

From the tally, find the empirical probability (to two decimal places) of each event.

46. no points   **47.** 1 point   **48.** 2 points

Simulation with Coin Tossing A coin was actually tossed 200 times, producing the following sequence of outcomes. Read from left to right across the top row, then from left to right across the second row, and so on, to the bottom row.

h h t h t	t h t h h	t t t t	t t h t t	h t h h h
t h t t h	h h h h h	h t t h h	h t h h t	h h t t h
t h t h t	h h t t t	h h h h h	t t h t h	t h t h h
t h h h h	h h h h t	t h t t h	t h h t h	t h h t t
t h h t t	t h t t t	h t h h t	t h h h t	h t t t t
h t t h h	h t t t t	t t t h t	t t t t h	t h t h h
h h h h h	t t h h t	t t h h t	h t h t t	h h h h t
h t h t t	h t t t t	h h t t h	t t t h h	t h t t h

Using this sequence, find the empirical probability of each of the following. Round to three decimal places.

49. two consecutive heads

50. two consecutive tails

51. three consecutive tosses of the same outcome

52. six consecutive tosses of the same outcome

Path of a Random Walk Using a Die and a Coin Exercises 52–59 of the fourth section of this chapter illustrated a simple version of the idea of a "random walk." Atomic particles released in nuclear fission also move in a random fashion. During World War II, John von Neumann and Stanislaw Ulam used simulation with random numbers to study particle motion in nuclear reactions. Von Neumann coined the name "Monte Carlo" for the methods used.

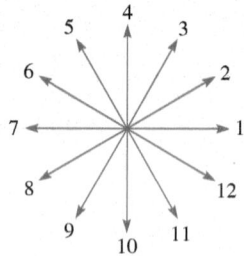

The figure suggests a model for random motion in two dimensions. Assume that a particle moves in a series of 1-unit "jumps," each one in a random direction, any one of 12 equally likely possibilities. One way to choose directions is to roll a fair die and toss a fair coin. The die determines one of the directions 1–6, coupled with heads on the coin. Tails on the coin reverses the direction of the die, so that the die coupled with tails gives directions 7–12. So 3h (meaning 3 with the die and heads with the coin) gives direction 3; 3t gives direction 9 (opposite to 3); and so on.

Solve each problem.

53. Simulate the motion described above with 10 rolls of a die and tosses of a coin. Draw the 10-jump path you get. Make your drawing accurate enough so you can estimate (by measuring) how far from its starting point the particle ends up.

54. Repeat the experiment of **Exercise 53** four more times. Measure distance from start to finish for each of the 5 "random trips." Add these 5 distances and divide the sum by 5, to arrive at an "expected net distance" for such a trip.

Random Walk *Consider another two-dimensional random walk governed by the following conditions.*

- *Start out from a given street corner, and travel one block north. At each intersection:*

- *Turn left with probability $\frac{1}{6}$.*

- *Go straight with probability $\frac{2}{6}\left(=\frac{1}{3}\right)$.*

- *Turn right with probability $\frac{3}{6}\left(=\frac{1}{2}\right)$.*
  (Never turn around.)

Solve each problem.

55. Use **Table 17** to simulate this random walk. For every 1 encountered in the table, turn left and proceed for another block. For every 2 or 3, go straight and proceed for another block. For every 4, 5, or 6, turn right and proceed for another block. Disregard all other digits— that is, 0s, 7s, 8s, and 9s. (Do you see how this scheme satisfies the probabilities given above?) This time, begin at the upper right corner of the table, running down the column 2, 6, 0, and so on, to the bottom. When this column of digits is used up, stop the "walk." Describe, in terms of distance and direction, where you have ended up relative to your starting point.

56. Explain how a fair die could be used to simulate this random walk.

Buffon's Needle Problem *The following problem was posed by Georges Louis Leclerc, Comte de Buffon (1707–1788) in his* Histoire Naturelle *in 1777. A large plane area is ruled with equidistant parallel lines, the distance between two consecutive lines of the series being "a". A thin needle of length*

$$\ell < a$$

is tossed randomly onto the plane. What is the probability that the needle will intersect one of these lines?

The answer to this problem is found using integral calculus, and the probability p is shown to be $p = \frac{2\ell}{\pi a}$. Solving for π gives the formula

$$\pi = \frac{2\ell}{pa},$$

which can be used to approximate the value of π experimentally. This was first observed by Pierre Simon de Laplace, and such an experiment was carried out by Johann Wolf, a professor of astronomy at Bern, in about 1850. (Source: Burton, David M., History of Mathematics: An Introduction. *Wm. C. Brown, 1995.)*

57. On a sheet of paper, draw a series of parallel lines evenly spaced across it. Then find a thin needle, or needlelike object, with a length less than the distance between adjacent parallel lines on the paper.

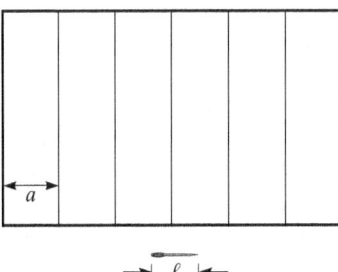

Carry out the following steps in order.

   (a) Measure and record the distance between lines (a) and the length of the needle (ℓ), using the same units for both.

   (b) Drop the needle onto the paper 100 times. Determine whether the needle "hits" a line or not, and keep a tally of hits and misses.

   (c) Now calculate

   $$p = \frac{\text{number of hits}}{100}.$$

   Is this probability value theoretical or empirical?

   (d) Enter the calculated value of p and the measured values of a and ℓ into the formula to obtain an approximation for π. Round this value to four decimal places.

   (e) The correct value of π, to four decimal places, is 3.1416. What value did your simulation give? How far off is it?

58. ***Dynamic Simulation***   See www.metablake.com/pi.swf for a dynamic illustration of the Buffon Needle Problem. (One of the authors of this text used 21,734 iterations, to get 13,805 intersections, for an empirical value of π to be 3.148569358927.)

FOR FURTHER THOUGHT

Expected Value of Games of Chance

Slot machines are a popular game for those who want to lose their money with very little mental effort. We cannot calculate an expected value applicable to all slot machines since payoffs vary from machine to machine. But we can calculate the "typical expected value."

A player operates a classic slot machine by pulling a handle after inserting a coin or coins. (We assume one coin in this discussion.) Reels inside the machine then rotate, and come to rest in some random order. (Most *modern* machines employ an electronic equivalent rather than actual rotating reels.) Assume that three reels show the pictures listed in **Table 18.** For example, of the 20 pictures on the first reel, 2 are cherries, 5 are oranges, 5 are plums, 2 are bells, 2 are melons, 3 are bars, and 1 is the number 7.

A picture of cherries on the first reel, but not on the second, leads to a payoff of 3 coins (*net* winnings: 2 coins); a picture of cherries on the first two reels, but not the third, leads to a payoff of 5 coins (*net* winnings: 4 coins). These and all other winning combinations are listed in **Table 19.**

Because, according to **Table 18,** there are 2 ways of getting cherries on the first reel, 15 ways of *not* getting cherries on the second reel, and 20 ways of getting anything on the third reel, we have a total of $2 \cdot 15 \cdot 20 = 600$ ways of getting a net payoff of 2.

This Cleveland Indians fan hit four 7s in a row on a progressive nickel slot machine at the Sands Casino in Las Vegas in 1988.

Table 18   Pictures on Reels

Pictures	Reels		
	1	2	3
Cherries	2	5	4
Oranges	5	4	5
Plums	5	3	3
Bells	2	4	4
Melons	2	1	2
Bars	3	2	1
7s	1	1	1
Totals	20	20	20

Table 19   Calculating Expected Loss on a Three-Reel Slot Machine

Winning Combinations	Number of Ways	Probability	Number of Coins Received	Net Winnings (in coins)	Probability Times Net Winnings
1 cherry (on first reel)	$2 \cdot 15 \cdot 20 = 600$	600/8000	3	2	1200/8000
2 cherries (on first two reels)	$2 \cdot 5 \cdot 16 = 160$	160/8000	5	4	640/8000
3 cherries	$2 \cdot 5 \cdot 4 = 40$	40/8000	10	9	360/8000
3 oranges	$5 \cdot 4 \cdot 5 = 100$	100/8000	10	9	900/8000
3 plums	$5 \cdot 3 \cdot 3 = 45$	45/8000	14	13	585/8000
3 bells	$\_ \cdot \_ \cdot \_ = \_$	____/8000	18	____	____/8000
3 melons (jackpot)	$\_ \cdot \_ \cdot \_ = \_$	____/8000	100		____/8000
3 bars (jackpot)	$\_ \cdot \_ \cdot \_ = \_$	____/8000	200	____	____/8000
3 7s (jackpot)	$\_ \cdot \_ \cdot \_ = \_$	____/8000	500	____	____/8000
Totals	__				6318/8000

Because there are 20 pictures per reel, there are a total of

$$20 \cdot 20 \cdot 20 = 8000 \text{ possible outcomes.}$$

Hence, the probability of receiving a net payoff of 2 coins is $\frac{600}{8000}$.

Table 19 takes into account all *winning* outcomes, with the necessary products for finding expectation added in the last column. However, since a *nonwinning* outcome can occur in

$$8000 - 988 = 7012 \text{ ways}$$

(with net winnings of -1 coin), the product

$$(-1) \cdot \frac{7012}{8000}$$

must also be included. Hence, the expected value of this particular slot machine is

$$\frac{6318}{8000} + (-1) \cdot \frac{7012}{8000} \approx -0.087 \text{ coin.}$$

On a machine costing one dollar per play, the expected *loss* (per play) is about

$$(0.087)(1 \text{ dollar}) = 8.7 \text{ cents.}$$

Actual slot machines vary in expected loss per dollar of play. But author Hornsby was able to beat a Las Vegas slot machine in 1988.

Table 20 comes from an article by Andrew Sterrett in *The Mathematics Teacher* (March 1967), in which he discusses rules for various games of chance and calculates their expected values. He uses expected values to find expected times it would take to lose $1000 if you played continually at the rate of $1 per play and one play per minute.

For Group or Individual Investigation

1. Explain why the entries of the "Net Winnings" column of **Table 19** are all one fewer than the corresponding entries of the "Number of Coins Received" column.

2. Find the 29 missing values in **Table 19** and verify that the final result,

$$\frac{6318}{8000},$$

is correct. (Refer to **Table 18** for the values in the "Number of Ways" column.)

3. In order to make your money last as long as possible in a casino, which game should you play? (Refer to **Table 20**.)

Table 20 Expected Time to Lose $1000

Game	Expected Value	Days	Hours	Minutes
Roulette (with one 0)	−$0.027	25	16	40
Roulette (with 0 and 00)	−$0.053	13	4	40
Chuck-a-luck	−$0.079	8	19	46
Keno (one number)	−$0.200	3	11	20
Numbers	−$0.300	2	7	33
Football pool (4 winners)	−$0.375	1	20	27
Football pool (10 winners)	−$0.658	1	1	19

FOR FURTHER THOUGHT

Assessing Randomness

Observe the two listings that follow, each of which simulates 200 tosses of a coin. Read from left to right across the top row, then from left to right across the second row, and so on, to the bottom row. One of the listings was obtained by actual tossing, and the other was made up by a student.

B

thhth	hthhh	thttt	ttttt	thhth
hhhtt	thtth	tthth	ttttt	hhtth
hthhh	ttthh	thhtt	tttth	thhth
thhht	hthhh	tttht	thhtt	hthht
hthht	hhhth	ththh	hthtt	tthht
htttt	thhhh	thhhh	hhthh	ththh
htthh	thhtt	tthth	ttttt	hthtt
tthht	ththh	hthhh	hthth	hthth

For Group or Individual Investigation

Make an educated guess about which is which. Consider the following fact of probability. When a fair coin is tossed 200 times, there is a 97% chance that there will be a run of six consecutive heads or six consecutive tails. Now look at the listings again, and based on this *almost* certain outcome, decide whether your earlier choice is correct. (The authors thank Marty Triola for his input into this For Further Thought feature.)

A

thhtt	thhtt	hthhh	tthth	htttth
hthht	thttt	thhth	tthht	hthhh
thhht	thttt	thhht	hthth	htthh
htthh	htthh	thttt	hhtht	thhht
thtth	hthhh	httht	hhtth	hthth
httth	hthhh	htthht	hhttt	thhth
htthh	hthtt	ttthht	hhtth	hhtth
thhtt	thhth	hthth	thhtt	htthh

StatCrunch Experiment: Flip Coin

Chapter **11** SUMMARY

KEY TERMS

11.1

deterministic
  phenomenon
random phenomenon
probability of an event
experiment
outcome
sample space
event
classical definition of
  probability

theoretical probability
empirical probability
law of large numbers (law
  of averages)
odds in favor
odds against

11.2

mutually exclusive events
random variable
probability distribution

11.3

conditional probability of
  B given A
independent events

11.4

Bernoulli trials
binomial probability
  distribution

11.5

expected value
  (mathematical expectation)
simulation
simulation methods
  (Monte Carlo methods)
random number generator

NEW SYMBOLS

$P(E)$ probability of event E

$P(B|A)$ probability of event B, given that event A has occurred

$E(x)$ expected value of random variable x

TEST YOUR WORD POWER

See how well you have learned the vocabulary in this chapter.

1. The **sample space** of an experiment is the
   A. set of all possible events of the experiment.
   B. set of all possible outcomes of the experiment.
   C. number of possible events of the experiment.
   D. number of possible outcomes of the experiment.

2. The **probability of an event** E
   A. must be a number greater than 1.
   B. may be a negative number.
   C. must be a number between 0 and 1, inclusive of both.
   D. cannot be 0.

3. If E is an event, the **probability of "not E"** must be
   A. equal to 0.
   B. equal to 1 plus the probability of E.
   C. equal to 1 minus the probability of E.
   D. less than the probability of E.

4. If the probability of event E is $\frac{a}{b}$, then the **odds in favor of E** are
   A. b to a.   B. $(b - a)$ to a.
   C. b to $(b - a)$.   D. a to $(b - a)$.

5. A **probability distribution**
   A. shows all possible values of a random variable, along with the probabilities that those values will occur.
   B. is a listing of empirical values of an experiment.
   C. is a number between 0 and 1, inclusive of both.
   D. applies the distributive property to probability.

ANSWERS

1. B **2.** C **3.** C **4.** D **5.** A

QUICK REVIEW

Concepts

Examples

11.1 BASIC CONCEPTS

Theoretical Probability Formula

If all outcomes in a sample space S are equally likely, and E is an event within that sample space, then the **theoretical probability** of event E is given by the following formula.

$$P(E) = \frac{\text{number of favorable outcomes}}{\text{total number of outcomes}} = \frac{n(E)}{n(S)}$$

The spinner shown here is spun twice. Find the probability that both spins yield the same number.

The sample space S is

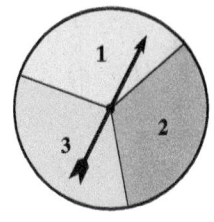

$$\{(1,1), (1,2), (1,3), (2,1), (2,2),$$
$$(2,3), (3,1), (3,2)(3,3)\}.$$

Of these, three indicate the same number:

$$E = \{(1,1), (2,2), (3,3)\}.$$

$$P(E) = \frac{n(E)}{n(S)} = \frac{3}{9} = \frac{1}{3}$$

Empirical Probability Formula

If E is an event that may happen when an experiment is performed, then an **empirical probability** of event E is given by the following formula.

$$P(E) = \frac{\text{number of times event } E \text{ occurred}}{\text{number of times the experiment was performed}}$$

A coin was actually tossed 100 times, producing the following sequence of outcomes. If E represents two consecutive tosses of heads (h), what is the empirical probability of E?

thhht thhht hthth thhhh thttt

ttttt tthht htthh tttth tttth

hhthh thttt thttt thtth thhht

hthhh ttthh hhhht ththt hhhht

Concepts | *Examples*

Law of Large Numbers (Law of Averages)
As an experiment is repeated more and more times, the proportion of outcomes favorable to any particular event will tend to come closer and closer to the theoretical probability of that event.

There are 99 cases of two consecutive tosses, and 24 of them show both heads.

$$P(E) = \frac{24}{99} \approx 0.242 \quad \text{Empirical probability}$$

In the experiment above, what is the theoretical probability that two consecutive heads will occur? How does the empirical result compare?

The theoretical probability is $\frac{1}{4} = 0.25$. This is very close to 0.242, the empirical value obtained above.

Odds
If all outcomes in a sample space are equally likely, a of them are favorable to the event E, and the remaining b outcomes are unfavorable to E, then

the **odds in favor** of E are a to b,

and   the **odds against** E are b to a.

Converting between Probability and Odds
Let E be an event.

• If $P(E) = \frac{a}{b}$, then the odds in favor of E are **a to $(b - a)$.**

• If the odds in favor of E are a to b, then $P(E) = \frac{a}{a + b}$.

A single card is drawn from a standard deck of 52 cards.

(a) What is the probability that the card is a queen?

There are 4 queens (and thus 48 non-queens).

$$P(\text{queen}) = \frac{4}{52} = \frac{1}{13} \quad \text{Probability of a queen}$$

(b) What are the odds in favor of drawing a queen?

4 to 48,   or   1 to 12   Odds in favor of a queen

(c) What are the odds against drawing a queen?

48 to 4,   or   12 to 1   Odds against a queen

11.2 EVENTS INVOLVING "NOT" AND "OR"

Properties of Probability
Let E be an event within the sample space S. That is, E is a subset of S. Then the following properties hold.

1. **$0 \le P(E) \le 1$** (The probability of an event is a number from 0 through 1, inclusive.)

2. **$P(\varnothing) = 0$** (The probability of an impossible event is 0.)

3. **$P(S) = 1$** (The probability of a certain event is 1.)

If 6 fair coins are tossed, what is the probability that at most 5 will show tails?

There are $2^6 = 64$ possible outcomes. Only one of them,

tttttt,

does not show at most 5 tails, so there are $64 - 1 = 63$ favorable outcomes.

$$P(\text{at most 5 tails}) = \frac{64 - 1}{64} = \frac{63}{64}$$

Probability of a Complement (for the Event "Not E")
The probability that an event E will *not* occur is equal to 1 minus the probability that it will occur.

$$P(\text{not } E) = 1 - P(E)$$

If a single card is drawn from a standard 52-card deck, find the probability that it is the following.

(a) a heart or a black card   **(b)** a heart or a 6

Concepts	Examples

Mutually Exclusive Events
Two events A and B are **mutually exclusive events** if they have no outcomes in common. Mutually exclusive events cannot occur simultaneously.

$P(\text{heart or black}) = P(\text{heart}) + P(\text{black})$ These are mutually exclusive.

$$= \frac{13}{52} + \frac{26}{52}$$

$$= \frac{39}{52}$$

$$= \frac{3}{4}$$

Addition Rule of Probability (for the Event "A or B")
If A and B are any two events, then the following holds.

$$P(A \text{ or } B) = P(A) + P(B) - P(A \text{ and } B)$$

If A and B are mutually exclusive, then the following holds.

$$P(A \text{ or } B) = P(A) + P(B)$$

$P(\text{heart or a 6}) = P(\text{heart}) + P(6) - P(\text{6 of hearts})$ These are not mutually exclusive.

$$= \frac{13}{52} + \frac{4}{52} - \frac{1}{52}$$

$$= \frac{16}{52}$$

$$= \frac{4}{13}$$

11.3 CONDITIONAL PROBABILITY AND EVENTS INVOLVING "AND"

Conditional Probability
The probability of event B, computed on the assumption that event A has happened, is the **conditional probability of B given A** and is denoted $P(B|A)$.

A single number is to be selected randomly from the following sample space.

$$S = \{1, 2, 3, 4, 5, 6, 7, 8, 9, 10, 11, 12\}$$

Find each probability given events A and B.

A: The selected number is even.

B: The selected number is a multiple of 3.

Conditional Probability Formula
The **conditional probability of B given A** is calculated as

$$P(B|A) = \frac{P(A \cap B)}{P(A)} = \frac{P(A \text{ and } B)}{P(A)}.$$

(a) $P(B)$

$B = \{3, 6, 9, 12\}$, so find $P(B)$ as follows.

$$P(B) = \frac{n(B)}{n(S)} = \frac{4}{12} = \frac{1}{3}$$

Independent Events
Two events A and B are **independent events** if knowledge about the occurrence of one of them has no effect on the probability of the other one occurring. A and B are independent if

$$P(B|A) = P(B), \quad \text{or, equivalently,} \quad P(A|B) = P(A).$$

(b) $P(A \text{ and } B)$

$A = \{2, 4, 6, 8, 10, 12\}$, so $A \cap B = \{6, 12\}$.

$$P(A \text{ and } B) = P(A \cap B) = \frac{2}{12} = \frac{1}{6}$$

Multiplication Rule of Probability (for the Event "A and B")
If A and B are *any two events,* then

$$P(A \text{ and } B) = P(A) \cdot P(B|A).$$

If A and B are *independent,* then

$$P(A \text{ and } B) = P(A) \cdot P(B).$$

(c) $P(B|A)$ Are events A and B independent?

In the reduced sample space of A, 2 of the 6 outcomes belong to B.

$$P(B|A) = \frac{2}{6} = \frac{1}{3}$$

Equivalently, $P(B|A) = \dfrac{P(A \cap B)}{P(A)} = \dfrac{\frac{1}{6}}{\frac{1}{2}} = \dfrac{1}{6} \cdot \dfrac{2}{1} = \dfrac{1}{3}.$

Because $P(B) = P(B|A)$, A and B are independent.

Concepts	*Examples*

11.4 BINOMIAL PROBABILITY

Binomial Probability Formula

When n independent repeated trials occur, where $p =$ probability of success and $q =$ probability of failure with p and q (where $q = 1 - p$) remaining constant throughout all n trials, the probability of exactly x successes is calculated as follows.

$$P(x) = {}_nC_x\, p^x q^{n-x}$$

$$= \frac{n!}{x!(n-x)!}\, p^x q^{n-x}$$

A die is rolled ten times. Find the probability that exactly four of the tosses result in a 3.

The probability of rolling a 3 on one roll is $p = \frac{1}{6}$, so the probability of not rolling a 3 is $q = 1 - \frac{1}{6} = \frac{5}{6}$.

$$P(\text{exactly four 3s}) = {}_{10}C_4\left(\frac{1}{6}\right)^4\left(\frac{5}{6}\right)^{10-4} \quad {\scriptstyle n = 10,\ x = 4}$$

$$= 210\left(\frac{1}{1296}\right)\left(\frac{15{,}625}{46{,}656}\right)$$

$$\approx 0.054$$

11.5 EXPECTED VALUE AND SIMULATION

Expected Value

If a random variable x can have any of the values $x_1, x_2, x_3, \ldots, x_n$, and the corresponding probabilities of these values occurring are $P(x_1), P(x_2), P(x_3), \ldots, P(x_n)$, then $E(x)$, the **expected value of x,** is calculated as follows.

$$E(x) = x_1 \cdot P(x_1) + x_2 \cdot P(x_2) + x_3 \cdot P(x_3)$$
$$+ \cdots + x_n \cdot P(x_n)$$

In a certain lottery, a player chooses four digits in a specific order. The leading digit may be 0, so a number such as 0460 is legitimate. A four-digit sequence is randomly chosen, and any player matching that selection receives a payoff of \$10,000. Find the expected gross winnings.

The probability of selecting the correct four digits is $\left(\frac{1}{10}\right)^4 = \frac{1}{10{,}000}$, and the probability of not selecting them is $1 - \left(\frac{1}{10}\right)^4 = \frac{9999}{10{,}000}$.

$$E(\text{gross winnings}) = \$10{,}000 \cdot \frac{1}{10{,}000} + \$0 \cdot \frac{9999}{10{,}000}$$

$$= \$1.00 \quad {\scriptstyle \text{A fair cost to play this game}}$$

Chapter 11 TEST

Numbers from Sets of Digits *Two numbers are randomly selected without replacement from the set*

$$\{1, 2, 3, 4, 5\}.$$

Find the probability of each event.

1. Both numbers are even.

2. Both numbers are prime.

3. The sum of the two numbers is odd.

4. The product of the two numbers is odd.

Days Off for Pizza Parlor Workers *The manager of a pizza parlor (which operates seven days a week) allows each of three employees to select one day off next week. Assuming the selection is done randomly and independently, find the probability of each event.*

5. All three select different days.

6. All three select the same day, given that all three select a day beginning with the same letter.

7. Exactly two of them select the same day.

Genetics of Cystic Fibrosis   *The chart represents genetic transmission of cystic fibrosis. C denotes a normal gene, and c denotes a cystic fibrosis gene. (Normal is dominant.) Both parents in this case are Cc, which means that they inherited one of each gene. Therefore, they are both carriers but do not have the disease.*

		Second Parent	
		C	c
First Parent	C		
	c		Cc

8. Complete the chart, showing all four equally likely gene arrangements.

9. Find the probability that a child of these parents will also be a carrier without the disease.

10. What are the odds that a child of these parents actually will have cystic fibrosis?

Drawing Cards   *A single card is chosen at random from a standard 52-card deck. Find the odds against its being each of the following.*

11. a heart

12. a red queen

13. an ace

14. a king or a black face card

Rolling Dice   *A pair of dice are rolled. Find the following.*

15. the probability of "doubles" (the same number on both dice)

16. the odds in favor of a sum greater than 2

17. the odds against a sum of "7 or 11"

18. the probability of a sum that is even and less than 5

Card Choices   *Two cards are drawn, without replacement, from a standard 52-card deck. Find the probability of each event.*

19. Both cards are red.

20. Both cards are the same color.

21. The second card is a queen, given that the first card is an ace.

22. The first card is a face card and the second is black.

Making Par in Golf   *Ted has a 0.78 chance of making par on each hole of golf that he plays. Today he plans to play just three holes. Find the probability of each event. Round answers to three decimal places.*

23. He makes par on all three holes.

24. He makes par on exactly two of the three holes.

25. He makes par on at least one of the three holes.

26. He makes par on the first and third holes but not on the second.

27. *Coin Tosses*   Five fair coins are tossed. Find the expected number of heads.

28. *Card Choices*   Two cards are drawn, with replacement, from a standard 52-card deck. Find the expected number of diamonds.

Selecting Committees   *A three-member committee is selected randomly from a group consisting of three men and two women.*

29. Let *x* denote the number of men on the committee, and complete the probability distribution table.

x	$P(x)$
0	0
1	
2	
3	

30. Find the probability that the committee members are not all men.

31. Find the expected number of men on the committee.

32. *Gender in Sequences of Babies*   Assuming boy and girl babies are equally likely, find the probability that a family with three children will have exactly two boys.

33. *Simulation of Pea Plant Reproduction with Coin Tossing*   Use the sequence of flower colors in **Example 10** of the fifth section of this chapter to approximate, to three decimal places, the probability that three successive offspring will all have white flowers.

Simulation with Coin Tossing   *A coin was actually tossed 100 times, producing the following sequence of outcomes. Read from left to right across the top row, then from left to right across the second row, and so on, to the bottom row.*

tttth	ththh	htttt	hhhth	ththh
hhhth	httht	hhhht	thttt	thtth
httth	hhhhh	hhhhh	tttht	ttttt
htthh	tttth	httth	tthhh	tthhh

Find, to three decimal places, the empirical probability of each event.

34. Two consecutive tails (tt) occur.

35. Three consecutive heads (hhh) occur.

12 Statistics

Suppose you're a psychological therapist who specializes in helping patients with various early-stage addictions. You have noted that under your treatment plan, 50% of your clients have been able to quit drinking (their cumulative probability of remission is 0.50) by 10 years after their dependency began. ("Remission" is defined as quitting for 1 year.)

Compare these results with the general situation, as reported in a long-term study and shown in the graph on the next page. Complete the following statements.

1. According to the long-term study, the more general cumulative probability of remission is 0.50 after about _____ years.

2. Your own results for 0.50 probability of remission are better than the average by about _____ years.

(Check your answers with those given four pages ahead.)

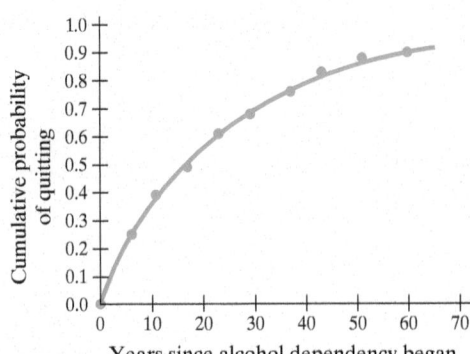

Data from G. Heyman, *Annual Review of Clinical Psychology.*

Depending on the comparison between the general results and your own results, you may want to either reconsider your treatment protocol or prepare a statistical report on your results for publication in a professional journal.

12.1 VISUAL DISPLAYS OF DATA

OBJECTIVES

1 Construct and interpret data displays.

2 Construct and interpret frequency and grouped frequency distributions.

3 Construct and interpret stem-and-leaf displays.

4 Construct and interpret bar graphs, circle graphs, and line graphs.

Basic Concepts

Governments collect and analyze an amazing quantity of "statistics." The word itself comes from the Latin *statisticus,* meaning "of the state."

A **population** includes *all* items of interest, and a **sample** includes *some,* but ordinarily not all, of the items in the population. See the Venn diagram in the margin.

To predict the outcome of an approaching presidential election, we may be interested in a population of many millions of voter preferences (those of all potential voters in the country). As a practical matter, however, even national polling organizations with considerable resources obtain only a relatively small sample, say 2000, of those preferences.

The study of statistics is divided into two main areas.

MAIN AREAS OF THE STUDY OF STATISTICS

1. **Descriptive statistics** consists of collecting, organizing, summarizing, and presenting data (information)—in short, "describing" the data. These descriptions can apply to populations or to samples.

2. **Inferential statistics** consists of drawing conclusions (making inferences) about populations on the basis of information from samples taken from those populations.

Population

Sample

Consider a population of 10,000 colored beads in a large bowl (see the first photo on the next page), 3000 of which (that is, 30%) are known to be green. From this description, we can make certain predictions by applying probability theory.

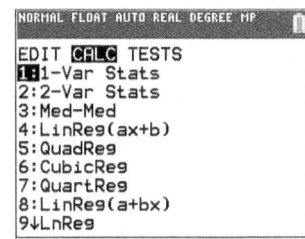

These screens give various **statistical options** on the TI-84 Plus C.

Most scientific and graphing calculators will perform all computations we describe in this chapter, including sorting of data, once they are entered. So your purpose in studying this material should be to understand the meanings of the various computed quantities. It is recommended that the actual computations, especially for sizable collections of data, be done on a calculator, or on a computer with suitable statistical software such as StatCrunch.

For example, a random sample of 25 beads from the bowl ought to contain *about*

$$30\% \text{ of } 25 = 0.30 \cdot 25 = 7.5 \text{ (say 7 or 8)}$$

green beads. Likewise, a random sample of 100 beads ought to contain *about*

$$30\% \text{ of } 100 = 0.30 \cdot 100 = 30$$

green beads. Of course, some variation from these sample "expected" proportions would be no surprise.

A population of 10,000

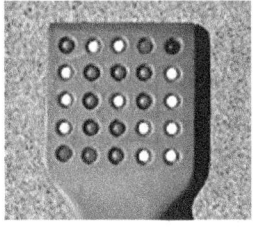
A random sample of 25

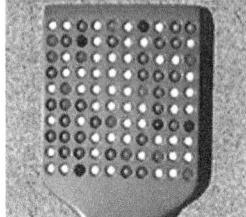
A random sample of 100

Now look at the second and third photos, which show *actual* random samples from the population of beads. If we did not know the actual proportions of green beads in the 10,000-bead population, then either of these samples could provide an "estimate" of that proportion. The first sample gives an estimate of

$$\frac{9}{25} = 0.36 = 36\% \text{ (about 3600 green beads)},$$

while the second sample gives

$$\frac{28}{100} = 0.28 = 28\% \text{ (about 2800 green beads)}.$$

As we would expect, the larger sample gives a more accurate estimate, because 28% is closer to the true proportion (30%) than is 36%.

Information that has been collected but not yet organized or processed is called **raw data.** It is often **quantitative** (or **numerical**), but it can also be **qualitative** (or **non-numerical**), as illustrated in **Table 1**.

Table 1 Examples of Raw Data

Quantitative data: The number of siblings in ten different families: 3, 1, 2, 1, 5, 4, 3, 3, 8, 2

Qualitative data: The makes of six different automobiles: Toyota, Ford, Kia, Toyota, Chevrolet, Honda

Quantitative data are generally more useful when they are **ranked,** or arranged in numerical order. In ranked form, the first list in **Table 1** appears as follows.

$$1, 1, 2, 2, 3, 3, 3, 4, 5, 8$$

Frequency Distributions

When a data set includes many repeated items, it can be organized into a **frequency distribution,** which lists only the distinct data values (x) along with their frequencies (f). The frequency designates the number of times the corresponding item occurred in the data set.

It is also helpful to show the **relative frequency** of each distinct item. This is the fraction, or percentage, of the data set represented by the item. If n denotes the total number of items, and a given item, x, occurred f times, then the relative frequency of x is

$$\frac{f}{n}.$$

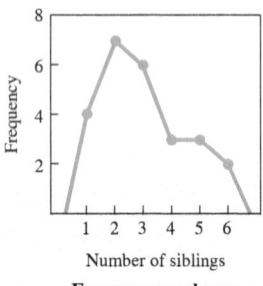

Number of siblings

Histogram

Figure 1

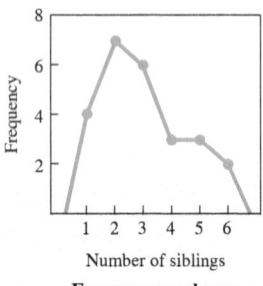

Number of siblings

Frequency polygon

Figure 2

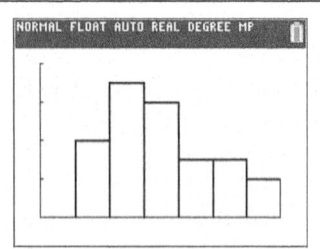

This histogram was generated with a graphing calculator using the data in **Table 2**. Compare with **Figure 1** above.

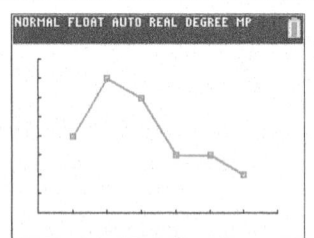

This line graph resembles the frequency polygon in **Figure 2** above. It was generated with a graphing calculator using the data in **Table 2**.

EXAMPLE 1 Constructing Frequency and Relative Frequency Distributions

The 25 members of a psychology class were polled about the number of siblings in their individual families. Construct a frequency distribution and a relative frequency distribution for their responses, which are shown here.

$$2, 3, 1, 3, 3, 5, 2, 3, 3, 1, 1, 4, 2, 4, 2, 5, 4, 3, 6, 5, 1, 6, 2, 2, 2$$

Solution

The data range from a low of 1 to a high of 6. The frequencies (obtained by inspection) and relative frequencies are shown in **Table 2**.

Table 2 Frequency and Relative Frequency Distributions for Numbers of Siblings

Number x	Frequency f	Relative Frequency $\frac{f}{n}$
1	4	$\frac{4}{25} = 16\%$
2	7	$\frac{7}{25} = 28\%$
3	6	$\frac{6}{25} = 24\%$
4	3	$\frac{3}{25} = 12\%$
5	3	$\frac{3}{25} = 12\%$
6	2	$\frac{2}{25} = 8\%$

Total: $n = 25$

Problem-Solving Strategy

You are already familiar with the three strategies "Make a table or chart," "Look for a pattern," and "Draw a sketch." The techniques of this section apply all three of these strategies to present raw data in more meaningful visual forms.

The numerical data of **Table 2** can be interpreted more easily with the aid of a **histogram.** A series of rectangles, whose lengths represent the frequencies, are placed next to one another as shown in **Figure 1**. On each axis, horizontal and vertical, a label and the numerical scale should be shown.

The information shown in the histogram in **Figure 1** can also be conveyed by a **frequency polygon,** as in **Figure 2**. Simply plot a single point at the appropriate height for each frequency, connect the points with a series of connected line segments, and complete the polygon with segments that trail down to the axis to the left of 1 and to the right of 6.

The frequency polygon is an instance of the more general *line graph,* discussed earlier in the text and later in this section.

Grouped Frequency Distributions

Data sets containing large numbers of items are often arranged into groups, or **classes.** All data items are assigned to their appropriate classes, and then a **grouped frequency distribution** can be set up and a graph displayed. Although there are no fixed rules for establishing the classes, most statisticians agree on a few general guidelines.

GUIDELINES FOR THE CLASSES OF A GROUPED FREQUENCY DISTRIBUTION

1. Make sure each data item will fit into one, and only one, class.

2. Try to make all classes the same width.

3. Make sure the classes do not overlap.

4. Use from 5 to 12 classes. Too few or too many classes can obscure the tendencies in the data.

StatCrunch Interpreting a Histogram

EXAMPLE 2 Constructing a Histogram and a Frequency Polygon

Forty students, selected randomly in the school cafeteria one morning, were asked to estimate the number of hours they had spent studying in the past week (including both in-class and out-of-class time). Their responses are recorded here.

These weekly "study times" data will be used to illustrate various concepts throughout the chapter.

18	60	72	58	33	15	12	36	16	29
26	41	45	25	32	24	22	55	30	31
55	39	29	44	29	14	36	31	45	62
36	52	47	38	36	23	33	44	17	24

Tabulate a grouped frequency distribution and a grouped relative frequency distribution, and construct a histogram and a frequency polygon for the given data.

Solution

The data range from a low of 12 to a high of 72—that is, over a range of

$$72 - 12 = 60 \text{ units.}$$

The widths of the classes should be uniform (by Guideline 2), and there should be from 5 to 12 classes (by Guideline 4). Five classes would imply a class width of about $\frac{60}{5} = 12$, while twelve classes would imply a class width of about $\frac{60}{12} = 5$. A class width of 10 will be convenient. We let our classes run from 10 through 19, from 20 through 29, and so on up to 70 through 79, for a total of seven classes. All four guidelines are met.

Next go through the data set, tallying each item into the appropriate class. The tally totals produce class frequencies, which in turn produce relative frequencies, as shown in **Table 3** below. The histogram is displayed in **Figure 3** to the left.

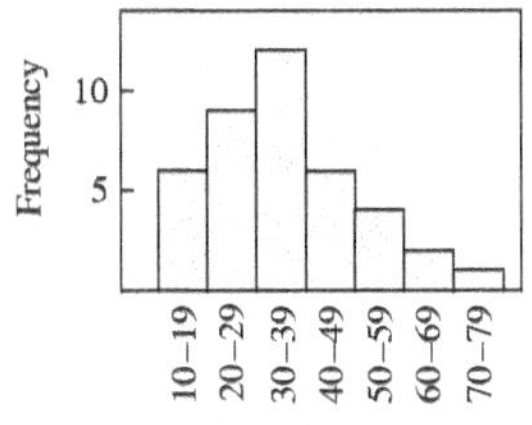

Weekly study times (in hours)
Grouped frequency histogram

Figure 3

Table 3 Grouped Frequency and Relative Frequency Distributions for Weekly Study Times

Class Limits	Tally	Frequency f	Relative Frequency $\frac{f}{n}$
10–19	⾏⾏ |	6	$\frac{6}{40} = 15.0\%$
20–29	⾏⾏ ||||	9	$\frac{9}{40} = 22.5\%$
30–39	⾏⾏ ⾏⾏ ||	12	$\frac{12}{40} = 30.0\%$
40–49	⾏⾏ |	6	$\frac{6}{40} = 15.0\%$
50–59	||||	4	$\frac{4}{40} = 10.0\%$
60–69	||	2	$\frac{2}{40} = 5.0\%$
70–79	|	1	$\frac{1}{40} = 2.5\%$
	Total:	$n = 40$	

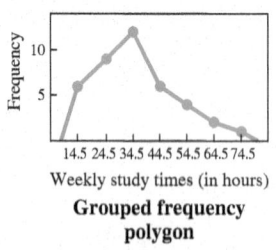

Grouped frequency polygon

Figure 4

Table 4 Grouped Frequency Distribution for Weekly Study Times

Class Limits	Frequency
10–19	6
20–29	9
30–39	12
40–49	6
50–59	4
60–69	2
70–79	1

The numbers 10, 20, 30, and so on are the **lower class limits,** while 19, 29, 39, and so on are the **upper class limits.** The common **class width** for the distribution is the difference of any two successive lower class limits (such as 30 − 20) or of any two successive upper class limits (such as 59 − 49). The class width for this distribution is 10, as noted earlier.

To construct a frequency polygon, notice that in a *grouped* frequency distribution, the data items in a given class are generally not all the same. We can obtain the "middle" value, or **class mark,** by adding the lower and upper class limits and dividing this sum by 2. We locate all the class marks along the horizontal axis and plot points above the class marks. The heights of the plotted points represent the class frequencies. The resulting points are connected just as for a nongrouped frequency distribution. The result is shown in **Figure 4.**

Stem-and-Leaf Displays

In **Table 3,** the tally marks give a good visual impression of how the data are distributed. In fact, the tally marks are almost like a histogram turned on its side. Nevertheless, once the tallying is done, the tally marks are usually dropped, and the grouped frequency distribution is presented as in **Table 4.**

The pictorial advantage of the tally marks is now lost. Furthermore, we cannot tell, from the grouped frequency distribution itself (or from the tally marks either, for that matter), what any of the original items were. For example, we know only that there were six items in the class 40–49. We do not know specifically what any of them were.

One way to avoid these shortcomings is to employ a tool of exploratory data analysis, the **stem-and-leaf display,** as shown in **Example 3.**

EXAMPLE 3   **Constructing a Stem-and-Leaf Display**

Present the study times data of **Example 2** in a stem-and-leaf display.

Solution

See **Example 2** for the original raw data. We arrange the numbers in **Table 5.** The tens digits, to the left of the vertical division line, are the "stems," while the corresponding ones digits are the "leaves." We have entered all items from the first row of the original data, from left to right, and then the items from the second row through the fourth row.

Table 5   Stem-and-Leaf Display for Weekly Study Times

1	8 5 2 6 4 7
2	9 6 5 4 2 9 9 3 4
3	3 6 2 0 1 9 6 1 6 8 6 3
4	1 5 4 5 7 4
5	8 5 5 2
6	0 2
7	2

Notice that the stem-and-leaf display of **Example 3** conveys at a glance the same pictorial impressions that the histogram in **Figure 3** conveys, without the need for constructing the drawing. It also preserves the exact data values.

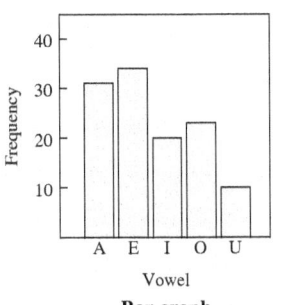

Bar graph

Figure 5

Bar Graphs, Circle Graphs, and Line Graphs

A frequency distribution of non-numerical observations can be presented in the form of a **bar graph,** which is similar to a histogram except that the rectangles (bars) usually are not touching one another and sometimes are arranged horizontally rather than vertically. The bar graph of **Figure 5** shows the frequencies of occurrence of the vowels A, E, I, O, and U in this paragraph.

EXAMPLE 4   Interpreting a Bar Graph

A 2017 study by Swiss researchers predicted a worldwide 2030 baseline of 200,000 tons of antibiotic use in food animals, with possible reductions if one or more of the following strategies were applied.

C: Capping use at 50 milligrams per kilogram of animal product

R: Reducing consumption of meat to 165 grams per day per person

T: Imposing a 50% tax on antibiotics

Refer to the bar graph in **Figure 6.** To the nearest percent, what reduction is predicted under each of the following intervention strategies?

(a) None   **(b)** C only   **(c)** C and T   **(d)** C, R, and T

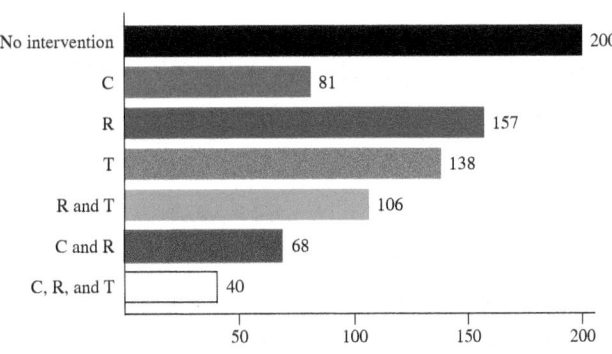

Estimated global antibiotic use in 2030 (1000 tons)

Data from www.sciencemag.org, September 28, 2017.

Figure 6

Solution

(a) $\dfrac{200 - 200}{200} = 0\%$ **(b)** $\dfrac{200 - 81}{200} \approx 60\%$

(c) We cannot tell from the graph.   **(d)** $\dfrac{200 - 40}{200} = 80\%$

A graphical alternative to the bar graph is the **circle graph,** or **pie chart,** which uses a circle to represent the total of all the categories and divides the circle into sectors, or wedges (like pieces of pie), whose sizes show the relative magnitudes of the categories. The angle around the entire circle measures 360°. So, for example, a category representing 20% of the whole should correspond to a sector whose central angle is 20% of 360°, that is,

$$0.20(360°) = 72°.$$

A circle graph shows, at a glance, the relative magnitudes of various categories.

Table 6 Portfolio Allocations

Investment Category	Percent of Total
Stocks	55%
Bonds	20
Cash	15
Metals	10

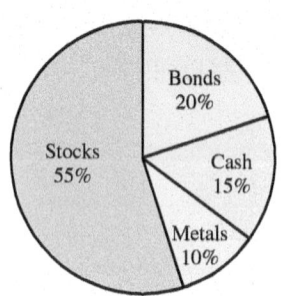

Investment allocations

Figure 7

EXAMPLE 5   **Constructing a Circle Graph**

Andre's investment advisor recommends the portfolio allocations shown in **Table 6.** Present this information in a circle graph.

Solution

The central angles for the sectors are as follows.

$$\text{Stocks:} \quad 0.55(360°) = 198°$$
$$\text{Bonds:} \quad 0.20(360°) = 72°$$
$$\text{Cash:} \quad 0.15(360°) = 54°$$
$$\text{Metals:} \quad 0.10(360°) = 36°$$

Draw a circle and mark off the angles with a protractor. (See **Figure 7.**)

To demonstrate how a quantity *changes,* say with respect to time, use a **line graph.** Connect a series of line segments that rise and fall with time, according to the magnitude of the quantity being illustrated. To compare the patterns of change for two or more quantities, we can even plot multiple line graphs together in a "comparison line graph." (A line graph looks somewhat like a frequency polygon, but the quantities graphed are not necessarily frequencies.)

EXAMPLE 6   **Interpreting a Comparison Line Graph**

Answer the following questions by referring to **Figure 8.**

(a) What time frame is reflected in the figure?

(b) In which years was resistance to ciprofloxacin greater than resistance to ceftriaxone?

(c) Over this time span, what was the highest resistance to ceftriaxone achieved, and when did it occur?

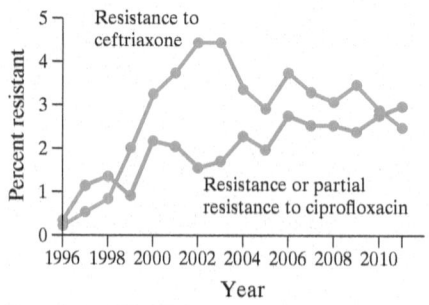

Data from U.S. CDC.

Antibiotic resistance of *Salmonella* infections in humans
Figure 8

Solution

(a) The figure represents the years 1996 through 2011.

(b) Resistance to ciprofloxacin was greater than resistance to ceftriaxone in the years 1996, 1997, 1998, and 2011.

(c) The highest resistance to ceftriaxone was about 4.5%, and it occured in 2002 and 2003.

WHEN WILL I **EVER** USE THIS?

Imagine that you are the therapist mentioned in the chapter opener and that, again, you want to compare the results you have achieved in your own practice with the broader results. In this case, you are interested in tobacco dependency and remission. The study that was referred to earlier produced the following data for smokers, where

x = years since dependency began, and

y = cumulative probability of quitting (for a year).

x	y	x	y	x	y
3.3	0.08	22.9	0.39	45.8	0.63
7.9	0.17	28.0	0.45	50.9	0.69
14.0	0.26	33.2	0.52	56.1	0.78
18.2	0.32	37.9	0.57	62.6	0.84

1. To clarify the meaning of the data, plot the twelve points and sketch a smooth curve that approximately fits the points.

2. Use the graph you have prepared to complete the following statement:

It takes about _____ years for 50% of smokers to quit the habit.

3. Compare with the graph in the chapter opener, and complete the same statement for drinkers:

It takes about _____ years for 50% of drinkers to quit the habit.

(Check your answers with those given below.)

4. Give some possible reasons for the discrepancy between the results for smokers and those for drinkers, as reflected in your answers to Questions 2 and 3 above.

Answers: **2.** 30 **3.** 17 **4.** Answers will vary.

12.1 EXERCISES

Generally, you may use statistical functions of your calculator or a statistics package such as StatCrunch, as directed by your instructor.

You can use **StatCrunch** *(available in MyLabMath) to solve exercises marked with the* **StatCrunch** .

CONCEPT CHECK *Write* true *or false for each statement.*

1. Usually a sample has fewer members than the population.

2. A frequency distribution and a relative frequency distribution are the same thing.

3. A grouped frequency distribution has the same number of groups as there are data items.

4. A grouped frequency distribution should have from 5 to 12 classes.

5. A histogram has numerical scales on both vertical and horizontal axes.

6. Some data sets are more accurately depicted in a circle graph than in a pie chart.

In each case, use the given data to do the following:

(a) *Construct frequency and relative frequency distributions, in a table similar to* **Table 2.**

(b) *Construct a histogram.*

(c) *Construct a frequency polygon.*

7. *Preparation for Summer* A magazine identified the following five "maintenance" activities often performed by people as summer approaches.

1. Prep the car for road trips.

2. Clean up the house or apartment.

3. Groom the garden.

4. Exercise the body.

5. Organize the wardrobe.

The following data are the responses of 30 people who were asked, on June 1, how many of the five activities they had accomplished.

```
1  1  3  1  0  3  0  0  2  1
2  2  0  0  5  3  4  0  1  0
4  2  0  2  0  1  0  1  2  3
```

8. *Favorite Numbers* The following data are the "favorite numbers" (single digits, 1 through 9) of the 24 pupils of a second grade class.

```
4  7  2  7  6  1
7  2  9  8  5  1
4  3  8  9  5  5
9  2  6  5  2  7
```

In each case, use the given data to do the following:

(a) *Construct grouped frequency and relative frequency distributions, in a table similar to* **Table 3.** *(Follow the suggested guidelines for class limits and class width.)*

(b) *Construct a histogram.*

(c) *Construct a frequency polygon.*

9. *Exam Scores* The scores of the 48 members of a sociology lecture class on a 50-point exam were as follows.

```
40  43  44  33  41  40  33  39
42  36  43  42  45  42  23  41
43  40  37  47  44  48  31  46
38  31  46  38  39  42  47  40
47  41  48  42  45  29  32  38
45  28  49  46  47  36  48  46
```

Use six classes with a uniform class width of 5 points, where the lower limit of the first class is 21 points.

10. *Charge Card Account Balances* The following raw data represent the monthly account balances (to the nearest dollar) for a sample of 50 brand-new charge card users.

```
51  175   46  138   79  118  90  163   88  107
75  154   85   60   42   54  62  128  114   73
108 119  116  145  129  130  81  105   96   71
83  145  117   60  125  130  94   88  136  112
62  165  118   84   74   62  81  110  108   71
```

Use seven classes with a uniform width of 20 dollars, where the lower limit of the first class is 40 dollars.

11. *Daily High Temperatures* The following data represent the daily high temperatures (in degrees Fahrenheit) for the month of June in a town in the Sacramento Valley of California.

```
79  84   88   96  102  104  99   97   92   94
85  92  100   99  101  104  97  108  106  106
90  82   74   72   83  107  98  102   97   94
```

Use eight classes with a uniform width of 5 degrees, where the lower limit of the first class is 70 degrees.

12. IQ Scores of College Freshmen The following data represent IQ scores of a group of 50 college freshmen.

121	109	118	92	130	112	114	117	122	115
121	107	108	113	124	112	111	106	116	118
104	107	118	118	110	124	115	103	100	114
96	124	116	123	104	135	113	126	116	111
127	134	98	129	102	103	107	113	117	112

Use nine classes with a uniform width of 5, where the lower limit of the first class is 91.

In each case, construct a stem-and-leaf display for the given data. Treat the ones digits as the leaves. For any single-digit data, use a stem of 0.

13. Games Won in the National Basketball Association Approaching midseason, the teams in the National Basketball Association had won the following numbers of games.

27	20	29	11	26	11	12	7	26	18
22	19	14	13	22	9	25	11	10	15
38	10	22	23	31	8	24	15	24	15

14. Accumulated College Units The students in a biology class were asked how many college units they had accumulated to date. Their responses follow.

22	4	13	12	21	33	15	17	12	24
32	42	26	11	53	62	42	25	13	8
54	18	21	14	19	17	38	17	20	10

15. Distances to School The following data are the daily round-trip distances to school (in miles) for 30 randomly chosen students attending a community college in California.

16	30	10	11	18	26	34	18	8	12
21	14	5	22	4	25	9	10	6	21
12	18	9	16	44	23	4	13	36	8

16. Yards Gained in the National Football League The following data represent net yards gained per game by National Football League running backs who played during a given week of the season.

25	19	36	73	37	88	67	33	54	79
19	39	45	22	58	7	30	43	24	36
65	43	33	55	40	29	112	60	86	62
52	29	18	25	41	3	49	16	32	46

Federal Revenue and Spending *The graph shows U.S. government receipts and outlays for 2006–2017. Refer to the graph for Exercises 17–21.*

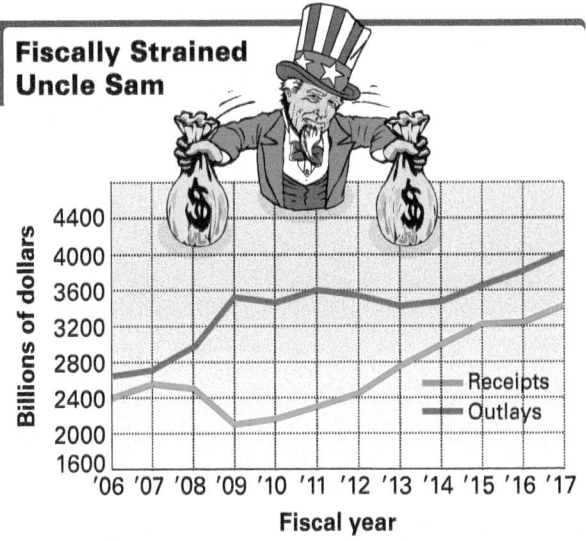

Data from Department of the Treasury, Office of Management and Budget.

17. For the period 2006–2017, list all years when receipts exceeded outlays.

18. Identify each of the following amounts, and indicate when it occurred.

   (a) the greatest one-year drop in receipts

   (b) the greatest one-year rise in outlays

   (c) the least deficit (outlays minus receipts)

19. In what years did receipts appear to climb faster than outlays?

20. About what was the greatest deficit (outlays minus receipts), and in what year did it occur?

21. Plot a point for each year and draw a line graph showing the federal surplus (+) or deficit (−) over the years 2006–2017.

Reading Bar Graphs of Economic Indicators *The bar graphs here show trends in several economic indicators over the period 2012–2017.*

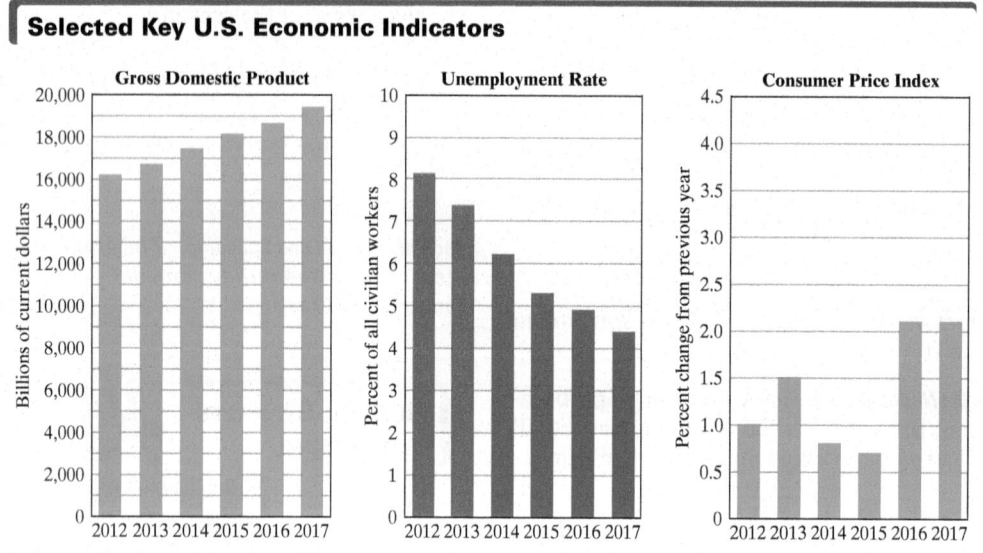

Selected Key U.S. Economic Indicators

Data from U.S. Department of Labor, Bureau of Labor Statistics, www.statista.com

Refer to the above graphs for Exercises 22–26.

22. About what was the gross domestic product in 2017?

23. Over the six-year period, about what was the greatest change in consumer price index, and when did it occur?

24. What was the greatest year-to-year change in the unemployment rate, and when did it occur?

25. Observing these graphs, what would you say was the most significant fact shown for 2016?

26. Explain why the gross domestic product would generally increase when the unemployment rate decreases.

Reading a Circle Graph of Job-Training Sources *The circle graph in the next column shows how a surveyed group of American workers were trained for their jobs. Refer to it for Exercises 27 and 28.*

27. What is the greatest single training source? To the nearest degree, what is the central angle of that category's sector?

28. What percent of job training was provided directly from employers, either formally or informally?

29. ***Sources of U.S. Budget Receipts*** The table at the right gives the major sources of government income for fiscal year 2016. Use the information to draw a circle graph.

Graph for Exercises 27 and 28

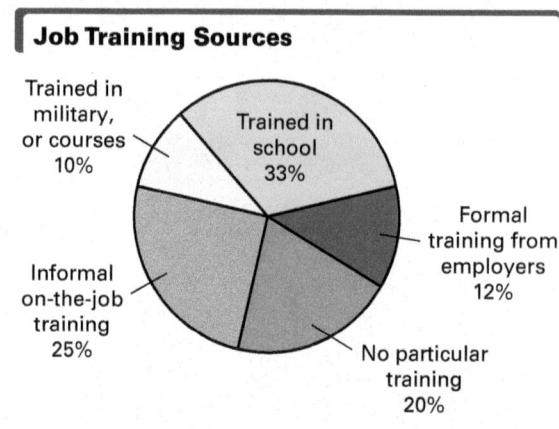

Job Training Sources

Table for Exercise 29

Source of Receipts	Approximate Percent of Total
Individual income taxes	47%
Corporate income taxes	9
Social insurance and retirement receipts	34
Excise taxes	3
Other receipts	7

Data from Congressional Budget Office; Office of Management and Budget.

30. *Correspondence between Education and Earnings* Data for 2016 showed that the median weekly earnings of American workers corresponded to educational level as shown in the accompanying table. Draw a bar graph that shows this information.

Educational Level	Median Weekly Earnings
Less than a high school diploma	$504
High school diploma	692
Some college, no degree	756
Associate's degree	819
Bachelor's degree	1156
Master's degree	1380
Doctoral degree	1664
Professional degree	1745

Data from Bureau of Labor Statistics.

Net Worth of Retirement Savings *Saura, wishing to retire at age 60, is studying the comparison line graph here, which shows (under certain assumptions) how the net worth of her retirement savings (initially $500,000 at age 60) will change as she gets older and as she withdraws living expenses from savings. Refer to the graph for Exercises 31–34.*

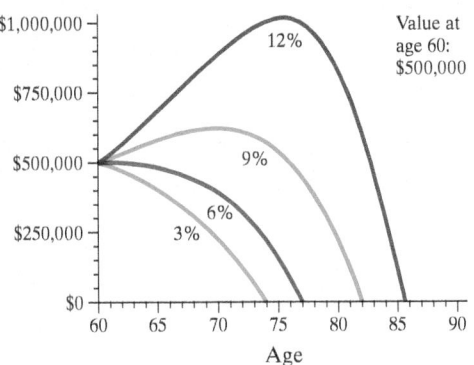

31. Assuming Saura can maintain an average annual return of 9%, how old will she be when her money runs out?

32. If she could earn an average of 12% annually, what maximum net worth would Saura achieve? At about what age would the maximum occur?

33. Suppose Saura reaches age 70 in good health, and the average annual return has proved to be 6%.

   (a) About how much longer can she expect her money to last?

   (b) What options might she consider in order to extend that time?

34. At age 71, about how many times more will Saura's net worth be if she averages a 12% return than if she averages a 3% return?

Sample Masses in a Geology Laboratory *Stem-and-leaf displays can be modified in various ways in order to obtain a reasonable number of stems. The following data, representing the measured masses (in grams) of thirty mineral samples in a geology lab, are shown in a **double-stem** display in* **Table 7.**

60.7	41.4	50.6	39.5	46.4
58.1	49.7	38.8	61.6	55.2
47.3	52.7	62.4	59.0	44.9
35.6	36.2	40.6	56.9	42.6
34.7	48.3	55.8	54.2	33.8
51.3	50.1	57.0	42.8	43.7

Table 7 Stem-and-Leaf Display for Mineral Sample Masses

(30–34)	3	4.7	3.8				
(35–39)	3	9.5	8.8	5.6	6.2		
(40–44)	4	1.4	4.9	0.6	2.6	2.8	3.7
(45–49)	4	6.4	9.7	7.3	8.3		
(50–54)	5	0.6	2.7	4.2	1.3	0.1	
(55–59)	5	8.1	5.2	9.0	6.9	5.8	7.0
(60–64)	6	0.7	1.6	2.4			

Write a short answer for each problem.

35. Describe how the stem-and-leaf display of **Table 7** was constructed.

36. Explain why **Table 7** is called a "double-stem" display.

37. In general, how many stems (total) are appropriate for a stem-and-leaf display? Explain your reasoning.

Solve each problem.

38. *Record Temperatures* According to the National Climatic Data Center, U.S. Department of Commerce, the highest temperatures (in degrees Fahrenheit) ever recorded in the 50 states (as of June 15, 2017) were as follows.

112	100	128	120	134	114	106	110	109	112
100	118	117	116	118	121	114	114	105	109
107	112	115	115	118	117	118	125	106	110
122	108	110	121	113	120	119	111	104	113
120	113	120	117	107	110	118	112	114	115

Present these data in a double-stem display.

39. *Letter Occurrence Frequencies in the English Language* The table below shows commonly accepted percentages of occurrence for the various letters in English language usage. (Code breakers have carefully analyzed these percentages as an aid in deciphering secret codes.)

For example, notice that E is the most commonly occurring letter, followed by T, A, O, N, and so on. The letters Q and Z occur least often. Referring to **Figure 5** in the text, would you say that the relative frequencies of occurrence of the vowels in the associated paragraph were typical or unusual? Explain your reasoning.

Letter	Percent		Letter	Percent
E	13		L	$3\frac{1}{2}$
T	9		C, M, U	3
A, O	8		F, P, Y	2
N	7		W, G, B	$1\frac{1}{2}$
I, R	$6\frac{1}{2}$		V	1
S, H	6		K, X, J	$\frac{1}{2}$
D	4		Q, Z	$\frac{1}{5}$

Frequencies and Probabilities of Letter Occurrence The percentages shown in Exercise 39 are based on a very large sampling of English language text. Since they are based on experiment, they are "empirical" rather than "theoretical." By converting each percent in that table to a decimal fraction, you can produce an **empirical probability distribution.**

For example, if a single letter is randomly selected from a randomly selected passage of text, the probability that it will be an E is 0.13. The probability that a randomly selected letter will be a vowel (A, E, I, O, or U) is

$$(0.08 + 0.13 + 0.065 + 0.08 + 0.03) = 0.385.$$

40. Rewrite the distribution shown in **Exercise 39** as an empirical probability distribution. Give values to three decimal places. Note that the 26 probabilities in this distribution—one for each letter of the alphabet—should add up to 1 (except for, perhaps, a slight round-off error).

41. (a) From your distribution in **Exercise 40,** construct an empirical probability distribution just for the vowels A, E, I, O, and U. (*Hint:* Divide each vowel's probability, from **Exercise 40,** by 0.385 to obtain a distribution whose five values add up to 1.) Give values to three decimal places.

(b) Construct an appropriately labeled bar graph from your distribution in part (a).

42. Based on the occurrences of vowels in the paragraph represented by **Figure 5,** construct a probability distribution for the vowels. Give probabilities to three decimal places. The frequencies are

A–31, E–34, I–20, O–23, U–10.

43. Is the probability distribution in **Exercise 42** theoretical or empirical? Is it different from the distribution in **Exercise 41?** Which one is more accurate? Explain your reasoning.

44. *Frequencies and Probabilities of Study Times* Convert the grouped frequency distribution in **Table 3** to an empirical probability distribution, using the same classes and giving probability values to three decimal places.

45. *Probabilities of Study Times* Recall that the distribution in **Exercise 44** was based on weekly study times for a sample of 40 students. Suppose one of those students was selected randomly. Using your distribution, find the probability that the study time in the past week for the student selected would have been in each of the following ranges.

(a) 20–29 hours   **(b)** 50–69 hours

(c) fewer than 40 hours   **(d)** at least 60 hours

Favorite Sports among Recreation Students *The 40 members of a recreation class were asked to name their favorite sports. The table shows the numbers who responded in various ways.*

Sport	Number of Class Members
Sailing	7
Archery	4
Snowboarding	10
Bicycling	8
Rock climbing	8
Rafting	3

Use this information in Exercises 46–48.

46. If a member of this class is selected at random, what is the probability that the favorite sport of the person selected is sailing?

47. (a) Based on the data in the table, construct a probability distribution, giving probabilities to three decimal places.

(b) Do you think the distribution in part (a) is theoretical, or empirical? Why?

48. Explain the distinction between a relative frequency distribution and a probability distribution.

FOR FURTHER THOUGHT

Expected and Observed Frequencies

When fair coins are tossed, the results on particular tosses cannot be reliably predicted. As more and more coins are tossed, however, the proportions of heads and tails become more predictable. This is a consequence of the "law of large numbers."

For example, if five coins are tossed, then the resulting number of heads, denoted x, is a "random variable," whose possible values are

0, 1, 2, 3, 4, and 5.

If the five coins are tossed repeatedly, say 64 separate times, then the binomial probability formula can be used to get **expected frequencies** (or **theoretical**

frequencies), as shown in the table. The first two columns of the table comprise the **expected frequency distribution** for 64 tosses of five fair coins.

In an actual experiment, we could obtain **observed frequencies** (or **empirical frequencies**), which would most likely differ somewhat from the expected frequencies. But 64 repetitions of the experiment should be enough to provide fair consistency between expected and observed values.

For Group or Individual Investigation

Toss five coins a total of 64 times, keeping a record of the number of heads obtained each time.

Number of Heads x	Expected Frequency e	Observed Frequency o
0	2	
1	10	
2	20	
3	20	
4	10	
5	2	

1. Enter your experimental results in the third column of the accompanying table, producing an **observed frequency distribution.**

2. Compare the second and third column entries.

3. Construct two histograms, one from the expected frequency distribution and one from your observed frequency distribution.

4. Compare the two histograms.

12.2 MEASURES OF CENTRAL TENDENCY

OBJECTIVES

1 Find and interpret means.
2 Find and interpret medians.
3 Find and interpret modes.
4 Infer central tendency measures from stem-and-leaf displays.
5 Discern symmetry in data sets.
6 Compare measures of central tendency.

A city issued the following numbers of building permits over a six-month period.

305, 285, 240, 376, 198, 264

A single number that is in some sense representative of this whole set of numbers, a kind of "middle" value, would be a **measure of central tendency.**

Mean

The most common measure of central tendency is the **mean** (or **arithmetic mean**). The mean of a sample is denoted $\bar{x}$ (read "x bar"), while the mean of a complete population is denoted μ (the lowercase Greek letter *mu*). For our purposes here, data sets are considered to be samples, so we use $\bar{x}$.

The mean of a set of data items is found by adding up all the items and then dividing the sum by the number of items. (The mean is what most people associate with the word "average.") Because adding up, or summing, a list of items is a common procedure in statistics, we use the symbol for "summation," Σ (the capital Greek letter *sigma*). Therefore, the sum of n items—say $x_1, x_2, \ldots, x_n$—can be denoted

$$\Sigma x = x_1 + x_2 + \cdots + x_n.$$

MEAN

The **mean** of n data items $x_1, x_2, \ldots, x_n$, is calculated as follows.

$$\bar{x} = \frac{\Sigma x}{n}$$

NORMAL FLOAT AUTO REAL DEGREE CL

mean({305,285,240,376,198, 264})
....278.

A calculator can find the mean of items in a list. This screen supports the text discussion of monthly building permits.

We find the central tendency of the monthly building permit figures as follows.

$$\text{Mean} = \bar{x} = \frac{\Sigma x}{n}$$

$$= \frac{305 + 285 + 240 + 376 + 198 + 264}{6} \quad \text{Add the monthly numbers.}$$
$$\text{Divide by the number of months.}$$

$$= \frac{1668}{6}, \quad \text{or} \quad 278$$

The mean value (the "average number of permits issued monthly") is 278.

EXAMPLE 1   Finding the Mean of a List of Sales Figures

Last year's annual sales in dollars for eight different online businesses were as follows.

NORMAL FLOAT AUTO REAL DEGREE CL

mean({374910,321872,242943 ,351147,382740,412111,3340 89,262900})
....335339.

This screen supports the result in **Example 1.**

374,910   321,872   242,943   351,147

382,740   412,111   334,089   262,900

Find the mean annual sales for the eight businesses.

Solution

$$\bar{x} = \frac{\Sigma x}{n} = \frac{2,682,712}{8} = 335,339 \quad \begin{array}{l}\text{Add the sales.}\\ \text{Divide by the number of businesses.}\end{array}$$

The mean annual sales amount is $335,339.

The following table shows the units and grades earned by one student last term.

Course	Grade	Units
Mathematics	A	3
History	C	3
Physics	B	5
Art	B	2
PE	A	1

Many calculators find the **mean** (as well as other statistical measures) automatically when a set of data items are entered. To recognize these calculators, look for a key marked $\bar{x}$, or perhaps μ, or look in a menu such as "LIST" for a listing of mathematical measures.

In one common method of defining **grade-point average,** an A grade is assigned 4 points, with 3 points for B, 2 for C, and 1 for D. Compute grade-point average as follows.

Step 1   Multiply the number of units for a course and the number assigned to each grade.

Step 2   Add these products.

Step 3   Divide by the total number of units.

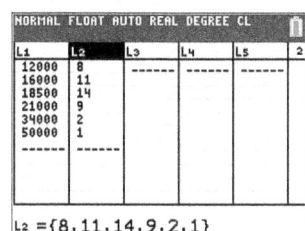

L2 ={8,11,14,9,2,1}

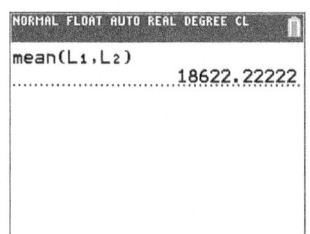

In these screens supporting **Example 2,** the salaries, and their frequencies, are first entered in List 1 and List 2, respectively.

Course	Grade	Grade Points	Units	(Grade Points) · (Units)
Mathematics	A	4	3	12
History	C	2	3	6
Physics	B	3	5	15
Art	B	3	2	6
PE	A	4	1	4
		Totals:	14	43

$$\text{Grade-point average} = \frac{43}{14} = 3.07 \ (\text{rounded})$$

The calculation of a grade-point average is an example of a **weighted mean,** because the grade points for each course grade must be weighted according to the number of units for the course. (For example, five units of A is better than two units of A.) The number of units is called the **weighting factor.**

WEIGHTED MEAN

The **weighted mean** of n numbers, $x_1, x_2, \ldots, x_n$ that are weighted by the respective factors $f_1, f_2, \ldots, f_n$ is calculated as follows.

$$\overline{w} = \frac{\Sigma(x \cdot f)}{\Sigma f}$$

In words, the weighted mean of a group of (weighted) items is the sum of all products of items times weighting factors, divided by the sum of all weighting factors.

The weighted mean formula is commonly used to find the mean for a frequency distribution. In this case, the weighting factors are the frequencies.

EXAMPLE 2   Finding the Mean of a Frequency Distribution of Salaries

Salary x	Number of Employees f
$12,000	8
$16,000	11
$18,500	14
$21,000	9
$34,000	2
$50,000	1

Find the mean salary for a small company that pays annual salaries to its employees as shown in the frequency distribution in the margin.

Solution

According to the weighted mean formula, we can set up the work as follows.

Salary x	Number of Employees f	Salary · Number $x \cdot f$
$12,000	8	$96,000
$16,000	11	$176,000
$18,500	14	$259,000
$21,000	9	$189,000
$34,000	2	$68,000
$50,000	1	$50,000
Totals:	45	$838,000

$$\text{Mean salary} = \frac{\Sigma(x \cdot f)}{\Sigma f} = \frac{\$838,000}{45} = \$18,622 \quad (\text{rounded})$$

For some data sets the mean can be a misleading indicator of average. Consider a small business that employs five workers at the following annual salaries.

$$\$16,500, \quad \$16,950, \quad \$17,800, \quad \$19,750, \quad \$20,000$$

The employees, struggling to survive on these salaries, approach the business owner, Aiguo, with the following computation.

$$\bar{x} = \frac{\$16,500 + \$16,950 + \$17,800 + \$19,750 + \$20,000}{5}$$

$$= \frac{\$91,000}{5}, \quad \text{or} \quad \$18,200 \quad \text{Mean salary (employees)}$$

Aiguo responds with his own calculations, for *all* workers (including his own salary of $188,000).

$$\bar{x} = \frac{\$16,500 + \$16,950 + \$17,800 + \$19,750 + \$20,000 + \$188,000}{6}$$

$$= \frac{\$279,000}{6}, \quad \text{or} \quad \$46,500 \quad \text{Mean salary (including Aiguo's)}$$

This mean is quite different, and it looks much more respectable.

The employees, of course, would argue that when Aiguo included his own salary in the calculation, it caused the mean to be a misleading indicator of average. This was so because Aiguo's salary is not typical. It lies a good distance away from the general grouping of the items (salaries). An extreme value like this is referred to as an **outlier.** Since a single outlier can have a significant effect on the value of the mean, we say that the mean is "highly sensitive to extreme values."

Median

Another measure of central tendency, which is not so sensitive to extreme values, is the **median.** This measure divides a group of numbers into two parts, with half the numbers below the median and half above it.

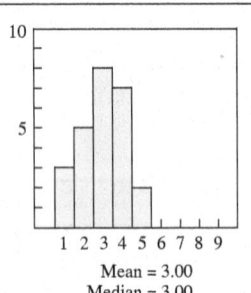

Mean = 3.00
Median = 3.00

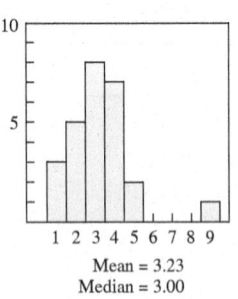

Mean = 3.23
Median = 3.00

The introduction of a single "outlier" above increased the mean by 8 percent but left the median unaffected.

Outliers should usually be considered as *possible* errors in the data.

MEDIAN

Find the **median** of a group of items as follows.

Step 1 Rank the items (that is, arrange them in numerical order from least to greatest).

Step 2 If the number of items is *odd*, the median is the middle item in the list.

Step 3 If the number of items is *even*, the median is the mean of the two middle items.

For Aiguo's business, all salaries (including his own), arranged in numerical order, are shown here.

$$\$16,500, \quad \$16,950, \quad \$17,800, \quad \$19,750, \quad \$20,000, \quad \$188,000$$

Thus, $$\text{median} = \frac{\$17,800 + \$19,750}{2} = \frac{\$37,550}{2} = \$18,775.$$

This figure is a representative average, based on all six salaries, that the employees would probably agree is reasonable.

EXAMPLE 3 Finding Medians of Lists of Numbers

Find the median of each list of numbers.

(a) 6, 7, 12, 13, 18, 23, 24

(b) 4, 7, 10, 13, 18, 23, 30, 35

(c) 17, 15, 9, 13, 21, 32, 41, 7, 12

(d) 147, 159, 132, 181, 174, 253

Solution

(a) This list is already ranked. The number of values in the list, 7, is odd, so the median is the middle value, or 13.

(b) The list is already ranked. Because the list contains an even number of items, namely 8, there is no single middle item. Find the median by taking the mean of the two middle items, 13 and 18.

$$\frac{13 + 18}{2} = \frac{31}{2} = 15.5 \leftarrow \text{Median}$$

(c) First rank the items.

$$7, 9, 12, 13, 15, 17, 21, 32, 41$$
$$\uparrow$$
$$\text{Median}$$

The middle number can now be identified. The median is 15.

(d) First rank the items.

$$132, 147, 159, 174, 181, 253$$

The two middle items are 159 and 174.

$$\frac{159 + 174}{2} = \frac{333}{2} = 166.5 \leftarrow \text{Median}$$

NORMAL FLOAT AUTO REAL DEGREE CL
median({6,7,12,13,18,23,24})
                            13
median({147,159,132,181,174,253})
                          166.5

The calculator can find the median of the entries in a list. This screen supports the results in **Examples 3(a) and 3(d).**

Locating the middle item (the median) of a frequency distribution is a bit more involved. First find the total number of items in the set by adding the frequencies ($n = \Sigma f$). Then the median is the item whose *position* is given by the following formula.

POSITION OF THE MEDIAN IN A FREQUENCY DISTRIBUTION

$$\text{Position of median} = \frac{n + 1}{2} = \frac{\Sigma f + 1}{2}$$

This formula gives only the position, and not the actual value, of the median.

EXAMPLE 4 **Finding Medians for Frequency Distributions**

Find the medians for the following distributions.

(a)

Value	1	2	3	4	5	6
Frequency	1	3	2	4	8	2

(b)

Value	2	4	6	8	10
Frequency	5	8	10	6	6

Solution

(a) Arrange the work as follows. Tabulate the values and frequencies, and the **cumulative frequencies,** which tell, for each different value, how many items have that value or a lesser value.

Value	Frequency	Cumulative Frequency	
1	1	1	1 item 1 or less
2	3	4	$1 + 3 = 4$ items 2 or less
3	2	6	$4 + 2 = 6$ items 3 or less
4	4	10	$6 + 4 = 10$ items 4 or less
5	8	18	$10 + 8 = 18$ items 5 or less
6	2	20	$18 + 2 = 20$ items 6 or less

Total: 20

Adding the frequencies shows that there are

20 items total.

(The final entry in the cumulative frequency column also shows the total number of items.)

$$\text{position of median} = \frac{20 + 1}{2} = \frac{21}{2} = 10.5$$

The median, then, is the average of the tenth and eleventh items. To find these items, make use of the cumulative frequencies. Since the value 4 has a cumulative frequency of 10, the tenth item is 4 and the eleventh item is 5, making the median

$$\frac{4 + 5}{2} = \frac{9}{2} = 4.5.$$

(b)

Value	Frequency	Cumulative Frequency
2	5	5
4	8	13
6	10	23
8	6	29
10	6	35

Total: 35

There are 35 items total.

$$\text{position of median} = \frac{35 + 1}{2} = \frac{36}{2} = 18$$

From the cumulative frequency column, the fourteenth through the twenty-third items are all 6s. This means the eighteenth item is a 6, so the median is 6.

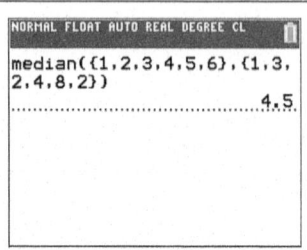

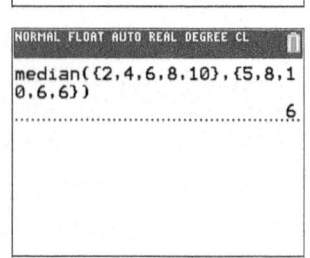

These two screens support the results in **Example 4**.

Mode

The third important measure of central tendency is the **mode.** Suppose ten students earned the following scores on a business law examination.

74, 81, 39, 74, 82, 80, 100, 92, 74, 85

Notice that more students earned the score 74 than any other score.

MODE

The **mode** of a data set is the value that occurs most often.

EXAMPLE 5 Finding Modes for Sets of Data

Find the mode for each set of data.

(a) 51, 32, 49, 49, 74, 81, 92

(b) 482, 485, 483, 485, 487, 487, 489

(c) 10,708, 11,519, 10,972, 17,546, 13,905, 12,182

(d)

Value	19	20	22	25	26	28
Frequency	1	3	8	7	4	2

Solution

(a) 51, 32, 49, 49, 74, 81, 92

The number 49 occurs more times than any other. Therefore, 49 is the mode. *The numbers do not need to be in numerical order when you look for the mode.*

(b) 482, 485, 483, 485, 487, 487, 489

Both 485 and 487 occur twice. This list is said to have *two* modes, or to be **bimodal.**

(c) No number here occurs more than once. This list has no mode.

(d)

Value	Frequency
19	1
20	3
22	8 ← Greatest frequency
25	7
26	4
28	2

The frequency distribution shows that the most frequently occurring value (and, thus, the mode) is 22.

It is traditional to include the mode as a measure of *central tendency,* because many important kinds of data sets do have their most frequently occurring values "centrally" located. And when the data items being studied are non-numerical, the mode is the only usable measure of central tendency. (The mean and median both fail to exist.)

Sometimes a distribution is bimodal, as in **Example 5(b).** In a large distribution, this term is commonly applied even when the two modes do not have exactly the same frequency. Three or more different items sharing the highest frequency of occurrence rarely offer useful information. We say that such a distribution has *no* mode.

Central Tendency from Stem-and-Leaf Displays

As seen in the previous section, data are sometimes presented in a stem-and-leaf display in order to give a visual impression of their distribution. We can also calculate measures of central tendency from a stem-and-leaf display. The median and mode are more easily identified when the "leaves" are ranked on their "stems."

In **Table 8** on the next page, we have ranked the leaves of **Table 5** in the previous section (which showed the weekly study times from **Example 2** of that section).

Table 8 Stem-and-Leaf Display for Weekly Study Times, with Leaves Ranked

1	2	4	5	6	7	8					
2	2	3	4	4	5	6	9	9	9		
3	0	1	1	2	3	3	6	6	6	6	8	9
4	1	4	4	5	5	7					
5	2	5	5	8							
6	0	2									
7	2										

EXAMPLE 6 Finding the Mean, Median, and Mode from a Stem-and-Leaf Display

For the data in **Table 8,** find the following.

(a) the mean   **(b)** the median   **(c)** the mode

Solution

(a) A calculator with statistical capabilities will automatically compute the mean. Otherwise, add all items (reading from the stem-and-leaf display) and divide by $n = 40$.

$$\text{mean} = \frac{12 + 14 + 15 + \cdots + 60 + 62 + 72}{40} = \frac{1414}{40} = 35.35$$

(b) In this case, $n = 40$ (an even number), so the median is the average of the twentieth and twenty-first items, in order. Counting leaves, we see that these will be the fifth and sixth items on the stem 3.

$$\text{median} = \frac{33 + 33}{2} = 33$$

(c) By inspection, we see that 36 occurred four times and no other value occurred that often. The mode is 36.

Symmetry in Data Sets

The most useful way to analyze a data set often depends on whether the distribution is **symmetric** or **nonsymmetric (or asymmetric).** In a "symmetric" distribution, as we move out from the central point, the pattern of frequencies is the same (or nearly so) to the left and to the right. In a "nonsymmetric" distribution, the patterns to the left and right are significantly different.

Figure 9 shows several types of symmetric distributions, while **Figure 10** shows some nonsymmetric distributions. A nonsymmetric distribution with a tail extending out to the left, shaped like a J, is **skewed to the left.** If the tail extends out to the right, the distribution is **skewed to the right.** Notice that a bimodal distribution may be either symmetric or nonsymmetric.

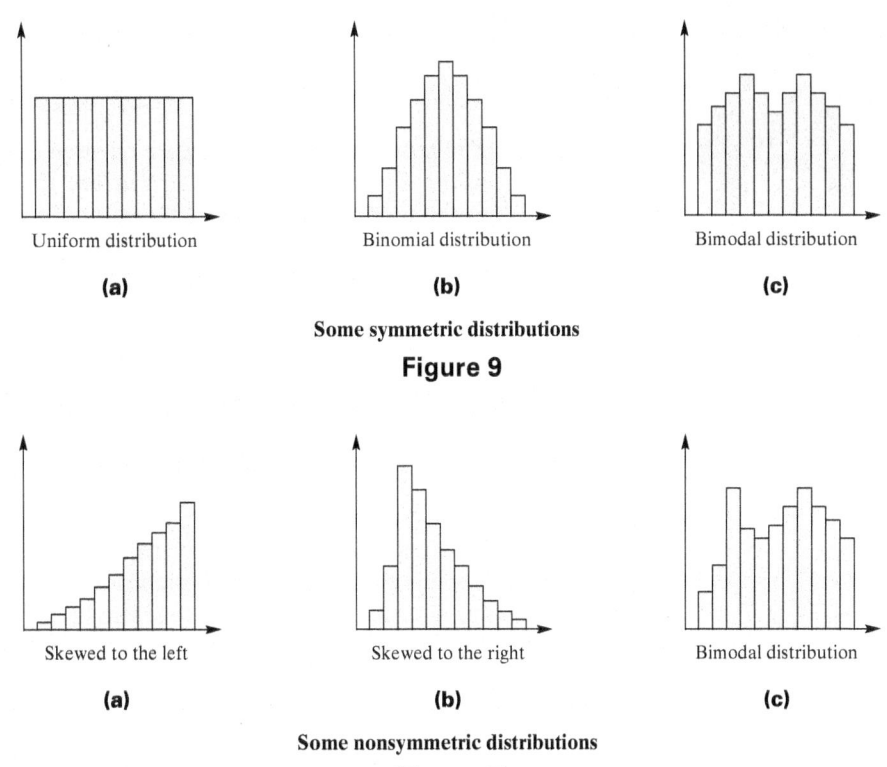

Some symmetric distributions

Figure 9

Some nonsymmetric distributions

Figure 10

Data Sets: Mean, Median, and Mode

Summary

We conclude this section with a summary of the measures presented and a brief discussion of their relative advantages and disadvantages.

SUMMARY OF THE COMMON MEASURES OF CENTRAL TENDENCY

The **mean** of a set of numbers is found by adding all the values in the set and dividing by the number of values.

The **median** is a kind of "middle" number. To find the median, first rank the values. For an *odd* number of values, the median is the middle value in the list. For an *even* number of values, the median is the mean of the two middle values.

The **mode** is the value that occurs with the greatest frequency. Some sets of numbers have two most frequently occurring values and are **bimodal.** Other sets have no mode at all (if no value occurs more often than the others or if more than two values occur most often).

Some helpful points of comparison follow.

1. For distributions of numerical data, the mean and median will always exist, while the mode may not exist. On the other hand, for non-numerical data, it may be that none of the three measures exists, or that only the mode exists.

2. Because even a single change in the data may cause the mean to change significantly, while the median and mode may not be affected at all, ***the mean is the most "sensitive" measure.***

3. In a symmetric distribution, the mean, median, and mode (if a single mode exists) will all be equal. In a nonsymmetric distribution, the mean is often unduly affected by relatively few extreme values and, therefore, may not be a good representative measure of central tendency. For example, distributions of salaries, family incomes, or home prices often include a few values that are much higher than the bulk of the items. In such cases, the median is a more useful measure.

4. ***The mode is the only measure covered here that must always be equal to one of the data items of the distribution.*** In fact, more of the data items are equal to the mode than to any other number. A fashion shop planning to stock only one hat size for next season would want to know the mode (the most common) of all hat sizes among its potential customers.

WHEN WILL I **EVER** USE THIS **?**

There is considerable controversy over how important professional intervention is, in general, for addiction remission. But it seems clear that for those who *do* need help with their follow-up recovery, a complete quarterly checkup is much more effective than just a brief interview in getting them back into treatment.

The graph reflects the results of an extensive study, *Drug and Alcohol Dependence 2012*, reported by M. L. Dennis and C. K. Scott. Conveying and interpreting information graphically is important in the work of psychologists and other social scientists.

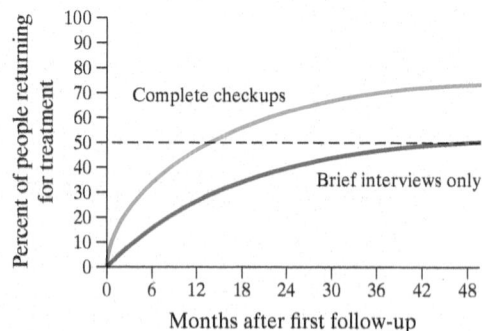

Effects of follow-up on readmission to treatment

Data from M. L. Dennis and C. K. Scott, *Drug and Alcohol Dependence 2012.*

The median time to readmission is the number of months by which half the people who needed help with their follow-up recovery had returned for treatment. Complete the following statements.

1. The median time for those having checkups was about _____ months.

2. The median time for those not having checkups was about _____ months.

3. For those who had complete quarterly checkups, the median time to readmission was reduced by about _____ months.

Answers: **1.** 14 **2.** 46 **3.** 32

12.2 EXERCISES

CONCEPT CHECK *Choose all that apply.*

    A. Mean   **B.** Median   **C.** Mode

1. This measure exists for every numerical data distribution.

2. This measure is found by first ranking the data.

3. This measure may not be a data item in the distribution.

4. Outliers have a disproportionate effect on this measure.

For each list of data, calculate

(a) *the mean (to the nearest tenth),*

(b) *the median, and*

(c) *the mode or modes (if any).*

5. 6, 9, 12, 14, 21

6. 20, 27, 42, 45, 53, 62, 62, 64

7. 218, 230, 196, 224, 196, 233

8. 26, 31, 46, 31, 26, 29, 31

9. 3.1, 4.5, 6.2, 7.1, 4.5, 3.8, 6.2, 6.3

10. 14,322, 16,959, 17,337, 15,474

Solve each problem.

11. *Gymnasts' Scores* An Olympic gymnast can earn an average score of 16.5 points for each of two vaults. Another can earn an average score of 15.5 points. The first gymnast falters, losing 2.375 points on her first vault and 1.575 points on her second vault. What average score must the second gymnast earn to win the gold? (*MT* calendar problem)

12. *Quiz Grades* The average of 5 quiz grades is 10. When the lowest grade is dropped and the new average is calculated, it turns out to be 11. What was the score of the dropped grade? (*MT* calendar problem)

Leading U.S. Businesses The table at the top of the next column shows the top five world companies, ranked by 2016 revenue.

Find each of the following quantities for these five companies.

13. the mean revenue

14. the median revenue

Business	Revenue (in billions of dollars)
Wal-Mart Stores, U.S.	$485.9
State Grid, China	315.2
Sinopec Group, China	267.5
China National Petroleum, China	262.6
Toyota Motor, Japan	254.7

Data from *The World Almanac and Book of Facts 2018; Fortune* magazine, July 2017.

Airline Fatalities in the United States The table pertains to scheduled commercial carriers over a decade that included 2001. Fatalities data include those on the ground except for the September 11, 2001, terrorist attacks. Use this information for Exercises 15–20.

U.S. Airline Safety, 1999–2008

Year	Departures (in millions)	Fatal Accidents	Fatalities
1999	10.9	2	12
2000	11.1	2	89
2001	10.6	6	531
2002	10.3	0	0
2003	10.2	2	22
2004	10.8	1	13
2005	10.9	3	22
2006	10.6	2	50
2007	10.7	0	0
2008	10.6	0	0

Data from *The World Almanac and Book of Facts 2010.*

For each category in Exercises 15–20, find

(a) *the mean (to the nearest tenth),*

(b) *the median, and*

(c) *the mode or modes (if any).*

15. departures (in millions) **16.** fatal accidents

17. fatalities

The year 2001 was clearly an anomaly. If the terrorist numbers are removed from the data for that year (4 fatal accidents and 265 fatalities), which of the three measures change and what are their new values for each of the following?

18. Exercise 16   **19. Exercise 17**

20. Following 2001, in what year did airline departures start to increase again?

Measuring Elapsed Times *While doing an experiment, a physics student recorded the following sequence of elapsed times (in seconds) in a lab notebook.*

$$2.16, \ 22.2, \ 2.96, \ 2.20, \ 2.73, \ 2.22, \ 2.39$$

21. Find the mean. **22.** Find the median.

When reviewing the calculations later, the student decided that the entry 22.2 should have been recorded as 2.22 and made that change in the listing.

23. Find the mean for the new list.

24. Find the median for the new list.

25. Which measure, the mean or the median, was affected more by correcting the error?

26. Find the mode(s), if any, for the original data and the changed data.

Scores on Management Examinations *Thao earned the following scores on her six management exams last semester.*

$$79, \ 81, \ 44, \ 89, \ 79, \ 90$$

27. Find the mean, median, and mode for Thao's scores.

28. Which of the three averages probably is the best indicator of Thao's ability?

29. Thao has a chance to replace her score of 44 by taking a "make-up" exam. What must she score on the make-up exam to get an overall average (mean) of 85?

For each of the following frequency distributions, find **(a)** *the mean (to the nearest tenth),* **(b)** *the median, and* **(c)** *the mode or modes (if any).*

30.

Value	Frequency
6	3
7	1
8	8
9	4

31.

Value	Frequency
615	13
540	7
605	9
579	14
586	7
600	5

32. Average Employee Salary A company has

15 employees with a salary of $21,500,

11 employees with a salary of $23,000,

17 employees with a salary of $25,800,

2 employees with a salary of $31,500,

4 employees with a salary of $38,900,

1 employee with a salary of $147,500.

Find the mean salary for the employees (to the nearest hundred dollars).

Grade-Point Averages *Find the grade-point average for each of the following students. Assume*

$$A = 4, \ B = 3, \ C = 2, \ D = 1, \ and \ F = 0.$$

Round to the nearest hundredth.

33.

Units	Grade
8	A
3	B
5	C

34.

Units	Grade
2	C
3	B
7	A
2	F

Federal Budget Totals *The table gives federal receipts and outlays for the years 2013–2017. (The 2017 figures are estimates.) Use this information for Exercises 35 and 36.*

Fiscal Year	Receipts (in billions of dollars)	Outlays (in billions of dollars)
2013	$2775.1	$3454.6
2014	3021.5	3506.1
2015	3249.9	3688.4
2016	3268.0	3852.6
2017	3459.7	4062.2

Data from *The World Almanac and Book of Facts 2018; Budget of the U.S. Government, Fiscal Year 2017,* Office of Management and Budget, Executive Office of the President.

Over the five-year period, find **(a)** *the mean and* **(b)** *the median for each of the following.*

35. receipts

36. outlays

World Cell Phone Use *In 2016, just the top six countries accounted for over 50% of cell phone subscriptions worldwide. Use the data in the table for Exercises 37 and 38.*

Country	Cell Phone Subscriptions (in millions)
China	1364.9
India	1127.8
United States	416.7
Indonesia	385.6
Russia	244.1
Brazil	231.4

Data from *The World Almanac and Book of Facts 2018*; International Telecommunications Union.

37. Find the approximate mean number of cell phone subscriptions for these six countries in 2016.

38. The United States accounted for about 5.547% of worldwide subscriptions. About how many subscriptions were active in the world that year?

39. Find the approximate mean number of cell phone subscriptions for the BRIC countries (Brazil, Russia, India, and China) in 2016.

40. Find the approximate mean number of cell phone subscriptions for the non-BRIC countries in the list in 2016.

Crew, Passengers, and Entertainers on Cruise Ships *The table shows, for four cruises, the numbers of crew members, passengers, and entertainers (not included as passengers). For each quantity in Exercises 41–43, find **(a)** the mean, and **(b)** the median.*

Cruise	Crew	Passengers	Entertainers
Alaska	185	1900	35
Mexico	223	3000	75
Scandinavia	175	1200	20
Mediterranean	215	2700	50

41. number of crew members per cruise

42. number of passengers per cruise

43. total number of persons per cruise

Olympic Medal Standings *The top ten medal-winning nations in the 2018 Winter Olympics at PyeongChang, South Korea, are shown in the table. Use the given information for Exercises 44–47.*

Medal Standings for the 2018 Winter Olympics

Place	Nation	Gold	Silver	Bronze	Total
1	Norway	14	14	11	39
2	Germany	14	10	7	31
3	Canada	11	8	10	29
4	United States	9	8	6	23
5	Netherlands	8	6	6	20
6	Korea	5	8	4	17
7	OAR	2	6	9	17
8	Switzerland	5	6	4	15
9	France	5	4	6	15
10	Sweden	7	6	1	14

Data from www.espn.com

Calculate the following for all nations shown.

44. the mean number of gold medals

45. the median number of bronze medals

46. the mode, or modes, for the number of silver medals

47. each of the following for the total number of medals
   (a) mean
   (b) median
   (c) mode or modes

*In Exercises 48 and 49, use the given stem-and-leaf display to identify **(a)** the mean, **(b)** the median, and **(c)** the mode or modes (if any) for the data represented.*

48. Online Sales The display here represents prices (to the nearest dollar) charged by 23 different online sellers for a new car alternator. Give answers to the nearest dollar.

9	9
10	2 3
10	5 8 9
11	1 2 3 3
11	5 6 7 8 8
12	0 2 4
12	5 6 7 9
13	4

49. *Scores on a Biology Exam* The display here represents scores achieved on a 100-point biology exam by the 34 members of a class.

4	7
5	1 3 6
6	2 5 5 6 7 8 8
7	0 4 5 6 7 7 8 8 8 8 9
8	0 1 1 3 4 5 5
9	0 0 0 1 6

50. *A Missing Test Score* Katie's Business professor lost his computer memory, which contained her five test scores for the course. A summary of the scores (each of which was an integer from 0 to 100) indicates the following:

The mean was 88.

The median was 87.

The mode was 92.

(The data set was not bimodal.) What is the least possible number among the missing scores?

51. Explain what an "outlier" is and how it affects measures of central tendency.

52. *Scores on a Math Quiz* The following are scores earned by 15 college students on a 20-point math quiz.

0, 0, 1, 14, 14, 15, 16, 16, 17, 17, 18, 18, 18, 19, 20

(a) Calculate the mean, median, and mode values.

(b) Which measure in part (a) is most representative of the data?

53. *Consumer Preferences in Food Packaging* A food processing company that packages individual cups of instant soup wishes to find out the best number of cups to include in a package. A survey of 22 consumers revealed that five prefer a package of 1, five prefer a package of 2, three prefer a package of 3, six prefer a package of 4, and three prefer a package of 6.

(a) Calculate the mean, median, and mode values for preferred package size.

(b) Which measure in part (a) should the food processing company use?

(c) Explain your answer to part (b).

*In Exercises 54–57, begin a list of the given numbers, in order, starting with the least one. Continue the list only until the median of the listed numbers is a multiple of 4. Stop at that point and find **(a)** the number of numbers listed, and **(b)** the mean of the listed numbers (to two decimal places).*

54. counting numbers

55. prime numbers

56. Fibonacci numbers

57. triangular numbers

Solve each problem.

58. Seven consecutive even whole numbers add up to 168. What is the result when their mean is subtracted from their median?

59. Bajir wants to include a fifth counting number, n, along with the numbers 2, 5, 8, and 9 so that the mean and median of the five numbers will be equal. How many choices does he have for the number n, and what are those choices?

60. If the mean, median, and mode are all equal for the set

$$\{70, 110, 80, 60, x\},$$

find the value of x.

For Exercises 61 and 62, refer to the grouped frequency distribution shown here.

Class Limits	Frequency f
21–25	5
26–30	3
31–35	8
36–40	12
41–45	21
46–50	38
51–55	35
56–60	20

61. Is it possible to identify, on the basis of the data shown in the table, any specific data items that occurred in this sample?

62. Is it possible to compute the actual mean for this sample?

Write a short answer for each problem.

63. Describe how you might approximate the mean for this sample. Justify your procedure.

64. *Average Employee Salaries* Refer to the salary data of **Example 2,** specifically the dollar amounts given in the salary column of the table. Explain what is wrong with simply calculating the mean salary by adding those six numbers and dividing the result by 6.

FOR FURTHER THOUGHT

Simpson's Paradox

A margin note in an earlier chapter showed how baseball player A can have a greater batting average than player B in each of two (or more) seasons, and yet have a lesser average than B overall when the individual seasons are combined.

This puzzling result is not confined to baseball. Indeed, it can arise in any area, including

- graduate school acceptance rates
- median wage rates
- medical research studies
- educational test scores
- hiring rates.

It is very important to interpret statistical measures carefully. This "paradox" was mentioned by researchers in 1899 and 1903 and was first described in a technical paper by Edward H. Simpson in 1951.

The table shows the results of an assessment of the effectiveness of a particular kidney stone treatment, as reported in a 1986 study.

For Group or Individual Investigation

1. Fill in the six blanks in the table (to the nearest tenth of a percent).

2. Which treatment was more effective for the 357 persons with small stones?

3. Which treatment was more effective for the 343 persons with large stones?

4. Which treatment was more effective for all 700 persons in the study?

5. If you were the attending physician, which treatment would you prescribe for your next patient with kidney stones?

	Treatment A			Treatment B		
	Treated	Effective	Percent Effective	Treated	Effective	Percent Effective
Small Stones	87	81		270	234	
Large Stones	263	192		80	55	
Combined	350	273		350	289	

Data from www.brookings.edu/search/simpson's

12.3   MEASURES OF DISPERSION

OBJECTIVES

1 Find the range of a data set.
2 Calculate the standard deviation of a data set.
3 Interpret measures of dispersion.
4 Calculate the coefficient of variation.

The mean is a good indicator of the central tendency of a set of data values, but it does not completely describe the data. Compare distribution A with distribution B in **Table 9**.

Both distributions have the same mean and the same median, and neither has a mode, so their central tendencies are identical. But in another way, the two distributions are quite different. In the first, 7 is a fairly typical value, but in the second, most of the values differ considerably from 7. What is needed here is some measure of the **dispersion,** or *spread*, of the data.

Table 9

	A	B
	5	1
	6	2
	7	7
	8	12
	9	13
Mean	7	7
Median	7	7

Range

The **range** of a data set is a straightforward measure of dispersion.

Table 9 (repeated)

	A	B
	5	1
	6	2
	7	7
	8	12
	9	13
Mean	7	7
Median	7	7

Table 10

Quiz	Gage	Allie
1	28	27
2	22	27
3	21	28
4	26	6
5	18	27
Mean	23	23
Median	22	27
Range	10	22

StatCrunch Mean/SD vs. Median/IQR

Once the data are entered, a calculator with statistical functions may actually show the range (among other things), or at least sort the data and identify the minimum and maximum items. (The associated symbols may be something like $\boxed{\text{MIN} \Sigma}$ and $\boxed{\text{MAX} \Sigma}$, or min$X$ and max$X$.) Given these two values, a simple subtraction produces the range.

> **RANGE**
>
> For any set of data, the **range** of the set is defined as follows.
>
> **Range = (greatest value in the set) − (least value in the set)**

EXAMPLE 1 Finding and Comparing Range Values

Find the ranges for distributions A and B in **Table 9,** and describe what they imply.

Solution

In distribution A, range = greatest value − least value = 9 − 5 = 4.

Similarly, in distribution B, range = 13 − 1 = 12.

Even though the two distributions have identical averages, distribution B exhibits three times more dispersion, or *spread,* than distribution A.

The range can be misleading if it is interpreted unwisely. For example, look at the points scored by Gage and Allie on five different quizzes, as shown in **Table 10.** The ranges for the two students make it tempting to conclude that Gage is more consistent than Allie. However, Allie is actually more consistent, with the exception of one very poor score. That score, 6, is an outlier which, if not actually recorded in error, must surely be due to some special circumstance. (Notice that the outlier does not seriously affect Allie's median score, which is more typical of her overall performance than is her mean score.)

Standard Deviation

One of the most useful measures of dispersion, the *standard deviation,* is based on *deviations* of the individual data values *from the mean.*

EXAMPLE 2 Finding Deviations from the Mean

Find the deviations from the mean for all data values in the following sample.

$$32, 41, 47, 53, 57$$

Solution

Add these values and divide by the total number of values, 5. The mean is 46. To find the deviations from the mean, subtract 46 from each data value.

Data value	32	41	47	53	57
Deviation	−14	−5	1	7	11

$$32 - 46 = -14 \qquad\qquad 57 - 46 = 11$$

To check your work, add the deviations. ***The sum of the deviations for a set of data is always 0.***

We cannot obtain a measure of dispersion by finding the mean of the deviations, because this number is always 0, since the positive deviations exactly cancel out the negative ones. To avoid this problem of positive and negative numbers canceling each other, we *square* each deviation.

The following chart shows the squares of the deviations for the data in **Example 2.**

Data value	32	41	47	53	57
Deviation	-14	-5	1	7	11
Square of deviation	196	25	1	49	121

$(-14) \cdot (-14) = 196$ $11 \cdot 11 = 121$

An average of the squared deviations could now be found by dividing their sum by the number of data values n (5 in this case), which we would do if our data values composed a population. However, since we are considering the data to be a sample, we divide by $n - 1$ instead.*

The average that results is itself a measure of dispersion, called the **variance,** but a more common measure is obtained by taking the square root of the variance. This compensates, in a way, for squaring the deviations earlier and gives a kind of average of the deviations from the mean, which is called the sample **standard deviation.** It is denoted by the letter s. (The standard deviation of a population is denoted σ, the lowercase Greek letter *sigma*.)

Continuing our calculations from the chart above, we obtain

$$s = \sqrt{\frac{196 + 25 + 1 + 49 + 121}{4}} = \sqrt{\frac{392}{4}} = \sqrt{98} \approx 9.90.$$

$$n - 1$$

The algorithm (process) described above for finding the sample standard deviation can be summarized as follows.

CALCULATION OF STANDARD DEVIATION

Let a sample of n numbers $x_1, x_2, \ldots, x_n$ have mean $\bar{x}$. Then the **sample standard deviation, s,** of the numbers is calculated as follows.

$$s = \sqrt{\frac{\Sigma(x - \bar{x})^2}{n - 1}}$$

The individual steps involved in this calculation follow.

Step 1 Calculate $\bar{x}$, the mean of the numbers.

Step 2 Find the deviations from the mean.

Step 3 Square each deviation.

Step 4 Sum the squared deviations.

Step 5 Divide the sum in Step 4 by $n - 1$.

Step 6 Take the square root of the quotient in Step 5.

NORMAL FLOAT AUTO REAL DEGREE CL

stdDev({32,41,47,53,57})
                    9.899494937
√(98)
                    9.899494937

This screen supports the text discussion. Note that the standard deviation reported agrees with the approximation for $\sqrt{98}$.

Most calculators find square roots, such as $\sqrt{98}$, to as many digits as you need using a key like $\boxed{\sqrt{x}}$. In this text, we normally give from two to four significant figures for such calculations.

---

*Although the reasons cannot be explained at this level, dividing by $n - 1$ rather than n produces a sample measure that is more accurate for purposes of inference. In most cases, the results using the two divisors are only slightly different.

The preceding description helps show why standard deviation measures the amount of spread in a data set. For actual calculation purposes, we recommend the use of a calculator, or a package such as StatCrunch, that does all the detailed steps automatically.

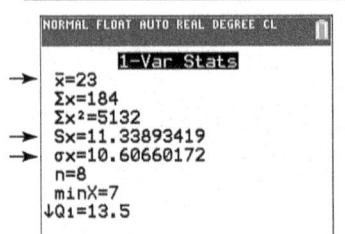

The sample in **Example 3** is stored in a list.

The arrows point to the mean and the sample and population standard deviations. See **Example 3**.

EXAMPLE 3   Finding a Sample Standard Deviation

Find the standard deviation of the following sample by using **(a)** the step-by-step process, and **(b)** the statistical functions of a calculator.

$$7, 9, 18, 22, 27, 29, 32, 40$$

Solution

(a) Carry out the six steps summarized above.

Step 1   Find the mean of the values.

$$\frac{7 + 9 + 18 + 22 + 27 + 29 + 32 + 40}{8} = 23$$

Step 2   Find the deviations from the mean.

Data value	7	9	18	22	27	29	32	40
Deviation	-16	-14	-5	-1	4	6	9	17

Step 3   Square each deviation.

Squares of deviations: 256   196   25   1   16   36   81   289

Step 4   Sum the squared deviations.

$$256 + 196 + 25 + 1 + 16 + 36 + 81 + 289 = 900$$

Step 5   Divide by $n - 1 = 8 - 1 = 7$

$$\frac{900}{7} \approx 128.57$$

Step 6   Take the square root

$$\sqrt{128.57} \approx 11.3$$

(b) Enter the eight data values in a list. Then calculate the 1-variable statistics for that list.

The result should again be 11.3.

If you *mistakenly* used the population standard deviation key, the result would be 10.6.

For data given in the form of a frequency distribution, you can enter the values and their frequencies in two separate lists and then calculate 1-variable statistics for those lists.

The following example is included only to strengthen your understanding of frequency distributions and standard deviation, not as a practical algorithm for calculating.

Table 11

Value	Frequency
2	5
3	8
4	10
5	2

```
NORMAL FLOAT AUTO REAL DEGREE CL
stdDev({2,3,4,5},{5,8,10,2
})
                .9073771726
√(19.76/24)
                .9073771726
```

The screen supports the result in **Example 4.**

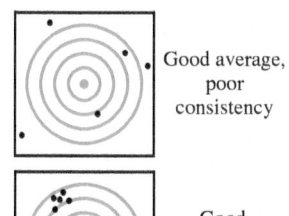

The more desirable basket depends on your objective.

Good average, poor consistency

Good consistency, poor average

In this case, good consistency (lesser dispersion) is more desirable than a good average (central tendency).

EXAMPLE 4 **Finding the Standard Deviation of a Frequency Distribution**

Find the sample standard deviation for the frequency distribution shown in **Table 11.**

Solution

Complete the calculations as shown in **Table 12** below. To find the numbers in the "Deviation" column, first find the mean, and then subtract the mean from the numbers in the "Value" column.

Table 12

Value	Frequency	Value Times Frequency	Deviation	Squared Deviation	Squared Deviation Times Frequency
2	5	10	−1.36	1.8496	9.2480
3	8	24	−0.36	0.1296	1.0368
4	10	40	0.64	0.4096	4.0960
5	2	10	1.64	2.6896	5.3792
Sums	25	84			19.76

$$\bar{x} = \frac{84}{25} = 3.36 \qquad s = \sqrt{\frac{19.76}{24}} \approx \sqrt{0.8233} \approx 0.91$$

Suppose tomatoes sell by the basket. Each basket costs the same, and each contains one dozen tomatoes. If you want the most fruit possible per dollar spent, you would look for the basket with the highest average weight per tomato (regardless of the dispersion of the weights). On the other hand, if the tomatoes are to be served on an hors d'oeuvre tray where "presentation" is important, you would look for a basket with uniform-sized tomatoes—that is, a basket with the lowest weight dispersion (regardless of the average of the weights). See the illustration at the side.

Another situation involves target shooting (also illustrated at the side). The five hits on the top target are, *on average,* very close to the bull's-eye. If the square grid is 10 by 10 units and the concentric circles have radii 2, 4, 6, and 8, then the hits shown are at

$$(-9, -7),\ (-5, 8),\ (2, -4),\ (6, 4),\ \text{and}\ (9, 2),$$

so the average position is the point

$$\left(\frac{-9 - 5 + 2 + 6 + 9}{5}, \frac{-7 + 8 - 4 + 4 + 2}{5}\right) = \left(\frac{3}{5}, \frac{3}{5}\right),$$

which is about 0.85 unit from the bull's-eye. However, the large dispersion (spread) of the hits—the lack of consistency—implies that improvement will require much effort on the part of the shooter. On the other hand, the bottom target exhibits a poorer *average.* (The hits in this case have an approximate average position at $(-3, 6)$, which is about 6.71 units from the bull's-eye.) But in this case the poor average is balanced by a small dispersion—that is, good consistency. Improvement will require only a minor adjustment of the gun sights. In general, consistent errors can be corrected more easily than more dispersed errors.

Data Sets: Mean, Median, and Mode

Interpreting Measures of Dispersion

A main use of dispersion measures is to compare the amounts of spread in two (or more) data sets as we did with distributions A and B earlier in this section.

EXAMPLE 5 Comparing Populations Based on Samples

Two companies, *A* and *B*, sell 12-ounce jars of instant coffee. Five jars of each were randomly selected from markets, and the contents were carefully weighed, with the following results.

$$A: \quad 12.02, \quad 12.08, \quad 11.99, \quad 11.96, \quad 11.99$$
$$B: \quad 12.40, \quad 12.21, \quad 12.36, \quad 12.22, \quad 12.27$$

Find **(a)** which company provides more coffee in its jars, and **(b)** which company fills its jars more consistently.

Solution

The mean and standard deviation values for both samples are shown in **Table 13**.

Table 13

Sample A	Sample B
$\bar{x}_A = 12.008$	$\bar{x}_B = 12.292$
$s_A = 0.0455$	$s_B = 0.0847$

(a) Since $\bar{x}_B$ is greater than $\bar{x}_A$, we *infer* that Company B most likely provides more coffee (greater mean) per jar.

(b) Since s_A is less than s_B, we *infer* that Company A seems more consistent (smaller standard deviation).

The conclusions drawn in **Example 5** are tentative, because the samples were small. We could place more confidence in inferences drawn from larger samples.

It is clear that a larger dispersion value means more "spread" than a smaller one. But it is difficult to express the exact meaning of dispersion. *It is impossible* (though it would be nice) to make a general statement like "Exactly half of the items of any distribution lie within one standard deviation of the mean. There is, however, one useful result that does apply to all data sets, no matter what their distributions are like. This result is named for the Russian mathematician Pafnuty Lvovich Chebyshev.

CHEBYSHEV'S THEOREM

For any set of numbers, regardless of how they are distributed, the fraction of them that lie within k standard deviations of their mean (where $k > 1$) is *at least*

$$1 - \frac{1}{k^2}.$$

Be sure to notice the words *at least* in the theorem statement. In certain distributions the fraction of items within k standard deviations of the mean may be more than $1 - \frac{1}{k^2}$, but in no case will it ever be less. The theorem is true for any value of k greater than 1 (integer or noninteger).

Pafnuty Lvovich Chebyshev
(1821–1894) was a Russian mathematician known mainly for his work on the theory of prime numbers. Chebyshev and French mathematician and statistician **Jules Bienaymé** (1796–1878) independently developed an important inequality of probability now known as the Bienaymé–Chebyshev inequality.

EXAMPLE 6 Applying Chebyshev's Theorem

What is the minimum percentage of the items in a data set that lie within 3 standard deviations of the mean?

Solution

With $k = 3$, $\quad 1 - \dfrac{1}{3^2} = 1 - \dfrac{1}{9} = \dfrac{8}{9} \approx 0.889 = 88.9\%$ ← Minimum percentage

$3^2 = 3 \cdot 3$, not $3 \cdot 2$

Coefficient of Variation

Look again at the top target pictured in the margin earlier. The dispersion, or spread, among the five bullet holes might not be especially impressive if the shots were fired from 100 yards, but it would be much more so at, say, 300 yards. There is another measure, the *coefficient of variation*, that takes this distinction into account. It is not strictly a measure of dispersion, as it combines central tendency and dispersion. It expresses the standard deviation as a percentage of the mean. ***Often this is a more meaningful measure than a straight measure of dispersion, especially when we are comparing distributions whose means are appreciably different.***

COEFFICIENT OF VARIATION

For any set of data, the **coefficient of variation** measures *relative dispersion*. It is calculated as follows.

$$V = \frac{s}{\bar{x}} \cdot 100\% \quad \text{for a sample} \qquad \text{or} \qquad V = \frac{\sigma}{\mu} \cdot 100\% \quad \text{for a population}$$

Table 14

Sample A	Sample B
$\bar{x}_A = 16.167$	$\bar{x}_B = 153.167$
$s_A = 3.125$	$s_B = 25.294$
$V_A = 19.3$	$V_B = 16.5$

EXAMPLE 7 Comparing Samples

Compare the dispersions in the two samples A and B.

$$A: 12, 13, 16, 18, 18, 20 \qquad B: 125, 131, 144, 158, 168, 193$$

Solution

Using a calculator, we apply the first formula of the previous definition to obtain the values shown in **Table 14**. From the calculated values, we see that sample B has a much larger dispersion (standard deviation) than sample A. But sample A actually has the larger *relative* dispersion (coefficient of variation). The dispersion within sample A is larger as a percentage of that sample's mean.

12.3 EXERCISES

CONCEPT CHECK *Complete each statement.*

1. The difference between the greatest data value and the least data value is the _____.

2. The _____ _____ is the square root of the variance.

3. Chebyshev's theorem states that, in any distribution, the fraction of items within 2 standard deviations of the mean is at least _____.

4. The coefficient of variation is a measure of _____ _____.

Solve each problem.

5. If your calculator finds both the sample standard deviation and the population standard deviation, which of the two will be a larger number for a given set of data? (*Hint:* Recall the difference in the ways the two standard deviations are calculated.)

6. If your calculator finds only one kind of standard deviation, explain how you could determine, without the calculator instructions, whether it is sample or population standard deviation.

Find **(a)** *the range, and* **(b)** *the standard deviation for each sample. If necessary, round answers to the nearest hundredth.*

7. 2, 4, 5, 8, 9, 11, 16

8. 16, 12, 10, 8, 19, 15, 22, 16, 5

9. 34, 27, 22, 41, 30, 15, 31

10. 62, 81, 57, 63, 75, 61, 88, 72, 65

11. 74.96, 74.60, 74.58, 74.48, 74.72, 75.62, 75.03, 75.10, 74.53

12. 311.8, 310.4, 309.3, 312.1, 312.5, 313.5, 310.6, 310.5, 311.0, 314.2

13.

Value	Frequency
9	2
12	5
8	6
3	4
1	2

14.

Value	Frequency
29	4
14	6
23	3
20	2
18	12
22	2
26	5

Find the least possible fraction of the numbers in a data set lying within the given number of standard deviations of the mean. Apply Chebyshev's theorem and give answers as common fractions reduced to lowest terms.

15. 2 **16.** 3 **17.** $\dfrac{5}{3}$ **18.** $\dfrac{7}{6}$

In a certain distribution of numbers, the mean is 50 and the standard deviation is 5. At least what fraction of the numbers are between the following pairs of numbers?

19. 40 and 60 **20.** 35 and 65

21. 30 and 70 **22.** 25 and 75

In a distribution with mean 80 and standard deviation 8, find the largest fraction of the numbers that could meet the following requirements.

23. less than 64 or more than 96

24. less than 62 or more than 98

25. less than 52 or more than 108

26. less than 60 or more than 100

Travel Accommodation Costs *Gabriel and Lucia took a road trip across the country. The room costs, in dollars, for their overnight stays are listed here.*

99	105	120	165	185	178
110	245	134	134	120	260

Use this distribution of costs for Exercises 27–30.

27. Find the mean of the distribution.

28. Find the standard deviation of the distribution.

29. How many of the cost amounts are within 1 standard deviation of the mean?

30. What does Chebyshev's theorem say about the number of the amounts that are within 2 standard deviations of the mean?

Samples *In each problem, two samples are given. In each case, (a) find both sample standard deviations, (b) find both sample coefficients of variation, (c) decide which sample has the higher dispersion, and (d) decide which sample has the higher relative dispersion.*

31. *A:* 3, 7, 4, 3, 8 *B:* 10, 8, 10, 6, 7, 3, 5

32. *A:* 65, 75, 69, 65, 71, 72, 68, 71, 67, 67

 B: 23, 35, 30, 32, 31, 36, 38, 29, 34, 33

Consider the following sample.

13, 14, 17, 19, 21, 22, 25

33. Compute the mean and standard deviation for the sample (each to the nearest hundredth).

34. Now add 5 to each item of the given sample, and compute the mean and standard deviation for the new sample.

35. Go back to the original sample. This time subtract 10 from each item, and compute the mean and standard deviation of the new sample.

36. Based on your answers for the previous three exercises, what happens to the mean and standard deviation when all items of a sample have the same constant k added or subtracted?

37. Go back to the original sample again. This time multiply each item by 3, and compute the mean and standard deviation of the new sample.

38. What happens to the mean and standard deviation when all items of a sample are multiplied by the same constant k.

Solve each problem.

39. Comparing Water Heater Lifetimes Two brands of electric water heaters, both carrying 6-year warranties, were sampled and tested under controlled conditions. Five of each brand failed after the numbers of months shown here.

 Brand A: 74, 65, 70, 64, 71

 Brand B: 69, 70, 62, 72, 60

 (a) Calculate both sample means.

 (b) Calculate both sample standard deviations.

 (c) Which brand apparently lasts longer?

 (d) Which brand has the more consistent lifetime?

Lifetimes of Engine Control Modules *Chin manages the service department of a trucking company. Each truck in the fleet utilizes an electronic engine control module. Long-lasting modules are desirable. A preventive replacement program also avoids costly breakdowns. For this purpose it is desirable that the modules be fairly consistent in their lifetimes, so that preventive replacements can be timed efficiently.*

Chin tested a sample of 20 Brand A modules, and they lasted 48,560 highway miles on the average (mean), with a standard deviation of 2116 miles. The listing below shows how long each of another sample of 20 Brand B modules lasted. Use these data for the following exercises.

44,660	51,300	45,680	48,840	47,510
61,220	49,100	48,660	47,790	47,210
48,050	49,920	47,420	45,880	50,110
52,910	47,930	45,800	46,690	49,240

40. According to the sampling, which brand of module has the longer average life (in highway miles)?

41. Which brand of module apparently has a more consistent (or uniform) length of life (in highway miles)?

42. If Brands A and B are the only modules available, which one should Chin purchase for the maintenance program? Explain your reasoning.

A Cereal-Marketing Study *A food distribution company conducted a study to determine whether a proposed premium to be included in boxes of its cereal was appealing enough to generate new sales. Four cities were used as test markets, where the cereal was distributed with the premium, and four cities as control markets, where the cereal was distributed without the premium. The eight cities were chosen on the basis of their similarity in terms of population, per capita income, and total cereal purchase volume. The results are shown in the following table.*

Percent Change in Average Market Share per Month

Test cities	1	+18
	2	+15
	3	+7
	4	+10
Control cities	1	+1
	2	−8
	3	−5
	4	0

43. Find the mean of the percent change in market share for the four control cities.

44. Find the mean of the percent change in market share for the four test cities.

45. Find the standard deviation of the percent change in market share for the control cities.

46. Find the standard deviation of the percent change in market share for the test cities.

47. Find the difference between the means of the test cities and the control cities. This difference represents the estimate of the percent change in sales due to the premium.

48. The two standard deviations from the test cities and the control cities were used to calculate an "error" of ± 7.95 for the estimate in **Exercise 47.** With this amount of error, what are the least and greatest estimates of the increase in sales?

(On the basis of the interval estimate of **Exercise 48,** the company decided to mass produce the premium and distribute it nationally.)

For Exercises 49 and 50, refer to the grouped frequency distribution shown below.

Class Limits	Frequency f
21–25	5
26–30	3
31–35	8
36–40	12
41–45	21
46–50	38
51–55	35
56–60	20

49. Is it possible to compute the actual standard deviation for this sample?

50. Describe how you might approximate the standard deviation for this sample. Justify your procedure.

FOR FURTHER THOUGHT

Measuring Skewness in a Distribution

The previous section included a discussion of "symmetry in data sets." Here we present a common method of measuring the amount of "skewness," or nonsymmetry, inherent in a distribution.

In a skewed distribution, the mean will be farther out toward the tail than the median, as shown here.

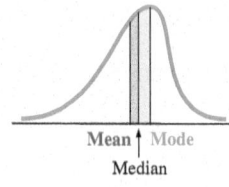 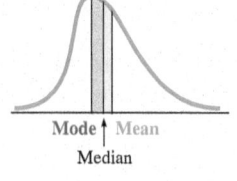

Mean ↑ Mode   Mode ↑ Mean
Median   Median
Skewed to the left   Skewed to the right

The degree of skewness can be measured by the **skewness coefficient SK**, which, like the coefficient of variation, involves both central tendency and dispersion and is calculated as follows.

$$SK = \frac{3 \cdot (\text{mean} - \text{median})}{\text{standard deviation}}$$

For Group or Individual Investigation

1. Under what conditions would the skewness coefficient be each of the following?

   (a) positive   **(b)** negative   **(c)** zero

2. Explain why the mean of a skewed distribution is always farther out toward the tail than the median.

3. In a skewed distribution, how many standard deviations apart are the mean and median in each case?

   (a) $SK = \frac{1}{2}$ **(b)** $SK = 1$ **(c)** $SK = 3$

4. Suppose $SK = -6$. Complete the following statement. The mean is _____ standard
   (how many?)
   deviations _____ the median.
   (above/below)

5. What is the median of a distribution with mean 5, standard deviation $\frac{7}{3}$, and skewness coefficient 1?

12.4   MEASURES OF POSITION

OBJECTIVES

1 Compute and interpret z-scores.

2 Compute and interpret percentiles.

3 Compute and interpret deciles and quartiles.

4 Construct and interpret box plots.

Measures of central tendency and measures of dispersion give us an effective way of characterizing an overall set of data. Central tendency indicates where, along a number scale, the overall data set is centered. Dispersion indicates how much the data set is spread out from the center point. And Chebyshev's theorem, stated in the previous section, sets a limit on what portions of the data set may be dispersed different amounts from the center point.

In some cases, we are interested in certain individual items within a data set, rather than in that set as a whole. So we would like to measure how an item fits into the collection, how its placement compares to those of other items in the collection. There are several common ways of creating such measures. Because they measure an item's position within the data set, we call them **measures of position.**

The z-Score

Each individual item in a sample can be assigned a **z-score,** which is the first measure of position that we consider. It is defined as follows.

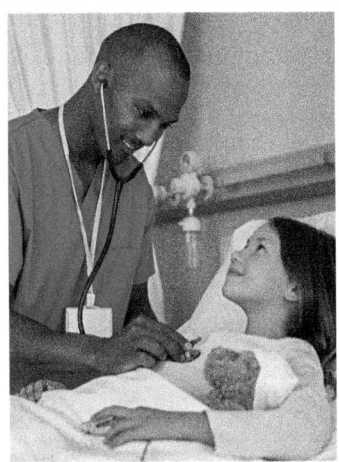

Lists of **the top ten jobs of 2018** may vary somewhat, depending on which organization's ratings are consulted. The list below is based generally on (a) job prospects, (b) salaries, (c) stress levels, and (d) "work-life balance." The dollar amount is the most recent median salary reported. See the Web site below for detailed information.

1.	Software developer	$100,080
2.	Dentist	$153,900
3.	Physician assistant	$101,480
4.	Nurse practitioner	$100,910
5.	Orthodontist	$208,000
6.	Statistician	$80,500
7.	Pediatrician	$168,990
8.	Obstetrician and gynecologist	$208,000
9.	Oral and maxillofacial surgeon	$208,000
10.	Physician	$196,380

(Data from www.money.usnews.com)

Visualizing z-Scores

THE *z*-SCORE

If x is a data item in a sample with mean $\bar{x}$ and standard deviation s, then the **z-score** of x measures the number of standard deviations by which x differs from $\bar{x}$. It is calculated as follows.

$$z = \frac{x - \bar{x}}{s}$$

Notice that z will be positive if x is greater than $\bar{x}$ but negative if x is less than $\bar{x}$. Chebyshev's theorem assures us that in any distribution whatsoever, at least 89% (roughly) of the items will lie within three standard deviations of the mean. That is, at least 89% of the items will have z-scores between -3 and 3. In fact, many common distributions, especially symmetric ones, have considerably more than 89% of their items within three standard deviations of the mean (as we will see in the next section). Hence, a z-score greater than 3 or less than -3 is rare.

EXAMPLE 1 Finding *z*-Scores

Suppose, for a given distribution, $\bar{x} = 340$ and $s = 22$. Find the z-scores (to the nearest hundredth) for these items within the distribution.

(a) $x = 349$ **(b)** $x = 300$ **(c)** $x = 382$

Solution

(a) $z = \frac{349 - 340}{22} = 0.41$ **(b)** $z = \frac{300 - 340}{22} = -1.82$

(c) $z = \frac{382 - 340}{22} = 1.91$

A common use of z-scores is to compare relative positions of items within two or more separate distributions.

EXAMPLE 2 Comparing Positions Using *z*-Scores

Two friends, Cyrus and Aisha, who take different history classes, had midterm exams on the same day. Cyrus's score was 86, and Aisha's was only 78. Based on z-scores, which student did relatively better, given the class data shown here?

	Cyrus	Aisha
Class mean	73	69
Class standard deviation	8	5

Solution
Calculate as follows.

$$\text{Cyrus: } z = \frac{86 - 73}{8} = 1.625 \qquad \text{Aisha: } z = \frac{78 - 69}{5} = 1.8$$

Because Aisha's z-score is higher, she was positioned relatively higher within her class than Cyrus was within his class.

Percentiles

When you take the Scholastic Aptitude Test (SAT), or any other standardized test taken by large numbers of students, your raw score usually is converted to a **percentile** score, which is defined as follows.

PERCENTILE

If approximately n percent of the items in a distribution are less than the number x, then x is the **nth percentile** of the distribution, denoted P_n.

For example, if you scored at the eighty-third percentile on the SAT, it means that you outscored approximately 83% of all those who took the test. (It does *not* mean that you got 83% of the answers correct.) Since the percentile score gives the position of an item within the data set, it is another "measure of position." The following example approximates percentiles for a fairly small collection of data.

EXAMPLE 3 Finding Percentiles

The following are the numbers of dinner customers served by a restaurant on 40 consecutive days. The numbers have been ranked least to greatest.

46	51	52	55	56	56	58	59	59	59
60	61	62	62	63	63	64	64	64	65
66	66	66	67	67	67	68	68	69	69
70	70	71	71	72	75	79	79	83	88

For this data set, find **(a)** the sixty-fifth percentile, and **(b)** the eighty-eighth percentile.

Solution

(a) The sixty-fifth percentile can be taken as the item below which 65% of the items are ranked. Since 65% of 40 is $0.65(40) = 26$, we take the twenty-seventh item, or 68, as the sixty-fifth percentile.

(b) Since 88% of 40 is $0.88(40) = 35.2$, we round *up* and take the eighty-eighth percentile to be the thirty-sixth item, or 75.

Technically, percentiles originally were conceived as a set of 99 values P_1, P_2, P_3, ..., P_{99} (not necessarily data items) along the scale that would divide the data set into 100 equal-sized parts. They were computed only for very large data sets. With smaller data sets, as in **Example 3,** dividing the data into 100 parts would necessarily leave many of those parts empty. However, the modern techniques of exploratory data analysis seek to apply the percentile concept to even small data sets. Thus, we use approximation techniques as in **Example 3.** Another option is to divide the data into a lesser number of equal-sized (or nearly equal-sized) parts, say ten parts, or just four parts, as we discuss next.

Deciles and Quartiles

Deciles are the nine values (denoted $D_1, D_2, \ldots, D_9$) along the scale that divide a data set into ten (approximately) equal-sized parts, and **quartiles** are the three values (Q_1, Q_2, and Q_3) that divide a data set into four (approximately) equal-sized parts. Since deciles and quartiles serve to position particular items within portions of a distribution, they also are "measures of position." We can evaluate deciles by finding their equivalent percentiles.

$$D_1 = P_{10}, \quad D_2 = P_{20}, \quad D_3 = P_{30}, \quad \ldots, \quad D_9 = P_{90}$$

EXAMPLE 4 Finding Deciles

Find the fourth decile for the dinner customer data of **Example 3.**

Solution

Refer to the ranked data table. The fourth decile is the fortieth percentile, and 40% of 40 is $0.40(40) = 16$. We take the fourth decile to be the seventeenth item, or 64.

Although the three quartiles also can be related to corresponding percentiles, notice that the second quartile, Q_2, also is equivalent to the median, a measure of central tendency. A common convention for computing quartiles goes back to the way we computed the median.

FINDING QUARTILES

For any set of data (ranked in order from least to greatest):

The **second quartile, Q_2,** is just the median, the middle item when the number of items is odd, or the mean of the two middle items when the number of items is even.

The **first quartile, Q_1,** is the median of all items below Q_2.

The **third quartile, Q_3,** is the median of all items above Q_2.

EXAMPLE 5 Finding Quartiles

Find the three quartiles for the data of **Example 3.**

Solution

Refer to the ranked data. The two middle data items are 65 and 66.

$$Q_2 = \frac{65 + 66}{2} = 65.5$$

The least 20 items (an even number) are all below Q_2, and the two middle items in that set are 59 and 60.

$$Q_1 = \frac{59 + 60}{2} = 59.5$$

The greatest 20 items are above Q_2.

$$Q_3 = \frac{69 + 70}{2} = 69.5$$

NORMAL FLOAT AUTO REAL DEGREE CL

1-Var Stats
↑Sx=8.463344796
σx=8.35688339
n=40
minX=46
Q₁=59.5
Med=65.5
Q₃=69.5
maxX=88

This screen supports the results of **Example 5.**

The Box Plot

A **box plot,** or **box-and-whisker plot,** involves the median (a measure of central tendency), the range (a measure of dispersion), and the first and third quartiles (measures of position), all incorporated into a simple visual display.

BOX PLOT

For a given set of data, a **box plot** (or **box-and-whisker plot**) consists of a rectangular box positioned above a numerical scale, extending from Q_1 to Q_3, with the value of Q_2 (the median) indicated within the box, and with "whiskers" (line segments) extending to the left and right from the box out to the minimum and maximum data items.

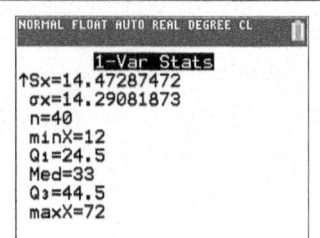

This screen supports the results of
Example 6.

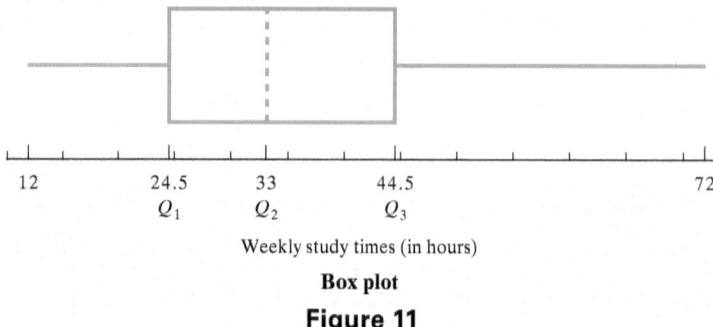

This box plot corresponds to the results
of **Example 6.** It indicates the median
in the display at the bottom. The TRACE
function of the TI-84 Plus C will locate the
minimum, maximum, and quartile values
as well.

EXAMPLE 6 **Constructing a Box Plot**

Construct a box plot for the forty weekly study times of **Example 2** in the first section
of this chapter.

Solution

To determine the quartiles and the
minimum and maximum values
more easily, we use the stem-and-
leaf display (with leaves ranked),
given in **Table 8.**

1	2 4 5 6 7 8
2	2 3 4 4 5 6 9 9 9
3	0 1 1 2 3 3 6 6 6 6 8 9
4	1 4 4 5 5 7
5	2 5 5 8
6	0 2
7	2

The median (determined earlier) is

$$\frac{33 + 33}{2} = 33.$$

From the stem-and-leaf display,

$$Q_1 = \frac{24 + 25}{2} = 24.5 \quad \text{and} \quad Q_3 = \frac{44 + 45}{2} = 44.5.$$

The minimum and maximum items, 12 and 72, are evident from the stem-and-
leaf display. The box plot is shown in **Figure 11.**

The box plot in **Figure 11** conveys the following important information:

1. central tendency (the location of the median);

2. the location of the middle half of the data (the extent of the box);

3. dispersion (the range is the extent of the whiskers); and

4. skewness (the nonsymmetry of both the box and the whiskers).

12.4 **EXERCISES**

CONCEPT CHECK *Write* true *or* false *for each statement.*

1. The z-score is the frequency of the item z in a frequency
distribution.

2. The nth percentile is the item that exceeds n percent of
the items in a distribution.

3. Every quartile can be found by computing a decile
instead.

4. Within any very large distribution, there are 100 percentiles, 10 deciles, and 4 quartiles.

5. A box plot specifically shows the minimum and maximum data values, as well as all of the quartiles.

6. A lack of symmetry in a distribution can show up in either the box or the whiskers of a box plot.

Suppose a distribution of 400 items has mean 125 and standard deviation 5.

7. Find the z-score of the item $x = 133$.

8. Find the z-score of the item $x = 113$.

9. How many items are less than Q_1?

10. How many items are greater than D_7?

Numbers of Restaurant Customers   *Refer to the dinner customers data of **Example 3**. Approximate each of the following. Use the methods illustrated in this section.*

11. the fifteenth percentile   **12.** the thirty-fifth percentile

13. the second decile   **14.** the eighth decile

In the following exercises, make use of z-scores.

15. **Relative Positions on Sociology Quizzes**   In a sociology class, Neil scored 5 on a quiz for which the class mean and standard deviation were 4.6 and 2.1, respectively. Janet scored 6 on another quiz for which the class mean and standard deviation were 4.9 and 2.3, respectively. Relatively speaking, which student did better?

16. **Relative Performances in Track Events**   In Saturday's track meet, Ramon, a high jumper, jumped 6 feet 5 inches. Conference high jump marks for the past season had a mean of 6 feet even and a standard deviation of 3.5 inches. Ric, Ramon's teammate, achieved 18 feet 8 inches in the long jump. In that event the conference season average (mean) and standard deviation were 16 feet 6 inches and 1 foot 10 inches, respectively. Relative to this past season in this conference, which athlete had a better performance on Saturday?

17. **Relative Lifetimes of Tires**   The lifetimes of Brand A tires are distributed with mean 45,000 miles and standard deviation 4500 miles, while Brand B tires last for only 38,000 miles on the average (mean) with standard deviation 2080 miles. Nicole's Brand A tires lasted 37,000 miles, and Yvette's Brand B tires lasted 35,000 miles. Relatively speaking, within their own brands, which driver got the better wear?

18. **Relative Ratings of Fish Caught**   In a certain lake, the trout average 12 inches in length with a standard deviation of 2.65 inches. The bass average 4 pounds in weight with a standard deviation of 0.9 pound. If Tobi caught an 18-inch trout and Katrina caught a 6-pound bass, then relatively speaking, which catch was the better trophy?

World's Largest Energy Producers and Consumers   *The table includes only countries in the top ten in 2013 for both production and consumption of energy. (Energy units are quadrillion Btu.) Population is for midyear 2017, in millions. Use this information for Exercises 19–30.*

Country	Population	Production	Consumption
China	1379	97.3	122.5
United States	327	76.4	97.1
Russia	142	54.0	30.5
Canada	36	15.8	14.4
India	1282	13.4	24.1

Data from *The World Almanac and Book of Facts 2018.*

Compute z-scores (accurate to one decimal place) for the following quantities.

19. Russia's population

20. U.S. production

21. China's consumption

22. India's consumption

In each case, determine which country occupied the given position.

23. the twenty-fifth percentile in population

24. the second decile in production

25. the third quartile in consumption

26. the first decile in production

Solve each problem.

27. Determine which was relatively higher: Canada in production or India in consumption.

28. Construct box plots for both production and consumption, one above the other in the same drawing.

29. What does your box plot of **Exercise 28** *for consumption* indicate about the following characteristics of the consumption data?

(a) the central tendency

(b) the dispersion

(c) the location of the middle half of the data items

30. Comparing your two box plots of **Exercise 28,** what can you say about energy among the world's top producers and consumers of 2013?

Write a short answer for each problem.

31. The text stated that for *any* distribution of data, at least 89% of the items will be within 3 standard deviations of the mean. Why couldn't we just move some items farther out from the mean to obtain a new distribution that would violate this condition?

32. Describe the basic difference between a measure of central tendency and a measure of position.

This chapter has introduced three major characteristics (central tendency, dispersion, and position) and has developed various ways of measuring them in numerical data. In each of the following exercises, a new measure is described. In each case, indicate which of the three characteristics you think it would measure, and explain why.

33. Midrange $= \dfrac{\text{minimum item} + \text{maximum item}}{2}$

34. Midquartile $= \dfrac{Q_1 + Q_3}{2}$

35. Interquartile range $= Q_3 - Q_1$

36. Semi-interquartile range $= \dfrac{Q_3 - Q_1}{2}$

Solve each problem.

37. The "skewness coefficient," defined in **For Further Thought** at the end of the previous section, is calculated as follows.

$$SK = \frac{3 \cdot (\bar{x} - Q_2)}{s}$$

Is this a measure of individual data items or of the overall distribution?

38. For the energy data preceding **Exercise 19,** calculate the skewness coefficient for the following.

(a) production

(b) consumption

39. From **Exercise 38,** how would you compare the skewness of production versus consumption?

40. In a national standardized test, Kimberly scored at the ninety-second percentile. If 67,500 individuals took the test, about how many scored higher than Kimberly?

41. Let the three quartiles (from least to greatest) for a large population of scores be denoted Q_1, Q_2, and Q_3.

(a) Is it necessarily true that

$$Q_2 - Q_1 = Q_3 - Q_2?$$

(b) Explain your answer to part (a).

In Exercises 42–45, answer yes *or* no *and explain your answer. (Consult **Exercises 33–36** for definitions.)*

42. Is the midquartile necessarily the same as the median?

43. Is the midquartile necessarily the same as the midrange?

44. Is the interquartile range necessarily half the range?

45. Is the semi-interquartile range necessarily half the interquartile range?

46. *Relative Positions on a Standardized Chemistry Test* Omer and Alessandro participated in the standardization process for a new statewide chemistry test. Within the large group participating, their raw scores and corresponding z-scores were as shown here.

	Raw Score	z-Score
Omer	60	0.69
Alessandro	72	1.67

Find the overall mean and standard deviation of the distribution of scores (to two decimal places).

Rating Passers in the National Football League *Passers (quarterbacks) in the National Football League are rated using the following formula.*

$$\text{Rating} = \frac{\left(250 \cdot \dfrac{C}{A}\right) + \left(1000 \cdot \dfrac{T}{A}\right) + \left(12.5 \cdot \dfrac{Y}{A}\right) + 6.25 - \left(1250 \cdot \dfrac{I}{A}\right)}{3},$$

where A = attempted passes,

C = completed passes,

T = touchdown passes,

Y = yards gained passing,

and I = interceptions.

A remarkable NFL family "dynasty" has been Archie Manning and his two sons, Peyton and Eli (all quarterbacks). The table shows the statistics for their best regular seasons, respectively.

NFL Passer	Archie	Peyton	Eli
Team	New Orleans Saints	Indianapolis Colts	New York Giants
Year	1980	2004	2015
A	509	497	618
C	309	336	387
T	23	49	____
Y	3,716	4,557	4,432
I	20	____	14
Rating points	____	121.1	93.6

Data from www.nfl.com

Find the missing entry in each passer's column.

47. Archie   **48.** Peyton   **49.** Eli

Archie Manning, father of NFL quarterbacks Peyton and Eli, signed this photo for author Hornsby's son, Jack.

The ratings for the ten leading passers in the league for 2017 regular season play are ranked in the following table.

Rank	NFL Passer	Rating Points
1	Carson Wentz, Philadelphia	75.9
2	Case Keenum, Minnesota	69.7
3	Tom Brady, New England	67.4
4	Dak Prescott, Dallas	66.7
5	Matt Ryan, Atlanta	63.7
6	Ben Roethlisberger, Pittsburgh	63.2
7	Matthew Stafford, Detroit	61.7
8	Alex Smith, Kansas City	61.6
9	Drew Brees, New Orleans	59.0
10	Russell Wilson, Seattle	58.3

Data from www.espn.com

Find the measures (to one decimal place) in Exercises 50–55.

50. the sixth decile

51. the three quartiles

52. the midrange (See **Exercise 33.**)

53. the ninety-fifth percentile

54. the interquartile range (See **Exercise 35.**)

55. the midquartile (see **Exercise 34.**)

56. Construct a box plot for the rating points data.

*Given our method of finding quartiles, **Example 5** shows that quartiles are not necessarily data items. For each data set in Exercises 57–60, **(a)** find all three quartiles, and **(b)** state how many of them are data items.*

57. 1 2 3 4 5 6 7 8

58. 1 2 3 4 5 6 7 8 9

59. 1 2 3 4 5 6 7 8 9 10

60. 1 2 3 4 5 6 7 8 9 10 11

61. The pattern of **Exercises 57–60** holds in general. Complete the following statements for any data set with n distinct items (no item occurs more than once).

   (a) If $n = 4k$ for some counting number k, then _____ of the quartiles are data items.

   (b) If $n = 4k + 1$ for some counting number k, then _____ of the quartiles are data items.

   (c) If $n = 4k + 2$ for some counting number k, then _____ of the quartiles are data items.

   (d) If $n = 4k + 3$ for some counting number k, then _____ of the quartiles are data items.

62. Suppose a data set contains n distinct items (no item occurs more than once). Determine how many of the three quartiles are data items in each case.

   (a) $n = 47$

   (b) $n = 128$

   (c) $n = 101$

   (d) $n = 166$

63. If exactly one quartile of a distribution is a data item, which quartile must it be?

64. If exactly two quartiles of a distribution are data items, which quartiles must they be?

12.5   THE NORMAL DISTRIBUTION

Discrete and Continuous Random Variables

Inferential statistics treats data distributions according to the nature of the random variable involved.

TWO KINDS OF RANDOM VARIABLES

- A **discrete random variable** can take on only certain fixed values.

  Example: The number of heads in five tosses of a fair coin may be only one of the following whole numbers.

  $$0, 1, 2, 3, 4, \text{ or } 5$$

- A **continuous random variable** can take on all real values over a range.

  Example: The diameter of mature camellia blossoms may be any real number in the range, say, of

  $$5 \text{ to } 25 \text{ centimeters.}$$

Most distributions discussed earlier in this chapter were *empirical* (based on observation). The distributions covered in this section are *theoretical* (based on theoretical probabilities). A knowledge of theoretical distributions enables us to identify when actual observations are inconsistent with stated assumptions.

The theoretical probability distribution for the discrete random variable "number of heads" when 5 fair coins are tossed is shown in **Table 15,** and **Figure 12** shows the corresponding histogram. The probability values can be found using the binomial probability formula or using Pascal's triangle, covered elsewhere in this text.

The normal curve was first developed by **Abraham De Moivre** (1667–1754), but his work went unnoticed for many years. It was independently redeveloped by Pierre de Laplace (1749–1827) and Carl Friedrich Gauss (1777–1855). Gauss found so many uses for this curve that it is sometimes called the *Gaussian curve.*

Table 15 Probability Distribution

x	$P(x)$
0	0.03125
1	0.15625
2	0.31250
3	0.31250
4	0.15625
5	0.03125
Sum:	1.00000

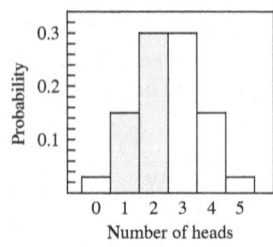

Figure 12

Each rectangle in **Figure 12** is 1 unit wide, so the *area* of the rectangle is also equal to the probability of the corresponding number of heads. The area, and thus the probability, for the event

$$\text{"1 head or 2 heads"}$$

is shaded in the figure. The graph consists of 6 distinct rectangles because "number of heads" is a *discrete* variable with 6 possible values. The sum of the 6 rectangular areas is exactly 1 square unit.

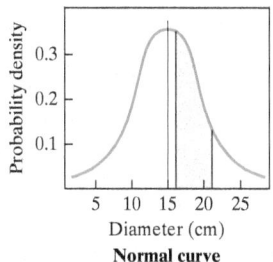

Normal curve

Figure 13

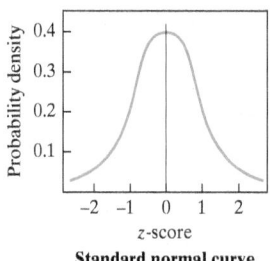

Standard normal curve

Figure 14

StatCrunch Distribution Demos

A probability distribution for camellia blossom diameters cannot be tabulated or graphed in quite the same way, because this variable is *continuous*. The graph would be smeared out into a "continuous" bell-shaped curve (rather than a set of rectangles) as shown in **Figure 13**. The vertical scale on the graph in this case shows what we call "probability density," the probability per unit along the horizontal axis.

Definition and Properties of a Normal Curve

The camellia blossom curve is highest at a diameter value of 15 cm, its center point, and drops off rapidly and equally toward a zero level in both directions. Such a symmetric, bell-shaped curve is called a **normal curve.** Any random variable whose graph has this characteristic shape is said to have a **normal distribution.**

The area under the curve along a certain interval is numerically equal to the probability that the random variable will have a value in the corresponding interval. The area of the shaded region in **Figure 13** is equal to the probability of a randomly chosen blossom having a diameter in the interval from the left extreme, say 16.4, to the right extreme, say 21.2. Normal curves are very important in the study of statistics because

> *a great many continuous random variables have normal distributions, and many discrete variables are distributed approximately normally.*

Each point on the horizontal scale of a normal curve lies some number of standard deviations from the mean (positive to the right, negative to the left). This number is the "standard score" for that point. It is the same as the z-score defined earlier in this chapter. By relabeling the horizontal axis, as in **Figure 14,** we obtain the **standard normal curve,** which we can use to analyze *any* normal (or approximately normal) distribution.

Figure 15 shows several of infinitely many possible normal curves. Each is completely characterized by its mean and standard deviation. Only one of these, the one marked S, is the *standard* normal curve. That one has mean 0 and standard deviation 1.

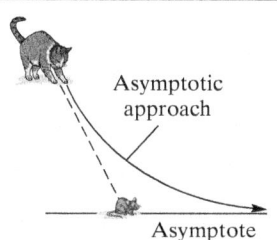

Close but Never Touching When a curve approaches closer and closer to a line, without ever actually meeting it (as a normal curve approaches the horizontal axis), the line is called an **asymptote,** and the curve is said to approach the line **asymptotically.** Imagine a mouse running in a straight line, at constant speed, and a cat approaching from the side at the same speed, always aiming at the mouse's current position.

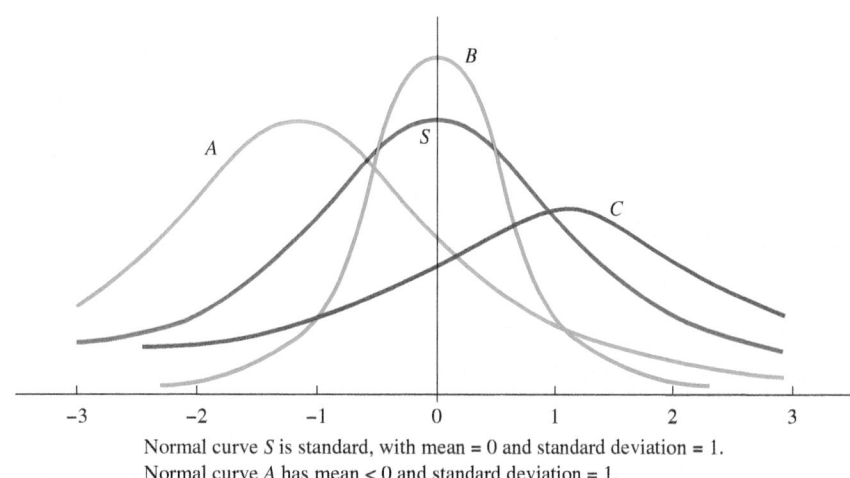

Normal curve S is standard, with mean = 0 and standard deviation = 1.
Normal curve A has mean < 0 and standard deviation = 1.
Normal curve B has mean = 0 and standard deviation < 1.
Normal curve C has mean > 0 and standard deviation > 1.

Figure 15

Several properties of normal curves are summarized on the next page and illustrated in **Figure 16**.

PROPERTIES OF NORMAL CURVES

The graph of a normal curve is bell-shaped and symmetric about a vertical line through its center.

The mean, median, and mode of a normal curve are all equal and occur at the center of the distribution.

Empirical Rule   The percentages of data values within given distances of the mean (in both directions) are approximately as follows.

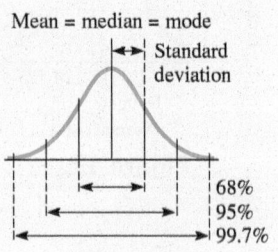

Figure 16

- 68% within 1 standard deviation
- 95% within 2 standard deviations
- 99.7% within 3 standard deviations

The empirical rule indicates that a very small percentage of the items in a normal distribution will lie more than 3 standard deviations from the mean (approximately 0.3%, divided equally between the upper and lower tails of the distribution). As we move away from the center, the curve *never* actually touches the horizontal axis. No matter how far out we go, there is always a chance of an item occurring even farther out. Theoretically then, the range of a true normal distribution is infinite.

EXAMPLE 1   **Applying the Empirical Rule**

Suppose that 300 pre-medical students take a midterm exam and the distribution of their scores can be treated as normal. Find the number of scores falling into each of the following intervals.

(a) Within 1 standard deviation of the mean

(b) Within 2 standard deviations of the mean

Solution

(a) A total of 68% of all scores lie within 1 standard deviation of the mean.

$$0.68(300) = 204 \quad \text{68\% = 0.68; There is a total of 300 scores.}$$

(b) A total of 95% of all scores lie within 2 standard deviations of the mean.

$$0.95(300) = 285 \quad \text{95\% = 0.95}$$

A Table of Standard Normal Curve Areas

Most questions we need to answer about normal distributions involve regions other than those within 1, 2, or 3 standard deviations of the mean. We might need the percentage of items within $1\frac{1}{2}$ or $2\frac{1}{5}$ standard deviations of the mean, or perhaps the area under the curve from 0.8 to 1.3 standard deviations above the mean.

In such cases, we refer to a table of area values, such as **Table 16** on the next page.

Although we utilize Table 16 in our examples here, statistics-capable calculators, and computer software packages such as StatCrunch, are also recommended, and they will usually give more precise answers.

The table gives the fraction of all scores in a normal distribution that lie between the mean and *z* standard deviations from the mean.

Because of the symmetry of the normal curve, the table can be used for values above the mean or below the mean.

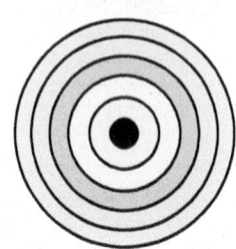

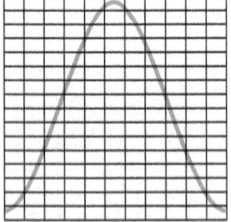

A normal distribution occurs in darts if the player, always aiming at the bull's-eye, tosses a fairly large number of times, and the aim on each toss is affected only by independent random errors.

The column under *A* gives the **fraction of the area under the entire curve** that is between *z* = 0 and a positive value of *z*.

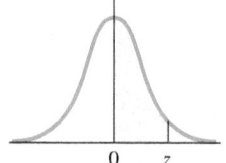

Because the curve is symmetric about the 0-value, the area between *z* = 0 and a *negative* value of *z* can be found by using the corresponding positive value of *z*.

StatCrunch Normal Calculator

Table 16 Areas under the Standard Normal Curve

z	A	z	A	z	A	z	A	z	A	z	A
.00	.000	.56	.212	1.12	.369	1.68	.454	2.24	.487	2.80	.497
.01	.004	.57	.216	1.13	.371	1.69	.454	2.25	.488	2.81	.498
.02	.008	.58	.219	1.14	.373	1.70	.455	2.26	.488	2.82	.498
.03	.012	.59	.222	1.15	.375	1.71	.456	2.27	.488	2.83	.498
.04	.016	.60	.226	1.16	.377	1.72	.457	2.28	.489	2.84	.498
.05	.020	.61	.229	1.17	.379	1.73	.458	2.29	.489	2.85	.498
.06	.024	.62	.232	1.18	.381	1.74	.459	2.30	.489	2.86	.498
.07	.028	.63	.236	1.19	.383	1.75	.460	2.31	.490	2.87	.498
.08	.032	.64	.239	1.20	.385	1.76	.461	2.32	.490	2.88	.498
.09	.036	.65	.242	1.21	.387	1.77	.462	2.33	.490	2.89	.498
.10	.040	.66	.245	1.22	.389	1.78	.462	2.34	.490	2.90	.498
.11	.044	.67	.249	1.23	.391	1.79	.463	2.35	.491	2.91	.498
.12	.048	.68	.252	1.24	.393	1.80	.464	2.36	.491	2.92	.498
.13	.052	.69	.255	1.25	.394	1.81	.465	2.37	.491	2.93	.498
.14	.056	.70	.258	1.26	.396	1.82	.466	2.38	.491	2.94	.498
.15	.060	.71	.261	1.27	.398	1.83	.466	2.39	.492	2.95	.498
.16	.064	.72	.264	1.28	.400	1.84	.467	2.40	.492	2.96	.498
.17	.067	.73	.267	1.29	.401	1.85	.468	2.41	.492	2.97	.499
.18	.071	.74	.270	1.30	.403	1.86	.469	2.42	.492	2.98	.499
.19	.075	.75	.273	1.31	.405	1.87	.469	2.43	.492	2.99	.499
.20	.079	.76	.276	1.32	.407	1.88	.470	2.44	.493	3.00	.499
.21	.083	.77	.279	1.33	.408	1.89	.471	2.45	.493	3.01	.499
.22	.087	.78	.282	1.34	.410	1.90	.471	2.46	.493	3.02	.499
.23	.091	.79	.285	1.35	.411	1.91	.472	2.47	.493	3.03	.499
.24	.095	.80	.288	1.36	.413	1.92	.473	2.48	.493	3.04	.499
.25	.099	.81	.291	1.37	.415	1.93	.473	2.49	.494	3.05	.499
.26	.103	.82	.294	1.38	.416	1.94	.474	2.50	.494	3.06	.499
.27	.106	.83	.297	1.39	.418	1.95	.474	2.51	.494	3.07	.499
.28	.110	.84	.300	1.40	.419	1.96	.475	2.52	.494	3.08	.499
.29	.114	.85	.302	1.41	.421	1.97	.476	2.53	.494	3.09	.499
.30	.118	.86	.305	1.42	.422	1.98	.476	2.54	.494	3.10	.499
.31	.122	.87	.308	1.43	.424	1.99	.477	2.55	.495	3.11	.499
.32	.126	.88	.311	1.44	.425	2.00	.477	2.56	.495	3.12	.499
.33	.129	.89	.313	1.45	.426	2.01	.478	2.57	.495	3.13	.499
.34	.133	.90	.316	1.46	.428	2.02	.478	2.58	.495	3.14	.499
.35	.137	.91	.319	1.47	.429	2.03	.479	2.59	.495	3.15	.499
.36	.141	.92	.321	1.48	.431	2.04	.479	2.60	.495	3.16	.499
.37	.144	.93	.324	1.49	.432	2.05	.480	2.61	.495	3.17	.499
.38	.148	.94	.326	1.50	.433	2.06	.480	2.62	.496	3.18	.499
.39	.152	.95	.329	1.51	.434	2.07	.481	2.63	.496	3.19	.499
.40	.155	.96	.331	1.52	.436	2.08	.481	2.64	.496	3.20	.499
.41	.159	.97	.334	1.53	.437	2.09	.482	2.65	.496	3.21	.499
.42	.163	.98	.336	1.54	.438	2.10	.482	2.66	.496	3.22	.499
.43	.166	.99	.339	1.55	.439	2.11	.483	2.67	.496	3.23	.499
.44	.170	1.00	.341	1.56	.441	2.12	.483	2.68	.496	3.24	.499
.45	.174	1.01	.344	1.57	.442	2.13	.483	2.69	.496	3.25	.499
.46	.177	1.02	.346	1.58	.443	2.14	.484	2.70	.497	3.26	.499
.47	.181	1.03	.348	1.59	.444	2.15	.484	2.71	.497	3.27	.499
.48	.184	1.04	.351	1.60	.445	2.16	.485	2.72	.497	3.28	.499
.49	.188	1.05	.353	1.61	.446	2.17	.485	2.73	.497	3.29	.499
.50	.191	1.06	.355	1.62	.447	2.18	.485	2.74	.497	3.30	.500
.51	.195	1.07	.358	1.63	.448	2.19	.486	2.75	.497	3.31	.500
.52	.198	1.08	.360	1.64	.449	2.20	.486	2.76	.497	3.32	.500
.53	.202	1.09	.362	1.65	.451	2.21	.486	2.77	.497	3.33	.500
.54	.205	1.10	.364	1.66	.452	2.22	.487	2.78	.497	3.34	.500
.55	.209	1.11	.367	1.67	.453	2.23	.487	2.79	.497	3.35	.500

Carl Friedrich Gauss (1777–1855) was one of the greatest mathematical thinkers of history. In his *Disquisitiones Arithmeticae,* published in 1798, he pulled together work by predecessors and enriched and blended it with his own into a unified whole. The book is regarded by many as the beginning of the modern theory of numbers.

   Of his many contributions to science, the statistical method of least squares is the most widely used today in astronomy, biology, geodesy, physics, and the social sciences. Gauss took special pride in his contributions to developing the method. Despite an aversion to teaching, he taught an annual course in the method for the last twenty years of his life.

   It has been said that Gauss was the last person to have mastered all of the mathematics known in his day.

All of the items in the table can be thought of as corresponding to the area under the curve. The total area is arranged to be 1.000 square unit, with 0.500 square unit on each side of the mean. The table shows that at 3.30 standard deviations from the mean, essentially all of the area is accounted for. Whatever remains beyond is so small that it does not appear in the first three decimal places.

EXAMPLE 2   Applying the Normal Curve Table

Use **Table 16** to find the percent of all scores that lie between the mean and the following values.

(a) 1 standard deviation above the mean

(b) 2.45 standard deviations below the mean

Solution

(a) Here $z = 1.00$ (the number of standard deviations, written as a decimal to the nearest hundredth). Refer to **Table 16**. Find 1.00 in the z column. The table entry is 0.341, so 34.1% of all values lie between the mean and 1 standard deviation above the mean.

   Another way of looking at this is to say that the area in color in **Figure 17** represents 34.1% of the total area under the normal curve.

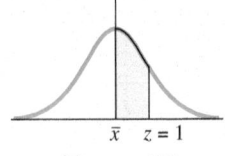

Figure 17

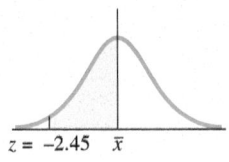

Figure 18

(b) Even though we go *below* the mean here (to the left), **Table 16** still works since the normal curve is symmetric about its mean. Find 2.45 in the z column. A total of 0.493, or 49.3%, of all values lie between the mean and 2.45 standard deviations below the mean. This region is colored in **Figure 18.**

EXAMPLE 3   Finding Probabilities of Phone Call Durations

The time lengths of phone calls placed through a certain company are distributed normally with mean 6 minutes and standard deviation 2 minutes. If one call is randomly selected from phone company records, what is the probability that it will have lasted more than 10 minutes?

Solution

Here 10 minutes is 2 standard deviations above the mean. The probability of such a call is equal to the area of the colored region in **Figure 19.**

   From **Table 16,** the area between the mean and 2 standard deviations above is 0.477 ($z = 2.00$). The total area to the right of the mean is 0.500. Find the area to the right of $z = 2.00$ by subtracting.

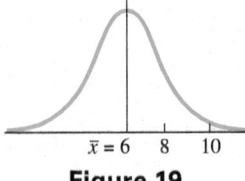

Figure 19

$$0.500 - 0.477 = 0.023$$

The probability of a call exceeding 10 minutes is 0.023, or 2.3%.

A Basic Consumer (T)issue   Brand X toilet paper, manufactured by Company Y, claims (on its wrapper) to contain 375 sheets per roll. But reporters on a local Midwest TV station found three randomly selected rolls to contain 360, 361, and 363 sheets, respectively.

   Shocked Company Y executives said that the **odds against** six rolls having fewer than 375 sheets each are 1 billion to 1 and a subsequent independent count by TV reporters agreed with Company Y.

   What happened the first time? Well, the reporters hadn't actually counted sheets but, rather, had measured rolls and divided the length of a roll by the length of one sheet. Small variations in length can be expected, which can add up over a roll, giving false results.

   This true story perhaps points out the distinction in probability and statistics between **discrete values** and **continuous values**.

 **Finding Areas under the Normal Curve**

Find the total area that is shaded in each of **Figures 20** and **21**.

Solution

For **Figure 20,** find the area from 1.45 standard deviations below the mean to 2.71 standard deviations above the mean. From **Table 16,** $z = 1.45$ gives $A = 0.426$, while $z = 2.71$ gives $A = 0.497$. The total area is the sum of these, or $0.426 + 0.497 = 0.923$.

Figure 20   Figure 21

   To find the area indicated in **Figure 21,** refer to **Table 16**. From the table, $z = 0.62$ gives $A = 0.232$, and $z = 1.59$ gives $A = 0.444$. To get the area between these two values of z, subtract the areas.

$$0.444 - 0.232 = 0.212$$

Interpreting Normal Curve Areas

Examples 2–4 emphasize the *equivalence* of three quantities, as follows.

MEANING OF NORMAL CURVE AREAS

In the standard normal curve, the following three quantities are equivalent.

1. **Percentage** (of total items that lie in an interval)
2. **Probability** (of a randomly chosen item lying in an interval)
3. **Area** (under the normal curve along an interval)

Which quantity we think of depends on how a particular question is formulated. They all can be evaluated by using A-values from **Table 16**.

   In general, when we use **Table 16,** z is the z-score of a particular data item x. When one of these values is known and the other is required, we use the formula

$$z = \frac{x - \bar{x}}{s}.$$

EXAMPLE 5 **Applying the Normal Curve to Driving Distances**

In one area, the distribution of monthly miles driven by motorists has mean 1200 miles and standard deviation 150 miles. Assume that the number of miles is closely approximated by a normal curve, and find the percent of all motorists driving the following distances.

(a) Between 1200 and 1600 miles per month
(b) Between 1000 and 1500 miles per month

Solution

(a) Start by finding how many standard deviations 1600 miles is above the mean. Use the formula for z.

$$z = \frac{1600 - 1200}{150} = \frac{400}{150} \approx 2.67$$

From **Table 16,** 0.496, or 49.6%, of all motorists drive between 1200 and 1600 miles per month.

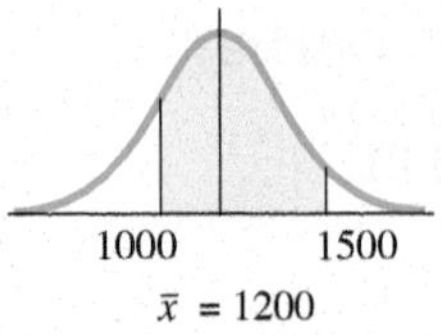

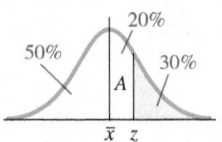

1000   1500

$\bar{x} = 1200$

Figure 22

(b) See **Figure 22.** Values of z must be found for both 1000 and 1500.

$$\text{For 1000:} \quad z = \frac{1000 - 1200}{150} = \frac{-200}{150} \approx -1.33$$

$$\text{For 1500:} \quad z = \frac{1500 - 1200}{150} = \frac{300}{150} = 2.00$$

From **Table 16,** $z = -1.33$ gives $A = 0.408$, and $z = 2.00$ gives $A = 0.477$. The total colored area in **Figure 22** is

$$0.408 + 0.477 = 0.885, \quad \text{or} \quad 88.5\%.$$

So 88.5% of all motorists drive between 1000 and 1500 miles per month.

EXAMPLE 6 Identifying a Data Value within a Normal Distribution

A particular normal distribution has mean $\bar{x} = 81.7$ and standard deviation $s = 5.21$. What data value from the distribution would correspond to $z = -1.35$?

Solution

$$\boxed{\text{Solve for } x.} \quad z = \frac{x - \bar{x}}{s} \qquad \text{z-score formula}$$

$$-1.35 = \frac{x - 81.7}{5.21} \qquad \text{Substitute the given values for } z, \bar{x}, \text{ and } s.$$

$$-1.35(5.21) = \frac{x - 81.7}{5.21}(5.21) \qquad \text{Multiply each side by 5.21 to clear the fraction.}$$

$$-7.0335 = x - 81.7 \qquad \text{Simplify.}$$

$$74.6665 = x \qquad \text{Add 81.7.}$$

Rounding to the nearest tenth, the required data value is 74.7.

EXAMPLE 7 Finding z-Values for Given Areas under the Normal Curve

Assuming a normal distribution, find the z-value meeting each condition.

(a) 30% of the total area is to the right of z.

(b) 80% of the total area is to the left of z.

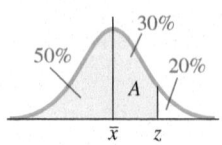

Figure 23

Figure 24

Solution

(a) Because 50% of the area lies to the right of the mean, there must be 20% between the mean and z. (See **Figure 23.**) In **Table 16,** $A = 0.200$ corresponds to $z = 0.52$ or 0.53, or we could average the two: $z = 0.525$.

(b) This situation is shown in **Figure 24.** The 50% to the left of the mean plus 30% additional makes up the 80%. From **Table 16,** $A = 0.300$ implies $z = 0.84$.

WHEN WILL I **EVER** USE THIS?

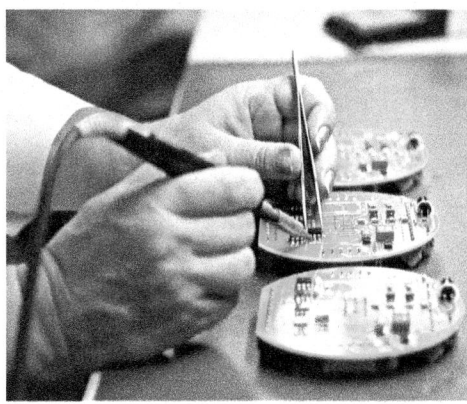

Suppose you are the human resources manager at an electronics assembly plant. By following current research, and also testing your own workers, you have determined that, for your plant environment,

- there is no discernible correlation between IQ score and manual dexterity measure, but that

- dexterity is closely correlated with the average time taken to assemble a particular cell phone (AT = assembly time).

An important new role for the plant will require persons with both high IQ (IQ > 120) and high dexterity—that is, low AT (specifically, AT < 12 minutes). You know that your current work force of 840 workers is distributed approximately normally in IQ and in AT score, and you know it is characterized by the values in the table below.

	Mean	Standard Deviation
IQ	105	8
AT	15.8 min	2.7 min

1. How many of your workers meet the IQ requirement for the new role?

2. How many meet the assembly time requirement?

3. Can you tell from the first two answers how many of your current workers qualify for the new role?

Answers: **1.** 25 **2.** 66 **3.** No, but it is at most 25. Personnel records on IQ and AT may identify which workers qualify.

(12.5) EXERCISES

Note: For problems requiring the calculation of z-scores or A-values, our answers are based on Table 16, *or the empirical rule when specified. By using a calculator or computer package, you will sometimes obtain a slightly more precise answer.*

CONCEPT CHECK *Write* true *or* false *for each statement.*

1. The range of a true normal distribution is about 6.7.

2. The mean, median, and mode are all equal in a normal distribution.

3. In a normal distribution, about 95% of all data values lie within 2 standard deviations of the mean.

4. Every continuous distribution has an infinite range.

Identify each variable quantity as discrete *or* continuous.

5. the number of heads in 30 rolled dice

6. the number of babies born in one day at a certain hospital

7. the average weight of babies born in a week

8. the heights of seedling aspen trees at two years of age

9. the time as shown on a digital watch

10. the time as shown on a watch with sweep hands

Measuring the Mass of Ore Samples *Suppose 100 geology students measure the mass of an ore sample. Due to human error and limitations in the reliability of the scale, not all the readings are equal. The results are found to closely approximate a normal curve, with mean 75 g and standard deviation 2 g.*

Use the symmetry of the normal curve and the empirical rule to estimate the number of students reporting readings in the following ranges.

11. more than 75 g

12. more than 77 g

13. between 71 and 79 g

14. between 69 and 73 g

Distribution of IQ Scores *On standard IQ tests, the mean is 100, with a standard deviation of 15. The results come very close to fitting a normal curve. Suppose an IQ test is given to a very large group of people. Use the* empirical rule *to find the percentage of people whose IQ scores fall into each category.*

15. more than 100   **16.** less than 85

17. between 70 and 130   **18.** less than 115

Find the percentage of area under a normal curve between the mean and the given number of standard deviations from the mean. (Note that positive indicates above the mean, while negative indicates below the mean.)

19. -1.68 **20.** 0.56 **21.** 1.60 **22.** -2.15

Find the percentage of the total area under a normal curve between the given values of z.

23. $z = -3.02$ and $z = 2.03$

24. $z = -1.56$ and $z = -1.09$

25. $z = 1.31$ and $z = 1.73$

26. $z = -1.42$ and $z = 0.98$

Find a value of z such that each condition is met.

27. 78% of the total area is to the left of z.

28. 33% of the total area is to the left of z.

29. 20% of the total area is to the right of z.

30. 58% of the total area is to the right of z.

Weights of Peaches *A fruit-packing company produced peaches last summer whose weights were normally distributed with mean 14 ounces and standard deviation 0.6 ounce. Among a sample of 1000 of those peaches, about how many could be expected to have weights as follows?*

31. more than 12 ounces

32. between 13.5 and 15.5 ounces

33. at least 14 ounces

34. between 14 and 15.2 ounces

35. between 14.9 and 15.5 ounces

36. less than 13.4 ounces

IQs of Employees *A large company employs workers whose IQs are distributed normally with mean 105 and standard deviation 12.5. Management uses this information to assign employees to projects that will be challenging, but not too challenging. What percent of the employees would have IQs that satisfy the following criteria?*

37. less than 90   **38.** more than 120

39. between 100 and 120   **40.** between 85 and 100

Net Weight of Cereal Boxes *A certain dry cereal is packaged in 12-oz boxes. The machine that fills the boxes is set so that, on the average, a box contains 12.4 oz. The machine-filled boxes have content weight that can be closely approximated by a normal curve. What is the probability that a randomly selected box will be underweight (net weight less than 12 oz) if the standard deviation is as follows?*

41. 0.5 oz   **42.** 0.4 oz   **43.** 0.3 oz   **44.** 0.2 oz

45. *Recommended Daily Vitamin Allowances* In nutrition, the recommended daily allowance of vitamins is a number set by the government to guide an individual's daily vitamin intake. Actually, vitamin needs vary dramatically from person to person, but the needs are closely approximated by a normal curve. To calculate the recommended daily allowance, the government first finds the average (mean) need for vitamins among people in the population and the standard deviation. The **recommended daily allowance** is then defined as the mean plus 2.5 times the standard deviation. What fraction of the population would receive adequate amounts of vitamins under this plan?

Recommended Daily Vitamin Allowances *Find the recommended daily allowance for each vitamin if the mean need and standard deviation are as follows. (See **Exercise 45**.)*

46. mean need $= 1600$ units;
standard deviation $= 140$ units

47. mean need $= 146$ units;
standard deviation $= 8$ units

*Assume the following distributions are all normal, and use the areas under the normal curve given in **Table 16** to find the appropriate areas.*

48. *Assembling Cell Phones* The times taken by workers to assemble a certain kind of cell phone are normally distributed with mean 18.5 minutes and standard deviation 3.8 minutes. Find the probability that one such phone will require less than 12.8 minutes in assembly.

49. *Finding Blood-Clotting Times* The mean clotting time of blood is 7.47 sec, with a standard deviation of 3.6 sec. What is the probability that an individual's blood-clotting time will be less than 7 sec or greater than 8 sec?

50. *Sizes of Fish* The average length of the fish caught in Vernal Lake is 11.6 in., with a standard deviation of 3.5 in. Find the probability that a fish caught there will be longer than 16 in.

51. *Size Grading of Eggs* To be graded extra large, an egg must weigh at least 2.2 oz. If the average weight for an egg is 1.5 oz, with a standard deviation of 0.4 oz, how many of five dozen randomly chosen eggs would you expect to be extra large?

Distribution of Student Grades *Professor Wang teaches a marketing course. He uses the following grading system.*

Grade	Score in Class
A	Greater than $\bar{x} + 1.5s$
B	$\bar{x} + 0.5s$ to $\bar{x} + 1.5s$
C	$\bar{x} - 0.5s$ to $\bar{x} + 0.5s$
D	$\bar{x} - 1.5s$ to $\bar{x} - 0.5s$
F	Below $\bar{x} - 1.5s$

What percent of his students receive the following grades?

52. A   **53.** B   **54.** C

55. Do you think that Professor Wang's system would be more likely to be fair in a large first-year lecture class or in a graduate seminar of five students? Why?

Normal Distribution of Student Grades *A teacher gives a test to a large group of students. The results are closely approximated by a normal curve. The mean is 72 with a standard deviation of 5. The teacher wishes to give As to the top 8% of the students and Fs to the bottom 8%. A grade of B is given to the next 15%, with Ds given similarly. All other students get Cs. Find the bottom point cutoffs (rounded to the nearest whole number) for the following grades.*

56. A   **57.** B   **58.** C   **59.** D

60. *Selecting Industrial Rollers* A manufacturer makes carts with industrial swivel rollers that must support a load of at least 125 pounds. Supplier A can provide rollers supporting 130 pounds on the average, with standard deviation 2.2 pounds, and Supplier B's rollers will support 135 pounds on the average, with standard deviation 4.2 pounds.

(a) If both distributions can be assumed normal, which supplier offers the smaller percentage of unsatisfactory rollers?

(b) Give your reasoning.

*A normal distribution has mean 76.8 and standard deviation 9.42. Follow the method of **Example 6** and find data values corresponding to the following values of z. Round to the nearest tenth.*

61. $z = 1.44$ **62.** $z = 0.65$

63. $z = -3.87$ **64.** $z = -1.89$

65. What percent of the items lie within 1.25 standard deviations of the mean

(a) in any distribution (using the results of Chebyshev's theorem)?

(b) in a normal distribution (by **Table 16**)?

66. Explain the difference between the answers to parts (a) and (b) of **Exercise 65**.

Chapter 12 SUMMARY

KEY TERMS

12.1

population
sample
descriptive statistics
inferential statistics
raw data
quantitative (numerical)
 data
qualitative
 (non-numerical) data
ranked data
frequency distribution
relative frequency
 distribution
histogram
frequency polygon
classes
grouped frequency
 distribution

lower class limits
upper class limits
class width
stem-and-leaf display
bar graph
circle graph
 (pie chart)
line graph
expected (theoretical)
 frequencies
observed (empirical)
 frequencies

12.2

measure of central
 tendency
mean
 (arithmetic mean)
weighted mean

weighting factor
outlier
median
cumulative frequency
mode
bimodal
symmetry in data sets
skewed to the left
skewed to the right
Simpson's paradox

12.3

dispersion (spread) of data
range
variance
standard deviation
Chebyshev's theorem
coefficient of variation
skewness coefficient (SK)

12.4

measures of position
z-score
percentile
decile
quartile
box plot
 (box-and-whisker plot)

12.5

discrete random variable
continuous random variable
normal curve
normal distribution
standard normal curve
asymptote
empirical rule

TEST YOUR WORD POWER

See how well you have learned the vocabulary in this chapter.

1. Using a **sample** to gain knowledge about a **population** is part of
   A. descriptive statistics.
   B. inferential statistics.
   C. qualitative data analysis.
   D. calculus.

2. A **class mark** is the
   A. average grade of all the class members.
   B. least item in a grouped frequency distribution.
   C. middle of a class in a grouped frequency distribution.
   D. number of items in the class.

3. A common property of **bar graphs, circle graphs, line graphs,** and **stem-and-leaf displays** is that they provide
   A. a visual representation of a data set.
   B. the range value of a data set.
   C. a way of comparing two data sets.
   D. clear divisions among the various classes.

4. The measure that, when it exists, is always equal to a data item is the
   A. mean.   B. standard deviation.
   C. median.   D. mode.

5. The **median** is an example of a
   A. weighted mean.
   B. measure of central tendency.
   C. measure of dispersion.
   D. measure of position.

6. For a frequency distribution, the formula $\frac{\Sigma f + 1}{2}$ gives
   A. the total number of data items.
   B. one of the measures of position.
   C. the position of the mean.
   D. the position of the median.

7. **Measures of dispersion** include
   A. mean, median, and mode.
   B. range, variance, and percentile.
   C. range, variance, and standard deviation.
   D. percentiles, quartiles, and deciles.

8. The **z-score** determines
   A. how many items must lie within 1 standard deviation of the mean.
   B. how many standard deviations from the mean a data item is.
   C. the ratio of the standard deviation to the mean.
   D. the percentage of items that lie below the given item.

9. A **bimodal** distribution will never be
   A. normal.
   B. skewed.
   C. symmetric.
   D. spread out more than 3 standard deviations from its mean.

10. In any approximately normal distribution, an item with a negative *z*-score will be
  A. negative.
  B. less than the distribution range.
  C. between the first and third quartiles.
  D. less than the distribution mean.

ANSWERS
1. B   **2.** C   **3.** A   **4.** D   **5.** B
6. D   **7.** C   **8.** B   **9.** A   **10.** D

QUICK REVIEW

Concepts   *Examples*

12.1 VISUAL DISPLAYS OF DATA

Qualitative raw data are non-numerical.

Consider the following data.

Category	Number of Positions Available	
Surgeons	3	
GP physicians	5	Qualitative data
Nurse practitioners	7	
Registered nurses	12	
Licensed vocational nurses	8	

Quantitative raw data are numerical.

32, 41, 18, 22, 32, 41, 32, 38, 18, 28, 41, 28, 38, 28, 41, 18, 22, 32, 28, 28   Quantitative data

Quantitative data can be **ranked** (arranged in numerical order).

18, 18, 18, 22, 22, 28, 28, 28, 28, 28, 32, 32, 32, 32, 38, 38, 41, 41, 41, 41   Quantitative data (ranked)

A **frequency distribution** makes a data presentation more concise.

A **relative frequency distribution** shows the fraction of the total number of items represented by each value.

x	f	$\frac{f}{n}$
18	3	0.15
22	2	0.10
28	5	0.25
32	4	0.20
38	2	0.10
41	4	0.20

Frequency distribution and relative frequency distribution

Total: $n = 20$

A **grouped frequency distribution** collects multiple items into separate classes.

Class limits are the least (or greatest) items that can occur in each class.

Class marks are the center points of the classes.

A **histogram** is a graphical presentation of a grouped frequency distribution.

A **frequency polygon** is a sequence of line segments representing a frequency distribution.

Histogram

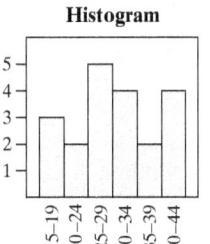

Frequency Polygon

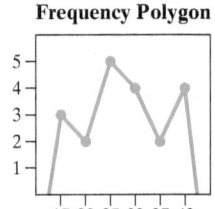

Lower class limits	15 20 25 30 35 40
Upper class limits	19 24 29 34 39 44
Class marks	17 22 27 32 37 42

Concepts	Examples

A **stem-and-leaf display** shows all data items in a visual arrangement.

$$
\begin{array}{ll}
(15\text{–}19)\ 1 & 8 \quad 8 \quad 8 \\
(20\text{–}24)\ 2 & 2 \quad 2 \\
(25\text{–}29)\ 2 & 8 \quad 8 \quad 8 \quad 8 \quad 8 \\
(30\text{–}34)\ 3 & 2 \quad 2 \quad 2 \quad 2 \\
(35\text{–}39)\ 3 & 8 \quad 8 \\
(40\text{–}44)\ 4 & 1 \quad 1 \quad 1 \quad 1
\end{array}
$$

A (double stem) stem-and-leaf display for the above data

12.2 MEASURES OF CENTRAL TENDENCY

The **mean,** the most common measure of central tendency, is computed using the formula

$$\bar{x} = \frac{\Sigma x}{n},$$

or, for frequency distributions,

$$\bar{x} = \frac{\Sigma(x \cdot f)}{\Sigma f}.$$

Find the mean for the distribution 2, 6, 3, 7, 4.

$$\bar{x} = \frac{\Sigma x}{n} = \frac{2 + 6 + 3 + 7 + 4}{5} = \frac{22}{5} = 4.4$$

Find the mean for the frequency distribution.

x	f	$x \cdot f$
3	2	6
4	5	20
5	3	15
Totals:	10	41

$$\bar{x} = \frac{\Sigma(x \cdot f)}{\Sigma f} = \frac{41}{10} = 4.1$$

A **weighted mean** is computed as

$$\bar{w} = \frac{\Sigma(x \cdot f)}{\Sigma f},$$

where f represents the "weighting factors." In the case of the mean of a frequency distribution, the mean is actually a weighted mean, where the weighting factors are the frequencies.

Find the grade-point average. (Grade-point average is a weighted mean, where the weighting factors are the numbers of units for which each grade is assigned.)

Grade Earned	Grade Points Assigned, x	Number of Units, f	$x \cdot f$
A	4	3	12
B	3	6	18
C	2	4	8
		Totals: 13	38

$$\text{Grade-point average} = \bar{w} = \frac{\Sigma(x \cdot f)}{\Sigma f} = \frac{38}{13} \approx 2.92$$

The **median** is the "center" item of a distribution, or the mean of the two center items if there are an even number of items.

The **mode,** when it exists, is the item that occurs in the distribution more times than any other. A distribution having two distinct items with higher frequencies is **bimodal.**

Find the median and mode of the data set

$$2, 6, 3, 7, 4, 5, 6, 3, 9.$$

First rank the data: 2, 3, 3, 4, 5, 6, 6, 7, 9.

The median is the middle item, 5.

This data set is bimodal. The modes are 3 and 6.

Concepts	*Examples*
It is possible to find measures of central tendency for a distribution presented in a stem-and-leaf display.	Consider this stem-and-leaf display.

$$
\begin{array}{c|ccccccc}
5 & 2 & 4 & 7 & 8 & 8 & & \\
6 & 1 & 2 & 5 & 5 & 6 & & \\
7 & 1 & 1 & 2 & 3 & 6 & 6 & 9 \\
8 & 2 & 2 & 4 & 5 & 8 & 8 & \\
9 & 0 & 0 & 0 & 2 & 4 & &
\end{array}
$$

Counting leaves gives $n = 28$.

$$\Sigma x = 52 + 54 + 57 + 58 + 58 + 61 + \cdots + 94 = 2081$$

$$\bar{x} = \frac{\Sigma x}{n} = \frac{2081}{28} \approx 74.3 \quad \text{Mean}$$

$$\frac{73 + 76}{2} = 74.5$$

The median is the mean of the 14th and 15th items.

Examining the leaves reveals that the mode is 90.

Symmetry and **skewness** are important properties of a distribution. • If the graph of a distribution drops off about equally in both directions from the center, it is symmetric. • If it trails off farther to the left (or right), it is "skewed" left (or right).	Symmetry and skewness are illustrated by these distributions. 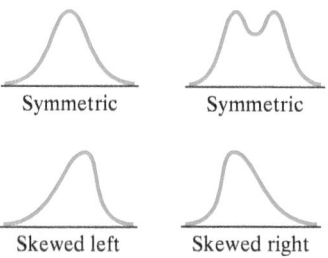

12.3 MEASURES OF DISPERSION

Concepts	*Examples*
Range is the distance from the least item to the greatest item of the data set.	Consider again the distribution

Range = (greatest value in the set)
− (least value in the set)

$$2, 6, 3, 7, 4, \quad \text{or, when ranked,} \quad 2, 3, 4, 6, 7.$$

$$\text{Range} = 7 - 2 = 5$$

Standard deviation s is a sort of "average" amount by which all items in a distribution differ from their mean.

$$s = \sqrt{\frac{\Sigma(x - \bar{x})^2}{n - 1}}$$

Since the mean is 4.4, the standard deviation is as follows.

$$s = \sqrt{\frac{(2 - 4.4)^2 + (3 - 4.4)^2 + (4 - 4.4)^2 + (6 - 4.4)^2 + (7 - 4.4)^2}{5 - 1}}$$

$$\approx 2.07$$

Chebyshev's Theorem

In any distribution, the fraction of all items that lie within k standard deviations of the mean (where $k > 1$) is *at least*

$$1 - \frac{1}{k^2}.$$

For $k = 2$, at least

$$1 - \frac{1}{2^2}, \quad \text{or} \quad 1 - \frac{1}{4} = \frac{3}{4},$$

of the items will lie within 2 standard deviations of the mean; that is, at least 3.75 items will lie between 0.26 and 8.54. In fact, in this case all 5 items lie within that range.

The **coefficient of variation** for a distribution expresses standard deviation as a percentage of the mean.

$$V = \frac{s}{\bar{x}} \cdot 100\%$$

Find the coefficient of variation for this distribution.

$$V = \frac{2.07}{4.4} \cdot 100\% \approx 47\%$$

Concepts	*Examples*

12.4 MEASURES OF POSITION

The **z-score** of an item gives the number of standard deviations that item is from the mean.

$$z = \frac{x - \bar{x}}{s}$$

Earlier, we saw a stem-and-leaf display for this data set.

52 54 57 58 58 61 62 65 65 66 71 71 72 73
76 76 79 82 82 84 85 88 88 90 90 90 92 94

The mean was 74.3. The standard deviation was 12.9.

Find the z-score for the item 84.

$$z = \frac{84 - 74.3}{12.9} \approx 0.75$$ 84 is three-fourths of a standard deviation *above* the mean.

Find the z-score for the item 62.

$$z = \frac{62 - 74.3}{12.9} \approx -0.95$$ 62 is $\frac{95}{100}$ of a standard deviation *below* the mean.

The **percentile** corresponding to an item tells roughly the percentage of all items that are below that item. There are ninety-nine percentiles, the first through the ninety-ninth.

For the above data, find the ninety-fifth percentile.

Since 95% of 28 (the number of items) is 26.6, we round up. The twenty-seventh item is 92.

For any numerical distribution, there are nine **deciles,** the first through the ninth.

Find the fourth decile (the same as the fortieth percentile).

40% of 28 is 11.2.

Therefore, take the twelfth item, which is 71.

For any numerical distribution, there are three **quartiles,** the first through the third.

Find the three quartiles for the data.

The median (Q_2) was found to be

74.5 (the mean of 73 and 76).

Q_1 is the median of the lower half of the items:

$$\frac{62 + 65}{2} = 63.5$$

Q_3 is the median of the upper half of the items:

$$\frac{85 + 88}{2} = 86.5$$

A **box plot** (or **box-and-whisker plot**) for a distribution is a visual representation that shows the least value, the greatest value, and all three quartiles.

Construct a box plot for this distribution of twenty-eight items.

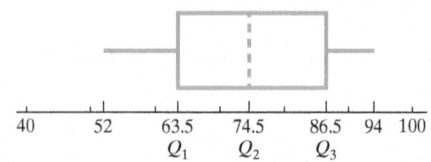

40	52	63.5	74.5	86.5 94	100
		Q_1	Q_2	Q_3	

Concepts	Examples

12.5 THE NORMAL DISTRIBUTION

Many quantities are **distributed approximately "normally"**—that is, according to a characteristic bell-shaped curve. Their distributions may have any mean and any (positive) standard deviation, but we can calculate their properties by relating them to the so-called standard normal curve, which has mean 0 and standard deviation 1. The total area under the standard normal curve, from $-\infty$ to $+\infty$, is 1 square unit.

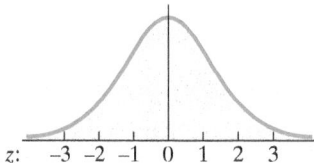

For any two items in a normal distribution, say x_1 and x_2, if we calculate their corresponding z-scores,

$$z_1 = \frac{x_1 - \bar{x}}{s} \quad \text{and} \quad z_2 = \frac{x_2 - \bar{x}}{s},$$

then the following three quantities are equivalent.

1. The **percentage** of items in the original distribution that are between x_1 and x_2

2. The **probability** that a randomly chosen item from the original distribution will lie between x_1 and x_2

3. The **area** under the standard normal curve between z_1 and z_2

Suppose the heights of the trainees on a large military base are normally distributed with mean 70 inches and standard deviation 5 inches.

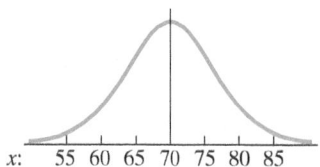

1. Find the *percentage* of trainees with heights between $x_1 = 70$ inches and $x_2 = 74$ inches.

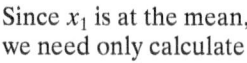

Since x_1 is at the mean, we need only calculate

$$z_2 = \frac{x_2 - \bar{x}}{s} = \frac{74 - 70}{5} = 0.8.$$

From the standard normal curve table,

$$A = 0.288, \quad \text{and the required percentage is} \quad 28.8\%.$$

2. Find the *probability* that a randomly selected trainee will have a height between $x_1 = 67$ inches and $x_2 = 77$ inches.

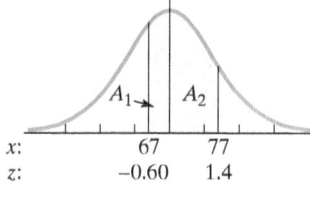

$$z_1 = \frac{67 - 70}{5} = -0.60, \quad \text{so} \quad A_1 = 0.226.$$

$$z_2 = \frac{77 - 70}{5} = 1.4, \quad \text{so} \quad A_2 = 0.419.$$

The required probability is

$$A_1 + A_2 = 0.226 + 0.419 = 0.645.$$

3. Find the *area* under the standard normal curve between the z-scores corresponding to trainee heights between 60 inches and 68 inches.

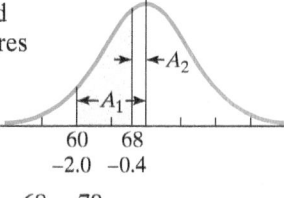

$$z_1 = \frac{60 - 70}{5} = -2.0 \quad \text{and} \quad z_2 = \frac{68 - 70}{5} = -0.4,$$

so $$A_1 = 0.477 \quad \text{and} \quad A_2 = 0.155.$$

The area under the curve between these two points is

$$A_1 - A_2 = 0.477 - 0.155 = 0.322.$$

Chapter 12 TEST

1. Top Five Social Networking Sites The table shows the number of persons age 2 and older in any U.S. location, in millions, who visited the listed Web sites at least once in June 2017.

Rank	Web Site	Visitors
1	Facebook and Messenger	202.0
2	Instagram	121.5
3	Twitter	110.3
4	LinkedIn	103.1
5	Snapchat	95.3

Data from comScore Inc.; comScore qSearch

Use this information to determine each of the following.

(a) the mean number of visitors per site

(b) the range

(c) the standard deviation

(d) the coefficient of variation

2. Cheaters Never Learn The table here shows the results of an educational study of university physics students, comparing exam scores with students' rates of copying others' homework. The numbers in the table approximate letter grades on a 4-point scale (4.0 is an A, 3.0 is a B, and so on). Answer the questions in terms of copy rate.

Copy Rate	Pretest	Exam 1	Exam 2	Exam 3	Final Exam
<10%	2.70	2.75	2.90	2.80	2.95
10% to 30%	2.50	2.45	2.35	2.40	2.30
30% to 50%	2.45	2.43	2.30	2.10	2.00
>50%	2.40	2.05	1.70	1.80	1.60

Data from research reported in *Physics Review–Special Topics–Physics Education* by David J. Palazzo, Young-Jin Lee, Rasil Warnakulasooriya, and David E. Pritchard of the Massachusetts Institute of Technology (MIT) physics faculty.

(a) Which students generally improved their exam performance over the course of the semester?

(b) Which students did better on exam 3 than on exam 2?

(c) Which students had lower scores consistently from one exam to the next throughout the semester?

(d) Do you think that copying homework is generally a *cause* of lower exam scores? Explain.

Champion Trees *The following table lists the 10 largest national champion trees (known as of May 5, 2017), based on the formula*

$$T = G + H + 0.25C,$$

where T = *total points,*
G = *girth (circumference of trunk 4.5 feet above the ground, measured in inches)*
H = *height, measured in feet*
and C = *average crown spread, measured in feet*

Tree Type	G (in.)	H (ft)	C (ft)	T	Location
Giant sequoia	1020	274	107	1321	Sequoia National Park, CA
Coast redwood	950	321	75	1290	Jedediah Smith Redwoods State Park, CA
Coast redwood	895	307	83	1223	Jedediah Smith Redwoods State Park, CA
Coast redwood	845	349	89	1216	Redwoods National Park, CA
Coast redwood	867	299	101	1191	Prairie Creek Redwoods State Park, CA
Western red cedar	761	159	45	931	Olympic National Park, WA
Sitka spruce	668	191	96	883	Olympic National Park, WA
Coast Douglas Fir	599	200	37	808	Olympic National Park, WA
Coast Douglas fir	505	281	71	804	Olympic National Forest, WA
Coast Douglas Fir	444	327	82	792	Coos County, OR

Data from *The World Almanac and Book of Facts 2018.*

Use this information for Exercises 3 and 4.

3. For the ten trees listed, find the following.

(a) the median height

(b) the first quartile in girth

(c) the seventh decile in total points

4. The eleventh-ranking tree in the country is a common bald cypress on Cat Island, Louisiana, with $G = 647$ inches, $H = 96$ feet, and $C = 74$ feet. For this tree, answer the following questions.

(a) Find its total points.

(b) Where would it have ranked based on girth alone?

(c) How much greater would its girth need to be to exceed the seventh-ranked girth figure?

(d) Assuming a roughly circular cross section at 4.5 feet above the ground, approximate the diameter of the trunk (to the nearest foot) at that height.

College Endowment Assets   *The table shows the top six U.S. colleges for 2016 endowment assets, in billions of dollars.*

Rank	College/University	Endowment Assets
1	Harvard University	34.5
2	Yale University	25.4
3	University of Texas System	24.2
4	Stanford University	22.4
5	Princeton University	22.2
6	Massachusetts Institute of Technology	13.2

Data from *World Almanac and Book of Facts 2018.*

5. Construct a bar graph for these data.

6. What percent (to one decimal place) of the total was held by Yale?

7. What was the median value?

Pediatrics Patients   *Casandra worked the 22 weekdays of last month in the pediatrics clinic and checked in the following numbers of patients.*

14	24	26	11	16	25	10	12	27	18	13
15	20	18	22	24	17	19	19	7	15	30

8. Construct grouped frequency and relative frequency distributions. Use five uniform classes of width 5 where the first class has a lower limit of 6. (Round relative frequencies to two decimal places.)

9. From your frequency distribution of **Exercise 8**, construct **(a)** a histogram and **(b)** a frequency polygon. Use appropriate scales and labels.

10. For the given data, how many uniform classes would be required if the first class had limits 7–9?

In Exercises 11–14, find the indicated measures for the following frequency distribution.

Value	9	11	13	15	17	19
Frequency	3	8	10	8	5	1

11. the mean

12. the median

13. the mode

14. the range

15. ***Exam Scores in a Criminal Justice Class***   The following data are scores achieved by 30 students on a criminal justice exam. Arrange the data into a stem-and-leaf display with leaves ranked.

52 43 84 65 77 70 79 61 66 80

55 86 68 78 71 38 54 64 67 73

33 49 61 91 84 77 50 67 79 72

Use the stem-and-leaf display below for Exercises 16–21.

```
2 | 1  2  4
2 | 6  7  9  9  9
3 | 0  1  1  2  2  3  3  4
3 | 6  6  7  8  8  9
4 | 1  1  2  4
4 | 5  6  9
5 | 2  4
5 | 8
6 | 0
```

Compute the measures required in Exercises 16–20.

16. the median

17. the mode(s), if any

18. the range

19. the third decile

20. the eighty-fifth percentile

21. Construct a box plot for the given data, showing values for the five important quantities on the numerical scale.

22. ***Test Scores in a Training Institute***   A certain training institute gives a standardized test to large numbers of applicants nationwide. The resulting scores form a normal distribution with mean 80 and standard deviation 5. Find the percent of all applicants with scores as follows. (Use the empirical rule.)

(a) between 70 and 90

(b) greater than 95 or less than 65

(c) less than 75

(d) between 85 and 90

Season Statistics in Major League Baseball *The tables below show the 2017 statistics on games won for all three divisions of both major baseball leagues. In each case,*

n = *number of teams in the division,*

$\bar{x}$ = *average (mean) number of games won,*

and s = *standard deviation of number of games won.*

American League

East Division	Central Division	West Division
$n = 5$	$n = 5$	$n = 5$
$\bar{x} = 83.0$	$\bar{x} = 79.6$	$\bar{x} = 82.4$
$s = 8.5$	$s = 15.3$	$s = 10.5$

National League

East Division	Central Division	West Division
$n = 5$	$n = 5$	$n = 5$
$\bar{x} = 76.4$	$\bar{x} = 80.8$	$\bar{x} = 83.8$
$s = 12.2$	$s = 9.4$	$s = 16.3$

Refer to the preceding tables for Exercises 23–26.

23. Overall, who had the greatest winning average, the East teams, the Central teams, or the West teams?

24. Overall, where were the teams the most "consistent" in number of games won: East, Central, or West?

25. Find (to the nearest tenth) the average number of games won for all East Division teams.

26. The Cleveland Indians, in the American League Central, won 102 games, while the L.A. Dodgers, in the National League West, won 104 games. Use z-scores to determine which of these two teams did relatively better within its own division of 5 teams.

Remission Data for Cocaine and Marijuana Addiction *The chapter opener and the* **When Will I Ever Use This?** *feature in the first section of this chapter showed research results for alcohol and tobacco addiction and recovery. (A 0.5 cumulative probability of remission required about 17 years for alcoholics, and about 30 years for smokers.) The table here gives data inferred from the same study for cocaine and marijuana addiction.*

Cocaine		Marijuana	
x	y	x	y
1.9	0.27	2.1	0.22
6.1	0.64	6.1	0.55
9.8	0.79	10.3	0.69
14.0	0.89	14.0	0.79
21.0	0.94	19.2	0.84

Data from G. Heyman, *Annual Review of Clinical Psychology 2013.*

27. Plot the data points from the table, draw two smooth curves, showing cocaine and marijuana addiction and recovery.

Then estimate the 0.5 cumulative probability of remission for:

(a) cocaine. **(b)** marijuana.

28. Make a conjecture about why remission periods are this much shorter for cocaine and marijuana than for alcohol and tobacco (about 17 and 30 years, respectively).

13 Personal Financial Management

Financial planners advise their clients in making decisions that will help them reach their financial goals. Many people find it beneficial to consult a professional who can offer a knowledgeable and objective perspective in this area. Financial planners can help people establish financial goals, increase their financial security, and even ask the right questions.

1. *Am I spending my income wisely?*
2. *Do I need a budget?*
3. *What are my financial priorities?*
4. *Should I buy a home, or is renting a better idea?*
5. *Am I saving enough for retirement or for my children's college education?*
6. *Am I taking advantage of available tax savings?*
7. *What are the best investment vehicles for me?*

Questions like these are within the purview of good financial planners, and many of the topics discussed in this chapter are related to them.

13.1 THE TIME VALUE OF MONEY

OBJECTIVES

1 Examine the meaning of interest.

2 Calculate simple interest.

3 Determine future value and present value.

4 Calculate compound interest.

5 Find the effective annual yield.

6 Use the rate of inflation to compare values over time.

Because the calculations in this chapter involve money, answers often will be rounded to the nearest $0.01. The TI-84 Plus C will display answers to the nearest 0.01 with the setting shown above.

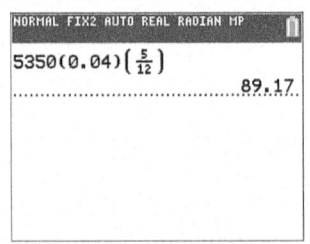

This is the computation required to solve **Example 1.**

Interest

To determine the value of money, we consider not only the amount (number of dollars) but also the particular point in time that the value is to be determined. If we borrow an amount of money today, we will repay a larger amount later. This increase in value is known as **interest.** The money *gains value over time.*

The amount of a loan or a deposit is called the **principal.** The interest is usually computed as a percent of the principal. This percent is called the **rate of interest** (or the **interest rate,** or simply the **rate**). The rate of interest is always assumed to be an annual rate unless otherwise stated.

Interest calculated only on principal is called **simple interest.** Interest calculated on principal plus any previously earned interest is called **compound interest.**

Simple Interest

Simple interest is calculated according to the following formula.

SIMPLE INTEREST

If P = principal, r = annual interest rate, and t = time (in years), then the **simple interest I** is calculated as follows.

$$I = Prt$$

EXAMPLE 1 **Finding Simple Interest**

Find the simple interest paid to borrow $5350 for 5 months at 4%.

Solution

$$I = Prt \qquad \text{Simple interest formula}$$

$$= \$5350(0.04)\left(\frac{5}{12}\right) \qquad \begin{array}{l}\text{5 months is } \frac{5}{12} \text{ of a year.}\\ P = \$5350, \ r = 4\% = 0.04\end{array}$$

$$= \$89.17 \qquad \text{Calculate.}$$

Future Value and Present Value

In **Example 1,** at the end of 5 months the borrower would have to repay

Principal   Interest

$5350   + $89.17 = $5439.17.

The total amount repaid is sometimes called the **maturity value** (or simply the **value**) of the loan. We will generally refer to it as the **future value,** or **future amount,** because when a loan is being set up, repayment will be occurring in the future. We use A to denote future amount (or value). The original principal, denoted P, can also be thought of as **present value.**

Future value depends on principal (present value) and interest as follows.

$$A = P + I = P + Prt = P(1 + rt)$$

FUTURE VALUE FOR SIMPLE INTEREST

If a principal P is borrowed at simple interest for t years at an annual interest rate of r, then the **future value** of the loan, denoted A, is calculated as follows.

$$A = P(1 + rt)$$

EXAMPLE 2   Finding Future Value for Simple Interest

Sergio took out a simple interest loan for \$210 to purchase textbooks and school supplies. If the annual interest rate is 4.5% and he must repay the loan after 6 months, find the future value (the maturity value) of the loan.

Solution

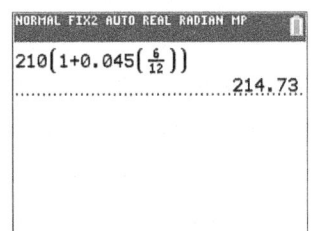

$$A = P(1 + rt) \qquad \text{Future value formula}$$

$$= \$210\left[1 + 0.045\left(\frac{6}{12}\right)\right] \qquad P = \$210,\ r = 4.5\% = 0.045,\ t = \frac{6}{12}$$

$$= \$214.73 \qquad \text{Calculate.}$$

This is the computation required to solve **Example 2.**

At the end of 6 months, Sergio will need to repay \$214.73.

Sometimes the future value is known, and we need to compute the present value. For this purpose, we solve the future value formula for P.

$$A = P(1 + rt) \qquad \text{Future value formula for simple interest}$$

$$P = \frac{A}{1 + rt} \qquad \text{Solve the formula for } P.$$

EXAMPLE 3   Finding Present Value for Simple Interest

Suppose that Sergio (**Example 2**) is granted a 6-month *deferral* of the \$210 payment rather than a loan. That is, instead of incurring interest for 6 months, he will have to pay just the \$210 at the end of that period. If he has extra money right now, and if he can earn 1% simple interest on savings, what lump sum must he deposit now so that its value will be \$210 after 6 months?

Solution

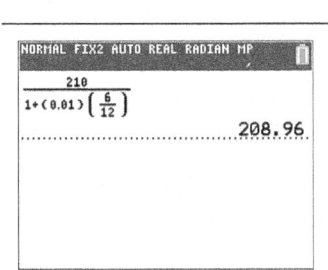

$$P = \frac{A}{1 + rt} \qquad \text{Future value formula solved for } P$$

$$P = \frac{\$210}{1 + (0.01)\left(\frac{6}{12}\right)} = \$208.96 \qquad \text{Substitute known values and calculate.}$$

This is the computation required to solve **Example 3.**

A deposit of \$208.96 now, growing at 1% simple interest, will grow to \$210 over a 6-month period.

Compound Interest

Interest paid on principal plus interest is called *compound interest*. After a certain period, the interest earned so far is *credited* (added) to the account, and the sum (principal plus interest) then earns interest during the next period.

EXAMPLE 4 **Comparing Simple and Compound Interest**

Compare simple and compound interest for a $1000 deposit at 2% interest for 5 years.

Solution

$$A = P(1 + rt) \qquad \text{Future value formula}$$

$$= \$1000(1 + 0.02 \cdot 5) \quad \text{\footnotesize $P = \$1000$, $r = 2\%$, $t = 5$}$$

$$= \$1100 \qquad \text{Calculate.}$$

After 5 years, $1000 grows to $1100 subject to 2% simple interest.

The result of compounding (annually) for 5 years is shown in **Table 1.**

Table 1 A $1000 Deposit at 2% Interest Compounded Annually

Year	Beginning Balance	Interest Earned $I = Prt$	Ending Balance
1	$1000.00	$1000.00(0.02)(1) = $20.00	$1020.00
2	$1020.00	$1020.00(0.02)(1) = $20.40	$1040.40
3	$1040.40	$1040.40(0.02)(1) = $20.81	$1061.21
4	$1061.21	$1061.21(0.02)(1) = $21.22	$1082.43
5	$1082.43	$1082.43(0.02)(1) = $21.65	$1104.08

Under annual compounding for 5 years, $1000 grows to $1104.08, which is $4.08 more than under simple interest.

Based on the compounding pattern of **Example 4,** we now develop a future value formula for compound interest. In practice, earned interest can be credited to an account at time intervals other than 1 year (usually more often). For example, it can be done semiannually, quarterly, monthly, or daily. This time interval is called the **compounding period** (or simply the **period**). Start with the following definitions:

P = original principal deposited, $\qquad r$ = annual interest rate,

n = number of periods per year, $\qquad t$ = number of years.

During each individual compounding period, interest is earned according to the simple interest formula, and as interest is added, the beginning principal increases from one period to the next. During the first period, the interest earned is given by

$$\text{Interest} = P(r)\left(\frac{1}{n}\right) \qquad \text{\footnotesize Interest = (Principal)(rate)(time),}$$
$$\text{\footnotesize one period} = \tfrac{1}{n} \text{ year}$$

$$= P\left(\frac{r}{n}\right). \qquad \text{\footnotesize Rewrite $(r)\left(\frac{1}{n}\right)$ as $\left(\frac{r}{n}\right)$.}$$

At the end of the first period, the account then contains

$$\text{Ending amount} = \overset{\text{\footnotesize Beginning amount}}{P} + \overset{\text{\footnotesize Interest}}{P\left(\frac{r}{n}\right)}$$

$$= P\left(1 + \frac{r}{n}\right). \qquad \text{\footnotesize Factor P from both terms.}$$

These screens support the Ending Balance figures in **Table 1.**

King Hammurabi tried to hold interest rates at 20% for both silver and gold, but moneylenders ignored his decrees.

Modern-day moneylenders ignore this cap as well. Penalty interest rates imposed on credit card balances after late payments can be nearly 30%.

Now during the second period, the interest earned is given by

$$\text{Interest} = \left[P\left(1 + \frac{r}{n}\right) \right](r)\left(\frac{1}{n}\right) \qquad \text{Interest} = [\,\text{Principal}\,](\text{rate})(\text{time})$$

$$= P\left(1 + \frac{r}{n}\right)\left(\frac{r}{n}\right),$$

so the account ends the second period containing

$$\text{Ending amount} = \overbrace{P\left(1 + \frac{r}{n}\right)}^{\text{Beginning amount}} + \overbrace{P\left(1 + \frac{r}{n}\right)\left(\frac{r}{n}\right)}^{\text{Interest}}$$

$$= P\left(1 + \frac{r}{n}\right)\left[1 + \frac{r}{n}\right] \qquad \text{Factor } P\left(1 + \frac{r}{n}\right) \text{ from both terms.}$$

$$= P\left(1 + \frac{r}{n}\right)^2. \qquad a \cdot a = a^2$$

Consider one more period, namely, the third. The interest earned is

$$\text{Interest} = \left[P\left(1 + \frac{r}{n}\right)^2 \right](r)\left(\frac{1}{n}\right) \qquad \text{Interest} = [\,\text{Principal}\,](\text{rate})(\text{time})$$

$$= P\left(1 + \frac{r}{n}\right)^2\left(\frac{r}{n}\right), \qquad \text{Multiply.}$$

so the account ends the third period containing

$$\text{Ending amount} = \overbrace{P\left(1 + \frac{r}{n}\right)^2}^{\text{Beginning amount}} + \overbrace{P\left(1 + \frac{r}{n}\right)^2\left(\frac{r}{n}\right)}^{\text{Interest}}$$

$$= P\left(1 + \frac{r}{n}\right)^2\left[1 + \frac{r}{n}\right] \qquad \text{Factor } P\left(1 + \frac{r}{n}\right)^2 \text{ from both terms.}$$

$$= P\left(1 + \frac{r}{n}\right)^3. \qquad a^2 \cdot a = a^3$$

Table 2 summarizes the preceding results.

Quantities like

$$\left(1 + \frac{r}{n}\right)^{nt}$$

can be evaluated using a key such as $\boxed{y^x}$ on a scientific calculator or $\boxed{\wedge}$ on a graphing calculator.

Table 2 Compound Amount

Period Number	Beginning Amount	Interest Earned During Period	Ending Amount
1	P	$P\left(\frac{r}{n}\right)$	$P\left(1 + \frac{r}{n}\right)$
2	$P\left(1 + \frac{r}{n}\right)$	$P\left(1 + \frac{r}{n}\right)\left(\frac{r}{n}\right)$	$P\left(1 + \frac{r}{n}\right)^2$
3	$P\left(1 + \frac{r}{n}\right)^2$	$P\left(1 + \frac{r}{n}\right)^2\left(\frac{r}{n}\right)$	$P\left(1 + \frac{r}{n}\right)^3$
$\vdots$	$\vdots$	$\vdots$	$\vdots$
nt	$P\left(1 + \frac{r}{n}\right)^{nt-1}$	$P\left(1 + \frac{r}{n}\right)^{nt-1}\left(\frac{r}{n}\right)$	$P\left(1 + \frac{r}{n}\right)^{nt}$

Compound Interest: Computing
Future Value

The lower right entry of **Table 2** provides the following formula.

FUTURE VALUE FOR COMPOUND INTEREST

If P dollars are deposited at an annual interest rate of r, compounded n times per year, and the money is left on deposit for t years, then the **future value, A** (the final amount on deposit), is calculated as follows.

$$A = P\left(1 + \frac{r}{n}\right)^{nt}$$

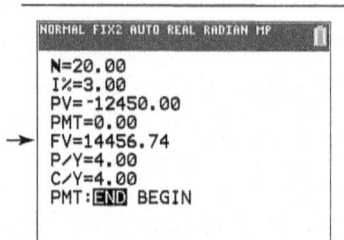

← Compound Interes...   ADVANCED

Principal Amount 3,419

Monthly Deposit 0

Period (month) 30

Annual Interest Rate% 2.1

Compounding Monthly

RESET   CALCULATE

Total Principal: **3,419.00**
Interest Amount: **184.13**
Maturity Value: **3,603.13**
APY: **2.1203%**

EMAIL   REPORT   TABLE

The screenshot above shows the Future (Maturity) Value and the Interest Amount calculated in **Example 5(a).**

Several **financial calculator apps** are available for smartphones. This one is the (free) Compound Interest Calculator from www.fncalculator.com.

EXAMPLE 5   Finding Future Value for Compound Interest

Find the future value (final amount on deposit) and the amount of interest earned for the following deposits.

(a) $3419 at 2.1% compounded monthly for 30 months

(b) $12,450 at 3% compounded quarterly for 5 years

Solution

(a) Here $P = \$3419$, $r = 2.1\% = 0.021$, $n = 12$, and $t = \frac{30}{12} = 2.5$.

$$A = P\left(1 + \frac{r}{n}\right)^{nt}$$

$$A = \$3419\left(1 + \frac{0.021}{12}\right)^{(12)(2.5)} = \$3603.13$$

Future value   Present value

$$\text{Interest earned} = \$3603.13 \ - \ \$3419$$

$$= \$184.13$$

(b) Here $P = \$12,450$, $r = 3\% = 0.03$, $n = 4$, and $t = 5$.

$$A = P\left(1 + \frac{r}{n}\right)^{nt}$$

$$A = \$12,450\left(1 + \frac{0.03}{4}\right)^{(4)(5)} = \$14,456.74$$

Future value   Present value

$$\text{Interest earned} = \$14,456.74 - \$12,450$$

$$= \$2006.74$$

```
NORMAL FIX2 AUTO REAL RADIAN MP

  N=20.00
  I%=3.00
  PV=-12450.00
  PMT=0.00
→ FV=14456.74
  P/Y=4.00
  C/Y=4.00
  PMT:END BEGIN
```

This screen shows the TVM solver, one of the financial applications available on the TI-84 Plus C. It enables the user to solve for a missing quantity related to the time value of money. The negative value in the Present Value (PV) field indicates a negative cash flow for the investor at the time of deposit. The arrow indicates the display that supports the Future Value (FV) answer in **Example 5(b).**

Compound interest problems sometimes require that we solve the formula for P to compute the present value when the future value is known.

$$A = P\left(1 + \frac{r}{n}\right)^{nt} \qquad \text{Future value for compound interest}$$

$$P = \frac{A}{\left(1 + \frac{r}{n}\right)^{nt}} \qquad \text{Solve the formula for } P.$$

The 1954 musical *The Pajama Game* was revived on Broadway twice, most recently in 2006 starring Harry Connick, Jr. The musical tells the story of the Sleeptite Pajama Factory workers attempting to secure a raise of $7\frac{1}{2}$ cents per hour.

The musical features a number called **"Seven and a Half Cents"** (after the novel that inspired the musical), with lyrics claiming that a small raise of $7\frac{1}{2}$ cents per hour will amount to $852.74 in 5 years, $1705.48 in 10 years, and $3411.96 in 20 years. Assuming that the figure for 5 years is correct, is the one for 10 years also correct? How about the one for 20 years? Oops! Judging by the first two figures, were the workers planning to invest their "extra" wages?

EXAMPLE 6   **Finding Present Value for Compound Interest**

Ashley will need $40,000 in 5 years to help pay for her college education. What lump sum, deposited today at 3% compounded quarterly, will produce the necessary amount?

Solution

This question requires that we find the present value P based on the following.

Future value:	$A = \$40{,}000$
Annual rate:	$r = 3\% = 0.03$
Periods per year:	$n = 4$
Number of years:	$t = 5$

$$P = \frac{A}{\left(1 + \frac{r}{n}\right)^{nt}} \qquad \text{Future value formula solved for } P$$

$$P = \frac{\$40{,}000}{\left(1 + \frac{0.03}{4}\right)^{(4)(5)}} \qquad \text{Substitute known values.}$$

$$P = \$34{,}447.59 \qquad \text{Calculate.}$$

Assuming interest of 3% compounded quarterly can be maintained, $40,000 can be attained 5 years in the future by depositing $34,447.59 today.

The next example uses logarithms (discussed in an earlier chapter). Consult your calculator manual if necessary.

EXAMPLE 7   **Finding the Time Required to Double a Principal Deposit**

In an investment account paying 3% interest, compounded daily, when will the amount in the account be twice the original principal?

Solution

The future value must equal two times the present value.

$$2P = P\left(1 + \frac{r}{n}\right)^{nt} \qquad \text{Substitute } 2P \text{ for } A \text{ in the future value formula.}$$

$$2 = \left(1 + \frac{r}{n}\right)^{nt} \qquad \text{Divide both sides by } P.$$

$$2 = \left(1 + \frac{0.03}{365}\right)^{365t} \qquad \text{Substitute values for } r \text{ and } n.$$

$$\log 2 = \log\left(1 + \frac{0.03}{365}\right)^{365t} \qquad \text{Take the logarithm of both sides.}$$

$$\log 2 = 365t \log\left(1 + \frac{0.03}{365}\right) \qquad \text{Use the power property of logarithms.}$$

$$t = \frac{\log 2}{365 \log\left(1 + \frac{0.03}{365}\right)} \qquad \text{Solve for } t.$$

$$t = 23.106 \qquad \text{Round to the nearest thousandth of a year.}$$

This equates to 23 years, 39 days (ignoring leap years).

The power of compound interest is being put to use in the town of Union City, Michigan. Eli Hooker, chairman of the local Bicentennial Committee in 1976, saw that there was not enough money to put on a proper celebration that year. So, in order to help his town prepare for the tricentennial in 2076, he collected twenty-five dollars apiece from 42 patriotic residents and deposited the money in a local bank. Compounded at 7%, that money would grow to a million dollars by 2076.

Unfortunately, the million dollars won't be worth as much then as we might think. If the community decides to hire people to parade around in historical costumes, it might have paid them $3 per hour in 1976. The going wage in 2076, assuming 3% annual inflation for a hundred years, would be nearly $60 per hour.

Example 8 shows the importance of starting early to maximize the long-term advantage of compounding.

EXAMPLE 8 Comparing Retirement Plans

Compare the results at age 67 for the following two retirement plans. Both plans earn 6% annual interest throughout the account-building period.

Plan A: Anika begins saving at age 20, deposits $2000 on every birthday from age 21 to age 35 (15 deposits, or $30,000 total), and thereafter makes no additional contributions.

Plan B: Barry waits until age 30 to start saving and then makes deposits of $2000 on every birthday from age 31 to age 67 (37 deposits, or $74,000 total).

Solution

Table 3 shows how both accounts build over the years. $30,000, deposited earlier, produces $45,882 more than $74,000, deposited later.

Table 3 Retirement Plans Compared

Age	Plan A	Plan B
20	0	0
25	11,274	0
30	26,362	0
35	46,552	11,274
40	62,297	26,362
45	83,367	46,552
50	111,564	73,571
55	149,298	109,729
60	199,795	158,116
65	267,371	222,870
67	300,418	254,536

In **Example 8,** the *early* deposits of Plan A outperformed the *greater number* of deposits of Plan B because the interest rate was high enough to result in a Plan A balance at age 35 that Plan B could never overtake, despite additional Plan B deposits from then on. However, if the interest rate had been 4% rather than 6%, Plan B would have overtaken Plan A (at age 58) and at age 67 would have come out ahead by $22,916 (to the nearest dollar).

Effective Annual Yield

Banks, credit unions, and other financial institutions often advertise two rates: first, the actual annualized interest rate, or **nominal rate** (the "named" or "stated" rate), and second, the rate that would produce the same final amount, or future value, at the end of 1 year if the interest being paid were simple rather than compound. This is called the "effective rate" or, more commonly, the **effective annual yield.** (It may be denoted **APY** for **"annual percentage yield."**) Because the interest is normally compounded multiple times per year, the effective annual yield will usually be somewhat higher than the nominal rate.

```
NORMAL FIX2 AUTO REAL RADIAN CL
▶Eff(2.50,4)
                        2.52
```

Financial calculators are programmed to compute effective interest rate. Compare to the result in **Example 9.**

EXAMPLE 9 Finding Effective Annual Yield

What is the effective annual yield of an account paying a nominal rate of 2.50%, compounded quarterly?

Solution

From the given data, $r = 0.025$ and $n = 4$. Suppose we deposited $P = \$1$ and left it for 1 year ($t = 1$). Then the compound future value formula gives

$$A = 1 \cdot \left(1 + \frac{0.025}{4}\right)^{4 \cdot 1} \approx 1.0252.$$

The initial deposit of $1, after 1 year, has grown to $1.0252.

$$\text{Interest earned} = 1.0252 - 1 \qquad \text{Interest = future value − present value}$$

$$= 0.0252, \quad \text{or} \quad 2.52\%$$

A nominal rate of 2.50% results in an effective annual yield of 2.52%.

Generalizing the procedure of **Example 9** gives the following formula.

EFFECTIVE ANNUAL YIELD

A nominal interest rate of r, compounded n times per year, is equivalent to the following **effective annual yield.**

$$Y = \left(1 + \frac{r}{n}\right)^n - 1$$

When shopping for loans or savings opportunities, a borrower should seek the least yield available, while a depositor should look for the greatest.

EXAMPLE 10 Comparing Savings Rates

Yolanda wants to deposit $2800 into a savings account and has narrowed her choices to the three institutions represented here. Which is the best choice?

Institution	Rate on Deposits of $1000 to $5000
Friendly Credit Union	2.08% annual rate, compounded monthly
Premier Savings	2.09% annual yield
Neighborhood Bank	2.05% compounded daily

Solution

Compare the effective annual yields for the three institutions.

Friendly: $\qquad Y = \left(1 + \dfrac{0.0208}{12}\right)^{12} - 1 = 0.0210 = 2.10\%$

Premier: $\qquad Y = 2.09\%$

Neighborhood: $\qquad Y = \left(1 + \dfrac{0.0205}{365}\right)^{365} - 1 = 0.0207 = 2.07\%$

The best of the three yields, 2.10%, is offered by Friendly Credit Union.

← Effective Rate Calculator

ANNUAL RATE   EFFECTIVE RATE   NOMINAL RATE

Nominal Annual Rate (%) 2.08

Compounding   Monthly (12/yr) ▾

Effective Annual Rate (%): **2.0999**

The screenshot shown here supports the effective annual yield value for the Friendly Credit Union in **Example 10.**

Table 4

Consumer Price Index*

Year	Average CPI-U	% Change in CPI-U
1980	82.4	13.5
1981	90.9	10.3
1982	96.5	6.2
1983	99.6	3.2
1984	103.9	4.3
1985	107.6	3.6
1986	109.6	1.9
1987	113.6	3.6
1988	118.3	4.1
1989	124.0	4.8
1990	130.7	5.4
1991	136.2	4.2
1992	140.3	3.0
1993	144.5	3.0
1994	148.2	2.6
1995	152.4	2.8
1996	156.9	3.0
1997	160.5	2.3
1998	163.0	1.6
1999	166.6	2.2
2000	172.2	3.4
2001	177.1	2.8
2002	179.9	1.6
2003	184.0	2.3
2004	188.9	2.7
2005	195.3	3.4
2006	201.6	3.2
2007	207.3	2.8
2008	215.3	3.8
2009	214.5	−0.4
2010	218.1	1.6
2011	224.9	3.2
2012	229.6	2.1
2013	233.0	1.5
2014	236.7	1.6
2015	237.0	0.1
2016	240.0	1.3
2017	245.1	2.1

Data from Bureau of Labor Statistics.

The period 1982 to 1984: 100

Inflation

Interest reflects how the dollar amount of an investment or loan *increases over time*. On the other hand, **price inflation** reflects how the value of each dollar *decreases over time* in terms of its ability to purchase goods or services.

The United State Bureau of Labor Statistics regularly publishes **consumer price index (CPI)** figures, which reflect the prices of common items necessary for living, such as food, clothing, housing, and transportation. **Table 4** gives values of the primary index representing "all urban consumers" (the CPI-U). The value shown for a given year is the average (arithmetic mean) of the twelve monthly figures for that year. For example, the average CPI-U for 2001 (177.1) was 2.8% greater than the average value for 2000 (172.2).

Deflation is a *decrease* in price levels from one year to the next. A brief period of minor deflation, along with a general economic slowdown, usually is called a **recession**. (For example, note the 2009 rate of −0.4 in **Table 4**).

Unlike the sudden and regular increases in account values under compound interest, price levels tend to fluctuate gradually over time. Thus, it is appropriate to use the formula for continuous compounding when making estimates of inflation.

FUTURE VALUE FOR CONTINUOUS COMPOUNDING

If an initial deposit of P dollars earns continuously compounded interest at an annual rate r for a period of t years, then the **future value, A,** is calculated as follows.

$$A = Pe^{rt}$$

EXAMPLE 11   Predicting Inflated Salary Levels

In 2017, the median salary for occupational therapists was \$83,200 (Data from Bureau of Labor Statistics, U.S. Dept. of Labor). Approximately what salary would a person in this field need in 2037 to maintain purchasing power, assuming the inflation rate were to persist at each of the following levels?

(a) 2% (approximately the 2017 level)

(b) 10% (approximately the 1981 level)

Solution

(a) In this case we can use the continuous compounding future value formula with $P = \$83,200$, $r = 0.02$, and $t = 20$ (since 20 years pass between 2017 and 2037), along with the $\boxed{e^x}$ key on a calculator.

$$A = Pe^{rt} = \$83,200\, e^{(0.02)(20)} = \$124,119.81$$

The required salary in 2037 would be about \$124,000.

(b) For this level of inflation, we have the following.

$$A = Pe^{rt} = \$83,200\, e^{(0.10)(20)} = \$614,769.47$$

The required salary in 2037 would be about \$615,000.

NORMAL FIX2 AUTO REAL RADIAN MP

83200e^0.02*20

                        124119.81

83200e^0.10*20

                        614769.47

These are the computations required to solve the two parts of **Example 11** on the previous page.

A proportion (introduced in an earlier chapter) is a statement that says two ratios are equal. To compare equivalent general price levels in any 2 years, we can use the proportion below.

INFLATION PROPORTION

For a given consumer product or service subject to average inflation, prices in two different years are related as follows.

$$\frac{\text{Price in year A}}{\text{Price in year B}} = \frac{\text{CPI in year A}}{\text{CPI in year B}}$$

EXAMPLE 12 Comparing a Tuition Increase to Average Inflation

The average cost of tuition and fees at U.S. institutions of higher learning increased from $1626 for the 1982–83 academic year to $12,219 for the 2016–17 academic year (Data from National Center for Education Statistics, U.S. Dept. of Education). Compare this increase to average inflation over the same period.

Solution

Since an academic year spans the last half of one calendar year and the first half of the next, use the average of the CPI-U figures from **Table 4** over each two-year period.

1982–83 academic year: $\dfrac{96.5 + 99.6}{2} = 98.05$ Average of CPI-U figures for 1982 and 1983

2016–17 academic year: $\dfrac{240.0 + 245.1}{2} = 242.55$ Average of CPI-U figures for 2016 and 2017

Let x represent what we would expect tuition and fees to cost in 2016–17 if their cost had increased at the average rate since the 1982–83 academic year.

$$\frac{\text{Cost in 2016–17}}{\text{Cost in 1982–83}} = \frac{\text{CPI for 2016–17}}{\text{CPI for 1982–83}}$$ Inflation proportion

$$\frac{x}{\$1626} = \frac{242.55}{98.05}$$ Substitute CPI values.

$$x = \frac{242.55}{98.05} \cdot \$1626$$ Solve for x.

$$x \approx \$4022$$

Now compare the actual 2016–17 cost of $12,219 with the expected figure, $4022.

$$\frac{\$12{,}219}{\$4022} = 3.04$$ 3.04 = 1 + 2.04 = 100% + 204%

Over the period of interest, the cost of tuition and fees at institutions of higher learning in the United States increased at a rate approximately 204% greater than the average CPI-U rate.

Monetary inflation devalues the currency just as **price inflation** does, but it does so through a direct increase in the money supply within the economy. Examples in the United States include the issuance in 2009 and 2010 of massive government debt and massive government spending in "stimulus" programs, as well as the Federal Reserve's purchasing of government and mortgage bonds through "quantitative easing," which continued into 2014.

When working with quantities, such as inflation, where continual fluctuations and inexactness prevail, we often develop rough "rules of thumb" for obtaining quick estimates. One example is the estimation of the **years to double,** which is the number of years it takes for the general level of prices to double for a given annual rate of inflation.

We can derive an estimation rule as follows.

$$A = Pe^{rt}$$ Future value formula

$$2P = Pe^{rt}$$ Prices are to double.

$$2 = e^{rt}$$ Divide both sides by P.

$$\ln 2 = rt$$ Take the natural logarithm of both sides.

$$t = \frac{\ln 2}{r}$$ Solve for t.

$$t = \frac{100 \ln 2}{100r}$$ Multiply numerator and denominator by 100.

$$\textbf{years to double} \approx \frac{\textbf{70}}{\textbf{annual inflation rate}}$$ 100 ln 2 ≈ 70 (rounded to the nearest 10)

(Because r is the inflation rate as a *decimal*, $100r$ is the inflation rate as a *percent*.) The result above usually is called the **rule of 70.** The value it produces, if not a whole number, should be rounded *up* to the next whole number of years.

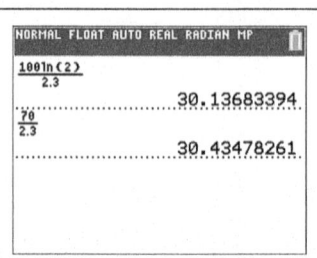

Because 100 ln 2 ≈ 70, the two results shown here are approximately equal. See **Example 13** and the preceding discussion.

EXAMPLE 13   Estimating Years to Double by the Rule of 70

Estimate the years to double for an annual inflation rate of 2.3%.

Solution

$$\text{Years to double} \approx \frac{70}{2.3} \approx 30.43 \quad \text{Rule of 70}$$

With a sustained inflation rate of 2.3%, prices would double in about 31 years.

13.1 EXERCISES

CONCEPT CHECK *Write* true *or* false *for each statement.*

1. Compound interest is calculated on principal only.

2. The interest paid on a loan is the difference between the maturity value and the original principal amount.

3. For a given annual interest rate, the shorter the compounding period is, the lower the interest earnings will be per year.

4. The nominal rate is normally higher than the effective annual yield.

5. Price inflation causes the value of currency to increase (inflate) over time.

6. Luxury items such as yachts and expensive jewelry are not included in the consumer price index.

In the following exercises, assume whenever appropriate that, unless otherwise known, there are 12 months per year, 30 days per month, and 365 days per year.

Find the simple interest owed for each loan.

7. $600 at 4% for 1 year   **8.** $5000 at 3% for 1 year

9. $920 at 3.5% for 9 months

10. $5400 at 5% for 4 months

11. $2675 at 5.4% for $2\frac{1}{2}$ years

12. $2620 at 4.82% for 32 months

Find the future value of each deposit if the account pays **(a)** *simple interest, and* **(b)** *interest compounded annually.*

13. $700 at 2% for 6 years

14. $4000 at 2.5% for 5 years

15. $2500 at 3% for 3 years

16. $3000 at 2% for 4 years

Solve each interest-related problem.

17. Simple Interest on a Late Property Tax Payment Darius was late on his property tax payment to the county. He owed $7500 and paid the tax 4 months late. The county charges a penalty of 5% simple interest. Find the amount of the penalty.

18. Simple Interest on a Loan for Equipment Alonso purchased new equipment for his yard maintenance business. He borrowed $2325 to pay for the equipment, and he agreed to pay off the loan in 9 months at 6% simple interest. Find the amount of interest he will owe.

19. Simple Interest on a Small Business Loan Tony opened a hot dog stand last April. He borrowed $6500 to pay for the stand and startup inventory, and he agreed to pay off the loan in 10 months at 7% simple interest. Find the *total amount* required to repay the loan.

20. Simple Interest on a Tax Overpayment Alvita is owed $530 by the Internal Revenue Service for overpayment of last year's taxes. The IRS will repay the amount at 4% simple interest. Find the *total amount* Alvita will receive if the interest is paid for 8 months.

Find the missing final amount (future value) and interest earned.

	Principal	Rate	Compounded	Time	Final Amount	Compound Interest
21.	$975	2%	quarterly	4 years	_____	_____
22.	$1150	2.5%	semiannually	6 years	_____	_____
23.	$7500	$3\frac{1}{2}$%	annually	25 years	_____	_____
24.	$3450	2.4%	semiannually	10 years	_____	_____

For each deposit, find the future value (that is, the final amount on deposit) when compounding occurs (a) annually, (b) semiannually, and (c) quarterly.

	Principal	Rate	Time
25.	$2000	4%	4 years
26.	$5000	2%	6 years
27.	$18,000	1%	3 years
28.	$10,000	3%	12 years

Occasionally a savings account may actually pay interest compounded continuously. For each deposit, find the interest earned if interest is compounded (a) semiannually, (b) quarterly, (c) monthly, (d) daily, and (e) continuously.

	Principal	Rate	Time
29.	$950	1.6%	4 years
30.	$2650	2.8%	33 months (In parts (d) and (e), assume 1003 days.)

31. Describe the effect of interest being compounded more and more often. In particular, how good is continuous compounding?

Solve each interest-related problem.

32. Finding the Amount Borrowed in a Simple Interest Loan Brynn takes out a 7% simple interest loan today that will be repaid 15 months from now with a payoff amount of $815.63. What amount is she borrowing?

33. Finding the Amount Borrowed in a Simple Interest Loan What is the maximum amount Dara can borrow today if it must be repaid in 4 months with simple interest at 8% and she knows that at that time she will be able to repay no more than $1500?

34. In the development of the future value formula for compound interest in the text, at least four specific problem-solving strategies were employed. Identify (name) as many of them as you can, and describe their use in this case.

Find the present value for each future amount.

35. $1000 (3% compounded annually for 5 years)

36. $14,000 (4% compounded quarterly for 2.5 years)

37. $9860 (3.5% compounded semiannually for 12 years)

38. $15,080 (5% compounded monthly for 6 years)

Finding the Present Value of a Compound Interest Retirement Account *Tanja wants to establish an account that will supplement her retirement income beginning 30 years from now. For each interest rate, find the lump sum she must deposit today so that $750,000 will be available at time of retirement.*

39. 5% compounded quarterly

40. 6% compounded quarterly

41. 5% compounded daily **42.** 6% compounded daily

Finding the Effective Annual Yield in a Savings Account *Suppose a savings and loan pays a nominal rate of 2% on savings deposits. Find the effective annual yield if interest is compounded as stated in Exercises 43–49. (Give answers to the nearest thousandth of a percent.)*

43. annually   **44.** semiannually

45. quarterly   **46.** monthly

47. daily   **48.** 1000 times per year

49. 10,000 times per year

50. Judging from **Exercises 43–49,** what do you suppose is the effective annual yield if a nominal rate of 2% is compounded continuously? Explain your reasoning.

Comparing CD Rates and Yields *A certificate of deposit (CD) is similar to a savings account, but deposits are generally held to maturity. They usually offer higher rates than savings accounts because they "tie up" money for a period of time and often require a minimum deposit. The table shows the two best 5-year CD rates available on a certain online listing on May 14, 2018. Use this information in Exercises 51 and 52.*

5-Year CD	Rate	Yield
Alaska USA FCU	3.000%	3.045%
First Internet Bank of Indiana	2.940%	2.984%

51. If you deposit $500 with Alaska USA FCU, how much will you have after 5 years?

52. How often does compounding occur in the First Internet Bank CD:

   semiannually, quarterly, monthly, or daily?

Solve each problem.

53. **Finding Years to Double** How long would it take to double your money in an account paying 4% compounded quarterly? (Answer in years plus days, ignoring leap years.)

54. **Comparing Principal and Interest Amounts** After what time period would the interest earned equal the original principal in an account paying 2% compounded daily? (Answer in years plus days, ignoring leap years.)

55. **Devising a Rate Formula** Solve the effective annual yield formula for r to obtain a general formula for nominal rate in terms of yield and the number of compounding periods per year.

56. **Finding the Nominal Rate of a Savings Account** Ridgeway Savings compounds interest monthly. The effective annual yield is 1.95%. What is the nominal rate?

57. **Comparing Bank Savings Rates** Bank A pays a nominal rate of 1.800% compounded daily on deposits. Bank B produces the same annual yield as Bank A but compounds interest only quarterly and pays no interest on funds deposited for less than an entire quarter.

   (a) What nominal rate does Bank B pay (to the nearest thousandth of a percent)?

   (b) Which bank should Melba choose if she has $2000 to deposit for 10 months? How much more interest will she earn than in the other bank?

   (c) Which bank should Tierra choose for a deposit of $5000 for one year? How much interest will be earned?

Estimating the Years to Double by the Rule of 70 *Use the rule of 70 to estimate the years to double for each annual inflation rate.*

58. 1%   59. 2%

60. 8%   61. 9%

Estimating the Inflation Rate by the Rule of 70 *Use the rule of 70 to estimate the annual inflation rate (to the nearest tenth of a percent) that would cause the general level of prices to double in each time period.*

62. 12 years

63. 7 years

64. 16 years

65. 22 years

66. Derive a rule for estimating the "years to triple"—that is, the number of years it would take for the general levels of prices to triple for a given annual inflation rate.

Estimating Future Prices for Constant Annual Inflation *The year 2018 prices of several items are given below. Find the estimated future prices required to fill the blanks in the chart. (Use the future value formula for continuous compounding and give a number of significant figures consistent with the 2018 price figures provided.)*

	Item	2018 Price	2024 Price 2% Inflation	2030 Price 2% Inflation	2024 Price 10% Inflation	2030 Price 10% Inflation
67.	1 lb of ground beef	$3.73	_____	_____	_____	_____
68.	House	$337,000	_____	_____	_____	_____
69.	New car	$36,300	_____	_____	_____	_____
70.	Gallon of gasoline	$2.88	_____	_____	_____	_____

Estimating Future Prices for Variable Annual Inflation *Assume that prices for the items below increased at the average annual inflation rates shown in* **Table 4.** *Use the inflation proportion to find the missing prices in the last column of the chart. Round to the nearest dollar.*

	Item	Price	Year Purchased	Price in 2017
71.	Evening dress	$175	2000	___
72.	Desk	$450	2002	___
73.	Lawn tractor	$1099	1996	___
74.	Designer puppy	$250	1998	___

Solve each interest-related problem.

75. ***Finding the Present Value of a Future Equipment Purchase*** An equipment rental company is planning to purchase 5 new forklifts in response to increasing demand. If the company is able to earn 6% compounded quarterly on invested money, what lump sum should be invested in order to purchase the forklifts 18 months later at a price of $12,000 each?

76. ***Comparing Medical Cost Increases to Overall Inflation*** The CPI-U for medical care expenses was 260.8 in 2000 and 484.0 in April 2018. The overall CPI-U was 172.2 in 2000 and 250.5 in April 2018.

(a) What would we expect the CPI-U for medical care expenses to be in April 2018 if medical care expenses had increased at the average rate of inflation since 2000?

(b) Compare actual increases in medical care expenses to the overall inflation rate for this time period.

13.2 CONSUMER CREDIT

OBJECTIVES

1 Identify the different types of consumer credit.

2 Determine the interest, payments, and cost for installment loans.

3 Determine account balances and finance charges for revolving loans.

Types of Consumer Credit

Consumer credit takes the form of loans extended to people who borrow money to finance the purchase of cars, furniture, appliances, jewelry, electronics, and other items. In this section, we will discuss two common types of consumer credit.

- An **installment loan** (or **closed-end credit**) involves borrowing a set amount up front and paying a series of equal installments (payments) until the loan is paid off. This type of credit commonly is used to finance the purchase of cars, furniture, and appliances.

- A **revolving loan** (or **open-end credit**) involves borrowing up to a **credit limit,** with no fixed number of payments—the consumer continues paying until no balance is owed. Additional credit often is extended (as long as the limit is not exceeded) before the initial amount is paid off. Examples of open-end credit include most department store charge accounts and bank charge cards such as MasterCard and VISA.

Installment Loans

Installment loans, set up under closed-end credit, often are based on **add-on interest.** This means that interest is calculated by the simple interest formula $I = Prt$, and we simply "add on" this amount of interest to the principal borrowed to arrive at the total debt (or amount to be repaid).

$$\text{Amount to be repaid} = \overset{\substack{\text{Amount} \\ \text{borrowed} \\ \downarrow}}{P} + \overset{\substack{\text{Interest} \\ \text{due} \\ \downarrow}}{Prt}$$

$$= P(1 + rt)$$

The total debt is then paid in equal periodic installments (usually monthly) to be made over the t years.

EXAMPLE 1 Repaying an Installment Loan

Ynez buys $5400 worth of furniture and appliances for her first apartment. She pays $1100 down and agrees to pay the balance at a 5% add-on rate for 2 years. Find

(a) the total amount to be repaid,   **(b)** the monthly payment, and

(c) the total cost of the purchases, including finance charges.

Solution

(a) Amount to be financed = $\overset{\text{Purchase amount}}{\$5400} - \overset{\text{Down payment}}{\$1100}$

$= \$4300$

Amount to be repaid $= P(1 + rt)$ Future value formula

$= \$4300(1 + 0.05 \cdot 2)$ Substitute known values.

$= \$4730$ Calculate.

(b) Monthly payment $= \dfrac{\$4730}{24}$ ← Amount to be repaid
← Number of payments

$= \$197.08$

(c) Total cost of purchases $= \overset{\text{Down payment}}{\$1100} + \overset{\text{Loan repayment}}{\$4730}$

$= \$5830$

Ynez will end up paying $5830, which is about 8% more than the price tag total.

Notice that the repayment amount in **Example 1,** $P(1 + rt)$, was the same as the final amount A in a savings account paying a simple interest rate r for t years. (See the previous section.) But in the case of savings, the bank keeps all of your money for the entire time period. In **Example 1,** Ynez did not keep the full principal amount for the full time period. She repaid it in 24 monthly installments. The 5% add-on rate turns out to be equivalent to a much higher "true annual interest" rate.

Revolving Loans

With a typical department store account, or bank card, a credit limit is established initially, and the consumer can make many purchases during a month (up to the credit limit). The required monthly payment can vary from a set minimum (which may depend on the account balance) up to the full balance.

At the end of each billing period (normally once a month), the customer receives an **itemized billing,** a statement listing purchases and cash advances, the total balance owed, the minimum payment required, and perhaps other account information. Any charges beyond cash advanced and cash prices of items purchased are called **finance charges.** Finance charges may include interest, an annual fee, credit insurance coverage, a time payment differential, or carrying charges.

Most revolving credit plans calculate finance charges by the **average daily balance method.** The average daily balance is the *weighted mean* of the various amounts of credit utilized for different parts of the billing period, with the weighting factors being the number of days that each credit amount applied.

Learning the following rhyme will help you in problems like the one found in **Example 2.**

Thirty days hath September,

April, June, and November.

All the rest have thirty-one,

Save February which has twenty-eight days clear,

And twenty-nine in each leap year.

EXAMPLE 2 **Using the Average Daily Balance Method**

The activity in Kaede's MasterCard account for one billing period is shown below. If the previous balance (on March 3) was $209.46, and the bank charges 1.3% per month on the average daily balance, find

(a) the average daily balance for the next billing (April 3),

(b) the finance charge to appear on the April 3 billing, and

(c) the account balance on April 3.

March 3	Billing date	
March 12	Payment	$50.00
March 17	Clothes	$28.46
March 20	Internet order	$31.22
April 1	Auto parts	$59.10

Solution

(a) First make a table that shows the beginning date of the billing period and the dates of all transactions in the billing period. Along with each of these, compute the running balance on that date.

Date	Running Balance
March 3	$209.46
March 12	$209.46 − $50 = $159.46
March 17	$159.46 + $28.46 = $187.92
March 20	$187.92 + $31.22 = $219.14
April 1	$219.14 + $59.10 = $278.24

Next, tabulate the running balance figures, along with the number of days until the balance changed. Multiply each balance amount by the number of days. The sum of these products gives the "sum of the daily balances."

Date	Running Balance	Number of Days Until Balance Changed	$\left(\text{Running Balance}\right) \cdot \left(\text{Number of Days}\right)$
March 3	$209.46	9	$1885.14
March 12	$159.46	5	$797.30
March 17	$187.92	3	$563.76
March 20	$219.14	12	$2629.68
April 1	$278.24	2	$556.48
		Totals: 31	$6432.36

$$\text{Average daily balance} = \frac{\text{Sum of daily balances}}{\text{Days in billing period}} = \frac{\$6432.36}{31} = \$207.50$$

Kaede will pay a finance charge based on the average daily balance of $207.50.

(b) The finance charge for the April 3 billing will be

$$1.3\% \text{ of } \$207.50 = 0.013 \cdot \$207.50 \quad \text{1.3\% = 1.3 × 0.01 = 0.013}$$
$$= \$2.70.$$

(c) The account balance on the April 3 billing will be the latest running balance plus the finance charge.

$$\$278.24 + \$2.70 = \$280.94$$

Other features of revolving credit accounts can be at least as important as the stated interest rate. For example:

1. Is an annual fee charged? If so, how much is it?
2. Is a special "introductory" rate offered? If so, how long will it last?
3. Are there other incentives, such as rebates, credits toward certain purchases, "free" airline miles, or return of interest charges for long-time use?

Credit cards and other revolving loans are relatively expensive credit. The monthly rate of 1.3% in **Example 2** is typical and is equivalent to an annual rate of 15.6%. A single card (like VISA or MasterCard), however, can be a great convenience.

The best practice is *not* to carry a balance from month to month, but to pay the entire new balance by the due date each month. Because purchases made during the month are not billed until the next billing date, and the payment due date may be 20 days or more after the billing date, items can often be charged without actually paying for them for nearly two months. To obtain this "free credit," resist the temptation to buy more than you can pay for by the next payment date.

13.2 EXERCISES

CONCEPT CHECK *Fill in the blank with the correct response.*

1. In the context of consumer credit, *installment* is another word for _____.

2. A car loan is an example of _____ credit.
(open-end/closed-end)

3. The _____ method is used to calculate finance charges for most credit card accounts.

4. The finance charges for a(n) _____ loan are determined when the loan account is opened and are added to the amount borrowed.

Round all monetary answers to the nearest cent unless directed otherwise. Assume that, unless otherwise known, there are 12 months per year, 30 days per month, and 365 days per year.

Financing an Appliance Purchase *Neema bought appliances costing $3795 at a store charging 6% add-on interest. She made a $1000 down payment and agreed to monthly payments over two years.*

5. Find the total amount to be financed.

6. Find the total interest to be paid.

7. Find the total amount to be repaid.

8. Find the monthly payment.

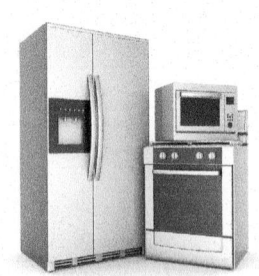

9. Find the total cost, for appliances plus interest.

10. What percent of the original price tag total did the financing cost?

Financing a New Car Purchase *Suppose you want to buy a new car that costs $32,500. You have no cash—only your old car, which is worth $4000 as a trade-in.*

11. How much do you need to finance to buy the new car?

12. The dealer says the interest rate is 4% add-on for 4 years. Find the total interest.

13. Find the total amount to be repaid.

14. Find the monthly payment.

15. Find your total cost, for the new car plus interest.

16. What percent of the original price tag total will the financing cost?

In Exercises 17–22, use the add-on method of calculating interest to find the total interest and the monthly payment.

	Amount of Loan	Length of Loan	Interest Rate
17.	$4500	4 years	4.5%
18.	$2700	2 years	4%
19.	$750	18 months	3.7%
20.	$2450	30 months	4.6%
21.	$535	16 months	5.1%
22.	$1725	29 months	3.4%

Solve each problem.

23. Finding the Monthly Payment for an Add-On Interest Furniture Loan Sanjay and Mira buy $8500 worth of furniture for their new home. They pay $3000 down. The store charges 10% add-on interest. The furniture is to be paid off in 30 monthly payments $(2\frac{1}{2}$ years$)$. Find the monthly payment.

24. Finding the Monthly Payment for an Add-On Interest Auto Loan Find the monthly payment required to pay off an auto loan of $15,780 over 3 years if the add-on interest rate is 3.9%.

25. Finding the Monthly Payment for an Add-On Interest Home Electronics Loan The total purchase price of a new home entertainment system is $14,240. If the down payment is $2900 and the balance is to be financed over 48 months at 6% add-on interest, what is the monthly payment?

26. *Finding the Monthly Payment for an Add-On Interest Loan*
What are the monthly payments that Holly pays on a loan of $1680 for a period of 10 months if 9% add-on interest is charged?

27. *Finding the Amount Borrowed for an Add-On Interest Car Loan*
Elon has misplaced the sales contract for his car and cannot remember the amount he originally financed. He does know that the add-on interest rate was 5.1% and the loan required a total of 36 monthly payments of $314.65 each. How much did Elon borrow (to the nearest dollar)?

28. *Finding an Add-On Interest Rate* Ella is making monthly payments of $259.06 to pay off a $3\frac{1}{2}$-year loan of $9400. What is her add-on interest rate (to the nearest tenth of a percent)?

29. *Finding the Term of an Add-On Interest Loan* How long (in years) will it take Makoto to pay off an $11,000 loan with monthly payments of $236.50 if the add-on interest rate is 5.8%?

30. *Finding the Number of Payments for an Add-On Interest Loan*
How many monthly payments must Jawann make on a $10,000 loan if he pays $417.92 per month and the add-on interest rate is 10.15%?

Finding Finance Charges *Find the finance charge for each charge account. Assume interest is calculated on the average daily balance of the account.*

	Average Daily Balance	Monthly Interest Rate
31.	$249.94	1.4%
32.	$350.75	1.5%
33.	$419.95	1.38%
34.	$450.21	1.26%
35.	$1073.40	1.425%
36.	$1320.42	1.375%

Finding Finance Charges and Account Balances Using the Average Daily Balance Method *For each credit card account, assume one month between billing dates (with the appropriate number of days) and interest of 1.3% per month on the average daily balance. Find* **(a)** *the average daily balance,* **(b)** *the monthly finance charge, and* **(c)** *the account balance for the next billing.*

37. Previous balance: $728.36

May 9	Billing date	
May 17	Payment	$200.00
May 30	Dinner	$46.11
June 3	Theater tickets	$64.50

38. Previous balance: $514.79

January 27	Billing date	
February 9	Candy	$11.08
February 13	Returns	$26.54
February 20	Payment	$59.00
February 25	Repairs	$71.19

39. Previous balance: $462.42

June 11	Billing date	
June 15	Returns	$106.45
June 20	Jewelry	$115.73
June 24	Car rental	$74.19
July 3	Payment	$115.00
July 6	Flowers	$68.49

40. Previous balance: $983.25

August 17	Billing date	
August 21	Internet order	$14.92
August 23	Returns	$25.41
August 27	Beverages	$31.82
August 31	Payment	$108.00
September 9	Returns	$71.14
September 11	Concert tickets	$110.00
September 14	Cash advance	$100.00

Finding Finance Charges *In Exercises 41 and 42, assume no purchases or returns are made.*

41. At the beginning of a 31-day billing period, Alicia has an unpaid balance of $720 on her credit card. Three days before the end of the billing period, she pays $600. Find her finance charge at 1.4% per month using the average daily balance method.

42. Vitaliy's VISA bill dated April 14 shows an unpaid balance of $1070. Five days before May 14, the end of the billing period, Vitaliy makes a payment of $900. Find his finance charge at 1.32% per month using the average daily balance method.

Analyzing a "90 Days Same as Cash" Offer *One version of the "90 Days Same as Cash" promotion was offered by a "major purchase card," which established an account charging 1.3875% interest per month on the account balance. Interest charges are added to the balance each month, becoming part of the balance on which interest is computed the next month. If you pay off the original purchase charge within 3 months, all interest charges are canceled. Otherwise, you are liable for all the interest. Suppose you purchase $3900 worth of carpeting under this plan.*

43. Find the interest charge added to the account balance at the end of

(a) the first month,

(b) the second month,

(c) the third month.

44. Suppose you pay off the account 1 day late (3 months plus 1 day). What total interest amount must you pay? (Do not include interest for the one extra day.)

45. Treating the 3 months as $\frac{1}{4}$ year, find the equivalent simple interest rate for this purchase (to the nearest tenth of a percent).

Various Charges of a Bank Card Account *Aimee's bank card account charges 1.3% per month on the average daily balance, as well as the following special fees:*

Cash advance fee:	*The greater of 5% and $10*
Late payment fee:	*$35*
Over-the-credit-limit fee:	*$25*

In the month of June, Aimee's average daily balance was $2846. She was on vacation during the month and did not get her account payment in on time, which resulted in a late payment and, as a consequence, charges accumulated to a sum above her credit limit. She also used her card for three $200 cash advances while on vacation. Find the following based on account transactions in that month.

46. interest charges to the account

47. special fees charged to the account

Write out your response to each of the following.

48. Is it possible to use a bank credit card for your purchases without paying anything for credit? If so, explain how.

49. Go online to research several different bank card programs. Compare their features (including those in fine print), and explain which deal would be best for you, and why.

50. Research and explain the difference, if any, between a "credit" card and a "debit" card.

51. Many charge card offers include the option of purchasing credit insurance coverage, whereby the insurer would make your monthly payments if you became disabled and could not work and/or would pay off the account balance if you died. Find out the details on at least one such offer, and discuss why you would or would not accept it.

52. Make a list of "special incentives" offered by bank cards you are familiar with, and briefly describe the pros and cons of each one.

53. One bank offers a credit card with "0% Intro APR—Balance Transfers" for 15 months, followed by a "Regular APR" ranging from 16.49% to 25.24%, and a "Balance Transfer Fee of 5%." Suppose you have a substantial balance on an existing card and are unable to make the next payment at this time. Would the new card give you a "free" 15-month extension on your balance if you transfered it from the old card? Why do you suppose there is a range of regular interest rates? Discuss why it would (or would not) be a good idea to apply for the second card.

54. Recall a car-buying experience you have had, or visit a new-car dealer and interview a salesperson. Describe the procedure involved in purchasing a car on credit.

Comparing Bank Card Accounts *Kiara is considering two bank card offers that are the same in all respects except for the following:*

> *Bank A charges no annual fee and charges monthly interest of 1.38% on the unpaid balance.*

> *Bank B charges a $30 annual fee and monthly interest of 1.21% on the unpaid balance.*

From her records, Kiara has found that the unpaid balance she tends to carry from month to month is quite consistent and averages $900.

55. Estimate her total yearly cost to use the card if she chooses the card from

   (a) Bank A.   **(b)** Bank B.

56. Which card is her better choice?

13.3   TRUTH IN LENDING

OBJECTIVES

1 Determine the annual percentage rate for different types of loans.

2 Calculate unearned interest.

Annual Percentage Rate (APR)

The Consumer Credit Protection Act, passed in 1968, has commonly been known as the **Truth in Lending Act.** The law addressed two major issues:

  1. How can we tell the true annual interest rate a lender is charging?

  2. How much of the finance charge are we entitled to save if we decide to pay off a loan sooner than originally scheduled?

Free consumer information is available from the Federal Trade Commission at www.consumer.ftc.gov. For example, you can find out about "Money & Credit," "Homes & Mortgages," "Jobs & Making Money," "Privacy & Identity," and many other consumer topics. You can also file consumer complaints and report identity theft using online forms.

Question 1 arose because lenders were computing and describing the interest they charged in several different ways. For example, how does 2% per month at Target compare to 9% per year add-on interest at a furniture store? Truth in Lending standardized the so-called true annual interest rate, or **annual percentage rate,** commonly denoted **APR.** All sellers (car dealers, stores, banks, insurance agents, credit card companies, and the like) must disclose the APR when asked, and the written contract must state the APR in all cases. This enables a borrower to more easily compare the true costs of different loans.

Theoretically, a borrower should not need to calculate APR, but it is possible to verify the value stated by the lender. Since the formulas for finding APR are quite involved, it is easiest to use a table or a financial calculator. **Table 5** identifies APR values to the nearest half percent from 4.0% to 12.0%, for loans requiring *monthly* payments and extending over the most common lengths for consumer loans from 6 to 60 months.

Table 5 relates the following three quantities.

APR = true annual interest rate (shown across the top)

n = total number of scheduled monthly payments (shown down the left side)

h = finance charge per $100 of amount financed (shown in the body of the table)

Table 5 Annual Percentage Rate (APR) for Monthly Payment Loans

Number of Monthly Payments (n)	Annual Percentage Rate (APR)																
	4.0%	4.5%	5.0%	5.5%	6.0%	6.5%	7.0%	7.5%	8.0%	8.5%	9.0%	9.5%	10.0%	10.5%	11.0%	11.5%	12.0%
	Finance Charge per $100 of Amount Financed (h)																
6	1.17	1.32	1.46	1.61	1.76	1.90	2.05	2.20	2.35	2.49	2.64	2.79	2.94	3.08	3.23	3.38	3.53
12	2.18	2.45	2.73	3.00	3.28	3.56	3.83	4.11	4.39	4.66	4.94	5.22	5.50	5.78	6.06	6.34	6.62
18	3.20	3.60	4.00	4.41	4.82	5.22	5.63	6.04	6.45	6.86	7.28	7.69	8.10	8.52	8.93	9.35	9.77
24	4.22	4.75	5.29	5.83	6.37	6.91	7.45	8.00	8.55	9.09	9.64	10.19	10.75	11.30	11.86	12.42	12.98
30	5.25	5.92	6.59	7.26	7.94	8.61	9.30	9.98	10.66	11.35	12.04	12.74	13.43	14.13	14.83	15.54	16.24
36	6.29	7.09	7.90	8.71	9.52	10.34	11.16	11.98	12.81	13.64	14.48	15.32	16.16	17.01	17.86	18.71	19.57
48	8.38	9.46	10.54	11.63	12.73	13.83	14.94	16.06	17.18	18.31	19.45	20.59	21.74	22.90	24.06	25.23	26.40
60	10.50	11.86	13.23	14.61	16.00	17.40	18.81	20.23	21.66	23.10	24.55	26.01	27.48	28.96	30.45	31.96	33.47

EXAMPLE 1 **Finding the APR for an Add-On Loan**

Recall that Ynez (in **Example 1** of the previous section) paid $1100 down on a $5400 purchase and agreed to pay the balance at a 5% add-on rate for 2 years. Find the APR for her loan.

Solution

As shown previously, the total amount financed was

Purchase price — Down payment

$5400 − $1100 = $4300.

The finance charge (interest) was

$$I = Prt = \$4300 \cdot 0.05 \cdot 2 = \$430.$$

Next find the finance charge per $100 of the amount financed. To do this, divide the finance charge by the amount financed, and then multiply by $100.

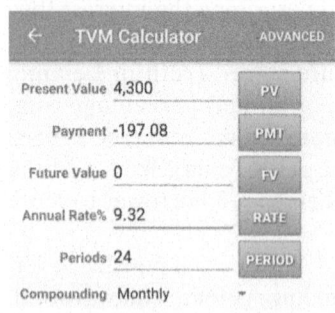

Financial calculator apps give another option (besides tables) for calculating APR. For **Example 1,** dividing the total amount to be repaid by the number of payments gives the monthly payment.

$$\frac{\$4300 + \$430}{24} = \$197.08$$

This amount can then be entered (as negative cash flow) into the financial calculator, and the rate calculated. We see here that the APR is 9.32%. (App used: TVM Calculator)

$$\left(\begin{array}{c}\text{Finance charge per}\\ \$100 \text{ financed}\end{array}\right) = \frac{\text{Finance charge}}{\text{Amount financed}} \cdot \$100$$

$$= \frac{\$430}{\$4300} \cdot \$100, \quad \text{or} \quad \$10$$

This amount, \$10, represents h, the finance charge per \$100 of the amount financed. Because the loan was to be paid over 24 months, look down to the "24 monthly payments" row of **Table 5** ($n = 24$). Then look across the table for the h-value closest to \$10.00, which is \$10.19. From that point, read up the column to find the APR, 9.5% (to the nearest half percent). In this case, a 5% add-on rate is equivalent to an APR of 9.5%.

The next example illustrates the use of a TVM (time value of money) Solver, an application available on many graphing calculators, including the TI-84 Plus C.

EXAMPLE 2 **Finding the APR for a Car Loan**

After making a down payment on her new car, Shanice still owed \$14,652. She agreed to repay the balance in 48 monthly payments of \$340.76 each. Find the APR of her loan to the nearest 0.01%.

Solution

Refer to the calculator screens in the margin. The first shows the APPS menu with the Finance option highlighted. Pressing ENTER will launch the second screen, which shows the financial applications menu. The TVM Solver is highlighted. Pressing ENTER again will launch the TVM solver. Enter the appropriate values as follows.

N = 48	Number of payments
PV = \$14,652	Present value of loan
PMT = −\$340.76	Monthly payment (negative cashflow)
P/Y = 12	Number of payments per year
C/Y = 12	Number of compoundings per year
PMT: END	Payments occur at the end of each month.

Finally, position the cursor to highlight the I% field, and press ALPHA SOLVE to solve for the APR.

APR ≈ 5.50%   I% is the annual interest rate.

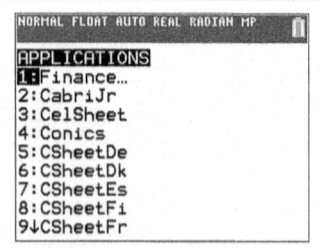

The arrow points to the APR value calculated in **Example 2.**

Financial calculators often are preferred to tables when more precision is desired, or when APR values or finance charge values fall between values in tables.

Unearned Interest

Question 2 at the beginning of this section arises when a borrower decides to pay off an installment loan earlier than originally scheduled. In such a case, the lender has not really "earned" the full finance charge originally disclosed. In such cases, the amount by which the original finance charge is reduced is the **unearned interest.** We will discuss two methods of calculating unearned interest, the **actuarial method** and the **rule of 78.** The Truth in Lending Act requires that the method for calculating this refund (or reduction) of finance charge (in case of an early payoff) be disclosed at the time the loan is initiated. Whichever method is used, the borrower may not, in fact, save all the unearned interest, since the lender is entitled to impose an **early payment penalty** to recover certain costs. A lender's intention to impose such a penalty in case of early payment also must be disclosed at initiation of the loan.

Rights and responsibilities apply to all credit accounts, and the consumer should read all disclosures provided by the lender. *The Fair Credit Billing Act* and *The Fair Credit Reporting Act* regulate, among other things, procedures for billing and for disputing bills, and for providing and disputing personal information on consumers.

If you have ever applied for a charge account, a personal loan, insurance, or a job, then information about where you work and live, how you pay your bills, and whether you've been sued or arrested or have filed for bankruptcy, appears in the files of Consumer Reporting Agencies (CRAs), which sell that information to creditors, employers, insurers, and other businesses.

For more detailed information on these and many other consumer issues, you may want to consult the Web site www.usa.gov/consumer.

UNEARNED INTEREST—ACTUARIAL METHOD

For an installment loan requiring *monthly* payments, which is paid off earlier than originally scheduled, let

R = regular monthly payment,

k = remaining number of scheduled payments (*after* current payment), and

h = finance charge per \$100, corresponding to a loan with the same APR and k monthly payments.

Then the **unearned interest, u,** is calculated as follows.

$$u = kR\left(\frac{h}{\$100 + h}\right)$$

Once the unearned interest u is calculated (by any method), the amount required to pay off the loan early is easily found. It consists of the present regular payment due, plus k additional future payments, minus the unearned interest.

PAYOFF AMOUNT

An installment loan requiring regular monthly payments R can be paid off early, along with the current payment. If the original loan had k additional payments scheduled (after the current payment), and the unearned interest is u, then, disregarding any possible prepayment penalty, the **payoff amount** is calculated as follows.

$$\text{Payoff amount} = (k + 1)R - u$$

EXAMPLE 3 Finding Early Payoff Amount (Actuarial Method)

Shanice got an unexpected pay raise and wanted to pay off her car loan of **Example 2** at the end of 3 years rather than paying for 4 years as originally agreed.

(a) Find the unearned interest (the amount she will save by retiring the loan early).

(b) Find the "payoff amount" (the amount required to pay off the loan at the end of 3 years).

Solution

(a) From **Example 2,** recall that $R = \$340.76$ and APR $= 5.5\%$. The current payment is payment number 36, so

$$k = 48 - 36 = 12.$$

Use **Table 5,** with 12 payments and APR 5.5%, to obtain $h = \$3.00$. Then use the actuarial method formula.

$$u = 12 \cdot \$340.76\left(\frac{\$3.00}{\$100 + \$3.00}\right) = \$119.10$$

By this method, Shanice will save \$119.10 in interest by retiring the loan early.

(b) The payoff amount is found by using the appropriate formula.

$$\text{Payoff amount} = (12 + 1)\$340.76 - \$119.10 = \$4310.78$$

The required payoff amount at the end of 3 years is \$4310.78.

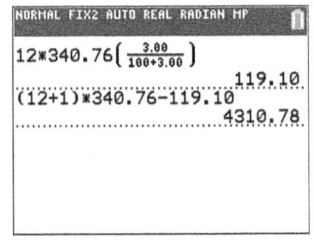

These are the computations required in the solution of **Example 3.**

The actuarial method, being based on the APR value, probably is the better method. However, some lenders historically used a second method, the **rule of 78.**

UNEARNED INTEREST—RULE OF 78

For a closed-end loan requiring *monthly* payments, which is paid off earlier than originally scheduled, let

F = original finance charge,

n = number of payments originally scheduled, and

k = remaining number of scheduled payments (*after* current payment).

Then the **unearned interest, *u*,** is calculated as follows.

$$u = \frac{k(k + 1)}{n(n + 1)} \cdot F$$

EXAMPLE 4   **Finding Early Payoff Amount (Rule of 78)**

Again assume that the loan in **Example 2** is paid off at the time of the thirty-sixth monthly payment. This time, however, instead of the actuarial method, use the rule of 78 to find

(a) the unearned interest, and

(b) the payoff amount.

Solution

(a) Begin by calculating the finance charge from **Example 2.**

$$\underset{\substack{\uparrow \\ \text{Total} \\ \text{payments}}}{} \quad \underset{\substack{\uparrow \\ \text{Amount} \\ \text{financed}}}{}$$

$$F = 48 \cdot \$340.76 - \$14,652 = \$1704.48$$

> Multiply first. Then subtract.

Use the formula with $k = 12$ and $n = 48$.

$$u = \frac{12(12 + 1)}{48(48 + 1)} \cdot \$1704.48 = \$113.05 \qquad \text{Rule of 78}$$

By the rule of 78, Shanice will save $113.05 in interest, which is $6.05 *less* than her savings by the actuarial method.

(b)   Payoff amount $= (12 + 1)\$340.76 - \$113.05 = \$4316.83$

The payoff amount is $4316.83, which is $6.05 *more* than the payoff amount calculated by the actuarial method in **Example 3.**

NORMAL FIX2 AUTO REAL RADIAN MP

48*340.76-14652
                        1704.48
$\frac{12(12+1)}{48(48+1)}$*1704.48
                         113.05
(12+1)340.76-113.05
                        4316.83

These are the computations required in the solution of **Example 4.**

When the rule of 78 was first introduced into financial law (by the Indiana legislature in 1935), loans were ordinarily written for 1 year or less, interest rates were relatively low, and loan amounts were less than they tend to be today. For these reasons the rule of 78 was acceptably accurate then. Today, however, with very accurate tables and/or calculators readily available, the rule of 78 is used much less often than previously.

Suppose we want to compute unearned interest accurately (so we don't trust the rule of 78), but the APR value, or the number of scheduled payments, or the number of remaining payments (or at least one of the three) is not included in **Table 5.** Then what? Actually, in the actuarial method, we can evaluate h (the finance charge per \$100 financed) using the same formula that was used to generate **Table 5.**

FINANCE CHARGE PER \$100 FINANCED

If an installment loan requires n equal monthly payments and APR denotes the true annual interest rate for the loan (as a decimal), then h, the **finance charge per \$100 financed,** is calculated as follows.

$$h = \frac{n \cdot \frac{\text{APR}}{12} \cdot \$100}{1 - \left(1 + \frac{\text{APR}}{12}\right)^{-n}} - \$100$$

EXAMPLE 5 Finding Unearned Interest and Early Payoff Amount

Hugo borrowed \$5000 to pay for music equipment for his band. His loan contract states an APR of 5.8% and stipulates 28 monthly payments of \$153.09 each. Hugo decides to pay the loan in full at the time of his nineteenth scheduled payment. Find

(a) the unearned interest, and

(b) the payoff amount.

Solution

(a) First find h from the finance charge formula just given.

Remember to use the remaining number of payments, $28 - 19 = 9$, as the value of n.

$$h = \frac{9\left(\frac{0.058}{12}\right)(\$100)}{1 - \left(1 + \frac{0.058}{12}\right)^{-9}} - \$100 = \$2.43$$

Next use the actuarial formula for unearned interest u.

Regular monthly payment: $R = \$153.09$

Remaining number of payments: $k = 28 - 19 = 9$

Finance charge per \$100: $h = \$2.43$

$$u = 9 \cdot \$153.09 \cdot \frac{\$2.43}{\$100 + \$2.43} = \$32.69$$

The amount of interest Hugo will save is \$32.69.

(b) Payoff amount $= (9 + 1)(\$153.09) - \$32.69 = \$1498.21.$

To pay off the loan at the time of his nineteenth scheduled payment, Hugo must pay \$1498.21.

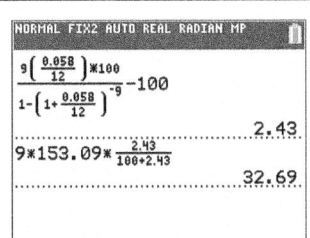

These are the computations required in the solution of **Example 5(a).**

13.3   EXERCISES

CONCEPT CHECK *Write* true *or* false *for each statement.*

1. The Truth in Lending Act requires lenders to penalize borrowers for early payoff of loans.

2. Unearned interest is the amount of interest *not* paid by the borrower when a loan is retired early.

3. The rule of 78 results in more accurate calculations of unearned interest than the actuarial method does.

4. When compared to the rule of 78, the actuarial method of calculation generally results in more savings for the borrower.

Round all monetary answers to the nearest cent unless otherwise directed.

Finding True Annual Interest Rate *Find the APR (true annual interest rate), to the nearest half percent, for each loan.*

	Amount Financed	Finance Charge	Number of Monthly Payments
5.	$1000	$38	12
6.	$1700	$81	24
7.	$6600	$750	30
8.	$5900	$943	60

Finding the Monthly Payment *Find the monthly payment for each loan.*

	Purchase Price	Down Payment	Finance Charge	Number of Monthly Payments
9.	$3500	$500	$275	24
10.	$4750	$450	$750	36
11.	$3850	$300	$800	48
12.	$9500	$1500	$1400	60

Finding True Annual Interest Rate *Find the APR (true annual interest rate), to the nearest 0.01%, for each loan.*

	Purchase Price	Down Payment	Add-On Interest Rate	Number of Payments
13.	$4190	$390	3.5%	12
14.	$3250	$750	7%	36
15.	$7480	$2200	5%	18
16.	$12,800	$4500	3.5%	48

Unearned Interest by the Actuarial Method *Each loan was paid in full before its due date.* **(a)** *Obtain the value of h from* **Table 5.** *Then* **(b)** *use the actuarial method to find the amount of unearned interest, and* **(c)** *find the payoff amount.*

	Regular Monthly Payment	APR	Remaining Number of Scheduled Payments after Payoff
17.	$346.70	4.0%	18
18.	$783.50	5.5%	12
19.	$595.80	7.0%	6
20.	$314.50	9.0%	24

Finding Finance Charge and True Annual Interest Rate *For each loan, find* **(a)** *the finance charge, and* **(b)** *the APR. In Exercises 21–24, use Table 5 and find the APR to the nearest 0.5%. In Exercises 25–28, use a TVM solver to find the APR to the nearest 0.01%. (Refer to* **Example 2***.)*

21. Joel financed a $1990 computer with 24 monthly payments of $91.50 each.

22. Margey bought a horse trailer for $6090. She paid $1240 down and paid the remainder at $175.50 per month for $2\frac{1}{2}$ years.

23. Rajesh still owed $2000 on his new garden tractor after the down payment. He agreed to pay monthly payments for 18 months at 4.6% add-on interest.

24. Amal paid off a $15,000 car loan over 3 years with monthly payments of $466.59 each.

25. Sebastian purchased a tablet for $650 and he will make 12 monthly payments of $55.87 each.

26. Tristan purchased a boat for $6995. He made a $1500 down payment, and he will make monthly payments of $142.59 for $3\frac{1}{2}$ years.

27. Jada borrowed $2895 to purchase a home entertainment system. She agreed to monthly payments of $81.06 for 40 months.

28. Kylie bought an elliptical exercise machine for $3999 and she will make 24 monthly payments of $175.08 each.

Comparing the Actuarial Method and the Rule of 78 for Unearned Interest *Each loan in Exercises 29–32 on the next page was paid off early. Find the unearned interest by* **(a)** *the actuarial method, and* **(b)** *the rule of 78.*

	Amount Financed	Regular Monthly Payments	Total Number of Payments Scheduled	Remaining Number of Scheduled Payments after Payoff
29.	$3310	$192.75	18	6
30.	$10,230	$237.91	48	12
31.	$29,850	$556.49	60	12
32.	$16,730	$516.57	36	18

Unearned Interest by the Actuarial Method *Each loan was paid in full before its due date.* **(a)** *Obtain the value of h from the appropriate formula. Then* **(b)** *use the actuarial method to find the amount of unearned interest, and* **(c)** *find the payoff amount.*

	Regular Monthly Payment	APR	Remaining Number of Scheduled Payments after Payoff
33.	$212	4.7%	4
34.	$575	5.9%	8

Comparing Loan Choices *Nadia needs to borrow $6000 to pay for NHL season tickets for her family. She can borrow the amount from a finance company (at 3.5% add-on interest for 3 years) or from the credit union (36 monthly payments of $183.12 each). Use this information for Exercises 35–39.*

35. Find the APR (to the nearest 0.01%) for each loan, and decide which one is Nadia's better choice.

36. Nadia takes the credit union loan. At the time of her thirtieth payment she pays it off. If the credit union uses the rule of 78 for computing unearned interest, how much will she save by paying in full now?

37. What would Nadia save in interest if she paid in full at the time of the thirtieth payment and the credit union used the actuarial method for computing unearned interest?

38. Under the conditions of **Exercise 37,** what amount must Nadia come up with to pay off her loan?

39. Suppose the credit union uses the rule of 78 and imposes an early payoff penalty of 1% of the original finance charge per month of early payment. For example, if Nadia pays the loan off 9 months early, then the penalty will be 9% of the original finance charge. Find the least value of k that would result in any savings for Nadia.

40. Describe why, in **Example 1,** the APR and the add-on rate differ. Which one is more legitimate? Why?

Approximating the APR of an Add-On Rate *To convert an add-on interest rate to its corresponding APR, some people recommend using the formula*

$$APR = \frac{2n}{n+1} \cdot r,$$

where r is add-on rate and n is total number of payments.

41. Apply the given formula to calculate the APR (to the nearest half percent) for the loan of **Example 1** ($r = 0.05, n = 24$).

42. Compare your APR value in **Exercise 41** to the value in **Example 1.** What do you conclude?

The Rule of 78 with Prepayment Penalty *A certain retailer's credit contract designates the rule of 78 for computing unearned interest and imposes a "prepayment penalty." In case of any payoff earlier than the due date, the lender will charge an additional 10% of the original finance charge. Find the least value of k (remaining payments after payoff) that would result in any net savings in each case.*

43. 24 payments originally scheduled

44. 60 payments originally scheduled

Write a short answer for each problem.

45. Why might a lender be justified in imposing a prepayment penalty?

46. Discuss reasons why a borrower might want to pay off a loan early.

47. Find out what federal agency you can contact if you have questions about compliance with the Truth in Lending Act.

48. Study the table below, which pertains to a 12-month loan. The column-3 entries are designed so that they are in the same ratios as the column-2 entries but will add up to 1 because their denominators are all equal to

$$1 + 2 + 3 + 4 + 5 + \ldots + 12 = \frac{12 \cdot 13}{2} = 78.$$

(This is the origin of the term "rule of 78.")

Month	Fraction of Loan Principal Used by Borrower	Fraction of Finance Charge Owed
1	12/12	12/78
2	11/12	11/78
3	10/12	10/78
4	9/12	9/78
5	8/12	8/78
6	7/12	7/78
7	6/12	6/78
8	5/12	5/78
9	4/12	4/78
10	3/12	3/78
11	2/12	2/78
12	1/12	1/78
		78/78 = 1

Suppose the loan is paid in full after eight months. Use the table to determine the unearned fraction of the total finance charge.

49. Find the fraction of unearned interest of **Exercise 48** by using the rule of 78 formula.

Understanding the Actuarial Method *The actuarial method of computing unearned interest assumes that, throughout the life of the loan, the borrower is paying interest at the rate given by the APR for money actually being used by the borrower. When contemplating complete payoff along with the current payment, think of k future payments as applying to a separate loan with the same APR and of h as being the finance charge per $100 of that loan. Refer to the following formula.*

$$u = kR \left(\frac{h}{\$100 + h} \right)$$

50. Describe in words the quantity represented by

$$\frac{h}{\$100 + h}.$$

51. Describe in words the quantity represented by kR.

52. Explain why the product of the two quantities above represents unearned interest.

13.4 THE COSTS AND ADVANTAGES OF HOME OWNERSHIP

OBJECTIVES

1 Explore the characteristics of fixed-rate mortgages.

2 Explore the characteristics of adjustable-rate mortgages.

3 Compute the total closing costs associated with mortgages.

4 Identify the recurring costs of home ownership.

Fixed-Rate Mortgages

For many decades, home ownership has been considered a centerpiece of the "American dream." For most people, a home represents the largest purchase of their lifetime, and it is certainly worth careful consideration.

A loan for a substantial amount, extending over a lengthy time interval (typically up to 30 years), for the purpose of buying a home or other property or real estate, and for which the property is pledged as security for the loan, is a **mortgage.** (In some areas, a mortgage may also be called a **deed of trust** or a **security deed.**)

- The time until final payoff is the **term** of the mortgage.
- The portion of the purchase price of the home that the buyer pays initially is the **down payment.**
- The **principal amount of the mortgage** (the amount borrowed) is found by subtracting the down payment from the purchase price.

With a **fixed-rate mortgage,** the interest rate remains constant throughout the term, and the initial principal balance, together with interest due on the loan, is repaid to the lender through regular (constant) periodic (we assume monthly) payments. This process is called **amortizing** the loan. The regular monthly payment needed to amortize a loan depends on the amount financed, the term of the loan, and the interest rate, according to the formula on the next page.

REGULAR MONTHLY PAYMENT

The **regular monthly payment** required to repay a loan of P dollars, together with interest at an annual rate r, over a term of t years, is calculated as follows.

$$R = \frac{P\left(\frac{r}{12}\right)}{1 - \left(\frac{12}{12 + r}\right)^{12t}}$$

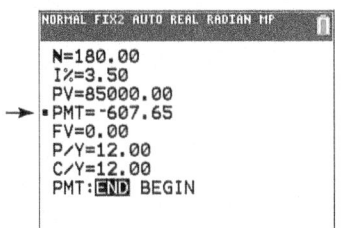

NORMAL FIX2 AUTO REAL RADIAN MP
N=180.00
I%=3.50
PV=85000.00
→ •PMT=-607.65
FV=0.00
P/Y=12.00
C/Y=12.00
PMT:**END** BEGIN

The arrow indicates a payment of $607.65, supporting the result of **Example 1.**

EXAMPLE 1 Using a Formula to Find a Monthly Mortgage Payment

Find the monthly payment necessary to amortize an $85,000 mortgage at 3.5% annual interest for 15 years.

Solution

$$R = \frac{\$85,000\left(\frac{0.035}{12}\right)}{1 - \left(\frac{12}{12 + 0.035}\right)^{(12)(15)}} = \$607.65$$

The amortization of the loan will require monthly payments of $607.65 for 15 years.

There are other methods for calculating monthly payments. The TVM solver of a financial calculator is one option, and tables provide another. **Table 6** gives payment values (per $1000 principal) for typical ranges of mortgage terms and interest rates.

Table 6 Monthly Payments to Repay Principal and Interest on a $1000 Mortgage

Annual rate (r)	Term of Mortgage (years) (t)						
	5	10	15	20	25	30	40
3.0%	17.96869	9.65607	6.90582	5.54598	4.74211	4.21604	3.57984
3.5%	18.19174	9.88859	7.14883	5.79960	5.00624	4.49045	3.87391
4.0%	18.41652	10.12451	7.39688	6.05980	5.27837	4.77415	4.17938
4.5%	18.64302	10.36384	7.64993	6.32649	5.55832	5.06685	4.49563
5.0%	18.87123	10.60655	7.90794	6.59956	5.84590	5.36822	4.82197
5.5%	19.10116	10.85263	8.17083	6.87887	6.14087	5.67789	5.15770
6.0%	19.33280	11.10205	8.43857	7.16431	6.44301	5.99551	5.50214
6.5%	19.56615	11.35480	8.71107	7.45573	6.75207	6.32068	5.85457
7.0%	19.80120	11.61085	8.98828	7.75299	7.06779	6.65302	6.21431
7.5%	20.03795	11.87018	9.27012	8.05593	7.38991	6.99215	6.58071
8.0%	20.27639	12.13276	9.55652	8.36440	7.71816	7.33765	6.95312
8.5%	20.51653	12.39857	9.84740	8.67823	8.05227	7.68913	7.33094
9.0%	20.75836	12.66758	10.14267	8.99726	8.39196	8.04623	7.71361
9.5%	21.00186	12.93976	10.44225	9.32131	8.73697	8.40854	8.10062
10.0%	21.24704	13.21507	10.74605	9.65022	9.08701	8.77572	8.49146
10.5%	21.49390	13.49350	11.05399	9.98380	9.44182	9.14739	8.88570
11.0%	21.74242	13.77500	11.36597	10.32188	9.80113	9.52323	9.28294

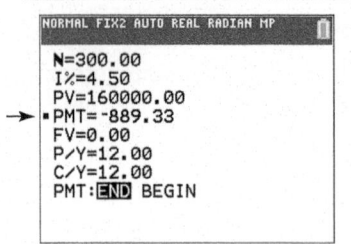

```
NORMAL FIX2 AUTO REAL RADIAN MP
  N=300.00
  I%=4.50
  PV=160000.00
▸■PMT=-889.33
  FV=0.00
  P/Y=12.00
  C/Y=12.00
  PMT:END BEGIN
```

Under the conditions of **Example 2,** the monthly payment is $889.33. Compare with the table method.

EXAMPLE 2 Using a Table to Find a Monthly Mortgage Payment

Find the monthly payment necessary to amortize a $160,000 loan at 4.5% for 25 years.

Solution

In **Table 6,** read down to the 4.5% row and across to the column for 25 years, to find the entry 5.55832. As this is the monthly payment amount needed to amortize a loan of $1000, and our loan is for $160,000, our required monthly payment is

$$160 \cdot \$5.55832 = \$889.33.$$

So that the borrower pays interest only on the money actually owed in a month, interest on real-estate loans is computed on the decreasing balance of the loan. Each equal monthly payment is first applied toward interest for the previous month. The remainder of the payment is then applied toward reduction of the principal amount owed.

Once the regular monthly payment has been determined, as in **Examples 1 and 2,** an **amortization schedule** (or **repayment schedule**) can be generated. It will show the allotment of payments for interest and principal, and the principal balance, for one or more months during the life of the loan. Tables showing these breakdowns are available from lenders or can be produced on a computer spreadsheet. The following steps demonstrate how the computations work.

Step 1 $\text{Interest for the month} = \left(\begin{array}{c}\text{Old balance}\\\text{of principal}\end{array}\right)\left(\begin{array}{c}\text{Annual}\\\text{interest rate}\end{array}\right)\left(\dfrac{1}{12}\text{year}\right)$

Step 2 $\text{Payment on principal} = \left(\begin{array}{c}\text{Monthly}\\\text{payment}\end{array}\right) - \left(\begin{array}{c}\text{Interest for}\\\text{the month}\end{array}\right)$

Step 3 $\text{New balance of principal} = \left(\begin{array}{c}\text{Old balance}\\\text{of principal}\end{array}\right) - \left(\begin{array}{c}\text{Payment on}\\\text{principal}\end{array}\right)$

This sequence of steps is done for each month. The new balance obtained in Step 3 becomes the Step 1 old balance for the next month.

Longer-term mortgages have become more popular in recent years. In fact, 40-year mortgages are not uncommon today. Extending the term may help a buyer "afford" a more expensive home. For example, the monthly payments for principal and interest on a $200,000 mortgage at 4% APR would be $954.83 for a 30-yr term, but only $835.88 for a 40-yr term. In some cases, this may be advantageous, but the ultimate effects of a longer-term loan should be considered. Over the lives of the two loans mentioned above, the total payments would be as follows.

    30-yr term: $343,739.01

    40-yr term: $401,220.93

Which loan would you prefer?

EXAMPLE 3 Preparing an Amortization Schedule

The Jacksons have a $120,000 mortgage with a term of 30 years and a 4.0% interest rate. Prepare an amortization schedule for the first 2 months of their mortgage.

Solution

First get the monthly payment. We use **Table 6.** (You could also use the formula.)

Mortgage amount in $1000s   Intersection of 4.0% row with 30-year column in **Table 6**

$$R = 120 \quad \cdot \quad \$4.77415 = \$572.90$$

Now apply Steps 1–3.

Step 1 $\text{Interest for the month} = \$120,000(0.04)\left(\dfrac{1}{12}\right) = \$400$

Step 2 $\text{Payment on principal} = \$572.90 - \$400 = \$172.90$

Step 3 $\text{New balance of principal} = \$120,000 - \$172.90 = \$119,827.10$

Starting with an old balance of $119,827.10, repeat the steps for the second month.

Step 1 Interest for the month = $119,827.10$(0.04)$\left(\frac{1}{12}\right)$ = \$399.42

Step 2 Payment on principal = \$572.90 − \$399.42 = \$173.48

Step 3 New balance of principal = \$119,827.10 − \$173.48 = **\$119,653.62**

These calculations are summarized in **Table 7.**

Table 7 Amortization Schedule

Payment Number	Interest Payment	Principal Payment	Balance of Principal
			\$120,000.00
1	\$400.00	\$172.90	\$119,827.10
2	\$399.42	\$173.48	**\$119,653.62**

Prevailing mortgage interest rates have varied considerably over the years. **Table 8** shows portions of the amortization schedule for the Jacksons' loan of **Example 3** and shows what the corresponding values would have been had their interest rate been 12.5%. (Rates that high have not been seen since the mid-1980s.) Notice how much interest is involved in this home mortgage. At the (low) 4.0% rate, \$86,243.41 in interest was paid, along with the \$120,000 principal. At a rate of 12.5%, the interest alone would total the huge sum of \$341,055.35, which is about 2.8 times the mortgage principal.

Table 8 Amortization Schedules for a $120,000, 30-Year Mortgage

Payment Number	4.0% Interest Monthly Payment: $572.90			Payment Number	12.5% Interest Monthly Payment: $1280.71		
	Interest Payment	Principal Payment	Balance of Principal		Interest Payment	Principal Payment	Balance of Principal
	Initially →		120,000.00		Initially →		120,000.00
1	400.00	172.90	119,827.10	1	1,250.00	30.71	119.969.29
2	399.42	173.48	119,653.62	2	1,249.68	31.03	119,938.26
3	398.85	174.05	119,479.57	3	1,249.36	31.35	119,906.91
12	393.55	179.35	117,886.76	12	1,246.29	34.42	119,609.63
60	362.49	210.41	108,537.02	60	1,224.11	56.60	117,458.11
152	287.12	285.78	85,851.63	152	1,133.87	146.84	108,704.57
153	286.17	286.73	85,564.90	153	1,132.34	148.37	108,556.20
230	202.43	370.47	60,358.56	230	951.19	329.52	90,984.42
231	201.20	371.70	59,986.86	231	947.75	332.96	90,651.46
294	114.50	458.40	33,890.53	294	641.10	639.61	60,905.68
295	112.97	459.93	33,430.60	295	634.43	646.28	60,259.40
296	111.44	461.46	32,969.13	296	627.70	653.01	59,606.39
359	3.80	569.10	571.00	359	26.27	1254.44	1267.51
360	1.90	571.00	0.00	360	13.20	1267.51	0.00
Totals:	86,243.41	120,000.00		**Totals:**	341,055.35	120,000.00	

Amortization Schedule			
No.	Amount	Interest	Principal	Balance
1	1,280.71	1,250.00	30.71	119,969.29
2	1,280.71	1,249.68	31.03	119,938.26
3	1,280.71	1,249.36	31.35	119,906.91

Amortizations to Go There are many free (and premium) apps for smartphones that will generate amortization tables. This screenshot is from the Loan Calculator feature of the fncalculator.com app. It agrees with the amortization of the 12.5% loan in **Table 8.**

A Note on Margin When determining the margin on a loan, the lender considers the creditworthiness of the borrower. Those with lower credit scores are a higher risk to the lender, so the margin is increased to cover that risk.

Credit scores are determined in part by comparing the amount of credit that borrowers have to the amount they use, as well as their record of making payments on time. The Fair Credit Reporting Act permits consumers to request a free credit report once per year from each of the three major credit-reporting agencies (Equifax, TransUnion, and Experian).

Potential homebuyers should establish good credit to qualify for lower interest rates. For sound advice on building good credit, check out www.fdic .gov/consumers/assistance/protection/ creditreport.html.

Table 8 also illustrates that payments made in the early years of a real-estate loan are mostly interest; only a small portion goes toward reducing the principal (especially when interest rates are high). Then, as principal decreases over the life of the loan, the amount of interest due each month also decreases, so larger portions of payments will apply toward principal.

Adjustable-Rate Mortgages

The lending industry uses many variations on the basic fixed-rate mortgage. An **adjustable-rate mortgage (ARM),** also known as a **variable-rate mortgage (VRM),** generally starts out with a lower rate than similar fixed-rate loans, but the rate changes periodically, reflecting changes in prevailing rates.

QUANTITIES GOVERNING ADJUSTABLE-RATE MORTGAGES (ARMS)

- **Adjustment period**—Time interval between rate adjustments (typically 1, 3, or 5 years)

- **Index**—Standard fluctuating average that is the basis for the new, adjusted rate (typically the 1-, 3-, or 5-year U.S. Treasury security rate or a national or regional "cost of funds" index)

- **Margin**—Additional amount added to the index by the lender (typically a few percentage points)

- **Discount**—Amount by which the *initial* rate may be less than the sum of the index and the margin (typically arranged between seller and lender)

- **Interest rate cap**—Limits on (interest) rate increases

- **Periodic cap**—Limit on rate increase per adjustment period (typically about 1% per 6 months or 2% per year)

- **Overall cap**—Limit on rate increases over the life of the loan (typically about 5% total)

- **Payment cap**—Limit on how much the payment can increase at each adjustment

- **Negative amortization**—An increasing loan principal (perhaps caused by a payment cap preventing the payment from covering a higher interest rate)

- **Convertibility feature**—A contractual ability to convert to a fixed-rate mortgage (usually at certain designated points in time)

- **Prepayment penalty**—Charges imposed by the lender if payments are made early

EXAMPLE 4 Comparing ARM Payments Before and After a Rate Adjustment

The Garcias pay $37,500 down on a $187,500 house and take out a 3/1 ARM (meaning that the initial rate stays fixed for the first three years and then adjusts each year after that to reflect the current index) for a 30-year term. The lender uses the 1-year Treasury index (at 2.2%) and a 2.8% margin.

(a) Find the monthly payment for the first three years.

(b) Suppose that after three years, the 1-year Treasury index has increased to 2.6%. Find the monthly payment for the fourth year.

Climate control is another cost that may get you involved with banks and interest rates after you finally get a roof over your head. The roof you see above does more than keep off the rain. It holds solar panels, part of the climate-control system in the building.

TVM Calculator	ADVANCED
Present Value 143,015.39	PV
Payment -839.58	PMT
Future Value 0	FV
Annual Rate% 5.4	RATE
Periods 324	PERIOD
Compounding Monthly	

This screen shows a monthly payment of $839.58, which supports the calculations of **Example 4(b).**

Solution

Cost of house   Down payment

(a) Mortgage amount = $187,500 − $37,500 = $150,000

The interest rate for the first three years will be

ARM interest rate = Index rate + Margin = 2.2% + 2.8% = 5.0%.

Now from **Table 6** (using 5.0% over 30 years) we obtain 5.36822.

Initial monthly payment = 150 · $5.36822 = $805.23

(b) During the first three years, some of the mortgage principal has been paid, so in effect we will now have a new "mortgage amount." (Also, the term will now be three years less than the original term.) The amortization schedule for the first three years (not shown here) yields a loan balance, after the thirty-sixth monthly payment, of $143,015.49. For the fourth year,

ARM interest rate = Index rate + Margin = 2.6% + 2.8% = 5.4%.

Because 5.4% is not included in **Table 6,** we use the regular monthly payment formula with the new mortgage balance and 27 years for the remaining term.

$$\text{Fourth-year monthly payment} = \frac{P\left(\frac{r}{12}\right)}{1 - \left(\frac{12}{12 + r}\right)^{12t}} \quad \text{Regular payment formula}$$

$$= \frac{\$143,015.49\left(\frac{0.054}{12}\right)}{1 - \left(\frac{12}{12 + 0.054}\right)^{(12)(27)}} \quad \text{Substitute known values.}$$

$$= \$839.58$$

The first ARM interest rate adjustment has caused the fourth-year monthly payment to rise to $839.58, which is an increase of $34.35 over the initial monthly payment.

A "seller buydown" occurs when the seller (a new-home builder, for example) pays the lender an amount in order to discount the buyer's loan. This reduces the initial rate and monthly payments, thereby giving the buyer a better chance to qualify for the loan. But it may be combined with higher initial fees or home price.

EXAMPLE 5 **Discounting a Mortgage Rate**

In **Example 4,** suppose that a seller buydown discounts the initial rate by 0.5%. Find the initial and fourth-year monthly payments.

Solution

The initial interest rate is discounted to 4.5% (rather than the 5% of **Example 4**), so the **Table 6** entry is 5.06685.

Initial monthly payment = 150($5.06685) = $760.03

The amortization schedule shows a balance at the end of three years of $142,401.79. The discount now expires, and the index has increased to 2.6%, so for the fourth year,

ARM interest rate = Index rate + Margin = 2.6% + 2.8% = 5.4%.

(This is just as in **Example 4.**)

Using the monthly payment formula, with $r = 5.4\%$ and $t = 27$,

$$\text{Fourth-year monthly payment} = \frac{\$142,401.79\left(\frac{0.054}{12}\right)}{1 - \left(\frac{12}{12 + 0.054}\right)^{(12)(27)}}$$

Substitute values in the monthly payment formula.

$$= \$835.98.$$

The initial monthly payment of $760.03 looks better than the $805.23 of **Example 4,** but at the start of year four, monthly payments jump by $75.95.

Closing Costs

Apart from principal and interest payments, buying a home involves a variety of one-time expenses called **closing costs,** or **settlement charges,** which are imposed when the loan is finalized (at "closing"). Often a buyer agrees to establish an **escrow account,** or **reserve account,** for the purpose of paying property taxes and home-owner's insurance premiums. In such cases, closing costs will include an "escrow deposit" to fund the escrow account.

Because escrow deposits increase closing costs, some buyers decide to pay property taxes and insurance premiums directly. However, some lenders lower the interest rate slightly for borrowers who use an escrow account, and many homeowners prefer the convenience of letting a third party process these payments. A buyer is entitled to a "good faith estimate" (GFE) of closing costs from the lender and, if desired, may shop for alternative providers of settlement services.

EXAMPLE 6 Computing Total Closing Costs

A buyer borrowed $130,000 at 5% interest to purchase a home. At the "closing" of the loan on June 19, the following closing costs were charged.

Loan origination fee (1% of mortgage amount)	$?
Broker loan fee	1625
Lender document and underwriting fees	534
Transfer taxes (for state and local fees on mortgages)	180
Title services and lender's title insurance	401
Title insurance	457
Government recording charges	55
Escrow deposit	1032
Daily interest charges for 12 days	?

Compute the total closing costs for this mortgage.

Solution

"Loan origination fees" are sometimes referred to as **points.** Each "point" amounts to 1% of the mortgage amount. By imposing points, the lender can effectively raise the interest rate without raising monthly payments (because points are normally paid at closing rather than over the life of the loan).

In this case, "one point" amounts to $1300. Because mortgage payments are typically made on the first of each month, and this loan closed on June 19, twelve days will pass before July 1. Interest on the $130,000 balance for these twelve days is

$$12 \cdot \$130,000 \cdot \frac{0.05}{365} \approx \$214.$$

Adding these amounts to those listed gives total closing costs of $5798.

Recurring Costs of Home Ownership

The primary considerations for most homebuyers are the following.

1. Accumulating the down payment

2. Having sufficient cash and income to qualify for the loan

3. Making the monthly mortgage payments

Three additional expenses of homeownership, listed below, should be anticipated realistically. These are **recurring costs** (as opposed to the one-time closing costs), and they can be significant.

1. **Property taxes** are collected by a county or other local government. Depending on the location and value of the home, taxes can range up to several thousand dollars annually. In most cases property taxes and mortgage interest are income tax deductible. Therefore, money expended for those items will decrease income taxes. This is one way in which the government has historically, through the tax code, encouraged home ownership.

2. **Homeowner's insurance** usually covers losses due to fire, storm damage, and other casualties. Some types, such as earthquake, flood, or hurricane coverage, may be unavailable or very expensive, depending on location.

3. All homes require **maintenance,** but these costs can vary greatly, depending mainly on the size, construction type, age, and condition of the home.

Insurance and maintenance expenses, though necessary to protect the home investment, are not generally tax deductible.

Government-backed mortgages, including FHA (Federal Housing Administration) and VA (Veterans Administration) loans, carry a government guarantee to protect the lender in case the borrower fails to repay the loan. Those who obtain a conventional loan, instead, are usually required to buy **private mortgage insurance (PMI),** unless they are able to make a down payment of at least 20% of the purchase price of the property.

This feature can be a surprise recurring expense to first-time home purchasers. It was introduced to protect lenders but indirectly also protects the buyers, who may lose a great deal if some catastrophe makes it impossible for them to make their mortgage payments.

EXAMPLE 7   **Determining the Affordability of Home Ownership**

The Jamiesons have household income that places them in a 30% combined federal and state tax bracket. They can afford a net average monthly expenditure of $1700 for a home. The home of their dreams is priced at $300,000. They have saved a 20% down payment and are prequalified for a 20-year, $240,000 fixed-rate mortgage at 4.5%. Property taxes are estimated to be $3900 annually, homeowner's insurance premiums would be $600 per year, and annual maintenance costs are estimated to be $1200. Can the Jamiesons afford their dream home?

Solution

Let's "do the math."

$$\text{Regular mortgage payment} = \underset{\substack{\text{Mortgage amount}\\\text{in \$1000s}}}{240} \cdot \underset{\substack{\text{Value from}\\\text{Table 6}}}{\$6.32649} \approx \$1518$$

$$\text{Monthly property taxes} = \frac{\$3900 \;\leftarrow\; \text{Annual taxes}}{12 \;\leftarrow\; \text{Months per year}}$$

$$= \$325$$

$$\text{Monthly insurance and maintenance cost} = \frac{\$600 + \$1200}{12} = \$150$$

$$\text{Total monthly expense} = \underset{\substack{\text{Mortgage}\\\text{payment}}}{\$1518} + \underset{\substack{\text{Property}\\\text{taxes}}}{\$325} + \underset{\substack{\text{Insurance and}\\\text{maintenance}}}{\$150}$$

$$= \mathbf{\$1993}$$

The **Tax Cuts and Jobs Act (TCJA) of 2017** limits deductions for state and local taxes (including property taxes and state income taxes) to $10,000 total, beginning in 2018 (for those who itemize deductions). Thus, the deductibility of property taxes will depend on how much a homeowner pays in state income tax and other state and local taxes. For example, the Jamiesons in **Example 7** will not be able to deduct the entire $3900 in property taxes if their other state and local taxes add up to more than $6100.

It seems that the Jamiesons cannot afford this home because

$$\$1993 > \$1700.$$

But wait—remember that mortgage interest and property taxes are both income tax deductible. Let's consider further.

$$\text{Monthly interest} = \$240{,}000(0.045)\left(\frac{1}{12}\right) \quad \text{\textit{Interest} = \textit{Prt}}$$

$$= \$900$$

Of the $1518 mortgage payment, $900 would be interest (initially).

$$\begin{array}{c} \text{Mortgage interest} \quad \text{Property tax} \\ \downarrow \qquad\qquad \downarrow \\ \text{Monthly deductible expenses} = \$900 \quad + \quad \$325 \quad = \quad \$1225 \end{array}$$

$$\begin{array}{c} \text{Deductible expense} \quad \text{Tax bracket} \\ \downarrow \qquad\qquad \downarrow \\ \text{Monthly income tax savings} = \$1225 \quad \cdot \quad 30\% \quad \approx \quad \$368 \end{array}$$

$$\begin{array}{c} \text{Gross cost} \quad \text{Tax savings} \\ \downarrow \qquad\qquad \downarrow \\ \text{Net monthly cost of home} = \$1993 \quad - \quad \$368 = \$1625 \end{array}$$

By considering the effect of taxes, we see that the net monthly cost of $1625 is indeed within the Jamiesons' $1700 affordability limit.

Example 7 raises some reasonable questions.

1. Is it wise to buy a house close to your spending limit, in light of future uncertainties?

2. What about the fact that the interest portion of mortgage payments, and therefore the tax savings, will decrease over time?

3. Won't taxes, insurance, and maintenance costs probably increase over time, making it more difficult to keep up the payments?

Some possible responses follow.

1. The Jamiesons may have built in sufficient leeway when they decided on their limit of $1700 per month.

2. **(a)** Look again at **Table 8** to see how slowly the interest portion drops.

   (b) Most people find that their income over time increases faster than expenses for a home that carries a fixed-rate mortgage. (A variable-rate mortgage is more risky and should have its initial rate locked in for as long as possible.)

   (c) Prevailing interest rates rise and fall over time. While rising rates will not affect a fixed-rate mortgage, falling rates may offer the opportunity to reduce mortgage expenses by refinancing. (For guidelines, see the margin note on refinancing.)

3. Here again, increases in income will probably keep pace. Although most things, including personal income, tend to follow inflation, the fixed-rate mortgage ensures that mortgage amortization, a major item, will stay constant.

Refinancing means initiating a new home mortgage and paying off the old one. In times of relatively low rates, lenders often encourage homeowners to refinance. Because setting up a new loan will involve costs, make sure you have a good reason before refinancing. Some possible reasons:

1. The new loan may be comparable to the current loan with a rate low enough to recoup the costs in a few years.

2. You may want to pay down your balance faster by switching to a shorter-term loan.

3. You may want to borrow out equity to cover other major expenses, such as education for your children.

4. Your present loan may have a large balloon feature that necessitates refinancing.

5. You may be uncomfortable with your variable (hence uncertain) ARM and want to convert to a fixed-rate loan.

WHEN WILL I **EVER** USE THIS?

Suppose you are a salesperson for a solar energy company. A couple who are potential customers contact you. To make your presentation more compelling, you provide cost comparisons of

"going solar"   and   "staying in the dark."

The customers give you permission to contact the power company and request a record of their power consumption over the last year, and they agree to meet the next day to go over the numbers. You explain that the government is providing a tax incentive in the form of a tax credit equal to 30% of the cost of the system. Your company adds to the incentive by offering a 12-month "same-as-cash" loan for that 30%, and a 15-year fixed-rate mortgage at 3.0% for the other 70% of the cost. This allows the customers to have zero "out-of-pocket" expense for the purchase and installation of the system. They can use their tax refund (which will include the solar tax credit) to pay off the 30% "same-as-cash" loan before any interest charges apply.

Analysis of this couple's power consumption for the last year shows that their average monthly bill was $200 (totaling $2400 for the year), and a $30,000 solar system would initially produce enough energy to match their consumption. Energy prices are projected to increase by 5% per year. A conservative estimate of solar panel degradation is 0.5% per year, and all equipment will be under warranty for 20 years. A cost comparison can now be provided for a 20-year period using the following calculations.

Initial principal amount of 3% loan = 0.7 · $30,000 = $21,000

$$\text{Monthly payment} = \frac{\$21{,}000\left(\dfrac{0.03}{12}\right)}{1 - \left(\dfrac{12}{12.03}\right)^{(12)(15)}} = \$145.02 \quad \text{Monthly payment formula}$$

First-year energy costs with solar ≈ 12 · $145 = $1740

First-year energy costs without solar = $2400 · 1.05 = $2520   5% price increase

Second-year energy costs without solar = $2400 · 1.05^2 = $2646   5% price increase

Second-year costs with solar = 0.005($2646) + $1740 ≈ $1753   0.5% degradation

Continuing these calculations results in the analysis of **Table 9.**

Cost of system:	$30,000	Interest rate:	3.0%
Same-as-cash loan:	$9,000	Term of loan (in years):	15
15-yr fixed-rate mortgage:	$21,000	Monthly payment:	$145.02

Table 9 Analysis of Cost/Savings of Solar Energy

Year	Annual Cost without Solar	Energy Bills		Monthly Loan Payments		Annual Cost with Solar
1	$2520	$0	+	$1740	=	$1740
2	2646	13	+	1740	=	1753
3	2778	28	+	1740	=	1768
4	2917	44	+	1740	=	1784
5	3063	61	+	1740	=	1801
6	3216	80	+	1740	=	1820
7	3377	101	+	1740	=	1841
8	3546	124	+	1740	=	1864
9	3723	149	+	1740	=	1889
10	3909	176	+	1740	=	1916
11	4105	205	+	1740	=	1945
12	4310	237	+	1740	=	1977
13	4526	272	+	1740	=	2012
14	4752	309	+	1740	=	2049
15	4989	349	+	1740	=	2089
16	5239	393	+	0	=	393
17	5501	440	+	0	=	440
18	5776	491	+	0	=	491
19	6065	546	+	0	=	546
20	6368	605	+	0	=	605
Totals:	$83,326					$30,723

20-Year Savings: $83,326 − $30,723 = $52,603

As you (and your customers) can see, the projected savings over the next 20 years are substantial.

WHEN WILL I **EVER** USE THIS?

Suppose you are a certified financial planner (CFP) with clients who need help deciding whether to purchase a home or rent. Their household income puts them in a 25% combined state/federal income tax bracket. If they purchased a home, they would stay in it for at least ten years. They are interested in a $280,000 home and are prequalified for a 30-year fixed mortgage at 5%. Comparable homes in the same area rent for $1,100 per month. Both home values and rents are projected to increase by 3% per year.

Should the clients buy or rent?

To answer this question, you lay out a side-by-side comparison of the costs. See **Table 10**.

Explanation of the Costs of Buying

To avoid purchasing private mortgage insurance, your clients want will make a down payment of $56,000 (20% of the $280,000 purchase price). Closing costs—including loan origination fees, discount points, documentation fees, and prorated interest and escrow deposits—typically come to about 3% of the purchase price, or $8,400.

Table 10 Comparing Buying and Renting over a Ten-Year Period

Buying		Renting	
Purchase Price: $280,000			
Up-front Costs of Purchasing		**Up-front Costs of Renting**	
Down payment	$56,000	Rental deposit (one month's rent)	$1,100
Closing costs (about 3% of purchase price)	$8,400		
Recurring Costs (30-yr fixed, 5%)		**Recurring Costs**	
Mortgage principal	$41,794	Rent (increasing 3% per year)	$151,323
Mortgage interest (incl. 25% tax deduction)	$76,878	Renter's insurance (1.32% of annual rent)	$1,997
Property Taxes (incl. 25% tax deduction)	$30,093		
Maintenance (0.5% of purchase price, rising with inflation)	$15,330		
Insurance (0.2% of home value)	$6,420		
Opportunity Costs		**Opportunity Costs**	
Down payment/closing costs	$33,244	Initial rent deposit	$568
Yearly costs	$41,749	Yearly costs	$35,292
Selling Costs		**Terminating Rental Agreement**	
Selling costs (6% of selling price)	$22,578	Return of initial deposit	−$1,100
Remaining principal	$182,206		
Proceeds from home sale	−$376,297		
Total Home Expenditures for Ten Years	$138,395	**Total Renting Expenditures for Ten Years**	$189,180

The recurring costs of home ownership include monthly principal and interest payments, property taxes (paid from an escrow account), maintenance costs, and homeowner's insurance. The amount of $41,794 for principal paid during the first ten years on the mortgage was calculated by amortizing the mortgage and adding the principal payments for 120 months. Your clients are in a 25% income tax bracket, so only 75% of the interest payments are considered "costs." Thus amortizing interest for the first 120 months, adding up these interest payments, and multiplying by 0.75 will yield the $76,878 figure for interest.

Property taxes in the county where the house is located are 1.25% of the value of the home.

$$\text{Year 1 property taxes} = 0.0125 \cdot \$280,000 = \$3,500$$

Property values are expected to increase by 3% per year.

$$\text{Year 2 property taxes} = \$3,500 \cdot 1.03 = \$3,605$$

Continuing this pattern yields the property tax amounts shown in **Table 11**. Property taxes are income tax deductible, so the actual expense for property taxes for the first ten years (considering the 25% tax bracket) is

$$(1 - 0.25) \cdot \$40,124 = 0.75 \cdot \$40,124 = \$30,093.$$

Maintenance costs are assumed to be 0.5% of the purchase price and grow at the rate of inflation, which is projected to be 2%.

$$\text{Year 1 maintenance costs} = 0.005 \cdot \$280,000 = \$1,400$$

$$\text{Year 2 maintenance costs} = \$1,400(1.02) = \$1,428$$

The cost increase due to inflation continues throughout the ten years, resulting in the amounts shown in **Table 12**.

Homeowner's insurance costs are assumed to be about 0.2% of the value of the home (which is projected to increase by 3% per year).

$$\text{Year 1 insurance costs} = 0.002 \cdot \$280,000 = \$560$$

$$\text{Year 2 insurance costs} = \$560(1.03) \approx \$577$$

Table 13 shows the insurance costs projected for the ten-year period.

Table 11

Year	Prop Tax
1	$3,500
2	3,605
3	3,713
4	3,825
5	3,939
6	4,057
7	4,179
8	4,305
9	4,434
10	4,567
Total	$40,124

Table 12

Year	Maintenance
1	$1,400
2	1,428
3	1,457
4	1,486
5	1,515
6	1,546
7	1,577
8	1,608
9	1,640
10	1,673
Total	$15,330

Table 13

Year	Insurance
1	$560
2	577
3	594
4	612
5	630
6	649
7	669
8	689
9	709
10	731
Total	$6,420

Opportunity cost is incurred when a choice is made to forgo an investment opportunity. For example, if your clients decide to purchase the home, they will be spending $56,000 on a down payment and $8,400 for closing costs, rather than investing it. For the purposes of your calculations, you have assumed that monies they have that are not spent on housing will instead be invested in the stock market at a 5% annual return. It is also assumed that the capital gains from these investments will be taxed at 15% (rather than the 25% income tax rate). Thus the "after-tax" effective growth rate will be

$$(1 - 0.15)(0.05) = 0.85(0.05) = 0.0425 = 4.25\%.$$

The net gain over ten years on the up-front costs of purchasing the home would be

$$\$64,400(1 + 0.0425)^{10} - \$64,400 \approx \$33,244.$$

To calculate the opportunity cost of the yearly expenses of home ownership, you begin with the monthly mortgage payments. The amortization of the 30-year fixed mortgage at 5% and beginning principal of $224,000 shows that the first month's principal payment will be $269 and the interest will be $933. Because the interest is tax deductible, the actual "cost" of the first month's mortgage payment would be

$$\$269 + 0.75(\$933) = \$969.$$

Investing this amount over ten years would yield an "after-tax" gain of

$$\$969(1 + 0.0425)^{10} - \$969 \approx \$500.$$

Performing similar calculations for the remaining 119 months of payments, as well as for the yearly costs of property taxes, maintenance, and insurance, yields the total opportunity cost of the yearly expenses of $41,749.

If your clients were to sell the home in ten years, the value would be

$$\$280,000(1.03)^{10} \approx \$376,297. \quad \text{Assume an increase of 3\% per year.}$$

This amounts to the proceeds from the sale. Closing costs (for realtor services, home inspection, etc.) can be expected to be about 6% of the selling price.

$$\text{Closing costs} = 0.06(\$376,297) = \$22,578$$

The remaining principal on the loan after ten years can be calculated as follows.

Original loan amount	Sum of principal payments	Remaining principal
↓	↓	↓
$224,000 −	$41,794 =	$182,206

Explanation of the Costs of Renting

Landlords typically charge a deposit equal to one month's rent. In your clients' case, this would be $1,100. The rent is expected to increase 3% per year over the next ten years. Rental agreements are often one-year leases, so the rent payment will be fixed for 12 months before it increases. All 120 projected rent payments are added to give the $151,323 figure for the rent payments over the ten-year period.

Renter's insurance in the area where your clients would like to live typically amount to 1.32% of the annual rent, which increases by 3% per year. Projected annual premiums for the ten years total $1,997.

Opportunity costs for rental expenses are calculated in a fashion similar to those for home ownership, taking into account the 15% tax rate on capital gains.

When all calculations are made, the difference between rental and ownership costs is

Rental costs	Ownership costs
↓	↓

$$\$189,180 - \$138,395 = \$50,786.$$

Your clients can save, on average, about $5,079 per year by buying rather than renting.

13.4 EXERCISES

CONCEPT CHECK *Fill in the blank with the correct response.*

Round all monetary answers in this exercise set to the nearest cent unless directed otherwise.

1. A(n) _____ is a listing of payments, showing the allotment of each payment toward principal and interest.

2. The amortization schedule for a(n) _____ rate mortgage may be completely determined at the initiation of the loan.

3. A(n) _____ may be established for the purpose of paying insurance premiums and/or property taxes.

4. Property taxes and maintenance expenses are examples of the _____ costs of home ownership.

5. The two factors determining interest rates for adjustable-rate mortgages are _____ and _____.

6. An adjustable-rate mortgage may be changed to a fixed-rate mortgage if it has a(n) _____ .

Monthly Payment on a Fixed-Rate Mortgage Find the monthly payment needed to amortize principal and interest for each fixed-rate mortgage. Use either the regular monthly payment formula or **Table 6,** as appropriate.

	Loan Amount	Interest Rate	Term
7.	$100,000	5.5%	20 years
8.	$23,000	8.0%	15 years
9.	$145,000	5.2%	25 years
10.	$95,000	6.3%	30 years
11.	$227,750	9.5%	25 years
12.	$195,450	5.0%	10 years
13.	$42,500	4.6%	5 years
14.	$200,000	3.5%	10 years

Amortization of a Fixed-Rate Mortgage Complete the first one or two months (as required) of each amortization schedule for a fixed-rate mortgage.

15. Mortgage: $69,500
Interest rate: 5.0%
Term of loan: 30 years

Amortization Schedule

Payment Number	Total Payment	Interest Payment	Principal Payment	Balance of Principal
1	(a) _____	(b) _____	(c) _____	(d) _____

16. Mortgage: $112,000
Interest rate: 5.0%
Term of loan: 20 years

Amortization Schedule

Payment Number	Total Payment	Interest Payment	Principal Payment	Balance of Principal
1	(a) _____	(b) _____	(c) _____	(d) _____

17. Mortgage: $143,200
Interest rate: 6.5%
Term of loan: 15 years

Amortization Schedule

Payment Number	Total Payment	Interest Payment	Principal Payment	Balance of Principal
1	(a) _____	(b) _____	(c) _____	(d) _____
2	(e) _____	(f) _____	(g) _____	(h) _____

18. Mortgage: $124,750
Interest rate: 4.50%
Term of loan: 25 years

Amortization Schedule

Payment Number	Total Payment	Interest Payment	Principal Payment	Balance of Principal
1	(a) _____	(b) _____	(c) _____	(d) _____
2	(e) _____	(f) _____	(g) _____	(h) _____

19. Mortgage: $113,650
   Interest rate: 5.75%
   Term of loan: 10 years

Amortization Schedule

Payment Number	Total Payment	Interest Payment	Principal Payment	Balance of Principal
1	(a) _____	(b) _____	(c) _____	(d) _____
2	(e) _____	(f) _____	(g) _____	(h) _____

20. Mortgage: $150,000
   Interest rate: 5.20%
   Term of loan: 16 years

Amortization Schedule

Payment Number	Total Payment	Interest Payment	Principal Payment	Balance of Principal
1	(a) _____	(b) _____	(c) _____	(d) _____
2	(e) _____	(f) _____	(g) _____	(h) _____

Finding Monthly Mortgage Payments   *Find the total monthly payment, including taxes and insurance.*

	Mortgage	Interest Rate	Term of Loan	Annual Taxes	Annual Insurance
21.	$62,300	5%	20 years	$610	$220
22.	$51,800	6%	10 years	$570	$145
23.	$89,560	3.5%	25 years	$915	$409
24.	$72,890	5.5%	15 years	$1850	$545

Comparing Mortgage Options   *Suppose you want to purchase a $300,000 home, and you have the required $60,000 down payment in savings. Complete the table below to compare different mortgage options presented by a mortgage broker (Data from Bankrate.com, accessed June 11, 2018).*

	Principal	Interest Rate	Term (years)	Payment	Total of Payments over Life of Loan	Total Interest Paid over Life of Loan
25.	$240,000	4.500%	30	_____	_____	_____
26.	$240,000	4.250%	20	_____	_____	_____
27.	$240,000	3.875%	15	_____	_____	_____
28.	$240,000	3.750%	10	_____	_____	_____

29. Compare the monthly payments, total of payments, and total interest paid for the loans in **Exercises 25 and 27.** Which loan option would you choose? Why?

30. Compare the monthly payments, total of payments, and total interest paid for the loans in **Exercises 26 and 28.** Which loan option would you choose? Why?

Long-Term Effect of Interest Rates   *You may remember seeing home mortgage interest rates fluctuate widely in a period of not too many years. The following exercises show the effect of changing rates. Refer to* **Table 8,** *which compared the amortization of a $120,000, 30-year mortgage for rates of 4.0% and 12.5%. Give values of each of the following for* **(a)** *a 4.0% rate, and* **(b)** *a 12.5% rate.*

31. monthly payments

32. the percentage of the first monthly payment that is principal

33. balance of principal after 1 year

34. the number of payments required to pay down half of the original principal

35. the first monthly payment that includes more toward principal than toward interest

36. amount of interest included in final monthly payment of mortgage

The Effect of Credit Rating on Mortgage Costs *Borrowers with higher credit scores are rewarded with lower interest rates on home mortgages. Suppose a loan of $200,000 is needed to purchase a home. Complete the table below to compare total interest paid over a 30-year term for borrowers with different credit scores.*

	Credit Score Range	Interest Rate	Total Interest Paid Over Life of Loan
37.	720 +	4.500%	_____
38.	700–719	4.625%	_____
39.	680–699	4.750%	_____
40.	660–679	4.875%	_____

The Effect of the Term on Total Amount Paid *Suppose a $60,000 mortgage is to be amortized at 5.2% interest. Find the total amount that would be paid for each term.*

41. 10 years   **42.** 20 years

43. 30 years   **44.** 40 years

The Effect of Adjustable Rates on the Monthly Payment *For each adjustable-rate mortgage, find* **(a)** *the initial monthly payment,* **(b)** *the monthly payment after the first adjustment, and* **(c)** *the change in monthly payment at the first adjustment.*

	Beginning Balance	Term	Initial Index Rate	Margin	Adjustment Period	Adjusted Index Rate	Adjusted Balance
45.	$175,000	20 years	2.5%	2.5%	1 year	4.0%	$169,772.24
46.	$244,500	30 years	3.2%	2.75%	3 years	2.6%	$234,840.87

(The "adjusted balance" is the principal balance at the time of the first rate adjustment. Assume no caps apply.)

The Effect of Rate Caps on Adjustable-Rate Mortgages *Millie has a 1-year ARM for $110,000 over a 20-year term. The margin is 2%, and the index rate starts out at 2.5% and increases to 5.0% at the first adjustment. The balance of principal at the end of the first year is $106,528.05. The ARM includes a periodic rate cap of 2% per adjustment period. Use this information for Exercises 47–50.*

47. Find **(a)** the interest owed and **(b)** the monthly payment due for the first month of the first year.

48. Find **(a)** the interest owed and **(b)** the monthly payment due for the first month of the second year. (Remember the periodic rate cap!)

49. What is the monthly payment adjustment at the end of the first year?

50. If the index rate has dropped slightly at the end of the second year, will the third-year monthly payments necessarily drop? Why or why not?

Closing Costs of a Mortgage *For Exercises 51–54, refer to the following list of closing costs for the purchase of a $175,000 house requiring a 20% down payment, and find each requested amount.*

Title insurance premium	$400
Document recording fee	60
Loan fee (two points)	____
Appraisal fee	400
Prorated property taxes	685
Prorated fire insurance premium	295

51. mortgage amount   **52.** loan fee

53. total closing costs

54. total amount of cash required of the buyer at closing (including down payment)

*Consider the scenario of **Example 7.** Recalling that mortgage interest is income tax deductible, find (to the nearest dollar) the additional initial net monthly savings resulting from each strategy. In each case, only the designated item changes. All other features remain the same.*

55. Change the mortgage term from 20 years to 30 years.

56. Change the mortgage from fixed at 4.5% to an ARM with an initial rate of 4.125%.

For Exercises 57–60, find each quantity for a $200,000 fixed-rate mortgage. (Give answers to the nearest dollar.)

(a) *Monthly mortgage payment (principal and interest)*

(b) *Monthly house payment (including property taxes and insurance)*

(c) *Initial monthly interest*

(d) *Income tax deductible portion of initial house payment*

(e) *Net initial monthly cost for the home (considering tax savings)*

	Term of Mortgage	Interest Rate	Annual Property Tax	Annual Insurance	Owner's Income Tax Bracket
57.	15 years	5.5%	$960	$480	20%
58.	20 years	6.0%	$840	$420	25%
59.	10 years	6.5%	$1092	$540	30%
60.	30 years	7.5%	$1260	$600	40%

Exercises 61 and 62 refer to the **When Will I Ever Use This?** *feature (concerning the certified financial planner) at the end of this section. Suppose your clients can afford to buy a $200,000 home and have the required 20% down payment in savings. Comparable rentals are available for $900 per month. Assume a combined income tax bracket of 25% and a return on investments of 4%, with a 15% tax rate on capital gains. Compare the costs of buying and renting under the given circumstances.*

61. Your clients plan to move in five years, so they would take out a 5/1 ARM at 4.0%, and then plan to sell the home and pay off the loan before the interest rate adjusts. If home values increase 3% per year and rents increase 2% per year, determine

(a) the opportunity costs that would result from the up-front costs of purchasing the home,

(b) the total costs of five years of ownership, and

(c) the difference between the costs of renting and the costs of owning over five years.

Buying		Renting	
Up-front Costs of Purchasing		**Up-front Costs of Renting**	
Down payment	$40,000	Rental deposit	$900
Closing costs	$6,000		
Recurring Costs		**Recurring Costs**	
Mortgage payments	$38,195	Rent	$56,204
Other	$17,685	Renter's insurance	$1,124
Opportunity Costs		**Opportunity Costs**	
Up-front costs	_____	Initial rent deposit	$164
Yearly costs	$5,278	Yearly costs	$5,094
Selling Costs		**Terminating Agreement**	
Selling costs	$13,911	Return of deposit	−$900
Remaining principal	$144,716		
Proceeds from sale	−$231,855		
Total Costs for 5 Years	_____	**Total Costs for 5 Years**	$62,586

62. Seeing the savings associated with home ownership, your clients now are considering staying in the area for 20 years. If they decided to purchase the home, they would take out a 30-year fixed mortgage at 4.75%. If home values increase by 3% per year, and rents also increase 3% per year, determine

(a) the proceeds from the sale of the same home in 20 years,

(b) the total costs of 20 years of ownership, and

(c) the difference between the costs of renting and the costs of owning over 20 years.

Buying		Renting	
Up-front Costs of Purchasing		**Up-front Costs of Renting**	
Down payment	$40,000	Rental deposit	$900
Closing costs	$6,000		
Recurring Costs		**Recurring Costs**	
Mortgage payments	$170,333	Rent	$290,200
Other	$87,425	Renter's insurance	$5,804
Opportunity Costs		**Opportunity Costs**	
Up-front costs	$43,778	Initial rent deposit	$857
Yearly costs	$106,350	Yearly costs	$112,388
Selling Costs		**Terminating Agreement**	
Selling costs	$21,673	Return of deposit	−$900
Remaining principal	$79,605		
Proceeds from sale	_____		
Total Costs for 20 Years	_____	**Total Costs for 20 Years**	$409,249

On the basis of material in this section, or your own research, give brief written responses to each problem.

63. Suppose your ARM allows conversion to a fixed-rate loan at each of the first five adjustment dates. Describe circumstances under which you would want to convert.

64. Should a home buyer always pay the smallest down payment that will be accepted? Explain.

65. Should a borrower always choose the shortest term available in order to minimize the total interest expense? Explain.

66. Under what conditions would an ARM probably be a better choice than a fixed-rate mortgage?

67. Why are fourth-year monthly payments less in **Example 5** than in **Example 4** even though the term and the interest rate are the same in both cases?

68. What tends to make ownership less expensive than renting, especially over many years?

69. Do you think the discount in **Example 5** actually makes the overall cost of the mortgage less? Explain.

70. Find out what is meant by each term and describe some of the features of each.

FHA-backed mortgage

VA-backed mortgage

Conventional mortgage

13.5 FINANCIAL INVESTMENTS

OBJECTIVES

1 Explore and analyze stock investments.

2 Explore and analyze bond investments.

3 Explore and analyze mutual fund investments.

4 Evaluate investment returns.

5 Explore strategies for building a nest egg.

Stocks

In a general sense, an *investment* is a way of putting resources to work so that (one hopes) they will grow. Our main emphasis here will be on the basic mathematical aspects of a restricted class of financial investments—namely, *stocks, bonds,* and *mutual funds.*

Buying stock in a corporation makes you a part owner of the corporation. You then share in any profits the company makes, and your share of profits is a **dividend.** If the company prospers (or if increasing numbers of investors believe it will prosper in the future), your stock will be attractive to others, so that you may, if you wish, sell your shares at a profit. The profit you make by selling for more than you paid is a **capital gain.** A negative gain, or **capital loss,** results if you sell for less than you paid. By **return on investment,** we mean the net difference between what you receive (including your sale price and any dividends received) and what you paid (your purchase price plus any other expenses of buying and selling the stock).

EXAMPLE 1 Finding the Return on Ownership of Stock

Etsuko bought 100 shares of stock in Company A on January 15, 2019, paying $30 per share. On January 15, 2020, she received a dividend of $0.50 per share, and the stock price had risen to $31.75 per share. (Ignore any costs other than the purchase price.) Find the following.

(a) Etsuko's total cost for the stock

(b) The total dividend amount

(c) Etsuko's capital gain if she sold the stock on January 15, 2020

(d) Etsuko's total return on her one year of ownership of this stock

(e) The percentage return

Solution

(a) Cost $= (\$30 \text{ per share}) \cdot (100 \text{ shares}) = \$3000$

She paid $3000 total.

(b) Dividend $= (\$0.50 \text{ per share}) \cdot (100 \text{ shares}) = \$50$

The dividend amount was $50.

(c)

$$\text{Capital gain} = \overset{\text{Change in price per share}}{(\$31.75 - \$30.00)} \cdot \overset{\text{Number of shares}}{100} = \$1.75 \cdot 100 = \$175$$

The capital gain was $175.

(d)

$$\text{Total return} = \overset{\text{Dividend}}{\$50} + \overset{\text{Capital gain}}{\$175} = \$225$$

Total return on the investment was $225.

(e) $\text{Percentage return} = \dfrac{\text{Total return}}{\text{Total cost}} \cdot 100\% = \dfrac{\$225}{\$3000} \cdot 100\% = 7.5\%$

The percentage return on the investment was 7.5%.

The **NASDAQ (National Association of Securities Dealers Automated Quotations),** unlike the NYSE and other exchanges that actually carry out trading at specific locations, is an electronic network of brokerages. The NASDAQ is known for trading high-tech company stocks, many of which tend to be newer and more volatile.

The price of a share of stock is determined by the law of supply and demand at institutions called **stock exchanges.** In the United States, the largest exchange is the New York Stock Exchange (NYSE), established in 1792 and located on Wall Street in New York City. Current prices and other information about particular stocks are published daily in many newspapers and on various Web sites, where information may be updated every few minutes. **Table 14** below shows examples of trading on June 15, 2018. The exchange opens at 9:30 A.M. and closes at 4:00 P.M., Eastern time.

Observe the column headings and the numbers for the first company listed, Allstate, with market symbol ALL. The first four numbers show that the first sale of Allstate shares after 9:30 A.M. was for $92.40 per share, the highest sale during the trading day was for $93.87, the lowest was for $92.37, and the last sale before 4:00 P.M. was for $93.81 per share.

Moving across to the right, the next two numbers show that the closing price for the day was $1.01 higher than that of the previous day—that is, 1.09% higher. The volume number indicates that 4,038,500 shares of Allstate stock were sold that day.

The next two numbers show that, over the last year, the greatest and least prices paid for ALL were $105.36 and $85.59, respectively. The following two numbers show that Allstate's current dividend is $1.84 per share and that this represents a current yield of 1.97% (of the closing price from the latest dividend date).

PE is the **price-to-earnings ratio,** the current price per share divided by the earnings per share over the past 12 months. For ALL, this ratio is 11.62. Finally, the year-to-date percent change means that Allstate's price per share is now 10.41% lower than it was at the start of this calendar year.

Table 14 Selected Stock Quotes (June 15, 2018)

Company	Symbol	Open	High	Low	Close	Net Change	% Change	Volume	52-Wk High	52-Wk Low	Annual Div	Dividend Yield	PE	YTD % Change
Allstate Corp	ALL	92.40	93.87	92.37	93.81	1.01	1.09%	4,038,500	105.36	85.59	1.84	1.97%	11.62	−10.41%
Bank of America	BAC	29.25	29.41	28.86	29.28	−0.22	−0.75%	105,079,500	33.05	22.73	0.48	1.61%	14.70	−0.81%
CarMax Inc	KMX	72.49	73.42	72.13	72.39	−0.83	−1.13%	3,371,800	77.64	57.05	0.00	0.00%	19.81	12.88%
Caterpillar Inc	CAT	150.85	151.00	148.43	150.02	−3.12	−2.04%	8,852,400	173.24	102.30	3.12	2.02%	18.37	−4.80%
Chevron Corp	CVX	127.10	127.12	123.84	124.04	−2.47	−1.95%	13,223,800	133.88	102.55	4.48	3.53%	27.81	−0.92%
Deere & Company	DE	148.57	150.20	145.69	148.75	−1.43	−0.95%	5,991,000	175.26	112.87	2.40	1.57%	19.19	−4.96%
Facebook Inc	FB	195.79	197.07	194.64	195.85	−0.96	−0.49%	21,851,000	197.28	147.80	0.00	0.00%	28.30	10.99%
FedEx Corp	FDX	262.51	264.99	260.84	264.56	1.08	0.41%	1,363,600	274.66	203.13	2.00	0.76%	19.25	6.02%
General Mills	GIS	44.35	45.51	44.35	45.43	0.92	2.07%	12,354,000	60.69	41.01	1.96	4.39%	14.63	−23.38%
Harley-Davidson	HOG	44.53	46.79	44.53	45.94	1.74	3.94%	4,921,600	56.95	39.34	1.48	3.34%	12.10	−9.71%
Hewlett-Packard	HPQ	23.70	23.70	23.44	23.59	−0.21	−0.88%	13,022,800	24.75	17.10	0.56	2.37%	12.86	12.28%
Pearson Plc	PSO	11.64	11.74	11.61	11.72	−0.02	−0.17%	479,600	12.51	7.62	0.33	2.83%	0.00	19.35%
Twitter Inc	TWTR	46.62	47.79	45.64	45.80	−0.96	−2.05%	51,489,500	47.79	15.67	0.00	0.00%	231.95	90.75%
Walt Disney	DIS	108.17	109.42	107.85	108.85	0.10	0.09%	16,169,200	113.19	96.20	1.68	1.58%	16.66	1.25%
Yum! Brands	YUM	83.22	83.44	81.59	82.62	−0.76	−0.91%	4,893,000	88.07	72.38	1.44	1.74%	25.65	1.24%

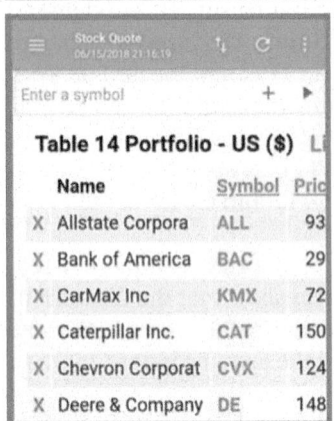

Pocket Portfolios   Today, investors can use smartphones to check their portfolios, issue buy and sell orders, and view the performance history of stocks on multiple exchanges. This screenshot (using the Stock Quote app from BISHINEWS) shows part of the portfolio from **Table 14.**

EXAMPLE 2   **Reading a Stock Table**

Use the stock table (**Table 14**) to find the required amounts below.

(a) The highest price for the last 52 weeks for CarMax Inc (KMX)

(b) The dividend for General Mills (GIS)

Solution

(a) Find the correct line from the stock table for CarMax.

The highest price for CarMax for the last 52 weeks was $77.64.

(b) The annual dividend value of $1.96 means that GIS paid a dividend of $1.96 per share.

EXAMPLE 3   **Finding the Cost of a Stock Purchase**

Find the cost for 50 shares of Deere (DE) stock, purchased at the low price for the day.

Solution

From **Table 14,** the low price for the day for Deere was $145.69 per share.

$$\underset{\substack{\text{Price per} \\ \text{share} \\ \downarrow}}{} \qquad \underset{\substack{\text{Number of} \\ \text{shares} \\ \downarrow}}{}$$

$$\text{Total cost} = \$145.69 \quad \cdot \quad 50 = \$7284.50$$

Fifty shares of this stock would cost $7284.50, plus any broker's fees.

Members of the public buy stocks through brokerage firms that have access to the exchange. Traditionally, clients would communicate directly with a stock broker, who would then contact a representative at the exchange to execute the buy or sell order. However, as technology has improved, it has become increasingly common, especially among "average" traders, for transactions to be conducted electronically. Brokers generally fall into one of two categories.

• **Full-service brokers** offer research, recommendations for stocks to buy and sell, and various other services, including wealth management, retirement planning, and tax advice.

• **Discount brokers** execute purchases and sales of stock for their clients and may offer some educational information to help clients make sound decisions, but they do not make specific stock recommendations to individual clients.

During the rapid growth of the stock market in the 1990s, several online companies appeared, offering much cheaper transactions than were available through conventional brokers. Subsequently, many conventional firms, as well as the major discount brokerages, introduced their own automated (online) services.

Brokerage firms charge a **commission** (the broker's fee) for executing an order. For full-service brokers, commissions are normally a percentage (usually between 1% and 2%) of the value (the **principal**) of the purchase or sale. Additional fees may be applied depending on whether the order is for a **round lot** of shares (a multiple of 100) or an **odd lot** (any portions of an order for fewer than 100 shares). A typical **odd-lot differential** is 12¢ per share.

It is possible to place a **limit order,** where the broker is instructed to execute a buy or sell if and when a stock reaches a predesignated price. An extra fee may be added to the commission on a limit order. Many discount brokerage firms do not charge these special fees; instead, they charge different commissions depending on whether the transactions are **automated trades** or **broker-assisted trades.**

Table 15 shows typical brokerage commissions for different transaction categories.

Table 15 Typical Discount Brokerage Commissions

	Automated (Online)	Automated (Phone)	Broker-Assisted
Brokerage Firm A	$5.95	$12.95	$32.95
Brokerage Firm B	$6.95	$34.99	$44.99
Brokerage Firm C	$4.95	$9.95	$29.95

Also, the Securities and Exchange Commission (SEC), a federal agency that regulates stock markets, supports its own activities by charging the exchanges, based on volume of transactions. Typically this charge is passed on to investors, through brokers, in the f orm of an **SEC fee** assessed on stock sales only (not on purchases). As of May 2018, this fee was $13.00 per million dollars of principal (rounded *up* to the next cent). For example, to find the fee for a sale of $1600, first divide $1600 by $1,000,000 and then multiply by $13.00.

$$\text{SEC fee} = \frac{\$1600}{\$1,000,000} \cdot \$13.00 = \$0.0208, \quad \text{which is rounded up to } \$0.03.$$

EXAMPLE 4 Finding Total Cost of Stock Purchases

Carina has a brokerage account with Brokerage Firm A, and on June 15, 2018, she executed a broker-assisted purchase of 200 shares of Harley-Davidson (HOG) stock at the average price of the day. She purchased 100 more shares online on September 12 for $48.12 per share. Find Carina's total cost for purchasing the 300 shares of HOG.

Solution

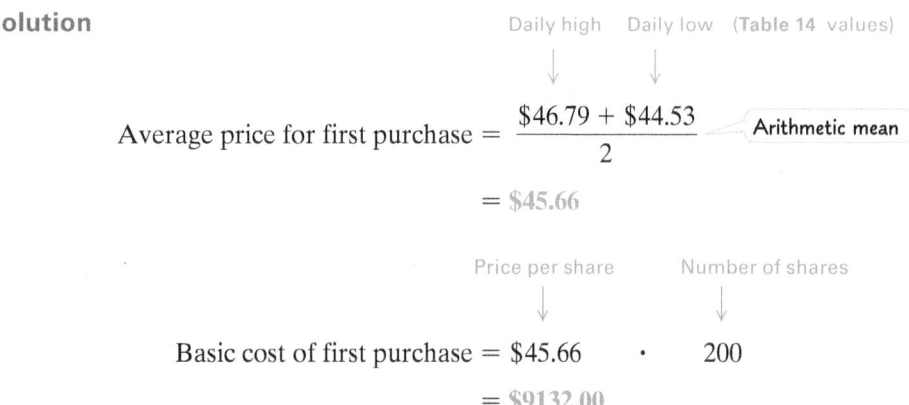

Daily high   Daily low   (**Table 14** values)

$$\text{Average price for first purchase} = \frac{\$46.79 + \$44.53}{2} \quad \text{Arithmetic mean}$$

$$= \$45.66$$

Price per share   Number of shares

$$\text{Basic cost of first purchase} = \$45.66 \quad \cdot \quad 200$$

$$= \$9132.00$$

Because the purchase was a broker-assisted transaction, Firm A charged a commission of $32.95 (as shown in **Table 15**).

Basic cost   Commission

$$\text{Total cost of first purchase} = \$9132.00 \quad + \quad \$32.95$$

$$= \$9164.95$$

Price per share   Number of shares

$$\text{Basic cost of second purchase} = \$48.12 \quad \cdot \quad 100$$

$$= \$4812.00$$

Paper trading lets beginners "pretend" to buy and sell stocks. Several brokerage firms offer virtual trading accounts with "play money." This enables clients to build a portfolio and test trading strategies before putting actual capital at risk. To see a list of brokers offering this option visit www.nerdwallet.com/blog/investing/virtual-trading-stock-market-simulators.

The second purchase was automated (online), so Firm A charged a commission of $5.95.

Basic cost   Commission

$$\text{Total cost of second purchase} = \$4812.00 + \$5.95$$

$$= \$4817.95$$

$$\text{Total cost for both transactions} = \$9164.95 + \$4817.95$$

$$= \$13,982.90 \quad \longleftarrow \text{Total cost}$$

EXAMPLE 5   **Finding the Return on a Stock Investment**

Suppose Carina (from **Example 4**) sold all 300 shares of HOG stock using an online transaction on December 19, 2018, at a price of $49 .20 per share. Find the return on the investment. (Assume that dividends were paid on September 5.)

Solution

The basic price of the stock sold is

$$(\$49.20 \text{ per share}) \cdot (300 \text{ shares}) = \$14,760.00.$$

Because Carina owned only 200 shares when dividends were paid, her dividend receipt (refer to **Table 14**) was

$$(\$1.48 \text{ per share}) \cdot (200 \text{ shares}) = \$296.00.$$

The sale was executed online, so Firm A charged a commission of $5.95 (see **Table 15**). Find the SEC fee as described before **Example 4.**

$$\frac{\$14,760.00}{\$1,000,000} \cdot \$13.00 = \$0.19188$$

Round up to obtain an SEC fee of $0.20. Then

Basic price   Commission   SEC fee

$$\text{Net proceeds from sale} = \$14,760.00 - \$5.95 - \$0.20$$

$$= \$14,753.85.$$

Finally, the return on investment will be the sum of dividends and proceeds from the sale, less the total cost of purchasing the stock (from **Example 4**).

$$\text{Return on investment} = \$296.00 + \$14,753.85 - \$13,982.90$$

$$= \$1066.95 \quad \longleftarrow \text{Final return}$$

Carina's return on the investment in HOG shares was $1066.95.

Bonds

Rather than using your capital (money) to buy shares of stock in a company, you may prefer merely to *lend* money to the company, receiving an agreed-upon rate of interest for the use of your money. In this case you would buy a **bond** from the company rather than purchase stock. The bond (loan) is issued with a stated term (life span), after which the bond "matures" and the principal (or **face value**) is paid back to you. Over the term, the company pays you a fixed rate of interest, which depends on prevailing interest rates at the time of issue (and to some extent on the company's credit rating), rather than on the underlying value of the company.

A corporate bondholder has no stake in company profits but is (quite) certain to receive timely interest payments. If a company is unable to pay both bond interest and stock dividends, the bondholders must be paid first.

EXAMPLE 6   Finding the Return on a Bond Investment

Justin invests $10,000 in a 5-year corporate bond paying 3.5% annual interest, paid semiannually. Find the total return on this investment, assuming Justin holds the bond to maturity (for the entire 5 years). Ignore any broker's fees.

Solution

By the simple interest formula,

$$I = Prt = \$10,000(0.035)(5) = \$1750.$$

Notice that this amount, $1750, is the *total* return, over the 5-year period, not the annual return. Also, the fact that interest is paid twice per year (1.75% of the face value each time) has no bearing here since compounding does not occur.

Mutual Funds

A **mutual fund** is a pool of money collected by an investment company from many individuals and/or institutions and invested in many stocks, bonds, or money market instruments. As of 2017, in the United States alone, nearly $19 trillion was invested in more than 9000 mutual funds. Over 57 million U.S. households held mutual funds, with the median mutual fund assets of these households approximately $120,000. Over a third ($8.8 trillion) of mutual fund assets were in retirement accounts.

MUTUAL FUNDS VERSUS INDIVIDUAL STOCKS AND BONDS

Advantages of Mutual Funds

1. *Simplicity*
   Let someone else do the work.

2. *Diversification*
   Being part of a large pool makes it easier to own interests in many different stocks, which decreases vulnerability to large losses in one particular stock or one particular sector of the economy.

3. *Access to New Issues*
   Initial public offerings (IPOs) can sometimes be very profitable. Whereas individual investors find it difficult to get in on these, the large overall assets of a mutual fund give its managers much more access.

4. *Economies of Scale*
   Large stock purchases usually incur smaller expenses per dollar invested.

5. *Professional Management*
   Professional managers may have the expertise to pick stocks and time purchases to better achieve the stated objectives of the fund.

6. *Indexing*
   With minimal management, an index fund can maintain a portfolio that mimics a popular index (such as the S&P 500). This makes it easy for individual fund investors to achieve returns at least close to those of the index.

The **Wall Street reform bill of 2010** put forth "the most sweeping set of changes to the financial regulatory system since the 1930s." Prompted by the failure of a number of large financial and other companies and a steep international recession beginning in 2007, the bill created a new Consumer Financial Protection Bureau within the Federal Reserve, headed by a presidentially appointed "independent" regulator, and with authority to write new rules to (theoretically) protect consumers from unfair or abusive practices in mortgages and credit cards. In fact, the bill creates vast bureaucratic control over virtually all financial areas discussed in this chapter.

It took over a year for the Democrat-controlled Congress to bring the 2010 bill to a vote. According to the watchdog group Public Citizen, over that period the financial industry spent nearly $600 million to hire some 1000 lobbyists to promote their interests. Debate raged about whether the bill was really to help consumers or to control business (or perhaps a combination of the two).

(*Source:* www.CNNMoney.com)

MUTUAL FUNDS VERSUS INDIVIDUAL STOCKS AND BONDS CONTINUED

Disadvantages of Mutual Funds

1. *Impact of One-Time Charges and Recurring Fees*
    Sales charges, management fees, "12b-1" fees (used to pay sales representatives), and fund expenses can mount and can be difficult to identify. And paying management fees does not guarantee getting *quality* management.

2. *Hidden Cost of Brokerage*
    Recurring fees and expenses are added to give the *expense ratio* of a fund. But commissions paid by the fund for stock purchases and sales are in addition to the expense ratio. (Since 1995, funds have been required to disclose their average commission costs in their annual reports.)

3. *Some Hidden Risks of Fund Ownership*

    (a) In the event of a market crash, getting out of the fund may mean accepting securities rather than cash, and these may be difficult to redeem for a fair price.

    (b) Managers may stray from the stated strategies of the fund.

    (c) Since tax liability is passed on to fund investors and depends on purchases and sales made by fund managers, investors may be unable to avoid inheriting unwanted tax basis.

    Source for advantages and disadvantages of mutual funds:
    Forbes Guide to the Markets, John Wiley & Sons, Inc.

The risks and rewards associated with stocks and bonds are reflected in the mutual funds that own shares of those stocks and bonds. Historically, the rate of return over most time periods has been greater for stocks than for bonds. Accordingly, stock mutual funds have more potential for higher returns (on average) than bond mutual funds. On the other hand, there are fewer uncertainties associated with bonds, and so investors holding shares of bond mutual funds are exposed to less risk than those invested in stock funds. "Mixed asset" (or blended) mutual funds are designed to help investors balance their portfolios by dealing in both stocks and bonds, thus increasing potential earnings with stock holdings, but reducing volatility by virtue of bond holdings.

Because a mutual fund owns many different stocks, each share of the fund represents a fractional interest in each of those companies. In an "open-end" fund (the most common type), new shares are issued to buyers while the fund company absorbs the shares of sellers. Similarly, a bond mutual fund can invest in many different bonds of varying maturities, buying and selling according to market conditions. Every stock or bond held by a fund contributes to the total value of the fund. Each day, the value of a share in the fund is called the *net asset value,* or **NAV.**

An abundance of information about **mutual funds** is available on the Internet. For example, the Investment Company Institute (ICI), the national association of the American investment company industry, provides material at the site www.ici.org/.

NET ASSET VALUE OF A MUTUAL FUND

If A = Total fund assets, L = Total fund liabilities, and N = Number of shares outstanding, then the **net asset value** is calculated as follows.

$$\text{NAV} = \frac{A - L}{N}$$

Robo-advisors have grown in popularity since the first one launched in 2008. They offer the advantages of 24/7 accessibility, low account minimums, efficient transactions, and extremely low cost compared to their human counterparts. Over $200 billion in assets were managed by these automated investment advisors in 2017, and this figure was projected to be $2 trillion by 2020. Mutual funds are a key component in robo-advisor portfolios.

EXAMPLE 7 Finding the Number of Shares in a Mutual Fund Purchase

Suppose, on a given day, a mutual fund has $500 million worth of stock, $500,000 in cash (not invested), and $300,000 in other assets. Total liabilities amount to $4 million, and there are 25 million shares outstanding. If Joanne invests $50,000 in this fund, how many shares will she obtain?

Solution

$$A = \$500,000,000 + \$500,000 + \$300,000$$

$$A = \$500,800,000$$

$$L = \$4,000,000$$

$$N = 25,000,000$$

$$\text{NAV} = \frac{A - L}{N} \qquad \text{Formula}$$

$$= \frac{\$500,800,000 - \$4,000,000}{25,000,000} \qquad \text{Substitute values of } A, L, \text{ and } N.$$

$$= \$19.872 \qquad \text{Calculate.}$$

Joanne's $50,000 investment will purchase.

$$\frac{\$50,000}{\$19.872} \approx 2516 \text{ shares.} \qquad \text{Nearest whole number}$$

Evaluating Investment Returns

Regardless of the type of investment you have or are considering (stocks, bonds, mutual funds, or others), it is important to be able to accurately evaluate and compare returns (profits or losses). While it is true that past performance is never a guarantee of future returns, it does provide information crucial to wise investing. A wealth of information on stock and bond markets and mutual funds is available in many publications (for example, the *Wall Street Journal, Investor's Business Daily*, and *Barron's Weekly*) and at numerous Web sites (for example, www.nasdaq.com, www.morningstar.com, and www.investopedia.com). Annual reports and other company filings for publicly traded companies and mutual funds are available at www.sec.gov/edgar/searchedgar/webusers.htm.

The most important measure of the performance of an investment is the **annual rate of return.** This is not *necessarily* the same as percentage return, as calculated in **Example 1,** because annual rate of return depends on the time period involved. Some commonly reported periods include daily, seven-day, monthly, month-to-date, quarterly, quarter-to-date, annual, year-to-date, 2-year, 3-year, 5-year, 10-year, 20-year, and "since inception" (since the fund or other investment vehicle was established).

The annual rate of return on an investment will depend on several factors, including the change in the value of each share during the investment period, whether the investment features dividends, how often dividends are distributed, and whether they are reinvested. **Examples 8 and 9** on the next page illustrate some of the many complications that may arise when one tries to evaluate and compare different investments.

EXAMPLE 8 Analyzing a Stock's Annual Rate of Return

Keshawn owns 50 shares of stock. His brokerage statement for the end of August showed that the stock closed that month at a price of $40 (per share), and the statement for the end of September showed a closing price of $40.19. For purposes of illustration, disregard any possible dividends, and assume that the money gained in a given month has no opportunity to earn additional returns. (Those earnings cannot be "reinvested.") Find the following.

(a) The value of these shares at the end of August

(b) Keshawn's monthly return on this stock

(c) The monthly percentage return   **(d)** The annual rate of return

Solution

(a)
$$\text{Value} = \overset{\text{Price per share}}{\$40} \cdot \overset{\text{Number of shares}}{50} = \$2000$$

(b)
$$\text{Return} = \overset{\text{Change in price per share}}{(\$40.19 - \$40)} \cdot \overset{\text{Number of shares}}{50}$$
$$= \$9.50$$

(c)
$$\text{Percentage return} = \frac{\text{Total return}}{\text{Total value}} \cdot 100\%$$
$$= \frac{\$9.50}{\$2000} \cdot 100\%$$
$$= 0.475\%$$

(d) Because monthly returns do not earn more returns, we use an "arithmetic" return here. Simply add 0.475% twelve times (or multiply 0.475% by 12) to obtain 5.7%. The annual rate of return is 5.7%.

EXAMPLE 9 Analyzing a Mutual Fund's Annual Rate of Return

Valerie owns 80 shares of a mutual fund with a net asset value of $10 on October 1. Assume that the only return that the fund earns is a dividend of 0.4% per month, which is automatically reinvested. Find the following.

(a) The value of Valerie's holdings in this fund on October 1

(b) The monthly return   **(c)** The annual rate of return

Solution

(a) Value = NAV $\cdot$ number of shares = $10 \cdot 80 = \$800$

(b) Monthly return = 0.4% of $800 = $3.20

(c) In this case, monthly returns get reinvested. Compounding occurs. Therefore, we use a "geometric" return rather than arithmetic. To make this calculation, first find the "return relative," which is 1 plus the monthly rate of return:

$$1 + 0.4\% = 1 + 0.004 = 1.004.$$

Then multiply this return relative 12 times (raise it to the 12th power) to get an annual return relative: $1.004^{12} \approx 1.049$. Finally, subtract 1 and multiply by 100 to convert back to a percentage rate. The annual rate of return is 4.9%.

A geometric return such as the one just found in **Example 9** often is called the "effective annual yield." (Notice that an arithmetic return calculation would have ignored the reinvestment compounding and would have understated the effective annual rate of return by 0.1%, because $12(0.4\%) = 4.8\%$.)

Building a Nest Egg

The most important function of investing, for most people, is to build an account for some future use, probably for retirement living. In **Example 8** of the first section of this chapter, we compared two retirement programs, Plan A and Plan B, to emphasize the importance of starting a retirement investment program early in life. The performance of investments over time is generally governed by formulas based on the summation of geometric sequences (see the margin note).

There are two major barriers to building a retirement nest egg (or whatever the goal is for accumulating wealth)—*inflation* and *(income) taxes.*

Table 4 (selected rows repeated at right) shows the actual historical inflation rates over 37 years (1980–2017). This reflects an average of about 3.0% per year. If we had invested for retirement over that time span, the inflation proportion indicates that we would have ended up, in 2017, paying about

Table 4   (repeated)
Consumer Price Index*
(Selected Years)

Year	Average CPI-U	% Change in CPI-U
1980	82.4	13.5
1985	107.6	3.6
1990	130.7	5.4
1995	152.4	2.8
2000	172.2	3.4
2005	195.3	3.4
2010	218.1	1.6
2015	237.0	0.1
2016	240.0	1.3
2017	245.1	2.1

Data from Bureau of Labor Statistics.
The period 1982 to 1984: 100

$$\frac{245.1}{82.4} \approx 3.0 \text{ times}$$

as much for goods and services as when we started, in 1980.

This inflationary effect, if unanticipated, could make a sizable dent in retirement living, but there is a fairly painless way to make provision for it. Periodically we can simply increase contributions by enough to keep pace with inflation.

The formula below allows us to compute the future values of an account where regular deposits are systematically adjusted for inflation. If the inflation rate is i, then the amount R is deposited at the end of year 1, $R(1 + i)$ is deposited at the end of year 2, and so on, with the deposit at the end of the year n, in general, being

$$R(1 + i)^{n-1}.$$

Putting off paying taxes for as long as possible, though often urged by financial advisors, is not always the best plan. The **Roth IRA,** available since 1998, is funded with taxed dollars, but all funds, both principal and earnings, are withdrawn later tax-free. For a young person, with many years to "grow" the investment, the prospect of avoiding taxes on that growth is attractive, especially if withdrawals will be made when the account holder is in a higher tax bracket than when contributions were made. Advantages also include less restriction on who qualifies, no forced withdrawals, and no restrictions on when withdrawals (of principal) can be made.

FUTURE VALUE OF AN INFLATION-ADJUSTED RETIREMENT ACCOUNT

Assume annual deposits into a retirement account, adjusted for inflation. If

i = annual rate of inflation,

R = initial deposit (at the end of year 1),

r = annual rate of return on money in the account,

and n = number of years deposits are made,

then the value V of the account at the end of n years is calculated as follows.

$$V = R\left[\frac{(1 + r)^n - (1 + i)^n}{r - i}\right]$$

EXAMPLE 10   **Adjusting a Retirement Account for Inflation**

In **Example 8** of the first section of this chapter, Anika executed Plan A, depositing $2000 on each birthday from age 21 to age 30, then stopped depositing, and at age 65 had $428,378. Barry, with Plan B, waited until age 31 to make the first $2000 deposit, then deposited $2000 every year to age 65, and came out with only $344,634.

Compute the final account value, at age 65, for Barry's Plan B, assuming that the 35 annual deposits had been adjusted for a 3.0% annual inflation rate.

Solution

Use the "inflation-adjusted" formula given in the box. The final account value is

$$V = \$2000\left[\frac{(1 + 0.08)^{35} - (1 + 0.030)^{35}}{0.08 - 0.030}\right] \approx \$478,859.$$

Let $R = \$2000$, $r = 0.08$, $i = 0.030$, and $n = 35$.

With inflation adjustment, Plan B, by age 65, accumulates $478,859.

The second barrier to building wealth is taxes. The more you invest and the more you earn, the higher the percentage our progressive tax system will claim of your earnings. An effective tool in softening this effect is to take full advantage of tax-deferred accounts, such as a TSA (Tax-Sheltered Annuity, or 403(b) Plan); an IRA (Individual Retirement Account, or 401(k) plan), especially if your employer will contribute matching funds; or, if self-employed, a SEP (Simplified Employee Pension) or a SIMPLE (Savings Incentive Match Plan for Employees) plan.

With most tax-deferred retirement accounts, the accumulated money is all taxed when it is withdrawn (presumably during retirement years). To demonstrate the decided advantage of tax deferral, we will compare the two situations using the following formulas. For a fair comparison, we look at both account values *after all taxes have been paid*.

TAX-DEFERRED VERSUS TAXABLE RETIREMENT ACCOUNTS

In both cases,

$R =$ amount withheld annually from current salary to build the retirement account,

$n =$ number of years contributions are made,

$r =$ annual rate of return on money in the account,

$t =$ marginal tax rate of account holder,

$V =$ final value of the account, **following all accumulations and payment of all taxes.**

Tax-Deferred Account
The entire amount R is contributed each year. All contributions, plus earnings, earn a return over the n years, at which time tax is paid on all money in the account. The final account value is calculated as follows.

$$V = \frac{(1 - t)R[(1 + r)^n - 1]}{r}$$

Taxable Account
Each amount R withheld from salary is taxed up front, decreasing the amount of the annual deposit to $(1 - t)R$. Furthermore, annual earnings are also taxed at the end of each year. But at the end of the accumulation period, no more tax will be due. The final account value is calculated as follows.

$$V = \frac{R[(1 + r(1 - t))^n - 1]}{r}$$

EXAMPLE 11 Comparing Retirement Accounts

Leah and Tyrell, both in a 22% marginal tax bracket, will each contribute $2000 annually for 20 years to retirement accounts that return 6% annually. Leah chooses a tax-deferred account, while Tyrell chooses a taxable account. Compare their final account values at the end of the accumulation period, after payment of all taxes.

Solution

We have $R = \$2000$, $n = 20$, $r = 0.06$, and $t = 0.22$.

Leah: $V = \dfrac{(1 - 0.22)\$2000[(1 + 0.06)^{20} - 1]}{0.06} \approx \$57{,}386$ Substitute values in the "tax-deferred" formula, and calculate.

Tyrell: $V = \dfrac{\$2000[(1 + 0.06(1 - 0.22))^{20} - 1]}{0.06} \approx \$49{,}872$ Substitute values in the "taxable" formula, and calculate.

$$\$57{,}386 - \$49{,}872 = \$7514$$

By deferring taxes, Leah ends up with $7514, or about 15%, more.

EXAMPLE 12 Comparing Retirement Accounts

Repeat **Example 11,** but this time let

$$R = \$5000, \quad n = 40, \quad r = 0.10, \quad \text{and} \quad t = 0.35.$$

Solution

Leah: $V = \dfrac{(1 - 0.35)\$5000[(1 + 0.10)^{40} - 1]}{0.10}$ Substitute values in the "tax-deferred" formula.

$\approx \$1{,}438{,}426$

Tyrell: $V = \dfrac{\$5000[(1 + 0.10(1 - 0.35))^{40} - 1]}{0.10}$ Substitute values in the "taxable" formula.

$\approx \$570{,}804$

With these higher parameters, Leah's advantage is dramatic:

$$1{,}438{,}426 - 570{,}804 = \$867{,}622, \quad \text{or about 152\%, more.}$$

Tax-deferred is not the same as "tax-free," which makes the result of **Example 12** less impressive than it may seem. This type of retirement income is taxed at the time it is withdrawn. People who sell tax-deferred plans stress that you will probably be in a lower tax bracket after retirement because of reduced income. But many find this is not so, especially compared to early career income, and especially if your retirement account investments have done very well, which could make your required withdrawals quite large. Besides, who can tell what general tax rates will be 30 years from now?

13.5 EXERCISES

CONCEPT CHECK *Fill in the blank with the correct response.*

1. Returns (profits) on stock investments come in the form of _____ and _____.

2. The NYSE and NASDAQ are examples of _____.

3. Full-service brokers provide more services than discount brokers, but they also charge higher _____.

4. _____ provide a way to invest in a company without owning a part of the company.

5. A(n) _____ is a popular investment vehicle that provides diversification through ownership of many different stocks and/or bonds.

6. Calculating the _____ on various investments enables one to compare their performances.

Refer to the stock table (**Table 14** *) for Exercises 7–34.*

Reading Stock Charts *Find each of the following.*

7. closing price for Deere & Company (DE)

8. 52-week high for Facebook Inc (FB)

9. percent change from the previous day for General Mills (GIS)

10. sales of the day for Twitter (TWTR)

11. year-to-date percentage change for Pearson Plc (PSO)

12. dividend for Chevron Corp (CVX)

13. the stock with the highest price-to-earnings ratio

14. the stock with the highest dividend yield

15. the stock with the highest volume (most shares sold) for the day

16. the stock with the highest opening price

Finding Stock Costs   *Find the basic cost (ignoring any broker fees) for each stock purchase, at the day's closing price.*

17. 300 shares of Allstate (ALL)

18. 200 shares of Walt Disney (DIS)

19. 400 shares of General Mills (GIS)

20. 500 shares of Facebook (FB)

Finding Stock Costs   *Find the cost, at the day's closing price, for each stock purchase. Include typical discount broker commissions as described in the text, using Firm B.*

	Stock Symbol	Number of Shares	Transaction Type
21.	FB	60	broker-assisted
22.	HPQ	70	broker-assisted
23.	BAC	355	phone
24.	TWTR	585	phone
25.	HOG	2500	online
26.	CVX	1500	online

Finding Receipts for Stock Sales   *Find the amount received by the sellers of each stock (at the day's closing prices). Deduct sales expenses as described in the text, using Firm C.*

	Stock Symbol	Number of Shares	Transaction Type
27.	ALL	400	broker-assisted
28.	KMX	600	broker-assisted
29.	CVX	500	phone
30.	DE	700	phone
31.	FDX	1350	online
32.	PSO	2740	online

Finding Net Results of Combined Transactions   *For each combined transaction (executed at the closing price of the day), find the net amount paid out or taken in. Assume typical expenses as outlined in the text, using Firm A.*

33. Genna bought 100 shares of Walt Disney and sold 120 shares of General Mills, both broker-assisted trades.

34. Jasmine bought 800 shares of Bank of America and sold 200 shares of Chevron, both online trades.

Costs and Returns of Stock Investments   *For each of the following stock investments, find*

(a) *the total purchase price,*

(b) *the total dividend amount,*

(c) *the capital gain or loss,*

(d) *the total return, and*

(e) *the percentage return. Ignore broker and SEC fees.*

	Number of Shares	Purchase Price per Share	Dividend per Share	Sale Price per Share
35.	40	$20.00	$2.00	$44.00
36.	20	$25.00	$1.00	$22.00
37.	100	$12.50	$1.08	$10.15
38.	200	$8.80	$1.12	$11.30

Total Return on Bond Investments   *For each of the following bond investments, find the total return.*

	Face Value	Annual Interest Rate	Term to Maturity
39.	$1000	5.5%	5 years
40.	$5000	6.4%	10 years
41.	$10,000	7.11%	3 months
42.	$50,000	4.88%	6 months

Net Asset Value of a Mutual Fund   *For each mutual fund investment, find* **(a)** *the net asset value, and* **(b)** *the number of shares purchased.*

	Amount Invested	Total Fund Assets	Total Fund Liabilities	Total Shares Outstanding
43.	$3500	$875 million	$36 million	80 million
44.	$1800	$643 million	$102 million	50 million
45.	$25,470	$2.31 billion	$135 million	263 million
46.	$83,250	$1.48 billion	$84 million	112 million

Finding Monthly and Annual Investment Returns *For each investment, assume that there is no opportunity for reinvestment of returns. In each case find* **(a)** *the monthly return,* **(b)** *the annual return, and* **(c)** *the annual percentage return.*

	Amount Invested	Monthly Percentage Return
47.	$645	1.3%
48.	$895	0.9%
49.	$2498	2.3%
50.	$4983	1.8%

Effective Annual Rate of Return of Mutual Fund Investments *Assume that each mutual fund investment earns monthly returns that are reinvested and subsequently earn at the same rate. In each case find* **(a)** *the beginning value of the investment,* **(b)** *the first monthly return, and* **(c)** *the effective annual yield.*

	Beginning NAV	Number of Shares Purchased	Monthly Percentage Return
51.	$9.63	125	1.5%
52.	$12.40	185	2.3%
53.	$11.94	350	1.83%
54.	$18.54	548	2.22%

Inflation-Adjusted Retirement Accounts *Find the future value (to the nearest dollar) of each inflation-adjusted retirement account. Deposits are made at the end of each year.*

	Annual Inflation Rate	Initial Deposit	Annual Rate of Return	Number of Years
55.	3%	$2000	5%	25
56.	2%	$1000	6%	20
57.	1%	$5000	6.5%	40
58.	4%	$2500	7%	30

The Effect of Tax Deferral on Retirement Accounts *Find the final value, after all taxes are paid, for each account if* **(a)** *taxes are deferred, or* **(b)** *taxes are not deferred. In both cases, deposits are made at the end of each year.*

	Marginal Tax Rate	Regular Deferred Contribution	Annual Rate of Return	Number of Years
59.	12%	$1000	5%	30
60.	10%	$500	7%	15
61.	35%	$1500	6%	10
62.	24%	$3000	10%	25

Required Annual Contributions to Reach Retirement Goal *The formulas for finding final values of tax-deferred and taxable retirement accounts can be solved for R to find the required annual contributions necessary to reach an amount V by the time you retire.*

63. (a) Solve the "tax-deferred" account formula for R.

   (b) Solve the "taxable" account formula for R.

Deciding How Much to Put Aside *Suppose that you plan to retire in 30 years and will need $1,000,000 (tax-paid) in your retirement account to do so. In Exercises 64 and 65, determine the required annual contribution toward retirement if your retirement account is* **(a)** *tax-deferred and* **(b)** *taxable. Assume that your marginal tax rate will remain at 25%.*

64. Suppose the annual rate of return on the money in your account is 6%.

65. Suppose the annual rate of return on the money in your account is 8%.

Solve each problem.

66. Monthly Deposits Adjusted for Inflation In an inflation-adjusted retirement account, let i denote annual inflation, let r denote annual return, and let n denote number of years, as in the text, but suppose deposits are made monthly rather than yearly with initial deposit R.

   (a) What expression represents the amount that would be deposited at the end of the second month?

   (b) Modify the future value formula for an inflation-adjusted account to accommodate the more frequent deposits.

   (c) Find (to the nearest dollar) the future value for monthly deposits of $100 (initially) for 20 years if $i = 0.03$ and $r = 0.06$.

67. Finding an Unknown Rate Given the compound interest formula

$$A = P\left(1 + \frac{r}{n}\right)^{nt},$$

do the following.

   (a) Solve the formula for r.

   (b) Find the annual rate r if $50 grows to $85.49 in 10 years with interest compounded quarterly. Give r to the nearest tenth of a percent.

68. Comparing Continuous with Annual Compounding It was suggested in the text that inflation is more accurately reflected by the continuous compounding formula than by periodic compounding. Compute the ratio

$$\frac{Pe^{rt}}{P(1 + r)^n} = \frac{e^{rt}}{(1 + r)^n}$$

to find how much (to the nearest tenth of a percent) continuous compounding exceeds annual compounding. Use a rate of 3% over a period of 30 years.

Spreading Mutual Fund Investments among Asset Classes Mutual funds (as well as other kinds of investments) normally are categorized into one of several "asset classes," according to the kinds of stocks or other securities they hold. Small capitalization funds are most aggressive, while cash is most conservative. Many investors, often with the help of an advisor, try to construct their portfolios in accordance with their stage of life. Basically, the idea is that a younger person can afford to be more aggressive (and assume more risk), while an older investor should be more conservative (assuming less risk). The investment diagrams here show typical percentage ranges that might be recommended by an investment advisor.

In Exercises 69–72, divide the given investor's money into the five categories so as to position them right at the middle of the recommended ranges.

Asset Classes

(a)	Aggressive Growth (small cap)
(b)	Growth
(c)	Growth & Income
(d)	Income
(e)	Cash

69. Alyssa, in her early investing years, with $20,000 to invest

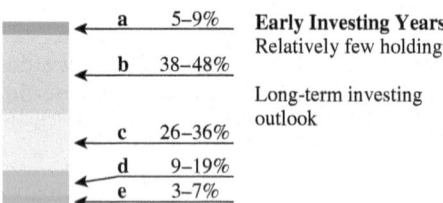

a 5–9% **Early Investing Years**
b 38–48% Relatively few holdings
c 26–36% Long-term investing outlook
d 9–19%
e 3–7%

70. Rohit, in his good earning years, with $250,000 to invest

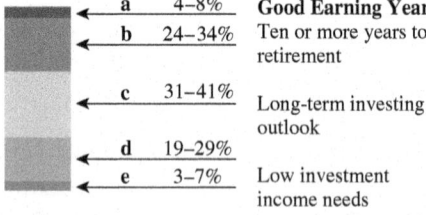

a 4–8% **Good Earning Years**
b 24–34% Ten or more years to retirement
c 31–41% Long-term investing outlook
d 19–29%
e 3–7% Low investment income needs

71. Akira, in his high-income/saving years, with $400,000 to invest

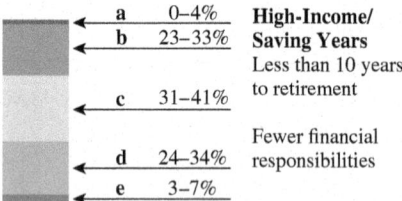

a 0–4% **High-Income/**
b 23–33% **Saving Years**
Less than 10 years to retirement
c 31–41%
Fewer financial
d 24–34% responsibilities
e 3–7%

72. Eve, retired, with $845,000 to invest

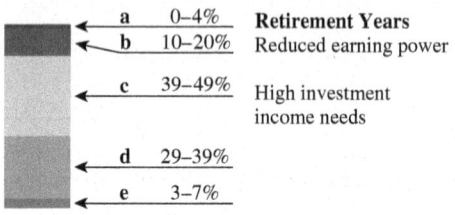

a 0–4% **Retirement Years**
b 10–20% Reduced earning power
c 39–49% High investment income needs
d 29–39%
e 3–7%

The Effect of Taxes on Investment Returns Income from municipal bonds generally is exempt from federal, and sometimes state, income tax. For an investor with a marginal combined state and federal tax rate of 35%, a taxable return of 6% would yield only 3.9% (65% of 6%). At that tax rate, a tax-exempt return of 3.9% is equivalent to a taxable return of 6%. Find the tax-exempt rate of return that is equivalent to the given taxable rate of return for each investor.

	Investor	Marginal Combined Tax Rate	Taxable Rate of Return
73.	Joshua	25%	5%
74.	Sherry	30%	7%
75.	Berna	35%	8%
76.	Tabari	40%	10%

Write responses to Exercises 77–84. Some research may be required.

77. Describe "dollar-cost averaging," and relate it to advantage number 1 of mutual funds as listed in the text.

78. From **Table 14** in the text, would you say that June 15, 2018, was a "good day" on Wall Street or a "bad day"?

79. Log onto www.napfa.org to research financial planners. Then report on what you would look for in a planner.

80. Discuss the advantages and disadvantages of using "robo-advisors," particularly in a volatile economy. Are you using, or would you consider using, a robo-advisor to build a nest egg? Explain.

81. Discuss the similarities and differences between mutual funds and exchange-traded funds. What advantages and/or risks are associated with each of these investment vehicles?

82. What is a *growth stock?* What is an *income stock?*

83. With respect to investing in corporate America, describe the difference between being an owner and being a lender.

84. Comment on the graph shown here.

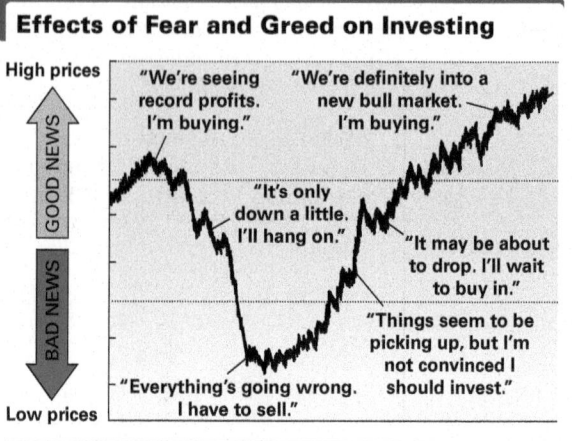

Effects of Fear and Greed on Investing

FOR FURTHER THOUGHT

Summary of Investments

The summary of various types of investments in **Table 16** is based on average cases and is intended only as a general comparison of investment opportunities. There are numerous exceptions to the characteristics shown. For example, although the table shows no selling fees for mutual funds, this is really true only for so-called no-load funds. The "loaded" funds do charge "early redemption fees" or other kinds of sales charges. This is not to say that a no-load fund is necessarily better. For example, the absence of sales fees may be offset by higher "management fees."

These various costs, as well as other important factors, will be clearly specified in a fund's prospectus and should be studied carefully before a purchase decision is made. Some expertise is advisable for mutual fund investing, just less so than with certain other options, like direct stock purchases, raw land, or collectibles.

To use the table, read across the columns to find the general characteristics of the investment.

"Liquidity" refers to the ability to get cash from the investment quickly.

For Group or Individual Investigation

List **(a)** some investment types you would recommend considering and **(b)** some you would recommend avoiding for a friend whose main investment objective is as follows.

1. Avoid losing money.
2. Avoid losing purchasing power.
3. Avoid having to learn about investments.
4. Avoid having to work on investments.
5. Be sure to realize a gain.
6. Be able to "cash in" at any time.
7. Get as high a return as possible.
8. Be able to "cash in" without paying fees.
9. Realize quick profits.
10. Realize growth over the long term.

Table 16 Summary of Investments

Investment	Protection of Principal	Protection against Inflation	Rate of Return (%)	Certainty of Return	Selling Fees	Liquidity	Long-term Growth	Requires Work from Investor	Expert Knowledge Required
Cash	excellent	none	0			excellent	no	no	no
Ordinary life insurance	good	poor	2–4	excellent	high	excellent	no	no	no
Savings bonds	excellent	poor	3–5	excellent	none	excellent	no	no	no
Bank savings	excellent	poor	2–4	excellent	none	excellent	no	no	no
Credit union	excellent	poor	3–4	excellent	none	excellent	no	no	no
Corporation bonds	good	poor	5–7	excellent	low	good	no	some	some
Tax-free municipal bonds	excellent	poor	2–5	excellent	medium	good	no	no	some
Corporation stock	good	good	3–12	good	medium	good	yes	some	some
Preferred stock	good	poor	4–8	excellent	medium	good	no	some	some
Mutual funds	good	good	3–12	good	none	good	yes	no	no
Your own home	good	good	4–10	good	medium	poor	yes	yes	no

Table 16 Summary of Investments (continued)

Investment	Protection of Principal	Protection against Inflation	Rate of Return (%)	Certainty of Return	Selling Fees	Liquidity	Long-term Growth	Requires Work from Investor	Expert Knowledge Required
Mortgages on the homes of others	fair	poor	5–8	fair	high	poor	no	yes	some
Raw land	fair	good	?	fair	medium	poor	yes	no	yes
Rental properties	good	good	4–8	fair	medium	poor	yes	yes	some
Stamps, coins, antiques	fair	good	0–10	fair	high	poor	yes	yes	yes
Your own business	fair	good	0–20	poor	high	poor	yes	yes	yes

Chapter 13 SUMMARY

KEY TERMS

13.1

interest
principal
interest rate
simple interest
compound interest
future (maturity) value
present value
period
nominal rate
effective annual yield
price inflation
consumer price index (CPI)
deflation
recession
years to double
rule of 70

13.2

consumer credit
installment loan

closed-end credit
revolving loan
open-end credit
credit limit
add-on interest
itemized billing
finance charges
average daily balance method

13.3

Truth in Lending Act
annual percentage rate (APR)
unearned interest
actuarial method
rule of 78
early payment penalty
payoff amount

13.4

mortgage
term

down payment
principal amount
fixed-rate mortgage
amortizing
amortization schedule
adjustable-rate mortgage (ARM)
closing costs
escrow account
points
private mortgage insurance
recurring costs
property taxes
homeowner's insurance
maintenance
opportunity cost

13.5

dividend
capital gain
capital loss

return on investment
stock exchanges
price-to-earnings ratio
commission
full-service broker
discount broker
principal
round lot
odd lot
odd-lot differential
limit order
automated trades
broker-assisted trades
SEC fee
bond
face value
mutual fund
net asset value (NAV)
annual rate of return

TEST YOUR WORD POWER

See how well you have learned the vocabulary in this chapter.

1. Compound interest is

A. calculated only on the original principal, and is generally more than simple interest.

B. calculated on principal and previously earned interest.

C. interest earned by two or more investments simultaneously.

D. generally less than simple interest, but more easily calculated.

2. An **installment loan** is
A. a type of open-ended credit loan with an undetermined number of payments.
B. a loan in which additional credit may be extended before the initial amount is paid off.
C. a type of closed-end credit that involves a predetermined number of equal payments.
D. a loan extended specifically for the installation of appliances or furniture.

3. Unearned interest is
A. the amount of income a taxpayer receives from investments as opposed to wages.
B. the penalty paid by the borrower for being delinquent on one or more monthly payments.
C. the amount by which the original finance charge is reduced when a loan is paid off early.
D. the interest paid for borrowing as opposed to that received from an investment.

4. An **escrow account** (or impound account) is
A. an account maintained by a lender and used to pay property taxes and insurance premiums.
B. an account that is terminated when documents are signed and escrow "closes."

C. always a requirement to obtain a mortgage.
D. a savings account from which principal and interest payments are made.

5. A **mortgage** is
A. an extension of unsecured credit, usually for a short period of time.
B. a form of unsecured debt, in which wages may be garnished if payments are missed.
C. the process of dividing principal and interest into equal periodic payments.
D. a loan for a substantial amount, and for which property is pledged as security.

6. The **capital gain** on an investment is
A. the overall return on the investment, including any dividends.
B. the profit resulting from selling the investment for more than the purchase price.
C. the annual increase in the value of the investment.
D. the difference between the purchase price and the selling price.

ANSWERS
1. B **2.** C **3.** C **4.** A **5.** D **6.** B

QUICK REVIEW

Concepts | *Examples*

13.1 THE TIME VALUE OF MONEY

Simple Interest
If principal P is borrowed for t years at an annual interest rate r, then (simple) interest is calculated as follows.

$$I = Prt$$

Future Value for Simple Interest
$$A = P(1 + rt)$$

Future Value for Compound Interest
Interest is compounded n times per year.

$$A = P\left(1 + \frac{r}{n}\right)^{nt}$$

If \$2000 is borrowed at 4.8% simple interest for 5 years, find the future value (at the end of the 5-year term).

$$A = P(1 + rt)$$
$$= \$2000[1 + 0.048(5)]$$
$$= \$2480$$

If the \$2000 is borrowed at 4.8% compounded monthly, find the future value (at the end of the 5-year term).

$$A = \$2000\left(1 + \frac{0.048}{12}\right)^{12(5)}$$
$$= \$2000(1.004)^{60}$$
$$= \$2541.28$$

Thus \$480 interest was paid with the simple interest loan, while \$541.28 was paid with compound interest.

Present Value for Compound Interest
Solve the future-value formula for P.

$$P = \frac{A}{\left(1 + \frac{r}{n}\right)^{nt}}$$

A company anticipates needing \$20,000 to purchase new equipment. How much must the company invest now at 6% annual interest, compounded daily, in order to have the necessary amount in 5 years?

$$P = \frac{\$20,000}{\left(1 + \frac{0.06}{365}\right)^{365(5)}} = \$14,816.73$$

Concepts	*Examples*
Effective Annual Yield The effective annual yield Y of an account paying a nominal annual rate r, compounded n times per year, is calculated as follows. $$Y = \left(1 + \frac{r}{n}\right)^n - 1$$	Find the effective annual yield of an account paying 3.000% annual interest, compounded quarterly. $$Y = \left(1 + \frac{0.03}{4}\right)^4 - 1 = 3.034\%$$
Future Value for Continuous Compounding $$A = Pe^{rt}$$	If \$12,500 is invested at 2.5% compounded continuously, find the account value after 7 years. $$A = \$12{,}500e^{0.025(7)} = \$14{,}890.58$$

13.2   CONSUMER CREDIT

Common Types of Consumer Credit **Installment loans** involve a set amount borrowed, to be paid off by a predetermined number of equal payments **(installments).** The total of the payments will be the loan amount plus the interest added on.	A business owner purchases \$5200 worth of office equipment by making a down payment of \$1000 and financing the remaining balance with an installment loan with an interest rate of 5.5% over 2 years. Find the total cost of the purchase. Amount financed: $P = \$5200 - \$1000 = \$4200$ Interest paid: $I = Prt$ $\qquad\qquad = \$4200(0.055)(2)$ $\qquad\qquad = \$462$ Amount to be repaid: $P + I = \$4200 + \$462 = \$4662$ Monthly payment: $\dfrac{\$4662}{24} = \$194.25$ Total cost: $\$1000 \;+\; \$4662 \;=\; \$5662$ $\qquad\qquad\quad\;\uparrow\qquad\qquad\;\uparrow$ $\qquad\qquad\;$Down$\qquad$Loan $\qquad\quad\;$payment$\;$ repayment
Revolving loans (such as credit card accounts) involve **open-end credit** up to a predetermined **credit limit.** Charges are made and then itemized at the end of each billing period (usually a month). **Finance charges** are generally determined by the **average daily balance** for a given billing period.	A bank charges 1.4% per month on average daily balances for credit cards it issues. A billing period opens on Sept. 3 with a balance of \$327.56. Transactions on the account for the period included a payment of \$100 on Sept. 13 and a purchase in the amount of \$158 on Sept. 25. Find the average daily balance, the finance charge, and the account balance for the Oct. 3 billing. The running balance was \$327.56 from Sept. 3 to Sept. 13 (for 10 days), \$227.56 from Sept. 13 to Sept. 25 (for 12 days), and \$385.56 from Sept. 25 to Oct. 3 (8 days). Average daily balance for the period: $$\frac{\$327.56(10) + \$227.56(12) + \$385.56(8)}{30} = \$303.03$$ Finance charge for Oct. 3 billing: $$0.014(\$303.03) = \$4.24$$ The account balance for the Oct. 3 billing statement will be the sum of the last running balance and the finance charge. $$\text{Ending balance} = \$385.56 + \$4.24$$ $$= \$389.80$$

Concepts　　　　　　　　　　　　　　　　　　　*Examples*

13.3　TRUTH IN LENDING

Lenders must (as required by the Truth in Lending Act) divulge the true cost of a loan by reporting the **annual percentage rate (APR)** to the borrower. **Table 5** in this section gives the APR for a loan with a given number of monthly payments and a given finance charge (per $100 financed).

Deon purchases a $10,200 car. He makes a $1500 down payment and agrees to pay 36 monthly payments of $270.62. The dealer claims a 7% annual interest rate.

　　Is this dealer being truthful in lending?

Amount financed:　$10,200 − $1500 = $8700

Cost of 36 monthly payments:　36 · $270.62 = $9742.32

Total finance charge:　$9742.32 − $8700 = $1042.32

Finance charge per $100 financed:

$$\frac{\$1042.32}{\$8700} \cdot \$100 = \$11.98$$

Table 5 shows that for 36 monthly payments, this finance charge per $100 financed corresponds to an APR of 7.5%. *The dealer is not being truthful in lending.*

Calculating Unearned Interest

When a borrower pays off a loan early, the lender has not "earned" the full amount of interest originally disclosed when the loan was initiated. The amount of **unearned interest** u is calculated in one of the following two ways.

Deon straightened things out with the dealer and financed the $8700 at 7% APR, which resulted in 36 monthly payments of $268.63. But he has decided to pay the loan off 12 months early.

　　Which method of calculating the payoff amount would be better for Deon?

Actuarial Method

$$u = kR\left(\frac{h}{\$100 + h}\right)$$

Here R is the regular monthly payment, k is the number of scheduled payments remaining, and h is the finance charge per $100 for a loan with the same APR and k monthly payments.

Actuarial method: We need to find h, the finance charge per $100 financed at 7% APR for 12 months. **Table 5** gives a value of $h = \$3.83$.

Unearned interest:

$$u = 12(\$268.63)\left(\frac{\$3.83}{\$100 + \$3.83}\right) = \$118.91$$

Payoff amount:

$$13 \cdot \$268.63 - \$118.91 = \$3373.28$$

Rule of 78

$$u = \frac{k(k+1)}{n(n+1)} \cdot F$$

Here F is the original finance charge, n is the number of payments originally scheduled, and k is the number of scheduled payments remaining.

Rule of 78: We need the corrected finance charge (using the smaller monthly payments).

$$F = 36 \cdot \$268.63 - \$8700 = \$970.68$$

Unearned interest:

$$u = \frac{12(13)}{36(37)} \cdot \$970.68 = \$113.68$$

Payoff Amount

To pay the loan off with k payments remaining (after the current payment), the borrower would need to pay the following.

$$\textbf{Payoff amount} = (k+1)R - u$$

Payoff amount:

$$13 \cdot \$268.63 - \$113.68 = \$3378.51$$

It is to Deon's advantage if the *actuarial method* is used, since the payoff amount would be $5.23 less.

Concepts	Examples

Finance Charge Per $100 Financed

If a loan is structured in such a way that its APR or number of remaining payments (or both) are not found in **Table 5,** then h, the finance charge per $100 financed, may be calculated using the following formula.

$$h = \frac{n \cdot \frac{APR}{12} \cdot \$100}{1 - \left(1 + \frac{APR}{12}\right)^{-n}} - \$100$$

Here n is the number of monthly payments required to pay off the loan.

Deon financed the $8700 at 7.8% interest and agreed to 42 monthly payments of $237.37. He decides to pay off the loan at the time of his 34th payment.

What is the finance charge per $100 financed?

$$h = \frac{8 \cdot \frac{0.078}{12} \cdot \$100}{1 - \left(1 + \frac{0.078}{12}\right)^{-8}} - \$100 = \$2.95 \qquad \text{Here } n = 42 - 34 = 8.$$

This figure can be used to calculate the unearned interest and the payoff amount by the actuarial method.

13.4 THE COSTS AND ADVANTAGES OF HOME OWNERSHIP

Amortizing a Mortgage

When a buyer borrows money to make a home purchase, the lender holds the property as security for the loan. Such an arrangement is a **mortgage.** A **fixed-rate mortgage** has regular periodic payments over the life **(term)** of the loan.

$$R = \frac{P\left(\frac{r}{12}\right)}{1 - \left(\frac{12}{12 + r}\right)^{12t}}$$

An **amortization schedule** shows all payments for the life of the loan, broken down into allocations for principal and interest for each payment.

Find the monthly payment on an $80,000 fixed-rate loan at 5.0% over 30 years.

$$R = \frac{\$80,000\left(\frac{0.05}{12}\right)}{1 - \left(\frac{12}{12 + 0.05}\right)^{12(30)}} = \$429.46$$

Amortization schedule for this loan (in part):

Payment Number	Interest Payment	Principal Payment	Balance of Principal
			$80,000.00
1	$333.33	$96.13	$79,903.87
2	$332.93	$96.53	$79,807.35
358	$5.31	$424.15	$851.35
359	$3.55	$425.91	$425.44
360	$1.77	$425.44	$0.00

An **adjustable-rate mortgage (ARM)** has a rate based on a fluctuating **index** and a **margin** determined by the lender. The introductory rate will stay fixed for a period of time, and then it will adjust according to the current value of the index.

Molly takes out a 5/1 ARM for $120,000, with a 30-year term. The introductory rate is determined by an index at 1.7% and a margin of 2.5% (a 4.2% rate). The initial payment amount for this loan is $586.82, and the amortization schedule for the first five years shows a principal balance of $108,883.79 after the 60th payment.

Payment Number	Interest Payment	Principal Payment	Balance of Principal
			$120,000.00
1	$420.00	$166.82	$119,833.18
2	$419.42	$167.40	$119,665.78
60	$381.81	$205.01	$108,883.79

Concepts	Examples

Concepts

When an adjustment occurs, the new payment amount is found using the above formula, replacing P with the principal balance at the end of the last adjustment period, replacing r with the new adjusted rate, and replacing t with the number of years remaining in the original term.

In addition to the **down payment** made on a mortgage, there are other one-time expenses involved in purchasing a home. These are the **closing costs.**

Recurring Costs of Homeownership
These include the following.
- **property taxes**
- **homeowner's insurance**
- **maintenance** of the property

Examples

At the end of five years, the index has increased to 2.4%. Adding the 2.5% margin gives an adjusted rate of 4.9% for the sixth year.

Find the new amount for payments 61–72.

$$R = \frac{\$108{,}883.79 \left(\frac{0.049}{12}\right)}{1 - \left(\frac{12}{12 + 0.049}\right)^{12(25)}} = \$630.20 \quad \text{Monthly payments jump \$43.38.}$$

Mary takes out a 20-year fixed rate mortgage at 4.5% for $116,000 to purchase a $145,000 home, and uses a "direct lender" to avoid broker fees. The loan closes eight days before the end of the month. Closing costs are shown.

Loan origination fee (2%)	$?
Document fees	525
Transfer taxes	140
Title services	350
Title insurance	600
Recording fees	90
Escrow deposit	1305
Daily interest charges	?

Compute the total closing costs for this mortgage.

Loan origination fee:

$$0.02(\$116{,}000) = \$2320$$

Daily interest charges (for 8 days on the beginning balance):

$$8 \cdot \$116{,}000 \cdot \frac{0.045}{365} \approx \$114$$

Add these amounts to those listed above.

$$\$5444 \quad \longleftarrow \text{Total closing costs}$$

Mary's combined tax bracket is 30%, and property taxes in her county amount to 1.2% of property value. Her homeowner's insurance rate is 0.3% of property value, and annual maintenance costs are estimated to be 0.4% of the purchase price.

Find the recurring costs (for the first year).

$$\text{Property taxes} = 0.012(\$145{,}000)(0.7)$$
$$= \$1218$$
$$\text{Insurance} = 0.003(\$145{,}000)$$
$$= \$435$$
$$\text{Maintenance} = 0.004(\$145{,}000)$$
$$= \$580$$

Concepts *Examples*

13.5 FINANCIAL INVESTMENTS

Classes of Investments

Investing puts capital to work for the investor.

Stocks

Purchasing company stock makes the investor a part owner of the company, so the investor shares in the profits (in the form of **dividends**) and in the growth (or decline) of the company's value over time.

An investor purchased 200 shares of FedEx Corp (FDX) stock at the closing price for June 15, 2018. He kept the stock for a year (during which one dividend was paid out) and then sold at a price of $275.25 per share. Assume the brokerage firm charged $5.95 for each transaction.

Find the return on investment for the FDX shares.

Table 14 in this section shows that the closing price of FDX stock on June 15, 2018 was $264.56.

$$\text{Basic cost:} \quad 200 \cdot \$264.56 = \$52,912.00$$

Table 14 shows the dividend to be $2.00 per share.

$$\text{Dividend for 200 shares:} \quad 200 \cdot \$2.00 = \$400.00$$

Proceeds from sale of the shares a year later:

$$200 \cdot \$275.25 = \$55,050.00$$

$$\text{SEC fee:} \quad \$55,050.00 \cdot \left(\frac{\$13.00}{\$1,000,000}\right) \approx \$0.71565 \rightarrow \$0.72$$

Expenses	
Cost of purchase	$52,912.00
Purchase commission	$5.95
Sale commission	$5.95
SEC fee	$0.72
Total:	**$52,924.62**

Income	
Proceeds from sale	$55,050.00
Dividend	$400.00
Total:	**$55,450.00**

Total income − Total expenses = Return on investment

$$\$55,450.00 - \$52,924.62 = \$2525.38$$

Bonds

When an investor purchases a bond, the issuing company agrees to pay a fixed rate of interest over the term of the bond, and then to pay back the original purchase price when the bond matures (at the end of the term).

Find the return on a $2000, 2-year corporate bond paying 4% annual interest.

$$I = Prt = \$2000(0.04)(2) = \$160$$

Mutual Funds

A mutual fund is a pool of money collected by an investment company to buy stocks, bonds, and other investments.

Find the net asset value of a mutual fund having $1.5 billion in assets, $280 million in liabilities, and 40 million shares outstanding.

Net Asset Value (NAV) of a Mutual Fund

$$NAV = \frac{A - L}{N}$$

$$NAV = \frac{\$1,500,000,000 - \$280,000,000}{40,000,000} = \$30.50$$

Find the number of shares that $10,000 would purchase.

Here A = total fund assets, L = total fund liabilities, and N = number of shares outstanding.

$$\frac{\$10,000}{\$30.50} \approx 328 \text{ shares}$$

Concepts	*Examples*
Annual Rate of Return This measure of the performance of an investment may change depending on the time period for which this rate is calculated.	Seung owns 1000 shares of a mutual fund with a net asset value of $15.00 on July 1. The fund pays a 0.2% dividend per month (reinvested in the fund), and on August 1, the NAV had increased to $15.12. Find the annual rate of return on his investment.

Holdings July 1: $1000 \cdot \$15.00 = \$15,000.00$

Dividend for July: $0.002(\$15,000.00) = \$30.00$

New value August 1: $1000 \cdot \$15.12 = \$15,120.00$

Holdings August 1: $\$15,120.00 + \$30.00 = \$15,150.00$

Rate of return for one month:

$$\frac{\$15,150 - \$15,000}{\$15,000} \cdot 100\% = 1.0\%$$

Annual rate of return for that month:

$$(1 + 0.01)^{12} - 1 \approx 0.127 = 12.7\%$$

Future Value of an Inflation-Adjusted Retirement Account

$$V = R\left[\frac{(1 + r)^n - (1 + i)^n}{r - i}\right]$$

where i = annual rate of inflation,

R = initial deposit (end of year 1),

r = annual rate of return on account,

n = number of years deposits are made.

The inflation rate is 3% per year, and $5000 is invested in a retirement account (with a 6% annual rate of return) at the end of year 1. Deposits are increased by 3% each year.
    Find the account value after 20 years.

$$V = \$5000\left[\frac{(1 + 0.06)^{20} - (1 + 0.03)^{20}}{0.06 - 0.03}\right]$$

$$= \$233,504.04$$

Retirement accounts can be tax-deferred (meaning that monies invested and earnings are not taxed until they are withdrawn in retirement), or taxable. If

R = amount withheld annually from salary to build the retirement account,

n = number of years contributions are made,

r = annual rate of return on account, and

t = marginal tax rate of account holder,

then the final value V of the account (including all earnings and payment of taxes) is calculated as follows.

$$V = \frac{(1 - t)R[(1 + r)^n - 1]}{r} \quad \text{Tax-deferred account}$$

$$V = \frac{R[(1 + r(1 - t))^n - 1]}{r} \quad \text{Taxable account}$$

Monique wants to contribute $10,000 annually to a retirement account and wishes to retire in 25 years. Her marginal tax rate is 30%. Assume an inflation rate of 2.5% over the next 25 years and an annual rate of return of 8% on investments.
    Should Monique invest in a tax-deferred account or a taxable account?

$$V = \frac{(1 - 0.30)(\$10,000)(1.08^{25} - 1)}{0.08} \quad \text{Tax-deferred account}$$

$$= \$511,741.58 \quad \leftarrow \text{Better option}$$

$$V = \frac{\$10,000[(1 + 0.08(0.7))^{25} - 1]}{0.08} \quad \text{Taxable account}$$

$$= \$363,099.06$$

The tax-deferred account is better by

$\$511,741.58 - \$363,099.06 = \$148,642.52$, or about 41%.

Chapter **13** TEST

Find all monetary answers to the nearest cent. Use tables and formulas from the chapter, or financial calculators, as necessary.

Finding the Future Value of a Deposit *Find the future value of each deposit.*

1. $1000 for 5 years at 4% simple interest

2. $500 for 2 years at 6% compounded quarterly

Solve each problem.

3. **Effective Annual Yield of an Account** Find the effective annual yield to the nearest hundredth of a percent for an account paying 3% compounded monthly.

4. **Years to Double by the Rule of 70** Use the rule of 70 to estimate the years to double at an inflation rate of 4%.

5. **Finding the Present Value of a Deposit** What amount deposited today in an account paying 4% compounded semi-annually would grow to $100,000 in 10 years?

6. **Finding Bank Card Interest by the Average Daily Balance Method** Jolene's MasterCard statement shows an average daily balance of $680. Find the interest due if the rate is 1.6% per month.

Analyzing a Consumer Loan *Julio buys a koi fish pond (and fish to put in it) for his wife on their anniversary. He pays $8000 for the pond with $2000 down. The dealer charges add-on interest of 3.5% per year, and Julio agrees to pay the loan with 36 equal monthly payments. Use this information for Exercises 7–10.*

7. Find the total amount of interest he will pay.

8. Find the monthly payment.

9. Find the APR value (to the nearest half percent).

10. Find **(a)** the unearned interest and **(b)** the payoff amount if he repays the loan in full with 12 months remaining. Use the most accurate method available.

11. **True Annual Interest Rate in Consumer Financing** Hot Tub Heaven wants to include financing terms in its advertising. If the price of a hot tub is $3000 and the finance charge with no down payment is $115 over a 12-month period (12 equal monthly payments), find the true annual interest rate (APR).

12. Explain what a mutual fund is, and discuss several of its advantages and disadvantages.

Analyzing the Cost of a Home Mortgage *A 20% down payment is made on a home purchased for $240,000, and a 30-year fixed mortgage at 5% interest is used to pay the balance. Annual property taxes are $3000, and annual insurance is $720. Use this information for Exercises 13–17.*

13. Find the monthly payment for principal and interest.

14. Find the total monthly payment, including taxes and insurance.

15. Amortize the first two months of the loan, showing principal payment, interest payment, and balance of principal.

Payment Number	Interest Payment	Principal Payment	Balance of Principal
			$192,000.00
1	_____	_____	_____
2	_____	_____	_____

16. How much interest will be paid over the life of the loan?

17. **Cost of Points in a Home Loan** If the lender charges two *points*, how much does that add to the cost of the loan?

Solve each problem.

18. Explain in general what *closing costs* are. Are they different from *settlement charges?*

19. **Adjusting the Rate in an Adjustable-Rate Mortgage** To buy your home, you obtain a 1-year ARM with a 2.25% margin and a 2% periodic rate cap. The index starts at 1.85% but has increased to 4.05% by the first adjustment date. What interest rate will you pay during the second year?

20. According to **Table 14** of this chapter, how many shares of Pearson Plc were traded on June 15, 2018?

Finding the Return on a Stock Investment *Jeong bought 1000 shares of stock at $12.75 per share. She received a dividend of $1.38 per share shortly after the purchase and another dividend of $1.02 per share one year later. Eighteen months after buying the stock, she sold the 1000 shares for $10.36 per share. (Disregard commissions and fees.)*

21. Find Jeong's total return on this stock.

22. Find her percentage return. Is this the *annual rate of return* on this stock transaction? Why or why not?

23. **The Final Value of a Retirement Account** $1800 is deposited at the end of each year in a tax-deferred retirement account. The account earns a 6% annual return, the account owner's marginal tax rate is 25%, and taxes are paid at the end of 30 years. Find the final value of the account.

24. What is meant by saying that a retirement account is "adjusted for inflation"?

14 Graph Theory

As a child, Ann Porter Wailes was often told that her name sounded like that of a writer. And as Ann grew up, she realized that her parents had aptly named her, because she absolutely loved writing. Even before entering college, Ann wanted to learn everything she could about England and its language. She was fortunate enough to be able to visit the United Kingdom several times before she graduated from high school. England was the land of her favorite literature, the study of British royalty fascinated her, and she visited many castles and courts with rich histories of the monarchy. John Keats became her favorite romantic poet.

Ann was enrolled in a mathematics survey course, and her interest was a bonus when she was assigned her first paper in an Introduction to English Literature class. The assignment was to analyze one of Keats's poems, and she chose his *Ode to Autumn*. The first stanza is shown on the next page, with the rhyme scheme illustrated by the letters A, B, C, D, and E.

Season of mists and mellow fruitfulness,	A
Close bosom-friend of the maturing sun;	B
Conspiring with him how to load and bless	A
With fruit the vines that round the thatch-eves run;	B
To bend with apples the moss'd cottage-trees	C
And fill all fruit with ripeness to the core;	D
To swell the gourd, and plump the hazel shells	E
With a sweet kernel; to set budding more,	D
And still more, later flowers for the bees,	C
Until they think that warm days will never cease,	C
For Summer has o'er brimm'd their clammy cells.	E

She learned in her Survey of Mathematics class that rhyme schemes can be analyzed by diagrams such as the following. This diagram is an example of a graph, the topic of this chapter.

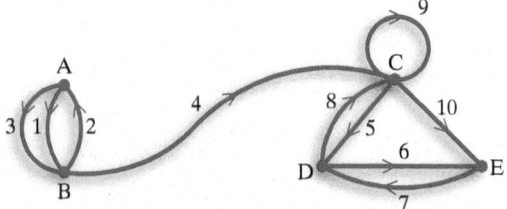

On one of Ann's visits to England, she toured Hampton Court and walked through its hedge maze, designed in 1690. In that same math class, she learned that the structure of mazes can also be analyzed by graphs. The figure on the left below is a diagram of the maze, and the one on the right offers a graphical interpretation, showing that the shortest path to the center is

$$\text{In} \rightarrow A \rightarrow B \rightarrow C \rightarrow D \rightarrow F \rightarrow G \rightarrow H \rightarrow \text{Center}.$$

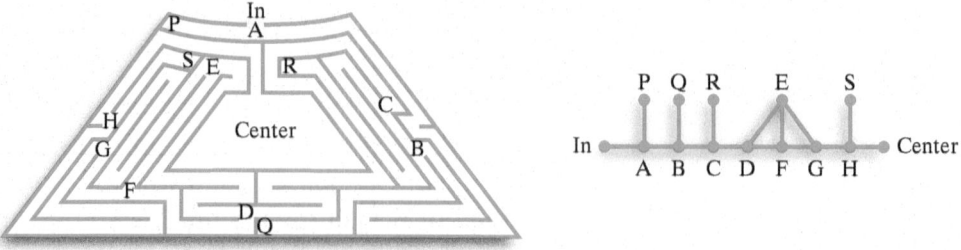

Ann soon realized that she had found her topic for a paper in yet *another* course (History of England). She had discovered mathematics in places she never dreamed it existed.

14.1 BASIC CONCEPTS

OBJECTIVES

1 Define and apply the meanings of the terms *graph, vertex, edge, simple graph,* and *degree of a vertex.*

2 Determine whether a graph is connected or disconnected, and identify its components.

3 Determine whether two graphs are isomorphic.

4 Find the sum of the degrees of a graph.

5 Define and apply the meanings of the terms *walk, path, circuit, weighted graph,* and *complete graph.*

6 Define and apply coloring and chromatic number of a graph and how these concepts are related to the four-color theorem.

7 Find the Bacon number of an actor by using a Web site.

Graphs

Suppose a preschool teacher has ten children in her class:

Andy, Claire, Dave, Erin, Glen, Katy, Joe, Mike, Sam, and Tim.

The teacher wants to analyze the social interactions among these children. She observes with whom the children play during recess over a 2-week period and records her observations.

Child	Played with
Andy	no one
Claire	Dave, Erin, Glen, Katy, Sam
Dave	Claire, Erin, Glen, Katy
Erin	Claire, Dave, Katy
Glen	Claire, Dave, Sam
Katy	Claire, Dave, Erin
Joe	Mike, Tim
Mike	Joe
Sam	Claire, Glen
Tim	Joe

It is not easy to see patterns in the children's friendships from this list. Instead, the information may be shown in a diagram, as in **Figure 1,** where each letter is the first letter of a child's name. Such a diagram is called a *graph.** Each dot is a **vertex** (plural **vertices**) of the graph. The lines between vertices are edges.

> ### GRAPH
>
> A **graph** is a collection of vertices (at least one) and edges. Each edge goes from a vertex to a vertex.

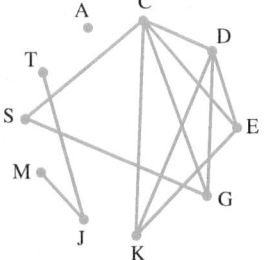

Figure 1

In **Figure 1,** each vertex represents one of the preschool children, and the edges show the relationship "play with each other." An edge must always begin and end at a vertex. We may have a vertex with no edges joined to it, such as A in **Figure 1.**

The graph in **Figure 1** has ten vertices and twelve edges. The only vertices of the graph are the dots. There is *no* vertex where edges TJ and SG intersect, because the intersection point does not represent one of the children. The positions of the vertices and the lengths of the edges have no significance. Also, the edges do not have to be drawn as straight lines. *Only the relationship between the vertices has significance, indicated by the presence or absence of an edge between them.*

Why not draw two edges for each friendship pair? For example, why not draw one edge showing that Tim plays with Joe and another edge showing that Joe plays with Tim? Since Tim and Joe play with each other, the extra edge would yield no additional information. A graph in which there is no more than one edge between any two vertices and in which no edge goes from a vertex to the same vertex is a **simple graph.** (See **Figure 2** on the next page.)

---

*Here *graph* will mean finite graph. A finite graph is a graph with finitely many edges and vertices.

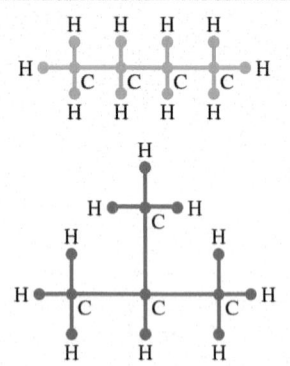

Why are they called graphs? The name was first used by chemist **A. Crum Brown.** He made a major breakthrough in chemistry in 1864 by inventing these now-familiar diagrams to show different molecules with the same chemical composition.

For example, both molecules illustrated above have chemical formula C_4H_{10}. Brown called his sketches **graphic formulae.** Mathematicians got involved in helping to classify all possible structures with a particular chemical composition and adopted the term "graph."

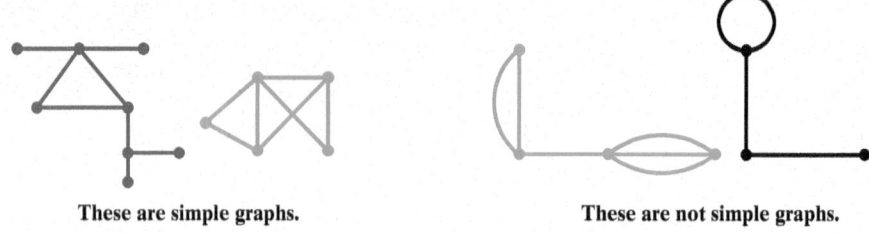

These are simple graphs.　　　These are not simple graphs.

Figure 2

In this chapter, the word **graph** *means simple graph, unless indicated otherwise.* (The meaning of *simple* here is different from its meaning in the phrase "simple curves" used in geometry.)

From the graph in **Figure 1,** we can determine the size of a child's friendship circle by counting the number of edges coming from that child's vertex. There are five edges coming from the vertex labeled C, indicating that Claire plays with five different children. The number of edges joined to a vertex is the **degree of the vertex.** In **Figure 1,** the degree of vertex C is 5, the degree of vertex M is 1, and the degree of vertex A is 0, as no edges are joined to A. (Here, the use of *degree* has nothing to do with its use in geometry in measuring the size of an angle.)

While **Figure 1** clarifies the friendship patterns of the preschoolers, the graph can be drawn in a more informative way, as in **Figure 3.** The graph still has edges between the same pairs of vertices. We think of **Figure 3** as one graph, even though it has three pieces, because it represents a single situation. By drawing the graph in this way, we make the friendship patterns more apparent.

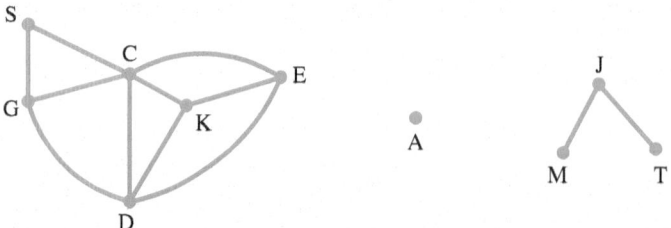

Figure 3

Although the graph in **Figure 3** looks different from that in **Figure 1,** it shows the same relationships between the vertices. The two graphs are said to be *isomorphic.*

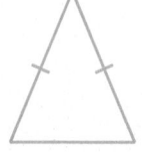

Isosceles triangle

The word **isomorphism** comes from the Greek words *isos,* meaning "same," and *morphe,* meaning "form." Other words that come from the Greek root *isos* include the following.

isobars: lines drawn on a map connecting places with the same barometric pressure;

isosceles: a triangle with two legs (*scelos*) of equal length; and

isomers: chemical compounds with the same chemical composition but different chemical structures.

ISOMORPHIC GRAPHS

Two graphs are **isomorphic** if there is a one-to-one matching between vertices of the two graphs, with the property that whenever there is an edge between two vertices of either one of the graphs, there is an edge between the corresponding vertices of the other graph.

A useful way to think about isomorphic graphs is to imagine a graph drawn with computer software that allows you to drag vertices without detaching any of the edges from their vertices. Any graph we can obtain by simply dragging vertices in this way will be isomorphic to the original graph.

Replacing a graph with an isomorphic graph sometimes can convey more about the relationships being examined. For example, it is clear from **Figure 3** that there are no friendship links between the group consisting of Joe, Tim, and Mike, and the other children. We say that the graph is *disconnected.*

CONNECTED AND DISCONNECTED GRAPHS

A graph is **connected** if we can move from each vertex of the graph to every other vertex of the graph *along edges of the graph*. If we cannot, the graph is **disconnected.** The connected pieces of a graph are called the **components** of the graph.

Figure 3 shows clearly that the friendship relationships among the children in the preschool class have three components. Although the graph in **Figure 1** also is disconnected and includes three components, the way in which the graph is drawn does not make this as clear.

Using the graph in **Figure 3** to analyze the relationships in her class, the teacher might make the following observations and decisions.

- Tim, Mike, and Joe do not interact with the other children.

- Andy's isolation is concerning.

- Claire seems to get along well with other children. As a result, encourage interaction between Claire and Tim, Mike, and Joe.

- Make a special attempt to encourage the children to include Andy in their games.

  Showing the friendship relationships as in **Figure 3** made it much easier to analyze them.

Problem-Solving Strategy

Using different colors for different components can help determine how many components a graph has. We choose a color, begin at any vertex, and color all edges and vertices that we can get to from our starting vertex along edges of the graph. If the whole graph is not colored, we then choose a different color and an uncolored vertex, and repeat the process, continuing until the whole graph is colored.

The **World Wide Web** is sometimes studied as a very large graph. The vertices are Web pages, and the edges are links between pages. This graph is unimaginably large.

Even so, because of the way the Web is connected, one can get from any randomly chosen Web page to any other, using hyperlinks, in an average of about 19 clicks. Even if the Web becomes 10 to 100 times larger, this will still be possible in only 20 or 21 clicks. Because of this, search engines such as Google can find keywords very quickly.

(*Source:* Antonio Gulli, Universit di Pisa, Informatica; Alessio Signorini, University of Iowa, Computer Science; R. Albert, H. Jeong, and A.-L. Barabsi, *Nature* 401, 130–131, 1999.)

EXAMPLE 1   **Determining Number of Components**

Is the graph in **Figure 4** connected or disconnected? How many components does the graph have?

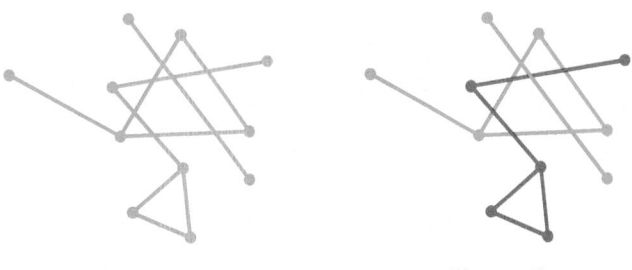

Figure 4   Figure 5

Solution

Coloring the graph helps to answer the question. The original graph in **Figure 4** is disconnected. The graph with color (**Figure 5**) shows three components.

James Joseph Sylvester (1814–1897) was the first mathematician to use the word *graph* for these diagrams in the context of mathematics. Sylvester spent most of his life in Britain, but he had an important influence on American mathematics from 1876 to 1884 as a professor at the newly founded Johns Hopkins University in Baltimore, Maryland. As a young man he worked as an actuary and lawyer in London, tutoring math on the side. One of his students was **Florence Nightingale.** His only published book, *The Laws of Verse*, was about poetry.

EXAMPLE 2   **Deciding Whether Graphs Are Isomorphic**

Are the two graphs in **Figure 6** isomorphic? Justify your answer.

(a)   (b)

Figure 6

Solution

These graphs are isomorphic. To show this, we label corresponding vertices with the same letters and color-code the matching edges in graphs (a) and (b) of **Figure 7.**

(a)   (b)

Figure 7

EXAMPLE 3   **Deciding Whether Graphs Are Isomorphic**

Are the two graphs in **Figure 8** isomorphic?

(a)   (b)

Figure 8

Solution

These graphs are not isomorphic. If we imagine graph (a) drawn with our special computer software, we can see that no matter how we drag the vertices, we cannot get this graph to look exactly like graph (b). For example, graph (b) has vertex X of degree 3 joined to vertex Y of degree 1. Neither of the vertices of degree 3 in graph (a) is joined to a vertex of degree 1.

EXAMPLE 4   **Summing Degrees**

For each graph in **Figure 9,** determine the number of edges and the sum of the degrees of the vertices. Note that graph (b) is not a simple graph.

(a)   (b)

Figure 9

Solution

Start by writing the degree of each vertex next to the vertex, as in **Figure 10.**

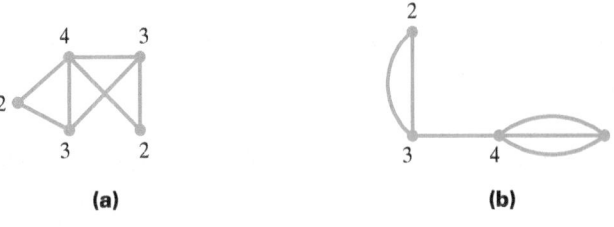

Figure 10

For graph (a):
Number of edges = 7
Sum of degrees of vertices
    = 2 + 3 + 2 + 3 + 4 = 14

For graph (b):
Number of edges = 6
Sum of degrees of vertices
    = 2 + 3 + 4 + 3 = 12

For each graph in **Example 4,** the sum of the degrees of the vertices is twice the number of edges. This is true for all graphs. Check that it is true for graphs in previous examples.

SUM-OF-THE-DEGREES THEOREM

In any graph, the sum of the degrees of the vertices equals twice the number of edges.

To understand why this theorem is true, we think of any graph and imagine cutting each edge at its midpoint (without adding any vertices to the graph, however). Now we have precisely twice as many half edges as we had edges in our original graph. For each vertex, we count the number of half edges joined to the vertex. This is, of course, the degree of the vertex. We add these answers to determine the sum of the degrees of the vertices. We know that the number of half edges is twice the number of edges in our original graph. So, the sum of the degrees of the vertices in the graph is twice the number of edges in the graph.

EXAMPLE 5 Using the Sum-of-the-Degrees Theorem

A graph has precisely six vertices, each of degree 3. How many edges does this graph have?

Solution

The sum of the degrees of the vertices of the graph is

$$3 + 3 + 3 + 3 + 3 + 3 = 18.$$

By the theorem above, this number is twice the number of edges, so the number of edges in the graph is

$$\frac{18}{2}, \quad \text{or} \quad 9.$$

EXAMPLE 6 **Planning a Tour**

Suppose we decide to explore the upper Midwest next summer. We plan to travel by Greyhound bus and would like to visit Grand Forks, Fargo, Bemidji, St. Cloud, Duluth, Minneapolis, Escanaba, and Green Bay. A check of the Greyhound Web site shows direct bus links between destinations as indicated.

City	Direct bus links with
Grand Forks	Bemidji, Fargo
Fargo	Grand Forks, St. Cloud, Minneapolis
Bemidji	Grand Forks, St. Cloud
St. Cloud	Bemidji, Fargo, Minneapolis
Duluth	Escanaba, Minneapolis
Minneapolis	Fargo, St. Cloud, Duluth, Green Bay
Escanaba	Duluth, Green Bay
Green Bay	Escanaba, Minneapolis

Draw a graph with vertices representing the destinations, and with edges representing the relation "there is a direct bus link." Is this a connected graph? Explain.

Solution

Begin with eight vertices representing the eight destinations, as in **Figure 11.** Use the bus link information to fill in the edges of the graph.

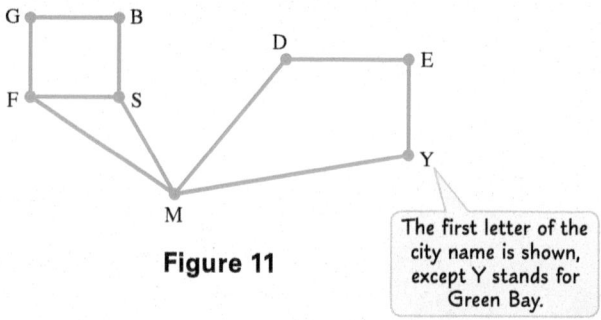

Figure 11

The first letter of the city name is shown, except Y stands for Green Bay.

Figure 11 is a connected graph. It shows that we can travel by bus from any of our chosen destinations to any other.

Walks, Paths, and Circuits

Example 6 suggests several important ideas. For example, what trips could we take among the eight destinations, using only direct bus links? If we do not mind riding the same bus route more than once, we could take the following route.

$$M \rightarrow S \rightarrow F \rightarrow S \rightarrow B$$

Note that we used the "St. Cloud to Fargo" link twice. We could not take the $M \rightarrow D \rightarrow S$ route, however, since there is no direct bus link from Duluth to St. Cloud. Each possible route on the graph is called a *walk*.

WALK

A **walk** in a graph is a sequence of vertices, each linked to the next vertex by a specified edge of the graph.

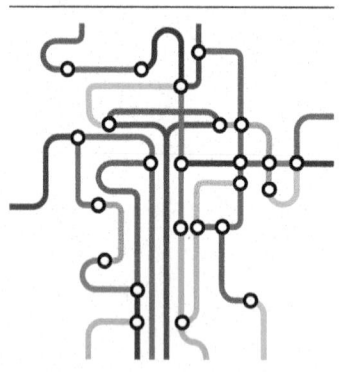

Maps of subway systems in large cities usually do not faithfully represent the positions of stations or the distances between them. They are essentially graphs with vertices representing stations, and edges showing subway links.

We can think of a walk as a route that we can trace with a pencil without lifting the pencil from the graph.

$$M \to S \to F \to S \to B \quad \text{and} \quad F \to M \to D \to E \quad \text{Walks in Figure 11}$$

$$M \to D \to S \quad \text{Not a walk (There is no edge between D and S in Figure 11.)}$$

Suppose that the travel time for our trip is restricted, and we do not want to use any bus link more than once. A *path* is a special kind of walk.

PATH

A **path** in a graph is a walk that uses no edge more than once.*

$$M \to S \to F \to M, \quad F \to M \to D \to E, \quad M \to S \to F \to M \to D \quad \text{Paths}$$

On the graph in **Figure 11,** trace these paths with a finger. Note that a path may use a *vertex* more than once.

$$B \to S \to F \to M \to S \to B \quad \text{Not a path (It uses edge BS more than once.)}$$

$$M \to D \to S \quad \text{Not a path (It is not even a walk.)}$$

Perhaps we want to begin and end our tour in the same city (and still not use any bus route more than once). A path such as this is known as a *circuit*. Two circuits that we could take are shown here.

$$M \to S \to F \to M, \quad M \to S \to F \to M \to D \to E \to Y \to M \quad \text{Circuits}$$

CIRCUIT

A **circuit** in a graph is a path that begins and ends at the same vertex.

Notice that a circuit is a kind of path and, therefore, is also a kind of walk.

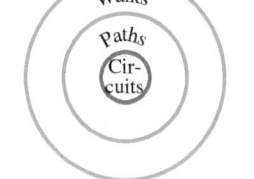

Walks
Paths
Circuits

This **Euler diagram** shows relationships among walks, paths, and circuits.

EXAMPLE 7 Classifying Walks

Using the graph in **Figure 12,** classify each sequence as a walk, a path, or a circuit.

(a) $E \to C \to D \to E$

(b) $A \to C \to D \to E \to B \to A$

(c) $B \to D \to E \to B \to C$

(d) $A \to B \to C \to D \to B \to A$

Solution

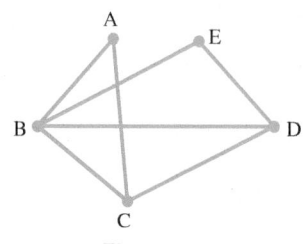

Figure 12

	Walk	Path	Circuit
(a)	No (no edge E to C)	No	No
(b)	Yes	Yes	Yes
(c)	Yes	Yes	No
(d)	Yes	No (edge AB is used twice)	No

If a sequence of vertices is not a walk, it cannot possibly be either a path or a circuit.

If a sequence of vertices is not a path, it cannot possibly be a circuit, because a circuit is defined as a special kind of path.

---

* Some texts call such a walk a *trail* and use the term *path* for a walk that visits no vertex more than once.

The 1980 movie *The Shining* concludes with a terrifying trip through a **maze.** Mazes are found on coins from ancient Knossos (Greece), in the sand drawings at Nazca in Peru, on Roman pavements, on the floors of Renaissance cathedrals in Europe, as the popular cornfield mazes accompanying Halloween pumpkin patches, and as hedge mazes in gardens. Graphs can be used to clarify the structure of mazes.

When planning our tour in **Example 6,** it would be useful to know the length of time the bus rides take. We write the time for each bus ride on the graph, as shown in **Figure 13.**

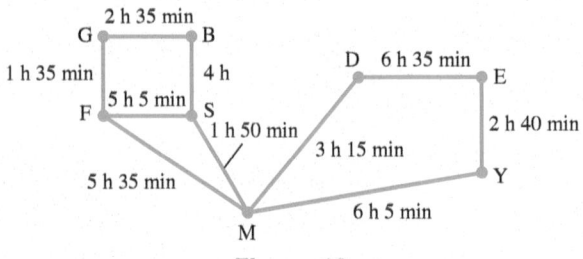

Figure 13

A graph with numbers on the edges, as in **Figure 13,** is a **weighted graph.** The numbers on the edges are **weights.**

Problem-Solving Strategy

Drawing a sketch is a useful problem-solving strategy. **Example 8** illustrates how drawing a sketch can simplify a problem that at first seems complicated.

EXAMPLE 8 Analyzing a Round-Robin Tournament

In a round-robin tournament, every contestant plays every other contestant. The winner is the contestant who wins the most games. Suppose six tennis players compete in a round-robin tournament. How many matches will be played?

Solution

Draw a graph with six vertices to represent the six players. Then, as in **Figure 14,** draw an edge for each match between a pair of contestants.

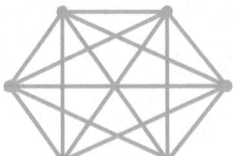

Figure 14

We can count the edges to find out how many matches must be played. Or we can reason as follows: There are 6 vertices each of degree 5, so the sum of the degrees of the vertices in this graph is $6 \cdot 5$, or 30. We know that the sum of the degrees is twice the number of edges. Therefore, there are 15 edges in this graph and, thus, 15 matches in the tournament.

Another way of solving this problem is to use combinations, covered earlier in the text. We must find $_6C_2$, the number of combinations of 6 things taken 2 at a time.

$$_6C_2 = \frac{6!}{2!(6-2)!} = \frac{6!}{2! \cdot 4!} = 15$$

Review:

$$_nC_r = \frac{n!}{r!\,(n-r)!}$$

Complete Graphs and Subgraphs

The graph for **Example 8** in **Figure 14** is an example of a *complete graph*.

COMPLETE GRAPH

A **complete graph** is a graph in which there is exactly one edge going from each vertex to each other vertex in the graph.

EXAMPLE 9 Deciding Whether a Graph Is Complete

Decide whether each of the graphs in **Figure 15** is complete. If a graph is not complete, explain why it is not.

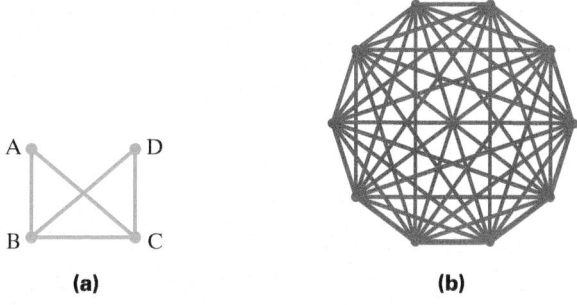

(a)   (b)

Figure 15

Solution

Graph (a) is not complete. There is no edge from vertex A to vertex D. Graph (b) is complete. (Check that there is an edge from each of the ten vertices to each of the other nine vertices in the graph.)

EXAMPLE 10 Finding a Complete Graph

Find a complete graph with four vertices in the preschoolers' friendship pattern depicted in **Figure 3,** repeated here.

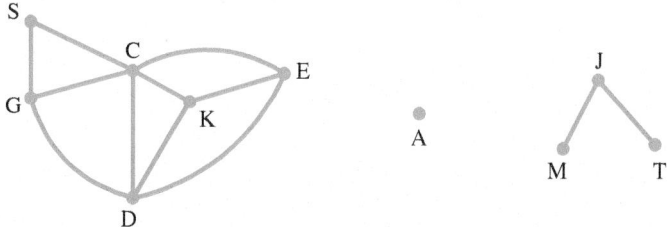

Figure 3 (repeated)

Solution

Figure 16 shows a portion of the friendship graph from **Figure 3.**

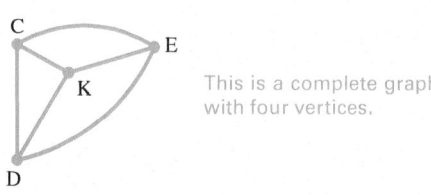

This is a complete graph with four vertices.

Figure 16

A good source for **real-life examples of graphs** is

Networks, Crowds and Markets: Reasoning about a Highly Connected World

by David Easley and Jon Kleinberg.

Figure 16 includes some complete graphs with three vertices. In general, a graph consisting of some of the vertices of the original graph and some of the original edges between those vertices is called a **subgraph.** Vertices or edges not included in the original graph cannot be in the subgraph.

A subgraph may include anywhere from one to all of the vertices of the original graph and anywhere from none to all of the edges of the original graph. Notice that a subgraph is a graph, and recall that in a graph, every edge goes from a vertex to a vertex. Therefore, no edge can be included in a subgraph without the vertices at both of its ends also being included.

- The subgraph shown in **Figure 16** is a complete graph, but a subgraph does not have to be complete. For example, the graphs in **Figures 17(a) and (b)** are both subgraphs of the graph in **Figure 3.**

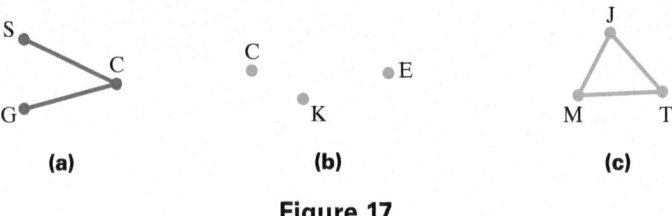

Figure 17

- The graph in **Figure 17(b)** is a subgraph even though it has no edges. (However, we could not form a subgraph by taking edges without vertices. In a graph, every edge goes from a vertex to a vertex.)

- The graph in **Figure 17(c)** is *not* a subgraph of the graph in **Figure 3,** because there was no edge between M and T in the original graph.

Graph Coloring

Tabatha schedules nurse trainees for the hospital departments listed on the left below. Trainee groups must be scheduled at different times if there are trainees working in both departments. The column on the right shows, for each department, which of the other departments have trainees in common with it. For example, the first row shows that each of the departments Radiology, Surgery, Obstetrics, and Pediatrics has one trainee, or more, also working in the ER.

Department	Other departments in which trainees in this department also work
Emergency (ER)	Radiology, Surgery, Obstetrics, Pediatrics
Obstetrics	Radiology, Surgery, Oncology, ER
Surgery	Pediatrics, Oncology, Geriatrics, ER, Obstetrics
Oncology	Pediatrics, Radiology, Geriatrics, Obstetrics, Surgery
Pediatrics	Radiology, ER, Surgery, Oncology
Radiology	Geriatrics, ER, Obstetrics, Oncology, Pediatrics
Geriatrics	Surgery, Oncology, Radiology

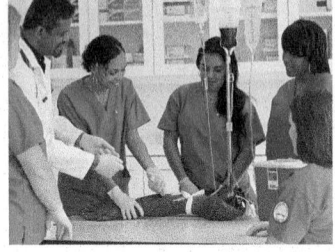

Tabatha's task is to figure out the least number of time slots needed for the seven trainee groups and to decide which (if any) can be scheduled at the same time. We can draw a graph, with a vertex for each department, and with an edge between two vertices *if there are trainees working in both departments.* See **Figure 18** on the next page.

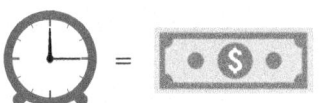

The Mathematics of Time Scheduling

Time is money.

This short sentence says a lot when it comes to productivity in the workplace. A branch of graph theory called **mathematical theory of scheduling** investigates how tasks can be arranged for a project to maximize efficiency. The terms *processors, tasks, processing time, precedence relations, project, critical time, finishing time, schedule, scheduling algorithms, priority list,* and *optimal schedule* are used in the theory of scheduling.

Tasks are represented by vertices, and precedence relations are represented by directed arrows from one vertex to another (similar to edges in a graph). The resulting diagram is called a directed graph (or digraph).

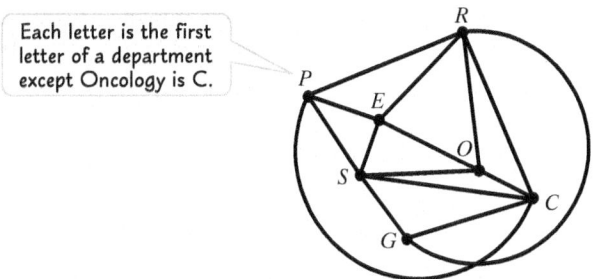

Each letter is the first letter of a department except Oncology is C.

Figure 18

If there is an edge between two vertices, then the corresponding departments have trainees in common, and the trainee groups must *not* be scheduled at the same time. To show this, we use a different color for each time slot for trainee groups. We will color the vertices with the different colors to show when the trainee groups could be scheduled.

To make sure that trainee groups are scheduled at different times if the departments have trainees in common, we must make sure that any two vertices joined by an edge have different colors. To ensure the least number of time slots, we use as few colors as possible. In **Figure 19,** we show one possible coloring of the vertices using three colors. We cannot color the graph as required with fewer than three colors.

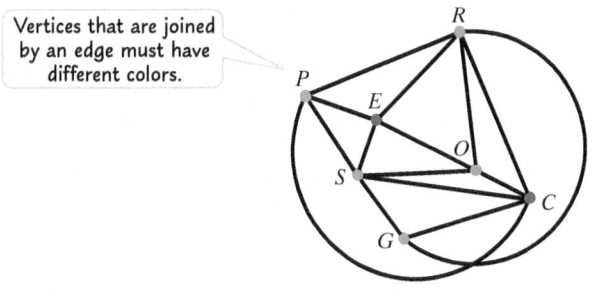

Vertices that are joined by an edge must have different colors.

Figure 19

Tabatha can schedule the required trainee groups in no fewer than three time slots. She can schedule Radiology and Surgery at one time, Pediatrics, Obstetrics, and Geriatrics at a second time, and Oncology and ER at a third time.

The result we achieved is called a *coloring** for the graph.

COLORING AND CHROMATIC NUMBER

A **coloring** for a graph is an assignment of a color to each vertex in such a way that vertices joined by an edge have different colors. The **chromatic number** of a graph is the least number of colors needed to achieve a coloring.

Using this terminology, Tabatha had to determine the chromatic number for the graph in **Figure 18.** The chromatic number was 3, which equaled the least number of time slots needed for the trainee groups.

The idea of using vertex coloring to solve this particular problem was straightforward. However, determining the chromatic number for a graph with many vertices or producing a coloring using the least possible number of colors can be difficult. No one has yet found an efficient method for finding the exact chromatic number for arbitrarily large graphs. (The term *efficient algorithm* has a specific, technical meaning for computer scientists.)

---

* This method is sometimes called a *vertex coloring* for the graph.

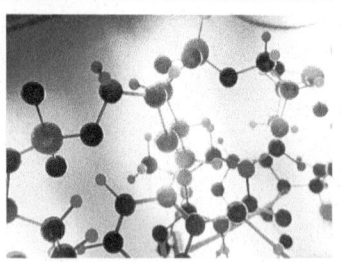

Networks are at the forefront of research in biology. Biologists have made enormous strides in understanding the structure of molecules in living organisms. It is now important to understand how these molecules interact with each other. This can be viewed as a network, or graph.

For example, current research aims to identify how proteins in cells interact with other proteins. Insights into such networks may help explain the growth of cancerous tumors. This may lead to better treatments for cancer. It was discovered that as graphs, these biological networks are surprisingly similar to the graphs representing the World Wide Web.

The following method for coloring a graph may not give a coloring with the least number of colors, but its result still may be helpful.

COLORING A GRAPH

Step 1 Choose a vertex with greatest degree, and color it. Use the same color to color as many vertices as you can without coloring two vertices the same color if they are joined by an edge.

Step 2 Choose a new color, and repeat what you did in Step 1 for vertices not already colored.

Step 3 Repeat Step 1 until all vertices are colored.

Graph coloring is used to solve many practical problems. It is useful in management science for solving scheduling problems, such as the "trainee groups" example. Graph coloring is also related to allocating transmission frequencies to TV and radio stations and cell phone companies.

The coloring of graphs has an interesting history involving maps. Maps need to be colored so that territories with common boundaries have different colors. How many colors are needed? For map makers, four colors have always sufficed. In about 1850, mathematicians started pondering this. For over a hundred years, they were able neither to prove that four colors are always enough nor to find a map that needed more than four colors. This was called the **four-color problem.**

In 1976 two mathematicians, Kenneth Appel and Wolfgang Haken, provided a proof that four colors are always enough. This proof created a controversy in mathematics, because it relied on computer calculations so lengthy (fifty 24-hour days of computer time) that no person could check the entire proof.

Figure 20

The postmark shown in **Figure 20** was used by the Urbana, Illinois, post office to commemorate the proof of the four-color theorem. (Appel and Haken were colleagues at the University of Illinois at Urbana-Champaign when they completed their proof.)

Bacon Numbers

The popular game *The Six Degrees of Kevin Bacon* was developed by three Albright College friends a number of years ago. The object of the game is to connect actor Kevin Bacon to another actor or actress using a sequence of movies, with six or fewer titles in the sequence. Thanks to the International Movie Data Base (www.imdb.com), this goal can be reached almost instantaneously. The minimum number of links necessary is called the **Bacon number** for the other performer. For example, Hayley Mills has a Bacon number of 2, because she can be connected to Kevin Bacon in 2 steps as follows:

• Hayley Mills was in *That Darn Cat!* (1965) with Roddy McDowall.

• Roddy McDowall was in *The Big Picture* (1989) with Kevin Bacon.

The process used in *The Six Degrees of Kevin Bacon* can be applied to other situations (Web pages, for example).

EXAMPLE 11 Determining a Bacon Number

Use the Web site

oracleofbacon.org

to determine the Bacon number for Mary Kaye Hebert, a lifelong friend of one of the authors of this text. She had a small speaking role in *Easy Rider* (1969). Use the default option of theatrical movies only.

Solution

According to the site, Mary Kaye Hebert has a Bacon number of 2.

• Mary Kaye Hebert was in *Easy Rider* (1969) with Carrie Snodgrass.

• Carrie Snodgrass was in *Wild Things* (1998) with Kevin Bacon.

It takes only 2 titles to connect Mary Kaye and Kevin. Her Bacon number is 2.

14.1 EXERCISES

CONCEPT CHECK *Complete each statement with the correct response(s).*

1. A graph is a collection of vertices (at least one) and _____, where each _____ goes from a vertex to a vertex.

2. Two graphs are _____ if there is a one-to-one matching between vertices of the two graphs, with the property that whenever there is an edge between two vertices of either one of the graphs, there is an edge between the corresponding vertices of the other graph.

3. A graph is _____ if we can move from each vertex of the graph to every other vertex of the graph along the edges of the graph. If not, the graph is _____ .

4. In any graph, the sum of the degrees of the vertices equals _____ times the number of edges.

5. A _____ in a graph is a sequence of vertices, each linked to the next vertex by a specified edge of the graph.

6. A _____ in a graph is a walk that uses no edge more than once.

7. A _____ in a graph is a path that begins and ends at the same vertex.

8. A _____ graph is a graph in which there is exactly one edge going from each vertex to each other vertex in the graph.

Vertices and Edges *Determine how many vertices and how many edges each graph has.*

9.

10.

11.

12.

13.

14.

Sum of Degrees *For each graph indicated, find the degree of each vertex in the graph. Then add the degrees to get the sum of the degrees of the vertices of the graph. What relationship do you notice between the sum of degrees and the number of edges?*

15. Exercise 9   **16. Exercise 10**

17. Exercise 11   **18. Exercise 12**

Isomorphic or Not Isomorphic? *Determine whether the two graphs are isomorphic. If so, label corresponding vertices of the two graphs with the same letters and color-code corresponding edges, as in **Example 2**. (Note that some exercises have more than one correct answer.)*

19.
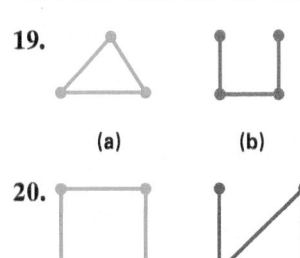

  (a)   (b)

20.

  (a)   (b)

21.

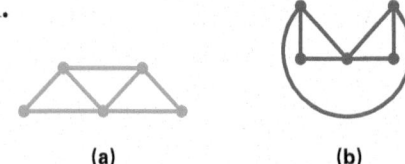

(a)   (b)

22.

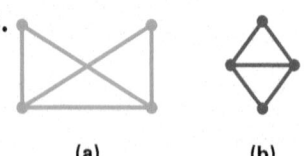

(a)   (b)

23.

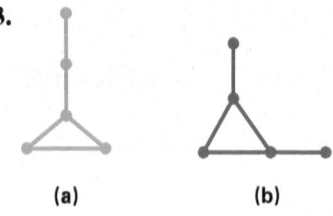

(a)   (b)

24.

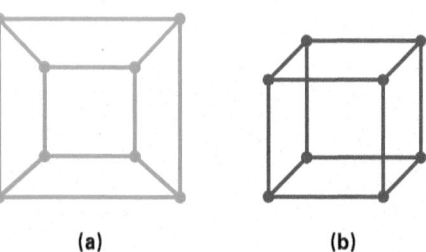

(a)   (b)

Connected or Disconnected? *Determine whether the graph is connected or disconnected. Then determine how many components the graph has.*

25.

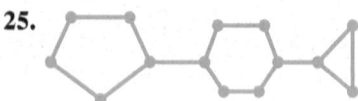

26.

27.

28.

29.    **30.**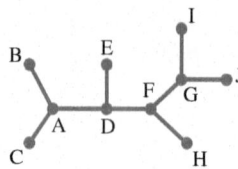

Number of Edges *Use the theorem that relates the sum of degrees to the number of edges to determine the number of edges in the graph (without drawing).*

31. A graph with 5 vertices, each of degree 4

32. A graph with 7 vertices, each of degree 4

33. A graph with 5 vertices, three of degree 1, one of degree 2, and one of degree 3

34. A graph with 8 vertices, two of degree 1, three of degree 2, one of degree 3, one of degree 5, and one of degree 6

Walks, Paths, Circuits *In Exercises 35–37, refer to the following graph.*

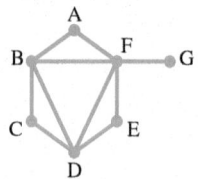

35. Which are walks in the graph? If not, why not?
(a) $A \rightarrow B \rightarrow C$
(b) $B \rightarrow A \rightarrow D$
(c) $E \rightarrow F \rightarrow A \rightarrow E$
(d) $B \rightarrow D \rightarrow F \rightarrow B \rightarrow D$
(e) $D \rightarrow E$
(f) $C \rightarrow B \rightarrow C \rightarrow B$

36. Which are paths in the graph? If not, why not?
(a) $B \rightarrow D \rightarrow E \rightarrow F$
(b) $D \rightarrow F \rightarrow B \rightarrow D$
(c) $B \rightarrow D \rightarrow F \rightarrow B \rightarrow D$
(d) $D \rightarrow E \rightarrow F \rightarrow G \rightarrow F \rightarrow D$
(e) $B \rightarrow C \rightarrow D \rightarrow B \rightarrow A$
(f) $A \rightarrow B \rightarrow E \rightarrow F \rightarrow A$

37. Which are circuits in the graph? If not, why not?
(a) $A \rightarrow B \rightarrow C \rightarrow D \rightarrow E \rightarrow F$
(b) $A \rightarrow B \rightarrow D \rightarrow E \rightarrow F \rightarrow A$
(c) $C \rightarrow F \rightarrow E \rightarrow D \rightarrow C$
(d) $G \rightarrow F \rightarrow D \rightarrow E \rightarrow F$
(e) $F \rightarrow D \rightarrow F \rightarrow E \rightarrow D \rightarrow F$

Walks and Paths *In Exercises 38 and 39, refer to the following graph.*

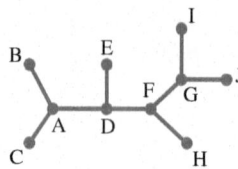

38. Which are walks in the graph? If not, why not?

(a) $F \to G \to J \to H \to F$

(b) $D \to F$

(c) $B \to A \to D \to F \to H$

(d) $B \to A \to D \to E \to D \to F \to H$

(e) $I \to G \to J$

(f) $I \to G \to J \to I$

39. Which are paths in the graph? If not, why not?

(a) $A \to B \to C$

(b) $J \to G \to I \to G \to F$

(c) $D \to E \to I \to G \to F$

(d) $C \to A$

(e) $C \to A \to D \to E$

(f) $C \to A \to D \to E \to D \to A \to B$

Walks, Paths, Circuits *In Exercises 40–45, refer to the following graph. In each case, determine whether the sequence of vertices is* **(i)** *a walk,* **(ii)** *a path,* **(iii)** *a circuit in the graph.*

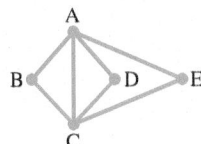

40. $A \to B \to C \to D \to E$

41. $A \to B \to C$

42. $A \to B \to C \to D \to A$

43. $A \to B \to A \to C \to D \to A$

44. $A \to B \to C \to A \to D \to C \to E \to A$

45. $C \to A \to B \to C \to D \to A \to E$

46. ***Graph Relationships*** *Explain why a graph that is a circuit must also be a walk.*

Complete or Not Complete? *Determine whether the graph is a complete graph. If not, explain why it is not complete.*

47.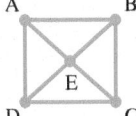

48.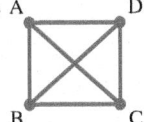

49.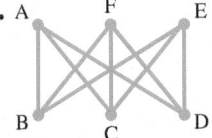

50.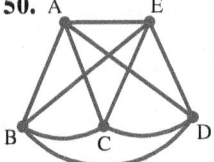

51.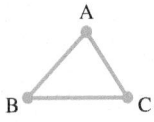

52.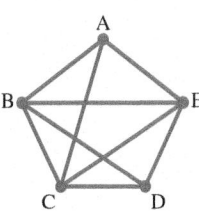

Solve each problem.

53. ***Chess Competition*** A chess master plays seven simultaneous games with seven other players. Draw a graph with vertices representing the players, and edges representing the chess games. How many games are being played?

54. ***Chess Competition*** Students from two schools compete in chess. Each school has a team of four students. Each student must play one game against each student on the opposing team. Draw a graph with vertices representing the students, and edges representing the chess games. How many games must be played in the competition?

55. ***Number of Handshakes*** There are six members on a hockey team (including the goalie). At the end of a hockey game, each member of the team shakes hands with each member of the opposing team. How many handshakes occur?

56. ***Dancing Partners*** At a party there were four males and five females. During the party each male danced with each female (and no female pair or male pair danced together). Draw a graph with vertices representing the people at the party, and edges of the graph showing the relationship

"danced with."

How many edges are there in the graph?

57. ***Spread of a Rumor*** A lawyer is preparing his argument in a libel case. He has evidence that a libelous rumor about his client was discussed in various telephone conversations among eight people. Two of the people involved had four telephone conversations in which the rumor was discussed, one person had three, four had two, and one had one such telephone conversation. How many telephone conversations were there in which the rumor was discussed among the eight people?

58. ***Number of Handshakes*** There are seven people at a business meeting. One of these shakes hands with four people, four shake hands with two people, and two shake hands with three people. How many handshakes occur? (*Hint:* Use the theorem stating that the sum of the degrees of the vertices in a graph is twice the number of edges.)

59. *Vertices and Edges of a Cube* Draw a graph with vertices representing the vertices (the corners) of a cube, and edges representing the edges of the cube. In your graph, find a circuit that visits four different vertices. What figure does your circuit form on the actual cube?

60. *Vertices and Edges of a Tetrahedron* Draw a graph with vertices representing the vertices (or corners) of a tetrahedron, and edges representing the edges of the tetrahedron. In the graph, identify a circuit that visits three different vertices. What figure does the circuit form on the actual tetrahedron?

61. *Students in the Same Class* Mary, Erin, Sue, Jane, Katy, and Brenda are friends at college. Mary, Erin, Sue, and Jane are in the same math class. Sue, Jane, and Katy take the same English composition class.

(a) Draw a graph with vertices representing the six students, and edges representing the relation "take a common class."

(b) Is the graph connected or disconnected? How many components does the graph have?

(c) In the graph, identify a subgraph that is a complete graph with four vertices.

(d) In the graph, identify three different subgraphs that are complete graphs with three vertices. (There are several correct answers.)

62. *Analyzing a Cube with a Graph* Draw a graph whose vertices represent the *faces* of a cube and in which an edge between two vertices shows that the corresponding faces of the actual cube share a common boundary. What is the degree of each vertex in the graph? What does the degree of any vertex in the graph reveal about the actual cube?

63. *Nonisomorphic Graphs* Draw two nonisomorphic (simple) graphs with 6 vertices, with each vertex having degree 3.

64. Here is another theorem about graphs: ***In any graph, the number of vertices with odd degree must be even.*** Explain why this theorem is true. (*Hint:* Use the theorem about the relationship between the number of edges and the sum of degrees.)

Coloring and Chromatic Numbers *Color the graph using as few colors as possible. Determine the chromatic number of the graph. (Hint: The chromatic number is fixed, but there may be more than one correct coloring.)*

65.

66.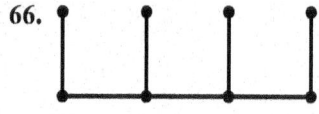

67.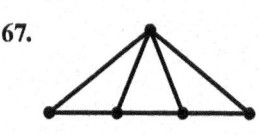

68.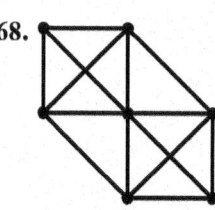

69.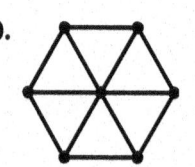

70.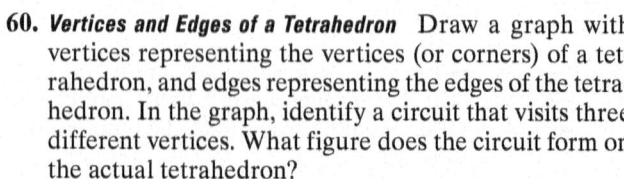

71. *Cycles* Color each graph using as few colors as possible. Use this to determine the chromatic number of the graph. (Graphs like these are called **cycles**.)

(a)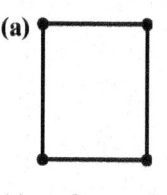

(b)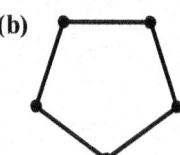

(c)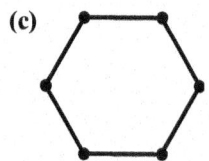

(d)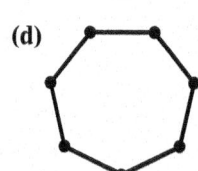

(e) Use the results from parts (a)–(d) to make a prediction about the chromatic number of a cycle. (*Hint:* Consider two cases.)

72. *Chromatic Number* Sketch a complete graph with the specified number of vertices, color the graph with as few colors as possible, and use this to determine the chromatic number of the graph.

(a) 3 vertices

(b) 4 vertices

(c) 5 vertices

(d) Write a general principle by completing this statement: A complete graph with n vertices has chromatic number ___.

(e) Why is the statement in part (d) true?

(f) Generalize further by completing this statement: If a graph has a *subgraph* that is a complete graph with n vertices, then the chromatic number of the graph must be at least ___.

Coloring and Chromatic Numbers *Color the vertices using as few colors as possible. Then state the chromatic number of the graph. (Hint: It might help to first identify the largest subgraph that is a complete graph.)*

73.

74.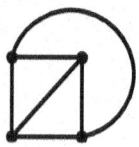

75. 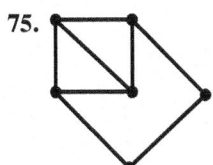   **76.**

Solve each problem.

77. *Scheduling Meeting Times* Campus Life must schedule weekly meeting times for the six organizations listed below so that organizations with members in common meet at different times.

Use graph coloring to determine the least number of different meeting times needed, and which organizations should meet at the same time.

Organization	Members also belong to
Choir	Caribbean Club, Dance Club, Theater, Service Club
Caribbean Club	Dance Club, Service Club, Choir
Service Club	Forensics, Choir, Theater, Caribbean Club
Forensics	Dance Club, Service Club
Theater	Choir, Service Club
Dance Club	Choir, Caribbean Club, Forensics

78. *Assigning Frequencies to Transmitters* Interference can occur between radio stations. To avoid this, transmitters that are less than 60 miles apart must be assigned different broadcast frequencies. The transmitters for 7 radio stations are labeled A through G. The information below shows, for each transmitter, which of the others are within 60 miles.

Use graph coloring to determine the least number of different broadcast frequencies that must be assigned to the 7 transmitters. Include a plan for which transmitters could broadcast on the same frequency to use this least number of frequencies.

Transmitters	Transmitter within 60 miles of this
A	B, E, F
B	A, C, E, F, G
C	B, D, G
D	C, F, G
E	A, B, F
F	A, B, D, E, G
G	B, C, D, F

79. *Inviting Colleagues to a Gathering* Several of Tiffany's colleagues do not get along with each other, as summarized below. Tiffany plans to organize several gatherings, so that colleagues who do not get along are not there at the same time.

Use graph coloring to determine the least number of gatherings needed, and which colleagues should be invited each time.

- Joe does not get along with Brad and Phil.
- Caitlin does not get along with Lindsay and Brad.
- Lindsay does not get along with Phil.
- Mary does not get along with Lindsay.
- Eva gets along with everyone.

Graph Coloring *Each graph shows the vertices and edges of the figure specified. Color the vertices of each graph in such a way that vertices with an edge between them have different colors. Use graph coloring to find a way to do this, and specify the least number of colors needed.*

80. Cube   **81.** Octahedron

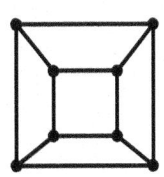   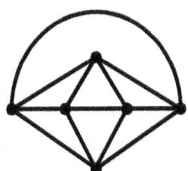

82. Dodecahedron

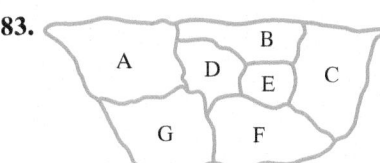

Graph Coloring *Draw a vertex for each region shown in the map. Draw an edge between two vertices if the two regions have a common boundary. Then find a coloring for the graph, using as few colors as possible.*

83.

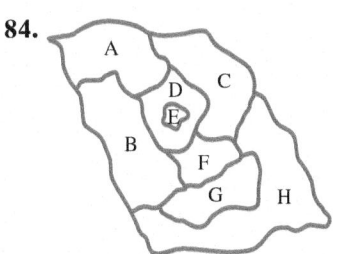

84.

85. *Map Coloring* Use graph coloring to determine the least number of colors that can be used to color the map so that areas with common boundaries have different colors.

Upstate New York

86. *Map Coloring* Use graph coloring to determine the least number of colors that can be used to color the map so that countries with common boundaries have different colors.

Countries of Southern Africa

Map Drawing *Draw a map with the specified number of countries that can be colored using the stated number of colors, and no fewer. (Countries may not consist of disconnected pieces.) If it is not possible to draw such a map, explain why not.*

87. 6 countries, 3 colors

88. 6 countries, 4 colors

89. 6 countries, 5 colors

90. *Four-Color Problem* Write a paper on the four-color problem. Include a discussion of the proof by Appel and Haken.

The Six Degrees of Kevin Bacon *Use the Web site*

oracleofbacon.org

to determine the Bacon number for each actor or actress. Use the default option of theatrical movies only.

91. Cameron Diaz   **92.** Drew Barrymore

93. Mae West   **94.** John Wayne

95. Theda Bara   **96.** Jackie Robinson (baseball player)

97. Donald J. Trump   **98.** Jennifer Lawrence (III)

99. Meryl Streep   **100.** Charlton Heston

Poetry Analysis *Graphs may be used to clarify the rhyme schemes in poetry. Vertices represent words at the end of a line, edges are drawn, and a path is used to indicate the rhyme scheme. See the chapter opener, where the first stanza of* Ode to Autumn *by John Keats has its rhyme scheme analyzed using a graph. Use a graph to analyze the rhyme scheme.*

101. Lines from *She Walks in Beauty,* by Lord Byron

She walks in beauty, like the night	A
Of cloudless climes and starry skies;	B
And all that's best of dark and bright	A
Meet in her aspect and her eyes;	B
Thus mellowed to that tender light	A
Which heaven to gaudy day denies.	B

102. Lines from *Annabel Lee,* by Edgar Allan Poe

It was many and many a year ago,	A
In a kingdom by the sea,	B
That a maiden there lived whom you may know	A
By the name of Annabel Lee;	B
And this maiden she lived with no other thought	C
Than to love and be loved by me.	B

103. Sonnet 3 of William Shakespeare

Look in thy glass, and tell the face thou viewest	A
Now is the time that face should form another;	B
Whose fresh repair if now thou not renewest,	A
Thou dost beguile the world, unbless some mother,	B
For where is she so fair whose unear'd womb	C
Disdains the tillage of thy husbandry?	D
Or who is he so fond will be the tomb	C
Of his self-love, to stop posterity?	D
Thou art thy mother's glass, and she in thee	D
Calls back the lovely April of her prime:	E
So thou through windows of thine age shall see	D
Despite of wrinkles this thy golden time.	E
But if thou live, remember'd not to be,	D
Die single, and thine image dies with thee.	D

104. Lines from *Ode to Graphs* by Vern E. Heeren

I still recall the day I first was told	A
That math could work on x's, y's, and z's.	B
The whole idea to me seemed more than bold,	A
But soon I drew my x-y graphs with ease.	B
And now you say before I should grow old,	A
I must perceive that graphs contain all these:	B
Walks, paths, and circuits with much more unfurled;	C
I think that graphs may hold the whole wide world!	C

14.2   EULER CIRCUITS AND ROUTE PLANNING

OBJECTIVES

1 State the Königsberg bridge problem and explain its historical relevance to graph theory.

2 State and apply Euler's theorem for Euler circuits.

3 Identify a cut edge of a graph.

4 Apply Fleury's algorithm to find an Euler circuit.

5 Minimize deadheading on a street grid by converting it to a graph with an Euler circuit.

Leonhard Euler (1707–1783) was a devoted father and grandfather. He had thirteen children and frequently worked on mathematics with children playing around him. He lost the sight in his right eye at age 31, a couple of years after writing the Königsberg bridge paper. He became completely blind at age 58 but produced more than half his mathematical work after that. Euler wrote nearly a thousand books and papers.

Königsberg Bridge Problem

In the early 1700s, the city of Königsberg was the capital of East Prussia. (Königsberg is now called Kaliningrad and is in Russia.) The river Pregel ran through the city in two branches with an island between the branches, and seven bridges joined various parts of the city as shown on the map in **Figure 21.**

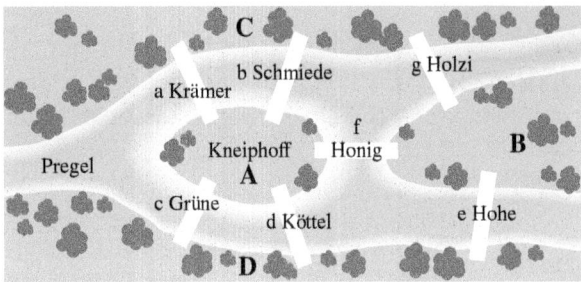

Figure 21

According to Leonhard Euler (pronounced "oiler"), the following problem (the **Königsberg bridge problem**) was well known in his time (around 1730).

> *Is it possible for a citizen of Königsberg to take a stroll through the city, crossing each bridge exactly once, and beginning and ending at the same place?*

Try to find such a route on the map in **Figure 21.** (Do not cross the river anywhere except at the bridges shown on the map.)

We can simplify the problem by drawing the map as a graph. See **Figure 22.** The graph focuses on the relations between the vertices. This is the relation "there is a bridge."

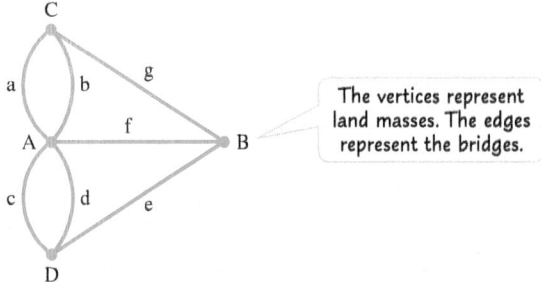

The vertices represent land masses. The edges represent the bridges.

Figure 22

In the graph in **Figure 22,** we have *two* edges between vertices A and C and *two* edges between vertices A and D. Thus, this graph is not a simple graph. We have labeled the edges (representing bridges) with lowercase letters. Now try to find a route as required in the problem.

Euler Circuits

The Königsberg bridge problem requires more than a path and more than a circuit. (Recall that a path is a walk that uses no edge more than once, and a circuit is a path that begins and ends at the same vertex.) The problem requires a circuit that uses *every* edge, *exactly* once. This is called an *Euler circuit.*

A university was founded in Königsberg in 1544. It was established as a Lutheran center of learning. Its most famous professor was the philosopher **Immanual Kant,** who was born in Königsberg in 1724. The university was completely destroyed during World War II.

EULER PATH AND EULER CIRCUIT

An **Euler path** in a graph is a path that uses every edge of the graph exactly once. An **Euler circuit** in a graph is a circuit that uses every edge of the graph exactly once.

EXAMPLE 1   Recognizing Euler Circuits

Consider the graph in **Figure 23.**

(a) Is $A \rightarrow B \rightarrow C \rightarrow D \rightarrow E \rightarrow F \rightarrow A$ an Euler circuit for this graph? Justify your answer.

(b) Does the graph have an Euler circuit?

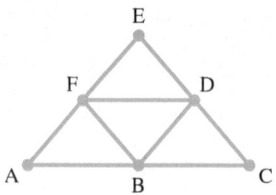

Figure 23

Solution

(a) $A \rightarrow B \rightarrow C \rightarrow D \rightarrow E \rightarrow F \rightarrow A$ is a circuit, but not an Euler circuit, because it does not use every edge of the graph. For example, the edge BD is not used in this circuit.

(b) In the graph in **Figure 23,** the circuit

$$A \rightarrow B \rightarrow C \rightarrow D \rightarrow E \rightarrow F \rightarrow D \rightarrow B \rightarrow F \rightarrow A$$

is an Euler circuit. Trace this path on the graph to check that it does use every edge of the graph exactly once.

Returning to the Königsberg bridge problem, it seemed that no route of the type required could be found. Why was this so? Euler published a paper in 1736 that explained why it is impossible to find a walk of the required kind, and he provided a simple way of deciding whether a given graph has an Euler circuit. Euler did not use the terms *graph* and *circuit,* but his paper was, in fact, the first paper on graph theory. In modern terms, Euler proved the first part of the theorem below. Part 2 was not proved until 1873.

EULER'S THEOREM

Suppose we have a connected graph.

1. If the graph has an Euler circuit, then each vertex of the graph has even degree.

2. If each vertex of the graph has even degree, then the graph has an Euler circuit.

What does Euler's theorem predict about the connected graph in **Figure 23?** The degrees of the vertices are as follows.

A: 2   B: 4   C: 2   D: 4   E: 2   F: 4

Each vertex has even degree, so it follows from part 2 of the theorem that the graph has an Euler circuit. (We already knew this, since we found an Euler circuit for this graph in **Example 1.**)

What does Euler's theorem suggest about the Königsberg bridge problem? Note that the degree of vertex C in **Figure 22** is 3, which is an odd number, and the graph is connected. So, by part 1 of the theorem, the graph cannot have an Euler circuit. (In fact, the graph has many vertices with odd degree, but the presence of *any* vertex of odd degree indicates that the graph does not have an Euler circuit.)

Why is Euler's theorem true? Consider the first part of the theorem:

If a connected graph has an Euler circuit, then each vertex has even degree.

Suppose we have a connected graph with an Euler circuit in the graph. Suppose the circuit begins and ends at a vertex A. Imagine traveling along the Euler circuit, from A all the way back to A, putting an arrow on each edge in our direction of travel. (Of course, we travel along each edge exactly once.)

Now consider any vertex B in the graph. It has edges joined to it, each with an arrow. Some of the arrows point toward B, some point away from B. The number of arrows pointing toward B is the same as the number of arrows pointing away from B, for every time we visit B, we come in on one unused edge and leave via a different unused edge. In other words, we can pair off the edges joined to B, with each pair having one arrow pointing toward B and one arrow pointing away from B. Thus, the total number of edges joined to B must be an even number, and therefore, the degree of B is even. A similar argument shows that our starting vertex, A, also has even degree.

The second part of Euler's theorem states the following:

If each vertex of a connected graph has even degree,
then the graph has an Euler circuit.

To show that this is true, we will provide a recipe (Fleury's algorithm) for finding an Euler circuit in any such graph. First, however, we consider more examples of how Euler's theorem can be applied.

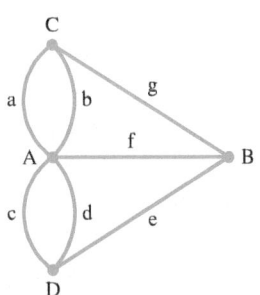

Figure 22 (repeated)

EXAMPLE 2   **Using Euler's Theorem**

Decide which of the graphs in **Figure 24** has an Euler circuit. Justify your answers.

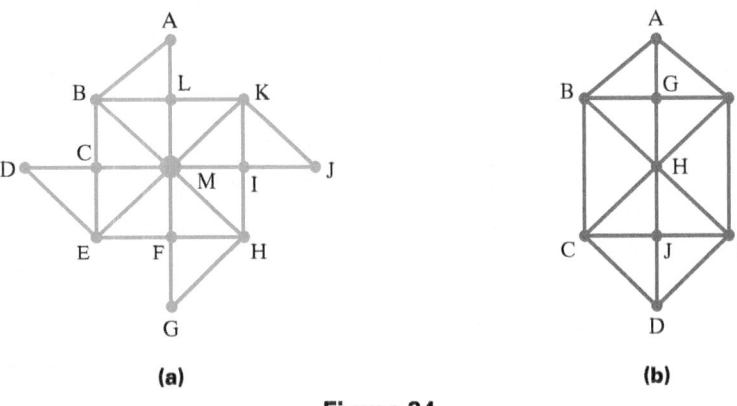

(a)   (b)

Figure 24

Solution

(a) The graph in **Figure 24(a)** is connected. If we write the degree next to each vertex on the graph, as in **Figure 25,** we see that each vertex has even degree. Since the graph is connected, and each vertex has even degree, it follows from Euler's theorem that the graph has an Euler circuit.

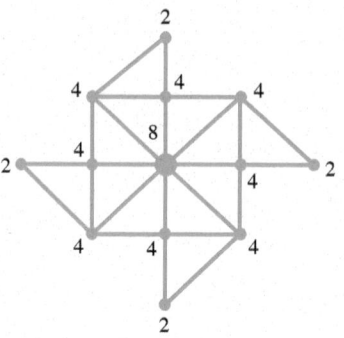

Figure 25

(b) In **Figure 24(b),** check the degrees of the vertices. Note that vertex A has degree 3, an odd degree. We need look no further. It follows from Euler's theorem that this connected graph does not have an Euler circuit. (All the other vertices in this graph except for D have even degree. However, if just one vertex has odd degree, then the graph does not have an Euler circuit.)

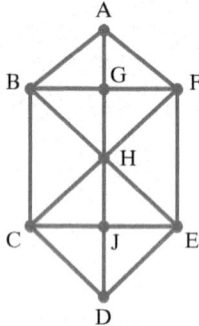

Figure 24(b) (repeated)

EXAMPLE 3 **Tracing Patterns**

Beginning and ending at the same place, is it possible to trace the pattern shown in **Figure 26** below without lifting the pencil off the page and without tracing over any part of the pattern (except isolated points) more than once?

Solution

The pattern in **Figure 26** is not a graph. (It has no vertices.) But what we are being asked to do is very much like being asked to find an Euler circuit. So we imagine that there are vertices at points where lines or curves of the pattern meet, obtaining the graph in **Figure 27.**

Figure 26

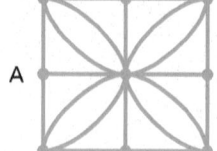

Figure 27

Now we can use Euler's theorem to decide whether we can trace the pattern. Vertex A has odd degree. Thus, by Euler's theorem, this graph does not have an Euler circuit. It follows that we cannot trace the original pattern in the manner required.

By 1875 a new bridge had been built in Königsberg, joining the land areas labeled C and D in **Figure 21.** It was now possible for a citizen of Königsberg to take a walk that used each bridge exactly once, provided that he or she started at A and ended at B, or vice versa.

The additional bridge did not change the answer for the Königsberg bridge problem. There were now paths (crossing each bridge exactly once), but still there was no circuit (beginning and ending at the same place).

The tracing problem in **Example 3** illustrates a practical problem that occurs in designing robotic arms to trace patterns. Automated machines that engrave identification tags for pets require such a mechanical arm.

Fleury's Algorithm

Fleury's algorithm can be used to find an Euler circuit in any connected graph in which each vertex has even degree. An algorithm is like a recipe—follow the steps and we achieve what we need. Before introducing Fleury's algorithm, we need a definition.

CUT EDGE

A **cut edge** in a graph is an edge whose removal disconnects a component of the graph.

We call such an edge a cut edge because removing the edge *cuts* a connected piece of a graph into two disconnected pieces.*

EXAMPLE 4   Identifying Cut Edges

Identify the cut edges in the graph in **Figure 28**.

Solution

This graph has only one component, so we must look for edges whose removal would disconnect the graph. DE is a cut edge. If we removed it, we would disconnect the graph into two components, obtaining the graph of **Figure 29(a)**. HG is also a cut edge. Its removal would disconnect the graph as shown in **Figure 29(b)**.

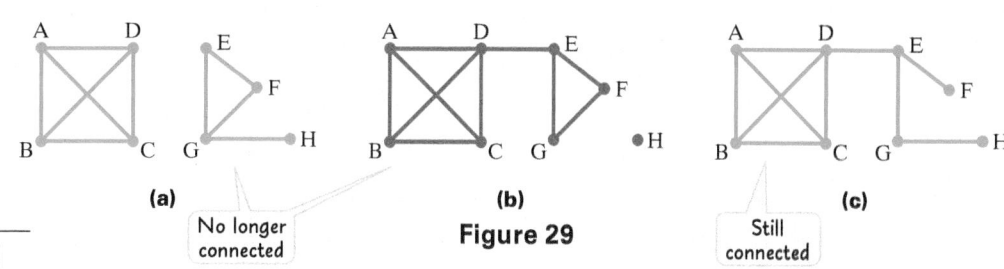

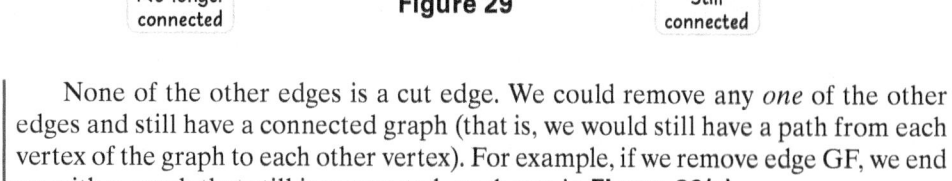

Figure 29

None of the other edges is a cut edge. We could remove any *one* of the other edges and still have a connected graph (that is, we would still have a path from each vertex of the graph to each other vertex). For example, if we remove edge GF, we end up with a graph that still is connected, as shown in **Figure 29(c)**.

Fleury's algorithm is used for finding an Euler circuit in a connected graph in which each vertex has even degree.

FLEURY'S ALGORITHM

Step 1   Start at any vertex. Go along any edge from this vertex to another vertex. *Remove this edge from the graph.*

Step 2   You are now on a vertex of the revised graph. Choose any edge from this vertex, subject to only one condition: Do not use a cut edge (*of the revised graph*) unless you have no other option. Go along your chosen edge. *Remove this edge from the graph.*

Step 3   Repeat Step 2 until you have used all the edges and returned to the vertex at which you started.

The graph on the left (Figure 28):

A   D   E

F

B   C   G   H

Figure 28

The word **algorithm** comes from the name of the Persian mathematician, **Abu Ja'far Muhammad ibn Musa Al-Khowârizmî,** who lived from about A.D. 780 to A.D. 850. One of the books he wrote was on the Hindu-Arabic system of numerals (the system we use today) and calculations using them. The Latin translation of this book is *Algoritmi de numero Indorum,* which means "Al-Khowârizmî on the Hindu Art of Reckoning."

---

* Some mathematicians use "bridge" instead of "cut edge" to describe such an edge.

EXAMPLE 5 **Using Fleury's Algorithm**

Find an Euler circuit for the graph in **Figure 30.**

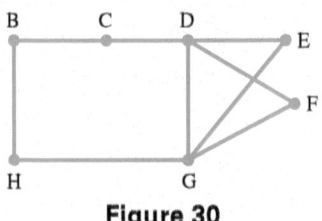

Figure 30

Solution

First we must check that this graph does indeed have an Euler circuit. A check shows that the graph is connected and that each vertex has even degree.

We can start at any vertex. We choose C. We could go from C to B or from C to D. We go to D, removing the edge CD. (We show it scratched out, but we must think of this edge as gone.) Our path begins: C → D. The revised graph is shown on the right in **Figure 31.**

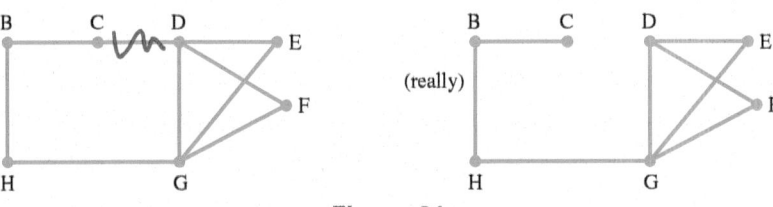

Figure 31

None of the edges DE, DF, and DG is a cut edge for our current graph, so we can choose to go along any one of these. We go from D to G, removing edge DG. So far, our path is C → D → G. (Keep a record of this path.) See **Figure 32.**

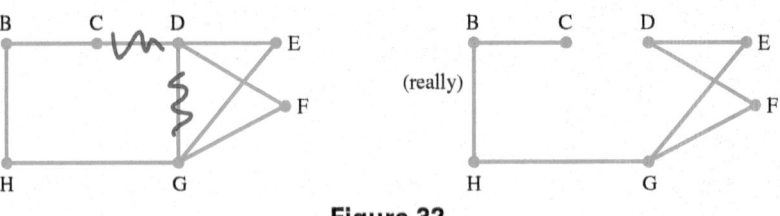

Figure 32

The remaining edges at G are GE, GF, and GH. Note, however, that GH is a cut edge for our revised graph, and we have other options. Thus, according to Fleury's algorithm, *we must not use GH at this stage.* (If we used GH now, we would never be able to get back to use the edges GF, GE, and so on.) We go from G to F.

Our path is C → D → G → F, and we show our revised graph in **Figure 33.**

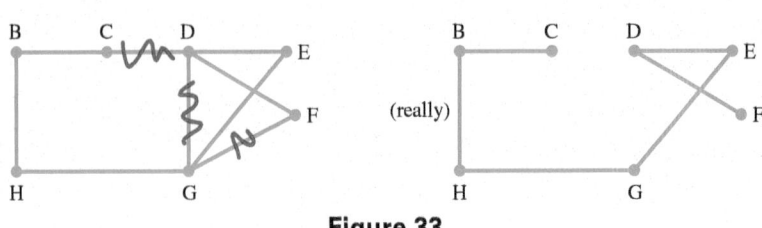

Figure 33

The **Tshokwe people of northeastern Angola** have a tradition involving continuous patterns in sand. The tracings are called *sona,* and they have ritual significance. They are used as mnemonics for stories about the gods and ancestors. Elders trace the drawings while telling the stories. The example shown here is called *skin of a leopard.*

(*Source:* Gerdes, Paul. *Geometry from Africa—Mathematical and Educational Explorations.* Mathematical Association of America, 1999, p. 171.)

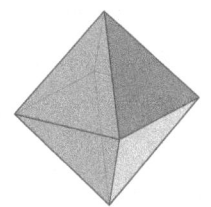

As seen earlier, there are only five **regular polyhedra**: the cube, the tetrahedron, the octahedron, the dodecahedron, and the icosahedron. There is only one which has the property that you can trace your finger along each edge exactly once, beginning and ending at the same vertex. It is the octahedron, because it is the only one with an even number of edges meeting each vertex.

Now we have only one edge from F, namely FD. This is a cut edge of the revised graph, but *since we have no other option,* we may use this edge. Our path is $C \rightarrow D \rightarrow G \rightarrow F \rightarrow D$, and our revised graph is shown in **Figure 34.**

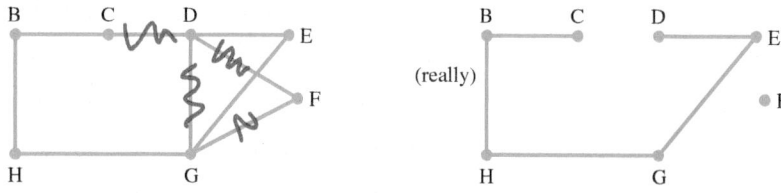

Figure 34

We have only one edge left at D, namely DE. It is a cut edge of the remaining graph, but since we have no other option, we may use it.

It is now clear how to complete the Euler circuit. (Be sure to return to the starting vertex.) The complete Euler circuit is as follows.

$$C \rightarrow D \rightarrow G \rightarrow F \rightarrow D \rightarrow E \rightarrow G \rightarrow H \rightarrow B \rightarrow C.$$

If we start with a connected graph with all vertices having even degree, then Fleury's algorithm will always produce an Euler circuit for the graph. ***Note that a graph that has an Euler circuit always has more than one Euler circuit.***

Even for a large graph, a computer can successfully apply Fleury's algorithm. This is not the case for all algorithms. (See **For Further Thought** on The Speed of Algorithms at the end of the next section.)

Route Planning

In planning routes for mail delivery, street sweeping, or snow plowing, the roads in the region to be covered usually do not have an Euler circuit. Unavoidably, certain stretches of road must be traveled more than once. This is referred to as **deadheading.** Planners then try to find a route on which the driver will spend as little time as possible deadheading.

In **Example 6,** for simplicity we use street grids for which the distances from corner to corner in any direction are the same.

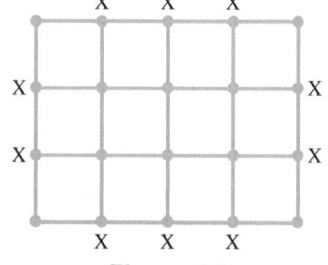

Figure 35

EXAMPLE 6   **Finding an Euler Circuit on a Street Grid**

The street grid in **Figure 35** has every vertex of odd degree marked with an X. Insert edges coinciding with existing roads to change it to a graph with an Euler circuit.

Solution

A certain amount of trial and error is needed to find the least number of edges that must be inserted to ensure that the street grid graph has an Euler circuit.

If we insert edges as shown in **Figure 36(a)** on the next page, we obtain a graph having an Euler circuit. (Check that all vertices now have even degree.) This solution introduces 11 additional edges, which turns out to be more deadheading than is necessary.

If we insert edges as shown in **Figure 36(b)** on the next page, with just 7 additional edges we obtain a graph with an Euler circuit. This is, in fact, the least number of edges that must be inserted (keeping to the street grid) to change the original street grid graph into a graph that has an Euler circuit. (There are different ways to insert the seven edges.)

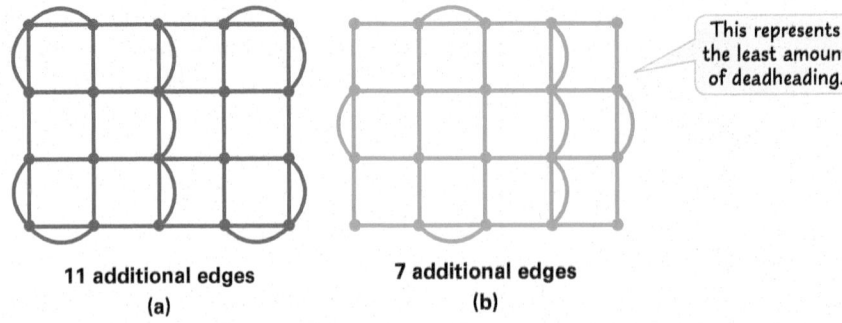

11 additional edges
(a)

7 additional edges
(b)

This represents the least amount of deadheading.

Figure 36

Nearest Neighbor Algorithm:
The Traveling Salesman Problem

WHEN WILL I **EVER** USE THIS?

Main road to post office

Imagine that you are in the management division of a small city post office. You have been assigned to redesign the rural mail carrier route, which is long out of date. Euler circuits and Fleury's algorithm can be used to create the most efficient route.

The figure shows a map of your rural district. In rural areas, mailboxes are placed along the same side of a road, so the mail delivery vehicle needs to travel along the road in one direction only. Suppose the roads off the main road in the figure show a mail delivery region.

It would be ideal if, in your new route, you could find an Euler circuit covering the entire route. If such a circuit existed, it would minimize the distance traveled for the delivery. A glance at the map (thinking of road intersections as vertices) reveals that there is no Euler circuit for this route.

As the next best thing, the delivery vehicle should retrace its path as little as possible. For simplicity, you assume that the time to travel any of the stretches of road when not delivering mail is roughly the same. In following the figure, you added red dashed edges to indicate stretches of road to be covered twice. For this delivery route, it is easy to see that you effectively inserted as few edges as possible to obtain a graph that has an Euler circuit.

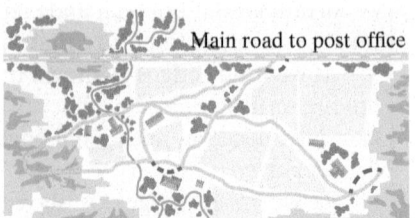

Main road to post office

Now you can use Fleury's algorithm to find an Euler circuit on the graph in the new figure, and this will give the most efficient route for mail delivery.

14.2 EXERCISES

CONCEPT CHECK *Complete each statement with the correct response(s).*

1. An Euler path (or circuit) in a graph is a path (or circuit) that uses every _____ of the graph exactly _____.

2. According to Euler's theorem, if a connected graph has an Euler circuit, then each vertex of the graph has _____ degree.

3. According to Euler's theorem, if each vertex of a connected graph has even degree, then the graph has a(n) _____ _____.

4. A(n) _____ _____ in a graph is an edge whose removal disconnects a component of the graph.

Euler Circuits *A graph is shown, and some sequences of vertices are specified. Determine which of these sequences show Euler circuits. For those that do not, explain why not.*

5.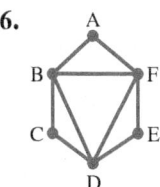

   (a) $A \to B \to C \to D \to A \to B \to C \to D \to A$

   (b) $C \to B \to A \to D \to C$

   (c) $A \to C \to D \to B \to A$

   (d) $A \to B \to C \to D$

6.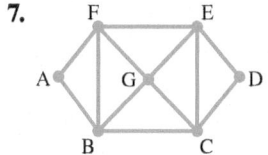

   (a) $A \to B \to C \to D \to E \to F \to A$

   (b) $F \to B \to D \to B$

   (c) $A \to B \to C \to D \to E \to F \to B \to D \to F \to A$

   (d) $A \to B \to F \to D \to B \to C \to D \to E \to F \to A$

7.

   (a) $A \to B \to C \to D \to E \to F \to A$

   (b) $A \to B \to C \to D \to E \to G \to C \to E \to$ $F \to G \to B \to F \to A$

   (c) $A \to B \to C \to D \to E \to C \to G \to E \to$ $F \to G \to E \to F \to A$

   (d) $A \to B \to G \to E \to D \to C \to G \to F \to$ $B \to C \to E \to F \to A$

Euler's Theorem *Use Euler's theorem to decide whether the graph has an Euler circuit. (Do not actually find an Euler circuit.) Justify each answer briefly.*

8.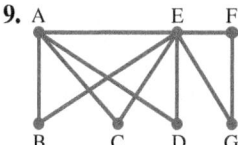

9.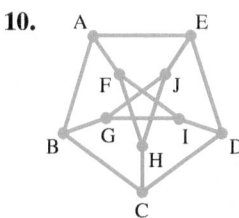

10.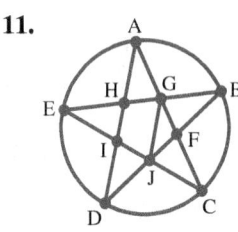

11.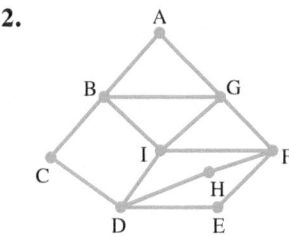

12.

Euler's Theorem *Use Euler's theorem to determine whether it is possible to begin and end at the same place, trace the pattern without lifting your pencil, and trace over no line in the pattern more than once.*

13.

14.

Euler's Theorem   *Use Euler's theorem to determine whether the graph has an Euler circuit, justifying each answer. Then determine whether the graph has a circuit that visits each vertex exactly once, except that it returns to its starting vertex. If so, write down the circuit. (There may be more than one correct answer.)*

15.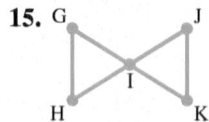

16.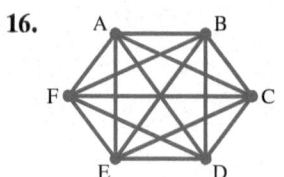

17.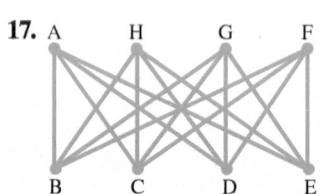

18.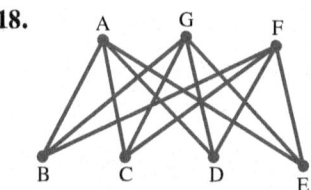

Floor Tilings   *Different floor tilings are shown. The material applied between tiles is called grout. For which of these floor tilings could the grout be applied beginning and ending at the same place, without going over any joint twice, and without lifting the tool? Justify each answer.*

19.

20.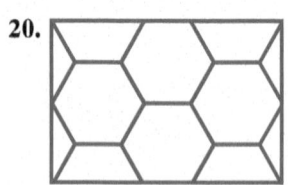

21.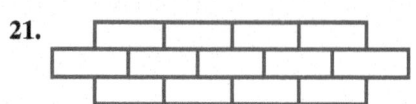

22.

Cut Edges   *Identify all cut edges in the graph. If there are none, say so.*

23.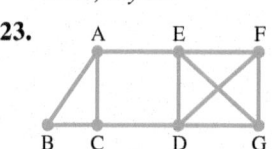

24.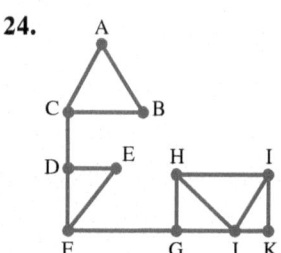

25.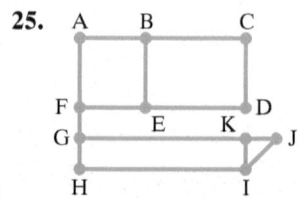

26.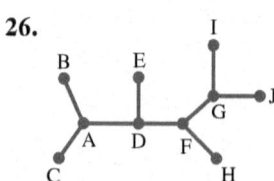

Fleury's Algorithm   *A graph is shown for which a student has been asked to find an Euler circuit starting at* A. *The student's revisions of the graph after the first few steps of Fleury's algorithm are shown, and in each case the student is now at* B. *For each graph, determine all edges that Fleury's algorithm permits the student to use for the next step.*

27.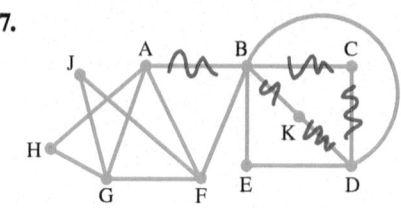

28.

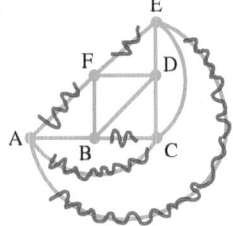

29.

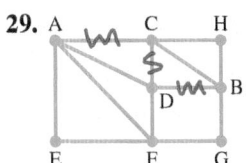

Fleury's Algorithm *Use Fleury's algorithm to find an Euler circuit for the graph, beginning and ending at* A. *(There are many different correct answers.)*

30.

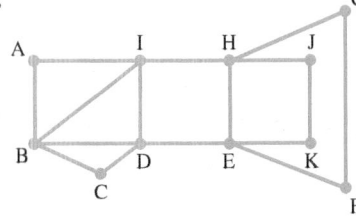

31.

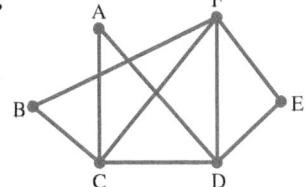

32.

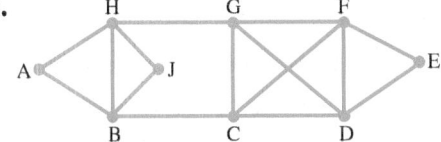

Euler's Theorem and Fleury's Algorithm *Use Euler's theorem to determine whether the graph has an Euler circuit. If not, explain why not. If the graph does have an Euler circuit, use Fleury's algorithm to find an Euler circuit for the graph. (There are many different correct answers.)*

33.

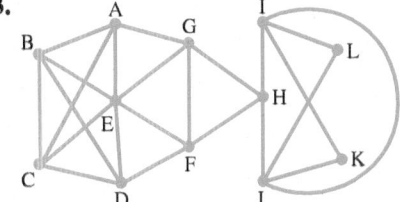

34.

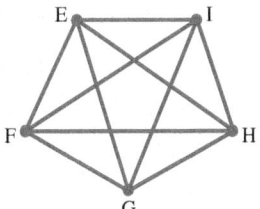

35.

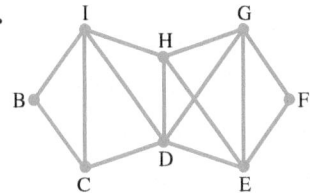

Solve each problem.

36. *Garden Design*   The graph below shows the layout of the paths in a botanical garden. The edges represent the paths.

(a) Has the garden been designed in such a way that it is possible for a visitor to find a route that begins and ends at the entrance to the garden (represented by the vertex A) and that goes along each path exactly once?

(b) If so, use Fleury's algorithm to find such a route.

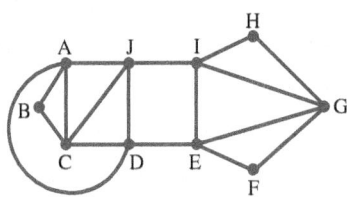

37. *Parking Pattern*   The map shows the roads on which parking is permitted at a national monument. This is a pay-and-display facility. A security guard has the task of periodically checking that all parked vehicles have a valid parking ticket displayed. He is based at the central complex, labeled A.

(a) Is there a route that he can take to walk along each of the roads exactly once, beginning and ending at A?

(b) If so, use Fleury's algorithm to find such a route.

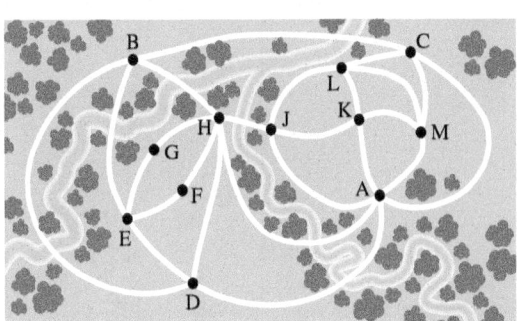

Floor Plans *The floor plan of a building is shown. For which of these is it possible to start outside, walk through each door exactly once, and end up back outside? Justify each answer. (Hint: Think of the rooms and "outside" as the vertices of a graph, and think of the doors as the edges of the graph.)*

38.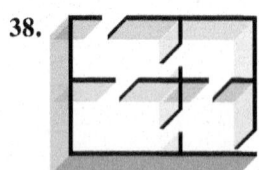

39.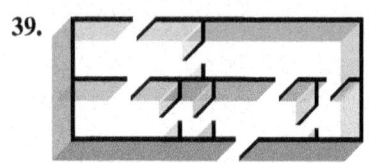

40.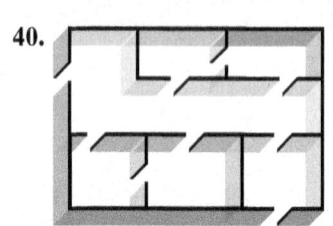

Consider the following theorem in the exercises below.

- **If a graph has an Euler path that begins and ends at different vertices, then these two vertices are the only vertices with odd degree. (All the rest have even degree.)**

- **If exactly two vertices in a connected graph have odd degree, then the graph has an Euler path beginning at one of these vertices and ending at the other.**

Determine whether the graph has an Euler path that begins and ends at different vertices. Justify your answer. If the graph has such a path, say at which vertices the path must begin and end.

41.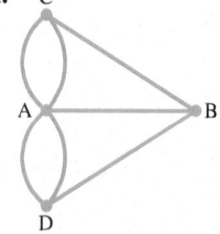

42.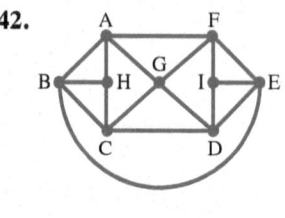

43.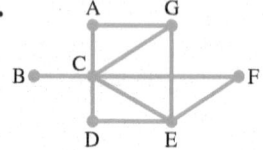

44.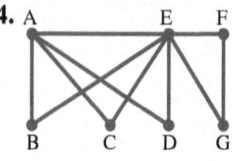

Floor Plans *Refer to the floor plan indicated, and determine whether it is possible to start in one of the rooms of the building, walk through each door exactly once, and end up in a different room from the one you started in. Justify each answer.*

45. Refer to the floor plan shown in **Exercise 38.**

46. Refer to the floor plan shown in **Exercise 39.**

47. Refer to the floor plan shown in **Exercise 40.**

New York City Bridges *The accompanying schematic map shows a portion of the New York City area, including tunnels and bridges.*

48. Is it possible to take a drive around the New York City area using each tunnel and bridge exactly once, beginning and ending in the same place?

49. Is it possible to take a drive around the New York City area using each tunnel and bridge exactly once, beginning and ending in different places? If so, where must the drive begin and end?

Circuits with Maximum Edges *The graph does not have an Euler circuit. For each graph find a circuit that uses as many edges as possible. (There is more than one correct answer in each case.) How many edges did you use in the circuit?*

50.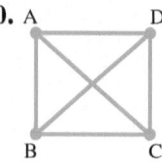

51.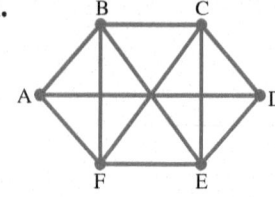

52.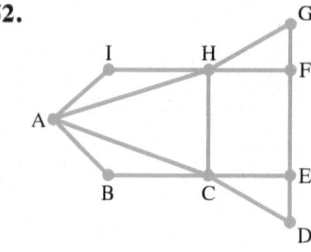

53. *Route Planning*   In the graph shown here, we have inserted just six edges in the original street grid from **Example 6.** Why is this not a better solution than the one in **Figure 36(b)?**

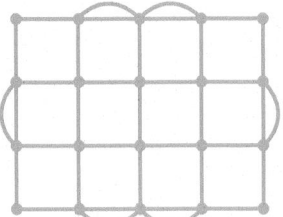

Route Planning   For each street grid, insert edges to obtain a street grid that has an Euler circuit. Try to insert as few edges as possible.

54.
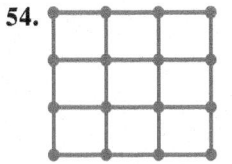

55.
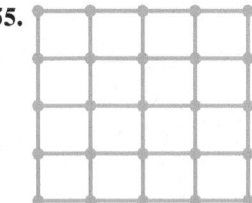

56.
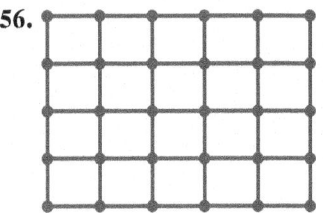

14.3   HAMILTON CIRCUITS AND ALGORITHMS

OBJECTIVES

1 Define and identify a Hamilton circuit in a graph.

2 Determine the number of Hamilton circuits in a complete graph with *n* vertices.

3 Define and apply the meanings of the terms *weighted graph* and *minimum Hamilton circuit.*

4 Apply the brute force algorithm for a complete, weighted graph.

5 Define and apply the meanings of the terms *efficient algorithm* and *approximate algorithm.*

6 Apply the nearest neighbor algorithm.

Hamilton Circuits

In this section we examine *Hamilton circuits* in graphs. The story of Hamilton circuits is a story of very large numbers and of unsolved problems in both mathematics and computer science. We start with a game, called the **Icosian game,** invented by Irish mathematician William Hamilton in the mid-nineteenth century. It uses a wooden board marked with the graph shown in **Figure 37.**

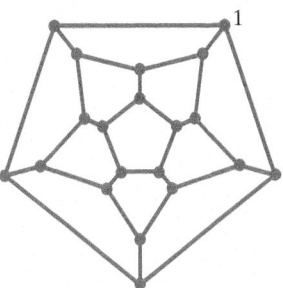

The Icosian Game

Figure 37

The game includes 20 pegs numbered 1 through 20. (The word *Icosian* comes from the Greek word for 20.) At each vertex of the graph there is a hole for a peg.

The simplest version of the game is as follows:

> *Put the pegs into the holes in order, following along the edges of the graph, in such a way that peg 20 ends up in a hole that is joined by an edge to the hole of peg 1.*

Try this now, numbering the vertices in the graph in **Figure 37,** starting at vertex 1.

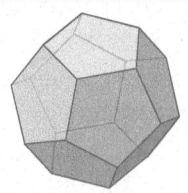

Dodecahedron

Hamilton sold the **Icosian game** idea to a games dealer for 25 British pounds. It went on the market in 1859, but it was not a huge success. A later version of the game, *A Voyage Around the World*, consisted of a regular dodecahedron with pegs at each of the 20 vertices. The vertices had names such as Brussels, Delhi, and Zanzibar. The aim was to travel to all vertices along the edges of the dodecahedron, visiting each vertex exactly once. We can show the vertices and edges of a dodecahedron precisely as in **Figure 37.**

The Icosian game asks that we find a circuit in the graph, but the circuit need not be an Euler circuit. The circuit must visit each *vertex* exactly once, except for returning to the starting vertex to complete the circuit. (The circuit may or may not travel all edges of the graph.) Circuits such as this are called *Hamilton circuits.*

HAMILTON CIRCUIT

A **Hamilton circuit** in a graph is a circuit that visits each vertex exactly once (returning to the starting vertex to complete the circuit).

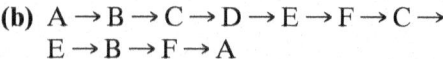 **Identifying Hamilton Circuits**

Which of the following are Hamilton circuits for the graph in **Figure 38?** Justify your answers briefly.

(a) $A \rightarrow B \rightarrow E \rightarrow D \rightarrow C \rightarrow F \rightarrow A$

(b) $A \rightarrow B \rightarrow C \rightarrow D \rightarrow E \rightarrow F \rightarrow C \rightarrow$
$E \rightarrow B \rightarrow F \rightarrow A$

(c) $B \rightarrow C \rightarrow D \rightarrow E \rightarrow F \rightarrow B$

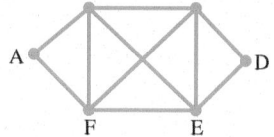

Figure 38

Solution

(a) $A \rightarrow B \rightarrow E \rightarrow D \rightarrow C \rightarrow F \rightarrow A$ is a Hamilton circuit for the graph. It visits each vertex of the graph exactly once and then returns to the starting vertex. (Trace this circuit on the graph to check it.)

(b) $A \rightarrow B \rightarrow C \rightarrow D \rightarrow E \rightarrow F \rightarrow C \rightarrow E \rightarrow B \rightarrow F \rightarrow A$ is not a Hamilton circuit, because it visits vertex B (and vertices C, E, and F) more than once. (This is, however, an Euler circuit for the graph.)

(c) $B \rightarrow C \rightarrow D \rightarrow E \rightarrow F \rightarrow B$ is not a Hamilton circuit because it does not visit all vertices in the graph.

Some graphs, such as the graph in **Figure 39,** do not have a Hamilton circuit. There is no way to visit all the vertices and return to the starting vertex without visiting vertex I more times than allowed.

It can be difficult to determine whether a particular graph has a Hamilton circuit, since there is no theorem that gives necessary and sufficient conditions for a Hamilton circuit to exist. This remains an unsolved problem in Hamilton circuits.

Fortunately, in many real-world applications of Hamilton circuits we are dealing with complete graphs, and any complete graph with three or more vertices does have a Hamilton circuit. For example,

$$A \rightarrow B \rightarrow C \rightarrow D \rightarrow E \rightarrow F \rightarrow A$$

is a Hamilton circuit for the complete graph shown in **Figure 40.** We could form a Hamilton circuit for any complete graph with three or more vertices in a similar way.

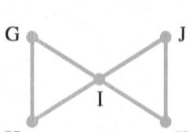

Figure 39

HAMILTON CIRCUITS FOR COMPLETE GRAPHS

Any complete graph with three or more vertices has a Hamilton circuit.

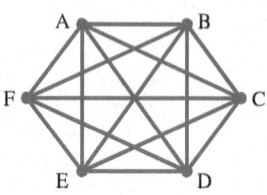

Figure 40

The graph in **Figure 40** has many Hamilton circuits. Try finding some that are different from the one given above.

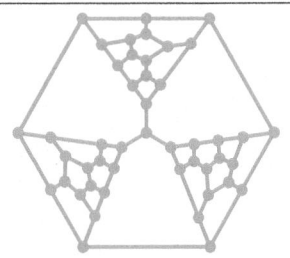

The Tutte graph, shown here, has no Hamilton circuit. This graph has a history connected to the four-color problem for map coloring. In 1880 Peter Tate provided a "proof" for the four-color theorem. It was based on the assumption that every connected graph with all vertices of degree 3, and which can be drawn with edges crossing nowhere except at vertices, has a Hamilton circuit. But in 1946, Tutte produced the graph shown here, showing that Tate's basic assumption was wrong and his proof was, therefore, invalid.

How many Hamilton circuits does a complete graph have? Before we count, we need an agreement about when two sequences of vertices will be considered to be *different* Hamilton circuits. For example, the Hamilton circuits

$$B \rightarrow C \rightarrow D \rightarrow E \rightarrow F \rightarrow A \rightarrow B \quad \text{and} \quad A \rightarrow B \rightarrow C \rightarrow D \rightarrow E \rightarrow F \rightarrow A$$

visit the vertices in essentially the same order in the graph in **Figure 40,** although the sequences start with different vertices. If we mark the circuit on the graph, we can describe it starting at any vertex. For our purposes, it is convenient to consider the two sequences above as representing the same Hamilton circuit.

WHEN HAMILTON CIRCUITS ARE THE SAME

Hamilton circuits that differ *only* in their starting points will be considered to be the same circuit.

Problem-Solving Strategy

Tree diagrams are often useful for counting.

We begin by counting the number of Hamilton circuits in a complete graph with four vertices. See **Figure 41.** We can use the same starting point for all the circuits we count. We choose A as the starting point. From A we can go to any of the three remaining vertices (B, C, or D). No matter which vertex we choose, we then have two unvisited vertices for our next choice. Then there is only one way to complete the Hamilton circuit. We illustrate this counting procedure in **Figure 42,** where the tree diagram shows that we have $3 \cdot 2 \cdot 1$ Hamilton circuits in the graph. We write this as 3! (read "3 factorial").

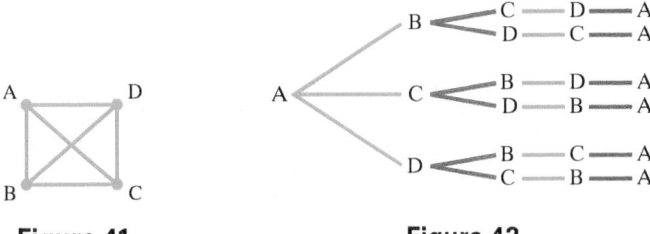

Figure 41   **Figure 42**

Returning to **Figure 40,** we can count the number of Hamilton circuits in the complete graph with 6 vertices. Start at vertex A. From A we can proceed to any of the five remaining vertices. No matter which we choose, we then have four choices for the next vertex along the circuit, three choices for the next, and two for the next. This means that we have

$$5 \cdot 4 \cdot 3 \cdot 2 \cdot 1 = 5! \quad \text{different Hamilton circuits in all.}$$

We would find that a complete graph with 10 vertices has 9! Hamilton circuits. The number of Hamilton circuits in a complete graph can be obtained by calculating the factorial of the number that is one less than the number of vertices.

NUMBER OF HAMILTON CIRCUITS IN A COMPLETE GRAPH

A complete graph with *n* vertices has the following number of Hamilton circuits.

$$(n - 1)!$$

This peg game is found throughout the United States at Cracker Barrel restaurants.

Traveling Salesman Problems and Archaeology Archaeologists often excavate sites in which there is no clear evidence to show which deposits were made earlier, and which later. Some archaeologists have used minimum Hamilton circuits to solve the problem. For example, if a site consists of a number of burials, they consider a complete graph with vertices representing the various burials, and weights on the edges corresponding roughly to how dissimilar the burials are. A minimum Hamilton circuit in this graph gives the best guess for the order in which the deposits were made. Dr. Edward C. Harris, in his *Principles of Archaeological Stratigraphy,* provided a system known today as **Harris matrices** (which are actually graphs) to assist archaeologists in determining how stratification occurred at a dig site.

As the number of vertices in a complete graph increases, the number of Hamilton circuits for that graph increases very quickly. Previously we considered how quickly exponential functions increase. Factorials increase quickly as well. For example,

$$25! = 15{,}511{,}210{,}043{,}330{,}985{,}984{,}000{,}000,$$

or approximately 1.6×10^{25}–more than a trillion trillion.

Minimum Hamilton Circuits

Consider that on a typical day, a UPS van might have to make deliveries to 100 different locations. For simplicity, assume that none are priority deliveries, so they can be made in any order. UPS wants to minimize the time to make these deliveries. Think of the 100 locations and the UPS distribution center as the vertices of a complete graph with 101 vertices.

In principle, the van can go from any of these locations directly to any other, shown by having an edge between each pair of vertices. Suppose we estimate the travel time between each pair of locations (of course, this would depend on distance and traffic conditions along the route). These estimates provide weights on the edges of the graph. The objective is to visit each location exactly once, to begin and end at the same place, and to take as little time as possible.

In graph theory terms, we need a Hamilton circuit for the graph with the *least possible total weight.* The **total weight** of a circuit is the sum of the weights on the edges in the circuit. We call such a circuit a *minimum Hamilton circuit* for the graph.

MINIMUM HAMILTON CIRCUIT

In a weighted graph, a **minimum Hamilton circuit** is a Hamilton circuit with the least possible total weight.

A problem whose solution requires us to find a minimum Hamilton circuit for a complete, weighted graph is often called a **traveling salesman problem** (or TSP). Think of the vertices of the complete graph as the cities that a salesperson must visit, and think of the weights on the edges as the cost of traveling directly between the cities. To minimize costs, the salesperson needs a minimum Hamilton circuit for the graph.

Traveling salesman problems are relevant to efficient routing of telephone calls and Internet connections. As another example, to manufacture integrated circuit silicon chips, many lines have to be etched on a silicon wafer. Minimizing production time involves deciding in which order to etch the lines: a traveling salesman problem. Likewise, to manufacture circuit boards for integrated circuits, laser-drilled holes must be made for connections. Again, the order in which the holes are drilled is a critical factor in production time.

Brute Force Algorithm

Suppose we are given a complete, weighted graph. How can we find a minimum Hamilton circuit for the graph? One way is to systematically list all the Hamilton circuits in the graph, find the total weight of each, and choose a circuit with least total weight. (In fact, it is sufficient to add up the weights on the edges for just half of the Hamilton circuits, since the total weight of a circuit is the same as the total weight of the circuit that uses the same edges in reverse order.)

The method just described is sometimes called the **brute force algorithm** because we find the solution by checking *all* the Hamilton circuits.

BRUTE FORCE ALGORITHM

Step 1   Choose a starting point.

Step 2   List all the Hamilton circuits with that starting point.

Step 3   Find the total weight of each circuit.

Step 4   Choose a Hamilton circuit with the least total weight.

EXAMPLE 2   Using the Brute Force Algorithm

Find a minimum Hamilton circuit for the complete, weighted graph shown in **Figure 43**.

Solution

Choose a starting point. For this example, we choose A. Now list all the Hamilton circuits starting at A. Because this is a complete graph with 4 vertices, there are

$$3! = 3 \cdot 2 \cdot 1 = 6 \text{ circuits.}$$

Thus, we must find 6 Hamilton circuits.

We need a systematic way of writing down all the Hamilton circuits. The "counting tree" of **Figure 42** shown earlier provides a guide. We start by finding all the Hamilton circuits that begin A → B, then include those that begin A → C, and finally include those that begin A → D, as shown below. Once all the Hamilton circuits are listed, pair those that visit the vertices in precisely opposite orders. (Circuits in these pairs have the same total weight. This will save some adding.) Finally, determine the sum of the weights in each circuit, as shown in **Table 1**.

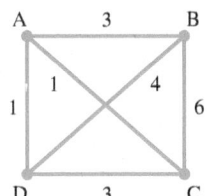

A   3   B
1   4   6
D   3   C

Figure 43

Table 1

Circuit	Total weight of the circuit
1. A → B → C → D → A	3 + 6 + 3 + 1 = 13
2. A → B → D → C → A	3 + 4 + 3 + 1 = 11 Minimum
3. A → C → B → D → A	1 + 6 + 4 + 1 = 12
4. A → C → D → B → A (opposite of 2)	11
5. A → D → B → C → A (opposite of 3)	12
6. A → D → C → B → A (opposite of 1)	13

We can now see that A → B → D → C → A is a minimum Hamilton circuit for the graph. The weight of this circuit is 11.

In principle, the brute force algorithm provides a way to find a minimum Hamilton circuit in any complete, weighted graph. In practice, it takes far too long, even for a complete, weighted graph with only 7 vertices. There are 6!, or 720, Hamilton circuits in such a graph, and we would have to calculate the total weight for half of these. That means we would have to do 360 separate calculations, in addition to listing the Hamilton circuits.

Most real-world problems involve a great deal more than 7 vertices. Manufacturing a large integrated circuit can involve a graph with almost one million vertices. Likewise, telecommunications companies routinely deal with graphs with millions of vertices. As the number of vertices in the graph increases, the task of finding a minimum Hamilton circuit using the brute force algorithm soon becomes too time-consuming for even our fastest computers. For example, if one of today's supercomputers had started using the brute force algorithm on a 100-vertex traveling salesman problem when the universe was created, it would not yet be done.

William Rowan Hamilton (1805–1865) spent most of his life in Dublin, Ireland. He was good friends with the poet William Wordsworth, whom he met while touring England and Scotland. Hamilton tried writing poetry, but Wordsworth tactfully suggested that his talents lay in mathematics rather than verse. Catherine Disney was the first great love of his life, but under pressure from her parents, she married another man, much wealthier than Hamilton. Hamilton never seemed to quite get over this.

In contrast, Fleury's algorithm for finding an Euler circuit in a graph does not take too long for our computers, even for rather large graphs. Algorithms that do not take too much computer time are called **efficient algorithms.** Computer scientists have so far been unable to find an efficient algorithm for the traveling salesman problem, and they suspect that it is simply impossible to create such an algorithm.

For the traveling salesman problem, there are some algorithms that do not take too much computer time and that give *reasonably good* solutions *most of the time.* Such algorithms are called **approximate algorithms,** because they give an approximate solution to the problem. The underlying idea is that from each vertex, we proceed to a "nearest" available vertex.

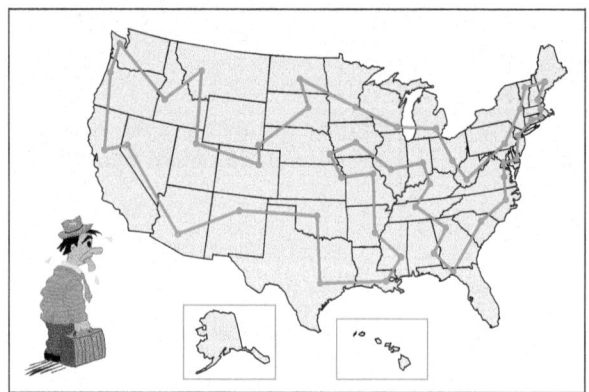

Figure 44

In 1985, researcher Shen Lin came up with the route shown in **Figure 44** for a salesman who wants to visit all capital cities in the forty-eight contiguous states, starting and ending at the same capital and traveling the shortest possible total distance. He could not prove that his 10,628-mile route was the shortest possible, but he offered $100 to anyone who could find a shorter one.

Nearest Neighbor Algorithm

One approximate algorithm for problems such as the traveling salesman problem is the **nearest neighbor algorithm.**

Nearest Neighbor Algorithm: The Traveling Salesman Problem

NEAREST NEIGHBOR ALGORITHM

Step 1 Choose a starting point for the circuit. Call this vertex A.

Step 2 Check all the edges joined to A, and choose one that has the least weight. Proceed along this edge to the next vertex.

Step 3 At each vertex you reach, check the edges from there *to vertices not yet visited.* Choose one with the least weight. Proceed along this edge to the next vertex.

Step 4 Repeat Step 3 until you have visited all the vertices.

Step 5 Return to the starting vertex.

EXAMPLE 3 **Using the Nearest Neighbor Algorithm**

A courier is based at the head office (A) and must deliver documents to four other offices (B, C, D, and E). The estimated time of travel (in minutes) between each of these offices is shown on the graph in **Figure 45** on the next page. The courier wants to visit the locations in an order that takes the least time. Use the nearest neighbor algorithm to find an approximate solution to this problem. Calculate the total time required to cover the chosen route.

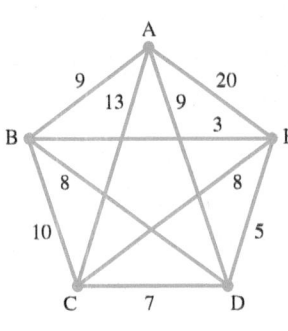

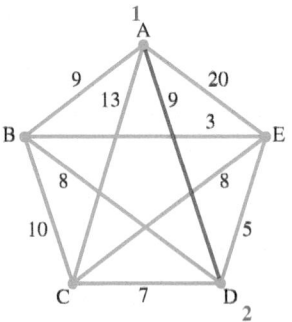

Figure 45   Figure 46

Solution

Step 1 Choose a starting point. Let's start at A.

Step 2 Choose an edge with least weight joined to A. Both AB and AD have weight 9, which is less than the weights of the other two edges joined to A. We choose AD* and keep a record of the circuit as we form it.

To help ensure that we do not visit a vertex twice, we number the vertices as we visit them and color the edges used, as in **Figure 46.**

Continue to number the vertices as we work through the rest of the solution.

Step 3 Now check edges joined to D, excluding DA (since we have already been to A). DC has weight 7, DB has weight 8, and DE has weight 5. DE has the least weight, so proceed along DE to E. Our circuit begins this way.

$$A \overset{9}{\to} D \overset{5}{\to} E$$

Step 4 Now repeat the process at E. Check all edges from E that go to a vertex not yet visited. EC has weight 8 and EB has weight 3. Proceed along EB to B.

$$A \overset{9}{\to} D \overset{5}{\to} E \overset{3}{\to} B$$

Step 5 We have one vertex not yet visited, C. We go next to C and now have visited all the vertices. We return to our starting point, A. Our Hamilton circuit is

$$A \overset{9}{\to} D \overset{5}{\to} E \overset{3}{\to} B \overset{10}{\to} C \overset{13}{\to} A.$$

Its total weight is $9 + 5 + 3 + 10 + 13 = 40$.

Our advice to the courier would thus be to visit the offices in the order shown in this circuit. This route will take about 40 minutes.

The circuit we found in **Example 3** is not the minimum Hamilton circuit for the graph in **Figure 45.** (We shall investigate this.) However, the point of an approximate algorithm is that a computer can implement it without taking too much time, and most of the time it will give a reasonably good solution to the problem.

If we had a computer performing the nearest neighbor algorithm for us, we could make a small adjustment that would give better results without taking too much longer. We could repeat the nearest neighbor algorithm for all possible starting points, and then choose from these the Hamilton circuit with the least weight. If we do this for the example above, we obtain the results in **Table 2** on the next page.

---

* If using a computer to do this, we would instruct the computer to make a random choice between AB and AD.

Table 2

Starting Vertex	Circuit Using Nearest Neighbor	Total Weight
A	$A \rightarrow D \rightarrow E \rightarrow B \rightarrow C \rightarrow A$	$9 + 5 + 3 + 10 + 13 = 40$
B	$B \rightarrow E \rightarrow D \rightarrow C \rightarrow A \rightarrow B$	$3 + 5 + 7 + 13 + 9 = 37$
C	$C \rightarrow D \rightarrow E \rightarrow B \rightarrow A \rightarrow C$	$7 + 5 + 3 + 9 + 13 = 37$
D	$D \rightarrow E \rightarrow B \rightarrow A \rightarrow C \rightarrow D$	$5 + 3 + 9 + 13 + 7 = 37$
E	$E \rightarrow B \rightarrow D \rightarrow C \rightarrow A \rightarrow E$	$3 + 8 + 7 + 13 + 20 = 51$

We would recommend that the courier use a circuit with total weight 37—for example,

$$D \rightarrow E \rightarrow B \rightarrow A \rightarrow C \rightarrow D.$$

This does not force him to start his journey at D. If he followed the route

$$A \rightarrow C \rightarrow D \rightarrow E \rightarrow B \rightarrow A,$$

he would be using the same edges as in $D \rightarrow E \rightarrow B \rightarrow A \rightarrow C \rightarrow D$, and his traveling time still would be 37 minutes.

Can we now be sure that we have found the minimum Hamilton circuit for the graph in **Figure 45?** No. This is still an *approximate* solution. If we check the total weights of all possible Hamilton circuits in the graph, we find the Hamilton circuit

$$A \overset{9}{\rightarrow} B \overset{3}{\rightarrow} E \overset{8}{\rightarrow} C \overset{7}{\rightarrow} D \overset{9}{\rightarrow} A$$

with total weight just 36 minutes.

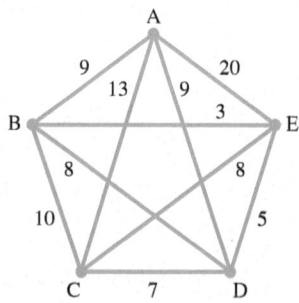

Figure 45 (repeated)

Nearest Neighbor Algorithm:
The Traveling Salesman Problem

The nearest neighbor algorithm simply will not find this minimum Hamilton circuit for the graph. However, all we expect of an approximate algorithm is that it give a reasonably good solution for the problem in a reasonable amount of time.

One method for generating solutions for the traveling salesman problem uses ant colony simulation. It models by creating paths observed in actual ant behavior to find the shortest paths between the ant nest and food sources. See **Figure 47.** There are many Web sites devoted to the Traveling Salesman Problem.

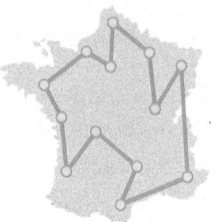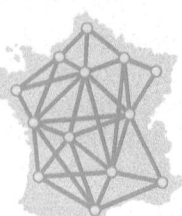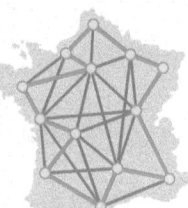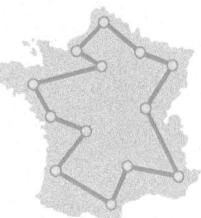

Figure 47

WHEN WILL I **EVER** USE THIS**?**

As a computer scientist, Len Testa was always fascinated with the idea of using computers to solve everyday problems. He was in graduate school at North Carolina A&T when he and his twin sister went on a trip to Walt Disney World in Florida. One day, they endured a two-hour wait for the Great Movie Ride, much of which was spent outside in the sweltering heat. In an interview with Coaster101.com, Testa said, "I thought to myself, 'there's got to be a better way to do this' where computers can help schedule all of this." He decided to write his master's degree thesis on minimizing wait times at Walt Disney World using the traveling salesman problem, one of the fundamental problems in mathematics and computer science.

The traveling salesman problem, or TSP, asks the following question:

Given a map of cities linked by roads, what is the most efficient way for a traveling salesman to visit every city once before returning home?

In Testa's research, the salesman is a guest at Walt Disney World, and the cities are attractions. In writing his thesis, Testa contacted Bob Sehlinger, author of *The Unofficial Guide to Walt Disney World*. Sehlinger, along with a colleague at MIT, had developed software that used mixed integer linear programming to create schedules, called "touring plans," that aimed to reduce guests' wait times. Testa thought he could improve on Sehlinger's software, and in 2003, after 3 years of collecting data at the Disney theme parks, he did just that. Sehlinger switched to using the new software, and in 2007, Testa become the co-author of *The Unofficial Guide to Walt Disney World*.

Len Testa is currently President of TouringPlans.com, which boasts a variety of tools to help people save time and money at Walt Disney World, as well as Disneyland Resort, Universal Orlando Resort, and Disney Cruise Line. TouringPlans.com features a crowd calendar, which allows users to see how crowded a particular theme park will be on any given day. Users can also create personalized touring plans, which are tailored to their preferences of attractions they wish to visit. Testa has truly made a career out of helping people save time and money at theme parks using computer science and mathematics.

Source: www.coaster101.com/2018/08/07/len-testa-and-touringplans-com-part-one-the-origins/

(14.3) EXERCISES

CONCEPT CHECK *Complete each statement with the correct response.*

1. A Hamilton circuit in a graph is a circuit that visits each vertex exactly _____ (returning to the starting vertex to complete the circuit).

2. Any complete graph with _____ or more vertices has a Hamilton circuit.

3. A complete graph with n vertices has _____ Hamilton circuits.

4. In a weighted graph, a minimum Hamilton circuit is a Hamilton circuit with the _____ _____ _____ weight.

Hamilton Circuits *A graph is shown, and some paths in the graph are specified. Determine which paths are Hamilton circuits for the graph. For those that are not, say why not.*

5.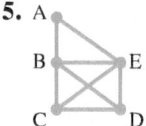

(a) $A \rightarrow E \rightarrow C \rightarrow D \rightarrow E \rightarrow B \rightarrow A$

(b) $A \rightarrow E \rightarrow C \rightarrow D \rightarrow B \rightarrow A$

(c) $D \rightarrow B \rightarrow E \rightarrow A \rightarrow B$

(d) $E \rightarrow D \rightarrow C \rightarrow B \rightarrow E$

6.

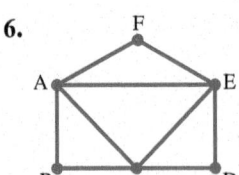

(a) $A \rightarrow B \rightarrow C \rightarrow D \rightarrow E \rightarrow C \rightarrow A \rightarrow E \rightarrow F \rightarrow A$

(b) $A \rightarrow C \rightarrow D \rightarrow E \rightarrow F \rightarrow A$

(c) $F \rightarrow A \rightarrow C \rightarrow E \rightarrow F$

(d) $C \rightarrow D \rightarrow E \rightarrow F \rightarrow A \rightarrow B$

Euler and Hamilton Circuits *Determine whether each sequence of vertices is a circuit, whether it is an Euler circuit, and whether it is a Hamilton circuit. Justify your answers.*

7.

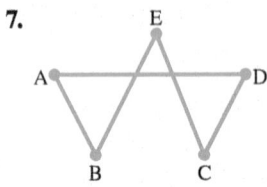

(a) $A \rightarrow B \rightarrow C \rightarrow D \rightarrow E \rightarrow A$

(b) $B \rightarrow E \rightarrow C \rightarrow D \rightarrow A \rightarrow B$

(c) $E \rightarrow B \rightarrow A \rightarrow D \rightarrow A \rightarrow D \rightarrow C \rightarrow E$

8.

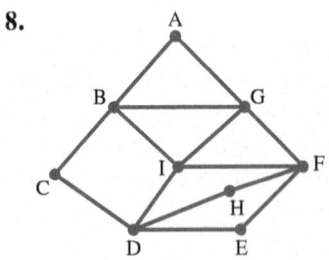

(a) $A \rightarrow B \rightarrow C \rightarrow D \rightarrow E \rightarrow F \rightarrow G \rightarrow A$

(b) $B \rightarrow I \rightarrow G \rightarrow F \rightarrow E \rightarrow D \rightarrow H \rightarrow F \rightarrow$ $I \rightarrow D \rightarrow C \rightarrow B \rightarrow G \rightarrow A \rightarrow B$

(c) $A \rightarrow B \rightarrow C \rightarrow D \rightarrow E \rightarrow F \rightarrow G \rightarrow H \rightarrow I \rightarrow A$

Hamilton Circuits *Determine whether the graph has a Hamilton circuit. If so, find one. (There may be many different correct answers.)*

9.

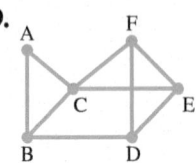

10.

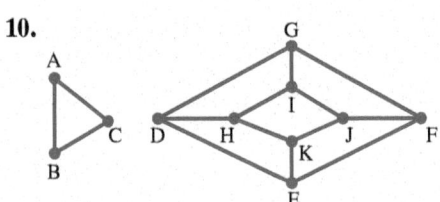

11.

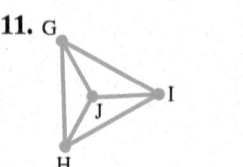

12.

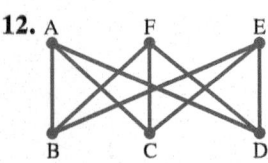

13.

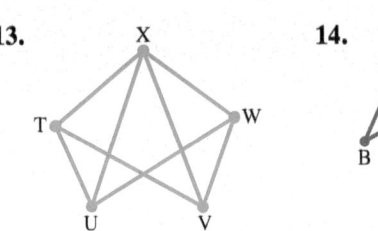

14.

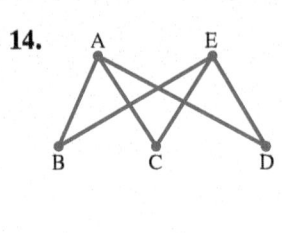

Hamilton and Euler Circuits *Solve each problem.*

15. Draw a graph that has a Hamilton circuit but no Euler circuit. Specify the Hamilton circuit, and explain why the graph has no Euler circuit. (There are many different correct answers.)

16. Draw a graph that has an Euler circuit but no Hamilton circuit. Specify an Euler circuit in your graph. (There are many different correct answers.)

17. Draw a graph that has both an Euler circuit and a Hamilton circuit. Specify these circuits. (There are many different correct answers.)

18. Decide whether each statement is true or false. If the statement is false, give an example to show that it is false.

(a) A Hamilton circuit for a graph must visit each vertex in the graph.

(b) An Euler circuit for a graph must visit each vertex in the graph.

(c) A Hamilton circuit for a graph must use each edge in the graph.

(d) An Euler circuit for a graph must use each edge in the graph.

(e) A circuit cannot be both a Hamilton circuit and an Euler circuit.

(f) An Euler circuit must visit no vertex more than once, except the vertex where the circuit begins and ends.

Hamilton and Euler Circuits *Determine whether an Euler circuit, a Hamilton circuit, or neither would solve the problem.*

19. Bandstands at a Festival The vertices of a graph represent bandstands at a festival, and the edges represent paths between the bandstands. A visitor wants to visit each bandstand exactly once, returning to her starting point when she is finished.

20. Relay Team Running Order The vertices of a complete graph represent the members of a five-person relay team. The team manager wants a circuit that will show the order in which the team members will run. (He will decide later who will start.)

21. Paths in a Botanical Garden The vertices of a graph represent places where paths in a botanical garden cross, and the edges represent the paths. A visitor wants to walk along each path in the garden exactly once, returning to his starting point when finished.

22. Western Europe The vertices of a graph represent the countries on the continent (Western Europe), and the edges represent border crossings between the countries. A traveler wants to travel over each border crossing exactly once, returning to the first country visited for his flight home to the United States.

23. Africa Vertices represent countries in sub-Saharan Africa, with an edge between two vertices if those countries have a common border. A traveler wants to visit each country exactly once, returning to the first country visited for her flight home to the United States.

24. X-Rays In using X-rays to analyze the structure of crystals, an X-ray diffractometer measures the intensity of reflected radiation from the crystal in thousands of different positions. Consider the complete graph with vertices representing the positions where measurements must be taken. The researcher must decide the order in which to take these readings, with the diffractometer returning to its starting point when finished.

Factorials *Use a calculator, if necessary, to find each value.*

25. 4!   **26.** 6!

27. 9!   **28.** 14!

Hamilton Circuits *Determine how many Hamilton circuits there are in a complete graph with this number of vertices. (Express answers using factorial notation.)*

29. 10 vertices   **30.** 15 vertices

31. 18 vertices   **32.** 60 vertices

33. List all Hamilton circuits in the graph that start at P.

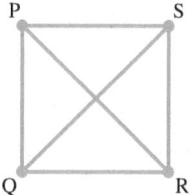

34. Hamilton Circuits List all Hamilton circuits in the graph that start at A.

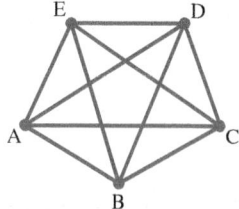

Hamilton Circuits *Refer to the following graph. List all Hamilton circuits in the graph that start with the indicated vertices.*

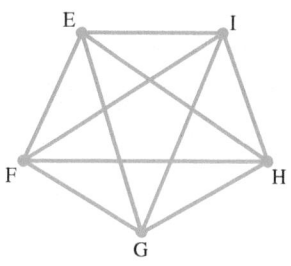

35. Starting E → H → I

36. Starting E → F → G

37. Starting E → F

38. Starting E → I → H

39. Starting E → G

40. Starting E → I

Brute Force Algorithm *Use the brute force algorithm to find a minimum Hamilton circuit for the graph. In each case determine the total weight of the minimum Hamilton circuit.*

41.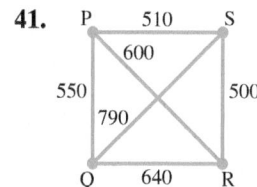

42.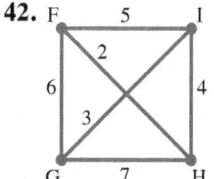

43.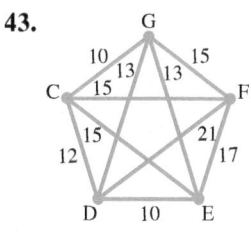

44.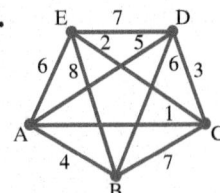

(b) Which of the circuits found in part (a) gives the best solution to the problem of finding a minimum Hamilton circuit for the graph?

(c) Just by looking carefully at the graph, find a Hamilton circuit in the graph that has lower total weight than any of the circuits found in part (a).

Nearest Neighbor Algorithm *Use the nearest neighbor algorithm starting at each of the indicated vertices to determine an approximate solution to the problem of finding a minimum Hamilton circuit for the graph. In each case, find the total weight of the circuit found.*

45.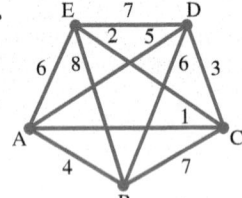

(a) Starting at A

(b) Starting at C

(c) Starting at D

(d) Starting at E

46.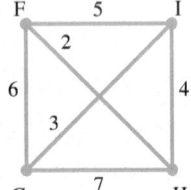

(a) Starting at F

(b) Starting at G

(c) Starting at H

(d) Starting at I

47.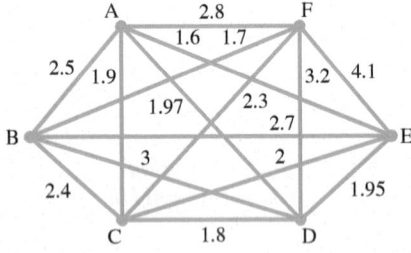

(a) Starting at A   **(d)** Starting at D

(b) Starting at B   **(e)** Starting at E

(c) Starting at C   **(f)** Starting at F

48. *Nearest Neighbor Algorithm* Refer to the accompanying graph. Complete parts (a)–(c) in order.

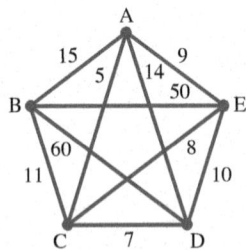

(a) Use the nearest neighbor algorithm starting at each of the vertices in turn to determine an approximate solution to the problem of finding a minimum Hamilton circuit for the graph. In each case, find the total weight of the circuit.

Hamilton Circuits *Find all Hamilton circuits in the graph that start at A. (Hint: Use counting trees.)*

49. 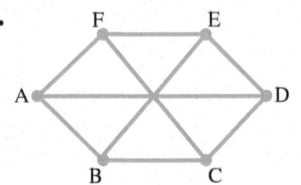   **50.**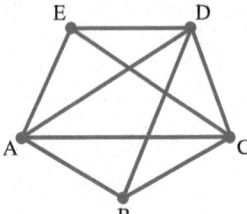

51.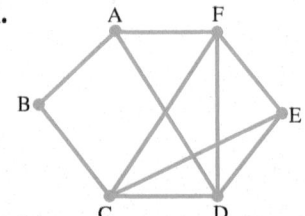

52.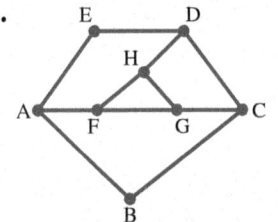

53. *Traveling Salesman Problem* The diagram represents seven cities, A through G, along with the distances in miles between many of the pairs.

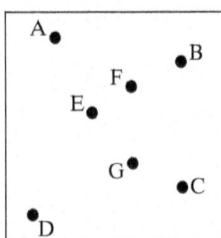

AB = 51   AF = 36   BF = 22   CG = 22   EF = 22

AD = 71   BC = 50   BG = 45   DE = 45   EG = 28

AE = 32   BE = 45   CD = 61   DG = 45   FG = 30

Determine what you believe to be the shortest possible distance that can be traveled so that the starting point and ending point are the same, and all cities have been visited.

The Icosian Game   *The graph below shows the Icosian game (described in the text) with the vertices labeled. Find a Hamilton circuit for the graph in the specified version of the game. (We suggest that you write numbers on the graph in the order in which you visit the vertices. There are different correct answers for these exercises.)*

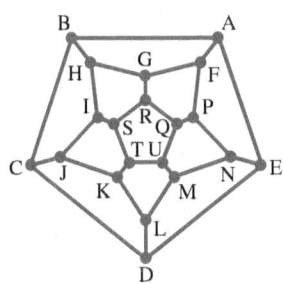

54. The circuit begins at A.

55. The circuit begins A → F.

56. The circuit begins A → B → H.

57. *Dirac's Theorem*   Paul A. M. Dirac proved the following theorem in 1952.

Suppose G is a (simple) graph with n vertices, n ≥ 3. If the degree of each vertex is greater than or equal to $\frac{n}{2}$, then the graph has a Hamilton circuit.

Refer to graphs (1)–(5) and answer parts (a)–(e) in order.

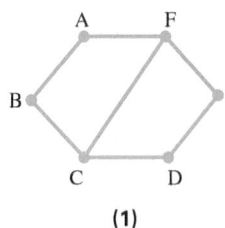

(1)

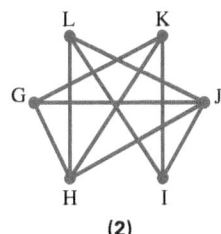

(2)

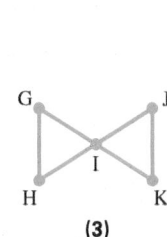

(3)

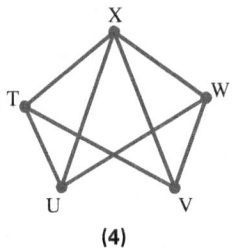

(4)

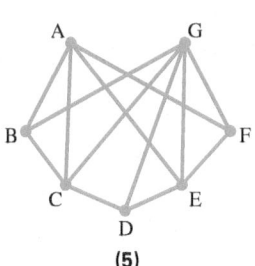

(5)

(a) Which graphs satisfy the condition that the degree of each vertex is greater than or equal to $\frac{n}{2}$?

(b) For which of the graphs can we conclude from Dirac's theorem that the graph has a Hamilton circuit?

(c) If a graph does *not* satisfy the condition that the degree of each vertex is greater than or equal to $\frac{n}{2}$, can we be sure that the graph does *not* have a Hamilton circuit? Justify your answer. (*Hint:* Study the accompanying graphs.)

(d) Is Dirac's theorem still true if $n < 3$? Justify your answer.

(e) Use Dirac's theorem to write a convincing argument that any complete graph with 3 or more vertices has a Hamilton circuit.

58. *Complete Bipartite Graph*   A graph is called a complete bipartite graph if the vertices can be separated into two groups in such a way that there are no edges between vertices in the same group, and there is an edge between each vertex in the first group and each vertex in the second group. An example is shown here.

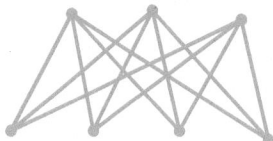

The notation we use for these graphs is $K_{m,n}$, where m and n are the numbers of vertices in the two groups. For example, the graph shown in this exercise is $K_{4,3}$ (or $K_{3,4}$). Complete parts (a)–(d) in order.

(a) Draw $K_{2,2}$, $K_{2,3}$, $K_{2,4}$, $K_{3,3}$, and $K_{4,4}$.

(b) For each of the graphs you have drawn in part (a), find a Hamilton circuit or say that the graph has no Hamilton circuit.

(c) Make a conjecture: What must be true about m and n for $K_{m,n}$ to have a Hamilton circuit?

(d) What must be true about m and n for $K_{m,n}$ to have an Euler circuit? Justify your answer.

The Speed of Algorithms

How do computer scientists classify the speed of algorithms?

First, they write the number of steps a computer using the algorithm takes to solve a problem as a function of the "size" of the problem. We can think of the size of the problem as the number of vertices in a graph.

Let's suppose that our graph has n vertices. For some algorithms the number of steps is a **polynomial function of n**—for example,

$$n^4 + 2n.$$

For other algorithms, the number of steps is an **exponential function of n**—for example,

$$2^n.$$

(See earlier chapters for more on polynomial and exponential functions.) As n increases, the number of steps increases much faster for exponential functions than for polynomial functions. Functions that involve factorials increase even faster.

In the table below we show values (some are approximate) for functions of the three types for different values of n. (We use scientific notation to make it easier to compare the sizes of the numbers.)

n	n^4	2^n	$n!$
5	6.3×10^2	3.2×10	1.2×10^2
10	1.0×10^4	1.0×10^3	3.6×10^6
15	5.1×10^4	3.3×10^4	1.3×10^{12}
20	1.6×10^5	1.0×10^6	2.4×10^{18}
25	3.9×10^5	3.4×10^7	1.6×10^{25}
30	8.1×10^5	1.1×10^9	2.7×10^{32}
40	2.6×10^6	1.1×10^{12}	8.2×10^{47}
50	6.3×10^6	1.1×10^{15}	3.0×10^{64}

In the first line of the table (where $n = 5$), n^4 gives the largest value. But as n gets larger, 2^n grows much faster than n^4, while $n!$ grows extremely fast. For example, if we double the size of n from 15 to 30, n^4 becomes a little more than 10 times as large, 2^n becomes about 3×10^4 (or 30,000) times as large, while $n!$ becomes approximately 2×10^{20} times as large. 10^{20} is more than a billion billion.

Algorithms whose time functions grow no faster than a polynomial function are called **polynomial time algorithms**. These are *efficient* algorithms—they do not take too much computer time.

Algorithms whose time functions grow faster than any polynomial function are called **exponential time algorithms**. These are *not* efficient algorithms. They are by nature too time-consuming for our computers. Our silicon chip computers are getting faster each year, but even this increase in speed hardly puts a dent in the time required for a computer to implement an exponential time algorithm for a very large graph.

Because the brute force algorithm for the traveling salesman problem has a time function involving $n!$, which grows faster than an exponential function of n, this algorithm is an exponential time algorithm.

For Group or Individual Investigation

Use the table to help answer these questions:

1. By what factor does n^4 grow if we double n from 10 to 20?

2. By what factor does 2^n grow if we double n from 10 to 20?

3. By what factor does $n!$ grow if we double n from 10 to 20?

4. Repeat Exercises 1–3 if we double n from 25 to 50.

5. Suppose it would take 2^{30} years using computers at their present speeds to solve a certain problem. (2^{30} is a little over one billion.) If we assume that computers double in speed each year, how long would we have to wait before we could solve the problem in 1 year?

14.4   TREES AND MINIMUM SPANNING TREES

Connected Graphs and Trees

Peggy wants to install an underground irrigation system in her garden. The system must connect the main valve (M) to outlets at various points, specifically the rose bed (R), perennial bed (P), daffodil bed (D), annuals bed (A), berry patch (B), vegetable patch (V), cut flower bed (C), and shade bed (S). These are the vertices of the graph in **Figure 48**.

Figure 48

The irrigation system will have pipes connecting these points. For the irrigation system to work well, there should be as few pipes as possible connecting all the outlets.

Think about this problem in terms of a graph with edges representing pipes. Peggy wants water to flow from the main valve (M) to each outlet. So we need to create a *connected* graph that includes all the vertices in **Figure 48**.

CONNECTED GRAPH (ALTERNATIVE DEFINITION)

A **connected graph** is one in which there is *at least one path* between each pair of vertices.

To see that this definition agrees with our definition in the first section of this chapter, observe that if we select any pair of vertices in the connected graph in **Figure 49** (for example, A and C), then there is at least one path between them. (In fact, there are many paths between A and C.) Two examples are

$$A \rightarrow B \rightarrow E \rightarrow C \quad \text{and} \quad A \rightarrow F \rightarrow C.$$

Peggy needs a *connected* graph. Also, she must use as few pipes as possible to connect all the outlets to the system. This means that the graph *must not contain any circuits*. If the graph had a circuit, we could remove one of the pipes in the circuit and still have a connected system of pipes.

Figure 50 shows one possible solution to Peggy's problem.

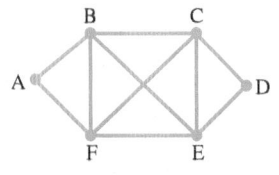

Figure 49

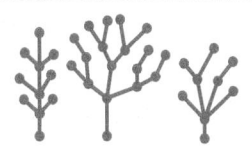

In graph theory, the botanical words go beyond trees. A **forest** is a graph with all components trees, as shown here. Later in this section you will find the words **root** and **leaf** applied to graphs. The terms **pruning** and **separating** arise in applications of trees to the computer analysis of images.

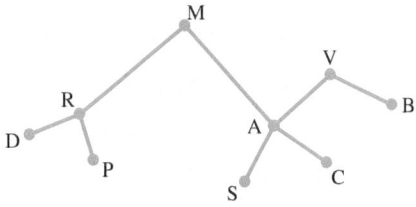

Figure 50

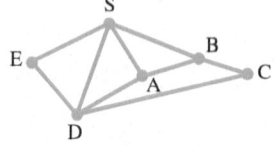

Benzene

It seems mathematician **Arthur Cayley** invented the term "tree" in graph theory in about 1857 in connection with problems related to counting all trees of certain types. Later he realized the relevance of his work to nineteenth-century organic chemistry and, in the 1870s, published a note on this. Of course, not all molecules have treelike structures. Friedrich Kekulé's realization that the structure of benzene is *not* a tree is considered one of the most brilliant breakthroughs in organic chemistry.

There is a name for graphs that are connected and contain no circuits.

TREE

A graph is a **tree** if the graph is *connected* and contains *no circuits*.

All five of the graphs shown in **Figure 51** are trees.

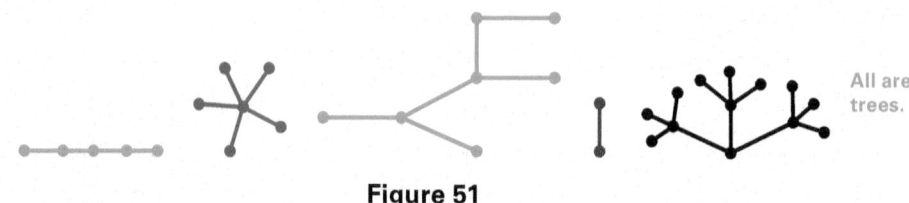

Figure 51

In contrast, none of the graphs shown in **Figure 52** is a tree. The graph in **Figure 52(a)** is not a tree, because it is not connected. Those in **Figures 52(b) and (c)** are not trees, because each contains at least one circuit.

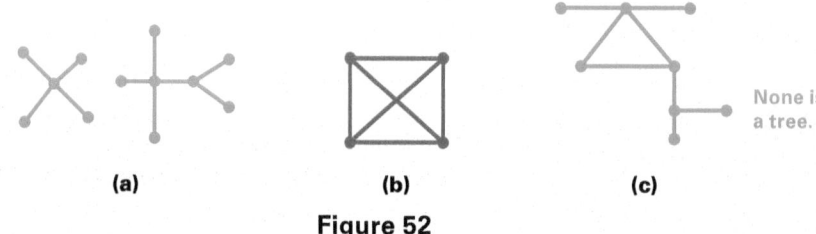

(a)   **(b)**   **(c)**

Figure 52

UNIQUE PATH PROPERTY OF TREES

In a tree, there is always **exactly one path** from each vertex in the graph to any other vertex in the graph.

For example, consider the vertices S and B in the tree in **Figure 50.** There is exactly one path from vertex S to vertex B, namely S → A → V → B.

This property follows from the definition of a tree. Because a tree is connected, there is always at least one path between each pair of vertices. And if the graph is a tree, there cannot be two different paths between a pair of vertices. If there were, these together would contain a circuit.

It is the unique path property of trees that makes them so important in real-world applications. This also is the reason why trees are not a very useful model for telephone networks. Each edge of a tree is a cut edge, so failure along one link would disrupt the service. Telephone companies rely on networks with many circuits to provide alternative routes for calls.

EXAMPLE 1   Renovating an Irrigation System

Joe has an old underground irrigation system in his garden, with connections at the source (S), outlets A through E in the garden, and existing underground pipes as shown in the graph in **Figure 53**. He wants to renovate this system, keeping only some of the pipes. Water must still be able to flow to each of the original outlets, and the graph of the irrigation system must be a tree. Design an irrigation system that meets Joe's objectives.

Figure 53

Just how fast are the fastest computers? Computer speeds are sometimes measured in **flops** (floating point operations per second). A speed of one gigaflop means that the computer can perform about one billion (10^9) calculations per second.

Solution

In graph theory terms, we need a *subgraph* of the graph in **Figure 53**. Because water still must be able to flow to each of the original outlets, we need a connected subgraph that *includes all the vertices of the original graph*. Finally, this subgraph must be a *tree*. Thus, we must remove edges (but no vertices) from the graph, without disconnecting the graph, to obtain a subgraph that is a tree. One way to achieve this is shown in **Figure 54.**

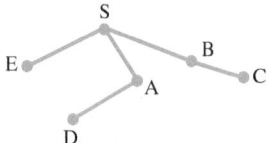

Figure 54

Spanning Trees

A subgraph of the type shown in **Figure 54** from **Example 1** is called a *spanning tree* for the graph. This kind of subgraph "spans" the graph in the sense that it connects all the vertices of the original graph into a subgraph.

SPANNING TREE

A **spanning tree** for a graph is a subgraph that includes every vertex of the original graph and is a tree.

We can find a spanning tree for any connected graph (think about that), and if the original graph has at least one circuit, it will have several different spanning trees.

Arthur Cayley (1821–1895) was a close friend of James Sylvester. They worked together as lawyers at the courts of Lincolns Inn in London and discussed mathematics with each other during work hours. Cayley considered his work as a lawyer simply a way to support himself so that he could do mathematics. In the fourteen years he worked as a lawyer, he published 250 mathematical papers. At age 42, he took a huge paycut to become a mathematics professor at Cambridge University.

EXAMPLE 2   **Finding Spanning Trees**

Consider the graph shown in **Figure 55**. Find two different spanning trees for the graph.

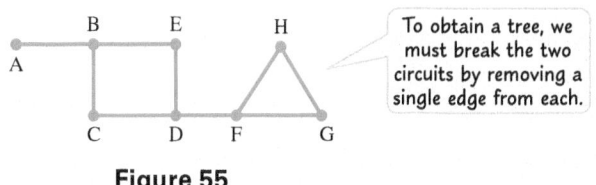

To obtain a tree, we must break the two circuits by removing a single edge from each.

Figure 55

Solution

There are several choices for which edges to remove. We could remove edges CD and FH, leaving the subgraph in **Figure 56(a)**. Alternatively, we could remove edges ED and FG, leaving the subgraph in **Figure 56(b)**. There are other correct solutions.

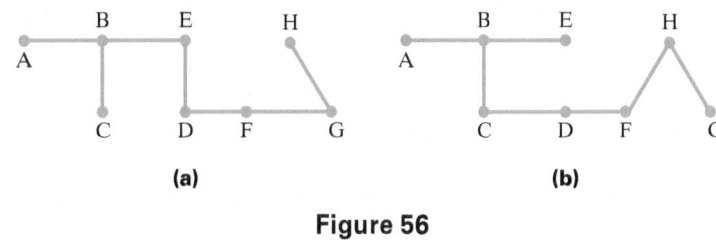

(a)   (b)

Figure 56

If two spanning trees use different edges from the original graph, we consider them to be *different* spanning trees even when they are isomorphic graphs. The subgraphs shown in **Figure 56** are different spanning trees, even though they happen to be isomorphic graphs. To decide whether they are isomorphic, change the labels in graph (a) as follows: Interchange C and E, and interchange G and H. By dragging vertices without detaching edges, we can obtain graphs that look alike.

Minimum Spanning Trees

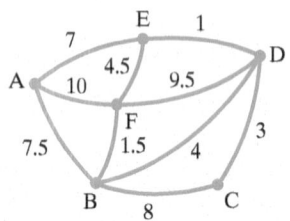

Figure 57

Suppose there are six villages in a rural district. At present all the roads in the district are gravel roads. There is not enough funding to pave all the roads, but there is a pressing need to have good paved roads so that emergency vehicles can travel easily between the villages. The vertices in the graph in **Figure 57** represent the villages in the district, and the edges represent the existing gravel roads. Estimated costs for paving the different roads are shown, in millions of dollars. (These estimates take into account distances, as well as features of the terrain.)

The council wants to pave enough roads so that emergency vehicles will be able to travel from any village to any other along paved roads, though possibly by a roundabout route. The council wants to achieve this at the *lowest possible cost*.

The problem calls for exactly one path from each vertex in the graph to any other, so it requires a spanning tree. This spanning tree should have minimum total weight. The total weight of a spanning tree is the sum of the weights of its edges.

MINIMUM SPANNING TREE

A spanning tree that has minimum total weight is a **minimum spanning tree** for the graph.

Kruskal's Algorithm

Fortunately for the district planners, there is a good algorithm for finding a minimum spanning tree for any connected, weighted graph. It is called **Kruskal's algorithm.** The algorithm is an efficient algorithm, because it does not take too much computer time for even very large graphs.

KRUSKAL'S ALGORITHM FOR FINDING A MINIMUM SPANNING TREE FOR ANY CONNECTED, WEIGHTED GRAPH

Choose edges for the spanning tree as follows.

Step 1 First edge: Choose any edge with minimum weight.

Step 2 Next edge: Choose any edge with minimum weight from *those not yet selected.* (At this stage, the subgraph may look disconnected.)

Step 3 Continue to choose edges of minimum weight from those not yet selected, except *do not select any edge that creates a circuit* in the subgraph.

Step 4 Repeat Step 3 until the subgraph connects all vertices of the original graph.

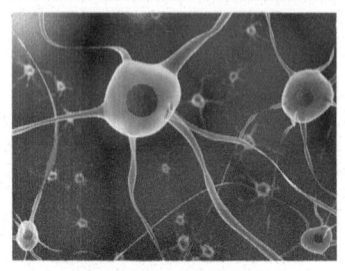

Computer analysis of images is a field of active research, particularly in medicine. (Computer analysis of images consists of having a computer detect relevant features of an image.) Some of this research uses minimum spanning trees. They have been used to identify protein fibers in photographs of cells. The computer detects some points lying on the fibers and then determines a minimum spanning tree to deduce the overall structure of the fiber. Some studies have used minimum spanning trees to analyze cancerous tissue.

Because Kruskal's algorithm gives a subgraph that is connected, includes all vertices of the original graph, and has no circuits, it certainly provides a spanning tree for the graph. Also, the algorithm always will give us a *minimum* spanning tree for the graph. The explanation for this fact is beyond the scope of this text. (Algorithms such as Kruskal's often are called *greedy* algorithms.)

EXAMPLE 3   Applying Kruskal's Algorithm

Use Kruskal's algorithm to find a minimum spanning tree for the graph in **Figure 57.**

Solution

Note that the graph is connected, so Kruskal's algorithm applies. Choose an edge with minimum weight. ED has the least weight of all the edges, so it is selected first. We color the edge a different color to show that it has been selected. See **Figure 58.**

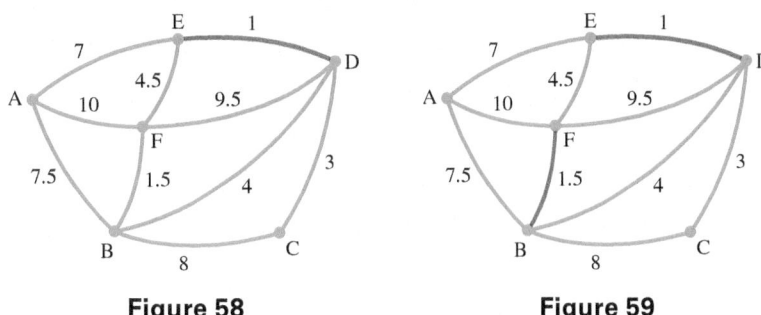

Figure 58   **Figure 59**

Of the remaining edges, BF has the least weight, so it is the second edge chosen, as shown in **Figure 59.** Note that it does not matter that at this stage the subgraph is not connected.

Continuing, we choose CD and then BD, as shown in **Figure 60.** Of the edges remaining, EF has the least weight. However, EF *would create a circuit* in the subgraph. We ignore EF and note that, of the edges remaining, AE has the least weight. Thus, AE is included in the spanning tree, as shown in **Figure 61.**

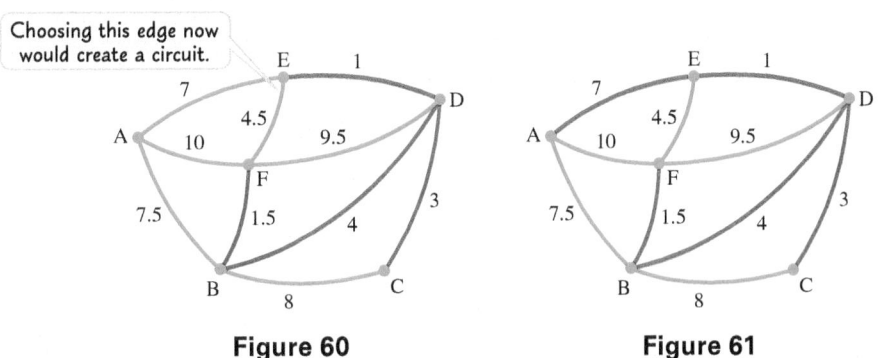

Choosing this edge now would create a circuit.

Figure 60   **Figure 61**

We now have a subgraph of the original that connects all vertices of the original graph into a tree. This subgraph is the required minimum spanning tree for the original graph. Its total weight is

$$1 + 1.5 + 3 + 4 + 7 = 16.5.$$

Thus, the district council can achieve its objective at a cost of

16.5 million dollars,

by paving only the roads represented by edges in the minimum spanning tree.

Number of Vertices and Edges

There is an interesting relationship between the number of vertices and the number of edges in a tree (sometimes called the **vertex/edge relationship**).

Minimum Spanning Trees in Astronomy   The structure of the universe may hold clues about its formation. The image shows the large-scale structure of the Milky Way against a background of many other galaxies. Galaxies are not uniformly distributed throughout the universe. Instead they seem to be arranged in clusters, which are grouped together into superclusters, which themselves seem to be arranged in vast chains called filaments. Some astronomers are using spanning trees to help map the filamentary structure of the universe.

Weather data are complex, involving numerous readings of temperature, pressure, and wind velocity over huge areas. Some current research uses minimum spanning trees to help computers interpret these data. For example, minimum spanning trees are used to identify developing cold fronts.

NUMBER OF VERTICES AND EDGES IN A TREE

If a graph is a tree, then the number of edges in the graph is one less than the number of vertices.

A tree with _n_ vertices has _n_ − 1 edges.

Consider the graphs shown in **Figure 62**. Information about these graphs is shown in **Table 3**. The three graphs in (a), (b), and (c) are not trees, and there is no uniform relation between the number of edges and the number of vertices for these graphs. In general, graphs may have more or fewer vertices than edges or an equal number of both. In contrast, the three graphs in (d), (e), and (f) are trees. The number of edges is indeed one less than the number of vertices for each of these trees.

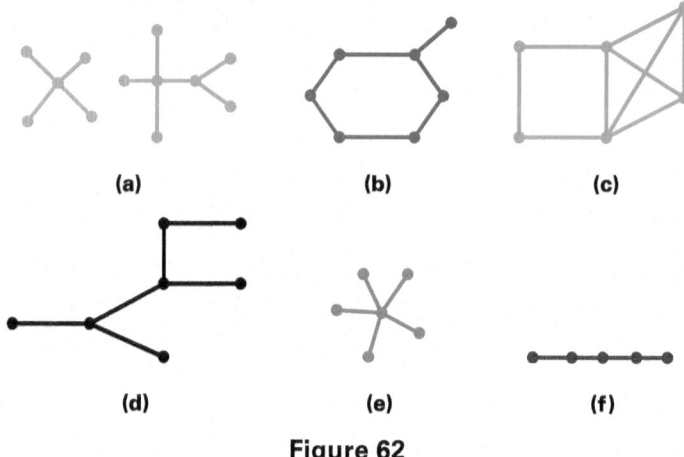

Figure 62

If we consider only connected graphs, then the converse of the theorem above is also true.

For a connected graph, if the number of edges is one less than the number of vertices, then the graph is a tree.

Table 3

Graph	Tree?	Number of Vertices	Number of Edges	
(a)	No	12	10	
(b)	No	7	7	
(c)	No	6	9	
(d)	Yes	7	6	6 = 7 − 1
(e)	Yes	6	5	5 = 6 − 1
(f)	Yes	5	4	4 = 5 − 1

Problem-Solving Strategy

Trial and error is a good problem-solving strategy. Choose an arbitrary proposed solution to the problem, and check whether it works. If it does not work, it may still yield insight into what to try next. (See **Example 5** on the next page.)

EXAMPLE 4 Using the Vertex/Edge Relationship

A chemist has synthesized a new chemical compound. She knows from her analyses that a molecule of the compound contains 54 atoms and that the molecule has a tree-like structure. How many chemical bonds are there in the molecule?

Solution

Think of the atoms in the molecule as the vertices of a tree, and think of the bonds as edges of the tree. This tree has 54 vertices. Since the number of edges in a tree is one less than the number of vertices, the molecule must have 53 bonds.

EXAMPLE 5 Finding a Graph with Specified Properties

Suppose a graph is a tree and has 15 vertices. What is the greatest number of vertices of degree 5 that this graph could have? Draw such a tree.

Solution

Because the graph is a tree with 15 vertices, it must have 14 edges. Recall that the sum of the degrees of the vertices of a graph is twice the number of edges. Therefore, the sum of the degrees of the vertices in the tree must be $2 \cdot 14$, which is 28.

Now we use trial and error. We want the greatest possible number of vertices of degree 5. Four vertices of degree 5 would contribute 20 to the total degree sum. This would leave $28 - 20 = 8$ as the degree sum of the remaining 11 vertices, but this would mean that some of those vertices would have no edges joined to them. The graph would not be connected and would not be a tree.

If we try 3 vertices of degree 5, this would contribute 15 to the total degree sum. It would leave $28 - 15 = 13$ as the degree sum for the remaining 12 vertices. We can draw a graph with these specifications, as in **Figure 63**. Thus, the greatest possible number of vertices of degree 5 in a tree with 15 vertices is 3.

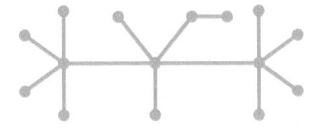

Figure 63

14.4 EXERCISES

CONCEPT CHECK *Complete each statement with the correct response(s).*

1. A connected graph is one in which there is at least one _____ between each pair of _____.

2. A graph is a tree if the graph is _____ and contains no _____.

3. A spanning tree for a graph is a subgraph that includes every _____ of the original graph and is a tree.

4. A spanning tree that has _____ _____ _____ is a minimum spanning tree for the graph.

Tree Identification *Determine whether the graph is a tree. If not, explain why it is not.*

5.

6.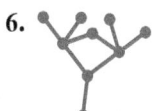

7.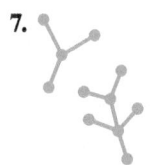

8.

9.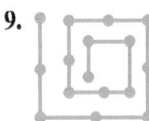

10. a complete graph with 6 vertices

11. a connected graph with all vertices of even degree

Tree Identification *Add edges (no vertices) to the graph to change the graph into a tree or explain why this is not possible. (Note: There may be more than one correct answer.)*

12. A B C E D G H F I

13.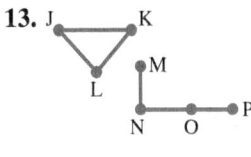

14. Q———R

S•   •T
       |
      •U

Tree or Not a Tree? *Determine whether the graph described must be a tree.*

15. *Spread of a Rumor* A sociologist is investigating the spread of a rumor. He finds that 18 people know of the rumor. One of these people must have started the rumor, and each must have heard the rumor for the first time from one of the others. He draws a graph with vertices representing the 18 people and edges showing from whom each person first heard the rumor.

16. *Internet Search Engine* You are using the Google search engine to search the Web for information on a topic for a paper. You start at the Google page and follow links you find, sometimes going back to a site you've already visited. To keep a record of the sites you visit, you show each site as a vertex of a graph. You draw an edge each time you connect for the *first* time to a *new* site.

17. *Tracking a Disease* A patient has contracted a mysterious disease. An employee of the Centers for Disease Control is trying to quarantine all people who have had contact, over the past week, either with the patient or with someone already included in the contact network. The employee draws a graph with vertices representing the patient and all people who have had contact as described. The edges of the graph represent the relation "the two people have had contact."

Trees and Cut Edges *Determine whether the statement is* true *or* false. *If the statement is false, draw an example to show that it is false.*

18. Every graph with no circuits is a tree.

19. Every connected graph in which each edge is a cut edge is a tree.

20. Every graph in which there is a path between each pair of vertices is a tree.

21. Every graph in which each edge is a cut edge is a tree.

Spanning Trees *Find three different spanning trees for each graph. (There are many different correct answers.)*

22.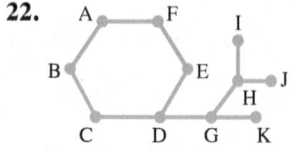

23.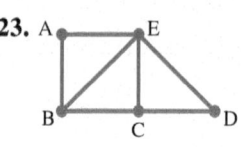

24.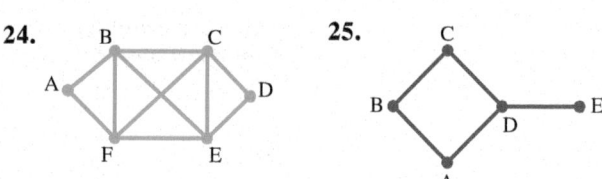

25.

Spanning Trees *Find all spanning trees for the graph.*

26. **27.**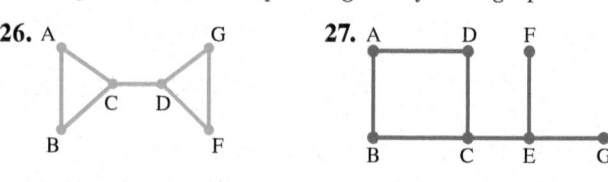

28. **29.**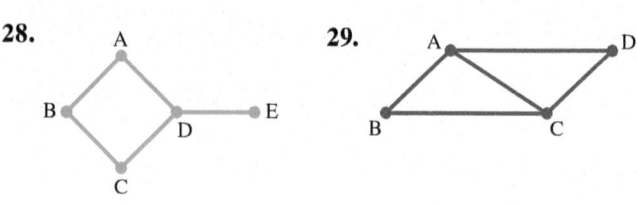

Spanning Trees *Determine how many spanning trees the graph has.*

30. **31.**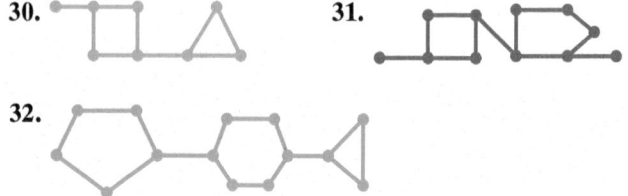

32.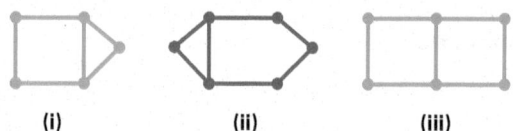

33. What is a general principle about the number of spanning trees in graphs such as those in **Exercises 30–32?**

34. *Spanning Trees* Complete the parts of this exercise in order.

(a) Find all the spanning trees of each of the following graphs.

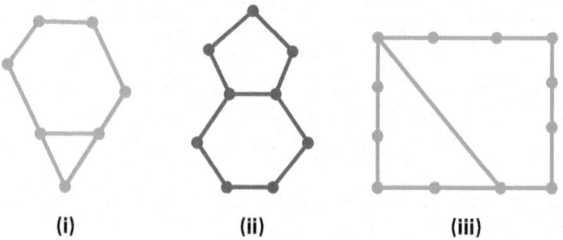

(i)   (ii)   (iii)

(b) For each of the following graphs, determine how many spanning trees the graph has.

(i)   (ii)   (iii)

(c) What is a general principle about the number of spanning trees in graphs of the kind shown in (a) and (b)?

Minimum Spanning Trees *Use Kruskal's algorithm to find a minimum spanning tree for the graph. Find the total weight of this minimum spanning tree.*

35.

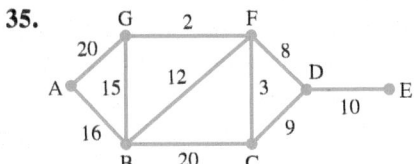

36.

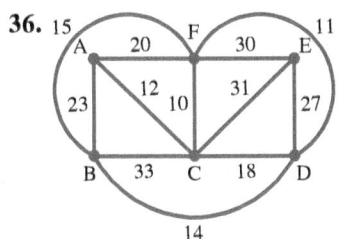

37.

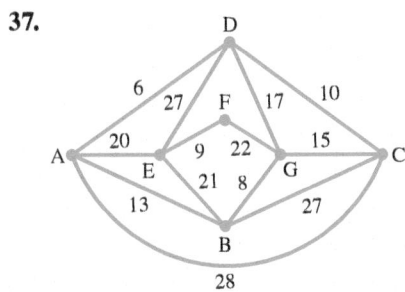

38.

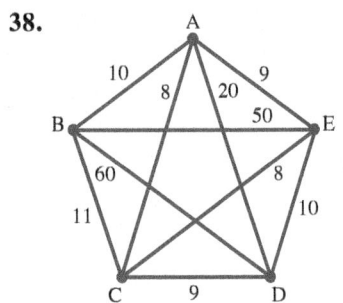

Kruskal's Algorithm *Solve each problem.*

39. A school consists of 6 separate buildings, represented by the vertices in the graph. There are paths, with lengths in feet, between some of the buildings as shown. Administrators want to cover some paths with roofs so that students can walk between buildings without getting wet when it rains. To minimize cost, they must select paths to cover such that the total covered length is as small as possible.

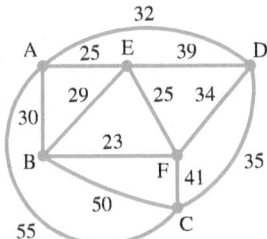

Use Kruskal's algorithm to determine which paths to cover and the total length of these pathways.

40. A town council is planning to provide town water to an area that previously relied on private wells. Water will be fed into the area at the point represented by the vertex labeled A in the graph below. Water must be piped to each of five main distribution points, represented by the vertices labeled B through F in the graph. Town engineers have estimated the cost of laying the pipes to carry the water between each pair of points in millions of dollars, as indicated in the graph. They must now select which pipes should be laid, so that there is exactly one route for the water to be pumped from A to any one of the five distribution points (possibly via another distribution point), and they want to achieve this at minimum cost.

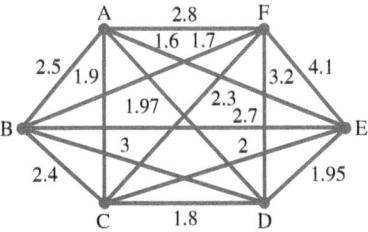

Use Kruskal's algorithm to decide which pipes should be laid. Find the total cost of laying the pipes.

Trees, Edges, and Vertices *Solve each problem.*

41. How many edges are there in a tree with 34 vertices?

42. How many vertices are there in a tree with 40 edges?

43. How many edges are there in a spanning tree for a complete graph with 63 vertices?

44. A connected graph has 27 vertices and 43 edges. How many edges must be removed to form a spanning tree for the graph?

45. We have said that we can always find at least one spanning tree for a connected graph—and usually more than one. Could it happen that different spanning trees for the same graph have different numbers of edges? Justify your answer.

46. Suppose we have a tree with 9 vertices.

(a) Determine the number of edges in the graph.

(b) Determine the sum of the degrees of the vertices in the graph.

(c) Determine the least possible number of vertices of degree 1 in this graph.

(d) Determine the greatest possible number of vertices of degree 1 in this graph.

(e) Answer parts (a) through (d) for a tree with n vertices.

47. Suppose we have a tree with 10 vertices.

   (a) Determine the number of edges in the graph.

   (b) Determine the sum of the degrees of the vertices in the graph.

   (c) Determine the least possible number of vertices of degree 4 in this graph.

   (d) Determine the greatest possible number of vertices of degree 4 in this graph.

   (e) Draw a graph to illustrate your answer in part (d).

48. Suppose we have a tree with 17 vertices.

   (a) Determine the number of edges in the graph.

   (b) Determine the sum of the degrees of the vertices in the graph.

   (c) Determine the greatest possible number of vertices of degree 5 in this graph.

   (d) Draw a graph to illustrate your answer in part (c).

49. *Computer Network Layout* A business has 23 employees, each with his or her own desk computer, all working in the same office. The managers want to network the computers. They need to install cables between individual computers so that every computer is linked into the network. Because of the way the office is laid out, it is not convenient to simply connect all the computers in one long line. Determine the least number of cables the managers need to install to achieve their objective.

50. *Design of a Garden* Maria has 12 vegetable and flower beds in her garden and wants to build flagstone paths between the beds so that she can get from each bed along flagstone paths to every other bed. She also wants a path linking her front door to one of the beds. Determine the minimum number of flagstone paths she must build to achieve this.

Tree or Not a Tree? *Exercises 51–53 refer to the following situation: We start with a tree and then draw in extra edges and vertices. For each, say whether it is possible to draw in the number of edges and vertices specified to end up with a connected graph. If it is possible, determine whether the resulting graph must be a tree, may be a tree, or cannot possibly be a tree. Justify each answer briefly.*

51. Draw in the same number of vertices as edges.

52. Draw in more vertices than edges.

53. Draw in more edges than vertices.

54. Starting with a tree, is it possible to draw in the same number of vertices as edges and end up with a disconnected graph? If so, give an example. If not, justify briefly.

Cayley's Theorem *Apply the following theorem, which was proved by Cayley in 1889:*

A complete graph with n vertices has n^{n-2} spanning trees.

55. How many spanning trees are there for a complete graph with 3 vertices? Draw a complete graph with 3 vertices, and find all the spanning trees.

56. How many spanning trees are there for a complete graph with 4 vertices? Draw a complete graph with 4 vertices, and find all the spanning trees.

57. How many spanning trees are there for a complete graph with 5 vertices?

Isomorphism *Solve each problem.*

58. Find all nonisomorphic trees with 4 vertices. How many are there?

59. Find all nonisomorphic trees with 5 vertices. How many are there?

60. Find all nonisomorphic trees with 6 vertices. How many are there?

61. Find all nonisomorphic trees with 7 vertices. How many are there?

62. *Vertex/Edge Relationship* In this exercise, we explore why the number of edges in a tree is one less than the number of vertices. Because the statement is clearly true for a tree with only one vertex, we will consider a tree *with more than one vertex.* Answer parts (a)–(h) in order.

   (a) How many components does the tree have?

   (b) Why must the tree have at least one edge?

   (c) Remove one edge from the tree. How many components does the resulting graph have?

   (d) You have not created any new circuits by removing the edge, so each of the components of the resulting graph is a tree. If the remaining graph still has edges, choose any edge and remove it. (You have now removed 2 edges from the original tree.) Altogether, how many components remain?

   (e) Repeat the procedure described in (d). If you remove 3 edges from the original tree, how many components remain? If you remove 4 edges from your original tree, how many components remain?

   (f) Repeat the procedure in (d) until you have removed all the edges from the tree. If you have to remove n edges, determine an expression involving n for the number of components remaining.

   (g) What *are* the components that remain when you have removed all the edges from the tree?

   (h) What can you conclude about the number of vertices in a tree with n edges?

FOR FURTHER THOUGHT

Binary Coding

The unique path property of a tree can be used to make a code. We illustrate with a **binary code**—that is, one that uses 0s and 1s. (Recall from earlier work that when we represent numbers in binary form, we use only the symbols 0 and 1.)

To set up the code we use a special kind of tree like that shown in the figure below. This tree is a **directed graph** (there are arrows on the edges). The vertex at the top (with no arrows pointing toward it) is called the **root** of the directed tree. The vertices with no arrows pointing away from them are called **leaves** of the directed tree. We label each leaf with a letter we want to encode. The diagram shown encodes only 8 letters, but we could easily draw a bigger tree with more leaves to represent more letters.

We now write a 0 on each branch that goes left and a 1 on each branch that goes right.

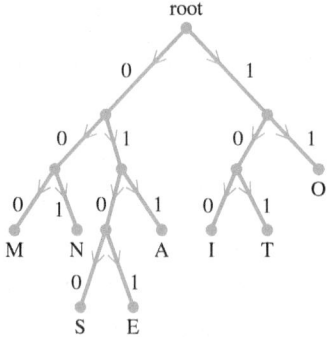

To show how the encoding works, we write the word MAT using the code. Follow the unique path from the root of the tree down to the appropriate leaf, noting in order the labels on the edges.

M is written 000.
A is written 011.
T is written 101.

Thus MAT is written 000011101.

We can easily translate this code using our tree. Let us see how we could decode 000011101. Referring to the tree, there is only one path from the root to a leaf that can give rise to those first three 0s, and that is the path leading to M. So we can begin to separate the code word into letters: 000–011101. Again following down from the root, the path 011 leads us unambiguously to A. So we have 000–011–101. The path 101 leads unambiguously to T.

The reason why we can translate the string of 0s and 1s back to letters without ambiguity is that no letter has a code the same as the first part of the code for a different letter.

For Group or Individual Investigation

1. Use the binary encoding tree to write the binary code for each of the following words:

   ANT, SEAT

2. Use the encoding tree to find the word represented by each of the following codes:

   0001111001, 0100011101

3. Decode the following messages (commas are inserted to show separation of words).

   (a) 0110011010100, 100001, 1010101001101

   (b) 0101011101, 0001000011010100, 100001, 101100001

   (c) 011, 0000101011001, 1010101011000

   (d) 00001010101101, 0000101, 011101, 1010101011

Directed Graphs

Chapter 14 SUMMARY

KEY TERMS

14.1
vertex
edge
simple graph
degree of a vertex
isomorphic graphs
connected graph
disconnected graph
components of a graph
walk
path
circuit

weighted graph
weights
complete graph
subgraph
coloring
chromatic number
four-color problem
Bacon number

14.2
Königsberg bridge problem
Euler path

Euler circuit
cut edge
Fleury's algorithm
deadheading

14.3
Icosian game
Hamilton circuit
total weight
minimum Hamilton circuit
traveling salesman problem
brute force algorithm

efficient algorithms
approximate algorithms
nearest neighbor algorithm

14.4
connected graph
tree
spanning tree
minimum spanning tree
Kruskal's algorithm
vertex/edge relationship

NEW SYMBOLS

n! *n* factorial

A → B a walk from vertex A to vertex B

TEST YOUR WORD POWER

See how well you have learned the vocabulary in this chapter.

1. Two graphs are **isomorphic** if
   A. they have the same number of edges.
   B. they have the same number of vertices.
   C. there is a one-to-one matching between vertices, with the property that whenever there is an edge between two vertices of either one of the graphs, there is an edge between the corresponding vertices of the other graph.
   D. each graph has the same chromatic number.

2. A **circuit** in a graph is
   A. a path that begins and ends with the same vertex.
   B. a path that begins and ends with different vertices.
   C. a path whose edges are assigned weights.
   D. a path that is not a walk.

3. An **Euler circuit** in a graph
   A. is a path that uses every edge of the graph exactly once.
   B. is a circuit that uses every edge of the graph exactly once.
   C. is a circuit that is composed of cut edges.
   D. has at least one vertex of odd degree.

4. A **Hamilton circuit** in a graph
   A. is a circuit that uses every edge exactly once (returning to the starting edge to complete the circuit).
   B. is a circuit that visits each vertex exactly once (returning to the starting vertex to complete the circuit).
   C. cannot exist if there are an odd number of vertices.
   D. cannot exist if there are an even number of vertices.

ANSWERS
1. C 2. A 3. B 4. B

QUICK REVIEW

Concepts

Examples

14.1 BASIC CONCEPTS

Graph

A **graph** is a collection of vertices (at least one) and edges. Each edge goes from a vertex to a vertex.

Isomorphic Graphs

Two graphs are **isomorphic** if there is a one-to-one matching between vertices of the two graphs, with the property that whenever there is an edge between two vertices of either one of the graphs, there is an edge between the corresponding vertices of the other graph.

Consider each graph.

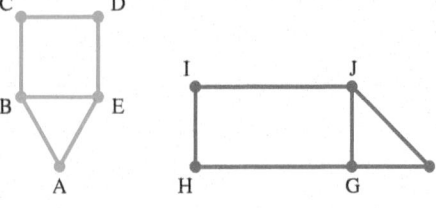

These two graphs are isomorphic.

Concepts	Examples
Connected and Disconnected Graphs A graph is **connected** if we can move from each vertex of the graph to every other vertex of the graph *along edges of the graph*. If not, the graph is **disconnected.** The connected pieces of a graph are the **components** of the graph.	The graph on the left is disconnected, as illustrated by the colors shown on the right. 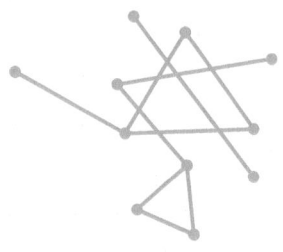 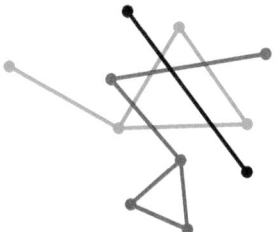
Sum-of-the-Degrees Theorem In any graph, the sum of the degrees of the vertices equals twice the number of edges.	Consider each graph. 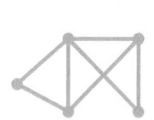  Number of edges = 7 Number of edges = 6 Sum of degrees of vertices = 14 Sum of degrees of vertices = 12
Walks, Paths, and Circuits A **walk** in a graph is a sequence of vertices, each linked to the next vertex by a specified edge of the graph. A **path** is a walk that uses no edge more than once. A **circuit** is a path that begins and ends at the same vertex. **Complete Graph** A **complete graph** is a graph in which there is exactly one edge going from each vertex to each other vertex.	Consider the graph shown here. 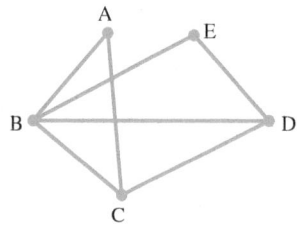 $E \rightarrow C \rightarrow D \rightarrow E$ is not a walk (and thus is neither a path nor a circuit). $B \rightarrow D \rightarrow E \rightarrow B \rightarrow C$ is a walk and a path, but not a circuit. $A \rightarrow C \rightarrow D \rightarrow E \rightarrow B \rightarrow A$ is all three: walk, path, and circuit.
Coloring and Chromatic Number A **coloring** for a graph is an assignment of a color to each vertex in such a way that vertices joined by an edge have different colors. The **chromatic number** of a graph is the least number of colors needed to achieve a coloring. **Coloring a Graph** *Step 1* Choose a vertex with greatest degree, and color it. Use the same color to color as many vertices as you can without coloring two vertices the same color if they are joined by an edge. *Step 2* Choose a new color, and repeat what you did in Step 1 for vertices not already colored. *Step 3* Repeat Step 1 until all vertices are colored.	This graph has chromatic number 3. 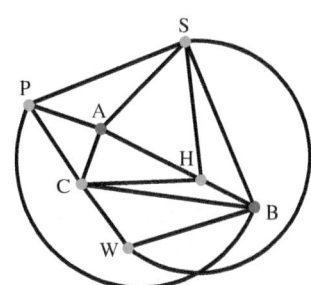

Concepts

Examples

14.2 EULER CIRCUITS AND ROUTE PLANNING

Euler Path and Euler Circuit
An **Euler path** in a graph is a path that uses every edge of the graph exactly once.

An **Euler circuit** in a graph is a circuit that uses every edge of the graph exactly once.

Euler's Theorem
Suppose we have a connected graph.

1. If the graph has an Euler circuit, then each vertex of the graph has even degree.

2. If each vertex of the graph has even degree, then the graph has an Euler circuit.

Cut Edge
A **cut edge** in a graph is an edge whose removal disconnects a component of the graph.

Fleury's Algorithm

Step 1 Start at any vertex. Go along any edge from this vertex to another vertex. *Remove this edge from the graph.*

Step 2 You are now on a vertex of the revised graph. Choose any edge from this vertex, subject to only one condition: Do not use a cut edge (*of the revised graph*) unless you have no other option. Go along your chosen edge. *Remove this edge from the graph.*

Step 3 Repeat Step 2 until you have used all the edges and returned to the vertex at which you started.

Consider each graph.

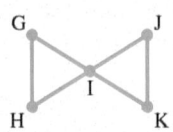

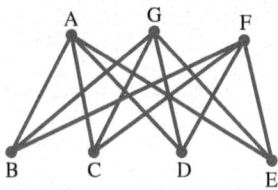

This graph has an Euler circuit, because all vertices have even degree.

This graph has no Euler circuit, because some vertices have odd degree.

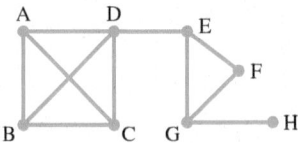

DE and HG are cut edges in this graph.

Fleury's algorithm can be used to show that

$$A \rightarrow C \rightarrow B \rightarrow F \rightarrow E \rightarrow D \rightarrow C \rightarrow F \rightarrow D \rightarrow A$$

is one of many Euler circuits for this graph.

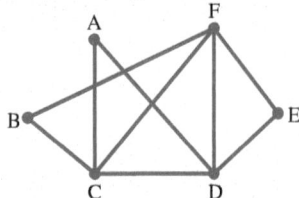

14.3 HAMILTON CIRCUITS AND ALGORITHMS

Hamilton Circuit
A **Hamilton circuit** in a graph is a circuit that visits each vertex exactly once (returning to the starting vertex to complete the circuit).

Any complete graph with three or more vertices has a Hamilton circuit. In this text, Hamilton circuits that differ *only* in their starting points are considered to be the same circuit.

Consider the following graph.

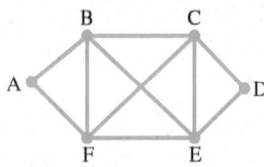

$$A \rightarrow B \rightarrow E \rightarrow D \rightarrow C \rightarrow F \rightarrow A \quad \text{Hamilton circuit}$$

$$A \rightarrow B \rightarrow C \rightarrow D \rightarrow E \rightarrow F \rightarrow C \rightarrow E \rightarrow B \rightarrow F \rightarrow A$$

and

$$B \rightarrow C \rightarrow D \rightarrow E \rightarrow F \rightarrow B$$

Not Hamilton circuits

Concepts	Examples

Number of Hamilton Circuits in a Complete Graph
A complete graph with n vertices has the following number of Hamilton circuits.

$$(n - 1)!$$

Consider the following graph.

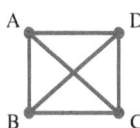

This complete graph has 4 vertices and thus has

$$(4 - 1)! = 3! = 3 \cdot 2 \cdot 1 = 6 \text{ Hamilton circuits.}$$

Minimum Hamilton Circuit
In a weighted graph, a **minimum Hamilton circuit** is a Hamilton circuit with the least possible total weight.

Consider the following graph.

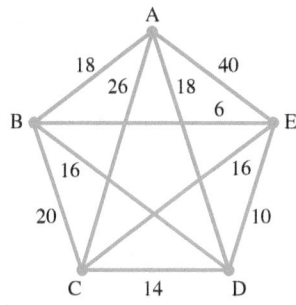

Nearest Neighbor Algorithm

Step 1   Choose a starting point for the circuit. Call this vertex A.

Step 2   Check all the edges joined to A, and choose one that has the least weight. Proceed along this edge to the next vertex.

Step 3   At each vertex you reach, check the edges from there *to vertices not yet visited*. Choose one with the least weight. Proceed along this edge to the next vertex.

Step 4   Repeat Step 3 until you have visited all the vertices.

Step 5   Return to the starting vertex.

The nearest neighbor algorithm shows that an approximate minimum Hamilton circuit is

$$A \rightarrow D \rightarrow E \rightarrow B \rightarrow C \rightarrow A$$

and has weight 80.

14.4   TREES AND MINIMUM SPANNING TREES

Connected Graph (Alternative Definition)
A **connected graph** is one in which there is *at least one path* between each pair of vertices.

Tree
A graph is a **tree** if the graph is *connected* and contains *no circuits*.

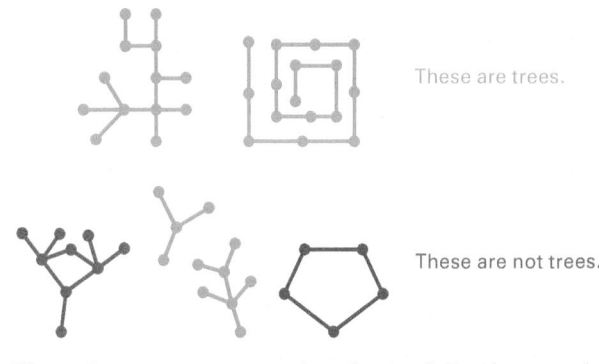

These are trees.

These are not trees.

Unique Path Property of Trees
In a tree, there is always **exactly one path** from each vertex in the graph to any other vertex in the graph.

If we choose any two vertices in the following graph, there is only one path from one to the other. See the red path as an illustration.

Spanning Tree and Minimum Spanning Tree
A **spanning tree** for a graph is a subgraph that includes every vertex of the original graph and is a tree.

A spanning tree that has minimum total weight is a **minimum spanning tree** for the graph.

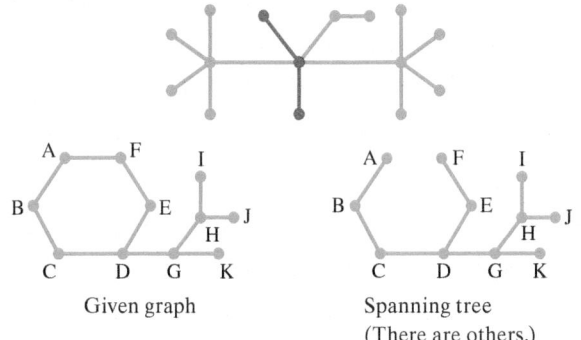

Given graph

Spanning tree
(There are others.)

Concepts	Examples

Kruskal's Algorithm for Finding a Minimum Spanning Tree for Any Connected, Weighted Graph
Choose edges for the spanning tree as follows.

Step 1 First edge: Choose any edge with minimum weight.

Step 2 Next edge: Choose any edge with minimum weight from *those not yet selected*. (At this stage, the subgraph may look disconnected.)

Step 3 Continue to choose edges of minimum weight from those not yet selected, except *do not select any edge that creates a circuit* in the subgraph.

Step 4 Repeat Step 3 until the subgraph connects all vertices of the original graph.

Consider the following graph.

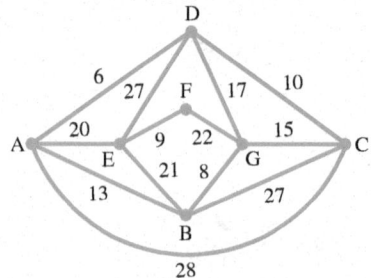

Kruskal's algorithm shows that this graph has the minimum spanning tree below, with total weight 66.

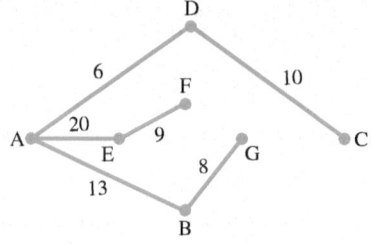

Number of Vertices and Edges in a Tree
If a graph is a tree, then the number of edges in the graph is one less than the number of vertices.

A tree with n vertices has $n - 1$ edges.

In the tree above, there are 7 vertices and

$$7 - 1 = 6 \text{ edges.}$$

Chapter 14 TEST

Basic Concepts *Refer to the following graph.*

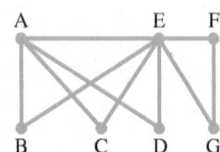

1. Determine how many vertices the graph has.

2. Determine the sum of the degrees of the vertices.

3. Determine how many edges the graph has.

4. Which of the following are paths in the graph? If not, why not?

(a) B → A → C → E → B → A

(b) A → B → E → A

(c) A → C → D → E

5. Which of the following are circuits in the graph? If not, why not?

(a) A → B → E → D → A

(b) A → B → C → D → E → F → G → A

(c) A → B → E → F → G → E → D → A → E → C → A

Solve each problem.

6. Components Draw a graph that has 2 components.

7. Sum of Degrees A graph has 10 vertices, 3 of degree 4, and the rest of degree 2. Use the theorem that relates the sum of degrees to the number of edges to determine the number of edges in the graph (without drawing the graph).

8. *Isomorphism*   Determine whether graphs (a) and (b) are isomorphic. If they are, justify this by labeling corresponding vertices of the two graphs with the same letters and color-coding the corresponding edges.

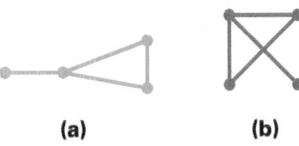

(a)   (b)

9. *Planning for Dinner*   Julia is planning to invite some friends for dinner. She plans to invite John, Adam, Bill, Tina, Nicole, and Rita. John, Nicole, and Tina all know each other. Adam knows John and Tina. Bill also knows Tina, and he knows Rita. Draw a graph with vertices representing the six friends whom Julia plans to invite to dinner and edges representing the relationship "know each other." Is your graph connected or disconnected? Which of the guests knows the greatest number of other guests?

10. *Chess Competition*   There are 8 contestants in a chess competition. Each contestant plays one chess game against every other contestant. The winner is the contestant who wins the most games. How many chess games will be played in the competition?

11. *Complete or Not Complete?*   Is the accompanying graph a complete graph? Justify your answer.

Coloring and Chromatic Number   *Color each graph using as few colors as possible. Use the result to determine the chromatic number of the graph.*

12.

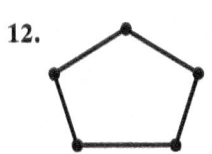

13.

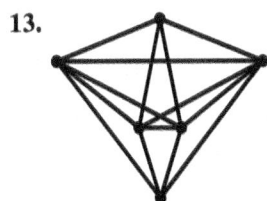

14. *Scheduling Exams*   A teacher at a high school must schedule exams for the senior class. In the table at the top of the next column, subjects in which there are exams are listed on the left. Several students must take more than one of the exams, as shown on the right. Exams in subjects with students in common must not be scheduled at the same time. Use graph coloring to determine the least number of exam times needed and to identify exams that can be given at the same times.

Exam	Students taking this must also take
History	English, Mathematics, Biology, Geography, Psychology
English	Mathematics, Psychology, History
Mathematics	Chemistry, Biology, History, English
Chemistry	Biology, Mathematics
Biology	Mathematics, Chemistry, History
Psychology	English, History, Geography
Geography	Psychology, History

15. *Euler Circuits*   Refer to the graph for **Exercises 1–5.** Which of the following are Euler circuits for the graph? For those that are not, why are they not?

(a) $A \to B \to E \to D \to A$

(b) $A \to B \to C \to D \to E \to F \to G \to A$

(c) $A \to B \to E \to F \to G \to E \to D \to A \to$
$E \to C \to A$

Euler's Theorem   *Use Euler's theorem to decide whether the graph has an Euler circuit. (Do not actually find an Euler circuit.) Justify your answer briefly.*

16.

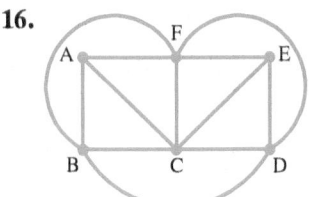

17.

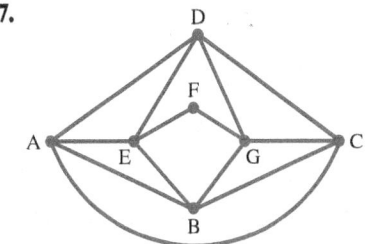

Solve each problem.

18. *Floor Plan*   The floor plan of a building is shown. Is it possible to start outside, walk through each door exactly once, and end up back outside? Justify your answer.

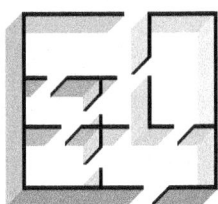

19. *Fleury's Algorithm* Use Fleury's algorithm to find an Euler circuit for the accompanying graph, beginning F → B.

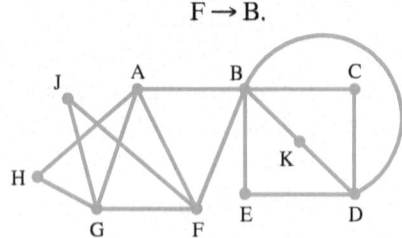

20. *Hamilton Circuits* Refer to the graph for **Exercises 1–5.** Which of the following are Hamilton circuits for the graph? If not, why not?

(a) A → B → E → D → A

(b) A → B → C → D → E → F → G → A

(c) A → B → E → F → G → E → D → A → E → C → A

21. *Hamilton Circuits* List all Hamilton circuits in the graph that start F → G. How many are there?

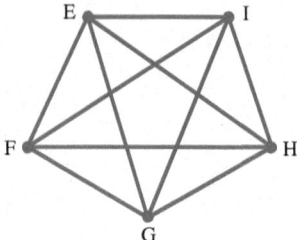

22. *Brute Force Algorithm* Use the brute force algorithm to find a minimum Hamilton circuit for the accompanying graph. Determine the total weight of the minimum Hamilton circuit.

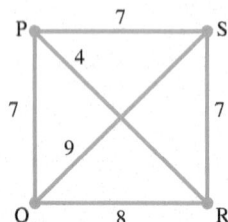

23. *Nearest Neighbor Algorithm* Use the nearest neighbor algorithm starting at A to find an approximate solution to the problem of finding a minimum Hamilton circuit for the graph below. Find the total weight of this circuit.

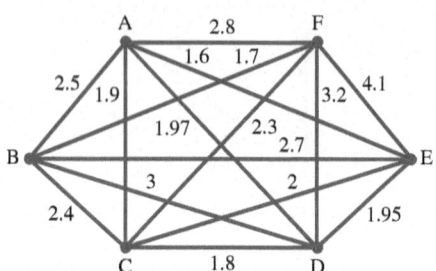

24. *Hamilton Circuits* How many Hamilton circuits are there in a complete graph with 25 vertices? (Leave your answer as a factorial.)

25. *Rock Band Tour Plan* The agent for a rock band based in Milwaukee is planning a tour for the band. He plans for the band to visit Minneapolis, Santa Barbara, Orlando, Phoenix, and St. Louis. Since the band members want to minimize the time they are away from home, they do not want to go to any of these cities more than once on the tour. Consider the complete graph with vertices representing the 6 cities mentioned. Does the problem of planning the tour require an Euler circuit, a Hamilton circuit, or neither for its solution?

26. *Nonisomorphic Trees* Draw three nonisomorphic trees with 7 vertices.

Trees *Decide whether the statement is* true *or* false.

27. Every tree has a Hamilton circuit.

28. In a tree, each edge is a cut edge.

29. Every tree is connected.

Solve each problem.

30. *Spanning Trees* Find all the different spanning trees for the following graph.

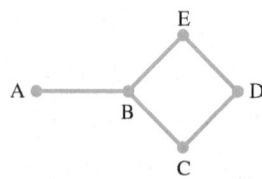

31. *Kruskal's Algorithm* Use Kruskal's algorithm to find a minimum spanning tree for the following graph.

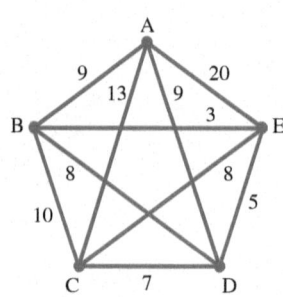

32. *Edges and Vertices* Determine the number of edges in a tree with 50 vertices.

15 Voting and Apportionment

Suppose that, as a political consultant, you mainly advise legislative and other bodies on the mathematical mechanics of making decisions. A nine-member committee is considering three options for addressing the problem of overcrowding in the state's prisons. The "voter profile" here shows how the members have ranked their options, which are

b: build more retention facilities,

r: grant some prisoners supervised release,

d: double the per-room occupancy rate.

Number of Committee Members	Ranking
4	r > b > d
2	b > d > r
3	d > b > r

1. Which option is ranked highest by the greatest number of members? (Would you judge this option to be the most desirable of the three?)

2. How many members prefer r to b? How many prefer b to r? Give the winner one point. How many members prefer r to d? How many prefer d to r? Give the winner a point. How many members prefer b to d? How many prefer d to b? Give this winner a point. Now which option has the greatest number of points? (Would you judge this option to be the most desirable among the three?)

3. Did the winning option in Question 1 have a *majority* of all first-place rankings? If not, eliminate the lowest-scoring option and reconsider the remaining two. Which of these is ranked higher by the greatest number of members? (Is this, then, the most desirable option?)

Turn two pages ahead for the answers. Questions such as these are the topic of this chapter.

15.1   THE POSSIBILITIES OF VOTING

OBJECTIVES

1. Apply the plurality method of voting.

2. Apply the pairwise comparison method.

3. Apply the Borda method.

4. Apply the Hare method.

Harold Washington won the 1983 Chicago Democratic mayoral primary with only 36% of the vote. He eventually became Chicago's first black mayor. He died of a heart attack on November 15, 1987, while in office.

Plurality Method

If a group of people want to make one choice from two alternatives, then each person indicates which of the two alternatives he or she prefers. The alternative awarded the majority of the votes is the choice of the group. Selecting one alternative from two is a straightforward process. The outcome, ignoring ties, is uncontestable.

Suppose a group of 100 people want to select one alternative from among three. The group decides each member should indicate his or her first preference, and the alternative with the most votes will be selected as the choice of the group. It is possible that none of the alternatives gets a **majority** (more than half) of the votes. For example, the outcome of the vote may be

$$\text{33 votes for Alternative a,}$$

$$\text{31 votes for Alternative b,}$$

and   36 votes for Alternative c.

Alternative c does *not* have a majority of the votes, but it does have a **plurality** (the greatest number) of the votes. Alternative c is declared the people's choice, even though 64 of 100 voters chose different alternatives. Whenever a group of people must select one alternative from three or more, an alternative favored by far fewer than a majority of the people may be selected.

Most voting methods require each voter to give a complete **ranking** of the candidates/alternatives (that is, an ordering of all candidates from highest to lowest preference). The number of possible rankings (or arrangements) of *n* candidates is *n*!—that is, *n* **factorial,** which was defined in an earlier chapter.

Problem-Solving Strategy
Many counting techniques covered earlier have direct application in the discussion of voting.

Suppose, for example, that there are twelve voters and three candidates/alternatives. Let the three candidates be a, b, and c. There are

$$3! = 3 \cdot 2 \cdot 1 = 6 \text{ possible ways to rank the three candidates.}$$

The notation a > b > c indicates the ranking in which Candidate a is the voter's first choice, Candidate b is the second choice, and Candidate c is the third choice. Suppose 3 of the 12 voters chose the ranking a > b > c, 3 voters chose the ranking c > a > b, 4 voters chose the ranking b > a > c, and 2 voters chose the ranking c > b > a. This **profile of voters** is displayed in **Table 1**. Although there are six possible rankings, only four are reflected in this voter profile (because no voters chose the other two possible rankings).

In our discussion of voting methods, we do not consider final election outcomes that result in ties.

The *plurality method* is the most common voting method. It is the process used by the 100 voters in the opening discussion and in nearly all state and federal elections in the United States.

Table 1

Number of Voters	Ranking
3	a > b > c
3	c > a > b
4	b > a > c
2	c > b > a

PLURALITY METHOD

In the **plurality method** of voting, each voter gives one vote to his or her top-ranked candidate. The candidate with the most votes, a plurality of the votes, wins the election.

The plurality method does not require a voter to rank the candidates completely. Only the voter's first choice is important in the vote tally.

When an election involves only two alternatives, the plurality of the votes is the majority of votes.

Table 2

Number of Voters	Ranking
6	b > p > c
5	c > p > b
4	p > c > b

EXAMPLE 1   Selecting a Banquet Entrée by the Plurality Method

The Culinary Arts Club is planning its annual banquet. Entrée choices are beef, chicken, and pork. The 15 members of the club rank the choices according to the profile in **Table 2**, where b represents beef, c chicken, and p pork. Determine the plurality winner among the three entrée choices.

Solution

Each voter's complete ranking is shown. *Only the top-ranked candidate matters in a plurality method election.* Beef receives 6 votes, chicken receives 5, and pork receives 4. The winner is beef because it receives a plurality of the votes.

One of the club members with the third ranking in **Table 2** notices that pork was ranked above beef by 5 + 4 = 9 members, while beef was ranked above pork by only 6 members. Therefore, the beef victory seems unfair to him, and he convinces a majority of the club members to push for some alternative to the plurality method of voting.

Pairwise Comparison Method

The *pairwise comparison method* requires voters to make a choice between every pair of candidates, and all possible pairs of candidates must be compared to find the outcome of the election. Elections decided by means of pairwise comparison are similar to round-robin tournaments in sports.

Recall from an earlier chapter that, in general,

$$_nC_r = \frac{n!}{r!(n-r)!}.$$

Example: $_3C_2 = \frac{3!}{2! \cdot 1!}$

$$= \frac{3 \cdot 2 \cdot 1}{2 \cdot 1 \cdot 1}$$

$$= 3$$

PAIRWISE COMPARISON METHOD

In the **pairwise comparison method** of voting, each voter gives a complete ranking of the candidates. For *each pair* of candidates a and b, the number of voters preferring Candidate a is compared with the number of voters preferring Candidate b.

- The candidate receiving more votes is awarded one point.
- If the two candidates receive the same number of votes, each is awarded half a point.
- After all pairs have been compared, the candidate with the most points wins.

If there are n candidates, then the number of pairwise comparisons is

$$_nC_2, \quad \text{the number of combinations of } n \text{ things taken 2 at a time.}$$

A voter's complete ranking of the candidates allows the direct comparison of any pair of candidates. All candidate positions in the rankings are ignored, except the positions of the two candidates being compared. The voter's preferred candidate of the pair is the candidate ranking higher in the order.

EXAMPLE 2 **Selecting a Banquet Entrée by the Pairwise Comparison Method**

Examine again the voter profile of **Table 2,** and determine the winning entrée by the pairwise comparison method.

Solution

The profile of how the 15 club members rank beef, pork, and chicken is repeated in **Table 3.** There are $n = 3$ candidates, so we must examine

Table 3

Number of Voters	Ranking
6	b > p > c
5	c > p > b
4	p > c > b

$$_nC_2 = {_3C_2} = 3 \text{ pairs,} \quad \text{here} \quad \text{p and c,} \quad \text{p and b,} \quad \text{c and b.}$$

p and c: 10 voters prefer p to c.   6 with the 1st ranking plus 4 with the 3rd ranking
              5 voters prefer c to p.   The 5 with the 2nd ranking

Pork beats chicken (10 to 5) and receives **1 point.**

p and b: 9 voters prefer p to b.   5 with the 2nd ranking plus 4 with the 3rd ranking
              6 voters prefer b to p.   The 6 with the 1st ranking

Pork beats beef (9 to 6) and receives **1 point.**

c and b: 9 voters prefer c to b.   5 with the 2nd ranking plus 4 with the 3rd ranking
              6 voters prefer b to c.   The 6 with the 1st ranking

Chicken beats beef (9 to 6) and receives **1 point.**

All three pairs have been considered, and the results are

pork with 2 points,   chicken with 1 point,   and   beef with 0 points.

The winner of the entrée election, using the pairwise comparison method, is pork.

Notice that different voting methods can lead to different outcomes. Beef is the plurality method selection of the Culinary Arts Club in **Example 1,** but **Example 2** shows that pork wins a pairwise comparison method election.

EXAMPLE 3 Selecting a Meeting Site by the Pairwise Comparison Method

The twenty-member executive board of the College Art Association (CAA) is meeting to decide the location of the next yearly meeting. They are considering four cities:

Atlanta a,   Boston b,   Chicago c,   and   Dallas d.

Each executive board member gives a complete ranking of the four cities. The profile is in **Table 4.** (Only 5 of the 4! = 24 possibilities appear as actual voter rankings.) Determine the winning city by the pairwise comparison method.

Solution

There are four candidates, so a total of $_nC_2 = {_4}C_2 = 6$ pairs of cities must be compared.

a and b,   a and c,   a and d,   b and c,   b and d,   c and d

Compare a and b. Atlanta is preferred to Boston by the 3 + 2 + 5 = 10 voters with the first three rankings. The remaining 6 + 4 = 10 voters with the last two rankings prefer Boston to Atlanta. The tie for this pair means Atlanta and Boston both receive $\frac{1}{2}$ point.

Atlanta wins the comparison with Chicago by a margin of

3 + 5 + 4 = 12   to   2 + 6 = 8.   Atlanta, 1 point

Dallas wins the comparison with Atlanta by a margin of

5 + 6 = 11   to   3 + 2 + 4 = 9.   Dallas, 1 point

Boston wins when compared with Chicago,

3 + 5 + 4 = 12   to   2 + 6 = 8.   Boston, 1 point

Boston also wins when compared with Dallas,

3 + 2 + 6 + 4 = 15   to   5.   Boston, 1 point

Chicago wins the comparison with Dallas,

3 + 2 + 6 = 11   to   5 + 4 = 9.   Chicago, 1 point

Thus, Atlanta has 1.5 points, Boston has 2.5 points, and Chicago and Dallas each have 1 point. The host city selected by the pairwise comparison method is Boston.

Borda Method

What is it about pork in the Culinary Arts Club profile that allows it to fare so well in the pairwise comparison method, while losing to beef in a plurality method election?

Every club member ranks pork first or second, and no one ranks it last.

The *Borda method* was first proposed in 1781 by Jean-Charles de Borda. Winners of many popular music awards and the winner of the annual Heisman Trophy are decided with versions of the Borda method. This method computes a weighted sum for each candidate to decide the outcome of the election.

Table 4

Number of Voters	Ranking
3	a > b > c > d
2	c > a > b > d
5	d > a > b > c
6	c > b > d > a
4	b > a > d > c

Jean-Charles de Borda (1733–1799), a Frenchman of diverse talents, was a military officer with skills on land and sea, serving as both a cavalry officer and a naval captain. In addition to **voting theory,** he studied and wrote about other topics in mathematics and physics.

Borda was fully aware that his voting method is easy for an insincere voter to manipulate. Such a voter could misrepresent his real ranking in order to help a favorite candidate. In reply to his detractors, Borda said he assumed his method would be used by honest men.

BORDA METHOD

In the **Borda method** of voting, each voter must give a complete ranking of the candidates. Let n be the number of candidates.

- Each first-place vote a candidate receives is worth $n - 1$ points.
- Each second-place vote earns $n - 2$ points.
- A third-place vote earns $n - 3$ points, a fourth-place vote earns $n - 4$ points, . . . , and a last-place vote earns $n - n = 0$ points.
- The candidate with the highest tally of points is the winner.

Notice that in a Borda method election with n candidates, *each voter* awards a total number of points given by

$$\underset{\substack{\text{to 1st}\\\text{choice}}}{(n-1)} + \underset{\substack{\text{to 2nd}\\\text{choice}}}{(n-2)} + \cdots + \underset{\substack{\text{to next-to-last}\\\text{choice}}}{(n-n+1)} + \underset{\substack{\text{to last}\\\text{choice}}}{(n-n)}.$$

That is,

$$(n - 1) + (n - 2) + \cdots + 1 + 0.$$

Dropping the 0 term and reversing the order of the terms, we obtain

$$1 + 2 + \cdots + (n - 2) + (n - 1).$$

Applying a special sum formula to this arithmetic sequence gives sum $= \frac{(n-1) \cdot n}{2}$. Now if there are v voters voting in the election, then the total number of Borda points awarded, *by all voters,* must be

$$\frac{(n-1) \cdot n \cdot v}{2}.$$

We can use this formula to check whether we have missed any points in our count.

EXAMPLE 4   Selecting a Banquet Entrée by the Borda Method

Determine the entrée selection for the Culinary Arts Club's annual banquet if they use a Borda method election.

Solution

Recall from **Examples 1 and 2** that there are $n = 3$ alternatives: beef, pork, and chicken. A first-place vote earns $n - 1 = 2$ points, a second-place vote earns 1 point, and a third-place vote earns 0 points. See **Table 5.**

Table 5

Number of Voters	Ranking	First Place Earns 2 Points	Second Place Earns 1 Point	Third Place Earns 0 Points
6	b > p > c	b	p	c
5	c > p > b	c	p	b
4	p > c > b	p	c	b

Beef earns $6 \cdot 2 = 12$ points from the six voters with the first ranking, $5 \cdot 0 = 0$ points from the five voters with the second ranking, and $4 \cdot 0 = 0$ points from the four voters with the third ranking.

The tally for beef is $12 + 0 + 0 = 12$ points.

Chicken earns $6 \cdot 0 = 0$ points from the six voters with the first ranking, $5 \cdot 2 = 10$ points from the five voters with the second ranking, and $4 \cdot 1 = 4$ points from the four voters with the last ranking.

The tally for chicken is $0 + 10 + 4 = 14$ points.

Pork earns $6 \cdot 1 = 6$ points from the six voters with the first ranking, $5 \cdot 1 = 5$ points from the five voters with the second ranking, and $4 \cdot 2 = 8$ points from the four voters with the third ranking.

The tally for **pork** is $6 + 5 + 8 = $ **19 points.**   ← Borda winner

Pork is the winner of the Borda method election. As a check, we counted $12 + 14 + 19 = 45$ total points, which agrees with the result from the formula.

$$\frac{(n-1) \cdot n \cdot v}{2} = \frac{2 \cdot 3 \cdot 15}{2} = 45 \quad \text{\small $n = 3$ and $v = 6 + 5 + 4 = 15$}$$

Pork won both the Borda method and the pairwise comparison method entrée elections. Both methods take into account the overall popularity of pork.

EXAMPLE 5   Selecting a Meeting Site by the Borda Method

Midwestern members of the executive board (see **Example 3**) are disappointed that Chicago was not selected as the site of the next yearly meeting, so they suggest a Borda method election. Determine the winning city by the Borda method.

Solution

The voter profile is repeated in **Table 6,** along with the Borda point values. The $n = 4$ alternatives (Atlanta, Boston, Chicago, and Dallas) mean each first-place vote earns 3 points, second place earns 2 points, third place earns 1 point, and fourth place earns 0 points.

Table 6

Number of Voters	Ranking	First Place (3 points)	Second Place (2 points)	Third Place (1 point)	Fourth Place (0 points)
3	$a > b > c > d$	a	b	c	d
2	$c > a > b > d$	c	a	b	d
5	$d > a > b > c$	d	a	b	c
6	$c > b > d > a$	c	b	d	a
4	$b > a > d > c$	b	a	d	c

Atlanta earns $(3 \cdot 3) + (2 \cdot 2) + (5 \cdot 2) + (6 \cdot 0) + (4 \cdot 2) = 31$ points.

Boston earns $(3 \cdot 2) + (2 \cdot 1) + (5 \cdot 1) + (6 \cdot 2) + (4 \cdot 3) = $ **37 points.**   ← Borda winner

Chicago earns $(3 \cdot 1) + (2 \cdot 3) + (5 \cdot 0) + (6 \cdot 3) + (4 \cdot 0) = 27$ points.

Dallas earns $(3 \cdot 0) + (2 \cdot 0) + (5 \cdot 3) + (6 \cdot 1) + (4 \cdot 1) = 25$ points.

The winner of the Borda method election is Boston.

Pork, the Borda method selection in **Example 4,** is also the pairwise comparison method selection in **Example 2.** Boston, the Borda method selection in **Example 5,** is also the pairwise comparison method selection in **Example 3.** *Do the Borda and pairwise comparison methods* always *select the same winner?* No.

Table 7

Number of Voters	Ranking
10	h > f > t > c
5	f > t > c > h

EXAMPLE 6 **Selecting a Class Pet by the Borda and Pairwise Comparison Methods**

Fifteen members of Kareem's kindergarten class are voting on a class pet. The alternatives are a hamster h, a fish f, a turtle t, and a canary c. The profile in **Table 7** shows remarkable agreement among the students. It uses only two of the 4! = 24 possible rankings. Compare the Borda and pairwise comparison results for this profile.

Solution

The number of class pet candidates is $n = 4$. If the Borda method is used, first place earns 3 points, second place 2 points, third place 1 point, and fourth place 0 points.

A hamster earns $(10 \cdot 3) + (5 \cdot 0) = 30$ points.

A fish earns $(10 \cdot 2) + (5 \cdot 3) = 35$ points.   ← Borda winner

A turtle earns $(10 \cdot 1) + (5 \cdot 2) = 20$ points.

A canary earns $(10 \cdot 0) + (5 \cdot 1) = 5$ points.

If the pairwise comparison method is used, a hamster wins each pairwise comparison with a fish, a turtle, and a canary, by a margin of 10 to 5. A fish wins each comparison with a turtle and a canary, by a margin of 15 to 0. A turtle wins the comparison with a canary, by a margin of 15 to 0.

A hamster earns 3 points.   ← Pairwise comparison winner

A fish earns 2 points.

A turtle earns 1 point.

A canary earns 0 points.

The Borda and pairwise comparison methods selected different winners.

Hare Method

The *Hare method* was first proposed in 1861 by Thomas Hare. It is a variation of the plurality method in which candidates are eliminated in sequential rounds of voting. Votes are transferred from eliminated candidates to remaining candidates. The president of France and public officials in Australia are selected by the Hare method.

Hare elections do not require the voters initially to rank the complete list of candidates. Instead, voters need to decide only which candidate they prefer most in each round. This is considered an advantage of the method.

Thomas Hare (1806–1891) was a British lawyer. His reason for devising his voting method was to secure proportional representation of all classes in the United Kingdom, including minorities, in the House of Commons and other elected groups.

The **Hare method** was considered to be "among the greatest improvements yet made in the theory and practice of government" by Hare's contemporary John Stuart Mill.

It also is known as the **plurality with elimination method** and as the **single transferable vote system**.

HARE METHOD

In the **Hare method** of voting, each voter gives one vote to his or her favorite candidate in the first round.

- If a candidate receives a *majority* of the votes, he or she is declared the winner.

- If no candidate receives a majority of the votes, the candidate (or candidates) with the fewest votes is (are) eliminated, and a second election is conducted, in the same way, on the remaining candidates.

- The rounds of voting continue to eliminate candidates until one candidate receives a *majority* of the votes. That candidate wins the election.

How many rounds of voting does it take to learn the winner of a Hare method election? It may take many rounds.

EXAMPLE 7 Selecting a Banquet Entrée by the Hare Method

Chicken entrée supporters of the Culinary Arts Club (see **Examples 1, 2, and 4**), having seen how pork was able to displace beef by appealing to different voting methods, decide to try the same strategy. They suggest a Hare method election. Does this method give them victory?

Solution

The 15-member club profile is repeated in **Table 8.** Club members vote for their favorite of the three entrées in the first round, with the following results.

> Beef gets 6 votes.
>
> Chicken gets 5 votes.
>
> Pork gets 4 votes.

No entrée receives a majority (8 or more) of the votes in the first round. Pork receives the fewest votes and is eliminated. (Chicken supporters are encouraged.)

In the second round of Hare method voting, club members vote for their favorite of the remaining entrées: beef or chicken. The 4 club members with the third ranking vote for chicken, because pork was eliminated in the first round and they prefer chicken to beef. The other 11 voters vote as they did in the first round.

> Beef gets 6 votes.
>
> Chicken gets 5 + 4 = 9 votes. ← Hare winner

Chicken receives a majority of the votes in the second round and indeed is declared the entrée selection by the Hare method.

Table 8

Number of Voters	Ranking
6	b > p > c
5	c > p > b
4	p > c > b

EXAMPLE 8 Selecting a Meeting Site by the Hare Method

The twenty-member executive board profile, repeated in **Table 9,** shows that Chicago would have won a plurality method election with 8 votes, but it has now lost both the pairwise comparison and the Borda method elections to Boston. (See **Examples 3 and 5.**) Now determine the winning city by the Hare method.

Solution

Refer to **Table 9.** First-round voting gives the following results.

> Atlanta gets 3 votes.
>
> Boston gets 4 votes.   No city receives a majority (11 or more) of the votes.
>
> Chicago gets 2 + 6 = 8 votes.   **Atlanta,** with the fewest votes, is eliminated.
>
> Dallas gets 5 votes.

Boston, Chicago, and Dallas compete in the second round. The 3 voters with the first ranking vote for Boston, because Atlanta was eliminated. Other voters vote as they did in the first round.

> Boston gets 4 + 3 = 7 votes.   Again no city receives a majority. **Dallas** has the fewest votes and is eliminated.
>
> Chicago gets 8 votes.
>
> Dallas gets 5 votes.

The third-round vote is between Boston and Chicago. The five voters with the middle ranking transfer their votes to Boston. The other 15 voters vote as they did in the second round.

> **Boston** gets 7 + 5 = 12 votes.. ← Hare winner
>
> Chicago gets 8 votes.

Boston is the winner of the Hare method election. Chicago loses out again.

Table 9

Number of Voters	Ranking
3	a > b > c > d
2	c > a > b > d
5	d > a > b > c
6	c > b > d > a
4	b > a > d > c

The voting methods introduced earlier are summarized in **Table 10.**

Table 10 Summary of Voting Methods

Voting Method	Mechanics of the Election Procedure
Plurality Method	Each voter gives one vote to his or her top-ranked candidate. The candidate with the most votes, a *plurality,* wins the election.
Pairwise Comparison Method	Each voter gives a complete ranking of the candidates. Each pair of candidates, a and b, are compared. • The one ranked higher by more voters receives one point. • If they get equal numbers of voters, each receives half a point. • After all pairs have been compared, the candidate with the most points wins the election.
Borda Method	Each voter gives a complete ranking of the n candidates. • A candidate receives $n - 1$ points for each first-place ranking by a voter, $n - 2$ points for each second-place ranking, . . . , and $n - n = 0$ points for each nth-place (last-place) ranking. • The candidate with the most points wins the election.
Hare Method	Each voter gives one vote to his or her top-ranked candidate. • If a candidate receives a *majority* of the votes, that candidate wins the election. • If no candidate receives a majority of the votes, the candidate (or candidates) with the fewest votes is (are) eliminated, and a second-round election is conducted, in the same way, on the remaining candidates. • Rounds continue to eliminate candidates until a candidate receives a majority of the votes. That candidate wins the election. *Note:* An option in the Hare method is to initially obtain complete rankings from the voters. This way, only one actual vote is required. If no candidate receives a majority of first-place rankings, subsequent rounds of voting, as required, can be simulated from the voter profile.

15.1 EXERCISES

CONCEPT CHECK *Write* true *or* false *for each statement.*

1. Usually, all voting methods presented in this section select the same winner in the end.

2. When there are three or more alternatives, a plurality winner may not receive a majority of the votes.

3. The discussion in this section ignored the possibility that in the preference a > b > c, it may be the case that a is *greatly preferred* to b, while b is only *slightly preferred* to c.

4. In a Borda method election with n candidates, each first-place vote that a candidate receives is worth n points.

5. In a pairwise comparison election with 4 candidates, the number of pairwise comparisons required is 6.

6. If no candidate receives a majority vote in the first round of a Hare method election with 5 candidates, then the top two candidates have a second-round runoff election.

Choosing a Poster Dog by the Plurality Method *A local animal shelter is choosing a poster dog. The choices are*

an Australian shepherd a,   a boxer b,   a cocker spaniel c,   a dalmatian d.

Thirteen staff members completely rank the four dogs on their ballots. The thirteen individual ballots are the columns in the table. In each of Exercises 7 and 8, complete the following.

(a) *How many different ways can a staff member complete his or her ballot?*

(b) *Write a voter profile for this election.*

(c) *Use the plurality method to determine the shelter's poster dog.*

7.

Voter	1	2	3	4	5	6	7	8	9	10	11	12	13
1st place	b	b	a	c	d	d	c	c	b	a	a	a	a
2nd place	c	c	d	d	c	a	a	a	c	c	d	b	b
3rd place	a	a	c	b	b	b	d	d	a	d	c	c	c
4th place	d	d	b	a	a	c	b	b	d	b	b	d	d

8.

Voter	1	2	3	4	5	6	7	8	9	10	11	12	13
1st place	a	b	d	c	a	a	b	c	d	a	a	a	a
2nd place	b	c	b	b	d	d	c	b	b	b	b	b	b
3rd place	c	a	c	a	c	c	d	d	c	c	d	d	c
4th place	d	d	a	d	b	b	a	a	a	d	c	c	d

Choosing a Poster Dog by Alternative Methods *For each voter profile of Exercises 9 and 10, determine the winning dog by the*

(a) *pairwise comparison method*

(b) *Borda method*

(c) *Hare method.*

9. The profile of voters found in **Exercise 7(b)**

10. The profile of voters found in **Exercise 8(b)**

Observing the Effect of the Number of Candidates *Answer each question in Exercises 11–16.*

11. If there are $n = 5$ candidates, then how many different rankings of the candidates are possible? What if there are $n = 7$ candidates?

12. If there are $n = 6$ candidates, how many different rankings of the candidates are possible? What if there are $n = 8$ candidates?

13. What are the implications of the number of different rankings for elections involving more than $n = 4$ candidates?

14. How many pairwise comparisons are needed to learn the outcome of an election involving $n = 5$ candidates? What if there are 7 candidates?

15. How many pairwise comparisons are needed to learn the outcome of an election involving $n = 6$ candidates? What if there are 8 candidates?

16. What are the implications of the number of pairwise comparisons needed to learn the outcome of an election involving more than $n = 4$ candidates?

Applying Four Voting Methods to a Voter Profile *For each of Exercises 17–24, determine the election winner using the*

(a) *plurality method*   (b) *pairwise comparison method*

(c) *Borda method*   (d) *Hare method.*

In Exercises 17 and 18, a 13-member committee is selecting a chairperson. The 3 candidates are

Amy a,   Byron b,   and   Chieko c.

Each committee member completely ranked the candidates on a separate ballot. Determine the selected chairperson.

17.

Number of Voters	Ranking
3	a > c > b
4	c > b > a
2	b > a > c
4	b > c > a

18.

Number of Voters	Ranking
4	a > b > c
2	b > c > a
4	b > a > c
3	c > a > b

In Exercises 19 and 20, a 13-member committee is selecting a new company logo from 3 alternatives:

a,   b,   *and*   c.

Each committee member completely ranked the possible logos on a separate ballot. Determine the new company logo.

19.

Number of Voters	Ranking
6	a > b > c
1	b > c > a
3	b > a > c
3	c > a > b

20.

Number of Voters	Ranking
2	a > c > b
3	c > b > a
4	b > a > c
4	b > c > a

In Exercises 21 and 22, a senator invited one person from each of the 21 counties in her state to a weekend workshop. The senator asked the attendees to rank the issues of

job creation j,

education e,

health care h,

and   gun control g

in order of importance to themselves and their counties. The 21 members each gave a complete ranking of the issues on a separate ballot. Determine which issue workshop members felt had the highest priority.

21.

Number of Voters	Ranking
6	h > j > g > e
5	e > g > j > h
4	g > j > h > e
3	j > h > g > e
3	e > j > h > g

22.

Number of Voters	Ranking
3	e > h > g > j
6	h > e > g > j
5	j > g > e > h
4	g > e > h > j
3	j > e > h > g

In Exercises 23 and 24, the 55 members of the Adventure Club are selecting one activity for their yearly vacation. Choices are

hiking in the desert h,

rock climbing c,

white-water kayaking k,

motorcycle touring m,

and   sitting by a mountain lake and tanning t.

Determine the selected activity.

23.

Number of Voters	Ranking
18	t > m > h > k > c
12	c > h > m > k > t
10	k > c > h > m > t
9	m > k > h > c > t
4	h > c > m > k > t
2	h > k > m > c > t

24.

Number of Voters	Ranking
18	t > k > h > m > c
12	c > h > k > m > t
10	m > c > h > k > t
9	k > m > h > c > t
4	h > c > k > m > t
2	h > m > k > c > t

Holding a Runoff Election *One common solution to an election in which no candidate receives a majority of first-place votes is to have a runoff between the two candidates with the most first-place votes (after eliminating all but those two). Determine the runoff selection in each of Exercises 25–28.*

25. The workshop voter profile in **Exercise 21**

26. The workshop voter profile in **Exercise 22**

27. The Adventure Club voter profile in **Exercise 23**

28. The Adventure Club voter profile in **Exercise 24**

A Different Kind of Runoff Election *Another solution to an election in which no candidate receives a majority of first-place votes is to eliminate all but the top three, have a runoff between the candidates who rank second and third, and eliminate the loser of that runoff. The winner of that election faces the candidate with the most first-place votes to decide the final outcome of the election. Use this process to select a winner in each of Exercises 29 and 30.*

29. The voter profile in **Exercise 23**

30. The voter profile in **Exercise 24**

The Pairwise Comparison Method   *Each table represents a pairwise comparison method election.*

(a) *Find the missing number of pairwise comparisons won.*

(b) *Identify the winning candidate and the number of pairwise points for the winner.*

31.

Candidate	a	b	c	d	e	f	g	h
Number of pairwise comparisons won	2	6	3	4	?	2	2	2

32.

Candidate	a	b	c	d	e	f	g	h
Number of pairwise comparisons won	4	5	2	?	3	1	2	3

33.

Candidate	a	b	c	d	e	f	g
Number of pairwise comparisons won	3	5	7	1	2	?	1

34.

Candidate	a	b	c	d	e	f	g
Number of pairwise comparisons won	4	6	3	1	5	1	?

The Borda Method   *Each table represents a Borda method election.*

(a) *Find the missing number of Borda points.*

(b) *Identify the winning candidate.*

35. Total number of voters: 15

Candidate	a	b	c
Number of Borda points	15	14	?

36. Total number of voters: 15

Candidate	a	b	c
Number of Borda points	17	14	?

37. Total number of voters: 20

Candidate	a	b	c	d	e
Number of Borda points	35	40	?	40	30

38. Total number of voters: 20

Candidate	a	b	c	d	e
Number of Borda points	?	50	25	40	55

The Coombs Method   *The **Coombs method** of voting is a variation of the Hare method. Voting takes place in rounds, but instead of the candidate with the fewest first-place votes, the candidate with the most last-place votes is eliminated. In each of Exercises 39 and 40, use the Coombs method to*

determine which issue the senator's workshop members felt had the highest priority.

39. The profile in **Exercise 21**

40. The profile in **Exercise 22**

41. Suppose that the pairwise comparisons for an election with the four alternatives a, b, c, and d have the following results.

> a beats b.   a beats c.
>
> b beats c.   b beats d.
>
> d beats a.   d beats c.

What is the overall outcome of the election? (This is considered a flaw of the pairwise comparison voting method.)

Write a short answer for each of Exercises 42–44.

42. Suppose only two candidates are running for office. Show that the pairwise comparison method, the Borda method, and the Hare method all select the same winner. Explain why all three methods reduce to a simple plurality vote for a top-ranked candidate.

43. Which voting method—plurality, pairwise comparison, Borda, or Hare—do you personally think is the best way to select one alternative from among three or more alternatives? Why do you think it is the best method?

44. Which voting method do you think would be best in an election with a very large number of candidates? Why?

Devising a Profile for Consistency among Voting Methods   *Given each 4-candidate voter profile, arrange a total of 21 voters in such a way that the plurality, pairwise comparison, Borda, and Hare methods all select candidate* d.

<table>
<tr><td>**45.**</td><td>**Number of Voters**</td><td>**Ranking**</td></tr>
<tr><td></td><td>?</td><td>a > b > d > c</td></tr>
<tr><td></td><td>?</td><td>b > c > d > a</td></tr>
<tr><td></td><td>?</td><td>c > d > a > b</td></tr>
<tr><td></td><td>?</td><td>d > a > b > c</td></tr>
<tr><td></td><td>?</td><td>a > d > c > b</td></tr>
</table>

<table>
<tr><td>**46.**</td><td>**Number of Voters**</td><td>**Ranking**</td></tr>
<tr><td></td><td>?</td><td>d > b > a > c</td></tr>
<tr><td></td><td>?</td><td>b > c > a > d</td></tr>
<tr><td></td><td>?</td><td>c > a > d > b</td></tr>
<tr><td></td><td>?</td><td>a > d > b > c</td></tr>
<tr><td></td><td>?</td><td>d > a > c > b</td></tr>
</table>

FOR FURTHER THOUGHT

Approval Method

The *approval method* was introduced in the 1970s. Like the Hare method, it is a variation of the plurality method. It is used to elect the secretary-general of the United Nations. Approval voting is useful for elections involving numerous candidates. The strength of the approval method is that voters do not need to rank the list of candidates completely. Instead, voters decide only which candidates they like and which they dislike.

If three candidates are in an approval method election, then each voter can decide to vote for his or her top-ranked candidate only, or to vote for both of his or her top-ranked candidates. (A vote for all three candidates would be meaningless, as it would fail to distinguish among the candidates.) The familiar profile for the Culinary Arts Club is repeated in the table with an additional column of information necessary for an approval method election simulation.

Approval method results of this profile can be generalized. Beef always receives 6 approval votes from the voters with the top ranking. Beef cannot get any additional votes because it is not ranked second by anyone. Chicken always receives 5 votes from the middle ranking. Chicken also can get up to $z = 4$ additional votes from the voters with the third ranking who rank it second. Pork always receives 4 votes from the third ranking. Pork also can get up to $x = 6$ and $y = 5$ additional votes from the 11 voters with the first and second rankings who rank pork second.

> **APPROVAL METHOD**
>
> In the **approval method** of voting, each voter is allowed to give one vote to as many candidates as he or she wishes. The candidate with the most approval votes wins the election.

<table>
<tr><td>**Number of Voters**</td><td>**Ranking**</td><td>**Number of Voters Voting for Two of the Three Candidates**</td><td></td></tr>
<tr><td>6</td><td>b > p > c</td><td>x, $0 \le x \le 6$</td><td>*x* represents how many vote for both b and p.</td></tr>
<tr><td>5</td><td>c > p > b</td><td>y, $0 \le y \le 5$</td><td>*y* represents how many vote for both c and p.</td></tr>
<tr><td>4</td><td>p > c > b</td><td>z, $0 \le z \le 4$</td><td>*z* represents how many vote for both p and c.</td></tr>
</table>

In summary:

Beef gets 6 approval votes.

Chicken gets $5 + z$ approval votes, $0 \le z \le 4$.

Pork gets $4 + x + y$ approval votes, $0 \le x \le 6$, $0 \le y \le 5$.

The outcome of the election depends on the values of the variables x, y, and z, which, in turn, depend on the individual voters' decisions whether to approve second-place candidates.

For Group or Individual Investigation

1. Determine the winner of this approval election under each of the following assumptions.

   (a) All the voters distrust the new system and vote only for their single top-ranked candidate.

   (b) Three of the six voters with the first ranking decide to approve of both beef and pork. All other voters approve of only their top-ranked candidate.

   (c) Two of the six voters with the first ranking decide to approve of both beef and pork, one of the five voters with the second ranking decides to approve of both chicken and pork, and three of the four voters with the third ranking decide to approve of both pork and chicken.

2. What other voting method is equivalent to the approval method under assumption 1(a) above?

3. Compare approval voting with the other four methods discussed in this section. Do you "approve" of the approval method?

4. The members of a soccer team are electing their captain using the approval method. Candidates include team members Jerica, Lori, Soo, and Alison. The 13 ballots have been marked as shown below. The symbol × indicates an approval vote for the candidate named in the first column.

   (a) Which candidate is selected as team captain by the approval method?

   (b) If the soccer team decides to name the two candidates with the most approval votes as their co-captains, which two candidates will be selected?

Candidate	Voter												
	1	2	3	4	5	6	7	8	9	10	11	12	13
Jerica	×	×	×		×	×	×	×	×				
Lori	×	×		×				×	×		×		×
Soo			×		×		×		×	×	×		
Alison	×		×	×		×						×	

15.2   THE IMPOSSIBILITIES OF VOTING

OBJECTIVES

1 Apply the majority criterion.

2 Apply the Condorcet criterion.

3 Apply the monotonicity criterion.

4 Apply the irrelevant alternatives criterion.

5 Explain Arrow's impossibility theorem.

Majority Criterion

Different voting methods applied to the same voter profile can show remarkable agreement or result in totally different outcomes. None of the voting methods presented in the previous section is perfect. Defects are revealed by considering the following four desirable attributes for any voting method.

1. the *majority criterion*

2. the *Condorcet criterion*

3. the *monotonicity criterion*

4. the *irrelevant alternatives* (IA) *criterion*

The first two criteria concern desirable qualities for a voting method when it is used a single time to determine a winner. The third and fourth criteria concern desirable qualities of a voting method when the method is to be used twice.

When a voter profile includes a candidate with a majority of first-place rankings—that is, a **majority candidate**—it seems reasonable to expect that candidate to be elected. This attribute is called the **majority criterion.**

MAJORITY CRITERION

If a candidate has a majority of first-place rankings in a voter profile, then that candidate should be the winner of the election.

If there are only two candidates, then a majority candidate wins the election no matter which method from the previous section is used. However, **Example 1** shows that a majority candidate in an election involving three or more candidates may not win the election if the Borda method is used.

EXAMPLE 1 Showing That the Borda Method Fails to Satisfy the Majority Criterion

The College Art Association (CAA) has a new executive board. (See **Examples 3, 5, and 8** of the previous section.) Its members are meeting to select a city to host the yearly meeting. The sites under consideration this time are Portland p, Boston b, Chicago c, and Orlando o. The voter profile for the new executive board shows how the 20 members rank the cities. (See **Table 11.**)

Board members agree to use the Borda method to decide the outcome of the vote. Determine the Borda winner.

Table 11

Number of Votes	Ranking	First Place (3 points)	Second Place (2 points)	Third Place (1 point)	Fourth Place (0 points)
8	c > p > b > o	c	p	b	o
4	c > b > o > p	c	b	o	p
4	b > p > o > c	b	p	o	c
4	b > o > p > c	b	o	p	c

Solution

Chicago supporters are delighted to discover they have a *majority* of first-place rankings. They are confident that, with 12 of 20 first-place votes, Chicago will be selected as the meeting site. The Borda tallies are calculated.

Portland earns $(8 \cdot 2) + (4 \cdot 0) + (4 \cdot 2) + (4 \cdot 1) = 28$ points.

Orlando earns $(8 \cdot 0) + (4 \cdot 1) + (4 \cdot 1) + (4 \cdot 2) = 16$ points.

Chicago earns $(8 \cdot 3) + (4 \cdot 3) + (4 \cdot 0) + (4 \cdot 0) = 36$ points.

Boston earns $(8 \cdot 1) + (4 \cdot 2) + (4 \cdot 3) + (4 \cdot 3) = 40$ **points**.   ← Borda winner

The Borda method selects Boston as the site for next year's meeting, even though Chicago had a majority of first-place rankings.

This failure to select the majority candidate is a defect in the Borda method.

No voting method that can result in a violation of the majority criterion can be considered completely satisfactory.

Do not misunderstand what it means when a voting method violates (or fails to satisfy) a criterion. For some voter profiles, if a candidate has a majority of first-place rankings, the Borda method *does* select that candidate; however, it takes only one contradictory profile to show that a voting system violates a criterion. Because the Borda method *does not* select the majority candidate for the CAA voter profile in **Example 1,** the Borda method does not satisfy the majority criterion.

The plurality method, the pairwise comparison method, and the Hare method do satisfy the majority criterion.

The Marquis de Condorcet (1743–1794) was a French aristocrat. His interests included mathematics, social science, economics, and politics. Sadly, Condorcet's political passions at the end of the French Revolution landed him in prison, where he died of questionable causes. It is hard to understand how France's counterpart to America's Thomas Jefferson could have come to such a tragic end.

Whatever the voter profile, if a majority candidate exists, these methods always select it. We can see this by reasoning as follows.

1. *Plurality method:*   Any candidate with a majority of first-place rankings automatically also has a plurality of first-place rankings and will, therefore, always be selected by the plurality method.

2. *Pairwise comparison method:*   Suppose Candidate x has a majority of first-place rankings in a field of n candidates. Every candidate is involved in $n - 1$ comparisons (once with each of the other candidates). Candidate x wins all $n - 1$ comparisons with the other candidates and receives $n - 1$ comparison points. Every other candidate loses the comparison with x and, therefore, receives at most $n - 2$ points. So the majority candidate will always be selected by the pairwise comparison method.

3. *Hare method:*   Any candidate with a majority of first-place rankings will win the election in the first round of voting and will, therefore, always be selected by the Hare method.

Condorcet Criterion

A candidate who can win a pairwise comparison with every other candidate is called a **Condorcet candidate,** in honor of the Marquis de Condorcet. It is not unusual for a voter profile to include a Condorcet candidate, but neither is it commonplace. Another standard of election fairness, the **Condorcet criterion,** calls for a Condorcet candidate, if one exists, to be selected as the choice of the group.

THE CONDORCET CRITERION

If a Condorcet candidate exists for a voter profile, then the Condorcet candidate should be the winner of the election.

Is the Condorcet criterion satisfied by the four main methods of the previous section? It is for any election involving only two candidates. However, it is not always satisfied when there are more than two candidates.

EXAMPLE 2   **Showing That the Plurality Method Fails to Satisfy the Condorcet Criterion**

Show that the plurality method of voting fails to satisfy the Condorcet criterion by applying it to the voter profile of the Culinary Arts Club of the previous section.

Solution

The voter profile, repeated in **Table 12,** shows that pork wins pairwise comparisons with both beef and chicken. The nine voters of the second and third rankings prefer pork to beef, while only the six voters of the first ranking prefer beef to pork. Also, the ten voters of the first and third rankings prefer pork to chicken, while only the five voters of the second ranking prefer chicken to pork.

If the plurality method is used by the club to select the entrée, then beef wins, because it is ranked first by six (a plurality) of the members. Beef is selected in spite of the popularity of pork displayed by the pairwise comparisons.

Because it has failed to select pork, the Condorcet candidate, the plurality method does not satisfy the Condorcet criterion.

Table 12

Number of Voters	Ranking
6	b > p > c
5	c > p > b
4	p > c > b

What about the other three methods of the previous section?

1. *Pairwise comparison method:* A Condorcet candidate, by definition, wins pairwise comparisons with every other candidate, earning $n-1$ pairwise points, while no other candidate can possibly earn more than $n-2$ points.

   The Condorcet candidate is always selected by the pairwise comparison method. (In fact, this method was created to satisfy the Condorcet criterion.)

2. *Borda method:* First observe that a majority candidate (having a majority of first-place rankings) automatically wins all pairwise comparisons with other candidates and is, therefore, also a Condorcet candidate. In **Example 1,** Chicago was a majority candidate, and therefore a Condorcet candidate, but was *not* selected by the Borda method. (Boston was the winner.)

   This counterexample establishes that the Borda method fails to satisfy the Condorcet criterion.

3. *Hare method:* We saw in **Example 2** that pork was a Condorcet candidate. However, in **Example 7** of the previous section, the Hare method selected chicken rather than pork for the same profile.

   The Hare method also fails to satisfy the Condorcet criterion.

While other voter profiles may not violate these criteria, a single counterexample is enough to establish that a given voting method fails to satisfy a criterion.

The majority criterion and the Condorcet criterion are desirable attributes of a voting method when it is used in a single election. The next two criteria involve a voting method's desirable attributes when a second election is held using the same method.

Monotonicity Criterion

Suppose the outcome of a first election is not binding, as in a straw poll. Voters may change their rankings before the next election. Suppose those who rearrange their rankings move the winner of the first election to the top of their rankings. It is reasonable to expect the winner of the first election, who now enjoys even more support, to win the second election also. This is the **monotonicity criterion.**

MONOTONICITY CRITERION

If Candidate x wins an election and, before a second election, the voters who rearrange their rankings move Candidate x to the top of their rankings, then Candidate x should win the second election.

How many of our four voting methods satisfy the monotonicity criterion? Unfortunately, only one of them does.

1. If candidate x has a plurality of votes in the first election and only x can get more first-place votes, then x must have a plurality in the second vote.

   The plurality method satisfies the monotonicity criterion.

2. Rearranging voters are required to move winning candidate x to the top of their rankings, but the criterion does not restrict shifts of the other candidates. This means that both Borda and pairwise comparison point tallies may increase for all candidates, which may prevent the original winner from winning the second election.

   Both the Borda method and the pairwise comparison method fail to satisfy the monotonicity criterion as it is stated.

3. Example 3, involving three rounds in each of two elections, establishes the following result.

The Hare method fails to satisfy the monotonicity criterion.

EXAMPLE 3 **Showing That the Hare Method Fails to Satisfy the Monotonicity Criterion**

An agency review board of 29 members is selecting a lead contracting company for some critical governmental security work. They will choose from a group of four companies: a, b, c, and d. They agree to use the Hare method to determine their selection. After an extensive review of perceived capabilities, past performance, and submitted proposals, the board decides to hold a nonbinding vote to see how they are tending in their judgments. The resulting voter profile is given in **Table 13.** Show that the monotonicity criterion is violated in this case.

Table 13

Number of Voters	Ranking
7	a > b > d > c
8	b > c > a > d
10	c > d > a > b
2	a > c > d > b
2	a > d > b > c

Solution

Round-one results: a gets 7 + 2 + 2 = 11 votes.

b gets 8 votes.

c gets 10 votes.

d gets 0 votes.

No company has a majority (15 or more) of the 29 votes. Company d has the fewest votes and is eliminated.

Round-two results: a gets 11 votes.

b gets 8 votes.

c gets 10 votes.

Again no company has a majority of the votes. Company b has the fewest votes and is eliminated. The eight voters with the second ranking vote for c in the next round, because b is eliminated.

Round-three results: a gets 11 votes.

c gets 10 + 8 = **18 votes.** ← Nonbinding winner

Company c is selected by the preliminary Hare method election.

That evening, supporters of Company c wine and dine the four voters with the two bottom rankings in **Table 13.** They convince these four voters to rearrange their rankings, placing c first.

Table 14 shows the voter profile for the official vote the next morning. The four voters with the bottom rankings have moved c into first place. The other 25 voters do not rearrange their rankings. Supporters of c believe the four additional first-place rankings will work in their favor.

Table 14

Number of Voters	Ranking
7	a > b > d > c
8	b > c > a > d
10	c > d > a > b
2	c > a > d > b
2	c > d > b > a

Round-one results: a gets 7 votes.

b gets 8 votes.

c gets 10 + 2 + 2 = 14 votes.

d gets 0 votes.

No company has a majority. Company d has the fewest votes and is eliminated.

Round-two results: a gets 7 votes.

b gets 8 votes.

c gets 14 votes.

No company has a majority of the votes. Company a has the fewest votes and is eliminated. The seven supporters of a with the top ranking will vote for b in the next round, because a is eliminated.

Round-three results: b gets 8 + 7 = **15 votes**.   ← Official winner

c gets 14 votes.

Company b wins the official Hare method election by a single vote in the final round. Company c loses the official vote, although c has more first-place votes than it had in the preliminary round. Because the winner of the first election, c, did not win the second election, this example violates the monotonicity criterion. Therefore, the Hare method does not satisfy the monotonicity criterion.

The conditions of the monotonicity criterion require the four voters who rearrange their rankings in **Example 3** to move c into first place. The voters also are permitted to reshuffle the other candidates in their rankings. To make the example obvious and dramatic, c was simply exchanged with the top-ranked candidate in the two bottom rankings, and the other candidates were not moved.

Irrelevant Alternatives Criterion

If a candidate who lost a first election drops out before a second vote, it is reasonable to expect the winner of the first election also to win the second. This voting method attribute is known as the **irrelevant alternatives criterion.**

IRRELEVANT ALTERNATIVES (IA) CRITERION

If Candidate x wins a first election, and one (or more) of the losing alternatives drops out before a second vote, the winner x of the first election should win the second election.

Table 15

Number of Voters	Ranking
6	b > p > c
5	c > p > b
4	p > c > b

1. Apply the plurality method to the Culinary Arts Club profile, repeated in **Table 15.** Beef, with six first-place votes, is selected as the entrée for the banquet. If pork is removed from the ballot, then in a second election between beef and chicken, the former pork supporters vote for chicken. Chicken receives a majority and wins the second plurality election. (If chicken is removed from the ballot, then pork wins the second decision between it and beef.)

The plurality method fails to satisfy the irrelevant alternatives (IA) criterion.

2. Now apply the Hare method to the same profile. Chicken is selected in the second round of the first Hare election. If beef, a loser, is then removed from the ballot, the second Hare election is reduced to a simple majority election between pork and chicken. The former beef supporters vote for pork, causing it to win a second Hare election.

The Hare method fails to satisfy the IA criterion.

3. Next, consider the Borda method. When an alternative is removed from a voter's ranking, the candidates formerly ranked below it move up in the ranking. Moving up in the rankings can allow an original Borda loser to gain points and defeat the original winner in a second election. For example, suppose a voter has the ranking

$$p > l > a > c > e.$$

Here, p is the original Borda winner. If l drops out before a second vote, p is still ranked first, a and c move up, and e remains ranked last:

$$p > a > c > e.$$

For every such voter, p earns 4 points before l drops out, but only 3 points after. On the other hand, a earns 2 points both before and after, and c earns 1 point both before and after. Thus the tallies for a and c are maintained, while the tally for p drops. This condition allows different winners in the first and second Borda elections.

The Borda method fails to satisfy the IA criterion.

4. Finally, consider the pairwise comparison method. The outcome of such an election depends on how many individual comparisons each candidate wins. When a candidate is dropped from an election, the pairwise comparisons involving that candidate are no longer considered in the voting process. **Example 4** will establish the following.

The pairwise comparison method also fails to satisfy the IA criterion.

EXAMPLE 4 **Showing That the Pairwise Comparison Method Fails to Satisfy the IA Criterion**

The conductor, the music director, and a small group of trustees are selecting a percussionist to replace their beloved timpani player, who is retiring from the world-famous American Orchestra. Five hopeful percussionists,

$$v, \quad w, \quad x, \quad y, \quad \text{and} \quad z,$$

perform audition pieces. The committee agrees to use the pairwise comparison method for their selection, because it allows direct, one-on-one comparisons of the five musicians. Only the results of the ten possible pairwise comparisons are shown—the actual voter profile is not given.

v ties with w.   v gets $\frac{1}{2}$ point; w gets $\frac{1}{2}$ point.

v beats x.   v gets 1 point.

y beats v.   y gets 1 point.

v beats z.   v gets 1 point.

x beats w.   x gets 1 point.

y beats w.   y gets 1 point.

z ties with w.   z gets $\frac{1}{2}$ point; w gets $\frac{1}{2}$ point.

x beats y.   x gets 1 point.

x beats z.   x gets 1 point.

z beats y.   z gets 1 point.

Before the conductor can call the selected percussionist, he gets an e-mail from percussionist y. Sensing she had not played well enough to win the audition, y accepted an offer to join a small European orchestra. Now that y no longer is a candidate, the conductor and the selection committee agree to vote again on the four remaining candidates (v, w, x, and z), using the pairwise comparison method. Show that the first and second elections yield different winners.

Solution

The pairwise comparison point total for each candidate is gathered from the ten individual results in the first election shown on the previous page.

x gets **3 points**.   ← Winner of first election

v gets $2\frac{1}{2}$ points.

y gets 2 points.

z gets $1\frac{1}{2}$ points.

w gets 1 point.

Percussionist x wins the first election by the pairwise comparison method.

The only change in the voter profile (not shown) for the second election is that the committee members remove y from their rankings for the second vote. Consequently, the results of the second round of comparisons agree with the previous results, but there are no longer any comparisons with percussionist y. Only six pairwise comparisons are needed for four candidates.

v ties with w.   v gets $\frac{1}{2}$ point; w gets $\frac{1}{2}$ point.

v beats x.   v gets 1 point.

v beats z.   v gets 1 point.

x beats w.   x gets 1 point.

z ties with w.   z gets $\frac{1}{2}$ point; w gets $\frac{1}{2}$ point.

x beats z.   x gets 1 point.

The results gathered from the six individual comparisons surprise the voters.

v gets $2\frac{1}{2}$ **points**.   ← Winner of second election

x gets 2 points.

w gets 1 point.

z gets $\frac{1}{2}$ point.

The first election selected percussionist x, but the second election selected percussionist v, violating the IA criterion. Therefore, we see that the pairwise comparison method fails to satisfy the IA criterion.

Arrow's Impossibility Theorem

The previous section introduced four ways to select one alternative from a set of alternatives. However, each of the methods fails to satisfy at least one of the four criteria considered desirable attributes of a selection process. **Table 16** on the next page summarizes which criteria are satisfied and which criteria are not satisfied by these voting methods.

FLORIDA

Palm Beach County

The following quote is from William C. Kimberling, deputy director of the FEC National Clearinghouse on Election Administration.

There have, in its 200 year history, been a number of critics and proposed reforms to the **Electoral College system**—most of them trying to eliminate it. But there are also staunch defenders of the Electoral College who, though perhaps less vocal than its critics, offer very powerful arguments in its favor.

Source: www.uselectionatlas.org

The 2000 presidential election focused on Palm Beach County, where confusion over the "butterfly" ballot may have resulted in enough voters mistakenly diverting votes away from Al Gore to cause Florida's 25 electoral votes to go to George W. Bush, giving him the national victory. The 2016 presidential election was one of several, over the years, that apparently gave one candidate the popular vote and the other candidate the electoral victory.

Still, many would argue that the Electoral College system is an essential part of the representative republic designed by the founding fathers, and that it helps ensure that minority interests have expression through the states.

Table 16 Summary of Desirable Criteria and Which Voting Methods Satisfy Them

Criterion	Voting Method			
	Plurality Method	**Pairwise Comparison Method**	**Borda Method**	**Hare Method**
Majority criterion	satisfied	satisfied	not satisfied (**Example 1**)	satisfied
Condorcet criterion	not satisfied (**Example 2**)	satisfied	not satisfied	not satisfied
Monotonicity criterion	satisfied	not satisfied	not satisfied	not satisfied (**Example 3**)
Irrelevant alternatives criterion	not satisfied	not satisfied (**Example 4**)	not satisfied	not satisfied

Kenneth J. Arrow (1921–2017) was an American economist. In 1972, he was awarded the Nobel Memorial Prize in Economic Sciences for his work on the theory of general economic equilibrium. His famous **impossibility theorem** about voting (1951) resulted from discussions he had while working at the RAND Corporation. Arrow's discovery of unavoidable imperfection ended the first era of voting theory. The focus has shifted to challenging questions such as what can be expected of a voting method and what can be done to help minimize the possibility that a particular paradox may occur.

(*Source:* From *Nobel Lectures, Economics 1969–1980,* Editor Assar Lindbeck, World Scientific Publishing Co., Singapore, 1992.)

Plurality voting considers only top-ranked candidates, so it manages to satisfy the majority criterion and the monotonicity criterion. The Hare method uses a series of plurality elections but satisfies only the majority criterion. The Borda method considers how a voter ranks a complete list of candidates, but although using it is easy, the Borda method fails to satisfy any of the four criteria.

The pairwise comparison method also considers a voter's complete ranking of the candidates and satisfies the majority and Condorcet criteria. This method does not satisfy either of the desired criteria for a second election but it is the only method to satisfy both desired criteria for a single election. ***This should make the pairwise comparison method the ideal choice when only one vote is taken.***

Unfortunately, the pairwise comparison method can be difficult to use if a large number of comparisons are required, as indicated by **Exercises 14–16** of the previous section. Furthermore, as shown by **Exercise 41** of the previous section, the pairwise comparison method may also fail to select a winner.

All of the voting methods presented in the previous section have shortcomings. The four fairness criteria may be reasonable individually, but together they are impossible to satisfy.

Economist Kenneth Arrow set out to invent a better voting system. He discovered his goal was mathematically impossible. A perfect voting system cannot exist. This famous result is known as **Arrow's Impossibility Theorem.** It is said that the theorem took Arrow less than a week to prove and ultimately ended the search for a perfect voting system that had started nearly 200 years earlier.

ARROW'S IMPOSSIBILITY THEOREM

For an election with more than two alternatives, there does not exist and never will exist any voting method that simultaneously satisfies the majority criterion, the Condorcet criterion, the monotonicity criterion, and the irrelevant alternatives criterion.

Arrow proved that for any voting method, there will be some voter profile for which one or more of the criteria are not satisfied. The comparison in **Table 16** illustrates this situation. Arrow's theorem also says that a contradictory profile of voters will exist for any new voting system yet to be invented.

WHEN WILL I **EVER** USE THIS**?**

Many citizens will find themselves on boards, councils, or other deliberative bodies, so it is wise to be familiar with the strengths and weaknesses of the decision-making methods in use—as well as with those of other available options. A good way to understand the workings of various methods is to examine what happens under different scenarios.

For instance, recall that **Example 3** established that the Hare method of voting violates the monotonicity criterion. Some have thought that the Hare method might satisfy monotonicity if that criterion had an additional stipulation:

Any voter moving x, the original winner, to the top of her or his rankings must leave all other candidates in the same relative positions as before.

1. How would the voter profile for the official vote in **Example 3** change under this new stipulation?

2. How many rounds of voting are required to determine a winner?

3. Would the final outcome now change, and if so, how?

Answers:
1. The two voters with the fifth ranking would now have a ranking of $c > a > d > b$.
2. Three rounds are required, just as before.
3. The final result would be the same as before. The Hare method still violates monotonicity, even with the additional stipulation.

15.2 EXERCISES

CONCEPT CHECK *Write* true *or* false *for each statement.*

1. Each of the four criteria for a desirable voting method is satisfied by exactly one of the four main voting methods of the previous section.

2. The monotonicity and irrelevant alternatives criteria apply only when more than one vote is to be taken.

3. The Borda voting method fails to satisfy any of the four criteria in this section.

4. The irrelevant alternatives criterion is satisfied by exactly two of the four voting methods.

Identifying Violations of the Majority Criterion *Answer each question for each voter profile of Exercises 5–8.*

(a) *Which alternative is a majority candidate?*

(b) *Which alternative is the Borda method winner?*

(c) *Does the Borda method violate the majority criterion for the voter profile?*

5.

Number of Voters	Ranking
6	$a > b > c$
3	$b > c > a$
2	$c > b > a$

6.

Number of Voters	Ranking
4	$b > c > a > d$
9	$c > a > d > b$
4	$a > c > b > d$
19	$d > c > b > a$

7.

Number of Voters	Ranking
16	$a > b > c > d > e$
3	$b > c > d > e > a$
5	$c > d > b > e > a$
3	$d > b > c > a > e$
3	$e > c > d > a > b$

8.

Number of Voters	Ranking
16	$b > e > c > d > a$
3	$a > c > b > d > e$
6	$c > a > e > b > d$
2	$d > a > c > e > b$
3	$e > c > a > d > b$

Identifying Violations of the Condorcet Criterion *Answer each question for each voting situation of Exercises 9–14.*

(a) *Which candidate is a Condorcet candidate?*

(b) *Which candidate is selected by the plurality method?*

(c) *Which candidate is selected by the Borda method?*

(d) *Which candidate is selected by the Hare method?*

(e) *Which voting method(s) — plurality, Borda, or Hare — violate(s) the Condorcet criterion for this voter profile?*

9. In **Exercise 17** of the previous section, a 13-member committee is selecting a chairperson. The three candidates are

Amy a, Byron b, and Chieko c.

The committee members ranked the candidates according to the following voter profile.

Number of Voters	Ranking
3	$a > c > b$
4	$c > b > a$
2	$b > a > c$
4	$b > c > a$

10. Repeat **Exercise 9** using the following voter profile (from **Exercise 18** of the previous section).

Number of Voters	Ranking
4	$a > b > c$
2	$b > c > a$
4	$b > a > c$
3	$c > a > b$

11. In **Exercise 21** of the previous section, a senator is holding a workshop. The senator asked the 21 workshop members to rank the following issues in order of importance.

job creation j, education e, health care h, and gun control g.

The workshop members rank the issues according to the following voter profile.

Number of Voters	Ranking
6	$h > j > g > e$
5	$e > g > j > h$
4	$g > j > h > e$
3	$j > h > g > e$
3	$e > j > h > g$

12. Repeat **Exercise 11** using the following voter profile (from **Exercise 22** of the previous section).

Number of Voters	Ranking
3	$e > h > g > j$
6	$h > e > g > j$
5	$j > g > e > h$
4	$g > e > h > j$
3	$j > e > h > g$

13. In **Exercise 23** of the previous section, members of the Adventure Club are selecting an activity for their yearly vacation. Choices are

hiking in the desert h, rock climbing c,

white-water kayaking k, motorcycle touring m,

and sitting by a mountain lake tanning t.

The 55 Adventure Club members ranked the choices according to the following voter profile.

Number of Voters	Ranking
18	$t > m > h > k > c$
12	$c > h > m > k > t$
10	$k > c > h > m > t$
9	$m > k > h > c > t$
4	$h > c > m > k > t$
2	$h > k > m > c > t$

14. Repeat **Exercise 13** using the following voter profile (from **Exercise 24** of the previous section).

Number of Voters	Ranking
18	$t > k > h > m > c$
12	$c > h > k > m > t$
10	$m > c > h > k > t$
9	$k > m > h > c > t$
4	$h > c > k > m > t$
2	$h > m > k > c > t$

Identifying Violations of the Monotonicity Criterion
Exercises 15–17 investigate the possibility of various voting methods violating the monotonicity criterion.

15. A 14-member committee is selecting a site for its next meeting. The choices are

Anaheim a, New Orleans n, Denver d, and Yakima y.

(a) The committee members decide to use the pairwise comparison method to select a site in a nonbinding decision. Prior to any discussion, the 14 members rank the choices according to the following voter profile.

Number of Voters	Ranking
5	a > n > d > y
4	y > d > n > a
3	y > d > a > n
2	n > a > d > y

Show that Anaheim is selected by the pairwise comparison method in the preliminary nonbinding decision.

(b) The 2 committee members with the bottom ranking in the table rearrange their ranking after listening to the discussions. The other 12 committee members stick with their original rankings of the cities. For the official vote, the 14 members rank the choices according to the following voter profile.

Number of Voters	Ranking
5	a > n > d > y
4	y > d > n > a
3	y > d > a > n
2	a > y > n > d

Use the pairwise comparison method to determine the site selection of the committee.

(c) Does the pairwise comparison method violate the monotonicity criterion in this selection process?

16. A 19-member committee is selecting a site for the next meeting. The choices are

Dallas d, Chicago c, Seattle s, and Boston b.

(a) The committee decides to use the Borda method to select a site in a nonbinding decision. Prior to any discussion, the 19 members rank the choices according to the following voter profile.

Number of Voters	Ranking
7	s > b > c > d
5	b > d > c > s
3	d > s > c > b
4	c > s > d > b

Show that the Borda method selects Seattle in the preliminary nonbinding decision.

(b) The 7 committee members with the bottom two rankings rearrange their rankings after listening to the discussions. The other 12 committee members stick with their original rankings of the cities. For the official vote, the 19 members rank the choices according to the following voter profile.

Number of Voters	Ranking
7	s > b > c > d
5	b > d > c > s
3	s > b > d > c
4	s > b > c > d

Use the Borda method to determine the site selection of the committee.

(c) Does the Borda method violate the monotonicity criterion in this selection process?

17. A 17-member committee is selecting a site for the next meeting. The choices are

Dallas d, Chicago c, Anchorage a, and Boston b.

(a) The committee decides to use the Hare method to select a site in a nonbinding decision. The 17 members rank the choices as follows.

Number of Voters	Ranking
6	a > c > b > d
5	b > d > a > c
4	c > d > b > a
2	c > a > b > d

Show that Anchorage is selected by the Hare method in the preliminary nonbinding decision.

(b) The 2 committee members with the bottom ranking rearrange their rankings after listening to the discussions. The other 15 committee members stick with their original rankings of the cities. For the official vote, the 17 members rank the choices according to the following voter profile.

Number of Voters	Ranking
6	a > c > b > d
5	b > d > a > c
4	c > d > b > a
2	a > c > b > d

Use the Hare method to determine the site selection of the committee.

(c) Does the Hare method violate the monotonicity criterion in this selection process?

Identifying Violations of the Irrelevant Alternatives Criterion
Exercises 18–20 investigate the possibility of various voting methods violating the irrelevant alternatives criterion.

18. Thirteen voters ranked three candidates a, b, and c according to the following voter profile.

Number of Voters	Ranking
6	a > b > c
5	b > c > a
2	c > b > a

(a) Show that Candidate a is selected if the plurality method is used to determine the outcome of the election.

(b) If losing Candidate c drops out, which candidate is selected by a second plurality method election?

(c) Does the plurality method violate the irrelevant alternatives criterion in this election process?

19. Four candidates, a, b, c, and d, are ranked by 175 voters according to the following voter profile.

Number of Voters	Ranking
75	a > c > d > b
50	c > a > b > d
30	b > c > d > a
20	d > b > c > a

(a) Show that Candidate a is selected if the plurality method is used to determine the outcome of the election.

(b) If losing Candidate b drops out, which candidate is selected by a second plurality vote?

(c) Does the plurality method violate the irrelevant alternatives criterion in this election process?

20. A subcommittee of the senator's workshop of **Exercises 21** and **22** of the previous section is asked to include the issue of military spending m in its breakout session discussion of the issues job creation j, education e, health care h, and gun control g. The 11 subcommittee members rank the five issues as shown in the following voter profile.

Number of Voters	Ranking
5	e > j > h > g > m
2	m > j > e > h > g
2	h > m > e > g > j
2	g > m > e > j > h

(a) If the pairwise comparison method is used, show that the subcommittee felt the issue of education had the highest priority.

(b) The workshop subcommittee decided to delete the losing issues of health care and gun control from further discussion. Use the pairwise comparison method to determine which of the remaining issues (education, job creation, or military spending) the senator's workshop subcommittee felt had the highest priority.

(c) Does the pairwise comparison method violate the irrelevant alternatives criterion in this selection process?

21. *Irrelevant Alternatives in a Pairwise Comparison Election* The conductor, the music director, and a small group of trustees select a new percussionist using the pairwise comparison method in **Example 4.** When Percussionist y drops out of the audition process, the pairwise comparison method violates the irrelevant alternatives criterion by selecting Percussionist v instead of original winner Percussionist x in a second pairwise comparison. If w, rather than y, had dropped out, would the irrelevant alternatives criterion have been violated in a second pairwise comparison?

22. *Irrelevant Alternatives in a Borda Method Election* Twenty-five voters rank three candidates a, b, and c according to the following voter profile.

Number of Voters	Ranking
13	c > b > a
8	b > a > c
4	b > c > a

(a) Show that Candidate b wins a Borda method election.

(b) Which candidate wins a second Borda method election if losing Candidate a drops out?

(c) Does the Borda method violate the irrelevant alternatives criterion in this election process?

23. *Irrelevant Alternatives in a Hare Method Election* Thirty-four voters rank three candidates a, b, and c according to the following voter profile.

Number of Voters	Ranking
12	a > c > b
10	b > a > c
8	c > b > a
4	c > a > b

(a) Show that Candidate a wins a Hare method election.

(b) If losing Candidate c drops out, which remaining candidate wins a Hare method election?

(c) Does the Hare method violate the irrelevant alternatives criterion in this election process?

24. *Irrelevant Alternatives in a Hare Method Election* For the profile of voters in **Exercise 17(a)**, Anchorage is selected by the Hare method in the preliminary nonbinding decision.

(a) If the losing city, Chicago, withdraws after the nonbinding vote, which city wins the official vote? Use the voter profile in **Exercise 17(a)** with Chicago deleted for the official Hare method election.

(b) Does the Hare method violate the irrelevant alternatives criterion in this selection process?

25. Explain why a violation of the majority criterion is an automatic violation of the Condorcet criterion.

26. *Irrelevant Alternatives in a Pairwise Comparison Election* What is it about the departure for Europe of Percussionist y in **Example 4** that hurts Percussionist x, the original winner, and benefits Percussionist v in the second pairwise comparison method selection?

Constructing Voter Profiles *Exercises 27–35 investigate the nature of voter profiles that cause different voting methods to violate criteria discussed in this section.*

27. Construct a voter profile for 19 voters and 6 candidates that has a majority candidate. Show that the majority candidate must win

5 pairwise comparison points,

but that another candidate can win

4 pairwise points.

(*Hint:* Only two of the 6! possible rankings are needed for the profile.)

28. Construct a voter profile for 40 voters and 4 candidates that shows the Borda method violates the majority criterion. Do this by assigning the remaining 19 voters to the rankings in the given incomplete voter profile in such a way that the majority candidate does not win a Borda method election.

Number of Voters	Ranking
21	d > b > c > a
?	b > c > a > d
?	c > b > d > a
?	a > c > b > d

29. Construct a voter profile for 13 voters and 4 candidates that has a Condorcet candidate who fails to be elected by the Borda method but is selected by both the Hare method and the plurality method.

30. Construct a voter profile for 13 voters and 4 candidates that has a Condorcet candidate that fails to be selected by the Borda method and the plurality method but is selected by the Hare method.

31. Construct a voter profile for 18 voters and 4 candidates that shows the pairwise comparison method violates the monotonicity criterion. The original voter profile for the 18 voters is given.

Number of Voters	Ranking
6	a > x > y > z
5	z > y > x > a
4	z > y > a > x
3	x > a > y > z

(a) Show that Candidate a wins the pairwise comparison method election for the original given profile.

(b) Rearrange the ranking of the 3 voters in the bottom row in such a way that for the altered voter profile, Candidate z wins a second pairwise comparison method election. Candidate a must be moved to first place in the new ranking.

32. Construct a voter profile for 14 voters and 4 candidates that shows the Borda method violates the monotonicity criterion. The original voter profile for the 14 voters is given.

Number of Voters	Ranking
5	m > b > s > c
4	b > s > c > m
3	s > m > c > b
2	c > m > s > b

(a) Show that Candidate m wins the Borda method election for the original given profile.

(b) Rearrange the rankings of the 5 voters in the bottom two rows in such a way that for the altered voter profile, Candidate b wins a second Borda method election. Candidate m must be moved to first place in the new ranking.

33. Delete exactly one of the losing candidates from the given voter profile in such a way that the plurality method violates the irrelevant alternatives criterion in a second election.

Number of Voters	Ranking
15	a > b > c > d
8	b > a > c > d
9	c > b > a > d
6	d > b > a > c

34. Delete two or three of the losing candidates from the voter profile below in such a way that the pairwise comparison method violates the irrelevant alternatives criterion in a second election.

Number of Voters	Ranking
15	a > c > b > d > e
6	e > c > a > b > d
6	b > e > a > d > c
6	d > e > a > c > b

35. Construct a voter profile for 41 voters and 3 candidates a, b, and c in such a way that the Borda method violates the irrelevant alternatives criterion in a second election. Arrange the voter profile so that b wins the first Borda election with the most Borda points, and c has the fewest Borda points. Delete c from the second Borda method vote.

Write a short answer for each of Exercises 36 and 37.

36. Now that you know the pros and cons of the plurality, pairwise comparison, Borda, and Hare methods, which do you think is the best method? Explain.

37. If you could have only two of the four criteria discussed in this section satisfied by a voting method, which would you choose? Why?

38. In **Exercise 41** of the previous section, there was a three-way tie among alternatives a, b, and d, with 2 points for each. Alternative c had no points. Would a pairwise comparison winner appear if c were eliminated and the points were recounted?

15.3 THE POSSIBILITIES OF APPORTIONMENT

OBJECTIVES

1 Investigate the issue of apportionment in the original United States House of Representatives.

2 Apply the Hamilton method.

3 Apply the Jefferson method.

4 Apply the Webster method.

5 Apply the Huntington-Hill method.

Introduction

The story of apportionment began as the dust of the American Revolution settled at the Constitutional Convention in 1787 in Philadelphia. America's founding fathers invented a system of government with three branches—executive, judicial, and legislative. The *Great Compromise,* found in Article I, Sections 2 and 3, created a legislative branch consisting of the Senate and the House of Representatives.

- Each state in the union would be represented by two senators in the Senate so that, within the Senate, all states have equal voting influence.

- The number of representatives a state has in the House of Representatives would be determined by the size of the state's population. Within the House, more populous states have greater voting influence than less populous states.

The allotment of House of Representative seats to the various states is an example of *apportionment.* In general, **apportionment** is a division or partition of identical, indivisible things according to some plan or proportion.

The apportionment methods discussed in this section are the following.

1. Hamilton method **2. Jefferson method**

3. Webster method **4. Huntington-Hill method**

Each of the first three was used for a portion of the first 150 years of U.S. national history and is named after its principal proponent, an important historical leader.

The Huntington-Hill method, used from 1942 to the present, is a bit more involved than the others mathematically. A fifth method, introduced in **Exercises 23–26,** is the **Adams method.** It was proposed by John Quincy Adams, the sixth U.S. president, but never was actually put into practice.

The **founding fathers,** when laying down the constitutional mandates concerning House of Representatives apportionment, were unable to fix the number of seats because of the country's rapidly growing and shifting population and the potential for new states beyond the first fifteen. They did require the following.

1. A census of the population every ten years (beginning in 1790)
2. Reapportionment based on the current census
3. Apportionment of seats to the states based on the states' populations
4. A population of at least 30,000 in any congressional district assigned a seat

The fact that no specific method of satisfying these requirements was designated gave rise to considerable debate and political controversy right from the start.

For example, after the first census in 1790, it was not until 1794 that Congress settled on the Jefferson method of apportionment. The Hamilton method, adopted first by Congress, was rejected by President Washington in the very first presidential veto. (**Examples 2 and 4** apply the Hamilton and Jefferson methods, respectively, to the fifteen states of 1794.)

Table 17 illustrates how apportionment applies to a variety of situations.

Table 17 Applications of Apportionment

Example	Things Being Apportioned	Entities Receiving Apportioned Things	Basis for Apportionment
U.S. House of Representatives	House seats	States	State populations
Parliamentary government	Parliamentary seats	Political parties	Percentage of total votes received by party
Rail transportation	Trains	Rail line	Passenger demand
Bus transportation	Buses	Bus route	Passenger demand
Air transportation	Terminals	Airports	Passenger demand
Colleges and universities	Faculty positions	Departments	Student enrollment
Colleges and universities	Teaching assistants	Professors	Student enrollment

In each application, to relate the situation to legislative apportionment, identify what quantities are analogous to House seats (the things being apportioned), the states (the entities receiving a share of the apportioned things), and the state populations (the basis for the apportionment). At times we may refer generically to "states," "state populations," and "seats," even though other corresponding meanings may apply.

Every apportionment method starts with the two preliminary steps shown next.

PRELIMINARY STEPS FOR APPORTIONMENT

Step 1 Compute the **standard divisor d.**

$$d = \frac{\text{total population}}{\text{total number of seats}}$$

Step 2 Compute the **standard quota Q** for each state.

$$Q = \frac{\text{state's population}}{d}$$

The first step determines the average number of people in the overall population per seat assigned. The second step produces the number of seats to which each state is entitled, based on its population. The standard quotas of Step 2 will always, by definition, sum to the number of seats to be apportioned. And if all the quotas happen to be integers, the two "preliminary" steps complete the apportionment. Nothing further is needed. However, this is extremely rare in practice, so our discussion will assume that some further adjustment is needed to obtain integer quotas that have the correct sum.

It is the mathematical details of this adjustment that distinguish one apportionment method from another.

Hamilton Method

The Hamilton method is sensible and easy to calculate. After the two preliminary steps already mentioned, the Hamilton method employs the following two additional steps.

HAMILTON METHOD

Step 1   Round each state's standard quota Q *down* to the nearest integer. Each state will get at least this many seats but must get at least one.

Step 2   Give any additional seats one at a time (until no seats are left) to the states with the largest fractional parts of their standard quotas.

Alexander Hamilton (1755–1804) was born in the British West Indies and came to New York in 1773. He was General Washington's aide-de-camp in the American Revolutionary War, an author of the influential *Federalist Papers,* and the first secretary of the treasury under President Washington. He is among the least understood of the founding fathers because of his aristocratic views and often is regarded as the patron saint of capitalism. Hamilton distrusted **Aaron Burr,** and the two argued publicly for years before their fateful duel in 1804 that took Hamilton's life.

Keep the following in mind:

- The first step achieves integer quotas and ensures that the total number of seats promised is no greater than the number of seats available.
- The second step is a pragmatic scheme for distributing any seats still available to the states Hamilton considered most deserving.

EXAMPLE 1   **Apportioning Computers to Schools Using the Hamilton Method**

The Highwood School District received a gift of 109 computers from a local manufacturer, who stipulated that the division of the identical machines must be based on the individual enrollments at the schools in the district:

The larger the school's enrollment, the more machines it is entitled to receive.

Naturally, each school must receive an integer number of computers. The district decides to apportion the computers by using the Hamilton method. The district's five schools and their enrollments are shown in the first two columns of **Table 18.** Complete the Hamilton apportionment.

Table 18

School	Enrollment	Standard Quota Q	Rounded-Down Q	Computers Apportioned
Applegate	335	16.14	16	16
Bayshore	456	21.96	21	**22**
Claypool	298	14.35	14	**15**
Delmar	567	27.31	27	27
Edgewater	607	29.24	29	29
Totals	2263	109	107	109

Solution

First complete the preliminary steps.

Step 1   Since the total enrollment for the district is

$$335 + 456 + 298 + 567 + 607 = 2263,$$

the standard divisor is $d = \dfrac{\text{total enrollment}}{\text{number of computers}} = \dfrac{2263}{109} = 20.761.$

The standard divisor represents the number of students there are for each of the 109 computers. There is one computer for every 20.761 students in the district.

Step 2 The standard quotas are given to two decimal places in the third column of **Table 18**. For example, for Applegate School,

$$Q = \frac{\text{Applegate enrollment}}{d} = \frac{335}{20.761} = 16.14.$$

Now complete the additional two steps of the Hamilton method.

Step 1 The standard quotas are rounded down to the nearest integer in the fourth column of the table. Each school will receive at least this number of computers.

Step 2 The total number of promised computers is only 107. Thus, two additional computers must be distributed. The Hamilton method distributes these two computers one at a time to the schools with the largest fractional parts of their standard quotas—hence, to Bayshore School, with fractional part 0.96, and to Claypool School, with fractional part 0.35.

The final apportionment of the 109 computers is shown in the final column of the table. The bold entries for Bayshore and Claypool Schools indicate that these schools received more computers than their rounded-down standard quota Q.

Consider Delmar School in **Example 1**. The fractional part of its standard quota is 0.31, which is not drastically smaller than the fraction (0.35) that gained Claypool School an additional computer. In practice, the fractional parts of the standard quotas for two "states" might need to be calculated to four or five decimal places to decide which is larger. Indeed, a "state" with a quota of $Q = 14.8678$ could get the last additional seat available (for a total of 15 seats), and a "state" with a quota of $Q = 14.8677$, if there was one, would get only its rounded-down quota of 14 seats. The Hamilton method has other, more serious and subtle problems.

EXAMPLE 2 **Applying the Hamilton Method to the 15 States of 1794**

The 1790 census determined that the population of the United States was 3,615,920. The Union had 15 states, and it was decided that the original House would have 105 seats. Use the Hamilton method to apportion the 105 seats.

Solution

Complete the two preliminary steps.

Step 1 $\quad\quad\quad d = \dfrac{\text{total population}}{\text{number of seats}} = \dfrac{3,615,920}{105} = 34,437.33$

Step 2 Virginia had the largest population with 630,560 residents. The standard quota for Virginia is

$$Q = \frac{\text{population of Virginia}}{d} = \frac{630,560}{34,437.33} = 18.310.$$

Using Q to three places is necessary for this example. Populations and standard quotas of the other states are shown in **Table 19** on the next page.

In the **history of apportionment,** as applied to U.S. House of Representatives seats, only four methods have actually been used. The reason for changing from one to another generally has been that the numerical results have seemed unfair from the perspective of one or more states. Apart from 1794, almost all reapportionments have been accomplished two years after the ten-year census. The four methods used, and their years of use, are summarized here.

Jefferson method	1794–1832
Hamilton method	1852–1892
Webster method	1842, and 1902–1932
Huntington-Hill method	1942– present

Table 19

State	Population	Standard Quota Q	Rounded-Down Q	Number of Seats
Virginia	630,560	18.310	18	18
Massachusetts	475,327	13.803	13	**14**
Pennsylvania	432,879	12.570	12	**13**
North Carolina	353,523	10.266	10	10
New York	331,589	9.629	9	**10**
Maryland	278,514	8.088	8	8
Connecticut	236,841	6.877	6	**7**
South Carolina	206,236	5.989	5	**6**
New Jersey	179,570	5.214	5	5
New Hampshire	141,822	4.118	4	4
Vermont	85,533	2.484	2	2
Georgia	70,835	2.057	2	2
Kentucky	68,705	1.995	1	**2**
Rhode Island	68,446	1.988	1	**2**
Delaware	55,540	1.613	1	**2**
Totals	3,615,920	105	97	105

Next, complete the Hamilton method steps.

Step 1 The values of Q are rounded down in the fourth column of the table. The total of this column, the number of seats promised, is 97. Eight more seats must be assigned.

Step 2 The apportionment of the seats by the Hamilton method is in the final column. The bold entries indicate the states that receive more seats than their rounded-down values of Q. They are the eight states with the largest fractional parts of their standard quotas.

Table 20

State	Standard Quota Q
Kentucky	1.**995**
South Carolina	5.**989**
Rhode Island	1.**988**
Connecticut	6.**877**
Massachusetts	13.**803**
New York	9.**629**
Delaware	1.**613**
Pennsylvania	12.**570**

The Hamilton apportionment of the original House would have given Delaware two seats, which is not constitutionally correct because the Constitution specifically requires that each seat represent at least 30,000 people. Delaware's population was 4460 people short of deserving two House seats.

The state of Virginia, where Thomas Jefferson made his home, would have received 18 seats in the first House if the Hamilton method of apportionment had been used. The actual 1794 House apportionment used the Jefferson method shown later in **Example 4.**

The theoretical Hamilton apportionment of the House in **Example 2** illustrates one of the more serious and subtle problems with the method. This is a problem of which President Washington was certainly aware when he vetoed the bill that would have made the Hamilton method the law of the land.

In **Example 2,** the additional eight seats are assigned based on fractional parts of the standard quotas, in the order shown in **Table 20.** It may be argued that these eight additional seats are not awarded based on population when the second step of the Hamilton method is executed.

The state with the next largest fractional part of its standard quota (after the eight states in Table 20) is Vermont, with $Q = 2.\textbf{484}$. The fraction 0.570 for Pennsylvania, the last state to receive an additional seat, is unarguably larger than Vermont's fraction of 0.484. However, simple division shows that the fractional part 0.484 of Vermont's value of Q is

$$\text{about } 19\% \left(\frac{0.484}{2.484} \right) \text{ of its total value of } Q = 2.484.$$

The fractional part 0.570 of Pennsylvania's value of Q is only

$$\text{about } 5\% \left(\frac{0.570}{12.570} \right) \text{ of its total value of } Q = 12.570.$$

Yet it is Pennsylvania, not Vermont, that gets an extra seat when Step 2 of Hamilton's method is used to assign the remaining eight seats.

The original detractors of the Hamilton method argued that this kind of situation showed that the method did not follow the constitutional mandate of apportionment by population. President Washington evidently felt their argument was valid.

Jefferson Method

The Jefferson method eliminates the pragmatic, yet problematic and perhaps unconstitutional, Step 2 of the Hamilton method. It does so by adjusting the standard divisor d to obtain a modified divisor md. This careful preliminary adjustment then means that the resulting rounded-down "modified quotas," mQ, for the various states will automatically sum to exactly the number of seats to be assigned. The Jefferson method is considered a **divisor method,** because it accomplishes the necessary adjustment by modifying the divisor rather than by changing some of the individual quotas after they are calculated.

Thomas Jefferson (1743–1826) is perhaps the most important man in American history. He shaped the very foundation of the country, helping to write both the Declaration of Independence and the Constitution of the United States. Many of the ideas in these important documents are the fruits of Jefferson's brilliant philosophical mind.

Jefferson was secretary of state under President Washington and later was elected president himself. Some of his other accomplishments include devising the decimal monetary system and nearly doubling the land area of the United States with the Louisiana Purchase in 1803.

JEFFERSON METHOD

Step 1 Compute **md,** the **modified divisor.**

Step 2 Compute **mQ,** the **modified quota** for each state.

$$mQ = \frac{\textbf{state's population}}{md}$$

Step 3 Round each state's modified quota **mQ** *down* to the nearest integer.

Step 4 Give each state this integer number of seats.

The preliminary steps have already produced the standard divisor d and the standard quotas Q for all states. Rounding these quotas *down* will (almost certainly) give values that sum to fewer than the number of available seats. (Why?) Step 1 of the Jefferson method is accomplished by decreasing d to obtain a modified divisor md. An optimum value of md (found, perhaps, by trial and error) will then produce, in Step 2, larger modified quotients mQ for the states that will, in turn, round *down* to integers that will automatically sum to the number of available seats. (The reason we *decrease* the value of d to obtain md is that a lesser divisor will give greater quotients.)

The values of md in the following examples were found by trial and error, slowly decreasing the value of the standard divisor d until the rounded values of mQ summed to the exact number of objects to be apportioned.

The U.S. Department of the Treasury issued commemorative quarters for each state starting in 1999. The quarters were phased into circulation in the same order in which the states they represent joined the Union. The next time you find a state quarter in your pocket, reflect on how the addition of that state changed the way the seats in the House of Representatives were apportioned and the mathematics involved in the process.

EXAMPLE 3 Apportioning New Sailboats to Caribbean Resorts

Sea Isle Vacations operates six resorts in the Caribbean. Company executives decide to use the Jefferson method to apportion thirty new sailboats, basing the apportionment on the number of rooms at each resort. The six individual resorts are named after hurricanes that spared them. The number of rooms at each location is in the second column of **Table 21**. The total number of rooms at the six resorts is 2013. Complete the Jefferson apportionment.

Table 21

Resort	Number of Rooms	Standard Quota Q	Rounded-Down Q	Modified Quota mQ	Rounded-Down mQ
Anna	345	5.14	5	5.75	5
Bob	234	3.49	3	3.90	3
Cathy	420	6.26	6	7.00	7
David	330	4.92	4	5.50	5
Ellen	289	4.31	4	4.82	4
Floyd	395	5.89	5	6.58	6
Totals	2013	30.00	27	33.55	30

Solution

First, complete the two preliminary steps.

Step 1
$$d = \frac{\text{total number of rooms}}{\text{number of new boats}} = \frac{2013}{30} = 67.1$$

The value of the standard divisor means that, theoretically, for every 67.1 rooms at a resort, that resort is entitled to one of the new boats.

Step 2 The standard quotas are in the third column of the table. Their rounded-down values, in the fourth column, sum to 27, which is less than 30.

Now complete the Jefferson method steps.

Step 1 The modified divisor md was found by slowly decreasing the value of d. For this example, when $md = 60$ the Jefferson method works. That is, when d is reduced to $md = 60$, the resulting rounded-down modified quotients sum to exactly 30. (In fact, any value of md from about 58.5 to 60 will work.)

Step 2 Using $md = 60$, we obtain the modified quota values in the fifth column. For example, the calculation for Resort Anna is as follows.

$$mQ = \frac{\text{Anna's number of rooms}}{md} = \frac{345}{60} = 5.75$$

Step 3 The modified quotas are rounded down in the final column of **Table 21**. The rounded-down values of mQ for the six resorts sum to exactly 30. This is guaranteed by the choice of $md = 60$.

Step 4 Each resort gets the number of new sailboats in the final column. The Jefferson method, being a divisor method, automatically apportions the exact number of new sailboats available. No boats remain unallocated.

If the Hamilton method is used to apportion the sailboats of **Example 3,** the smallest resort, Bob, receives four boats rather than three, while the largest resort, Cathy, receives six boats rather than seven. This is no coincidence. The Hamilton method always favors smaller "states," while the Jefferson method always favors larger "states." This bias helped Jefferson's large home state of Virginia in 1794.

EXAMPLE 4 Applying the Jefferson Method to the 15 States of 1794

Use the Jefferson method to apportion the 1794 House of Representatives based on the 1790 census. (The total population of the existing 15 states was 3,615,920, and the number of seats had been set at 105. The standard divisor $d = 34,437.33$ was computed in **Example 2.**)

Solution

Step 1 The modified divisor used by Congress for the actual Jefferson method apportionment was $md = 33,000$. Considering the hours of manual calculations it probably required to test each guess of the value of md, it must have been a welcome surprise that when $md = 33,000$, the Jefferson method works.

Step 2 The modified quota for Virginia is $\frac{630,560}{33,000} = 19.108$. The modified quotas for all the states are shown in **Table 22.**

Table 22

State	Population	Modified Quota mQ	Rounded-Down mQ
Virginia	630,560	19.108	19
Massachusetts	475,327	14.404	14
Pennsylvania	432,879	13.118	13
North Carolina	353,523	10.713	10
New York	331,589	10.048	10
Maryland	278,514	8.440	8
Connecticut	236,841	7.177	7
South Carolina	206,236	6.250	6
New Jersey	179,570	5.442	5
New Hampshire	141,822	4.298	4
Vermont	85,533	2.592	2
Georgia	70,835	2.147	2
Kentucky	68,705	2.082	2
Rhode Island	68,446	2.074	2
Delaware	55,540	1.683	1
Totals	3,615,920	109.576	105

Step 3 The rounded-down modified quotas mQ in the final column sum to the required 105 seats because of the choice of the value of md.

Step 4 Each state gets the number of seats shown in the final column of the table, that is, the integer part of its modified quota mQ.

Traditional rounding (*up* to the next higher integer for fractional parts 0.5 or greater, *down* to the next lower integer for fractional parts less than 0.5) may seem preferable to Hamilton's "rounding down" scheme. However, the resulting rounded quotas sometimes would still require adjustments, and some quotas may be too large. States, therefore, could not be *promised* at least the number of seats given by their quotas. Once a set of rounded quotas is announced, it is more appealing psychologically to grant some states additional seats than it is to "take away" seats from other states.

The numbers in the final column represent the way the House of Representatives was actually apportioned in 1794. The largest state, Jefferson's home state of Virginia, received 19 seats in the 1794 apportionment of the House of Representatives. Delaware, the smallest state in the Union, received only one seat when the Jefferson method was used.

When President Washington rejected the Hamilton method on constitutional grounds, supporters of other apportionment methods questioned whether the modified divisor in the Jefferson method was constitutionally sound. Jefferson argued that the Constitution required only that the House apportionment be based on population and not on a standard divisor. Jefferson method supporters convinced skeptics that the modified divisor *md* still produced quotients that reflect state populations and that, because the same divisor is used for all states, the method was on firm constitutional ground.

Webster Method

The Jefferson method was used for decades without incident, until the 1820s. The apportionment of the House based on the 1820 census revealed a major flaw in this method that is discussed later. The flaw might have been dismissed as a freak incident if the same anomaly had not occurred again following the 1830 census. The Jefferson method was replaced by another divisor method proposed and championed by Daniel Webster. No one suspected it was flawed in the same way. The Webster method was used to apportion the House following the 1840 census, then again from 1900 until it was replaced by the Huntington-Hill method in 1941.

The appeal of the Webster method is simple. Standard quotas are not rounded down to the nearest integer. Instead, they are rounded using a traditional rounding scheme. A quota with a fractional part greater than or equal to 0.5 rounds up, and a quota with a fractional part less than 0.5 rounds down.

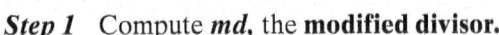

WEBSTER METHOD

Step 1 Compute *md,* the **modified divisor.**

Step 2 Compute *mQ,* the **modified quota** for each state.

$$mQ = \frac{\text{state's population}}{md}$$

Step 3 Round each state's modified quota *mQ up* to the nearest integer if its fractional part is greater than or equal to 0.5 and *down* to the nearest integer if its fractional part is less than 0.5.

Step 4 Give each state this integer number of seats.

The Webster-modified divisor *md* is a bit harder to find than the modified divisor for the Jefferson method because the rounding of *mQ* values can go either way, up or down, in the traditional rounding scheme of the Webster method. The Jefferson method always rounds the values of *mQ* down to the nearest integer. Rounding the values of *mQ* down allows the modified divisor for the Jefferson method to be found by slowly decreasing the value of the standard divisor *d*, to produce slightly greater modified quotas with a greater sum. Traditional rounding of the standard quota values of *Q* can cause their sum to be less than, equal to, or even greater than the actual number of seats to be apportioned.

Daniel Webster (1782–1852) was an important constitutional lawyer, a brilliant speaker, and a folk hero. He served in the House and the Senate and was secretary of state under Presidents Harrison, Tyler, and Fillmore. Webster spent his life keeping a watchful eye on the power of individual states and was an ardent supporter of the national government.

EXAMPLE 5 Apportioning Legislative Seats Using the Webster Method

A small island nation in the Pacific adopted a constitution modeled on the U.S. Constitution. Unlike the United States, the new nation, Timmu, decided to have just one legislative assembly with 131 seats. Like the U.S. House of Representatives, the seats of Timmu's legislature are to be apportioned based on the populations of the nation's individual states. The Timmu Constitution does not specify an apportionment method, but the leadership of the island nation voted to use the Webster method.

A census found that the population of Timmu is currently 47,841. The standard divisor is found as follows.

$$d = \frac{47,841}{131} = 365.20 \quad \text{To five significant figures, which is consistent with the total population figure}$$

The small nation is divided into six states. **Table 23** shows, in its second, third, and fourth columns, the state populations, the values of the standard quotas to three places, and the traditionally rounded values of the standard quotas.

Complete the Webster apportionment to find an appropriate modified divisor and modified quotients, and to allocate exactly 131 seats among the six states.

Solution

Step 1 The sum of the traditionally rounded values of Q is 132, which is greater than the number of seats in the Timmu house. The value of the modified divisor md needed for the Webster method is found by slowly *increasing* the value of the standard divisor $d = 365.20$, which produces smaller quotas with a smaller sum. We find that when $md = 366.00$, the Webster method works.

Step 2 The modified quota for Abo is found by dividing its population by the value of the modified divisor.

$$mQ = \frac{5672}{366.00} = 15.497 \quad \text{To three places}$$

Modified quotas to three places are given in the fifth column of **Table 23**.

Table 23

State	Population	Standard Quota Q	Traditionally Rounded Q	Modified Quota mQ	Traditionally Rounded mQ
Abo	5672	15.531	16	15.497	15
Boa	8008	21.928	22	21.880	22
Cio	2400	6.572	7	6.557	7
Dao	7289	19.959	20	19.915	20
Ekko	4972	13.614	14	13.585	14
Foti	19,500	53.395	53	53.279	53
Totals	47,841		132		131

Step 3 The six values of mQ are traditionally rounded in the sixth column of the table. The value of $md = 366.00$ guarantees that the traditionally rounded values of mQ sum to 131 exactly.

Step 4 Each state's number of seats is shown in the final column of the table.

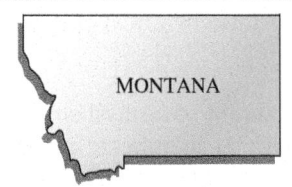

Huntington-Hill Method

Congress currently uses the Huntington-Hill method to apportion the U.S. House of Representatives. This method is also a divisor method. The method uses a modified divisor to produce modified quotas that, when rounded by the unique Huntington-Hill rounding scheme, sum to exactly the number of seats being apportioned.

Executing the Huntington-Hill method of apportionment is very similar to executing the Webster method.

1. First, the standard quotas Q are rounded by the Huntington-Hill rounding scheme and summed. This sum can be less than, equal to, or greater than the number of seats being apportioned.

2. Based on the sum of the rounded values of Q, a modified divisor for the Huntington-Hill method is selected. The modified divisor is used to produce modified quotas that are rounded according to the Huntington-Hill rounding scheme and summed.

3. If the sum of modified quotas is equal to the number of seats being apportioned, then the apportionment is complete and each state receives its rounded integer number of seats.

4. If the sum is not equal to the number of seats being apportioned, the modified divisor is tweaked until the sum is equal to the number of seats being apportioned.

The Huntington-Hill unique rounding scheme uses the *geometric mean* of two numbers.

> The *geometric mean* of two nonnegative numbers a and b is the square root of the product of the two numbers, $\sqrt{a \cdot b}$.

The Huntington-Hill method uses a geometric mean of $a =$ the integer part of a modified quota and $b = a + 1$ as the cutoff point for rounding the modified quota up to the nearest integer, instead of down to the nearest integer.

For example, suppose a state has a modified quota of $mQ = 6.49$. The Huntington-Hill method cutoff point for rounding this modified quota up to 7, the nearest integer, is the geometric mean of 6 and $6 + 1 = 7$. The geometric mean of 6 and 7 is equal to

$$\sqrt{6 \cdot 7} = \sqrt{42} = 6.481.$$

The value $mQ = 6.49$ is greater than 6.481, the geometric mean of 6 and 7, so the modified quota is rounded up to 7. Note that the modified quota $mQ = 6.49$ would not have been rounded up to 7 using a traditional rounding scheme but is rounded up to 7 by the Huntington-Hill rounding scheme.

HUNTINGTON-HILL METHOD

Step 1   Compute *md,* the **modified divisor.**

Step 2   Compute *mQ,* the **modified quota** for each state.

Step 3   Round each state's modified quota *mQ* to the nearest integer using the Huntington-Hill rounding scheme.

Step 4   Give each state this integer number of seats.

| EXAMPLE 6 | Apportioning Council Seats to Clubs Using the Huntington-Hill Method |

Three student clubs are sponsored by the Science Department of a small college: the physics club, with 29 members; the biology club, with 31 members; and the chemistry club, with 22 members. Use the Huntington-Hill method to apportion the 13 available club council seats to these clubs, based on their memberships.

Solution

The given data are shown in the first two columns of **Table 24.**

Table 24

Club	Number of Members	Standard Quota Q	Rounded Q	Modified Quota mQ	Rounded mQ
Physics	29	4.60	5	4.53	5
Biology	31	4.91	5	4.84	5
Chemistry	22	3.49	4	3.44	3
Totals	82		14		13

First complete the preliminary steps, as follows.

Step 1
$$d = \frac{\text{total number of members}}{\text{total number of seats}} = \frac{82}{13} = 6.31$$

Step 2 The standard quotas are in the third column of the table. For example,

$$Q = \frac{\text{physics club members}}{d} = \frac{29}{6.31} = 4.60.$$

Now complete the steps of the Huntington-Hill method.

Step 1 The Huntington-Hill rounding scheme is applied to the standard quotas in the third column to produce the values in the fourth column.

$$\sqrt{4 \cdot 5} = \sqrt{20} = 4.472, \quad \text{and} \quad 4.60 > 4.472 \quad \text{Physics club}$$

The standard quota 4.60 rounds *up* to 5. All three standard quotas round up, and the fourth column sums to 14, which is greater than the number of available seats. To obtain the required, *smaller* sum of 13, we *increase* the divisor d from 6.31 to at least 6.36. Any value from 6.36 to 6.48 will produce the required sum. For example, we can use $md = 6.4$.

Step 2 The modified quotas are shown in the fifth column of the table.

$$mQ = \frac{29}{6.4} = 4.53 \quad \text{Physics club}$$

Step 3 All modified quotas are rounded using the Huntington-Hill rounding scheme in the final column. The rounded mQ values sum exactly to 13.

Step 4 See the final column. The Huntington-Hill method, like any divisor method, automatically apportions exactly the number of seats available.

WHEN WILL I **EVER** USE THIS?

Suppose that you have become the superintendent of Highwood School District just in time to oversee the distribution of the 109 donated computers among your five schools. Your board has suspended the decision to apportion the computers using the Hamilton method because of a distrust of rounding all quotas *down*, and they have also rejected the Jefferson method for the same reason. You realize that the board members will be more comfortable with *traditional* rounding of quotas (that is, up for fractional parts 0.5 or greater, and down for fractional parts less than 0.5). Therefore, you decide to recommend that the Webster method be applied.

The table shows the school enrollments and standard quotas from **Example 1**.

Apportionment of Computers by the Webster Method

School	Enrollment	Standard Quota Q	Traditionally Rounded Q	Modified Quota mQ	Traditionally Rounded mQ
Applegate	335	16.14			
Bayshore	456	21.96			
Claypool	298	14.35			
Delmar	567	27.31			
Edgewater	607	29.24			
Totals	2263	109			

1. Confirm that the standard divisor is $d = 20.761$.
2. Confirm the five standard quota values in the table.
3. Fill in the traditionally rounded quotas in the fourth column, and confirm that they sum to less than 109.
4. Assume you calculate that the Webster method will work for a modified divisor of $md = 20.6$. Use this value to fill in the modified quotas in the fifth column (each to two decimal places).
5. Round the mQ values traditionally to fill in the sixth column, and confirm that this column now sums to 109.
6. Describe the difference between the Webster result and the Hamilton result in the final column of **Table 18**.

Answers:
3. $16 + 22 + 14 + 27 + 29 = 108 < 109$ **4.** 16.26, 22.14, 14.47, 27.52, 29.47
5. $16 + 22 + 14 + 28 + 29 = 109$ **6.** With the Webster method, Claypool loses one and Delmar gains one.

15.3 EXERCISES

CONCEPT CHECK *Write* true *or* false *for each statement.*

1. Among the Hamilton, Jefferson, Webster, and Huntington-Hill methods of apportionment, all are divisor methods except the Hamilton method.

2. Every method discussed in this section begins with computing the standard divisor d and the standard quota Q for each state.

3. If the rounded standard quotas sum to fewer than the number of seats to be apportioned, then the standard divisor d must be adjusted upward.

4. The Hamilton and Jefferson methods both round quotas down, while the Webster and Huntington-Hill methods both round quotas up.

Find each quantity (to the nearest whole number) for the decades of U.S. history designated in Exercises 5 and 6.

(a) *The average number of people represented per seat in the House of Representatives*

(b) *The average number of House seats assigned per state*

5. The 1790s
(The required data are given in **Example 2.**)

6. The 2000s
(A total of 435 seats were apportioned to the fifty states. The 2000 census determined the total population to be 281,424,177.)

7. What was the practical consequence of Washington's veto of the Hamilton method after the 1790 census?

8. Refer to the first four columns of **Table 21** in **Example 3.** Use the Hamilton method to apportion the sailboats.

Solve each problem.

9. **New Trees for Wisconsin Parks** The schoolchildren of Wisconsin began a statewide effort one year to raise money to purchase new trees for five state parks. Altogether, the students raised enough money to purchase 239 trees. A committee decided that the trees should be apportioned based on the amount of land in each of the five parks, as shown.

State park	a	b	c	d	e
Acres	1429	8639	7608	6660	5157

(a) Find the total number of acres in the five state parks, and compute the standard divisor for the apportionment of the 239 trees.

(b) Use the Hamilton method to apportion the trees.

(c) Use the Jefferson method to apportion the trees. As always, the modified divisor needed for the Jefferson apportionment is found by slowly decreasing the value of the standard divisor computed in part (a). Use $md = 122$.

(d) Round the values of all of the parks' standard quotas, using a traditional rounding scheme. Is the sum of the traditionally rounded values less than, equal to, or greater than the number of trees to be apportioned?

(e) Should the value of the modified divisor for the Webster method apportionment be less than, equal to, or greater than the standard divisor computed in part (a)? Why?

(f) Use the Webster method to apportion the trees. Use $md = 124$.

(g) Round the values of all of the parks' standard quotas, using the Huntington-Hill rounding scheme. Is the sum of these rounded values less than, equal to, or greater than the number of trees to be apportioned?

(h) Should the value of the modified divisor for the Huntington-Hill method apportionment be less than, equal to, or greater than the standard divisor computed in part (a)? Why?

(i) Use the Huntington-Hill method to apportion the trees. Use $md = 124$.

(j) Describe any differences among the four apportionments.

10. **Apportioning Computers to Schools** Enrollments for the Highwood School District, the standard divisor, and the Hamilton apportionment of its 109 new computers are in **Example 1.** Use the Jefferson method to apportion the 109 computers the Highwood School District received. Use $md = 20.25$.

11. *Assigning Faculty to Courses* The English department at Oaks College has the faculty to offer 11 sections in any combination of these four courses: Fiction Writing, Poetry, Short Story, and Great Books. The number of sections apportioned to each course is to be based on enrollments in the courses, as shown

Course	Fiction Writing	Poetry	Short Story	Great Books
Enrollment	56	35	78	100

(a) Find the total enrollment for the four courses and the standard divisor for the apportionment of the 11 sections.

(b) Use the Hamilton method to apportion the sections.

(c) Use the Jefferson method to apportion the sections. Use $md = 20$.

(d) Round the values of all of the courses' standard quotas, using a traditional rounding scheme. Is the sum of the traditionally rounded values less than, equal to, or greater than the number of sections to be apportioned?

(e) Should the value of the modified divisor for the Webster method apportionment be less than, equal to, or greater than the standard divisor computed in part (a)? Why?

(f) Use the Webster method to apportion the sections. Use $md = 23$.

(g) Round the values of all of the courses' standard quotas, using the Huntington-Hill rounding scheme. Is the sum of the Huntington-Hill rounded values less than, equal to, or greater than the number of sections to be apportioned?

(h) Should the value of the modified divisor for the Huntington-Hill method apportionment be less than, equal to, or greater than the standard divisor computed in part (a)? Why?

(i) Use the Huntington-Hill method to apportion the sections.

(j) Describe any differences among the four apportionments.

(k) If you were one of the 100 students enrolled in the Great Books course, which method would you hope was used to apportion the sections? Why?

(l) If you were one of the 35 students enrolled in the Poetry course, which method would you hope was *not* used to apportion the sections? Why?

12. *Apportioning Sailboats to Resorts* The number of rooms at each Sea Isle resort and the standard divisor are shown in **Example 3.**

(a) Use the Webster method to apportion the 30 new sailboats. Follow the process in **Exercise 9(d) and (e)** to find *md*.

(b) Use the Huntington-Hill method to apportion the 30 new sailboats. Follow the process in **Exercise 9 (g) and (h)** to find *md*.

(c) The Jefferson apportionment of the sailboats is shown in **Example 3.** The Hamilton apportionment results follow that example (or consult your Hamilton results for **Exercise 8**). Describe any differences among the four apportionments.

13. The information for apportioning the Timmu House of Representatives and the Webster apportionment of the 131 seats are shown in **Example 5.**

(a) Use the Hamilton method to apportion the seats.

(b) Use the Jefferson method to apportion the seats. Use $md = 355$.

(c) Use the Huntington-Hill method to apportion the seats. Use $md = 367$.

(d) Describe any differences among the four apportionments.

14. Show that the Webster method apportionment of the 1794 U.S. House of Representatives agrees with the Hamilton method apportionment in **Example 2.**

15. *Assigning Nurses to Hospitals* Forty nurses have volunteered to help at 5 hospitals treating the victims of an earthquake in a mountainous region of Mexico. The Mexican government decides to apportion the nurses based on the number of beds at each of the hospitals.

Hospital	A	B	C	D	E
Number of beds	137	237	337	455	555

(a) Find the total number of beds at the 5 hospitals and compute, to three decimal places, the standard divisor for the apportionment of the 40 nurses.

(b) Use the Hamilton method to apportion the nurses.

(c) Use the Jefferson method to apportion the nurses. Use $md = 40$.

(d) Round the values of the hospitals' standard quotas, using a traditional rounding scheme. Is the sum less than, equal to, or greater than the number of nurses to be apportioned?

(e) Should the value of the modified divisor for the Webster method apportionment of the nurses be less than, equal to, or greater than the standard divisor computed in part (a)? Why?

(f) Use the Webster method to apportion the nurses. Use $md = 43.1$.

(g) Use the Huntington-Hill method to apportion the nurses. Use $md = 43.3$.

(h) How do the four apportionments compare?

16. *Raising Money for Trees* The governor of Wisconsin postponed the purchase and apportionment of trees and challenged the schoolchildren to raise money for an additional 200 trees. The children met his challenge in only six months. The trees were then apportioned based on the number of acres at the parks, as given in **Exercise 9.**

(a) Find the total number of acres in the parks, and compute the standard divisor for the apportionment of 439 trees.

(b) Use the Hamilton method to apportion the trees.

(c) Use the Jefferson method to apportion the trees.

(d) Round the parks' standard quota values using a traditional rounding scheme. Is the sum of these rounded values less than, equal to, or greater than the number of trees to be apportioned?

(e) Should the value of the modified divisor for the Webster method apportionment be less than, equal to, or greater than the value of the standard divisor computed in part (a)? Why?

(f) Use the Webster method to apportion the trees.

(g) Round the parks' standard quota values using the Huntington-Hill method. Is the sum of these rounded values less than, equal to, or greater than the number of trees to be apportioned?

(h) Should the value of the modified divisor for the Huntington-Hill method apportionment be less than, equal to, or greater than the standard divisor computed in part (a)? Why?

(i) Use the Huntington-Hill method to apportion the trees.

(j) How do the four apportionments of the 439 trees compare? Comment on the political wisdom of Wisconsin's governor.

17. Find the Huntington-Hill method cutoff point for rounding a modified quota of $mQ = 5.470$ up to 6, the nearest integer.

18. Find the Huntington-Hill cutoff point for rounding up $mQ = 56.499$.

19. *Creating a Profile of School Bus Riders* Create a ridership profile for 5 school bus routes that must service 949 students for which the Hamilton, Jefferson, and Webster method apportionments of 16 school buses to the 5 routes all are different.

20. *Creating a Profile of Course Enrollments* Create an enrollment profile for 4 science courses with a total enrollment of 198 students for which all of the following conditions are true: (1) The Hamilton method and Webster method apportionments of 9 teaching assistants to the 4 courses are the same. (2) The Jefferson method apportionment is different. (3) The modified divisor for the Webster apportionment equals the standard divisor.

21. *Creating a Profile of State Populations* Create a population profile for 5 states with a total population of 1000 for which both of the following conditions are true: (1) The Hamilton, Jefferson, and Webster method apportionments of 100 legislative seats all agree. (2) The modified divisors for the Jefferson and Webster apportionments equal the standard divisor.

22. *Designing a Profile to Achieve a Desired Result* Suppose the total population is 1000 and there are 100 seats to be apportioned to 5 states. What simple condition guarantees that the Hamilton, Jefferson, and Webster apportionments are all the same and the modified divisors for the Jefferson and Webster methods equal the standard divisor? Explain why the condition guarantees what it does.

The Adams Method *A fifth apportionment method, considered by Congress but never adopted for use, was proposed by John Quincy Adams, the sixth president of the United States. The* **Adams method** *is another divisor method. It uses a modified divisor selected to produce modified quotas that, when rounded* up *to the nearest integer, produce a sum that is equal to the number of seats being apportioned.*

	Adams Method
Step 1	Compute *md*, the **modified divisor.**
Step 2	Compute *mQ*, the **modified quota** for each state.
Step 3	Round each state's modified quota *mQ* up to the nearest integer.
Step 4	Give each state this integer number of seats.

The standard quotas rounded up to the nearest integer can automatically sum to the number of seats to be apportioned. In that case, the Adams modified divisor md equals d, the standard divisor, and the apportionment is done. Otherwise, the modified divisor needed for an Adams method apportionment is found by slowly increasing the value of the standard divisor.

23. (a) Use the Adams method to apportion the 40 volunteer nurses to the 5 hospitals in Mexico. See **Exercise 15.** The standard divisor, computed in **Exercise 15(a)**, was $d = 43.025$. Use $md = 46$.

(b) How does the Adams apportionment compare to the other four methods applied in **Exercise 15?**

24. Explain why the modified divisor needed for an Adams method apportionment is always found by slowly *increasing* the value of the standard divisor.

25. Having now considered five different apportionment methods, which do you think is best? Give reasons to support your choice.

26. What advantages does the Jefferson method have over the Adams method? Explain.

FOR FURTHER THOUGHT

Redistricting

The U.S. Constitution mandates that Congressional seats be apportioned on the basis of states' populations. Exactly how the individual states determine their representatives, both to Congress and to their own state legislative bodies, was left to the states. This led, from the start, to some politically and racially based maneuverings. The shady political strategy usually was for the party in power (which was drawing the lines) to concentrate the opposition party's voters in one district, or a few districts, and leave most districts dominated by majority voters.

In 1812, Massachusetts governor Elbridge Gerry signed into law a redistricting plan that favored his Republican Party over the Federalists. The caricature in the illustration suggests that the name for the questionable practice, *gerrymandering*, combines "*gerry*" and "sala*mander*."

Over the years, both federal and state laws have sought to restrict such abuses. Here is a typical requirement list.

1. Districts must (or should) be contiguous (in one piece).

2. Districts must utilize existing geographic or political boundaries (city, county, . . .).

3. Districts must (or should) not divide communities of common interest.

4. Districts must (or should) be connected by transportation links.

5. Districts must (or should) not be drawn for the purpose of favoring a political party, incumbent, or other person.

A U.S. Supreme Court action in 2018 sent redistricting challenges by Democratic Wisconsin voters and Republican Maryland voters back to the lower courts, but another challenge from North Carolina was expected to be heard in 2019.

Imagine a small state (pictured below) with four congressional seats (hence, four districts) whose population is concentrated almost entirely in six metropolitan regions within six corresponding counties. Data tracking has determined the voting trends and the populations of the six metro regions A through F and the "other" sparsely populated rural areas (all shown in the table).

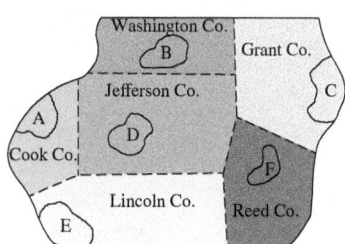

Jurisdiction	Democratic	Republican	Total
A	12,400	61,300	
B	11,100	10,700	
C	14,300	13,300	
D	27,200	25,800	
E	17,100	89,400	
F	12,700	78,900	
Other	600	800	

Totals:

For Group or Individual Investigation

1. Compute the totals at the right and the bottom of the table.

2. Determine (to the nearest percent) the Democrat portion of the state's electorate.

3. Imagine you were on the state's Democrat-appointed redistricting commission, and the requirement list above *did not* include requirement number 5. How could you draw lines for the 4 districts to essentially turn the representation of the people "upside down"?

4. How many of the state's representatives could you almost guarantee would be Democrats? What percentage is that?

5. What happens if the following requirement were added to the list?

   All districts must have approximately equal populations (within 5%).

6. Research the meaning of "compactness" when applied to congressional districts.

7. Who is responsible for drawing and/or approving district lines in your state or jurisdiction.

15.4 THE IMPOSSIBILITIES OF APPORTIONMENT

Introduction

The Hamilton, Jefferson, and Webster methods of apportionment have all been used to apportion the seats in the U.S. House of Representatives.

Over time, why did all three of these methods fall from favor?

In this section we address that question by considering the potential pitfalls of each apportionment method.

Quota Rule

The Jefferson method apportionment of the House of Representatives following the 1820 census caused concern. New York had a standard quota of $Q = 32.503$, which rounds (up) to 33. However, when the seats were apportioned, New York received 34 seats, because the modified divisor used for the Jefferson method gave New York a modified quota mQ that rounded (down) to 34. This caused some alarm (because New York received more seats than its value of Q rounded up to the nearest integer), but not enough to cause rejection of the Jefferson method.

According to the 1830 census, New York had a standard quota of $Q = 38.593$, which, when rounded up to the nearest integer, is 39. The Jefferson apportionment calculated a modified quota mQ for New York that rounded (down) to 40. Thus, two consecutive House apportionments using the Jefferson method had awarded the state of New York more House seats than many people believed was fair.

The Jefferson method was known to naturally favor large states such as New York. The method's partiality for large states alone had not been enough to condemn it. However, the second time the Jefferson method apportioned more seats to a state than the state's value of Q rounded up to the nearest integer, it was rejected and replaced by the Webster method. The Jefferson method failed to pass the basic test of fairness known as the **quota rule.**

Note that an apportionment method satisfies the **quota rule** only if there is no possible circumstance under which the required condition would be violated.

QUOTA RULE

The integer number of objects apportioned to each state must always be either the standard quota Q rounded down to the nearest integer, or the standard quota Q rounded up to the nearest integer.

A **quota method** of apportionment uses the standard divisor d and apportions seats to each state equal to the standard quota Q either rounded up or rounded down to the nearest integer. *The quota rule is satisfied by any quota method of apportionment by definition.*

The Hamilton method is a quota method. The values of Q are rounded down, and each state is guaranteed at least this number of seats. Any additional seats are awarded one at a time to the states with the largest fractional parts of their values of Q. A state cannot receive more than one additional seat in the second step of the Hamilton method because there are never more additional seats than states.

Historical House of Representatives apportionments of the 1820s and 1830s reveal that the Jefferson method does not satisfy the quota rule, because a state can receive more seats than its value of Q rounded up to the nearest integer. A state can also receive fewer seats than its value of Q rounded down to the nearest integer. Replacement of the Jefferson method with the Webster method was ironic, because the Webster method also fails to satisfy the quota rule.

The Jefferson and Webster methods are **divisor methods** of apportionment. They use modified divisors, not the standard divisor.

All divisor methods of apportionment fail to satisfy the quota rule.

EXAMPLE 1   **Checking a Jefferson Apportionment for Quota Rule Violation**

The Nexus Motor Company currently produces 20,638 automobiles per week at six different factories, as shown in **Table 25.** The CEO and the board want to apportion 500 new employees, using the Jefferson method, according to the weekly automobile production at each facility. Legislatively speaking, the newly hired employees are seats, the six factories are states, and the production at each factory is a state population. Complete the apportionment. Is the quota rule satisfied?

Table 25

Factory	Cars Produced per Week
A	1429
B	3000
C	1642
D	9074
E	4382
F	1111
Total	20,638

Solution

The preliminary steps are as follows.

Step 1 $d = \dfrac{\text{total weekly production}}{\text{number of new employees}} = \dfrac{20{,}638}{500} = 41.276$

Step 2   The standard quotas are shown in the third column of **Table 26.**

Table 26

Factory	Cars Produced per Week	Standard Quota Q	Rounded-Down Q	Modified Quota mQ	Rounded-Down mQ
A	1429	34.621	34	34.807	34
B	3000	72.681	72	73.073	73
C	1642	39.781	39	39.995	39
D	9074	**219.837**	**219**	221.021	**221**
E	4382	106.163	106	106.735	106
F	1111	26.916	26	27.061	27
Totals	20,638	500	496		500

In 1842, the **U.S. House of Representatives** actually decreased in size. The apportionment of 1832 had divided 240 seats; the apportionment of 1842 divided only 223 seats. The House has had 435 seats since 1912.

Notwithstanding a direct constitutional mandate, the House was not reapportioned following the 1920 census. Technically, this means that none of the business conducted by the House in the 1920s was completely honorable.

The Huntington-Hill method, adopted in 1941 and currently used to apportion the House, is named for **Joseph Hill** and **Edward V. Huntington.** Hill was chief statistician of the Bureau of the Census. He proposed the method in 1911. Huntington, a professor of mathematics at Harvard, gave it his endorsement.

The method was studied by the National Academy of Sciences. The Academy concluded in 1929 that the Huntington-Hill method was more neutral than the other method under consideration in its treatment of both large and small states. In 1980, Balinski and Young (profiled later in this section) showed that the Academy's conclusion was false. The method does have a slight bias for small states, because of its liberal rounding scheme. Nonetheless, Huntington-Hill remains the law.

Next, complete the Jefferson steps.

Step 1 The rounded-down Q values (in the fourth column) sum to 496, which is less than 500. The value of the modified divisor md is found by trial and error, slowly decreasing the value of the standard divisor d. Here, $md = 41.055$ works in the Jefferson method.

Step 2 The modified quotas for the six factories appear in the fifth column of the table. (The mQ values are calculated by dividing the weekly production of each factory by the modified divisor md.)

Step 3 The rounded-down modified quotas are shown in the final column of the table. They sum to exactly 500.

Step 4 Each factory gets the number of new employees shown in the final column of the table.

The Jefferson method violates the quota rule in this apportionment, and therefore, the Jefferson method does *not* satisfy the quota rule. The standard quota of the largest factory, Factory D, is

$$Q = 219.837.$$

The quota rule states that the number of employees apportioned to Factory D must be either 219 or 220. The apportionment of 221 new employees to Factory D violates the quota rule.

The large weekly production of Factory D in **Example 1** is naturally favored by Jefferson's method. Decreasing the value of the standard divisor to find the value of md allows the modified quota of the large production factory to grow more rapidly than the modified quotas of the other factories. The same kind of paradoxical apportionment is also possible when the Webster method is used. It is this possibility of a quota rule violation that taints both the Jefferson and Webster methods.

Alabama Paradox

The required tweaking of the standard divisor in any divisor method is an open invitation for quota violations, especially if one state's population is large (or small) in relation to other state populations. The Hamilton method and all quota methods automatically satisfy the quota rule by their definition. The Hamilton method would be the method of choice if the quota rule were the sole test of fairness for an apportionment method, but it is not the only such benchmark.

As the population of the United States grew and additional states joined the union, the size of the U.S. House of Representatives slowly increased from the original 105 seats to the current 435 seats. Preparations for increasing the size of the House following the 1880 census revealed a surprising flaw in the Hamilton method. Government workers calculated the different Hamilton method apportionments for all the different House sizes from 275 to 350 seats.

The sample apportionments showed that if the House had 299 seats, then Alabama would receive 8 seats. However, if the House size were 300, then Alabama would receive only 7 seats. This odd result is known as the *Alabama paradox*.

ALABAMA PARADOX

The situation in which an increase in the number of objects being apportioned actually forces a state to lose one of those objects is known as the **Alabama paradox.**

Table 27

School	Enrollment
Applegate	335
Bayshore	456
Claypool	298
Delmar	567
Edgewater	607
Total	**2263**

Hayes or Tilden? Congress decided to divide 283 seats following the 1870 census. They selected this number of seats because with it, the official apportionment done by the Hamilton method agreed with the theoretical outcome of the Webster method. A power grab on the floor of the House resulted in the apportionment of 292 seats, without using the lawful Hamilton method.

It was this unconstitutional House that gave **Rutherford B. Hayes** enough electoral votes to become president. If the Hamilton method had been used, Samuel J. Tilden, who had won a majority of the national votes, would have had enough electoral votes to become president.

EXAMPLE 2 Checking a Hamilton Apportionment for Occurrence of the Alabama Paradox

A longtime resident of Highwood has given one additional computer to the school district. The gift brings the total number of computers to be apportioned by the Hamilton method to 110. School enrollments are repeated in **Table 27.** Complete the apportionment, and observe whether the Alabama paradox has occurred.

Solution

Total enrollment in the school district is 2263, and now 110 computers are to be apportioned. The preliminary steps are as follows.

Step 1
$$d = \frac{\text{total enrollment}}{\text{number of computers}} = \frac{2263}{110} = 20.573$$

Step 2 Standard quotas for each school are computed by dividing the enrollment for the school by d, as shown below for Applegate School. Standard quotas for all five schools are shown in the third column of **Table 28.**

$$Q = \frac{\text{Applegate enrollment}}{d} = \frac{335}{20.573} = 16.28$$

Table 28

School	Enrollment	Standard Quota Q	Rounded-Down Q	110 Computers Apportioned	109 Computers Apportioned
Applegate	335	16.28	16	16	16
Bayshore	456	22.16	22	22	22
Claypool	298	14.49	14	14	15
Delmar	567	27.56	27	28	27
Edgewater	607	29.50	29	30	29
Totals	2263	110	108	110	109

Next, complete the Hamilton method steps.

Step 1 The standard quotas are rounded down to the nearest integer in the fourth column. Each school receives at least this integer number of computers. The values of Q sum to 108, so two additional computers are to be apportioned in the second step of Hamilton's method.

Step 2 The two additional computers are apportioned one at a time to the two schools with the largest fractional parts of their standard quotas. Delmar School gets the first additional computer, and Edgewater School gets the other. The final apportionment of the 110 computers is shown in the fifth column of the table. For comparison, the last column of the table shows the previous Hamilton apportionment of 109 computers from **Example 1** of the previous section.

Comparing the last two columns of the table reveals the following:

The gift of the additional computer has caused the Alabama paradox to occur.

With the Hamilton method, Claypool School receives 15 computers when 109 computers are apportioned and only 14 when 110 are apportioned. The increase in the number of computers being apportioned actually forces Claypool School to forfeit one of its machines.

The paradoxical apportionment in **Example 2** is caused by the pragmatic second step in the Hamilton method. Additional seats are distributed to the states with the greatest fractional parts of their standard quotas. And if the number of items to be apportioned changes, these fractional parts can shift erratically, possibly causing strange phenomena such as the Alabama paradox.

Population Paradox

The quota rule is an important measure of fairness for an apportionment method. The Hamilton method always satisfies the quota rule, even if it does allow the possibility of the Alabama paradox. The Alabama paradox can occur only when the number of objects being apportioned *increases*. If the number of objects being apportioned is permanently *fixed,* then the Hamilton method appears to shine again.

The size of the U.S. House of Representatives has been fixed at 435 seats for many years, so the following questions naturally arise.

1. Why are the seats still apportioned with a divisor method that invites quota rule violations?

2. What keeps the Hamilton method from really shining as the apportionment method of choice?

The Hamilton method allows for paradoxical outcomes other than the Alabama paradox. One of them, discovered around 1900, is known as the *population paradox*.

POPULATION PARADOX

The **population paradox** occurs when, based on updated population figures, a reapportionment of a fixed number of seats causes a state to lose a seat to another state. This occurs even though the percent increase in the population of the state that loses the seat is greater than the percent increase in the population of the state that gains the seat.

Table 29

School	Number of Interviews
Lincoln	55
St. Francis	125
Marshall	190
Total	370

Table 30

School	Revised Number of Interviews
Lincoln	60
St. Francis	145
Marshall	210
Total	415

EXAMPLE 3   **Checking a Hamilton Apportionment for Occurrence of the Population Paradox**

Eleven students in a graduate school education class plan to interview 370 educators at three different elementary schools, as shown in **Table 29.** They are interested in how art and music education impact scores on standardized mathematics exams. The students decide to apportion themselves into three interview teams using the Hamilton method, based on the number of educators to be interviewed at each school. In legislative lingo, the interviewers are comparable to House seats; they are what is being apportioned. The three schools are states, and the number of interviews to be conducted at each school is a state population.

A month before the interviews are to take place, the graduate students decide to include additional interviews with retired educators from the three schools. Five of the retired educators are from Lincoln, 20 are from St. Francis, and 20 are from Marshall. The interviewers must now talk with

$$370 + 45 = 415 \text{ people.}$$

They will now reapportion themselves using the Hamilton method, based on the revised numbers shown in **Table 30.**

Complete both apportionments, and observe whether the population paradox has occurred.

Solution

The initial Hamilton apportionment proceeds as follows, with the two preliminary steps done first.

Step 1 Eleven graduate students are to interview 370 educators. The standard divisor is as follows.

$$d = \frac{\text{total number of interviews}}{\text{number of interviewers}} = \frac{370}{11} = 33.64$$

Step 2 Standard quotas are computed by dividing the number of interviews at each school by $d = 33.64$. The values of Q are shown in the third column of **Table 31.**

Table 31 Hamilton Apportionment for a Total Population of 370

School	Number of Interviews	Standard Quota Q	Rounded-Down Q	Interviewers Apportioned
Lincoln	55	1.635	1	1
St. Francis	125	3.716	3	**4**
Marshall	190	5.648	5	**6**
Totals	370	11	9	11

Now complete the Hamilton steps.

Step 1 The values of Q are rounded down in the fourth column. At least this many graduate students will conduct the interviews at each school. The sum of the rounded Q values is 9, so two additional interviewers are to be assigned in the next step of the Hamilton apportionment.

Step 2 The two schools with the largest fractional parts of their values of Q are St. Francis (0.716) and Marshall (0.648). The final apportionment of the interviewers appears in the last column of the table. Bold numbers in the last column indicate that St. Francis and Marshall get more interviewers than their value of Q rounded down.

With additional interviews, the Hamilton reapportionment proceeds as follows. Again, the preliminary steps are done first.

Step 1
$$d = \frac{415}{11} = 37.73$$

Step 2 The values of Q are recalculated by dividing the new number of interviews at each school by the new standard divisor $d = 37.73$. See the third column of **Table 32.**

Table 32 Hamilton Apportionment for a Total Population of 415

School	Number of Interviews	Standard Quota Q	Rounded-Down Q	Interviewers Apportioned
Lincoln	60	1.59	1	**2**
St. Francis	145	3.84	3	**4**
Marshall	210	5.57	5	5
Totals	415	11	9	11

Now complete the Hamilton steps.

Step 1 The values of Q are rounded down in the fourth column. They sum to 9, leaving two additional interviewers to be apportioned in the final step.

Step 2 The two schools with the largest fractional parts of their Q values are St. Francis (0.84) and Lincoln (0.59). Bold entries in the last column indicate these two schools get more interviewers than their values of Q rounded down. The revised apportionment of the interviewers is in the last column.

The original Hamilton apportionment gives one additional interviewer to each of the St. Francis and Marshall interview teams. An additional interviewer still is assigned to the St. Francis team by the revised apportionment, but the second additional interviewer is assigned to the Lincoln interview team, rather than to the Marshall team.

Comparing the growth rates of the number of interviews at Lincoln and Marshall reveals the following:

The team of graduate student interviewers has suffered the population paradox.

The addition of the retired educators to the interview list causes the number of interviews at Marshall to increase by

$$\frac{210 - 190}{190} = 10.53\%$$

and the number of interviews at Lincoln to increase by

$$\frac{60 - 55}{55} = 9.09\%.$$

The percent increase in the number of interviews to be conducted at Marshall is greater than the percent increase in the number of interviews at Lincoln, yet in the final Hamilton method apportionment, the Marshall interview team loses a member to the Lincoln interview team.

New States Paradox

The Hamilton method of apportionment allows the population paradox for the same reason it allows the Alabama paradox. Both paradoxes are triggered by the pragmatic second step of the method, which awards the unapportioned seats based on the fractional parts of the values of Q, the standard quotas. Paradoxes occur because the changes in the fractional parts of Q, due to an increased number of seats or revised populations, are erratic. If the surplus seats are distributed based on erratic values, then unsettling results are bound to occur.

Another paradox triggered by the pragmatic second step of the Hamilton method is called the *new states paradox*. The new states paradox was discovered by Balinski and Young, discussed later in this section.

The **new states paradox** occurred in 1907, when Oklahoma joined the Union as the 46th state. Remember this fact when you find a commemorative quarter for Oklahoma (issued in 2008) among the coins in your pocket.

NEW STATES PARADOX

The **new states paradox** occurs when a reapportionment of an increased number of seats, necessary because of the addition of a new state, causes a shift in the apportionment of the original states.

EXAMPLE 4 Checking a Hamilton Apportionment for Occurrence of the New States Paradox

Table 33

Community	Population
Original	8500
Annexed	1671
Total	10,171

Several years ago the town of Red Bluff annexed a small unincorporated subdivision. The population of the original town was 8500 and the annexed subdivision had a population of 1671. (See **Table 33**.) The new town was governed by 100 council members. The Hamilton method was used to apportion council seats to the new town.

During the following years, council members from the annexed subdivision lobbied intensely and tirelessly. They finally convinced the town to annex a second small unincorporated subdivision with a population of 545. Populations of the original town and the first annexed subdivision did not change. The town council voted to increase the number of council seats from 100 to 105. They reasoned that if the established standard divisor of

$$d = \frac{\text{total population}}{\text{number of seats}} = \frac{10{,}171}{100} = 101.71$$

was used, then a population of 545 would have a standard quota of

$$Q = \frac{545}{101.71} = 5.36, \quad \text{which rounds down to 5.}$$

Thus 5 was the logical number of new seats to add.

The new situation is reflected in **Table 34**. Complete the Hamilton apportionment after the first annexation and then again after the second annexation. Observe whether the new states paradox has occurred.

Table 34

Community	Population
Original	8500
1st Annexed	1671
2nd Annexed	545
Total	10,716

Solution

The apportionment following the first annexation proceeds as follows with the preliminary steps done first. The total population of the new community was

$$8500 + 1671 = 10{,}171,$$

and 100 council seats were available. This means the standard divisor is as already shown above.

Step 1 $\qquad\qquad\qquad\qquad d = 101.71$

Step 2 The standard quotas for the original and annexed populations are shown in the third column of **Table 35**.

Table 35

Community	Population	Standard Quota Q	Rounded-Down Q	Council Seats Apportioned
Original	8500	83.57	83	84
Annexed	1671	16.43	16	16
Totals	10,171	100	99	100

Now complete the Hamilton steps.

Step 1 The rounded-down values of Q, in the fourth column, sum to 99, leaving one seat to be apportioned.

Step 2 The additional council seat is awarded to the original town. Its fractional part of Q is 0.57, which is greater than 0.43, the annexed population's fractional part of Q. The final apportionment appears in the final column.

An updated Hamilton method apportionment of the 105 council seats shows why the first annexed subdivision lobbied so intensely for the annexation of another subdivision. The updated apportionment was done with a revised value of the standard divisor *d,* reflecting the increased total population of Red Bluff and the increased number of council seats.

The updated total population was $8500 + 1671 + 545 = 10{,}716.$

Step 1
$$d = \frac{10{,}716}{105} = 102.057$$

Step 2 The updated standard quotas are in the third column of **Table 36.**

Now complete the Hamilton steps.

Step 1 The rounded-down values of *Q,* in the fourth column, sum to 104, leaving one additional seat to be apportioned.

Step 2 The fractional part of *Q* for the first annexed subdivision is 0.37, which is greater than the fractional parts of the values of *Q* for the other two populations. It receives the additional available seat. The reapportionment of the 105 council seats is in the final column of the table.

Table 36

Community	Population	Standard Quota Q	Rounded-Down Q	Council Seats Apportioned
Original	8500	83.29	83	83
1st Annexed	1671	16.37	16	17
2nd Annexed	545	5.34	5	5
Totals	10,716	105	104	105

Comparing the two apportionments shows the following:

> ***Annexing a second subdivision with a population of 545 and adding new seats to the council causes the new states paradox to occur.***

The first apportionment splits 100 council seats, giving 84 seats to the original town and 16 seats to the first annexed subdivision. The second apportionment again splits 100 seats between the original town and the first annexed subdivision, but the apportionment shifts, giving only 83 seats to the original town and 17 seats to the first annexed subdivision.

Balinski and Young Impossibility Theorem

The Jefferson, Webster, Huntington-Hill, and Adams methods are all examples of divisor methods. They are subject to the possibility of a quota rule violation, because they depend on a modified standard divisor–not the standard divisor. Modifying the standard divisor forces the modified quotas, rounded according to the prescribed scheme, to sum to exactly the number of seats being apportioned.

The Hamilton method is an example of a quota method. This method uses the standard divisor, so it avoids quota rule violations. However, it allows puzzling apportionments. The final step of the Hamilton method permits the possibility of the Alabama, population, or new states paradox. In other words, increases in the number of seats to be apportioned, changes in population figures, or increases in the number of states can cause the Hamilton method to go awry.

Controversial presidential elections have occurred a number of times in U.S. history. A recent case, emphasizing the significance of apportionment methods, was the 2000 face-off between George W. Bush and Albert Gore. Bush defeated Gore by 271 to 266 electoral votes. If the Jefferson method (of apportionment) had been used by Congress following the 1990 census, Gore would have won with 271 electoral votes. And if the Hamilton method had been used, the Electoral College vote would have been tied at 269 each, invoking the Twelfth Amendment requirement that the House of Representatives elect the president if no candidate receives a majority of the electoral vote. That duty has fallen to the House only once (in 1824) since the Twelfth Amendment was ratified in 1804.

An unexpected 1980 discovery by Michel Balinski and Peyton Young is hinted at in **Table 37,** which is a summary of fairness standards relative to the five methods we have discussed.

Table 37   Summary of Fairness Standards and Which Apportionment Methods Meet Them

Method	Quota Rule	Fairness Standard		
		Alabama Paradox	Population Paradox	New States Paradox
Hamilton	Satisfied	Can occur	Can occur	Can occur
Jefferson	Not satisfied	Cannot occur	Cannot occur	Cannot occur
Webster	Not satisfied	Cannot occur	Cannot occur	Cannot occur
Huntington-Hill	Not satisfied	Cannot occur	Cannot occur	Cannot occur
Adams	Not satisfied	Cannot occur	Cannot occur	Cannot occur

H. Peyton Young, Senior Fellow at the Brookings Institution (Washington DC), in 1982 coauthored the book *Fair Representation: Meeting the Ideal of One Man, One Vote* with **Michael L Balinski**, a mathematician at CNRS and École Polytechnique (Paris). In addition to framing their important impossibility theorem and deriving key insights into the Huntington-Hill method, they maintained that the Webster method is the only divisor method of apportionment that is not biased to either large or small states.

Balinski and Young themselves devised an ingenious "quota method" that also avoids the Alabama paradox, but they rejected its use, partly because it still allows the population paradox to occur.

Recall that Kenneth Arrow was in search of a flawless voting method when, instead, he discovered that no such method can possibly exist. Much the same thing occurred in the theory of apportionment. Scholars and politicians had always expected that mathematicians would eventually find the perfect apportionment method. Instead, mathematicians Balinski and Young showed that a perfectly satisfactory method is impossible.

BALINSKI AND YOUNG IMPOSSIBILITY THEOREM

Any apportionment method that satisfies the quota rule will, by its nature, permit the possibility of a paradoxical apportionment. Likewise, any apportionment method that does not permit the possibility of a paradoxical apportionment will, by its nature, fail to satisfy the quota rule.

Like Kenneth Arrow's work, the **Balinski and Young Impossibility Theorem** answered an age-old question of great social importance. Unfortunately, even very talented mathematicians cannot find solutions that do not exist. Fortunately, they can tell the world when perfection does not exist and can begin developing ways of coping with unavoidable imperfection.

EXERCISES

CONCEPT CHECK *Write* true *or* false *for each statement.*

1. Of the five apportionment methods discussed in this chapter, only the Hamilton method satisfies the quota rule.

2. Of the four fairness standards discussed in this section, only the quota rule is satisfied by the Hamilton method.

3. The Balinski and Young impossibility theorem shows that the U.S. representative system is inferior to other forms of government.

4. An apportionment method may someday be devised that will satisfy all four fairness standards.

Quota Rule Violations with the Jefferson Method   In each case, show that if the Jefferson method of apportionment is used, then a violation of the quota rule occurs.

5. 132 seats are apportioned. Use *md* = 595.

State	a	b	c	d
Population	17,179	7500	49,400	5824

6. 290 seats are apportioned. Use *md* = 48.4.

State	a	b	c	d	e
Population	2567	1500	8045	950	1099

Alabama Paradox with the Hamilton Method *In each case, show that if the Hamilton method of apportionment is used, then the Alabama paradox occurs.*

7. Seats increase from 204 to 205.

State	a	b	c	d
Population	3462	7470	4265	5300

8. Seats increase from 126 to 127.

State	a	b	c	d	e
Population	263	808	931	781	676

Population Paradox with the Hamilton Method *In each case, show that the population paradox occurs when the Hamilton method is used for both initial and revised populations. (Populations are given in thousands.)*

9. 11 seats are apportioned.

State	a	b	c
Initial population	55	125	190
Revised population	61	148	215

10. 13 seats are apportioned.

State	a	b	c
Initial population	930	738	415
Revised population	975	750	421

New States Paradox with the Hamilton Method *In each case, use the Hamilton method to apportion legislative seats to two states with the given populations. Show that the new states paradox occurs if a new state with the indicated population and the appropriate number of additional seats is included in a second Hamilton method apportionment. The appropriate number of seats to add for the second apportionment is the state's standard quota of the new state, computed using the original two-state standard divisor, rounded down to the nearest integer. The populations are given in hundreds.*

11. 75 seats are apportioned originally.

State	Original State a	Original State b	New State c
Population	3184	8475	330

12. 83 seats are apportioned originally.

State	Original State a	Original State b	New State c
Population	7500	9560	1500

Violations of the Quota Rule? *For each apportionment situation, determine whether the quota rule is violated by* **(a)** *the Huntington-Hill method and* **(b)** *the Adams method (explained before* ***Exercise 23*** *of the previous section). Whenever a violation occurs, note whether the method apportions a state too many or too few seats.*

13. 100 seats are apportioned. The Huntington-Hill method works with the standard divisor $d = 1000$. The Adams method works with $md = 1025$.

State	a	b	c	d
Population	86,875	4215	5495	3415

14. 100 seats are apportioned. The Huntington-Hill method works with $md = 990$. The Adams method works with $md = 1025$.

State	a	b	c	d
Population	86,915	4325	5400	3360

15. 220 seats are apportioned. The Huntington-Hill method works with $md = 204$. The Adams method works with $md = 208$.

State	a	b	c	d	e
Population	1720	3363	6960	24,223	8800

16. 219 seats are apportioned. The Huntington-Hill method works with the standard divisor $d = 205.781$. The Adams method works with $md = 209$.

State	a	b	c	d	e
Population	1720	3363	6960	24,223	8800

17. ***Playing Politics with Apportionment*** In **Example 4,** council members from the first subdivision annexed to the town of Red Bluff lobbied for the annexation of a second unincorporated subdivision with a population of 545 because they realized this would cause the new states paradox to occur. The first annexed subdivision took one of the original 100 council seats away from the original town when 5 new seats were included in an updated Hamilton method apportionment. If the population of the second subdivision had been only 531, why would it have been prudent for the first subdivision to wait until a baby was born and the population increased to 532?

18. Consider the situation described in **Exercise 17.** If, by strange coincidence, the population of the second subdivision had been 548, why would it have been urgent for the first subdivision to act before a baby was born and the population increased to 549?

19. Knowing the pros and cons of the Hamilton, Jefferson, Webster, Huntington-Hill, and Adams methods of apportionment, concerning possible violations of the fairness standards, which method do you think is the best? Explain.

20. Which outcome related to the impossibility of apportionment—a quota rule violation, the Alabama paradox, the population paradox, or the new states paradox—do you find most disturbing? Explain.

21. Refer to **Exercise 20.** Which outcome related to the impossibility of apportionment do you find least disturbing? Explain.

22. The Jefferson and Adams methods are both subject to quota rule violations. But when violations do occur, one of these methods always apportions too many seats, and the other always apportions too few seats. Which is which?

23. Describe your impression of the challenges of apportionment in a representative republic. Is there a simple solution for these challenges?

24. Were you aware, before studying this chapter, of the mathematics involved in apportionment?

Chapter 15 SUMMARY

KEY TERMS

15.1
majority
plurality
ranking
n factorial
profile of voters
plurality method
pairwise comparison
  method
Borda method
Hare method

15.2
majority candidate
majority criterion
Condorcet candidate
Condorcet criterion
monotonicity
  criterion
irrelevant alternatives
  criterion
Arrow's Impossibility
  Theorem

15.3
apportionment
Hamilton method
Jefferson method
Webster method
Huntington-Hill method
Adams method
standard divisor
standard quota
divisor method
modified divisor

modified quota
geometric mean

15.4
quota rule
quota method
divisor method
Alabama paradox
population paradox
new states paradox
Balinski and Young
  Impossibility Theorem

NEW SYMBOLS

a > b > c	voter's ranking	***d***	standard divisor
Q	standard quota	***md***	modified divisor
mQ	modified quota		

TEST YOUR WORD POWER

See how well you have learned the vocabulary in this chapter.

1. A **profile of voters** is
   A. a demographic summary of the major constituency groups in an election.
   B. a list of eligibility requirements for voting in an election.
   C. a table showing how all voters ranked all the alternatives in an election.
   D. a table showing how all voters ranked only the top three candidates in an election.

2. A **majority candidate** is
   A. the winner of the election under the voting method employed.
   B. an alternative with more than half of all first-place rankings.
   C. a candidate with more first-place rankings than any other candidate.
   D. a candidate with a majority of all points awarded.

3. In general, **apportionment** is a process for
   A. assigning the same number of objects to all recipients.
   B. dividing objects into fractional parts so that all recipients receive a fair share.
   C. assigning weights to objects so that each recipient receives the same total weight.
   D. dividing identical, indivisible objects among two or more recipients.

4. In the apportionment methods described in this chapter, the **final state quotas**
   A. always sum to the number of "seats" to be apportioned.
   B. are commonly the same as the standard quotas.
   C. are always within 0.5 of their respective standard quotas.
   D. always result from dividing the original quotas by the modified divisor.

5. A **divisor method** of apportionment involves
   A. dividing the quotas by the standard divisor to obtain modified quotas.
   B. dividing the populations by the modified divisor to obtain modified quotas.
   C. dividing the populations by the standard divisor to obtain modified quotas.
   D. dividing the populations by modified quotas to obtain the modified divisor.

6. A **quota method** of apportionment
   A. can fail to satisfy the quota rule.
   B. always has initial rounded quotas that do not need to be modified.
   C. may allow a paradoxical apportionment.
   D. always favors large states over small states.

ANSWERS
1. C   2. B   3. D   4. A   5. B   6. C

QUICK REVIEW

| Concepts | Examples |

15.1   THE POSSIBILITIES OF VOTING

A voter's **ranking** is an arrangement of all candidates in order of preference.

A **profile of voters** shows how many voters chose the various rankings. If there are n candidates, there are

$$n! \text{ possible rankings,}$$

but some may not be chosen by any voters.

The ranking $a > b > c$ indicates that a is the voter's first choice, b is the second choice, and c is the third choice.

Consider the following voter profile.

Number of Voters	Ranking
3	$a > b > c$
3	$c > a > b$
4	$b > a > c$
2	$c > b > a$

Only 4 of the $3! = 6$ possible rankings are included.

Voting Methods

Plurality Method
Each voter gives one vote to his or her top-ranked candidate, and the candidate with the most votes wins.

In the voter profile above,

   a gets 3 votes,   b gets 4 votes,   c gets $3 + 2 = 5$ votes.

Since 5 is greater than 3 or 4, c is the plurality winner.

Pairwise Comparison Method
Candidates are compared in pairs. The candidate preferred by more voters receives one point. (If they tie, each receives a half point.) If there are n candidates,

$$_nC_2$$

comparisons are made, and that many points are awarded. The candidate with the most points wins.

In the voter profile above,

   a ties with b, 6 to 6;   a beats c, 7 to 5;   b beats c, 7 to 5.

The $_3C_2 = 3$ total points are awarded as follows:

$$1\tfrac{1}{2} \text{ to a,}\quad 1\tfrac{1}{2} \text{ to b,}\quad 0 \text{ to c.}$$

No candidate's points exceed all the others, so in this case no winner is selected. (This possibility is a disadvantage of the pairwise comparison method.)

Concepts	*Examples*

Borda Method

If there are n candidates, then candidates receive points at the rate of

$n - 1$ for each first-place vote

$n - 2$ for each second-place vote

$\vdots$

$n - n = 0$ for each nth-place vote.

The candidate with the most points wins. The total number of points will be

$$[1 + 2 + \cdots + (n - 1)] \cdot v,$$

where $v =$ number of voters.

Consider the following voter profile.

Number of Voters	Ranking
9	a > c > b > d
8	b > d > a > c
7	c > d > b > a
5	c > a > b > d

(Only 4 of the 4! = 24 possible rankings are utilized.)

Points earned: a: $9 \cdot 3 + 8 \cdot 1 + 5 \cdot 2 = 45$

b: $9 \cdot 1 + 8 \cdot 3 + 7 \cdot 1 + 5 \cdot 1 = 45$

c: $9 \cdot 2 + 7 \cdot 3 + 5 \cdot 3 = 54$ ← Winner

d: $8 \cdot 2 + 7 \cdot 2 = 30$

The sum $45 + 45 + 54 + 30 = 174$ agrees with the following.

$$[1 + 2 + \cdots + (n - 1)] \cdot v = 6 \cdot 29 = 174$$

Hare Method

Each candidate gives one vote to his or her favorite candidate.

- A candidate wins only when he or she gets a *majority* of the votes.
- If that does not happen in the first round of voting, the candidate (or candidates) with the fewest votes is (are) eliminated, and another election is conducted on the remaining candidates.
- The process continues until one candidate gets a majority of the votes. That candidate wins the election.

Consider the above voter profile. It takes more than half of 29 votes to win—that is, at least 15. In the first round,

 a gets 9;   b gets 8;   c gets 12;   d gets 0   (all less than 15).

So d is eliminated. In the second round, the vote is still

 9 for a;   8 for b;   12 for c   (all less than 15).

So b is eliminated. In the third round, the vote is

 c: 12;   a: $9 + 8 = 17$.   ← Winner in the third round

15.2 THE IMPOSSIBILITIES OF VOTING

Different voting methods may result in the same winner or in different winners.

No method is perfect. The following four attributes are desirable but cannot always be guaranteed.

Majority Criterion

A majority candidate (one with a majority of first-place rankings in a voter profile) should win the election.

Condorcet Criterion

A Condorcet candidate (one who can win a pairwise comparison with every other candidate) should win the election.

Consider the following voter profile.

Number of Voters	Ranking
14	a > b > c
7	b > c > a
6	c > b > a

Candidate a is a majority candidate, but a Borda method election results in the point count

 a: 28;   b: 34;   c: 19.   ← b wins the Borda election.

Consider the following voter profile.

Number of Voters	Ranking
3	c > b > a
4	b > a > c
2	a > c > b
4	a > b > c

Candidate b is a Condorcet candidate, but Candidate a wins a plurality method election.

Concepts	*Examples*

Monotonicity Criterion

Suppose the outcome of a first election is not binding. If Candidate x wins in the first election and, before a second election, the voters who rearrange their rankings move Candidate x to the top of their rankings, then Candidate x should win the second election.

Consider the following voter profile.

Number of Voters	Ranking
7	q > r > s > t
5	s > t > r > q
6	q > r > t > s
8	t > s > r > q

In a nonbinding first election using the pairwise comparison method, the point count is

$$q: 1\tfrac{1}{2}; \quad r: 1\tfrac{1}{2}; \quad s: 1; \quad t: 2.$$

So t wins the first election. However, if the five voters with the second ranking change their ranking to

$$t > q > s > r,$$

then q wins the second election with $2\tfrac{1}{2}$ points to 2 points for t.

Irrelevant Alternatives Criterion

If one or more of the losing alternatives of a first election drop out before a second election, then the winner of the first election should also win the second election.

Consider the following voter profile.

Number of Voters	Ranking
4	d > c > a > b
5	b > c > d > a
7	a > b > c > d

Arrow's Impossibility Theorem

For an election with more than two alternatives, no possible voting method could simultaneously satisfy the

1. majority criterion,
2. Condorcet criterion,
3. monotonicity criterion, and
4. irrelevant alternatives criterion.

Candidate b wins a first Borda method election, but if d then drops out, Candidate a wins the second Borda method election.

If Hare method elections are conducted, Candidate a wins the first election, but if b then drops out, Candidate c wins the second election.

Each voting method violates at least one of the four criteria.

15.3 THE POSSIBILITIES OF APPORTIONMENT

Preliminary Steps for Apportionment

Step 1 Compute the standard divisor, *d*.

$$d = \frac{\text{total population}}{\text{total number of seats}}$$

Step 2 Compute the standard quota *Q* for each state,

$$Q = \frac{\text{state's population}}{d}$$

Round standard quotas to integers. If rounded quotas sum to the available number of seats, the apportionment is done. If not, adjustments are needed.

Consider the following apportionment.

Suppose 195 seats are to be apportioned to five states with populations (in thousands) as shown here.

State	a	b	c	d	e
Population	1429	8639	7608	6660	1671

Then $d = \dfrac{26{,}007}{195} = 133.37$. The states' standard quotas are

$$10.715, \quad 64.775, \quad 57.045, \quad 49.937, \quad 12.529.$$

Concepts	Examples

Additional Apportionment Steps

Hamilton method (a quota method):
Round all standard quotas *down*.

Add seats, one by one, to those states with the greatest fractional parts in their standard quotas, until the resulting quotas have the appropriate sum.

In the apportionment of the 195 seats (on the preceding page), the rounded-down quotas are

$$10, \quad 64, \quad 57, \quad 49, \quad 12.$$

These sum to 192, which is 3 less than the 195 available seats.

Divisor methods:
All remaining methods are **divisor methods.** Convert the standard divisor to a **modified divisor** *md* to obtain **modified quotas** *mQ* that will automatically sum to the number of available seats.

The greatest fractional parts are in the standard quotas of state d, then b, then a.

$$11, \quad 65, \quad 57, \quad 50, \quad 12 \quad \text{Final state quotas}$$

Adjust the modified divisor downward (upward) until the modified quotas produce the correct sum.

Jefferson method: Round quotas down.

For the five states above, adjust the divisor *d* downward to *md* = 131.

$$10, \quad 65, \quad 58, \quad 50, \quad 12 \quad \text{Modified state quotas}$$

Webster method: Round quotas traditionally (up for fractional part 0.5 or greater, down for fractional part less than 0.5).

Traditional rounding produces rounded quotas that sum to 196. Adjust *d* upward to *md* = 133.8.

$$11, \quad 65, \quad 57, \quad 50, \quad 12 \quad \text{Modified state quotas}$$

Huntington-Hill method: Round up or down based on the geometric mean of the two integers immediately below and above the modified quotas.

Rounded quotas sum to 196. Adjusting *d* upward to *md* = 133.8 also works in this case. The rounded modified quotas *mQ* are

$$11, \quad 65, \quad 57, \quad 50, \quad 12. \quad \text{Modified state quotas}$$

Adams method: Round quotas up.

Rounded-up quotas sum to 197. Adjusting *d* upward to *md* = 135 produces the rounded-up modified quotas

$$11, \quad 64, \quad 57, \quad 50, \quad 13. \quad \text{Modified state quotas}$$

 15.4 THE IMPOSSIBILITIES OF APPORTIONMENT

Quota Rule
Each state's apportioned seats should equal the state's standard quota, rounded either up or down.

All five methods applied to the apportionment situation for **Section 15.3** satisfied the quota rule (for that case). However, all divisor methods are vulnerable to violations of the quota rule.

Alabama Paradox
An increase in the number of available seats results in a state losing a seat.

The Hamilton method, a quota method, is vulnerable to the Alabama paradox.

Population Paradox
Based on updated population figures, a reapportionment of a fixed number of seats causes a state to lose a seat to another state. This occurs even though the percent increase in the population of the state that loses the seat is greater than the percent increase in the population of the state that gains the seat.

The Hamilton method is vulnerable to the population paradox.

Concepts	Examples
New States Paradox A reapportionment of an increased number of seats, necessitated by the addition of a new state, causes a shift in the apportionment of the original states.	The Hamilton method is vulnerable to the new states paradox.
Balinski and Young Impossibility Theorem Any apportionment method that satisfies the quota rule will permit the possibility of a paradoxical apportionment. Likewise, any method that does not permit the possibility of a paradoxical apportionment will fail to satisfy the quota rule.	The Hamilton method, a quota method, is of the first type. The Jefferson, Webster, Huntington-Hill, and Adams methods (all divisor methods) are of the second type.

Chapter 15 TEST

1. How many different complete rankings are possible in an election with 7 candidates?

2. How many comparisons must be considered in a pairwise comparison election with 10 alternatives?

3. What is the majority criterion?

4. What is the Condorcet criterion?

5. Who was Kenneth J. Arrow? Explain in your own words what Arrow's Impossibility Theorem says about voting methods.

6. Why is the irrelevant alternatives criterion an important measure of fairness in a two-stage election process?

Use the given voter profile for Exercises 7–14.

Number of Voters	Ranking
16	a > b > c
8	b > c > a
7	c > b > a

7. Which candidate, if any, is a majority candidate?

8. Which candidate, if any, is a Condorcet candidate?

9. In a Borda method election, what total number of Borda points will be awarded?

10. Determine the Borda winner.

11. Is the majority criterion violated for this Borda election? Why, or why not?

12. Is the Condorcet criterion violated for this Borda election? Why, or why not?

13. Suppose that after the initial Borda election, the seven voters with the last ranking rearrange their ranking to be b > c > a, and a second Borda election is conducted. Is the monotonicity criterion violated? Why, or why not?

14. Suppose that after the initial Borda election, Candidate a drops out and a second Borda election is conducted. Is the irrelevant alternatives criterion violated? Why, or why not?

Use the given voter profile for Exercises 15–17.

Number of Voters	Ranking
8	c > m > b > s
7	s > b > m > c
6	s > b > c > m
5	m > c > b > s

15. Use the pairwise comparison method to determine the preferred alternative.

16. Find the pairwise comparison winner again for the case where the five voters with the last ranking rearrange their ranking to be c > s > m > b.

17. What do the outcomes of **Exercises 15 and 16** show about the pairwise comparison method of voting?

Use the given voter profile for Exercises 18–20.

Number of Voters	Ranking
10	a > b > c
7	b > c > a
5	c > b > a

18. Use the plurality method to determine the preferred alternative.

19. Find the plurality winner in a second election if Candidate c drops out.

20. What do the outcomes of **Exercises 18 and 19** show about the plurality method of voting?

21. Explain what Balinski and Young discovered about apportionment methods in general.

One hundred seats are to be apportioned to 4 states as shown below. Use this information for Exercises 22–28.

State	a	b	c	d
Population	2354	4500	5598	23,000

22. Find the standard divisor to two decimal places.

23. Find the standard quota, to two decimal places, for each state.

a: _____   b: _____   c: _____   d: _____

24. Find the *initial rounded quota,* for each state, for each of the following apportionment methods.

(a) Hamilton   a: ___   b: ___   c: ___   d: ___

(b) Jefferson   a: ___   b: ___   c: ___   d: ___

(c) Webster   a: ___   b: ___   c: ___   d: ___

(d) Huntington-Hill

a: ___   b: ___   c: ___   d: ___

25. Complete the Hamilton apportionment.

State	a	b	c	d
Number of seats	___	___	___	___

26. (a) For the Jefferson apportionment, should the modified divisor be less than, equal to, or greater than the standard divisor?

(b) Use a whole number modified divisor about 6.5 units different from the standard divisor, and complete the Jefferson apportionment.

State	a	b	c	d
Number of seats	___	___	___	___

27. (a) For the Webster apportionment, should the modified divisor be less than, equal to, or greater than the standard divisor?

(b) Use a whole number modified divisor about 2.5 units different from the standard divisor, and complete the Webster apportionment.

State	a	b	c	d
Number of seats	___	___	___	___

28. (a) For the Huntington-Hill apportionment, should the modified divisor be less than, equal to, or greater than the standard divisor?

(b) Use a whole number modified divisor about 2.5 units different from the standard divisor, and complete the Huntington-Hill apportionment.

State	a	b	c	d
Number of seats	___	___	___	___

29. What is the quota rule?

30. Explain the Alabama paradox.

31. Explain the population paradox.

32. Explain the new states paradox.

CHAPTER 1 THE ART OF PROBLEM SOLVING

1.1 Exercises

1. 1, 2, 3, 4, 5 **3.** 16 **5.** 81 **7.** inductive **9.** deductive
11. deductive **13.** inductive **15.** deductive
17. deductive **19.** inductive **21.** inductive
23. Answers will vary. **25.** 21 **27.** 3072 **29.** 63
31. $\frac{11}{12}$ **33.** 216 **35.** 52
37. One such list is 10, 20, 30, 40, 50, ...
39. $(98{,}765 \times 9) + 3 = 888{,}888$
41. $3367 \times 15 = 50{,}505$
43. $11{,}111 \times 11{,}111 = 123{,}454{,}321$
45. $3 + 6 + 9 + 12 + 15 = \frac{15(6)}{2}$
47. $5(6) + 5(36) + 5(216) + 5(1296) + 5(7776) = 6(7776 - 1)$
49. $\frac{1}{2} + \frac{1}{4} + \frac{1}{8} + \frac{1}{16} + \frac{1}{32} = 1 - \frac{1}{32}$ **51.** 20,100 **53.** 320,400
55. 15,400 **57.** 2550 **59.** 1 (These are the numbers of chimes a clock rings, starting with 12 o'clock, if it rings the number of hours on the hour, and 1 chime on the half-hour.)
61. (a) Answers will vary. **(b)** The middle digit is always 9, and the sum of the first and third digits is always 9 (considering 0 as the first digit if the difference has only two digits). **(c)** Answers will vary.
63. 142,857; 285,714; 428,571; 571,428; 714,285; 857,142. Each result consists of the same six digits, but in a different order; $142{,}857 \times 7 = 999{,}999$

1.2 Exercises

1. 11 **3.** 6 + 1, or 7 **5.** arithmetic; 56
7. geometric; 1215 **9.** neither **11.** geometric; 8
13. neither **15.** arithmetic; 22 **17.** 79 **19.** 450
21. 4032 **23.** 32,758 **25.** 57; 99
27. $(4321 \times 9) - 1 = 38{,}888$
29. $999{,}999 \times 4 = 3{,}999{,}996$
31. $21^2 - 15^2 = 6^3$ **33.** $5^2 - 4^2 = 5 + 4$
35. $1 + 5 + 9 + 13 = 4 \times 7$ **37.** 45,150 **39.** 228,150
41. 2601 **43.** 250,000 **45.** $S = n(n + 1)$
47. Answers will vary. **49.** *row 1:* 28, 36; *row 2:* 36, 49, 64; *row 3:* 35, 51, 70, 92; *row 4:* 28, 45, 66, 91, 120; *row 5:* 18, 34, 55, 81, 112, 148; *row 6:* 8, 21, 40, 65, 96, 133, 176
51. $8(1) + 1 = 9 = 3^2$; $8(3) + 1 = 25 = 5^2$; $8(6) + 1 = 49 = 7^2$; $8(10) + 1 = 81 = 9^2$
53. The pattern is 1, 0, 1, 0, 1, 0, ...

55.

$$25 = 10 + 15 \qquad 36 = 15 + 21$$

57. 256 **59.** 117 **61.** 235 **63.** $N_n = \frac{n(7n - 5)}{2}$
65. a square number **67.** a perfect cube **69.** 42
71. 419 **73.** $\frac{101}{2}$ **75.** 2048 **77.** $\frac{1}{2048}$ **79.** $\frac{5}{2048}$
81. To find any entry within the body of the triangle, add the two entries immediately to the left and to the right in the row above it. For example, in the sixth row, $10 = 4 + 6$. The next three rows of Pascal's triangle are as follows.

		1		6		15		20		15		6		1	
	1		7		21		35		35		21		7		1
1		8		28		56		70		56		28		8		1

83. The sums along the diagonals are 1, 1, 2, 3, 5, 8. These are the first six terms of the Fibonacci sequence.

1.3 Exercises

1. 18 (in the year 2031)
3. 49 (each amount from 1 cent to 49 cents)
5. 11 **7.** 50 **9.**

11. IN **13.** 41
15. 15 **17.** 32

19. 0 (The product of the first and last digits is the two-digit number between them.) **21.** 11 **23.** 48.5 in.
25. 42 **27.** 6 **29.** If you multiply the two digits in the numbers in the first row, you will get the second row of numbers. The second row of numbers is a pattern of two numbers (8 and 24) repeating. **31.** A **33.** You should choose a sock from the box labeled *red and green socks*. Because it is mislabeled, it contains only red socks or only green socks, determined by the sock you choose. If the sock is green, relabel this box *green socks*. Since the other two boxes were mislabeled, switch the remaining label to the other box and place the label that says *red and green socks* on the unlabeled box. No other choice guarantees a correct relabeling because you can remove only one sock. **35.** D

37. Here is one solution.   **39.** Here is one solution.

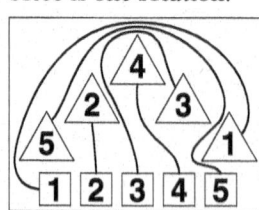

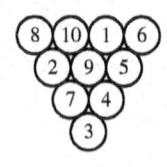

41. D   **43.** $\frac{1}{3}$ **45.** 90   **47.** 55 mph   **49.** 07   **51.** 437
53. 3   **55.** 3 socks   **57.** 6   **59.** the nineteenth day
61. 1967   **63.** Eve has $5, and Adam has $7.

65.
$$
\begin{array}{r}
4\ 0\ 2 \\
\times\quad 3\ 9 \\
\hline
1\ 5,\ 6\ 7\ 8
\end{array}
$$

67.

6	12	7	9
1	15	4	14
11	5	10	8
16	2	13	3

69. 25 pitches (The visiting team's pitcher retires 24
consecutive batters through the first eight innings, using
only one pitch per batter. His team does not score either.
Going into the bottom of the ninth tied 0–0, the first
batter for the home team hits his first pitch for a home
run. The pitcher threw 25 pitches and loses the game by a
score of 1–0.)   **71.** Q

73. Here is one solution.

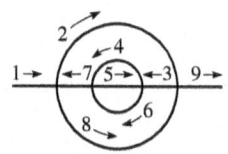

75. 6   **77.** 86 cm   **79.** Dan
(36) is married to Jessica
(29); James (30) is married
to Cathy (31).

81. 6

	X	X
X		X
X	X	

One of the
possibilities

1.4 Exercises
1. positive   **3.** 0   **5.** 43.8   **7.** 2.3589   **9.** 7.48
11. 7.1289   **13.** 6340.338097   **15.** 1
17. 1.061858759   **19.** 2.221441469   **21.** 3.141592653
23. 0.7782717162   **25.** 1   **27.** the same as
29. negative   **31.** yes   **33.** Answers will vary.
35. Answers will vary.   **37.** 63   **39.** 14   **41.** B
43. A   **45.** D   **47.** 13%   **49.** 12 million
51. 2012, 2013, 2014, 2015, and 2016

53. 2010: about 193 billion lb; 2016: about 213 billion lb
55. 2009 and 2010   **57.** 2010: 9.5%; 2015: 5%; 4.5%

Chapter 1 Test
1. inductive   **2.** deductive
3. $65,359,477,124,183 \times 68 = 4,444,444,444,444,444$
4. 351   **5.** 31,375   **6.** 65; $65 = 1 + 7 + 13 + 19 + 25$
7. 1, 8, 21, 40, 65, 96, 133, 176; The pattern is 1, 0, 1, 0,
1, 0, 1, 0, ...   **8.** The first two terms are both 1. Each
term after the second is found by adding the two previous
terms. The next term is 34.   **9.** $\frac{1}{4}$ **10.** 31   **11.** 9   **12.** 35
13. 3   **14.** 108 in., or 9 ft.   **15.** 3   **16.** The sum of the
digits is always 9.   **17.** 9.907572861 (Answers may vary due
to the model of calculator used.)   **18.** 34.328125   **19.** D
20. (a) The percent was decreasing.   **(b)** 2010: 39.5%;
2013: 36%   **(c)** 3.5%

CHAPTER 2 THE BASIC CONCEPTS OF SET THEORY

2.1 Exercises
1. true   **3.** false   **5.** F   **7.** E   **9.** B   **11.** H
13. $\{1, 2, 3, 4, 5, 6\}$ **15.** $\{0, 1, 2, 3, 4\}$
17. $\{6, 7, 8, 9, 10, 11, 12, 13, 14\}$
19. $\{2, 4, 8, 16, 32, 64, 128, 256\}$
21. $\{0, 2, 4, 6, 8, 10\}$ **23.** $\{$Lake Erie, Lake Huron,
Lake Michigan, Lake Ontario, Lake Superior$\}$
25. $\left\{1, \frac{1}{2}, \frac{1}{3}, \frac{1}{4}, \frac{1}{5}, \cdots\right\}$

**In Exercises 27 and 29, there are other ways to describe
the sets.**   **27.** $\{x \mid x$ is a rational number$\}$
29. $\{x \mid x$ is an odd natural number less than 76$\}$
31. the set of single-digit integers   **33.** the set of states of
the United States   **35.** finite   **37.** infinite   **39.** infinite
41. 8   **43.** 500   **45.** 26   **47.** 39   **49.** 28   **51.** Answers
will vary.   **53.** well defined   **55.** not well defined
57. $\notin$ **59.** $\in$ **61.** $\in$ **63.** false; Answers will vary.
65. false   **67.** true   **69.** true   **71.** false   **73.** true
75. true   **77.** false   **79.** false   **81.** true   **83.** false
85. true   **89.** Answers will vary.   **91.** $\{2\}$ and $\{3, 4\}$
(Other examples are possible.)   **93.** $\{a, b\}$ and $\{a, c\}$
(Other examples are possible.)   **95. (a)** $\{$Brock, Heather,
Boris$\}$, $\{$Brock, Heather, Natalie$\}$, $\{$Brock,
Taylor, Boris$\}$, $\{$Brock, Taylor, Natalie$\}$, $\{$Heather,
Taylor, Boris$\}$, $\{$Heather, Taylor, Natalie$\}$
(b) $\{$Brock, Boris, Natalie$\}$, $\{$Heather, Boris, Natalie$\}$,
$\{$Taylor, Boris, Natalie$\}$ **(c)** $\{$Brock, Heather, Taylor$\}$

2.2 Exercises

1. true **3.** false **5.** D **7.** B **9.** $\not\subseteq$ **11.** $\subseteq$ **13.** $\subseteq$
15. $\not\subseteq$ **17.** both **19.** $\subseteq$ **21.** both **23.** neither
25. true **27.** false **29.** true **31.** false **33.** true
35. false **37.** true **39.** false **41. (a)** 64 **(b)** 63
43. (a) 32 **(b)** 31 **45.** $\varnothing$ **47.** $\{5, 7, 9, 10\}$
49. {Higher cost, Lower cost, Educational, More time to
see the sights in Florida, Less time to see the sights in
Florida, Cannot visit friends along the way, Can visit
friends along the way} **51.** {Higher cost, More time to
see the sights in Florida, Cannot visit friends along
the way} **53.** $\varnothing$ **55.** {A, B, C, D, E} (All are present.)
57. $\{A, B, C\}, \{A, B, D\}, \{A, B, E\}, \{A, C, D\},$
$\{A, C, E\}, \{A, D, E\}, \{B, C, D\}, \{B, C, E\}, \{B, D, E\},$
$\{C, D, E\}$ **59.** $\{A\}, \{B\}, \{C\}, \{D\}, \{E\}$ **61.** 32
63. $2^{25} - 1 = 33{,}554{,}431$ **65. (a)** 15 **(b)** 16; It is now
possible to select *no* bills. **67. (a)** s **(b)** s **(c)** $2s$
(d) Adding one more element will always double
the number of subsets, so the expression 2^n is true in
general.

2.3 Exercises

1. true **3.** true **5.** B **7.** A **9.** E **11.** $\{a, c\}$
13. $\{a, b, c, d, e, f\}$ **15.** $\{b, d, f\}$ **17.** $\{d, f\}$
19. $\{a, b, c, e, g\}$ **21.** $\{e, g\}$ **23.** $\{a\}$
25. $\{e, g\}$ **27.** $\{d, f\}$

**In Exercises 29 and 31, there may be other acceptable
descriptions.** **29.** the set of all elements that either
are in A, or are not in B and not in C **31.** the set
of all elements that are in C but not in B, or are in A
33. $\{e, h, c, l, b\}$ **35.** $\{e, h, c, l, b\}$ **37.** the set
of all tax returns filed in 2018 without itemized
deductions **39.** the set of all tax returns with itemized
deductions or showing business income, but not
selected for audit **41.** always true **43.** not always
true **45. (a)** $\{1, 3, 5, 2\}$ **(b)** $\{1, 2, 3, 5\}$
(c) For any sets X and Y, $X \cup Y = Y \cup X$.
47. (a) $\{1, 3, 5, 2, 4\}$ **(b)** $\{1, 3, 5, 2, 4\}$ **(c)** For any
sets X, Y, and Z, $X \cup (Y \cup Z) = (X \cup Y) \cup Z$.
49. $A \times B = \{(d, p), (d, i), (d, g), (o, p), (o, i),$
$(o, g), (g, p), (g, i), (g, g)\}; B \times A = \{(p, d), (p, o),$
$(p, g), (i, d), (i, o), (i, g), (g, d), (g, o), (g, g)\}$
51. $n(A \times B) = 210; n(B \times A) = 210$ **53.** 6

55.
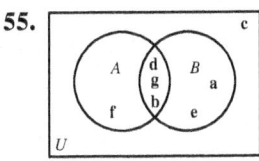

57.
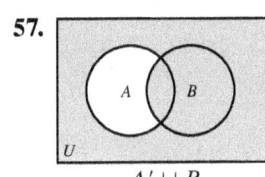

$A' \cup B$

59.
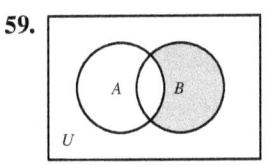

$B \cap A'$

61.
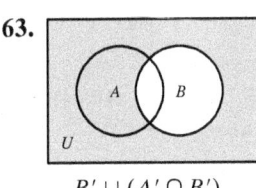

$B' \cap B = \varnothing$

63.
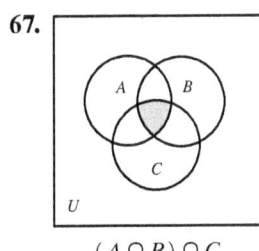

$B' \cup (A' \cap B')$

65.
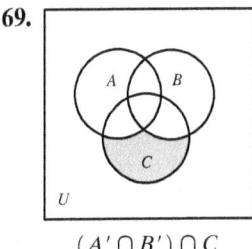

67.
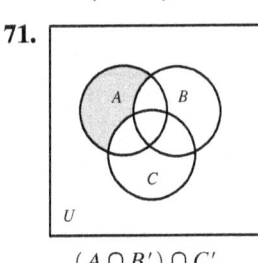

$(A \cap B) \cap C$

69.
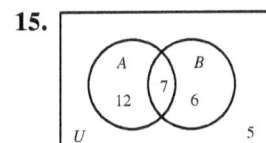

$(A' \cap B') \cap C$

71.
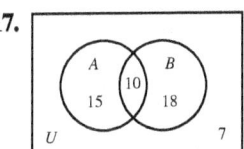

$(A \cap B') \cap C'$

73. $A' \cap B'$, or $(A \cup B)'$
75. $(A \cup B) \cap (A \cap B)'$,
or $(A \cup B) - (A \cap B)$,
or $(A - B) \cup (B - A)$
77. $(A \cap B) \cup (A \cap C)$,
or $A \cap (B \cup C)$

79. $A \cap B = \varnothing$ **81.** This statement is true for any set A.
83. $B \subseteq A$ **85.** always true **87.** not always true
89. Answers will vary.

2.4 Exercises

1. false **3.** true **5. (a)** 5 **(b)** 7 **(c)** 0 **(d)** 2 **(e)** 8
7. (a) 1 **(b)** 3 **(c)** 4 **(d)** 0 **(e)** 2 **(f)** 8 **(g)** 2
(h) 6 **9.** 21 **11.** 7 **13.** 35

15.
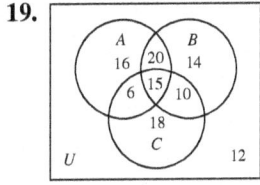

17.
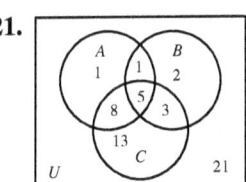

19.

21.

23. 90 **25. (a)** 2 **(b)** 4 **27. (a)** 25 **(b)** 30 **(c)** 2
(d) 10 **(e)** 19 **(f)** 57 **29. (a)** 8 **(b)** 26 **(c)** 60
31. maximum: 37; minimum: 31 **33. (a)** 31 **(b)** 24
(c) 11 **(d)** 45 **(e)** 52 **(f)** 2 **35. (a)** 6 **(b)** 473
(c) 835 **(d)** 50 **(e)** 297 **(f)** 386 **37.** Answers will vary.

Chapter 2 Test

1. {a, b, c, d, e} **2.** {a, b, d} **3.** {c, f, g, h} **4.** {a, c}
5. false **6.** true **7.** true **8.** true **9.** false **10.** false
11. true **12.** true **13.** 16 **14.** 15

Answers may vary in Exercises 15–18. **15.** the set of
odd integers between -4 and 10 **16.** the set of days of
the week **17.** $\{x \mid x$ is a negative integer$\}$
18. $\{x \mid x$ is a multiple of 8 between 20 and 90$\}$ **19.** $\subseteq$
20. neither

21.
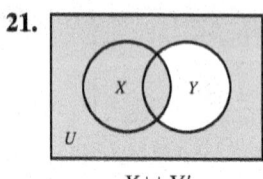
$X \cup Y'$

22.
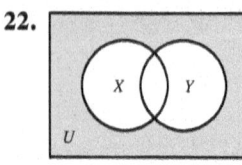
$X' \cap Y'$

23.
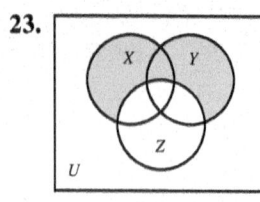
$(X \cup Y) - Z$

24.
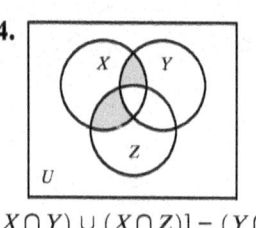
$[(X \cap Y) \cup (X \cap Z)] - (Y \cap Z)$

25. Answers will vary. **26.** {Electric razor}
27. {Adding machine, Baking powder, Pendulum clock,
Thermometer} **28.** {Telegraph, Zipper} **29. (a)** 22
(b) 12 **(c)** 28 **30. (a)** 16 **(b)** 32 **(c)** 33 **(d)** 45
(e) 14 **(f)** 26

CHAPTER 3 INTRODUCTION TO LOGIC

3.1 Exercises

1. compound statement **3.** existential **5.** statement
7. not a statement **9.** statement **11.** statement
13. statement **15.** not a statement **17.** not compound
19. compound **21.** compound **23.** not compound
25. Her aunt's name is not Hermione. **27.** No book
is longer than this book. **29.** At least one computer
repairman can play blackjack. **31.** Someone does not
love somebody sometime. **33.** $x \leq 10$ **35.** $x < 2$
37. Answers will vary. **39.** She does not have green eyes.
41. She has green eyes and he is 60 years old.
43. She does not have green eyes or he is 60 years old.

45. She does not have green eyes or he is not 60 years old.
47. It is not the case that she does not have green eyes and
he is 60 years old. **49.** $p \wedge {\sim}q$ **51.** ${\sim}p \vee q$
53. ${\sim}(p \vee q)$ or, equivalently, ${\sim}p \wedge {\sim}q$
55. Answers will vary. **57.** C **59.** A, B **61.** A, C
63. B **65.** true **67.** false **69.** true **71.** true
73. false **75.** Answers will vary. **77.** Every person here
has made mistakes before. **79.** $(\forall c)({\sim}f)$

3.2 Exercises

1. disjunction; false; false **3.** $6 > 2$ **5.** 4 **7.** false
9. true **11.** They must both be false. **13.** F **15.** T
17. T **19.** T **21.** F **23.** T **25.** T **27.** F **29.** T
31. T **33.** F **35.** F **37.** T **39.** T **41.** T **43.** 4
45. 16 **47.** 128 **49.** six **51.** FFTF **53.** TTTT
55. TFFF **57.** FFFFTFFF **59.** FTFTTTTT
61. TTTTTTTTTTTFTTT **63.** You can't pay me now
and you can't pay me later. **65.** It is not summer or there
is snow. **67.** I did not say yes or she did not say no.
69. $6 - 1 \neq 5$ or $9 + 13 = 7$ **71.** T **73.** F

75.

p	q	$p \underline{\vee} q$
T	T	F
T	F	T
F	T	T
F	F	F

77. F **79.** T
81. He was looking at his own son.
83. The negation of a conjunction is
equivalent to the disjunction of the
negations.

3.3 Exercises

1. false **3.** false **5.** is not **7.** series **9.** If you just
believe, then you can do it. **11.** If it is an even integer
divisible by 5, then it is divisible by 10. **13.** If it is a
grizzly bear, then it does not live in California. **15.** If
they are surfers, then they can't stay away from the beach.
17. true **19.** true **21.** false **23.** F **25.** T **27.** T
29. Answers will vary. **31.** If they do not collect
classics, then he fixes cars. **33.** If she sings for a living,
then they collect classics and he fixes cars. **35.** If he does
not fix cars, then they do not collect classics or she sings
for a living. **37.** $c \to g$ **39.** $g \wedge (s \to {\sim}c)$ **41.** $g \to s$
43. T **45.** F **47.** T **49.** T **51.** Answers will vary.
53. TTTF **55.** TTFT **57.** TTTT; tautology
59. TTTTTTFT **61.** TTTFTTTTTTTTTTTT
63. one **65.** That is an authentic diamond and I am
not surprised. **67.** You talk in your sleep and you
mention my name. **69.** The bullfighter doesn't get
going and doesn't get gored. **71.** You do not call or I
will answer.

73. They do not turn the ball over one more time or they'll lose.   **75.** They are not champions or they have had their challenges.   **77.** equivalent   **79.** equivalent   **81.** equivalent   **83.** $(p \wedge q) \vee (p \wedge \sim q)$; The statement simplifies to p.   **85.** $p \vee (\sim q \wedge r)$ **87.** $\sim p \vee (p \vee q)$; The statement simplifies to T.   **89.** The statement simplifies to $p \wedge q$.

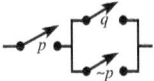

91. The statement simplifies to F.

93. The statement simplifies to $(r \wedge \sim p) \wedge q$.

95. The statement simplifies to $p \vee q$.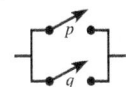

97. $525.60

3.4 Exercises

1. (a) contrapositive   **(b)** converse   **(c)** inverse
3. contrapositive   **5. (a)** If you were an hour, then beauty would be a minute.   **(b)** If beauty were not a minute, then you would not be an hour.   **(c)** If you were not an hour, then beauty would not be a minute.
7. (a) If you don't fix it, then it ain't broke.   **(b)** If it's broke, then you fix it.   **(c)** If you fix it, then it's broke.
9. (a) If he comes, then you built it.   **(b)** If you don't build it, then he won't come.   **(c)** If he doesn't come, then you didn't build it.   **11. (a)** If it is dangerous to your health, then you walk in front of a moving car.   **(b)** If you do not walk in front of a moving car, then it is not dangerous to your health.   **(c)** If it is not dangerous to your health, then you do not walk in front of a moving car.
13. (a) If they flock together, then they are birds of a feather.   **(b)** If they are not birds of a feather, then they do not flock together.   **(c)** If they do not flock together, then they are not birds of a feather.
15. (a) $\sim q \rightarrow p$ **(b)** $\sim p \rightarrow q$ **(c)** $q \rightarrow \sim p$
17. (a) $\sim q \rightarrow \sim p$ **(b)** $p \rightarrow q$ **(c)** $q \rightarrow p$
19. (a) $(q \vee r) \rightarrow p$ **(b)** $\sim p \rightarrow (\sim q \wedge \sim r)$
(c) $(\sim q \wedge \sim r) \rightarrow \sim p$ **21.** Answers will vary.
23. If it has legs of 3 and 4, then it has a hypotenuse of 5.
25. If a number is a whole number, then it is a rational number.   **27.** If the graffiti are to be covered, then two coats of paint must be used.   **29.** If I do logic puzzles, then I am driven crazy.   **31.** If a number is a whole number, then it is an integer.   **33.** If employment

improves, then the economy recovers.   **35.** If their pitching improves, then the Cubs will win the pennant.
37. If the figure is a rectangle, then it is a parallelogram with perpendicular adjacent sides.   **39.** If a three-digit number whose units digit is 5 is squared, then the square will end in 25.   **41.** If a triangle has two perpendicular sides, then it is a right triangle.   **43.** D   **45.** Answers will vary.   **47.** true   **49.** false   **51.** false   **53.** contrary
55. contrary   **57.** consistent
59. (1) $p \rightarrow (p \rightarrow q)$ **(2)** $(p \rightarrow q) \rightarrow p$

3.5 Exercises

1. false   **3.** false   **5.** valid   **7.** invalid   **9.** valid
11. invalid   **13.** invalid   **15.** yes
17. All people with blue eyes have blond hair.
    Erin does not have blond hair.
    ——————————————
    Erin does not have blue eyes.
19. valid   **21.** invalid   **23.** valid   **25.** invalid
27. invalid   **29.** valid

3.6 Exercises

1. premises; conclusion   **3.** false   **5.** conjunction
7. valid by reasoning by transitivity   **9.** valid by modus ponens   **11.** fallacy by fallacy of the converse   **13.** fallacy by fallacy of the inverse   **15.** valid by modus tollens
17. valid by disjunctive syllogism   **19.** valid   **21.** invalid
23. valid   **25.** invalid   **27.** valid   **29.** invalid
31. Every time something squeaks, I need WD-40.
    Every time I need WD-40, I go to the hardware store.
    ——————————————————————
    Every time something squeaks, I go to the hardware store.

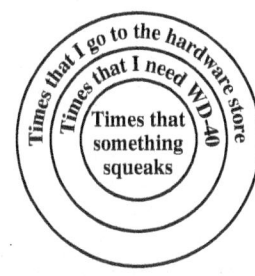

33. valid   **35.** invalid   **37.** valid   **39.** valid
41. (a) When the cable company keeps you on hold, your dad gets punched over a can of soup.   **(b)** The cable company doesn't keep you on hold.   **43.** If it is my poultry, then it is a duck.   **45.** If it is a guinea pig, then it is hopelessly ignorant of music.   **47.** If I can read it, then I have not filed it.   **49.** If it is a teachable kitten, then it does not have green eyes.   **51. (a)** $p \rightarrow \sim s$ **(b)** $r \rightarrow s$

(c) $q \rightarrow p$ (d) None of my poultry are officers.

53. (a) $r \rightarrow \sim s$ (b) $u \rightarrow t$ (c) $\sim r \rightarrow p$ (d) $\sim u \rightarrow \sim q$

(e) $t \rightarrow s$ (f) All pawnbrokers are honest.

55. (a) $r \rightarrow w$ (b) $\sim u \rightarrow \sim t$ (c) $v \rightarrow \sim s$ (d) $x \rightarrow r$

(e) $\sim q \rightarrow t$ (f) $y \rightarrow p$ (g) $w \rightarrow s$ (h) $\sim x \rightarrow \sim q$

(i) $p \rightarrow \sim u$ (j) I can't read any of Brown's letters.

Chapter 3 Test

1. $6 - 3 \neq 3$ **2.** Some roses are not red.

3. No members of the class went on the field trip.

4. I fall in love and it will not be forever.   **5.** She did not apply or she got a student loan.   **6.** $\sim p \rightarrow q$

7. $p \rightarrow q$ **8.** $\sim q \leftrightarrow \sim p$ **9.** You won't help me and I will help you.   **10.** It is not the case that you will help me or I will not help you. (Equivalently: You won't help me and I will help you.)   **11.** T   **12.** T   **13.** F   **14.** T

15. Answers will vary.   **16.** (a) The antecedent must be true and the consequent must be false.   (b) Both component statements must be true.   (c) Both component statements must be false.   (d) Both component statements must have the same truth value.   **17.** FTFF   **18.** TTTT (tautology)   **19.** false   **20.** true

Wording may vary in the answers for Exercises 21–25.

21. If the number is an integer, then it is a rational number.

22. If a polygon is a rhombus, then it is a quadrilateral.

23. If a number is divisible by 4, then it is divisible by 2.

24. If she digs dinosaur bones, then she is a paleontologist.

25. (a) If the graph helps me understand it, then a picture paints a thousand words.   (b) If a picture doesn't paint a thousand words, then the graph won't help me understand it.   (c) If the graph doesn't help me understand it, then a picture doesn't paint a thousand words.   **26.** (a) $(q \wedge r) \rightarrow \sim p$ (b) $p \rightarrow (\sim q \vee \sim r)$

(c) $(\sim q \vee \sim r) \rightarrow p$ **27.** valid   **28.** (a) A   (b) F   (c) C

(d) D   **29.** valid   **30.** invalid

CHAPTER 4 NUMERATION SYSTEMS

4.1 Exercises

1. A   **3.** A, B, C   **5.** C   **7.** 12,034   **9.** 7,610,729

11.

13.

15.

17.

19. 173   **21.** 14,000,000   **23.** MMDCCCLXI

25. $\overline{\text{XXV}}$DCXIX   **27.** 936   **29.** 2009

31.   **33.**   **35.** to   **37.** to

39. 406   **41.** 113   **43.** 53,601   **45.** 2021   **47.** 903

49. 1504   **51.** 8128   **53.** 622,500 shekels   **55.** Answers will vary.   **57.** Answers will vary.   **59.** 99,999   **61.** 3124

63. $10^d - 1$ **65.** 9999

4.2 Exercises

1. A, B, C   **3.** A, C   **5.** C   **7.** Mayan; 12

9. Babylonian; 32   **11.** Greek; 234   **13.** Babylonian; 2601   **15.** Greek; 15,149   **17.** Mayan; 2640   **19.** Mayan; 59,954   **21.** Greek; 99,999   **23.** Babylonian; 80,474

25.   **27.**   **29.**   **31.**

33.   **35.**   **37.**   **39.**   **41.**   **43.**

45. $\lambda\theta$ **47.** $\upsilon\iota\beta$ **49.** $,\beta\psi\xi\theta$ **51.** $\overset{\epsilon}{\text{M}},\delta\psi\kappa\varsigma$ **53.** $\chi\pi\alpha$

55. $\omega\pi\eta$

4.3 Exercises

1. true   **3.** false   **5.** $(7 \cdot 10^1) + (3 \cdot 10^0)$

7. $(8 \cdot 10^3) + (3 \cdot 10^2) + (3 \cdot 10^1) + (5 \cdot 10^0)$

9. $(3 \cdot 10^3) + (6 \cdot 10^2) + (2 \cdot 10^1) + (4 \cdot 10^0)$

11. $(1 \cdot 10^7) + (4 \cdot 10^6) + (2 \cdot 10^5) + (0 \cdot 10^4) +$ $(6 \cdot 10^3) + (0 \cdot 10^2) + (4 \cdot 10^1) + (0 \cdot 10^0)$

13. 75   **15.** 4380   **17.** 70,401,009   **19.** 79   **21.** 43

23. 109   **25.** 733   **27.** 6   **29.** 206   **31.** 242   **33.** 49,801

35. 10,256   **37.** 63,259

39.   **41.**

43. 1885   **45.** 16,425   **47.** 3,035,154   **49.** 496

51. 217,204   **53.** 460   **55.** 32,798   **57.** Answers will vary.   **59.** fours complement; Answers will vary.

61. 1; Answers will vary.

4.4 Exercises

1. A   **3.** A, D, F   **5.** 1, 2, 3, 4, 5, 6, 10, 11, 12, 13, 14, 15, 16, 20, 21, 22, 23, 24, 25, 26   **7.** 1, 2, 3, 4, 5, 6, 7, 8, 10, 11, 12, 13, 14, 15, 16, 17, 18, 20, 21, 22   **9.** 13_{five}; 20_{five}

11. B6E$_{\text{sixteen}}$; B70$_{\text{sixteen}}$ **13.** least: 1000$_{\text{five}}$ = 125;
greatest: 4444$_{\text{five}}$ = 624   **15.** 956   **17.** 874   **19.** 28,854
21. 139   **23.** 4312   **25.** 93$_{\text{sixteen}}$ **27.** 2131101$_{\text{five}}$
29. 1001001011$_{\text{two}}$ **31.** 111134$_{\text{six}}$ **33.** 102112101$_{\text{three}}$
35. 32$_{\text{seven}}$ **37.** 1031321$_{\text{four}}$ **39.** 11110111$_{\text{two}}$ **41.** 467$_{\text{eight}}$
43. 10101100$_{\text{two}}$ **45.** 2D$_{\text{sixteen}}$ **47.** 37$_{\text{eight}}$ **49.** 1427
51. Answers will vary.   **53.** 01000011$_{\text{two}}$ **55.** 01101011$_{\text{two}}$
57. BASE   **59.** 4E6577$_{\text{sixteen}}$ **61.** 1 and 15   **63. (a)** The
binary ones digit is 1.   **(b)** The binary twos digit is 1.
(c) The binary fours digit is 1. **(d)** The binary eights digit is 1.
(e) The binary sixteens digit is 1. **65.** 6   **67.** yes
69. yes   **71.** yes   **73.** yes   **75.** Answers will vary.
77. no   **79.** yes   **81.** Append a 0 on the right.   **83.** 0
85. Answers will vary.   **87.** 20120011$_{\text{three}}$ **89.** 25657$_{\text{nine}}$
91. 11$_{\text{two}}$ = 3   **93.** 1111$_{\text{two}}$ = 15   **95.** 30$_{\text{four}}$ = 12

Chapter 4 Test
1. Egyptian; 1534   **2.** Roman; 10,474   **3.** Chinese; 385
4. Babylonian; 155,540   **5.** Mayan; 50,511   **6.** Greek;
53,524   **7.** 1035   **8.** 23,862   **9.** 13,227   **10.** 1476
11. 37   **12.** 73   **13.** 230   **14.** 48,879   **15.** 110001$_{\text{two}}$
16. 43210$_{\text{five}}$ **17.** 256$_{\text{eight}}$ **18.** E8D$_{\text{sixteen}}$ **19.** 1105
20. 101110101101$_{\text{two}}$ **21.** There is less repetition of
symbols.   **22.** Place values are understood by position.
23. There are fewer symbols to learn.   **24.** There are
fewer digits in the numerals.   **25.** Answers will vary.
26. 7$_{\text{nine}}$ = 21$_{\text{three}}$, 6$_{\text{nine}}$ = 20$_{\text{three}}$, and 5$_{\text{nine}}$ = 12$_{\text{three}}$. So
765$_{\text{nine}}$ = 212012$_{\text{three}}$.

CHAPTER 5 NUMBER THEORY

5.1 Exercises
1. false   **3.** true   **5.** true   **7.** true   **9.** 1, 2, 3, 4, 6, 12
11. 1, 2, 4, 7, 14, 28   **13. (a)** no   **(b)** yes   **(c)** no
(d) no   **(e)** no   **(f)** no   **(g)** no   **(h)** no   **(i)** no
15. (a) yes   **(b)** yes   **(c)** yes   **(d)** yes   **(e)** yes
(f) yes   **(g)** yes   **(h)** yes   **(i)** yes   **17.** Answers will vary.
19. square root; square root; square root   **21.** 101, 103,
107, 109, 113, 127, 131, 137, 139, 149, 151, 157, 163, 167,
173, 179, 181, 191, 193, 197, 199; There are 21.   **23.** 179
and 193, 181 and 197, 191 and 197, 193 and 199   **25.** 2, 3; no
27. It must be 0.
29.

$1320 = 3 \cdot 2^3 \cdot 5 \cdot 11$
$= 2^3 \cdot 3 \cdot 5 \cdot 11$

31. $2 \cdot 3^2 \cdot 7$ **33.** $7 \cdot 13^2$ **35.** $2 \cdot 3 \cdot 5 \cdot 19$ **37.** no
39. yes   **41.** no   **43.** no   **45.** yes   **47.** yes   **49.** The
number must be divisible by both 3 and 5. That is, the sum
of the digits must be divisible by 3, and the last digit must
be 5 or 0.   **51.** 0, 2, 4, 6, 8   **53.** 0, 4, 8   **55.** 0, 6   **57.** 8
59. 27   **61.** 64   **63.** leap year   **65.** not a leap year
67. Answers will vary.   **69.** Answers will vary.
71. 1981 and 1987   **73.** 2041

5.2 Exercises
1. false   **3.** true   **5.** true
7. $37 = 1^2 + 6^2$; $41 = 4^2 + 5^2$; $53 = 2^2 + 7^2$
9. 6   **11.** 6   **13. (a)** 65,537   **(b)** 251   **15.** Answers will
vary.   **17.** Answers will vary.   **19.** Answers will vary.
21. composite; 30,031 = 59 · 509   **23.** 63   **25.** $2^p - 1$
27. 3 and 31   **29.** Answers will vary.

5.3 Exercises
1. true   **3.** true   **5.** true   **7.** false   **9.** false
11. The sum of the proper divisors is 496:
$1 + 2 + 4 + 8 + 16 + 31 + 62 + 124 + 248 = 496.$
13. 8191 is prime; 33,550,336   **15.** $85 = 6^2 + 7^2 = 2^2 + 9^2$
17. deficient   **19.** abundant   **21.** 12, 18, 20, 24
23. $1 + 3 + 5 + 7 + 9 + 15 + 21 + 27 + 35 + 45 + 63 +$
$105 + 135 + 189 + 315 = 975$, and $975 > 945$, so 945 is
abundant.   **25.** $1 + 2 + 4 + 8 + 16 + 32 + 37 + 74 +$
$148 + 296 + 592 = 1210$ and $1 + 2 + 5 + 10 + 11 + 22 +$
$55 + 110 + 121 + 242 + 605 = 1184$ **27.** $5 + 7$
29. $11 + 29$ **31. (a)** Let $a = 5$ and $b = 3$; $11 = 5 + 2 \cdot 3$
(b) $17 = 3 + 2 \cdot 7 = 7 + 2 \cdot 5 = 11 + 2 \cdot 3 = 13 + 2 \cdot 2$
33. 59 and 61   **35.** 4   **37.** 58 and 85   **39. (a)** $3^4 - 1 = 80$
is divisible by 5.   **(b)** $2^6 - 1 = 63$ is divisible by 7.
41. $5^2 + 2 = 27 = 3^3$ **43.** Answers will vary.
45. False; For the first six, the sequence is 6, 8, 6, 8, 6, 6.
47. one; not happy   **49.** both; happy   **51.** B   **53.** 7; yes
55. 15; no   **57.** 24; 23; 25; yes; no   **59.** Answers will vary.
61. B   **63.** $20 = 1 + 4 + 5 + 10$
65. $18 = 1 + 2 + 6 + 9 = 3 + 6 + 9$
67. $1 + 2 + 5 + 7 + 10 + 14 + 35 = 74 > 70$
69. $2^2 \cdot 11 \cdot 19$ **71.** 11; 1, 2, 4, 11, 19, 22, 38, 44, 76, 209,
418; 844   **73.** 8; no
75. 1,000,000,000,000,066,600,000,000,000,001   **77.** 13

5.4 Exercises
1. true   **3.** false   **5.** false   **7.** true   **9.** false   **11.** 28
13. 11   **15.** 17   **17.** 10   **19.** 30   **21.** 6   **23.** 6   **25.** 18
27. 45   **29.** 240   **31.** 405   **33.** 300   **35.** 108   **37.** 693
39. 2160   **41.** 180   **43.** 180   **45.** 9450

47. (a) $p^a q^a r^b$ (b) $p^b q^b r^c$ **49.** 15 **51.** 12
53. p and q are relatively prime. **55.** 144th
57. 48 (5 stacks of 48 pennies, 6 stacks of 48 nickels) **59.** $600; 25 books **61.** 16 inches

5.5 Exercises

1. true **3.** false **5.** false **7.** 987
9. o, o, e, o, o, e, o, o, e, o, o, e **11.** $\frac{1 + \sqrt{5}}{2}$
13. $1 + 2 + 5 + 13 + 34 = 55$; Each expression is equal to 55.
15. $1 + 1 + 2 + 3 + 5 = 13 - 1$; Each expression is equal to 12. **17.** $1 - 2 + 5 - 13 + 34 = 5^2$; Each expression is equal to 25. **19.** $13^2 - 5^2 = 144$; Each expression is equal to 144. **21.** (There are other ways to do this.)
(a) $39 = 34 + 5$ (b) $59 = 55 + 3 + 1$ (c) $99 = 89 + 8 + 2$
23. (a) The greatest common factor of 10 and 4 is 2, and the greatest common factor of $F_{10} = 55$ and $F_4 = 3$ is $F_2 = 1$. (b) The greatest common factor of 12 and 6 is 6, and the greatest common factor of $F_{12} = 144$ and $F_6 = 8$ is $F_6 = 8$. (c) The greatest common factor of 14 and 6 is 2, and the greatest common factor of $F_{14} = 377$ and $F_6 = 8$ is $F_2 = 1$. **25.** (a) $8^2 - 3 \cdot 21 = 1$ (b) $2 \cdot 34 - 8^2 = 4$
(c) $8^2 - 1 \cdot 55 = 9$ (d) The difference will be 25, because we are obtaining the squares of the terms of the Fibonacci sequence. $13^2 - 1 \cdot 144 = 25 = 5^2$. **27.** 123
29. Each sum is 2 less than a Lucas number.
31. (a) $8 \cdot 18 = 144$; Each expression is equal to 144.
(b) $8 + 21 = 29$; Each expression is equal to 29.
(c) $8 + 18 = 2 \cdot 13$; Each expression is equal to 26.
(d) $18 + 47 = 5 \cdot 13$; Each expression is equal to 65.
33. 3, 4, 5 **35.** 105, 208, 233 **37.** The sums are 1, 1, 2, 3, 5, 8, 13. They are terms of the Fibonacci sequence.
39. $\frac{1 + \sqrt{5}}{2} \approx 1.618033989$ and $\frac{1 - \sqrt{5}}{2} \approx -0.618033989$.
After the decimal point, the digits are the same.
41. negative **43.** 987

Chapter 5 Test

1. true **2.** false **3.** true **4.** true **5.** false **6.** true
7. (a) yes (b) yes (c) no (d) yes (e) yes (f) no
(g) no (h) yes (i) no **8.** (a) composite (b) prime
(c) neither **9.** $2^2 \cdot 3 \cdot 5 \cdot 7 \cdot 13$ **10.** Answers will vary.
11. (a) abundant (b) deficient (c) perfect **12.** B
13. 71 and 73 **14.** 9 **15.** 2002 **16.** Monday
17. 28,657 **18.** $55 - (5 + 8 + 13 + 21) = 8$; Each expression is equal to 8. **19.** B **20.** 68
21. The process will yield 4 for any term chosen. **22.** A

CHAPTER 6 THE REAL NUMBERS AND THEIR REPRESENTATIONS

6.1 Exercises

1. 5 **3.** 0 **5.** $\sqrt{18}$ (There are others.) **7.** true
9. true **11.** (a) 3, 7 (b) 0, 3, 7 (c) $-9, 0, 3, 7$
(d) $-9, -1\frac{1}{4}, -\frac{3}{5}, 0, 3, 5.9, 7$ (e) $-\sqrt{7}, \sqrt{5}$
(f) All are real numbers. **13.** 2,592,000
15. 16,162 **17.** -665.75 **19.** 5436 **21.** $-220°$
23. Pacific Ocean, Indian Ocean, Caribbean Sea, South China Sea, Gulf of California **25.** true
27. **29.**
31. (a) A (b) A (c) B (d) B **33.** (a) 2 (b) 2
35. (a) -6 (b) 6 **37.** (a) -3 (b) 3 **39.** (a) 0
(b) 0 **41.** -12 **43.** -8 **45.** 3 **47.** $|-3|$, or 3
49. $-|-6|$, or -6 **51.** $|5 - 3|$, or 2 **53.** true
55. true **57.** true **59.** false **61.** true **63.** false
65. (a) Austin; an increase of 19.93%
(b) Chicago; an increase of 0.35%
67. Transportation (because $|-4.2| > |0.9|$)

6.2 Exercises

1. positive; $18 + 6 = 24$ **3.** greater; $-14 + 9 = -5$
5. the number with the greater absolute value is subtracted from the one with the lesser absolute value; $5 - 12 = -7$ **7.** additive inverses; $4 + (-4) = 0$
9. negative; $-5(15) = -75$ **11.** -20 **13.** -4
15. -11 **17.** 9 **19.** 20 **21.** 24 **23.** -1296 **25.** 6
27. -6 **29.** 0 **31.** -6 **33.** 27 **35.** undefined **37.** -1
39. -4 **41.** 7 **43.** 13 **45.** A, B, C **47.** commutative property of addition **49.** inverse property of addition
51. identity property of multiplication **53.** identity property of addition **55.** associative property of multiplication
57. closure property of multiplication **59.** associative property of addition **61.** distributive property
63. identity **65.** (a) messing up your room
(b) spending money (c) decreasing the volume on your telephone **67.** -81 **69.** 81 **71.** -81 **73.** -81
75. (a) -11; The returns decreased 11 percent. (b) -13; The returns decreased 13 percent. (c) 13; The returns increased 13 percent. (d) -1; The returns decreased 1 percent. **77.** 50,395 ft **79.** 1345 ft **81.** 136 ft
83. -12 **85.** -6 **87.** $+5$ degrees **89.** -2 degrees
91. 0 degrees (no change) **93.** $+5$ degrees
95. 43; 34; 29; 36; 38; 36; 37; 41 **97.** 27 feet
99. $-285 - (-495) = 210$ **101.** 469 B.C. **103.** $-60°$F
105. 116°F

6.3 Exercises

1. rational number   **3.** 1   **5.** 459; 221; 5967
7. $\frac{7}{12}$; $\frac{4}{3}$ **9.** 6 or 4 (Either answer is acceptable.)
11. A, C, D   **13.** C   **15.** $\frac{1}{3}$ **17.** $-\frac{3}{7}$ **Answers will vary**
in Exercises 19 and 21.   **19.** $\frac{6}{16}, \frac{9}{24}, \frac{12}{32}$ **21.** $-\frac{10}{14}, -\frac{15}{21}, -\frac{20}{28}$
23. (a) $\frac{1}{3}$ (b) $\frac{1}{4}$ (c) $\frac{2}{5}$ (d) $\frac{1}{3}$ **25.** the dots in the
intersection of the triangle and the rectangle as a part of
the dots in the entire figure   **27.** (a) Callie (b) Jessica
(c) Jessica (d) Terry (e) Johnny and Rocky; $\frac{1}{2}$ **29.** $\frac{1}{2}$
31. $\frac{43}{48}$ **33.** $-\frac{5}{24}$ **35.** $\frac{23}{56}$ **37.** $\frac{27}{20}$ **39.** $\frac{5}{12}$ **41.** $\frac{1}{9}$ **43.** $\frac{3}{2}$
45. $\frac{3}{2}$ **47.** $\frac{13}{3}$ **49.** $\frac{29}{10}$ **51.** $6\frac{3}{4}$ **53.** $6\frac{1}{8}$ **55.** $-17\frac{7}{8}$
57. $1\frac{5}{16}$ **59.** $37\frac{7}{8}$ **61.** $7\frac{3}{8}$ **63.** $4\frac{1}{8}$ **65.** $\frac{9}{16}$ inch
67. $30\frac{1}{4}$ in.   **69.** (a) $1\frac{1}{8}$ in. (b) $1\frac{7}{8}$ in.   **71.** $11\frac{55}{64}$ in.
73. 8 cakes (There will be some sugar left over.)
75. $16\frac{5}{8}$ yd   **77.** $9638   **79.** $\frac{5}{8}$ **81.** $\frac{19}{30}$ **83.** $-\frac{3}{4}$
85. repeating   **87.** terminating   **89.** terminating
91. 0.75   **93.** 0.1875   **95.** $0.\overline{27}$ **97.** $0.\overline{285714}$ **99.** $\frac{2}{5}$
101. $\frac{17}{20}$ **103.** $\frac{467}{500}$ **105.** $\frac{67}{99}$ **107.** $\frac{7}{165}$ **109.** $\frac{1}{90}$
111. (a) $0.\overline{3}$, or 0.333 . . . (b) $0.\overline{6}$, or 0.666 . . .
(c) $0.\overline{9}$, or 0.999 . . . (d) $1 = 0.\overline{9}$

6.4 Exercises

1. true   **3.** true   **5.** false   **7.** rational   **9.** irrational
11. rational   **13.** rational   **15.** irrational   **17.** rational
19. irrational   **21.** (a) $0.\overline{8}$ (b) irrational; rational
**The number of digits shown will vary among calculator
models in Exercises 23–29 and 49–59.**
23. 6.244997998   **25.** 3.885871846   **27.** 29.73213749
29. 1.060660172   **31.** 6   **33.** 2.4 m^2 **35.** 5.4 ft
37. 2.5 sec   **39.** 1.7 amps   **41.** 71 mph   **43.** 54 mph
45. 392,000 mi^2 **47.** The area and the perimeter are both
numerically equal to 36.   **49.** $5\sqrt{2}$; 7.071067812
51. $5\sqrt{3}$; 8.660254038   **53.** $12\sqrt{2}$; 16.97056275
55. $\frac{5\sqrt{6}}{6}$; 2.041241452   **57.** $\frac{\sqrt{7}}{2}$; 1.322875656   **59.** $\frac{\sqrt{21}}{3}$;
1.527525232   **61.** $3\sqrt{17}$ **63.** $4\sqrt{7}$ **65.** $10\sqrt{2}$
67. $3\sqrt{3}$ **69.**

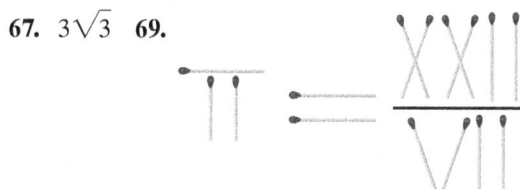

71. The result is 3.1415929, which agrees with the first
seven digits in the decimal for π.   **73.** 3   **75.** 4
77. The computer will forever be occupied by this task
"to the exclusion of all else" and thus will no longer be
able to interfere.   **79.** ϕ is positive, and its conjugate is
negative. The units digit of ϕ is 1, and the units digit of its
conjugate is 0. The decimal digits agree.   **81.** If a 9 were
under the radical, the number would equal the "nice"
integer value 3. $\sqrt{3}$ is irrational, and its decimal is not so
nice.   **83.** It is just a coincidence that 1828 appears back-to-
back early in the decimal. There is no repetition indefinitely,
which would be indicative of a rational number.
85. 4   **87.** 7   **89.** 6   **91.** 1   **93.** 4   **95.** 8   **The number
of decimal digits shown will vary among calculator
models in Exercises 97–103.**   **97.** 3.50339806
99. 5.828476683   **101.** 10.06565066   **103.** 5.578019845

6.5 Exercises

1. true   **3.** true   **5.** true   **7.** false   **9.** 11.315
11. −4.215   **13.** 0.8224   **15.** 47.5   **17.** 31.6   **19.** $30.13
21. (a) $466.02 (b) $190.68   **23.** $1229.09
25. (a) 50,000 (b) 55,000 (c) 54,800 (d) 54,790
27. (a) 0.89245 (b) 0.8925 (c) 0.892 (d) 0.89 (e) 0.9
(f) 1   **29.** (a) E (b) D (c) B (d) F (e) C
(f) A   **31.** (a) $33\frac{1}{3}$% (b) 25% (c) 40% (d) $33\frac{1}{3}$%
33. 42%   **35.** 36.5%   **37.** 0.8%   **39.** 210%   **41.** 0.96
43. 0.0546   **45.** 0.003   **47.** 4.00   **49.** 0.005   **51.** 0.035
53. 20%   **55.** 1%   **57.** $37\frac{1}{2}$%   **59.** 150%   **61.** Answers
will vary.   **63.** 124.8   **65.** 2.94   **67.** 150%   **69.** 600
71. 1.4%   **73.** 8%   **75.** 3.2%   **77.** 25.1%   **79.** 740%
81. about 122%   **83.** 21.2%   **85.** $895 million   **87.** A
89. C   **91.** No, the price is $96.00.   **93.** 0.79 m^2
95. 0.44 m^2 **97.** 116 mg   **99.** 129 mg
101. (a) $14.7 - 40 \cdot 0.13$ (b) 9.5 (c) 8.075; walking
(5 mph)   **103.** (a) 0.031 (b) 0.035   **105.** 297
107. (a) .574 (b) .463 (c) .562 (d) .469   **109.** $4.50
111. $0.75   **113.** $12.00   **115.** $36.00   **117.** three (and
you would have 0.01¢ left over)   **119.** $0.06, or 6¢
121. 1000   **123.** 25,000%

Chapter 6 Test

1. (a) 12 (b) 0, 12 (c) −4, 0, 12 (d) $-4, -\frac{3}{2}, -0.5$,
0, 4.1, 12 (e) $-\sqrt{5}, \sqrt{3}$ (f) $-4, -\sqrt{5}, -\frac{3}{2}, -0.5$,
0, $\sqrt{3}$, 4.1, 12   **2.** (a) C (b) B (c) D (d) A
3. (a) false (b) true (c) true (d) false   **4.** 4
5. 10   **6.** 3   **7.** −43   **8.** 213.8°F   **9.** 5296 ft   **10.** (a) E
(b) A (c) B (d) D (e) F (f) C   **11.** (a) Mary and
Janie (b) Priya and Charlene (c) Mary (d) Jackie
and Mabel; $\frac{2}{5}$ (e) Janie   **12.** $\frac{11}{16}$ **13.** $\frac{57}{160}$ **14.** $-\frac{2}{5}$
15. $\frac{4}{9}$ **16.** (a) 0.45 (b) $0.41\overline{6}$ **17.** (a) $\frac{18}{25}$ (b) $\frac{58}{99}$
18. (a) irrational (b) rational (c) rational (d) rational
(e) irrational (f) irrational   **19.** (a) 12.247448714 (b) $5\sqrt{6}$

20. (a) 4.913538149 **(b)** $\frac{13\sqrt{7}}{7}$ **21. (a)** -45.254834
(b) $-32\sqrt{2}$ **22.** 0.08, or 8% **23. (a)** 13.81
(b) -0.315 **(c)** 38.7 **(d)** -24.3 **24. (a)** 350
(b) 346.04 **(c)** 346.045 **25.** \$54.00 **26.** D
27. (a) 8% **(b)** 81% **(c)** 4,412,000
28. (a) $26\frac{2}{3}\%$ **(b)** $66\frac{2}{3}\%$ **29.** 1656; 1008; 16%; 6%
30. 75 mg

CHAPTER 7 THE BASIC CONCEPTS OF ALGEBRA

7.1 Exercises

1. A **3.** Two equations are equivalent if they have the same solution set. **5.** 30 **7.** $\{-1\}$ **9.** $\{3\}$
11. $\{-7\}$ **13.** $\{0\}$ **15.** $\left\{-\frac{5}{3}\right\}$ **17.** $\left\{-\frac{1}{2}\right\}$ **19.** $\{2\}$
21. $\{-2\}$ **23.** $\{7\}$ **25.** $\{2\}$ **27.** $\{4\}$ **29.** $\{0\}$
31. $\{-2\}$ **33.** $\{2000\}$ **35.** $\{25\}$ **37.** $\{40\}$
39. contradiction; $\varnothing$ **41.** conditional; $\{-8\}$
43. conditional; $\{0\}$ **45.** identity; {all real numbers}
47. $A = \frac{150p}{w}$ **49.** $\ell = \frac{v \cdot ACH}{60}$ **51.** $t = 2\ell - 2a - 2b$
53. $\ell = \frac{r_1 x - r_2 x}{r_2}$, or $\ell = \frac{r_1 x}{r_2} - x$ **55.** $r = \frac{\ell + h^2}{2h}$
57. $\mathscr{A} = \frac{20TS - dT}{4}$ **59.** $t = \frac{d}{r}$ **61.** $b = \frac{\mathscr{A}}{h}$ **63.** $a = P - b - c$
65. $b = \frac{2\mathscr{A}}{h}$ **67.** $h = \frac{S - 2\pi r^2}{2\pi r}$, or $h = \frac{S}{2\pi r} - r$
69. $h = \frac{3V}{\pi r^2}$ **71.** 59 **73.** -4 **75.** 5 **77.** -25

7.2 Exercises

1. D; There cannot be a fractional number of cars.
3. A; Distance cannot be negative. **5.** expression
7. equation **9.** expression **11.** $x - 12$
13. $(x - 6)(x + 4)$ **15.** $\frac{25}{x}$ $(x \neq 0)$ **17.** 3 **19.** 6
21. -3 **23.** Beyoncé: \$169 million; Guns N' Roses:
\$131 million **25.** wins: 67; losses: 15 **27.** Democrats: 46;
Republicans: 52 **29.** shortest piece: 15 inches; middle
piece: 20 inches; longest piece: 24 inches **31.** gold: 26;
silver: 18; bronze: 26 **33.** 70 milliliters **35.** \$250
37. \$24.85 **39.** 4 liters **41.** 5 liters **43.** 1 gallon
45. \$4000 at 1.5%; \$8000 at 2% **47.** \$10,000 at 2.25%;
\$19,000 at 1.5% **49.** Pennies: 17; dimes: 17; quarters: 10
51. students: 305; non students: 105 **53.** Row 1 Seats: 54;
Row 2 Seats: 51 **55.** 328 miles **57.** No. The distance is
$55\left(\frac{1}{2}\right) = 27.5$ miles. **59.** 3.218 hours **61.** 1.715 hours
63. 9.18 meters per second **65.** 10.11 meters per second
67. 18 miles **69.** $1\frac{3}{4}$ hours **71.** 11:00 A.M. **73.** 8 hours

7.3 Exercises

1. compare: A, D **3. (a)** C **(b)** D **(c)** B **(d)** A

5. $\frac{5}{8}$ **7.** $\frac{1}{4}$ **9.** $\frac{2}{1}$ **11.** $\frac{3}{1}$ **13.** 10-lb size; \$0.429
15. 32-oz size; \$0.093 **17.** 128-oz size; \$0.051
19. 36-oz size; \$0.049 **21.** $\{35\}$ **23.** $\{-1\}$ **25.** $\left\{-\frac{27}{4}\right\}$
27. 2 tablets **29.** 0.8 milliliter **31.** 910 milligrams
33. 7.5 milliliters **35.** 12 milligrams **37.** \$44.55
39. 12,500 fish **41.** $25\frac{2}{3}$ inches **43.** 2.0 inches
45. $2\frac{5}{8}$ cups **47. (a)** 30 minutes **(b)** Yes
49. 20 pounds **51.** about 302 pounds **53.** \$450
55. 100 pounds per square inch **57.** 144 feet
59. 448.1 pounds **61.** 24 **63.** 42 **65.** 6.2 pounds
67. \$252 **69.** \$273 **71.** about 267 square miles

7.4 Exercises

1. $>$, $<$ (or $<$, $>$); $\geq$, $\leq$ (or $\leq$, $\geq$)
3. $(0, \infty)$ **5.** D **7.** B **9.** F

11. $[7, \infty)$

13. $[5, \infty)$

15. $(7, \infty)$

17. $(-4, \infty)$

19. $(-\infty, -40]$

21. $(-\infty, 4]$

23. $\left(-\infty, -\frac{15}{2}\right)$

25. $\left[\frac{1}{2}, \infty\right)$

27. $(3, \infty)$

29. $(-\infty, 4)$

31. $\left(-\infty, \frac{23}{6}\right]$

33. $(1, 11)$

35. $[-14, 10]$

37. $[-5, 6]$

39. $\left[-\frac{14}{3}, 2\right]$

41. $\left[-\frac{1}{2}, \frac{35}{5}\right]$

43. $\left(-\frac{1}{3}, \frac{1}{9}\right]$

45. after 40 miles
47. 2 miles
49. 80 or greater
51. 90 or greater
53. 71 or greater
55. $-0.15 \leq x - 18 \leq 0.15$; between and inclusive of
17.85 and 18.15 ounces **57.** $-0.75 \leq x - 52.5 \leq 0.75$;
between and inclusive of 51.75 and 53.25 milliliters

59. (a) 140 to 184 pounds   **(b)** Answers will vary.
61. 26 DVDs

7.5 Exercises

1. (a) C   **(b)** A   **(c)** B   **(d)** D   **3. (a)** 6; 4; 6.3; 4
(b) 5; 2; 5.71; -2 **5.** 16   **7.** 625   **9.** -32 **11.** -8
13. -81 **15.** $\frac{1}{8}$ **17.** $\frac{1}{49}$ **19.** $-\frac{1}{49}$ **21.** $\frac{16}{5}$ **23.** 125
25. $\frac{25}{16}$ **27.** $\frac{9}{20}$ **29.** 1   **31.** 1   **33.** 0   **35.** 10^{16} **37.** $\frac{1}{10}$
39. $\frac{1}{5}$ **41.** 10^2 **43.** 10^3 **45.** 10^1, or 10   **47.** 5^2 **49.** 10^{11}
51. $\frac{1}{10^6}$ **53.** $\frac{4}{10^{14}}$ **55.** 2.3×10^2 **57.** 8.1×10^9
59. 1.4×10^{12} **61.** 2×10^{-2} **63.** 5.623×10^{-1}
65. 4.4×10^{-6} **67.** 6500   **69.** 28,000,000   **71.** 15,000,000,000
73. 0.42   **75.** 0.0152   **77.** 0.00000833   **79.** 6×10^5
81. 2×10^5 **83.** 2×10^5 **85. (a)** 3.174×10^8
(b) $\$1.727 \times 10^{13}$ **(c)** \$54,410   **87.** \$109.02
89. approximately 9.5×10^{-7} parsec   **91.** 300 seconds
93. approximately 5.87×10^{12} miles   **95. (a)** 5.95×10^2
people per square mile   **(b)** 999 square miles   **97.** When
Kirk said "one to the fourth power," he indicated that the
sound would *not* be changed at all, because $1^4 = 1$.

7.6 Exercises

1. B   **3.** C   **5.** $x^2 - x + 3$ **7.** $9x^2 - 4x + 4$
9. $6x^4 - 2x^3 - 7x^2 - 4x$ **11.** $-2x^2 - 13x + 11$
13. $x^2 - 5x - 24$ **15.** $28x^2 + x - 2$ **17.** $12x^5 - 20x^3 + 4x^2$
19. $4x^2 - 9$ **21.** $x^2 + 14x + 49$ **23.** $25x^2 - 30x + 9$
25. $10x^4 + 34x^3 + 7x^2 - 13x + 6$ **27.** $12(x + 3)$
29. $2x(4x^2 - 2x + 3)$ **31.** $2x^2(4x^2 + 3x - 6)$
33. $(x - 5)(x + 3)$ **35.** $(x + 7)(x - 5)$
37. $6(x - 10)(x + 2)$ **39.** $3x(x + 1)(x + 3)$
41. $(3x - 2)(2x + 3)$ **43.** $(5x + 3)(x - 2)$
45. $(7x + 2)(3x - 1)$ **47.** $2x^2(4x - 1)(3x + 2)$
49. $(x - 4)^2$ **51.** $(x + 7)^2$ **53.** $(3x - 2)^2$
55. $2(4x - 3)^2$ **57.** $(2x + 7)^2$ **59.** $(x + 6)(x - 6)$
61. $(10 + x)(10 - x)$ **63.** $(3x + 4)(3x - 4)$
65. $(5x^2 + 3)(5x^2 - 3)$ **67.** $(x^2 + 25)(x + 5)(x - 5)$
69. $13x(x^2 + 3x + 4)$ **71.** $(6x - 7)(2x + 5)$
73. $(2x + 7)^2$ **75.** $(x^4 + 9)(x^2 + 3)(x^2 - 3)$

7.7 Exercises

1. cannot   **3.** two (2)   **5.** $\{-3, 9\}$ **7.** $\left\{\frac{7}{2}, -\frac{1}{5}\right\}$
9. $\{-7, 0\}$ **11.** $\{-3, 4\}$ **13.** $\{-7, -2\}$ **15.** $\left\{-\frac{5}{2}, 1\right\}$
17. $\left\{-\frac{1}{2}, \frac{1}{6}\right\}$ **19.** $\{-2, 4\}$ **21.** $\{\pm 8\}$ **23.** $\{\pm 2\sqrt{6}\}$
25. $\varnothing$ **27.** $\{1, 7\}$ **29.** $\{4 \pm \sqrt{3}\}$ **31.** $\left\{\frac{5 \pm \sqrt{13}}{2}\right\}$
33. $\left\{\frac{-3 \pm \sqrt{37}}{2}\right\}$ **35.** $\left\{\frac{1 \pm \sqrt{3}}{2}\right\}$ **37.** $\left\{\frac{1 \pm \sqrt{5}}{2}\right\}$
39. $\left\{\frac{-1 \pm \sqrt{2}}{2}\right\}$ **41.** $\left\{\frac{1 \pm \sqrt{29}}{2}\right\}$ **43.** $\varnothing$ **45.** eastbound

ship: 80 miles; southbound ship: 150 miles   **47.** 120 feet
49. 412.3 feet   **51.** 8, 15, 17   **53.** 5 centimeters,
12 centimeters, 13 centimeters   **55. (a)** 1 second and
8 seconds   **(b)** 9 seconds after it is projected
57. 0.7 second and 4.0 seconds   **59.** 1 foot
61. length: 26 meters; width: 16 meters

Chapter 7 Test

1. $\{2\}$ **2.** $\{4\}$ **3.** identity; {all real numbers}
4. $v = \frac{S + 16t^2}{t}$, or $v = \frac{S}{t} + 16t$ **5.** Hawaii: 4021 square
miles; Maui: 728 square miles; Kauai: 551 square miles
6. 5 liters   **7.** 2.2 hours   **8.** 16 slices for \$4.38
9. 3 tablets   **10.** 2300 miles   **11.** 200 amps
12. $(-\infty, 4]$ ⟵┼┼┼╌┤┼┼┼┼⟶   **13.** $(-2, 6]$ ⟵┼╀┼╌┤┼⟶
　　　　　　　　　0　　4　　　　　　　　　　　　　　　-2 0　　6
14. 82 or greater   **15.** $\frac{16}{9}$ **16.** -64 **17.** $\frac{64}{27}$ **18.** 0   **19.** 1
20. 10^1, or 10   **21. (a)** 693,000,000   **(b)** 0.000000125
22. 3×10^{-4} **23.** about 255 minutes   **24.** $4x^2 + 6x + 10$
25. $15x^2 - 14x - 8$ **26.** $16x^4 - 9$ **27.** $x^2 + 12x + 36$
28. $(x + 7)(x - 4)$ **29.** $(2x + 9)(x + 5)$
30. $(5x + 7)(5x - 7)$ **31.** $(x - 9)^2$
32. $2x(x - 5)(x - 4)$ **33.** $\left\{-\frac{3}{2}, \frac{1}{3}\right\}$ **34.** $\{\pm\sqrt{13}\}$
35. $\left\{\frac{1 \pm \sqrt{29}}{2}\right\}$ **36.** 0.87 second

CHAPTER 8 GRAPHS, FUNCTIONS, AND SYSTEMS OF EQUATIONS AND INEQUALITIES

8.1 Exercises

1. y **3.** $(0, -4)$ **5.** 5   **7.** $(0, 0)$; 7   **9.** $(4, -7)$; 3
11. (a) I   **(b)** III   **(c)** II   **(d)** IV   **(e)** none
13.–21.

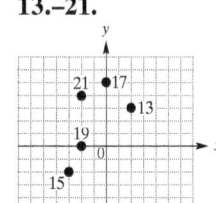

23. (a) $\sqrt{34}$ **(b)** $\left(\frac{1}{2}, \frac{5}{2}\right)$
25. (a) $\sqrt{61}$ **(b)** $\left(\frac{1}{2}, 1\right)$
27. (a) $\sqrt{146}$ **(b)** $\left(-\frac{1}{2}, \frac{3}{2}\right)$
29. (a) $\sqrt{40} = 2\sqrt{10}$
(b) $(-1, 4)$

31. B   **33.** D   **35.** $x^2 + y^2 = 36$
37. $(x + 1)^2 + (y - 3)^2 = 16$ **39.** $x^2 + (y - 4)^2 = 3$
41. $(-2, -3)$; 2   **43.** $(-5, 7)$; 9   **45.** $(2, 4)$; 4

47.

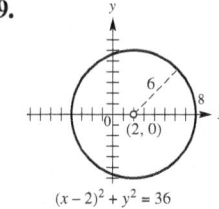

$x^2 + y^2 = 36$

49.

$(x - 2)^2 + y^2 = 36$

51.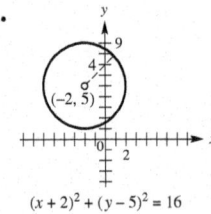

$(x+2)^2 + (y-5)^2 = 16$

53.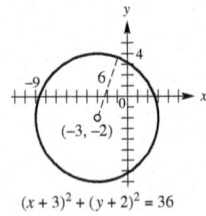

$(x+3)^2 + (y+2)^2 = 36$

55. (a) 26.1%   **(b)** This is very close to the actual figure.   **57.** \$23,439   **59.** Answers will vary.

61. The epicenter is $(-2, -2)$.

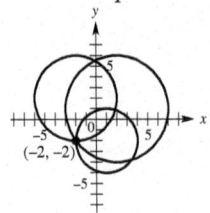

8.2 Exercises

1. $\frac{3}{10}$ **3.** $\frac{3}{2}$ **5. (a)** positive   **(b)** undefined   **(c)** negative   **(d)** zero

7. $(0, -4), (4, 0), (2, -2),$
$(3, -1)$

9. $(0, 5), \left(\frac{5}{2}, 0\right), (1, 3),$
$(2, 1)$

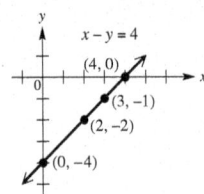

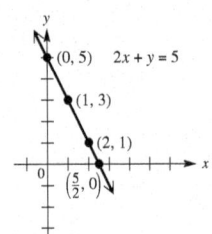

11. $(0, 4), (5, 0), \left(3, \frac{8}{5}\right), \left(\frac{5}{2}, 2\right)$

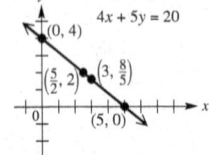

13.

x	y
0	4
$\frac{8}{3}$	0
2	1
4	-2

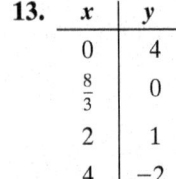

15.

x	y
0	3
3	2
9	0
12	-1

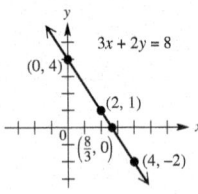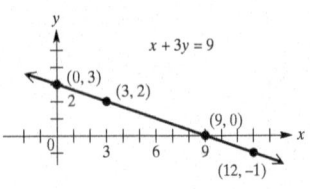

17. $(4, 0)$; $(0, 6)$

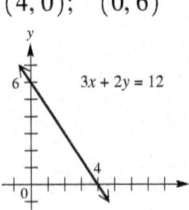

19. $(2, 0)$; $\left(0, \frac{5}{3}\right)$

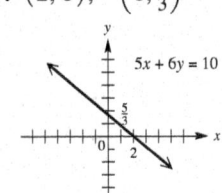

21. $\left(\frac{5}{2}, 0\right)$; $(0, -5)$

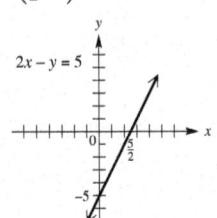

23. $(2, 0)$; $\left(0, -\frac{2}{3}\right)$

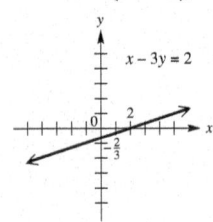

25. $(0, 0)$; $(0, 0)$

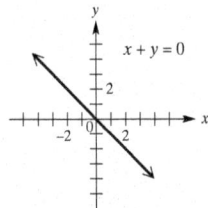

27. $(0, 0)$; $(0, 0)$

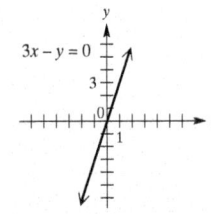

29. $(2, 0)$;   none

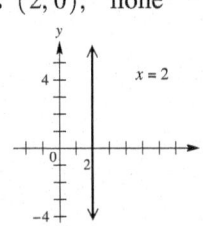

31. none; $(0, 4)$

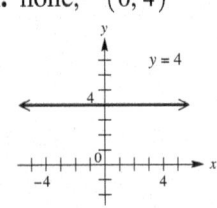

33. C   **35.** A   **37.** D   **39.** B   **41.** 8   **43.** $-\frac{5}{6}$ **45.** 0

47. undefined

49.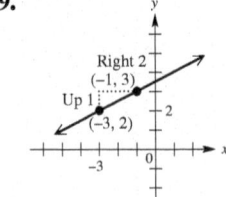

51.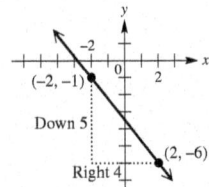

53.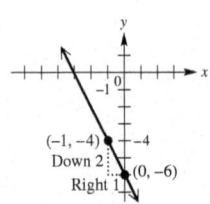

55.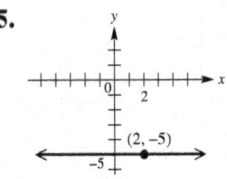

57.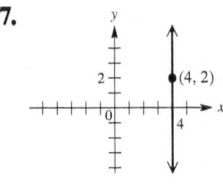

59. parallel   **61.** perpendicular   **63.** neither parallel nor perpendicular   **65.** $\frac{7}{10}$ **67.** $-\$4000$ per year; The value of the machine is decreasing by an average of $\$4000$ per year during those years.

69. (a) In 2016, there were 396 million wireless subscriber connections in the United States.   **(b)** 16   **(c)** The number of subscribers increased by an average of 16 million per year from 2011 to 2016.

71. (a) -9 theaters per year   **(b)** The negative slope means that the number of drive-in theaters *decreased* by an average of 9 per year from 2010 to 2017.   **73.** $\$0.05$ per year; The price of a gallon of gasoline increased by an average of $\$0.05$ per year from 1996 to 2016.

75. -16 million cameras per year; The number of digital cameras shipped worldwide decreased by an average of 16 million per year from 2010 to 2016.

8.3 Exercises

1. $-\frac{1}{3}$ **3.** -4 **5.** D   **7.** B   **9.** A   **11.** C   **13.** H
15. B   **17.** $y = 3x - 3$ **19.** $y = -x + 3$ **21.** $y = -2x + 18$
23. $y = -\frac{3}{4}x + \frac{5}{2}$ **25.** $y = \frac{1}{2}x + \frac{13}{2}$ **27.** $y = 4x - 12$
29. $y = 5$ **31.** $x = 9$ **33.** $x = 0.5$ **35.** $y = 8$
37. $y = 2x - 2$ **39.** $y = -\frac{1}{2}x + 4$ **41.** $y = 5$ **43.** $x = 7$
45. $y = -3$ **47.** $y = 5x + 15$ **49.** $y = -\frac{2}{3}x + \frac{4}{5}$
51. $y = \frac{2}{5}x + 5$ **53. (a)** $y = -x + 12$ **(b)** -1 **(c)** $(0, 12)$
55. (a) $y = -\frac{5}{2}x + 10$ **(b)** $-\frac{5}{2}$ **(c)** $(0, 10)$
57. (a) $y = \frac{2}{3}x - \frac{10}{3}$ **(b)** $\frac{2}{3}$ **(c)** $\left(0, -\frac{10}{3}\right)$ **59.** $y = 3x - 19$
61. $y = \frac{1}{2}x - 1$ **63.** $y = -\frac{1}{2}x + 9$ **65.** $y = 7$
67. $C = \frac{5}{9}F - \frac{160}{9}$ **69.** -40 degrees

8.4 Exercises

1. set of ordered pairs   **3.** range; dependent   **5.** linear; line   **7.** 5   **9.** 2   **11.** -1 **13.** -13 **15.** 3   **17.** 0

19.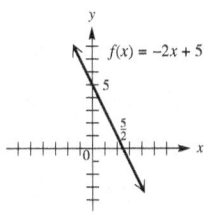

domain and range: $(-\infty, \infty)$

21.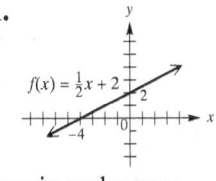

domain and range: $(-\infty, \infty)$

23.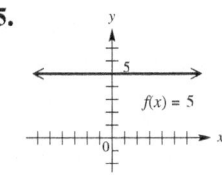

domain and range: $(-\infty, \infty)$

25.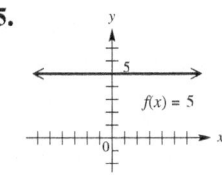

domain: $(-\infty, \infty)$; range: $\{5\}$

27. 4   **29.** 16   **31. (a)** $C(x) = 0.02x + 200$ **(b)** $R(x) = 0.04x$ **(c)** 10,000   **(d)**

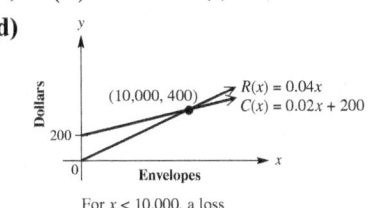

For $x < 10{,}000$, a loss
For $x > 10{,}000$, a profit

33. (a) $C(x) = 3.00x + 2300$ **(b)** $R(x) = 5.50x$ **(c)** 920   **(d)**

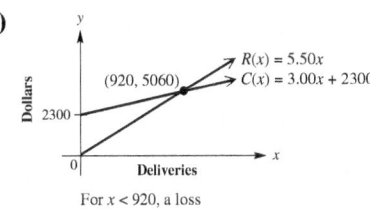

For $x < 920$, a loss
For $x > 920$, a profit

35. (a) $f(x) = 329.2x + 7132$ **(b)** $\$9107$
37. (a) $f(x) = 0.996x + 34.3$ **(b)** 89.1%
39. (a) $f(x) = 76.9x$ **(b)** 30,760 kilometers per second
41. (a) $\$0$; $\$2.50$; $\$5.00$; $\$7.50$ **(b)** $2.50x$

(c)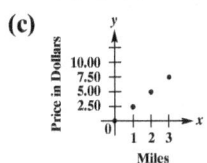

43. (a) $\$160$ **(b)** 70 mph   **(c)** 66 mph   **(d)** for speeds more than 80 mph   **45. (a)** 500 cubic centimeters   **(b)** 90° C

8.5 Exercises

1. $(0, 4)$ **3.** $x = -3$ **5.** narrower   **7.** F   **9.** C   **11.** E
13. $(0, 0)$ **15.** $(0, 4)$ **17.** $(1, 0)$ **19.** $(-3, -4)$
21. downward; narrower   **23.** upward; wider

25.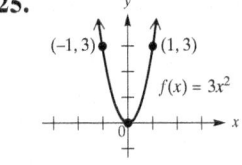

27.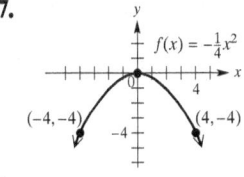

29.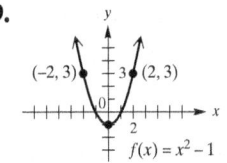

31.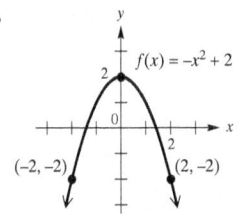

33.

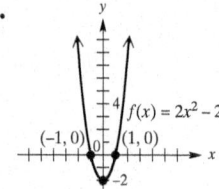

35.

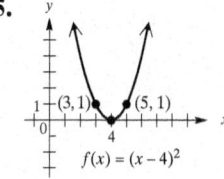

37.

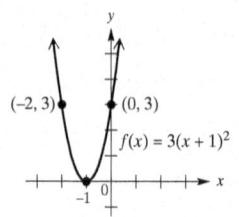

39.

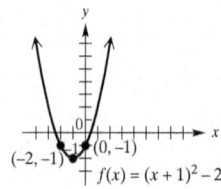

41.

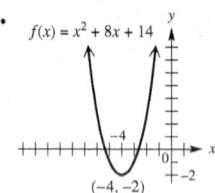

43.

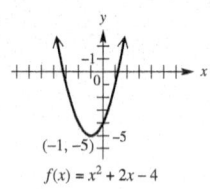

45.

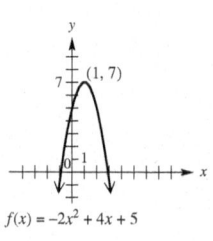

47. 25 meters; 625 square meters **49.** 16 feet; 2 seconds **51.** 4.1 seconds; 81.6 meters **53.** $f(45) = 161.5$; This means that when the speed is 45 mph, the stopping distance is 161.5 feet.

8.6 Exercises

1. rises; falls **3.** does not **5.** rises; falls **7.** does not **9.** 2.56425419972 **11.** 1.25056505582 **13.** 7.41309466897 **15.** 0.0000210965628481

17.

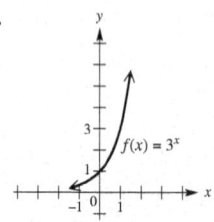

19.

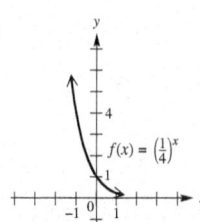

21. 20.0855369232 **23.** 0.018315638889 **25.** $2 = \log_4 16$ **27.** $-3 = \log_{2/3}\left(\frac{27}{8}\right)$ **29.** $2^5 = 32$ **31.** $3^1 = 3$ **33.** 1.38629436112 **35.** -1.0498221245

37.

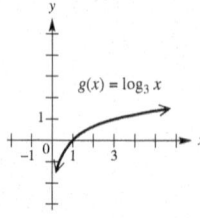

39.

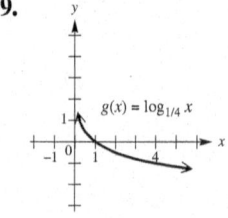

41. (a) $20,812.08 **(b)** $20,814.14 **43. (a)** $28,196.16 **(b)** $28,196.17 **45.** Plan A is better by $121.81.

47. (a) 2% compounded quarterly **(b)** $807.19 **49.** 27.73 years **51.** 54.06 years **53. (a)** 8.18 billion **(b)** 2046 **55.** about 9000 years **57. (a)** 440 grams **(b)** 387 grams **(c)** 264 grams **(d)** 21.66 years **59.** 1611.97 years **61.** 4 hours **63. (a)** 0.6°C **(b)** 0.3°C **65. (a)** 1.4°C **(b)** 0.5°C

8.7 Exercises

1. 3; -6 **3.** D; The ordered-pair solution must be in quadrant IV. **5.** yes **7.** no **9.** B **11.** A

13. $\{(2,2)\}$

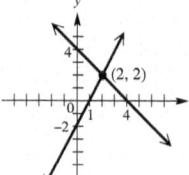

15. $\{(3,-1)\}$ **17.** $\{(2,-3)\}$ **19.** $\left\{\left(\frac{3}{2},-\frac{3}{2}\right)\right\}$ **21.** $\left\{\left(\frac{6-2y}{7},y\right)\right\}$ **23.** $\varnothing$ **25.** $\{(2,-4)\}$ **27.** $\{(1,2)\}$ **29.** $\left\{\left(\frac{22}{9},\frac{22}{3}\right)\right\}$ **31.** $\{(2,3)\}$ **33.** $\{(5,4)\}$ **35.** $\varnothing$

37. $\left\{\left(\frac{2-y}{3},y\right)\right\}$ **39.** $\left\{\left(-5,-\frac{10}{3}\right)\right\}$

41. $\{(2,6)\}$ **43.** Netflix **45.** $\{(4.4,48.7)\}$ **47.** $80x - 7y = -420$ **49.** $\{(4.375,110)\}$

8.8 Exercises

1. (a) 12 ounces **(b)** 30 ounces **(c)** 48 ounces **(d)** 60 ounces **3.** $4.29x **5. (a)** $(10-x)$ mph **(b)** $(10+x)$ mph **7.** $0.10x + 0.25y$ **9.** wins: 102; losses: 60 **11.** length: 94 feet; width: 50 feet **13.** dark clay: $5 per kilogram; light clay: $4 per kilogram **15.** square: 12 cm; triangle: 8 cm **17.** cappuccino: $3.59; caffe latte: $4.19 **19.** ribeye: $30.30; salmon: $21.60 **21.** New York: $549; Washington: $462 **23.** AT&T: $163.8 billion; Verizon: $125.1 billion **25.** Boston: $360.66; Cleveland: $179.44 **27.** 15% solution: $26\frac{2}{3}$ liters; 33% solution: $13\frac{1}{3}$ liters **29.** 3 liters **31.** 50% juice: 150 liters; 30% juice: 50 liters **33.** $1.20 candy: 100 pounds; $2.40 candy; 60 pounds **35.** 4%: $10,000; 3%: $5000 **37.** 1%: $1000; 2%: $2000 **39.** train: 60 mph; plane: 160 mph **41.** boat: 21 mph; current: 3 mph

8.9 Exercises

1. solid **3.** cannot

5.

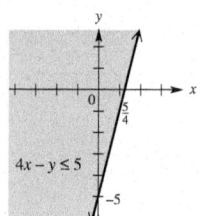

7.

9.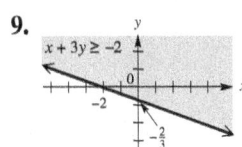

11.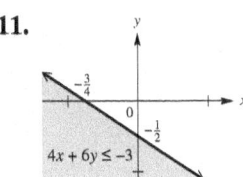

13.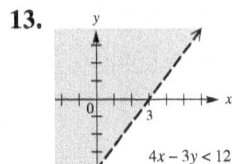

15.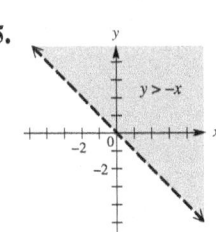

17. C   **19.** B

21.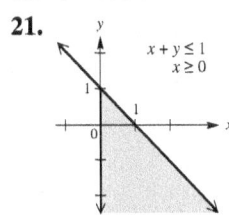

23.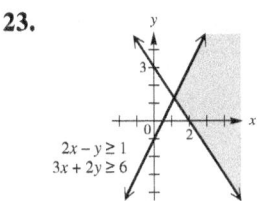

25.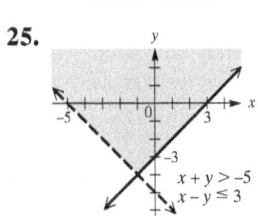

27. maximum of 65 at $(5, 10)$; minimum of 8 at $(1, 1)$ **29.** $(1, 1)$; 7

31. $\left(\frac{17}{3}, 5\right)$; $\frac{49}{3}$

33. Ship 20 to A and 80 to B, for a minimum cost of $1040.
35. Take 3 red pills and 2 blue pills, for a minimum cost of $0.70 per day.   **37.** Make 3 batches of cakes and 6 batches of cookies, for a maximum profit of $210.

Chapter 8 Test
1. $\sqrt{41}$ **2.** $(x + 1)^2 + (y - 2)^2 = 9$

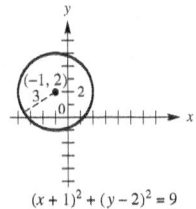

3. The midpoint is $(2014, 361)$, so there were approximately 361 million subscribers in 2014.

4. x-intercept: $\left(\frac{8}{3}, 0\right)$;
y-intercept: $(0, -4)$

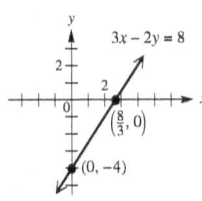

5. $\frac{2}{7}$ **6. (a)** $y = -\frac{2}{5}x + \frac{13}{5}$ **(b)** $y = -\frac{1}{2}x - \frac{3}{2}$
(c) $y = -\frac{1}{2}x + 2$ **7.** B   **8. (a)** $f(x) = 46x + 852$
(b) $1128 billion   **9.** 2.5 million subscribers per year

10. (a) $y = 0.05x + 0.50$ **(b)** $(1, 0.55), (5, 0.75),$
$(10, 1.00)$ **11.** $y = \frac{2}{3}x + 1$ **12. (a)** -5 **(b)** 2
13. 500 units; $30,000

14. axis: $x = -3$; vertex: $(-3, 4)$

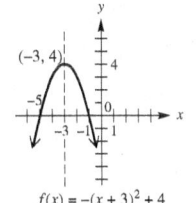

15. 80 feet by 160 feet; 12,800 square feet
16. (a) 2116.31264888
(b) 0.157237166314   **(c)** 3.15955035878
17. (a) $12,740.13   **(b)** $12,742.04
18. (a) 1.62 grams   **(b)** 1.18 grams
(c) 0.69 gram   **(d)** 2.00 grams

19. $\{(4, -2)\}$ **20.** $\varnothing$ **21.** Arizona: $59; Baltimore: $75
22. $6-per-lb nuts: 30 lb; $3-per-lb candy: 70 lb

23.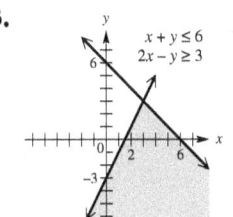

24. Raise 4 pigs and 12 geese, for a maximum profit of $1120.

CHAPTER 9 GEOMETRY

9.1 Exercises

(The art here is not to scale with the exercise art.)
1. 90   **3.** equal   **5.** true   **7.** false   **9.** true
11. false   **13.** true   **15.** false
17. (a) $\overleftrightarrow{AB}$ **(b)** [segment diagram A to B]
19. (a) $\overrightarrow{CB}$ **(b)** [ray diagram A, B, C]
21. (a) $\overrightarrow{BC}$ **(b)** [ray diagram B, C, D]
23. (a) $\overrightarrow{BA}$ **(b)** [ray diagram A, B]

25. F   **27.** D   **29.** B   **31.** E

There may be other correct forms of the answers in Exercises 33–39.   **33.** $\overleftrightarrow{MO}$ **35.** $\overleftrightarrow{NO}$ **37.** $\varnothing$ **39.** $\overrightarrow{OP}$
41. 52°   **43.** 1°   **45.** $(90 - x)°$ **47.** $(10 - x)°$ **49.** 48°
51. 154°   **53.** $(180 - y)°$ **55.** $(x + 90)°$ **57.** 40°
59. 52°   **61.** $\angle CBD$ and $\angle ABE$; $\angle CBE$ and $\angle DBA$
63. (a) 52°   **(b)** 128°   **65.** 107° and 73°   **67.** 75° and 75°
69. 139° and 139°   **71.** 35° and 55°   **73.** 49° and 49°
75. 48° and 132°   **77. (a)** 3   **(b)** 6   **(c)** 7
(d) 7; exterior   **79. (a)** 180   **(b)** 180   **(c)** 180; 180
(d) 0   **(e)** 0   **(f)** 3

9.2 Exercises

1. chord   **3.** equilateral (or equiangular)   **5.** false
7. false   **9.** true   **11.** Answers will vary.   **13.** both

15. closed   **17.** closed   **19.** neither   **21.** convex

23. convex   **25.** not convex   **27.** right, scalene

29. acute, equilateral   **31.** obtuse, scalene

33. obtuse, isosceles   **35.** right, scalene

37. An isosceles right triangle is a triangle having a 90° angle and two perpendicular sides of equal length.

39. $A = 50°$; $B = 70°$; $C = 60°$ **41.** $A = B = C = 60°$

43. $A = B = 55°$; $C = 70°$ **45.** 155°   **47.** 360°

49. 360°   **51.** (a) O (b) $\overrightarrow{OA}$, $\overrightarrow{OC}$, $\overrightarrow{OB}$, $\overrightarrow{OD}$
(c) $\overleftrightarrow{AC}$, $\overleftrightarrow{BD}$ (d) $\overleftrightarrow{AC}$, $\overleftrightarrow{BD}$, $\overleftrightarrow{BC}$, $\overleftrightarrow{AB}$ (e) $\overleftrightarrow{BC}$, $\overleftrightarrow{AB}$
(f) $\overleftrightarrow{AE}$ **53.** With the radius of the compass greater than one-half the length PQ, place the point of the compass at P and swing arcs above and below line r. Then, with the same radius and the point of the compass at Q, swing two more arcs above and below line r. Locate the two points of intersection of the arcs above and below, and call them A and B. With a straightedge, join A and B. AB is the perpendicular bisector of PQ.

55. With the radius of the compass greater than the distance from P to r, place the point of the compass at P and swing an arc intersecting line r in two points. Call these points A and B. Swing arcs of equal radius to the left of line r, with the point of the compass at A and at B, intersecting at point Q. With a straightedge, join P and Q. PQ is the perpendicular from P to line r.

57. With any radius, place the point of the compass at P and swing arcs to the left and right, intersecting line r in two points. Call these points A and B. With an arc of sufficient length, place the point of the compass first at A and then at B, and swing arcs either both above or both below line r, intersecting at point Q. With a straightedge, join P and Q. PQ is perpendicular to line r at P.

59. With any radius, place the point of the compass at A and swing an arc intersecting the sides of angle A at two points. Call the point of intersection on the horizontal side B and call the other point of intersection C. Draw a horizontal working line, and locate any point A' on this line. With the same radius used earlier, place the point of the compass at A' and swing an arc intersecting the working line at B'. Return to angle A, and set the radius of the compass equal to BC. On the working line, place the point of the compass at B' and swing an arc intersecting the first arc at C'. Now draw line $A'C'$. Angle A' is equal to angle A.   **61.** Answers will vary.

9.3 Exercises

1. false   **3.** true

5.

STATEMENTS	REASONS
1. $AC = BD$	**1.** Given
2. $AD = BC$	**2.** Given
3. $AB = AB$	**3.** Reflexive property
4. $\triangle ABD \cong \triangle BAC$	**4.** SSS congruence property

7.

STATEMENTS	REASONS
1. $\angle BAC = \angle DAC$	**1.** Given
2. $\angle BCA = \angle DCA$	**2.** Given
3. $AC = AC$	**3.** Reflexive property
4. $\triangle ABC \cong \triangle ADC$	**4.** ASA congruence property

9.

STATEMENTS	REASONS
1. $\overleftrightarrow{DB}$ is perpendicular to $\overleftrightarrow{AC}$.	**1.** Given
2. $AB = BC$	**2.** Given
3. $\angle ABD = \angle CBD$	**3.** Both are right angles by definition of perpendicularity.
4. $DB = DB$	**4.** Reflexive property
5. $\triangle ABD \cong \triangle CBD$	**5.** SAS congruence property

11. 108°   **13.** 67°, 67°   **15.** Answers will vary.

17. $\angle H$ and $\angle F$; $\angle K$ and $\angle E$; $\angle HGK$ and $\angle FGE$; $\overrightarrow{HK}$ and $\overrightarrow{FE}$; $\overrightarrow{GK}$ and $\overrightarrow{GE}$; $\overrightarrow{HG}$ and $\overrightarrow{FG}$

19. $\angle A$ and $\angle P$; $\angle C$ and $\angle R$; $\angle B$ and $\angle Q$; $\overrightarrow{AC}$ and $\overrightarrow{PR}$; $\overrightarrow{CB}$ and $\overrightarrow{RQ}$; $\overrightarrow{AB}$ and $\overrightarrow{PQ}$

21. $\angle P = 76°$; $\angle M = 45°$; $\angle A = \angle N = 59°$

23. $\angle T = 74°$; $\angle Y = 28°$; $\angle Z = \angle W = 78°$

25. $\angle T = 20°$; $\angle V = 64°$; $\angle R = \angle U = 96°$

27. $a = 12$; $b = 9$ **29.** $a = 6$; $b = \frac{15}{2}$ **31.** $x = 6$

33. $x = 165$ **35.** $c = 111\frac{1}{9}$ **37.** $r = \frac{108}{7}$ **39.** 60 m

41. 250 m, 350 m   **43.** 112.5 ft   **45.** 10 cups   **47.** $c = 17$

49. $a = 13$ **51.** $c = 50$ m   **53.** $a = 20$ in.   **55.** The sum of the squares of the two shorter sides of a right triangle is equal to the square of the longest side.

57. $(3, 4, 5)$ **59.** $(7, 24, 25)$ **61.** Answers will vary.

63. $(3, 4, 5)$ **65.** $(7, 24, 25)$ **67.** Answers will vary.

69. $(4, 3, 5)$ **71.** $(12, 35, 37)$ **73.** Answers will vary.

75. 24 m   **77.** 18 ft   **79.** 4.55 ft   **81.** 19 ft, 3 in.

83. 28 ft, 10 in.   **85.** $4\sqrt{2}$ **87.** 18   **89.** 55°

91. $\frac{\sqrt{6} + \sqrt{2}}{2}$ **93.** Answers will vary.   **95.** Answers will vary.

9.4 Exercises

1. 18   **3.** 8   **5.** perimeter   **7. (a)** 28 cm   **(b)** 48 cm²

9. (a) $12\frac{2}{3}$ cm   **(b)** 10 cm²   **11. (a)** 13 in.   **(b)** 8 in.²

13. (a) 101.2 mm   **(b)** 418 mm²   **15. (a)** 12.8 cm
(b) 8 cm²   **17. (a)** 6.28 cm   **(b)** 3.14 cm²
19. (a) 113.04 m   **(b)** 1017.36 m²   **21.** 4 m
23. 300 ft, 400 ft, 500 ft   **25.** 50 ft   **27.** 23,800.10 ft²
29. 14,600 mi²   **31.** 12 in., 12π in., 36π in.²
33. 5 ft, 10π ft, 25π ft²   **35.** 6 cm, 12 cm, 36π cm²
37. 10 in., 20 in., 20π in.   **39.** $\frac{20}{\pi}$ yd, $\frac{40}{\pi}$ yd, 40 yd
41. 17   **43.** 7   **45.** 5.7   **47.** 6   **49.** 5   **51.** 2.4
53. (a) 20 cm²   **(b)** 80 cm²   **(c)** 180 cm²   **(d)** 320 cm²
(e) 4   **(f)** 3; 9   **(g)** 4; 16   **(h)** n^2 **55.** n^2 **57.** \$800
59. 80   **61.** 76.26   **63.** 132 ft²   **65.** 5376 cm²
67. 145.34 m²   **69.** 16-in. pizza   **71.** 16-in. pizza
73. $\frac{1}{2}(a+b)(a+b)$
75. $\frac{1}{2}(a+b)(a+b) = \frac{1}{2}ab + \frac{1}{2}ab + \frac{1}{2}c^2$
77. $\frac{25\pi}{4} - 12$ **79.** $r^2\sqrt{3}$ **81.** 26 in.²   **83.** $(4-\pi)r^2$

85. 36   **87.** 5 in.

9.5 Exercises

1. true   **3.** true   **5.** false   **7. (a)** $22\frac{1}{2}$ m³   **(b)** $49\frac{1}{4}$ m²
9. (a) 33,493.33 ft³   **(b)** 5024 ft²   **11. (a)** 197.82 cm³
(b) 188.4 cm²   **13. (a)** 65.94 m³   **(b)** 100.00 m²
15. 168 in.³   **17.** 1969.10 cm³   **19.** 427.29 cm³
21. 381.51 in.³   **23.** 1,694,000 m³   **25.** 0.52 m³
27. 288π in.³, 144π in.²   **29.** 2 cm, 16π cm²
31. 1 m, $\frac{4}{3}\pi$ m³   **33.** volume   **35.** 270 ft³   **37.** $\sqrt[3]{2}\,x$
39. \$8100   **41.** \$37,500   **43.** 65.7%   **45.** 2.5   **47.** 6
49. 6 cm   **51.** 300 mi   **53.** 6   **55.** $S = \frac{3\sqrt{3}}{2} + 9; V = \frac{9}{4}$
57. Answers will vary.   **59.** 4, 4, 6, 2   **61.** 8, 6, 12, 2
63. 20, 12, 30, 2

9.6 Exercises

(The answers are given in blue for this section.)
1. false   **3.** true   **5.** true

7.

9.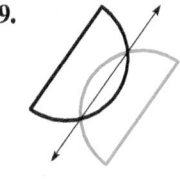

11.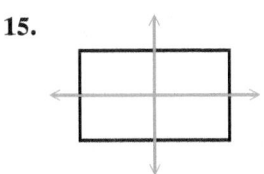

13. The figure is its own reflection image.

15.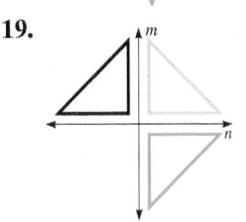

17.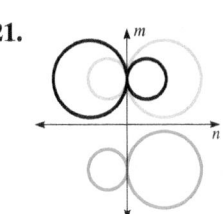

19.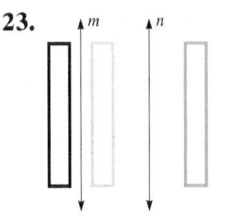

21.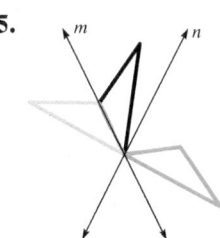

23.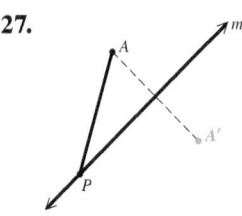

25.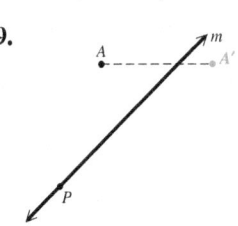

27. **29.**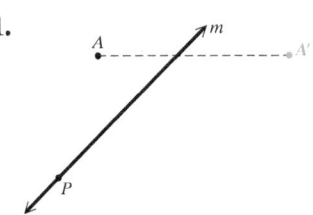

31.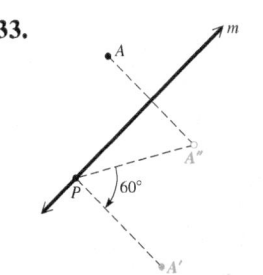

33. **35.** **37.**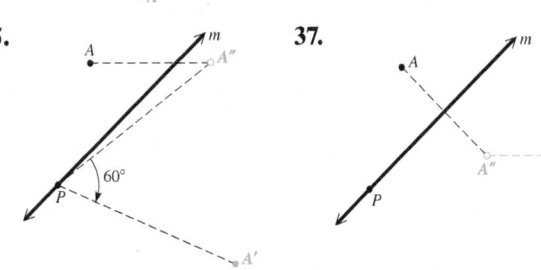

39. no

41.

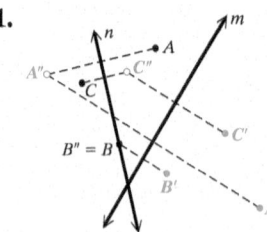

43.

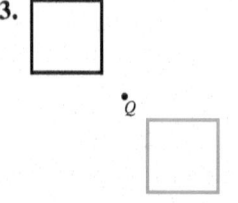

45.

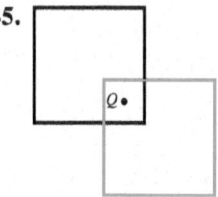

47.

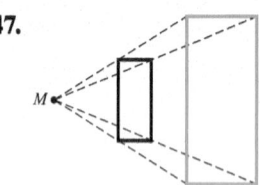

49.

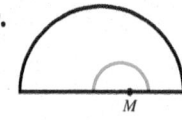

51.

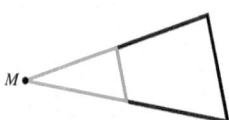

53. 6   **55.** $R_P{}^2 \cdot C_P{}^{6n+2}$

57. (a)   **(b)**

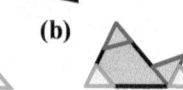

(c) The sum of the interior angles is 180°.

59.

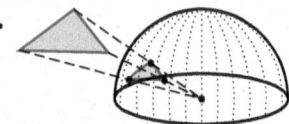
$\mathcal{A} = \frac{1}{2}h(b_1 + b_2)$

9.7 Exercises

1. Euclidean   **3.** Lobachevskian   **5.** greater than
7. Riemannian   **9.** Euclidean   **11.** N
13. $C = 2\pi < 2\pi r = \pi^2$
15.

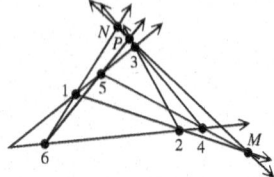

17. Yes; Any two distinct lines have at least one point in common.   **19.** Yes; Any two points in a plane have at least one line of the plane in common.   **21.** Yes; Every point is contained by at least three lines of the plane.
23. no   **25.** yes   **27.** C   **29.** A, E   **31.** B, D   **33.** A, E
35. 1   **37.** 3   **39.** 1
41. (a)–(g)

(h) Suppose that a hexagon is inscribed in an angle. Let each pair of opposite sides be extended so as to intersect. Then the three points of intersection thus obtained will lie in a straight line.

9.8 Exercises

1. false   **3.** false   **5.** false   **7.** 0.842, 0.452, 0.842, 0.452, ... The two attractors are approximately 0.842 and 0.452.   **9.** 4   **10.** 4   **11.** 2   **12.** $\frac{2}{1} = 2$ **13.** $\frac{4}{1} = 4$
14. $\frac{3}{1} = 3; \frac{9}{1} = 9$ **15.** $\frac{4}{1} = 4; \frac{16}{1} = 16$ **16.** 4, 9, 16, 25, 36, 100   **17.** Each ratio in the bottom row is the square of the scale factor in the top row.   **18.** 4   **19.** 4, 9, 16, 25, 36, 100   **20.** Each ratio in the bottom row is again the square of the scale factor in the top row.   **21.** Answers will vary. Some examples are: $3^d = 9$, thus $d = 2$; $5^d = 25$, thus $d = 2$; $4^d = 16$, thus $d = 2$.   **22.** 8   **23.** $\frac{2}{1} = 2; \frac{8}{1} = 8$
24. 8, 27, 64, 125, 216, 1000   **25.** Each ratio in the bottom row is the cube of the scale factor in the top row.
26. Since $2^3 = 8$, the value of d in $2^d = 8$ must be 3.
27. $\frac{3}{1} = 3$ **28.** 4   **29.** 1.262, or $\frac{\ln 4}{\ln 3}$ **30.** $\frac{2}{1} = 2$
31. 3   **32.** It is between 1 and 2.   **33.** 1.585, or $\frac{\ln 3}{\ln 2}$
35. 1.262, or $\frac{\ln 4}{\ln 3}$

Chapter 9 Test

1. (a) 48°   **(b)** 138°   **(c)** acute   **2.** 40°, 140°
3. 45°, 45°   **4.** 30°, 60°   **5.** 30°, 45°, 105°   **6.** 130°, 50°
7. 117°, 117°   **8.** C   **9.** Answers will vary.
10. neither   **11.** both   **12.** 72 cm²   **13.** 60 in.²   **14.** 68 m²
15. 180 m²   **16.** 57 cm²   **17.** 24π in.   **18.** 1978 ft
19.

STATEMENTS	REASONS
1. $\angle CAB = \angle DBA$	1. Given
2. $DB = CA$	2. Given
3. $AB = AB$	3. Reflexive property
4. $\triangle ABD \cong \triangle BAC$	4. SAS congruence property

20. 64 ft   **21.** 29 m
22.

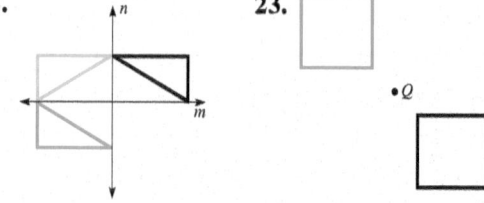

23.

24. (a) 904.32 in.³   **(b)** 452.16 in.²   **25. (a)** 864 ft³
(b) 552 ft²   **26. (a)** 1582.56 m³   **(b)** 753.60 m²
27. Answers will vary.   **28. (a)** yes   **(b)** no   **29.** no
30. The only attractor is approximately 0.524.

CHAPTER 10 COUNTING METHODS

10.1 Exercises

1. $6 \cdot 7 = 42$ **3.** 5   **5.** *AB, AC, AD, AE, BA, BC, BD,*
BE, CA, CB, CD, CE, DA, DB, DC, DE, EA, EB, EC, ED;
20 ways   **7.** *AB, AD, BA, BD, CE, DA, DB, EC;* 8 ways
9. *ACE, AEC, BCE, BEC, DCE, DEC;* 6 ways
11. *ABC, ABD, ABE, ACD, ACE, ADE, BCD, BCE,*
BDE, CDE; 10 ways   **13.** *CEA, CEB, CED;* 3 ways
15. 1   **17.** 3   **19.** 5   **21.** 5   **23.** 3   **25.** 1   **27.** 18
29. 15   **31.** 30   **33.** 27   **35.** 15

37.

	2	4	6
2	22	24	26
4	42	44	46
6	62	64	66

39. 22, 33, 55, 77   **41.** 23, 37, 53, 73

43.

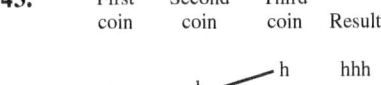
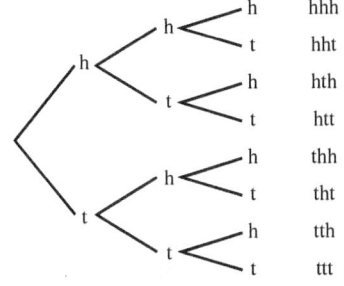

	First coin	Second coin	Third coin	Result

(a) hhh, hht, hth, thh   (b) hhh   (c) hht, hth, htt, thh,
tht, tth, ttt   (d) htt, tht, tth, ttt
45. 16   **47.** 36   **49.** 17   **51.** 72   **53.** 12   **55.** 10   **57.** 6
59. 20   **61.** 9   **63.** $4 \cdot 5 \cdot 5 = 100$ **65.** 49   **67.** 21
69. $2 \cdot 25 \cdot 24 \cdot 23 = 27{,}600$ **71.** 16   **73.** 13   **75.** 4
77. (a) 1600   (b) $4k + 1$ for all positive integers k
(c) $4k^2$ **79.** 3   **81.** (a) 3   (b) 3   (c) none

10.2 Exercises

1. 720   **3.** $4 \cdot 3 \cdot 2 \cdot 1 = 24$ **5.** $1 \cdot 2 \cdot 1 = 2$ **7.** 24   **9.** 72
11. 20   **13.** 1225   **15.** 28   **17.** 10   **19.** (a) 720   (b) 120
21. (a) 306   (b) 153   **23.** 3,628,800   **25.** 3,991,680
27. 11,440   **29.** 980,179,200   **31.** 134,596   **33.** 60
35. 2,162,160   **37.** $2^3 = 8$ **39.** Answers will vary.
41. $5! = 120$ **43.** $3 \cdot 2 = 6$ **45.** $3 \cdot 3 = 9$
47. $3 \cdot 2 \cdot 1 = 6$ **49.** $5 \cdot 2 \cdot 4 = 40$ **51.** $2 \cdot 3 \cdot 4 = 24$
53. $2^6 = 64$ **55.** $5^{10} = 9{,}765{,}625$ **57.** $2 \cdot 3 \cdot 4 \cdot 5 = 120$
59. $2 \cdot 3 \cdot 4 \cdot 3 = 72$ **61.** $2 \cdot 3 \cdot 1 \cdot 3 = 18$ **63.** $6^3 = 216$
65. $9 \cdot 10^4 = 90{,}000$ **67.** $2 \cdot 4 \cdot 6 = 48$ **69.** $5! = 120$
71. 800   **73.** (a) 6   (b) 5   (c) 4   (d) 3   (e) 2
(f) 1; 720   **75.** (a) 3   (b) 3   (c) 2   (d) 2   (e) 1
(f) 1; 36   **77.** 516,243   **79.** 636   **81.** 48

10.3 Exercises

1. 3   **3.** 6   **5.** 5   **7.** 504   **9.** 95,040   **11.** 330   **13.** 45
15. 220   **17.** 116,280   **19.** 43,680   **21.** 126   **23.** 792
25. $1.805037696 \times 10^{11}$ **27.** 225,792,840   **29.** permutation
31. combination   **33.** permutation   **35.** permutation
37. $_8P_5 = 6720$ **39.** $_{12}P_2 = 132$ **41.** $_{25}P_5 = 6{,}375{,}600$
43. $_6P_3 = 120$ **45.** (a) $_6C_3 = 20$ (b) $_6C_2 = 15$
47. $_{18}C_5 = 8568$ **49.** (a) $_{13}C_5 = 1287$ (b) $_{26}C_5 = 65{,}780$
(c) 0 (impossible)   **51.** $_9C_3 = 84$
53. $_{20}C_3 \cdot _{17}C_5 \cdot _{12}C_{12} = 7{,}054{,}320 \left(\text{or } \dfrac{20!}{12! \cdot 5! \cdot 3!} \right)$
55. $_{26}P_3 \cdot _{10}P_3 \cdot _{26}P_3 = 175{,}219{,}200{,}000$ **57.** $2 \cdot _{25}P_3 = 27{,}600$
59. $7 \cdot _{12}P_8 = 139{,}708{,}800$ **61.** (a) $6! = 720$
(b) $2 \cdot 4! = 48$ (c) $4! = 24$
63. $_{15}C_1 \cdot _{14}C_2 \cdot _{12}C_3 \cdot _9C_4 \cdot _5C_5 = 37{,}837{,}800$
65. $\dfrac{_8C_3 \cdot _5C_3 \cdot _2C_2}{2!} = 280$ **67.** (a) $_{13}C_4 \cdot 39 = 27{,}885$
(b) $_{12}C_2 \cdot _{40}C_3 = 652{,}080$ (c) $_{26}C_2 \cdot _{13}C_2 \cdot 13 = 329{,}550$
69. $4 \cdot _{13}C_5 = 5148$ **71.** (a) $_7P_2 = 42$ (b) $3 \cdot 6 = 18$
(c) $_7P_2 \cdot 5 = 210$ **73.** $_{20}C_3 = 1140$ **75.** $_8P_3 = 336$
77. $_9C_2 \cdot _7C_3 \cdot _4C_4 \cdot 2 \cdot 3 \cdot 4 = 30{,}240$
79. (a) $_6C_2 \cdot _6C_3 \cdot _6C_4 = 4500$
(b) $3! \cdot _6C_2 \cdot _6C_3 \cdot _6C_4 = 27{,}000$ **81.** (a) $6! = 720$
(b) 745,896   **83.** Each equals 220.   **85.** $_4C_3 = 4$
87. $_{13}P_2 = 156$ **89.** 624; 5108; 123,552

10.4 Exercises

1. A   **3.** F   **5.** D   **7.** 6   **9.** 20   **11.** 56   **13.** 36
15. $_7C_1 \cdot _3C_3 = 7$ **17.** $_7C_3 \cdot _3C_1 = 105$ **19.** $_8C_3 = 56$
21. $_8C_5 = 56$ **23.** $_9C_4 = 126$ **25.** $1 \cdot _8C_3 = 56$
(or $126 - 70$)   **27.** $x^6 + 6x^5 + 15x^4 + 20x^3 + 15x^2 + 6x + 1$
29. $y^4 - 16y^3 + 96y^2 - 256y + 256$
31. $16x^4 + 160x^3 + 600x^2 + 1000x + 625$
33. 1   **35.** 10   **37.** 5   **39.** 32   **41.** the even-numbered
rows   **43.** (a) All are multiples of the row number.
(b) The same pattern holds.   (c) The same pattern holds.
Each entry is a multiple of 11.   **45.** $\dots 8, 13, 21, 34, \dots$;
A number in this sequence is the sum of the two preceding
terms. This is the Fibonacci sequence.   **47.** Row 8
49. The sum of the squares of the entries across the top
row equals the entry at the bottom vertex.
Wording may vary in the answers for Exercises 51 and 53.
51. sum = N; Any entry in the array equals the sum of the
column of entries from its immediate left upward to the top of
the array.   **53.** sum = $N - 1$; Any entry in the array equals
1 more than the sum of all entries whose cells make up the
largest rectangle entirely to the left and above that entry.

10.5 Exercises

1. $2^7 = 128$; $2^7 - 1 = 127$ **3.** $52 - (13 + 4 - 1) = 36$
5. $2^4 - 1 = 15$ **7.** $2^8 - 1 = 255$ **9.** $2^7 - 1 = 127$
11. 120 **13.** $36 - 6 = 30$ **15.** $6 + 6 - 1 = 11$
17. $52 - 1 = 51$ **19.** $52 - 4 = 48$ **21.** $90 - 9 = 81$
23. $29 + 9 - 2 = 36$ **25.** (a) $_{10}C_3 = 120$ (b) $_9C_3 = 84$
(c) $120 - 84 = 36$ **27.** $_7C_3 - _5C_3 = 25$
29. (a) $_8P_4 - _5P_4 = 1560$ (b) $_8C_4 - _5C_4 = 65$
31. $_{10}P_3 - _7P_3 = 510$ **33.** $2 \cdot 26^2 + 2 \cdot 26^3 = 36{,}504$
35. $_{12}C_4 - _8C_4 = 425$ **37.** $13 + 4 - 1 = 16$
39. $12 + 26 - 6 = 32$ **41.** $9 + 8 - 5 = 12$
43. $11 + 5 = 16$ **45.** $2{,}598{,}960 - _{13}C_5 = 2{,}597{,}673$
47. $2{,}598{,}960 - _{40}C_5 = 1{,}940{,}952$
49. $_{10}C_0 + _{10}C_1 + _{10}C_2 = 56$ **51.** $2^{10} - 56 = 968$
53. 56 **55.** $26^2 \cdot 10^3 - _{26}P_2 \cdot _{10}P_3 = 208{,}000$
57. $1 \cdot 12 + 3 \cdot 13 = 51$ **59.** $_4C_3 + _3C_3 + _5C_3 = 15$
61. $_{12}C_3 - _4C_1 \cdot _3C_1 \cdot _5C_1 = 160$ **63.** Answers will vary.
65. Answers will vary.

Chapter 10 Test

1. $6 \cdot 7 \cdot 7 = 294$ **2.** $6 \cdot 7 \cdot 3 = 126$ **3.** $6 \cdot 6 \cdot 5 = 180$
4. $6 \cdot 5 \cdot 1 = 30$ end in 0; $5 \cdot 5 \cdot 1 = 25$ end in 5;
$30 + 25 = 55$ **5.** 720 **6.** 56 **7.** 1320 **8.** 56 **9.** 13

10.

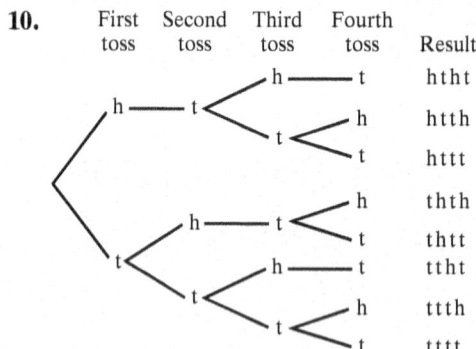

First toss	Second toss	Third toss	Fourth toss	Result
				htht
				htth
				httt
				thth
				thtt
				ttht
				ttth
				tttt

11. $4! = 24$ **12.** $2 \cdot 3 \cdot 4! = 144$ **13.** $_{26}P_5 = 7{,}893{,}600$
14. $32^5 = 33{,}554{,}432$ **15.** $_7P_2 = 42$ **16.** $5! = 120$
17. $_{10}C_4 = 210$ **18.** $_{10}C_2 \cdot _8C_2 = 1260$ **19.** $_{10}C_5 \cdot _5C_5 = 252$
20. $\frac{_{10}C_4 \cdot _6C_4}{2!} = 1575$ **21.** $2^{10} - (_{10}C_0 + _{10}C_1 + _{10}C_2) = 968$
22. $2^5 = 32$ **23.** $2^3 = 8$ **24.** $2 \cdot 2^3 = 16$ **25.** 13 **26.** 2
27. $32 - (1 + 5) = 26$ **28.** $1 \cdot _6C_2 = 15$
29. $1 \cdot 1 \cdot _5C_1 = 5$ **30.** $2 \cdot _5C_2 = 20$
31. $_5C_3 + _5C_2 \cdot _2C_1 = 30$ **32.** $x^4 + 8x^3 + 24x^2 + 32x + 16$
33. $\frac{6!}{2! \cdot 3!} = 60$ **34.** $_{60}C_3 \cdot _{40}C_2 = 26{,}691{,}600$
35. $_9C_4 = 126$ **36.** the counting numbers

CHAPTER 11 PROBABILITY

11.1 Exercises

1. $\frac{1}{6}$ **3.** $\frac{1}{2}$ **5.** 5 to 1 **7.** (a) $\frac{1}{3}$ (b) $\frac{1}{3}$ (c) $\frac{1}{3}$ **9.** (a) $\frac{1}{2}$
(b) $\frac{1}{3}$ (c) $\frac{1}{6}$ **11.** (a) $\{1, 2, 3\}$ (b) 2 (c) 1 (d) 3 (e) $\frac{2}{3}$
(f) 2 to 1 **13.** (a) $\{11, 12, 13, 21, 22, 23, 31, 32, 33\}$
(b) $\frac{2}{3}$ (c) $\frac{1}{3}$ (d) $\frac{1}{3}$ (e) $\frac{4}{9}$ **15.** (a) $\frac{1}{50}$ (b) $\frac{2}{50} = \frac{1}{25}$
(c) $\frac{3}{50}$ (d) $\frac{4}{50} = \frac{2}{25}$ (e) $\frac{5}{50} = \frac{1}{10}$ **17.** (a) $\frac{1}{36}$ (b) $\frac{2}{36} = \frac{1}{18}$
(c) $\frac{3}{36} = \frac{1}{12}$ (d) $\frac{4}{36} = \frac{1}{9}$ (e) $\frac{5}{36}$ (f) $\frac{6}{36} = \frac{1}{6}$ (g) $\frac{5}{36}$
(h) $\frac{4}{36} = \frac{1}{9}$ (i) $\frac{3}{36} = \frac{1}{12}$ (j) $\frac{2}{36} = \frac{1}{18}$ (k) $\frac{1}{36}$
19. $\frac{36}{2{,}598{,}960} \approx 0.00001385$ **21.** $\frac{624}{2{,}598{,}960} \approx 0.00024010$
23. $\frac{1}{4} \cdot \frac{5108}{2{,}598{,}960} \approx 0.00049135$ **25.** $\frac{175}{200} = 0.875$
27. (a) $\frac{5}{9}$ (b) $\frac{49}{144}$ (c) $\frac{5}{48}$ **29.** $\frac{1}{4}$ **31.** $\frac{1}{4}$ **33.** (a) $\frac{3}{4}$
(b) $\frac{1}{4}$ **35.** $\frac{1}{250{,}000} = 0.000004$ **37.** $\frac{1}{4}$ **39.** $\frac{1}{4}$ **41.** $\frac{2}{4} = \frac{1}{2}$
43. $\frac{1}{500} = 0.002$ **45.** 160 **47.** $\frac{2}{4} = \frac{1}{2}$ **49.** 7 to 4
51. 9 to 2 **53.** 4 to 7 **55.** 4 to 1 **57.** 2 to 3 **59.** 37 to 63
61. $\frac{12}{31}$ **63.** (a) 0 (b) no (c) yes
65. (a) $3 \cdot 1 \cdot 2 \cdot 1 \cdot 1 \cdot 1 = 6$ (b) $\frac{6}{720} = \frac{1}{120} \approx 0.0083$
67. (a) $4 \cdot 3! \cdot 3! = 144$ (b) $\frac{144}{720} = \frac{1}{5} = 0.2$
69. $\frac{1}{_{26}P_3} \approx 0.000064$ **71.** $\frac{1}{9}$ **73.** $\frac{2}{_7C_2} = \frac{2}{21} \approx 0.095$
75. $\frac{_5C_3}{_{12}C_3} = \frac{10}{220} \approx 0.045$ **77.** $\frac{1}{_5P_3} = \frac{1}{60} \approx 0.017$
79. (a) $\frac{8}{9^2} = \frac{8}{81} \approx 0.099$ (b) $\frac{4}{_9C_2} = \frac{1}{9} \approx 0.111$
81. 1 **83.** $\frac{5}{31}$ **85.** $\frac{9}{9 \cdot 10} = \frac{1}{10}$ **87.** $\frac{1}{15}$

11.2 Exercises

1. $\frac{1}{3}$ **3.** $\frac{8}{15}$ **5.** $\frac{1}{2}$ **7.** $\frac{5}{6}$ **9.** $\frac{2}{3}$ **11.** (a) $\frac{2}{13}$ (b) 2 to 11
13. (a) $\frac{11}{26}$ (b) 11 to 15 **15.** (a) $\frac{9}{13}$ (b) 9 to 4 **17.** $\frac{2}{3}$
19. $\frac{7}{36}$ **21.** $\frac{5}{12}$ **23.** $\frac{2}{3}$ **25.** yes **27.** 0.76 **29.** 0.92 **31.** 0.04
33. 0.005365 **35.** 0.971285

37.

x	$P(x)$
3	0.1
4	0.1
5	0.2
6	0.2
7	0.2
8	0.1
9	0.1

39. Answers will vary.
41. Answers will vary.
43. $n(A') = s - a$
45. $P(A) + P(A') = 1$
47. 180 **49.** 60 **51.** 1 **53.** $\frac{1}{4}$

11.3 Exercises

1. $\frac{4}{12} = \frac{1}{3}$ **3.** $\frac{2}{12} \cdot \frac{1}{11} = \frac{1}{66}$ **5.** $\frac{7}{15}$ **7.** $\frac{2}{5}$ **9.** $\frac{1}{15}$ **11.** 1

13. $\frac{1}{7}$ **15.** $\frac{1}{8}$ **17.** $\frac{2}{4} = \frac{1}{2}$ **19.** $\frac{2}{6} = \frac{1}{3}$ **21.** $\frac{2}{4} = \frac{1}{2}$ **23.** $\frac{1}{2}$

25. independent   **27.** not independent   **29.** independent

31. $\frac{52}{100} = \frac{13}{25}$ **33.** $\frac{69}{100}$ **35.** $\frac{14}{31}$ **37.** $\frac{4}{7} \cdot \frac{4}{7} = \frac{16}{49}$ **39.** $\frac{2}{7} \cdot \frac{1}{7} = \frac{2}{49}$

41. $\frac{4}{7} \cdot \frac{3}{6} = \frac{2}{7}$ **43.** $\frac{1}{6}$ **45.** 0   **47.** $\frac{12}{51} = \frac{4}{17}$ **49.** $\frac{12}{52} \cdot \frac{11}{51} = \frac{11}{221}$

51. $\frac{4}{52} \cdot \frac{11}{51} = \frac{11}{663}$ **53.** $\frac{26}{52} \cdot \frac{26}{51} = \frac{13}{51}$ **55.** $\frac{1}{3}$ **57.** 1

59. $\frac{3}{10}$ (the same)   **61.** $\frac{1}{2} \cdot \frac{1}{2} \cdot \frac{1}{2} \cdot \frac{1}{2} \cdot \frac{1}{2} \cdot \frac{1}{2} = \frac{1}{64}$ **63.** 0.490

65. 0.027   **67.** 0.95   **69.** 0.23   **71.** $\frac{1}{20}$ **73.** $\frac{1}{10}$ **75.** $2^6 = 64$

77. 0.1479   **79.** 0.400   **81.** 0.080   **83.** $(0.90)^4 = 0.6561$

85. $_4C_2 \cdot (0.10) \cdot (0.20) \cdot (0.70)^2 = 0.0588$ **87.** 0.2704

89. 0.2496   **91.** 0.30   **93.** 0.49   **95.** $\frac{9}{13}$ **97.** 0

99. approximately 0.90355   **101.** $\frac{5}{108}$ **103.** $\frac{1}{3}$

105. $\frac{7}{2210}$ **107.** $\frac{27}{55}$ **109.** $\frac{3}{5}$ **111.** $\frac{1}{5}$ **113.** $\frac{1}{136}$

115. $\frac{8}{11!}$, or $\frac{8}{39,916,800}$ **117.** $\frac{5}{12}$ **119.** $\frac{1}{4}$

11.4 Exercises

1. constant   **3.** $\frac{15}{64}$, or 0.234375   **5.** $\frac{1}{8}$ **7.** $\frac{3}{8}$ **9.** $\frac{3}{4}$ **11.** $\frac{1}{2}$

13. Answers will vary.   **15.** $\frac{1}{128}$ **17.** $\frac{21}{128}$ **19.** $\frac{35}{128}$ **21.** $\frac{7}{128}$

23. $\frac{125}{216}$ **25.** $\frac{5}{72}$ **27.** 0.041   **29.** 0.268   **31.** .302

33. Answers will vary.   **35.** 0.228   **37.** 0.299   **39.** 0.032

41. 0.259   **43.** 0.167   **45.** 0.010   **47.** $6p^2(1-p)^2$ **49.** 0.883

51. 0.073   **53.** $\frac{1}{1024} \approx 0.001$ **55.** $\frac{45}{1024} \approx 0.044$

57. $\frac{210}{1024} = \frac{105}{512} \approx 0.205$ **59.** $\frac{772}{1024} \approx 0.754$

11.5 Exercises

1. $\frac{1}{4}$ **3.** $\frac{4}{16} = \frac{1}{4}; \frac{3}{4}$ **5.** 2   **7.** \$1   **9.** \$0.50   **11.** no (expected net winnings: $-\frac{3}{4} ¢$)

13. 1.69   **15. (a)** $-\$60$ **(b)** \$36,000   **(c)** \$72,000

17. \$0.46   **19.** \$2700   **21.** 2.7   **23.** a decrease of 50

25. Project *B*   **27.** Project *A*   **29.** \$2200   **31.** \$81,000

33. Do not purchase the insurance (because $\$86,000 > \$81,000$).   **35.** \$1500; \$3000; \$17,500; \$27,000

37. \$56,000   **39.** 48.7%   **41.** $\frac{15}{47} \approx 0.319$ **43.** $\frac{18}{50} = 0.36$; This is quite close to 0.375, the theoretical value.

45. no   **47.** $\frac{6}{50} = 0.12$ **49.** $\frac{48}{199} \approx 0.241$ **51.** $\frac{48}{198} \approx 0.242$

53. Answers will vary.   **55.** The walk ends 12 blocks north of the starting point.   **57. (a)** Answers will vary.   **(b)** Answers will vary.   **(c)** The value of *p* will vary; empirical   **(d)** Answers will vary.
(e) Answers will vary.

Chapter 11 Test

1. $\frac{_2C_2}{_5C_2} = \frac{1}{10}$ **2.** $\frac{_3C_2}{_5C_2} = \frac{3}{10}$ **3.** $\frac{6}{10} = \frac{3}{5}$ **4.** $\frac{3}{10}$ **5.** $\frac{7}{7} \cdot \frac{6}{7} \cdot \frac{5}{7} = \frac{30}{49}$

6. $\frac{7}{19}$ **7.** $1 - \left(\frac{30}{49} + \frac{1}{49}\right) = \frac{18}{49}$ **8.** row 1: CC; row 2: cC, cc

9. $\frac{1}{2}$ **10.** 1 to 3   **11.** 3 to 1   **12.** 25 to 1   **13.** 12 to 1

14. 11 to 2   **15.** $\frac{6}{36} = \frac{1}{6}$ **16.** 35 to 1   **17.** 7 to 2

18. $\frac{4}{36} = \frac{1}{9}$ **19.** $\frac{25}{102}$ **20.** $\frac{25}{51}$ **21.** $\frac{4}{51}$ **22.** $\frac{3}{26}$

23. $(0.78)^3 \approx 0.475$ **24.** $_3C_2 \cdot (0.78)^2 \cdot (0.22) \approx 0.402$

25. $1 - (0.22)^3 \approx 0.989$

26. $(0.78) \cdot (0.22) \cdot (0.78) \approx 0.134$ **27.** $\frac{5}{2}$ **28.** $\frac{1}{2}$

29. $\frac{3}{10}; \frac{6}{10}; \frac{1}{10}$ **30.** $\frac{9}{10}$ **31.** $\frac{18}{10} = \frac{9}{5}$ **32.** $\frac{3}{8}$

33. $\frac{1}{24} \approx 0.042$ **34.** $\frac{29}{99} \approx 0.293$ **35.** $\frac{18}{98} \approx 0.184$

CHAPTER 12 STATISTICS

12.1 Exercises

1. true   **3.** false   **5.** true

7. (a)

x	f	$\frac{f}{n}$
0	10	$\frac{10}{30} \approx 33\%$
1	7	$\frac{7}{30} \approx 23\%$
2	6	$\frac{6}{30} = 20\%$
3	4	$\frac{4}{30} \approx 13\%$
4	2	$\frac{2}{30} \approx 7\%$
5	1	$\frac{1}{30} \approx 3\%$

(b) 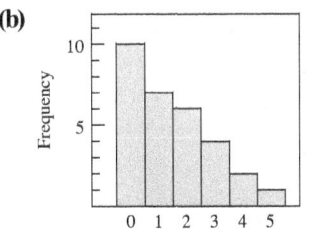   **(c)**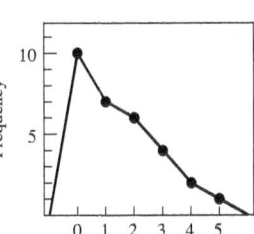

9. (a)

Class Limits	Tally	Frequency f	Relative Frequency $\frac{f}{n}$
21–25		1	$\frac{1}{48} \approx 2.1\%$
26–30	||	2	$\frac{2}{48} \approx 4.2\%$
31–35	|||||	5	$\frac{5}{48} \approx 10.4\%$
36–40	||||| ||||| ||	12	$\frac{12}{48} = 25.0\%$
41–45	||||| ||||| |||||		16	$\frac{16}{48} \approx 33.3\%$
46–50	||||| ||||| ||	12	$\frac{12}{48} = 25.0\%$

Total: $n = 48$

9. (b) 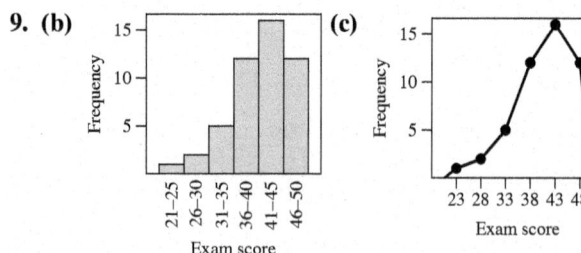 **(c)**

11. (a)

Class Limits	Tally	Frequency f	Relative Frequency $\frac{f}{n}$
70–74	‖	2	$\frac{2}{30} \approx 6.7\%$
75–79	∣	1	$\frac{1}{30} \approx 3.3\%$
80–84	‖∣	3	$\frac{3}{30} = 10.0\%$
85–89	‖	2	$\frac{2}{30} \approx 6.7\%$
90–94	‖‖	5	$\frac{5}{30} \approx 16.7\%$
95–99	‖‖ ‖	7	$\frac{7}{30} \approx 23.3\%$
100–104	‖‖ ∣	6	$\frac{6}{30} = 20.0\%$
105–109	‖‖	4	$\frac{4}{30} \approx 13.3\%$

Total: $n = 30$

(b) 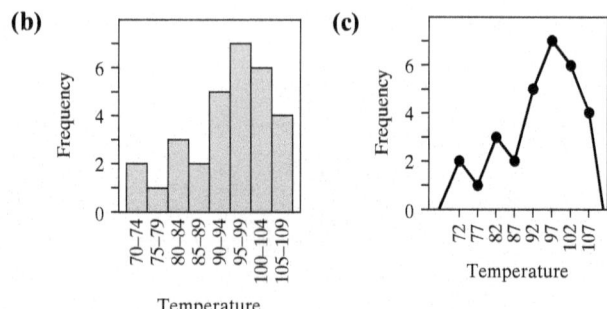 **(c)**

13.
```
0 | 7 9 8
1 | 1 1 2 8 9 4 3 1 0 5 0 5 5
2 | 7 0 9 6 6 2 2 5 2 3 4 4
3 | 8 1
```

15.
```
0 | 8 5 4 9 6 9 4 8
1 | 6 0 1 8 8 2 4 0 2 8 6 3
2 | 6 1 2 5 1 3
3 | 0 4 6
4 | 4
```

17. none **19.** 2007, 2010, 2012, 2013, 2014, and 2015

21.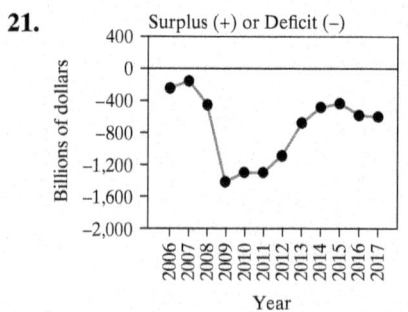

23. about 2.1% in 2016 and 2017
25. Answers will vary.
27. school; 119°
29.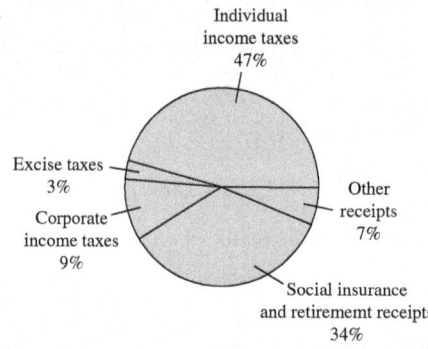

31. about 82 years **33. (a)** about 7 years
(b) Answers will vary. **35.** Answers will vary.
37. Answers will vary. **39.** Answers will vary.
41. (a)

Letter	Probability
A	0.208
E	0.338
I	0.169
O	0.208
U	0.078

(b)

43. Answers will vary. **45. (a)** 0.225 **(b)** 0.150
(c) 0.675 **(d)** 0.075
47. (a)

Sport	Probability
Sailing	0.175
Archery	0.100
Snowboarding	0.250
Bicycling	0.200
Rock climbing	0.200
Rafting	0.075

(b) empirical; Answers will vary.

12.2 Exercises

1. A, B **3.** A, B **5. (a)** 12.4 **(b)** 12 **(c)** none
7. (a) 216.2 **(b)** 221 **(c)** 196 **9. (a)** 5.2 **(b)** 5.35
(c) 4.5 and 6.2 **11.** more than 14.525 **13.** $317.2 billion
15. (a) 10.7 **(b)** 10.65 **(c)** 10.6

17. (a) 73.9 **(b)** 17.5 **(c)** 0 **19.** mean = 47.4; median and mode remain the same

21. 5.27 seconds **23.** 2.41 seconds **25.** the mean

27. mean = 77; median = 80; mode = 79 **29.** 92

31. (a) 589.6 **(b)** 586 **(c)** 579 **33.** 3.19

35. (a) $3154.8 billion **(b)** $3249.9 billion

37. about 628.4 million **39.** about 742.1 million

41. (a) 199.5 **(b)** 200 **43. (a)** 2444.5 **(b)** 2542.5

45. 6 **47. (a)** 22 **(b)** 18.5 **(c)** 15 and 17

49. (a) 74.8 **(b)** 77.5 **(c)** 78

51. Answers will vary.

53. (a) mean = 3; median = 3; mode = 4 **(b)** mode

(c) Answers will vary. **55. (a)** 4 **(b)** 4.25 **57. (a)** 6

(b) 9.33 **59.** three choices: 1, 6, 16 **61.** no

63. Answers will vary.

12.3 Exercises

1. range **3.** $\frac{3}{4}$ **5.** the sample standard deviation

7. (a) 14 **(b)** 4.74 **9. (a)** 26 **(b)** 8.38 **11. (a)** 1.14

(b) 0.37 **13. (a)** 11 **(b)** 3.89 **15.** $\frac{3}{4}$ **17.** $\frac{16}{25}$ **19.** $\frac{3}{4}$

21. $\frac{15}{16}$ **23.** $\frac{1}{4}$ **25.** $\frac{4}{49}$ **27.** $154.58 **29.** nine

31. (a) $s_A = 2.35$; $s_B = 2.58$ **(b)** $V_A = 46.9\%$; $V_B = 36.9\%$

(c) sample B ($s_B = 2.58 > 2.35$) **(d)** sample A

($V_A = 46.9\% > 36.9\%$) **33.** 18.71; 4.35

35. 8.71; 4.35 **37.** 56.14; 13.04

39. (a) $\bar{x}_A = 68.8$; $\bar{x}_B = 66.6$ **(b)** $s_A = 4.21$; $s_B = 5.27$

(c) brand A ($\bar{x}_A > \bar{x}_B$) **(d)** brand A ($s_A < s_B$)

41. Brand A ($s_B = 3539 > 2116$) **43.** −3.0 **45.** 4.2

47. 15.5 **49.** no

12.4 Exercises

1. false **3.** false **5.** true **7.** 1.6 **9.** 100 **11.** 58

13. 59 **15.** Janet (because $z = 0.48 > 0.19$) **17.** Yvette

(because $z = -1.44 > -1.78$) **19.** −0.8 **21.** 1.3 **23.** Russia

25. United States **27.** India in consumption (India's consumption z-score was −0.7, Canada's production z-score was −1.0, and −0.7 > −1.0.) **29. (a)** The median is 30.5 quadrillion Btu. **(b)** The range is $122.5 - 14.4 = 108.1$ quadrillion Btu. **(c)** The middle half of the items extend from 19.3 to 109.8 quadrillion Btu.

31. Answers will vary. **33.** Answers will vary.

35. Answers will vary. **37.** the overall distribution

39. Production is skewed left, and consumption is skewed right. **41. (a)** no **(b)** Answers will vary.

43. no; Answers will vary.

45. yes; Answers will vary. **47.** 81.8 **49.** 35

51. $Q_1 = 61.6$, $Q_2 = 63.5$, $Q_3 = 67.4$ **53.** $P_{95} = 75.9$

55. 64.5 **57. (a)** 2.5, 4.5, 6.5 **(b)** none **59. (a)** 3, 5.5, 8

(b) two **61. (a)** none **(b)** one **(c)** two **(d)** three **63.** Q_2

12.5 Exercises

1. false **3.** true **5.** discrete **7.** continuous

9. discrete **11.** 50 **13.** 95 **15.** 50% **17.** 95%

19. 45.4% **21.** 44.5% **23.** 97.8% **25.** 5.3% **27.** 0.77

29. 0.84 **31.** 1000 **33.** 500 **35.** 61 **37.** 11.5%

39. 54.0% **41.** 0.212 **43.** 0.092 **45.** 0.994, or 99.4%

47. 166 units **49.** 0.888 **51.** about 2 eggs **53.** 24.2%

55. Answers will vary. **57.** 76 **59.** 65 **61.** 90.4

63. 40.3 **65. (a)** at least 36% **(b)** 78.8%

Chapter 12 Test

1. (a) 126.4 million **(b)** 106.7 million **(c)** 43.3 million

(d) 34.3% **2. (a)** those who copied less than 10%

(b) those with copy rates from 10% to 30% or greater than 50% **(c)** those with a copy rate from 30% to 50%

(d) Answers will vary. **3. (a)** 290 feet **(b)** 599 inches

(c) 1223 **4. (a)** 762 **(b)** 8th **(c)** 22 inches

(d) 17 feet

5.

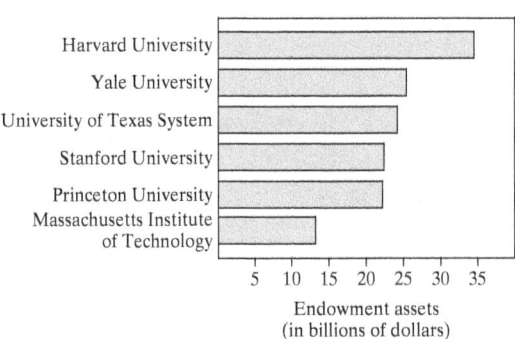

Top Six U.S. College Endowment Funds

6. 17.9% **7.** $23.3 billion

8.

Class Limits	Frequency f	Relative Frequency $\frac{f}{n}$
6–10	2	$\frac{2}{22} \approx 0.09$
11–15	6	$\frac{6}{22} \approx 0.27$
16–20	7	$\frac{7}{22} \approx 0.32$
21–25	4	$\frac{4}{22} \approx 0.18$
26–30	3	$\frac{3}{22} \approx 0.14$

9. (a) 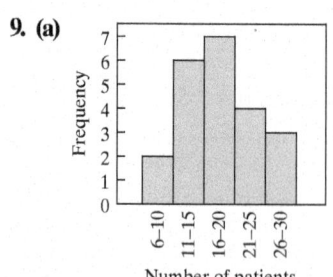   **(b)**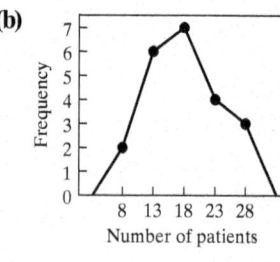

10. 8   **11.** 13.4   **12.** 13   **13.** 13   **14.** 10

15.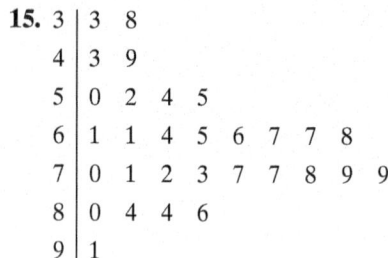

16. 36   **17.** 29   **18.** 39   **19.** 31   **20.** 49

21.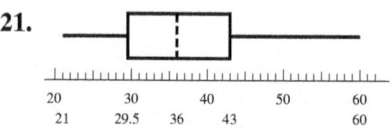

22. (a) about 95%   **(b)** about 0.3%   **(c)** about 16%
(d) about 13.5%   **23.** West   **24.** East   **25.** 79.7
26. Indians (because $z = 1.46 > 1.24$)

27.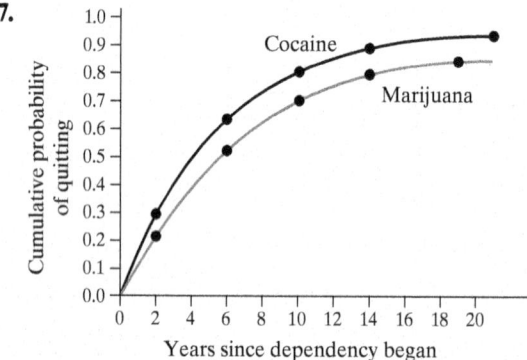

(a) about 4 years   **(b)** about 6 years   **28.** Answers will
vary.

CHAPTER 13 PERSONAL FINANCIAL MANAGEMENT

13.1 Exercises

1. false   **3.** false   **5.** false   **7.** $24   **9.** $24.15
11. $361.13   **13. (a)** $784   **(b)** $788.31   **15. (a)** $2725
(b) $2731.82   **17.** $125   **19.** $6879.17   **21.** $1055.99;
$80.99   **23.** $17,724.34; $10,224.34   **25. (a)** $2339.72
(b) $2343.32   **(c)** $2345.16   **27. (a)** $18,545.42
(b) $18,546.80   **(c)** $18,547.49   **29. (a)** $62.53
(b) $62.66   **(c)** $62.74   **(d)** $62.79   **(e)** $62.79

31. Answers will vary.   **33.** $1461.04   **35.** $862.61
37. $6502.06   **39.** $168,910.81   **41.** $167,364.81
43. 2.000%   **45.** 2.015%   **47.** 2.020%   **49.** 2.020%
51. $580.90   **53.** 17 years, 152 days
55. $r = n\left[(Y + 1)^{1/n} - 1\right]$ **57. (a)** 1.804%
(b) Bank A; $2.63   **(c)** no difference; $90.81
59. 35 years   **61.** 8 years   **63.** 10.0%   **65.** 3.2%
67. $4.21; $4.74; $6.80; $12.40   **69.** $40,900; $46,100;
$66,100; $121,000   **71.** $249   **73.** $1717   **75.** $54,872.53

13.2 Exercises

1. payment   **3.** average daily balance   **5.** $2795
7. $3130.40   **9.** $4130.40   **11.** $28,500   **13.** $33,060
15. $37,060   **17.** $810.00; $110.63   **19.** $41.63; $43.98
21. $36.38; $35.71   **23.** $229.17   **25.** $292.95   **27.** $9824
29. 5 years   **31.** $3.50   **33.** $5.80   **35.** $15.30
37. (a) $607.33   **(b)** $7.90   **(c)** $646.87
39. (a) $473.96   **(b)** $6.16   **(c)** $505.54   **41.** $9.27
43. (a) $54.11   **(b)** $54.86   **(c)** $55.62   **45.** 16.9%
47. $90.00   **49.–53.** Answers will vary.   **55. (a)** $149.04
(b) $160.68

13.3 Exercises

1. false   **3.** false   **5.** 7.0%   **7.** 8.5%   **9.** $136.46
11. $90.63   **13.** 6.40%   **15.** 9.27%   **17. (a)** $3.20
(b) $193.51   **(c)** $6393.79   **19. (a)** $2.05   **(b)** $71.81
(c) $4098.79   **21. (a)** $206   **(b)** 9.5%   **23. (a)** $138
(b) 8.5%   **25. (a)** $20.44   **(b)** 5.75%   **27. (a)** $347.40
(b) 6.78%   **29. (a)** $20.00   **(b)** $19.59   **31. (a)** $159.70
(b) $150.86   **33. (a)** $0.98   **(b)** $8.23   **(c)** $1051.77
35. finance company APR: 6.60%; credit union APR:
6.22%; choose credit union (because 6.22 < 6.60)
37. $19.64   **39.** 13   **41.** 9.6%   **43.** 8
45.–47. Answers will vary.   **49.** $\frac{5}{39}$
51. Answers will vary.

13.4 Exercises

1. amortization schedule   **3.** escrow account or
reserve account   **5.** index; margin   **7.** $687.89
9. $864.64   **11.** $1989.84   **13.** $794.26   **15. (a)** $373.09
(b) $289.58   **(c)** $83.51   **(d)** $69,416.49   **17. (a)** $1247.43
(b) $775.67   **(c)** $471.76   **(d)** $142,728.24
(e) $1247.43   **(f)** $773.11   **(g)** $474.32   **(h)** $142,253.92
19. (a) $1247.53   **(b)** $544.57   **(c)** $702.96
(d) $112,947.04   **(e)** $1247.53   **(f)** $541.20   **(g)** $706.33
(h) $112,240.71   **21.** $480.31   **23.** $558.69   **25.** $1216.04;
$437,774.40; $197,774.40   **27.** $1760.25; $316,845.00;
$76,845.00   **29.** Answers will vary.   **31. (a)** $572.90
(b) $1280.71   **33. (a)** $117,886.76   **(b)** $119,609.63

35. (a) payment 153   **(b)** payment 295   **37.** $164,813.20
39. $175,584.40   **41.** $77,072.40   **43.** $118,609.20
45. (a) $1154.92   **(b)** $1298.51   **(c)** an increase of $143.59
47. (a) $412.50   **(b)** $695.91   **49.** $118.88   **51.** $140,000
53. $4640   **55.** $302   **57. (a)** $1634   **(b)** $1754
(c) $917   **(d)** $997   **(e)** $1555   **59. (a)** $2271   **(b)** $2407
(c) $1083   **(d)** $1174   **(e)** $2055   **61. (a)** $8370
(b) $42,300   **(c)** $20,286   **63.–69.** Answers will vary.

13.5 Exercises

1. dividends; capital gains   **3.** commissions
5. mutual fund   **7.** $148.75   **9.** 2.07% higher
11. 19.35% higher   **13.** TWTR   **15.** BAC   **17.** $28,143.00
19. $18,172.00   **21.** $11,795.99   **23.** $10,429.39
25. $114,856.95   **27.** $37,493.56   **29.** $62,009.24
31. $357,146.40   **33.** $5499.38 net paid out   **35. (a)** $800
(b) $80   **(c)** $960   **(d)** $1040   **(e)** 130%   **37. (a)** $1250
(b) $108   **(c)** −$235   **(d)** −$127   **(e)** −10.16%
39. $275.00   **41.** $177.75   **43. (a)** $10.49   **(b)** 334
45. (a) $8.27   **(b)** 3080   **47. (a)** $8.39   **(b)** $100.68
(c) 15.6%   **49. (a)** $57.45   **(b)** $689.40   **(c)** 27.6%
51. (a) $1203.75   **(b)** $18.06   **(c)** 19.56%   **53. (a)** $4179
(b) $76.48   **(c)** 24.31%   **55.** $129,258   **57.** $993,383
59. (a) $58,466.19   **(b)** $52,785.51   **61. (a)** $12,851.28
(b) $11,651.81   **63. (a)** $R = \dfrac{rV}{(1-t)[(1+r)^n - 1]}$
(b) $R = \dfrac{rV}{(1 + r(1-t))^n - 1}$ **65. (a)** $11,769.91   **(b)** $16,865.22
67. (a) $r = n\left[\left(\dfrac{A}{P}\right)^{1/(nt)} - 1\right]$ **(b)** 5.4%

		Growth &		
Aggressive Growth	**Growth**	**Income**	**Income**	**Cash**
69. $1400	$8600	$6200	$2800	$1000
71. $8000	$112,000	$144,000	$116,000	$20,000

73. 3.75%   **75.** 5.2%   **77.–83.** Answers will vary.

Chapter 13 Test

1. $1200   **2.** $563.25   **3.** 3.04%   **4.** 18 years
5. $67,297.13   **6.** $10.88   **7.** $630   **8.** $184.17
9. 6.5%   **10. (a)** $75.97   **(b)** $2318.24   **11.** 7.0%
12. Answers will vary.   **13.** $1030.70   **14.** $1340.70

15. Payment Number	**Interest Payment**	**Principal Payment**	**Balance of Principal**
			$192,000.00
1	$800.00	$230.70	$191,769.30
2	$799.04	$231.66	$191,537.64

16. $179,052.00   **17.** $3840   **18.** Answers will vary.
19. 6.1%   **20.** 479,600   **21.** $10.00   **22.** 0.078%; no;
The percentage return was not for a 1-year period.
23. $106,728.55   **24.** Answers will vary.

CHAPTER 14 GRAPH THEORY

14.1 Exercises

1. edges; edge   **3.** connected; disconnected   **5.** walk
7. circuit   **9.** 7 vertices, 7 edges   **11.** 10 vertices, 9 edges
13. 6 vertices, 9 edges   **15.** Two have degree 3. Three
have degree 2. Two have degree 1. Sum of degrees is 14.
This is twice the number of edges.   **17.** Six have degree 1.
Four have degree 3. Sum of degrees is 18. This is twice the
number of edges.   **19.** not isomorphic   **21.** isomorphic;
Corresponding edges should be the same color. AB should
match AB, etc.

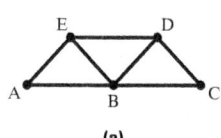 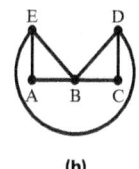

        (a)   **(b)**

23. not isomorphic   **25.** connected, 1 component
27. disconnected, 3 components   **29.** disconnected,
2 components   **31.** 10   **33.** 4   **35. (a)** yes   **(b)** No,
because there is no edge from A to D.   **(c)** No, because
there is no edge from A to E.   **(d)** yes   **(e)** yes   **(f)** yes
37. (a) No, because it does not return to the starting vertex.
(b) yes   **(c)** No, because there is no edge from C to F.
(d) No, because it does not return to the starting vertex.
(e) No, because the edge from F to D is used more than
once.   **39. (a)** No, because there is no edge from B to C.
(b) No, because the edge from I to G is used more than once.
(c) No, because there is no edge from E to I.   **(d)** yes
(e) yes   **(f)** No, because the edge from A to D is used more
than once.   **41.** It is a walk and a path, not a circuit.
43. It is a walk, not a path, not a circuit.   **45.** It is a walk
and a path, not a circuit.   **47.** No. For example, there is
no edge from A to C.   **49.** No. For example, there is no
edge from A to F.   **51.** yes

53. 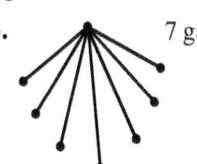   7 games   **55.** 36 handshakes

57. 10 telephone conversations
59. A → B → C → D → A corresponds
to tracing around the edges of a single
face. (The circuit is a square.)

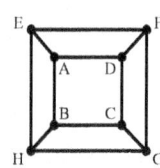

61. (a) 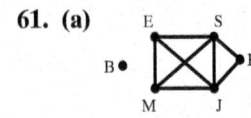 **(b)** disconnected, two components

(c) 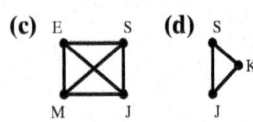 **(d)**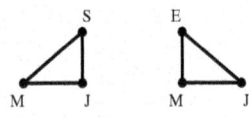

63. one answer: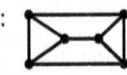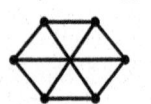

65. 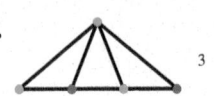 **67.**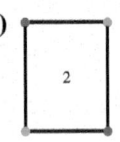

69.  **71. (a)**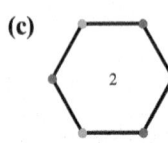

(b) 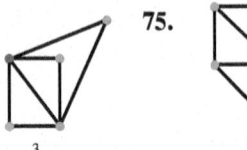 **(c)**  **(d)**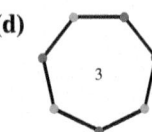

(e) A cycle with an odd number of vertices has chromatic number 3. A cycle with an even number of vertices has chromatic number 2.

73. 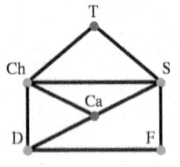 **75.**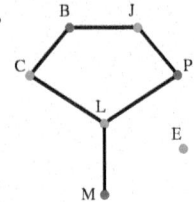

77. 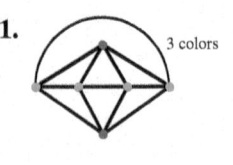 Three times: Choir and Forensics, Service Club and Dance Club, and Theater and Caribbean Club. (There are other possible groupings.)

79.  Three gatherings: Brad and Phil and Mary, Joe and Lindsay, and Caitlin and Eva. (There are other possible groupings.)

81. 3 colors **83.** 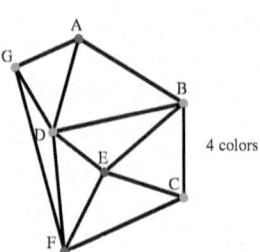 4 colors

85. 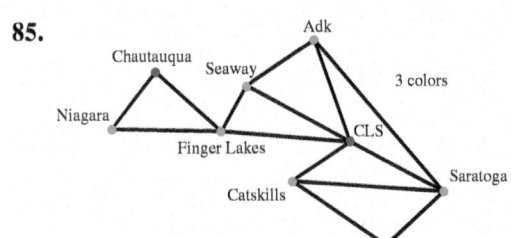 3 colors

In Exercise 87, there are many different ways to draw the map.

87.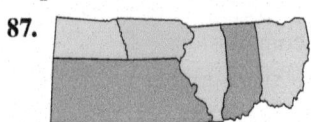

89. By the four-color theorem, this is not possible.

91. 2 **93.** 2 **95.** 3 **97.** 2 **99.** 1

101.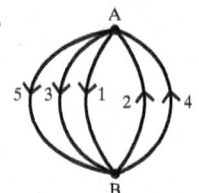

103.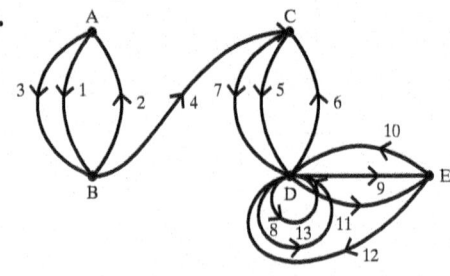

14.2 Exercises

1. edge; once **3.** Euler circuit **5. (a)** No, it is not a path. **(b)** yes **(c)** No, it is not a walk. **(d)** No, it is not a circuit. **7. (a)** No, some edges are not used (e.g., $B \to F$). **(b)** yes **(c)** No, it is not a path. **(d)** yes **9.** Yes, all vertices have even degree. **11.** No, some vertices (e.g., G) have odd degree. **13.** yes **15.** It has an Euler circuit. All vertices have even degree. No circuit visits each vertex exactly once. **17.** It has an Euler circuit. All vertices have even degree. $A \to B \to H \to C \to G \to D \to F \to E \to A$ visits each vertex exactly once. **19.** No, some vertices have odd degree. **21.** No, some vertices have odd degree. **23.** none **25.** FG **27.** $B \to E$ or $B \to D$ **29.** $B \to C$ or $B \to H$ **31.** $A \to C \to B \to F \to E \to D \to C \to F \to D \to A$ **33.** The graph has an Euler circuit. $A \to G \to H \to J \to I \to L \to J \to K \to I \to H \to F \to G \to E \to F \to D \to E \to C \to D \to B \to C \to A \to E \to B \to A$

35. The graph does not have an Euler circuit. Some vertices (e.g., C) have odd degree.   **37.** There is such a route:
A→D→B→C→A→H→D→E→B→H→G→
E→F→H→J→L→C→M→A→K→M→L→
K→J→A
39. It is not possible. Some vertices (e.g., room at upper left) have odd degree.   **41.** No. There are more than two vertices with odd degree.   **43.** yes; B and G   **45.** It is possible. Exactly two of the rooms have odd numbers of doors.   **47.** It is not possible. All rooms have even numbers of doors.   **49.** yes; Manhattan and Randall's Island.
There are other correct answers in Exercise 51.
51. A→B→C→D→E→C→F→B→E→F→A; There are 10 edges in any such circuit.
53. In the graph shown, not all vertices have even degree. For example, the vertex marked X is of degree 5. So this graph still does not have an Euler circuit.

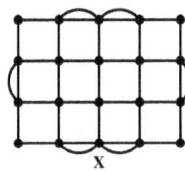

55. (There are other ways of inserting the additional edges.)

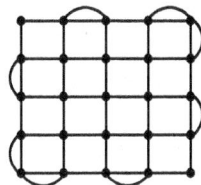

8 additional edges

14.3 Exercises
1. once   **3.** $(n-1)!$ **5. (a)** No, because it visits E twice.   **(b)** yes   **(c)** No, because it does not visit C.
(d) No, because it does not visit A.   **7. (a)** None, because there is no edge from B to C.   **(b)** all three
(c) none; Edge AD is used twice.
9. A→B→D→E→F→C→A
11. G→H→J→I→G   **13.** X→T→U→W→V→X
15. Hamilton circuit: A→B→C→D→A. The graph has no Euler circuit, because at least one of the vertices has odd degree. (In fact, all have odd degree.)

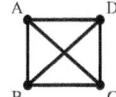

17. A→B→C→D→E→F→A
is both a Hamilton and an Euler circuit.

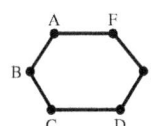

19. Hamilton circuit   **21.** Euler circuit

23. Hamilton circuit   **25.** 24   **27.** 362,880
29. 9!   **31.** 17!
33. P→Q→R→S→P   P→Q→S→R→P
P→R→Q→S→P   P→R→S→Q→P
P→S→Q→R→P   P→S→R→Q→P
35. E→H→I→F→G→E   E→H→I→G→F→E
37. E→F→G→H→I→E   E→F→G→I→H→E
E→F→H→G→I→E   E→F→H→I→G→E
E→F→I→G→H→E   E→F→I→H→G→E
39. E→G→F→H→I→E   E→G→F→I→H→E
E→G→H→F→I→E   E→G→H→I→F→E
E→G→I→F→H→E   E→G→I→H→F→E
41. P→Q→R→S→P; Weight is 2200.
43. C→D→E→F→G→C; Weight is 64.
45. (a) A→C→E→D→B→A; Total weight is 20.
(b) C→A→B→D→E→C; Total weight is 20.
(c) D→C→A→B→E→D; Total weight is 23.
(d) E→C→A→B→D→E; Total weight is 20.
47. (a) A→E→D→C→F→B→A; Total weight is 11.85.
(b) B→F→C→D→E→A→B; Total weight is 11.85.
(c) C→D→E→A→B→F→C; Total weight is 11.85.
(d) D→C→A→E→B→F→D; Total weight is 12.9.
(e) E→A→C→D→B→F→E; Total weight is 14.1.
(f) F→B→C→D→E→A→F; Total weight is 12.25.
49. A→B→C→D→E→F→A
A→B→C→F→E→D→A
A→B→E→D→C→F→A
A→B→E→F→C→D→A
A→D→E→F→C→B→A
A→D→E→B→C→F→A
A→D→C→B→E→F→A
A→D→C→F→E→B→A
A→F→E→B→C→D→A
A→F→E→D→C→B→A
A→F→C→D→E→B→A
A→F→C→B→E→D→A
51. A→B→C→D→E→F→A
A→B→C→E→D→F→A
A→B→C→E→F→D→A
A→B→C→F→E→D→A
A→D→E→F→C→B→A
A→D→F→E→C→B→A
A→F→D→E→C→B→A
A→F→E→D→C→B→A

53. 252 miles

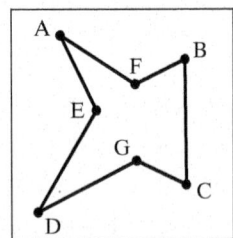

55. $A \to F \to G \to R \to S \to T \to U \to Q \to P \to N \to$
$M \to L \to K \to J \to I \to H \to B \to C \to D \to E \to A$

57. (a) graphs (2) and (4)   **(b)** graphs (2) and (4)
(c) No. Graph (1) provides a counterexample.
(d) No. If $n < 3$, the graph will have no circuits at all.
(e) The degree of each vertex in a complete graph with
n vertices is $(n - 1)$. If $n \geq 3$, then $(n - 1) > \frac{n}{2}$. So we
can conclude from Dirac's theorem that the graph has a
Hamilton circuit.

14.4 Exercises

1. path; vertices   **3.** vertex   **5.** tree
7. No, because it is not connected.   **9.** tree
11. No, because it has a circuit.
13. It is not possible, because the graph has a circuit.
15. tree   **17.** not necessarily a tree   **19.** true
21. false

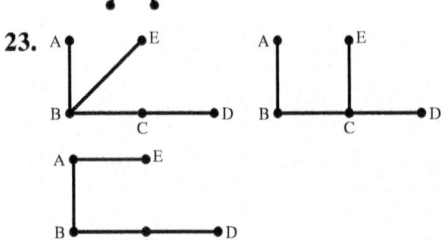

23.

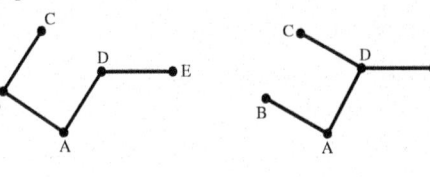

25.

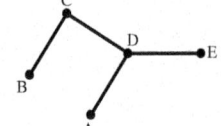

27.

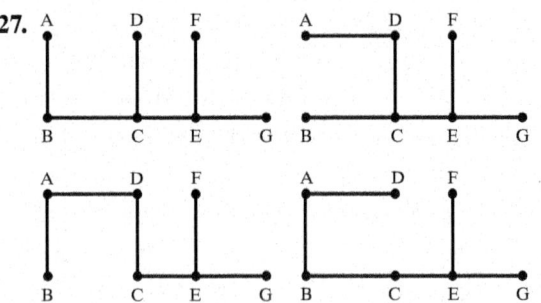

29.

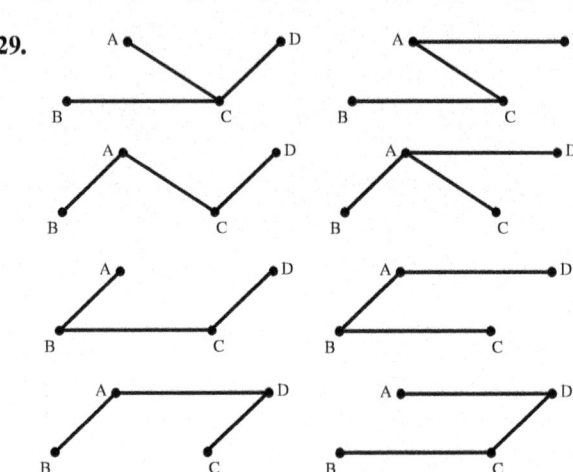

31. 20   **33.** If a connected graph has circuits, none of
which have common edges, then the number of spanning
trees for the graph is the product of the numbers of edges
in all the circuits.

35. Total weight is 51.

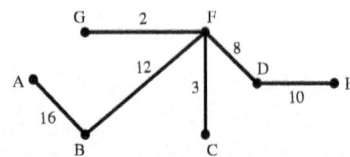

37. Total weight is 66.

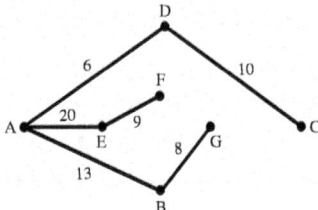

39. Total length to be covered is 140 ft.

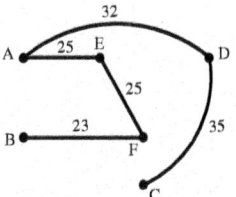

41. 33   **43.** 62   **45.** Different spanning trees must have
the same number of edges. The number of vertices in the
tree is the number of vertices in the original graph, and
the number of edges has to be one less than this.

47. (a) 9 **(b)** 18 **(c)** 0 **(d)** 2 **(e)**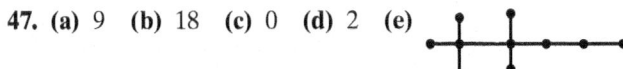

49. 22 cables   **51.** This is possible. The graph must be a tree because it has one fewer edges than vertices.

53. This is possible. The graph cannot be a tree, because it would have at least as many edges as vertices.

55. 3: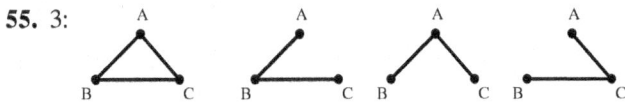

57. 125

59. 3 nonisomorphic trees:

61. 11 nonisomorphic trees:

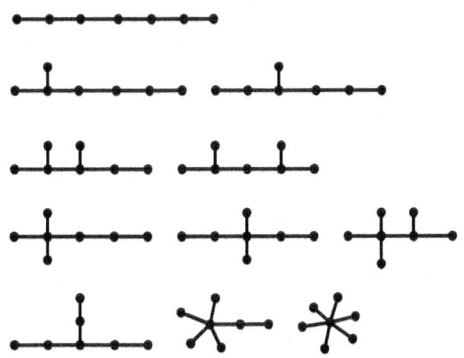

Chapter 14 Test

1. 7 **2.** 20 **3.** 10 **4. (a)** No, because edge AB is used twice. **(b)** yes **(c)** No, because there is no edge from C to D. **5. (a)** yes **(b)** No, because (for example) there is no edge from B to C. **(c)** yes

6. For example: 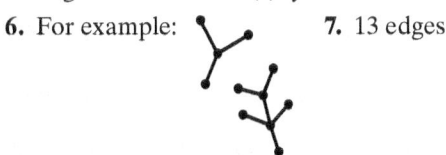   **7.** 13 edges

8. The graphs are isomorphic. Corresponding edges should be the same color. AB should match AB, etc.

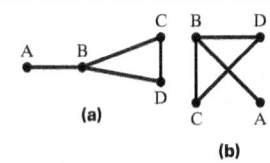

9. 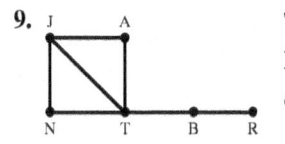 The graph is connected. Tina knows the greatest number of other guests.

10. 28 games   **11.** Yes, because there is an edge from each vertex to each of the remaining 6 vertices.

12.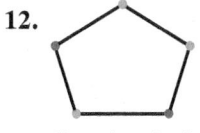

Chromatic number: 3

13.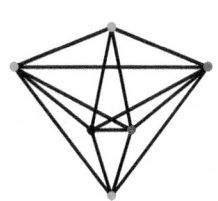

Chromatic number: 5

14. 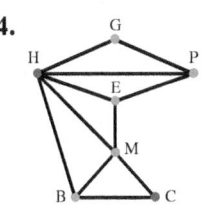 The chromatic number is 3, so three separate exam times are needed. Exams could be at the same times as follows: Geography, Biology, and English; History and Chemistry; Mathematics and Psychology.

15. (a) No, because it does not use all the edges. **(b)** No, because it is not a circuit (for example, there is no edge from B to C). **(c)** yes   **16.** No, because some vertices have odd degree.   **17.** Yes, because all vertices have even degree.   **18.** No, because two of the rooms have odd numbers of doors.

19. F → B → E → D → B → C → D → K → B → A → H → G → F → A → G → J → F

20. (a) No, because it does not visit all vertices. **(b)** No, because it is not a circuit (for example, there is no edge from B to C). **(c)** No, because it visits some vertices twice before returning to the starting vertex.

21. F → G → H → I → E → F   F → G → H → E → I → F
F → G → I → H → E → F   F → G → I → E → H → F
F → G → E → H → I → F   F → G → E → I → H → F
There are 6 such Hamilton circuits.

22. P → Q → S → R → P; Total weight is 27.

23. A → E → D → C → F → B → A; Total weight is 11.85.

24. 24! **25.** Hamilton circuit

26. Any three of these:

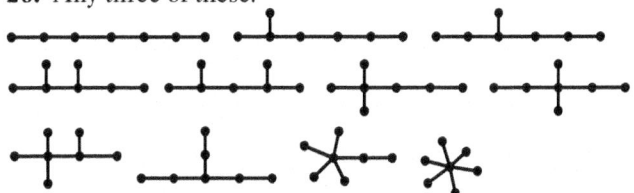

27. false **28.** true **29.** true

30. There are 4 spanning trees.

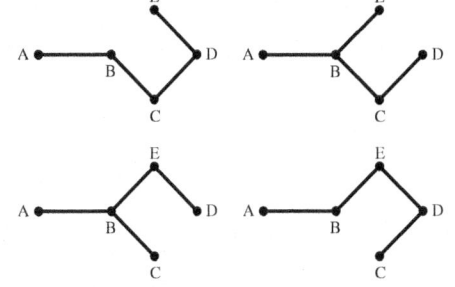

31. Weight is 24.

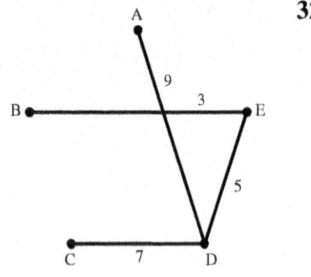

32. 49

CHAPTER 15 VOTING AND APPORTIONMENT

15.1 Exercises

1. false   **3.** true   **5.** true   **7. (a)** 4! = 24

(b)

Number of Voters	Ranking
3	b > c > a > d
2	a > d > c > b
1	c > d > b > a
1	d > c > b > a
1	d > a > b > c
2	c > a > d > b
1	a > c > d > b
2	a > b > c > d

(c) Australian shepherd   **9. (a)** c, with 3 (pairwise) points   **(b)** a, with 24 (Borda) points   **(c)** c
11. 5! = 120; 7! = 5040   **13.** For any value of $n \geq 5$, there are at least 120 possible rankings of the candidates. This means a voter must select one ranking from a huge number of possibilities. It also means the mechanics of any election method, except the plurality method, are difficult to manage.   **15.** $_6C_2 = 15$; $_8C_2 = 28$ **17. (a)** b, with 6 first-place votes   **(b)** c, with 2 (pairwise) points   **(c)** b, with 16 (Borda) points   **(d)** c   **19.** Logo a is selected by all the methods.   **21. (a)** e, with 8 first-place votes   **(b)** j, with 3 (pairwise) points   **(c)** j, with 40 (Borda) points   **(d)** h   **23. (a)** t, with 18 first-place votes   **(b)** h, with 4 (pairwise) points   **(c)** m, with 136 (Borda) points   **(d)** k   **25.** h beats e, 13 to 8.   **27.** c beats t, 37 to 18.   **29.** In a runoff between k and c, activity k is selected. Activity k faces activity t, and k is selected.
31. (a) 7   **(b)** e, with 7 pairwise points   **33. (a)** 2
(b) c, with 7 pairwise points   **35. (a)** 16   **(b)** c, with 16 Borda points   **37. (a)** 55   **(b)** c, with 55 Borda points
39. j   **41.** No winner is selected. Alternatives a, b, and d tie.   **43.** Answers will vary.   **45.** Answers will vary. One possible arrangement of the voters is 2, 4, 5, 7, 3.

15.2 Exercises

1. false   **3.** true   **5. (a)** Alternative a (6 of 11 first-place votes)   **(b)** b   **(c)** yes   **7. (a)** Alternative a (16 of 30 first-place votes)   **(b)** b   **(c)** yes   **9. (a)** c   **(b)** b
(c) b   **(d)** c   **(e)** plurality and Borda methods
11. (a) j   **(b)** e   **(c)** j   **(d)** h   **(e)** plurality and Hare methods   **13. (a)** h   **(b)** t   **(c)** m   **(d)** k   **(e)** all three methods   **15. (a)** a has 2 pairwise points, y and d have $1\frac{1}{2}$ pairwise points each, and n has 1 pairwise point.
(b) y has $2\frac{1}{2}$ pairwise points, a has 2 pairwise points, d has 1 pairwise point, and n has $\frac{1}{2}$ pairwise point, so y is selected.   **(c)** Yes; the rearranging voters moved a, the winner of the nonbinding election, to the top of their ranking, but y wins the official selection process.
17. (a) d drops out after one round of votes; b drops out after the second round; in the third vote, a is preferred to c by a margin of 11 to 6.   **(b)** d drops out after one round of votes; c drops out after the second round; in the third vote, b is preferred to a by a margin of 9 to 8.
(c) Yes; the rearranging voters moved a, the winner of the nonbinding election, to the top of their ranking, but b wins the official selection process.   **19. (a)** Candidate a has 75 first-place votes.   **(b)** Candidate c has 80 first-place votes.   **(c)** yes   **21.** No; however, the second pairwise comparison would result in a tie.   **23. (a)** Round one eliminates b; a is preferred to c in the second round, by a margin of 22 to 12.   **(b)** b   **(c)** yes   **25.** Answers will vary.
27. Answers will vary. One possible voter profile is given.

Number of Voters	Ranking
10	a > b > c > d > e > f
9	b > f > e > c > d > a

Candidate a is a majority candidate and wins all of her or his pairwise comparisons by a margin of 10 to 9, earning 5 pairwise points. Candidate b wins all of her or his comparisons except the one with a, earning 4 pairwise points.   **29.** Answers will vary. One possible profile is the voter profile for the animal shelter poster dog contest in **Exercise 8** of the previous section.   **31. (a)** Candidate a has 2 pairwise points.   **(b)** Answers will vary. Two options: the 3 bottom-row voters all switch to either a > z > x > y or a > z > y > x.   **33.** Delete Candidate c.
35. Answers will vary. One possible profile is given.

Number of Voters	Ranking
21	a > b > c
12	b > c > a
8	b > a > c

37. Answers will vary.

15.3 Exercises

1. true   **3.** false   **5. (a)** 34,437   **(b)** 7

7. Virginia received 19 seats rather than 18. Delaware received only 1 seat rather than 2.   **9. (a)** 29,493; 123.40

(b)

State park	a	b	c	d	e
Number of trees	11	70	62	54	42

(c)

State park	a	b	c	d	e
Number of trees	11	70	62	54	42

(d) The traditionally rounded values of Q sum to 240, which is greater than the number of trees to be apportioned.

State park	a	b	c	d	e
Traditionally rounded Q	12	70	62	54	42

(e) The value of md for the Webster apportionment should be greater than $d = 123.40$, because greater divisors make lesser modified quotas with a lesser total sum.

(f)

State park	a	b	c	d	e
Number of trees	12	70	61	54	42

(g) The Huntington-Hill rounded values of Q sum to 240, which is greater than the number of trees to be apportioned.

(h) The value of md for the Huntington-Hill apportionment should be greater than $d = 123.40$, because greater divisors make lesser modified quotas with a lesser total sum.

(i)

State park	a	b	c	d	e
Number of trees	12	70	61	54	42

(j) The Hamilton and Jefferson apportionments are the same. The Webster and Huntington-Hill apportionments are the same, but they differ from the Hamilton and Jefferson apportionments.

11. (a) 269; $d = 24.45$

(b)

Course	Fiction Writing	Poetry	Short Story	Great Books
Number of sections	2	2	3	4

(c)

Course	Fiction Writing	Poetry	Short Story	Great Books
Number of sections	2	1	3	5

(d) The traditionally rounded values of Q sum to 10, which is less than the number of sections to be apportioned.

Course	Fiction Writing	Poetry	Short Story	Great Books
Traditionally rounded Q	2	1	3	4

(e) The value of md for the Webster apportionment should be less than $d = 24.45$ because lesser divisors make greater modified quotas with a greater total sum.

(f)

Course	Fiction Writing	Poetry	Short Story	Great Books
Number of sections	2	2	3	4

(g) The Huntington-Hill rounded values of Q sum to 11, which is equal to the number of sections to be apportioned.

(h) Let $md = d$, because the correct number of sections is already apportioned.

(i) $md = d = 24.45$

Course	Fiction Writing	Poetry	Short Story	Great Books
Enrollment	2	2	3	4

(j) The Hamilton, Webster, and Huntington-Hill apportionments all agree. The Jefferson apportionment is different from the other three.

(k) The Jefferson method apportions 5 sections to Great Books, compared to only 4 sections apportioned by each of the other methods. With 5 sections, average class size would be 20 rather than 25.

(l) The Jefferson method apportions only 1 section to Poetry, for a class size of 35, whereas any of the other methods apportions 2 sections, allowing for two classes of, say, 17 and 18 students.

13. (a)

State	Abo	Boa	Cio	Dao	Ekko	Foti
Number of seats	15	22	7	20	14	53

(b)

State	Abo	Boa	Cio	Dao	Ekko	Foti
Number of seats	15	22	6	20	14	54

(c) The Huntington-Hill apportionment is the same as the Hamilton apportionment in part (a).

(d) The Hamilton, Webster, and Huntington-Hill methods all agree. The Jefferson method apportions one more seat to the largest state, Foti, and one fewer to the smallest state, Cio.

15. (a) 1721; $d = 43.025$

(b)

Hospital	A	B	C	D	E
Number of nurses	3	5	8	11	13

15. (c) $md = 40$

Hospital	A	B	C	D	E
Number of nurses	3	5	8	11	13

(d) The traditionally rounded values of Q sum to 41, which is greater than the number of nurses to be apportioned.

Hospital	A	B	C	D	E
Traditionally rounded Q	3	6	8	11	13

(e) The value of md for the Webster apportionment should be greater than $d = 43.025$, because greater divisors make lesser modified quotas with a lesser total sum.

(f) Hospital	A	B	C	D	E
Number of nurses	3	5	8	11	13

(g) The Huntington-Hill method also agrees with the Hamilton method in part (b).

(h) In this case, all four methods agree.

17. $\sqrt{5 \cdot 6} = 5.477$

19. Answers will vary. One possible ridership profile is given.

Bus route	a	b	c	d	e	Total
Number of riders	131	140	303	178	197	949

21. Answers will vary. One possible population profile is given.

State	a	b	c	d	e	Total
Population	50	230	280	320	120	1000

23. (a)

Hospital	A	B	C	D	E
Number of nurses	3	6	8	10	13

(b) The Hamilton, Jefferson, Webster, and Huntington-Hill methods all agreed, but the Adams method gives a different apportionment.   **25.** Answers will vary.

15.4 Exercises

1. true   **3.** false

5.

State	a	b	c	d
Q rounded down/up	28/29	12/13	**81/82**	9/10
Number of seats	28	12	**83**	9

7.

State	a	b	c	d
Number of seats if $n = 204$	35	74	42	53
Number of seats if $n = 205$	34	75	43	53

9.

State	a	b	c
Initial number of seats	1	4	6
Percent growth	10.91%	18.40%	13.16%
Revised number of seats	2	4	5

11. Two additional seats are added for the second apportionment.

State	Original State a	Original State b	New State c
Initial number of seats	20	55	*****
Revised number of seats	21	54	2

13. (a) With the Huntington-Hill method, no violation occurs.   **(b)** The Adams method violates the quota rule. State a, with standard quota $Q = 86.875$, should receive 86 or 87 seats but instead receives 85, which is too few.

15. (a) Wth the Huntington-Hill method, no violation occurs.   **(b)** The Adams method violates the quota rule. State d, with standard quota $Q = 118.250$, should receive 118 or 119 seats but instead receives 117, which is too few.

17. The new states paradox does not occur if the new population is 531, but it does occur if the new population is 532.   **19.** Answers will vary.   **21.** Answers will vary.   **23.** Answers will vary.

Chapter 15 Test

1. $7! = 5040$ **2.** $_{10}C_2 = 45$ **3.** Answers will vary.   **4.** Answers will vary.   **5.** Answers will vary.   **6.** Answers will vary.   **7.** Candidate a   **8.** Candidate a   **9.** 93   **10.** Candidate b   **11.** Yes, a is a majority candidate, but b wins the election.   **12.** Yes, a is a Condorcet candidate, but b wins the election.   **13.** No; b won initially, and b still wins.   **14.** No; b won initially, and b still wins.   **15.** Candidate c   **16.** Candidate s   **17.** The pairwise comparison method can violate the monotonicity criterion.   **18.** Candidate a   **19.** Candidate b   **20.** The plurality method can violate the irrelevant alternatives criterion.   **21.** The Balinski and Young Impossibility Theorem says that any apportionment method devised will either violate the quota rule or permit the possibility of a paradoxical apportionment (the Alabama paradox, the population paradox, or the new states paradox).   **22.** 354.52   **23.** a: 6.64; b: 12.69; c: 15.79; d: 64.88   **24. (a)** a: 6; b: 12; c: 15; d: 64   **(b)** a: 6; b: 12; c: 15; d: 64   **(c)** a: 7; b: 13; c: 16; d: 65   **(d)** a: 7; b: 13; c: 16; d: 65   **25.** a: 6; b: 13; c: 16; d: 64   **26. (a)** less than   **(b)** a: 6; b: 12; c: 16; d: 66   **27. (a)** greater than   **(b)** a: 7; b: 13; c: 16; d: 64   **28. (a)** greater than   **(b)** a: 7; b: 13; c: 16; d: 64   **29.** Answers will vary.   **30.** Answers will vary.   **31.** Answers will vary.   **32.** Answers will vary.

CREDITS

Text Credits

CHAPTER 1: **7** Howard Eves, *In Mathematical Circles*, Prindle, Weber & Schmidt, Inc. **20** *How to Solve it: A New Aspect of Mathematical Method*, George Pólya, Princeton University Press, 1945. **21** Fibonacci, L., & Sigler, L. E. (2002). *Fibonacci's Liber Abaci: A Translation into Modern English of Leonardo Pisano's Book of Calculation.* **26–30** Exercises 1–49, 51: From the monthly calendar of *Mathematics Teacher.* Copyright © by National Council of Teachers of Mathematics. Used by permission of National Council of Teachers of Mathematics. **32** Exercises 77, 80, 82: From the monthly calendar of *Mathematics Teacher.* Copyright © by National Council of Teachers of Mathematics. Used by permission of National Council of Teachers of Mathematics. **33** Screenshots courtesy of Texas Instruments. **35** Excerpt from *Innumeracy: Mathematical Illiteracy and Its Consequences* by John Allen Paulos. Published by Macmillan & Company, © 1988. **45** Summary Example: From the monthly calendar of *Mathematics Teacher.* Copyright © by National Council of Teachers of Mathematics. Used by permission of National Council of Teachers of Mathematics. **46–47** Exercises 9–15: From the monthly calendar of *Mathematics Teacher.* Copyright © by National Council of Teachers of Mathematics. Used by permission of National Council of Teachers of Mathematics.

CHAPTER 2: **62** Exercise 63: From the monthly calendar of *Mathematics Teacher.* Copyright © by National Council of Teachers of Mathematics. Used by permission of National Council of Teachers of Mathematics. **79–80** Exercises 23, 71: From the monthly calendar of *Mathematics Teacher.* Copyright © by National Council of Teachers of Mathematics. Used by permission of National Council of Teachers of Mathematics.

CHAPTER 3: **89–90** Screenshots courtesy of Texas Instruments. **95–97** Screenshots courtesy of Texas Instruments. **104** Problem from *The Lady or the Tiger and Other Logic Puzzles* by Raymond M. Smullyan. Copyright © 2009 by Dover Publications. Used by permission of Dover Publications; Smullyan, R. M. (1978). *What Is the Name of this Book?: The Riddle of Dracula and Other Logical Puzzles.* Englewood Cliffs, Prentice-Hall, Inc. **105** Ebert, R. (2009). *Roger Ebert's Four Star Reviews: 1967–2007.* Andrews McMeel Publishing. **107** Screenshots courtesy of Texas Instruments. **134** DIRECTV.

CHAPTER 4: **150** Ascher, M. (1991). *Ethnomathematics: A Multicultural View of Mathematical Ideas.* New York: Chapman & Hall. **167** Screenshots courtesy of Texas Instruments. **171** Nicolas Queijo; © Google.

CHAPTER 5: **184** Quoted in Gauss zum Gedächtniss (1856) by Wolfgang Sartorius von Waltershausen. **186–187** Program Primes written by Charles W. Gantner and provided courtesy of Texas Instruments; Screenshots courtesy of Texas Instruments. **196** Screenshots courtesy of Texas Instruments. **197** Exercises 9–12: From the monthly calendar of *Mathematics Teacher.* Copyright © by National Council of Teachers of Mathematics. Used by permission of National Council of Teachers of Mathematics. **204** The Prime Pages. **207–208** Screenshots courtesy of Texas Instruments. **210–211** Screenshots courtesy of Texas Instruments. **215** Springer International Publishing. **216** Texas Instruments Incorporated.

CHAPTER 6: **228** Leopold Kronecker. **231–233** Screenshots courtesy of Texas Instruments. **238** Screenshots courtesy of Texas

Instruments. **240–242** Screenshots courtesy of Texas Instruments. **251–253** Screenshots courtesy of Texas Instruments. **255** Screenshots courtesy of Texas Instruments. **258–259** Screenshots courtesy of Texas Instruments. **266–267** Screenshots courtesy of Texas Instruments. **276** Screenshots courtesy of Texas Instruments.

CHAPTER 7: **333** Archimedes, *The Works of Archimedes*, University Press, 1897. **343** Screenshots courtesy of Texas Instruments. **345** Screenshots courtesy of Texas Instruments. **348–349** Screenshots courtesy of Texas Instruments.

CHAPTER 8: **380** Screenshots courtesy of Texas Instruments. **385–386** Screenshots courtesy of Texas Instruments. **403–405** Screenshots courtesy of Texas Instruments. **413–416** Screenshots courtesy of Texas Instruments. **418** Screenshots courtesy of Texas Instruments. **421–424** Screenshots courtesy of Texas Instruments. **426–427** Screenshots courtesy of Texas Instruments. **432** Screenshots courtesy of Texas Instruments. **434–435** Screenshots courtesy of Texas Instruments. **445** Screenshots courtesy of Texas Instruments.

CHAPTER 9: **478** Exercises 47–50: From the monthly calendar of *Mathematics Teacher.* Copyright © by National Council of Teachers of Mathematics. Used by permission of National Council of Teachers of Mathematics; Exercise 52: "The Wizard of Oz," 1939, Metro-Goldwyn-Mayer; **489** Exercises 85–92: From the monthly calendar of *Mathematics Teacher.* Copyright © by National Council of Teachers of Mathematics. Used by permission of National Council of Teachers of Mathematics. **502–503** Exercises 77–79, 81–83, 85 : From the monthly calendar of *Mathematics Teacher.* Copyright © by National Council of Teachers of Mathematics. Used by permission of National Council of Teachers of Mathematics. **512** Exercises 49, 52–55: From the monthly calendar of *Mathematics Teacher.* Copyright © by National Council of Teachers of Mathematics. Used by permission of National Council of Teachers of Mathematics. **534–535** Screenshots courtesy of Texas Instruments. **538–539** Exercises 9–34: From the monthly calendar of *Mathematics Teacher.* Copyright © by National Council of Teachers of Mathematics. Used by permission of National Council of Teachers of Mathematics. **547** Screenshots courtesy of Texas Instruments.

CHAPTER 10: **558** Exercises 59–62: From the monthly calendar of *Mathematics Teacher.* Copyright © by National Council of Teachers of Mathematics. Used by permission of National Council of Teachers of Mathematics. **565–566** Screenshots courtesy of Texas Instruments. **568–570** Exercises 64, 71, 77–82: From the monthly calendar of *Mathematics Teacher.* Copyright © by National Council of Teachers of Mathematics. Used by permission of National Council of Teachers of Mathematics. **570** "Twinkle, Twinkle, Little Star" Lyric by Jane Taylor, 1806. **572** Screenshots courtesy of Texas Instruments. **574–575** Screenshots courtesy of Texas Instruments. **580** Screenshots courtesy of Texas Instruments. **581** Exercises 51, 53, 54: From the monthly calendar of *Mathematics Teacher.* Copyright © by National Council of Teachers of Mathematics. Used by permission of National Council of Teachers of Mathematics. **584** Exercise 82: From the monthly calendar of *Mathematics Teacher.* Copyright © by National Council of Teachers of Mathematics. Used by permission of National Council of Teachers of Mathematics. **589** Exercise 40: From the monthly calendar of *Mathematics*

Teacher. Copyright © by National Council of Teachers of Mathematics. Used by permission of National Council of Teachers of Mathematics. **592–593** Screenshots courtesy of Texas Instruments. **596–597** Exercises 33, 36, 53, 54, 56: From the monthly calendar of *Mathematics Teacher.* Copyright © by National Council of Teachers of Mathematics. Used by permission of National Council of Teachers of Mathematics. **602** Exercise 35: From the monthly calendar of *Mathematics Teacher.* Copyright © by National Council of Teachers of Mathematics. Used by permission of National Council of Teachers of Mathematics.

CHAPTER 11: **614** Davis, P. J., & Hersh, R. (1986). *Descartes' Dream: The World According to Mathematics.* San Diego: Harcourt Brace Jovanovich. **615–616** Exercises 71–72, 81–84: From the monthly calendar of *Mathematics Teacher.* Copyright © by National Council of Teachers of Mathematics. Used by permission of National Council of Teachers of Mathematics. **624** Exercises 51, 53–54: From the monthly calendar of *Mathematics Teacher.* Copyright © by National Council of Teachers of Mathematics. Used by permission of National Council of Teachers of Mathematics. **630** Sagan, C. (2000). *Carl Sagan's Cosmic Connection: An Extraterrestrial Perspective.* Cambridge: Cambridge University Press. **635–637** Exercises 62, 95–119: From the monthly calendar of *Mathematics Teacher.* Copyright © by National Council of Teachers of Mathematics. Used by permission of National Council of Teachers of Mathematics. **641–642** Screenshots courtesy of Texas Instruments. **644** Exercise 47: From the monthly calendar of *Mathematics Teacher.* Copyright © by National Council of Teachers of Mathematics. Used by permission of National Council of Teachers of Mathematics.

CHAPTER 12: **669–670** Screenshots courtesy of Texas Instruments. **682–683** Screenshots courtesy of Texas Instruments. **685–686** Screenshots courtesy of Texas Instruments. **691** Exercises 11–12: From the monthly calendar of *Mathematics Teacher.* Copyright © by National Council of Teachers of Mathematics. Used by permission of National Council of Teachers of Mathematics. **697–699** Screenshots courtesy of Texas Instruments. **707–708** Screenshots courtesy of Texas Instruments.

CHAPTER 13: **732–734** Screenshots courtesy of Texas Instruments. **736** Screenshots courtesy of Texas Instruments and Financial Calculators, www.fncalculator.com **739** Screenshots courtesy of Texas Instruments and Financial Calculators, www.fncalculator.com **741–742** Screenshots courtesy of Texas Instruments. **752** Screenshots courtesy of Texas Instruments and Financial Calculators, www.fncalculator.com **753–755** Screenshots courtesy of Texas Instruments. **759–760** Screenshots courtesy of Texas Instruments. **763** Screenshots courtesy of Financial Calculators, www.fncalculator.com

CHAPTER 14: **802** *Ode to Autumn* by John Keats. **820** *She Walks in Beauty* by Lord Byron; *Annabel Lee* by Edgar Allan Poe; "Sonnet 3" of William Shakespeare; *Ode to Graphs* by Vern E. Heeren.

Photo Credits

COVER: Ammentorp/123RF; Andrey Armyagov/123RF; Andrii Bicher/123RF; Andriy Popov/123RF; Antonio Balaguer Soler/123RF; Antonio Diaz/123RF; Ariadna de raadt/123RF; Atic12/123RF; Auremar/123RF; Belchonock/123RF; Blend Images/123RF; Carla Zagni/123RF; Cathy Yeulet/123RF; Chalermpon Poungpeth/123RF; Comaniciu Dan/123RF; Denis Ismagilov/123RF; Dmitriy Shironosov/123RF; Dmitry Kalinovsky/123RF; Dmytro Gilitukha/123RF; Dolgachov/123RF; Dotshock/123RF; Erwin Purnomo Sidi/123RF; Evgeniy Shkolenko/123RF; Franckreporter/Getty Images; Galina Barskaya/123RF; Goran Bogicevic/123RF; Gstockstudio/123RF; Guruxox/123RF; Iakov Filimonov/123RF; Ian Allenden/123RF; Iofoto/123RF; Jaromír Chalabala/123RF; Jk21/123RF; Juan Carlos Tinjaca Rodriguez/123RF; Karel Miragaya/123RF; Katarzyna Bialasiewicz/123RF; Konstantin Pelikh/123RF; Kurhan/123RF; Kzenon/123RF; Mark Adams/123RF; Mark Agnor/123RF; Matthias Ziegler/123RF; Mihai Blanaru/123RF; NejroN/123RF; Nenad Aksic/123RF; Noraisman Sahran/123RF; Nyul/123RF; Olegdudko/123RF; Patcharaporn Fuwiroj/123RF; Paulphoto/123RF; Petro/123RF; Photovs/123RF; Pornprasit Raksaman/123RF; Rades6/123RF; Rawpixel.com/Shutterstock; Rawpixel/123RF; Robert Przybysz/123RF; Rommel Canlas/123RF; Scott Griessel/123RF; Scyther5/123RF; Seoterra/123RF; Sergey Mironov/123RF; Shannon Fagan/123RF; Sheeler/123RF; Stockbroker/123RF; Stylephotographs/123RF; Tatsiana Yatsevich/123RF; Terry Schmidbauer/123RF; Thesupe87/123RF;Tyler Olson/123RF; Viktoriia Hnatiuk/123RF; Volodymyr Melnyk/123RF; Wang Tom/123RF; Wavebreak Media Ltd/123RF; Wong yu liang/123RF; Zlikovec/123RF; Zoomteam/123RF

FRONT MATTER: **xxiv** (top) Carole Heeren, (center) courtesy of John Hornsby, (bottom) r.levy photo/rhlphoto.com.

CHAPTER 1: **1** Courtesy of Terry Krieger **4** Jules Selmes/Pearson Education, Inc. **5** Pearson Education, Inc. **7** (top) Leoks/Shutterstock, (bottom) Snehasis Panja/Shutterstock. **14** Andrii Zhezhera/123RF **20** Pearson Education, Inc. **21** AP Images **22** From the monthly calendar of *Mathematics Teacher.* Copyright © by National Council of Teachers of Mathematics. Used by permission of National Council of Teachers of Mathematics. **23** Pearson Education, Inc. **24** Jakob Metzger/Shutterstock **26** Courtesy of John Hornsby **27** Olena Kachmar/123RF **30** (top) Mdorottya/123RF, (bottom left) George Yong/Shutterstock, (bottom right) Cheuk-king Lo/Pearson Education, Inc. **33** (top): Adrian825/iStock/Gettyimages, (bottom) Image used with permission by Texas Instruments, Inc. **34** Everett Collection **35** Ron Tarver/Krt Hfo/Newscom **36** Monkey Business Images/Shutterstock **37** (top) Dream Pictures/Blend Images/Alamy Stock Photo, (bottom) Elena Schweitzer/Fotolia **39** Monkey Business Images/Shutterstock **41** (left) Melody Mulligan/Shutterstock, (right) Jupiterimages/Photos.com/Getty Images

CHAPTER 2: **49** Cathy Yeulet/123RF **50** Pearson Education, Inc. **53** Beth Anderson/Pearson Education, Inc. **57** Paul Fearn/The History Collection/Alamy Stock Photo **62** (left) Nimon/Shutterstock; Robynrg/Shutterstock, (right) Beth Anderson/Pearson Education, Inc. **71** Belchonock/123RF **72** Andriy Popov/123RF **78** Sturti/E+/Getty Images **79** Everett Collection **80** Lifestyle pictures/Alamy Stock Photo **81** (top) Debby Wong/Shutterstock, (bottom) MC2 Eric C. Tretter/Defenseimagery.mil **86** belchonock/123RF

CHAPTER 3: **87** Rawpixel.com/Shutterstock **88** Pearson Education, Inc. **89** Pearson Education, Inc. **90** Scala/Art Resource **91** JST-Photography/Shutterstock **93** Monkey Business Images/Shutterstock **94** Pearson Education, Inc. **98** Pearson Education, Inc. **99** Pearson Education, Inc. **100** Pearson Education, Inc. **104** (top) Wavebreak Media Ltd/123RF, (bottom) Dover Publications (courtesy of Christopher Heeren) **105** (left) Courtesy of John Hornsby, (right) Pearson Education, Inc. **106** Radius Images/Alamy Stock Photo **109** Pirita/Shutterstock **111** Denys Prykhodov/Fotolia **112** (top) Urosr/Fotolia, (bottom) Courtesy of Christopher Heeren **113** David Pineda Svenske/Shutterstock **115** Pearson Education, Inc. **116** Pearson Education, Inc. **117** Pictorial Press Ltd/Alamy Stock Photo **118** Pearson Education, Inc. **120** Pearson Education, Inc. **121** Pearson Education, Inc. **125** Crush Rush/Shutterstock **128** Python Pictures/Ronald Grant Archive/Alamy Stock Photo **129** Aisyaqilumaranas/Shutterstock **131** RTRO/Alamy Stock Photo **132** Cathy Yeulet/123RF **134** The Chinese Gardener/Shutterstock

CHAPTER 4:   **143** Milles Studio/Shutterstock **144** Science & Society Picture Library/GettyImages **145** (top) Art Collection 2/Alamy Stock Photo, (bottom) Leemage/Universal Images Group/Getty Images **146** Joserpizarro/123RF **147** Believeinme33/123RF **148** EQRoy/Shutterstock **150** Little Monster/Stockimo/Alamy Stock Photo **156** World History Archive/Alamy Stock Photo **158** Canada [Addison Wesley Canada]/Pearson Education, Inc **161** Payless Images/Shutterstock **162** (top) Pearson Education, Inc., (bottom) NYPL/Science Source **167** Suchatbky/Shutterstock **168** Photos.com/Getty Images **171** Courtesy of Christopher Heeren **172** Courtesy of John Hornsby **174** Wavebreak Media Ltd/123RF **177** Image Flow/Shutterstock

CHAPTER 5:   **183** B Christopher/Alamy Stock Photo **190** Courtesy of Vern Heeren **191** Pearson Education, Inc. **192** Andriy Bezuglov/123RF **193** Pearson Education, Inc. **194** National Institute on Aging **196** Tim Brakemeier/picture-alliance/dpa/AP Images **199** Pearson Education, Inc. **201** TriStar Pictures/Everett Collection **202** AF archive/Alamy Stock Photo **203** Georgios Kollidas/Alamy Stock Photo **204** Pearson Education, Inc. **207** Dover Publications (courtesy of Vern E. Heeen) **209** Pearson Education, Inc. **212** (top) Fotofermer/Shutterstock, (bottom): ZealPhotography/Alamy Stock Photo **213** Monkey Business Images/Shutterstock **217** (left) Kathathep/Shutterstock, (right) Alex Staroseltsev/Shutterstock **218** Vvr/Fotolia **219** (left) Scala/Art Resource, New York, (right) KMNPhoto/Shutterstock **222** Everett Collection

CHAPTER 6:   **227** 279photo Studio/Shutterstock **234** Jules Selmes/Pearson Education, Inc. **236** Galyna Andrushko/123RF **239** Pearson Education, Inc. **244** Scanrail/123RF **247** Bizoo_n/Fotolia **254** (top, bottom) U.S. Mint **255** Auremar/123RF **256** EMG Network/Pearson Education, Inc. **257** (left) Pearson Education, Inc., (right) Marcel Derweduwen/123RF **259** Rick Seeney/Shutterstock **260** Terry Putman/123RF **261** U.S. Mint **262** Dolgachov/123RF **264** Yaroslaff/Shutterstock **267** (top) Kurhan/Shutterstock, (bottom) Courtesy of Charles D. Miller **269** Pearson Education, Inc. **273** Jason Maehl/Shutterstock **274** Courtesy of Terry McGinnis **278** GaudiLab/Shutterstock **279** (top) Stylephotographs/123RF, (bottom) Moodboard/Alamy Stock Photo **280** (top, top center, bottom) Courtesy of John Hornsby, (bottom center) Courtesy of Terry McGinnis **281** Iakov Filimonov/Shutterstock **282** Jacob Lund/Shutterstock **283** Elnur/Shutterstock **285** (top) Daniel D Malone/Shutterstock, (bottom) U.S. Mint **286** (top) Uwimages/Fotolia, (bottom) Nd3000/Shutterstock **287** Action Sports Photography/Shutterstock **288** (left) Pearson Education, Inc., (top, middle right) Courtesy of John Hornsby **296** Cheryl Casey/Shutterstock

CHAPTER 7:   **297** Courtesy of John Hornsby **299** Jax10289/Shutterstock **300** Cobalt88/Shutterstock **301** Pearson Education, Inc. **302** Pearson Education, Inc. **303** PhotoStock10/Shutterstock **304** Sofya Apkalikova/123RF **308** Leonard Zhukovsky/Shutterstock **309** Levent Konuk/Shutterstock **312** Courtesy of John Hornsby **313** Historic Collection/Alamy Stock Photo **314** Alexandr Shevchenko/Shutterstock **316** White House Photo/Alamy Stock Photo **317** Courtesy of John Hornsby

318 (top) John Green/Cal Sport Media/Alamy Stock Photo, (bottom) Simon Balson/Alamy Stock Photo **320** Matka_Wariatka/Shutterstock **322** Linza/Shutterstock **323** Emily Barker/Shutterstock **324** Courtesy of John Hornsby **325** Joggie Botma/Shutterstock **326** (top) Andrew Fielding/Southcreek Global/ZUMA Press, Inc./Alamy Stock Photo, (bottom) Valerio Pardi/123RF **327** Ximagination/123RF **328** Cathy Yeulet/123RF **329** Stephen VanHorn/Shutterstock **330** (top) Courtesy of John Hornsby, (bottom) Kakisnow/123RF **331** Michal_K/Shutterstock **332** Vincent Noel/Shutterstock **333** Pearson Education, Inc. **334** Courtesy of Charles D. Miller **335** Pearson Education, Inc. **338** Courtesy of John Hornsby **339** Konstantin Pelikh/123RF **340** Monkey Business Images/Shutterstock **341** Tupungato/Shutterstock **342** William Perugini/123RF **350** Catwalker/Shutterstock **352** Klaus Lang/All Canada Photos/Alamy Stock Photo **363** Pearson Education, Inc. **364** Pearson Education, Inc. **372** Martin Froyda/Shutterstock **373** Maderla/Shutterstock

CHAPTER 8:   **375** Courtesy of John Hornsby **377** Courtesy of Charles D. Miller **378** Courtesy of Charles D. Miller **381** Ritchie B. Tongo/Epa/Shutterstock **384** Courtesy of John Hornsby **388** Courtesy of John Hornsby **396** Pearson Education, Inc. **399** Barry Blackburn/Shutterstock **402** Courtesy of John Hornsby **403** Nadezda Murmakova/Shutterstock **405** Africa Studio/Shutterstock **406** Sam Edwards/OJO Images Ltd/Alamy Stock Photo **407** Tatagatta/Shutterstock **410** (top) Cristovao31/Fotolia, (bottom) Igor Borisenko/123rf **411** Macor/123RF **412** Titelio/Shutterstock **415** Everett Historical/Shutterstock **416** PPL/Shutterstock **419** (left) Aaron Kohr/Shutterstock, (right) S Curtis/Shutterstock **420** (top) Pearson Education, Inc., (bottom)Tsuneo/123RF **425** Chronicle/Alamy Stock Photo **426** Holger Hollemann/Epa/Shutterstock **427** Chitsanupong Chuenthananont/123RF **428** (top) Cathy Yeulet/123RF, (bottom) Deepspace/Shutterstock **430** (top) Elenabsl/123RF, (center) ChameleonsEye/Shutterstock, (bottom) Juriah Mosin/Shutterstock **431** Wyatt Rivard/Shutterstock **433** Derek Bayes/Lebrecht Music & Arts/Alamy Stock Photo **438** Salvatore Massara/123RF **439** Mitch Gunn/Shutterstock **443** (top) Antonio Diaz/123RF, (bottom) Nadia Zagainova/Shutterstock **448** Wu Hong/Epa/Shutterstock **450** P72/Shutterstock **459** Studio 8/Pearson Education Ltd

CHAPTER 9:   **461** Dino Fracchia/Alamy Stock Photo **462** Universal Art Archive/Alamy Stock Photo **463** Oksana Perkins/Fotolia **464** Dmitry Kalinovsky/Shutterstock **467** Arisia/Shutterstock **475** (top, bottom): Pearson Education, Inc. **476** Moviestore collection Ltd/Alamy Stock Photo **479** V J Matthew/Shutterstock **480** Chonrawit Boonprakob/Shutterstock **482** World History Archive/Alamy Stock Photo **483** Courtesy of John Hornsby **484** Oksana.Perkins/Shutterstock **488** Courtesy of Charles D. Miller **490** From the monthly calendar of *Mathematics Teacher*. Copyright © by National Council of Teachers of Mathematics. Used by permission of National Council of Teachers of Mathematics. **491** Doniyor Mamanov/Shutterstock **492** GeniusMinus/Fotolia **496** Steve Tulley/Alamy Stock Photo **497** Pearson Education, Inc. **498** (top) Tbradford/iStock/Getty Images, (bottom) Wavebreak Media Ltd/123RF **500** (top) Courtesy of John Hornsby, (bottom): Pearson Education, Inc. **502** Pearson Education, Inc. **504** Dincer.agin/Shutterstock **506** Courtesy of Christopher Heeren **506** Courtesy of David Honda **509** Dino Fracchia/Alamy Stock Photo **513** Christian Heisch/Shutterstock **514** Scaliger/Fotolia **515** Brainmaster/E+/Getty Images **524** Pearson Education, Inc. **525** Courtesy of Tom Lehrer **526** VimilkVimin/Shutterstock **527** Vatchara katpakong/123RF **528** Pearson Education, Inc. **532** Dan Thornberg/Shutterstock **535** Sabina Louise Pierce/AP Images

INDEX OF APPLICATIONS

INDEX

applications of, 234–235, 244–246
complex numbers and, 269
as decimals, 230
division of, 240–241
explanation of, 52, 92, 230
irrational numbers and, 265–271
multiplication of, 239–240
operations with, 237–246
operations with decimals and, 275–278
operations with percent and, 277–281
order and, 231–232
properties of addition and multiplication of, 243–244
rational numbers and, 250–260
sets of, 228–231
subtraction of, 238
Reasoning
circular, 124
deductive. *See* Deductive reasoning
inductive. *See* Inductive reasoning
post hoc, 125
Reasoning by transitivity, 129
Recession, 740
Reciprocal, 243
Recommended daily allowance, 721
Recorde, Robert, 231
Rectangles
area of, 493
explanation of, 471
Golden, 218, 270, 491
perimeter of, 491
Rectangular coordinate system, 376–377
Rectangular parallelepiped, 504–506
Recurring costs, 765
Recursion formula, 215
Red herring, 125
Redistricting, 909
Reductio ad absurdum, 265
Refinancing, 766
Reflection image, 513–514
Reflection(s)
examples of, 514
explanation of, 513–515, 518
glide, 517, 518
point, 516
product of two, 515–516
Reflection transformation, 514
Reflexive axiom, 306
Region of feasible solutions, 448
Regular monthly payment
explanation of, 759
formula for, 759
table for finding, 759–760
Regular polygons, 470, 504
Regular polyhedra, 504–505, 827
Regular tessellation, 522
Relations
equivalence, 306
explanation of, 401

functions and, 375, 401–402
Relative frequency, 669
Relatively prime numbers, 208
Repayment schedule, 760–762
Repeating decimals, 258–260, 265
Reserve account, 764
Residue, 196
Retirement accounts, 738, 785–787
Return on investment, 776
Revenue models, 405–406
Revolving loans, 746–748
Rhind papyrus, 145
Rhombus, 471
Richter, Charles F., 428
Richter, P. H., 536
Richter scale, 428
Riemann, Georg Friedrich Bernhard, 191, 526–528
Riemann Hypothesis, 191
Riemann sum, 528
Riese, Adam, 239
Right angles, 464, 471
Right circular cones, 504–505, 507
Right circular cylinders
example of, 506
explanation of, 506
volume and surface area of, 506
Right triangles. *See also* Pythagorean theorem
examples of, 483–484
explanation of, 471
legs of, 482
Risk, 608
Rodricks, J., 608
Rogers, Christopher Eric, 407–408
Roman numerals, 147–148, 229, 239
Root, of a directed tree, 847
Rotation
explanation of, 515–517, 518
magnitude of, 515
Roth IRA, 785
Roulette, 647
Rounding, 900
Round lot of shares, 778
Route planning, 827–828
Rubik's Cube, 1–2, 555
Rule of 70, 742
Rule of 78, 752, 754–755
Rule of False Position, 300
Russell, Bertrand, 108, 115, 116, 117
Russian peasant method, 163

S

Saccheri, Girolamo, 524–525, 527
Sagan, Carl, 630
Samples, 668
Sample space, 604
Sample standard deviation, 698
Savings Incentive Match Plan for Employees (SIMPLE), 786
Scale factor, 538

Scalene triangles, 471
Scaling (self-similarity), 535
Scheduling, mathematical theory of, 813
Scientific calculators. *See also* Calculators; Graphing calculators
for converting bases, 171
explanation of, 32
future value on, 736
powers of *e* on, 422
statistics using, 669
Scientific notation
conversion from, 349–350
conversion to, 349
examples of, 350
explanation of, 33, 348
Search for Extraterrestrial Intelligence (SETI), 194, 630–631
Secant lines, 474
SEC fee, 779
Second component in an ordered pair, 66
Sectors, 497
Securities and Exchange Commission (SEC), 779
Security deeds, 758. *See also* Mortgages
Segments, line, 378–379, 463
Sehlinger, Bob, 841
Seife, Charles, 229
Self-similarity (scaling), 535
Seller buydown, 763
Semicircles
angles inscribed in, 474
explanation of, 474
Semi-interquartile range, 710
Semiperimeter, 273, 496
Semi-regular tessellations, 523
Separating, in graph theory, 847
Set-builder notation, 50, 229
Set equality, 53, 55, 58
Set(s)
cardinal numbers of, 52
Cartesian product of, 67–68
complement of, 56–57, 68
De Morgan's laws and, 70
designation for, 50–51
difference of, 65–66, 68
disjoint, 63
elements of, 50, 59
empty, 50
equal, 53, 55, 58
equivalent, 53, 55
explanation of, 50–51, 573
finite, 52
infinite, 52
intersection of, 62–63, 68
members of, 50
null, 50
operations on, 68
proper subset of, 58, 60
of real numbers, 228–231